6th International Conference on Fracture Fatigue and Wear (FFW 2017)

Journal of Physics: Conference Series Volume 843

Porto, Portugal
26 – 27 July 2017

ISBN: 978-1-5108-4187-1
ISSN: 1742-6588

Printed from e-media with permission by:

Curran Associates, Inc.
57 Morehouse Lane
Red Hook, NY 12571

Some format issues inherent in the e-media version may also appear in this print version.

Printed by Curran Associates, Inc. (2017)

For permission requests, please contact the Institute of Physics
at the address below.

Institute of Physics
Dirac House, Temple Back
Bristol BS1 6BE UK

Phone: 44 1 17 929 7481
Fax: 44 1 17 920 0979

techtracking@iop.org

Additional copies of this publication are available from:

Curran Associates, Inc.
57 Morehouse Lane
Red Hook, NY 12571 USA
Phone: 845-758-0400
Fax: 845-758-2633
Email: curran@proceedings.com
Web: www.proceedings.com

Table of contents

Volume 843

6th International Conference on Fracture Fatigue and Wear
26–27 July 2017, Porto, Portugal

Accepted papers received: 9 May 2017
Published online: 31 May 2017

Preface

Fracture

Fatigue

Wear

IOP Conf. Series: Journal of Physics: Conf. Series **843** (2017) 011001 doi:10.1088/1742-6596/843/1/011001

Proceedings of 6th International Conference on Fracture Fatigue and Wear FFW 2017, Porto, Portugal, 26-27 July 2017

Editorial

This volume contains the proceedings of the 6th International Conference on Fracture Fatigue and Wear FFW 2017, Porto, Portugal, 26-27 July 2017. Fracture Fatigue and Wear 2017 is the 6th FFW conference. Previous conferences were celebrated in Jinan (China, 2010), Kitakyushu (Japan, 2013), Kitakyushu (Japan, 2014), Gent (Belgium, 2015) and Kitakyushu (Japan, 2016).

The sixth international conference on Fracture, Fatigue and Wear will be held in the city of Porto, Portugal. The overall objective of the conference is to bring together international scientists and engineers in academia and industry in fields related to Fracture Mechanics, Fatigue of Materials, Tribology and Wear of Materials. The conference covers industrial engineering applications of the above topics including theoretical and analytical methods, numerical simulations and experimental techniques. One of the aims of conference is to promote cooperation between international scientists and engineers from a large number of disciplines, who are involved in research related to Fracture, Fatigue and Wear. The presentations of FFW 2017 are divided into 3 main sessions, namely 1) Fracture, 2) Fatigue and 3) Wear.

The organising committee is grateful to keynote speakers; Professor Hong Liang, Mechanical Engineering, Texas A & M University, for her keynote lecture entitled 'Direct and in situ Observation of Initiation of Wear ' and Professor Xudong Qian, Department of Civil and Environmental Engineering, National University of Singapore, for his keynote lecture entitled 'Ductile Fracture Assessment and Representation in Large-Scale Structures'.

Special thanks go to members of the Scientific Committee of FFW 2017 for reviewing the articles published in this volume and for judging their scientific merits. Based on the comments of reviewers and the scientific merits of the submitted manuscripts, the articles were accepted for publication in the conference proceedings and for presentation at the conference venue. The accepted papers are of a very high scientific quality and contribute to advancement of knowledge in all research topics relevant to FFW conference.

Finally, the organising committee would like to thank all authors, who have contributed to this volume and presented their research work at the confrence venue in Porto.

Professor Magd Abdel Wahab

Chairman of FFW 2017

Peer review statement

All papers published in this volume of *Journal of Physics: Conference Series* have been peer reviewed through processes administered by the proceedings Editors. Reviews were conducted by expert referees to the professional and scientific standards expected of a proceedings journal published by IOP Publishing.

IOP Conf. Series: Journal of Physics: Conf. Series **843** (2017) 012001 doi:10.1088/1742-6596/843/1/012001

The prediction of residual stress and its influence on the mechanical properties of weld joint

J. Ni[1] and M. A. Wahab[1]

[1] Ghent University, Laboratory Soete, Belgium

E-mail: magd.abdelwahab@ugent.be

Abstract. A three-dimensional metallo-thermo-mechanical analysis of bead on plate welding is performed in this work. This coupled model enables to capture the microstructural development and temperature history at local region. As a result, the residual stress is evaluated based on the temperature-dependent mechanical properties computed by the mixture of individual phase. Isotropic hardening is assumed in the finite element (FE) analysis. At the same time, the distribution of residual stress is also predicted by treating the mechanical properties as integral values of sheet metal. The two simulated fields of stress and strain after welding are analysed and compared. Moreover, as it is known that welding changes the mechanical properties of the original material, especially in fusion zone (FZ) and heat affected zone (HAZ), the stress and strain data at interested areas (HAZ and FZ) are subtracted for comparison. The predicted stress and strain fields are imported to subsequent simulation of standard tensile test. The stress-strain curves are compared with the one of base material. It is found that residual stress has significant influence on the structural performance of weld joints.

1. Introduction

Residual stress in welded structure or component can result in increasing degradation of mechanical performance. In existence of residual stress, the corrosion is accelerated and fatigue cycle is shortened [1]. Therefore, understanding the generation of residual stress and predicting its distribution are of the utmost importance for welding process.

The welding simulation is always performed in finite element (FE) code as an efficient tool for prediction of residual stress and strain [2-4]. However, some of the researchers [3, 5] simply treated the steel as an overall material and adopted temperature-dependent properties of specific steel for simulation. Other authors [2, 6] included the effect of phase transformation in analysis but only the main product martensite is evaluated. To predict residual stress in a more reasonable way, the development of all transformation products need to be tracked [7]. The most common metallurgical model is the Johnson-Mehl-Avrami-Kolmogorov (JMAK) equation [8], which describes the isothermal transition and is often implemented with the additivity law [9-11] for anisothermal process. One flexible model is proposed by Leblond et al. [12] in which the volume fraction is expressed in an incremental function of equilibrium volume fraction and characteristic time. The function is capable of predicting microstructural development in arbitrary thermal history. Although widely used, parameters in both models need to be calibrated with time-temperature-transformation (TTT) and/or continuous cooling transformation (CCT) diagrams. Differently, Bhadeshia et al. [13-16] published a series of research works which amount to a metallurgical model that is independent of metallurgical diagrams. In the present work, the model of Bhadeshia et al. [13-16] is chosen to track the microstructural evolution.

As the microstructure is determined, the expansion coefficient, elastic modulus, Poisson's ratio and yield stress are interpolated linearly according to the volume proportion. Two plastic models, i. e. perfectly plastic and linear hardening, are performed. To analyze the effect of phase transformation, simulation which treats the material as an overall unit is also conducted. The resultant residual stress of all the three simulations and the distribution of product phases are extracted and presented for comparison. Finally, the residual stress is imported to another model to conduct tensile test simulation.

2. Experimental setup

Bead-on-plate specimen is manufactured from low-alloyed sheet metal S700 by Gas Metal Arc Welding (GMAW). The thickness of the plate is 8.0 mm and the length 800 mm. Although the metallurgical model is independent of continuous cooling transformation (CCT) diagram, the chemical compositions of plate and filler material (FM) are needed, which are listed in Table 1. The welding speed is controlled at 300 mm/min so that continuous bead is obtained.

Table 1. Chemical composition of S700 and filler material

	Main alloying elements (wt.%)			
	C	Si	Mn	Ni + Mo + Cr
Plate	0.06	0.05	1.9	0.7
FM	0.1	0.9	1.5	

3. Finite element analysis

Finite element analysis (FEA) should be well conducted and may require experiences and skills. FEA has been widely used in the literature for many engineering applications that have been recently published [17-34]. The commercial FE code ABAQUS is used to conduct the coupled thermo-metallo-mechanical analysis. Figure 1 shows the bead geometry after welding predicted by another commercial software. It is seen that the resultant bead geometry is not symmetric. Therefore, the whole dimension model rather symmetric model is built for simulation.

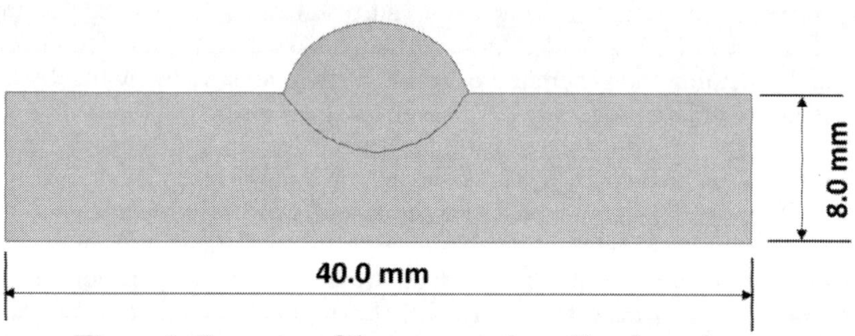

Figure 1. Geometry of the cross-section of bead-on-plate

An approximated heat source proposed by Goldak et. al. [35] is used to simulate the development of temperature field during the whole process. The power distribution q in front quadrant is expressed as [35]:

$$q(x,y,z,t) = \frac{6\sqrt{3} f_f Q_{inp}}{a_f b c \pi \sqrt{\pi}} e^{-3[x+v(\tau-t)]^2/a_f^2} e^{-3y^2/b^2} e^{-3z^2/c^2} \quad (1)$$

The fraction of front ellipsoid is denoted by f_f. Parameters a_f, b and c are the lengths of ellipsoid semi-axes. Q_{inp} is heat input rate, v is the welding speed and τ a lag factor needed to define the position of the heat source at $t = 0$. The distribution in rear quadrant is written in the same way, but with different fraction f_r and semi-axis a_r. The calibrated parameters are listed in Table 2. The heat source is applied in ABAQUS by writing user subroutine DFLUX.

Table 2. Calibrated parameters in heat source

a_f (mm)	a_r (mm)	b (mm)	c (mm)	f_f	Q_{inp} (W)
7.0	19.5	4.8	3.2	1.5	5474

Moreover, the temperature-dependent thermal properties are used in order to precisely capture the temperature. The value of the specific heat (***c***) and heat conductivity (***k***) are shown in Figure 2. The thermal properties at 1400°C will be used for situation where temperature exceeds 1400°C.

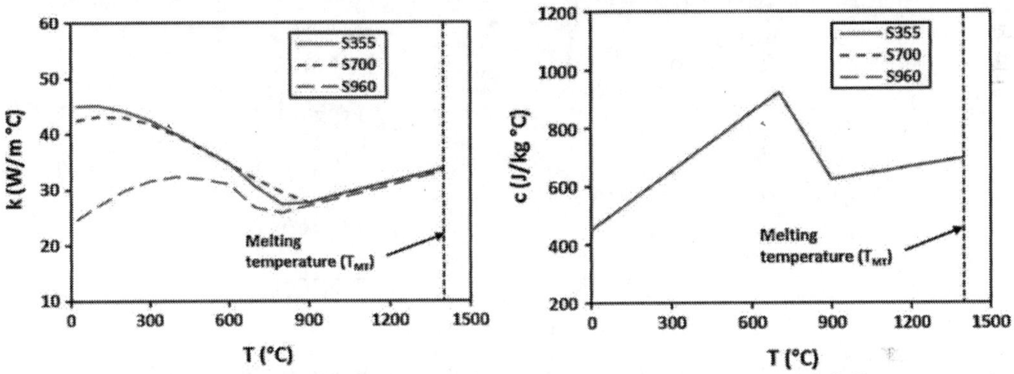

Figure 2. Temperature dependent thermal properties ***c*** and ***k*** [3]

To analyze the influence of microstructure, the volume fractions of each phase or micro-constituent are predicted by the metallurgical framework of Bhadeshia [13-16]. In his model, the start and finish transformation temperatures are estimated based on thermodynamic theories rather than empirical function. For transforming kinetics, the model assumes the reconstructive transformation to occur at grain boundaries but also includes the possibility of inclusion nucleation. The nucleation rates were treated as temperature-dependent and the growth rates were determined by solving carbon diffusion equations [13]. The initiation of bainite growth was assumed to be displacive with a strain energy of 400 J/mol [36]. The advantages of this framework are: 1) it omits the necessity of determining parameters with known results and 2) it requires only the basic information of the chemical composition, the temperature history and the austenite grain size for input. Totally, five products, allotriomorphic ferrite (α), Widmannstätten ferrite (α_w), pearlite (α_p), bainite (α_b) and martensite (α') are considered in the simulation. The sequence in which the individual phase is arranged is listed in Table 3.

Table 3. The sequence of field variables and corresponding phase

FV1	FV2	FV3	FV4	FV5	FV6
α	α_w	α_p	α_b	α'	γ

IOP Conf. Series: Journal of Physics: Conf. Series **843** (2017) 012001 doi:10.1088/1742-6596/843/1/012001

The material property are assigned in two ways. As the base material and the beam are treated as single unit, the whole model is applied unique temperature-dependent mechanical properties, which are shown in Figure 3. The Poisson's ratio is considered to be constant value of 0.29 over whole temperature range.

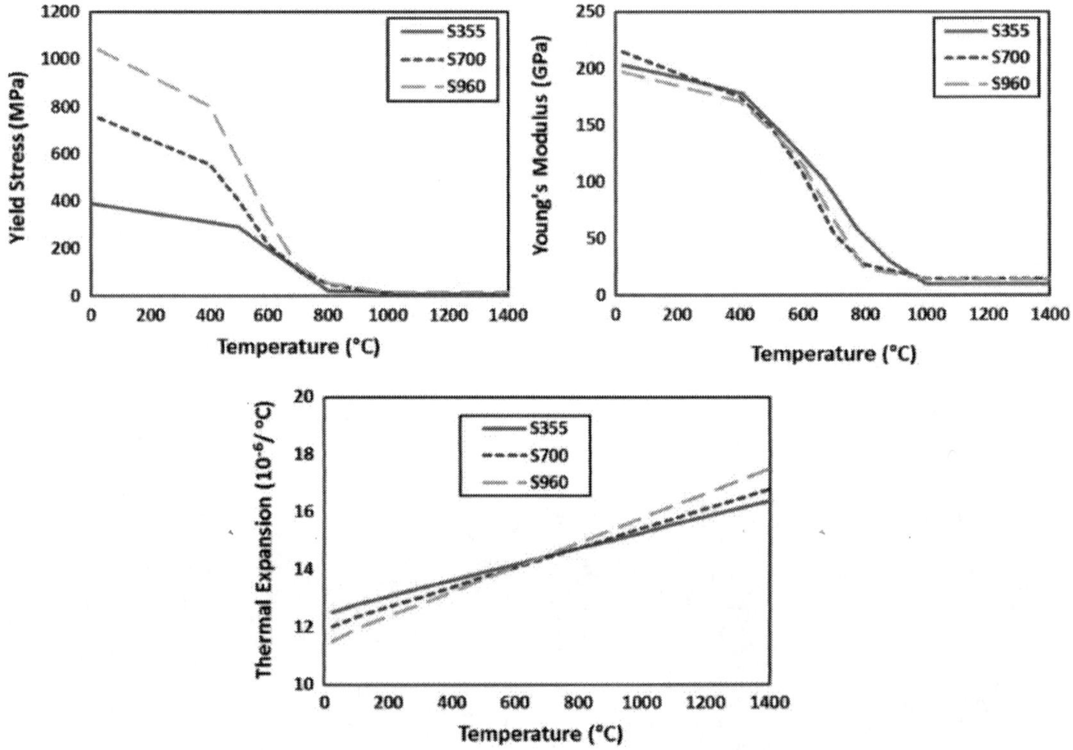

Figure 3. Temperature dependent yield stress, Young's modulus and thermal expansion [3]

The second method, which treats the material as an assembly of phases or constituents, applies mechanical properties to individual microstructure. The mechanical properties are listed in Table 4. The overall mechanical properties are obtained by linearly interpolating the individual properties with corresponding volume fractions. The property of α_W is not listed in Table 4 but it is considered the same as α and α_p. With interpolation, two hardening cases, i. e. perfectly plastic and isotropic hardening (the tangent modulus of which is assumed to be 2000 MPa) are simulated.

Table 4. The sequence of field variables and corresponding phase [37]

		Temperature (°C)			
Property	Phase	0	300	600	800
Elastic modulus, GPa	γ	200	175	150	124
	α and α_p	210	193	165	120
	α_b	210	193	165	120
	α'	200	185	168	--
Poisson's ratio	γ	0.29	0.31	0.33	0.35
	α and α_p	0.28	0.30	0.31	0.33
	α_b	0.28	0.30	0.31	0.33
	α'	0.28	0.30	0.31	--

IOP Conf. Series: Journal of Physics: Conf. Series **843** (2017) 012001 doi:10.1088/1742-6596/843/1/012001

Expansion coefficient, K^{-1}	γ		2.1×10^{-5}		
	α and α_p		1.4×10^{-5}		
	α_b		1.4×10^{-5}		
	α'		1.3×10^{-5}		
Yield strength, MPa	γ	190	110	30	20
	α and α_p	360	230	140	30
	α_b	440	330	140	30
	α'	1600	1480	1260	--

The main algorithm is summarized in Figure 4. The thermal analysis is run before metallurgical and mechanical analyses in order to calculate the cooling rate. Therefore, the proposed model is sequential coupled, which ignores the effects of latent heat and plastic work. However, it is reasonable since the amount of energy produced by these two terms is far less than the welding power [38, 39]. The initial temperature of the whole structure is assigned to 20°C and the convection coefficient is 30W/m²°C.

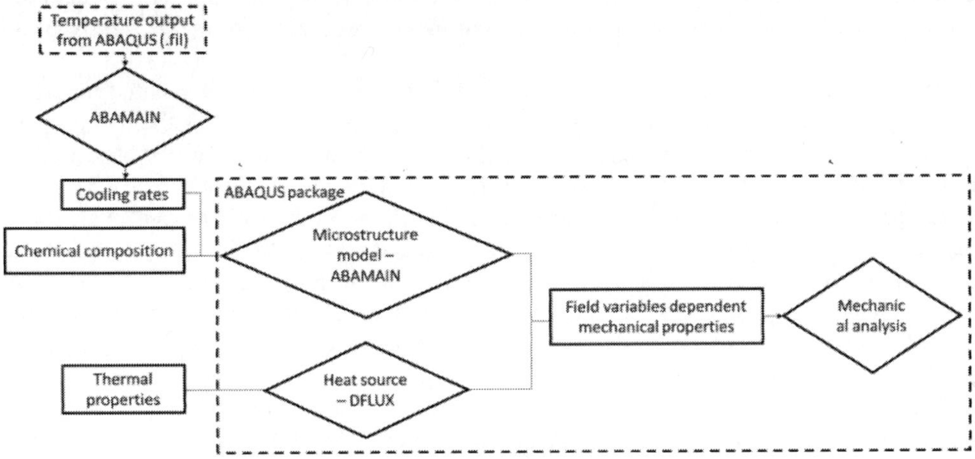

Figure 4. Flow chart of the fully coupled model

After the simulation of welding process, the structure is further subjected to a tensile load along the welding direction. The stress and strain curves are plotted in different region to show the influence of residual stress on mechanical performance of weld structure.

4. Results and discussion

The temperature histories of two nodes, A and B as shown in Figure 5, are plotted. Node A is selected inside melt region and Node B lies at the boundary between FZ and HAZ. It is seen that the maximum temperature that Node B experiences is about 1400°C, which is in agreement with the experimental data [3]. This accordance shows the reasonability of the calibrated heat source model. The distribution of α_W (Widmannstätten ferrite) at the end of simulation is also presented. Since the temperature at A drops faster than B and the transformation to α_W is displacive, the volume fraction of α_W at A should be higher than the value at B. This conclusion is validated by the contour plot of FV2 in Figure 5.

IOP Conf. Series: Journal of Physics: Conf. Series **843** (2017) 012001 doi:10.1088/1742-6596/843/1/012001

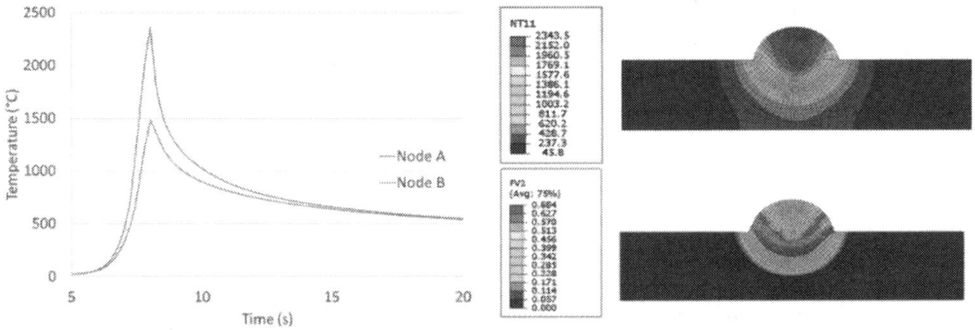

Figure 5. Temperature histories of nodes and contour plots of temperature and α_W

As it is mentioned in section 3. three different approaches are used to apply mechanical properties. The first approach is to interpolate the values corresponding to phase volume fraction at various temperature and assumes a linear hardening tangent modulus of 2000 MPa (IPLH). The second way interpolates the mechanical properties similarly but assumes perfectly plastic (IPPP). The third approach simply uses overall mechanical properties as shown in Figure 3 and assumes perfectly plastic as well (OPPH).

The transverse residual stress at the plate surface of middle section is presented in Figure 6. In all three cases, the transverse residual stress is tensile. This is reasonable since as the weld cools down, the material of this region begins to shrink. However, the tensile stress is not equally distributed. It increases at first as it approaches to the centreline and drops to minimum value at the boundary between HAZ and FZ. All the three simulation results share the same distribution pattern. The peak value of transverse stress in linear hardening case is highest among the three analysis, which means that the tensile stress caused by cooling is large enough to yield the material. When the material is assumed to be perfectly plastic, the transverse stress in FZ calculated by interpolating properties is lower than using overall properties. In this region, the stress state is not sufficient to cause a global yield and as a result, the influence of individual phase becomes clear. Since α_W and α' possess higher specific volume than γ, the transition from γ to α_W and α' will lead to an extra expansion. This expansion releases the tensile stress partially, which explains the lower value of transverse stress when using interpolated properties.

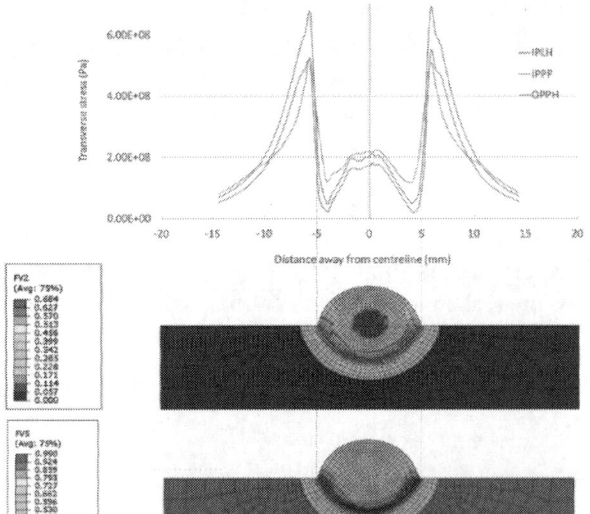

Figure 6. Transverse residual stress at middle cross section and volume fractions of α_W and α'

Finally, a homogenous load is applied to the welded structure in longitudinal direction (or welding direction). The stress-strain curves are plotted at three positions, A, B and C. It can be seen that in existence of residual stress (A and B), the relation between stress and strain is no longer linear. The bead is subjected to tensile stress in longitudinal direction in initial, which leads to a faster yielding. In contrast, the material in HAZ is subjected to compressive state.

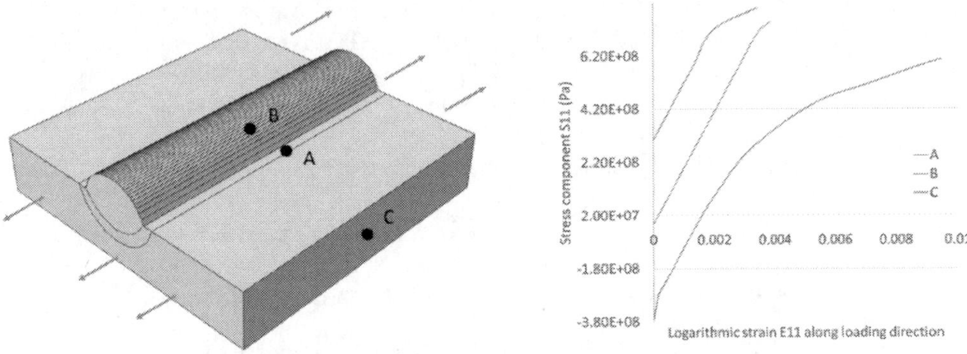

Figure 7. The stress-strain curves of selected points

5. Conclusion

A fully coupled thermo-metallo-mechanical model is developed to predict the distribution of residual stress and analyse its influence on mechanical behaviour. The metallurgical model proposed by Bhadeshia et al. [13-16] proves an efficient way to track the microstructural development. To evaluate the influence of phase transformation on residual stress, mechanical properties are interpolated with the phase volume fraction. By comparing with the result from an overall model, it is found that the volume expansion due to transition will affect the magnitude and distribution of residual stress, especially in FZ. In the zone where material is already yielded, the difference of residual stress is not so significant. Moreover, a tensile test simulation is performed subsequently after the welding simulation. The stress-strain curves of three typical regions show that the material in FZ and HAZ does not behaves linearly even in elastic region. As the welded structure is subjected to external load, the residual stress will accelerate yielding in some areas while retard yielding in other areas, which makes the prediction of structure failure even more complicated.

6. Acknowledgements

The authors would like to acknowledge the MaDurOS project and the support from SIM and VLAIO.

7. References

[1] Barsoum Z and Barsoum I 2009 *Engineering Failure Analysis* **16**(1) 449-467
[2] Deng D and Murakawa H 2008 *Computational Materials Science* **43**(4) 681-695
[3] Bhatti AA, Barsoum Z, Murakawa H and Barsoum I 2015 *Materials & Design* **65** 878-889
[4] Junyan Ni and Wahab MA 2017 *Computers & Structures* **186** 35–49
[5] Teng TL and Lin CC 1998 *International Journal of Pressure Vessels and Piping* **75**(12) 857-864
[6] Deng D 2009 *Materials & Design* **30**(2) 359-366
[7] Borjesson L and Lindgren LE 2001 *Journal of Engineering Materials and Technology-Transactions of the Asme* **123**(1) 106-111
[8] Avrami M 1939 *The Journal of Chemical Physics* **7**(12) 1103-1112
[9] Cahn JW 1956 *Acta Metallurgica* **4**(6) 572-575

[10] Deng D, Ma N and Murakawa H 2012 *Transactions of JWRI* **2011** 79-82
[11] Inoue T 1996 *Transactions of JWRI* **25**(2) 69-87
[12] Leblond JB and Devaux J 1984 *Acta Metallurgica* **32**(1) 137-146
[13] Jones SJ and Bhadeshia HKDH 1997 *Acta Materialia* **45**(7) 2911-2920
[14] Takahashi M and Bhadeshia HKDH 1991 *Materials Transactions Jim* **32**(8) 689-696
[15] Khan SA and Bhadeshia HKDH 1990 *Materials Science and Engineering a-Structural Materials Properties Microstructure and Processing* **129**(2) 257-272
[16] Rees GI and Bhadeshia HKDH 1992 *Materials Science and Technology* **8**(11) 985-993
[17] X. Nguyen H, N. Nguyen T, Abdel Wahab M, Bordas SPA, Nguyen-Xuan H and P. Voa T 2017 *Computer Methods in Applied Mechanics and Engineering* **313** 904-940
[18] Tran Vinh L, Lee J, Nguyen-Van H, Nguyen-Xuan H and Abdel Wahab M 2015 *International Journal of Non-Linear Mechanics* **72** 42-52
[19] Tran LV, Phung-Van P, Lee J, Wahab MA and Nguyen-Xuan H 2016 *Composite Structures* **140** 655-667
[20] Thai CH, Ferreira AJM, Abdel Wahab M and Nguyen-Xuan H 2016 *Acta Mechanica* **227**(5) 1225-1250
[21] Thai C, Zenkour AM, Abdel Wahab M and Nguyen-Xuan H 2016 *Composite Structures* **139** 77-95
[22] Phung-Van P, Tran LV, Ferreira AJM, Nguyen-Xuan H and Abdel-Wahab M 2016 *Nonlinear Dynamics* 1-16; doi:10.1007/s11071-11016-13085-11076
[23] Phung Van P, Nguyen LB, Tran Vinh L, Dinh TD, Thai CH, Bordas SPA, Abdel Wahab M and Nguyen-Xuan H 2015 *International Journal of Non-Linear Mechanics* **76** 190-202
[24] Phung Van P, De Lorenzis L, Thai CH, Abdel Wahab M and Nguyen-Xuan H 2015 *Computational Materials Science* **96** 495-505
[25] Phung Van P, Abdel Wahab M, Liew KM, Bordas SPA and Nguyen-Xuan H 2015 *Composite Structures* **123** 137-149
[26] Yue T and Abdel Wahab M 2017 *Tribology International* **107** 274-282
[27] Yue T and Abdel Wahab M 2016 *Materials* **9** 597; doi:510.3390/ma9070597
[28] Resende Pereira KdF, Bordas S, Tomar S, Trobec R, Depolli M, Kosec G and Abdel Wahab M 2016 *Materials* **9** 639; doi:610.3390/ma9080639
[29] Noda N-A, Chen X, Sano Y, Wahab MA, Maruyama H, Fujisawa R and Takase Y 2016 *MATERIALS & DESIGN* **96** 476-489
[30] Ferjaoui A, Yue T, Abdel Wahab M and Hojjati-Talemi R 2015 *International Journal of Fatigue* **73** 66-76
[31] Tran Vinh L, Lee J, Ly HA, Abdel Wahab M and Nguyen-Xuan H 2015 *International Journal of Mechanical Sciences* **96-97** 65-78
[32] Yue T and Abdel Wahab M 2016 *Materials* **9**(7)
[33] Phung-Van P, Qui LX, Nguyen-Xuan H and Wahab MA 2017 *Composite Structures* **166** 120–135
[34] Phung-Van P, Ferreira AJM, Nguyen-Xuan H and Abdel Wahab M 2017 *Composites Part B: Engineering* **118** 125-134
[35] Goldak JA and Akhlaghi M 2006 Computational welding mechanics ed editors. Springer Science & Business Media
[36] Bhadeshia HKDH 1981 *Acta Metallurgica* **29**(6) 1117-1130
[37] Totten GE 2002 Handbook of Residual Stress and Deformation of Steel ed editors. ASM International
[38] Kirkaldy JS and Sharma RC 1982 *Scripta Metallurgica* **16**(10) 1193-1198
[39] Dai H 2012 Modelling residual stress and phase transformations in steel welds ed editors. INTECH Open Access Publisher

IOP Conf. Series: Journal of Physics: Conf. Series **843** (2017) 012002 doi:10.1088/1742-6596/843/1/012002

Detailed investigation of the analysis conditions in the evaluation of bonded joints by cohesive zone models

R.J.B. Rocha[a], R.D.S.G. Campilho[a,b,*]

[a] Departamento de Engenharia Mecânica, Instituto Superior de Engenharia do Porto, Instituto Politécnico do Porto, Rua Dr. António Bernardino de Almeida, 431, 4200-072 Porto, Portugal

[b] INEGI – Pólo FEUP, Rua Dr. Roberto Frias, s/n, 4200-465 Porto, Portugal

* raulcampilho@gmail.com

Abstract. Cohesive Zone Models (CZM) are widely used for the strength prediction of adhesive joints. This work studies the influence of different conditions used in CZM simulations to model a thin adhesive layer in single-lap joints (SLJ) under a tensile loading, for an estimation of their influence on the strength prediction under diverse geometrical and material conditions. Adhesives ranging from brittle to highly ductile and overlap lengths (L_O) between 12.5 and 50 mm were considered. Several damage initiation and growth criteria were tested. The analysis carried out in this work allowed to conclude that CZM is a powerful technique for strength prediction of bonded joints, provided that the modelling conditions are properly defined.

1. Introduction

Adhesive technology has earned interest in the major fields of industrial applications, since adhesively-bonded joints have many advantages over conventional mechanical fasteners. Among other benefits, this joining method preserves the integrity of parent materials, gives the possibility of join different materials and provides more uniform stress distributions. Adhesive bonds also promote good strength-weight and cost-effectiveness ratios. On the other hand, limitations of bonded joints include the disassembly difficulty without causing damage, low resistance to temperature and humidity, the requirement of a surface treatment and joint design orientated towards the elimination of peel stresses [1]. The strength prediction of bonded joints began with analytical techniques, firstly performed by Volkersen [2]. With these techniques, bonded joints are quickly analysed, although with simplifying assumptions regarding the material behaviour (e.g. neglecting the adhesives' plasticity), loading and boundary conditions. Numerical techniques (i.e. finite elements (FE)) are more adequate. Barenblatt [3] and Dugdale [4] proposed the concept of cohesive zone to describe damage under static loads with no need of an initial crack. Since then, CZM were improved and tested to simulate crack initiation and growth. CZM are based on cohesive elements [5], connecting solid elements of structures and can be easily implemented in FE software to model the fracture behaviour. This method simulates damage along a predefined crack path by the establishment of traction-separation laws that correlate the cohesive tractions (t_n for tension and t_s for shear) with the relative displacements (δ_n for tension and δ_s for shear). Different criteria can be used to assess damage initiation and growth. Gustafson and Waas [6] have investigated the influence of the constitutive parameters of the adhesive for the CZM analysis of several tests commonly used to describe the joints' performance. The Double-Cantilever Beam (DCB), the End-Notched Flexure (ENF) test and the SLJ were evaluated. It was found that the DCB test is only sensitive to the fracture toughness in tension (G_{IC}), which makes this test suitable to define this parameter for

CZM simulations. Oppositely, the ENF and SLJ results are influenced by several parameters, such as G_{IC}, fracture toughness in shear (G_{IIC}) and the shear cohesive strength (t_s^0). The tensile cohesive strength (t_n^0) promotes a negligible effect in the studied tests and, thus, it can be estimated independently to those tests. CZM is a precise technique for bonded joints, provided that the modelling conditions are those known to work well for this type of structures (for example using a quadratic stress criterion for damage initiation and a linear power law energetic criterion for damage propagation).

This work evaluates the influence of the different damage initiation and growth criteria on the P_m estimation for SLJ, considering different geometrical and material conditions. Validation with experimental data is considered. Adhesives ranging from brittle to highly ductile and L_O between 12.5 and 50 mm were considered.

2. Experimental work

The high strength and ductile aluminium alloy AA6082 T651 was chosen for the adherends. The tensile mechanical properties of this material were obtained in the work of Campilho et al. [7]: Young's modulus (E) of 70.07±0.83 GPa, tensile yield stress (σ_y) of 261.67±7.65 MPa, tensile strength (σ_f) of 324±0.16 MPa and tensile failure strain (ε_f) of 21.70±4.24%. Three structural adhesives were tested: the brittle epoxy Araldite® AV138, the ductile epoxy Araldite® 2015 and the ductile polyurethane Sikaforce® (Sika®, Baar, Switzerland) 7752, characterized in previous works regarding the most relevant properties [7-9]. The tensile mechanical properties (E, σ_y, σ_f and ε_f) were obtained by bulk tests on specimens with dogbone shape (NF 76-142 standard). The shear mechanical properties estimated with Thick Adherend Shear Tests (TAST) considering the ISO 11003-2 standard. The G_{IC} and G_{IIC} were taken from DCB and ENF tests, respectively, using beam theories. The DCB tests were carried out according to the ISO 15024-1 standard. No standard is available for ENF tests; thus, common practices for composite toughness determination were followed. The properties of the adhesives are detailed in Table 1.

Table 1 – Properties of the adhesives Araldite® AV138, Araldite® 2015 and Sikaforce® 7752 [7-9].

Property	AV138	2015	7752
Young's modulus, E [GPa]	4.89±0.81	1.85±0.21	0.49±0.09
Poisson's ratio, ν	0.35 [a]	0.33 [a]	0.30 [a]
Tensile yield stress, σ_y [MPa]	36.49±2.47	12.63±0.61	3.24±0.48
Tensile strength, σ_f [MPa]	39.45±3.18	21.63±1.61	11.48±0.25
Tensile failure strain, ε_f [%]	1.21±0.10	4.77±0.15	19.18±1.40
Shear modulus, G [GPa]	1.56±0.01	0.56±0.21	0.19±0.01
Shear yield stress, τ_y [MPa]	25.1±0.33	14.6±1.3	5.16±1.14
Shear strength, τ_f [MPa]	30.2±0.40	17.9±1.8	10.17±0.64
Shear failure strain, γ_f [%]	7.8±0.7	43.9±3.4	54.82±6.38
Toughness in tension, G_{IC} [N/mm]	0.20 [b]	0.43±0.02	2.36±0.17
Toughness in shear, G_{IIC} [N/mm]	0.38 [b]	4.70±0.34	5.41±0.47

[a] manufacturer's data
[b] estimated in reference [7]

The SLJ geometry and characteristic dimensions are presented in Figure 1. The tested geometrical parameter in this work is L_O. The dimensions of the specimens are: joint length between grips L_T=170 mm, adherends' thickness t_P=3 mm, width B=25 mm, adhesive thickness t_A=0.2 mm and L_O=12.5, 25, 37.5 and 50 mm. Joint preparation before bonding consisted of sandblasting and cleaning with acetone, which resulted in full cohesive failures of the adhesive layer in all specimens. Assembly was undertaken in a steel mould and calibrated nylon wire was placed between the adherends to produce the desired value of t_A. Curing was carried out at room temperature for one week. In order to perform the experimental tests at room temperature, a Shimadzu® (Shimadzu, Kyoto, Japan) AG-X 100 testing machine with a 100 kN load cell was used. The displacement rate was 1 mm/min. Five samples were considered for each joint configuration.

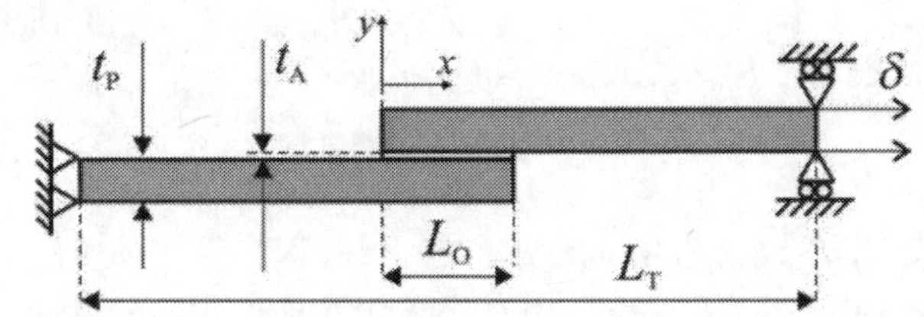

Figure 1 – Geometry and characteristic dimensions of the SLJ specimens [8]

3. Numerical work

The software ABAQUS® (Dassault Systèmes, Vélizy-Villacoublay, France) was used for P_m prediction by CZM. The adherends were modelled as elastic-plastic isotropic materials, and by 4-node plane-strain elements (CPE4 from ABAQUS®). The adhesive layer was modelled by 4-node cohesive elements. Both simulations consisted of a two-dimensional (2D) and geometrically non-linear analysis. In order to reduce the computational effort, mesh grading was considered by using the bias effect. CZM are based on a relationship between stresses and relative displacements (in tension or shear) connecting paired nodes of cohesive elements, to simulate the elastic behaviour up to t_n^0 in tension or t_s^0 in shear and subsequent softening up to failure. The areas under the traction-separation laws in tension or shear are equalled to G_{IC} or G_{IIC}, by the respective order. The triangular law assumes an initial linear elastic behaviour followed by linear degradation. Damage initiation can be specified by different criteria, either stress or strain based. The available stress-based damage initiation criteria are the maximum nominal stress (MAXS), quadratic nominal stress (QUADS) and maximum principal stresses (MAXPS) criteria expressed, by the same order, as [10]

$$\max\left\{\frac{\langle t_n\rangle}{t_n^0},\frac{t_s}{t_s^0}\right\}=1 \quad ; \quad \left\{\frac{\langle t_n\rangle}{t_n^0}\right\}^2+\left\{\frac{t_s}{t_s^0}\right\}^2=1 \quad ; \quad \left\{\frac{\langle\sigma_{max}\rangle}{\sigma_{max}^0}\right\}=1. \tag{1}$$

$\langle\ \rangle$ are the Macaulay brackets, emphasizing that a purely compressive stress state does not initiate damage [11]. σ_{max} and σ_{max}^0 represent the current and the allowable maximum principal stress. The strain-based maximum nominal strain (MAXE), quadratic nominal strain criterion (QUADE) and maximum principal strain (MAXPE) criteria are described as (by the same order) [10]

$$\max\left\{\frac{\langle\varepsilon_n\rangle}{\varepsilon_n^0},\frac{\varepsilon_s}{\varepsilon_s^0}\right\}=1 \ ; \ \left\{\frac{\langle\varepsilon_n\rangle}{\varepsilon_n^0}\right\}^2+\left\{\frac{\varepsilon_s}{\varepsilon_s^0}\right\}^2=1 \ ; \ \left\{\frac{\langle\varepsilon_{max}\rangle}{\varepsilon_{max}^0}\right\}=1, \tag{2}$$

where ε_n and ε_s are the current tensile and shear strain, respectively. ε_n^0 and ε_s^0 are corresponding peak strains. ε_{max} and ε_{max}^0 represent the current and the allowable maximum principal strain. By the fulfilment of the above mentioned criteria, the material stiffness initiates a degradation process. Complete separation and failure are predicted by a damage evolution law. The power law criterion states that failure under mixed-mode conditions is governed by a power law interaction of the energies required to cause failure in the individual (normal and shear) modes. It is given by the expression [10]

$$\left\{\frac{G_I}{G_{IC}}\right\}^\alpha+\left\{\frac{G_{II}}{G_{IIC}}\right\}^\alpha=1, \tag{3}$$

where α is the power law parameter, and G_I and G_{II} relate to the work done by the traction and corresponding relative displacements in the normal and shear directions, respectively. The Benzeggagh-Kenane (BK) [12] fracture criterion is particular useful when the critical fracture energies during deformation purely along the first and the second shear directions are the same; i.e., G_{IIC}=G_{IIIC}. It is given by

$$G_{IC}+\left(G_{IIC}+G_{IC}\right)\left\{\frac{G_S}{G_T}\right\}^\eta=G_C, \tag{4}$$

where G_S=G_{II} + G_{III}, G_T=G_I + G_S and η is a characteristic material parameter. The relevant CZM parameters used in this work were taken from the data of Table 1.

4. Results

4.1. Experimental and numerical joint strength
The average experimental values of P_m and respective deviation for the joints bonded with the three adhesives are depicted in Figure 2. The CZM predictions using the common modelling conditions for bonded joints are also presented (quadratic stress criterion damage initiations and linear power law criterion for growth).

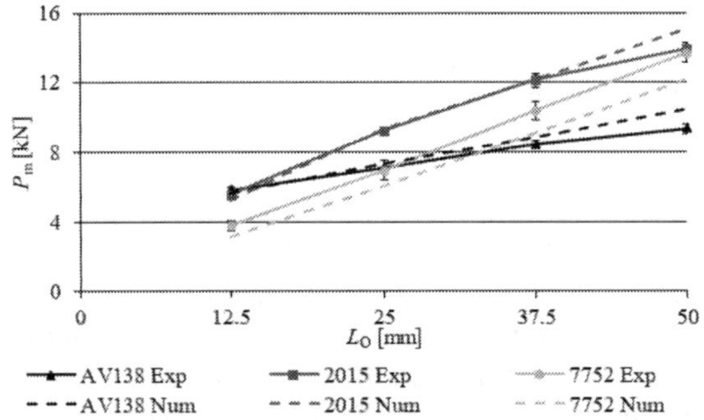

Figure 2 – Experimental and CZM values of P_m for each value of L_O and adhesive type

Comparing the experimental and numerical results for the adhesive Araldite® AV138 (Figure 2), these were very similar except for L_O=50 mm, with a maximum relative deviation of 11.8%. Due to the brittleness of the adhesive Araldite® AV138, P_m is attained with minimum plasticization at the overlap edges, justifying the good results obtained with a triangular CZM. A good correlation was found for the joints bonded with the adhesive Araldite® 2015, except for L_O=50 mm, where P_m was underestimated by 8.0%. Figure 2 presents numerically underestimated values of P_m for all L_O for the joints bonded with the adhesive Sikaforce® 7752. The under prediction of the P_m values is related to this adhesive's large plasticity, which is not accurately modelled by the triangular CZM.

4.2. Different conditions of the numerical simulations

4.2.1. Damage initiation criterion. First, the stress-based damage initiation criteria were evaluated by comparison between the experimental data, the prediction performed in Section 4.1 (QUADS criterion), and the MAXS and MAXPS criteria. Figure 3 shows the comparative results for the three adhesives. The results for the adhesive Araldite® AV138 (Figure 3 a) show close results to the experimental ones for the MAXS and QUADS criteria, contrarily to MAXPS. Actually, the MAXPS criterion gives identical P_m values irrespectively of L_O, since failure took place by tensile net failure of the adherends. Between the other two criteria, the QUADS criterion gives more accurate results compared to the experimental data (maximum deviation of 16.9% for the MAXS and 11.8% for the QUADS, both considering L_O=50 mm). Figure 3 b, related to the Araldite® 2015, presents once more closer results for the MAXS and QUADS criteria. With the MAXPS criterion failure took place by tensile net failure of the adherends. For the presented adhesive, the QUADS criterion gave the closest results to the experiments. The maximum error regarding the experimental data was 8.8% (MAXS), and 8.0% (QUADS) for L_O=50 mm. The strength prediction performed for the adhesive Sikaforce® 7752 shows that the MAXS and QUADS criteria give P_m under predicted results. The MAXPS criterion revealed, once more, to be a poor choice for P_m. However, evaluation data places the MAXS criterion as the best choice to perform P_m estimation for the joints bonded with this adhesive, since smaller under predictions were attained for all L_O values. The maximum relative deviations were 17.1% (QUADS), and 15.7% (MAXS), found for L_O=12.5 mm.

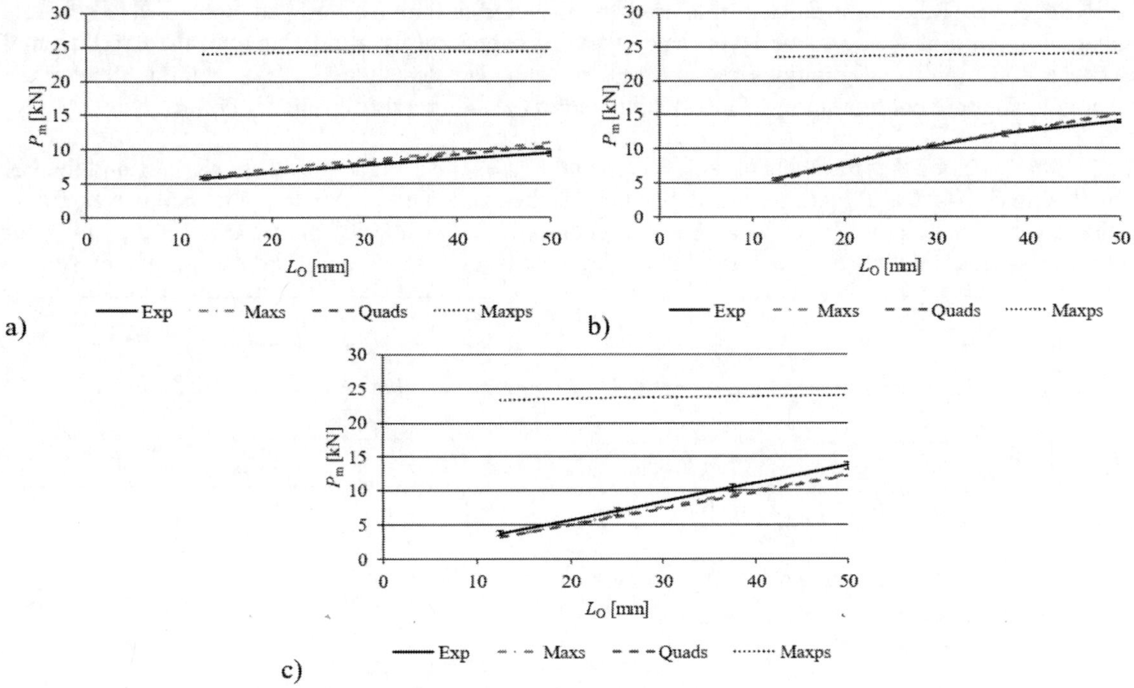

Figure 3 – Comparison between the experimental data and different damage initiation stress criteria for the joints bonded with the adhesive Araldite® AV138 (a), Araldite® 2015 (b) and Sikaforce® 7752 (c)

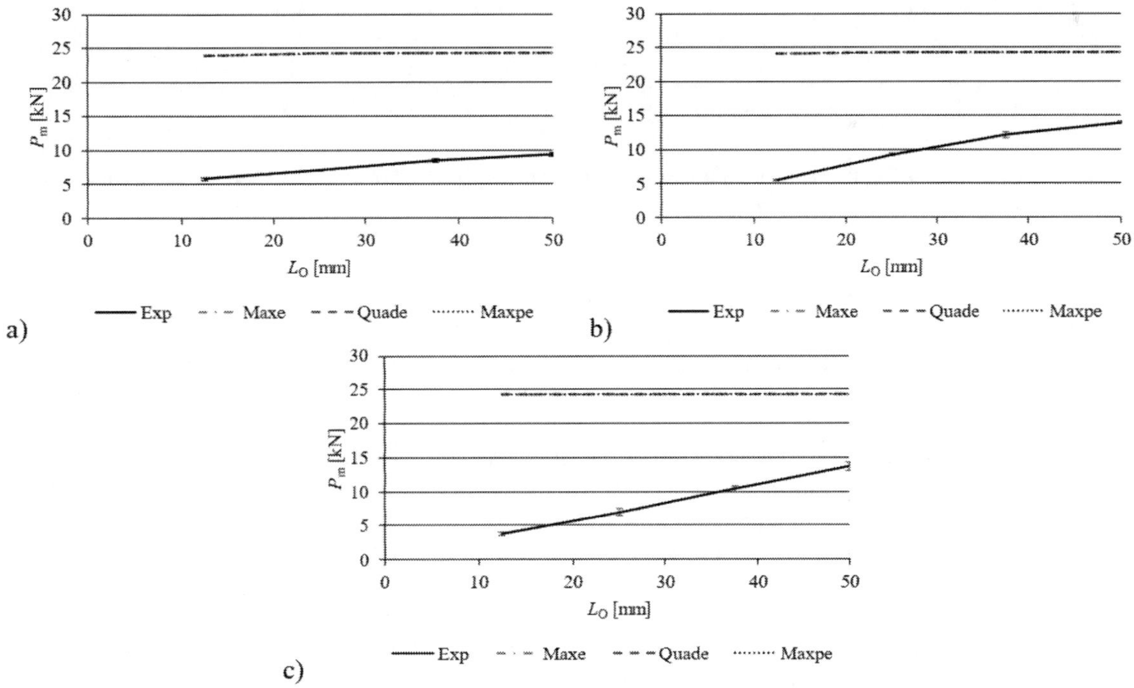

Figure 4 – Comparison between the experimental data and different damage initiation strain criteria for the joints bonded with the adhesive Araldite® AV138 (a), Araldite® 2015 (b) and Sikaforce® 7752 (c)

The evaluation of the strain-based initiation criteria was carried out by comparison between the experimental data, the MAXE, the QUADE and the MAXPE criteria. Figure 4 presents this evaluation for all adhesives. Regardless the adhesive used, strain-based initiation criteria are not suited to simulate damage initiation in the adhesive layer, since they all overshoot by a large amount the experimental results. In some joint configurations, damage initiated at a load close to the tensile net failure load of the adherends. In other configurations, P_m corresponded to the adherends' tensile net failure.

4.2.2. Damage growth criterion. This sub-Section addresses the accuracy of the power law and the BK growth criteria for the P_m prediction of the three adhesives. The power law evaluation was firstly performed by varying the power parameter α (equation (3)), considering values 0.5, 1, 1.5, and 2, for further comparison with the experimental data. The results are presented in Figure 5 for all adhesives.

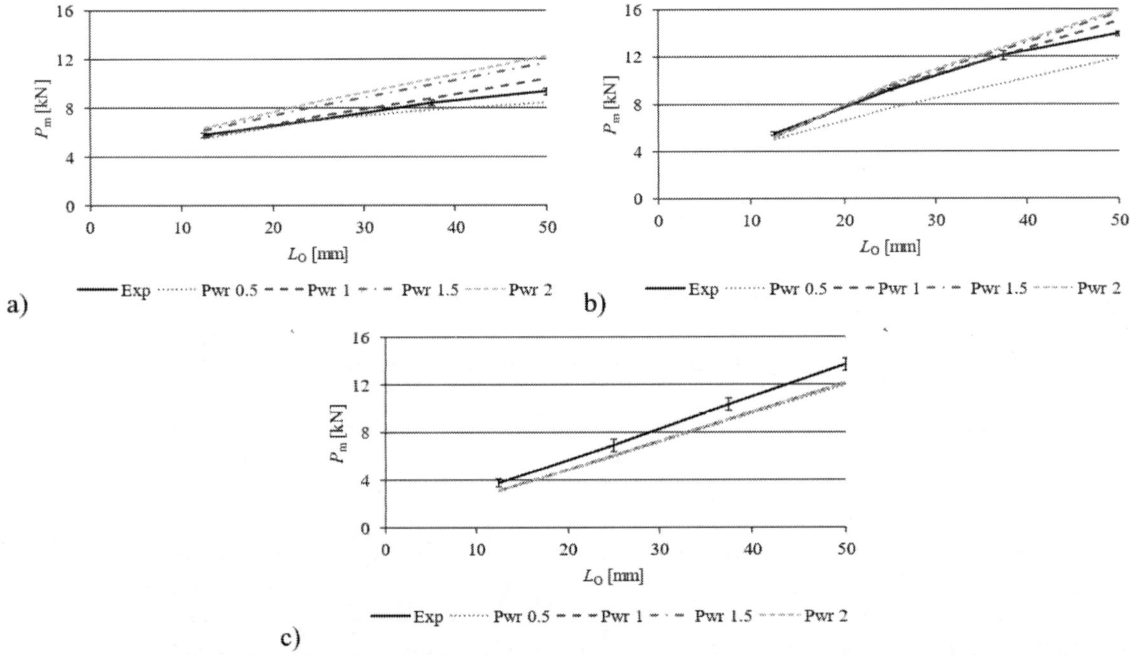

Figure 5 – Comparison between different parameter values of the power law grow criteria for P_m prediction for the joints bonded with the adhesive Araldite® AV138 (a), Araldite® 2015 (b) and Sikaforce® 7752 (c)

Figure 5 a, related to the adhesive Araldite® AV138, reports that the Power 0.5 suits best for this adhesive. Actually, an average deviation of 4.9% was achieved for all L_O values, followed by Power 1, with an average difference of 5.9%. Power 1.5 over predicts P_m by the average value of 16.2%, while Power 2 gives the worst P_m predictions. As L_O increases, deviations also grow, regardless the value of the power parameter. The comparison shown in Figure 5 b reveals Power 1 as the best for the joints bonded with the adhesive Araldite® 2015. Power 1.5 and Power 2 predicted P_m with an average deviation of 6.6 and 7.3%, respectively. The Power 0.5 criterion, which was the best choice for the strength prediction of the previous adhesive, presents now the worst P_m estimation. The increase of L_O promotes the rise of relative deviations, by comparison with the experimental data, equally to what was observed for the Araldite® AV138. The evaluation depicted in Figure 5 c, comparing different parameters of the power law criterion for the adhesive Sikaforce® 7752, provides identical P_m under estimation values, whatsoever were the chosen α or L_O values.

The evaluation of the BK growth criterion for P_m prediction was performed by comparison with the Power law criterion (α=1), and the experimental data. For the characteristic parameter η of equation (4) the values 0.5, 1, and 2.5 were used. It is shown that the condition that provides the most accurate P_m prediction for the adhesive Araldite® AV138 was BK 2.5 (Figure 6 a), with an average deviation of

IOP Conf. Series: Journal of Physics: Conf. Series **843** (2017) 012002 doi:10.1088/1742-6596/843/1/012002

4.1%, close to that obtained with Power 1 (4.3%). Nonetheless, a more detailed analysis reveals P_m under estimations for the lower L_O values: -7.5, and -1.3% for L_O=12.5, and 25 mm, and P_m over estimations for higher L_O values: 0.6, and 7.1% for L_O=37.5, and 50 mm, respectively. In Figure 6 b (Araldite® 2015), it is shown that the BK criterion under predicts P_m for the lowest L_O value, by an average error of 4.3% for the three values of η. Oppositely, for higher L_O, big deviations by excess were found: P_m was over predicted by values up to 48.3% (η=0.5), 47.2% (η=1), and 16.7% (η=2.5), attained with L_O=50 mm. In view of those deviations, the power law criteria gives the most accurate P_m predictions for this adhesive. The comparison depicted in Figure 6 c, regarding the evaluation of the damage growth criteria for the adhesive Sikaforce® 7752, shows that all criteria underestimate P_m, with very similar average deviations of -17.1, -12.9, -12.2, and -11.3% for L_O=12.5, 25, 37.5, and 50 mm, respectively.

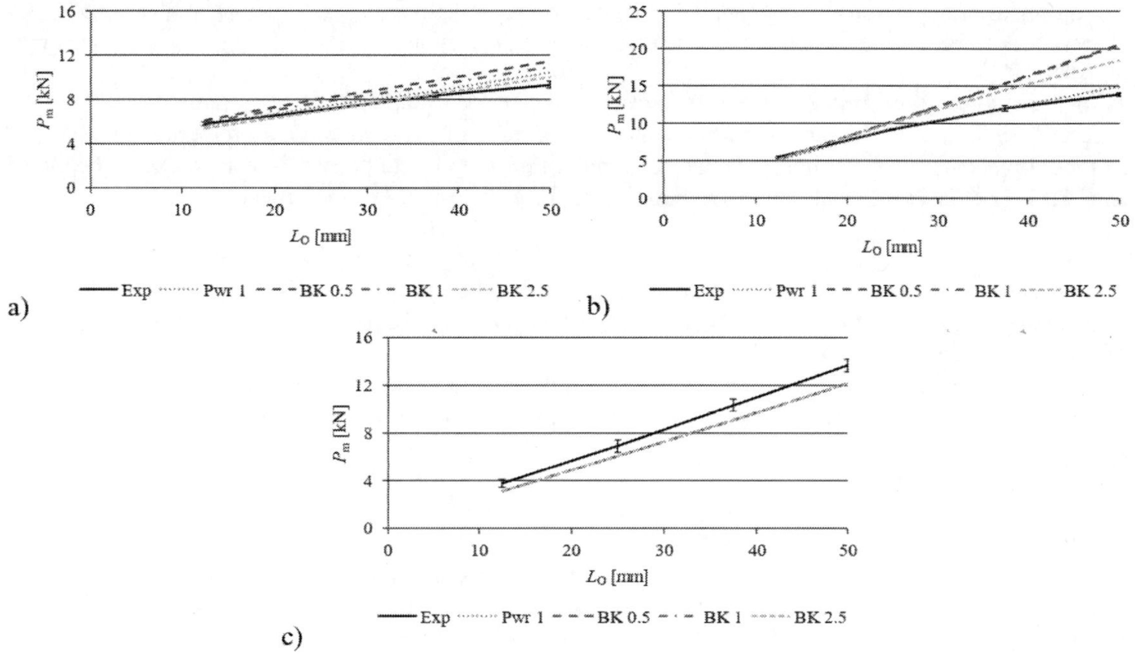

Figure 6 – Comparison between power law and BK growth criteria for P_m prediction for the joints bonded with the adhesive Araldite® AV138 (a), Araldite® 2015 (b) and Sikaforce® 7752 (c)

The evaluation of the BK growth criterion for P_m prediction was performed by comparison with the Power law criterion (α=1), and the experimental data. For the characteristic parameter η of equation (4) , the values 0.5, 1, and 2.5 were used. It is shown that the condition that provides the most accurate P_m prediction for the adhesive Araldite® AV138 was BK 2.5 (Figure 6 a), with an average deviation of 4.1%, close to that obtained with Power 1 (4.3%). In Figure 6 b (Araldite® 2015), it is shown that the BK criterion under predicts P_m for the lowest L_O value, by an average error of 4.3% for the three values of η. Oppositely, for higher L_O, big deviations by excess were found. The comparison depicted in Figure 6 c, regarding the evaluation of the damage growth criteria for the adhesive Sikaforce® 7752, shows that all criteria underestimate P_m, with very similar average deviations.

5. Conclusions

This work aimed at evaluating different simulation conditions in the CZM analysis for an accurate strength prediction of SLJ. Strength prediction by CZM was performed, using reliable CZM conditions for bonded joints (triangular mixed-mode CZM law, QUADS initiation criterion and linear CZM law). For the adhesives Araldite® AV138 and Araldite® 2015, the P_m prediction was accurate. For the adhesive Sikaforce® 7752, P_m was under predicted by 13.4% (average for all L_O), due to the large plasticity that this adhesive can endure, which is not correctly modelled by the triangular CZM. After this study, different modelling conditions were tested, whose main conclusions were as follows:

- **Damage initiation criterion**: The different strain-based initiation damage criteria (MAXE, QUADE and MAXPE) showed that, notwithstanding the type of adhesive, all criteria provided large errors. The stress-based MAXPS criterion also attained large P_m deviations. Oppositely, the QUADS and MAXS stress-based criteria presented better accuracy for the adhesives Araldite® AV138 and Araldite® 2015 with the QUADS criterion (average deviations of 5.9% for the Araldite® AV138 and 3.8% for the adhesive Araldite® 2015). The Sikaforce® 7752 results were not satisfactory, yet the MAXS criterion presented the closer predictions.
- **Damage growth criterion**: For the adhesive Araldite® AV38, P_m results showed that the best criterion was the BK with η=2.5 (average error of 4.1%) despite, for the limit L_O values, P_m had been wrongly estimated by ≈7.3%. P_m for the moderate ductile adhesive Araldite® 2015 was best predicted by a Power law criterion with α=1 (average error of 3.79%), yet it also presented an average P_m deviation of 6.8% for the two limit L_O values. Regarding the strength prediction of the ductile adhesive Sikaforce® 7752, none of the tested criteria was able to accurately evaluate it. P_m was under predicted by an average of ≈13% for all simulations performed.

Former studies addressed the effect of the CZM law shape [8] and cohesive parameters [13] on the strength prediction of SLJ which, together with the present analysis, constitute an integrated analysis of this modelling technique applied to bonded joints. Upon the conclusions of these works, it becomes easier to choose the best set of numerical conditions for an accurate P_m prediction.

References

[1] E.W. Petrie, Handbook of adhesives and sealants, 2nd ed ed., McGraw-Hill, New York, 1999.
[2] O. Volkersen, Die Niektraftverteilung in zugbeanspruchten mit konstanten laschenquerschritten, Luftfahrtforschung 15 (1938) 41-47.
[3] G.I. Barenblatt, The formation of equilibrium cracks during brittle fracture. General ideas and hypotheses. Axially-symmetric cracks, Journal of Applied Mathematics and Mechanics 23(3) (1959) 622-636.
[4] D.S. Dugdale, Yielding of steel sheets containing slits, J. Mech. Phys. Solids 8(2) (1960) 100-104.
[5] P. Feraren, H.M. Jensen, Cohesive zone modelling of interface fracture near flaws in adhesive joints, Eng. Fract. Mech. 71(15) (2004) 2125-2142.
[6] P.A. Gustafson, A.M. Waas, The influence of adhesive constitutive parameters in cohesive zone finite element models of adhesively bonded joints, Int. J. Solids. Struct. 46(10) (2009) 2201-2215.
[7] R.D.S.G. Campilho, M.D. Banea, A.M.G. Pinto, L.F.M. da Silva, A.M.P. de Jesus, Strength prediction of single- and double-lap joints by standard and extended finite element modelling, Int. J. Adhes. Adhes. 31(5) (2011) 363-372.
[8] R.D.S.G. Campilho, M.D. Banea, J.A.B.P. Neto, L.F.M. da Silva, Modelling adhesive joints with cohesive zone models: effect of the cohesive law shape of the adhesive layer, Int. J. Adhes. Adhes. 44 (2013) 48-56.
[9] T.M.S. Faneco, R.D.S.G. Campilho, F.J.G. da Silva, R.M. Lopes, Strength and fracture characterization of a novel polyurethane adhesive for the automotive industry, J. Test. Eval. (2016) DOI: 10.1520/JTE20150335.
[10] ABAQUS® Documentation, Dassault Systèmes, Vélizy-Villacoublay, 2009.
[11] J. Jing, F. Gao, J. Johnson, F.Z. Liang, R.L. Williams, J. Qu, Simulation of dynamic fracture along solder–pad interfaces using a cohesive zone model, Eng. Fail. Anal. 16(5) (2009) 1579-1586.
[12] M.L. Benzeggagh, M. Kenane, Measurement of mixed-mode delamination fracture toughness of unidirectional glass/epoxy composites with mixed-mode bending apparatus, Composites Science and Technology 56(4) (1996) 439-449.
[13] R.D.S.G. Campilho, M.D. Banea, J.A.B.P. Neto, L.F.M. da Silva, Modelling of single-lap joints using cohesive zone models: Effect of the cohesive parameters on the output of the simulations, J. Adhesion 88(4-6) (2012) 513-533.

eXtended Finite Element Method applied to the strength prediction of adhesively-bonded joints

J.T.S. Xará[a], R.D.S.G. Campilho[a,b,*]

[a] Departamento de Engenharia Mecânica, Instituto Superior de Engenharia do Porto, Instituto Politécnico do Porto, Rua Dr. António Bernardino de Almeida, 431, 4200-072 Porto, Portugal
[b] INEGI – Pólo FEUP, Rua Dr. Roberto Frias, s/n, 4200-465 Porto, Portugal
* raulcampilho@gmail.com

Abstract. High-strength composites are widely used in several industries, such as aeronautical, automotive and naval, and they can be combined with metals to provide significant advantages in structural design. The application of adhesive bonding to these assemblies supposes the existence of reliable design tools to accurately analyse the joints' behaviour. In this context, the eXtended Finite Element Method (XFEM) is a recent possibility to predict bonded' joints fracture behaviour. This work aims to study by XFEM single-*L* adhesive joints between aluminium components and carbon-epoxy composites under peeling loads, considering the variation of the *L*-shaped adherend's thickness (t_{P2}) and adhesives of distinct ductility. The XFEM analysis was either based on stress or strain criteria for assessment of damage initiation, and different damage law types for crack propagation. Validation was undertaken with experimental data. The XFEM analysis revealed that this method is very accurate when using the stress-based quadratic initiation criterion and the triangular propagation law. It was shown that the *L*-shaped adherend's geometry and the adhesive type are the most important parameters affecting the joints' strength.

1. Introduction

Structural bonding using adhesives has been increasing in recent years because of the advantages that this joining method presents over more traditional joining methods such as fastening, welding and riveting. The aeronautical industry is the one that most contributed to the use and development of adhesives. Structural components can be bonded using various types of adhesive joints: single and double-lap joints, scarf joints or single-*L* joints. The single-lap joint is the most common and studied due to its easy fabrication and because the adhesive is mainly loaded with shear (τ_{xy}) stresses [1]. However, this type of joint performs poorly in comparison with other joint configurations due to the eccentric transmission of load resulting in high normal (σ_y) peak stresses at the overlap edges. The double-lap joint solves this problem due to its symmetric configuration [2, 3]. Thus, it is possible to obtain a joint more than twice as strong as a single-lap joint for the same overlap length (L_O) [4]. The use of scarf joints promotes the reduction of peak stresses in the adhesive layer, which results in a higher strength per bonded area. Single-*L* joints have some similarities with *T*-joints but, instead of having only two *L*-shaped adherends facing each other and bonded to a flat adherend, only one *L*-part is considered. Li et al. [5] studied the stress distributions in the adhesive layer of a single-*L* joint using a Finite Element Method (FEM) analysis. The joint was analysed under three loading conditions, two linear loads and one bending moment (M). The results allowed to conclude that the loading in the y-direction (P_y) is the most harmful to the adhesive layer by resulting in higher peak stresses in comparison with the loading in the *x*-direction (P_x) and *M*. It was also shown in a parametric study on L_O that, as this parameter increased, the peak peel stresses gradually diminished.

Failure prediction in adhesive joints can be undertaken using two alternatives: analytical and numerical methods. Analytical methods consist of closed-form analyses. Numerical methods, such as the FEM, are more suitable to model joints with complex geometries and elaborate material models. Continuum mechanics modelling by FEM is based on a stress or strain limit state and give good results provided that the failure criteria are properly selected. Damage mechanics modelling, based either on stress intensity factors or energetic approaches, induce damage in the models by the onset and propagation of cracks. This method allows the possibility to simulate fracture and step-by-step damage at a pre-defined crack path or arbitrarily with a finite region up to complete structural failure. Cohesive Zone Modelling (CZM) is a damage mechanics-based technique that can be incorporated in FEM analyses to promote static or fatigue damage in structures. The XFEM is an improvement of the FEM standard modelling procedure to induce damage growth. XFEM damage modelling excels CZM by enabling cracks to grow freely through continuum-based finite elements without the requirement of introducing special purpose elements. Campilho et al. [6] evaluated the CZM and XFEM techniques for the strength prediction of single and double-lap bonded joints with aluminium adherends and a brittle adhesive (Araldite® AV138). The CZM results were accurate but damage growth in the adhesive could not be simulated by the XFEM because of the mode-mixity. Mubashar et al. [7] studied adhesive joints by combining CZM and XFEM. The XFEM was used for crack modelling in the adhesive fillets and bond length, and the CZM was used for the interfaces between the fillet/adhesive bond and the adherends. It was concluded that the XFEM could predict the crack trajectory in the adhesive region of an adhesive joint. This hybrid model proved to be efficient in predicting the crack initiation and growth and also in estimating the strength of adhesive joints. Curiel Sosa and Karapurath [8] studied the application of the XFEM in modelling delamination damage in Fibre Metal Laminates (FML) under mode I loading by the Double-Cantilever Beam (DCB) configuration. In order to validate the XFEM analysis, the load-displacement (P-δ) curves were compared with the experimental results and with a CZM simulation from another work (Airoldi et al. [9]), and a good match was found.

This work aims to study by XFEM single-*L* adhesive joints between aluminium components and carbon-epoxy composites under peeling loads, considering the variation of t_{P2} and adhesives of distinct ductility. The XFEM analysis was either based on stress or strain criteria for assessment of damage initiation, and different damage law types for crack propagation. Validation was undertaken with experimental data and also CZM results from a previous work [10].

2. Experimental work

The experimental work considered a carbon-fibre reinforced plastic (CFRP) pre-preg (SEAL® Texipreg HS 160 RM) with a ply thickness of 0.15 mm for the flat adherends of the single-*L* joint, to produce a unidirectional laminate ($[0]_{20}$). The fabrication of the composite adherends consisted of manually stacking 20 layers of CFRP and curing for 1 hour at 130 °C and pressure of 2 bar in a hot plates press. Table 1 presents the elastic orthotropic properties of a unidirectional carbon-epoxy ply aligned in the fibres direction to be used in the numerical simulations [10].

Table 1 – Elastic orthotropic properties of a unidirectional carbon-epoxy ply aligned in the fibres direction (*x*-direction; *y* and *z* are the transverse and through-thickness directions, respectively) [10].

E_x=1.09E+05 MPa	ν_{xy}=0.342	G_{xy}=4315 MPa
E_y=8819 MPa	ν_{xz}=0.342	G_{xz}=4315 MPa
E_z=8819 MPa	ν_{yz}=0.380	G_{yz}=3200 MPa

The *L*-shaped adherends are made of a high-strength aluminium alloy sheet (AA6082 T651), whose relevant mechanical properties are as follows: Young's modulus (*E*) of 70.1±0.8 GPa, tensile yield stress

(σ_e) of 261.7±7.7 MPa, tensile failure strength (σ_f) of 324.0±0.2 MPa and tensile failure strain (ε_f) of 21.7±4.2%.

The adhesives considered in the experimental work are the following: the Araldite® AV138, a brittle epoxy, and the SikaForce® 7752, a highly ductile polyurethane. The relevant properties to be obtained for the adhesives are E, the shear modulus (G), the strengths in tension (t_n^0) and shear (t_s^0), G_{IC} and the shear fracture toughness (G_{IIC}). Table 2 presents all data obtained for the two adhesives used in this work.

Table 2 – Properties of the adhesives Araldite® AV138 and Sikaforce® 7752 [6, 11, 12].

Property	**AV138**	**7752**
Young's modulus, E [GPa]	4.89±0.81	0.49±0.09
Poisson's ratio, ν	0.35 *	0.30 [a]
Tensile yield strength, σ_y [MPa]	36.49±2.47	3.24±0.48
Tensile failure strength, σ_f [MPa]	39.45±3.18	11.48±0.25
Tensile failure strain, ε_f [%]	1.21±0.10	19.18±1.40
Shear modulus, G [GPa]	1.56±0.01	0.19±0.01
Shear yield strength, τ_y [MPa]	25.1±0.33	5.16±1.14
Shear failure strength, τ_f [MPa]	30.2±0.40	10.17±0.64
Shear failure strain, γ_f [%]	7.8±0.7	54.82±6.38
Toughness in tension, G_{IC} [N/mm]	0.20 [a]	2.36±0.17
Toughness in shear, G_{IIC} [N/mm]	0.38 [a]	5.41±0.47

* manufacturer's data

[a] estimated in reference [6], [b] estimated from Hooke's law

The geometry and respective dimensions of the single-L joints are represented in Fig. 1. The dimensions considered are (in mm): L_O=25, out-of-plane width of the joint b=25, specimen length L_T=80, flat adherend thickness t_{P1}=3, t_{P2}=1, 2, 3 and 4, curved element free length L_A=60, curved element radius R=5 and t_A=0.2.

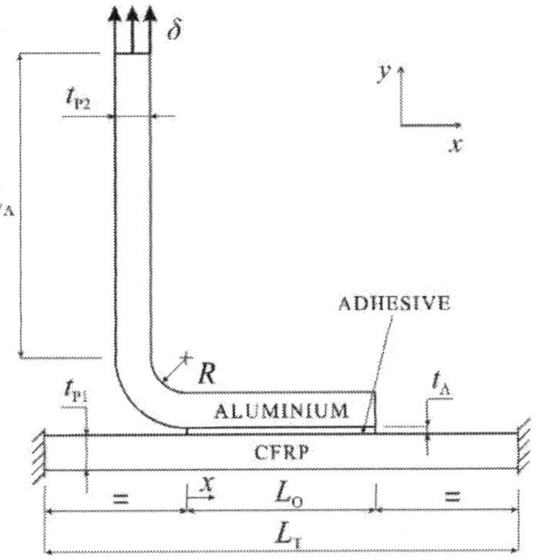

Fig. 1 – Geometry and dimensions of the single-L joint.

The fabrication of the single-L joints consisted of several steps. First, CFRP plates were manufactured using an unidirectional CFRP pre-preg in order to produce the base adherends of the joint. The fabrication of the composite plates was performed by manual stacking followed by curing in a hot plates press under specified pressure and heat conditions. Bending of the aluminium adherends was executed using a manual bending machine after cutting to the final dimensions from a laminated aluminium plate. Surface preparation of both aluminium and CFRP adherends consisted of manual abrasion with fine mesh sandpaper and cleaning with acetone. After, the adherends were bonded using a steel mould in order to guarantee a correct alignment and using calibrated spacers to ensure the stipulated t_A value. The adhesive layer curing lasted for one week and the removal of the excess adhesive at the specimens' side was performed by precision milling. The tensile tests were executed in an Instron® 3367, which is an electro-mechanical testing machine, with a 30 kN load cell. These tests were carried out at room temperature and constant velocity of 0.5 mm/min. Five specimens were tested for each joint configuration.

3. Numerical work

The geometrically non-linear and two-dimensional XFEM analysis was undertaken in the FEM software Abaqus®. Both the adherends and adhesive were regarded as elasto-plastic with solid FEM elements (plane-strain elements with the Abaqus® reference CPE4) Apart from this, an enriched XFEM formulation was considered for the adhesive layer's solid elements. Fig. 2 gives an example of mesh refinement for the XFEM simulations.

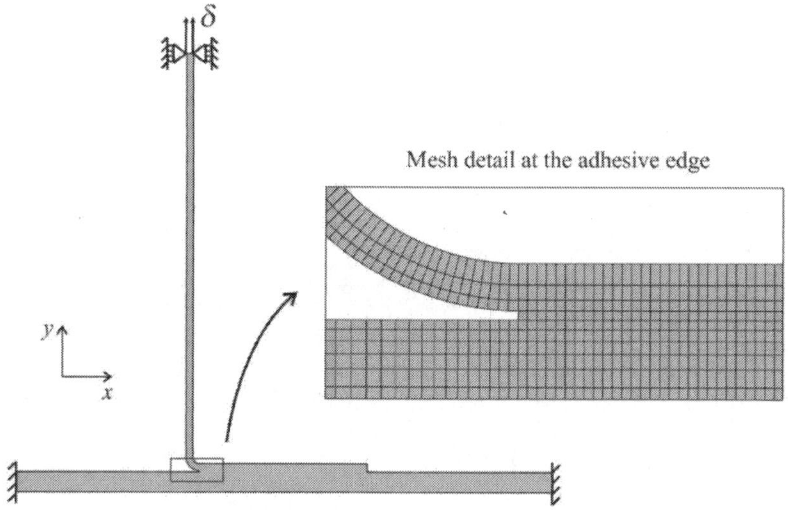

Fig. 2 – Boundary conditions and mesh detail at the adhesive edge (example for joint with t_{P2}=1 mm).

Table 3 – Parameters of the Araldite® AV138 and SikaForce® 7752 for XFEM modelling.

Property	AV138	7752
E [GPa]	4.89	0.49
G [GPa]	1.56	0.19
σ_{max}^0 [MPa]	39.45	11.48
ε_{max}^0 [%]	1.21	19.18
t_n^0 [MPa]	39.45	11.48
t_s^0 [MPa]	30.2	10.17
ε_n^0 [%]	1.21	19.18
ε_s^0 [%]	7.8	54.82

All meshes were constructed taking advantage of bias effects to increase the computational efficiency. The finite element's size was graded across L_O to provide a higher refinement near the overlap edges, vertically in the adherends towards the adhesive layer, and also in the unbonded portion of the adherends in the direction of the bond. The quadrilateral elements' size in the adhesive layer was defined as 0.2×0.2 mm^2. To be consistent with the experimental tests, the boundary conditions consisted of clamping the base CFRP adherend at both sides, and horizontal restraining and tensile loading the upper edge of the *L*-part (Fig. 2). Table 3 summarizes the parameters introduced in Abaqus® for XFEM modelling, obtained from the data of Table 2. For a full description of the XFEM formulation, the reader can refer to the Abaqus® documentation [13].

4. Results

4.1. Experimental strength

Fig. 3 shows the experimental maximum load (P_m) values for the joints with the Araldite® AV138 and Sikaforce® 7752 as a function of t_{P2}.

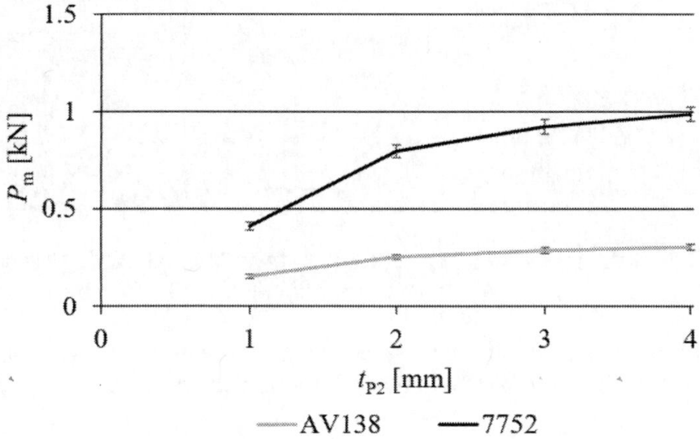

Fig. 3 – Experimental P_m values for the joints with the Araldite® AV138 and Sikaforce® 7752.

The joints with the Araldite® AV138 revealed a major P_m increase between t_{P2}=1 and 2 mm (64.5%), but this improvement gradually diminishes for higher t_{P2} (compared to t_{P2}=1 mm, improvement of 86.3% for t_{P2}=3 mm and 97.5% for t_{P2}=4 mm). The P_m improvement was found to be related to the increased stiffness of the *L*-part, which in turn reduced σ_y peak peel stresses and enabled spreading σ_y stresses across a larger length in the adhesive. Compared to these joints, for the joints with the Sikaforce® 7752 P_m is much higher, although this adhesive is less strong (Table 2). It was found that, depending on t_{P2}, P_m for this adhesive can exceed 3 times P_m of the Araldite® AV138, even though the Sikaforce® 7752 has practically ¼ of the tensile strength. However, the Sikaforce® 7752 is much less stiff (Table 2), leading to flatter σ_y stress plots in the vicinity of the damage initiation site (pull-out edge of the adhesive layer). For this adhesive, the P_m increase from the joint configuration with t_{P2}=1 mm is 93.5% (t_{P2}=2 mm), 124.3% (t_{P2}=3 mm) and 140.1% (t_{P2}=4 mm). Compared to the Araldite® AV138, the evolution of P_m with t_{P2} is higher, because of the increased capacity of the adhesive to plasticize further within the adhesive layer.

4.2. Strength prediction

4.2.1. Study of the damage initiation criterion. A triangular growth law was used in all simulations of this Section. The linear energetic propagation criterion was used. Fig. 4 gives two examples of crack growth in the adhesive layer for the MAXS (maximum nominal stress), MAXE (maximum nominal strain), QUADS (quadratic nominal stress) and QUADE (quadratic nominal strain) criteria (a) and MAXPS (maximum principal stress) and MAXPE (maximum principal strain) criteria (b), in both cases considering a joint with t_{P2}=1 mm.

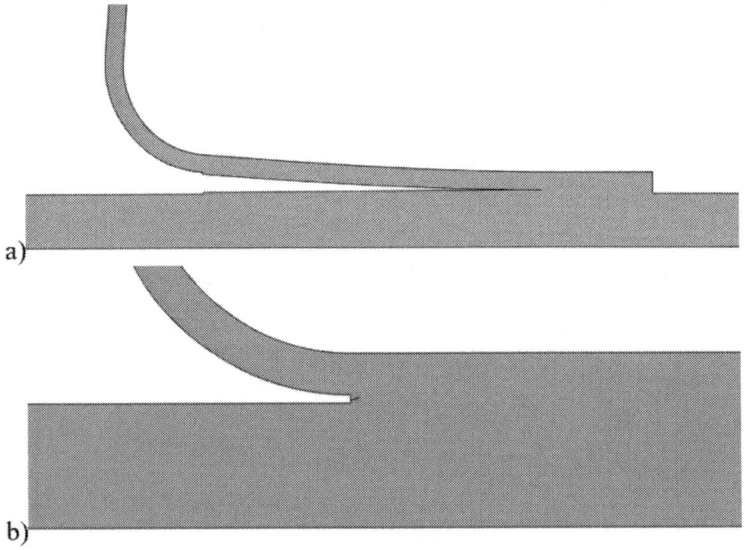

Fig. 4 – Example of crack growth in the adhesive layer for the MAXS, MAXE, QUADS and QUADE criteria (a) and MAXPS and MAXPE criteria (b).

Fig. 5 shows a comparison between the experimentally obtained P_m values and the respective numerical predictions of the different XFEM damage initiation criteria, for the two tested adhesives and as a function of t_{P2}.

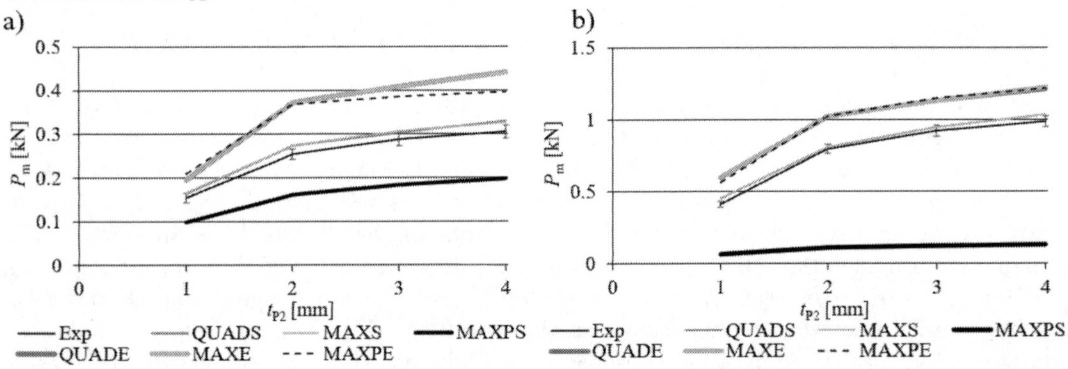

Fig. 5 – Experimental and numerical P_m values with different damage initiation criteria for the joints with the Araldite® AV138 (a) and Sikaforce® 7752 (b).

For the Araldite® AV138 (Fig. 5 a), by comparison with the experiments, the QUADS and MAXS criteria clearly provide the best P_m predictions. Nonetheless, for most t_{P2} values the numerical P_m points are slightly above the experimental range defined by the respective scatter. Compared to the experimentally obtained values, for the QUADS criterion and the four values of t_{P2} (1, 2, 3 and 4 mm), the percentile variations are 6.9%, 7.7%, 5.4% and 8.2%, respectively. Considering the MAXS criterion, the percentile variations are 7.5%, 7.8%, 6.4% and 7.2%, by the same order. The other stress-based criterion, MAXPS, has proved not to work well in the technique that it was applied, giving a maximum percentile variation to the experiments of -59.9%. The strain-based criteria (QUADE, MAXE and MAXPE) all overestimated the values of P_m. The maximum percentile deviations were 46.8%, 46.8% and 45.3% for the QUADE, MAXE and MAXPE criteria, respectively (in all cases for t_{P2}=2 mm). The accuracy of the predictions by applying the same six damage initiation criteria for the joints bonded with the SikaForce® 7752 (Fig. 5 b) is qualitatively identical to those obtained with the Araldite® AV138, although P_m have a different magnitude. The QUADS and MAXS damage initiation criteria were again the most accurate and, for this adhesive, the numerical values were mostly within the experimental range. In the case of the QUADS criterion, the percentile variation to the average experimental values was 9.7%, 1.6%, 2.7% and 4.8% for increasing t_{P2} between 1 and 4 mm. For the MAXS criterion, the same deviation was 10.1%, 1.6%, 2.7% and 4.8%, for t_{P2}=1, 2, 3 and 4 mm, respectively. These two criteria behave similarly, resulting in similar % differences. The MAXPS results revealed the highest offset to the experimental data, presenting a maximum percentile deviation of -86.8%. The predictions with the three initiation criteria based on strains (QUADE, MAXE and MAXPE) presented relatively approximate values among them but significantly offset to the experimental P_m values. Based on the experimental results, the maximum percentile deviations were 44.6%, 45.9 and 36.6% for the QUADE, MAXE and MAXPE criteria, respectively (in all cases for t_{P2}=1 mm).

4.2.2. Study of the law shape and damage growth criterion. As a complement of the former analysis on the damage initiation criterion, a comparison was also carried out regarding the softening law shape during damage growth. The triangular law results obtained in the previous analysis are thus compared with numerical predictions using an exponential law and power law exponents of 0.5, 1 and 2. All numerical simulations considered the QUADS damage initiation criterion. Fig. 6 compares P_m regarding these conditions and the experimental data for the joints bonded with the adhesives Araldite® AV138 (a) and SikaForce® 7752 (b).

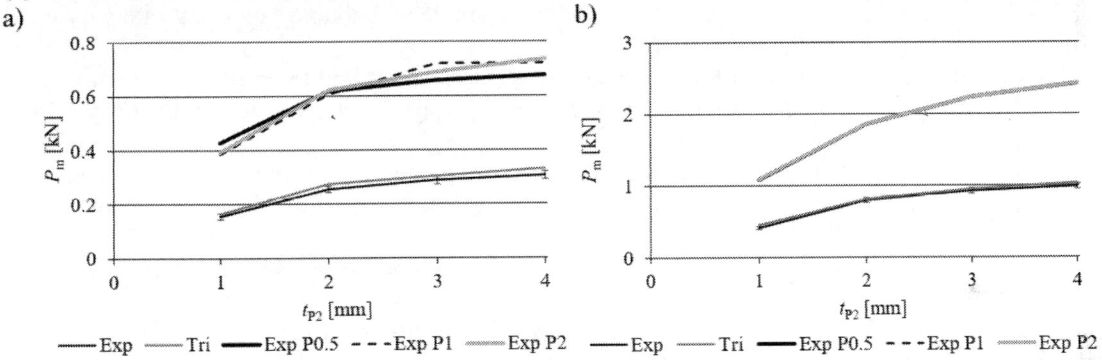

Fig. 6 – Experimental and numerical P_m values with different law shapes and damage growth criteria for the joints with the Araldite® AV138 (a) and Sikaforce® 7752 (b).

For the joints bonded with the Araldite® AV138, the exponential propagation law greatly overestimates P_m, for the three tested values of α. A maximum percentile different to the average experimental values of 160.9%, 137.6% and 139.4% was attained for α=0.5, 1 and 2, respectively. Identically, for the joints bonded with the Sikaforce® 7752, the exponential propagation law resulted in a large deviation of P_m compared to the experiments. A maximum percentile variation between the simulations with an exponential law and the experimental tests of 139.5%, 139.5% and 141.9% was found for α=0.5, 1 and 2, respectively. For both adhesives, this large difference is justified by the higher failure displacements resulting from the exponential damage laws, which artificially enlarge the damage length in the adhesive.

5. Conclusions

This work aimed at studying the behaviour of hybrid single-*L* bonded joints between aluminium and CFRP adherends, considering different geometric conditions (t_{P2} values) and adhesives of different strength and ductility. The experimental analysis showed that the Sikaforce® 7752 adhesive gives the best results in this particular joint configuration for all t_{P2} values tested, showing that a less strong but ductile adhesive behaves better in this joint configuration than a stronger but more brittle adhesive. The percentile P_m improvement for the joints bonded with the Sikaforce® 7752 over those bonded with the Araldite® AV138 was 166.3%, 213.4%, 220.7% and 223.9% for increasing t_{P2} between 1 and 4 mm.

The geometric parameter t_{P2} largely influences the joints' strength, with advantage of higher values. A detailed study was then undertaken to validate the XFEM technique for strength prediction of these joints. The study of the effect of the damage initiation criterion showed that the QUADS and MAXS criteria give the most accurate results. Considering the joints bonded with the Araldite® AV138, for the QUADS criterion the maximum error compared to the average experimental values was 8.2% (t_{P2}=4 mm),while for the MAXS criterion it was 7.8% (t_{P2}=2 mm). The maximum errors for the joints bonded with the Sikaforce® 7752 were 9.7% (t_{P2}=1 mm) for the QUADS criterion and 10.1% (t_{P2}=1 mm) for the MAXS criterion. The strain-based criteria (QUADE, MAXE and MAXPE) largely overestimated P_m for both adhesives. For both adhesives, the MAXPS criterion proved to be the most misfit criterion. The study of the propagation law showed that the triangular law is much more accurate than the exponential law, which is not considered suitable to model the behaviour of adhesive joints. It is thus concluded that the XFEM, using the aforementioned criterion and growth law, is a very precise tool for the strength prediction of single-*L* joints bonded with adhesives ranging from brittle to ductile.

References

[1] R.D.F. Moreira, R.D.S.G. Campilho, Parametric Study of the Reinforcement Geometry on Tensile Loaded Scarf Adhesive Repairs, J. Adhesion 92(7-9) (2016) 586-609.
[2] S.L.S. Nunes, R.D.S.G. Campilho, F.J.G. da Silva, C.C.R.G. de Sousa, T.A.B. Fernandes, M.D. Banea, L.F.M. da Silva, Comparative failure assessment of single and double-lap joints with varying adhesive systems, J. Adhesion 92 (2016) 610-634.
[3] U.T.F. Carvalho, R.D.S.G. Campilho, Application of the direct method for cohesive law estimation applied to the strength prediction of double-lap joints, Theor. Appl. Fract. Mec. 85, Part A (2016) 140-148.
[4] K.C. Shin, J.J. Lee, Bond Parameters to Improve Tensile Load Bearing Capacities of Co-Cured Single and Double Lap Joints with Steel and Carbon Fiber-epoxy Composite Adherends, J. Compos. Mater. 37(5) (2003) 401-420.
[5] W. Li, L. Blunt, K.J. Stout, Analysis and design of adhesive-bonded tee joints, Int. J. Adhes. Adhes. 17(4) (1997) 303-311.
[6] R.D.S.G. Campilho, M.D. Banea, A.M.G. Pinto, L.F.M. da Silva, A.M.P. de Jesus, Strength prediction of single- and double-lap joints by standard and extended finite element modelling, Int. J. Adhes. Adhes. 31(5) (2011) 363-372.
[7] A. Mubashar, I.A. Ashcroft, A.D. Crocombe, Modelling damage and failure in adhesive joints using a combined XFEM-cohesive element methodology, J. Adhesion 90(8) (2014) 682-697.
[8] J.L. Curiel Sosa, N. Karapurath, Delamination modelling of GLARE using the extended finite element method, Composites Science and Technology 72(7) (2012) 788-791.
[9] A. Airoldi, M. Vesco, S. van der Zwaag, A. Baldi, G. Sala, Damage in GLARE laminates under indentation loads: experimental and numerical results, Proceedings of the 17th international conference on composite materials, Edinburgh, UK, 2009.
[10] N.R.E. Domingues, R.D.S.G. Campilho, R.J.C. Carbas, L.F.M. da Silva, Experimental and numerical failure analysis of aluminium/composite single-L joints, Int. J. Adhes. Adhes. 64 (2016) 86-96.
[11] R.D.S.G. Campilho, M.D. Banea, J.A.B.P. Neto, L.F.M. da Silva, Modelling adhesive joints with cohesive zone models: effect of the cohesive law shape of the adhesive layer, Int. J. Adhes. Adhes. 44 (2013) 48-56.
[12] T. Faneco, R. Campilho, F. Silva, R. Lopes, Strength and Fracture Characterization of a Novel Polyurethane Adhesive for the Automotive Industry, J. Test. Eval. 45(2) (2017) n.a.
[13] Abaqus®, Documentation, in: D. Systèmes (Ed.) Vélizy-Villacoublay, 2013.

IOP Conf. Series: Journal of Physics: Conf. Series **843** (2017) 012004 doi:10.1088/1742-6596/843/1/012004

Theoretical prediction of energy release rate for interface crack initiation by thermal stress in environmental barrier coatings for ceramics

E Kawai[1], Y Umeno[1]

[1] Institute of Industrial Science, The University of Tokyo, 4-6-1, Komaba Meguro-ku, Tokyo 153-8505, Japan

E-mail: kawai@ulab.iis.u-tokyo.ac.jp

Abstract. As weight reduction of turbines for aircraft engines is demanded to improve fuel consumption and curb emission of carbon dioxide, silicon carbide (SiC) fiber reinforced SiC matrix composites (SiC/SiC) are drawing enormous attention as high-pressure turbine materials. For preventing degradation of SiC/SiC, environmental barrier coatings (EBC) for ceramics are deposited on the composites. The purpose of this study is to establish theoretical guidelines for structural design which ensures the mechanical reliability of EBC. We conducted finite element method (FEM) analysis to calculate energy release rates (ERRs) for interface crack initiation due to thermal stress in EBC consisting of Si-based bond coat, Mullite and Ytterbium (Yb)-silicate layers on a SiC/SiC substrate. In the FEM analysis, the thickness of one EBC layer was changed from 25 μm to 200 μm while the thicknesses of the other layers were fixed at 25 μm, 50 μm and 100 μm. We compared ERRs obtained by the FEM analysis and a simple theory for interface crack in a single-layered structure where ERR is estimated as nominal strain energy in the coating layers multiplied by a constant factor (independent of layer thicknesses). We found that, unlike the case of single-layered structures, the multiplication factor is no longer a constant but is determined by the combination of consisting coating layer thicknesses.

1. Introduction

To improve fuel consumption and curb emission of carbon dioxide, the weight reduction of turbines for aircraft is demanded [1-3]. Silicon carbide (SiC) fiber reinforced SiC matrix composites (SiC/SiC) are attractive materials as high-pressure turbine materials because this composites have the superior heat resistance and the density of it is approximately one-third of Nickel-based super alloys which are used for turbine materials [4, 5].

At high temperature, SiC/SiC composites react with oxygen to form an oxide layer, which hinders degradation of SiC/SiC [6, 7]. However, under the high temperature and water vapor environments where high-pressure turbines are used, the oxide layer reacts with water vapor to form a volatile hydroxide layer [6, 7]. This leads to wall thinning or disappearance of the oxide layer and degradation of SiC/SiC. In order to maintain the property of SiC/SiC, development of environmental barrier coatings (EBC) for SiC/SiC is essential. EBC is fabricated by depositing several layers on a SiC/SiC substrate at high temperature and cooling it to room temperature. During the fabrication process, delamination cracks are initiated due to the thermal stress which occurs by the difference in coefficients of thermal expansion (CTEs) of the coating layers and the substrate. For sophisticated design to prevent the interface crack initiation, it is necessary to theoretically identify critical

IOP Conf. Series: Journal of Physics: Conf. Series **843** (2017) 012004 doi:10.1088/1742-6596/843/1/012004

thicknesses of the coating layers. The aim of this study is to establish theoretical guidelines for the condition of the coating layer thicknesses that ensures the mechanical reliability of EBC.

2. Theoretical framework of energy release rate for interface crack initiation in multi-layered structure

Assuming a biaxial stress state, the energy release rate (ERR) for interface crack initiation in single-layered isotropic elastic structure is evaluated by the following equation [8],

$$G = Z\frac{\sigma^2 h_{\mathrm{film}}}{\overline{E}_{\mathrm{film}}},\ \overline{E}_{\mathrm{film}} = \frac{E_{\mathrm{film}}}{1-\nu_{\mathrm{film}}^2}, \tag{1}$$

where G is ERR, Z is a dimensionless constant depending on the cracking pattern and elastic mismatch (Z = 1.028 for crack initiation at interface with vanishing elastic mismatch [8]). E_{film}, ν_{film} and h_{film} are Young's modulus, Poisson's ratio and thickness of the film, respectively. The thermal stress, σ, which occurs due to cooling from high temperature (T_{h}) to room temperature (T_{r}), is written as

$$\sigma = (\alpha_{\mathrm{film}} - \alpha_{\mathrm{sub}})(T_{\mathrm{h}} - T_{\mathrm{r}})\frac{E_{\mathrm{film}}}{1-\nu_{\mathrm{film}}}. \tag{2}$$

Note that α_{film} and α_{sub} are CTEs of the film and substrate, respectively.

To predict ERR for interface crack initiation in multi-layered structure, we regard the coating layers above the objective interface as one film and the other layers below the interface as a substrate. Then, ERR is given by

$$G = Z\sum_{i=1}^{n}\frac{\sigma_i^2 h_i}{\overline{E}_i},\ \overline{E}_i = \frac{E_i}{1-\nu_i^2}. \tag{3}$$

Here, i = 1,···, n indicate the layers of the objective structure. Note that Z in eq. (3) is no longer a constant determined by elastic properties of the components but depends on thicknesses of the coating layers and substrate because ERR should be governed by the 'effective' mismatch between the layers over and below the interface in question. To examine the significance of the dependence of Z on layer thicknesses, in Sec.3 we calculate ERR using numerical calculations to be compared with G_{T}, which is calculated assuming Z is a constant ($Z = Z_{\mathrm{T}} \equiv 1.028$).

3. Analytical procedure

The thermal stress finite element method (FEM: ABAQUS, ver6.14.5) analysis was conducted to calculate ERR (G_{F}) for each interface crack initiation in multi-layered structure and consider the dependence of Z on the layer thicknesses by comparing G_{F} and ERR obtained by eq. (3) with $Z = Z_{\mathrm{T}} \equiv$ 1.028 (G_{T}). The temperature difference upon cooling ($\Delta T = T_{\mathrm{h}} - T_{\mathrm{r}}$) was 1375 K. Figure 1 shows the FEM analysis model and boundary conditions. The axisymmetric model is adopted for EBC consisting of Yb-silicate (YbS), Mullite (Mu) and Si-based bond coat (BC) layers on a SiC/SiC substrate. All material were treated as isotropic elastic materials. In this paper, we refrain from disclosing the material properties. The region near the interfaces (YbS/Mu, Mu/BC and BC/(SiC/SiC)), where the stress concentration is expected, was divided into a finer mesh. In this analysis, the thickness of one layer in EBC was changed from 25 μm to 200 μm while the thicknesses of the other layers were fixed at 25 μm (Type A), 50 μm (Type B) and 100 μm (Type C). The displacement constraint in the y direction was imposed on the bottom end of the model. It was assumed that the interface crack (length: 900 μm) was initiated from the right side of the model.

G_{F} was obtained by

$$G_F = -\frac{\Pi_{\mathrm{C}} - \Pi_{\mathrm{NC}}}{\Delta A}, \tag{4}$$

where Π_{C} and Π_{NC} are nominal strain energies of the FEM models with and without crack, respectively, and ΔA is crack propagation area.

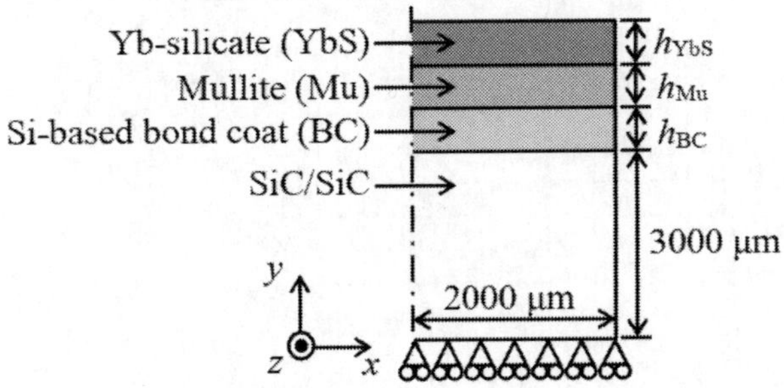

Fig. 1 Schematic illustration of FEM analysis model and boundary conditions.

4. Results and discussion

4.1. Theoretical prediction of ERRs for interface crack initiation in EBC

4.1.1. Si-based bond coat/(SiC/SiC) substrate interface. Figure 2 (a) shows the ratio of factor Z_T/Z determined by G_T/G_F for BC/(SiC/SiC) interface - thickness of the layers (h_i) relationship of Type A. As shown in Fig. 2 (a), Z_T/Z was changed by increasing the thickness of the layers. The relationships between Z_T/Z and h_i are given by

$$\frac{Z_T}{Z} = -1.00\times10^{-4} h_{YbS}^2 + 0.0327 h_{YbS} + 5.72, \tag{5}$$

$$\frac{Z_T}{Z} = -5.00\times10^{-4} h_{Mu}^2 + 0.135 h_{Mu} + 3.41, \tag{6}$$

$$\frac{Z_T}{Z} = 2.00\times10^{-4} h_{BC}^2 - 0.0506 h_{BC} + 6.44. \tag{7}$$

Here, h_{YbS}, h_{Mu} and h_{BC} are thicknesses of YbS, Mu and BC layers, respectively.

Figures 2 (b) and (c) show the Z_T/Z-h_i relationships of Type B and Type C, respectively. In both cases, Z_T/Z increases with the increase of h_{YbS} and h_{Mu} while it decreases with increasing h_{BC}. The relationships between Z_T/Z and h_i of these types are written as follows:

Type B

$$\frac{Z_T}{Z} = -3.00\times10^{-4} h_{YbS}^2 + 0.111 h_{YbS} + 1.63, \tag{8}$$

$$\frac{Z_T}{Z} = -2.00\times10^{-4} h_{Mu}^2 + 0.0940 h_{Mu} + 2.30, \tag{9}$$

$$\frac{Z_T}{Z} = 3.00\times10^{-4} h_{BC}^2 - 0.0784 h_{BC} + 9.63, \quad 50 \le h_{BC} \le 200 \tag{10}$$

Type C

$$\frac{Z_T}{Z} = -4.00\times10^{-5} h_{YbS}^2 + 0.0601 h_{YbS} + 1.57, \tag{11}$$

$$\frac{Z_T}{Z} = -8.00\times10^{-5} h_{Mu}^2 + 0.0449 h_{Mu} + 3.42, \tag{12}$$

$$\frac{Z_T}{Z} = 2.00\times10^{-3} h_{BC}^2 - 0.628 h_{BC} + 52.8, \quad 50 \le h_{BC} \le 200 \tag{13}$$

Clearly, Z_T/Z-h_i relationship is determined by the combination of the film thicknesses, meaning that the theoretical prediction of ERR for BC/(SiC/SiC) interface crack initiation in EBC is described as

$$G_{BC/(SiC/SiC)} = Z_{BC/(SiC/SiC)}(h_{YbS}, h_{Mu}, h_{BC})\left(\frac{\sigma_{YbS}^2 h_{YbS}}{\overline{E}_{YbS}} + \frac{\sigma_{Mu}^2 h_{Mu}}{\overline{E}_{Mu}} + \frac{\sigma_{BC}^2 h_{BC}}{\overline{E}_{BC}}\right),\ \overline{E}_i = \frac{E_i}{1-\nu_i^2},\ i = \text{YbS, Mu, BC}. \tag{14}$$

Here, $Z_{BC/(SiC/SiC)}(h_{YbS}, h_{Mu}, h_{BC})$ is a factor dependent on the combination of the YbS, Mu and BC layers.

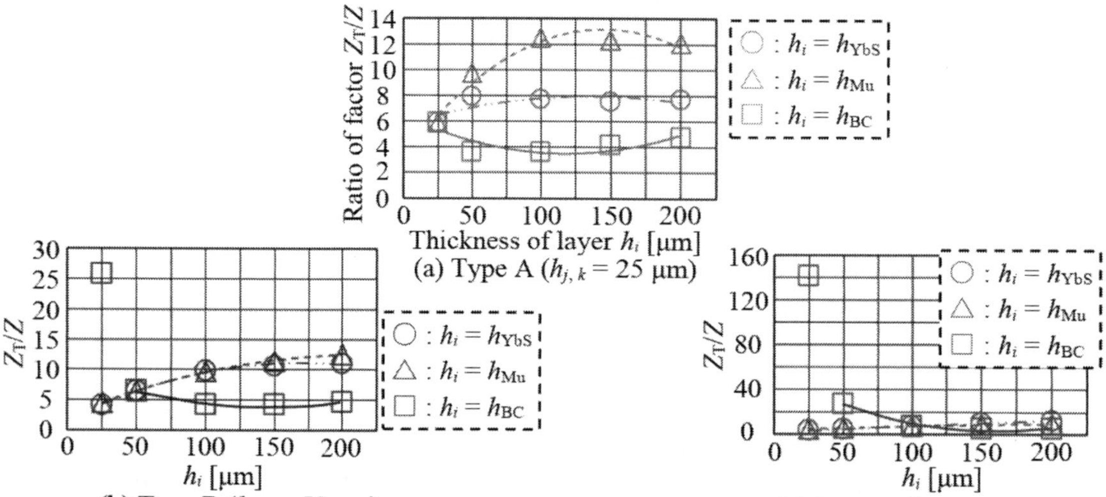

(a) Type A ($h_{j,k}$ = 25 μm)

(b) Type B ($h_{j,k}$ = 50 μm)

(c) Type C ($h_{j,k}$ = 100 μm)

Fig. 2 Ratio of factor (Z_T/Z) for Si-based bond coat/(SiC/SiC) substrate interface – thickness of layers (h_i) relationships of (a) Type A, (b) Type B and (c) Type C.

4.1.2 Mullite/Si-based bond coat interface. Figure 3 shows the Z_T/Z-h_i relationships for Mu/BC interface. In all types, Z_T/Z increases with the increase of h_{YbS} while it decreases with increasing h_{Mu} and h_{BC}. The relationships between Z_T/Z and h_i for Mu/BC interface are written as

Type A

$$\frac{Z_T}{Z} = 0.0811 h_{YbS} + 0.754, \tag{15}$$

$$\frac{Z_T}{Z} = 6.00\times10^{-5} h_{Mu}^2 - 0.0143 h_{Mu} + 2.58, \tag{16}$$

$$\frac{Z_T}{Z} = 1.00\times10^{-4} h_{BC}^2 - 0.0316 h_{BC} + 2.93, \tag{17}$$

Type B

$$\frac{Z_T}{Z} = 4.00\times10^{-5} h_{YbS}^2 + 0.0197 h_{YbS} + 0.560, \tag{18}$$

$$\frac{Z_T}{Z} = 8.00\times10^{-5} h_{Mu}^2 - 0.0217 h_{Mu} + 2.75, \tag{19}$$

$$\frac{Z_T}{Z} = 1.00\times10^{-4} h_{BC}^2 - 0.0380 h_{BC} + 3.53, \tag{20}$$

Type C

$$\frac{Z_T}{Z} = 6.40\times10^{-3} h_{YbS} + 0.483, \tag{21}$$

$$\frac{Z_T}{Z} = 8.00\times10^{-5} h_{Mu}^2 - 0.0241 h_{Mu} + 2.70, \tag{22}$$

$$\frac{Z_T}{Z} = 1.00\times10^{-4} h_{BC}^2 - 0.0436 h_{BC} + 4.15. \tag{23}$$

These results suggest that, similarly to the case of BC/(SiC/SiC) interface, Z_T/Z for Mu/BC interface is dependent on the combination of the coating layer thicknesses in EBC.

The ERR for Mu/BC interface crack initiation in EBC is described as

$$G_{Mu/BC} = Z_{Mu/BC}\left(h_{YbS}, h_{Mu}, h_{BC}\right)\left(\frac{\sigma_{YbS}^2 h_{YbS}}{\bar{E}_{YbS}} + \frac{\sigma_{Mu}^2 h_{Mu}}{\bar{E}_{Mu}}\right), \bar{E}_i = \frac{E_i}{1-\nu_i^2}, i = \mathrm{YbS, Mu}. \quad (24)$$

Here, $Z_{Mu/BC}(h_{YbS}, h_{Mu}, h_{BC})$ is a factor dependent on the thicknesses of the layers in EBC.

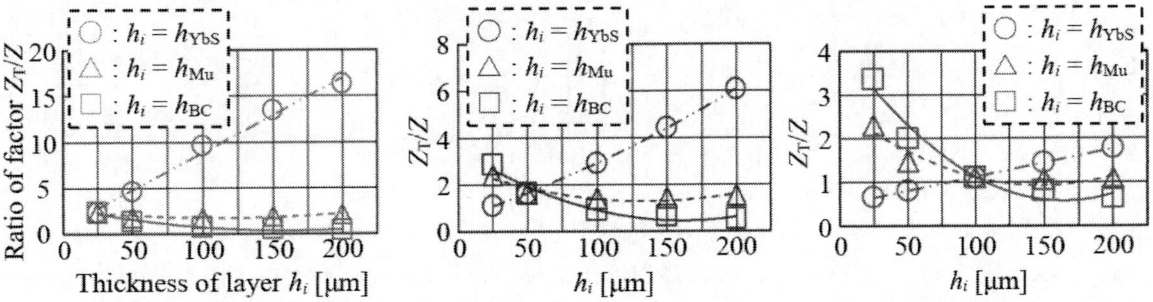

(a) Type A ($h_{j,k}$ = 25 μm) (b) Type B ($h_{j,k}$ = 50 μm) (c) Type C ($h_{j,k}$ = 100 μm)

Fig. 3 Ratio of factor (Z_T/Z) for Mullite/Si-based bond coat interface – thickness of layers (h_i) relationships of (a) Type A, (b) Type B and (c) Type C.

4.1.3 Yb-silicate/Mullite interface. Figure 4 shows the Z_T/Z-h_i relationships for YbS/Mu interface. As well as the cases of BC/(SiC/SiC) and Mu/BC interfaces, (Z_T/Z)s of each type for YbS/Mu interface were changed by increasing the coating layer thicknesses. The relationships between Z_T/Z and h_i for YbS/Mu interfaces are written as

Type A

$$\frac{Z_T}{Z} = 0.0116h_{YbS} + 2.23, \quad (25)$$

$$\frac{Z_T}{Z} = 1.00\times10^{-4} h_{Mu}^2 - 0.0322h_{Mu} + 3.07, \quad (26)$$

Type B

$$\frac{Z_T}{Z} = 5.00\times10^{-5} h_{YbS}^2 + 3.00\times10^{-4} h_{YbS} + 3.00, \quad (27)$$

$$\frac{Z_T}{Z} = 3.00\times10^{-4} h_{Mu}^2 - 0.0945h_{Mu} + 7.70, \quad (28)$$

$$\frac{Z_T}{Z} = 0.0480h_{BC} + 1.06, \quad (29)$$

Type C

$$\frac{Z_T}{Z} = 3.00\times10^{-5} h_{YbS}^2 + 5.4\times10^{-3} h_{YbS} + 1.89, \quad (30)$$

$$\frac{Z_T}{Z} = 7.00\times10^{-4} h_{Mu}^2 - 0.226h_{Mu} + 19.3, \quad 50 \le h_{Mu} \le 200 \quad (31)$$

$$\frac{Z_T}{Z} = 0.0219h_{BC} + 0.597. \quad (32)$$

Here, the Z_T/Z-h_{BC} relationship of Type A cannot be defined in this analysis.

These equations indicate that the Z_T/Z-h_i relationship for YbS/Mu interface is also determined by the combination of the film thicknesses. Therefore, the theoretical equation of ERR for YbS/Mu interface crack initiation in EBC is modified as following equation:

$$G_{YbS/Mu} = Z_{YbS/Mu}\left(h_{YbS}, h_{Mu}, h_{BC}\right)\frac{\sigma_{YbS}^2 h_{YbS}}{\bar{E}_{YbS}}, \bar{E}_i = \frac{E_i}{1-\nu_i^2}, i = \mathrm{YbS} \quad (33)$$

Here, $Z_{YbS/Mu}(h_{YbS}, h_{Mu}, h_{BC})$ is a factor dependent on the thicknesses of the layers in EBC.

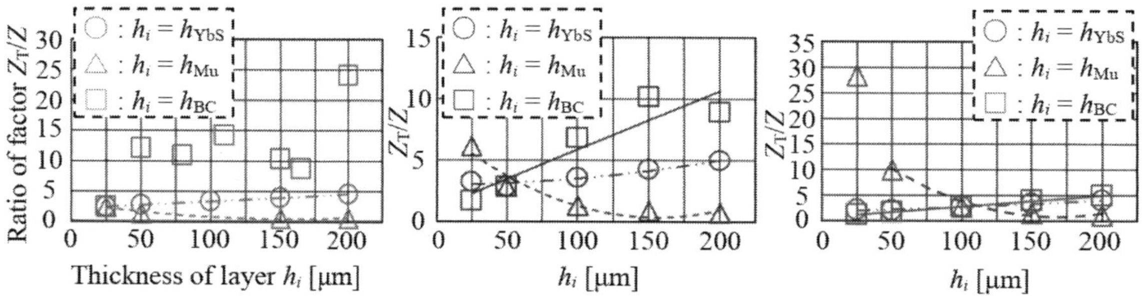

(a) Type A ($h_{j,k}$ = 25 μm) (b) Type B ($h_{j,k}$ = 50 μm) (c) Type C ($h_{j,k}$ = 100 μm)

Fig. 4 Ratio of factor (Z_T/Z) for Yb-silicate/Mullite interface – thickness of layers (h_i) relationships of (a) Type A, (b) Type B and (c) Type C.

4.2. Condition of acceptable EBC layer thicknesses

Figures 5 (a) ~ (c) schematically show conditions on EBC layer thicknesses such that interface crack initiation is avoided for Type A. Following the Griffith theory, crack initiation occurs if ERR exceeds a critical value (G_c) determined by the interface fracture toughness. If Z in eq. (3) was a constant value (independent of layer thicknesses), ERR would be a linear function of thickness(es) of layer(s) above the interface in question. This means that there would be a plane that gives the critical layer thickness(es) and that the region surrounded by the plane and three axes would give acceptable layer thickness(es) to prevent interface crack initiation (see solid lines in Figs. 5 (a) ~ (c)). In fact, the shape of the region of acceptable EBC layer thicknesses becomes more complex because of the dependence of the factor on the layer thicknesses (eqs. (14), (24) and (33)), as is schematically shown as the hatched solids in Figs. 5 (a) ~ (c). Note that the region is depicted in a simple fashion; i.e., the actual shape of the boundaries between acceptable and unacceptable thicknesses may be complicated. The abovementioned Z_T/Z-h_i relationships (eqs. (5) ~ (7), (15) ~ (17) and (25) ~ (26)) gives the actual shape.

As shown in Fig. 2 (a), Z_T/Z is greater than 1 ($G_T > G_F$) regardless of h_i. This means that the actual ERR is smaller than that estimated by eq. (3) with $Z = Z_T \equiv 1.028$. Thus, the condition on the coating layer thicknesses to prevent the BC/(SiC/SiC) interface crack initiation is loosened as indicated by the blue hatched area in Fig. 5 (a). On the other and, because Z_T/Z becomes smaller than 1 with the increase of h_{BC} as shown in Fig. 3 (a), the region of permissible layer thicknesses for Mu/BC interface is contracted as indicated by the green hatched area in Fig. 5 (b).

As shown in Fig. 5 (d), the region overlapped with the conditions for all interfaces gives permissible EBC layer thicknesses to prevent any interface crack initiation due to thermal stress during the fabrication process.

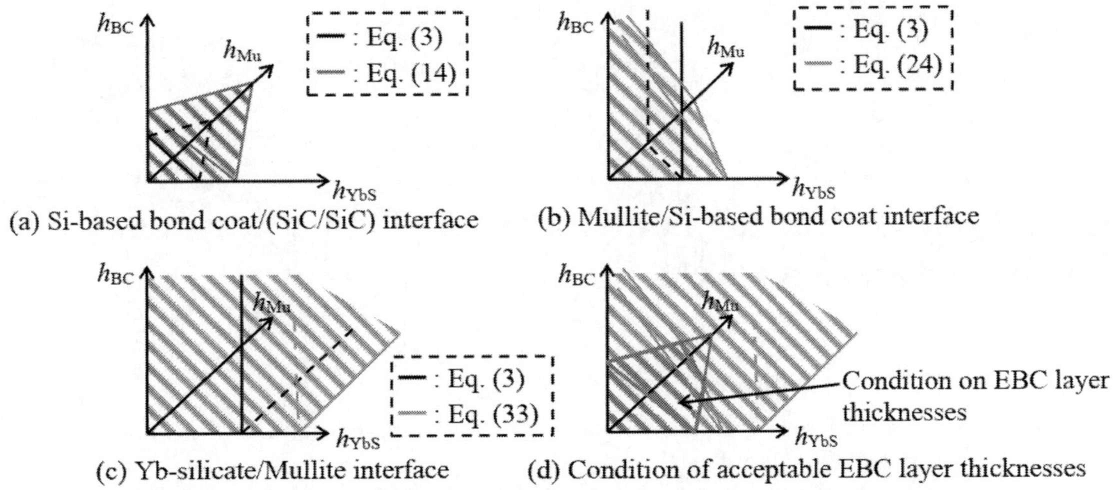

Fig. 5 Condition on EBC layer thicknesses for Type A to prevent the (a) Si-based bond coat/(SiC/SiC) interface, (b) Mullite/Si-based bond coat interface, (c) Yb-silicate/Mullite interface crack initiation due to thermal stress during fabrication process and (d) condition of acceptable EBC layer thicknesses for Type A which is obtained by superposing (a) ~ (c).

5. Conclusion

To establish theoretical prediction for acceptable EBC layer thicknesses that ensure the mechanical reliability of EBC, we conducted FEM analysis to calculate ERRs for interface crack initiation due to thermal stress in EBC. The results were compared with a simple theory for interface crack in a single-layered structure where ERR is estimated as nominal strain energy in the coating layers multiplied by a constant factor. It was shown that the factor is no longer a constant but depends on layer thicknesses in EBC. We obtained the factor as a function of EBC layer thicknesses and presented condition of EBC layer thicknesses based on the results.

References

[1] David L M, Tito T S and James A D, 1986, *J.Mater.Eng*, **8** 1, 80-91
[2] Anand A and Alta Y N C, 1997, *J.Mater.Process.Tech.*, **63**, 384-94
[3] Joosung J L, Stephen P L, Ian A W and Andreas S, 2001, *Annu.Rev.Energy Environ*, **26**, 167-200
[4] Irene S and Jim S, 2004, *Int.J.Appl.Ceram.Technol*, **1** 4, 291-301
[5] Tresa M P and Sammy T, 2006, *J.Propul.Power*, **22** 2, 361-74
[6] Kang N L and Robert A M, 1996, *Surf.Coat.Tech.*, **86-87**, 142-8
[7] Sivakumar R, Surendra N T, Kang N L, Ramakrishna T B and Dennis S F, 2011, *Surf.Coat.Tech.*, **205**, 3578-81
[8] John W H and Zhigang S 1991 *Advances in Applied Mechanics* vol 29 ed John W H and Theodore Y Wu (Amsterdam: Elsevier) pp 63-191

Acknowledgment

This work was supported by Council for Science, Technology and Innovation (CSTI), Cross-ministerial Strategic Innovation Promotion Program (SIP), "Structural Materials for Innovation" (Funding agency: JST, Japan Science and Technology Agency).

Liquid phase diffusion bonding of A1070 by using metal formate coated Zn sheet

K Ozawa[1], **S Koyama**[2] **and I shohji**[2]

[1]Graduate School of Science and Technology, Gunma University, Tenjin-cho, Kiryu, Gunma 376-8515, Japan
[2]Faculty of Science and Technology, Gunma University, Tenjin-cho, Kiryu, Gunma 376-8515, Japan

koyama@gunma-u.ac.jp

Abstract. Aluminium alloy have high strength and easily recycle due to its low melting point. Therefore, aluminium is widely used in the manufacturing of cars and electronic devices. In recent years, the most common way for bonding aluminium alloy is brazing and friction stir welding. However, brazing requires positional accuracy and results in the formation of voids by the flax residue. Moreover, aluminium is an excellent heat radiating and electricity conducting material; therefore, it is difficult to bond together using other bonding methods. Because of these limitations, liquid phase diffusion bonding is considered to the suitable method for bonding aluminium at low temperature and low bonding pressure. In this study, the effect of metal formate coating processing of zinc surface on the bond strength of the liquid phase diffusion bonded interface of A1070 has been investigated by SEM observation of the interfacial microstructures and fractured surfaces after tensile test. Liquid phase diffusion bonding was carried out under a nitrogen gas atmosphere at a bonding temperature of 673 K and 713 K and a bonding load of 6 MPa (bonding time: 15 min). As a result of the metal formate coating processing, a joint having the ultimate tensile strength of the base aluminium was provided. It is hypothesized that this is because metallic zinc is generated as a result of thermal decomposition of formate in the bonded interface at lower bonding temperatures.

1. Introduction

In the past, methods such as brazing, friction stir welding and laser welding have been used for bonding aluminum alloy [1, 2]. However, the adverse effect of the flux has on the environment is also something need to be concerned [3]. In addition, these methods have some shortcomings: (1) a high heat input is required to compensate for high heat radiation and (2) micro cracks are developed owing to the softening of the weld zone. Moreover, tools used for friction stir welding have short lifespan and this translates to higher running costs. Hence, in this study, liquid phase diffusion bonding method is considered to the most suitable method for bonding of A1070. In liquid phase diffusion bonding, materials are bonded together by applying heat and pressure to promote interdiffusion without significant deformation. Moreover, since liquid phase diffusion bonding involves only low bonding temperature, there is little damage to a component, thus making this technique suitable for bonding precision assemblies. However, a bonding surface has various defects such as surface irregularities, a native oxide film and a processing layer, which act as inhibitors to successful bonding.

In this study, liquid phase diffusion bonding of A1070 with Zn sheet is carried out on Zn sheet surfaces treated for metal salt coating processing using formic acid. The bonded surfaces were then examined to determine the effectiveness of metal salt coating processing.

2. Experimental details

As shown in Figure 1(a), the pure Zn sheet specimen (thickness: 0.1, 0.8 and 2.0 mm, 99.9% purity) with dimensions of 11 × 11 mm were used. As shown in Figure 1(b), the aluminum specimen is a cylindrical bar cut from A1070 (Table 1). The bonding surface is polished by a #1200 emery paper. In addition, the reason why a Zn sheet was chosen as an insert metal is because a liquid phase is generated by a reaction with the base aluminum and it is supposed that the molten metal fills the gap of the bonding interface. Zn sheet surfaces were modified by boiling in formic acid for optimum time for metal salt coating on the bonding surface. Furthermore, grazing-angle incidence reflection-absorption infrared (GIRAS-IR) spectroscopy was employer to obtain IR spectra at a nanometer-scale depth from the modified zinc surface.

Liquid phase diffusion bonding is performed under a nitrogen gas (99.99% purity) atmosphere using the following conditions; 6 MPa for bonding load, 15 min for bonding time and 673 K and 713 K for bonding temperature. In addition, the effect of the thickness and the modification time of the Zn sheet on the tensile strength of joint is investigated.

After liquid phase diffusion bonding, the joint strength is evaluated using a tensile tester. The tensile test is carried out using INSTRON 5567 universal test machine at room temperature at a displacement speed of 10 mm/min.

Further, a specimen showing the vertical cross-section of the bonded interface was also prepared and was observed using scanning electron microscopy (SEM) and energy-dispersive X-ray spectroscopy (EDX) performed at an acceleration voltage of 15 kV.

3. Results and discussion

3.1. Optimization of the Zn sheet thickness and metal formate coating processing time

To examine the effect of the Zn sheet thickness has on the bond strength of joint, tensile test is carried out after liquid phase diffusion bonding is performed using different Zn sheet thickness ranging from 0.1 ~ 2.0 mm. The bonding temperature is fixed at 673 K and 713 K and the relationships between Zn sheet thickness and tensile strength are shown in Figure 2. As shown in Figure 2, the highest bond strength joints were obtained when the Zn sheet thickness was 0.8 mm at both bonding temperatures.

Table 1. Chemical compositions of A1070 aluminium used in this study (UACJ Corporation, Japan).

Elements	Si	Fe	Cu	Mn	Zn	Ti	V	Al
mass%	0.05	0.13	0	0	0	0	0.02	Bal.

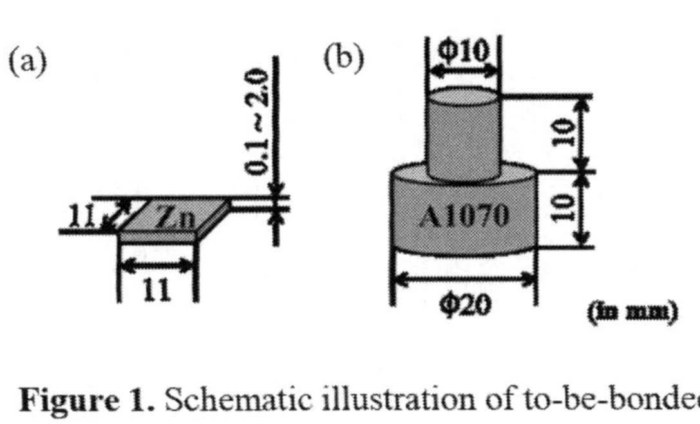

Figure 1. Schematic illustration of to-be-bonded specimen: (a) pure Zn sheet and (b) A1070.

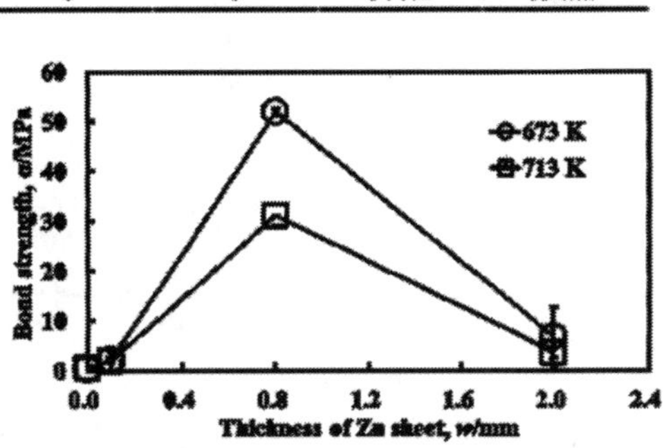

Figure 2. Relationship between bond strength and thickness of Zn sheet.

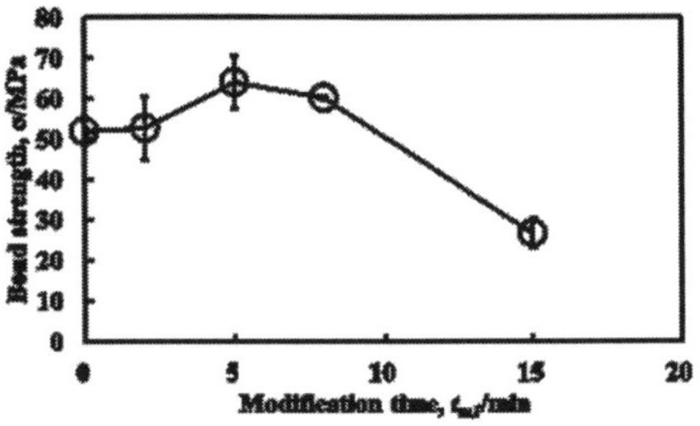

Figure 3. Relationship between bond strength and modification time (T= 673 K).

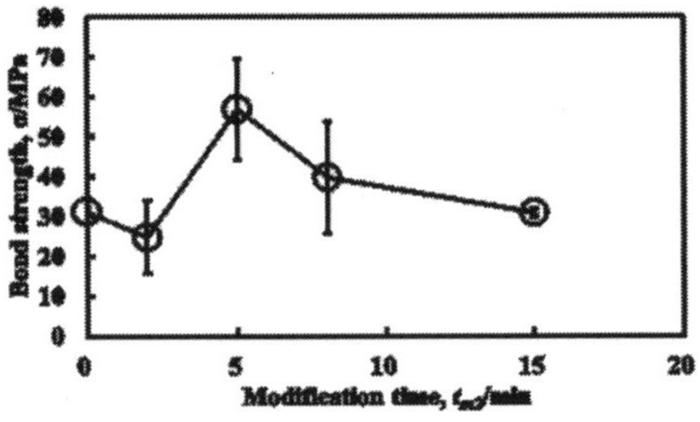

Figure 4. Relationship between bond strength and modification time (T= 713 K).

From this examination results, when the Zn sheet thickness is thin, the roughness of the bonding surface was not filled; and when the thickness is the optimum thickness, Zn-Al eutectic structure (tensile strength is higher than A1070) was remained in the bonded interface. On the other hand, when the Zn sheet thickness is thick, solidification crack occurred at the bonded interface, and the bond strength decreased. From these reasons, it was determined that the optimum Zn sheet thickness to use in this study is 0.8 mm.

Figures 3 and 4 shows the relationship between the modification time and bond strength. The highest bond strength joint were obtained when the metal formate coating processing time was 5 min at both bonding temperature. From these examination results, when the metal formate coating processing time is short, the zinc oxide film is not substituted to a zinc formate; and when the time is long, an excessive quantity of the formate (oxide and gas is generated at the time of thermolysis) is generated. Therefore, it was determined that the optimum modification time to use in this study is 5 min.

3.2. Effect of metal formate coating processing on the bond strength

Figure 5 shows the relationship between the bond strength of the joint and the bonding temperature when the optimum conditions discussed in section 3.1 were applied. In order to illustrate the effect of metal formate coating processing, the corresponding relationship for a non-modified joint are also shown. As shown in Figure 5, the bond strength decreased with bonding temperature irrespective of metal formate coating processing. However, the bond strength of the joint has about 2 times higher the strength than that of non-modified joint and has the ultimate tensile strength of the base A1070 aluminum at the bonding temperature of 713 K. Furthermore, all the joints broken at the bonded interface and necking was observed in all of them.

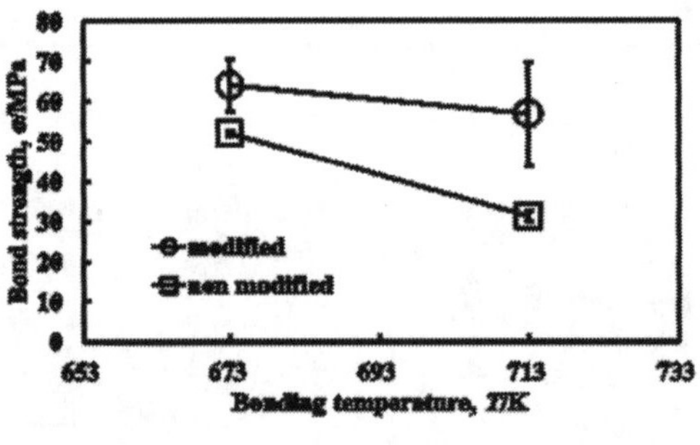

Figure 5. Effect of metal formate coating processing on relationships between bond strength and bonding temperature.

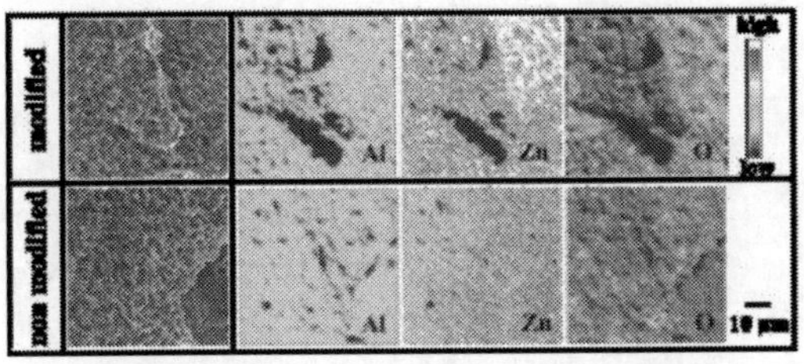

Figure 6. SEM micrographs and EDX analysis of fractured surfaces of joints after tensile test (T= 673 K).

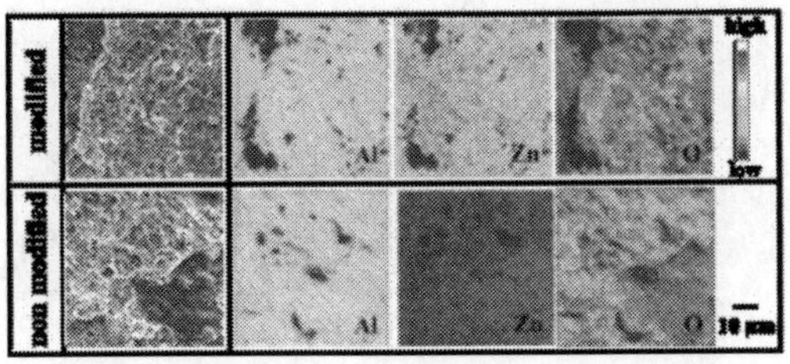

Figure 7. SEM micrographs and EDX analysis of fractured surfaces of joints after tensile test (T= 713 K).

3.3. Observation of fractured surface

Figures 6 and 7 shows the fractured joint surface after the tensile test for specimens bonded at 673 K and 713 K. As shown in Figure 6, when the bonding temperature of 673 K, ductile fracture characteristics were observed irrespective of metal formate coating processing. However, when the metal formate coating processing was not applied, brittle fracture characteristics were observed, although it was not observed when the metal formate coating processing was applied. In addition, when the metal formate coating processing was applied, Zn was detected in more areas compared with the case that the coating processing was not applied. It is therefore thought that a high-bond strength joint was obtained at a same bonding temperature with metal formate coating processing because zinc formate in the bond interface underwent a thermal decomposition reaction and eutectic reaction after close contact between zinc and aluminum was achieved to mention it later.

With a rise in bonding temperature of 713 K, the bond strength were decreased irrespective of metal formate coating processing. The reason for this is because molten zinc was exhausted to the circumference of the junction by a zinc sheet reaching a melting point at this bonding temperature. In addition, when the metal formate coating processing was not applied, the bond strength was very low. The reason for this is because an oxide film of the molten zinc surface inhibited a eutectic reaction with an aluminum base metal.

3.4. Observation of bonded interface

To determine the relationship between interfacial microstructure and bond strength, the bonded interface was observed by SEM. Figure 8 shows SEM micrographs of the bonded interface. When the bonding temperature of 673 K, a lamellar structure with a thickness of about 5 μm was observed irrespective of metal formate coating processing. As a result of the EDX analysis, this lamellar structure revealed that it was a eutectic structure containing about 30 mass% Zn. When the metal formate coating processing was not applied, black-colored thin layer with a thickness of about 1 μm were observed between lamellar structure and an aluminum base metal. As a result of the EDX analysis, this black-colored thin layer revealed that it was zinc oxide, namely a zinc oxide film.

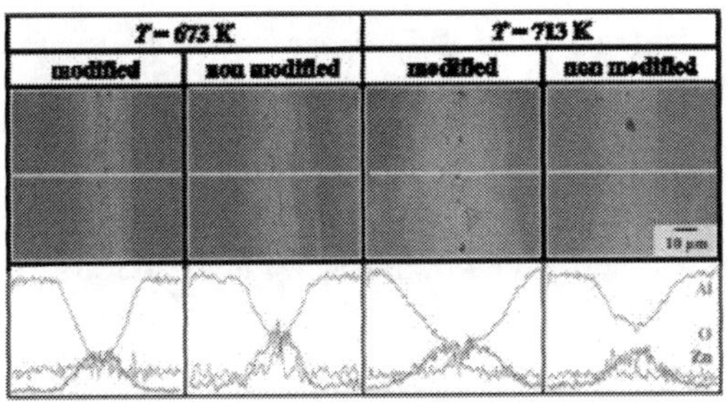

Figure 8. SEM micrographs and EDX analysis of bonded interfaces.

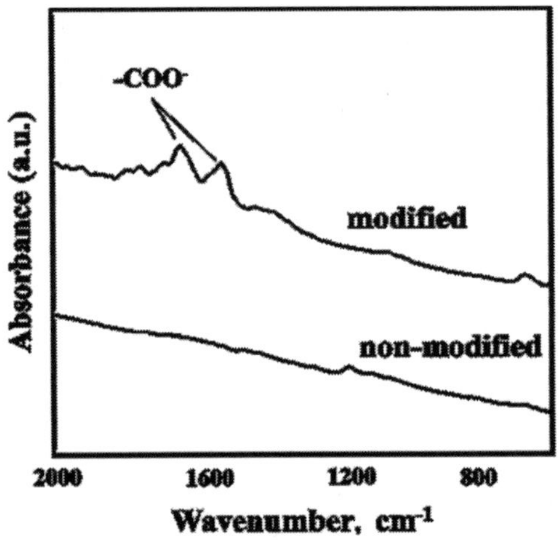

Figure 9. FT-IR analysis of Zn surfaces with/without metal formate coating processing.

With a rise in bonding temperature of 713 K, when the metal formate coating processing was applied, a divided lamellar structure was observed. However, when the metal formate coating processing was not applied, almost none of the lamellar structure was observed. Therefore, it is inferred that the promotion of the eutectic reaction with an aluminum base metal by applying the metal formate coating processing greatly had an influence on the improvement of the bond strength.

3.5. Chemical analysis of metal formate coating surface

Figure 9 shows the FT-IR analysis result of the Zn sheet surface before and after metal formate coating processing. As shown in Figure 9, the absorption bands of about 1637 and 1504 cm^{-1} correspond to a carboxyl ligand. When the thermal decomposition of Zinc formate were occurred in a bonding process, the compound thermally decomposed in to nano particles of zinc oxide and several kind of gases at temperature of 560 K or more [4, 5]. It is therefore thought that a high-bond-strength joint was obtained at a lower bonding temperature with metal formate coating processing because zinc formate underwent a thermal decomposition reaction after close contact between A1070 and Zn atomic plane was achieved.

4. Conclusions

The conclusions of this study are summarized as follows.

(1) From the FT-IR spectrum, it is revealed that zinc formate, for which the decomposition reaction occurs at a low temperature is formed by metal formate coating processing on the zinc surface.
(2) A higher bond strength can be achieved at the same bonding temperature through the use metal formate coating processing.
(3) The increase surface area of the metallic-Zn promotes the eutectic reaction between Zn sheet and A1070 base metal contributes to increase of bond strength of joint.

Acknowledgement
This work was supported by JSPS KAKENHI Grant Number JP26820124.

References

[1] Hao H L, Ni D R, Huang H, Wang D, Xiao B L, Nie Z R, Ma Z Y 2013 *Mater. Sci. Eng.* A **559** 889

IOP Conf. Series: Journal of Physics: Conf. Series **843** (2017) 012005 doi:10.1088/1742-6596/843/1/012005

[2] Dai W, Xue S, Lou J, Wang S, 2014 *Mater. Des.* **42** 395
[3] Xu W F, Liu J H, Chen D L, Luan G H, Yao J S 2012 *Mater. Sci. Eng.* A **548** 89
[4] Viertelhaus M, Anson C E, Powell A K 2005 *Z. Anorg. Allg. Cem.* **631** 2365
[5] Boruah S, Mustafiza S, Saikia D, Saikia H J, Saikia P P, Baruah M K 2016 *American Chem. Sci. J.* **11** 1

How to obtain the adhesive strength for double lap joint by using single lap joint

N A Noda[1], R Li[1], Y Sano[1], Y Takase[1], R Takaki[1], T Miyazaki[2]

[1] Department of Mechanical Engineering, Kyushu Institute of Technology, 1-1 Sensui-cho Tobata-ku, Kitakyushu-shi 804-8550, Japan
[2] Department of Mechanical Systems Engineering, University of the Ryukyus, 1 Senbaru, Nishihara-cho, Nakagami-gun, Okinawa 903-0213, Japan

E-mail: noda@mech.kyutech.ac.jp

Abstract. The testing method of adhesive strength of lap joint is prescribed in Japanese Industrial Standard (JIS K6850). However, it has been reported that the strength of double lap joint (DLJ) is about two times larger than the one of single lap joint (SLJ). Therefore, suitable testing method has been required from industries. In this study, the equivalent conditions of strength for SLJ and DLJ are investigated in terms of the intensity of singular stress field (ISSF) appearing at the interface end. First, in order to minimize the bend effect for SLJ, the effect of the specimen geometry on ISSF and deformation angle at the interface corner is considered under the same adhesive geometry and load P. It is found that the minimum ISSF of SLJ can be obtained when the adherend thickness t_1 is large enough, and the deformation angle at interface corner is also smallest when adherend thickness t_1 is large enough. Therefore, it is necessary to use the specimen with thicker adherend thickness. Then, the equivalent conditions of strength for SLJ and DLJ is investigated by changing adherend thickness. The results show that the strength of the DLJ in JIS (t_1=1.5mm) can be obtained by using the SLJ with adherend thickness t_1=7mm. When the adherend thickness $t_1 \geq$25mm, the strength of SLJ is nearly equal to that of DLJ.

1. Introduction

Since adhesively joints are economical, practical and easy to be used, they have been widely used in a variety of industries and a number of studies of adhesive joints have been made so far[1-4]. The authors investigated the adhesive butt joint strength in Figure 1 by changing the adhesive thickness and material combination [5]. It is found that the adhesive strength of butt joint can be expressed as the critical intensity of singular stress field (ISSF) $K_{\sigma c}$ =const [5]. The adhesive strength of single lap joint (SLJ) also can be expressed as $K_{\sigma c}$ =const [6],[7].

The testing methods for the adhesive strength of lap joints are standardized by Japanese Industrial standard (JIS) [8]. Compared with double lap joint (DLJ), SLJ can be used conveniently. However, the shear strength of DLJ in Figure 2(a) is about two times larger than the one of SLJ in Figure 2(b) (see Figure 3) [9]. Therefore, in this study, the equivalent conditions of strength for SLJ and DLJ are investigated in terms of the ISSF.

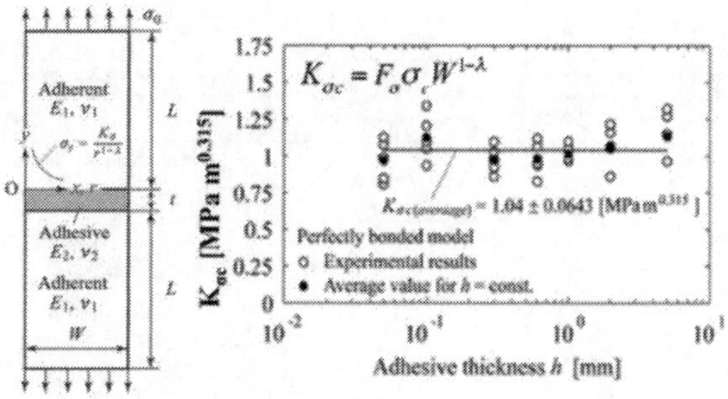

Figure 1. Adhesive strength expressed as $K_{\sigma c}$ =const for butt joint.

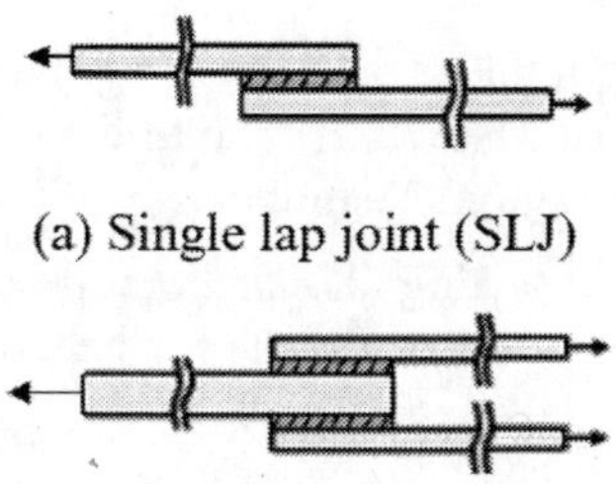

(a) Single lap joint (SLJ)

(b) Double lap joint (DLJ)

Figure 2. Two kinds of lap joint specimens.

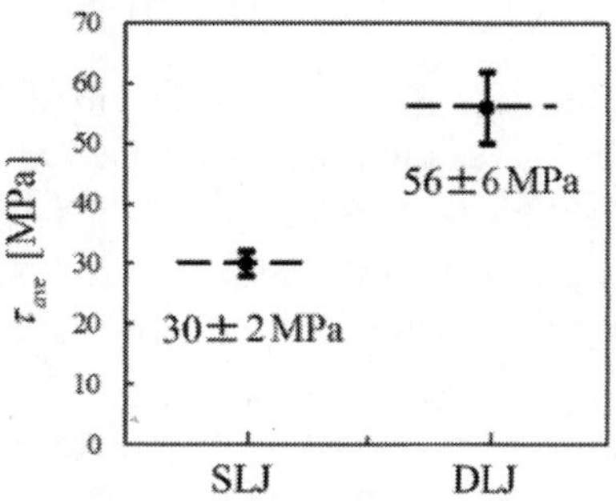

Figure 3. Adhesive shear strength for SLJ and DLJ (Adherend: S45C, Adhesive: Epoxy).

2. Pure shear testing to minimize K_{σ,λ_1}

Figure 4 shows the schematic illustration of the analysis models. It has been reported that the singularity exists near the interface corner, and the singularity depending on the singular indexes λ_1 and λ_2 at the interface. In this paper, λ_1=0.6062, λ_2=0.9989. The stress σ_θ at r direction (θ=0) can be expressed as follows. The notation r denotes the radial distance away from the corner singular point O_1.

$$\sigma_\theta = \frac{K_{\sigma,\lambda_1}}{r^{1-\lambda_1}} + \frac{K_{\sigma,\lambda_2}}{r^{1-\lambda_2}} \cong \frac{K_{\sigma,\lambda_1}}{r^{1-\lambda_1}}\left(1 + C_\sigma r^{\lambda_2-\lambda_1}\right) \tag{1}$$

Here, K_{σ,λ_1} and K_{σ,λ_2} are ISSFs. The values of K_{σ,λ_1} and K_{σ,λ_2} can be obtained by using the method presented in [6-7]. The ISSF can be represented with only K_{σ,λ_1} since C_σ is almost constant expressed as $C_\sigma = -5.321 \pm 0.338$.

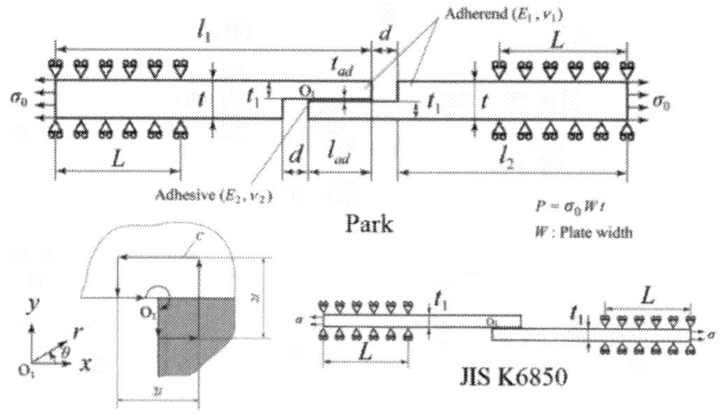

Figure 4. Analysis model and boundary condition.

The butt joint in Figure 1 is used to obtain the adhesive strength under pure tension [5] and the SLJ in Figure 2 is used to obtain the adhesive strength under pure shear. However, due to the deformation of SLJ during testing, the peeling force is applied to the adhesive region. Then the K_{σ,λ_1} at the interface corner is affected by the peeling force. Since the single lap joint testing should be done under pure shear loading, smaller K_{σ,λ_1} is desirable. Therefore, in order to minimize the K_{σ,λ_1}, the effect of the specimen geometry is considered under the same adhesive geometry and load P based on the specimen used by Park [10] in Figure 4 (Adherend: Aluminum alloy 6061-T6, Adhesive: FM73M epoxy). The total length of the specimen is 225mm, adhesive length l_{ad} =25mm, adhesive thickness t_{ad} =0.15mm, d =10mm, P =14.15N. L is fixed boundary length, t_1 is adherend thickness.

Figure 5 shows the deformation near the interface corner. In this study, the deformation angle at interface corner C will be considered to explain the deformation. Here, the deformation angle θ_C is considered by using the maximum θ_C since the value of maximum θ_C is almost constant independent of element sizes.

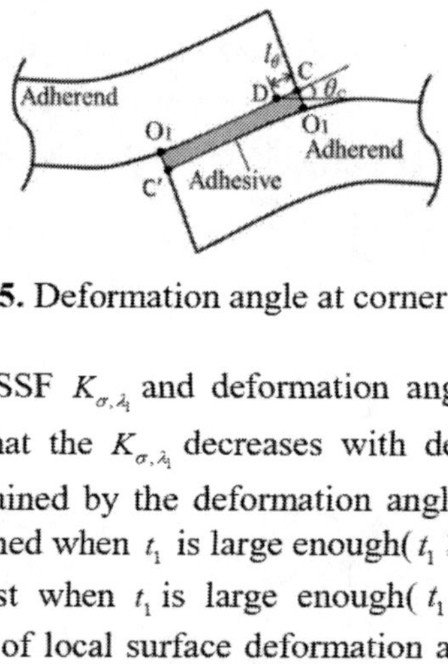

Figure 5. Deformation angle at corner edge C.

The relationship between the ISSF K_{σ,λ_1} and deformation angle θ_C for three types of models is shown in Figure 6. It is found that the K_{σ,λ_1} decreases with decreasing θ_C. This means that the changing of the ISSF can be explained by the deformation angle at the interface corner. Here, the minimum ISSF of SLJ can be obtained when t_1 is large enough($t_1 \geq$25mm), and the deformation angle at interface corner is also smallest when t_1 is large enough($t_1 \geq$ 25mm). The possible reason of minimum $K_{\sigma,\lambda_1} \neq 0$ is the existence of local surface deformation at the interface corner even for very large thickness. Therefore, it is necessary to use thick adherend.

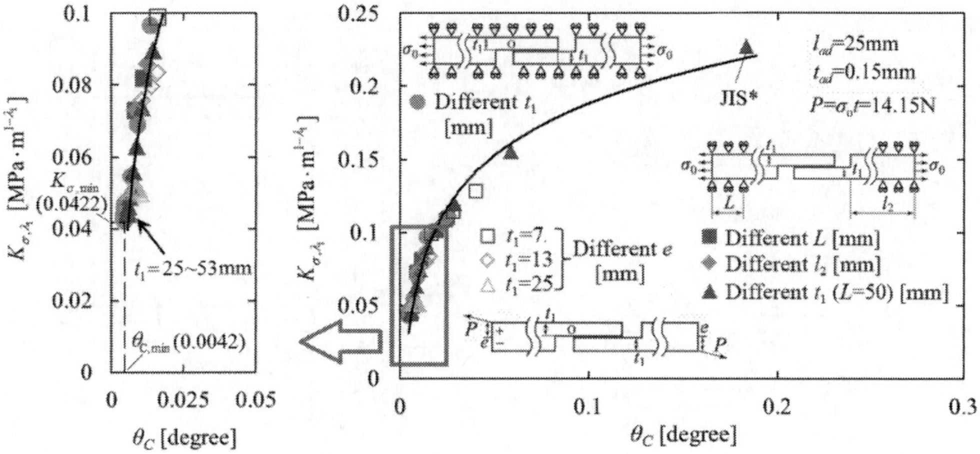

Figure 6. Relationship between K_{σ,λ_1} and θ_C.

3. Equivalent conditions of strength for lap joints in terms of ISSF

The experimental results show that the strength of DLJ is about two times larger than the one of SLJ (see Figure 7(a)) [9]. However, the critical ISSF $K_{\sigma c}$ of SLJ is almost the same as the $K_{\sigma c}$ of DLJ (see Figure 7(b)). Therefore, in this chapter, the equivalent conditions of strength for the DLJ and SLJ in Figure 8 are investigated in terms of the ISSF K_{σ,λ_1}. Here, based on the conclusions in chapter 2, the effect of the adherend thickness on the K_{σ,λ_1} is considered. Since end tab is often bonded at the ends of experimental specimens to reduce bend effect when loaded, the influence of the tab on K_{σ,λ_1} is also considered in this chapter. The same material as adherend is used for tab. The total length of the specimen is 225mm, tab length is 90mm, adhesive length l_{ad}=25mm, adhesive thickness t_{ad}=0.15mm, P=14.15N.

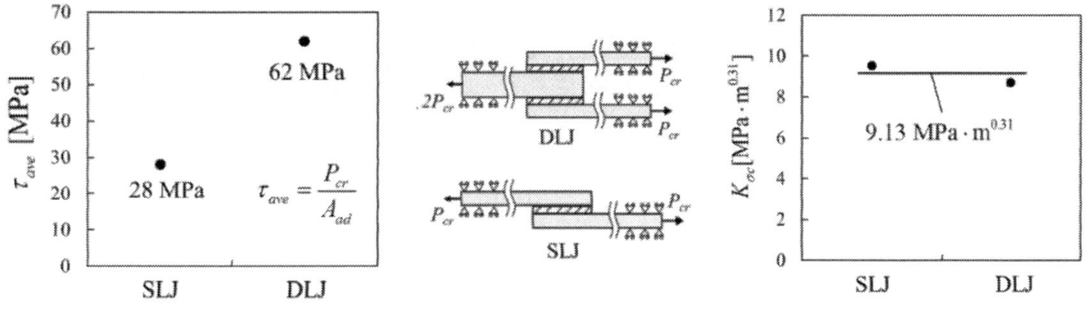

Figure 7. (a) Average shear strengths of SLJ and DLJ, (b) $K_{\sigma c}$ of SLJ and DLJ (Adherend: S45C, Adhesive: Epoxy B).

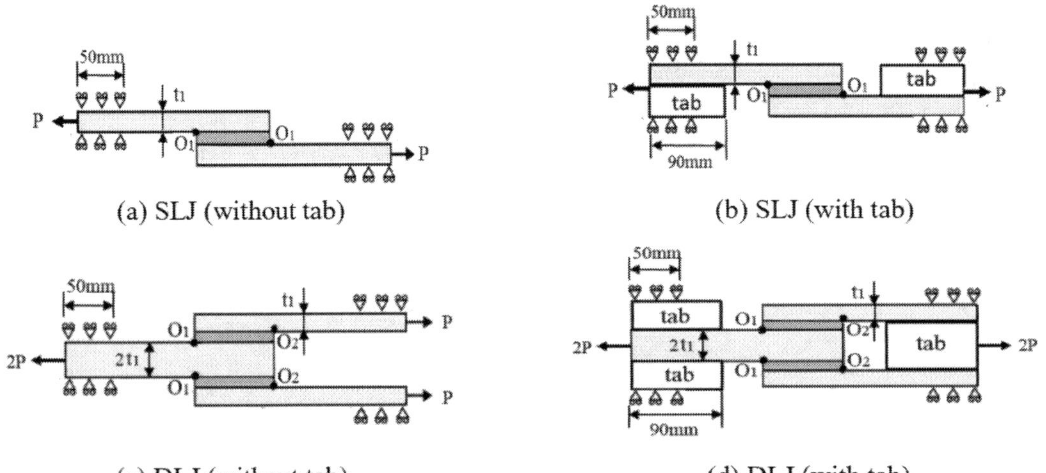

(a) SLJ (without tab) (b) SLJ (with tab)

(c) DLJ (without tab) (d) DLJ (with tab)

Figure 8. Models of SLJ and DLJ.

For DLJ, the K_{σ,λ_1} at interface corner O_1 is not equal to the K_{σ,λ_1} at interface corner O_2. Figure 9 shows the results of K_{σ,λ_1} at interface corners O_1 and O_2. It is found that the K_{σ,λ_1} for the specimen with tab is nearly equal to the K_{σ,λ_1} for the specimen without tab, the K_{σ,λ_1} at corner O_1 is large than that at corner O_2. Therefore, the fracture may occur at corner O_1 during testing. For this reason, the equivalent conditions of strength for SLJ and DLJ will be considered by using the K_{σ,λ_1} at corner O_1.

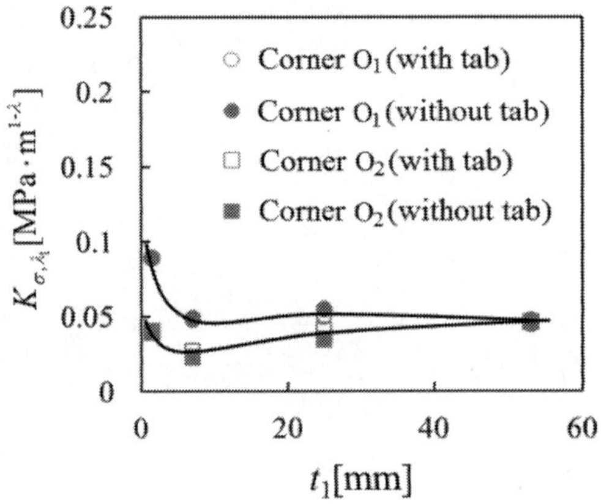

Figure 9. Results of DLJ (see Figure 8(c),(d)).

Figure 10 shows the results of K_{σ,λ_1} at interface corner O_1 with different adherend thickness t_1 for SLJ and DLJ. It is found that the K_{σ,λ_1} for the specimen with tab is nearly equal to the K_{σ,λ_1} for the specimen without tab, the K_{σ,λ_1} decreases with increasing t_1. When $t_1 \geq 25$mm, the minimum K_{σ,λ_1} can be obtained and the minimum $K_{\sigma,\lambda_1} \approx 0$. At that time, the bend effect is minimized, the possible reason of minimum $K_{\sigma,\lambda_1} \neq 0$ is the existence of local surface deformation at the interface corner even for very large thickness. In JIS, the adherend thickness t_1=1.5mm (see Figure 4). The strength of DLJ with t_1=7mm is nearly equal to that of DLJ with t_1=1.5mm (JIS) since the K_{σ,λ_1} of SLJ and DLJ are nearly the same. For the same reason, the strength of SLJ is nearly equal to that of DLJ when $t_1 \geq 25$mm .

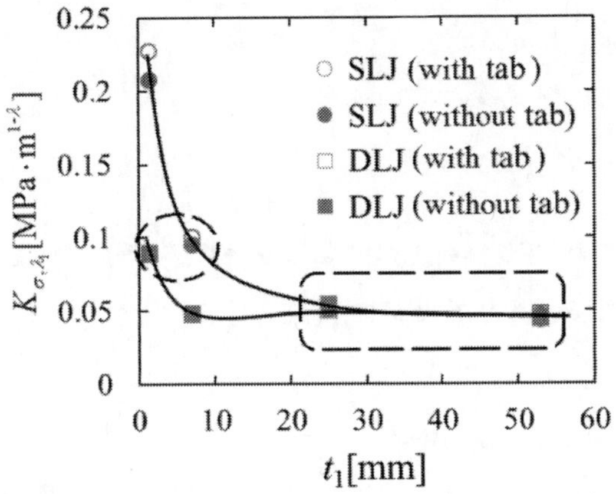

Figure 10. Comparison of SLJ and DLJ at corner O_1(see Figure 8(a)-(d)).

4. Conclusion

(1) The minimum ISSF of SLJ can be obtained when t_1 is large enough, and the deformation angle at interface corner is also smallest when t_1 is large enough.

(2) The strength of SLJ with t_1=7mm is nearly equal to that of DLJ with t_1=1.5mm (JIS) since the ISSFs of SLJ and DLJ are nearly the same. For the same reason, the strength of SLJ is nearly equal to that of DLJ when $t_1 \geq$25mm .

References

[1] Jen YM, Ko CW 2010 *Int J of Fatigue* **32** 330.
[2] Tsai MY, Morton J 2010 *Int J Solids Struct* **47** 3317.
[3] Lee-Sullivan LiGP, Thring RW 1999 *Compos Struct* **46** 395.
[4] Imanaka M, Ishii K, Nakayama H 1999 *Engng Fract Mech* **62** 409.
[5] Noda NA, Miyazaki T, Li R, Uchikoba T and Sano Y 2015 *Int J Adhes Adhes* **61** 46.
[6] Miyazaki T, Noda NA, Uchikoba T, Li R and Sano Y 2014 *Soc Automob Eng Jpn* **45** 895.
[7] Miyazaki T, Noda NA, Li R, Uchikoba T and Sano Y 2013 *J Jpn Inst Electron Packag* **16** 143.
[8] JIS K6850:1999.
[9] Ikegami K, Kyogoku H, Kawagoe H, Sugibayashi T, Nono K., Fujii T, Motoie K and Yoshida F 1997 *Jpn Soc Mech Eng A* **63** 830.
[10] Park JH, Choi JH and Kweon JH 2010 *Compos Struct* **92** 2226.

Fracture evaluation of thermally sprayed coatings in dependence on cohesive strength

J Schubert[1] and Z Česánek[1]

[1]Senior researcher, Výzkumný a zkušební ústav Plzeň, Pilsen, CZ

E-mail: schubert@vzuplzen.cz

Abstract. Measuring the cohesive strength of thermally sprayed coatings is relatively difficult matter, which can be accessed in many directions. This issue is nowadays solved by use of Scratch test method. This method is not completely sufficient for the cohesive strength testing because the coating is under load of combined stresses during the Scratch test. The reason to develop this method was need for exact measurement of tensile cohesion toughness of thermally sprayed coatings, which could provide results as close to a classic tensile test as possible. Another reason for development of this method was the impossibility of direct comparison with results obtained by other methods. Tested coatings were prepared using HP / HVOF (Stellite 6, NiCrBSi, CrC-NiCr and Hastelloy C-276). These coatings were selected as commonly used in commercial sector and also on because of rising customer demand for ability to provide such coating characteristics. The tested coatings were evaluated in terms of cohesive strength (method based on tensile strength test). Final fractures were evaluated by optical microscopy together with scanning electron microscopy and EDS analysis. As expected higher cohesive strength showed metallic coatings with top results of coating Stellite 6. Carbide coatings showed approximately third of the cohesion strength in comparison with metal based coating.

1. Introduction

This paper is based on research which was conducted with aim to develop simple and fast method allowing to measure cohesive strength of coatings. Another condition was that the method has to be able provide reputable results with low costs. The study is part of more complex research focused on influence of coating on mechanical properties of substrate material (mainly fatigue life).

Reason for development of such method was demand of customers which asks for precise specification of mechanical properties of coatings used commercially.

The method described further in this paper proved to fulfil all requirement and is able to provide all necessary data together with sufficient repeatability. There are few drawbacks related to measurement of carbide based coatings which leads to a slightly higher volatility of measured results. The reasons for this deviation are described father in this paper.

This issue is nowadays solved by use of Scratch test method [1] which is not completely sufficient for such kind of testing. The demand was for test as close to standard tensile testing as possible. As it was mentioned above, this method will be used with combination with other mechanical properties evaluation processes. This complete overview of properties is necessary to determine the influence of coatings on fatigue life of coating/substrate system. As it is clear, there are many factors affecting fatigue life. One of them is residual stress mainly located around interface between coating and substrate material. This factor can also play role in affecting nucleation of cracks in case of cohesive (tensile) stress test [2,3,4].

Furthermore, cohesive strength is one of at the most influencing factor determining wear mechanism of coated parts and it is possible to find correlation between them [5].

2. Experiment
The following paragraphs describes design of experiment samples and the method they were used.

2.1. Experimental equipment
Selected coatings were tested on cohesion strength properties using newly constructed device. The method is based two cylindric arts inserted in to each other (shown in Figure 1). During the spray process both parts of the sample are tightened together by threaded bolt. Dividing plane are coated with appropriate coating in 40 mm width and with 400 μm thickness. Afterwards the sample is subjected to tensile test (in this case provided by device ZWICK Z250 in laboratories of VZÚ Plzeň). Design of this sample together with appropriate sample mounting should provide elimination of shear force during the test procedure.

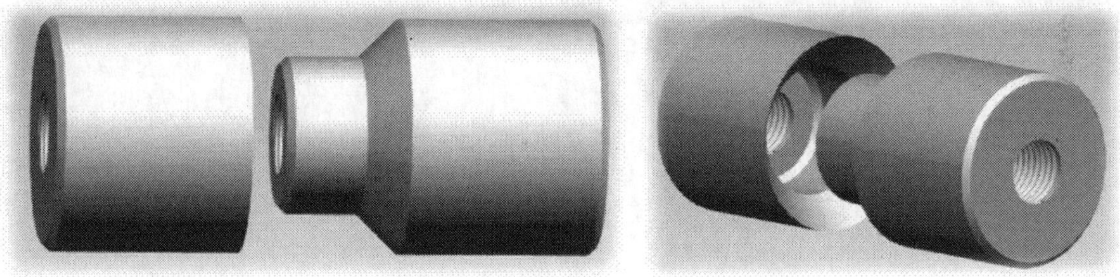

Figure 1. Design of test sample.

The coatings were deposited by system HP/HVOF TAFA JP5200 in laboratories of VZÚ Plzeň using previously optimized parameters. Material of substrate samples was 316L stainless steel. Before coating process samples were degreased and grit blasted with brown corundum (with particle distribution F22).

After the tensile tests the samples were further examined. First of all, the microstructure of each coating was captured and evaluated by scanning electron microscopy (SEM – FEI Quanta 200). The fracture surface was examined using Dyno-Lite AM7000 portable microscope.

2.2. Measurement conditions

The measurement conditions during tensile test were as follows:

- Preload 10 MPa
- Load sped 0,0001 1/s

3. Results and discussion
Each of the selected coatings were sprayed using standard pre-optimized parameters which are used for commercial contract and thus are the most representative for further testing and properties evaluation. Powders used for deposition of selected coatings are commercially available and this paper uses the designation according to company Flame Spray Technologies.

3.1. Scanning electron microscopy evaluation

IOP Conf. Series: Journal of Physics: Conf. Series **843** (2017) 012007 doi:10.1088/1742-6596/843/1/012007

The reason to analyse microstructure was to prove that the samples were prepared under same conditions as commercially produced coatings, and to prove that there is no difference between test sample microstructure and commercially produced microstructure of coating. Difference in microstructure (pores, particle distribution, etc.) would lead to not corresponding results. All tested coatings prepared for this experiment shown satisfactory comparable microstructure with our standards. Examples of individual microstructure in cross-section is shown in following figures (figure 2-5).

3.1.1. Cr_3C_2-NiCr coating
Commercially available carbide based powder (FST 588.071) was used for deposition of this coating. Following figures (figure 2) represent microstructure captured by SEM microscopy. The resulting microstructure was acknowledged as suitable and identical to commercially produced coatings.

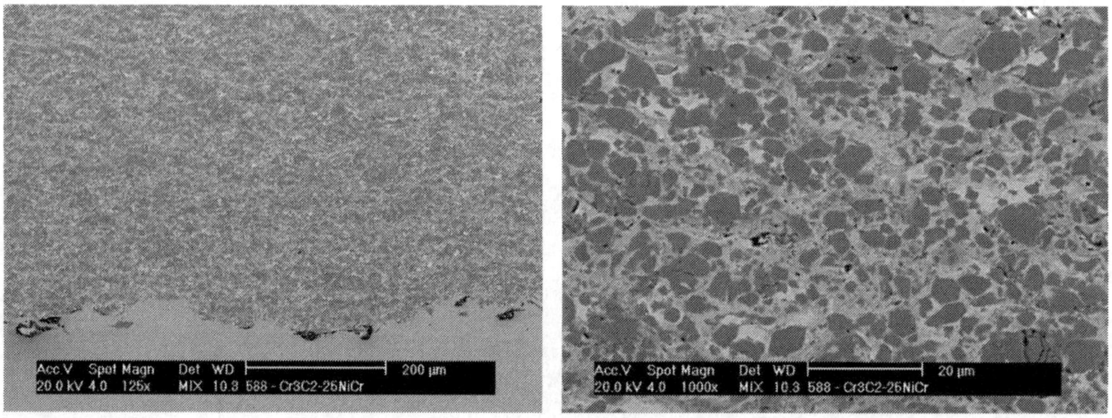

Figure 2. CrC-NiCr coating microstructure cross-section; b) 125 magnification, a) 1000 magnification.

3.1.2. Hastelloy C-276 coating
Commercially available alloy based powder (FST 341.33) was used for deposition of this coating. Following figures (figure 3) represent microstructure captured by SEM microscopy. The resulting microstructure was acknowledged as suitable and identical to commercially produced coatings.

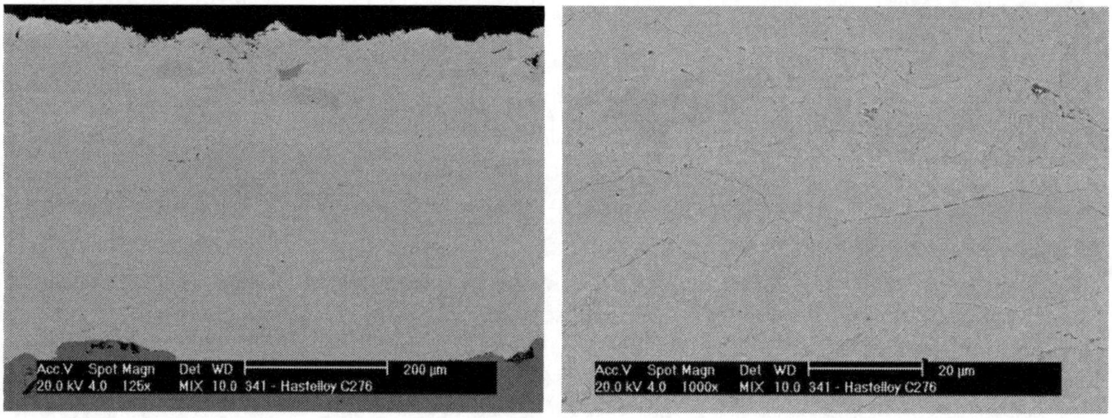

Figure 3. Hastelloy C-276 coating microstructure cross-section; b) 125 magnification, a) 1000 magnification.

3.1.3. *NiCrBSi coating*

Commercially available alloy based powder (FST 771.33) was used for deposition of this coating. Following figures (figure 4) represent microstructure captured by SEM microscopy. The resulting microstructure was acknowledged as suitable and identical to commercially produced coatings.

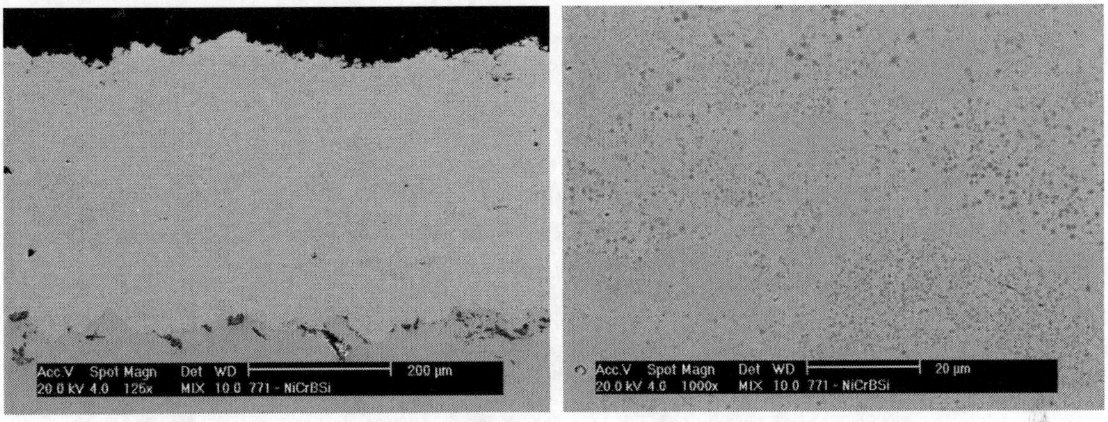

Figure 4. NiCrBSi coating microstructure cross-section; b) 125 magnification, a) 1000 magnification..

3.1.4. *Stellite 6 coating*

Commercially available alloy based powder (FST 484.33) was used for deposition of this coating. Following figures (figure 5) represent microstructure captured by SEM microscopy. The resulting microstructure was acknowledged as suitable and identical to commercially produced coatings.

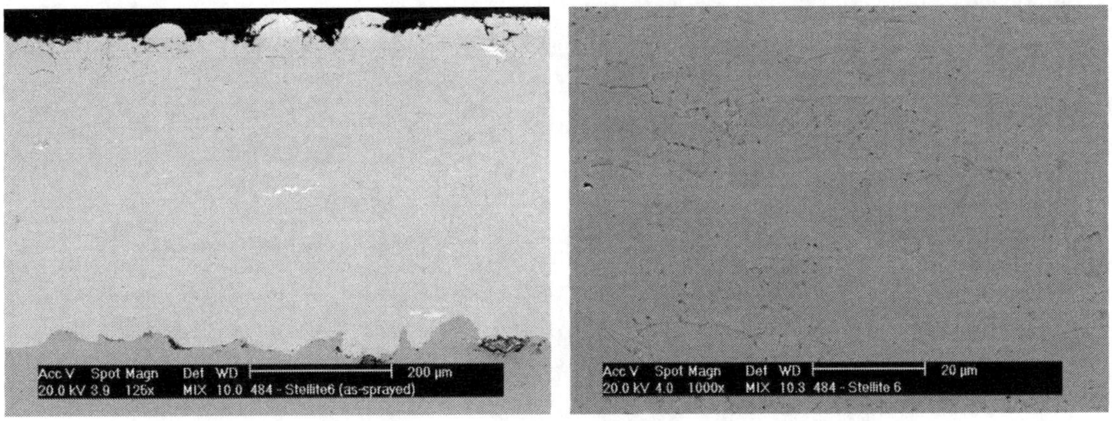

Figure 5. Stellite 6 coating microstructure cross-section; b) 125 magnification, a) 1000 magnification.

3.2. *Cohesion strength test*

Following section of this paper represents results of cohesion strength testing conducted according to previously mentioned parameters. Figures 6-9 represents results of cohesion strength testing of each selected coating separately. Results of average values for each coating are provided in table 1 and figure 10. Each coating was tested seven times to ensure relevant results. The lowest and the biggest

IOP Conf. Series: Journal of Physics: Conf. Series **843** (2017) 012007 doi:10.1088/1742-6596/843/1/012007

value obtained for each and singe coating was excluded. This procedure should ensure more accurate results. Results of coating CrC-NiCr (see figure 5) were the most volatile from all measurements. Also, this coating shown lowest cohesive strength from all tested materials. The reason of this results is the most inhomogeneous microstructure of all tested coatings (as shown in figure 2). Carbide particles embedded in metallic matrix acts as tension concentrator and serves as initiator of fracture cracking.

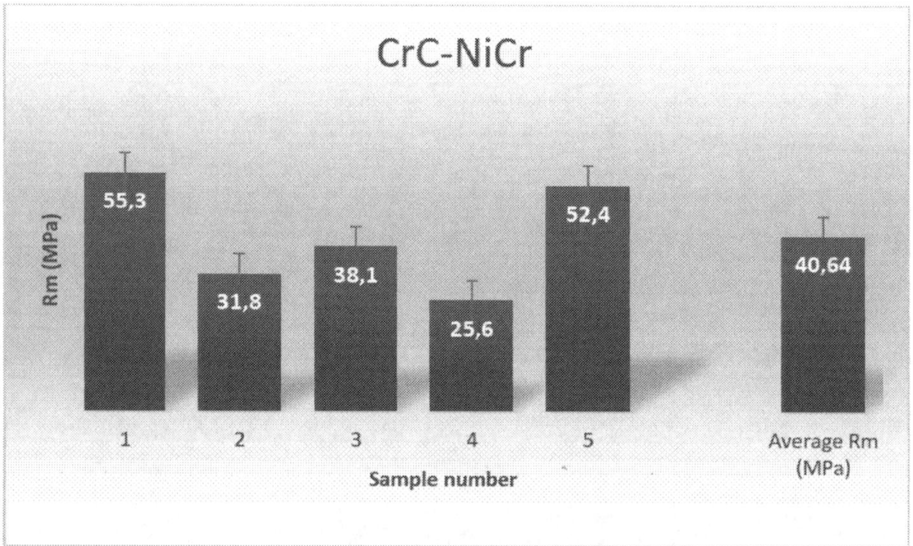

Figure 6. Cohesive strength results of coating C_3r_2C-NiCr.

Coating Hastelloy C-276 (results shown in figure 6) provided lower results than expected but with low deviation. Expectations ranged to be somewhere around Stellite 6 results (displayed further in this paper). The reason for this behaviour is still unclear because even the microstructure (figure 3) is fine and homogenous. Causes of this behaviour will be further examined.

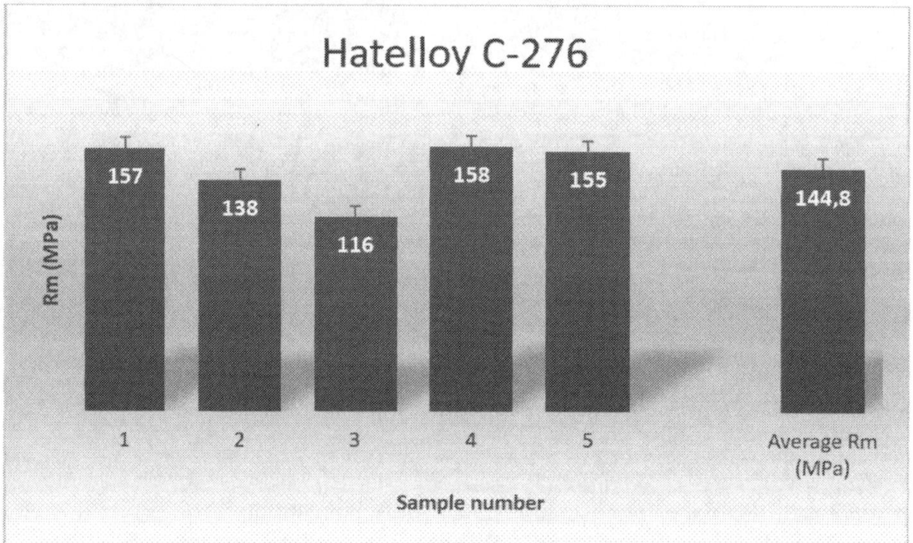

Figure 7. Cohesive strength results of coating Hastelloy C-276.

Alloy based coating NiCrBSi (figure 7) showed average values between all tested coatings with decent result deviation. It is expected that reason for lower values of this coating is formation of new structures in microstructure (figure 4).

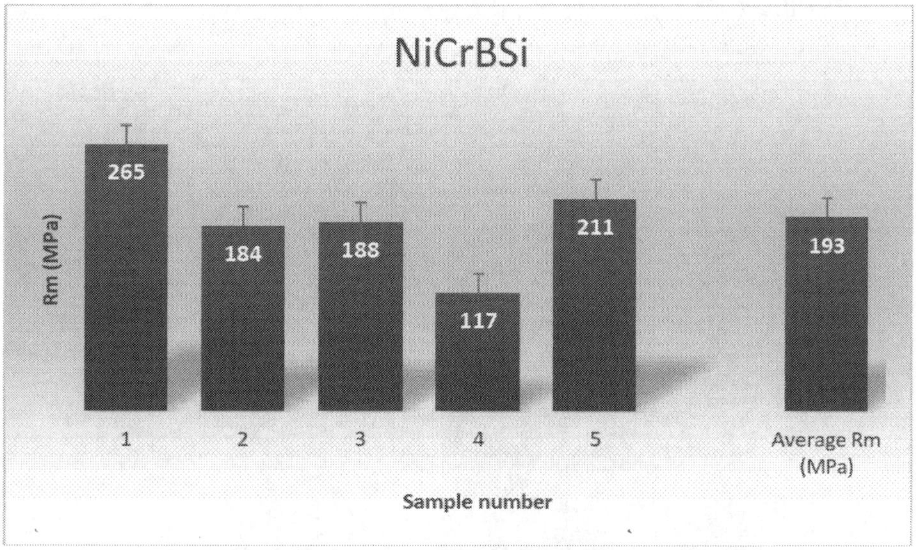

Figure 8. Cohesive strength results of coating NiCrBSi.

Last but not least alloy based coating Stellite 6 (figure 9) provided best results according to expectance. Low deviation of values ensures reproducibility of the test. This results also points at well done optimisation of thermal spray process ensuring fine microstructure with low share of flaws and defects.

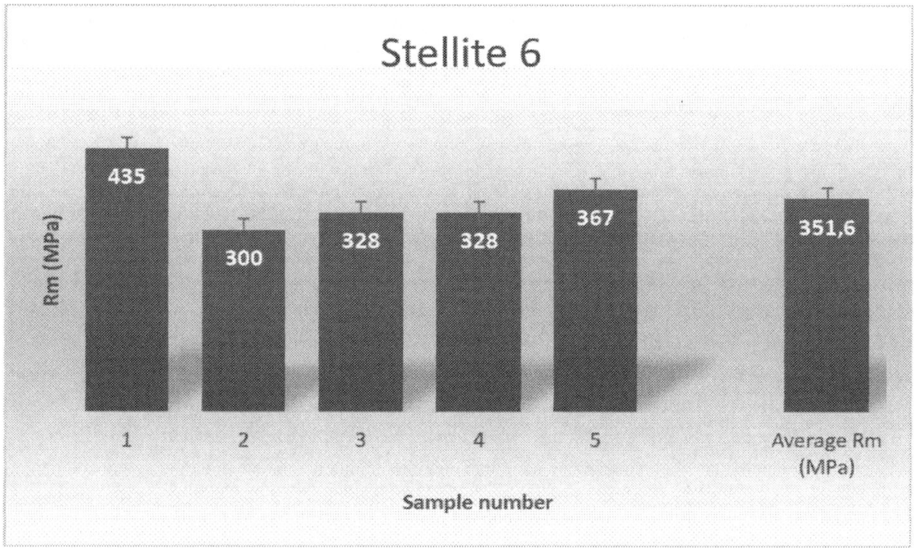

Figure 9. Cohesive strength results of coating Stellite 6.

Table 1. Cohesive tensile strength average values.

Coating	Rm (MPa)	Standard deviation
NiCrBSi	193	23,8
Stellite 6	351,6	23,4
CrC-NiCr	40,6	5,7
Hastelloy C-276	144,8	8,0

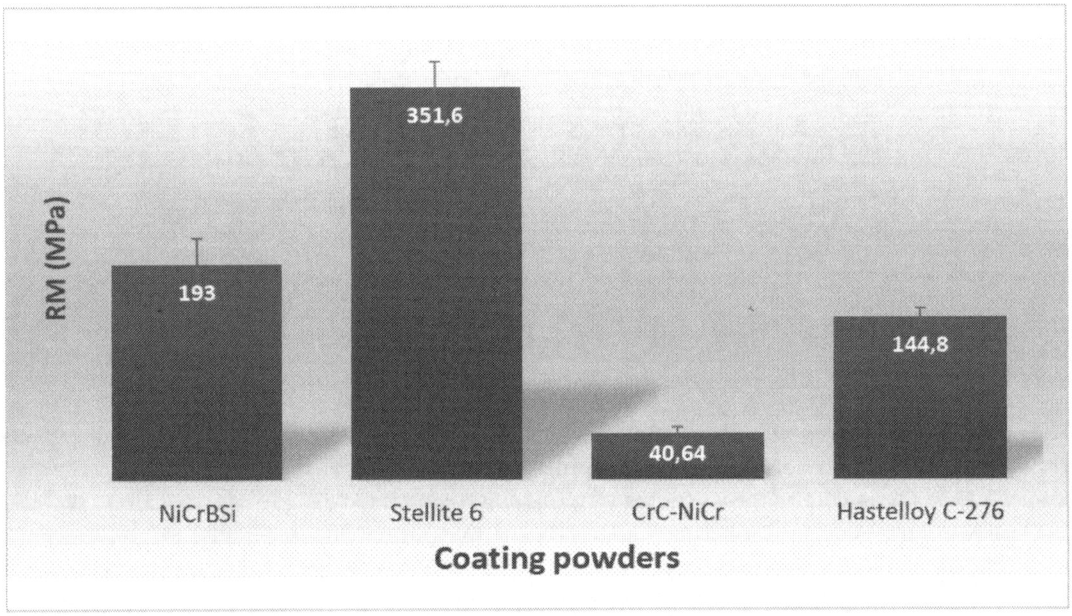

Figure 10. Comparison of average cohesive strength of tested coatings.

3.3. Fracture surface

Following paragraph is focused on example of characteristic representatives of two fracture surfaces. Figure 11a shows fracture surface of alloy based coating Hastelloy C-276. Each of tested alloy based materials exhibited same fracture surfaces and it is redundant to display all of them. The picture also shows fragile fracture mechanism with small plastic areas. This behaviour is based on microstructure of coating which is mainly consisted from splats mechanically bonded together.

On the other hand, carbide based coating CrC-NiCr is great representative of this category of coatings. Figure 11b shows fracture surface of this coating and it is clear, that the mechanism of cracking is plainly based on fragile cracking together with splatation and delamination.

Figure 11. Fracture surfaces (230 magnification); a) Hastelloy C-276, b) CrC-NiCr.

4. Conclusion

Results obtained by this research proved, that there are still new ways how to evaluate mechanical properties of thermally sprayed coatings. Quick, cheap and reliable determination of cohesive strength of coatings is important from many perspectives. Excluding the customers demand to know exact properties of delivered coating there are other useful applications for this method.

It is possible to correlate cohesive strength with mechanism of wear which allows to predict the behaviour of specific coating under defined stress conditions. On the other hand, this method can be used to correlate fatigue properties of coated system.

Results indicate that this method is primarily suitable for alloy based coatings where it provides replicable measurements. Alloy based coatings exhibit low volatility in measured values with low standard deviation in proportion to measured values. For this reason, this method seems to be reliable for such use.

Results obtained measuring carbide based coatings ale less convincing but still the 12.5% deviation is acceptable.

The research will continue as mentioned above and will focus on other carbide based coatings to prove or disproof usefulness of such method for this type of coatings. Furthermore, the correlation between cohesive strength and wear mechanism will be examined together with influence of residual stress influence.

References

[1] Vencl A Arostegui S Favaro G Zivic F Mrdak M Mitrović S Popovic V 2011 *Evaluation of adhesion/cohesion bond strength of the thick plasma spray coatings by scratch testing on coatings cross-sections,* (Tribology International/Elsevier) pp 1281-1288

[2] Yang Y C Chang E 2001 *Influence of residual stress on bonding strength and fracture of plasma-sprayed hydroxyapatite coatings on Ti–6Al–4V substrate*, (Biomaterials / Elsevier) pp 1827–1836

[3] Kadolkar P Dahotre N B 2003 *Effect of processing parameters on the cohesive strength of laser surface engineered ceramic coatings on aluminum alloys (Materials Science and Engineering: A)* (Elsevir) pp 183–191

[4] Ivosevic M Knight R Kalidindi S R Palmese G R Sutter K L 2005 *Adhesive/cohesive properties of thermally sprayed functionally graded coatings for polymer matrix composites* (Journal of Thermal Spray Technology) pp 45–51

[5] Kim H J Kweon Y G Chang R W 1994 *Wear and erosion behavior of plasma-sprayed WC-Co coatings* (Journal of Thermal Spray Technology) pp 169–178

Acknowledgement
The paper was prepared thanks to the project of Technology Agency of the Czech Republic no. TE01020068.

Fracture toughness in Mode I (G_{IC}) for ductile adhesives

P Gálvez[1], RJC Carbas[2], RDSG Campilho[3], J Abenojar[1], MA Martinez[1] and LFM da Silva[2]

[1] Department of Material Science and Engineering, Universidad Carlos III de Madrid, Avenida de la Universidad 30, 28911, Leganés, Spain.

[2] Departamento de Engenharia Mecânica, Faculdade de Engenharia da Universidade do Porto, Portugal.

[3] Departamento de Engenharia Mecânica, Instituto Superior de Engenharia do Porto, Instituto Politécnico do Porto, Portugal.

pgalvez@ing.uc3m.es

Abstract. Works carried out in this publication belong to a project that seeks the replacement of welded joints by adhesive joints at stress concentration nodes in bus structures. Fracture toughness in Mode I (G_{IC}) has been measured for two different ductile adhesives, SikaTack Drive and SikaForce 7720. SikaTack Drive is a single-component polyurethane adhesive with high viscoelasticity (more than 100%), whose main use is the car-glass joining and SikaForce 7720 is double-component structural polyurethane adhesive. Experimental works have been carried out from the test called Double Cantilever Beam (DCB), using two steel beams as adherents and an adhesive thickness according to the problem posed in the Project, of 2 and 3 mm for SikaForce 7720 and SikaTack Drive, respectively. Three different methods have been used for measuring the fracture toughness in mode I (G_{IC}) from the values obtained in the experimental DCB procedure for each adhesive: Corrected Beam Theory (CBT), Compliance Calibration Method (CCM) and Compliance Based Beam Method (CBBM). Four DCB specimens have been tested for each adhesive. Dispersion of each G_{IC} calculation method for each adhesive has been studied. Likewise variations between the three different methods have been also studied for each adhesive.

1. Introduction

In recent years, the use of adhesives for structural applications [1] is growing due to the benefits and solutions that this technology is capable of providing to different industries. Among the potential beneficiaries of adhesive technology are Aerospace and Automotive Industries.

Using adhesives enable improving the performance obtained with traditional joints, allowing the joint of dissimilar materials, reducing the structure weight and manufacturing costs, improving the resistance to dynamic and static loads, reaching quasi homogeneous stress distribution and better damage tolerance [2].

Actually, traditional joining techniques (welding, rivets, screws…) continue being used mostly, but problems related to these joints make necessary to develop other joining technologies. It is necessary to know all the characteristics of any material designed for be used in structural applications, so that it is possible to predict its behavior for the raised solicitations. Finite element software are increasingly widespread in adhesive joints, being able to simulate the behavior of the joint. To obtain a correct operation of the program it is necessary to know precisely all the properties of the involved materials in the adhesive joint. Thus, mechanical characterization of adhesive is mandatory prior to its use.

Fracture resistance values of structural adhesives can be calculated with different tests, widely studied in the literature [3,4,5]. In this work, Double Cantilever Beam (DCB) test have been carried out in order to obtain the Fracture Toughness in mode I (G_{IC}) value of two ductile polyurethane adhesives, SikaTack Drive and Sika Force 7720.

Most common methodologies for analysis of G_{IC} are based on linear elastic fracture mechanics (LEFM), including Compliance Calibration Method (CCM) and Corrected Beam Theory (CBT), but recently developed method such Compliance-Based Beam Method (CBBM) is based on the crack equivalent concept, depending only on the specimen`s compliance during the test [6]. DCB test allows obtaining the resistance to crack initiation and propagation, being able to calculate the R-curve (Resistance Curve), plotting G_{IC} versus crack length [7].

1.1. DCB Data Analysis

To avoid the accumulation of data, in this work only the formulas are exposed, being able to find more complete information in [6] and [8]:

-Compliance Calibration Method (CCM):

$$G_{IC} = \frac{P^2}{2b}\frac{dC}{da}$$

P is the load and b the specimen width. The compliance (C) is calculated by $C = \frac{\delta}{P}$ being δ the specimen displacement. $C = f(a)$ curves are adjusted by cubic polynomials ($C = C_3a^3 + C_2a^2 + C_1a + C_0$).

-Corrected Beam Theory (CBT):

$$G_{IC} = \frac{3P\delta}{2b(a + |\Delta|)}$$

Corrected Beam Theory (CBT) is based on elementary beam theory including the effects of crack tip rotation and deflection, being Δ crack length correction for crack tip rotation and deflection. For calculating Δ, a linear regression analysis of $(C)^{1/3}$ versus a data is carried out.

-Compliance-Based Beam Method (CBBM) [8,9]:

$$G_{IC} = \frac{6P^2}{b^2h}\left(\frac{2a_{eq}^2}{h^2E_f} + \frac{1}{5G}\right)$$

This method, based on the crack equivalent concept, depends only on the compliance of the specimen during the test. a_{eq} is the equivalent crack length, estimated from the experimental compliance. $a_{eq} = a + |\Delta| + \Delta a_{FPZ}$, being a real crack length, Δ the root rotation correction for the initial crack length and Δa_{FPZ} the correction because the fracture process zone (FPZ) is considered. h is the specimen height; E_f is a corrected flexural modulus, which takes into account some phenomena that may affect the $P - \delta$ curves, such as the stiffness variability between different specimens; and G is the substrate´s shear modulus.

2. Experimental Procedure

2.1. *Double Cantilever Beam Test (DCB)*

Different materials can be used as substrate for developing DCB Test, normally metals such as Steel and Aluminum, but it is possible find other works carried out with substrates made of unidirectional fiber-composite materials [10]. Two steel substrates have been used in this work for manufacturing each specimen. So that each pair of substrates are bonded together with the adhesive. A cohesive failure is essential to calculate correctly the fracture toughness of the studied adhesives.

Tests are carried out at room conditions on a universal testing machine through an opening process by pin loading at the beam-ends, with 1mm/min of test velocity. Tensile testing machine reproduces for each test the curves of applied load versus displacement (P-δ curves). The crack propagation is measured over the length of the adhesive with a calibrated ruler using a digital photography system, taking photos every five seconds until the process is finalized (Figure 1).

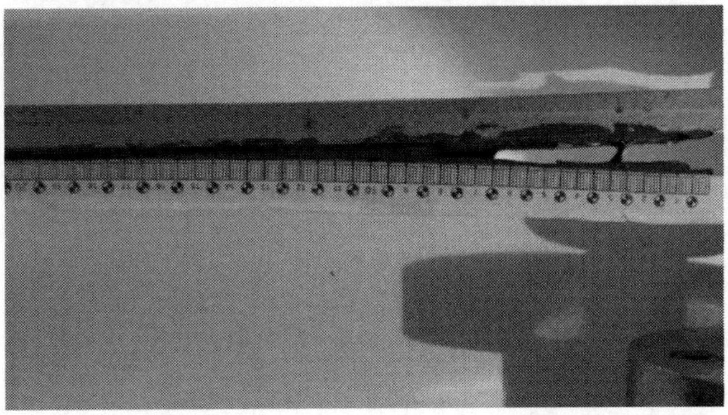

Figure 1. Crack propagation measured over the adhesive length.

2.2. *Test Geometry*

The test specimen sketch is shown in Figure 2, being F_c the applied force, whose values are obtained from the collected data during the test developed by the tensile testing machine. d= substrate thickness, being a constant value of 12.5mm. a=crack length, result of the sum of initial crack length and measured length values. w_{med}=substrate width, being a constant value of 25mm. The thickness of the adhesive layer is different on both adhesives, being 3 mm for Sikatack Drive and 2 mm for SikaForce 7720.

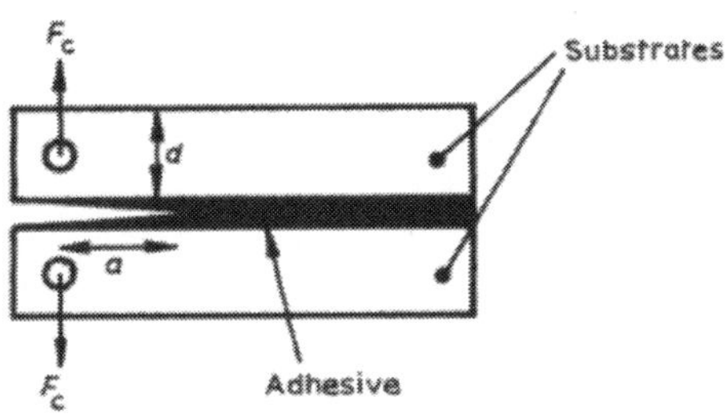

Figure 2. Measurements of DCB Specimen used [3].

IOP Conf. Series: Journal of Physics: Conf. Series **843** (2017) 012008 doi:10.1088/1742-6596/843/1/012008

2.3. Materials
Two different polyurethane structural adhesives supplied by Sika ® are used for the development of this work. One single-component adhesive with high viscoelasticity (more than 100%) whose trade name is SikaTack Drive. And other double-component ductile adhesive, called SikaForce 7720.

3. Results
From the DCB tests, $P - \delta$ curves were obtained for both adhesives. Representative curves of the studied adhesives are shown in Figure 3. As shown in the figure, the double-component adhesive (SikaForce 7720) presents less elongation at break due to its lower elastoplasticity. In the case of SikaTack, a greater elongation for higher values of rupture is obtained. In terms of elastic-plastic bahavior, it is important to remark that the most ductile adhesive, SikaTack Drive, is able to continue deforming for a longer time before reaching the minimum breaking value.

Figures 4 and 5 show representative curves of Resistance (R-Curve) of both adhesives. G_{IC} values are obtained from these graphs. The horizontal part of the curves represents G_{IC} values, so that, after a first peak of maximum G_{IC} value, this value stabilizes. This stabilization zone shows a set of similar values, where the sought value is obtained. G_{IC} values for both adhesives are shown in Table 1.

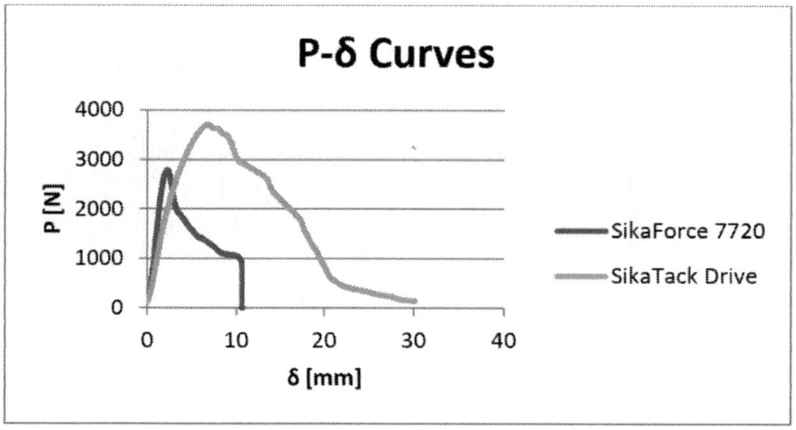

Figure 3. P-δ curves for SikaForce 7720 and SikaTack Drive. As the graph shows, SikaTack Drive is more ductile than SikaForce 7720, being also greater the maximum breaking value.

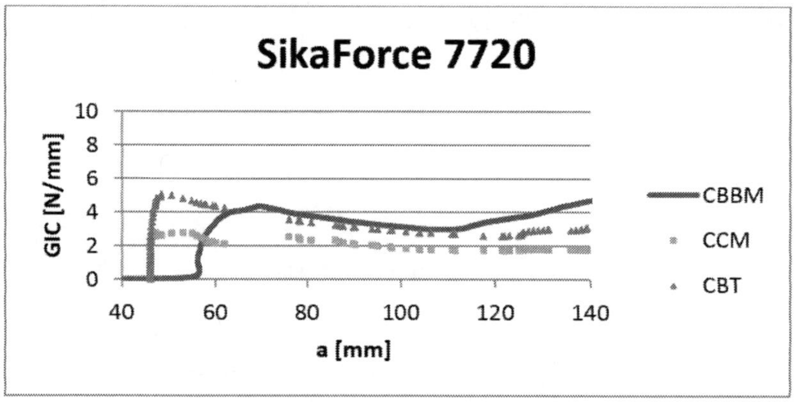

Figure 4. Representative Resistance Curve (R-Curve) for SikaForce 7720.

IOP Conf. Series: Journal of Physics: Conf. Series **843** (2017) 012008 doi:10.1088/1742-6596/843/1/012008

As Figure 4 shown (SikaForce 7720), CBBM and CBT curves are similar, being the curve for CCM similar but with lower G_{IC} values. Figure 5 (SikaTack Drive) shows similar curves for CBBM and CCM models, being different the CBT curve. CBBM model depends only on the compliance of the specimen during the test, with G_{IC} not being affected by errors due to the crack propagation measuring over the length of the adhesive with a calibrated ruler.

The CBT model includes a crack length correction for crack tip rotation and deflection (Δ). So better behavior is obtained in CBT model for Sikaforce 7720 than for SikaTack Drive. In the case of CCM model, the opposite happens. Better elastic-plastic behavior of SikaTack Drive allows better adjustment of $C = f(a)$ curves by cubic polynomials ($C = C_3a^3 + C_2a^2 + C_1a + C_0$).

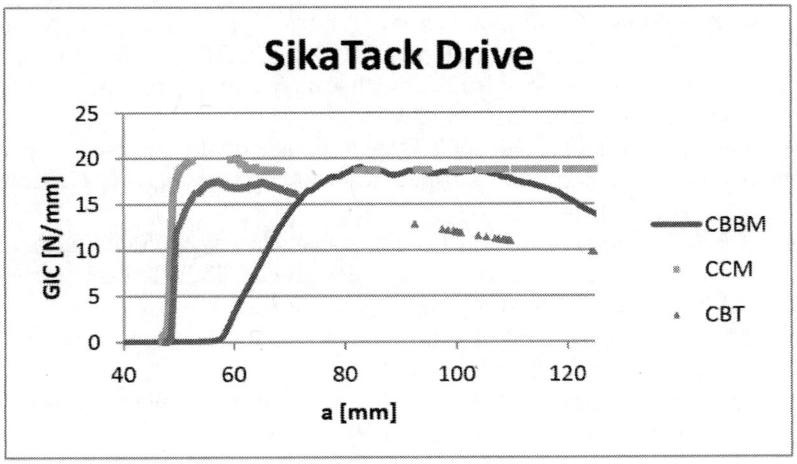

Figure 5. Representative Resistance Curve (R-Curve) for SikaTack Drive.

Table 1. G_{IC} values (in N/mm) for both studied adhesives with the three employed models.

	SikaTack Drive	SikaForce 7720
CBBM (N/mm)	20.63±1.56	3.83±0.72
CCM (N/mm)	18.35±1.40	2.18±0.28
CBT (N/mm)	17.45±2.77	3.24±0.86

4. Conclusions

Recent works have demonstrated the validity of using CBBM model in adhesives, being probably the most reliable method to obtain G_{IC}. This is due to the difficulty of measuring experimentally crack growth with an optical system, which is aggravated by high elastic-plastic behavior of both studied adhesives. The CBBM is only based on the crack equivalent concept, depending only on the compliance of the specimen during the test.

The CBT model shows better results for SikaForce 7720 and CCM model shows better results for SikaTack Drive. This is due to the different properties of each adhesive, together with the different adjustment of each model.

IOP Conf. Series: Journal of Physics: Conf. Series **843** (2017) 012008 doi:10.1088/1742-6596/843/1/012008

5. References

[1] Higgins A 2000 Adhesive bonding of aircraft structures *International Journal of Adhesion and Adhesives* **20** 367-76
[2] Banea M D and da Silva L F M 2009 Adhesively bonded joints in composite materials: an overview *Proceedings of the Institution of Mechanical Engineers, Part L: Journal of materials design and applications* **223**
[3] Kinloch A J 1987 Fracture mechanics of adhesive joints *Adhesion and Adhesives* chapter 7 pp 264-338
[4] Blackman B R K 2011 Fracture Tests *Handbook of Adhesion Technology* pp 474-501
[5] Lee R J, Butler J R, Yates T and Davidson R 1997 Test methods for adhesive fracture properties *MTS adhesives project 2 failure modes and criteria* Report No. 5 Annex 1
[6] Banea M D, da Silva L F M and Campilho R D S G 2012 Effect of temperature on tensile strength and mode I fracture toughness of a high temperature epoxy adhesive *Journal of adhesion science and technology* **26** 939-53
[7] Lamut M, Turk R and Torkar M 2008 Determination of the adhesive fracture energy G_C of structural adhesives using DCB and peel tests *RMZ-Materials and Geoenvironment* **55** 476-89
[8] de Moura M F S F, Campilho R D S G and Gonçalves J P M 2008 Crack equivalent concept applied to the fracture characterization of bonded joints under pure mode I loading *Composites science and technology* **68** 2224-30
[9] de Moura M F S F, Gonçalves J P M, Chousal J A G and Campilho R D S G 2008 *International Journal of adhesion and Adhesives* **28** 419-26
[10] Blackman B, Dear J P, Kinloch A J and Osiyemi S 1991 The calculation of adhesive fracture energies from double-cantilever beam test specimens *Journal of materials science letters* **10** 253-56

Acknowledgments

Spanish Ministry of Education and Science through the Project TRA2014-56471-C4-2-R, Institute of Materials Alvaro Alonso Barba (IAAB) and Sika®.

IOP Conf. Series: Journal of Physics: Conf. Series **843** (2017) 012009 doi:10.1088/1742-6596/843/1/012009

Failure Analysis of Fencing Blades

D Kibaroglu[1,a], M Baydogan[1,2,b], H Cimenoglu[1,c], B Bas[1], C Yagsi[1], N Aliyeva[1]

[1] Department of Metallurgical and Materials Engineering, Istanbul Technical University, 34469 Maslak – Istanbul / TURKEY

[2] Prof. Dr. Adnan Tekin Material Sciences and Production Technologies Applied Research Centre, Istanbul Technical University, 34469 Maslak – Istanbul / TURKEY

[a]dilaykibaroglu@gmail.com, [b]baydogan@itu.edu.tr, [c]cimenogluh@itu.edu.tr

Abstract. This study deals with the failure analysis of broken fencing blades (one épée and one foil). For the characterization of the broken blades, metallographic examinations, chemical analysis, hardness measurements, fracture surface examinations and tensile tests were performed. Maximum stress occurred at the outer fibres of the blades was estimated as high as 1456 MPa and 1298 MPa for épée and foil, respectively. Results showed that failure of the blades was initiated from a notch, which has been formed as the result of an impact action during training, or from the groove machined along the blade for inserting an electrical wire. In order to increase resistance of the blades against such failures, alternative blade material, modified blade geometry and a surface hardening treatment were proposed.

1. Introduction

Fencing is an art of assault and defence with a sword or similar weapon in which two athletes fight indirectly, through their weapons, and physical contact is forbidden. Modern fencing is a competitive sport of which rules are based on the old-time duelling [1]. There are three main blades used in modern fencing, which are épée, foil, and sabre. There are specific requirements and properties which are determined by FIE (International Fencing Federation) for these three type of blades. Each of them has different technique and target area. Therefore, shapes, geometries, cross-sections, and flexibilities of the blades are different from each other. For instance, foil is rectangular while épée is triangular in cross section. Total lengths of the swords are also different from each other. The maximum length of foil and épée swords is 110 cm each while sabre has a length of 105 cm. In addition to these specifications, there are several testing methods, which are performed on the blades used in Olympic Games and international professional competitions, to determine various properties including ductility, fracture toughness, corrosion resistance, fatigue strength and traction as well as eddy current inspection is also required [2]. Since the blades are exposed to continuous bending during fencing games, they are expected to have superior mechanical properties to ensure a longer life. That is why, it is commonly desired that the blades are resistant to fracture and fatigue [3]. To meet these requirements, maraging steel is mostly preferred material and is used in professional competitions due to its ultra-high strength (up to 2500 MPa), high toughness and fracture toughness. Maraging steel blades are manufactured by forging and are then subjected to a suitable heat treatment, which is an essential step for the blade to gain superior mechanical properties [4, 5]. Heat treatment of maraging steels is a demanding and high cost process which is why the maraging steel blades are very expensive. This leads to the emergence of a need for a cheaper alternative and spring steels stand out because of their bending resistance and ability to return to their original shape without deforming. Spring steel is started to be used in amateur fencing competitions or trainings as a lower quality alternative to maraging steel. Nonetheless, it has failed at possessing the requirements even for the inferior fencing blades. The blades exhibit poor durability and are broken easily after a short period of usage. Moreover, the lack of standardization for production of lower quality fencing blades makes it difficult to obtain desired properties.

In this study, fencing blades which are used in amateur competitions or trainings were investigated. Fracture analysis was performed on the broken blades to understand what causes the fracture and to offer a solution to this problem.

2. Experimental Procedure

Swords which have been taken from ITU Fencing Club were examined. Some of the swords were broken during trainings or competitions as shown in figure 1. Duration until breakage can vary depending on the quality of blades. The higher the quality, the longer will be the durability of the weapon.

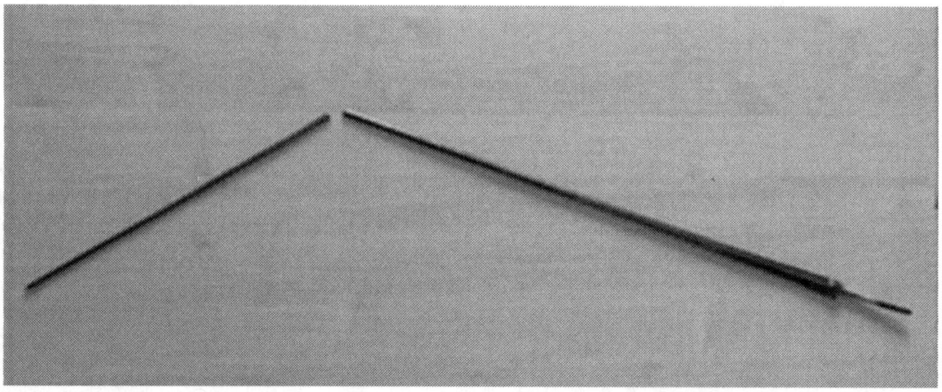

Figure 1. General view of a broken blade (épée).

In order to determine the properties of the swords and to obtain detailed information; chemical, microstructural, and mechanical analysis were performed on épée and foil blades. Firstly, chemical analysis was done for both épée and foil blades by optical emission spectroscopy (GNR Metal Lab Plus, Italy) to identify the blade material. Before chemical analysis, one sample for each blade was cut from the close-handle sections of the blades and they were sequentially cleaned in boiling water and alcohol. For microstructural analysis, two samples for each blade were cut from the tip and handle sections of the blades and were mounted by epoxy resin at 160°C under a pressure of 98 MPa. The samples were gently ground and polished by using SiC papers (with successive use of 400, 600, 800, 1200, 2500 grid emery papers) and alumina suspensions (with 1 μm and 0.3 μm particle size and 200 rpm rotating speed). The mirror finished samples were examined by an optical microscope (Leica ICC 50 HD, Germany) before etching to determine whether there are inclusions. Then, etching was applied to the samples by using 5% nital solution. When microstructural analysis was completed, hardness measurements of the samples were carried out by hardness tester (Zwick/Roell ZHR, Germany). The hardness scale was selected as Rockwell C using a normal load of 150 kgf (1471 N) and a diamond cone indenter. The measurements were repeated as five times for each sample and the values were averaged. Surface defects and fracture surfaces of the broken blades (an épée and a foil) were examined by a digital microscope (Cooling Tech, China) and a scanning electron microscope (Hitachi TM 1000, Japan) in back scattered mode. Stress analysis was performed on an épée and a foil blade to reveal the stresses occurred on the blades. For this purpose, bending stresses exerted on the outer fibers of the blades were estimated by stress – deflection equations. Tensile tests were carried out to determine tensile strength of the blades and to compare it with the results of the stress analysis. One specimen with approximately 20 cm in length for each blade was cut from the close-tip sections. Test was performed once for each blade with a constant cross head speed of 2 mm per minute. Tensile strength was calculated by dividing the maximum load to the cross sectional area of the blade at the point of fracture.

3. Results

3.1. Chemical Analysis and Structural Characterization

Chemical analysis by optical emission spectroscopy revealed that blade materials are 56Si7 quality spring steel according to EN 10132-4 standard [6]. Figure 2 shows as-polished optical micrographs of the samples. It is clear that blade microstructures contain some non-metallic inclusions which are spherical in shape and possibly belong to oxide inclusions. Etched optical micrographs of the samples were shown in figure 3. Both épée and foil blades have almost the same characteristic features in their micrographs. They have a ferritic–pearlitic microstructure close to the handle, while contain martensite phase close to the tip of the blades which are frequently exposed to impact during competitions. It clearly suggests that a quenching heat treatment has been applied to the tip of these blades.

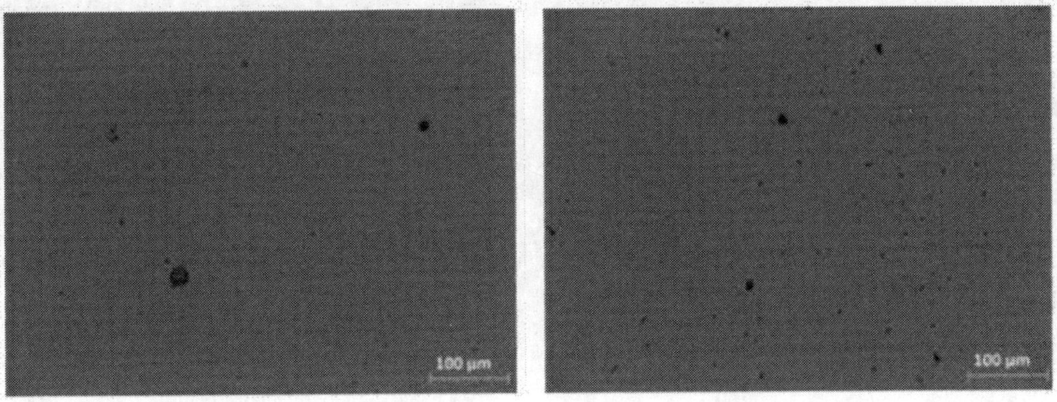

(a) (b)

Figure 2. As-polished optical micrographs of the blade materials (a) épée, (b) foil.

Hardness values, measured from the handle and the tip sections of épée were 25 and 56 HRC, respectively. Similarly, the foil has a hardness of 33 HRC at its handle and 59 HRC at its tip. Measured hardness values from different sections of the blades were in well agreement with the microstructural characteristics observed in the corresponding sections of the blades shown in figure 3a and b.

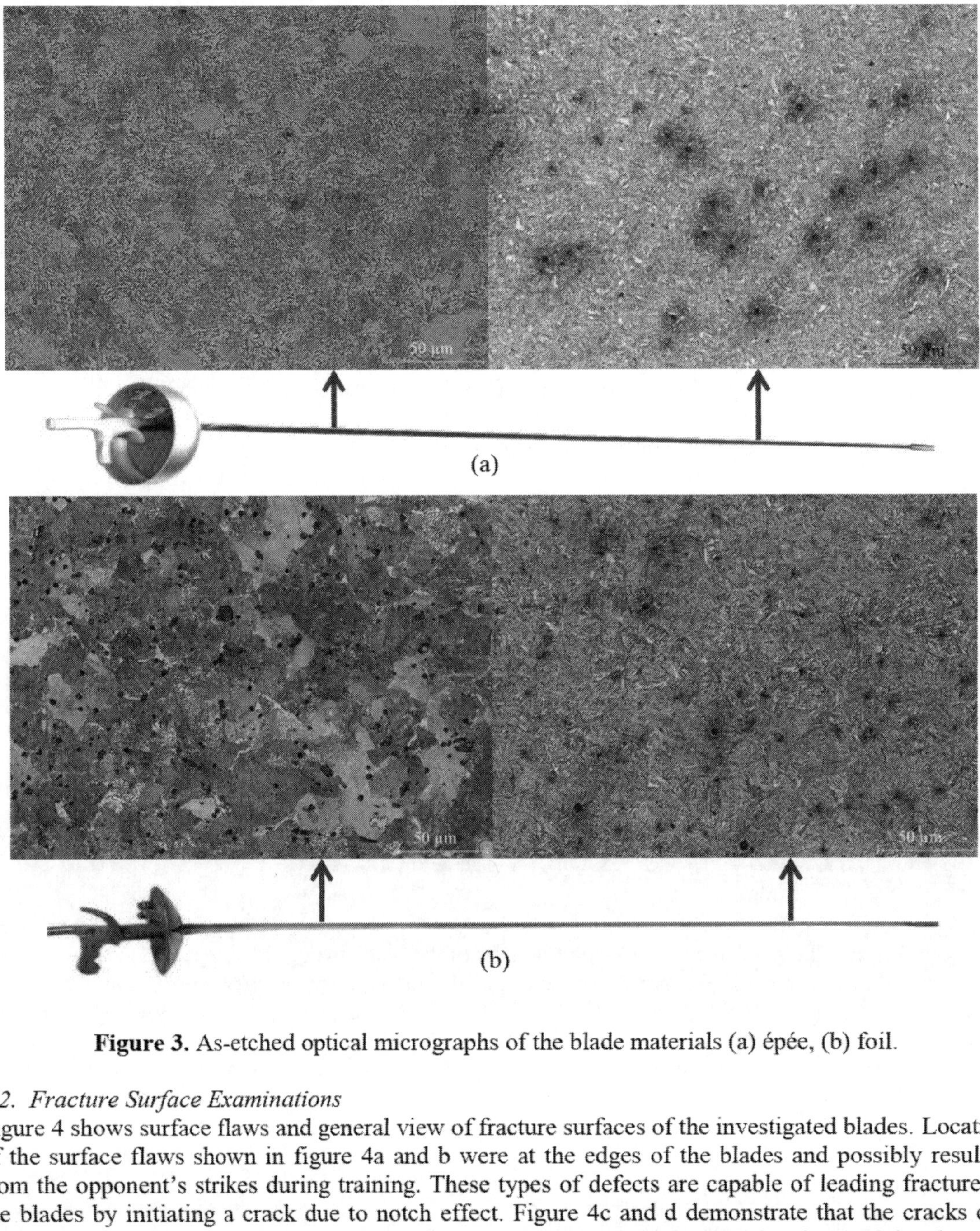

Figure 3. As-etched optical micrographs of the blade materials (a) épée, (b) foil.

3.2. Fracture Surface Examinations

Figure 4 shows surface flaws and general view of fracture surfaces of the investigated blades. Location of the surface flaws shown in figure 4a and b were at the edges of the blades and possibly resulted from the opponent's strikes during training. These types of defects are capable of leading fracture of the blades by initiating a crack due to notch effect. Figure 4c and d demonstrate that the cracks can also be originated from the grooves, where stress concentration is high. It can be observed that fracture characteristics of both épée and foil were brittle in nature without exhibiting any significant plastic deformation as shown in figure 4c and d. SEM fractographs shown in figure 5a and b revealed that general characteristics of the fracture surface were mixed mode fracture which contains intergranular brittle fracture and dimple areas.

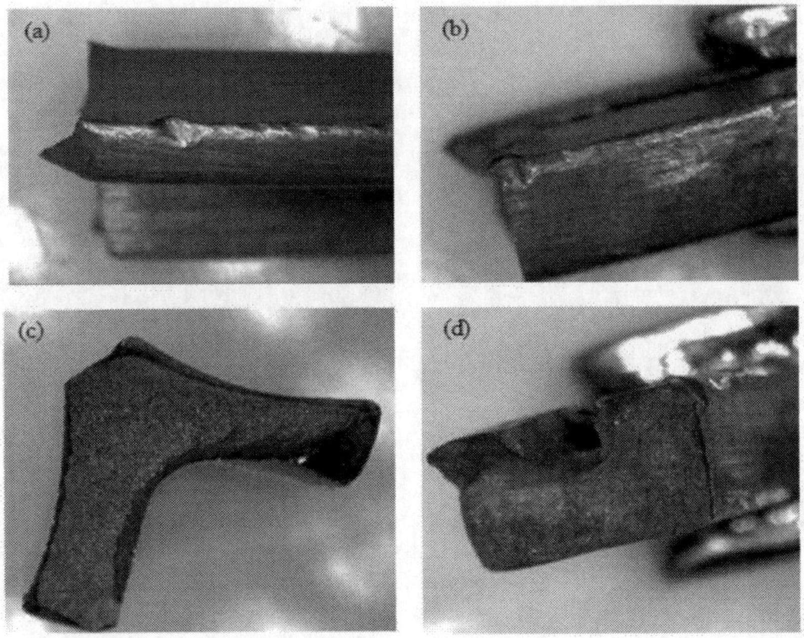

Figure 4. Surface flaws on the surface of (a) épée, and (b) foil, and general view of fracture surface of (c) épée and (d) foil.

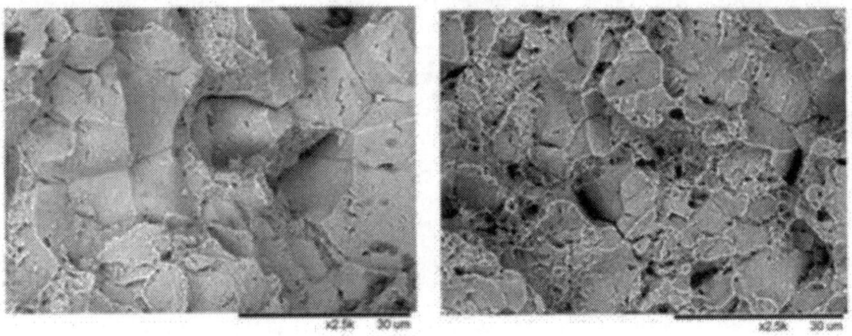

(a) (b)

Figure 5. SEM images of the fracture surfaces of (a) épée, and (b) foil.

3.3. Estimation of Stress Exerted on the Blades

In order to estimate maximum stress exerted on the blades during competition, the blades were first forced to be bent by a fencer as much as a real competition (figure 6). In this loaded condition, maximum deflection of the blades was measured. Assuming that deflection of the blades in this loaded configuration is symmetrical along the length of the blades, bend radius was calculated and the corresponding tensile stress at the outermost fibers of the blades was estimated according to equation (1)

$$\sigma = \frac{E \cdot Y}{R} \tag{1}$$

where σ is tensile stress, Y is distance from the neutral axis, E is modulus of elasticity (210 GPa) and R is bend radius. Based on this approximation, maximum stresses which are exerted on the outermost fibers of the blades were estimated as 1456 MPa and 1298 MPa for épée and foil, respectively [7].

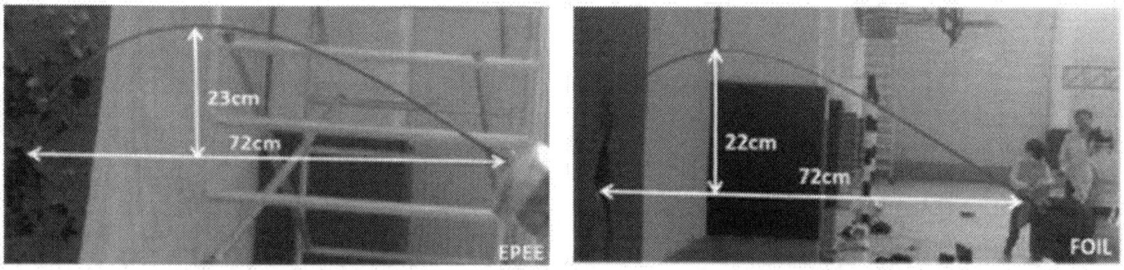

(a) (b)

Figure 6. Bending of blades during attack, (a) épée and (b) foil.

It is clearly shown by the above calculations that very high stresses are exerted on the surface of the blades during a real competition. In order to resist such high stresses, the blades should be made of a material with a very high strength, as described in the previous sections. Tensile strength of the blades were determined as 1538 MPa and 1939 MPa for épée and foil, respectively. When these values are compared with estimated stress values exerted on the blades, it is clear that maximum stress is close to the tensile strength of the blades. It also dictates that it is critical to choose a very high strength material to prevent a sudden failure during fencing.

4. Discussion

As presented in the previous sections, very high stresses, which are comparable to tensile strength of the material, can be produced during a fencing attack. It is therefore likely to be occurred a sudden fracture of the blades, which might be dangerous for the fencers. In order to prevent occurrence of such a failure, a previous research suggests a steel – steel composite (martensite fibers in a tough austenite phase) [3]. In this study, on the other hand, three approaches - selecting a very high strength material, modifying the geometry of the blade or modifying the surface of the blade materials - were developed to eliminate the problem. Details of these suggestions were given in the following subsections.

4.1. Alternative Material

Precipitation hardened stainless steels, which are a group of very high strength steels, were suggested to be an alternative material for high strength fencing blades. Well known member of this alloy is 17-4 PH whose name stems from the additions of 17% Cr and 4% Ni. This steel also contains 0.3% Nb and 4% Cu. Mechanical, chemical and physical properties of precipitation hardening stainless steels are similar to those of maraging steels [8]. That is why, they can meet the requirements for a high strength fencing blade. Nevertheless, 17-4 PH steels are almost as expensive as maraging steels since they have similar heat treatment process, yet they are still considerable alternative despite the high cost.

4.2. Modifying the Blade Geometry

As stated in Section 3.2, cracks in the fencing blades can be originated by the notches on the surface or from the groove machined along the blades. Therefore, modifying the blade geometry by changing the depth of the groove, through which the electric wire passes, was suggested to prevent occurrence of high stress concentration on the blades. This can help reducing the stress concentration factor (K_t). Typical cross sectional view of a foil is shown in figure 7. According to equation (2) it can be clearly seen that h value considerably affects K_t value [9]. When groove depth is decreased from 1 mm to 0.4

mm, K_t value decreases from 2.79 to 1.94. This change corresponds to 30% decrease in stress concentration factor and minimizes the possibility of crack initiation from the groove.

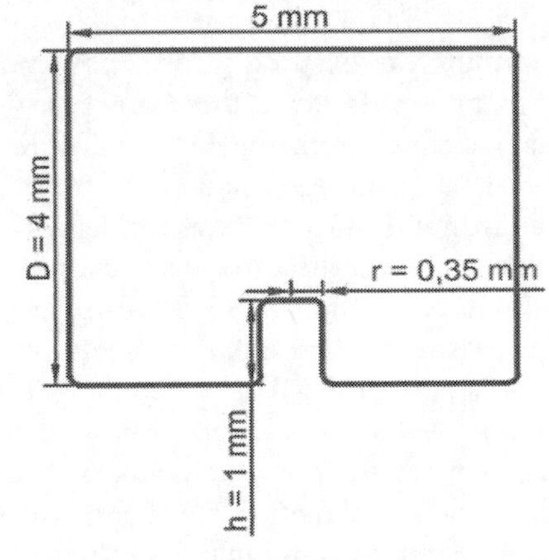

Figure 7. Schematic cross-sectional view of the foil.

$$K_t = C_1 + C_2\left(\frac{h}{D}\right) + C_3\left(\frac{h}{D}\right)^2 + C_4\left(\frac{h}{D}\right)^3 \quad (2)$$

h is groove depth, D is thickness of the foil and C_1, C_2, C_3, and C_4 are parameters determined by h/r ratio. Before the modification, since h/r ratio (2.86) was in between 2 and 20, C values are calculated through equation (2.1-2.4);

$$C_1 = 2.966 + 0.5\left(\frac{h}{r}\right) - 0.09\left(\frac{h}{r}\right)^2 \quad (2.1)$$

$$C_2 = -6.475 - 1.126\left(\frac{h}{r}\right) + 0.019\left(\frac{h}{r}\right)^2 \quad (2.2)$$

$$C_3 = 8.023 + 1.253\left(\frac{h}{r}\right) - 0.02\left(\frac{h}{r}\right)^2 \quad (2.3)$$

$$C_4 = -3.572 - 0.634\left(\frac{h}{r}\right) - 0.01\left(\frac{h}{r}\right)^2 \quad (2.4)$$

After the modification, since h/r ratio (1.14) was in between 0.5 and 2, C values are calculated through equation (2.5-2.8);

$$C_1 = 1795 + 1.481\left(\frac{h}{r}\right) - 0.211\left(\frac{h}{r}\right)^2 \quad (2.5)$$

$$C_2 = -3.544 - 3.677\left(\frac{h}{r}\right) + 0.578\left(\frac{h}{r}\right)^2 \quad (2.6)$$

$$C_3 = 5.459 + 3.691\left(\frac{h}{r}\right) - 0.565\left(\frac{h}{r}\right)^2 \quad (2.7)$$

$$C_4 = -2.678 - 1.531\left(\frac{h}{r}\right) + 0.205\left(\frac{h}{r}\right)^2 \quad (2.8)$$

4.3. Surface Hardening

Another solution to prevent failures of the blades can be surface hardening of the blades by a proper heat treatment to increase their resistance to fencer's strikes. For this purpose, the proposed heat treatment is nitriding without formation of a brittle white layer. It can be carried out by forming nitride layers composed of S-phase which is a metastable nitrogen supersaturated phase on the surface and it is possible to apply it to spring steel with either gas or plasma nitriding methods [10]. By means of

IOP Conf. Series: Journal of Physics: Conf. Series **843** (2017) 012009 doi:10.1088/1742-6596/843/1/012009

nitriding, formation of notches on the surface of the blades becomes more unlikely and fracture possibility decreases.

5. Conclusion

This study was undertaken to contribute the existing information and to provide some metallurgical insights on the blades used in fencing sports since there are few works available in the literature concerning material characteristics of such equipment. Hardness of the blades in near tip sections were determined in the range of 56-59 HRC while near handle sections had a hardness of 25-33 HRC. Failure of the blades was originated from the surface flaws formed as a result of fencer's strike or from the grooves machined along the blades for inserting the electrical wire. Tensile stresses exerted on the blades were estimated as high as 1456 MPa and 1298 MPa for épée and foil, respectively, based on maximum deflection which can be possible during a real competition. Tensile strength values of épée and foil were determined as 1538 MPa and 1939 MPa, respectively. Since maximum stresses occurred on the blades were very high fraction of tensile strength, it was therefore concluded that very high strength materials such as 17-4 PH stainless steels should be used for manufacturing the blades. On the other hand, it was proposed that blade geometry can be modified in such a way that stress concentration around the groove can be minimized. Finally a proposed heat treatment, namely S-phase nitriding without the formation of a brittle white layer, might be performed to increase resistance of the surface against the formation of crack inducing flaws during fencing.

Acknowledgements

Authors greatly acknowledge the supports provided by Istanbul Technical University (ITU) Fencing Club during this study.

References

[1] Fare M 2014 History of Fencing, British Academy of Fencing. Retrieved January 26, 2017 from http://www.baf-fencing.com/history-of-fencing.html

[2] FIE 2014 Book 3 FIE Material Rules for Competitions. Retrieved January 28, 2017 from http://fie.org/uploads/3/18830-weapons%20and%20 equipment %20 control.pdf

[3] Smallman R E and Bishop R J 1999 Modern Physical Metallurgy and Materials Engineering Science, Process, Applications. (6th Eds.), Materials for sports, (Oxford: Butterworth-Heinemann)

[4] Kubesh K, McNeil N and Bellotto K 2007 *Fencing: Ancient to modern.* California: In the Hands of a Child

[5] Heat-treatment for fencing weapons 2014 IPE Magazine Retrieved on January 28, 2017 from http://www.ipesearch.co.uk/page_521688.asp

[6] British Standards Institution. Cold rolled narrow steel strip for heat treatment – Technical delivery conditions - Part 4: Spring steels and other applications. EN 10132-4:2000

[7] Kibaroglu D et al. 2016 *Design of Fencing Weapons* (Bachelor's thesis, Istanbul Technical University, Istanbul, Turkey)

[8] AK Steel 2007 17-4 PH stainless steel Retrieved on January 30, 2017 from http://www.aksteel.com/pdf/markets_products/stainless/precipitation/17-4_ph_data_sheet.pdf

[9] Young W C and Budynas R G 2002 *Roark's Formulas for Stress and Strain.* (7th Eds.), Stress Concentration, (New York: McGraw-Hill Companies)

[10] Baranowska J et al 2009 The Properties of Hard Coating Composed of S-Phase Obtained by PVD Method (Koszalin: Koszalin University of Technology) **33** 4

IOP Conf. Series: Journal of Physics: Conf. Series **843** (2017) 012010 doi:10.1088/1742-6596/843/1/012010

Stress intensity factor for small edge crack in interfacial singularity region

Kazuhiro Oda[1], **Masato Kawakita**[2], **Noriko Tsutsumi**[1]

1 Department of Mechanical and Energy Systems Engineering, Oita University, 700 Dunnoharu, Oita, Japan
2 Graduate school of Oita University, 700 Dunnoharu, Oita, Japan

E-mail: oda-kazuhiro@oita-u.ac.jp

Abstract. Fracture of the bonded dissimilar materials generally initiates near the interface, or just from the interface edge due to the edge stress singularity. In this study, the stress intensity factor of an edge crack close to the interface between the dissimilar materials is considered. The small edge crack is strongly dominated by the singular stress field near the interface edge. The analysis of stress intensity factor of small edge crack near the interface in bi-material and butt joint plates is carried out by changing the length and the location of the crack and the region dominated by the interface edge is examined. It is found that the dimensionless stress intensity factor of small crack, normalized by the singular stress at the crack tip point in the bonded plate without the crack, is equal to 1.12, independent of the material combination and adhesive thickness, when the relative crack length with respect to the crack location is less than 0.01. Therefore, the stress intensity factor of the small edge crack can be simply evaluated not only in the bi-material plate but also in the butt joint plate.

1. Introduction

Adhesive bonded joints subjected to thermal or mechanical loading have the singular stress at the interface edge, and the interfacial debonding often occurs. Therefore, a number of studies about the bonded strength have been made experimentally and numerically by using the intensity of the singular stress at the interface edge in jointed dissimilar materials. The singular stress field near the interface edge can be presented by the stress singularity index and the intensity of the singular stress.

Recently, Chen and Nisitani [1], and Kubo and Ohji [2] investigated the singular stress field near the interface edge again by the eigenvalue expansion method. To evaluate the intensity of the singular stress field near the interface edge, Munz and Yang [3] proposed the numerical extrapolation method for single stress singularity problems, and Xu et al. [4] developed the numerical method to determine both multiple stress singularities and the related stress intensity coefficients by using conventional numerical results. Noda et al. [5, 6] showed the accurate method to analyze the intensity of the singular stress at the interface edge by using the stress ratio at the interface edge calculated by the finite element method. Moreover, Noda et al. [7-10] reported that the stress intensity factor of the small interface crack in bonded materials, as well as the intensity of the singular stress at the interface corner, is effective for evaluating the adhesive strength.

On the other hand, the problem of cracks occurring in the vicinity of the interface edge has also been studied. Fett et al. [11, 12] investigated the edge singularity induced crack by using the weight functions. Xu et al. [13] provided the approximate solution to calculate the stress intensity factors of

IOP Conf. Series: Journal of Physics: Conf. Series **843** (2017) 012010 doi:10.1088/1742-6596/843/1/012010

small crack near the interface for any material combinations. In this study, we focus on a small edge crack near the interface between dissimilar materials. The stress intensity factor of a small edge crack close to the interface is considered to be strongly dominated by the singular stress field at the interface edge. By using the stress intensity factor of small crack, it is possible to represent the intensity of the singular stress field near interface edge as an alternative parameter. The analysis of stress intensity factor of small edge crack will be performed by changing the length and position of the crack near the interface and the region dominated by the interface edge will be examined.

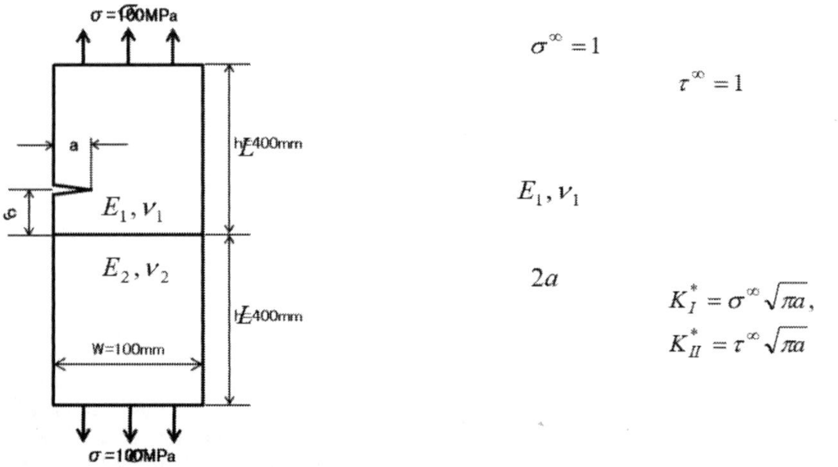

(a) Edge crack near the interface in bi-material plate (b) Single crack in an infinite plate
Figure 1. Analysis model of (a) unknown and (b) reference problems.

2. Method of Analysis

We consider an edge crack parallel to the interface of bonded dissimilar materials, as shown in Figure 1, where 'c' is the distance between the interface and the crack, and 'a' is the crack length. It is assumed here that the crack is induced by the singular stress field near the interface edge. In the FEM analysis, the versatile program MSC.Marc is used and the four-node-quadrilateral element is selected. The FEM mesh pattern near the crack tip is made fine systematically. The stress intensity factor of the edge crack is analyzed by the crack tip stress method [14, 15].

The crack tip stress method is a highly accurate method to calculate the stress intensity factor of a crack problem using the finite element method, and has been proposed by Nisitani [14, 15]. According to the principle of the crack tip stress method, the ratio of the stress intensity factors of the two strips with a crack subjected to tension is nearly equal to the ratio of the y-directional stresses at the crack tips obtained by FEM. This is the reason why the singular stress fields are represented by the crack tip stress $\sigma_{y0,\mathrm{FEM}}$ under the condition of the same mesh pattern [14, 15]. Based on the above fact, equation (1) is obtained. In this relation, K_I stands for the stress intensity factor, $\sigma_{y0,\mathrm{FEM}}$ stands for the y-directional stress at the crack tip calculated by FEM and the asterisk * means the values of the reference problem as shown in Figure 1b.

$$\frac{K_I}{\sigma_{y0,EFM}} \fallingdotseq \frac{K_I^*}{\sigma^*_{y0,FEM}} \tag{1}$$

If the stress intensity factor K_I^* of the reference problem is known, we can easily obtained the K_I of the unknown problem from the relation (1) because the stresses $\sigma^*_{y0,\mathrm{FEM}}$ and $\sigma_{y0,\mathrm{FEM}}$ of the reference and the unknown problems can be easily calculated by FEM. In this analysis, the exact solution of a single crack in an infinite plate is used as the reference problem (Figure 1b).

3. Numerical results and discussion

3.1. Comparison of numerical results

The stress intensity factors of edge crack illustrated in Figure 1a have been analyzed by Xu et al. [13] using the boundary element method. The present results are compared with the results of Xu et al. [13] when the crack length a=0.1 mm, the distance between crack and interface c=2 mm, the plate width W=100 mm and the plate length L=400 mm. Material combinations used in this analysis are shown in Table 1. The numerical results of the stress intensity factors for three types of material combinations are shown in Table 2. It is found that the present results are in good agreement with the results of Xu et al. [13].

Table 1. Material combinations and constants [13].

Material combinations	Young's modulus E (GPa)	Poisson's ratio ν	Singularity order 1-λ
Si_3N_4	304	0.27	0.0852
Cu	108	0.33	
Si_3N_4	304	0.27	0.0145
S45C	206	0.30	
S45C	206	0.30	0.2823
Epoxy	4.93	0.33	

Table 2. Comparison of the numerical results.

Material combinations	Present		Xu et al.[13]	
	$K_I / \sigma\sqrt{\pi a}$	$K_{II} / \sigma\sqrt{\pi a}$	$K_I / \sigma\sqrt{\pi a}$	$K_{II} / \sigma\sqrt{\pi a}$
Si_3N_4 / Cu	1.695	0.00413	1.701	0.0039
Si_3N_4 / S45C	1.268	0.00057	1.259	0.0005
S45C / Epoxy	3.510	0.0294	3.478	0.0286

3.2. Normalized stress intensity factors of small edge crack near the interface in bonded plate

In order to examine the effect of the interface edge singularity, the stress intensity factors of edge crack are calculated when the crack length and location are changed systematically ($10^{-4}<a<10$ mm, $0.5<c<2.0$ mm). In Figure 2, the relation between the normalized stress intensity factors and the relative crack length for the material combination of Si3N4-Cu are indicated. The normalized SIF is based on the remote stress σ. As shown in Figure 2, the values of K_I converge to constant value and the values of K_{II} approach zero when the relative crack length $a/c<0.01$. The values of K_I for small crack are larger as the distance from interface decreases because of the edge singularity. Similar relations can be obtained for other material combinations. Therefore, it is considered that the stress intensity factor of small crack is controlled by the singular stress field near the interface edge.

Next, we examine the other normalized stress intensity factor based on the singular stress near the interface edge. The newly defined normalized stress intensity factor can be expressed as

$$F_I = \frac{K_I}{\sigma_{y,eq}\sqrt{\pi a}}, \quad F_{II} = \frac{K_{II}}{\sigma_{y,eq}\sqrt{\pi a}}. \tag{2}$$

Here, $\sigma_{y,eq}$ is the singular stress in y-direction at the crack tip position in the same bonded plate without the crack, as shown in Figure 3.

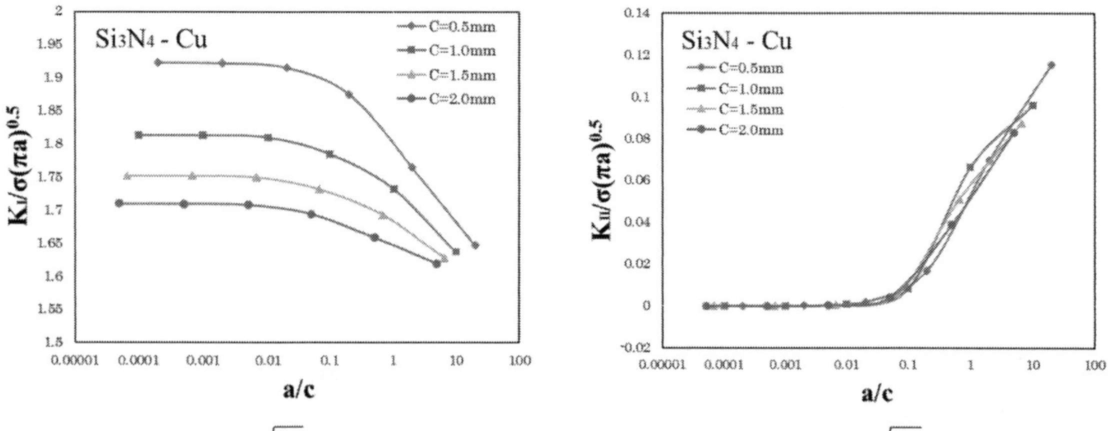

(a) $K_I / \sigma\sqrt{\pi a}$ vs. a/c (b) $K_{II} / \sigma\sqrt{\pi a}$ vs. a/c

Figure 2. Relation between the normalized stress intensity factor and the relative crack length a/c.

The singular stress $\sigma_{y,eq}$ near the interface edge in material '1' can be given as follows [13]

$$\sigma_{y,eq} = \frac{\rho(\sin\theta)^{1-\lambda}}{4}\{[(\lambda+2-\cos\lambda\pi)d + b\sin\lambda\pi]\cos(\lambda-1)\theta \\ -[(\lambda+2+\cos\lambda\pi)b + d\sin\lambda\pi]\sin(\lambda-1)\theta + (\lambda-1)[d\cos(\lambda-3)\theta - b\sin(\lambda-3)\theta]\} \tag{3}$$

Here,

$$\rho = \frac{\lambda c^{\lambda-1}K_\sigma}{\lambda^2 - \sin^2(\lambda\pi/2)}, \quad \theta = \tan^{-1}(c/a), \\ b = 2F - \frac{1}{\lambda}[\sin\lambda\pi - F(1-\cos\lambda\pi)], \quad d = 2 - \frac{1}{\lambda}(1-\cos\lambda\pi + F\sin\lambda\pi), \\ F = \frac{\lambda D + \sin(\lambda\pi/2)\cos(\lambda\pi/2) - D\sin^2(\lambda\pi/2)}{\lambda + \sin^2(\lambda\pi/2) + D\sin(\lambda\pi/2)\cos(\lambda\pi/2)}, \\ D = \frac{2\beta[\lambda^2 + (2\lambda+1)\sin^2(\lambda\pi/2)] - 2\alpha[\lambda^2 - \lambda\cos(\lambda\pi)]}{\sin(\lambda\pi)[1-2\lambda(\beta-\alpha)]} \tag{4}$$

In the above equations, K_σ is the parameter describing the intensity of the singular stress field near the interface edge, and α, β are Dundurs' dissimilar materials parameters.

$$\alpha = \frac{\Gamma(\kappa_1+1)-(\kappa_2+1)}{\Gamma(\kappa_1+1)+(\kappa_2+1)}, \quad \beta = \frac{\Gamma(\kappa_1-1)-(\kappa_2-1)}{\Gamma(\kappa_1+1)+(\kappa_2+1)}, \\ \Gamma = \frac{G_2}{G_1}, \quad \kappa_j = \begin{cases}(3-\nu_j)/(1+\nu_j) & \text{plane stress} \\ 3-4\nu_j & \text{plane strain}\end{cases} \tag{5}$$

The singularity index 1-λ of the singular stress field near the interface edge can be determined by solving the following eigenequation

$$\lambda^2(\lambda^2-1)\alpha^2 + \left(\lambda^2 - \sin^2\frac{\lambda\pi}{2}\right)^2\beta^2 + 2\lambda^2\alpha\beta\left(\sin^2\frac{\lambda\pi}{2}\right) + \frac{1}{4}\sin^2(\lambda\pi) = 0. \tag{6}$$

The intensity K_σ of the singular stress field near the interface edge usually can be determined by the numerical extrapolation method as shown in equation (7).

$$K_\sigma = \lim_{r \to 0} \frac{\sigma_y}{r^{1-\lambda}}, \quad at\ \theta = 0 \tag{7}$$

Figure 3 shows the relation between the normalized stress intensity factor F_I and the relative crack length a/c. Figure 4 also indicates the F_I-a/c relations for different material combinations. As illustrated in these figures, regardless of the crack length and the distance from interface, the normalized factors can be represented by one curve and is a function of a/c. Especially, it is very interesting that the values of F_I equal to 1.12 irrespective of the material combinations when $a/c<0.01$. At this case, the F_{II}-values are very small. Therefore, the stress intensity factor of small crack near the interface edge can be expressed as $K_I = 1.12\sigma_{y,eq}\sqrt{\pi a}$. In addition, from equations (3) and (4), it can be seen that the stress intensity factor K_I of the small edge crack and the intensity K_σ of the singular stress field strength near the interface edge are one-to-one correspondence.

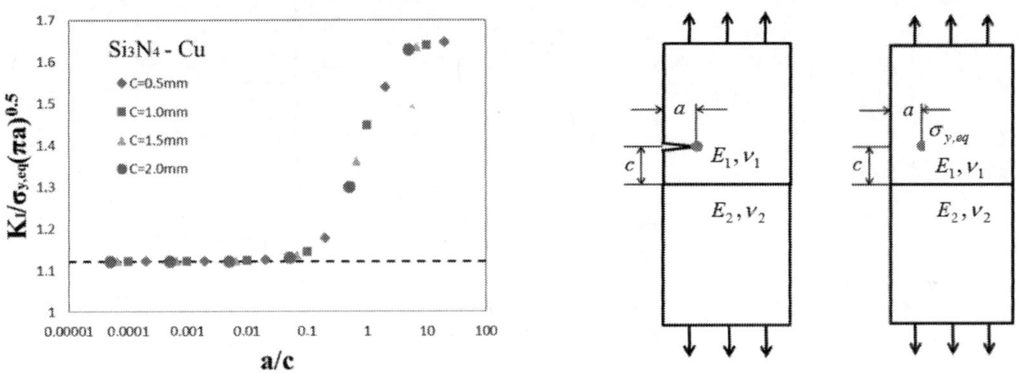

Figure 3. Relation between F_I and t a/c for Si_3N_4-Cu.

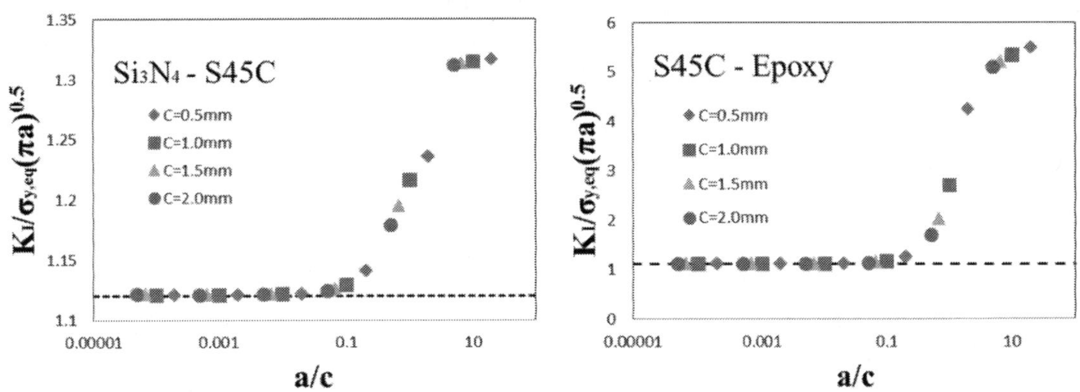

(a) Si_3N_4-S45C (b) S45C-Epoxy

Figure 4. Relation between F_I and t a/c for different material combination.

3.3. Normalized stress intensity factors of small edge crack near the interface in butt joint

Next, we consider the stress intensity factors of edge crack near the interface in butt joint specimen. As shown in Figure 5, the adherent is material 1 and the adhesive is material 2. The stress intensity factors of the small edge crack are analyzed by the crack tip stress method when the adhesive layer thickness is h=1 mm. In Figure 5, the normalized factors are shown for various crack lengths and locations. In this case, the stress value calculated by FEM is used as the value of $\sigma_{y,eq}$. As shown in Figure 5, the normalized stress intensity factor is 1.12 when a/c<0.01.

Figure 6 represents the relation between the dimensionless stress intensity factors and the relative crack length a/h with respect to the adhesive thickness when the adhesive thickness h=1 and 2 mm, respectively. As shown in Figure 6, the normalized stress intensity factor is also 1.12 when a/h<0.01. Therefore, even in the case of the butt joint, the stress intensity factor of the small edge crack in the edge singularity region can be represented by $K_I = 1.12\sigma_{y,eq}\sqrt{\pi a}$.

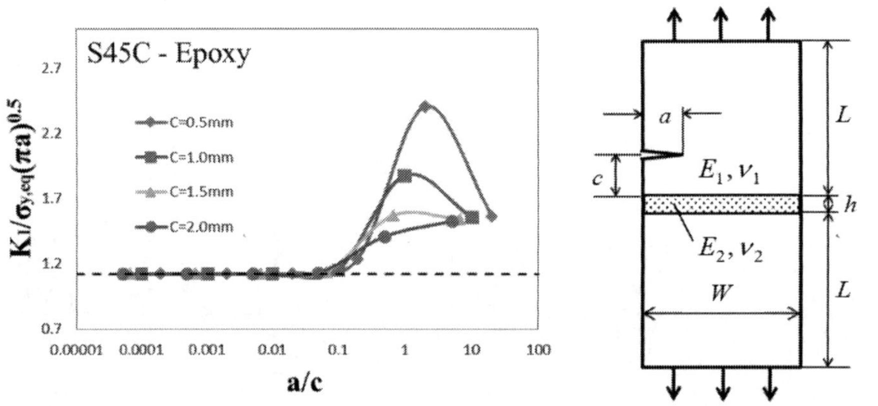

Figure 5. Relation between F_I and a/c for S45C-Epoxy butt joint when h= 1mm.

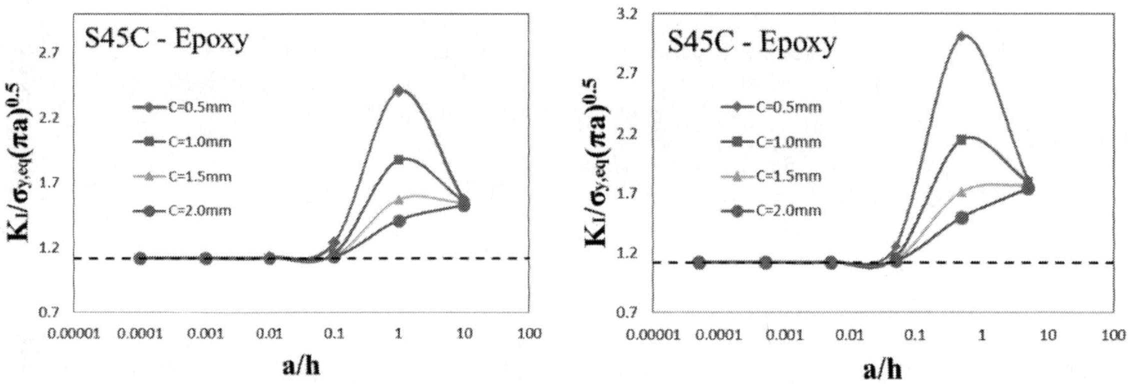

(a) h=1 mm (b) h=2 mm

Figure 6. Relation between F_I and a/h for butt joint with different adhesive thickness.

4. Conclusion

In this study, the stress intensity factor of the edge crack near the interface between dissimilar materials were analyzed and the region affected by the singular stress at the interface edge were examined. The numerical results of edge crack near the interface were normalized by using the singular stress $\sigma_{y,eq}$ at the crack tip point in the bonded plate without the crack. The normalized stress

IOP Conf. Series: Journal of Physics: Conf. Series **843** (2017) 012010 doi:10.1088/1742-6596/843/1/012010

intensity factors converged with the constant value 1.12 irrespective of the material combination when the relative crack length with respect to the distance from the interface a/c<0.01. Therefore, the stress intensity factor of the small edge crack near the interface can be simply expressed as $K_I = 1.12\sigma_{y,eq}\sqrt{\pi a}$. The expression can be also applied to the butt joint problem in the case that the relative crack length with respect to the adhesive thickness a/h<0.01.

References

[1] Chen D H and Nisitani H 1992 Singular stress field in two bonded wedges *Transactions of the Japan Society of Mechanical Engineers* **58** 457–464 (in Japanese)

[2] Kubo S and Ohji K 1991 Geometrical conditions of no free-edge stress singularities in edge-bonded elastic dissimilar wedges *Transactions of the Japan Society of Mechanical Engineers* **57** 632–636 (in Japanese).

[3] Munz D and Yang Y Y 1993 Stresses near the edge of bonded dissimilar materials described by two stress intensity factors *International Journal of Fracture* **60** 169–177.

[4] Xu J Q, Liu Y H and Wang X G 1999 Numerical methods for the determination of multiple stress singularities and related stress intensity coefficients *Engineering Fracture Mechanics* **63** 775–790.

[5] Noda N-A, Zhang Y, Lan X and Takaishi K 2011 Strength analysis for the adhesive layer on the basis of intensity of singular stress *Procedia Engineering* **10** 722–727.

[6] Noda N-A, Zhang Y, Takaishi K, Lan X and Oda K 2012 Intensity of singular stress for single-lap joints *Transactions of the Japan Society of Mechanical Engineers* **78** 651–655 (in Japanese)

[7] Noda N-A, Lan X, Michinaka K, Zhang Y and Oda K 2010 Stress Intensity Factor of an Edge Interface crack in a Bonded Semi-infinite Plate *Transactions of the Japan Society of Mechanical Engineers* **76** 1270–1277 (in Japanese)

[8] Oda K, Lan X, Noda N-A and Michinaka K 2012 Effect of arbitrary bi-material combination and bending loading conditions on stress intensity factors of an edge interface crack *International Journal of Structural Integrity* **3** 457 – 475.

[9] Noda N-A, Miyazaki T, Uchikoba T, Li R, Sano Y and Takase Y 2014 Convenient debonding strength evaluation based on the intensity of singular stress for adhesive joints, *Journal of the Japan Institute of Electronics Packaging* **17** 132-142 (in Japanese).

[10] Noda N-A, Misyazaki T, Li R, Uchikoba T, Sano Y and Takase Y 2015 Debonding strength evaluation in terms of the intensity of singular stress at the interface corner with and without fictitious crack *International Journal of Adhesion & Adhesives* **61** 46–64.

[11] Fett T, Tilscher M and Munz D 1997a Weight functions for cracks near the interface of a bimaterial joint and application to thermal stresses *Engineering Fracture Mechanics* **56** 87–100.

[12] Fett T, Tilscher M and Munz D 1997b Weight functions for sub-interface cracks *International Journal of Solids and Structures* **34** 393–400.

[13] Xu J-Q, Wang X-G and Mutoh Y 2001 Stress intensity factors of a surface crack near an interface end *International Journal of Fracture* **111** 251–264.

[14] Nisitani H, Kawamura T, Fujisaki W and Fukuda T 1999 Determination of highly accurate values of stress intensity factor or stress concentration factor of plate specimen by FEM *Transactions of the Japan Society of Mechanical Engineers* **65A** 26–31 (in Japanese).

[15] Nisitani H and Teranishi T 2004 K_I of a circumferential crack emanating from an ellipsoidal cavity obtained by the crack tip stress method in FEM *Engineering Fracture Mechanics* **71** 579–585.

IOP Conf. Series: Journal of Physics: Conf. Series **843** (2017) 012011 doi:10.1088/1742-6596/843/1/012011

Safety assessment of Cracked K-joint Structure Based on Fracture Mechanics

Wang Xin [1], Yan Pengyu[1], Du Jianwei[1] and Cai Fuhai[2]

[1] School of Mechanical Engineering, Dalian University of Technology, Dalian, 116024, China
[2] Changzhou YILIY Engineering machinery co. Ltd., Changzhou, 213100, China

E-mail:wangxbd21@163.com

Abstract. The K-joint is the main bearing structure of lattice jib crane. During frequent operation of the crane, surface cracks often occur at its weld toe, and then continue to expand until failure. The safety of the weak structure K-joint of the crane jib can be evaluated by BS7910 failure assessment standard in order to improve its utilization. The finite element model of K-joint structure with cracks is established, and its mechanical properties is analyzed by ABAQUS software, the results show that the crack depth has a great influence on the bearing capacity of the structure compared with the crack length. It is assumed that the K-joint with the semi-elliptical surface crack under the action of the tension propagate stably under the condition that the c/a (ratio of short axis to long axis of ellipse) is about 0.3. The safety assessment of K-joint with different lengths crack is presented according to the 2A failure assessment diagram of BS7910, and the critical crack of K-joint under different loads can be obtained.

1. Introduction

Large tonnage cargo lifting machinery is widely used in engineering, and the bearing structure tends to produce fatigue damage due to fluctuating loads, which greatly influence the safety of the machinery. K-joint structure is an important bearing structure of lattice jib crane, and the cracks often occur in the weld toe, and then expand during the working process, which is a great threat to the safety of the crane. A safe and economical life assessment method for fatigue strength analysis can provide a reliable basis to determine whether it is necessary to repair or immediately stop operation or even to for scrap [1]. At present, there are two fatigue life assessment methods for welded structures: traditional fatigue analysis and fatigue analysis, the former combines with S-N curve and cumulative damage theory, many scholars [2-3] use this kind of method to assess the life of the engineering machinery. But the method is only suitable for the prediction of fatigue crack initiation life because it does not consider the initial imperfection of the structure. The latter method provides a greater reference value for the safety evaluation of crane based on fracture mechanics to consider the initial imperfection of structure. Literature [4-5] shows the use of fracture mechanics to evaluate the residual life of crane, but there is a lack of a widely used standard for the life assessment of the crane with defects.

In this paper, it is put forward the residual life assessment method and process of structure with crack by combining with fitness for purpose assessment and fatigue life assessment. Taken typical load bearing structure K-joint of 25t crawler crane as a model, the safety assessment of K-joint with different crack lengths is analyzed based on BS7910. And then, the critical crack lengths of K-joint under different stress amplitude are obtained. By the Paris formula in fracture mechanics, the residual life of the K-joint can be calculated from the initial crack to the critical crack. The proposed

IOP Conf. Series: Journal of Physics: Conf. Series **843** (2017) 012011 doi:10.1088/1742-6596/843/1/012011

assessment method and process can provide theoretical basis for the structural optimization design, crack detection and making maintenance cycle, and greatly reduce the cost of design and maintenance.

2. K-joint model with cracks

In this paper, the research object is the K-joint of 25t crawler crane in Figure.1, whose parameters are shown as table 1 and table 2. During the annual inspection, 20% main welding seam should be detected by magnetic particle inspection and their sizes and shape are inspected. K-joint is a typical bearing structure of crane, and its surface cracks tend to be initiated in the weld toe stress zone, then expand along the length direction and the depth direction. When the crack extends to the limit lengths, the structure will be suddenly unstable or even broken. According to the correlation statistics of the actual crack location and analysis of the paper [6], the crack is established to the stress concentration zone at weld toe named crown in the finite element analysis of K-joint in Figure.2. The surface crack at the toe of the K-joint structure can be simplified into a semi elliptical surface crack in Figure.3, where 3(a) is a 3D crack along the weld toe line, the length of which is $2a$, the depth of which is c; 3(b) is the 2D projection of the 3D crack on the chord section, the length of which is $2a'$, the depth of which is c'.

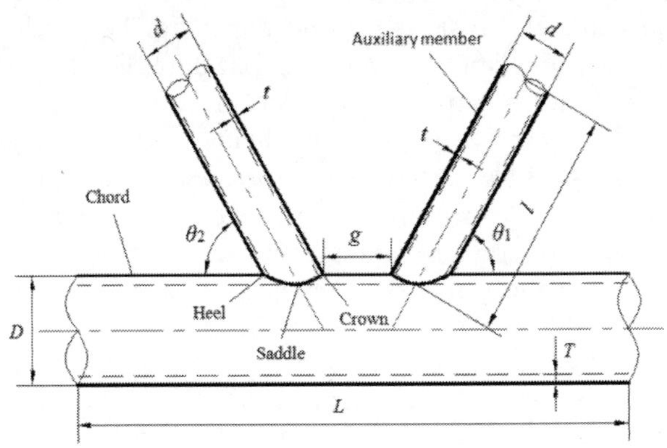

Figure 1. Main view of the K-joint

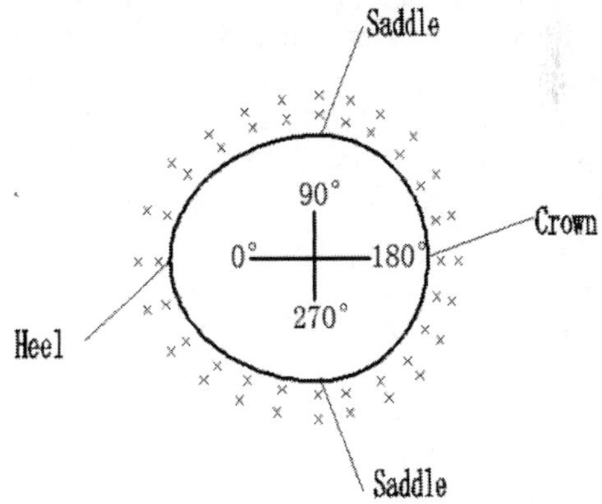

Figure 2. Weld toe profile

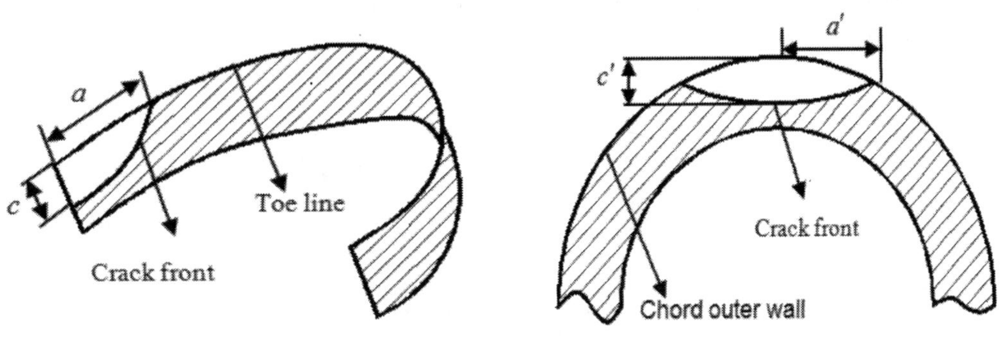

(a) (b)

Figure 3. (a) 3D crack surface (b) 2D crack surface

Table 1. Material Properties.

Yield strength (MPa)	Tensile strength (MPa)	Elongation rate (%)	Impact energy,20°C (J)
345	573	23.45	76

Table 2. Model Parameters.

D (mm)	d (mm)	L (mm)	l (mm)	T (mm)	t (mm)	g (mm)	θ_1 (°)	θ_2 (°)
76	38	600	250	6	4	30	60	60

It is necessary to use the FEM to do mechanics analysis of K-joint in order to obtain the evaluation parameters for its safety assessment. The ABAQUS is one kind of commercial finite element software, and it can well handle the crack problem. In the software, the linear elastic analysis and elastic-plastic analysis can be provided for finite element calculation. In the linear elastic analysis, the structure is considered elastic material, and the parameter stress intensity factor K representing brittle fracture of structure can be obtained. In the elastic-plastic analysis, the true stress-strain data of the material needs to be input into the software, the structure is considered plastic material, and the parameter plastic collapse load P_c representing plastic damage bearing capacity of structure can be obtained.

According to above-mentioned crack location, the crack tip is established at the corresponding position in the software. The precision of finite element calculation is determined by the mesh generation around the crack tip. In this paper, the high order singular element that degenerate hexahedral element is used to simulate the displacement singularity at the crack tip, hexahedral element to simulate 1/4 node prism element outside near the crack tip, and the four pyramid element and tetrahedral element can be used to simulate the crack tip and crack tip away from the transition region, and hexahedral element to simulate area away from the crack tip. In the linear elastic analysis, the strain singularity should be $1/r^{1/2}$ at the crack tip (r is the distance from the crack tip to near location). In the elastic-plastic analysis, the plastic zone is formed at the crack tip, the strain singularity should be $1/r$. The crack tip singularity can be well simulated by the corresponding degradation treatment of the crack tip element, and the evaluation parameters such as stress intensity factor K can be obtained accurately.

The finite element model of K-joint with crack is established according to above-mentioned geometric parameters as shown in Figure.4. In the process of the crawler crane loading and unloading, the tension acting on the auxiliary member plays a decisive role to cause the crack propagation. Therefore, in the finite element model, the loading and boundary condition are showed in Figure.4(a), where P_a is the tension acting on one auxiliary member and the six degrees of freedom are all restricted in end face of two chords and another auxiliary member.

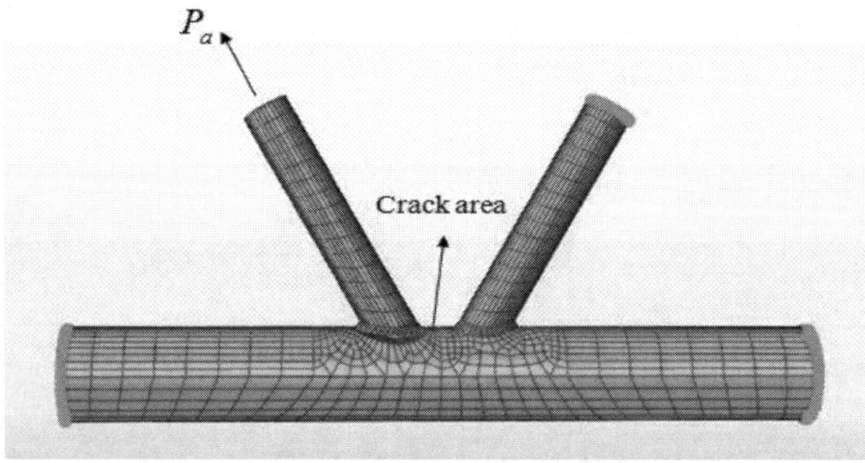

(a)

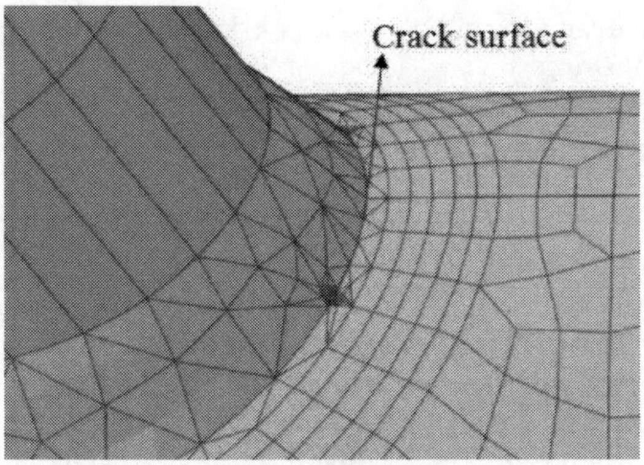

(b)

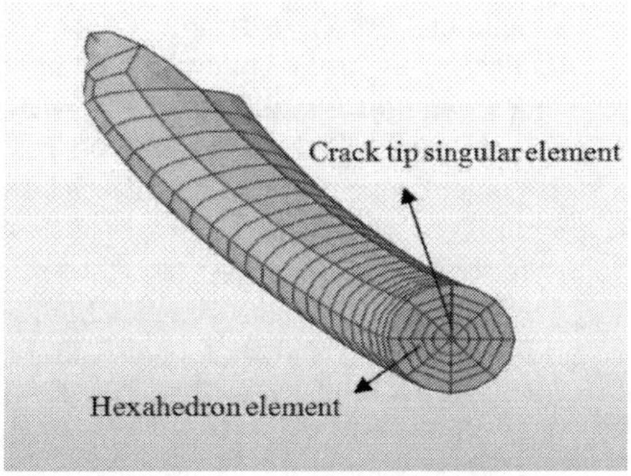

(c)

Figure 4. (a) Finite element model of K-joint with crack (b) Crack tip region (c) Crack tip lateral element

3. BS7910 standard

BS7910 standard [7] is a quantitative evaluation of the ability to satisfy the design function on the basis of ensuring the safety of the defective structure. It is based on the evaluation method of fracture mechanics, and plays a complementary role to the traditional quality control standards. It has brought significant economic benefits in the field of pressure vessels and oil pipelines. The failure assessment diagram (FAD) in Figure.5 considers both failure modes of plastic failure and brittle fracture, and the transverse axis parameter L_r reflects the plastic damage performance of structure, the vertical axis parameter K_r represents the brittle fracture performance of structure. With the increase of the load or crack length, the loading curve is close to the failure assessment curve (FAC), and the point of intersection between the two curves is called the critical point. The load or crack size before this critical point is considered to be safe, after which the load or crack size is considered to be not acceptable.

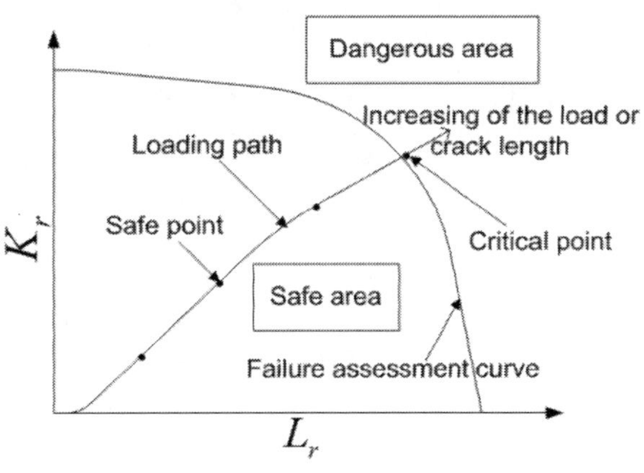

Figure 5. Failure assessment diagram

The BS7910 standard contains three levels of evaluation, with the increase in the level of assessment, assessment accuracy will increase, but the data will be increased. The three evaluation levels are as follows:

- The first level is a simple evaluation method, divided into 1A and 1B two kinds of assessment methods. In 1A level, the horizontal axis and vertical axis are limit fixed point in FAD. In 1B level, it is based on the empirical formula calculation of failure assessment. The first level evaluation criterion is simple and conservative, and it can be used as a primary criterion when there are only limited material data or applied stress.

- The second level is the most widely evaluation method, divided into 2A and 2B two kinds of assessment methods. Each method has its own failure curve equations and cut-off lines. The 2A level does not need the material stress-strain data, the 2B level needs the true stress and strain data of the material to construct the failure curve, and the second level evaluation criterion can be taken into account for the structure of the medium and low toughness.
- The third level evaluation is suitable for ductile materials, and considers the use of this method if the material is high strain hardening index or the need for analysis of the crack stable tearing fracture. It divided into 3A, 3B, 3C three kinds of assessment methods, 3A and 3B is the promotion of 2A and 2B, 3C is an assessment level for the structure with specific materials and specific geometry. The third level evaluation criterion is suitable for high toughness and high strength structure with cracks.

Although the BS7910 provides a higher level of safety assessment, the higher the accuracy of the assessment, but more complex assessment process and more structural parameters are required in a higher level of assessment. The K-joint structure is regarded as research object in this paper, according to its structure and material properties, it is recommended to use more economical and conservative 2A level failure assessment in engineering practice. The failure curve of 2A level is represented by an equation:

$$K_r = \left(1 - 0.14L_r^2\right)\left[0.3 + 0.7\exp\left(-0.65L_r^2\right)\right] \quad (1)$$

Where $L_r = P_a/P_c$, P_a is applied load, P_c is plastic collapse load; in linear elastic analysis, $K_r = K_I/K_{mat}$, K_I is stress intensity factor, K_{mat} is fracture toughness of steel, according to the formula provided by BS7910, the value of K_{mat} is 3037.7 $\mathrm{MPa}\cdot\mathrm{mm}^{1/2}$.

4. Mechanical analysis of K-joint structure with cracks

4.1. Fracture strength analysis

The surface crack of K-joint structure often occur near the top of the chord, and the crack can extend along the direction of length and depth at the same time when the structure is subjected to fluctuating

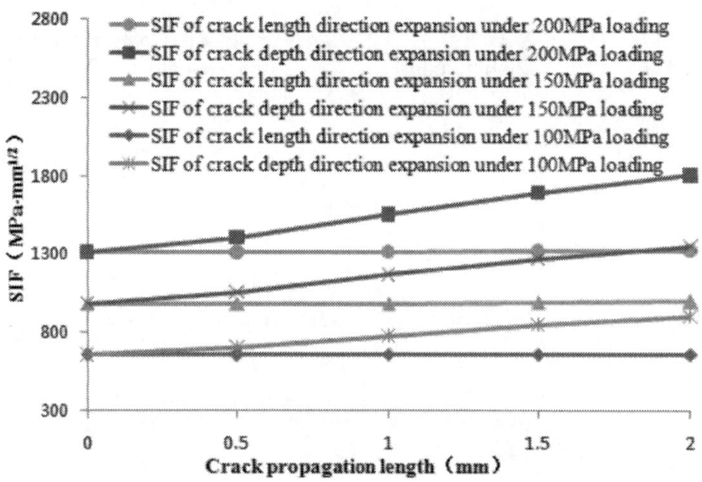

Figure 6. Method to acquire plastic collapse load

load. In the ABAQUS software, the surface load of 100MPa, 150MPa and 200MPa are applied in K-joint respectively in linear elastic analysis. At present, the finite element software can't solve the stress intensity factor of the dynamic crack. In this paper, the crack propagation process is simulated by extending the 0.5mm solution in the direction of crack length and depth. The variation of the stress intensity factor of the initial crack at c =3mm and $2a$ =20mm in the length direction and depth direction is showed in Figure.6. It can be seen that under the same load, the stress intensity factor will increase with the crack growth, and the influence of the stress intensity factor on the crack depth direction is larger than the length direction.

4.2. Bearing capacity analysis

In practice, the plastic deformation often occurs near the crack region of K-joint, and the plastic damage will accumulate with the increase of load amplitude. It is necessary to conduct elastic-plastic analysis to obtain the plastic bearing capacity of the K-joint structure for safety assessment. When the load on the component or structure made of ideal elastic-plastic material reaches a certain value, even if the load is not increased, the plastic deformation on them will continue to increase, this state is called the collapse state, and the corresponding load is called the plastic collapse load. The plastic collapse load can express the bearing capacity of structural plastic damage. The load-strain curves of the welded joint of K-joint at different loads were fitted with the help of ABAQUS software of the K-joint elastic-plastic finite element analysis, and the plastic collapse load is obtained by using two elastic slope criterion($\tan\alpha = 2\tan\theta$) [6] in Figure.7. According to the above method, the plastic collapse load of K-joint structure without crack can be obtained for 313MPa.

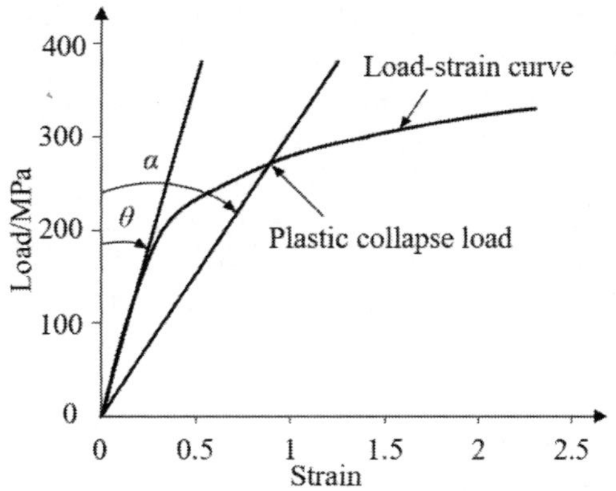

Figure 7. Method to acquire plastic collapse load

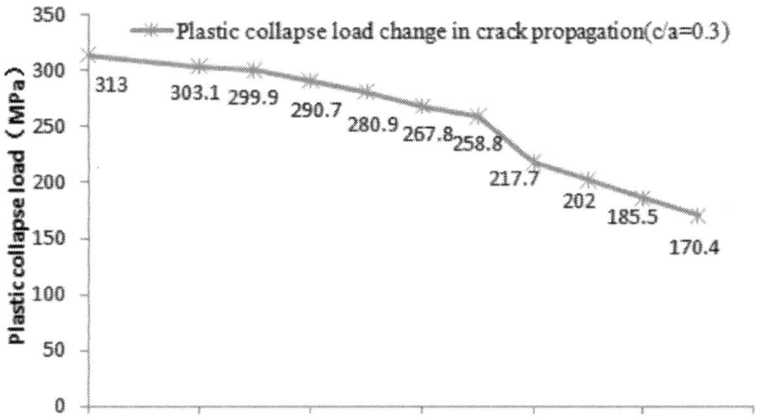

Figure 8. Plastic collapse load in crack propagation

IOP Conf. Series: Journal of Physics: Conf. Series **843** (2017) 012011 doi:10.1088/1742-6596/843/1/012011

The semi-elliptical crack expansion in 0.3 short axis ratio (c/a =0.3) of plastic collapse load change of K-joint is showed in Figure.8. The horizontal axis represents the crack length is $2a$, the vertical axis represents the K-joint's plastic collapse load under different crack length. With the expansion of the crack, the structural defects are more obvious, and the plastic bearing capacity of K-joint structure decrease more obviously. When the crack length is extended to 40mm, the surface crack penetrates the chord wall, the plastic collapse load of the K-joint decline significantly, namely the K-joint structure plastic bearing capacity decrease rapidly. As a result, the penetration crack has a great influence on the bearing capacity of the structure.

5. Critical crack prediction of K-joint with cracks

When a crane works a period of time, surface crack can initiate on the surface of K-joint, as the load cycle, the crack can be extended to the crack depth and the length direction at the same time. When the crack penetrates the wall thickness, the crack can continue to expand steadily along the length direction, and the crack can be considered penetrating crack. When the crack length is extended to a critical value, the structure will accelerate the expansion, and soon failure or even fracture. It is assumed that the K-joint with the semi-elliptical surface crack under the action of the tension will propagate stably under the condition that the c/a is about 0.3.

In this paper, the failure of K-joint with different cracks is evaluated by 2A level provided by BS7910, and the critical crack length of K-joint under different loads can be obtained [8]. In this paper, the safety of K-joint under different loads is evaluated through increasing crack length under short axis ratio of 0.3 in the propagation path. At a is 20mm, the crack penetrates the chord, and then the extended crack can be simplified as a penetrating crack. The stress intensity factor of the K-joints in the corresponding load is obtained by using the linear elastic analysis of finite element software ABAQUS, and the safety assessment point of K-joint with different crack is obtained in FAD using the above formula. The K-joint with growing crack safety assessment diagram is shown in Figure.9. The intersection point of the crack propagation path curve and FAC is the critical crack assessment point of the K-joint under the load. The crack size corresponding to this point is named critical crack. With the increase of the applied load, the evaluation point will be closer to the failure assessment curve. When the crack penetrates the chord, the safety of the structure will decrease obviously. The critical crack length $2a_c$ of the K-joint structure under different loads is showed in Table 3.

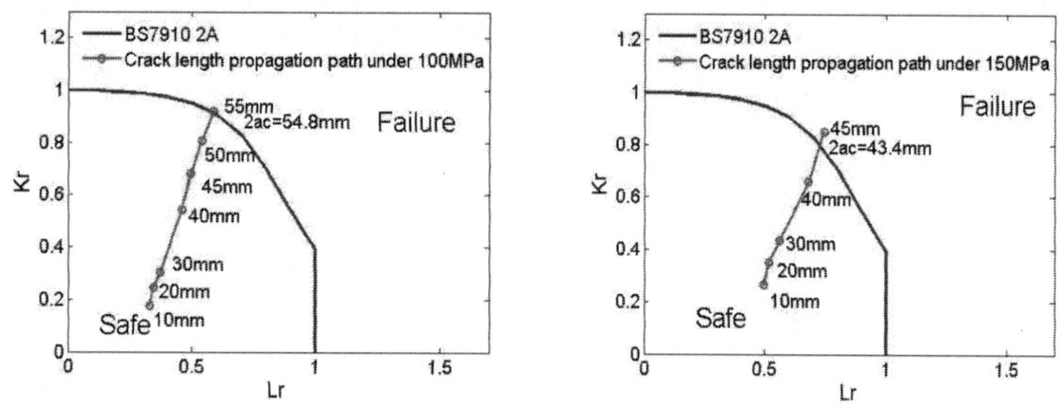

(a) (b)

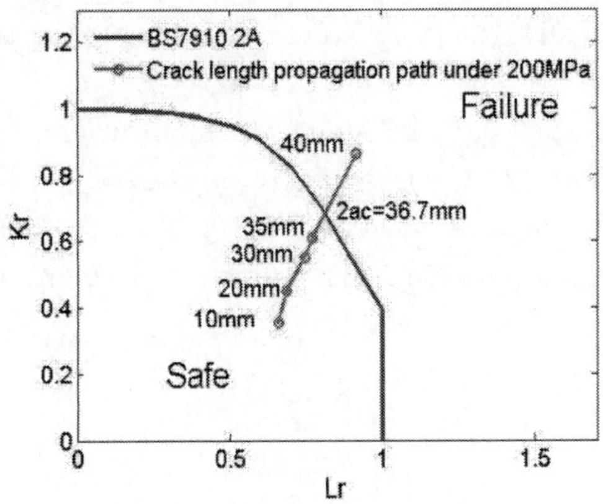

(c)

Figure 9.(a) Safety assessment under 100MPa (b) Safety assessment under 150MPa(c) Safety assessment under 200MPa

Table 3 Critical crack size under different loads

P_a (MPa)	100	150	200
$2a_c$ (mm)	54.8	43.4	36.7

6. Conclusions

The stress intensity factor of crack front line node can be computed by using the finite element software ABAQUS mechanical analysis of K-joint with cracks, and the plastic collapse load of K-joint with cracks can be obtained with two elastic slope criterion. With cracks extension, the mechanical strength of the K-joint structure decreases. The result shows that the crack depth has a great influence on the bearing capacity of the K-joint structure by comparing the mechanical properties of the crack propagation along depth and length direction and should avoid crack propagation along depth direction in practical engineering. The critical crack length of K-joint with cracks under different loads can be obtained based on the 2A level evaluation provided by BS7910 and it can provide a theoretical basis for the safe use of K-joint structure.

Acknowledgments

The work was supported by the National Science and technology support program of China (Grant No. 2015BAF07B01).

References

[1] Li Yugang, Lin Yan and Ji Zhuoshang 1994 *China Offshore Platform* **14**, Study on fatigue crack control of offshore oil platform 5-6

[2] Zheng Yuqi, Wang Sanmin and Wang Yanping 2013 *Mechanical Strength* **35**, A type of truck crane should evaluate the stress and strain fatigue life of the boom 850-854

[3] Wu Tianmin, YanTingjun and Wang Dongshen 2014 *Petroleum Machinery* **42**, Study on fatigue life assessment of offshore platform crane 66-69

[4] Zhou Taiquan and Chen Hongtian 2009 *Ship Mechanics* **13**, Fatigue crack propagation and fatigue life calculation of steel members with initial crack 91-99

[5] Wang Aihong, XuGening and Gao Youshan 2012 *Chinese Journal of Mechanical Engineering* **18**, Random stress spectrum acquisition and fatigue residual life estimation of bridge crane 192-198

IOP Conf. Series: Journal of Physics: Conf. Series **843** (2017) 012011 doi:10.1088/1742-6596/843/1/012011

[6] Shao Yongbo, Song Shengzhi and Li Tao 2013 *Engineering Mechanics* **30**, Study on safety of circular tubular T-joints containing fatigue crack based on failure assessment diagram (FAD)184-193

[7] BS7910:2005. Guide to method for assessing the acceptability of flaws in metallic structures.

[8] Li Yongzheng and Hu Gang 2015 *Proceedings of the 2015 Conference on Ship Hydrodynamics* Study on critical crack length of typical welded structures based on FAD failure assessment diagram.(Harbin:Ship mechanics academic committee/China shipbuilding newsroom) pp 205-211

The effect of pearlite on the hydrogen-induced ductility loss in ductile cast irons

T Matsuo[1, 2]

[1]Department of Mechanical Engineering, Fukuoka University, 8-19-1 Nanakuma, Jonan-ku, Fukuoka, 814-0180, Japan
[2]Institute of Material Science and Technology, Fukuoka University, 8-19-1 Nanakuma, Jonan-ku, Fukuoka, 814-0180, Japan

E-mail: tmatsuo@fukuoka-u.ac.jp

Abstract. Hydrogen energy systems, such as a hydrogen fuel cell vehicle and a hydrogen station, are rapidly developing to solve global environmental problems and resource problems. The available structural materials used for hydrogen equipments have been limited to only a few relatively expensive metallic materials that are tolerant for hydrogen embrittlement. Therefore, for the realization of a hydrogen society, it is important to expand the range of materials available for hydrogen equipment and thereby to enable the use of inexpensive common materials. Therefore, ductile cast iron was, in this study, focused as a structural material that could contribute to cost reduction of hydrogen equipment, because it is a low-cost material having good mechanical property comparable to carbon steels in addition to good castability and machinability. The strength and ductility of common ductile cast irons with a ferritic-pearlitic matrix can be controlled by the volume fraction of pearlitic phase. In the case of carbon steels, the susceptibility to hydrogen embrittlement increases with increase in the pearlite fraction. Toward the development of ferritic-pearlitic ductile cast iron with reasonable strength for hydrogen equipment, it is necessary to figure out the effect of pearlite on the hydrogen embrittlement of this cast iron. In this study, the tensile tests were conducted using hydrogen-precharged specimens of three kinds of ferritic-pearlitic ductile cast irons, JIS-FCD400, JIS-FCD450 and JIS-FCD700. Based on the results, the role of pearlite in characterizing the hydrogen embrittlement of ductile cast iron was discussed.

1. Introduction

A new energy network utilizing hydrogen has been developed as a solution for increasing global warming and energy resource problems. Although a lot of efforts are undertaken for a development of hydrogen energy equipments such as a fuel cell vehicle and a hydrogen station, there are still many obstacles to overcome. From the viewpoint of material strength, hydrogen embrittlement (i.e. the degradation of strength properties due to hydrogen penetration into a metallic material) needs to be solved to ensure the integrity of hydrogen energy society. At present, to address this problem, only a few relatively expensive metallic materials which are less susceptible to hydrogen embrittlement (e.g. austenitic stainless steel containing high Ni content) are used for hydrogen equipments in Japan [1].

This is a reason why such hydrogen equipments involve a significant cost. Toward the realization of a hydrogen energy society, it is essential to expand the range of materials available for hydrogen equipments to include inexpensive, common materials. Therefore, ductile cast iron (DCI) was focused in this study as a candidate of low cost, common material available for hydrogen equipments.

DCI is characterized by good mechanical properties (strength, toughness and wear resistance) and therefore, it is used for a variety of applications such as a crankshaft and a gear in cars and pipeline, etc. One noticeable advantage of casting manufacturing is its geometric freedom for product design. In other words, casting manufacturing offers an inexpensive, easy production of monolithic structural components of complex geometry. Therefore, the high production cost for an intricately shaped component and multiple part assemblies is expected to be reduced by shift from conventional steel-based fabrication technique (e.g. machining, mechanical joint and weld) to DCI-based one. If the resistant design to hydrogen embrittlement for DCI is established, casting can be a new option for manufacturing the hydrogen equipments, which will contribute to a cost reduction.

A wide range of mechanical properties of DCIs mainly depends on the matrix structure (e.g. ferrite, pearlite, ferrite-pearlite, martensite and austenite) obtained by the control of chemical composition and heat treatment. In commonly used ferritic-pearlitic DCIs, the strength and toughness are controlled by the volume fraction of pearlitic phase. Regarding carbon steels, it has been reported that a pearlitic phase is susceptible to hydrogen embrittlement [2-4]. Nishiguchi et al. [2] investigated the effect of hydrogen and pearlite on the tensile properties of several kinds of carbon steels, and reported that the ductility loss due to hydrogen was pronounced with increase in the volume fraction of pearlitic phase. Accordingly, it is important to elucidate the pearlite effect on the hydrogen embrittlement of common ferritic-pearlitic DCIs as well. However, there are only a few researches [5, 6] on the hydrogen embrittlement of DCIs and the effect of pearlite has not been reported.

In this study, the tensile tests were carried out using hydrogen-precharged specimens of three kinds of ferritic-pearlitic DCIs: JIS-FCD400, JIS-FCD450 and JIS-FCD700. The effect of pealite on hydrogen-induced ductility loss was investigated in terms of strain rate dependency.

2. Expermental procedure

2.1. Material and specimen

Three kinds of ferritic-pearlitic DCIs with different volume fraction of pearlite were used in this study: JIS-FCD400 with an almost fully ferritic matrix, JIS-FCD450 with a ferritic matrix with sparsely distributed pearlitic phase and JIS-FCD700 with a pearlitic matrix containing a thin ferritic envelope surrounding graphite. Microstructures of these DCIs are shown in Fig. 1. The chemical compositions are listed in Table 1. The tensile specimen shown in Fig. 2 was machined from as-cast blocks and then polished by a #2000 emery paper and paste of alumina particles with a diameter of 1 μm.

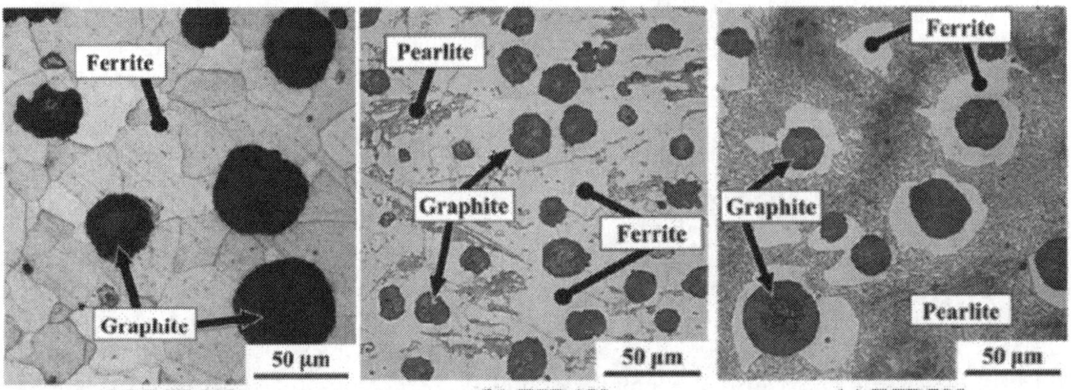

(a) FCD400 (b) FCD450 (c) FCD700

Figure 1. Microstructures.

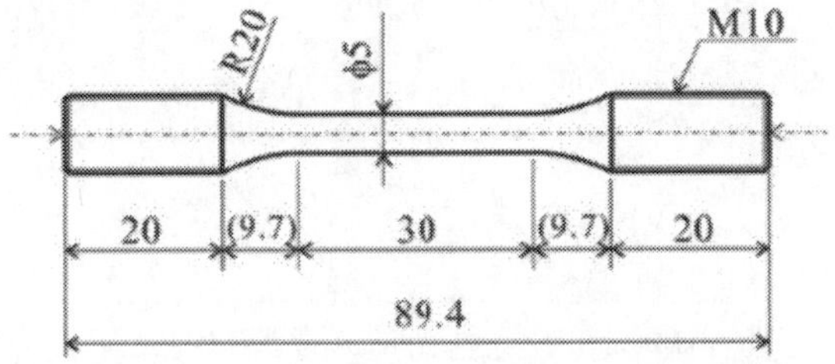

Figure 2. Shape and dimensions of tensile specimen in mm.

Table 1. Chemical compositions.

	C	Si	Mn	P	S	Cu	Mg
FCD400	3.68	2.16	0.29	0.024	0.010	0.01	0.035
FCD450	3.70	1.78	0.34	0.058	0.007	-	-
FCD700	3.77	2.99	0.44	0.023	0.110	-	-

2.2 Hydrogen charging and tensile test

In order to charge hydrogen in a specimen, the specimen was soaked in a 20 mass% of aqueous solution of NH_4SCN at 40 °C prior to tensile test. The hydrogen charging times were 24 h for FCD450 and 168 h for FCD400 and FCD700. These specimens were hereafter called 24h-charged specimen and 168h-charged specimen, respectively, whereas the specimen without hydrogen charging was called uncharged specimen.

The tensile tests for both of charged and uncharged specimens were conducted in air at ambient temperature. To elucidate the strain rate dependency on hydrogen-induced ductility loss, the tests were conducted at different cross head speeds (CHSs) of 0.02 mm/min, 1 mm/min (only for FCD450) and 50 mm/min.

2.3 Analysis of hydrogen storage and fracture morphology

The amount of hydrogen adsorption into a specimen after hydrogen charging was measured by a thermal desorption spectroscope (TDS) in which temperature was raised up to 800 °C at a rate of 0.5 °C/sec.

Fracture surface was observed by a scanning electron microscope (SEM). The fracture morphology inside the material, e.g. void growth and coalescence and crack initiation, was examined by observing the longitudinal cross section of tensile-fractured specimen.

3. Results and discussion

3.1 Hydrogen content

The hydrogen contents of hydrogen-charged specimens obtained by TDS measurement are listed in Table 2 together with the mean graphite diameters and the area fractions of ferrite, pearlite and graphite. Regardless of area fraction of pearlite, a much greater amount of hydrogen was stored in DCI than in ordinary carbon steel. In general, the amount of hydrogen storage of hydrogen-charged carbon steel is less than or comparable to 1 mass ppm [7]. Takai et al. [8] reported that most of hydrogen in hydrogen-charged ferritic-pearlitic DCI exists in the graphite-matrix interface. Ogi et al. [9] also reported the similar results for ferritic DCI. Matsunaga et al. [5] figured out that a large amount of hydrogen was stored not only at the graphite-matrix interface but also inside the graphite nodule by using the hydrogen microprint technique. The hydrogen content of DCI was much larger compared with the increase in hydrogen content of carbon steel when the fraction of pearlitic phase was increased [2]. Therefore, it is concluded that the graphite-matrix interface and the inside of graphite nodule are a major hydrogen storage site for ferritic-pearlitic DCI.

Table 2. Mean graphite diameter, area fraction of each phase and hydrogen content.

Material	Mean graphite diameter (μm)	Area fraction (%)			Hydrogen content (mass ppm)
		Ferrite	Pearlite	Graphite	
FCD400	38	86	2	12	28.1
FCD450	19	72	15	13	4.4
FCD700	28	22	64	14	7.0

3.2 Relationship between ductility loss and crosshead speed (CHS)

Figure 3 shows the stress-strain curves obtained by the tensile tests of FCD400 and FCD700. The strength properties (i.e. flow stress and ultimate tensile stress) were slightly affected by hydrogen, but its effect was not remarkable. On the other hand, the ductility was severely deteriorated by hydrogen. The relative reduction of area (RRA), i.e. the ratio of reduction of area of hydrogen-charged specimen to that of uncharged specimen, is often employed as a measure of the degree of hydrogen embrittlement [10-12]. The RRAs were plotted as a function of CHS in Fig.4. It should be noted that the effect of CHS on the RRA depends on the fraction of pearlite. In FCD450 and FCD700 that contain a certain amount of pearlite, the RRA decreased with decrease in CHS. Namely, decreasing CHS intensified the hydrogen effect on the ductility. This phenomenon is understandable and is consistent with a conventional remark that low strain rate causes a more pronounced degradation owing to increased hydrogen accumulation at the fracture site. In ferritic DCI (FCD400), however, the RRA remained the almost same level despite a decrease in CHS.

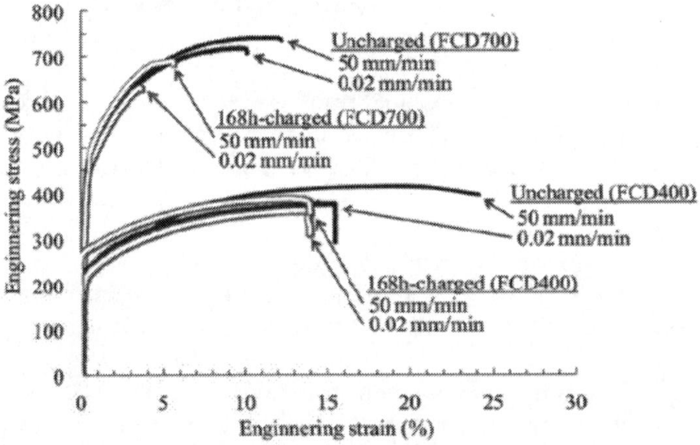

Figure 3. Stress-strain curves of FCD400 and FCD700.

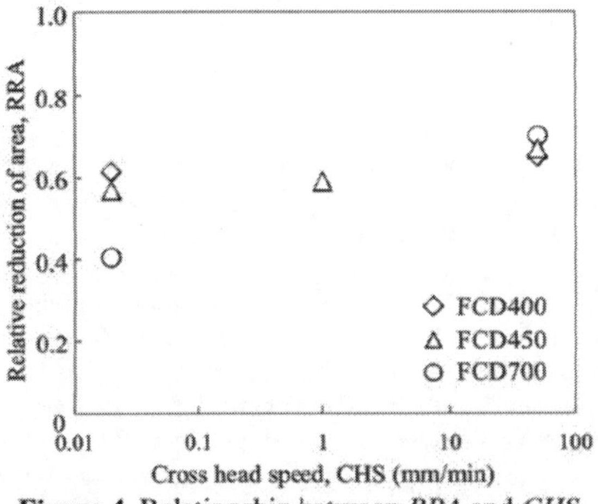

Figure 4. Relationship between *RRA* and *CHS*.

3.3 Fracture morphology

Figure 5 and 6 shows the fracture surfaces of uncharged and 168h-charged FCD400 and FCD700 that are tensile-fractured at CHS of 0.02 mm/min. Regarding FCD400, the fracture surface was covered by dimples originated from graphite nodules in uncharged specimen, whereas the quasi-cleavage fracture was observed partially in 168h-charged specimen. On the other hand, in FCD700, there is no significant difference in between the fracture surfaces of uncharged and 168h-charged specimens. Both of them were mostly covered by brittle-like fracture surfaces and flat fracture surfaces were observed around graphite nodules (see Fig. 6(c)).

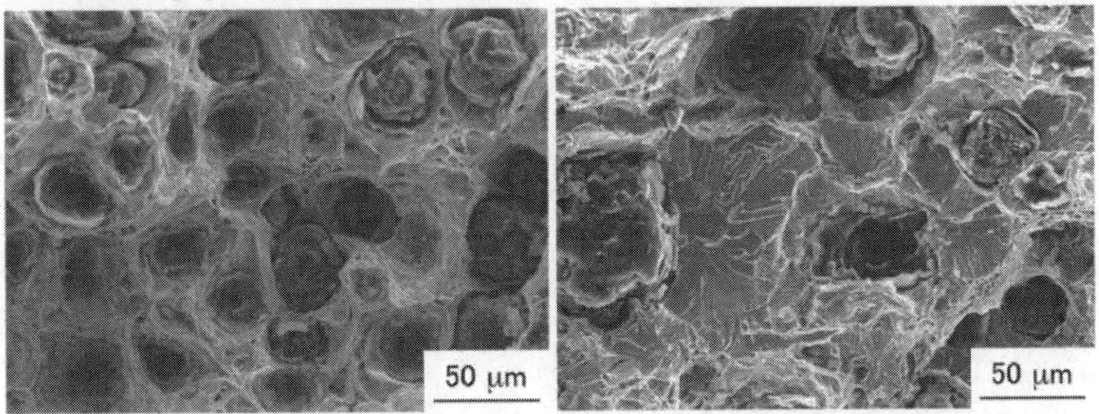

(a) Uncharged (b) 168h-charged

Figure 5. Fracture surfaces of FCD400 at *CHS*=0.02 mm/min.

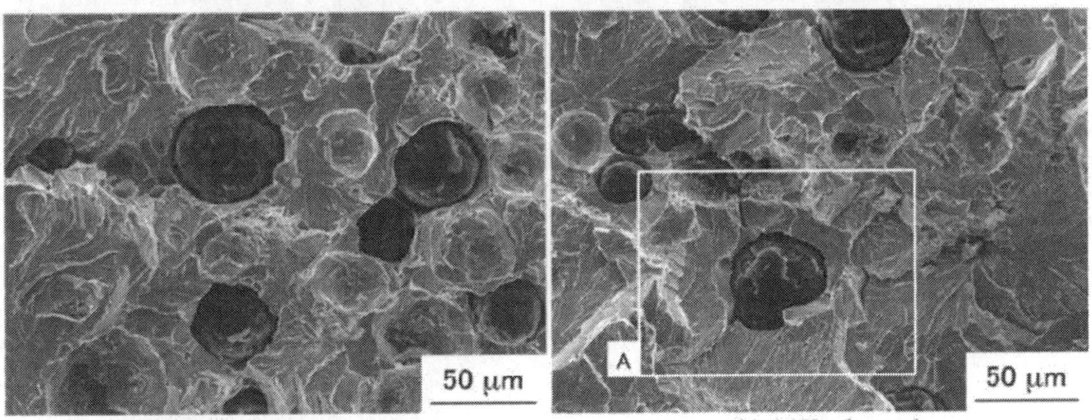

(a) Uncharged (b) 168h-charged

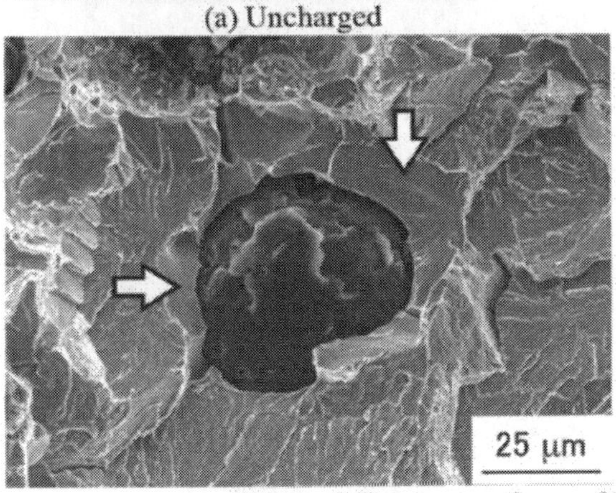

(c) Magnification of A in (b). Arrows indicate the flat fracture surfaces.

Figure 6. Fracture surfaces of FCD700 at *CHS*=0.02 mm/min.

IOP Conf. Series: Journal of Physics: Conf. Series **843** (2017) 012012 doi:10.1088/1742-6596/843/1/012012

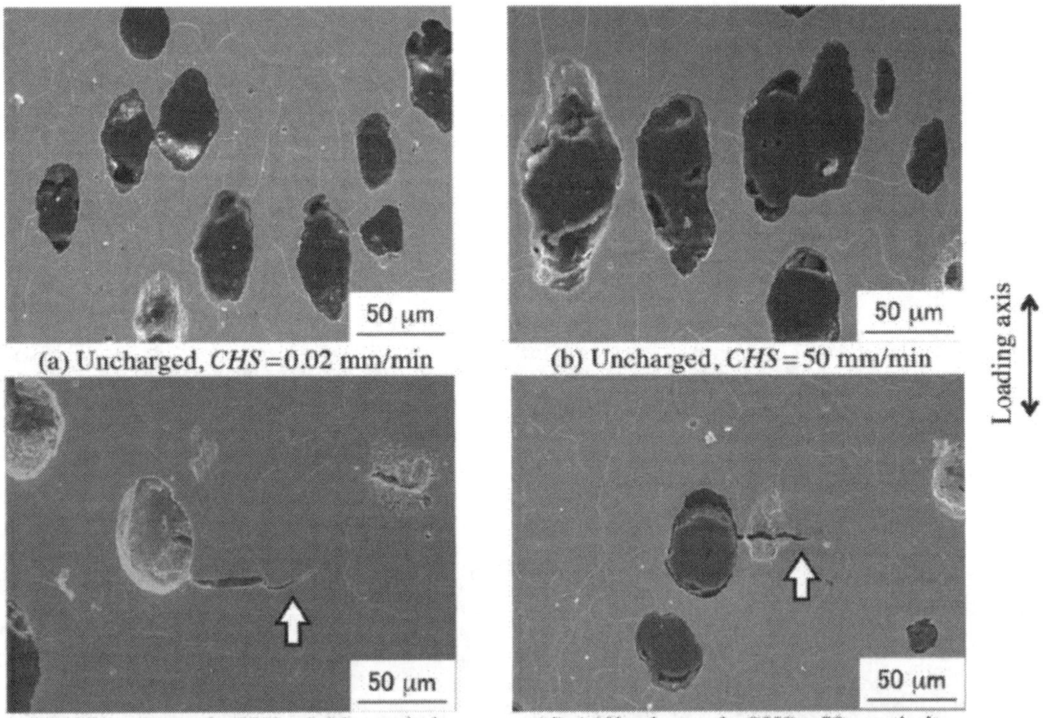

(a) Uncharged, *CHS* = 0.02 mm/min (b) Uncharged, *CHS* = 50 mm/min

(c) 168h-charged, *CHS* = 0.02 mm/min (d) 168h-charged, *CHS* = 50 mm/min

Figure 7. Fracture morphologies observed on the longitudinal cross sections of FCD400.

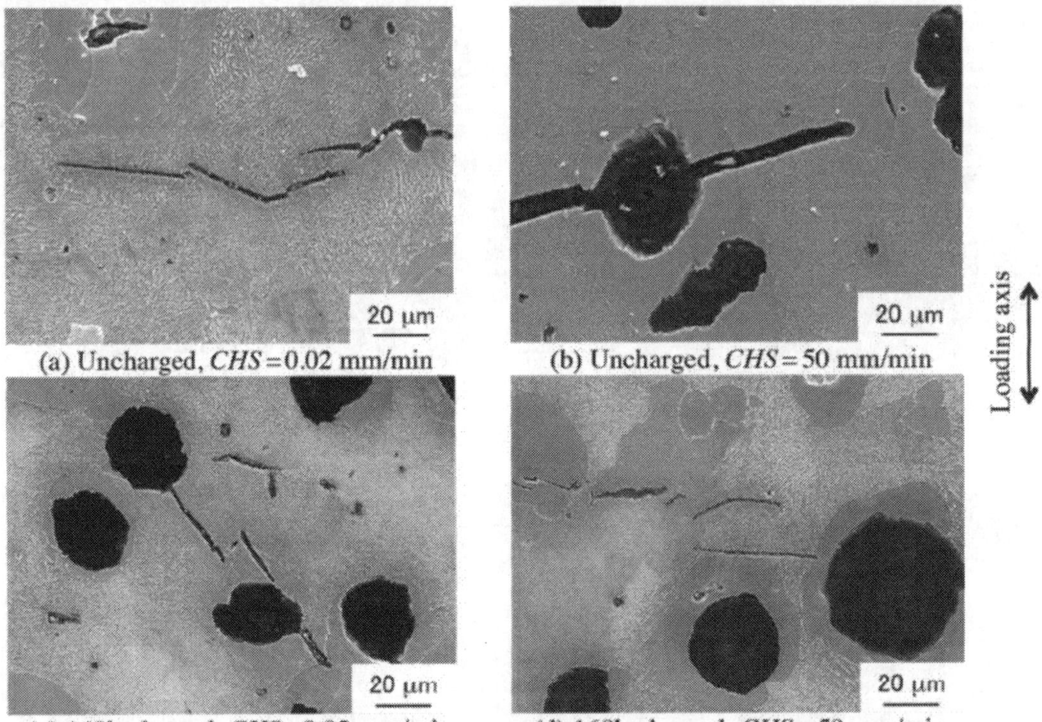

(a) Uncharged, *CHS* = 0.02 mm/min (b) Uncharged, *CHS* = 50 mm/min

(c) 168h-charged, *CHS* = 0.02 mm/min (d) 168h-charged, *CHS* = 50 mm/min

Figure 8. Fracture morphologies observed on the longitudinal cross sections of FCD700.

In order to elucidate the fracture process under each condition, the longitudinal cross section of the specimen after tensile rupture were observed and the typical images of SEM observation are shown in Figs. 7 and 8. In uncharged FCD400, voids originated from graphite nodules were elongated in the loading direction and then they coalesced to each other (see Fig. 7 (a), (b)). In 168h-charged FCD400, several cracks emanated from some of graphite nodules and grew perpendicular to the loading direction (indicated by arrows in Fig. 7 (c), (d)). These cracks resulted in the quasi-cleavage fracture observed on the fracture surface, which was considered to be a cause of ductility loss in FCD400. On the other hand, in both of uncharged and 168h-charged FCD700, several cracks were observed at the pearlite (see Fig. 8). The brittle-like fracture on the most of fracture surface of FCD700 was attributed to this pearlite cracking. Moreover, the flat fracture around graphite nodules is thought to be due to cracking of the ferrite layer surrounding graphite nodules.

3.4 Effect of pearlite and CHS on the fracture mechanism

Based on the observation of fracture morphology, the fracture mechanism of each condition is considered as schematically shown in Fig. 9. In FCD400 with a fully ferritic matrix, a graphite-matrix debonding is the prologue of fracture [13]. As mentioned above, a great amount of hydrogen exists at the graphite-matrix interface and the inside of graphite nodules. This graphite-supplied hydrogen (at least hydrogen trapped at the interface) fills immediately the interspace between graphite and matrix after debonding [5]. It is well known that hydrogen enhances a localized slip deformation called Hydrogen Enhanced Localized Plasticity, HELP [14, 15], which causes the quasi-cleavage cracking instead of the ductile dimple development. Accordingly, hydrogen supplied from the interspace to crack tip affects the fracture process, resulting in the growth of quasi-cleavage cracking in FCD400. It should be noted that the interspace is filled with hydrogen after debonding, and thereby, the sufficient amount of hydrogen can be supplied to the fracture site even at high CHS (i.e. 50 mm/min). At low CHS (i.e. 0.02 mm/min), although a small amount of hydrogen from the matrix diffuses to the crack

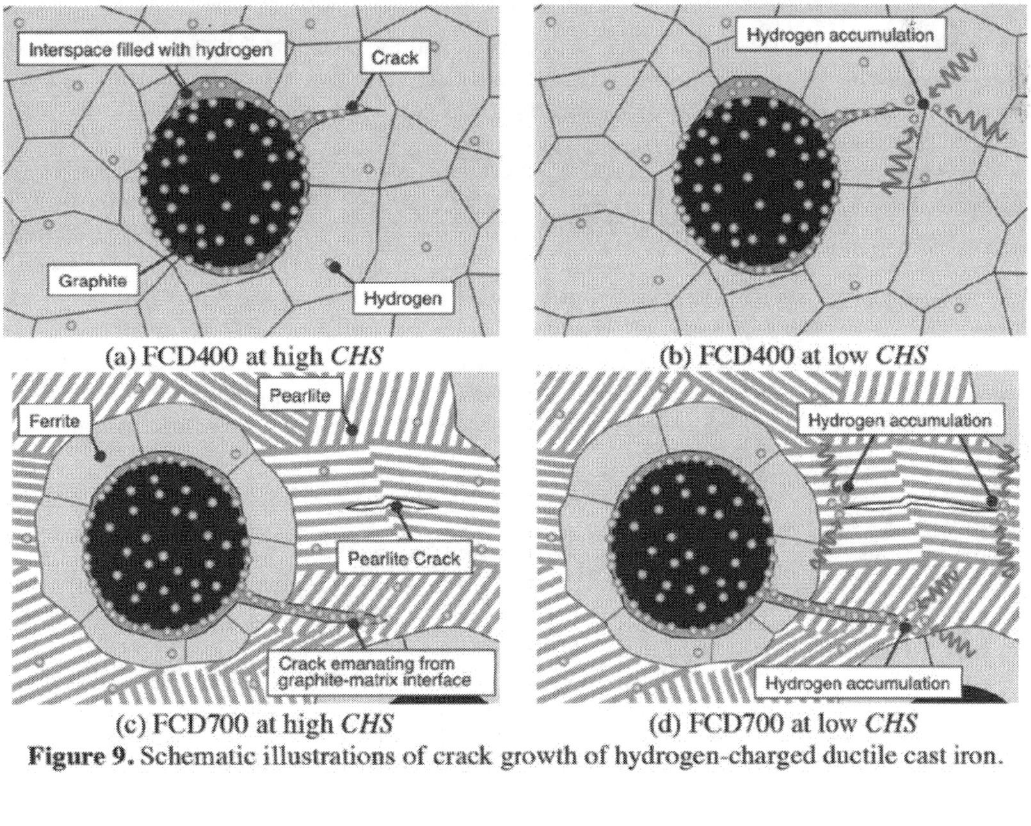

Figure 9. Schematic illustrations of crack growth of hydrogen-charged ductile cast iron.

tip, the effect of this hydrogen is very small compared with the hydrogen supplied directly from the interspace. In the case of FCD400, as a consequence, the hydrogen-induced ductility loss was not affected by CHS.

In FCD700, the pearlite cracking occurs in both uncharged and 168h-charged specimens. It has been reported that the growth of pearlite cracking in carbon steel is facilitated by the accumulation of solute hydrogen at the crack tip [2]. Therefore, the cracking initiated inside the pearlite is facilitated with decrease in the CHS, because the hydrogen supplied from the matrix becomes pronounced owing to the longer loading time. However, the cracking activated by the graphite-matrix debonding is not affected by the CHS in common with the case of FCD400.

In summary, only hydrogen that is supplied from graphite nodules plays a key role for cracking and thereby degrading ductility in the case of high CHS. In addition to this mechanism, in the case of low CHS, the hydrogen accumulated at the crack tip via the matrix is also influential for cracking inside the pearlite in FCD700 and as a result, assists the extension of fracture surface, reducing ductility. This is probably the reason why the hydrogen-induced ductility loss of ferritic-pearlitic DCIs becomes more susceptible to CHS with increase in the volume fraction of pearlite.

4. Conclusions

The tensile tests of hydrogen-charged ferritic-pearlitic DCIs with different volume fraction of pearlite were conducted. The effect of pearlite on the hydrogen-induced ductility loss of DCI was investigated in terms of strain rate dependency and the following conclusions were obtained:

1. The amount of hydrogen stored in ferritic-pearlitic DCI was much greater than that in ordinary carbon steel and it poorly correlated with the volume fraction of pearlite.
2. In FCD450 and FCD700 that contain a certain amount of pearlite, the reduction of area (RRA) decreased as the crosshead speed (CHS) decreased and this tendency was more pronounced in FCD700 than in FCD450. In FCD400, on the other hand, the RRA remained almost the same level despite a decrease in CHS.
3. In FCD400, the void growth and coalescence were mainly observed for the fracture process of uncharged material, while the quasi-cleavage cracking occurred partially in hydrogen-charged material. In FCD700, regardless of hydrogen-charging, pearlite cracking accounted for the majority of fracture process.
4. The difference in strain rate dependency of hydrogen-induced ductility loss in between ferritic-pearlitic DCIs can be explained as follows. At high CHS, only hydrogen that is supplied directly from interspace due to the graphite-matrix debonding plays a key role for degradation of ductility. At low CHS, additionally, the small amount of hydrogen supplied from the matrix accumulates at the crack tip. This hydrogen accumulation can not work for cracking from graphite nodule in FCD400 because a large amount of hydrogen inside the crack affects its growth dominantly, resulting in no strain rate dependency. On the other hand, in FCD700, the hydrogen supply from the matrix assists cracking inside the pearlite together with cracking from graphite that is facilitated by hydrogen supplied from graphite itself, and consequently it contributes further degradation of ductility with decrease in CHS.

Acknowledgment

The author would like to thank Mr. T. Usuda for his extensive experimental works.

References

[1] Yamada T and Kobayashi H 2012 *Koatsu gas* **49** 885-93
[2] Nishiguchi H, Fukushima Y, Matsuoka S and Murakami Y 2008 *Trans. Jpn. Soc. Mech. Eng.* A **74** 1016-25
[3] Nishiguchi H, Fukushima Y, Matsuoka S and Murakami Y 2010 *Trans. Jpn. Soc. Mech. Eng.* A **76** 1459-68

IOP Conf. Series: Journal of Physics: Conf. Series **843** (2017) 012012 doi:10.1088/1742-6596/843/1/012012

[4] Costa J E and Thompson A W 1982 *Metall. Trans.* A **13** 1315-18
[5] Matsunaga H, Usuda T, Yanase K and Endo M 2014 *Metall. Mater. Trans.* A **45** 1315-26
[6] Matsunaga H, Nakashima T, Yamada K, Matsuo T, Yamabe J and Matsuoka S 2016 *ASME. Pressure Vessels and Piping Conference* **6B** PVP2016-63536
[7] Matsuo T, Homma N, Matsuoka S and Murakami Y 2008 *Trans. Jpn. Soc. Mech. Eng.* A **74** 1164-73
[8] Takai K, Chiba Y, Noguchi K and Nozue A 2002 *Metall. Mater. Trans.* A **33** 2659–65
[9] Ogi K, Hagi H, Tahara A, Sawamoto A, Ikeda H and Hayashi Y 1992 *J. Jpn. Found. Eng. Soc.* **64** 186–91
[10] *Safety Standard for Hydrogen and Hydrogen Systems* 2005 NASA NSS 1740
[11] Michler T and Naumann J 2010 *Int. J. Hydrogen* **35** 821-32
[12] San Marchi C, Michler T, Nibur K A and Somerday B P 2010 *Int. J. Hydrogen* **35** 9736-45
[13] Dong M J, Prioul C and François D 1997 *Metall. Mater. Trans.* A **28** 2245-54
[14] Beachem C D 1972 *Metall. Trans.* **3** 437-51
[15] Birnbaum H K and Sofronis P 1994 *Mater. Sci. Eng.* A **176** 191-202

IOP Conf. Series: Journal of Physics: Conf. Series **843** (2017) 012013 doi:10.1088/1742-6596/843/1/012013

Anti-plane interface cracks between two dissimilar orthotropic layers

M Ayatollahi[1], S Varasteh[1]

[1]Faculty of Engineering, University of Zanjan, P.O.Box 45195-313, Zanjan, Iran

Corresponding author: E-mail address: mo.ayatollahy@gmail.com

Abstract. The problem of an anti-plane crack situated in the interface between two bonded dissimilar orthotropic layers is considered. Fourier transforms are used to reduce the problem to a system of singular integral equations with simple Cauchy kernel. The integral equations are solved numerically by converting to a system of linear algebraic equations and by using a collocation technique. The equations are solved for various crack length, material combinations, and external loads. The numerical results given in the paper include the stress intensity factors of several interfacial cracks

1. Introduction

The purpose of this study is to consider the so-called interface crack problem. From microelectronics to structural engineering many physical applications require joining of two dissimilar materials. Layered materials are extensively used in various products and devices to improve structural performance such as strength and durability. Thermal barrier coatings, for example, are used in turbine blades to protect the core material from thermal fatigue [1]. The increasing concern in recent years with the mechanical failure initiating at the interfacial regions in many technologically important multiphase materials requires a better understanding of the interaction between the cracks that may exist in these regions. The modern composites, diffusion bonded materials used in microelectronics, and the thermal barrier coating of superalloys coated by ceramics used in jet engines may be mentioned as some examples for such multiphase materials. Wide applications of adhesive bonds in aerospace industry for both aluminum and composite structures are well documented. Failures of these layered structures are often attributed to interfacial damage in the form of microcracks or debonded zones, which are either preexisting or developed during service due to mechanical/thermal fatigue and environmental degradations. If a crack appears at an interface between two dissimilar materials, the stress intensity factors should be solved to ensure that catastrophic fracture of composite materials does not occur. Stresses and displacements for an interface crack between two dissimilar elastic half-planes are solved by reducing the boundary conditions to a Hilbert problem [2]. For non-interacting cracks at the interface between two dissimilar materials, Pecorari and Kelly [3] have obtained an explicit expression of the effective normal spring stiffness and have shown that their expression reduces to that of Baik and Thompson [4] when elastic moduli difference and crack area fraction approach zero. While crack interactions have been investigated in regard to overall effective material properties [5]. Yang et al. [6] solved the fracture problems near the interface crack tip of double dissimilar orthotropic composite materials. The asymmetrical dynamic propagation problem on the edges of mode III interface crack subjected to superimpose loads was investigated by Lu et al. [7]. Zhao et al. [8] obtained the solution of interface crack in the bi-material composite by means of complex variable function.

The objective of the present paper is to investigate the problem of multiple interface cracks in two dissimilar orthotropic layers under the condition of anti-plane deformation. To solve the proposed problem, the Fourier transform is employed together with the distributed dislocation technique to the derivation of a singular integral equation with a Cauchy kernel. The results turn out that the cracks interaction strongly affect the magnitude of stress intensity factors. Also, the influence of the material properties and crack position on the stress intensity factors are presented.

2. Formulation of the problem

Consider the anti-plane elasticity problem for dissimilar orthotropic layers weakened by screw dislocation shown in Fig. 1. The superscripts 1 and 2 correspond to the bottom and top orthotropic materials. Referring to Fig. 1, the dislocation is situated in the origin of Cartesian coordinates along the interface.

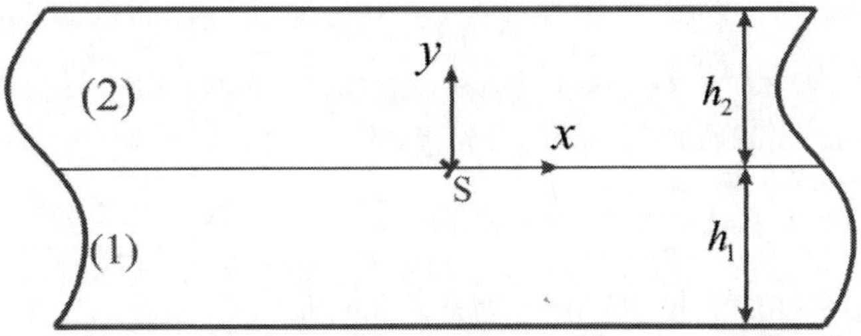

Fig. 1. Schematic view of the medium with screw dislocation.

The x- and y-axes are in the directions of principal material orthotropy of both layers. The only nonzero displacement component is the out of plane component $w(x,y)$. Consequently, the constitutive equations of the layers in a Cartesian coordinate system are:

$$\sigma_{zx1}(x,y)=G_{zx1}\frac{\partial w_1(x,y)}{\partial x},\qquad \sigma_{zy1}(x,y)=G_{zy1}\frac{\partial w_1(x,y)}{\partial y},\qquad -h_1\le y\le 0$$

$$\sigma_{zx2}(x,y)=G_{zx2}\frac{\partial w_2(x,y)}{\partial x},\qquad \sigma_{zy2}(x,y)=G_{zy2}\frac{\partial w_2(x,y)}{\partial y},\qquad 0\le y\le h_2 \tag{1}$$

The equilibrium equations in terms of displacement result in

$$\frac{\partial^2 w_1}{\partial y^2}+g_1^2\frac{\partial^2 w_1}{\partial x^2}=0,\qquad -h_1\le y\le 0$$

$$\frac{\partial^2 w_2}{\partial y^2}+g_2^2\frac{\partial^2 w_2}{\partial x^2}=0,\qquad 0\le y\le h_2 \tag{2}$$

where $g_i^{\,2}=G_{zxi}/G_{zyi}$ $(i=1,2)$. Note that body forces are not considered in the present work. For this dislocation problem, the upper and lower surfaces of structure are stress free. The problem will be solved under the following boundary and continuity conditions:

$$\sigma_{zy2}(x,h_2)=0$$

$$\sigma_{zy1}(x,-h_1)=0$$

$$\sigma_{zy2}(x,0^-)=\sigma_{zy2}(x,0^+)$$

$$w_2(x,0^+)-w_1(x,0^-)=b_zH(x)\qquad |x|<\infty \tag{3}$$

where $H(x)$ is the Heaviside-step function, the former relation, denotes the multivaluedness of displacement. By taking Fourier transforms in x, the general solution of Eqs.(2) may be expressed as:

$$w_1^*(s,y) = A_1(s)e^{g_1|s|y} + A_2(s)e^{-g_1|s|y} \qquad -h_1 \le y \le 0$$
$$w_2^*(s,y) = A_3(s)e^{g_2|s|y} + A_4(s)e^{-g_2|s|y} \qquad 0 \le y \le h_2 \qquad (4)$$

The functions $A_i(s), i = 1,2,3,4$ are unknown. The application of conditions (3) gives the unknown coefficients. After determining coefficients, referring to (4) and using the constitutive equations with the aid of inverse Fourier transform, the stress field associated with a single dislocation in the interface of two dissimilar orthotropic layers can be written as:

$$\sigma_{zy1}(x,y) = -\frac{g_1 g_2 G_{zy1} G_{zy2} b_z}{\pi}\int_0^\infty \frac{\sinh(g_2 s h_2)\sinh(g_1 s(y+h_1))}{M_1(s)}\sin(sx)ds \qquad (5)$$

where

$$M_1(s) = g_2 G_{zy2}\sinh(g_2 s h_2)\cosh(g_1 s h_1) + g_1 G_{zy1}\sinh(g_1 s h_1)\cosh(g_2 s h_2) \qquad (6)$$

By observing that for large values of s, from (5) we find:

$$\sigma_{zy1\infty}(x,y) = \lim_{s\to\infty}\sigma_{zy1}(x,y) = -\frac{g_1 g_2 G_{zy1} G_{zy2} b_z}{\pi(g_1 G_{zy1} + g_2 G_{zy2})}\int_0^\infty e^{syg_1}\sin(sx)ds \qquad (7)$$

The singular parts can be evaluated by the use of the following identities:

$$\int_0^\infty e^{sy}\sin(sx)ds = \frac{x}{x^2+y^2}, \qquad y<0$$
$$\int_0^\infty e^{sy}\cos(sx)ds = -\frac{y}{x^2+y^2}, \qquad y<0 \qquad (8)$$

After performing the appropriate asymptotic analysis and separating the singular parts of the kernels, we obtain:

$$\sigma_{zy1}(x,y) = -\frac{g_1 g_2 G_{zy1} G_{zy2} b_z}{\pi}\{\int_0^\infty [\frac{\sinh(g_2 s h_2)\sinh(g_1 s(y+h_1))}{M_1(s)} - \frac{e^{syg_1}}{g_1 G_{zy1} + g_2 G_{zy2}}]\sin(sx)ds$$
$$+ \frac{1}{g_1 G_{zy1} + g_2 G_{zy2}}\frac{x}{x^2 + g_1^2 y^2}\} \qquad (9)$$

There may be observed that stress componenty exhibit the familiar Cauchy type singularity at the dislocation location.

3. Formulation of multiple interface cracks

In this section, the basic concepts of the technique is introduced, such as the dislocation. In this part, the method will be expanded to deal with multiple interface cracks by considering two dissimilar orthotropic layers weakened by N interface cracks. The cracks configuration are presented in parametric form as:

$$x_i = x_i(s)$$
$$y_i = y_i(s) \qquad i \in \{1,2,\dots,N\} \qquad -1 \le s \le 1 \qquad (10)$$

Two orthogonal coordinate systems *s-n* are chosen on the i-th crack such that their origin is located on the crack while the *s*-axis remains tangent to the crack surface. Suppose dislocations with unknown dislocation density $B_{zj}(t)$ are distributed on the infinitesimal segment $\sqrt{[x_j'(t)]^2+[y_j'(t)]^2}dt$ on the surface of the *j*-th crack. The traction components on the surface of the i-th crack in the presence of dislocations distribution on the surfaces of all *N* cracks mentioned above, yield:

$$\sigma_{zn}(x_i(s),y_i(s))=\sum_{j=1}^{N}\int_{-1}^{1}K_{ij}(s,t)B_{zj}(t)\sqrt{[x_j'(t)]^2+[y_j'(t)]^2}dt \tag{11}$$

where $\sigma_{zn}(x_i,y_i)$ is the anti-plane traction on the surface of the i-th crack in terms of stress. The kernel of integral equations (11) takes the following form

$$K_{ij}(\mathrm{s},\mathrm{t})=-\frac{g_1g_2G_{zy1}G_{zy2}}{\pi}\{\int_0^{\infty}(\frac{\sinh(g_2sh_2)\sinh[g_1s((y_i-y_j)+h_1)]}{M_1(s)}-\frac{e^{s(y_i-y_j)g_1}}{g_1G_{zy1}+g_2G_{zy2}})$$

$$\times\sin(s(x_i-x_j))ds+\frac{1}{g_1G_{zy1}+g_2G_{zy2}}\frac{x_i-x_j}{(x^2+g_1^{\,2}(y_i-y_j)^2)}\} \tag{12}$$

The left hand side of the Eqs. (11) represent stress component at the presumed location of the cracks with negative sign. Employing the definition of dislocation density functions, the equations for the crack opening displacement across *j*th crack become:

$$w_j^+(s)-w_j^-(s)=\int_{-1}^{s}B_{zj}(t)\sqrt{[x_j'(t)]^2+[y_j'(t)]^2}dt \qquad j=1,2,...,N \tag{13}$$

Since the crack is an embedded crack, from the condition of single-valuedness of displacements it follows that:

$$\int_{-1}^{1}B_{zj}(t)\sqrt{[x_j'(t)]^2+[y_j'(t)]^2}dt=0, \qquad j=1,2,...,N \tag{14}$$

The Cauchy singular integral equations (11) and (14) are to be solved simultaneously, to obtain the dislocation density. This is accomplished by means of Gauss–Chebyshev uadrature scheme developed in [9]. The stress fields near a crack tips having square-root singularity, can be expressed as:

$$B_{zj}(t)=\frac{g_{zj}(t)}{\sqrt{1-t^2}}, \quad -1\le t\le 1,\ j=1,2,...,N \tag{15}$$

Substituting (15) into (11) and (14) and discretizing the domain, $-1\le t\le 1$, the integral equations reduced to the system of linear algebraic equations. The function $g_{zj}(t)$ are obtained via solution of the system of equations. The stress intensity factors for the i-th interface crack may be defined and evaluated as follows:

$$K_{Rj}^{M}=-\frac{\mu_1}{2}\left[[x_j'(1)]^2+[y_j'(1)]^2\right]^{\frac{1}{4}}g_{zj}(1)$$

$$K_{Lj}^{M}=\frac{\mu_1}{2}\left[[x_j'(-1)]^2+[y_j'(-1)]^2\right]^{\frac{1}{4}}g_{zj}(-1) \qquad j=1,2,...,N \tag{16}$$

The details of the derivation of fields intensity factors to reach (16) are not given here.

4. Results and discussions

The formulation described above can be easily implemented numerically. In the first example, the effect of ratios of moduli of elasticity of the orthotropic layers on the stress intensity factors is studied. The symmetry of the problem with respect to *y*-axis implies that the stress intensity factors at the crack tips are identical. Comparing stress intensity factors in orthotropic layers reveals that material orthotropy g_2 enhances K_{III}/K_0 but material orthotropy g_1 attenuates normalized stress intensity factor.

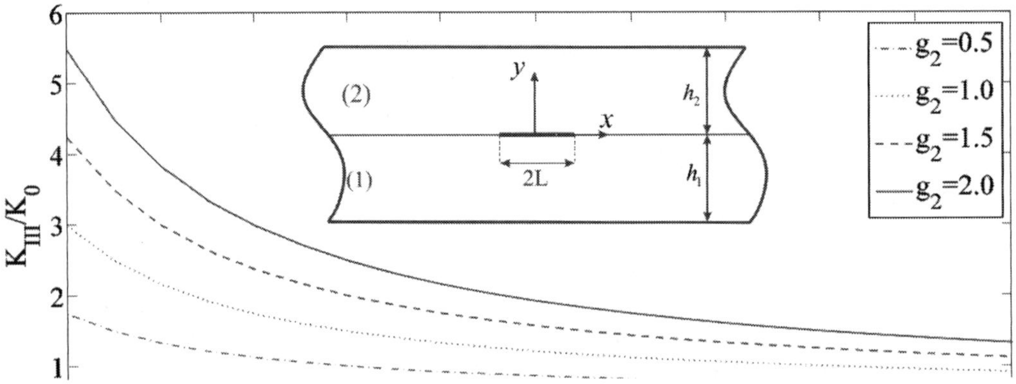

Fig. 2. Normalized stress intensity factor of crack tip versus the ratios of material orthotropy.

In the next example, two interface cracks L_1R_1 and L_2R_2 with equal length $2L$ are considered. The dimensionless stress intensity factor with g_2/g_1 is shown in Fig. 3 . It can be found that increasing the value of the parameter g_2/g_1 results increasing the value of stress intensity factor. the ratios of material orthotropy of the layers becomes more significant.

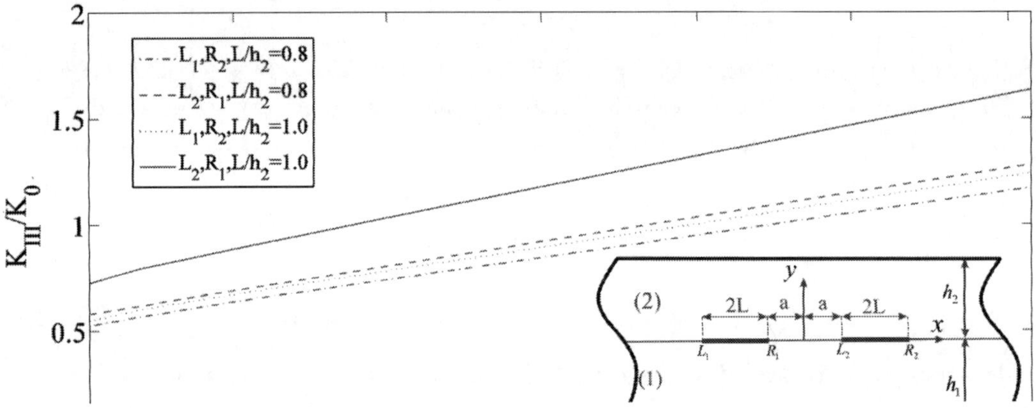

Fig. 3. Dimensionless stress intensity factors of two interface cracks with g_2/g_1 .

Fig. 4 shows the effect of varying the crack tip distance a on the stress intensity factors. The lengths of cracks remain fixed while the cracks location are changing with the same rate. As it may be observed K_{III}/K_0 , is decreased by growing the cracks distance.

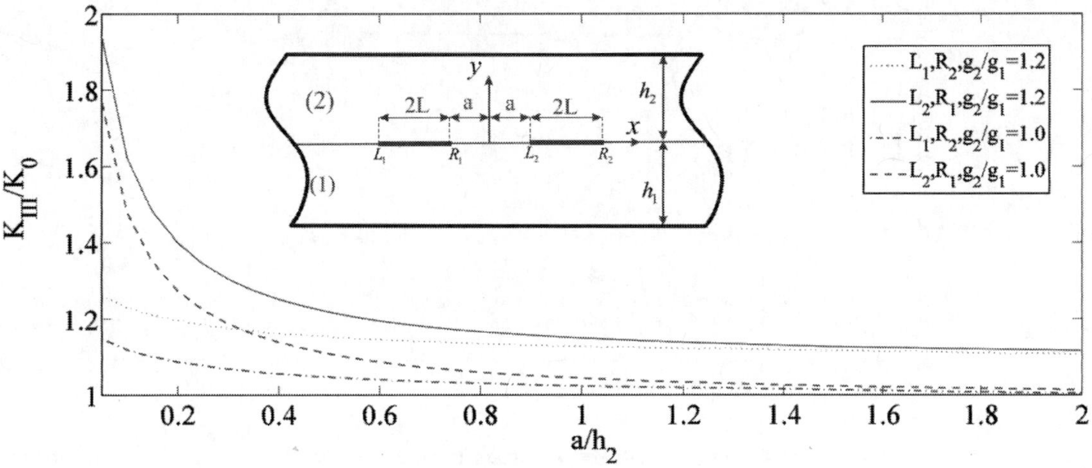

Fig. 4. Dimensionless stress intensity factors of two interface cracks with a/h_2 for different values of the g_2/g_1 .

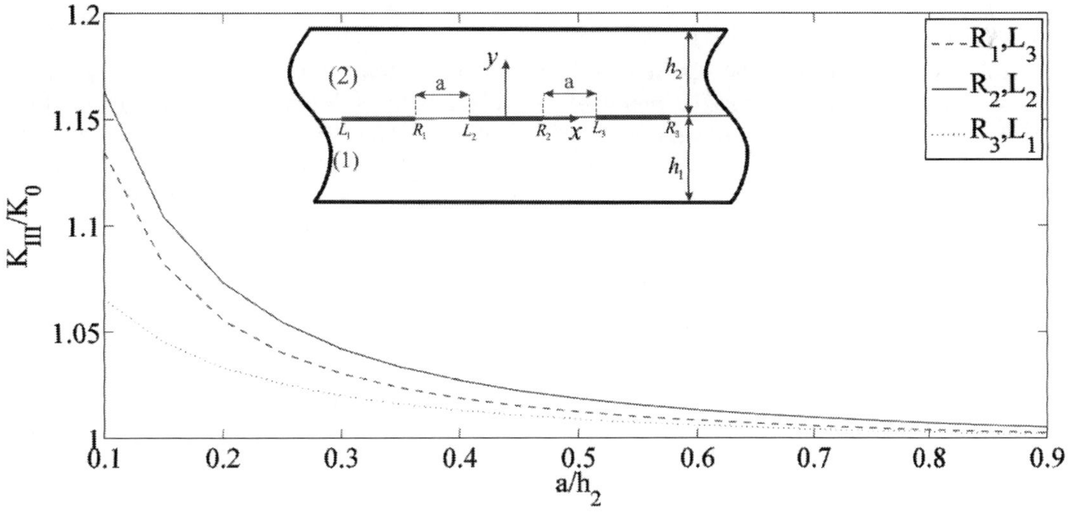

Fig. 5. Dimensionless stress intensity factors of three interface cracks with a/h_2 .

We study the intraction of three identical interface cracks. Fig. 5. Shows the influence of the dimensionless crack tip distance a/h_2 on the normalized stress intensity factors. It is evident from this figure that increasing the crack distance leads to a reduction of the stress intensity factors.

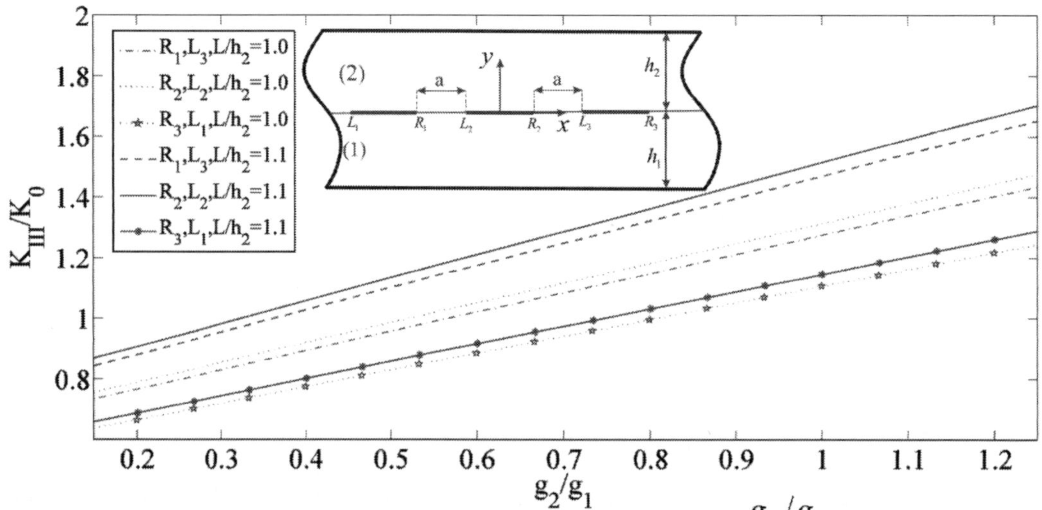

Fig. 6. Dimensionless stress intensity factors of three interface cracks with g_2/g_1 for different values of the L/h_2 .

In the last example, The effect of the ratios of material orthotropy of the layers on the stress intensity factors of three equal lengths cracks is examined. As it may be observed, the stress intensity factors increase rapidly as the crack lengths and ratios of material orthotropy increase.

5. Conclusions

Analytic solution for the stress fields caused by the Volterra type dislocation in the interface of two dissimilar orthotropic layers is obtained.The dislocation solution is utilized to perform integral equations for a medium weakened by interfacial cracks. These equations are of Cauchy singular type solvable by numerical methods to obtain dislocation density on the cracks surfaces. Several examples are solved and stress intensity factors are determined for interacting cracks. The interaction of multiple cracks shows that the stress intensity factors of the approaching crack tips intensify. Moreover, the stress intensity factor increases by increasing the crack length.

References

[1] Miller, R.A., 1987. Current status of thermal barrier coatings: an overview.Surface Coating Technology 30, 1–11.

[2] England, A.H., A crack between dissimilar media, ASME J. Appl. Mech. 32 (1965) 400–402.

[3] Pecorari, C., Kelly, P.A., 2000. The quasi-static approximation for a cracked interface between a layer and a substrate. Journal of the Acoustical Society of America 107 (5), 2454–2461.

[4] Baik, J., Thompson, R.B., 1984. Ultrasonic scattering from imperfectinterfaces: a quasi-static model. Journal of Nondestructive Evaluation 4 (3/4), 177–196.

[5] Nemat-Nasser, S., Yu, N., Hori, M., 1993. Solids with periodically distributed cracks. International Journal of Solids and Structures 30 (15), 2071–2095.

[6] Yang, W.-y., Zhang, S.-q., Li, J.-l. and Ma, Y.-l. 2009. Interface crack problems for mode II of double dissimilar orthotropic composite materials, Applied Mathematics and Mechanics, 30(5), 585-594.

[7] Lu, N.-C., Li, X.-G., Cheng, Y.-H. and Cheng, J., 2013. Asymmetrical Dynamic Propagation Problem on the Edges of Mode III Interface Crack Subjected to Superimpose Loads, Journal of Mechanics, 29(02), pp. 319-326.

[8] Zhao, W.B., Zhang X.X., Cui, X.C., Yang W.Y., 2014. Analysis of stress intensity factor in orthotropic bi-material mixed interface crack. Applied Mathematics and Mechanics 35(10), 1271–1292.

[9] F. Erdogan, G.D. Gupta and T.S. Cook, Numerical solution of singular integral equations, in: G.C. Sih (Ed.) Methods of analysis and solutions of crack problems: Recent developments in fracture mechanics Theory and methods of solving crack problems, Springer Netherlands, (Dordrecht, 1973, 368-425).

Analysis of multiple edge cracks in a non-homogeneous piezoelectric layer

M. Nourazar, M. Ayatollahi
Faculty of Engineering, University of Zanjan, P. O. Box 45195-313, Zanjan, Iran
Corresponding author's: mo.ayatollahy@gmail.com

Abstract. This study is concerned with the treatment of the several edge cracks in a functionally graded piezoelectric (FGP) layer under anti-plane mechanical and in-plane electrical loading. The edge crack is assumed to be either electrically impermeable or permeable. The problem is formulated by using distributed dislocation technique. The integral equations are constructed for the analysis of a FGP layer, in which the unknown variables are dislocation densities. By use of the dislocation densities, the field intensity factors are calculated. Numerical examples are provided to show the effect of the location and orientations of edge crack upon the stress intensity factors.

1. Introduction

Piezoelectric materials generate an electric field when subjected to strain fields and undergo deformation when an electric field is applied. This inherent electromechanical coupling is widely used in electronic industry. The technical applications include wave guides, sensors, phase invertors, transducers. Studies on the properties of piezoelectric composites have been carried out by numerous investigators. In particular, there is a growing interest among researchers in solving fracture mechanics problems. Because piezoelectric composite materials are very brittle and cracked piezoelectric composite materials usually contain multiple cracks, the interaction between cracks may significantly affect their fracture behavior.

There has been significant progress in the study of electroelastic fields disturbed by cracks in piezoelectric materials recently. The problem of a finite crack in a strip of FGPM was analyzed by Li and Weng [1]. Wang [2] studied a mode-III crack in FGPMs. A FGPM strip with eccentric crack under anti-plane shear was analyzed by Shin and Kim [3]. Hu et al. [4] and Yong and Zhou [5] studied both the impermeable and permeable cracks in a FGPM layer bonded to two dissimilar homogeneous piezoelectric half spaces. Zhou and Wang [6] solved an anti-plane shear crack in FGPMs using non-local theory. Jiang [7] investigated the fracture behavior of FGPMs with dielectric cracks. Zhou and Chen [8] studied the interaction of two parallel mode-I limited-permeable cracks in a FGPM. Interaction between an electrically permeable crack and the imperfect interface in FGPM was solved by Li and Lee [9]. Yan and Jiang [10] investigated the parallel cracks in FGPMs.

In the present study, the interactions of multiple edge cracks in a FGP strip investigated using the distributed dislocation technique. The Fourier transform was employed to reduce boundary value problem to the integral equations. Numerical results are provided to show the effects of the length and distance of cracks on the field intensity factors.

2. Formulation of the problem

Consider FGP layer with the poling axis z occupies the region, and is thick enough in the z-direction to allow a state of anti-plane shear. The FGP layer containing a Volterra-type screw dislocation with the line of dislocation parallel to the edges of the layer is analyzed. The piezoelectric boundary value problem is simplified considerably if we consider only the out-of-plane mechanical and the in-plane electric fields such that:

$$u = 0, v = 0, w = w(x, y)$$

$$E_x = E_x(x,y),\ E_y = E_y(x,y),\ E_z = 0. \tag{1}$$

In this case, the constitutive equations for the piezoelectric material can be expressed as

$$\sigma_{xz} = c_{44}(y)\frac{\partial w}{\partial x} + e_{15}(y)\frac{\partial \varphi}{\partial x}$$

$$\sigma_{yz} = c_{44}(y)\frac{\partial w}{\partial y} + e_{15}(y)\frac{\partial \varphi}{\partial y}$$

$$D_x = e_{15}(y)\frac{\partial w}{\partial x} - \varepsilon_{11}(y)\frac{\partial \varphi}{\partial x}$$

$$D_y = e_{15}(y)\frac{\partial w}{\partial y} - \varepsilon_{11}(y)\frac{\partial \varphi}{\partial y} \tag{2}$$

where $c_{44}(y)$ is the elastic stiffness measured in a constant electric field, $e_{15}(y)$ is the piezoelectric constant and $\varepsilon_{11}(y)$ is the dielectric measured at a constant strain. The elastic displacement and electric potential must satisfy the governing equations as follows:

$$c_{44}(y)\nabla^2 w + e_{15}(y)\nabla^2\varphi + \frac{\partial c_{44}(y)}{\partial y}\frac{\partial w}{\partial y} + \frac{\partial e_{15}(y)}{\partial y}\frac{\partial \varphi}{\partial y} = 0$$

$$e_{15}(y)\nabla^2 w - \varepsilon_{11}(y)\nabla^2\varphi + \frac{\partial e_{15}(y)}{\partial y}\frac{\partial w}{\partial y} - \frac{\partial \varepsilon_{11}(y)}{\partial y}\frac{\partial \varphi}{\partial y} = 0 \tag{3}$$

The material properties of the FGP layer are adopted as follows:

$$c_{44}(y) = c_{440}e^{2\lambda y},\quad e_{15}(y) = e_{150}e^{2\lambda y}$$
$$\varepsilon_{11}(y) = \varepsilon_{110}e^{2\lambda y},\quad \rho(y) = \rho_0 e^{2\lambda y} \tag{4}$$

where λ is the non-homogeneous material constant and $[c_{440}, e_{150}, \varepsilon_{110}, \rho_0]$ are values at the plane $y = 0$. The above assumption is unrealistic for all the material properties, however it would allow us to shed some light on the influence of the material gradient upon the stress intensity factors. Under the above consideration, the governing equations can be simplified to the following forms:

$$c_{440}\nabla^2 w + e_{150}\nabla^2\varphi + 2\lambda c_{440}\frac{\partial w}{\partial y} + 2\lambda e_{150}\frac{\partial \varphi}{\partial y} = 0$$

$$e_{150}\nabla^2 w - \varepsilon_{110}\nabla^2\varphi + 2\lambda e_{150}\frac{\partial w}{\partial y} - 2\lambda\varepsilon_{110}\frac{\partial \varphi}{\partial y} = 0. \tag{5}$$

With the use of Bleustein function

$$\psi(x,y) = \varphi - \alpha w(x,y) \tag{6}$$

In which $\alpha = e_{150}/\varepsilon_{110}$, Eqs. (5) can be written as follows:

$$\nabla^2 w + 2\lambda\frac{\partial w}{\partial y} = 0,$$

$$\nabla^2\psi + 2\lambda\frac{\partial \psi}{\partial y} = 0. \tag{7}$$

where $\tilde{c}_{44} = c_{440} + (e_{150})^2/\varepsilon_{110}$ is the piezoelectric stiffened elastic constant. The constitutive equation (2) can be written as follows

$$\sigma_{xz}(\mathrm{x,y}) = [(c_{440}+\alpha e_{150})\frac{\partial w}{\partial x} + e_{150}\frac{\partial \psi}{\partial x}]\exp(2\lambda y)$$

$$\sigma_{yz}(\mathrm{x,y}) = [(c_{440}+\alpha e_{150})\frac{\partial w}{\partial y} + e_{150}\frac{\partial \psi}{\partial y}]\exp(2\lambda y)$$

$$D_x(\mathrm{x,y}) = -\varepsilon_{110}\exp(2\lambda y)\frac{\partial \psi}{\partial x}$$

$$D_y(\mathrm{x,y}) = -\varepsilon_{110}\exp(2\lambda y)\frac{\partial \psi}{\partial y} \tag{8}$$

From the physical viewpoint, two electrical crack boundary conditions have been commonly used to describing the fracture behavior in the piezoelectric materials, permeable and impermeable ones. Of these extreme cases for the crack boundary conditions where permittivity of these cracks are assumed to be infinite and zero for a slit-like crack,respectively. For impermeable case, by the using equation (6), the following conditions for screw dislocation are considered,

$$\sigma_{yz}(x,0)=0,\quad \sigma_{yz}(x,h)=0,\quad \sigma_{yz}(x,\zeta^-)=\sigma_{yz}(x,\zeta^+),$$
$$D_y(x,0)=0,\quad D_y(x,h)=0,\quad D_y(x,\zeta^-)=D_y(x,\zeta^+),$$
$$w(x,\zeta^-)-w(x,\zeta^+)=b_{mz}H(x-\eta),$$
$$\psi(x,\zeta^-)-\psi(x,\zeta^+)=(b_p-\alpha b_{mz})H(x-\eta),\ \lim_{|x|\to\infty} w=0,\ \lim_{|x|\to\infty}\psi=0, \tag{9}$$

where in Eq. (9) b_{mz} and b_p designate dislocation Burgers vectors. Although the jump in the electric potential is not a type of dislocation, it is referred here as electric dislocation for convenience. To solve the problem stated above, it is convenient to employ Fourier transforms to Eqs. (7) and (9). Let the solutions of Eqs. (7) be given by:

$$w(x,y)=\frac{b_{mz}e^{-\lambda(y-\xi)}}{4\pi}\int_{-\infty}^{+\infty}\frac{\sinh[\beta(\xi-h)]}{\beta\sinh(\beta h)}(\pi\delta(\omega)-i/\omega)\left(\beta\cosh\beta y+\lambda\sinh\beta y\right)e^{i\omega x}d\omega$$
$$0<y<\xi$$

$$w(x,y)=\frac{b_{mz}e^{-\lambda(y-\xi)}}{4\pi}\int_{-\infty}^{+\infty}\frac{\sinh(\beta\xi)}{\beta\sinh(\beta h)}(\pi\delta(\omega)-i/\omega)(\beta\cosh[\beta(y-h)]+\lambda\sinh[\beta(y-h)])e^{i\omega(x-\eta)}d\omega$$
$$\xi<y<h$$

$$\psi(x,y)=\frac{(b_p-\alpha b_{mz})e^{-\lambda(y-\xi)}}{4\pi}\int_{-\infty}^{+\infty}\frac{\sinh[\beta(\xi-h)]}{\beta\sinh(\beta h)}(\pi\delta(\omega)-i/\omega)\left(\beta\cosh\beta y+\lambda\sinh\beta y\right)e^{i\omega(x-\eta)}d\omega$$
$$0<y<\xi$$

$$\psi(x,y)=\frac{(b_p-\alpha b_{mz})e^{-\lambda(y-\xi)}}{4\pi}$$
$$\times\int_{-\infty}^{+\infty}\frac{\sinh(\beta\xi)}{\beta\sinh(\beta h)}(\pi\delta(\omega)-i/\omega)[\beta\cosh(\beta(y-h))+\lambda\sinh(\beta(y-h))]e^{i\omega(x-\eta)}d\omega$$
$$\xi<y<h \tag{10}$$

where $,\beta=\sqrt{\lambda^2+\omega^2}$. In this case, with the aid of constitutive equations the stress and electric displacement components are as follows:

$$\sigma_{zx}(x,y)=-\frac{e^{\lambda(\xi+y)}}{2\pi}(c_{440}b_{mz}+e_{150}b_p)$$
$$\times\int_0^\infty\frac{\sinh[\beta(h-\xi)]}{\beta\sinh(\beta h)}(\beta\cosh\beta y+\lambda\sinh\beta y)\cos(\omega(x-\eta))d\omega,\qquad 0<y<\xi$$
$$\sigma_{zx}(x,y)=\frac{e^{\lambda(\xi+y)}}{2\pi}(c_{440}b_{mz}+e_{150}b_p)$$
$$\times\int_0^\infty\frac{\sinh(\beta\xi)}{\beta\sinh(\beta h)}(\beta\cosh[\beta(y-h)]+\lambda\sinh[\beta(y-h)])\cos(\omega(x-\eta))d\omega,\quad \xi<y<h$$

$$\sigma_{zy}(x,y)=(c_{440}+\alpha e_{150})\frac{e^{\lambda(\xi+y)}b_{mz}}{\pi}\int_0^\infty\frac{\omega\sinh[\beta(\xi-h)]}{\beta\sinh(\beta h)}\sinh(\beta y)\sin(\omega(x-\eta))d\omega$$
$$+\frac{e_{150}(b_p-\alpha b_{mz})e^{\lambda(\xi+y)}}{\pi}\int_0^\infty\frac{\omega\sinh[\beta(\xi-h)]}{\beta\sinh(\beta h)}\sinh(\beta y)\sin(\omega(x-\eta))d\omega\qquad 0<y<\xi$$
$$\sigma_{zy}(x,y)=(c_{440}+\alpha e_{150})\frac{e^{\lambda(\xi+y)}b_{mz}}{\pi}\int_0^\infty\frac{\omega\sinh(\beta\xi)}{\beta\sinh(\beta h)}\sinh(\beta(y-h))\sin(\omega(x-\eta))d\omega$$
$$+\frac{e_{150}(b_p-\alpha b_{mz})e^{\lambda(\xi+y)}}{\pi}\int_0^\infty\frac{\omega\sinh(\beta\xi)}{\beta\sinh(\beta h)}\sinh(\beta(y-h))\sin(\omega(x-\eta))d\omega\qquad \xi<y<h$$
$$D_x(x,y)=\frac{\varepsilon_{110}(b_p-\alpha b_{mz})e^{\lambda(\xi+y)}}{2\pi}\int_0^\infty\frac{\sinh[\beta(h-\xi)]}{\beta\sinh(\beta h)}(\beta\cosh\beta y+\lambda\sinh\beta y)\cos(\omega(x-\eta))d\omega,\qquad 0<y<\xi$$
$$D_x(x,y)=-\frac{\varepsilon_{110}(b_p-\alpha b_{mz})e^{\lambda(\xi+y)}}{2\pi}$$
$$\times\int_0^\infty\frac{\sinh(\beta\xi)}{\beta\sinh(\beta h)}(\beta\cosh[\beta(y-h)]+\lambda\sinh[\beta(y-h)])\cos(\omega(x-\eta))d\omega,\quad \xi<y<h$$

$$D_y(x,y)=-\frac{\varepsilon_{110}(b_p-\alpha b_{mz})e^{\lambda(\xi+y)}}{\pi}\int_0^\infty\frac{\omega\sinh(\beta(\xi-h))}{\beta\sinh(\beta h)}\sinh(\beta y)\sin(\omega(x-\eta))d\omega\qquad 0<y<\xi$$
$$D_y(x,y)=-\frac{\varepsilon_{110}(b_p-\alpha b_{mz})e^{\lambda(\xi+y)}}{\pi}\int_0^\infty\frac{\omega\sinh(\beta\xi)}{\beta\sinh(\beta h)}\sinh(\beta(y-h))\sin(\omega(x-\eta))d\omega\qquad \xi<y<h$$
(11)

It may be seen that the integrals in (11) are bounded at $\omega=0$. If we further observe that the integrands in (11) are continuous functions of ω, it is then clear that any singularity the kernels must be due to the asymptotic behavior of the integrands as $|\omega|$ approaches infinity. By adding and subtracting the asymptotic expressions of the integrands, we find:

$$\sigma_{zx}(x,y) = -\frac{e^{\lambda(\xi+y)}(c_{440}b_{mz}+e_{150}b_p)}{2\pi}\left\{\frac{y-\xi}{(x-\eta)^2+(y-\xi)^2}\right.$$
$$\left.-\int_0^\infty[\frac{(1-\mathrm{e}^{-2\beta(\mathrm{h}-\xi)})\left(\beta(1+\mathrm{e}^{-2\beta y})+\lambda(1-\mathrm{e}^{-2\beta y})\right)}{\beta(1-\mathrm{e}^{-2\beta h})}-1]\mathrm{e}^{\omega(y-\xi)}\cos(\omega(x-\eta))d\omega\right\}, \qquad 0<y<\xi$$

$$\sigma_{zx}(x,y) = -\frac{e^{\lambda(\xi+y)}(c_{440}b_{mz}+e_{150}b_p)}{2\pi}\left\{\frac{y-\xi}{(x-\eta)^2+(y-\xi)^2}\right.$$
$$\left.+\int_0^\infty[\frac{(1-\mathrm{e}^{-2\beta\xi})\left(\beta(1+\mathrm{e}^{2\beta(y-h)})+\lambda(\mathrm{e}^{2\beta(y-h)}-1)\right)}{\beta(1-\mathrm{e}^{-2\beta h})}-1]\mathrm{e}^{-\omega(y-\xi)}\cos(\omega(x-\eta))d\omega\right\}, \qquad \xi<y<h$$

$$\sigma_{zy}(x,y) = \frac{(c_{440}b_{mz}+e_{150}b_p)e^{\lambda(\xi+y)}}{2\pi}\left\{\frac{x-\eta}{(x-\eta)^2+(y-\xi)^2}\right.$$
$$\left.+\int_0^\infty[\frac{\omega(\mathrm{e}^{2\beta(\xi-h)}-1)(1-\mathrm{e}^{-2\beta y})}{\beta(1-\mathrm{e}^{-2\beta h})}-1]\mathrm{e}^{\omega(y-\xi)}\sin(\omega(x-\eta))d\omega\right\} \qquad 0<y<\xi$$

$$\sigma_{zy}(x,y) = \frac{(c_{440}b_{mz}+e_{150}b_p)e^{\lambda(\xi+y)}}{2\pi}\left\{\frac{x-\eta}{(x-\eta)^2+(y-\xi)^2}\right.$$
$$\left.-\int_0^\infty[\frac{\omega(1-\mathrm{e}^{-2\beta\xi})(\mathrm{e}^{2\beta(y-h)}-1)}{\beta(1-\mathrm{e}^{-2\beta h})}+1]\mathrm{e}^{-\omega(y-\xi)}\sin(\omega(x-\eta))d\omega\right\} \qquad \xi<y<h$$

$$D_y(x,y) = \frac{\varepsilon_{110}(\alpha b_{mz}-b_p)e^{\lambda(\xi+y)}}{2\pi}\left\{\frac{x-\eta}{(x-\eta)^2+(y-\xi)^2}\right.$$
$$\left.+\int_0^\infty[\frac{\omega(\mathrm{e}^{2\beta(\xi-h)}-1)(1-\mathrm{e}^{-2\beta y})}{\beta(1-\mathrm{e}^{-2\beta h})}-1]\mathrm{e}^{\omega(y-\xi)}\sin(\omega(x-\eta))d\omega\right\} \qquad 0<y<\xi$$

$$D_y(x,y) = \frac{\varepsilon_{110}(\alpha b_{mz}-b_p)e^{\lambda(\xi+y)}}{2\pi}\left\{\frac{x-\eta}{(x-\eta)^2+(y-\xi)^2}\right.$$
$$\left.-\int_0^\infty[\frac{\omega(1-\mathrm{e}^{-2\beta\xi})(\mathrm{e}^{2\beta(y-h)}-1)}{\beta(1-\mathrm{e}^{-2\beta h})}+1]\mathrm{e}^{-\omega(y-\xi)}\sin(\omega(x-\eta))d\omega\right\} \qquad \xi<y<h$$

$$D_x(x,y) = \frac{(e_{150}b_{mz}-\varepsilon_{110}b_p)e^{\lambda(\xi+y)}}{2\pi}\left\{-\frac{y-\xi}{(x-\eta)^2+(y-\xi)^2}\right.$$
$$\left.+\int_0^\infty[\frac{(1-\mathrm{e}^{-2\beta(\mathrm{h}-\xi)})\left(\beta(1+\mathrm{e}^{-2\beta y})+\lambda(1-\mathrm{e}^{-2\beta y})\right)}{\beta(1-\mathrm{e}^{-2\beta h})}-1]\mathrm{e}^{\omega(y-\xi)}\cos(\omega(x-\eta))d\omega\right\}, \qquad 0<y<\xi$$

$$D_x(x,y) = -\frac{(e_{150}b_{mz} - \varepsilon_{110}b_p)e^{\lambda(\xi+y)}}{2\pi}\left\{\frac{y-\xi}{(x-\eta)^2+(y-\xi)^2}\right.$$
$$\left.+\int_0^{\infty}\left[\frac{(1-e^{-2\beta\xi})\left(\beta(1+e^{2\beta(y-h)})+\lambda(e^{2\beta(y-h)}-1)\right)}{\beta(1-e^{-2\beta h})}-1\right]e^{-\omega(y-\xi)}\cos(\omega(x-\eta))d\omega\right\},\qquad \xi<y<h \tag{12}$$

We may observe that stress components exhibit the familiar Cauchy-type singularity at dislocation location.

The dislocation solutions accomplished in the preceding section may be employed to analyze FGP layer containing several edge cracks with any arrangement. A crack configuration with respect to coordinate system x, y may be described in parametric form as:

$$x_i = x_i(s)$$
$$y_i = y_i(s) \qquad i=1,2,\ldots,N \qquad -1\le s\le 1 \tag{13}$$

Next, covering the crack surfaces by dislocations, the principle of superposition is invoked to obtain the tractions on a given crack surface. The anti-plane traction and electric potential components on the face of the ith crack due to the presence of distribution of the above-mention dislocations on all N cracks yields.

$$\sigma_{yz}(x_i(s),y_i(s)) = \sum_{j=1}^{N}\int_{-1}^{1}[K_{ij}^{11}(s,t)B_{mzj}(t) + K_{ij}^{12}(s,t)B_{pj}(t)]l_j\,dt$$
$$D_y(x_i(s),y_i(s)) = \sum_{j=1}^{N}\int_{-1}^{1}[K_{ij}^{21}(s,t)B_{mzj}(t) + K_{ij}^{22}(s,t)B_{pj}(t)]l_j\,dt \tag{14}$$

By virtue of Bueckner's principle (see, e.g., Hills et al., 1996), the left hand side of Eqs. (14) are stress components and the electric displacement at the presumed location of the cracks with negative sign, which implies impermeable crack boundary conditions. From the definition of density function, the equation for the crack opening displacement and electric potential across jth crack become

$$w_j^-(s) - w_j^+(s) = \int_{-1}^{s} l_j B_{mj}(t)dt$$
$$\phi_j^-(s) - \phi_j^+(s) = \int_{-1}^{s} l_j B_{pj}(t)dt \tag{15}$$

For the edge cracks, taking the embedded crack tip at $t=-1$, the dislocation densities may be expressed as:

$$B_{kj}(t) = g_{kj}(t)\sqrt{1-t/1+t},\quad -1\le t\le 1,\ k\in\{m,p\} \tag{16}$$

The parameters $g_{kj}(t)$ are obtained by solving the system of equations (14, 16). The stress and electric intensity factors for edge crack take the forms

$$(K_{III}^m)_{Li} = c_{44}(y_{Li})\left[[(x_i'(-1)]^2+[y_i'(-1)]^2\right]^{\frac14} g_{mi}(-1) + e_{15}(y_{Li})\left[[(x_i'(-1)]^2+[y_i'(-1)]^2\right]^{\frac14} g_{pi}(-1)$$
$$(K_{III}^D)_{Li} = e_{15}(y_{Li})\left[[(x_i'(-1)]^2+[y_i'(-1)]^2\right]^{\frac14} g_{mi}(-1) - \varepsilon_{11}(y_{Li})\left[[(x_i'(-1)]^2+[y_i'(-1)]^2\right]^{\frac14} g_{pi}(-1) \tag{17}$$

For brevity, the details of the derivation of fields intensity factors are not given here.

IOP Conf. Series: Journal of Physics: Conf. Series **843** (2017) 012014 doi:10.1088/1742-6596/843/1/012014

3. Numerical results and discussion

The analysis developed in the preceding section allows the consideration a FGP layer with any number of cracks with different shape and orientations. In the computational procedure, we consider the PZT-4 piezoelectric ceramic of which material properties are given as follows:

$$c_{44} = 2.56\times10^{10}\frac{N}{m^2}, e_{15} = 12.7\frac{C}{m^2}, \varepsilon_{11} = 64.6\times10^{-10}\frac{C}{Vm}, \rho_0 = 7.5\times10^3\frac{kg}{m^3}$$

where N is the force in Newton, C is the charge in Coulomb and V is the electric potential in Volt. The stress intensity factor is normalized by $K_0 = \tau_0\sqrt{L}$ and electromechanical coupling factor that we used is defined by $ecf = D_0 e_{15}/\tau_0\varepsilon_{11}$.

The first example is considered the problem of a functionally graded piezoelectric layer weakened by an edge crack. The applied loads are constant anti-plane mechanical ($\sigma_{xz}=\sigma_0$) and in-plane electrical loading with magnetiude (D_0), while $D_0 e_{15}/\tau_0\varepsilon_{11} = 1.0$. The plot of the normalized stress intensity factor versus crack length for different values of the non-homogeneous parameter (λh=0,1) is shown in Fig (1). Increasing the non-homogeneous parameter results in lower stress intensity factor. Fig. (1) also shows that the stress intensity factor increases with an increase in crack length for two different electric boundary conditions.

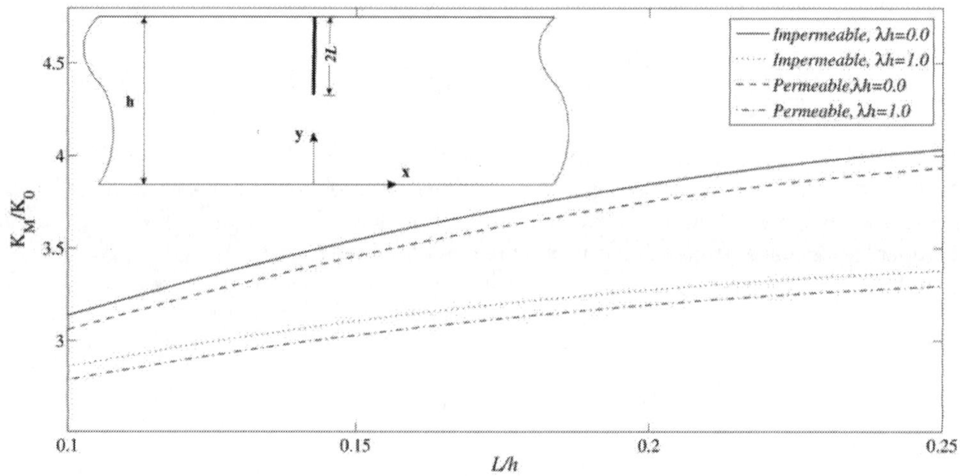

Fig. 1. Variation of stress intensity factors for an edge crack versus L/h.

Fig. (2) shows the normalized stress intensity factor, K_M/K_0 as function of the non-homogeneous parameter λL and for different crack orientations. As it may be observed, the stress intensity factor decreases rapidly as λL increases. The similar trend may be noticed for the two different electric boundary conditions.

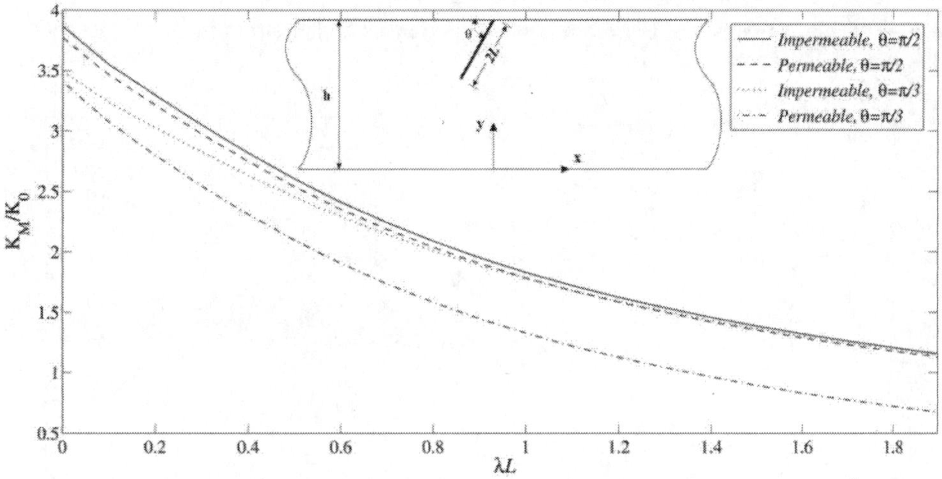

Fig. 2. Variation of stress intensity factors for an edge crack versus λL.

The effect of crack orientation on the stress intensity factor is examined by considering an edge crack with constant length (Fig. 3). As the crack angle increases, the stress intensity factors increase. The maximum stress intensity factor for the crack tips occur when the crack surface traction is maximum. The trend of variation of stress intensity factor remains the same by changing the FGM constant.

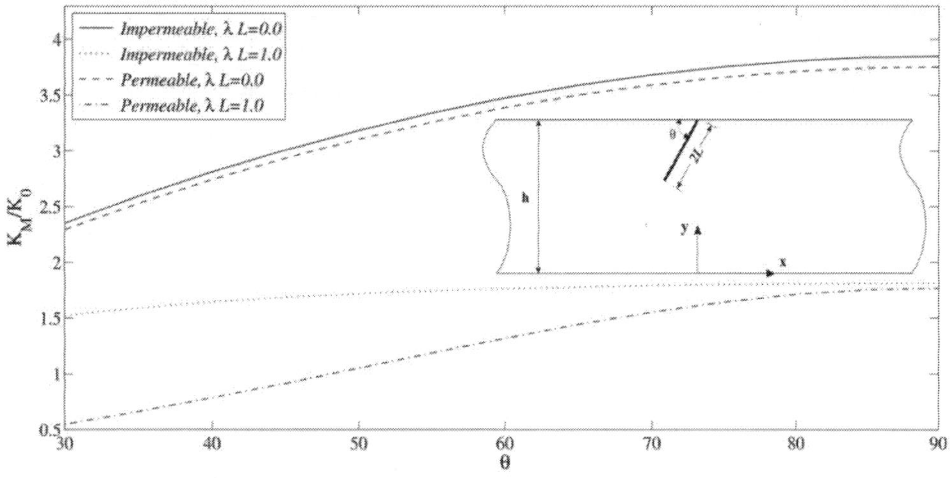

Fig. 3. Variation of stress intensity factors for a rotating edge crack.

IOP Conf. Series: Journal of Physics: Conf. Series **843** (2017) 012014 doi:10.1088/1742-6596/843/1/012014

In the next example, the FGP layer containing two parallel cracks with equal length is considered. The variations of K_M / K_0 with crack length is depicted in Fig(4) for three different crack distances. When the distance between the crack tips is minimum obviously experiences higher stress intensity factors.

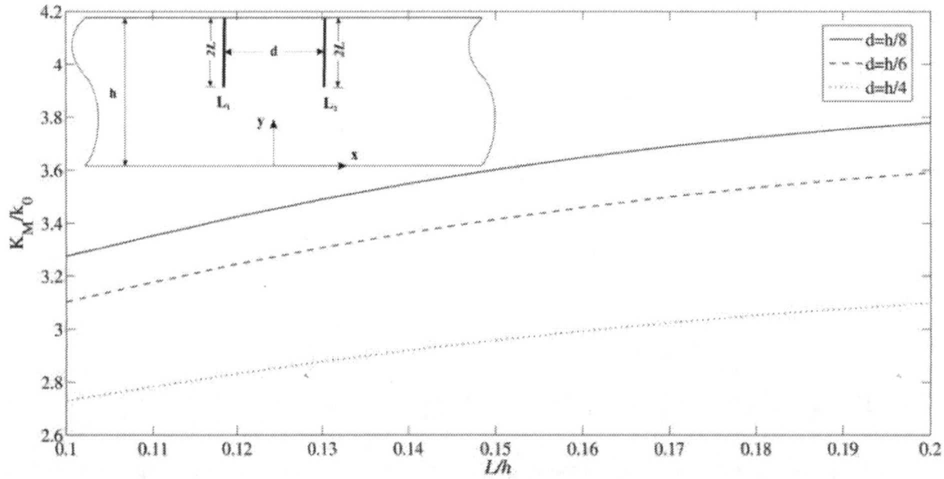

Fig. 4. Variation of stress intensity factors of two parallel edge crack.

In the last example, the interaction between two equal-length edge cracks is examined. The dimensionless stress intensity factors versus the crack orientation is shown in Fig. (5). As it was expected, the variation of stress intensity factor of rotating crack tip L_2 is more pronounced than that of the tip L_1.

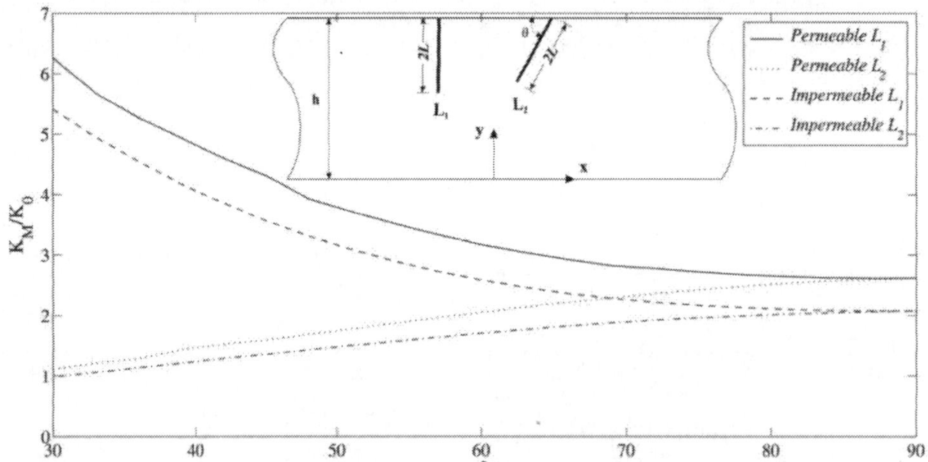

Fig. 5. Variation of stress intensity factors of two intracting edge crack.

4. Conclusions

The present work deals with the behavior of a FGP layer containing multiple edge cracks subjected to electro-mechanical loads. The dislocation solution is utilized to perform integral equations for FGP layer with multiple edge cracks. These equations are of Cauchy singular type solvable by numerical methods to obtain dislocation density on the crack surfaces. Numerical calculations are carried out to study the effect of the geometry of the several edges cracks and the material properties on the resulting stress intensity factor.

References

[1] Li C and Weng GJ, 2002a Anti-plane crack problem in functionally gradedpiezoelectric materials, ASME J. Appl. Mech. **69** 481–488

[2] Wang BL 2003 A mode-III crack in functionally graded piezoelectric materials, Mech. Res. Commun. **30** 151–159

[3] Shin JW and Kim TU 2003. Functionally graded piezoelectric strip with eccentriccrack under anti-plane shear, J. Mech. Sci. Technol. **17** 854–859

[4] Hu KQ, Zhong Z and Jin B 2005 Anti-plane shear cracking functionally gradientpiezoelectric layer bonded to dissimilar half spaces, Int. J. Mech. Sci. **47** 82–93

[5] Yong HD and Zhou YH 2007 A mode-III crack in a functionally graded piezoelectricstrip bonded to two dissimilar piezoelectric half-planes, Compos. Struct. **79** 404–410

[6] Zhou ZG and Wang B 2006 Non-local theory solution for an anti-plane shearpermeable crack in functionally graded piezoelectric materials, Appl. Compos. Mater. **13** 345–367

[7] Jiang LY 2008 The fracture behavior of functionally graded piezoelectric materialswith dielectric cracks, Int. J. Fract. **149** 87–104

[8] Zhou Z and Chen ZT 2008 The interaction of two parallel mode-I limitedpermeable cracks in a functionally graded piezoelectric material, Euro. J. Mech. A Solids **27** 824–846

[9] Li YD and Lee KY 2009a Interaction between an electrically permeable crack andthe imperfection interface in a functionally graded piezoelectric sensor, Int. J. Eng. Sci. **47** 363–371

[10] Yan Z and Jiang LY 2010 Interaction of parallel dielectric cracks in functionallygraded piezoelectric materials, Acta Mech. **211** 251–269

[11] Hills DA, Kelly PA, Dai DN and Korsunsky AM 1996. Solution of Crack Problems: The Distributed Dislocation Technique, Kluwer: Academic Publishers

Response of an orthotropic half-plane subjected to transient anti-plane loading with multiple edge cracks

M. Ayatollahi[1]
[1] Faculty of Engineering, University of Zanjan, P.O. Box 45195-313, Zanjan, Iran

Corresponding author's: mo.ayatollahy@gmail.com

Abstract. This study deals with the problem of multiple edge cracks in an elastic orthotropic half-plane under transient loading. The dislocation solution is utilized to derive integral equations for multiple interacting edge cracks in an orthotropic half-plane. These equations are solved numerically thereby obtaining the dislocation density function on the crack faces and stress intensity factors at crack tips. Numerical results are obtained to illustrate the variation of the dynamic stress intensity factors as a function of crack length and material properties.

1. Introduction

With the growing use of composites in great variety of engineering structures, in recent years the problems regarding their structured intensity and failure have been studied extensively. In the case of dynamic loads, two loading cases are of interest, harmonic loading and impact loading [1]. Various attempts have been made to analyze half-plane weakened by cracks under harmonic load. Impact loads applied on cracked structures may cause catastrophic failure. It is therefore, of great importance to investigate the transient response of cracked materials. The stationary semi-infinite crack under uniform step loading in the crack faces was considered first by Maue [2]. Loeber and Sih [3] investigated the dynamic stress intensity factor for a finite crack in the infinite plane under anti-plane deformation. Ma and Hou [4] investigated the response of an elastic solid containing a crack subjected to suddenly concentrated point loads acting at a finite distance from the crack tip. Wang et al. [5] obtained the dynamic stress intensity factor for mode III loading. They analyzed the anti-plane response of a non-homogeneous composite material containing several cracks subjected to dynamic impact. The failure behavior of fiber reinforced composites involving cracked matrix and imperfectly bounded fibers under dynamic anti-plane loading was investigated by Meguid et al. [6]. Results show the effect of the interaction between a main crack and a completely debounded fiber upon the dynamic stress intensity factors. The transient dynamic stress intensity factor was determined for an interface crack between two dissimilar isotropic viscoelastic bodies under impact loading by Wei et al. [7]. The transient response of an infinite orthotropic material with finite crack under point loading was studied by Rubio et al. [8]. Chen et al. [9] investigated the transient response of the internal crack in a functionally graded orthotropic strip. Results show the effects of the material parameter, the crack configuration and the orthotropic property on the dynamic stress intensity factors. Shul and Lee [10] considered a subsurface crack in a functionally graded coating layer on the layered half-space subjected to an anti-plane impact load. Lira-Vergaraand Rubio-Gonzalez [11] analyzed the transient response of an non-homogeneous orthotropic material with an interfacial finite crack under shear impact load. The problem of a homogeneous linear elastic body containing multiple collinear cracks under anti-plane dynamic load was considered by Wu et al. [12]. However, those previous solutions only considered the embedded or edge crack with simple shape and patterns. Monfared and Ayatollahi [13], investigated the scattering of anti-plane harmonic stress waves by multiple cracks in an orthotropic half-plane.

In this paper, we consider the problem of an orthotropic half-plane with multiple edge cracks The main purpose is toprovide an analytical treatment to investigate the transient behavior of orthotropic half-plane under transient load, which is important for the design and numerical simulation of composite structures. Based on the use of integral transforms, the problem is reduced to a singular integral equation, which can be solved using Chebyshev polynomial expansions. Numerical results are provided to show the effect of cracks interaction and material property on the stress intensity factors

IOP Conf. Series: Journal of Physics: Conf. Series **843** (2017) 012015 doi:10.1088/1742-6596/843/1/012015

2. Problem statements

We consider an orthotropic half-plane containing a screw dislocation under anti-plane deformation. A Cartesian coordinate is assumed in such a way that the x and y axes are taken as directions of principal material orthotropy and y-axis is along dislocation path. The constitutive equations for anti-plane problem of the orthotropic material are as follows:

$$\sigma_{zx}(x,y,t) = G_{zx}\frac{\partial w(x,y,t)}{\partial x},\ \sigma_{zy}(x,y,t) = G_{zy}\frac{\partial w(x,y,t)}{\partial y}. \tag{1}$$

where G_{zx} and G_{zy} are elastic constants. In the absence of body force the equation of motion is:

$$\frac{\partial^2 w(x,y,t)}{\partial x^2} + \frac{1}{f^2}\frac{\partial^2 w(x,y,t)}{\partial y^2} = s_l^2\frac{\partial^2 w(x,y,t)}{\partial t^2}. \tag{2}$$

where s_l^2 is wave slowness defined by $s_l = 1/c_l$ and c_l being the shear wave speed, $f = \sqrt{G_{zx}/G_{zy}}$, $c_l = \sqrt{G_{zx}/\rho}$ and ρ is the mass density. The conditions representing a Volterra-type screw dislocation located at the positive part of the y-axis in an orthotropic half-plane are

$$w(0^+,y,t) - w(0^-,y,t) = b_z(t)H(y), \sigma_{zx}(0^+,y,t) = \sigma_{zx}(0^-,y,t), \sigma_{zy}(x,h,t) = 0., \tag{3}$$

where, $H(.)$ is the Heaviside step function, $b_z(t)$ is the time dependent Burgers vector. The second Eq. (3)impliesthe continuity of traction crossing the dislocation line. Employing the symmetry of the problem with respect to the y-axis, Equation (2)may be solved in the x> 0. The boundary conditions for the half-plane are:

$$\sigma_{zy}(x,h,t) = 0,\ w(x,0^+,t) = w(x,0^-,t),\ \sigma_{zy}(x,0^+,t) = \sigma_{zy}(x,0^-,t),\ w(0,y,t) = \frac{1}{2}b_z(t)H(y). \tag{4}$$

The solution to equation (2) is achieved by means of integral transforms. The Laplace and Fourier sine transforms are applied to Eq. (2), assuming that the orthotropic half-plane is initially stationary, leads to:

$$\frac{d^2\bar{w}^*(\zeta,y,p)}{dy^2} - f^2(\zeta^2 + s_l^2 p^2)\bar{w}^* = 0., \tag{5}$$

The general solution of (5) is:

$$\bar{w}_1^*(\zeta,y,p) = A_1(\zeta,p)e^{-yf\beta} + B_1(\zeta,p)e^{yf\beta} + \frac{\zeta b_z(p)}{2\beta^2}, \qquad 0<y<h$$

$$\bar{w}_2^*(\zeta,y,p) = B_2(\zeta,p)e^{yf\beta}. \qquad y<0 \tag{6}$$

where $\beta = \sqrt{\zeta^2 + s_l^2 p^2}$, the functions $A_1(\zeta,p)$, $B_1(\zeta,p)$ and $B_2(\zeta,p)$are unknown functions. By using the inversion of the Fourier sine transform of Eqs. (6), the anti-plane displacement in the Laplace domain can be obtained as:

$$\bar{w}_1(x,y,p) = -\frac{b_z(p)}{2\pi}\int_0^{+\infty}\frac{\zeta\sin(\zeta x)}{\beta^2}(e^{-yf\beta} + e^{-(2h-y)f\beta} - 2)d\zeta, \qquad 0<y<h$$

$$\bar{w}_2(x,y,p) = \frac{b_z(p)}{2\pi}\int_0^{+\infty}\frac{\zeta\sin(\zeta x)}{\beta^2}(e^{yf\beta} - e^{-(2h-y)f\beta})d\zeta. \qquad y<0 \tag{7}$$

From Eqs. (1) and (7), the stress components in Laplace domain may be written as:

$$\bar{\sigma}_{zx}(x,y,p) = -\frac{G_{zx}b_z(p)}{2\pi}\int_0^{+\infty}\frac{\zeta^2\cos(\zeta x)}{\beta^2}(e^{-yf\beta} + e^{-(2h-y)f\beta} - 2)d\zeta,$$

$$, \qquad 0<y<h$$

$$\bar{\sigma}_{zy}(x,y,p) = \frac{fG_{zy}b_z(p)}{2\pi}\int_0^{+\infty}\frac{\zeta\sin(\zeta x)}{\beta}(e^{-yf\beta} - e^{-(2h-y)f\beta})d\zeta,$$

$$\bar{\sigma}_{zx}(x,y,p)=\frac{G_{zx}b_z(p)}{2\pi}\int_0^{+\infty}\frac{\zeta^2\cos(\zeta x)}{\beta^2}(e^{yf\beta}-e^{-(2h-y)f\beta})d\zeta$$

$$\bar{\sigma}_{zy}(x,y,p)=\frac{fG_{zy}b_z(p)}{2\pi}\int_0^{+\infty}\frac{\zeta\sin(\zeta x)}{\beta}(e^{yf\beta}-e^{-(2h-y)f\beta})d\zeta \qquad y<0$$

(8)

In order to specify the singular behavior of the stress components, the asymptotic behavior of the integrands in (8) should be examined. The singular parts of the kernels in the first relation of Eq. (8) can be separated after performing the appropriate asymptotic analysis and use the table of integral transforms [14]. Eq. (8) may be recast to more appropriate forms:

$$\bar{\sigma}_{zx}(x,y,p)=-\frac{G_{zx}b_z(p)}{2\pi}\left[\pi s_l p e^{-s_l p|x|}+\frac{yf}{x^2+y^2f^2}+\int_0^{+\infty}(\frac{\zeta^2}{\beta^2}(e^{-yf\beta}+e^{-(2h-y)f\beta})-e^{-yf\beta})\cos(\zeta x)d\zeta\right],$$

$$\bar{\sigma}_{zy}(x,y,p)=\frac{fG_{zy}s_l xpb_z(p)}{2\pi}\left[\frac{K_1(s_l p\sqrt{x^2+y^2f^2})}{\sqrt{x^2+y^2f^2}}-\frac{K_1(s_l p\sqrt{x^2+(2h-y)^2f^2})}{\sqrt{x^2+(2h-y)^2f^2}}\right]. \; 0<y<h$$

$$\bar{\sigma}_{zx}(x,y,p)=-\frac{G_{zx}b_z(p)}{2\pi}\left[\frac{yf}{x^2+y^2f^2}-\int_0^{+\infty}(\frac{\zeta^2}{\beta^2}(e^{yf\beta}-e^{-(2h-y)f\beta})-e^{yf\zeta})\cos(\zeta x)d\zeta\right],$$

$$\bar{\sigma}_{zy}(x,y,p)=\frac{fG_{zy}s_l xpb_z(p)}{2\pi}\left[\frac{K_1(s_l p\sqrt{x^2+y^2f^2})}{\sqrt{x^2+y^2f^2}}-\frac{K_1(s_l p\sqrt{x^2+(2h-y)^2f^2})}{\sqrt{x^2+(2h-y)^2f^2}}\right]. \; y<0. \qquad (9)$$

The details of the analysis will not be given in this paper. We observe that stress components are Cauchy singular at the dislocation position, which is a well-known feature of the stress fields caused by Volterra-type dislocations.

3. Orthotropic half-plane weakened by multiple edge cracks

The dislocation solutions accomplished in the foregoing section is extended to analyze orthotropic half-plane with several edge cracks. Cracks configurations are presented in parametric form as

$$x_i(\eta)=x_i+l_i\eta, \; -1\le\eta\le1$$
$$y_i(\eta)=y_i, \quad i=1,2,....,N. \qquad (10)$$

To derive the integral equations for the crack problem, covering the crack faces by dislocations with unknown densities $b_z(\eta,p)$ in an orthotropic half-plane. By use of Buckner principle [15] the integral equation for N edge cracks is

$$\sigma_{zx}(x_i(\eta),y_i,p)=\sum_{j=1}^{N}l_i\int_{-1}^{1}K_{ij}(s,\eta,p)B_{zj}(\eta,p)d\eta \quad -1\le\eta\le1, \; i=1,2,...,N. \qquad (11)$$

Employing the definition of dislocation density function, the equations for the crack opening displacement across the jth crack reads:

$$w_j^-(s,p)-w_j^+(s,p)=\int_{-1}^{s}l_jB_{zj}(\eta,p)d\eta, \; j=1,2,...,N. \qquad (12)$$

The numerical inversion of Laplace transform is carried out via Stehfest's method [16].

$$\sigma_{zxi}(x_i(\eta),y_i,\frac{\ln2}{t}n)=\sum_{j=1}^{N}l_i\int_{-1}^{1}K_{ij}(s,\eta,\frac{\ln2}{t}n)B_{zj}(\eta,\frac{\ln2}{t}n)d\eta, \qquad (13)$$

Stress fields for the cracks in orthotropic materials are singular at crack tips with a square-root singularity; Hence, the dislocation density functions for edge cracks can be represented by

$$B_{zj}(\eta,\frac{\ln2}{t}n)=g_{zj}(\eta,\frac{\ln2}{t}n)\sqrt{\frac{1-\eta}{1+\eta}}, \quad -1\le\eta\le1, \; j=1,...,N. \qquad (14)$$

We substitute Eq. (14) into the Eq. (13) and apply the numerical technique developed by Erdogan [17] for the solution of integral equations to solve the resultant equations. The inverse Laplace transform of the solution becomes

$$g_{zj}(\eta,t)=\frac{\ln 2}{t}\sum_{n=1}^{M} v_n G_{zj}(\eta,\frac{\ln 2}{t}n),\ -1\le\eta\le 1,\ j=1,...,N. \tag{15}$$

where $v_n=(-1)^{\frac{M}{2}+n}\sum_{k=[0.5(n+1)]}^{\min(\frac{M}{2},n)}\frac{k^{\frac{M}{2}}(2k)!}{(\frac{M}{2}-k)!k!(k-1)!(n-k)!(2k-n)!}$. (16)

The stress intensity factor for *j*th edge crack is [18]:

$$(k_{III}(t))_j=\sqrt{l_j G_{zx} G_{zy}}\ g_{zj}(-1,t), \qquad j=1,2,...,N. \tag{17}$$

4. Numerical results and discussion

In all the proceeding examples, The quantities of interest are the dimensionless stress intensity factors, $k(t)/k_0$. For convenience, the dynamic stress intensity factors are normalized by $k_0=\tau_0\sqrt{a/2}$,except example 2 where $k_0=\tau_0\sqrt{a_1/2}$,for uniform shear traction.

In the first example, an orthotropic half-plane under constant shear traction $\tau_{zy}=\tau_0 H(t)$ weakened by an edge crack with length $a=0.5h$ is analyzed, Fig. 1. The deviser for normalizing time is $t_0=a/2s_l$, where a is the crack length. It is seen that, the dynamic stress intensity factor increases very quickly with time, reaching a pick then decreases in magnitude and tends to the quasi static solution for sufficiently large normalized time. From Fig. 1, we may conclude that dynamic stress intensity factor of the edge crack exhibit small variations versus $f=\sqrt{G_{zx}/G_{zy}}$.

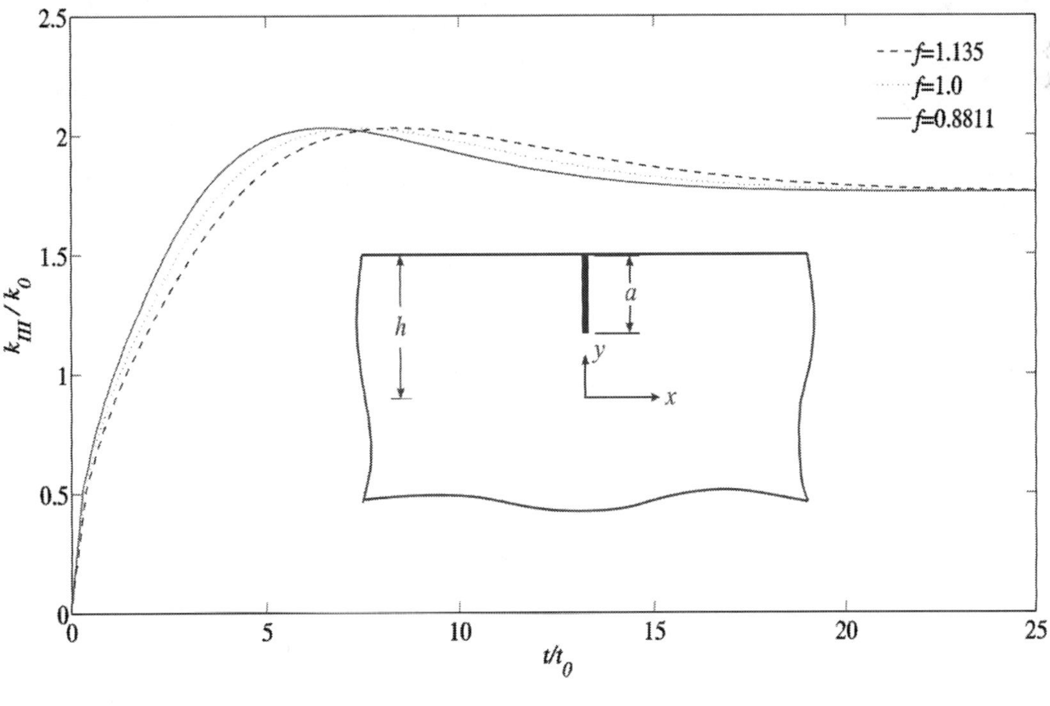

Fig. 1. Variation of normalized dynamic stress intensity factor of an edge crack.

In the next example, let us consider an orthotropic half-plane weakened by two unequal length edge cracks under anti-plane impact load, we study effects of the cracks location on dynamic stress intensity factors of crack tips Fig. 2. Note that, the variation of stress intensity factors of the tips L_2 is significant and the times at which the maxima of dynamic stress intensity factor occur for the tip L_1 are shorter than that the tip L_2.

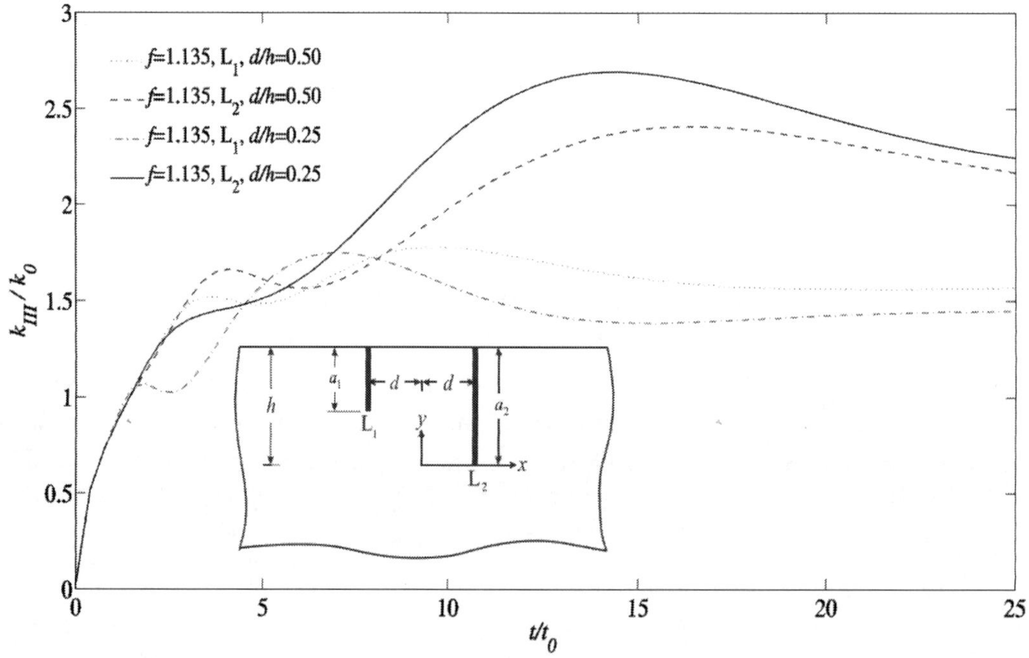

Fig . 2. Variation of normalized dynamic stress intensity factor of two parallel edge cracks.

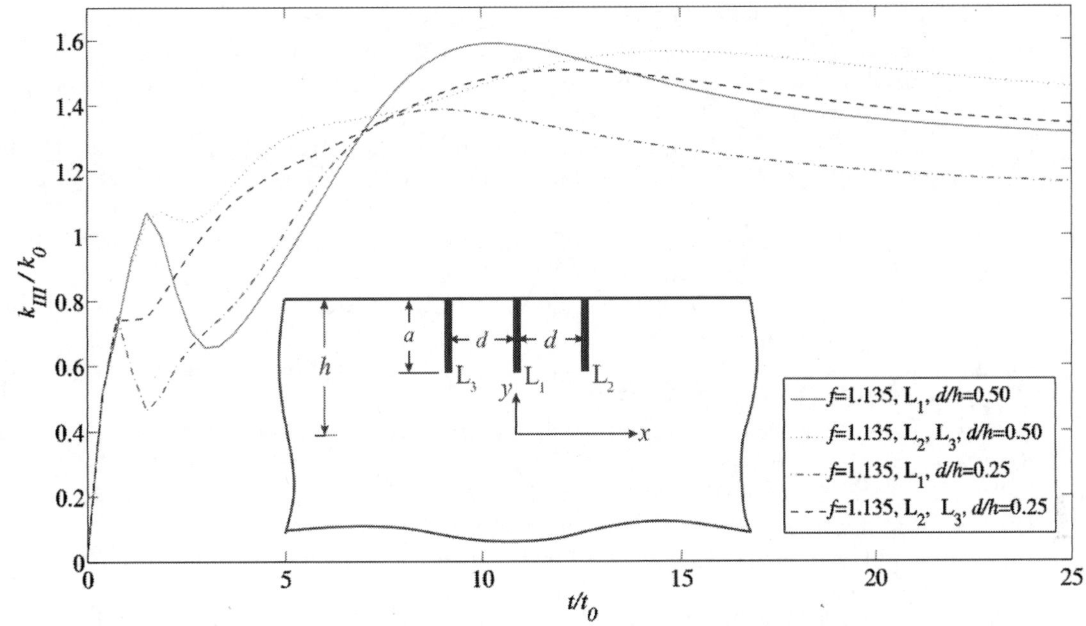

Fig. 3. Variation of normalized dynamic stress intensity factor of three identical parallel edge cracks.

Three equal-length cracks which are normal to the boundary are shown in Fig. 3. The problems are symmetric with respect to the *y*-axis. For other values of crack distance similar plots with different values of $k(t)/k_0$ may be obtained. As it was expected, the variation of stress intensity factors of three interacting crack tip namely L_1 is more pronounced than that of the tips L_2.

5. Conclusions

The dynamic response of an orthotropic half-plane with multiple edge cracks is studied analytically under the action of anti-plane impact loads. It may be observed that, once a solution is available for boundary loading with step function time dependence, extension to boundary loading with other time dependence is quite simple. According to the calculation results, it is found that the parameters such as the time, crack length and material properties have significant influences on the dynamic stress intensity factors. To show the applicability of the procedure more examples are solved and the stress intensity factors for multiple edge cracks are obtained.

References

[1] Freund LB 1998 Dynamic fracture mechanics, Cambridge University Press, New York
[2] Maue AW 1953 Die beuguhgelastischerwellenan der halbebene, ZAMM **33** 1–10
[3] Loeber JF and Sih GC 1968 Diffraction of anti-plane shear waves by a finite crack, J. Acoust. Soc. Am **44** 90–98
[4] Ma CC and Hou YC 1990 Theoretical analysis of the transient response for a stationary in plane-crack subjected to dynamic impact loading, Int. Engng. Sci. **28** 1321-1329
[5] Wang BL, Han JC and Du SY 1999 Multiple crack problem in nonhomogeneous composite materials subjected to dynamic antiplane shearing, Int. J. Fract. **100** 343-353
[6] Meguid SA and Wang XD 1999 Wave scattering from cracks and imperfectly bounded in homogeneities in advanced materials, Mech. Mater. **31** 187-195.

[7] Wei PJ, Zhang SY, Wu YL and Li KY 2000 Dynamic SIF of interface crack problem between two dissimilar viscoelastic bodies under impact loading, Int. J. Fract. **105**127-136

[8] Rubio-Gonzalez C and Mason JJ 2001 Greens function for the stress intensity factor evolution in finite cracks in orthotropic materials, Int. J. Fract. **108** 317-336

[9] Chen J, Liu Z and Zou Z 2002 Transient internal crack problem for a nonhomogeneous orthotropic strip, Int. J. Eng. Sci. **40** 1761-1774

[10] Shul CW and Lee KY 2002 Subsurface eccentric crack in a functionally graded coating layer on the layered half-space under an anti-plane shear impact load, Int. J. Solids Struct. **39** 2019-2029

[11] Lira-VergaraEand, Rubio-Gonzalez C 2005 Dynamic stress intensity factor of interfacial finite cracks in orthotropic materials. Int. J. Fract. **135** 285-309

[12] Wu KC and Chen JC 2011 Transient analysis of collinear cracks under anti-plane dynamic loading, Eng. procedia**10** 924-929

[13] Monfared MM and Ayatollahi M 2012 Elastodynamic analysis of a cracked orthotropic half-plane, Appl. Math. Model.**36** 2350-2359

[14] Jeffrey A and Zwillinger D 2000 Table of integrals, series and products, sixth ed., Academic press

[15] Hills DA, Kelly PA, Dai DN and Korsunsky AM 1996 Solution of Crack Problems: The Distributed Dislocation Technique, Kluwer Academic Publishers

[16] Cohen AM 2007 Numerical Methods for Laplace Transform Inversion. Springer, Berlin

[17] Erdogan F, Gupta GD and Cook TS1973 Numerical Solution of Singular Integral Equations, Method of Analysis and Solution of Crack Problems, Edited by G.C. Sih, Noordhoof, Leyden, Holland

[18] Liebowitz H 1968 Fracture mechanics, Academic Press, New York

Buckling analysis of nanoplates using IGA

P Phung-Van[1], M Abdel-Wahab[1], H Nguyen-Xuan[2]

1 Department of Electrical Energy, Metals, Mechanical Constructions and Systems, Faculty of Engineering and Architecture, Ghent University, Belgium
2 Center for Interdisciplinary Research in Technology, Hutech University, Ho Chi Minh City, Vietnam

E-mail: Phuc.PhungVan@UGent.be

Abstract. Isogeometric analysis (IGA) based on HSDT is used to simulate buckling analysis of nanoplates. The material properties of nanoplates based on the Mori–Tanaka schemes and the rule of mixture are used. The differential nonlocal equations with size effect are utilized. The nonlocal governing equations are approximated according to IGA, that satisfies naturally the higher-order derivatives continuity requirement in weak form of nanoplates. Several numerical results are presented to demonstrate the reliability of the proposed method.

1. Introduction

There has recently been a fast growth in applications of nanoscale structures, which are primarily concerned with fabrication of FGMs. This is considered a new revolutionary of materials with enhanced functionality. FGM nanoplates, as specific nanostructures, have been applied in the engineering and technology sectors such as mechanical engineering, aeronautic manufacturing industry, nuclear engineering, etc. Therefore, the understanding of the behaviors of the FGM nanoplates is essential for the development of nanostructures because of their huge potential applications in the real life.

To carry out numerical analysis of nanostructures, three general types of modelling approaches can be listed as follows: (i) hybrid atomistic continuum mechanic, (ii) atomistic and (iii) continuum mechanics. In continuum mechanics, one of the well-known theories is the nonlocal continuum theory of Eringen [1]. Finite Element method (FEM) is a very well know numerical techniques and has been used for a wide range of applications [2-15]. Using FEM, Free vibration analysis of FG size-dependent nanobeams was studied by Alshorbagy et al. [16]. The free vibration and buckling analyses of FG nanoplate using Navier solution subjected to thermal load were reported in Ref. [17]. Size-dependent thermal stability analysis of embedded FG annular nanoplates resting on an elastic foundation under various types of thermal using an exact analytical solution was introduced in Ref. [18]. The mechanical behaviour for homogenous nonlinear microbeams [19], FG nonlinear microbeams [20]. A nonlinear microbeams model based on the strain gradient elasticity is introduced in Refs. [21, 22]. Farokhi et al. [23] studied nonlinear dynamics of a geometrically imperfect microbeams with using Hamilton's principle for the nonlinear differential equation of motion for an initially curved beam. free vibration of FGM circle plates using the analytical solution based on the first-order shear deformation theory (FSDT) and the nonlocal theory was derived in Ref [24]. Size-dependent free analysis of FGM square plate based on FSDT was reported by Natarajan et al. [25]. Size-dependent analysis of FG nanoplates using IGA based on quasi-3D theory is recently examined in Ref [26]. Recently, Phung-Van et al. studied static and free vibration analyses of FG-CNTRC nanoplates [27] and nonlinear transient analysis of FGM nanoplates [28]. Furthermore, mechanical behaviours of FGM composite plates based on the local continuum theory have recently been published in Refs. [29-46].

As one may note, almost cited references deal with the modelling of micro/nano-beams. A very limited literature is available for nano-scale FGM structures. This paper hence aims to fill in this gap by developing a size-dependent geometrically nonlinear transient analysis of FGM nanoplates by a combination of IGA and the nonlocal continuum theory based on HSDT. In particular, we show that IGA based on HSDT fulfilling C2-continuity requirements can easily achieve the higher-order derivatives in the framework of the nonlocal continuum theory, which is of interest in this study. Size effects based on Eringen [1] in the differential nonlocal equations are performed. The effect of nonlocal approach on buckling analysis of the FGM nanoplates with various volume fraction exponents are discussed in details.

2. Functional graded materials

A functionally graded material nanoplate made of ceramic and metal is considered in this research. The properties materials based on the rule of mixture can be given by:

$$P(z)=\left(P_c-P_m\right)V_c+P_m \; ; \; V_c=\left(\frac{1}{2}+\frac{z}{t}\right)^n \; (n\geq 0) \; ; \; V_m=1-V_c \tag{1}$$

where m and c represent the metal and ceramic constituents, respectively, P refers to the effective material properties including the thermal conductivity k, v Poisson's ratio, ρ density, E Young's modulus and α thermal expansion. V_m and V_c are the volume fraction of the metal and ceramic, respectively, z is the thickness coordinate of plate and varies from $-t/2$ to $t/2$ and n is the volume fraction exponent.

The effective bulk and shear moduli based on the Mori-Tanaka scheme [47] can be defined as:

$$\frac{K_e-K_m}{K_c-K_m}=\frac{V_c}{1+V_m\frac{K_c-K_m}{K_m+4/3\mu_m}} \; ; \; \frac{\mu_e-\mu_m}{\mu_c-\mu_m}=\frac{V_c}{1+V_m\frac{\mu_c-\mu_m}{\mu_m+f_1}} \tag{2}$$

where $f_1=\frac{\mu_m(9K_m+8\mu_m)}{6(K_m+2\mu_m)}$. The Young's modulus and Poisson's ratio are now expressed by:

$$E_e=\frac{9K_e\mu_e}{3K_e+\mu_e} \; ; \; v_e=\frac{3K_e-2\mu_e}{2(3K_e+\mu_e)} \tag{3}$$

3. Theoretical formulation

3.1 Nonlocal elasticity theory

In the nonlocal elasticity theory proposed by Eringen [1], the stress field depends on all points of the considered body, and an equivalent form of differential equations of nonlocal stress at any points **x** can be expressed as follows:

$$\left(1-\mu\nabla^2\right)\sigma_{ij}=t_{ij} \tag{4}$$

where $\mu=e_0 l, 0\leq\mu\leq 4$ is the small-scale effect; l is an internal characteristic length; e_0 is material constant and $\nabla^2=\partial^2/\partial x^2+\partial^2/\partial y^2$ is the Laplace operator.

3.2 Displacement field

The displacement field can be expressed as follows [48]:

$$\begin{aligned} u(x,y,z)&=u_0+z\beta_x+cz^3\left(\beta_x+w_{,x}\right) \\ v(x,y,z)&=v_0+z\beta_y+cz^3\left(\beta_y+w_{,y}\right), \quad (-h/2\leq z\leq h/2) \\ w(x,y,z)&=w_0 \end{aligned} \tag{5}$$

where $c = 4/3h^2$.

The Green strain-displacement relations are now given as:

$$\boldsymbol{\varepsilon}=[\varepsilon_{xx}\ \varepsilon_{yy}\ \gamma_{xy}]^T=\boldsymbol{\varepsilon}_m+z\boldsymbol{\kappa}_1+z^3\boldsymbol{\kappa}_2\ ;\ \boldsymbol{\gamma}=\begin{bmatrix}\gamma_{xz} & \gamma_{yz}\end{bmatrix}^T=\boldsymbol{\varepsilon}_s+z^2\boldsymbol{\kappa}_s \tag{6}$$

where

$$\boldsymbol{\varepsilon}_m=\begin{bmatrix}u_{0,x}\\ v_{0,y}\\ u_{0,y}+v_{0,x}\end{bmatrix};\ \boldsymbol{\kappa}_1=\begin{bmatrix}\beta_{x,x}\\ \beta_{y,y}\\ \beta_{x,y}+\beta_{y,x}\end{bmatrix};\ \boldsymbol{\kappa}_2=c\begin{bmatrix}\beta_{x,x}+w_{0,xx}\\ \beta_{y,y}+w_{0,yy}\\ \beta_{x,y}+\beta_{y,x}+2w_{0,xy}\end{bmatrix};\ \boldsymbol{\varepsilon}_s=\begin{bmatrix}\beta_x+w_{0,x}\\ \beta_y+w_{0,y}\end{bmatrix};\ \boldsymbol{\kappa}_s=3c\begin{bmatrix}\beta_x+w_{0,x}\\ \beta_y+w_{0,y}\end{bmatrix} \tag{7}$$

The nonlocal equations in Eq. (4) can be rewritten as:

$$\begin{bmatrix}\sigma_{xx}\\ \sigma_{yy}\\ \sigma_{zz}\\ \tau_{xy}\\ \tau_{xz}\\ \tau_{yz}\end{bmatrix}-\mu\nabla^2\begin{bmatrix}\sigma_{xx}\\ \sigma_{yy}\\ \sigma_{zz}\\ \tau_{xy}\\ \tau_{xz}\\ \tau_{yz}\end{bmatrix}=\begin{bmatrix}Q_{11} & Q_{12} & Q_{13} & 0 & 0 & 0\\ Q_{21} & Q_{22} & Q_{23} & 0 & 0 & 0\\ Q_{31} & Q_{32} & Q_{33} & 0 & 0 & 0\\ 0 & 0 & 0 & Q_{66} & 0 & 0\\ 0 & 0 & 0 & 0 & Q_{55} & 0\\ 0 & 0 & 0 & 0 & 0 & Q_{44}\end{bmatrix}\begin{bmatrix}\varepsilon_{xx}\\ \varepsilon_{yy}\\ \varepsilon_{zz}\\ \gamma_{xy}\\ \gamma_{xz}\\ \gamma_{yz}\end{bmatrix} \tag{8}$$

where the material constants are given by:

$$\begin{aligned}&Q_{11}=Q_{22}=\frac{E_e}{1-\nu_e^2},\quad Q_{33}=Q_{44}=Q_{55}=Q_{66}=\frac{E_e}{2(1+\nu_e)},\\ &C_{12}=C_{21}=\frac{E_e\nu_e}{1-\nu_e^2},\ Q_{13}=Q_{31}=Q_{23}=Q_{32}=0\end{aligned} \tag{9}$$

3.3 Isogeometric analysis

The displacement field of the plate can be expressed as:

$$\mathbf{u}^h(\xi,\eta)=\sum_{I=1}^{m\times n}R_I(\xi,\eta)\mathbf{d}_I \tag{10}$$

where $\mathbf{d}_I=\begin{bmatrix}u_{0I} & v_{0I} & \beta_{xI} & \beta_{yI} & w_I\end{bmatrix}^T$ is the vector of degrees of freedom associated with the control point I, and R_I is the shape function as defined in Ref [49].

The in-plane and shear strains can be expressed as:

$$\begin{bmatrix}\boldsymbol{\varepsilon}_0 & \boldsymbol{\kappa}_1 & \boldsymbol{\kappa}_2 & \boldsymbol{\kappa}_s\end{bmatrix}^T=\sum_{I=1}^{m\times n}\begin{bmatrix}\mathbf{B}_I^m & \mathbf{B}_I^{b1} & \mathbf{B}_I^{b2} & \mathbf{B}_I^s\end{bmatrix}^T\mathbf{d}_I=\sum_{I=1}^{m\times n}\begin{bmatrix}\mathbf{B}_I^{mb} & \mathbf{B}_I^s\end{bmatrix}^T\mathbf{d}_I \tag{11}$$

where

$$\begin{aligned}&\mathbf{B}_I^m=\begin{bmatrix}R_{I,x} & 0 & 0 & 0 & 0\\ 0 & R_{I,y} & 0 & 0 & 0\\ R_{I,y} & R_{I,x} & 0 & 0 & 0\end{bmatrix},\quad \mathbf{B}_I^{b1}=-\begin{bmatrix}0 & 0 & R_{I,xx} & 0 & 0\\ 0 & 0 & R_{I,yy} & 0 & 0\\ 0 & 0 & 2R_{I,xy} & 0 & 0\end{bmatrix},\\ &\mathbf{B}_I^{b2}=\begin{bmatrix}0 & 0 & 0 & R_{I,x} & 0\\ 0 & 0 & 0 & 0 & R_{I,y}\\ 0 & 0 & 0 & R_{I,y} & R_{I,x}\end{bmatrix},\quad \mathbf{B}_I^s=\begin{bmatrix}0 & 0 & 0 & R_I & 0\\ 0 & 0 & 0 & 0 & R_I\end{bmatrix}\end{aligned} \tag{12}$$

The governing algebraic equations for buckling analysis can be obtained as follows

$$\left(\mathbf{K}-\lambda\mathbf{K}_g\right)\mathbf{d}=\mathbf{0} \tag{13}$$

where

$$\begin{aligned}&\mathbf{K}=\int_\Omega\left[\left(\mathbf{B}^{mb}\right)^T\mathbf{D}_{mb}\mathbf{B}_I^{mb}+\left(\mathbf{B}^s\right)\mathbf{D}_s\mathbf{B}^s\right]d\Omega\\ &\mathbf{K}_g=\int_\Omega\left[\left(\mathbf{B}^g\right)^T-\lambda\nabla^2\left(\mathbf{B}^g\right)^T\right]\mathbf{N}_0\mathbf{B}^g d\Omega\end{aligned} \tag{14}$$

in which

$$\mathbf{D}_{mb} = \begin{bmatrix} \mathbf{A} & \mathbf{B} & \mathbf{E} \\ \mathbf{B} & \mathbf{D} & \mathbf{F} \\ \mathbf{E} & \mathbf{F} & \mathbf{H} \end{bmatrix} \; ; \; \mathbf{B}_I^s = \begin{bmatrix} 0 & 0 & R_{I,x} & 0 & 0 \\ 0 & 0 & R_{I,y} & 0 & 0 \end{bmatrix}$$
$$A_{ij}, B_{ij}, D_{ij}, E_{ij}, F_{ij}, H_{ij} = \int_{-h/2}^{h/2} \left(1, z, z^2, f(z), zf(z), f^2(z)\right) C_{ij} \mathrm{d}z \;, \quad i, j = 1,2,6 \tag{15}$$
$$D_{sij} = \int_{-h/2}^{h/2} \left[f'(z)\right]^2 C_{ij} \mathrm{d}z \;, \qquad i, j = 4,5$$

4. Numerical results

In this study, the critical buckling loads $\bar{P}_{cr} = P_{cr}R^2 / D_m$ of Al/Al2O3 circular plate (radius R = 10, h = 0.34) are performed. Table 1 shows Non-dimensional buckling load of simply supported and clamped circle nanoplate. It can be observed that the present results match well with the reference solution [26].

Table 1. Material properties of FGM plates.

BCs	μ	Model	n				
			0	1	2	5	10
SSSS	0	RPT-1 [26]	22.7572	9.6391	8.1921	6.9000	6.0401
		RPT [26]	22.7567	9.6389	8.1920	6.9002	6.0399
		Quasi 3D [26]	23.1059	10.0048	8.5094	7.0907	6.1583
		IGA	22.7565	9.6263	8.1811	6.8955	6.0381
	1	RPT-1 [26]	21.8371	9.2241	7.8382	6.6039	5.7855
		RPT [26]	21.8367	9.2257	7.8396	6.6053	5.7863
		Quasi 3D [26]	22.1473	9.5755	8.1448	6.7891	5.8991
		IGA	21.8371	9.2172	7.8328	6.6045	5.7871
	4	RPT-1 [26]	19.4782	8.1913	6.9602	5.8774	5.1565
		RPT [26]	19.4779	8.1920	6.9608	5.8784	5.1572
		Quasi 3D [26]	19.7025	8.4927	7.2273	6.0387	5.2522
		IGA	19.4795	8.1788	6.9501	5.8754	5.1572
CCCC	0	RPT-1 [26]	79.3355	31.5217	26.7612	23.1849	20.7162
		RPT [26]	79.3304	31.5202	26.7602	23.1835	20.7141
		Quasi 3D [26]	82.3402	33.5902	28.5574	24.4165	21.6023
		IGA	79.3204	31.5156	26.7567	23.1807	20.7108
	1	RPT-1 [26]	69.1774	27.4857	23.3347	20.2163	18.0637
		RPT [26]	69.1730	27.4844	23.3338	20.2151	18.0619
		Quasi 3D [26]	71.7287	29.2705	24.8837	21.2721	18.8188
		IGA	69.1652	27.4808	23.3311	20.2129	18.0593
	4	RPT-1 [26]	49.9794	19.8579	16.8589	14.6059	13.0507
		RPT [26]	49.9762	19.8569	16.8582	14.6050	13.0494
		Quasi 3D [26]	51.6664	21.1039	17.9361	15.3215	13.5514
		IGA	49.9809	19.8586	16.8593	14.1510	12.7227

5. Conclusions

In this paper, size-dependent buckling analysis of nanoplates using using IGA based on HSDT is investigated. The material properties of FGM based on the Mori–Tanaka schemes and the rule of mixture are considered. The differential nonlocal equations are utilized to take into account effect of the size-dependent. The buckling nonlocal governing equations of motion is approximated according to IGA based on HSDT, which satisfies naturally the 3rd derivatives of displacement field. The effects of volume fraction exponent and small scale parameter on buckling analysis of nanoplates are also performed. Numerical results proved high accuracy and reliability of the present method in comparison with other available numerical approaches.

Acknowledgements
P Phung-Van would like to acknowledge the financial support of the Special Research Fund (BOF) of Ghent University (BOF16/PDO/88).

References
[1] Eringen AC 1983 *Journal of Applied Physics* **54**(9) 4703-4710
[2] Yue T and Abdel Wahab M 2017 *Tribology International* **107** 274-282
[3] Pereira K, Yue T and Abdel Wahab M 2017 *Tribology International* **110** 222-231
[4] Martínez JC, Vanegas Useche LV and Wahab MA 2017 *International Journal of Fatigue* **100, Part 1** 32-49
[5] Kumar D, Biswas R, Poh LH and Abdel Wahab M 2017 *Tribology International* **109** 124-132
[6] Bhatti NA and Abdel Wahab M 2017 *Tribology International* **109** 552-562
[7] Yue T and Abdel Wahab M 2016 *Materials* **9** 597; doi:510.3390/ma9070597
[8] Resende Pereira KdF, Bordas S, Tomar S, Trobec R, Depolli M, Kosec G and Abdel Wahab M 2016 *Materials* **9** 639; doi:610.3390/ma9080639
[9] Ferjaoui A, Yue T, Abdel Wahab M and Hojjati-Talemi R 2015 *International Journal of Fatigue* **73** 66-76
[10] Yue T and Abdel Wahab M 2014 *Wear* **321** 53-63
[11] Noda N-A, Chen X, Sano Y, Wahab MA, Maruyama H, Fujisawa R and Takase Y 2016 *MATERIALS & DESIGN* **96** 476-489
[12] Gadala I, Abdel Wahab M and Alfantazi A 2016 *MATERIALS & DESIGN* **97** 287-299
[13] Wang C, Sun Q, Abdel Wahab M, Zhang X and Xu L 2015 *Waste Management* **43** 19-27
[14] Vanegas-Useche L, Abdel Wahab M and Parker G 2015 *Waste Management* **43** 28-36
[15] Junyan Ni and Wahab MA 2017 *Computers & Structures* **186** 35–49
[16] Alshorbagy AE, Eltaher MA and Mahmoud FF 2011 *Applied Mathematical Modelling* **35**(1) 412-425
[17] Ansari R, Ashrafi M, Pourashraf T and Sahmani S 2015 *Acta Astronaut* **109** 42-51
[18] Ashoori AR, Sallari E and Sadough Vanini SA 2016 *International Journal of Mechanical Sciences* **119** 369-411
[19] Asghari M, Kahrobaiyan MH and Ahmadian MT 2010 *International Journal of Engineering Science* **48**(12) 1749-1761
[20] Ke LL, Wang YS, Yang J and Kitipornchai S 2012 *International Journal of Engineering Science* **50**(1) 256-267
[21] Rajabi F and Ramezani S 2012 *Archive of Applied Mechanics* **82** 363-376
[22] Ghayesh MH, Amabili M and Farokhi H 2013 *International Journal of Engineering Science* **63** 52-60
[23] Farokhi F, Ghayesh MH and Amabili M 2013 *International Journal of Engineering Science* **68** 11-23
[24] Hosseini-Hashemi S, Bedroud M and Nazemnezhad R 2013 *Composite Structures* **103** 108-118
[25] Natarajan S, Chakraboty S, Thangavel M, Bordas SPA and Rabczuk T 2012 *Computational Materials Science* **65** 74-80
[26] Nguyen NT, Hui D, Lee J and Nguyen-Xuan H 2015 *Computer Methods in Applied Mechanics and Engineering* **297** 191-218
[27] Phung-Van P, Lieu X Q, Nguyen-Xuan H and Abdel-Wahab M 2017 *Composite Structures* **166** 120-135
[28] Phung-Van P, Ferreira AJM, Nguyen-Xuan H and Abdel-Wahab M 2017 *Composites Part B Engineering*

IOP Conf. Series: Journal of Physics: Conf. Series **843** (2017) 012016 doi:10.1088/1742-6596/843/1/012016

[29] Phung-Van P, Nguyen-Thoi T, Luong-Van H and Lieu-Xuan Q 2014 *Computer Methods in Applied Mechanics and Engineering* **270** 15-36

[30] Phung-Van P, Nguyen-Thoi T, Dang-Trung H and Nguyen-Minh N 2014 *Composite Structures* **111** 553-565

[31] Nguyen-Thoi T, Bui-Xuan T, Phung-Van P, Nguyen-Hoang S and Nguyen-Xuan H 2014 *KSCE Journal of Civil Engineering* **18**(4) 1072-1082

[32] Phung-Van P, Nguyen-Thoi T, Luong-Van H, Thai-Hoang C and Nguyen-Xuan H 2014 *Computer Methods in Applied Mechanics and Engineering* **272** 138-159

[33] Phung-Van P, Thai CH, Nguyen-Thoi T and Nguyen-Xuan H 2014 *Composites Part B-Engineering* **60** 227-238

[34] Luong-Van H, Nguyen-Thoi T, Liu GR and Phung-Van P 2014 *Engineering Analysis with Boundary Elements* **42** 8-19

[35] Nguyen-Thoi T, Luong-Van H, Phung-Van P, Rabczuk T and Tran-Trung D 2013 *International Journal for Composite Materials* **3(A)** 19-27

[36] Phung-Van P, Luong-Van H, Nguyen-Thoi T and Nguyen-Xuan H 2014 *International Journal for Numerical Methods in Engineering* **98**(13) 988-1014

[37] Phung-Van P, Nguyen-Thoi T, Le-Dinh T and Nguyen-Xuan H 2013 *Smart Materials and Structures* **22**(9)

[38] Nguyen-Thoi T, Rabczuk T, Lam-Phat T, Ho-Huu V and Phung-Van P 2014 *Theoretical and Applied Fracture Mechanics* **72** 150-163

[39] Phung-Van P, Nguyen-Thoi T, Tran LV and Nguyen-Xuan H 2013 *Computational Materials Science* **79** 857-872

[40] Phung-Van P, Tran LV, Ferreira AJM, Nguyen-Xuan H and Abdel-Wahab M 2017 *Nonlinear Dynamics* **87**(2) 879-894

[41] Phung-Van P, Nguyen LB, Tran LV, Dinh TD, Thai CH, Bordas SPA, Abdel-Wahab M and Nguyen-Xuan H 2015 *International Journal of Non-Linear Mechanics* **76** 190-202

[42] Phung-Van P, De Lorenzis L, Thai CH, Abdel-Wahab M and Nguyen-Xuan H 2015 *Computational Materials Science* **96** 495-505

[43] Nguyen-Thoi T, Luong-Van H, Phung-Van P, Rabczuk T and Tran-Trung D 2013 *International Journal of Composite Materials* **3**(A) 19-27

[44] Tran LV, Phung-Van P, Lee J, Wahab MA and Nguyen-Xuan H 2016 *Composite Structures* **140** 655-667

[45] Phung-Van P, Nguyen-Thoi T, Bui-Xuan T and Lieu-Xuan Q 2015 *Computational Materials Science* **96** 549-558

[46] Nguyen-Thoi T, Phung-Van P, Thai-Hoang C and Nguyen-Xuan H 2013 *International Journal of Mechanical Sciences* **74** 32-45

[47] Mori T and Tanaka K 1973 *Acta Metallurgica* **21**(5) 571-574

[48] Reddy JN 1984 *Journal of Applied Mechanics-Transactions of the Asme* **51**(4) 745-752

[49] Cottrell JA, Hughes TJR and Bazilevs Y 2009 Isogeometric analysis, towards integration of CAD and FEA ed editors. Wiley

IOP Conf. Series: Journal of Physics: Conf. Series **843** (2017) 012017 doi:10.1088/1742-6596/843/1/012017

Impact resistance of metal skin-carbon fiber reinforced polymer laminate structures for the production of lightweight vehicles body frame

Luca Quagliato[1], Dongwook Kim[2], Changsoon Jang[2], Naksoo Kim[2] and Seokmoo Hong[3]

[1] Department of Management and Engineering, University of Padua, Vicenza, I-36100, Italy
[2] Department of Mechanical Engineering, Sogang University, Seoul, 04107, Republic of Korea
[3] Department of Metal Mold Design Engineering, Kongju National University, Cheonan, 32588, Republic of Korea

E-mail: nskim@sogang.ac.kr

Abstract. The research presented in this paper deals, in the first part, with the material characterization of an innovative class of material, made of steel skin and sheet molding compound core, the last one realized with vinyl-ester resin and chopped carbon fibers. Utilizing the results of the material characterization, the case of an automotive B-pillar is taken into account in order to test and prove the reliability of the utilization of this innovative typology of material, instead of the standard steel. The steel skin – SMC hybrid laminate structures are realized by means of stamping process under warm forming condition, in order to favorite the curing of the SMC layers. The result presented along the paper will show that the steel skin – SMC material not only have better specific resistance that those of both standard and high strength steel but also better crashworthiness properties, in comparison to the considered steel. Based on the results of the numerical simulations, a useful way thought which to estimate the crash worthiness property is proposed in this paper and the results will show how the B-pillar realized with steel skin – SMC has a higher impact resistance, in comparison the steel one with the same weight and reduced thickness.

1. Introduction

In the nowadays automotive industry, the vehicles weight reduction is becoming a critical issue, driven by environmental growing concerns as well by the shortage of fossil combustibles. However, the utilization of lightweight material has shown not always to comply with the requirements of high mechanical properties required for some parts of the vehicles body frame, such as in case of comparing the mechanical properties of aluminum and steel. To this aim, lightweight hybrid structures made of steel skin and carbon fiber reinforced polymer core are a concrete innovative option for the lightening of vehicles body frame and, if well designed and manufactured, can also improve the mechanical properties of the relevant components. In the literature, different authors presented results concerning the material characterization of the sheet molding compound parts, both from the material and from the numerical point of view, among them. Kim et al. [1] studied the basic of the SMC process by means of numerical simulations, taking into account both deformation and curing phases of the process. Katayama et al. [2] developed a CAE model for the prediction of the stiffness in parts with ribs made of SMC, showing how the material flow prediction is more complex in presence of ribs. Evan et al. [3] studied the mechanical properties of carbon fiber molding compounds, showing how they can be processed by utilizing the conventional SMC glass fiber process. In addition to that, Lauter et al. [4] studied a manufacturing method for the realization of a vehicle B-pillar structure by utilizing a metal skin – CFRP (carbon fiber reinforced polymer) made of layups of unidirectional fibers reinforced prepregs. Moreover, Liu et al. [5] studied feasible design method for the fabrication of lightweight vehicles by utilizing fabrics of carbon fibers.

In addition, concerning the impact test for the determination for the crash worthiness of composite material, some contributions are available in the literature, among them: Meo et al. [6] developed a numerical simulation for the analysis of the influence of low-velocity impacts on the aircraft sandwich panels. Bandi et al. [7] developed a design method for the realization of crashworthy structures where the energy abortion is chosen a-priori as a project input. Finally, Lewis et al. [8] proposed an interesting method to correlate impact strength, calculated from the results of the impact test, and the fracture toughness. The present research work is subdivide into two different parts. In the first part, the materials composing the steel skin – sheet molding compound (SMC) laminate structure, the SK5 steel and the original SMC, are tested by means of simple tension test. In addition, since also the adhesive utilized between steel skins and SMC core is vital for the load transfer, also a shear strength test has been conducted. Finally, the mechanical properties of the steel skin – SMC laminate structure has been tested by means of 3-points bending test. In the second part of the paper, focusing the attention on a vehicle B-pillar simplified shape, an impact test resistance simulation has been set-up considering both the original steel and the laminate structure made of steel skin and SMC core. From the results of the numerical models, the principle stress components are exported and utilized for the calculation of the strain energy density by utilizing the Beltrami equation. The comparison between the amount of strain energy density for each element of the mesh and the critical strain energy density allows calculating the number of elements those have reached, or passed, the critical value hence to compare the crashworthiness of steel and laminate steel skin – SMC products. Following this procedure it is possible to estimate the overall amount of damaged element out of the total and compare the impact resistance of the steel skin – SMC laminate structure with that of the original steel.

2. Material characterization and model constants determination

In order to enhance the strength of the original SK5 steel, quenching and tempering operation at 780°C 120°C, respectively, have been carried out. This procedure allows obtaining higher yield strength values at the expenses of part of the elongation to fracture. The rationale behind this choice is given by the fact that, as it will be shown in this section, the SMC workpiece manifests a brittle behavior characterized by a high value for the break stress thus, a high strength steel shall promote a more uniform behavior of the composite steel-SMC structure.

The simple tension test has been conducted at the speed of 2mm/min utilizing the INSTRON 3367 test machine, with a maximum load of 3tons. The size of the specimen, the ASTM E8 type 1, are reported in Figure 1 whereas the fitting of the plastic part of the stress-strain curve, operated by utilizing the Swift model, is shown in Figure 2.

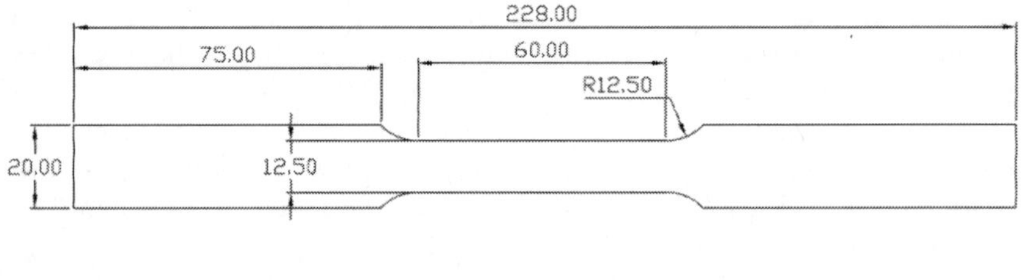

Figure 1. SK5 simple tension test specimen

In addition to that, the material properties derived from the material characterization of the SK5 steel are reported in Table 1.

Table 1. Material properties for the SK5 steel

Parameter	**Description**
Young's modulus	190GPa
Yield strength	1450MPa
UTS	1796MPa
Density	7.8744 kg/dm^3

Concerning the material testing for the SMC, simple tension test have been carried out by utilizing the specimen shown in Figure 3a, whose size are reported in in Figure 3b.

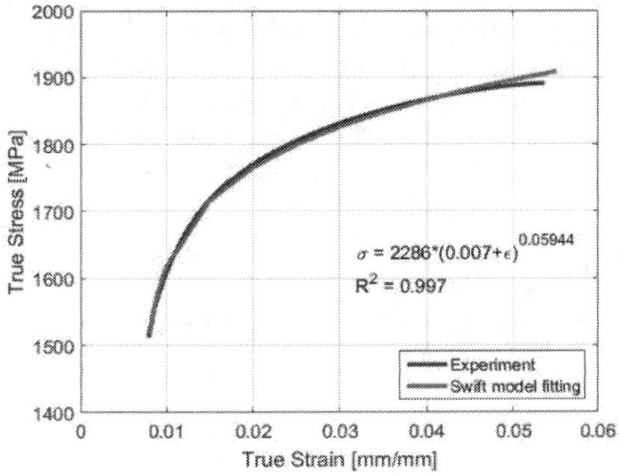

Figure 2. SK5 simple tension test specimen

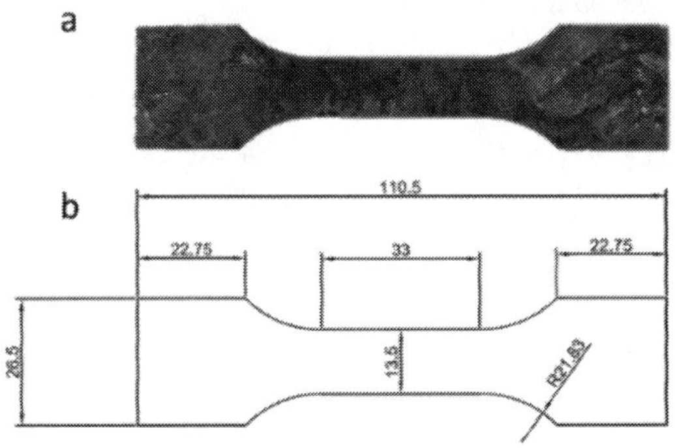

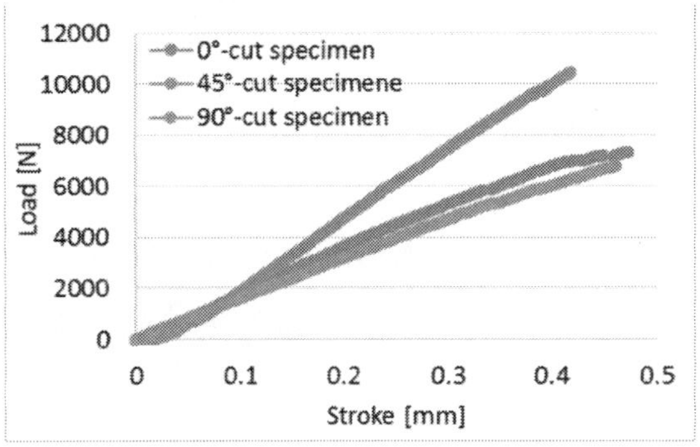

Figure 3. (a) SMC specimen and (b) dimensions

Figure 4. Load-stroke for SMC specimens along 0°, 45° and 90° directions

The simple tension test specimens have been derived from a single plate realized by compressing three pre-cured SMC layers, each one with the initial thickness of 2.5±0.5mm, by means of stamping process. In order to favorite the curing of the SMC, both top and bottom die have been heated up to 130°C and the plate has been kept under pressure for 2 minutes after the stamping process in order to assure a complete curing. At the end of the stamping process, the SMC plate has a thickness of 3.2±0.05mm. In order to account for possible anisotropies in the material, due to the random orientation of the carbon fibers in the sheet molding compound process, simple tension specimens have been cut along 0°, 45° and 90° directions in the stamped plate and the results, in terms of load-stroke curves, are presented in Figure 4. The density of the SMC has been calculated in 1.8431kg/dm^3 and the material properties, relevant for the three tested directions, are reported in Table 2.

Table 2. Material properties for the SMC

Parameter	Young's modulus [GPa]	Break stress [MPa]
0°-cut specimen	20.91	184.19
45°-cut specimen	29.30	259.65
90°-cut specimen	18.79	169.35

The last material property to be tested is the shear strength of the adhesive utilized to bond the steel skins to the SMC core, which is of vital importance for the load transfer between these two different materials. However, the adhesive can also represent a weak point in the laminate structure since it represent a discontinuity in the stress field between the two materials and can generate strong stress concentrations. The utilized adhesive is a vinyl-ester polymer, whose composition is shown in following Table 3.

Table 3. Chemical composition of the vinyl-ester adhesive

Component	Percentage	Component	Percentage
Epoxy	35 ~ 43%	Toning agent	0.5 ~ 1%
Hardener	6 ~ 8%	Additive	4.5 ~ 6.5%
Filler	29 ~ 39%	Thinner	5 ~ 7%
Accelerator	1.5 ~ 2.5%	Glass bead	4 ~ 6%

The shear strength test specimens have been realized by applying a 0.1mm layer of adhesive on a 0.4mm thick metal skin and, on the top of it, a 3.2mm piece of SMC has been placed, as reported in Figure 5, where also the fracture area after the test is shown.

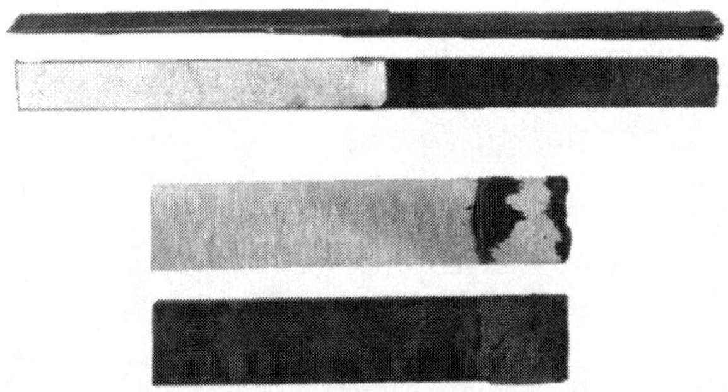

Figure 5. Shear strength specimen and fracture area after the test

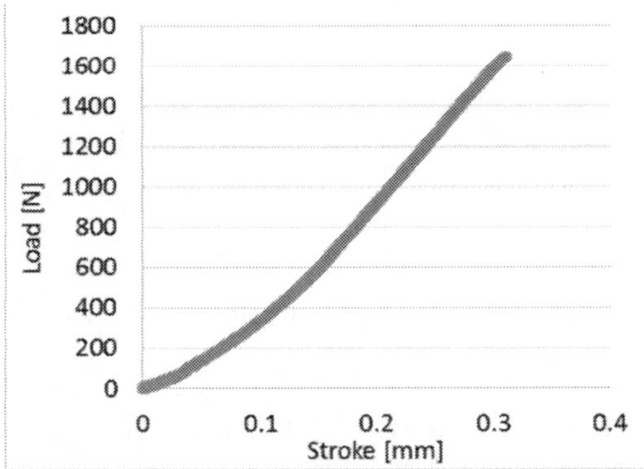

Figure 6. Load-stroke curve of the adhesive shear strength test

The bonding area measures 13mmx13mm and the bonded pieces have been blocked into the right position during the curing time (20minute) and the cooling down time (12 hours).

The load-stroke curve relevant for the shear strength of the utilized adhesive is shown in following Fig. 5 where a slightly non-linear behavior can be seen from the results and where the break load is identified in 1642.7N.

Finally, in order to test the material properties of the laminate structure made of steel skins and SMC core, bending test on specimens with size shown in Figure 7 have been carried out. In Figure 8 both the experiment settings and the specimen after failure are presented.

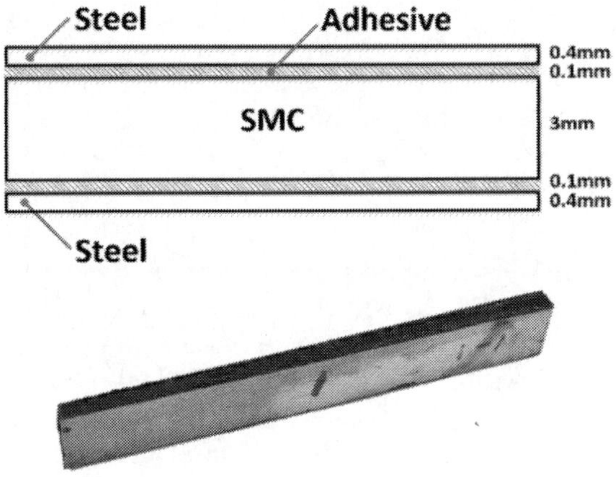

Figure 7. Steel skin – SMC specimen and relevant dimensions

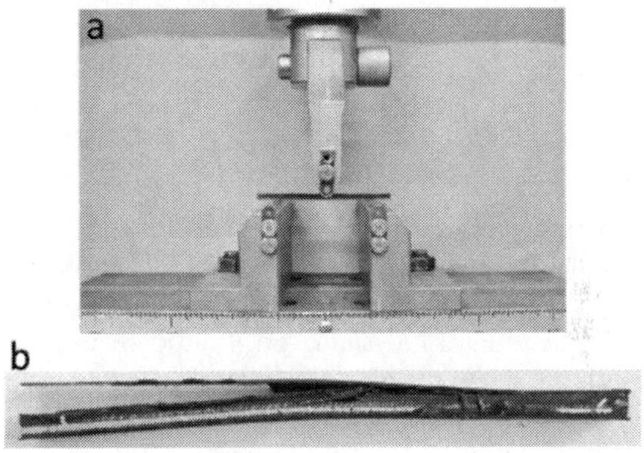

Figure 8. Steel skin – SMC 3-points bending test setting and (b) specimen after failure

The 3-point bending tests have been conducted on the Universal Test Machine TO-100-IC, which has a load limit of 10tons, with a test speed of 2mm/min. The load-stroke and engineering stress-strain curves, relevant for the bending test of the steel skin – SMC specimens are shown in Figure 9 and Figure 10, respectively, where the results are presented with those of the only SMC, showing an improvement in both Young's modulus and break stress.

By gathering together all the results, it is possible to draw a first conclusion concerning the specific strength of the steel skin – SMC laminate structure, in comparison with standard steel. Considering a generic steel, with an yield strength of 600MPa and the SK5 steel utilized in this paper, which as a yield strength of 1350MPa, the specific strength is identified in 0.76MPa/(g/cm^3) and 1.72 MPa/(g/cm^3), respectively, whereas that of the steel skin – SMC hybrid material is calculated in 2.28MPa/(g/cm^3), showing the improvement of the proposed material in comparison to standard one.

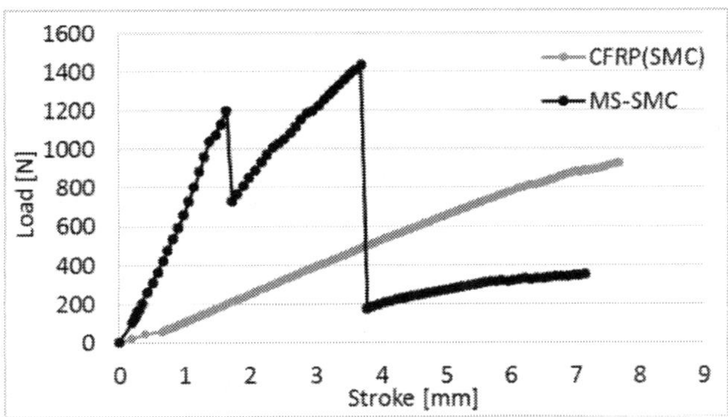

Figure 9. Load-stroke curves for steel skin – SMC and only SMC

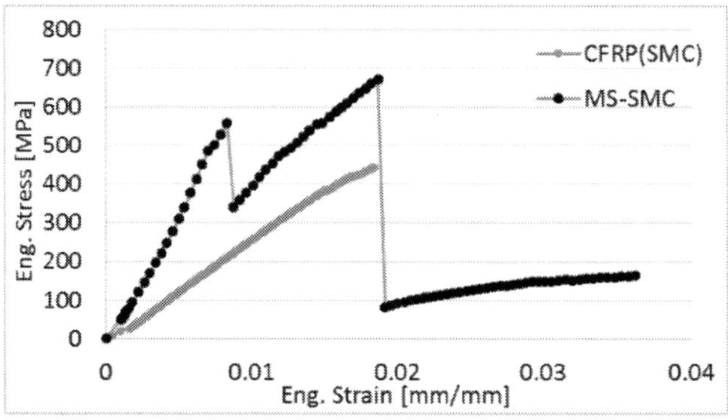

Figure 10. Load-stroke curves for steel skin – SMC and only SMC

3. Numerical model implementation

In order to prove the crashworthiness properties of laminate structures made of steel skin and SMC core, a numerical simulation has been implemented in ABAQUS/Explicit simulating the impact of an object of cylindrical shape on the top of the simplified B-pillar. In order to be able to compare the results in a fair way, the thickness of the B-pillar realized with steel has been set to 1.5mm whereas that of steel skin – SMC to 4mm, resulting in both products to have the same weight. The implementation of the simulation is shown in Figure 11 and Figure 12, for the steel product and for the laminate product, respectively.

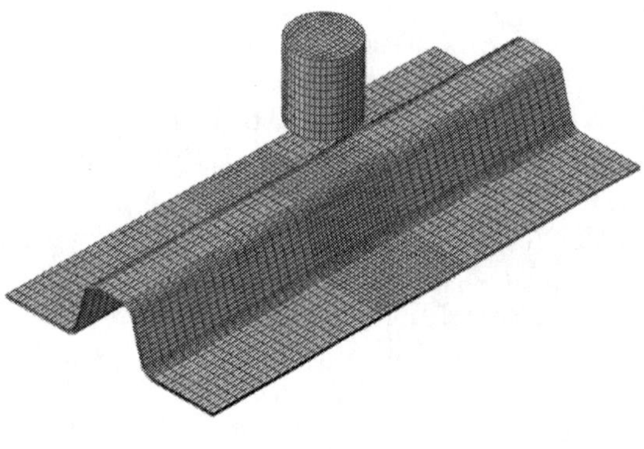

Figure 11. ABAQUS impact resistance simulation (steel)

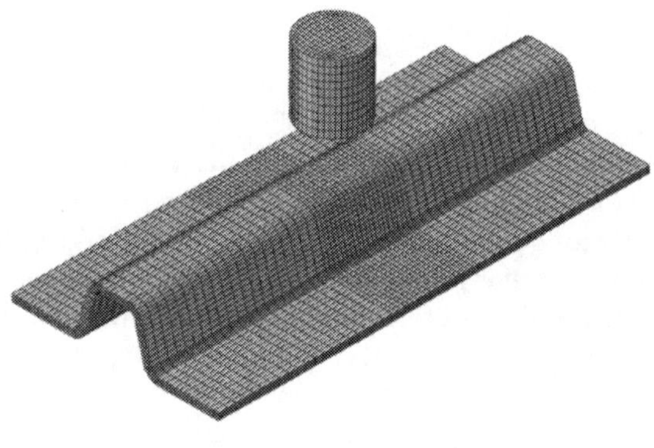

Figure 12. ABAQUS impact resistance simulation (laminate steel skin - SMC)

In the numerical model, the C3D8R element with reduced integration has been utilized and both models have been meshed with a fine mesh size of 2.5mm side and a coarse mesh size of 5mm side. Concerning the boundary conditions applied to the model, the link between the B-pillar and the rest of the body frame of the vehicle has been simulated by imposing four fully constrained zones on the four corners of the low areas of the part, in order to simulate the relevant rivets, as shown in Figure 13, where the fixed nodes have been highlighted in red.

The cylinder, which represents the impactor is, moved, till the impact with the B-pillar surface, with a constant speed of 27.8m/s and has a mass of 400kg, which represent a total amount of 155kJ transferred to the part.

IOP Conf. Series: Journal of Physics: Conf. Series **843** (2017) 012017 doi:10.1088/1742-6596/843/1/012017

Figure 13. Constraint boundary conditions

4. Results and discussion

As a first result of the numerical simulations, the von Mises yield stress distribution, results of both steel and laminate products, in shown in Figure 14a and Figure 14b, respectively.

The more spread stress field in the laminate structure, with a lower maximum value for the von Mises equivalent stress, is a first confirmation of the capability of this laminate structure material to better redistribute the stresses when loaded. The steel part instead presents an higher value for the von Mises equivalent stress and the area where the load is redistributed is small, which is a typical behaviour for high strength steel, as the case of the hardened SK5.

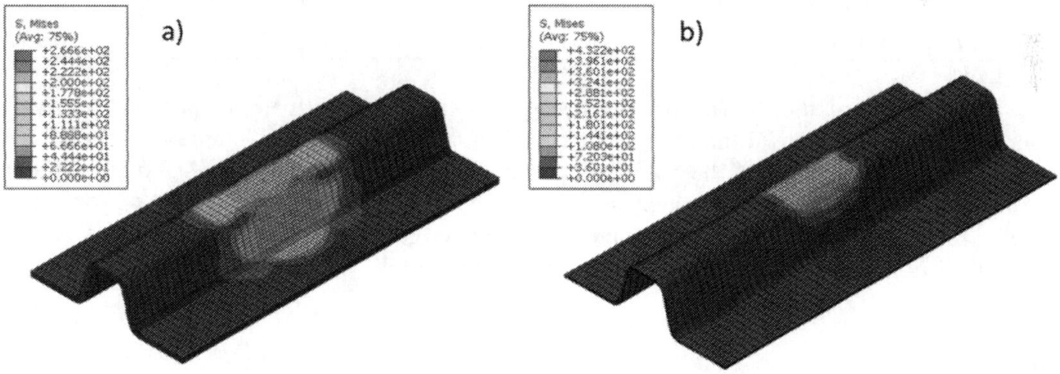

Figure 14. (a) von Mises stress distribution for the laminate steel skin – SMC B-pillar and (b) for the SK5 steel B-pillar

In order to have a more clear comparison between the crashworthiness properties of the laminate steel skin – SMC laminate structure and those of the SK5 steel, the strain energy density has been calculated by utilizing the Beltrami Eq. (1).

$$U_t = \frac{1}{2E}\left[\sigma_1^2 + \sigma_2^2 + \sigma_3^2 - 2\nu\left(\sigma_1\sigma_2 + \sigma_2\sigma_3 + \sigma_1\sigma_3\right)\right] \tag{1}$$

By exporting, from both simulations, the principle stress tensor and inputting the relevant components into Eq. (1), it is possible to determine the total amount of strain energy density in the fine mesh portion of the B-pillar. In addition to that, by considering the critical value for the strain energy density, Eq. (2), the amount of failed elements can be determined and used for the comparison of the crashworthiness properties. In Eq (2), σ_R stands for the break stress under bending loading conditions for the relevant material. The flexural strength of the SK5 has been derived from the previously presented result of the uniaxial tension test in order to carry out a proper comparison with the flexural strength determined for the laminate structure in the 3-points bending test.

$$W_c = \frac{1}{2E}\sigma_R^2 \tag{1}$$

For comparison, in Table 2, the data of the total strain energy density on all the elements in the fine mesh portion of the B-pillar are reported along with the number of failed elements.

Table 1. Strain energy density results comparison

Parameter	Steel skin – SMC	SK5 steel
Total strain energy density	394.40485	119.6991
Flexural strength	671.6MPa	300MPa
Young's modulus	63.93GPa	190GPa
Critical strain energy density	3.53	0.24
N° of failed elements	0	64
Volume of failed elements	$0mm^3$	$1000mm^3$
% of failed elements	0%	1.05%

As a consequence of the higher specific strength of the laminate material in comparison to the original steel, also the crashworthiness properties of the steel skin – SMC laminate structures have shown to be higher than those of the SK5 steel. Especially, the capability of the laminate structure to better redistribute the stresses consequent to the impact is one of the main reason to explain why, being same the weight, no failed element have been detected in the steel skin – SC product simulation whereas 1.05% of the elements in the steel one have shown to fail.

5. Conclusion

The preliminary results shown in this paper, concerning the material characterization and the crashworthiness properties of laminate structures made of steel skin and SMC core, have proved the capability of this new typology of replacing standard ones, made of traditional material, such as steel.

In addition to that, the reduced density makes the laminate structure to be of very interest for the automotive industry, where the weight reduction is a key aspect in the design of the vehicles of the future. Finally, the utilization of an energy criterion for the comparison of the crashworthiness properties between different materials has proved to be a quick and practical approach, easy to implement in all the FEM structural program.

Although the results presented in this paper are based on material characterization and preliminary application on test-products, the positive outcome of the research encourages further effort in this direction for a better understanding of fundamental, processing and possible applications for this new class of hybrid laminate materials.

References

[1] Kim N S 1992 *Conf. Kor. Soc. Tech. Plast.* **3** 177–199
[2] Katayama T, Shinohara M, Hakotami M, Kitade A abd Kono D 2001 *J. Mat. Proc. Tech.* **119** 237–243
[3] Evans A D, Qian C C, Turner T A, Harper L T and Warrio N A 2016 *Comp. Part A* **90** 1–12
[4] Lauter C, Niewel J, Siewers B, Zanft B and Tröster T 2015 *Int. Conf. Exp. Mech.* **2690** 1–7
[5] Liu Q, Lin Y, Zong Z, Sun G and Li Q 2013 *Comp. Str.* **97** 231–238
[6] Meo M, Morris A J, Vignjevic R and Marengo G 2003 *Comp. Str.* **62** 353–360
[7] Bandi P, Schmiedeler J P and Tovar A 2013 *J. Mech. Des.* **135** 1–11
[8] Lewis G and Mladsi S 2000 *Biom.* 755–781

IOP Conf. Series: Journal of Physics: Conf. Series **843** (2017) 012018 doi:10.1088/1742-6596/843/1/012018

Formability prediction for AHSS materials using damage models

R. Amaral[1], Abel D. Santos[2], José César de Sá[2] and Sara Miranda[1]

[1] Institute of Science and Innovation in Mechanical and Industrial Engineering, Rua Dr. Roberto Frias, 4200-465 Porto, Portugal

[2] Faculty of Engineering, University of Porto, Rua Dr. Roberto Frias 400, 4200-465 Porto, Portugal

E-mail: ramaral@inegi.up.pt, abel@fe.up.pt, cesarsa@fe.up.pt, smiranda@inegi.up.pt

Abstract. Advanced high strength steels (AHSS) are seeing an increased use, mostly due to lightweight design in automobile industry and strict regulations on safety and greenhouse gases emissions. However, the use of these materials, characterized by a high strength to weight ratio, stiffness and high work hardening at early stages of plastic deformation, have imposed many challenges in sheet metal industry, mainly their low formability and different behaviour, when compared to traditional steels, which may represent a defying task, both to obtain a successful component and also when using numerical simulation to predict material behaviour and its fracture limits. Although numerical prediction of critical strains in sheet metal forming processes is still very often based on the classic forming limit diagrams, alternative approaches can use damage models, which are based on stress states to predict failure during the forming process and they can be classified as empirical, physics based and phenomenological models. In the present paper a comparative analysis of different ductile damage models is carried out, in order numerically evaluate two isotropic coupled damage models proposed by Johnson-Cook and Gurson-Tvergaard-Needleman (GTN), each of them corresponding to the first two previous group classification. Finite element analysis is used considering these damage mechanics approaches and the obtained results are compared with experimental Nakajima tests, thus being possible to evaluate and validate the ability to predict damage and formability limits for previous defined approaches.

1. Introduction

Sheet metal formability can be defined as the ability of metal to deform without necking or fracture into a desired shape. Every sheet metal can be deformed without failing only up to a certain limit, which is normally known as forming limit curve (FLC). Usually it is experimentally determined for defined linear strain paths, by using Marciniak tests, where a sheet metal sample is strained by a flat-bottomed cylindrical punch having a frictionless in-plane deformation, or by using Nakajima tests, in which the sample is deformed by an hemispherical punch (out-of-plane) and by varying the sample width, different linear strain paths can be obtained.

The experimental determination of FLC for sheet metal is time consuming and requires expensive equipment. Because of the expenses involved on the experimental procedure, many mathematical criteria for plastic instability have been proposed, empirically as well as theoretically [1, 5], to predict the forming limits of the metallic material, making these methods very attractive. However, since these models don't take into consideration the variation of the strain path and microscopic damage, some authors proposed modifications, by considering such variation during the deformation [6, 7], or by introducing new models [8], like a new path independent failure criterion to predict necking in sheet metal forming.

Another possible approach is considering damage models based on stress state to predict failure during the forming process. These damage models can be classified as empirical, physics based and phenomenological [9, 10].

In this work two different ductile damage models were selected, Johnson-Cook and GTN, each of them corresponding to the previous group classification, to predict the forming limit curve of a dual-phase steel (DP780). The selected models, namely GTN and Johnson-Cook, are already available in finite element program Abaqus.

To study the previous damage models, a dual-phase DP780 steel with a nominal thickness of 0.8 mm was selected and the corresponding experimental forming limit curve was obtained. A detailed description of the procedure will be presented in further section.

Also in this work, finite element simulations of Nakajima test were carried out and the obtained results were compared with experimental FLD, as well as, the punch force vs. displacement curves, in order to evaluate the implemented ductile damage models and their ability to predict the material forming limit.

2. Damage models

Gurson Tvergaard Needleman (GTN) damage model is used to determine the forming limit curve of anisotropic sheet metals. The original model proposed by Gurson [11] was further extended by Tvergaard and Needleman [12, 13] replacing the variable f by a modified parameter f^* in order to take into account the accelerated void coalescence process after a critical void volume fraction f_c is reached. The yield function of the GTN damage model is given by:

$$\phi = \left(\frac{\overline{\sigma}}{\sigma_Y}\right)^2 + 2q_1 f^* cosh\left(\frac{3}{2}q_2\frac{\sigma_H}{\overline{\sigma}}\right) - \left(1 + q_3 f^{*2}\right) \tag{1}$$

where q_1, q_2 and q_3 are the model parameters, σ_Y the flow stress and σ_H and $\overline{\sigma}$ the hydrostatic stress and equivalent von Mises stress, respectively. The modified porosity parameter f^* is calculated by:

$$f^* = \begin{cases} f & f \leq f_c \\ f_c + \dfrac{1/q_1 - f_c}{f_F - f_c}(f - f_c) & f > f_c \end{cases} \tag{2}$$

The evolution law of the void volume fraction $\dot{f} = \dot{f_G} + \dot{f_N}$ take into account the void growth and nucleation, both variables expressed, respectively, by:

$$\begin{aligned} \dot{f_G} &= (1 - f)\, tr\, \dot{\boldsymbol{\varepsilon}}_{\boldsymbol{p}} \\ \dot{f_N} &= \frac{f_N}{S_N\sqrt{2\pi}}\, exp\left[-\frac{1}{2}\left(\frac{\dot{\varepsilon}_p - \varepsilon_N}{S_N}\right)^2\right]\dot{\varepsilon}_p \end{aligned} \tag{3}$$

where f_N, S_N and ε_N are material parameters and $\dot{\varepsilon}_p$ is the plastic strain rate tensor.

The model proposed by Johnson and Cook [14] has material degradation under plastic deformation expressed by damage indicators, which, besides the postulation of a phenomenological strain rate and temperature dependent hardening law, defines the fracture strain ε_f as a function of stress triaxiality η, plastic strain rate $\dot{\varepsilon}_p$ and temperature θ as:

$$\varepsilon_f = \left(d_1 + d_2 \cdot e^{d_3\eta}\right)\left[1 + d_4 ln\left(\frac{\dot{\varepsilon}_p}{\dot{\varepsilon}_0}\right)\right](1 + d_5\theta) \tag{4}$$

where d_1, d_2, d_3, d_4 and d_5 are material parameters.

3. Material mechanical characterization

The material used in this work is a dual-phase DP780 steel sheet with a thickness of 0.8 mm. Scanning electron microscopy (SEM) observations were performed on the initial material (figure 1) showing that the microstructure is defined by the presence of hard martensite particles (lighter areas) dispersed in the soft ferritic matrix [15, 16].

The average size of ferritic grains (D) and the volume percentage of martensite (V_M), obtained from microstructural observations is presented in table 1.

Figure 1: SEM images of DP780 steel [15, 16].

Table 1: Average size of ferritic grains (D) and volume percentage of martensite (V_M) in DP780 steel [15, 16].

Material	D (μm)	V_M (%)
DP780	6	25

Uniaxial tensile tests were performed at room temperature using a defined 0.0016 s^{-1} initial strain rate. Specimens have been machined according to ASTM E 8M-04 standard and the tests were conducted for three different directions relative to the rolling direction (0°, 45° and 90°).

Results were averaged from three experiments for each direction, in order to test and ensure repeatability. Figure 2 shows the true stress vs. strain hardening curves for DP780, according to the defined loading directions. Additionally, hydraulic biaxial bulge tests of DP780 were also performed, being the biaxial stress vs. strain curve presented in Fig. 3.

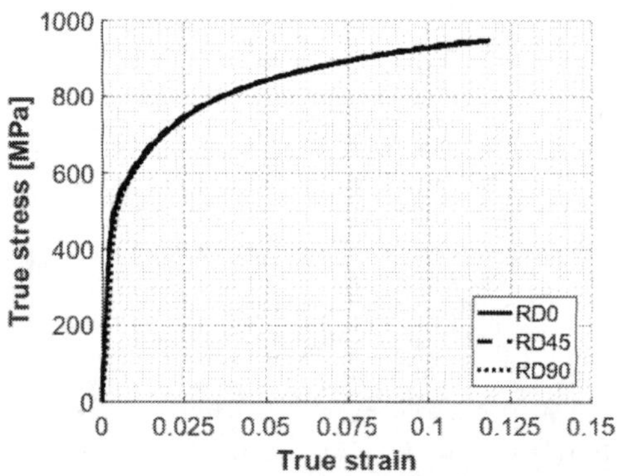

Figure 2: Tensile true stress vs. strain curves of DP780 for different loading directions.

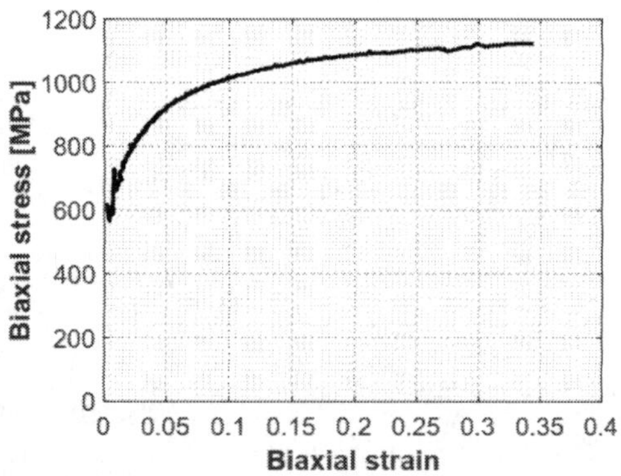

Figure 3: Biaxial stress vs. strain curve from bulge test of DP780.

The mechanical properties acquired from tensile tests are presented in table 2, namely yield stress (YS), ultimate tensile strength (UTS), elongation at yield point (e_0), uniform elongation (e_u) and total elongation (et), as well as, the anisotropic coefficients measured for the same evaluated loading directions.

Table 2: Mechanical properties for DP780 under current study, along rolling direction.

Direction	YS [MPa]	UTS [MPa]	e_0 [%]	e_u [%]	e_t [%]	r-value
0°	526.18	843.10	0.47	12.53	17.96	0.70
45°	537.72	845.87	0.49	11.78	18.10	1.05
90°	517.78	840.95	0.49	12.35	18.04	0.88

In order to describe the hardening behaviour of the DP780, the Swift Law was used. The identified parameters are presented also in table 3.

Table 3: Swift law parameters for DP780 steel.

$\sigma = K\left(\varepsilon_0 + \varepsilon^p\right)^n$	
K [MPa]	1319.2
ε_0	0.0015
n	0.149

3.1. Parameters of Damage models

For the Johnson-Cook damage model, strain rate and temperature effects were not considered, corresponding to parameters d_4 and d_5, respectively. In these model the failure strain depends the stress triaxility so it's important the determination of parameters d_1, d_2 and d_3.

To characterize the damage behaviour of the material DP780 using the selected damage models, the corresponding parameters were identified through an inverse optimization FE methodology. The procedure uses as reference the tensile and biaxial bulge test data to find the optimum fitting parameters, which represents the best correlation between numerical and experimental curves.

Since the strain rate and temperature effects of Johnson-Cook damage model were not considered, the parameters d_4 and d_5 were not included in the identification scheme.

In case of GTN, the standard deviation S_N was set to 0.2 which is often used by other authors in literature. Moreover, the values of parameters q_1 and q_2

Furthermore, the values set for parameters q_1 and q_2 were based on characterizations conducted by others, in order to reduce the number of variables to be identified.

The parameters used for Lemaitre damage model has been identified by Niazi et al [16]. Table 4 presents the parameters of each ductile damage model.

Table 4: Implemented damage model parameters (DP780).

GTN damage model								
Parameters	q_1	q_2	q_3	f_c	f_F	ε_N	S_N	f_N
Values	1.5	1	2.25	0.021	0.04	0.3	0.1	0.001

Johnson-Cook damage model				
Parameters	d_1	d_2	d_3	d_4, d_5
Value	0.94	-0.29	6.48	0

4. FLD experimental determination

Nakajima et al. [17] proposed to use metal sheet strips with different widths, stretched by a hemispherical punch (figure 4), in order to obtain the material forming limit curve. By varying the sample width, different strain paths can be achieved (figure 5).

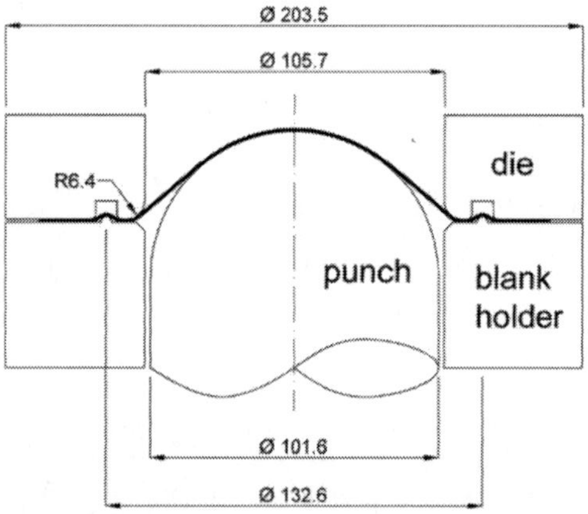

Figure 4: Tool dimensions and FLD setup used in Nakajima tests.

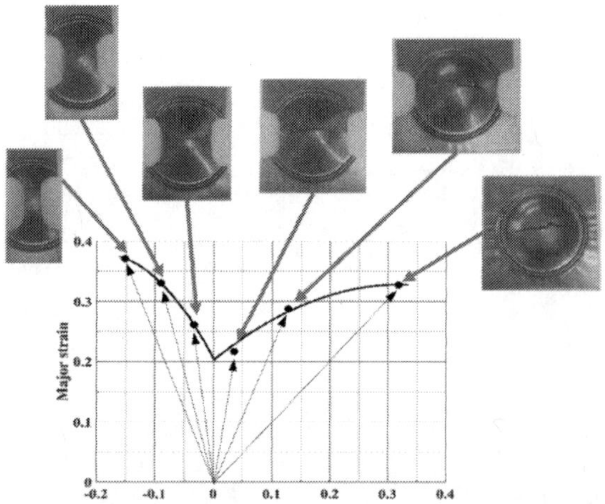

Figure 5: Obtained FLC points for different specimen widths.

The Nakajima testing procedure uses a hemispheric punch of 101.6 mm diameter and drawbeads in tools, to restrain the blank between the die and blank-holder and also to avoid stress concentration at the transition from strip section to the blank. Nakajima experiments were performed according to the standard ISO 12004-2 [18], to measure the DP780 formability, as well as to determine the corresponding forming limit curve. To obtain a wide range on forming limit curve the specimen widths are varied and then different strain paths and values are obtained. The defined specimens for testing have the following widths: 25, 50, 75, 100, 125 and 175 mm [19, 20].

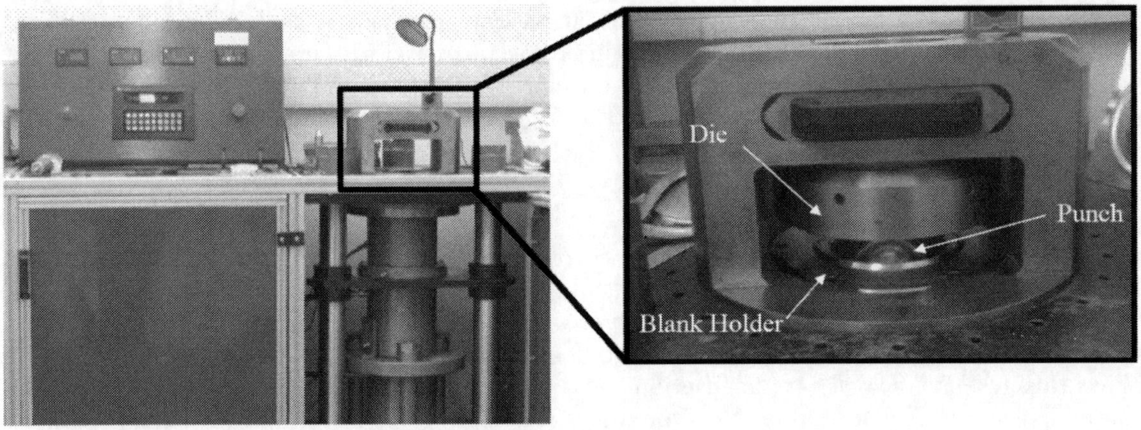

Figure 6: Universal testing machine used for Nakajima test and FLD determination.

All experimental tests were conducted on a hydraulic testing machine, as seen in figure 6. The tooling setup is interchangeable depending on the type of experiments. The friction between punch and blank was minimized by applying a teflon foil at the top of the punch, in addition to a lubricant at the sheet surface (Ferrocoat N 6130). Necking or fracture visualization determined the moment to stop the test.

Strain measurements on samples can be done through several methods, but in this paper the method used was a variant of Bragard method [21], which it is applicable to perform measurements on grids of tangled circles, obtained in this work by an electrochemical etching process. For interpolation, it was followed the recommendation of using the circles whose deformation differences of adjacent circles is less than 5%.

The obtained experimental data points are represented by fitting curve that has been adjusted, using the least square approximation approach. This methodology performs an adjustment for the drawing region and another for the stretching region. The figure 8 shows the forming limit curve of the studied dual-phase material and corresponding experimental points.

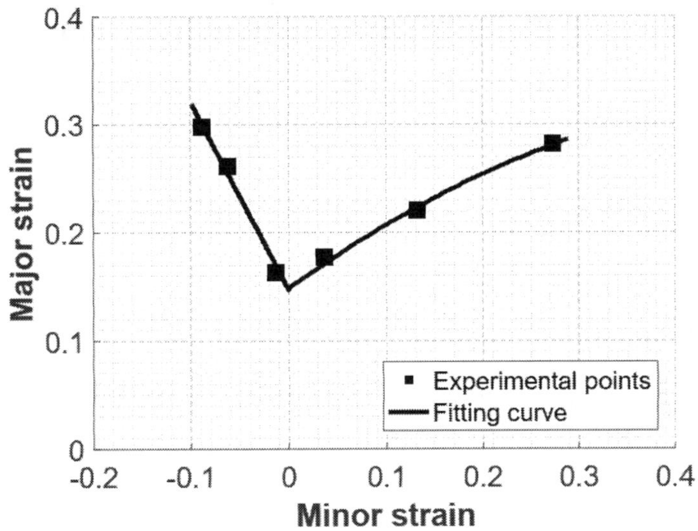

Figure 7: DP780 experimental points and corresponding fitting curve.

IOP Conf. Series: Journal of Physics: Conf. Series **843** (2017) 012018 doi:10.1088/1742-6596/843/1/012018

The FLC_0 point can be determined experimentally from a tensile test of a large width specimen or using a smaller range of Nakajima specimens widths in plane strain region. A mathematical expression (equation 5) based on material thickness (t) and hardening coefficient (n) can be used to obtain a reference FLC_0 value [22, 23].

$$FLC_0 = ln\left[1 + \left(\frac{13.3 + 14.13 \cdot t}{100}\right)\frac{n}{0.21}\right] \quad for\ n \leq 0.21 \Leftrightarrow FLC_0 = 0.16 \qquad (5)$$

The experimental FLC_0 point was obtained by the interception of the fitting curves and its value is 0.152, showing a good agreement with the analytical result.

Figure 8 shows the corresponding specimens fractured during the experimental Nakajima tests. The necking and fracture occur at the dome thus showing that friction was minimized. Since this test is affected by the friction between the punch head and the specimen when contacted, the sheet deformation could be limited to the regions in contact with the punch.

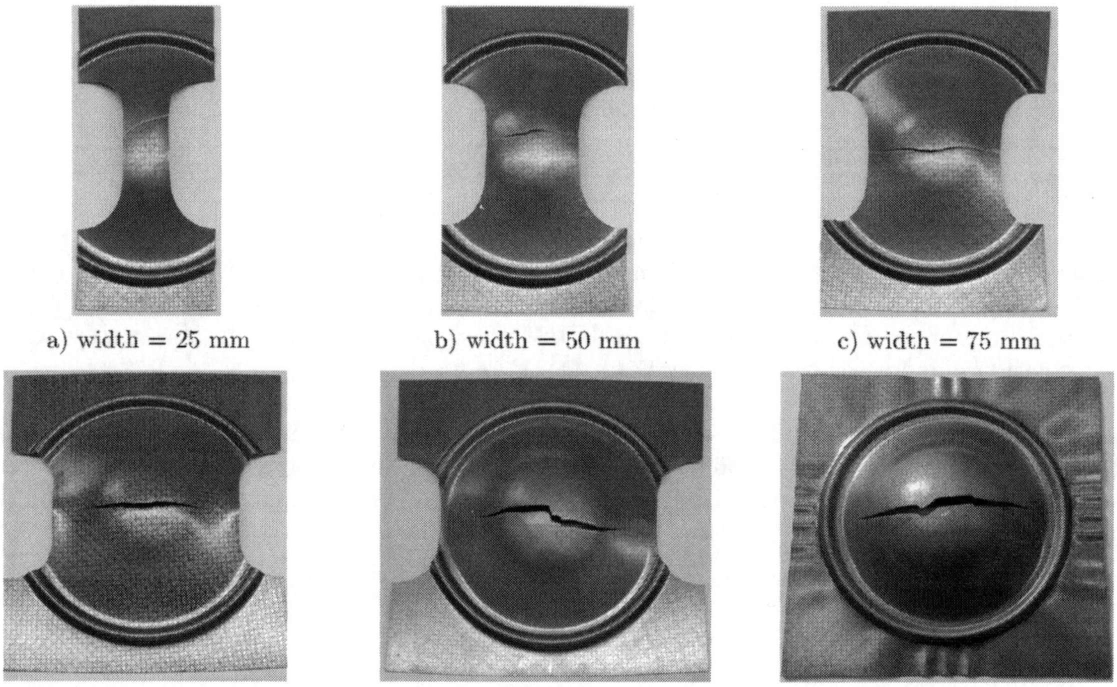

a) width = 25 mm b) width = 50 mm c) width = 75 mm

d) width = 100 mm e) width = 125 mm f) width = 175 mm

Figure 8: Obtained FLC points for different specimen widths.

5. Finite Element Analysis

Numerical simulations were performed using a 3D finite element explicit analysis with Abaqus/Explicit code and defining a full model for the Nakajima test (figure 9).

The blank was discretized by four node shell elements with reduced integration (S4R type from Abaqus Library) and tools were discretizated by analytical rigid surfaces. A constant friction coefficient of 0.05 was defined for the interacting surfaces. As for the material mechanical modelling, the sheet blank is defined as an elasto-plastic material with isotropic hardening defined by the constitutive model already presented in section 3. Johnson-Cook and GTN ductile damage models are available in FE program.

Figure 9: Nakajima setup used in finite element model.

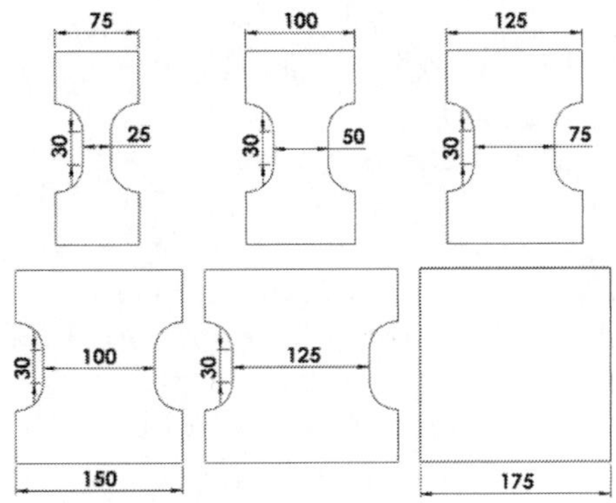

Figure 10: Sample dimensions used in Nakajima numerical simulations.

6. Results and Discussion

Limit strain values from the numerical simulation were obtained using the same experimental methodology. Each ductile damage model FLC was formed by a fitted curve to the strains points acquired from different sample widths (as seen in figure 10) using the least-squares method. The comparison of DP780 forming limit curves, determined experimentally and using the ductile damage models is shown in figure 11.

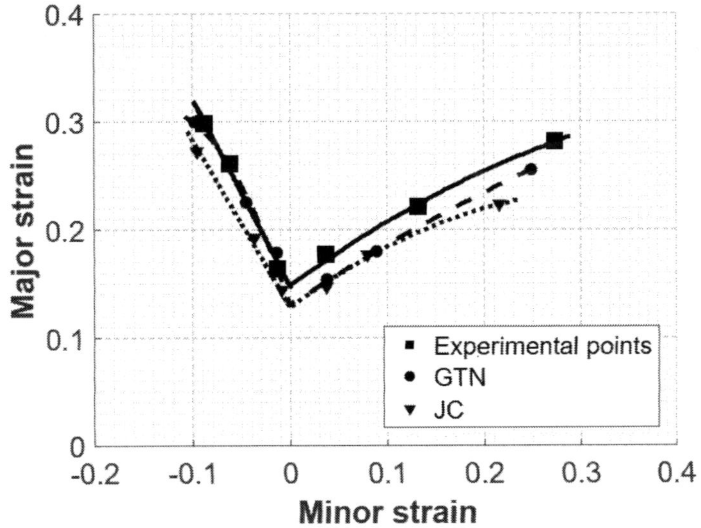

Figure 11: Numerical model of deep drawing cylindrical cup test.

Globally, the forming limit curves obtained from numerical simulation using the damage models and presented in figure 11 show a similar tendency compared to the experimental data. The GTN criterion prediction of forming limit curve gives closer results when compared with experimental data. In case of Johnson-Cook damage model, the numerical points are more dispersed.

The Johnson-Cook and GTN damage models seem to have a lower formability in plane strain state and also in the stretching region, being more evident in equibiaxial loading case. This trend, more evident in Jonhson-Cook damage model, which shows closer results to experimental points in left FLD region and an underestimation of forming limits in right side, can be adjusted by the user, since this model expression is an exponential based formula. Due to the influence of experimental reference data used to identify the model parameters, a closer fitting to some part of FLC could be made.

By comparison of the formability prediction of DP780 using these two ductile damage models with experimental results, it can be concluded that the GTN model gives a better approximation and shows a closer overall prediction to current experimental data. However, with a smaller initial relative density, the formability prediction of GTN can be increased and the FLC could be more closer in plane strain and stretching region (right side of FLD) but a possible overestimation in drawing region (left side of FLD) could be happen.

Additionally to this study, a comparison of punch force vs. displacement curves were performed, in order to evaluate the numerical and experimental data, being also a complement to the FLD analysis.

Figure 12 shows the curves of punch force vs. displacement evolution during deformation for each sample width defined in this work. Numerical results for current damage models and experimental data are compared on the same graph.

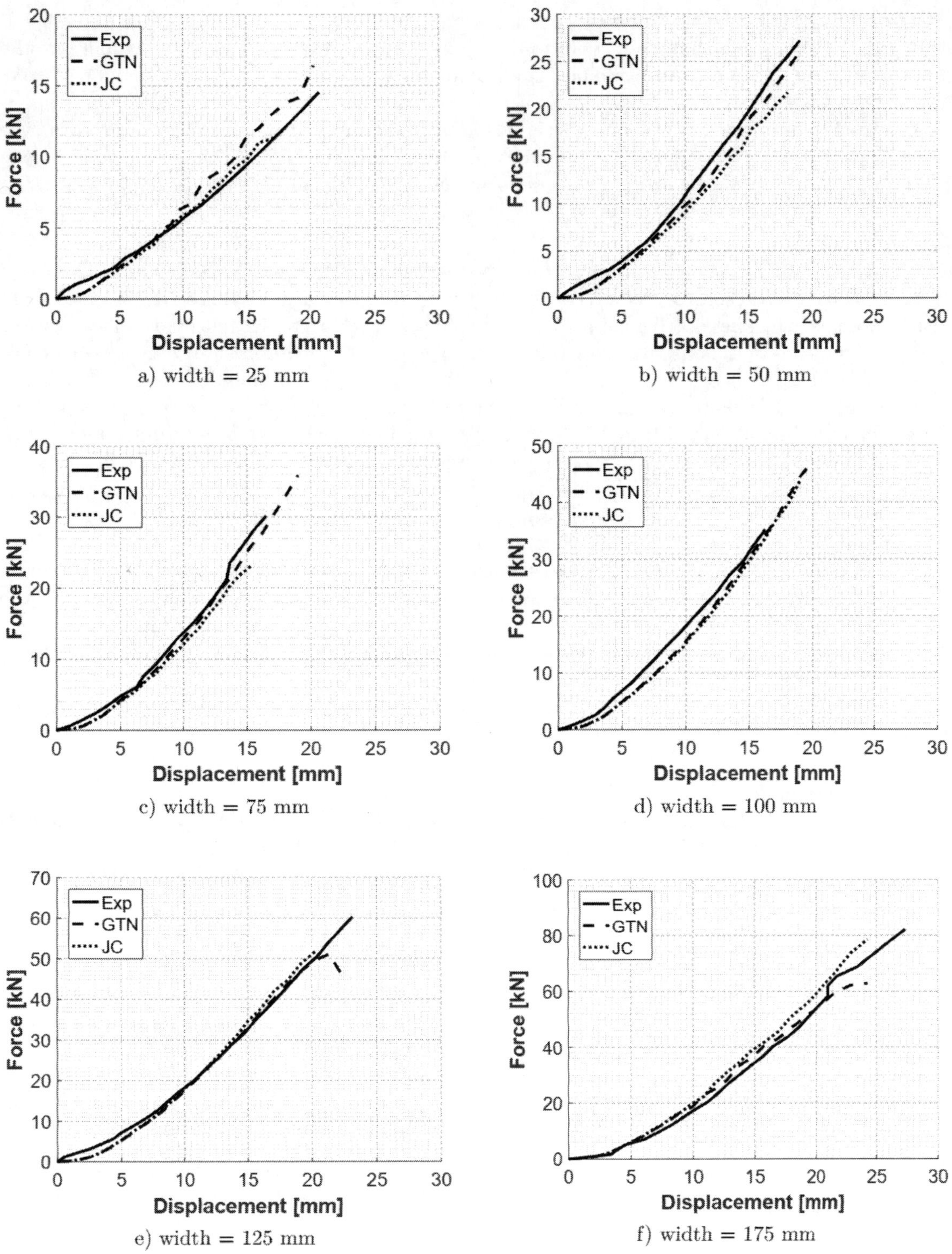

a) width = 25 mm
b) width = 50 mm
c) width = 75 mm
d) width = 100 mm
e) width = 125 mm
f) width = 175 mm

Figure 12: Comparison of experimental punch force evolution for different specimen widths, with the numerical results from the damage models.

Figure 12 presents the evolution of the punch force vs. displacement during forming process where experimental data and numerical results are compared.

It can be observed that the evolution of punch force predicted is very similar to the experimental punch force. Additionally, this comparison allows the evaluation of punch stroke at which fracture occurs.

Since GTN damage model has a closer forming limit curve to the experimental data, the punch stroke is in a very good agreement with experimental observation. In contrast, the Johnson-Cook model shows a smaller value of punch stroke for all of defined sample widths.

7. Conclusions

In this paper, two different ductile damage models, Gurson-Tvergaard-Needleman (GTN) and Johnson-Cook, were used to predict the forming limit curve of a dual-phase DP780 steel sheet with a nominal thickness of 0.8 mm. Their ability to reproduced the material behaviour was analysed and validated using the Nakajima experimental test, for different specimen geometries, as a reference.

The FLD of DP780 using the damage models was formed through a set of Nakajima finite element simulations, which GTN numerical prediction showed a good agreement with the experimental results. In general, the forming limits predicted by the GTN model were more closer than those obtained with the Johnson-Cook damage model. The parameters used for Johnson-Cook damage model provided smaller forming limits of the DP780 steel and consequently an underestimation of material formability.

A comparison of punch force vs. displacement was made between the finite element results obtained from the two models and the experimental data. The analysis of results shows that the punch force evolution is very close to the experiment, for all tested specimen widths. However, the Johnson-Cook damage model predicted a smaller punch force, since the damage of material happened at a earlier stage and consequently a smaller punch stroke.

Further studies will include the influence of damage parameters identification and also the friction model considered in numerical simulation results.

7.1. Acknowledgments

Authors gratefully acknowledge the funding of SciTech, R&D project NORTE-01-0145-FEDER-000022 cofinanced by NORTE2020, through FEDER and the financial support of the Portuguese Foundation for Science and Technology (FCT) under project PTDC/EMS-TEC/6400/2014. The first author is also grateful to the FCT for the Doctoral grant SFRH/BD/119362/2016.

References

[1] A. Col: "FLCs: Past, present and future", IDDRG 2002 Conference, Nagoya, Japan, 107-25 (2002).
[2] D. Banabic: "Sheet Metal Forming Processes - Constitutive Modelling and Numerical Simulation", Springer-Verlag GmbH, (2010).
[3] A. Hrivnak, E. Evin: "Formability of steel sheets", Kosice: Elfa, (2004).
[4] Morteza Nurcheshmeh: "Numerical prediction of sheet metal forming limits", Dissertation, University of Windsor, (2011).
[5] A. Pradeau, S. Thuillier and J.W. Yoon, Prediction of failure in bending of an aluminium sheet alloy, International Journal of Mechanical Sciences, **119**, 23-35 (2016).
[6] P. Hora, L. Tong, J. Reissner: "A prediction method for ductile sheet metal failure in FE-simulation", Proceedings of the Numisheet96 Conference (Dearborn/Michigan), 252-56 (1996).
[7] Z. Marciniak, K. Kuczynski: "Limit strains in the processes of stretch forming sheet metal", International Journal of Mechanical Sciences, **9**(9), 609-612 (1967).
[8] T.B. Stoughton and J.W. Yoon, A new approach for failure criterion for sheet metals, International Journal of Plasticity, **27**(3), 440-59 (2011).
[9] Y. Bai and T. Wierzbicki, A comparative study of three groups of ductile fracture loci in the 3D space, Engineering Fracture Mechanics, **135**, 147-67 (2015).

IOP Conf. Series: Journal of Physics: Conf. Series **843** (2017) 012018 doi:10.1088/1742-6596/843/1/012018

[10] T. Wierzbicki, Y. Bao, Y.W. Lee, Y. Bai, Calibration and evaluation of seven fracture models, International Journal of Mechanical Sciences, **47**(4-5), 719-43 (2005).

[11] L. Gurson,Continuum Theory of Ductile Rupture by Void Nucleation and Growth: Part I-Yield Criteria and Flow Rules for Porous Ductile Media, Journal of Materials Processing Technology, **99**, 2-15 (1977).

[12] V. Tvergaard, On localization in ductile materials containing spherical voids, International Journal of Fracture, **18**, 237-52 (1982).

[13] V. Tvergaard and A. Needleman, Analysis of the cup-cone fracture in a round tensile bar, Acta Metallurgica, **32**, 157-69 (1984).

[14] G.R. Johnson, W.H. Cook. Engineering Fracture Mechanics, **21**, 31-48 (1985).

[15] R. Amaral, H. Campos, J.A. Sousa, A.B. Lopes, A.D. Santos: "Inovação em Materiais: Aços Bifásicos - DP500, DP600, DP780, caracterização mecânica e microestrutural", TECNOMETAL (AIMMAP - Associação dos Industriais Metalúrgicos, Metalomecânicos e afins de Portugal), **218**, 4-9 (2015).

[16] R. Amaral, A.D. Santos, J.A. Sousa, A.B. Lopes : "The Influence of Microstructure on the Mechanical Behaviour of Dual Phase Steels", Materials Design and Applications, **65**(3) (2017).

[17] K Nakajima, T Kikuma, and K Hasaku: "Study of the formability of steel sheets". Yawata Technical Report No.264, 141-54 (1968).

[18] ISO 12004 standard: Metallic materials guidelines for the determination of forming limit diagrams, (2008).

[19] M. Parente, R. Safdarim, A.D. Santos, A. Loureiro, P. Vilaca, R.M Natal Jorge: A study on the formability of aluminum tailor welded blanks produced by friction stir welding, International Journal of Advanced Manufacturing Technology, **83**, 2129-41 (2016).

[20] R. Safdarian, H. Moslemi Naeini, R.M. Natal Jorge, A.D. Santos, M.P.L Parente: A comparative study of forming limit diagram prediction of tailor welded blanks, International Journal of Material Forming, 8(2) (2014).

[21] A. Bragard, J.C. Baret, H. Bonnarens, Metallurgical Rap. CRM, 33, 53-63 (1972).

[22] S.P. Keeler, W.G. Brazier, Relationship between laboratory material properties and press shop formability, Proceedings of conference on microalloy, **75**, 517-28 (1977).

[23] S.K. Paul: Theoretical analysis of strain- and stress-based forming limit diagrams, Journal of Strain Analysis for Engineering Design, 48(3), 177-88 (2013).

Mechanics and complications of reverse shoulder arthroplasty: morse taper failure analysis and prospective rectification

HLD Hoskin[1], E Furie[1, 3], W Collins[2], TM Ganey[1] and DR Schlatterer[1]

[1] Department of Orthopeadics, Wellstar Atlanta Medical Center, Atlanta Georgia, 30312, USA

[2]Material Analysis Center, Georgia Tech Research Institute, Atlanta, Georgia, 30332 USA

[3]Global Orthopedics Atlanta, Georgia, 30312, USA

E-mail: Timothy.ganey@wellstar.org

Abstract. Since Sir John Charnley began his monumental hip arthroplasty work in 1958, clinical researchers have been incrementally improving longevity and functionality of total joint systems, although implant failure occurs on occasion. The purpose of this study is to report the fracture of the humeral tray Morse taper of a reverse total shoulder system (RTSS), which to date has not been reported with metallurgic analysis for any RTSS. There was no reported antecedent fall, motor vehicle collision, or other traumatic event prior to implant fracture in this case. Analysis was performed on the retrieved failed implant by Scanning Electron Microscopy (SEM) and Electron Dispersion Spectroscopy (EDS) in an attempt to determine the failure method, as well as to offer improvements for future implants. At the time of revision surgery all explants were retained from the left shoulder of a 61-year old male who underwent a non-complicated RTSS 4 years prior. The explants, particularly the cracked humeral tray, were processed as required for SEM and EDS. Analysis was performed on the failure sites in order to determine the chemical composition of the different parts of the implant, discover the chemical composition of the filler metal used during the electron beam welding process, and to detect any foreign elements that could suggest corrosion or other evidence of failure etiology. Gross visual inspection of all explants revealed that implant failure was a result of dissociation of the taper from the humeral tray at the weld, leaving the Morse taper embedded in the humeral stem while the tray floated freely in the patient's shoulder. SEM further confirmed the jagged edges noted grossly at the weld fracture site, both suggesting failure due to torsional forces. EDS detected elevated levels of carbon and oxygen at the fracture site on the taper only and not on the humeral tray. In order to determine the origin of the high levels of C and O, it was considered that in titanium alloys, C and O are used as stabilizers that help raise the temperature at which titanium can be cast. Since the presence of stabilizers reduces ductility and fatigue strength, all interstitial elements are removed after casting. Considering this, the presence of C and O suggests that not all of the interstitials were removed during the manufacturing process, resulting in decreased fatigue strength. Further destructive analytical testing would verify weld quality and failure mode. RTSSs are quite successful in select patients not amenable to traditional shoulder arthroplasty options. This case report highlights how an implant may function well for several years and then suddenly fail without warning. SEM and EDS analysis suggest that residual C and O in the taper lowered the metal implant's integrity, leading to torsional cracking at the weld junction of the humeral tray and the taper. The elevated levels of C and O measured at fracture sites on both the tray and the taper suggest poor quality filler metal

IOP Conf. Series: Journal of Physics: Conf. Series **843** (2017) 012019 doi:10.1088/1742-6596/843/1/012019

or failure to remove all interstitial elements after casting. In both cases, the results would be decreased fatigue strength and overall toughness, leading to mechanical failure. A manufacturer's recall of all implants soon followed the reporting of this implant failure; subsequently, the metal materials were changed from Ti_6Al_4V to both titanium alloy and cobalt-chrome alloy (Co-Cr-Mo). Time will tell if the alterations were sufficient.

1. Introduction

Total joint arthroplasty has been regarded as one of the most efficacious orthopedic interventions of its time. Though inhibiting patients' ability to participate in most high-impact activities post operation, joint replacement hip and total shoulder systems provide significant pain relief and increased mobility for patients suffering from osteoarthritis or other instances of cartilaginous degeneration [1]. Analysis of the mechanics and complications of total hip arthroplasty has led to experimentation with various materials, including titanium, cobalt-chrome, polymers, and ceramics [2]. This wide range of material options has allowed surgeons to offer patients implants specific to patients' ages, individual lifestyles, and activity levels. A significant amount of total hip arthroplasty research focuses on optimizing implant materials to withstand pressure, loading, and friction that the acetabular joints undergo on a day to day basis [3]. As newer shoulder procedures are developed and perfected, previous research on total hip devices serves as a predicate in creating safe, effective systems. However, as the acetabular and glenohumeral joints vary in more than basic anatomy, different methods of testing are needed for total shoulder systems and RTSSs.

The reverse shoulder design incorporates a socket where the head of the humerus is normally located and a ball where the glenoid is normally situated; by switching the components of a normal shoulder, greater stability is added. The configuration allows the deltoid to elevate the shoulder in place of the rotator cuff muscles, helping patients to overcome shoulder pseudo-paralysis and carry out basic shoulder functions on a daily basis [4]. Although both total shoulder and RTSSs are generally successful, cases of both non-mechanical and mechanical failure have been reported. Previously recorded issues include glenoid component disengagement (**Fig. 1**), loosening of humeral stem/metaphysis resulting in interfragmentary strain (**Fig. 2**), and Morse taper dissociation (**Fig. 3**). In nearly all cases, the failures were not reportedly caused by single incidents [5].

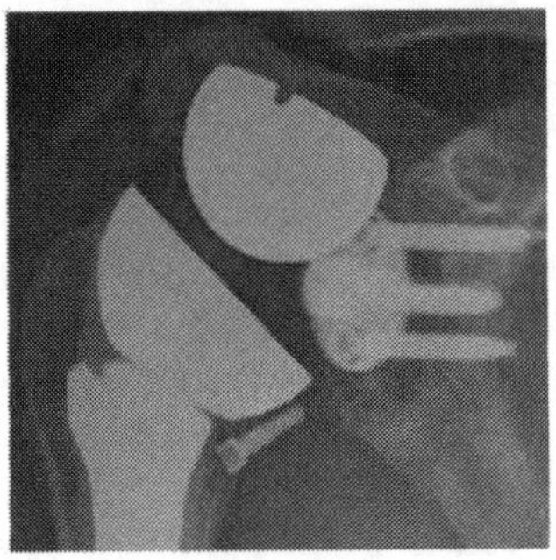

Fig. 1 Glenoid component disengagement reported with various designs of reverse total shoulders (Ref. 18).

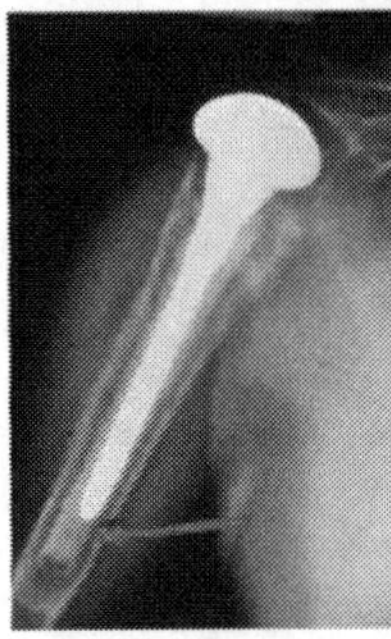

Fig. 2 Loose infected humeral component. AP radiograph shows lucencies about the cement; distal cement fracture (arrow) caused by stem loosening (Ref. 3).

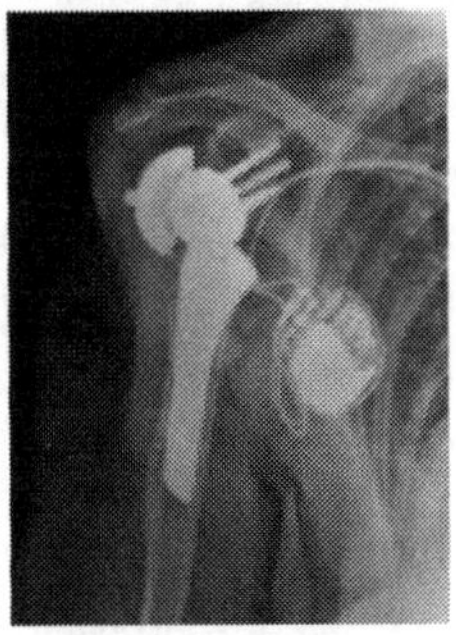

Fig. 3 X-Ray displaying dissociation of the Morse taper from the humeral stem (Ref. 10).

Examples of mechanical failure bring into question the integrity of the metal. In one case, a patient had a RTSS implanted for about 3 years before developing pain and "clunking" noises. Radiographs revealed glenosphere disengagement as well as detachment of the humeral tray from the stem (**Fig. 4**) [5]. The taper dissociated from the humeral tray at the weld, leaving the taper embedded in the humeral stem while the tray floated freely in the patient's shoulder (**Fig. 5**). Note jagged edges at the fracture site, which suggest failure due to torsional forces (**Fig. 6**).

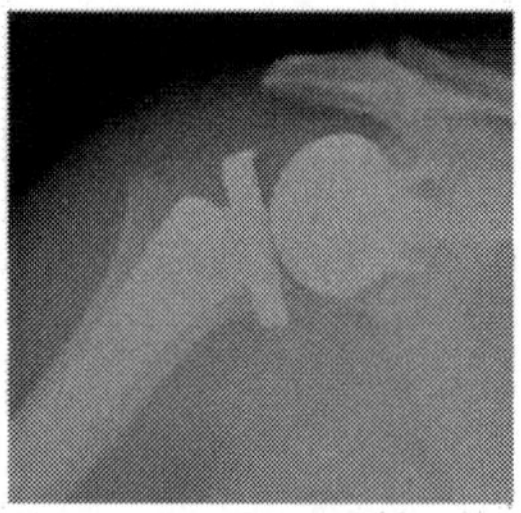

Fig. 4 Post-operative film of completed, fully functional reverse total shoulder system (Ref. 11).

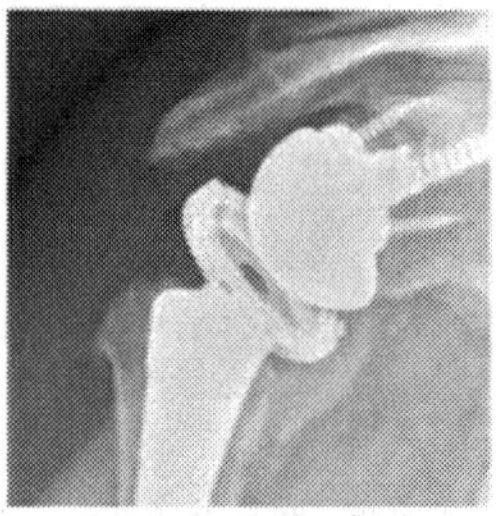

Fig. 5 Radiographic findings reveal disengaged glenoid component and dissociation of the humeral tray (Ref. 11).

Fig. 6 Separated tray (still attached to polyethylene cup) and taper removed from patient. Note jagged edges of facture site (Ref.11).

FDA testing requirements for total hip and total shoulder procedures differ mostly in that the shoulder is not usually subjected to the greater axial loading and torsional forces the hip experiences on a daily basis [6]. Testing processes consider glenohumeral joint flexibility, reduced loading, and greater range of motion (higher tendency for instability and/or dislocation) than the acetabular joint [7]. RTSSs are newer and not as thoroughly researched as total hip procedures; revisions to FDA testing and requirements for RTSS implants are needed. Fatigue strength, biocompatibility, and other requirements must be verified as adequate for the new RTSSs, as there is considerable difference in forces on the shoulder and hip. The recent development of RTSS as a surgical intervention has been coupled with greater incidence of morbidity in these new systems, as is common in the early phases of the innovation of any new technique. Failure of the original RTSS system has guided research to refine and implement improvements to the designs, materials, instrumentation systems, and the procedures themselves.

Author EF offers a similar case that prompted this review and eventually became the foundation for this case report. In part, he summarized, "*A 61-year-old male was evaluated 4 years following a non-complicated left RTSS. The patient was doing well until the sudden onset of pain, "heaviness" and limited motion after light lifting, coming in approximately 3 weeks after the onset of symptoms. Physical exam revealed no obvious deformity. Motion was limited to 90 degrees flexion, 75 degrees abduction and 45 degrees external rotation passively with pain. No evident open wounds, swelling, ecchymosis or erythema. Radiographic evaluation revealed a separation of the humeral tray at the trunnion with retention of the reverse Morse taper humeral component within the stem. Despite aggressive efforts to remove the taper, it remained within the humeral stem.*" EF also noted no antecedent trauma prior to the catastrophic RTSS failure. **Fig. 7** displays the radiograph of the separated humeral tray/taper component. Also pictured is the stem/taper and tray once removed from the patient; note how the edges of the fracture site are relatively smooth compared to those viewed in the former example (**Fig. 8**).

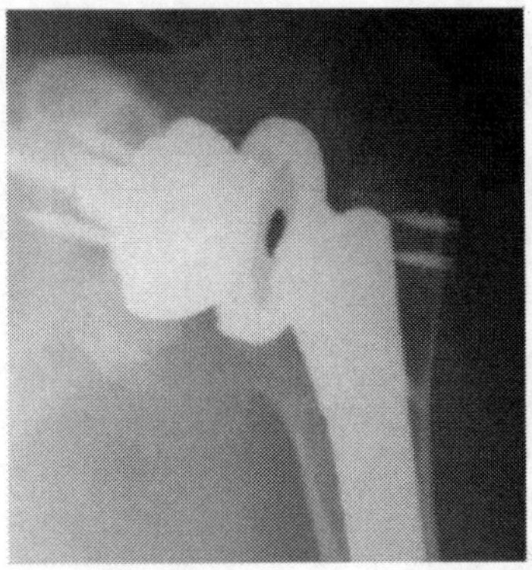

Fig. 7 (left) Radiographic evaluation revealing failure of the humeral tray-Morse taper component.

Fig. 8 (right) Humeral tray/polyethylene cup component and Morse taper retained within humeral stem. Note smooth edges of fracture site.

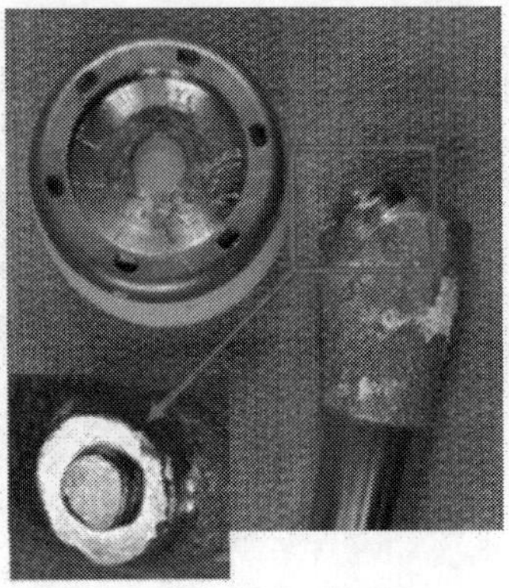

2. Materials & Methods

At the time of revision surgery all explants were retained from the left shoulder of a 61-year old male who underwent a non-complicated RTSS 4 years prior. The explants were processed as required for SEM and EDS. Analysis was performed on the failure sites in order to determine the chemical composition of the different parts of the implant, discover the chemical composition of the filler metal used during the electron beam welding process, and to detect any foreign elements that could suggest corrosion or other evidence of the cause of failure.

Analysis of the failed implant began with separation of the Morse taper from the humeral stem; the taper was so firmly inserted that the insertion site at the superior end of the stem had to be cut out with a Dremel handheld rotary tool (**Fig. 9**). During the procedure to fix the failed RTSS, surgeons aggressively attempted to remove the dissociated taper from the stem using pliers or a similar device. The removal attempts badly damaged the surface of the trunnion at the tray/taper connection, making taper analysis more difficult than that performed on the tray (**Fig. 10**).

SEM imaging was executed on both the fracture site along the rim of the humeral tray and the trunnion of the taper, as shown in **Figs. 11-16**. Electron Dispersion Spectroscopy (EDS) was performed on the failure sites in order to determine the chemical composition of the implant, discover the composition of the filler metal used during the electron beam welding process, and to detect any foreign elements that could suggest corrosion or other evidence of the cause of failure [8]. EDS of the inner rim of the humeral tray revealed unusually high levels of C and O. EDS was also performed on both compromised and uncompromised sections of the trunnion of the Morse taper (**Fig. 18**). EDS was executed on the smooth shaft, uncompromised fracture site, and a compromised area, which provided chemical composition data of three separate areas for comparison. Slightly elevated but insignificant levels of C and O were detected at the fracture site on the taper, and significant levels were detected on the humeral tray.

Fig. 9 Result of the removal of the embedded taper using a rotary tool. Every effort was made to ensure no damage was inflicted on the taper during removal.

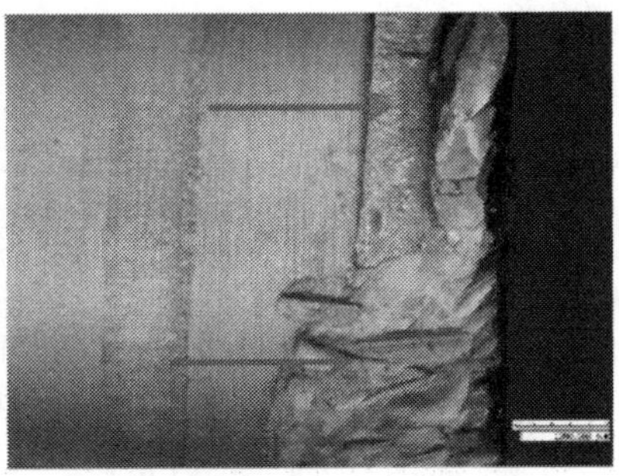

Fig. 10 HIROX imaging provided evidence of uncompromised areas on the taper. Here, the difference between uncompromised (red arrow) and compromised (blue arrow) areas is displayed.

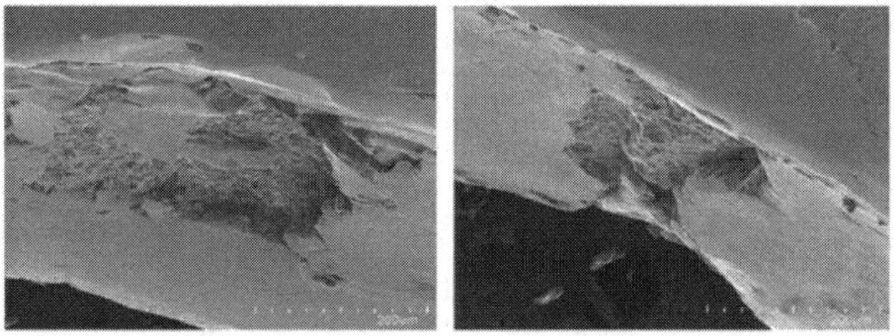

Figs. 11-12 SEM imaging of different areas of the inner rim of the humeral tray reveal similar fracture patterns in different locations.

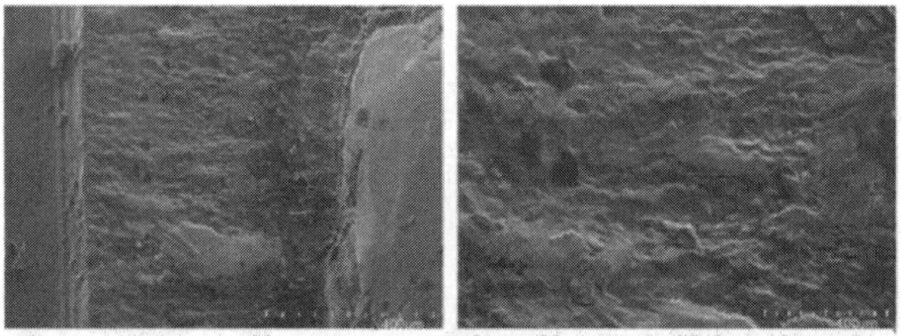

Figs. 13-14 SEM imaging of the same uncompromised area of the inner rim of the humeral tray, 120x (**left**) and 350x (**right**).

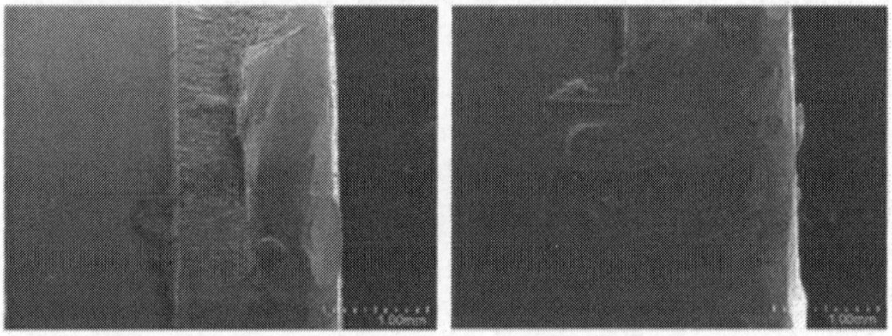

Fig. 15 uncompromised areas such as that shown here yielded acceptable fracture data (red arrow).
Fig. 16 Compromised areas of the trunnion such as this one yielded no usable data.

3. Results

Gross visual inspection of all explants revealed that implant failure was a result of Morse taper dissociation from the humeral tray at the weld. SEM further confirmed the likelihood of torsional failure due to the presence of jagged edges at the fracture site. EDS detected elevated levels of C and O at the edge transition of the humeral tray and not on the taper. **Fig. 17 (area 3)** indicates the presence of various elements, transitioning from the planar area of the tray to the inner rim. Comparison of the presence of elements on the planar area of the tray, the edge of the inner rim, and on the inner rim itself enabled determination of failure method. The presence of other trace elements (Na, P, Si, Cl, etc.) was assumed to be due to handling during testing.

IOP Conf. Series: Journal of Physics: Conf. Series **843** (2017) 012019 doi:10.1088/1742-6596/843/1/012019

Fig. 17 EDS of the humeral tray. The planar area of the humeral tray (1) revealed the presence of mostly titanium and aluminum, which is ordinary considering that the composition of the alloy is Ti6Al4V. The rim of the tray (2) showed similar elemental findings. However, the fracture site (3) displays elevated levels of carbon and oxygen, which provided significant information regarding how the implant failed.

Image Name:
Humeral Tray ID Edge Transition - 1

- Image Resolution: 512 by 384
- Image Pixel Size: 0.71 µm
- Acc. Voltage: 15.0 kV
- Magnification: 350

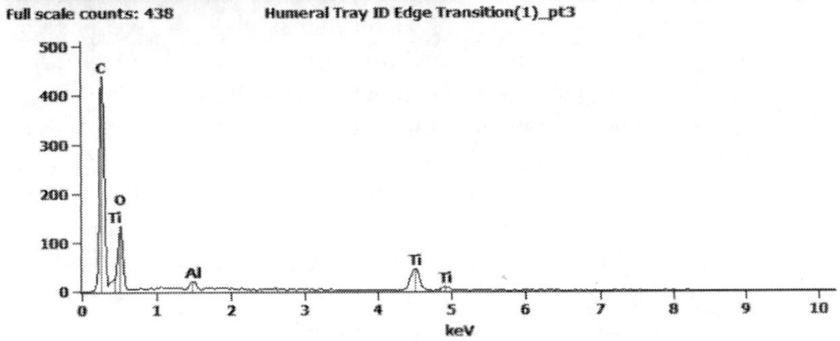

EDS was performed on both compromised and uncompromised sections of the trunnion of the Morse taper (**Fig. 18**). EDS was executed on the smooth shaft, uncompromised fracture site, and a compromised area. Insignificant levels of C and O were detected at the fracture site on the taper, and lack of foreign elements suggests that either no filler metal was used to weld the two pieces together, or that the filler metal is also composed of Ti_6Al_4V .

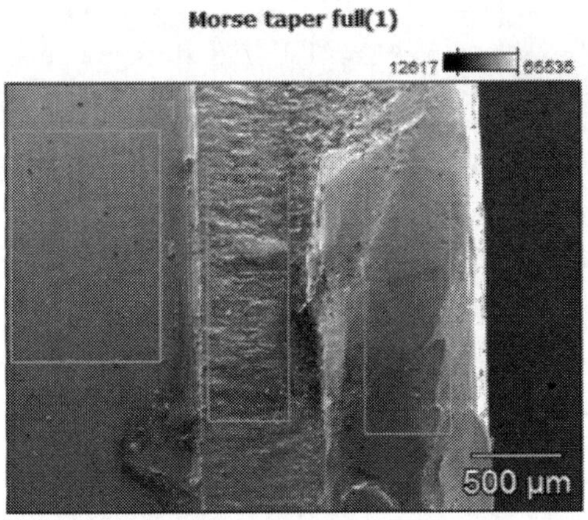

Fig. 18 EDS of the Morse taper at the failure site. The smooth shaft of the taper (1) revealed the presence of mostly titanium and aluminum, which is ordinary considering that the composition of the alloy is Ti6Al4V. The uncompromised area of the fracture site (2) showed slightly increased levels of carbon, oxygen, and nitrogen, which are all interstitial elements. The compromised area (3) displays traces of iron and cobalt, which most likely originated from tools with which surgeons attempted to remove the taper from the stem.

Image Name: Morse Taper Full – 1

- Image Resolution: 512 by 384
- Image Pixel Size: 6.22 µm
- Acc. Voltage: 15.0 kV
- Magnification: 40

4. Discussion

Fatigue testing for acetabular and glenohumeral joints is designed for each joint based on primary function. The contact between the acetabulum and femoral head is much broader and closer than the contact between the glenoid and the humeral head; in the shoulder, this implies greater range of motion but decreased stability [9]. In order to determine the respective joint loads and muscle forces during active motion, one study applied five advanced biomechanical techniques to obtain clinically relevant data regarding joint loading. Techniques included in vivo instrumented implants to capture joint forces and movements, load sensors to assess external forces such as ground reaction forces, and synchronized data of joint loads, external loads, and skeletal motion as applied to a musculoskeletal computer model for the calculation of muscle forces and muscle activation patterns, among other testing methods. In the hip, peak loading reaches 270% of body weight during walking and 520% during running. Climbing stairs induces similar peak forces with an 80% increase in rotational movements. Glenohumeral contact forces range up to only 130% of body weight when performing daily activities, although concentration of forces varies as the center of gravity is altered [3]. Activities inducing forces of 150% of body weight or more such as doing a handstand are not typically executed by patients who have undergone total shoulder arthroplasty [10]. Successful total joint arthroplasty in the shoulder must be preceded by careful consideration of dependency on ligaments and thorough biomechanical analysis of joint constraints and motion; incorrect placement or poor balancing can result in undue joint loading and wear, as well as instability [9].

In titanium alloys, carbon and oxygen are used as stabilizers that help raise the temperature at which titanium can be cast. Since the presence of stabilizers reduces ductility and fatigue strength, the interstitial elements are removed after casting [11]. Considering this, the presence of C and O could suggest that not all of the interstitials were removed during the manufacturing process, resulting in decreased fatigue strength. The presence of C and O might also be attributed to various bodily deposits. The lack of significant traces of foreign elements suggests either no filler metal was used or a Ti_6Al_4V filler metal. According to the American Filler Metals Company, a titanium alloy surgical implant should be Grade 23 Ti_6Al_4V with a filler metal of equal composition and grade. Grade 23 titanium alloy has lower levels of Al, C, O, and other interstitial elements to improve toughness, fabricability, and weldability [11]. Additional destructive metallurgical examination is required to evaluate the nature and location of residual C, O, and other metal processing aids. To eliminate residual biological deposits, a combination of solvent cleaning and high temperature ashing procedures could be undertaken.

In the summer of 2011, the FDA posted a recall notice for three sizes of the Biomet standard RTSS humeral tray with locking ring, due to possible faulty locking ring mechanism on the humeral trays [12]. The 510(k) summary report submitted several months later, dated January 11, 2012, acknowledged product improvement as the reason for the submission [6]. The materials composing the humeral tray were changed from Ti_6Al_4V to both titanium alloy and cobalt-chrome-molybdenum alloy (Co-Cr-Mo). Using the Vicker's hardness values for measurement, CoCrMo has a higher hardness value than Ti6Al4V; conclusively, CoCrMo is more fracture resistant than Ti_6Al_4V due to its greater interfacial adhesion strength and stronger surface oxide [13]. CoCrMo is less biocompatible than titanium alloy but much stronger [9], but in view of the elderly age of most patients in which the new system is utilized, negative effects such as metallosis would most likely take more time than the patient's remaining lifespan to set in. Taking into consideration the health and well-being of the patient, the sacrifice of some biocompatibility for the sake of producing stronger implants was a decision that positively influenced total shoulder and RTSS development.

5. Conclusions

RTSSs have reportedly been quite successful in select patients not amenable to traditional shoulder arthroplasty options. This case report highlights how an implant may function well for several years and then suddenly fail without warning. New information guides both development and improvement of total shoulder and RTSSs and surgical techniques. The total hip is barely 125 years old; technological advancements since then have led to better-engineered implants, extended life due to reduced wear properties, and better personalization of anatomical replacements to sustain normal joint movement.

Based on the results gathered from SEM and EDS analysis, it is likely that several factors contributed to the mechanical failure of the implant. The striations observed on the uncompromised areas of the taper (**Figs. 13, 14, 15, 18**) are an indication of brittle-like failure, most likely caused by fatigue. The combination of rotational and toggle-like forces acting on the weld site produced wave movements [14]. This wave-like motion would have occurred any time the patient engaged in activities requiring flexion, abduction, and/or external rotation, and constant cyclic loading over the course of nearly four years ultimately resulted in failure of the implant. Once enough brittle-like failure had occurred over time,

IOP Conf. Series: Journal of Physics: Conf. Series **843** (2017) 012019 doi:10.1088/1742-6596/843/1/012019

the trunnion of the taper pulled away from the humeral tray, fully separating the pieces by means of ductile rupture (**Fig. 19**) [15].

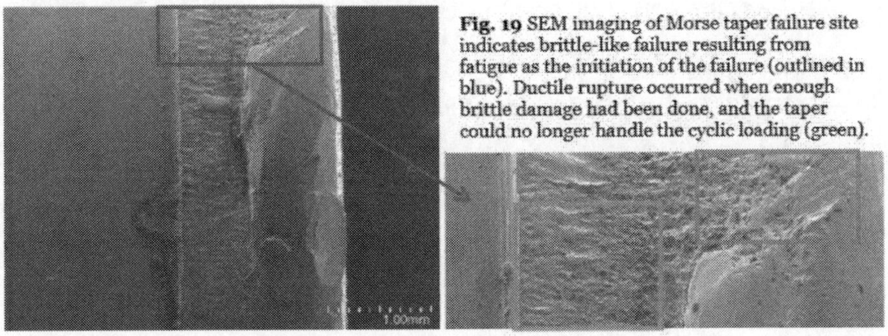

Fig. 19 SEM imaging of Morse taper failure site indicates brittle-like failure resulting from fatigue as the initiation of the failure (outlined in blue). Ductile rupture occurred when enough brittle damage had been done, and the taper could no longer handle the cyclic loading (green).

Despite the cyclic loading and forces placed on the shoulder during everyday activities, the main issue surrounding the failure of this implant is the poor design of the system. The entire system is undersized considering the actions the shoulder is expected to be able to carry out; the short length of the Morse taper, thinness of the humeral tray, weaker metal system, and small surface area of the weld site between the taper and tray all pose probable causes of failure. Fortunately, the design flaws of this particular system were quickly recognized, and engineers took action to recall the titanium alloy implants, recasting the pieces using stronger materials. Time will tell if the alterations were sufficient.

6. References

[1] Virani NA, et al 2012. Reverse shoulder arthroplasty components and surgical techniques that restore glenohumeral motion. *Journal of Shoulder and Elbow Surgery*. 2012.

[2] Berquist, T H 2007. Orthopedic Appliances and Prostheses. *Musculoskeletal Imaging Companion. 2nd ed.* N.p.: Lippincott Williams & Wilkins, 2007. Pg. 998. Print.

[3] Bottlang M, et al 2011. Musculoskeletal Biomechanics: Principles of Orthopedics. *Orthopaedic Knowledge Update*. Vol. 10. American Academy of Orthopaedic Surgeons. 2011. Pgs. 59-72.

[4] Biomet Orthopedics, 2014. *Comprehensive reverse shoulder system: surgical technique*. Warsaw, IN: Biomet Orthopedics, 2012. *www.biomet.com*. Biomet, Inc. Web. 13 June 2014.

[5] Matsen, FA III. 2014. Failings of a Reverse Total Shoulder. *Web log post. Www.shoulderarthritis.blogspot.com. Blogspot, Frederick A. Matsen III, M.D.*, 24 Nov. 2012. Web. 23 July 2014.

[6] U. S. Food and Drug Administration. 2014. *Comprehensive Reverse Shoulder Humeral Tray: A summary of 510(k) safety and effectiveness information in accordance with the requirements of 21 CFR 807.92.* www.accessdata.fda.gov. Food and Drug Administration, n.d., 2012. Web. 14 July 2014.

[7] Zumstein, et al 2011. Problems, complications, reoperations, and revisions in reverse total shoulder arthroplasty: a systematic review. *Journal of Shoulder and Elbow Surgery*, 2011; 20(1): 146-157[19] *www.ncbi.nlm.nih.gov*. Web. 23 July 2014.

[8] Olson DL et al 1993 Procedure Development and Practice Considerations for Electron-Beam Welding. *Welding, Brazing and Soldering*: *ASM Handbook*, Vol. 6. Eds. ASM International: 1993. pgs. 851–73.

[9] Jaffar AA, et al 2006. Biomechanical Aspects of Shoulder and Hip Articulations: A Comparison of Two Ball and Socket Joints. *Al-Khwarizmi Engineering Journal,* 2006; 2(1): 1-14.

[10] Graichen F, et al 2007. In vivo glenohumeral contact forces: Measurements in the first patient 7 months post-operatively. *Journal of Biomechanics,* 2007; 40(10): 2139-49.

[11] American Filler Metals Company 2014. American filler metals: quality welding alloys. Houston, TX: *American Filler Metals, n.d. www.amfiller.com*. American Filler Metals Company. Web. 9 July 2014.

[12] U. S. Food and Drug Administration. 2014. *Class 2 Device Recall Shoulder Humeral Tray.* www.accessdata.fda.gov. Food and Drug Administration, n.d. Web. 14 July 2014.

[13] Goldberg JR and Gilbert JL, 2004. The electrochemical and mechanical behavior of passivated and TiN/A1N-coated CoCrMo and Ti6Al4V alloys. *Science Direct: Biomaterials 25. Elsevier Ltd.* 2004. Pgs. 851-864.

[14] Liao Y, et al 2013. CoCrMo Metal-on-Metal Hip Replacements. *Physical chemistry chemical physics :* PCCP, 2013. 15(3):10.1039/c2cp42968c. doi:10.1039/c2cp42968c. https://www.ncbi.nlm.nih.gov/pmc/articles/PMC3530782/. Web. 14 Feb 2017.

[15] Escobedo JP, et al 2013. Effect of Shock-Wave Profile on Brittle Failure. *ArXiv & Los Alamos National Laboratory, NM, USA*. arXiv, 2013. https://arxiv.org/pdf/1303.0393. Web. 14 February 2017.

IOP Conf. Series: Journal of Physics: Conf. Series **843** (2017) 012020 doi:10.1088/1742-6596/843/1/012020

Experimental Study on Failure Mechanism of Single Lap-shear Bond Joint with Dissimilar Materials

S H Chung[1], B C Park[1], H J Chun[1] and J C Park[2]

[1]Department of Mechanical Engineering, Yonsei University, Seoul, Republic of Korea
[2]Commercial Vehicle CAE Team, Hyundai Motor Group, Hwaseong, Republic of Korea

E-mail: hjchun@yonsei.ac.kr

Abstract. The use of bond joints has increased recently, owing to a number of advantages they have over mechanical fasteners. However, most relevant studies have focused on bond joints with the same adherends. In this study, a quasi-static tensile test was conducted to investigate the failure mechanism of bond joints, considering different overlap lengths and dissimilar adherends. In the experiments, two modes of failure were observed: the interfacial and cohesive failure modes. The experiments showed that the length of separation of an overlap area through the interfacial failure was almost the same for different specimens.

1. Introduction

Adhesive joints are being increasingly used in the aerospace, automotive, and other industries where conventional fasteners, such as bolts, rivets, and welding are typically used. Although adhesive joints need extensive surface preparation and pre-treatment, they have a number of advantages over traditional fasteners, including savings in terms of time and cost, higher strength-to-weight ratio, corrosion and fatigue resistance, crack retardance, and damping characteristics, among others. [1, 2] Adhesive joints also allow the avoidance of machining the adherends and stress concentration in the vicinity of the resultant holes, which affect the overall load-carrying capacity of the joints. In other words, bond joints transfer loads more efficiently than mechanically fastened joints. In spite of these many advantages, it is difficult to predict the failure of bond. This is because the failure load and failure mode of bond joints are different from various other bonding methods and parameters. [3, 4]
Many researchers have attempted to predict the failure load of bond joint and different models have been proposed. Sheppard et al. proposed a damage zone model based on a critical damage zone size and strain-based failure criteria to predict the failure load of bond joints. [5] In the damage zone model, joint failure regions appeared to be insensitive to finite element mesh refinement and the regions could be predicted. Apalak and Engin studied the initiation and propagation of damaged zones in the adherends and adhesive layer of adhesive single and double lap joints. [6] In their study, non-linearity of both the adhesive and adherends was considered. After interfacial failure was occurred, damaged zone growth was observed in the end of adhesive regions because of lateral straining.
In the fracture mechanics approaches, a cohesive zone model was suggested to predict the failure load. Fracture mechanics uses an energy-based failure criterion, such as the J-integral or critical energy release rate. However, the cohesive zone model is appropriate for analyzing the fracture behavior of bond joints when the adherends considered only to have elastic deformation.

Hart-Smith first suggested the elasto-plasticity model. [7] The growth of the plastic deformation area in the elasto-plasticity adhesive layer of different types of adhesive bonded joints was examined, and the plastic deformation areas appearing at the free ends of the adhesive layer were shown. In the elasto-plasticity model, stress-strain curve of adhesive should be obtained through the coupon test. However, in the elasto-plasticity model, Hart-Smith neglected the normal stresses acting on the adhesive layer which might be the principal reason of joint failure.
There have been many studies on the failure mechanism and failure prediction of bond joints. However, most of them have focused on bond joints using the same adherends. Thus, in this study, a quasi-static tensile test was conducted to investigate the failure mechanism of single lap-shear bond joints with dissimilar adherends.

2. Material and methods

2.1. Material properties
Al6061-T6 aluminum and SS400 steel were used as adherends, while Araldite AV138/HV998 was used as the adhesive in the bond joints in this study. Metal adherends were tested to obtain mechanical properties and the test method was in accordance with ASTM E8/E8M-13, which is the standard test method for tension of metallic materials. Adhesive bulk tensile test specimens were manufactured and tested. The test method was in accordance with ASTM D638, which is the standard test method for tensile properties of plastics. A vacuum desiccator was used before the curing process to remove the air bubbles within the adhesive. Additionally, a vacuum bag was used to remove the air bubbles when the adhesive was cured at high temperature.
A universal testing machine (MTS 810) was used for the tensile test. Load was applied by stroke control at a rate of 2 mm/min for the adherends and 0.3 mm/min for the adhesive. A strain gage was attached to the center of the specimens to measure Young's modulus and Poisson's ratio. The strain signal was measured using a dynamic strain meter and the data was recorded on a PC through an A/D converter. The mechanical properties of the adhesive and adherends are shown in Table. 1.

Table 1. Mechanical properties of adhesive and metals.

	Young's modulus (GPa)	Ultimate strength (MPa]	Poisson's ratio
AV138/HV998	4.5	41.04	0.35
Al6061-T6	68.9	310	0.33
SS400	190	400	0.26

2.2. Specimens and experimental procedures of bond joint
In this study, the bond joints were designed for single lap-shear testing according to ASTM D1002, which is the standard test method for apparent shear strength of single-lap-joint adhesively bonded metal specimens. The dimensions of the bond joints were changed when the joints were made to observe the change of failure mode and failure load. The geometry of the bond joint specimens is shown in Figure 1. The specimens of bond joint are shown in Figure 2.
A universal testing machine (MTS 810) was used and load was applied by stroke control at a rate of 0.5 mm/min, which is the standard for ASTM D1002. The experimental setup is shown in Figure 3. Three specimens were tested for each test condition and the mean value was taken.

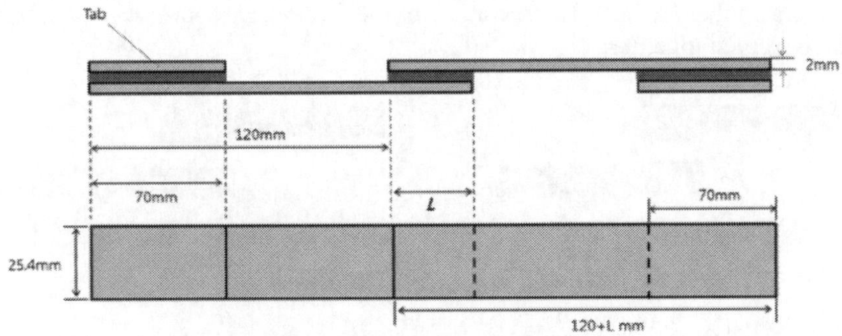

Figure 1. Specimen geometry of bond joint.

Figure 2. Specimen of dissimilar metal bond joint; overlap length: 20 mm, 40 mm, and 60 mm.

Figure 3. Experimental setup of single lap-shear test.

3. Results and discussion

3.1. Failure modes of single lap-shear bond joints

The failure modes of single-lap bond joints could be divided into cohesive failure and interfacial failure, as shown in Figure 4. The failure modes depend on the mechanical properties of the adhesive and adherends. The cohesive failure occurred when the adhesive is separated to both adherends. This failure mode was very radical and the crack initiation could not be observed. It can be inferred that the crack initiation and final failure occurred at the same time. Interfacial failure was a progressive damage occurring with plastic deformation of the adhesive along the bonded boundary. In this failure mode, the crack initiated in both ends of the overlap region and propagated along the interface between adhesive and adherend. [8] However, in our experiment, cohesive and interfacial failure modes are usually observed together.

(a) Cohesive failure (b) Interfacial failure

Figure 4. Failure modes of adhesive joints.

3.2. Single lap-shear bond joints with dissimilar metal adherends

In order to investigate the failure load and failure mode of the adhesive joints of dissimilar materials, Al6061-T6 aluminum and SS400 steel adherends, as well as AV138 / HV998 adhesive were used. From the experimental results, a boundary line was observed in the failure surface, as shown in the Figure 5. As mentioned before, the two modes of failure were observed with the interfacial mode occurring first. Additionally, experiments showed that the length of separation of an overlap area through the interfacial failure were almost the same for different specimens.

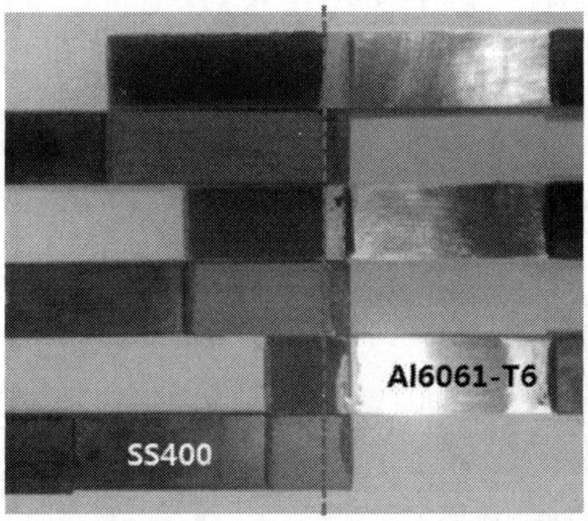

Figure 5. Surfaces of the failed specimens.

As P.N.B Reis et al. studied the effects of asymmetric properties of bond joints, the flexural rigidity of adherends played a major role in the performance of the bond joints. [8] In this study, SS400 steel adherend, which has approximately three times the elastic modulus of Al6061-T6 aluminum, was used and compared with joints between aluminum adherends. Compared to the same adherends joints, in this case Al6061-T6/Al6061-T6, using a stiffer material in a dissimilar joint, in our case SS400/Al6061-T6, resulted in larger overall flexural rigidity of the joint, which caused smaller bending rotation as well. Consequently, the smaller bending rotation of the joint causes the lower out of plane stress at the end of the adhesive joint. Thus, as shown in Figure 6, the failure loads of SS400/Al6061-T6 joints are larger than the failure loads of Al6061-T6/Al6061-T6 joints.

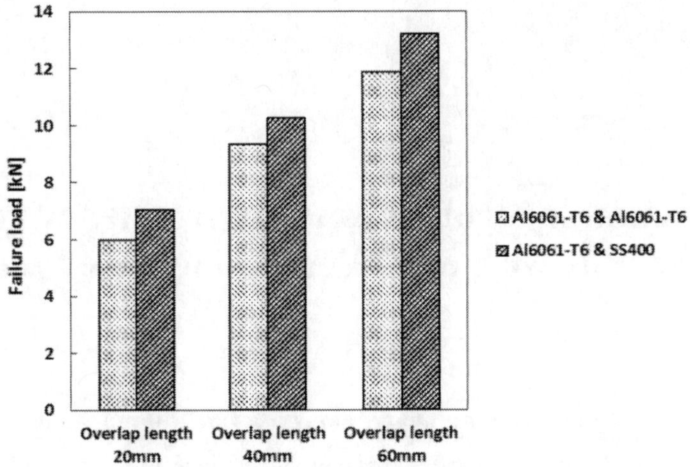

Figure 6. Failure loads of bond joints with same and dissimilar adherends.

4. Conclusion

The results of this study confirmed that the failure loads of bond joints with dissimilar adherends, in our case SS400/Al6061-T6, were larger than the failure loads of bond joints with same adherends, in this case Al6061-T6/Al6061-T6. Experimental investigations were conducted to elucidate the failure mechanism of single lap-shear bond joints, while taking into account the effects of the elastic modulus of adherends. From the experimental results, cohesive and interfacial failure modes of single lap-shear bond joint were observed together. The length of separation of an overlap area through the interfacial failure were almost the same for different specimens.

References

[1] Patrick R L (Ed.) 1976 *Structural adhesives with emphasis on aerospace applications* vol 4 (New York: Marcel Dekker Inc.)
[2] Kuno J K 1979 *Proc. Conf. on Structural Adhesives Bonding* (El Segundo)
[3] Mallick P K 1988 *Fiber-reinforced composite* (New York: Marcel Dekker Inc.)
[4] Choi J H and Lee D G 1996 *J. Adhesion* **55** pp 245-260.
[5] Sheppard A *et al* 1998 *Int. J. Adhesion Adhesives* **18** pp 385-400.
[6] Apalak M K *et al* 2003 *J. Adhesion Sci. Technol.* **17** pp 1889-1921
[7] Hart-Smith L J 1973 *Adhesive-Bonded Single Lap Joints* (Hampton VA: Langley Research Center) p 116
[8] Reis P N B *et al* 2011 *Int. J. of Adhesion and Adhesives* **31** pp 193-201

Acknowledgement

This work was supported by the Commercial Vehicle CAE Research Lap of the Hyundai Motor Group of Republic of Korea.

IOP Conf. Series: Journal of Physics: Conf. Series **843** (2017) 012021 doi:10.1088/1742-6596/843/1/012021

Damage detection in structures using frequency response functions ensemble with extended cosine based indicator

Yun-Lai Zhou[1], Xudong Qian[1], Hongyou Cao[1], Magd Abdel Wahab[2]

[1] Department of Civil and Environmental Engineering, National University of Singapore, 2 Engineering Drive 2, Singapore 117576

[2] Soete Laboratory, Faculty of Engineering and Architecture, Ghent University, Technologiepark Zwijnaarde 903, B-9052 Zwijnaarde, Belgium

E-mail: Magd.AbdelWahab@UGent.be

Abstract. Although extensive investigations exist for structural damage detection, which often solely considers the stiffness reduction, or specific damage type like concrete spalling, this study aims to address a general novelty detection procedure for detecting the novelty, or so-called outlier from structural dynamic analysis with applying cosine based indicator in frequency response functions. This study also addresses a short review for the cosine indicator and modal assurance criterion, and their interrelation as well. The eight degree-of-freedom system is considered for the verification, and proves well performance in novelty detection.

1. Introduction

In-service structures undergo various forms of occasions that can affect their dynamic responses, such as temperature, bolt loosening, concrete spalling and so on. These changes in comparison with the initial structural state cannot be always considered as damages, but might be better called as novelty, or so-called outlier. The detection of these novelties makes structural health monitoring (SHM) of high significance, and also make it arduous since various factors can induce novelty. The accurate detection of damage rather than environmental change shall be afterwards challenging. In contrast to defining a damage-caused stiffness reduction, this study considers all response changes as novelty.

Over the past few years, various inspection techniques have been developed, such as vibration based, impedance based, probability based, and so on. The techniques can involve magnetic particle testing, acoustic testing, guided wave, new sensors like fiber bragg gratting (FBG) and so on. For large-scale structures, vibration based techniques still occupy a large portion of applications due to their simple conduction, and effective work [1]. Vibration based techniques can be summarized into two categories: experimental modal analysis (EMA) and operational modal analysis (OMA) [2]. EMA usually applies to the structures in laboratory testing, and measures as much data as possible especially for the excitation and response, later to use frequency response functions (FRFs) and modal parameters such as modal damping, modal frequencies, mode shapes and their extensions like modal strain energy (MSE) in identifying the potential structural damages, or more generally, novelties. OMA behaves as an alternative for EMA by avoiding the measurement of excitation, and tries to achieve the same functions as EMA, which implies further applications for real engineering structures. For OMA, transmissibility,

IOP Conf. Series: Journal of Physics: Conf. Series **843** (2017) 012021 doi:10.1088/1742-6596/843/1/012021

as reference free technique, raised for decades ago, has been applied in several directions, such as damage detection [2-18], damage quantification [8, 10, 19], and so on. Review for transmissibility are available in [2, 10, 12].

Even alternatives can be found for novelty detection in terms of specific damage, like weight change can apply for spalling. It would be desirable to develop a strategy for general detection of novelty. This study tries to use extended cosine based indicator in the detection of novelty. Remaining work is organized as follows: section 2 gives the background of cosine indicator and its relation with modal assurance criterion (MAC) [19], and further applies in the FRFs to extract indicators for novelties; section 3 uses a eight degree-of-freedom system as verification, and section 4 addresses the concluding remarks.

2. Theoretical background

2.1. Cosine indicator and modal assurance criterion (MAC)

Consine and MAC have been widely used separately, and recent advancement combined these two indicators, and addressed their kernel agreeing well, and detailed discussion can refer to [19]. For cosine indicator, which can be expressed as:

$$\cos(X,Y)=\frac{X^T Y}{|X||Y|} \tag{1}$$

where X, Y mean two vectors, $()^T$ means the transpose of ().

Then, MAC can be illustrated as:

$$MAC(X,Y)=\frac{\left(X^T Y\right)}{\left(X^T X\right)^{\frac{1}{2}}\left(Y^T Y\right)^{\frac{1}{2}}}=\left(\frac{X^T Y}{|X||Y|}\right)^2=\left(\cos(X,Y)\right)^2 \tag{2}$$

This relation unveils the interrelation between cosine indicator and MAC, and also explains why MAC can be applied as an objective function [19].

2.2. Frequency response functions

For a single loaded linear elastic structural system, the damped vibration can be illustrated with the second order differential equation:

$$M\ddot{x}(t)+C\dot{x}(t)+Kx(t)=f(t) \tag{3}$$

where M, C, and K denote the mass, damaping, and stiffness matrices of the structure, respectively. $f(t)$ represents all the possible time dependent loading. x(t) means the responses of the structure.

Then, the *FRF* can also be expressed as:

$$H_{(i,r)}=\frac{X_i}{F_r} \tag{4}$$

where X_i, and F_r refer to the frequency spectrum of response x_i and excitation f_r. FRF can also expressed as the accumulation of discrete modes,

$$H_{(i,r)}(\omega)=\sum_{p=1}^{n}\frac{\phi_p^i\phi_p^r}{k_p-\omega^2 m_p+j\omega c_p} \tag{5}$$

where p denotes the pth mode, n means the number of modes considered. k_p, m_p and c_p mean modal stiffness, mass and damping, respectively. From this equation, it is possible to see that the modal parameters are hidden in the FRFs, and therefore possible to extract these modal parameters from FRFs through some algorithms, like expectation maximum algorithm.

2.3. Novelty indicator

From the discussion above, novelty indicators can be defined as the extension of cosine indicator [19],

$$CI = \left(\cos(FRF^{u}, FRF^{d})\right)^{Q} \tag{6}$$

where $()^u$, $()^d$ denote the value under undamaged and damaged states, respectively. This cosine extension indicator has also been applied in transmissibility for damage detection [19], and it proves to be effective in detecting damages, and can measure the relative damage.

3. Experimental verification

An eight degree-of-freedom system shown in Figure 1 [20] is taken into consideration for unveiling the feasibility of the proposed approach in detecting the novelty. Detailed information for this experiment can be found in [16]. Bolts, springs, translating masses form the system. Different types of damage are simulated, linear and non-linear. Excitation of the system is driven by the voltage supplied. In this study, only the 14% damage at location 5 is considered, and the excitation is random, and changing voltage with 3V, 4V, 5V and 6V. Each response is recorded with an average of 40 measurements, and hanning window is used [20]. Further information about this experimental data analysis can also refer to [21].

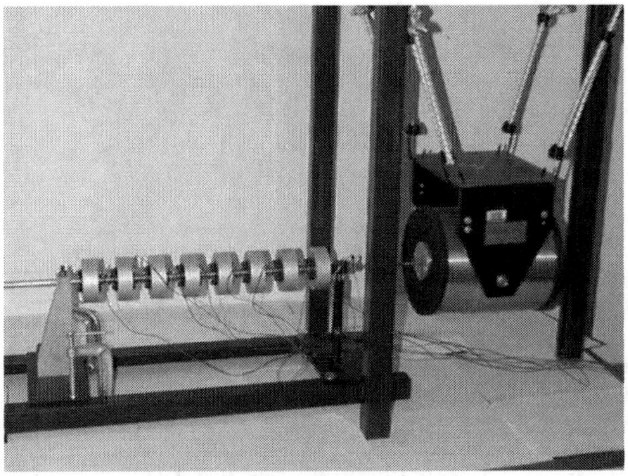

Figure 1. Experimental setup.

4. Results analysis

Results are computed and discussed in this section as follows in order to draw out the feasibility of the proposed novelty detection procedure. Figure 2 demonstrates the FRF (9, 2) for the structural system with and without damage under 6V excitation, and one can clearly find the natural frequency shift leftward, which implies the stiffness decrease. Certainly herein, it is possible to use the frequency change to estimate the stiffness reduction, since the change in the stiffness matrix leads to a corresponding change in the eigenvalue.

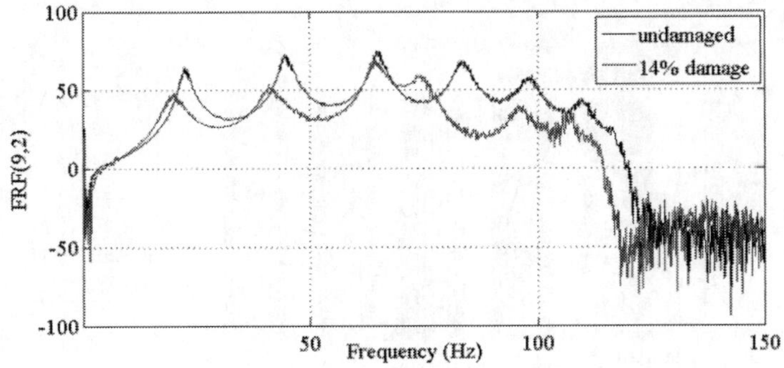

Figure 2. FRF(9, 2) for the structure with and without damage under 6V excitation.

Figure 3 shows the FRF (9, 2) for the structure without damage under 6V, 5V, 4V, and 3V excitation. From this figure, clear differences exist between these FRFs in the light of different voltage excitation. Similar phenomenon can be found in Figure 4 for the case with 14% damage. In theory, FRFs should be stable without change in the existing damage and nonlinearity. Herein, these difference can attribute to small nonlinearities, operational errors, or systematic errors. In general, these factors are all summarized as novelty in this study. Further investigation is indispensable in unveiling the kernel.

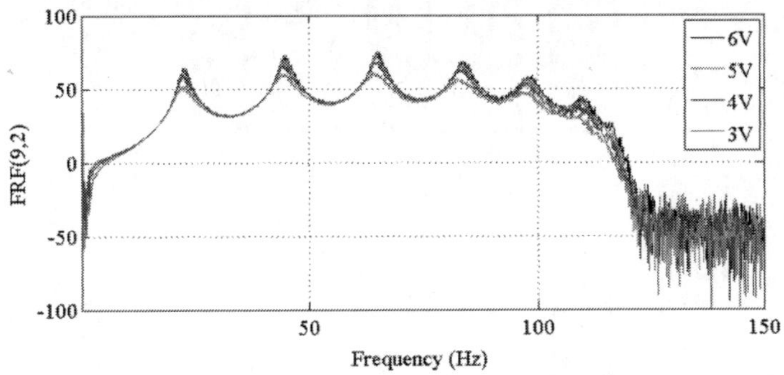

Figure 3. FRF(9, 2) for the structure without damage under 6V, 5V, 4V, and 3V excitation.

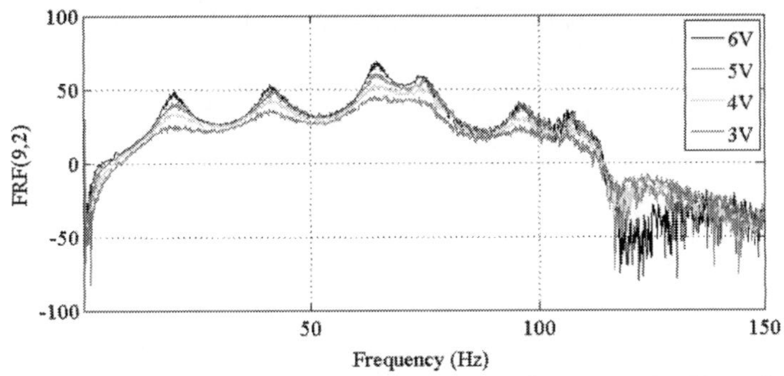

Figure 4. FRF(9, 2) for the structure with damage under 6V, 5V, 4V, and 3V excitation.

Figures 5, 6, and 7 demonstrate the CI (Q=1), CI (Q=2), and CI (Q=4) for undamaged and damaged scenarios, under 6V, 5V, 4V and 3V excitation. All novelties are successfully detected, like damaged scenarios for these three CIs, and as Q increases from 1 to 2, and 4, the CI performances become better as the difference of CI between undamaged and damaged scenarios increases. For the novelty induced between different voltage excitations, as the excitation voltage decreases from 6V, 5V, 4V, to 3V, the CIs increase. For different FRFs, the CIs show small variations without changing the general trend.

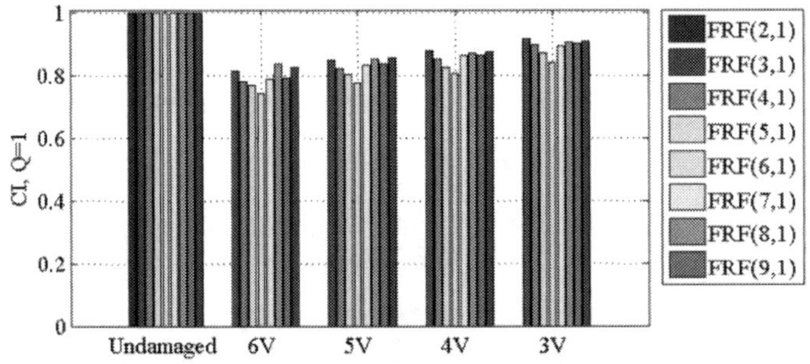

Figure 5. CI (Q=1) for the structure for undamaged and damaged, under 6V, 5V, 4V, and 3V excitation.

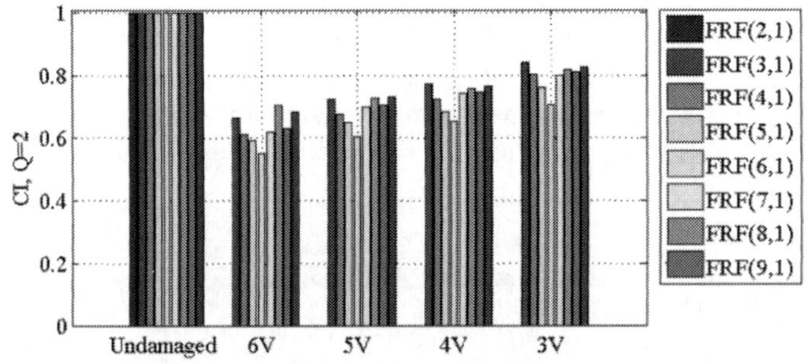

Figure 6. CI (Q=2) for the structure for undamaged and damaged, under 6V, 5V, 4V, and 3V excitation.

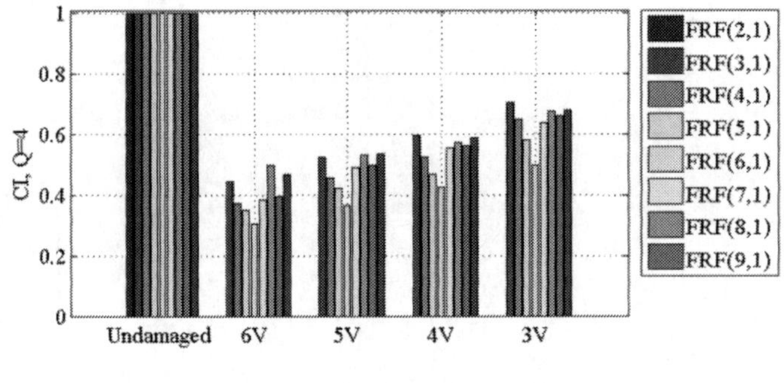

Figure 7. CI (Q=4) for the structure for undamaged and damaged, under 6V, 5V, 4V, and 3V excitation.

5. Conclusions

This study applies the cosine extended indicator in FRFs for detecting the novelties. Compared with an experimental investigation on an eight degree-of-freedom structure, this study demonstrates that the proposed novelty detection procedure performs well in identifying the novelties, regardless of the type of damages. The sensitivity of CI depends on the Q, which relies heavily on the engineering experience.

Acknowledgement

The first author thanks CWO (Commissie Wetenschappelijk Onderzoek), Faculty of Engineering and Architecture, Ghent University for providing financial support for a research stay at Soete Laboratory.

ReferenceUncategorized References

[1] Sohn H, Farrar CR, Hemez FM, Shunk DD, Stinemates SW, Nadler BR and Czarnecki JJ 2004 *Technical Report LA-13976-MS* Los Alamos National Laboratory
[2] Zhou Y-L, Maia NMM and Wahab MA 2016 *Journal of Vibration and Control* (In press)
[3] Zhou YL, Perera R and Sevillano E 2012 *Proceedings of the 6th European Workshop on Structural Health Monitoring* Dresden, Germany, July 2012
[4] Zhou YL and Perera R 2013 *Proceedings of the 5th European-American Workshop on Reliability of NDE* Berlin, Germany, October 2013
[5] Zhou YL, Figueiredo E, Maia N, Sampaio R and Perera R 2014 *Proceedings of International Conference on Noise and Vibration Engineering (Isma2014) and International Conference on Uncertainty in Structural Dynamics (Usd2014)* 527-535
[6] Zhou YL and Perera R 2014 *Proceedings of the 9th European Conference on Structural Dynamics (EURODYN 2014)* Oporto, Portugal, June 2014
[7] Zhou YL, Figueiredo E, Maia N, Sampaio R and Perera R 2015 *Structural Control & Health Monitoring* **22**(10) 1209-1222
[8] Zhou Y-L, Figueiredo E, Maia NMM and Perera R 2015 *Shock and Vibration* **2015**(Article ID 290714)
[9] Zhou Y-L, Wahab MA and Perera R 2015 *International Journal of Fracture Fatigue and Wear* **3** 254-259
[10] Zhou YL 2015 *PhD thesis*
[11] Zhou Y-L and Wahab MA 2016 *Journal of Vibroengineering* **18**(7) 4491-4499
[12] Zhou Y-L, Maia NMM, Sampaio R and Wahab MA 2016 *Structural health monitoring - An International Journal* (In press)
[13] Zhou Y-L, Maia NMM, Sampaio R, Qian XD and Wahab MA 2016 *Proceedings of the 2016 Leuven Conference on Noise and Vibration Engineering (ISMA 2016)* **Leuven, Belgium, September 2016.**
[14] Zhou Y-L, Yang XB and Wahab MA 2016 *International Journal of Fracture Fatigue and Wear* **4** 200-203
[15] Khatir S, Belaidi I, Serra R, Abdel Wahab M and Khatir T 2016 *Journal of Vibroengineering* **18**(1) 202-213
[16] Khatir A, Tehami M, Khatir S and Abdel Wahab M 2016 *Journal of Vibroengineering* **18**(8) 5063-5073
[17] Gillich G-R, Praisach Z-I, Abdel Wahab M, Gillich N, Mituletu IC and Nitescu C 2016 *Shock and Vibration* **2016**(Article ID 2086274) 10 pages; http://dx.doi.org/10.1155/2016/2086274
[18] Khatir S, Belaidi I, Serra R, Abdel Wahab M and Khatir T 2015 *Mechanika* **21**(6) 472-479
[19] Zhou Y-L and Wahab MA 2017 *Engineering Structures* **2017**(141) 175-183
[20] http://www.lanl.gov/projects/national-security-education-center/engineering/ei-software-download/thanks.php
[21] Worden K, Allen DW, Sohn H, Stinemates DW and Farrar CR 2002 *Los Alamos National Laboratory Report LA-13903-MS*

A proposal for unification of fatigue crack growth law

V. Kobelev

University of Siegen, D-57076, Siegen, Germany

kobelev@imr.mb.uni-siegen.de

Abstract. In the present paper, the new fractional-differential dependences of cycles to failure for a given initial crack length upon the stress amplitude in the linear fracture approach are proposed. The anticipated unified propagation function describes the infinitesimal crack length growths per increasing number of load cycles, supposing that the load ratio remains constant over the load history. Two unification fractional-differential functions with different number of fitting parameters are proposed. An alternative, threshold formulations for the fractional-differential propagation functions are suggested. The mean stress dependence is the immediate consequence from the considered laws. The corresponding formulas for crack length over the number of cycles are derived in closed form.

1. Introduction

The functional relation between the loading properties, stress gradient and the physical time-depending characteristics of materials are essential for the statistical fatigue analysis and analytical evaluation number of the cycles to failure [1-2]. The traditional evaluations concepts based on the Palmgren-Miner's rule of damage accumulation, rain-flow counting of time-dependent loads, Wöhler curve and Basquin equation, Paris-Erdogan law and diverse extensions of the fracture and damage mechanics approaches [3]. The book [4] reviews the numerical treatment of fatigue microscopic crack propagation together with their implementation in fatigue-life prediction models. It was recently pointed out that fracture and fatigue are not scale invariant [5]. An attempt to account the scale-invariance in the traditional fatigue approach is performed in this paper. The approach is based on the introduction of fractal length that divides the crack propagation into two characteristic scales.

There are three stages of fatigue fracture commonly distinguished: initiation, propagation, and final rupture [6-7]. The first stage (stage I) of fatigue is referred to as initiation. Initiation is probably the most complex stage of fatigue fracture. The most significant factor about the initiation stage of fatigue fracture is that the irreversible alterations in the metal are caused by cyclic shear stresses. The accumulation of microscopic faults over a large number of load applications, leads to cumulative damage. At the location of a severe stress concentration, the number depends on the geometry of the part as well as on environmental, stress, metallurgical, and strength conditions, as will become apparent. During the stage I, the spread of fatigue crack per unit cycle decelerates smoothly with number of cycles and the process approaches its second stage.

The second stage of fatigue is known as crack propagation (stage II). The propagation stage of fatigue causes the microscopic crack to change direction and grow perpendicular to the tensile stress. The second, or propagation, stage of fatigue is typically the most readily recognizable area of a fatigue fracture. The Paris law describes the stage II propagation law [8]. The traditional form of Paris law pronounces the spread of fatigue crack per unit cycle as a power function of the range of stress intensity factor. Towards the end of the stage II, the spread of fatigue crack per unit cycle begins to accelerate smoothly with number of cycles and the process turns to the third stage.

The final, third stage of fatigue is the final rupture (stage III). As the propagation of the fatigue crack endures, progressively sinking the cross-sectional area of the test specimen, it eventually deteriorates the part so significantly that final, broad fracture occurs with a couple of load cycles. The fracture mode may be either ductile (with a dimpled fracture surface) or brittle (with a cleavage, or intergranular, fracture surface). Occasionally occur the combinations of these modes, depending upon the metal concerned, the stress level and the environment. In the course of the stage III, the spread of fatigue crack per unit cycle progressively accelerates with each cycle.

Numerous load damage models, which extend the celebrated Paris and Erdogan propagation law, were proposed since the 1970s. The article [9] reviews the articles focused on the prediction of fatigue properties of structures under variable amplitude loading. The reliability and accuracy of prediction models and the physical concept of fatigue damage was discussed. The reviewed fatigue life models were based on the scientific and engineering knowledge about fatigue of material and structures under constant and variable amplitude loading.

The near-threshold deviation of the common propagation law was suggested by [10]. The observed rapidly increasing growth towards ductile tearing was accounted. The brittle fracture correction of the power law was proposed in the cited article in the analytical form.

In the work [11] several combinations of high and low amplitude of stress intensity factors values were studied. A similar approach of for transition region between high and low amplitude of stress intensity factors was proposed in [12].

A more general "unified law" accounts certain deviations from the power-law regime [13]. The implemented extension of the Paris law for crack propagation results in the generalized Wöhler fatigue curves.

2. The propagation laws for cyclic loads

The crack propagation is calculated as a function of the range of the stress intensity factor:

$$K = K_{max} - K_{min} .$$

The factors in this equation are

K_{max} the maximum stress intensity factor and

K_{min} the minimum stress intensity factor per cycle.

The common form of Paris' law quantifies the fatigue life of a specimen for a given particular crack size a. The range of stress intensity factor reads:

$$K = Y\sigma\sqrt{\pi a} .$$

The range of stress intensity factor depends on:

$\sigma = \sigma_{max} - \sigma_{min}$ the stress range,

σ_{max} the maximum stress,

σ_{min} the minimum stress per cycle

Y a dimensionless parameter that reflects the geometry.

The parameter Y possesses the value 1 for a center crack in an infinite sheet. Hereafter, this value is assumed for briefness: $Y = 1$.

In the present paper the traditional form for propagation law is implemented:

$$c\frac{dn}{da} = U(K) \qquad \text{or} \qquad c\frac{dn}{da} = U\left(\sigma\sqrt{\pi a}\right). \tag{1}$$

The coefficients in Eq. (1) are:

$c = c(R)$ the material constant for a given stress ratio R,

IOP Conf. Series: Journal of Physics: Conf. Series **843** (2017) 012022 doi:10.1088/1742-6596/843/1/012022

$R = \frac{K_{min}}{K_{max}} < 1$ the stress ratio of cyclic load,

$K_m \equiv \frac{K_{max} + K_{min}}{2} = \frac{1+R}{1-R}\frac{K}{2}$ the mean value of stress intensity factor.

The *propagation function* $U(K)$ describes the infinitesimal crack length growths per increasing number of load cycles n, supposing that the load ratio R remains constant over the load history. It is presumed that the cracks with the initial length δ exist in the material. The cycle count starts with the beginning of crack propagation from its initial length:

$$n\big|_{a=\delta} = 0. \tag{2}$$

Thus, the range K and the mean value K_m of stress intensity factor remain constant. The variations of range K and the mean value K_m of stress intensity factor over the load history could be accounted by applications of damage accumulation theories [14-15].

3. Fractal crack equation (type I)

The *propagation function* $U(K)$ of the first type

$$\begin{aligned} L_1[n(a)] &\equiv \frac{dn}{da} + \kappa^{\alpha-1}\frac{d^\alpha n}{da^\alpha} \\ L_1[n(a)] &= \frac{U(K)}{c}, \qquad U(K) = K^{-p}, \end{aligned} \tag{3}$$

where the stress intensity factor, corresponds to the transition to VHCF regime is

$$\kappa = \sqrt{\pi\lambda}\ \sigma, \tag{4}$$

λ is the characteristic transition length.

The method is used to calculate the fractional derivative using the Davison-Essex definition [19], that is, first differentiate n times, and then integrate $n-\alpha$ times:

$$\frac{d^\alpha n(a)}{da^\alpha} = \frac{1}{\Gamma(n-\alpha)}\int_0^a (a-t)^{n-\alpha-1}\frac{d^n n(t)}{dt^n}dt, \tag{5}$$

where n smallest integer greater than or equal to a number α , $\alpha \le n$. This definition handles differentiation orders $\alpha > -1$.

At first assume that $0 < \alpha < 1$. The definition of the fractional derivative in the equation (3) leads to the integral-differential equation:

$$\frac{dn}{da} + \frac{(\pi\lambda\sigma^2)^{\alpha-1}}{\Gamma(1-\alpha)}\int_0^a (a-t)^{-\alpha}\frac{d^n n(t)}{dt^n}dt - \frac{(\pi\sigma^2 a)^{-p/2}}{c} = 0 \tag{6}$$

Application of the Laplace transform to the equation (6) leads to the following linear algebraic equation

$$\mathcal{L}\left\{\frac{dn}{da}+\frac{\left(\pi\lambda\sigma^2\right)^{\alpha-1}}{\Gamma(1-\alpha)}\int_0^a (a-t)^{-\alpha}\frac{d^n n(t)}{dt^n}dt-\frac{\left(\pi\sigma^2 a\right)^{-p/2}}{c}\right\}= \tag{7}$$
$$=s^{\frac{p}{2}-1}\left(\pi\sigma^2\right)^{\frac{p}{2}}\Gamma\left(1-\frac{p}{2}\right)+c\,n_0\left[1+\left(s\pi\lambda\sigma^2\right)^{\alpha-1}\right]-c\left[s+\left(\pi\lambda\sigma^2\right)^{\alpha-1}s^{\alpha}\right]\tilde{n}(s)=0$$

in terms of a Laplace transform of the function $\tilde{n}(s)=\mathcal{L}\{n(a)\}$ with an unknown constant n_0. The solution of the equation (7) delivers the transformed function:

$$\tilde{n}(s)=\frac{1}{c}\frac{s^{\frac{p}{2}-1}\left(\pi\sigma^2\right)^{\frac{p}{2}}\Gamma\left(1-\frac{p}{2}\right)}{s+\left(\pi\lambda\sigma^2\right)^{\alpha-1}s^{\alpha}}+\frac{n_0}{s}. \tag{8}$$

The application of Bromwich integral inverse Laplace transform of the function (8) results the solution of fractional differential equation (3):

$$n_{1,1}(a)=\mathcal{L}^{1}\{n(s)\}=\frac{\left(\pi\lambda\sigma^2\right)^{1-p}}{c\,\Gamma\left(-\frac{p}{2}+\frac{3}{2}\right)}\exp\left(\frac{a}{\pi\lambda\sigma^2}\right)\cdot$$
$$\left[\Gamma\left(-\frac{p}{2}+\frac{3}{2},\frac{a}{\pi\lambda\sigma^2}\right)\Gamma\left(-\frac{p}{2}+1\right)-\Gamma\left(-\frac{p}{2}+1,\frac{a}{\pi\lambda\sigma^2}\right)\Gamma\left(-\frac{p}{2}+\frac{3}{2}\right)\right]+n_0, \tag{9}$$
$$N_{1,1}(a_1,a_2)=n_{1,1}(a_1)-n_{1,1}(a_2)$$

One limit case corresponds to an infinite scaling parameter λ:

$$\kappa^{\alpha-1}\frac{d^{\alpha}n}{da^{\alpha}}-\frac{\left(\pi\sigma^2 a\right)^{-p/2}}{c}\equiv\frac{\left(\pi\lambda\sigma^2\right)^{\alpha-1}}{\Gamma(1-\alpha)}\int_0^a (a-t)^{-\alpha}\frac{d^n n(t)}{dt^n}dt-\frac{\left(\pi\sigma^2 a\right)^{-p/2}}{c}=0. \tag{10}$$

The application of the Laplace transform to the Eq. (10) and leads to the transformed function

$$\tilde{n}(s)=\frac{1}{c}\left[s^{\frac{p}{2}-1-\alpha}\left(\pi\sigma^2\right)^{\frac{p}{2}+1-\alpha}\lambda^{1-\alpha}\Gamma\left(1-\frac{p}{2}\right)\right]+\frac{n_0}{s}. \tag{11}$$

The inverse Laplace transform of (11) provides the expression for the asymptotic limit of the function (9) for an infinite scaling parameter λ:

$$N_{1,2}(a_1,a_2)=\frac{\left(\pi\sigma^2\right)^{\frac{1-p}{2}}\Gamma\left(-\frac{p}{2}+1\right)}{c\,\Gamma\left(-\frac{p}{2}+\frac{3}{2}\right)}\sqrt{\lambda}\left[a_1^{\frac{1-p}{2}}-a_2^{\frac{1-p}{2}}\right]. \tag{12}$$

The other limit case corresponds to an infinitesimally small scaling parameter λ:

$$\frac{dn}{da}-\frac{\left(\pi\sigma^2 a\right)^{-p/2}}{c}=0 \tag{13}$$

with an elementary solution

$$N_{1,3}(a_2,a_1)=\frac{2(\pi\sigma^2)^{\frac{p}{2}}}{c(2-p)}\left[a_2^{1-\frac{p}{2}}-a_1^{1-\frac{p}{2}}\right]. \tag{14}$$

4. Fractal crack equation (type II)

The second formulation presumes the application of the fractional differentiation operator L_2 to the right side of the Eq. (1):

$$\frac{dn}{da}=L_2\left[\frac{K^{-p}}{c}\right], \qquad L_2\left[\frac{K^{-p}}{c}\right]=\frac{K^{-p}}{c}-\kappa^{-\alpha}\frac{d^\alpha}{da^\alpha}\left(\frac{K^{-p}}{c}\right). \tag{15}$$

The definition of the fractional derivative leads an ordinary inhomogeneous differential equation of the first order:

$$\frac{dn}{da}-\frac{(\pi\sigma^2 a)^{-p/2}}{c}+\lambda^\alpha\frac{\Gamma(1-p/2)}{\Gamma(\alpha+1-p/2)}\frac{(\pi\sigma^2 a)^{\alpha-p/2}}{c}=0. \tag{16}$$

The Laplace transform (16) brings the following expression

$$\tilde{n}(s)=\frac{\Gamma\left(1-\frac{p}{2}\right)}{c}\left[-s^{\frac{p}{2}-2-\alpha}(\pi\sigma^2)^{\frac{p}{2}-\alpha}\lambda^{-\alpha}+s^{\frac{p}{2}-2}(\pi\sigma^2)^{\frac{p}{2}}\right]+\frac{n_0}{s}, \tag{17}$$

which results in the formula for the number of cycles

$$n_{2,1}(a)=\frac{\Gamma\left(1-\frac{p}{2}\right)}{c}\left[\frac{(\pi\sigma^2)^{-\frac{p}{2}}a^{1-\frac{p}{2}}}{\Gamma\left(2-\frac{p}{2}\right)}-\frac{(\pi\sigma^2)^{\frac{\alpha-p}{2}}a^{1+\frac{\alpha-p}{2}}}{\Gamma\left(2+\alpha-\frac{p}{2}\right)}\lambda^\alpha\right]+n_0, \tag{18}$$

$$N_{2,1}(a_2,a_1)=n_{2,1}(a_2)-n_{2,1}(a_1).$$

The limit case corresponds to an infinite scaling parameter λ:

$$\frac{dn}{da}=\lambda^\alpha\frac{\Gamma(1-p/2)}{\Gamma(\alpha+1-p/2)}\frac{(\pi\sigma^2 a)^{\alpha-p/2}}{c}=0 \tag{19}$$

with the corresponding solution

$$N_{2,2}(a_1,a_2)=\frac{\Gamma\left(1-\frac{p}{2}\right)}{c\Gamma\left(2+\alpha-\frac{p}{2}\right)}(\pi\sigma^2)^{\frac{\alpha-p}{2}}\lambda^\alpha\left[a_1^{1+\frac{\alpha-p}{2}}-a_2^{1+\frac{\alpha-p}{2}}\right]. \tag{20}$$

The other limit case corresponds to an infinitesimally small scaling parameter λ leads once again to the equation (13) with an already known solution
$N_{1,3}(a_2,a_1)=N_{2,3}(a_2,a_1)$.

5. Effect of stress ratio

The load ratio is defined as the ratio of the algebraically minimum over the maximum load, Eq. (1). The experiments demonstrate that the load ratio affects the fatigue crack growth and threshold behaviour. Namely, the fatigue crack propagation rate and threshold value vary with the applied load ratio (Walker, 1970). If the load ratio is positive, the experiments reveal that the necessary stress intensity factor range for growth decreases with increasing positive values. In the region of the negative load ratio the required stress intensity factor range for growth (threshold stress intensity range) decreases as load ratio decreases. Reaching a definite negative value, known as saturation point, the required stress intensity factor range for fatigue crack growth stabilizes.

Consider now the variation of load ratio for the proposed propagation laws. The closed form solution for the cycles to failure for a given initial crack length upon the stress amplitude could be found for given load ratio.

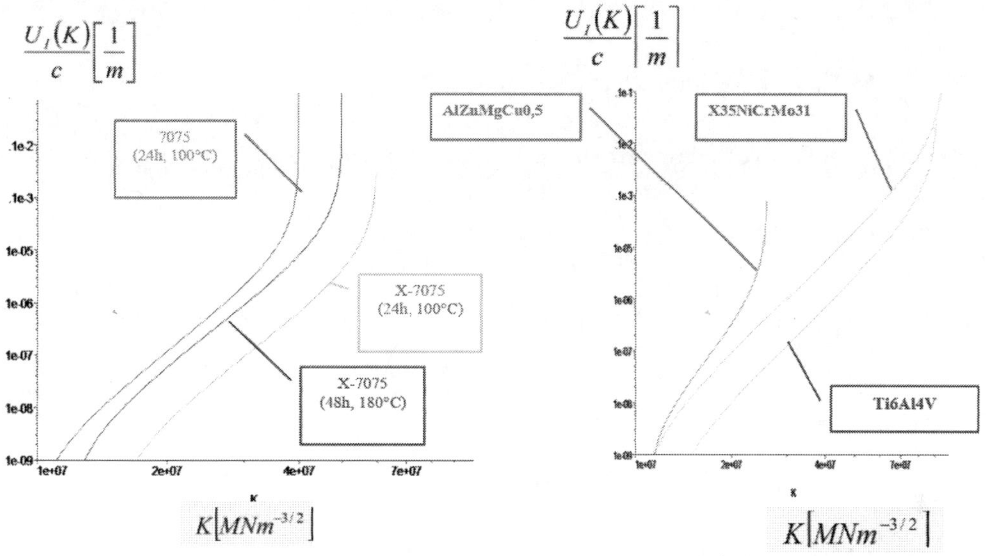

Fig. 1 The plot of the relations **between the crack growth rate** $dn/da = U_I(K)/c$ **and the range of stress intensity factor** K **for different materials [17,18]**

Instead of an empirical approach, in this paper we use the equations to derive the influence of the stress ratio on the crack growth rate explicitly. For this purpose consider two harmonically varying loads with equal amplitude, but with the different mean values. The mean values and amplitudes are assumed to be positive, such that:

$$K_m > 0, \qquad K_m^* > 0, \qquad K > 0.$$

Both loads lead to the harmonically varying stress intensity factors. The harmonically varying stress intensity factors correspondingly are:

$$K(t) = K_m + \tfrac{1}{2}K\sin t, \qquad K^*(t) = K_m^* + \tfrac{1}{2}K\sin t. \tag{21}$$

For both considered cyclic loads the ratios are equal correspondingly to:

$$R = \frac{K_m - \frac{1}{2}K}{K_m + \frac{1}{2}K}, \qquad R^* = \frac{K_m^* - \frac{1}{2}K}{K_m^* + \frac{1}{2}K}. \tag{22}$$

Consider at first the positive load cycles

$$K(t) > 0, \qquad K^*(t) > 0.$$

For the positive load cycles the stress ratios are in the ranges:

$$-1 < R < 1, \qquad -1 < R^* < 1. \tag{23}$$

The damage caused by the varying load depends linearly upon the stress intensity factor in the power p along the cycle. Consequently, the relation between fatigue coefficients, which correspond to the different stress ratios $c(R)$ and $c(R^*)$, reduces to:

$$\Lambda\left[R, R^*\right] = \frac{c(R)}{c(R^*)},$$

$$c(R) = \frac{c_0}{2\pi}\int_0^{2\pi} \left(K_m + \tfrac{1}{2}K\sin t\right)^p dt, \tag{24}$$

$$c(R^*) = \frac{c_0}{2\pi}\int_0^{2\pi} \left(K_m^* + \tfrac{1}{2}K\sin t\right)^p dt.$$

The integrals in (19) are expressed in term of the Meijer G- function [20]:

$$\frac{1}{2\pi}\int_0^{2\pi} \left(K_m + \tfrac{1}{2}K\sin t\right)^p dt = \mathbf{G}_{pq}^{mn}\left(\left(\frac{K}{2K_m}\right)^2 \middle| \begin{matrix} \frac{1-p}{2}, -\frac{p}{2} \\ 1 \end{matrix}\right) K_m^p.$$

The parameters of Eq. (24) are:

$$\frac{1-R}{1-R} = \frac{K}{2K_m}, \qquad \frac{1-R^*}{1-R^*} = \frac{K}{2K_m^*}. \tag{25}$$

Consequently the relation between fatigue coefficients for two different load ratios is:

$$\Lambda\left[R, R^*\right] = \frac{\lambda[R]}{\lambda[R^*]}, \tag{26}$$

where

$$\lambda[R] = \mathbf{G}_{pq}^{mn}\left(\frac{(1-R)^2}{(1+R)^2} \middle| \begin{matrix} \frac{1-p}{2}, -\frac{p}{2} \\ 1 \end{matrix}\right) \frac{(1+R)^p}{(1-R)^p}.$$

6. Threshold-limited formulations for propagation functions

For some alloys there is a theoretical value for stress amplitude below which the material will not fail for any number of cycles. In this paper the method of representation of crack propagation functions through appropriate elementary functions is employed. The choice of the elementary functions is motivated by the phenomenological data and covers broad region possible parameters. With the introduced crack propagation functions differential equations describing the crack propagation are solved rigorously. The resulting closed form solutions allow the evaluation of crack propagation histories on one side, and the effects of stress ratio on crack propagation, on the other side.

Two new unified functions (type I and type II) were proposed in the paper (Kobelev, 2017). The functions were suggested in the form that incorporates the three commonly accepted stages of fatigue. The advantage of the newly proposed functions is the closed form solution of crack propagation.

The simplified representation of unified propagation function for the damage growths per cycle accounts the Paris propagation law together with transition regions at high and low amplitudes of stress intensity factors:

$$U(K) = K^{-p} \frac{1-\left(\frac{K}{K_2}\right)^m}{1-\left(\frac{K_1}{K}\right)^m}. \tag{27}$$

$m_2 > 1$ is the exponent at the short-term limit,

$m_1 > 1$ is the endurance limit exponent.

The function (27) delivers the closed form solution of the ordinary differential equation with the initial condition (2):

$$n = n(a, \delta, \sigma). \tag{28}$$

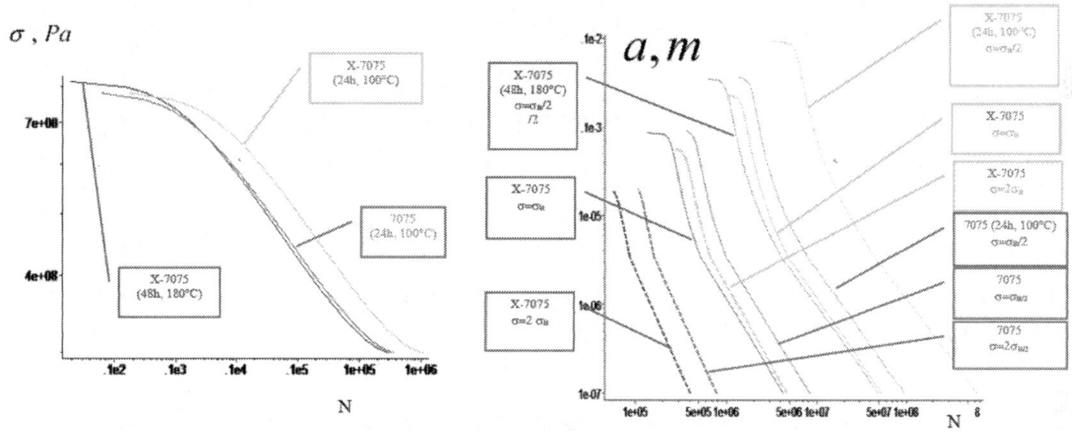

Fig. 2 The dependencies between the cycles to failure N for the given initial crack lengths upon the stress amplitude σ (s-N-curve, left) and cycles to failure N for the given stresses upon the length of crack (a-N-curve, right)

Precisely, the function $n(a, \delta, \sigma)$ delivers the number of cycles for growth of the crack length from the initial value δ to the given value a, assuming that the stress range σ and mean stress

$$\sigma_m = \frac{\sigma_{\max} + \sigma_{\min}}{2} \tag{29}$$

remain constant.

From the equation (1) two ultimate crack sizes could be determined immediately. Firstly, the value:

$$\pi a_2 = K_2^2 / \sigma^2$$

is the critical crack length at which instantaneous fracture will occur. Secondly, the value:

$$\pi a_1 = K_1^2 / \sigma^2$$

is the initial crack length at which fatigue crack growth starts for the given stress range. The number of cycles to failure at $a_1 = \delta$ is infinite. For a finite number of cycles before fracture the initial crack length δ must satisfy the condition:

IOP Conf. Series: Journal of Physics: Conf. Series **843** (2017) 012022 doi:10.1088/1742-6596/843/1/012022

$$a_1 < \delta < a_2 \ .$$

The relations between the crack growth rate

$$\frac{dn}{da} = \frac{U(K)}{c} \quad \text{with} \quad K = \sqrt{\pi a}\ \sigma$$

and the range of stress intensity factor K for simulated materials are shown on the **Fig.1**. The endurance and short time threshold exponents were experimentally acquired [17], alloys AlZnMgCu0,5, Ti6Al4V, X35NiCrMo31) and [18], aluminum alloys X-7075 (180°), X-7075 (100°), 7075 (100°).

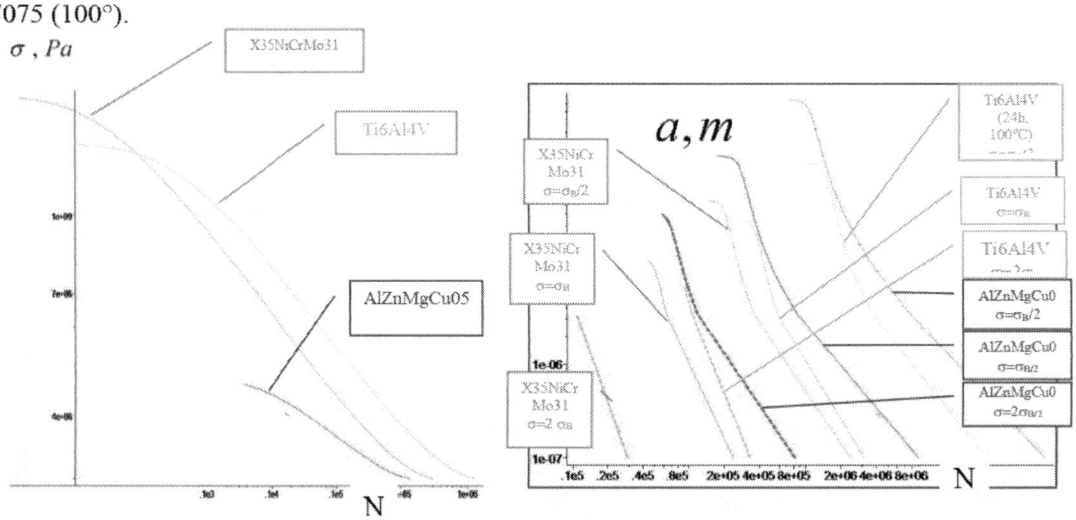

Fig. 3 The dependencies between the cycles to failure N for the given initial crack lengths upon the stress amplitude σ (s-N-curve, left) and cycles to failure N for the given stresses upon the length of crack (a-N-curve, right)

The solution of the ordinary differential equation (27) with the corresponding initial condition delivers closed form analytical expression for the remaining number of cycles to fracture:

$$n(a,\delta,\sigma) = \tilde{n}(a,\sigma) - \tilde{n}(\delta,\sigma), \tag{30}$$

where the auxiliary functions are:

$$\tilde{n}(a,\sigma) = \frac{2aK^{-p}}{c\,m}\left[\frac{K^m}{K_2^m}\mathbf{G}^{mn}_{pq}\left(\frac{K^m}{K_1^m}\middle|\begin{matrix}0, & \dfrac{p-2}{m}\\ 0, & \dfrac{-m+p-2}{m}\end{matrix}\right) - \mathbf{G}^{mn}_{pq}\left(\frac{K^m}{K_1^m}\middle|\begin{matrix}0, & \dfrac{m+p-2}{m}\\ 0, & \dfrac{p-2}{m}\end{matrix}\right)\right], \tag{31}$$

In the expression (31) appears the Meijer G- function [19]. The following example illustrates the behaviour of the metallic material with the unified propagation function (27). The diagrams that express the dependence of cycles to failure, S-N curves are presented on the pictures on left side of **Fig.2** and **Fig.3**. The same initial length of the crack was assumed for definiteness in all cases $\delta = 10^{-4} m$. The stress intervals for calculation depend on threshold stress intensity factors:

$$\sigma_1 < \sigma < \sigma_2, \qquad \sigma_1 = K_1\sqrt{\pi\delta}\ , \qquad \sigma_2 = K_2\sqrt{\pi\delta}\ .$$

The dependencies of crack lengths on cycles, a-N curves are drawn on the right pictures. The number of cycles was calculated as the function of length of the crack in the interval

$$10^{-7} m < a < 10^{-2} m .$$

The NSa surface N-S and curves and a-N curves are shown on the **Fig.2** for aluminium alloys X-7075 (180°), X-7075 (100°), 7075 (100°) from [17]. The S-N curves and a-N curves are displayed on the **Fig.3** for aluminium alloy AlZnMgCu0,5, titanium alloy Ti6Al4V, and steel X35NiCrMo31 [18].

7. Conclusion

The paper introduces the closed form analytical expression for crack length over number of cycles. Two new fractional-differential functional dependences express the damage growth per cycle are introduced. These functions allow the unification of different fatigue laws in a single expression. The unified fatigue law provides closed form analytical solutions for crack length upon the mean value and range of cyclic variation of stress intensity factor. The solution expresses the number of cycles to failure as the function of the initial size of the crack and eliminates the solution of nonlinear ordinary differential equation of the first order. The explicit formulas for stress against the number of cycles to failure are delivered for both proposed unified fatigue laws.

The different common expressions, which account the influence of the stress ratio, are immediately applicable. For the proposed unified propagation functions, the ranges for the stress load factor are extended using the introduction of effective stress intensity. The solution leads to the effective stress intensity factor range, effective mean value of stress intensity factor and effective stress ratio.

The proposed methods are applicable for the modeling of tribological effects on fatigue. Fretting fatigue is an outcome involving different influences. The major factors that effect fretting fatigue are slip amplitude, coefficient of friction, shear load and loading phase difference [21-24]. In the recent paper [25] study, three numerical models were introduced for simulation the effect of both in phase and out of phase loading on contact stresses and damage initiation locations. In the paper [25] was pointed out that phase difference affects the shear traction and tensile stress profiles at the contact interface, whereas no significant effect is observed on convergence efficiency. The methods, which proposed in the currect article, could be applied for the simulation of the fretting fatigue processes.

The proposed method is also appropriate for investigation of fatigue behaviour compression springs at a very high number of cycles [26]. The reported variation of the slope in S-N diagram could be explained by the transition to a fractal fracture regime [27,28]. The corresponding models of fracture under static load conditions were proposed in [29]

References

[1] Carpinteri A., Ed. 1994 *Handbook of Fatigue Crack Propagation in Metallic Structures*, Elsevier Science B.V. Philadelphia, PA
[2] Suresh S. 1998 *Fatigue of materials*, Cambridge University Press, Cambridge
[3] Richard H.A., Sander M. 2012 *Ermüdungsrisse*, ISBN 9 78-3-8348-1594-1, DOI 10.1007/978-3-8348-8663-7, Springer-Vieweg, Berlin
[4] Krupp U. 2007 *Fatigue Crack Propagation in Metals and Alloys: Microstructural Aspects and Modelling Concepts*, ISBN: 978-3-527-31537-6, Wiley-VCH, Berlin
[5] Turcotte D. 1997*, Fractals And Chaos in Geology and Geophysics*, Cambridge University Press
[6] Totten G. 2008 Fatigue crack propagation, *Advanced Materials & Processes*, May 2008, p. 39-41
[7] Leese G.E., Socie D., Eds. 1989 *Multiaxial Fatigue, Analysis & Experiments*, SAE Pub. AE-14, Warrendale, PA
[8] Paris P., Erdogan F. 1963 A critical analysis of crack propagation laws. *J. Basic Eng. Trans. ASME*, 528–534
[9] Beden S. M., Abdullah S., Ariffin A. K. 2009 Review of Fatigue Crack Propagation Models for Metallic Components, *European Journal of Scientific Research*, ISSN 1450-216X Vol.28 No.3, pp.364-397

[10] Donahue, R. J. Clark, H. M., Atanmo, P., Kumble, R., McEvily, A. J. 1972 Crack opening displacement and the rate of fatigue crack growth. *Int. J. Fract. Mech.* 8, 209–219, DOI 10.1007/BF0070388

[11] Kanninen M.F., Popelar C.H. 1985 *Advanced Fracture Mechanics*. Oxford University Press, New York

[12] McEvily A.J., Groeger J. 1977 On the threshold for fatigue-crack growth. *Fourth International Conference on Fracture*, vol. 2. University of Waterloo Press, Waterloo, Canada, pp. 1293–1298

[13] Xue L. 2008 A unified expression for low cycle fatigue and extremely low cycle fatigue and its implication for monotonic loading, *International Journal of Fatigue*, 30, 1691–1698

[14] Pugno N., Cornetti P., Carpinteri A. 2007 New unified laws in fatigue: From the Wöhler's to the Paris' regime, *Engineering Fracture Mechanics*, 74, 595–601

[15] Wheeler, O. E. 1972 Spectrum loading and crack growth, *J. Basic Engng*. 94, 181–186

Noroozi A.H., Glinka G., Lambert S. 2005 A two parameter driving force for fatigue crack growth analysis, *International Journal of Fatigue*, Volume 27, Issues 10–12, October–December 2005, Pages 1277–1296

[16] Sorensen A.1969 A General Theory of Fatigue Damage Accumulation, *J. Basic Eng* 91(1), 1-14, doi:10.1115/1.3571021

[17] Schwalbe K.-H. 1980 *Bruchmechanik metallischer Werkstoffe*, Carl Hanser Verlag, München, Wien

[18] Albrecht J., Martin J.W.R., Lüthering G., Martin J.W. 1976 Influence of micromechanisms on fatigue crack propagation rate of Al-alloys, *4th Int. Conf. on the Strength of Metals and Alloys*, Conference Proceedings Nancy, 30 August-3 September 1976, E.N.S.M.I.M, Nancy, 1408 P.

[19] Davison, M., Essex, G. C., 1998, Fractional Differential Equations and Initial Value Problems, *The Mathematical Scientist*, (December 1998): 108-116

[20] Prudnikov, A. P.; Brychkov, Yu; and Marichev, O. 1990 *Integrals and Series, Volume 3: More Special Functions*. New York: Gordon and Breach Science Publishers, 1990.

[21] Kumar D, Biswas R, Poh LH and Abdel Wahab M 2017 Fretting fatigue stress analysis in heterogeneous material using direct numerical simulations in solid mechanics Tribology International 109 124-132

[22] Resende Pereira KdF, Bordas S, Tomar S, Trobec R, Depolli M, Kosec G and Abdel Wahab M 2016 On the Convergence of Stresses in Fretting Fatigue Materials 9 639

[23] Ferjaoui A, Yue T, Abdel Wahab M and Hojjati-Talemi R 2015 Prediction of fretting fatigue crack initiation in double lap bolted joint International Journal of Fatigue 73 66-76

[24] Bhatti NA and Wahab MA 2017 Finite element analysis of fretting fatigue under out of phase loading conditions Tribology International 109 552-562

[25] Hojjati Talemi R and Abdel Wahab M 2013 Fretting fatigue crack initiation lifetime predictor tool: using damage mechanics approach Tribology International 60 176-186

[26] Pyttel B., Schwerdt D., Berger C. 2011 Very high cycle fatigue – Is there a fatigue limit? Int. J. Fatigue, 33, 49–58.

[27] Pyttel B., Brunner I., Kaiser B., Berger C., Mahendran M. 2014 Fatigue behaviour of helicalcompression springs at a very high number of cycles – Investigation of various influences. Int J Fatigue; 60:101-109.

[28] Korn M., Rohm T., Lang K.-H. 2014, Very High Cycle Fatigue Resistance ($N>10^7$) of Middle- and High Strength Heat Treatment Conditions of 42CrMo4, Proc. 6th International Conference on VHCF, October 15–18, 2014, Chengdu, China

[29] Mosolov A.B. (1993) Mechanics of Fractal Cracks in Brittle Solids. Europhys. Lett., 24 (8), pp. 673-678

[30] Yavari A., Sarkani S., Moyer E.T. 2002 The mechanics of self-similar and self-affine fractal cracks International Journal of Fracture 114: 1–27

IOP Conf. Series: Journal of Physics: Conf. Series **843** (2017) 012023 doi:10.1088/1742-6596/843/1/012023

Plasma immersion ion implantation on 15-5PH stainless steel: influence on fatigue strength and wear resistance

R Bonora[1], M O H Cioffi[1] and H J C Voorwald[1]

Materials and Technology Department, Fatigue and Aeronautical Materials Research Group, Engineering Faculty of Guaratinguetá, FEG/UNESP – São Paulo State University, Av. Ariberto Pereira da Cunha, 333 – Guaratinguetá – SP, 12516-410, Brazil.

hjvoorwald@gmail.com

Abstract. Surface improvement in steels is of great interest for applications in industry. The aim of this investigation is to study the effect of nitrogen ion implantation on the axial fatigue strength and wear resistance of 15-5 PH stainless steel. It is well know that electroplated coatings, which are used to improve abrasive wear and corrosion properties, affects negatively the fatigue strength. It is also important to consider requirements to reduce the use of coated materials with electroplated chromium and cadmium, that produce waste, which is harmful to health and environment. The HVOF (High velocity oxygen fuel) process provides hardness, wear strength and higher fatigue resistance in comparison to electroplated chromium. Plasma immersion ion implantation has been used to enhance the hardness, wear, fatigue and corrosion properties of metals and alloys. In the present research the fatigue life increased twice for 15-5 PH three hours PIII treated in comparison to base material. From the abrasive wear tests a lower pin mass reduction was observed, associated to the superficial treatments. The improvement of fatigue and mechanical performance is attributed to a combination of nitrides phase structure and compressive residual stresses during the PIII treatment.

1. Introduction

Surface treatments in mechanical components are considered necessary to improve wear and corrosion properties of applied materials. It is well known that wear and corrosion processes on aircraft structures are controlled by electroplated chromium and cadmium, hard anodizing and, more recently, HVOF thermal spray coatings [1-3]. The decrease in fatigue strength of coated parts is associated to microcracks density, high tensile residual stresses and to the presence of high density of porous and oxide inclusions into the coating, that commonly forms during the process [4-5]. Compressive residual stresses induced in the surface layers of the material after controlled shot peening processes, delay nucleation and propagation of fatigue cracks, increasing materials performance [6-7]. Rotating bending fatigue experimental results indicate that electroplated aluminum and IVD aluminum coatings may replace cadmium electroplating on AISI 4340 steel [8].

The PIII process avoids common problems like microcracks and poor adhesion coating/substrate. The plasma immersion ion implantation and deposition (PIII D) technique was responsible for the good adhesion strength between the nitrogen doped diamond-like carbon (N-DLC) coating and NiTi substrate. The corrosion resistance of NiTi alloys was also improved [9]. The effect of temperature on the surface mechanical properties of H13 steel implanted with nitrogen ions using plasma immersion ion implantation was to improve hardness and wear resistance by increasing the PIII treatment temperature. An opposite behavior was observed in the presence of carbon (C) ion [10]. Results were attributed to nitride formation and lattice expansion. A relatively constant concentration of nitrogen with depth was observed in experiments using AISI 304 stainless steel plates subjected to nitrogen ion implantations by plasma immersion [11]. Experiments performed with AISI 304 stainless steel treated by plasma immersion ion implantation showed that high nitrogen concentration and/or thicker layers improved hardness and wear resistance, whereas simultaneously reduced the surface fatigue strength [12].

Improvement in the mechanical behavior and in the rolling contact fatigue strength of nitrogen PIII treated AISI 52100 bearing steel was attributed to a combination of nitrides phase structure and compressive residual stresses during the process [13].

In the present research, fatigue and wear resistance of 15-5 PH stainless steel nitrogen PIII treated for one, two and three hours was evaluated Axial S-N curves were obtained in fatigue tests. In order to study the effect of residual stresses on the fatigue strength, the stress field was measured by an X-ray tensometry after the treatments. Scanning electron microscopy was used to investigate the fatigue crack initiation points. Wear resistance was evaluated using pin-on-disk apparatus.

2. Experimental procedures

2.1. Material and mechanical properties
The 15-5 PH martensitic stainless steel, which combines high mechanical strength and good corrosion resistance, has the chemical composition according to table 1.

Table 1, Chemical composition (wt%). 15-5 PH stainless steel.

C	Mn	P	S	Si	Cr	Ni	Mo	Nb	Cu
0.07 max	1.0 max	0.03 max	0.015 max	1.0 max	14.0 – 15.5	3.5 – 5.5	0.5 max	0.45 max	2.5 – 4.5

The desired mechanical properties were obtained by a precipitation hardening treatment at 552°C for four hours. Ultimate tensile strength: 1125 MPa and Hardness: 39.5 HRc.

2.2. Plasma immersion ion implantation
A vacuum chamber composes the PIII system with a specimen holder, plasma source and a high voltage pulse modulator. The specimen to be treated is placed in the chamber in which a plasma containing the ions to be implanted is generated. To start the cleaning step with argon gas, the chamber was pumped to a base pressure of 2.0×10^{-2} mbar. The cleaning procedure was concluded in 15 minutes. The nitrogen implantation process was performed by applying negative high voltage pulses of 10.0kV, pulse length 50μs and bombardment frequency 1.5KHz. The average working temperature was 396°C and treatment times were chosen as one, two and three hours. The experimental setup is represented in figure 1.

IOP Conf. Series: Journal of Physics: Conf. Series **843** (2017) 012023 doi:10.1088/1742-6596/843/1/012023

Figure 1. PIII system experimental setup.

The PIII treated specimen surfaces were analyzed by atomic force microscopy. Nanohardness tests were performed with a nanoidenter XP System, with 400mN maximum load. Distribution of indentations occurred in a matrix 5x5, separated by a distance equal to 50μm. To investigate the fatigue crack initiation in the specimen groups and possible correlation with treatment times, scanning electron microscopy was used. Vickers microhardness measurements were performed according to ASTM E384.

2.3. *Axial fatigue tests*

Fatigue tests were conducted to complete fracture or 10^6 load cycles, using a sinusoidal load of 10Hz frequency and load ratio R=-1.0 at room temperature. The test specimen used for the axial fatigue tests, in figure 2, was prepared according to standard ASTM E466.

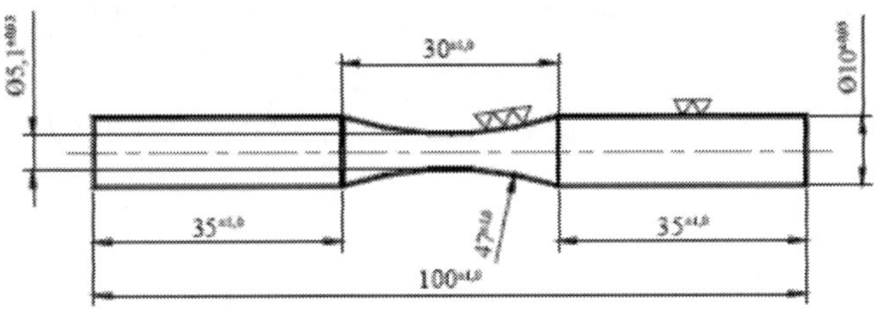

Figure 2. Axial fatigue test specimen.

Four groups of axial fatigue specimens were prepared to obtain S-N curves:

- 12 specimens/15-5 PH stainless steel;
- 12 specimens/15-5 PH stainless steel/ 1 hour PIII treated;
- 12 specimens/15-5 PH stainless steel/ 2 hours PIII treated;
- 12 specimens/15-5 PH stainless steel/ 3 hours PIII treated.

2.4 *Residual stress measurement*

The X-ray diffraction method was used to determine the residual stress field induced by the PIII treatments. The accuracy of the stress measurement was $\Delta\sigma=\pm 10$MPa.

2.5 Abrasive wear test

Dry sliding wear tests were performed on 15-5PH (one, two and three hours PIII treatment)/aluminum-bronze 630 tribological pair, using a pin-on-disk equipment according to ASTM G99 standard. Tribological characterization was conducted with a normal force of 5N and a speed of 0.5m/s. Parameters were collected to measure mass loss of the metal pin and disk using an analytical balance. At every 400m route, specimens were cleaned in ultrasonic bath and weighted. The total distance for each tribological pair was 2000m.

3. Results and discussion

Residual stresses for base metal and 15-5PH stainless steel PIII treated for one, two and three hours are represented in table 2.

Table 2. 15-5 PH stainless steel. Residual stresses.

Specimen	Depth (mm) 0.00	0.01	0.1
Material	**Stress (MPa)**		
15-5 PH	+36	+110	+110
15-5 PH PIII/1H	-115	+30	+43
15-5 PH PIII/2H	-230	+12	+78
15-5 PH PIII/3H	-270	-60	+25

Tensile residual stresses were obtained for 15-5PH stainless steel at surface and 0.01mm/0.1mm depth, respectively 36MPa, 110MPa and 110MPa. Increase in compressive residual stresses at specimen surface occurred associated to increase in the PIII treatment times, respectively -115MPa, -230MPa and -270MPa for one, two and three hours treatments.

For one and two hours PIII treatment, tensile residual stresses are observed for 0.01mm and 0.1mm from surface, but still lower than residual stresses for base metal at the same depth. In the case three hours PIII treatment, -60MPa and 25MPa residual stresses are related to 0.01mm and 0.1mm, respectively.

The axial S-N curves for 15-5PH stainless steel base metal and PIII treated for one, two and three hours, are indicated in figure 3.

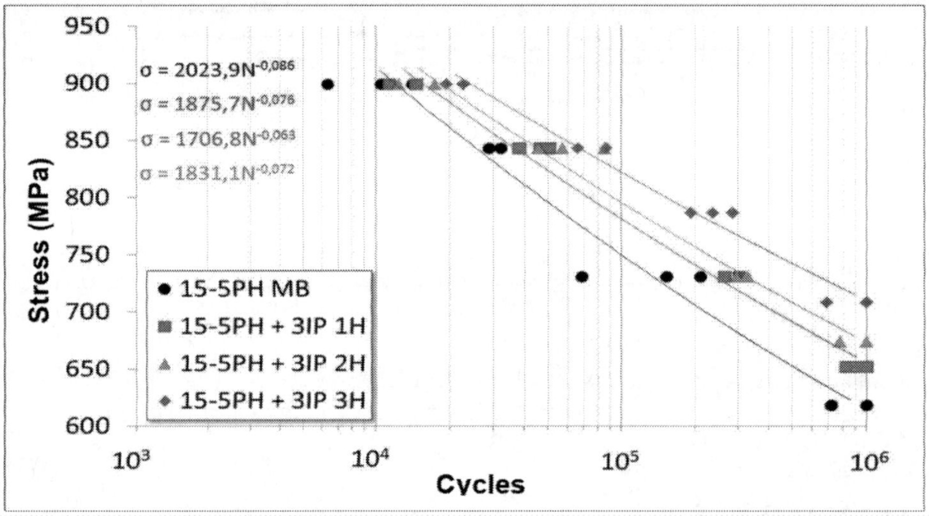

Figure 3. Axial S-N curves. Base metal and 15-5PH stainless steel PIII treated for one, two and three hours.

In table 3, the average number of cycles to failure for 15-5PH stainless steel and base metal PIII treated for one, two and three hours are N, N1, N2 and N3, respectively.

Table 3. Number of cycles to failure. 15-5PH stainless steel.

Cycles Stress (MPa)	15-5PH N	15-5PH/1H N1	15-5PH/2H N2	15-5PH/3H N3
900	10358	13046	14960	21029
843	30409	45021	62773	76029
787	-	-	-	237735
730	144017	283071	315348	-
710	-	-	-	688274
675	-	-	783129	-
650	-	857727	-	-
620	718405	-	-	-

According to figure 3 and table 3, the PIII treatment increased the low and high cycle axial fatigue resistance of 15-5PH stainless steel. Table 3 indicates, for maximum applied stress 900MPa, 10358, 13046, 14960 and 21029 average cycles to failure for 15-5PH stainless steel and base material PIII treated for one, two and three hours, respectively. An improvement in fatigue life twice for PIII/3H was observed in comparison to 15-5PH stainless steel. With respect to the high cycle fatigue regime, table 3 shows fatigue strength for number of cycles to failure near to 10^6, equal to 620 MPa, 650 MPa, 675 MPa and 710 MPa for base material and 15-5PH PIII treated, respectively, for one, two and three hours. As the improvement in fatigue strength is correlated to the compressive residual stresses, which may avoid fatigue crack nucleation or even retard propagation, an increase in fatigue life is expected with higher compressive residual stresses from PIII treatments. According to table 2, higher compressive residual stresses at surface and 0.01mm depth were obtained for 15-5PH stainless steel PIII treated for three hours. Table 3 shows higher number of cycles to failure for this treatment condition. Improvement in the rolling contact fatigue of AISI 52100 bearing steel occurred by nitrogen plasma immersion ion implantation technique [13]. The maximum microhardness and the effective elastic modulus value also increased.

In table 4, increase in the fatigue strength related to the PIII treatments is represented by the ratios N1/N, N2/N and N3/N.

Table 4. Ratios Ni/N.

Ni/N Stress (MPa)	N1/N	N2/N1	N2/N	N3/N1	N3/N	N3/N2
900	1.26	1.15	1.44	1.61	2.03	1.4
843	1.48	1.39	2.06	1.69	2.5	1.21
730	1.96	1.11	2.19	-	-	-

From table 4, for maximum applied stresses 900MPa, 843MPa and 730MPa an increase in the ratio Ni/N is observed with increase in the PIII treatment times one, two and three hours. As an example, for 900MPa, values for Ni/N are 1.26, 1.44 and 2.03, respectively for 15-5PH PIII treated for one, two and three hours respectively. An improvement in Ni/N is observed with the increase in PIII treatment times. The same behavior is indicated in table 4 for 843MPa and 730MPa. The influence of maximum applied stress on Ni/N for each PIII treatment temperature studied is also represented in table 4. In case one hour PIII treatment, values for N1/N are 1.26, 1.48 and 1.96 for 900MPa, 843MPa and 730MPa, respectively. The same tendency occurs for N2/N and N3/N. The conclusion is that improvement in Ni/N is also associated to the decrease in maximum applied stress, from low to high cycle fatigue.

IOP Conf. Series: Journal of Physics: Conf. Series **843** (2017) 012023 doi:10.1088/1742-6596/843/1/012023

SEM photomicrographs for 15-5 PH stainless steel and base metal PIII treated for 1, 2 and 3 hours are shown in figures 4a, 4b, 4c and 4d, respectively.

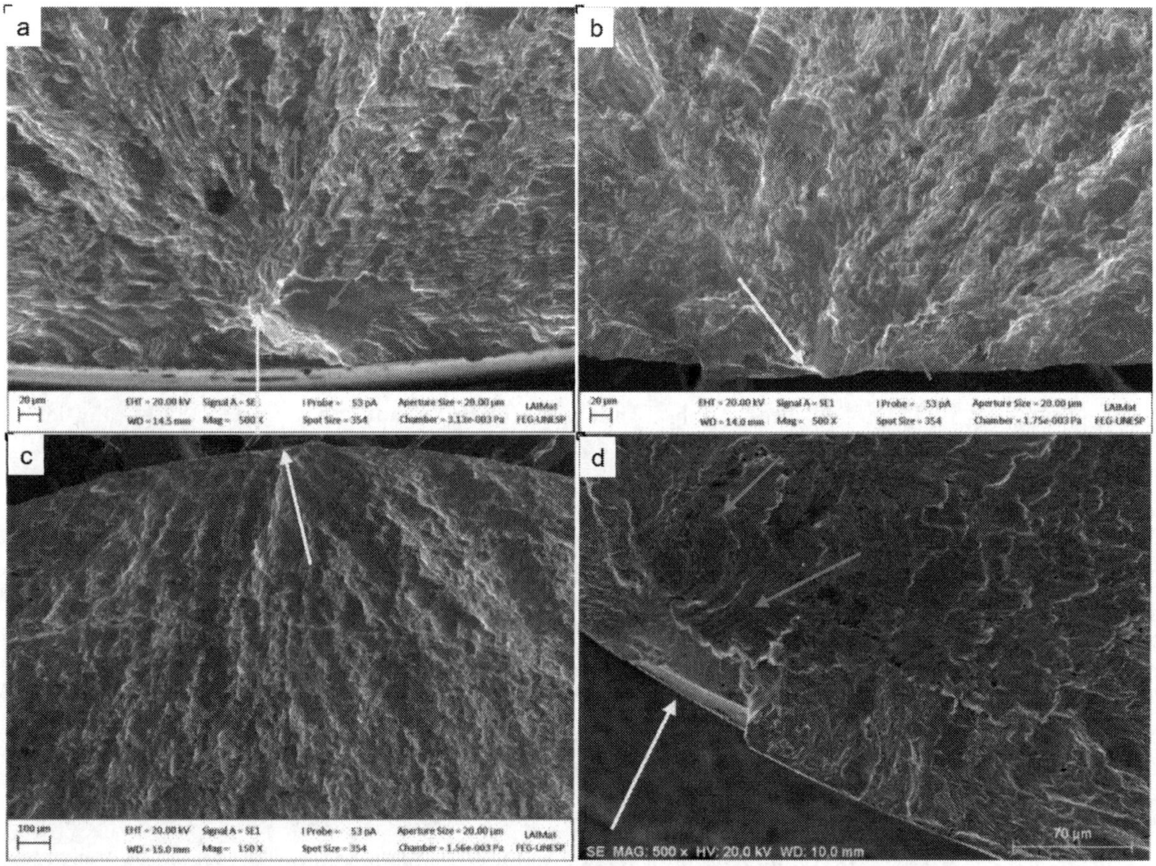

Figure 4. Fatigue fracture surface. a) Base metal, b) PIII/1H, c) PIII/2H, d) PIII/3H.

Figure 4 indicates fatigue crack initiation at surface, for all conditions studied. Results from nanohardness test are indicated in figure 5.

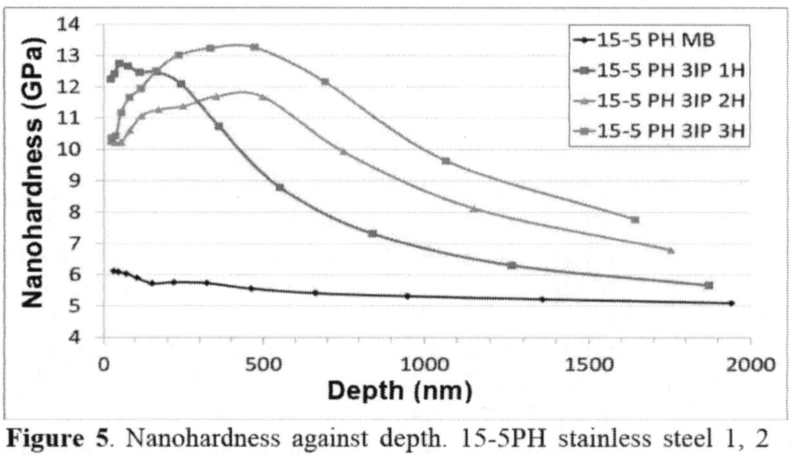

Figure 5. Nanohardness against depth. 15-5PH stainless steel 1, 2 and 3 hours PIII treatments.

An increase in hardness close to the surface is attributed to the specimens preparation. For 500nm depth, considering the base material nanohardness equal to 5GPa, experimental data in figure 4 shows for PIII treatments, 9GPa, 12GPa and 13GPa for one, two and three hours, respectively.

It can be seen in figure 6, that the PIII treatment applied on the 15-5PH disk decreased the pin mass loss in comparison to the base material. Moreover, the highest reduction was obtained with one hour treatment. It can be inferred that the decrease in mass loss is related to two changes caused by the disk PIII treatments: increase in roughness and hardness both in nanometric scale, according to figures 5 and 7. Improved tribological properties were observed for N-PIII into SS304 samples, as a consequence of a surface layer composed of oxygen and nitrogen [12]. Experimental results also indicate improvement in hardness and wear resistance of H13 steel due to the PIII treatment [10].

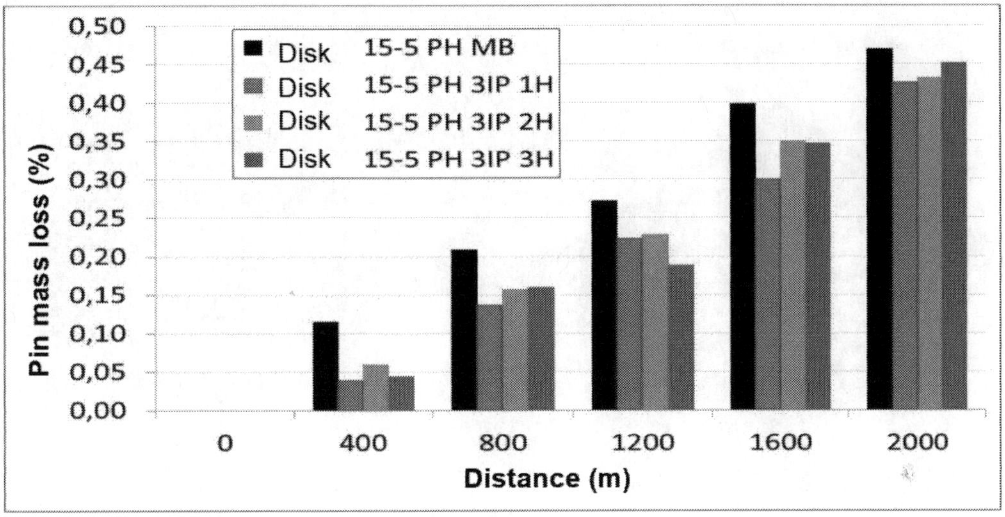

Figure 6. Pin mass loss.

Figure 7 shows atomic force microscopy images of the 15-5PH stainless steel before and after the implantation process.

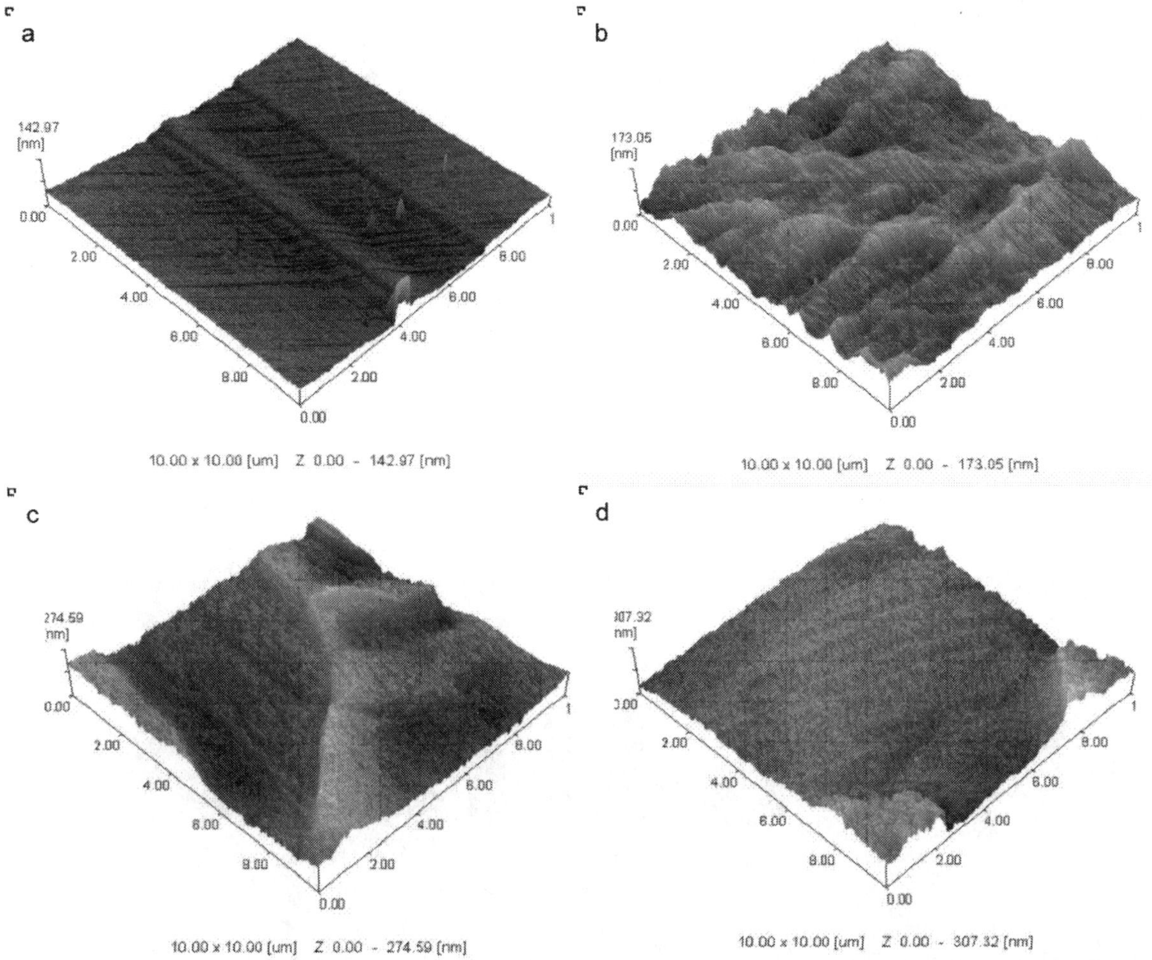

Figure 7. Atomic force microscopy. a) 15-5PH stainless steel, b) PIII/1H, c) PIII/2H, d) PIII/3H.

A characteristic feature observed in the untreated 15-5PH stainless steel is the presence of two regions on the sample surface, maybe originated by the polishing. Possible interference in the measured roughness, 142.9nm, occurred. Surface roughness for PIII treatments were 173nm, 274.6nm and 307.3nm for 1,2 and 3 hours, respectively.

As already mentioned, the increase in surface roughness and material nanohardness associated to the PIII treatments, played an important role on the pair pin-disk abrasive wear results obtained.

4. Conclusions

1. Increase on the 15-5 PH stainless steel axial fatigue strength occurred with PIII treatments for 1, 2 and 3 hours. Higher number of cycles to failure is directly related to the increase in treatment time.
2. Tensile residual stresses are present at base material surface and 0.01/0.1mm depth. Increase in compressive residual stresses at surface was observed for PIII treatments, -115MPa; -230MPa and -270MPa, respectively for 1, 2 and 3 hours.
3. The plasma immersion ion implantation increased, significatively the 15-5 PH stainless steel nanohardness at 500nm depth. Higher nanohardness resulted from longer treatment times.

4. The PIII treatment applied on the 15-5PH disk decreased the pin mass loss in comparison to the base material. Increase in roughness and hardness in nanometric scale, are associated to this behavior.

Acknowledgements

The author H.J.C. Voorwald is grateful to CNPq - National Council for Scientific and Technological Development, process number 303832/2014-2.

References

[1] Souza RC, Voorwald HJC and Cioffi MOH Fatigue strength of HVOF sprayed Cr3C2-25NiCr and WC-10Ni on AISI 4340 steel 2008 *Surf. Coat. Int.* **203** 191-198

[2] Costa MYP, Venditti MLR, Voorwald HJC, Cioffi MOH and Cruz TG 2009 Effect of WC-10%Co-4%Cr coating on the Ti-6Al-4V alloy fatigue strength *Mater. Sci. Eng. A* **507** 29-36

[3] Voorwald HJC, Rocha PCF, Cioffi MOH and Costa MYP 2007 Residual stress influence on fatigue lifetimes of electroplated AISI 4340 high strength steel *Fatigue Fract. Eng. Mater. Struct.* **30** 1084-1097

[4] Vieira LFS, Voorwald HJC and Cioffi MOH 2015 Fatigue performance of AISI 4340 steel Ni-Cr-B-Si-Fe HVOF thermal spray coated *Procedia Eng.* **114** 606-612

[5] Voorwald HJC, Souza RC, Pigatin WL and Cioffi MOH Evaluation of WC-17Co and WC-10Co-4Cr thermal spray coatings by HVOF on the fatigue and corrosion strength of AISI 4340 steel *Surf. Coat. Int.* **190** 155-164

[6] Gagliano M and Vergani L 2004 An approach for prediction of fatigue strength of shot peened components *Eng. Fract. Mech.* **71** 501-512

[7] Curtis S, De Los Rios ER, Rodopoulos LA and Levers A 2003 Analysis of the effects of controlled shot peening on fatigue damage of high strength aluminum alloys *Int. J. Fatigue* **25** 59-66

[8] Voorwald HJC, Peres MPP, Costa MYP and Cioffi MOH 2010 Evaluation on fatigue strength of AISI 4340 steel aluminum coated by electroplating and IVD processes *J. Mater. Sci.* **45** 6094-6100

[9] Yan-Ling P and Yang L 2013 Surface modification of NiTi alloys using nitrogen doped diamond-like carbon coating fabricated by plasma immersion ion implantation and deposition *J. Alloys Compd.* **581** 873-876

[10] Mursen S 2007 Characterization of surface mechanical properties of H13 steel implanted by plasma immersion ion implantation *J. Mater. Process. Technol.* **189** 367-373

[11] Castro-Colin M, Durrer W, López JA and Ramirez-Homs E 2016 Surface modification by nitrogen plasma immersion ion implantation on austenitic AISI 304 stainless steel *J. Iron Steel Res. Int.* **23(4)** 380-384

[12] Fernandes BB, Mändl S, Oliveira RM and Veda M 2014 Mechanical properties of nitrogen-rich surface layers on SS 304 treated by plasma immersion ion implantation *Appl. Surf. Sci.* **310** 278-283

[13] Liu H, Tang B, Wang L, Wang X and Jiang B 2007 Fatigue life and mechanical behaviors of bearing steel by nitrogen plasma immersion ion implantation *Surf. Coat. Technol.* **201** 5273-5277

Effects of friction coefficient and shape of impact object on the fatigue life of specimens with impact damage

Zhan Zhixin[1,*], Hu Weiping[1] and Meng Qingchun[1]

[1]School of Aeronautics Science and Engineering, Beihang University, Beijing, 100191, China

[*]zzxupc@163.com

Abstract. This paper proposes a new method based on the continuum damage mechanics to analyse the effects of friction coefficient and shape of impact object on the fatigue life, which considers the impact damage, elastic-plastic damage and coupled effects between the stress field and damage field. Firstly, the impact process is simulated and the initial impact damage is calculated. Secondly, the damage-coupled elastic-plastic constitutive equations and fatigue damage evolution equations are used to calculate the fatigue damage of material. Thirdly, the finite element implementation of theoretical models is conducted to conduct fatigue life calculation. At last, effects of friction coefficient and shape of impact object on the fatigue life are discussed.

1. Introduction

In the flight environment, some objects such as birds and metal debris tend to be ingested into the air flow channel and hit on the aircraft turbine engines, which will induce adverse effect on the performance of engine and reduce the fatigue life of structures markedly [1-2]. However, if all of the structures with impact damage are considered to be rejected products, it will be a kind of waste because some defective structures are still enabled to sustain cyclic loading and meet the fatigue life requirements [3]. Thus it's necessary to present a method to evaluate the effect caused by the impact damage and to calculate the fatigue life for structures with impact defect.

Fatigue life prediction for components with impact damage is an important issue in engineering application. In fact, many methods [4-8] such as the stress equivalents, the local stress strain method, the critical plane method and the continuum damage mechanics based method have been proposed to predict the fatigue life of structures. Ferjaoui A et al. [9] used the continuum damage mechanics to conduct the prediction of fretting fatigue crack initiation in double lap bolted joint. Hu P et al. [10] proposed a continuum damage mechanics approach coupled with an improved pit evolution model to predict the fatigue life of the corrosion fatigue of aluminum alloy. Dezecot S et al. [11] conducted the 3D characterization and modeling of low cycle fatigue damage mechanisms at high temperature in a cast aluminum alloy. Mei J et al. [12] presented a comprehensive investigation into non-proportional loading induced multi-axial fatigue damage in aluminum alloys using a recently developed multi-axial fatigue damage parameter. Kumara D et al. [13] conducted the fretting fatigue stress analysis in heterogeneous material using direct numerical simulations in solid mechanics. Bhattia NA et al. [14] used three numerical models to model the effect of both in phase and out of phase loading on contact stresses and damage initiation locations. Pereira K et al. [15] investigated the singularity's presence in fretting fatigue stresses distributions at a contact interface. However, the methods applied to plain fatigue problems cannot be used directly for failure analyses of defected structures due to the complex effects, such as the impact damage and the redistribution of stress field around the impact pit.

In the previous work [16], the method to calculate the fatigue life for a specimen with an impact pit considering impact damage, residual stress relaxation and elastic-plastic fatigue damage was proposed, but the effects of friction coefficient and shape of impact object on the fatigue life were not discussed. The objective of this study is to research the effects of the two aspects on fatigue life.

2. Theoretical models

2.1. *Initial impact damage analysis*

In this paper, the quasi-static numerical method is applied to simulate the foreign object impact process. The Armstrong-Frederick elastic-plastic model is used to simulate the constitutive relationship of the structure material.

The deterioration of material induced by the impact can be represented by the damage variable in the framework of continuum damage mechanics. For isotropic materials, the damage variable D [17] is used to represent the stiffness deterioration of RVE (Representative Volume Element), which is expressed by

$$D = \frac{(E - E_D)}{E} \tag{1}$$

where E is Young's Modulus of material and E_D is the effective Young's Modulus of RVE with damage. The size of RVE is about 0.1mm for metals [18]. As E_D ranges from E to 0, D varies between 0 and 1.

After the completion of the residual stress analysis, the initial damage D_0 induced by the plastic deformation can be calculated according to Lemaitre's plasticity damage model [18]:

$$D_0 = [\frac{\sigma_{eq}^2 R_v}{2ES}]^m \Delta p \tag{2}$$

where σ_{eq} is the equivalent stress and Δp is the accumulated plastic strain. S and m are material parameters. R_v is the triaxiality function.

2.2. *Fatigue damage analysis*

2.2.1. *Damage-coupled elastic–plastic constitutive model*

In the frame work of small deformation, total strain ε_{ij} can be divided as

$$\varepsilon_{ij} = \varepsilon_{ij}^e + \varepsilon_{ij}^p \tag{3}$$

where ε_{ij}^e and ε_{ij}^p are elastic strain and plastic strain, respectively. The elastic strain takes the form

$$\varepsilon_{ij}^e = \frac{1+\upsilon}{E}(\frac{\sigma_{ij}}{1-D}) - \frac{\upsilon}{E}(\frac{\sigma_{kk}\delta_{ij}}{1-D}) \tag{4}$$

where E, v and σ_{ij} are Young's Modulus, Poisson's ratio and stress components, respectively. The evolution of plastic strain is defined as

$$\dot{\varepsilon}^p = \dot{\lambda}\frac{\partial f}{\partial \sigma_{ij}} = \frac{3}{2}\frac{\dot{\lambda}}{1-D}\frac{S_{ij}\cdot(1-D)^{-1} - \alpha_{ij}}{(S_{ij}\cdot(1-D)^{-1} - \alpha_{ij})_{eq}} \tag{5}$$

$$\dot{p} = \sqrt{\frac{2}{3}\dot{\varepsilon}_{ij}^p\dot{\varepsilon}_{ij}^p} = \frac{\dot{\lambda}}{1-D} \tag{6}$$

where $\dot{\lambda}$ is the plastic multiplier plastic multiplier and S_{ij} is the deviatoric part of the stress, α_{ij} is the deviatoric part of the back stress. The nonlinear kinematic hardening model is used to represent the kinematic hardening behavior

$$\alpha_{ij} = \sum_{k=1}^{M}\alpha_{ij}^{(k)},\ \dot{\alpha}_{ij}^{(k)} = (1-D)(\frac{2}{3}C_k\dot{\varepsilon}_{ij}^p - \gamma_k\alpha_{ij}^{(k)}\dot{p}) \tag{7}$$

where C_k and γ_k are material constants that can be determined from experimental tests.

2.2.2. Fatigue damage evolution model

Two fatigue damage models are employed to calculate the elastic damage and plastic damage, and the damage evolution rate $\dot{D}$ can be obtained by

$$\dot{D} = \dot{D}_e + \dot{D}_p \tag{8}$$

For the multiaxial cyclic loading, the elastic damage evolution equation [19] of the non-linear continuous damage model can be written as

$$\dot{D}_e = \frac{dD_e}{dN} = [1-(1-D)^{\beta+1}]^{\alpha} \cdot [\frac{A_{II}}{M_0(1-3b_2\sigma_{H,m})(1-D)}]^{\beta} \tag{9}$$

where N is the number of cycles until failure, β , M_0 and b_2 are material constants that are determined by fatigue tests. A_{II} is the amplitude of the octahedral shear stress. $\sigma_{H,m}$ is the mean hydrostatic stress. α is defined: $\alpha = 1 - a\langle [A_{II} - \sigma_{l0}(1-3b_1\sigma_{H,m})]/(\sigma_u - \sigma_{e,\max})\rangle$, where $\sigma_{e,\max}$ is the maximum equivalent stress over a loading cycle. a and b_2 are material constants. The plastic damage evolution equation can be written as

$$\dot{D}_p = (\frac{\sigma_{eq}^2 R_v}{2ES(1-D)^2})^m \dot{p} \tag{10}$$

3. Numerical algorithm and finite element model

3.1. Numerical algorithm

In this study, the damage mechanics finite element numerical method [20-21] is used to calculate the fatigue life of a specimen with impact damage. The entire numerical computation includes simulation of the impact process, calculation of the residual stress, computation of the initial impact damage, and calculation of the elastic-plastic fatigue damage. The detailed steps of the calculation are:

1) Initialize all the parameters.
2) Conduct a quasi-static simulation analysis of the impact process to obtain the plastic strain field and residual stress field.
3) Calculate the initial damage field caused by plastic deformation during the impact process.
4) Modify the material properties for each element according to the initial damage. Then the cyclic stress and the accumulated plastic strain are calculated under cyclic loading.
5) The increment of fatigue damage can be calculated after ΔN loading cycles, which is given as

$$\Delta D^{(i+1)} = \Delta N \cdot \Delta D_e^{(i+1)} + \Delta N \cdot \Delta D_p^{(i+1)} \tag{11}$$

Then, the total extent of damage corresponding to the total loading cycle of N is obtained as

$$D^{(i+1)} = D^{(i)} + \Delta D^{(i+1)} \tag{12}$$

6) If the accumulation of damage at any element reaches 1, a fatigue crack initiates at this element and the corresponding number of cycles is the fatigue crack initiation life, or else, a recalculation of the material properties of each element is conducted using the following equation

$$E^{(i+1)} = E^{(i)}(1-D^{(i+1)}) \tag{13}$$

Then, the analysis of the stress field and the damage field is repeated until the accumulation of damage at any element reaches 1.

3.2. Finite element model

The material of the specimens in the experiments is TC4 titanium alloy. The method of parameters identification for the above models is presented in the literature [8] and the parameters for TC4 are listed in Table 1. Considering the symmetry of the model, only 1/4 of the model is built and the symmetry boundary conditions are applied at the plane of symmetry. The foreign object is set as rigid and the structure is set as elastoplastic. A 3-D 8-node solid element is used, which is defined by 8 nodes having three degrees of freedom per node. The mesh density is verified and the convergence analysis of stresses is conducted in order to obtain a converged solution. In total, 54372 elements and 50800 nodes are created. The schematic diagram of model and the established finite element model is shown in figure 1. Cyclic loading is applied in the longitudinal direction with a maximum applied stress of 500 MPa and a stress ratio of 0.1. In the previous research [16], we have validated the CDM approach and we just analyze the effects of friction coefficient and shape of impact object on the fatigue life in this study.

Table 1. The calibrated material parameters.

Elastic damage model	β	M_0	b_1	b_2	a
	1.95	114339	0.001	0.0004	0.72
Plastic damage model	S/MPa			m	
	14429			0.2976	

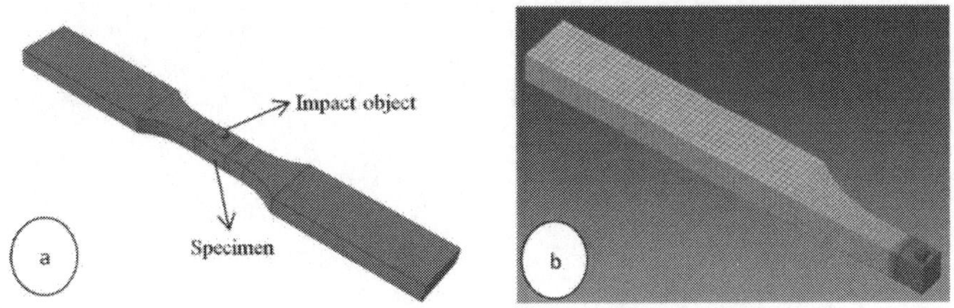

Figure 1. (a) The schematic diagram of whole model; (b) the established finite element model.

4. Computed results and discussions

4.1. *Effect of the friction coefficient*

The friction coefficient *f* between the impact object and specimen is an important factor influencing the elastic-plastic damage and fatigue life. The friction coefficient studied here are 0, 0.1 and 0.2, respectively. The radius of impact object is 1.0mm and the depth of impact is 0.43mm.

After conducting the quasi-static simulation of the impact process, the residual stress field can be obtained. The distribution of longitudinal residual stress around the impact pit is shown in figure 2 and the longitudinal residual stress along the distance from the bottom of the impact pit is shown in figure 3. It is clear that the distributions and changing trends of residual stress are similar and the maximum longitudinal compressive residual stress increases with the increase of friction coefficient, which is favourable to increase the fatigue life.

The calculated fatigue lives corresponding to the different friction coefficients are 133175, 181975 and 202475, respectively. It is clear that the fatigue life increases with the increase of friction coefficient. Curves of total damage, initial impact damage, plastic damage and elastic damage versus the number of cycles corresponding to the dangerous element are shown in figure. 4. The changing trend of damage is similar, that is the damage increases slowly in the beginning of loading cycle and faster and faster afterward and the subsequent plastic damage is the dominating compared with the elastic damage.

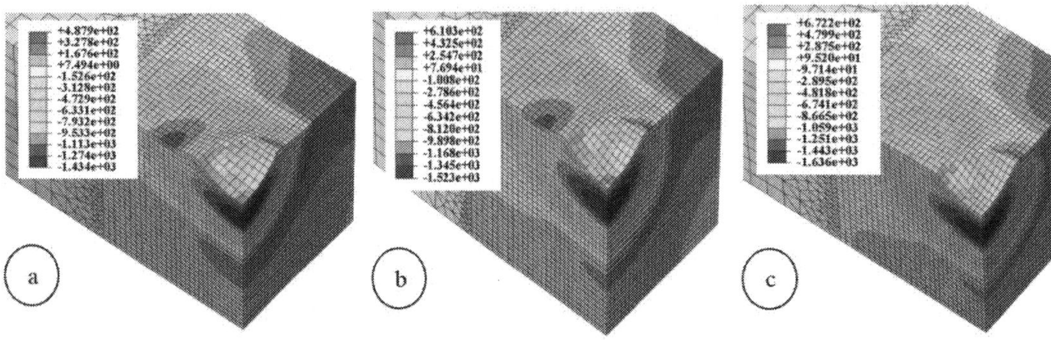

Figure 2. Distribution of longitudinal residual stress with respect to different friction coefficient: (a) f=0; (b) f=0.1; (c) f=0.2.

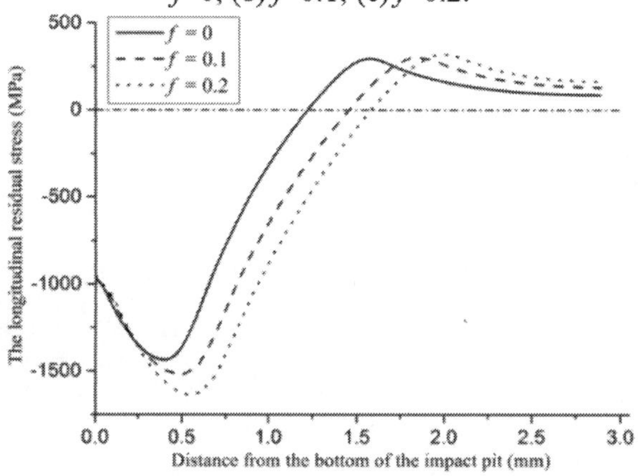

Figure 3. Longitudinal residual stress along the distance from the bottom of the impact pit with respect to different friction coefficient.

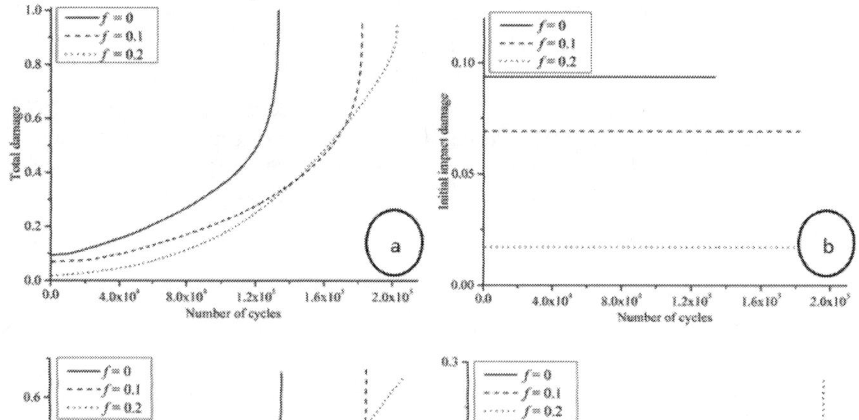

Figure 4. Curves of damage versus the number of cycles with respective to different friction coefficient: (a) total damage, (b) initial impact damage, (c) plastic damage, (d) elastic damage.

4.2. Effect of the shape of impact object

The shape of impact object is another important factor influencing the elastic-plastic damage and fatigue life. The friction coefficient between the impact object and specimen is 0. The depth of impact is 0.43mm. Three different shapes of impact object are discussed. Case 1: spheroid with a=1, b=2, c=1; Case 2: sphere with R=1; Case 3: spheroid with a=2, b=1, c=1.

After conducting the quasi-static simulation of the impact process, the residual stress field can be obtained. The distribution of longitudinal residual stress around the impact pit is shown in figure 5 and the longitudinal residual stress along the distance from the bottom of the impact pit is shown in figure 6. It is clear that the maximum longitudinal tensile residual stress of case 1 is much larger than other two cases, which can be adverse to the fatigue life of specimen. We can see that the distributions and changing trends of residual stress along the distance from the bottom of the impact pit with respect to the different shapes of impact object are similar.

The calculated fatigue lives corresponding to the different cases are 119675, 133175 and 163375, respectively. The curves of total damage, initial impact damage, plastic damage and elastic damage versus the number of cycles corresponding to the dangerous element are shown in figure. 7. The changing trend of damage is similar, that is the damage increases slowly in the beginning of loading cycle and faster and faster afterward and the subsequent plastic damage is the dominating compared with the elastic damage.

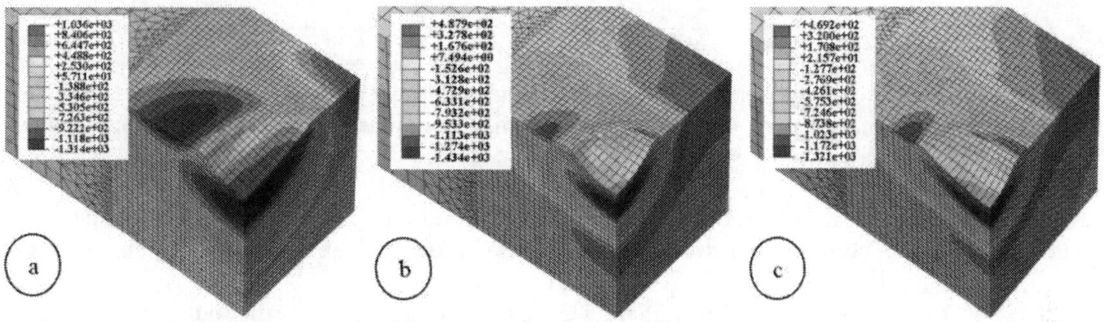

Figure 5. Distribution of longitudinal residual stress with respect to the different shape of impact object: (a) case 1; (b) case 2; (c) case 3.

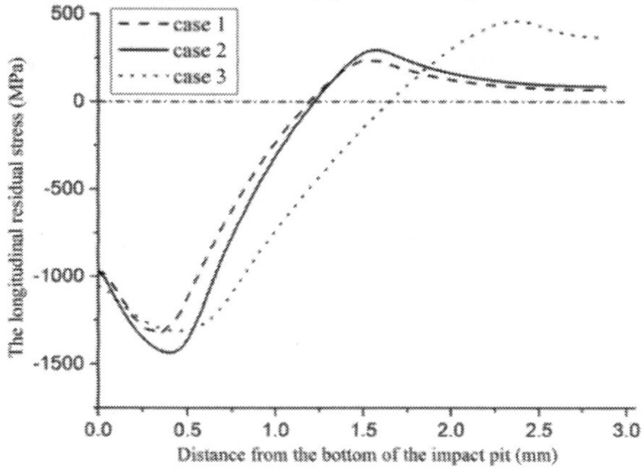

Figure 6. Longitudinal residual stress along the distance from the bottom of the impact pit with respect to the different shape of impact object.

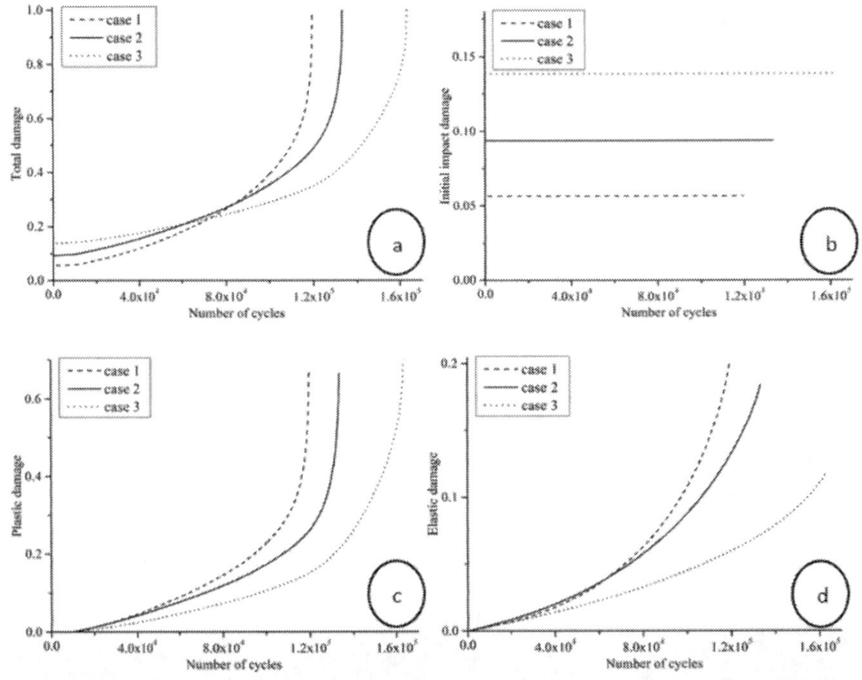

Figure 7. Curves of damage versus the number of cycles with respective to the different shape of impact object: (a) total damage, (b) initial impact damage, (c) plastic damage, (d) elastic damage.

5. Conclusions

In this study, the effects of friction coefficient and shape of impact object on the fatigue life are discussed by using a continuum damage mechanics based method. The following conclusions can be made:

(1) The impact simulation is conducted and the residual stress and plastic strain fields are obtained, and the initial impact damage is calculated according to the plasticity damage model. Then the fatigue damage evolution equations are used to calculate the fatigue damage of material under cyclic loadings.

(2) The damage mechanics finite element numerical algorithm is presented which can considers the coupling effect of stress and strain field, initial impact damage and fatigue damage of material.

(3) Effects of friction coefficient and shape of impact object on the fatigue life are analysed. Firstly, the distributions and changing trends of residual stress are similar for all the cases; secondly, the longitudinal compressive residual stress increases with the increase of friction coefficient, which is favourable for the fatigue life; thirdly, the damage increases slowly in the beginning of loading cycle and faster and faster afterward and the subsequent plastic damage is the dominating compared with the elastic damage for all the cases.

References

[1] Nowell D, Duo P, Stewart I F. Prediction of fatigue performance in gas turbine blades after foreign object damage. International journal of fatigue 2003; 25(9): 963-969.

[2] Peters J O, Ritchie R O. Influence of foreign-object damage on crack initiation and early crack growth during high-cycle fatigue of Ti–6Al–4V. Engineering Fracture Mechanics 2000; 67(3): 193-207.

[3] Zhan Z, Hu W, Zhang M, et al. Experimental method for and theoretical research on defect tolerance of fixed plate based on damage mechanics. Chinese Journal of Aeronautics 2013; 26(5): 1195-1201.

[4] C.A. Gonçalves, J.A. Araújo, E.N. Mamiya. Multiaxial fatigue: a simple stress based criterion for

hard metals. International Journal of Fatigue 2005; 27(2): 177-187.
[5] R. Seshadri. The Generalized Local Stress Strain (GLOSS) Analysis—Theory and Applications. Journal of Pressure Vessel Technology 1991; 113(2): 219-227.
[6] Karolczuk A, Macha E. A review of critical plane orientations in multiaxial fatigue failure criteria of metallic materials. International Journal of Fracture 2005; 134(3-4): 267-304.
[7] Chaboche J-L. Continuous damage mechanics—a tool to describe phenomena before crack initiation. Nuclear Engineering and Design 1981; 64(2): 233-247.
[8] Zhan Z, Hu W, Zhang M, et al. Revised damage evolution equation for high cycle fatigue life prediction of aluminum alloy LC4 under uniaxial loading. Applied Mathematics and Mechanics 2015; 36(9): 1185-1196.
[9] Ferjaoui A, Yue T, Wahab M A, et al. Prediction of fretting fatigue crack initiation in double lap bolted joint using Continuum Damage Mechanics. International Journal of Fatigue 2015; 73: 66-76.
[10] Hu P, Meng Q, Hu W, et al. A continuum damage mechanics approach coupled with an improved pit evolution model for the corrosion fatigue of aluminum alloy. Corrosion Science 2016; 113: 78-90.
[11] Dezecot S, Maurel V, Buffiere J Y, et al. 3D characterization and modeling of low cycle fatigue damage mechanisms at high temperature in a cast aluminum alloy. Acta Materialia 2017; 123: 24-34.
[12] Mei J, Dong P. Modeling of path-dependent multi-axial fatigue damage in aluminum alloys. International Journal of Fatigue 2017; 95: 252-263.
[13] Kumar D, Biswas R, Poh L H, et al. Fretting fatigue stress analysis in heterogeneous material using direct numerical simulations in solid mechanics. Tribology International 2017; 109: 124-132.
[14] Bhatti N A, Wahab M A. Finite element analysis of fretting fatigue under out of phase loading conditions. Tribology International 2017; 109: 552-562.
[15] Pereira K, Bordas S, Tomar S, et al. On the Convergence of Stresses in Fretting Fatigue. Materials 2016; 9(8): 639.
[16] Zhan Z, Hu W, Shen F, et al. Fatigue life calculation for a specimen with an impact pit considering impact damage, residual stress relaxation and elastic-plastic fatigue damage. International Journal of Fatigue 2017; 96: 208–223.
[17] Lemaitre J. Mechanics of solid materials: Cambridge university press; 1994.
[18] Lemaitre J, Desmorat R. Engineering damage mechanics: ductile, creep, fatigue and brittle failures: Springer; 2005.
[19] Marmi A K, Habraken A M, Duchene L. Multiaxial fatigue damage modelling at macro scale of Ti–6Al–4V alloy. International Journal of Fatigue 2009; 31(11): 2031-2040.
[20] Zhan Z, Hu W, Zhang M, et al. The fatigue life prediction for structure with surface scratch considering cutting residual stress, initial plasticity damage and fatigue damage. International Journal of Fatigue 2015; 74: 173-182.
[21] Zhan Z, Hu W, Meng Q, et al. Continuum damage mechanics-based approach to the fatigue life prediction for 7050-T7451 aluminum alloy with impact pit. International Journal of Damage Mechanics 2016; 25(7): 943-966.

Investigation of fatigue assessments accuracy for beam weldments considering material data input and FE-mode type

Yevgen Gorash*, Tugrul Comlekci and Donald MacKenzie

University of Strathclyde, MAE, James Weir Bld, 75 Montrose St, Glasgow G1 1XJ, UK

E-mail: yevgen.gorash@strath.ac.uk

Abstract. This study investigates the effects of fatigue material data and finite element types on accuracy of residual life assessments under high cycle fatigue. The bending of cross-beam connections is simulated in ANSYS Workbench for different combinations of structural member shapes made of a typical structural steel. The stress analysis of weldments with specific dimensions and loading applied is implemented using solid and shell elements. The stress results are transferred to the fatigue code nCode DesignLife for the residual life prediction. Considering the effects of mean stress using FKM approach, bending and thickness according to BS 7608:2014, fatigue life is predicted using the Volvo method and stress integration rules from ASME Boiler & Pressure Vessel Code. Three different pairs of S-N curves are considered in this work including generic seam weld curves and curves for the equivalent Japanese steel JIS G3106-SM490B. The S-N curve parameters for the steel are identified using the experimental data available from NIMS fatigue data sheets employing least square method and considering thickness and mean stress corrections. The numerical predictions are compared to the available experimental results indicating the most preferable fatigue data input, range of applicability and FE-model formulation to achieve the best accuracy.

1. Introduction

Connection by welding is the most effective fabrication process, which is used for a relatively fast manufacturing of big assemblies using simple structural members. Welded joints between metal parts are produced by causing fusion, which includes melting the the base metal and adding the filler material. The phase transformation of even a small amount of the structural material usually results in significant residual stresses, heat effected zone with weaker mechanical characteristics, rugged surface geometrical features and various welding defects – cracks, distortion, inclusions, incomplete penetration, etc. – refer to [1] for more details. In general, the nature of the welding process means that weldments have a lower fatigue strength than the base material of the parts, which are joined together. The negative effect of welding on the integral strength of the structure is usually minimised during the design process. For example, welded joints need to be kept away from highly stressed areas, since they increase the stresses even more. An infinite fatigue life can be theoretically provided for the base material by identification of the fatigue strength limit, which can be used as a stress limit in the design analysis. Thus, by a proper positioning of weldments the main loading can be carried out primarily by the base material providing an infinite fatigue life. However, even in well-designed

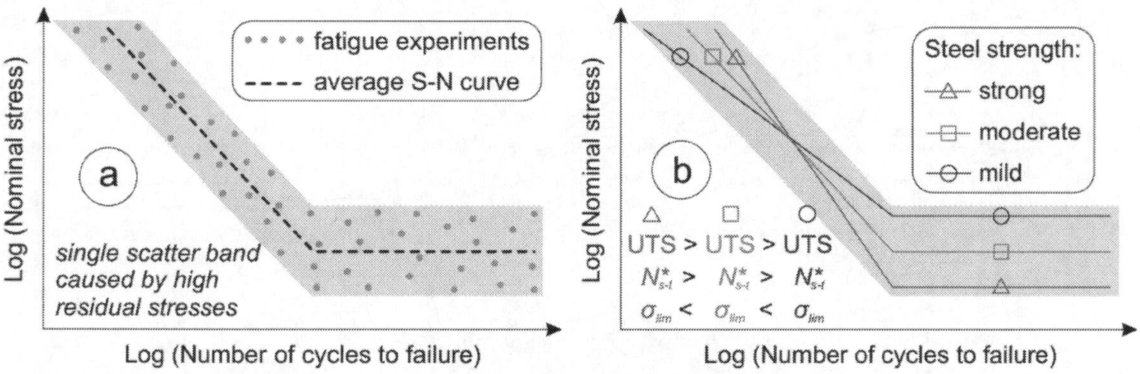

Figure 1. Fatigue strength of weldments in relation to the strength of base metal: a) conventional approach to S-N curve, b) investigated observations.

structures, where the weldments are placed away from the load path, the fatigue failures are typically found in weldments [2]. Therefore, the residual life prediction for welded structure should be based in the first instance upon fatigue analysis of weldments [1].

The fatigue behaviour of weldments has been studied in terms of the geometry of the members, the stresses to which they are subjected, and the materials of which they are fabricated [3]. In regard to the choice of base material, there are a few experimental observations, which may explain its relation to the fatigue strength of corresponding weldments. Within the conventional approach, when steels of widely differing grades are welded, the resulting S-N curves tend to fall within a single scatter band, as shown in Fig. 1a. The principal reason for this is that superior fatigue strength of high-strength steels as base material is eliminated by the high residual stresses in the welds, which usually approaches the yield strength. However, closer examination of the fatigue curve slopes reveals that low-strength steels (with lower σ_u) tend to have better fatigue resistance in long-term domain under low loads while high-strength steels (with higher σ_u) tend to have better fatigue strength in short-term domain (N^*_{s-t}) under high loads, as shown in Fig. 1b. The detailed discussion of this observation is available in [4, 5], which conclude that the fatigue strength of weldments is not completely independent of the base material strength, it is rather inversely proportional to the σ_u of the base material. Therefore, provision of more specific S-N curves for different groups of steels (e.g. mild, moderate and hard) may increase the quality of fatigue assessments. This paper addresses the comparison of specific S-N curve and generic S-N curves for investigation of accuracy of residual life predictions for welded structures.

The most effective way of fatigue assessment is a postprocessing of FEA results of a structural analysis in the form of stress / strain fields (geometry input) in combination with input of load history and fatigue material data. There is a variety of fatigue assessment tools available for FEA from basic tools in form of add-ins or modules for CAD/CAE products to advanced stand-alone or integrated fatigue postprocessors [6]. The fatigue code nCode DesignLife embedded in ANSYS Workbench environment has been chosen for this study, since a number of advanced features have been implemented in it to facilitate the effective fatigue analysis of welds. The main methods implemented in nCode DesignLife with theoretical background on fatigue of welds and validation cases are outlined in [2, 7].

It should be noted that in total majority of published studies the specific modules for weld fatigue analysis were not used, the investigators preferred conventional approaches like hot-spot stress or notch stress methods which required structural stress from FEA as input. In contrast to previous works, this study investigates the robustness of the available seam weld fatigue analysis module of nCode DesignLife which is based on nominal stress method in application

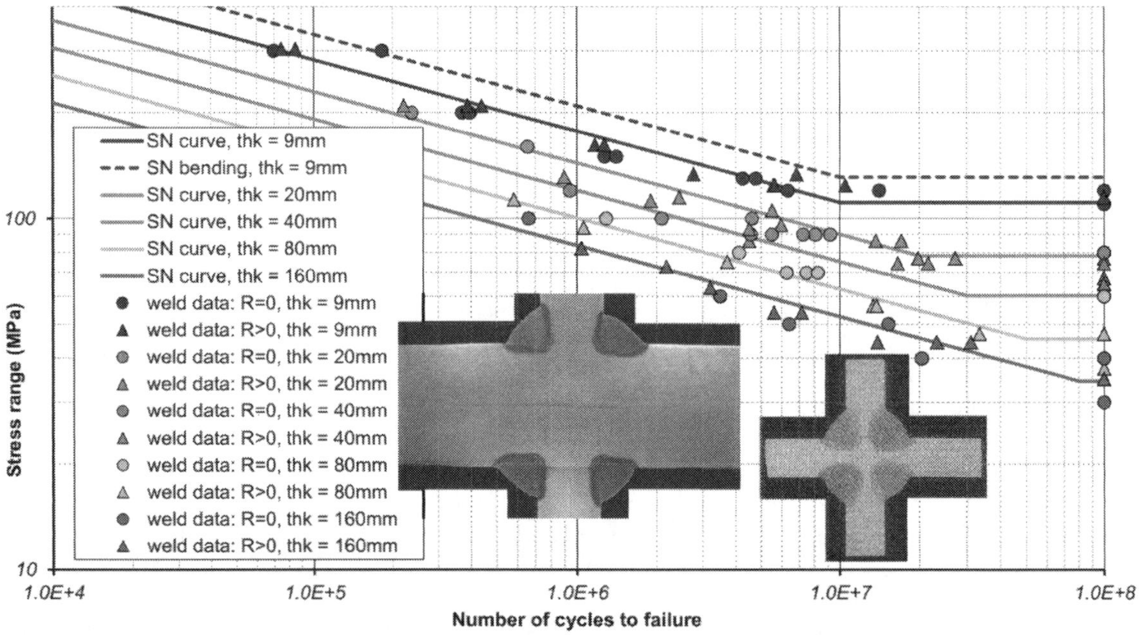

Figure 2. Experimental and fitted S-N curves of cruciform welded joints made of SM490B steel for different thicknesses [10] and normalised to $R = 0$.

to solid and shell FE-models. This paper presents a numerical comparative study in order to validate not only available analysis facilities in nCode DesignLife for weld models in solid and shell formulation, but also the significance of fatigue material data input. This work is built on the outcomes of the initial study [4] and develops it further considering a proper induction of the mean stress correction and finalising the fatigue assessments for solid FE-models of all weldment configurations. The experimental studies of welded thin-walled cross-beam connections under cyclic bending by Mashiri *et al.* [8, 9] have been chosen as a validation in this numerical study.

2. Handling the fatigue properties of weldments

The specimens in experiments [8, 9] were manufactured from cold-formed high-strength steel of grades C350LO (σ_y = 350 MPa and σ_u = 430 MPa) and C450LO (σ_y = 450 MPa and σ_u = 500 MPa) according to the Australian Standard AS3678. These two grades represent the lower (Grade C350) and upper (Grade C450) bounds of the big international group of weldable, general-purpose, high-strength structural steels, which includes the following grades [11]: *Grade 50 (A, B, C, D)* from British Standard BS4360; *St52-3* from German Standard DIN17100; *G3106-SM490 (A, B, C)* from Japanese Standard JIS; *Fe510 (B, C, D)* from International Standard ISO630; *A572-345 (-415)* from American Standard ASTM; *S355 (JR, J0, etc.)* from European Standard EN10025.

These steels are roughly equivalent in chemical composition and have similar elastic properties with elastic modulus of $E = 2 \cdot 10^5$ MPa and Poisson's ratio of $\nu = 0.3$. Since specific fatigue curves for the weldments made of grades C350 or C450 are unavailable in the nCode DesignLife material database, an equivalent fatigue data input is required. The principal aspect in fatigue of weldments is availability of the appropriate experimental data for a long-term strength domain. The most suitable fatigue datasets of this kind are provided by National Institute for Materials Science (Tsukuba, Japan) for the Japanese equivalent from the list above – steel SM490B. The datasets are presented by 5 NIMS Fatigue Data Sheets [10] available in 5 parts for cruciform

Table 1. Fatigue constants for three variants of the material, different types of amplitude, values of stress ratio R and reference thickness t_{ref}.

material	No.	amplitude	t_{ref} [mm]	n	R	bending	SRI [MPa]	b	slope (1/b)
SM490B steel welds	1a	CA	9	0.26	0	stiff	2914.8	0.2027	4.93
						flex	3439.5	0.2027	4.93
	1b	CA	1		0	stiff	5160.7	0.2027	4.93
						flex	6089.7	0.2027	4.93
generic SN curves from nCode DL	2a	VA	1	0.16667	-1	stiff	18000	0.3333	3.00
						flex	36000	0.3333	3.00
	2b	CA	1		-1	stiff	25960	0.3333	3.00
						flex	51920	0.3333	3.00
	2c	CA	1		0	stiff	20768	0.3333	3.00
						flex	41536	0.3333	3.00
generic SN curves from nCode FEF	3a	VA	3	0.16667	-1	stiff	8569	0.2632	3.80
						flex	11478	0.219	4.57
	3b	CA	3		-1	stiff	13090	0.2632	3.80
						flex	17534	0.219	4.57
	3c	CA	1		-1	stiff	13090	0.2632	3.80
						flex	17534	0.219	4.57
	3d	CA	1		0	stiff	10472	0.2632	3.80
						flex	14027	0.219	4.57

weldments of 5 different thicknesses (9 mm, 20 mm, 40 mm, 80 mm, 160 mm), which provide the details of fatigue tests with stress ratio R from 0 to 1 and duration of experiments up to 10^8 cycles, as illustrated in Fig. 2. This experimental data needs to be properly fitted by an S-N curve equation, which is adapted to fatigue analysis capabilities implemented in nCode DesignLife, and corresponding fatigue parameters need to be accurately identified.

An advantage of nCode DesignLife as a fatigue postprocessor in the availability of effective approaches for fatigue analysis of weldments in both shell and solid elements formulations. The important feature of these approaches is that they are relatively non-sensitive to the quality of finite element mesh. The "Volvo" Method [12], used for coarse shell models, is suitable for making FE-based fatigue assessments of welded joints with the minimum of user intervention being required [2, 7]. Weld fatigue analysis in application to the solids requires more efforts compared to shells as discussed in [5, 2, 7]. The key element of this approach is the stress integration method proposed in ASME BPVC Code [13], when the stress is extracted at several points through thickness, and then extrapolated to produce membrane and bending components. Recently the fatigue analysis in solid formulation has been drastically improved and accelerated with the release of an ACT extension for ANSYS Workbench titled nCode Weldline [14], which brings a dramatic progress into the analysis of complex assemblies with a free-form geometry and arbitrary location of welded joints [5].

An essential feature of fatigue analysis is the mean stress effect, which needs to be carefully considered in this work. Weld fatigue assessment is based on a mean stress correction using the FKM approach [15], in which the mean stress sensitivity is defined by 4 factors in 4 regimes: $R > 1$, $-\infty \leq R < 0$, $0 \leq R < 0.5$, $0.5 \leq R < 1$. The idea of FKM method is usually illustrated in the form of a constant life or Haigh diagram, where the factors M_{1-4} correspond to the slopes in coordinates of mean stress σ_{m} and stress amplitude σ_{a}. In contrast to classical mean stress correction approaches (like Goodman, Gerber or Soderberg), the FKM approach is not related to material characteristics σ_{y} and σ_{u}. It also allows to determine the equivalent stress amplitude σ_{a} at a particular material R-ratio, but it has better applicability for characterisation of environmental (corrosion / erosion) and technological (welding) effects on fatigue life. In

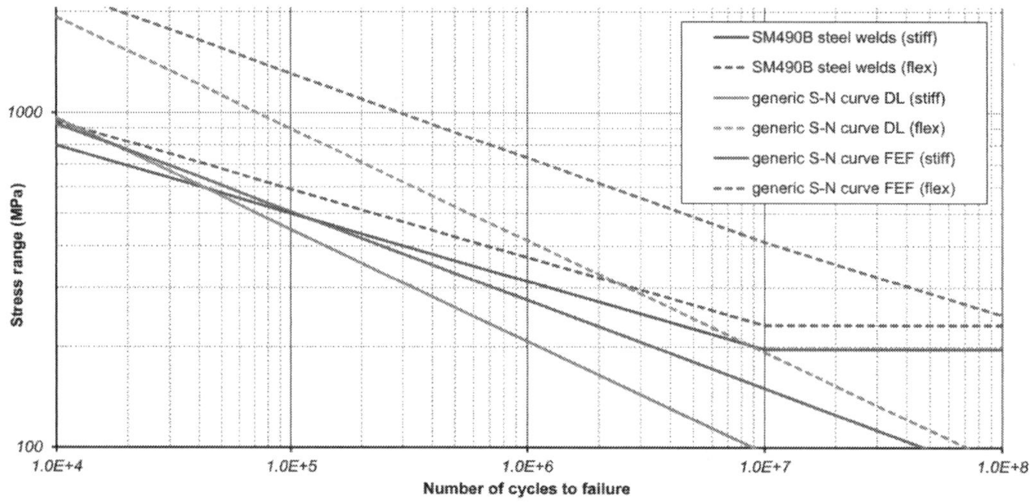

Figure 3. Comparison of three pairs of S-N curves to be used in weldments fatigue analysis, which are normalised to $t_{\rm ref} = 1$ mm of and $R = 0$.

definition of the generic S-N curves used for fatigue analysis of weldments, the nCode guidelines [2, 7] recommend the following values of the FKM factors: $M_1 = 0$, $M_2 =$ -0.25 and $M_3 = M_4$ = -0.1. For the specific steel SM490B, these factors may have different values and need to be identified using the available experiments [10] in the range $0 < R < 1$.

Other important effects directly related to the fatigue of welds are bending and thickness effects, which are included in fatigue analysis according to the British Standard BS7608 [16]. If a reference thickness is exceeded, the fatigue strength is reduced by a correction factor:

$$k_{\rm tb} = \left(\frac{t_{\rm ref}}{t}\right)^n \left[1 + 0.18\,\Omega^{1.4}\right], \tag{1}$$

where t – thickness of the welded components, $t_{\rm ref}$ – reference thickness, n – thickness exponent.

In notation (1), the fatigue strength increases with increasing bending component (defined by bending ratio Ω) for a decreasing stress range gradient through the thickness. However, the design S-N curves relate to applied loading conditions that produce predominantly membrane stresses. So the S-N curve corresponding to pure bending condition can be obtained from a basic membrane S-N curve by setting the bending ratio $\Omega = 1$. The potentially detrimental effect of increased thickness but beneficial effect from applied bending are combined by the application of the correction factor $k_{\rm tb}$ using Eq. (1) to the stress ranges $\Delta\sigma_{\rm b}$ obtained from the relevant basic S-N curve [16]:

$$\Delta\sigma = k_{\rm tb}\,\Delta\sigma_{\rm b}, \tag{2}$$

where $\Delta\sigma$ is a nominal stress range in the structural component under consideration of bending and thickness correction. The basic S-N curve is fitted using the standard nCode DesignLife definition, where the curve consists of 3 linear segments on a log-log plot. The central and long-term domains are defined by the formula [7]:

$$\Delta\sigma_{\rm b} = \begin{cases} I_{\Delta\sigma 1}\, N_*^{-b_1} & \text{if } \; N_* < N_{C1} \\ I_{\Delta\sigma 1}\, N_{C1}^{(b_2 - b_1)}\, N_*^{-b_2} & \text{otherwise} \end{cases}, \tag{3}$$

where N_* – number of cycles to failure; $I_{\Delta\sigma 1}$ – stress range intercept (MPa); b_1 – first fatigue strength exponent; N_{C1} – transition life; b_2 – second fatigue strength exponent. Transition life

N_{C1} defines the point on the curve, where it transitions to the second slope b_2. If b_2 is set to zero, this acts as a fatigue limit.

For the purpose of accurate identification of fatigue parameters, the original set of experimental data [10] for steel SM490B welds available for the range of thicknesses t and stress ratios R needs to be normalised to the equality condition with $t_{\mathrm{ref}} = 9$ mm and $R = 0$. The R-normalisation of the stress range $\Delta\sigma_{\mathrm{R0}} = 2\,\sigma_{\mathrm{a0}}$ is obtained using the idea of FKM mean stress correction as explained in [5]. The t-normalisation of the stress range $\Delta\sigma_{\mathrm{t9}}$ is done using the thickness correction in form (1) as

$$\Delta\sigma_{\mathrm{t9}} = \Delta\sigma\ (t_{\mathrm{ref}}\ /\ t)^{-n}\,, \tag{4}$$

where the thickness exponent n is unknown.

The resultant set of experimental data [10] after normalisation is fitted with Eq. (3) as illustrated in Fig. 4. The fitting accuracy is characterised by the coefficient of determination R^2, which needs to be maximised for the best result. A manual optimisation procedure based on enumeration with the goal to maximise R^2 and determine corresponding values of M_3 and n was implemented in MS Excel. Using the step of 0.0001 for M_3 in the range from 0.015 to 0.03 and the step of 0.001 for n in the range from 0.2 to 0.3 the global maximum value of $R^2 = 0.9484959$ has been achieved with corresponding values of $M_3 = -0.0283$ and $n = 0.26$. The least squares fit of the power function to the normalised experimental data as shown in Fig. 4 results into the constants set No. 1a for constant amplitude (CA), which is listed in Table 1 with additional constants for long-term domain as $b_2 = 0$ and $N_{C1} = 10^7$, as recommended in BS7608 [16]. This constants set for steel SM490B welds is complemented by the remaining FKM factors as recommended in [15]: $M_1 = 0$, $M_2 = 3 \cdot M_3 =$ -0.0849 and $M_4 = M_3 =$ -0.0283.

The result of fitting with Eq. (3) is illustrated in Fig. 2 with solid blue line. The result of bending correction for $\Omega = 1$ using Eqs (1) and (2) is shown with the dashed blue line. The result of thickness correction using Eqs (1) and (2) is shown with solid lines for thicknesses 20, 40, 80, 160 mm. Since $t_{\mathrm{ref}} = 9$ mm for the constants set No. 1a, it produces very conservative fatigue life predictions for the welded components with $t < 9$ mm. Thus, the constants need to be extrapolated up to $t_{\mathrm{ref}} = 1$ mm to become suitable for fatigue analysis of structural members used in experiments by Mashiri *et al.* [8, 9] having thicknesses as $t = 1.4$ mm, 3 mm and 4 mm. This transformation is done combining Eqs (1)-(3) as

$$I^{t}_{\Delta\sigma1} = I_{\Delta\sigma1}\,(t_{\mathrm{ref}}\ /\ t^{new}_{\mathrm{ref}})^{n}\,, \tag{5}$$

where t^{new}_{ref} is a new reference thickness and $I^{t}_{\Delta\sigma1}$ is a new corresponding stress range intercept. Using the value of $t^{\mathrm{new}}_{\mathrm{ref}} = 1$ mm in Eq. (5), the constant set No. 1a is transformed into the constant set No. 1b, which is listed in Table 1 and illustrated in Fig. 3 with solid and dashed blue S-N curves. This constants set is used in nCode DesignLife for the fatigue analysis of weldments as a specific user fatigue data input characterising the material of weldments.

The advantage of specific fatigue data input needs to be confirmed by comparing it to the generic S-N curves for seam welds made of weldable structural steels available in nCode software databases. DesignLife material database contains a pair of generic S-N curves with the standard slope $1/b = 3$, $t_{\mathrm{ref}} = 1$ mm and constants given in Table 1 under No. 2a for the assumption of variable amplitude (VA). This assumptions means that $N^{CA}_{*} = 3\ N^{VA}_{*}$, because the quality of production is not as good as specimens resulting into Miner's sum effectively reducing to 1/3. Mathematically transformation from VA to CA is expressed by increase of the stress range intercept as

$$I^{CA}_{\Delta\sigma1} = I^{VA}_{\Delta\sigma1}\,(1/3)^{-b_1}\,, \tag{6}$$

where $I^{CA}_{\Delta\sigma1}$ and $I^{VA}_{\Delta\sigma1}$ are the stress range intercepts for CA and VA correspondingly. Using Eq. (6), the constant set No. 2a for VA is transformed into the constant set No. 2b for CA,

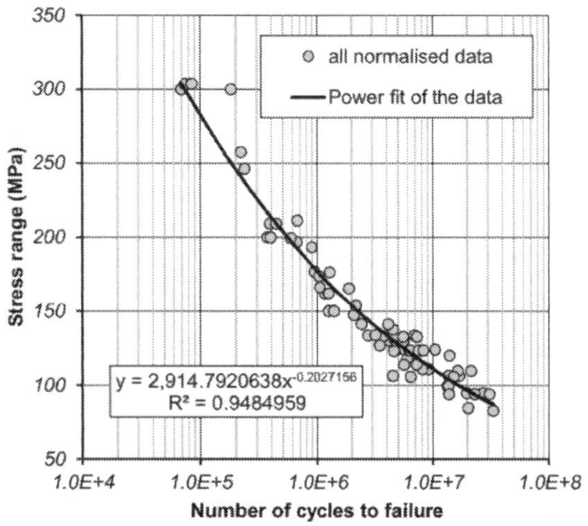

Figure 4. All fatigue data from [10] normalised to $t_{\mathrm{ref}} = 9$ mm and $R = 0$.

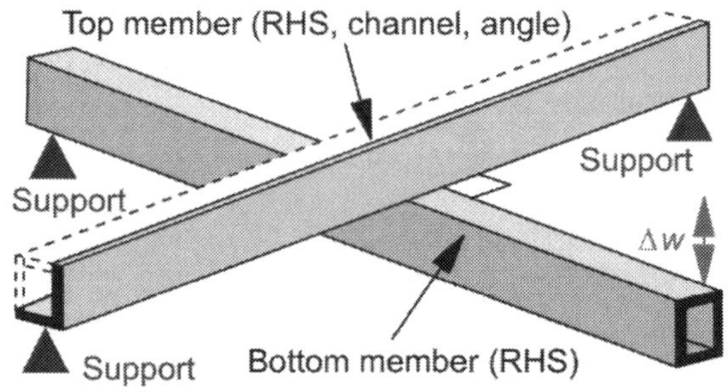

Figure 5. The fatigue tests for RHS / Channel / Angle to RHS welded cross-beam connections [8, 9] are addressed for simulations in current study.

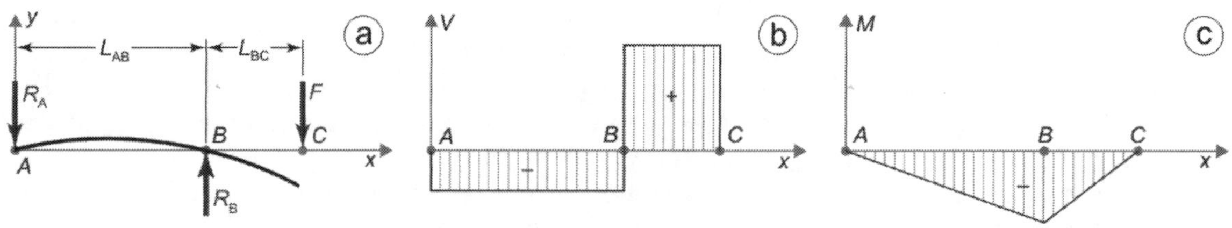

Figure 6. Diagrams of (a) vertical deflection Y, (b) shear force V, (c) bending moment M for overhanging load applied to simply supported beam [17].

which is listed in Table 1. This constants set is used in nCode DesignLife for the fatigue analysis of weldments as a recent formulation of generic fatigue data input, which has been revised to meet the requirement for standard S-N curve slope of 3.

An alternative and older generic fatigue data input considered in this works is a pair of generic S-N curves for seam welds with slopes different from standard (3.8 and 4.57) from nCode FE-Fatigue, a legacy fatigue postprocessor from nCode, predecessor of DesignLife. These S-N curves are described by the constants set No. 3a for $t_{\mathrm{ref}} = 3$ mm and VA, which is given in Table 1 and historically preceded the set No. 2a. In order to be compared to the sets No. 1b and 2b, this constants set No. 3a requires a transformation. Firstly, Eq. (6) is used to do a VA-CA transformation resulting into the set No. 3b. Secondly, Eq. (5) is used to do reduce t_{ref} from 3 mm to 1 mm resulting into the set No. 3c listed in Table 1.

For a proper graphical comparison, all three sets of fatigue constants should correspond to the same stress ratio R, e.g. 0 as in case of the set No. 1b. Hence the sets No. 2b and 3c need to be converted from $R = -1$ to $R = 0$ using FKM mean stress correction as explained in [5]. The obtained sets No. 2c and 3d are shown in Fig. 3 with green and red S-N curves correspondingly and compared to the set No. 1b (blue curves). The visual comparison clearly indicates that the nCode DL set (2c) is going to provide the most conservative fatigue life predictions with the nCode FEF set (3d) giving non-conservative predictions, while the SM490B set (1b) will be somewhere in between nCode DL and FEF outputs with moderate fatigue life predictions.

Table 2. Dimensions of 6 variants of the beam cross-sections [mm] according to [18, 19] and corresponding area moments of inertia [mm^4].

Square Hollow Sections				Rectangular Hollow Sections	
50x50x3 SHS	50x50x1.6 SHS	35x35x3 SHS	35x35x1.6 SHS	50x25x3 RHS	50x25x1.6 RHS
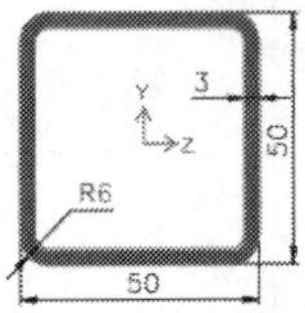	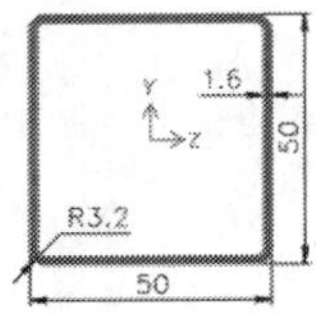	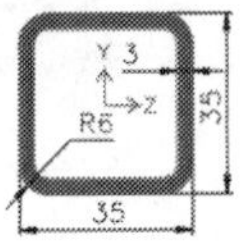	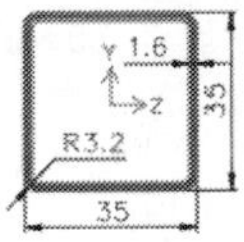	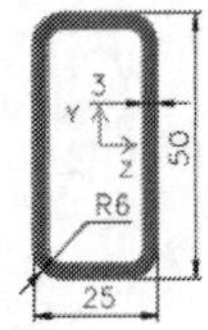	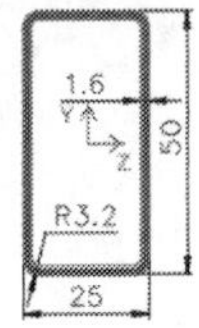
194671.36	117050.41	59483.1	37886.23	111721.36	70182.15

3. FEA-based fatigue assessment of weldments

The specimens in experiments [8, 9] had three ends (left bottom and two top supports) constrained to ground using the hinge coupling and one end left free, as illustrated in Fig. 5. This unconstrained right bottom support has an out-of-plane orthogonal displacement w applied cyclicly, which corresponds to a particular nominal stress range $\Delta\sigma$. The values of $\Delta\sigma$ for each experiment having particular cross-beam connection are reported in [4, 5]. Using the stress ratio $R = 0.1$ from the fatigue experiments [8, 9], the nominal stress range $\Delta\sigma$ can be converted into the maximum nominal stress σ_{max} using the following equation:

$$\sigma_{\text{max}} = \sigma_{\text{ave}} + \sigma_{\text{a}}, \tag{7}$$

with

$$\sigma_{\text{ave}} = \sigma_{\text{a}} \frac{1+R}{1-R} \quad \text{and} \quad \sigma_{\text{a}} = \frac{\Delta\sigma}{2}, \tag{8}$$

where σ_{ave} is an average nominal stress and σ_{a} is a nominal stress amplitude. The values of σ_{max} reported in [4, 5] are required for the assessment of corresponding out-of-plane displacement w applied to the unconstrained beam end.

For the assessment of w, the case of overhanging load applied to simply supported beam available from [17] and shown in Fig. 6, is used as a structural equivalent to the cross-beam connection. In this case, locations A and B on a beam from the simplified model are vertically constrained, and the location C has the orthogonal force applied to it. In experiments [8, 9], location A corresponds to the left constrained end of the bottom beam, location B – toe of the weld connecting the beams, and location C – right unconstrained end of the bottom beam. The analytical solution for a structural analysis problem with overhanging load applied to a simply supported beam is provided in [17] including equations and diagrams for vertical deflection Y, shear force V and bending moment M, which are shown in Fig. 6. The most relevant for this study is the equation for deflection of the beam in location C, which is further used as w:

$$Y_{\text{C}} = -\frac{F\, L_{\text{BC}}^2\, (L_{\text{AB}} + L_{\text{BC}})}{3\, E\, I_{\text{Z}}}, \tag{9}$$

where E is an elastic modulus for the material of beam taken from Sec. 2, and I_{Z} is an area moments of inertia [mm^4] about the neutral axis Z for the beam cross-section. The particular values of I_{Z} for six variants of the beam profiles for the bottom member are reported in Table 2. These values are calculated using the real geometry of beam profiles in CAD-software SolidWorks with dimensions from Australian / New Zealand standard [18] and technical specification [19], which are shown in Table 2. The beam profile is considered to be located in $Y - Z$ plane with

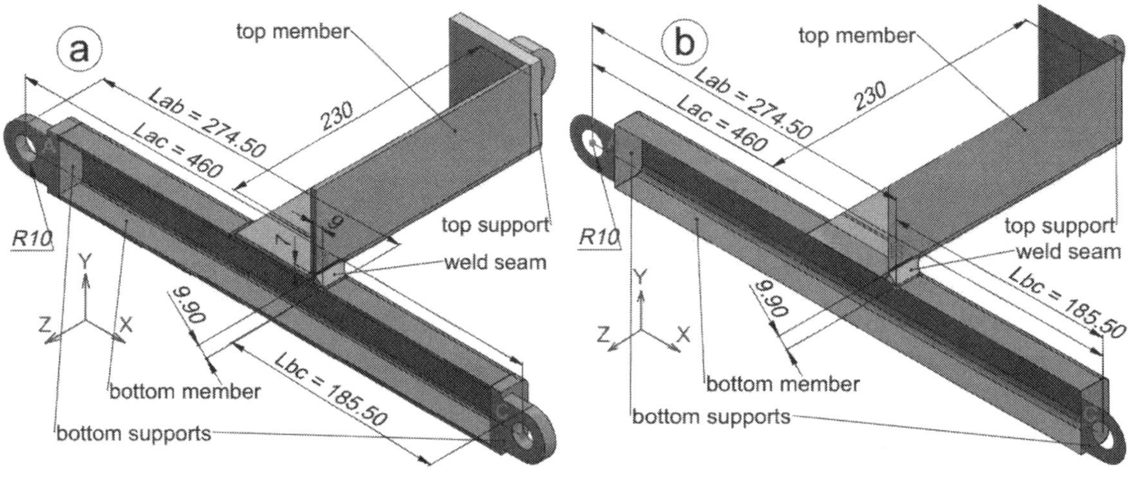

Figure 7. Geometry, components definition and dimensions (mm) for 75x75x4 CA to 50x50x3 SHS connection using (a) solids and (b) surfaces.

the neutral axis Z going through the profile centre. Therefore, a beam representing the bottom member bends around the axis Z. In notation (9), the applied bending force F is estimated using the assumption of maximum bending stress being the maximum nominal stress $\sigma_{\max}$ in experiments. Referring to the design guide [20], in experiments [8, 9] the nominal stress is caused by the basic load, which was the "bending moment in the bottom member". So the bending force F is obtained from the classic formula for determining the maximum bending stress in the outermost layer of the beam under simple bending as:

$$\sigma_{\max} = \frac{M\,H_{\mathrm{Y}}}{I_{\mathrm{Z}}} = \frac{F\,L_{\mathrm{BC}}\,H_{\mathrm{Y}}}{I_{\mathrm{Z}}} \quad \Longrightarrow \quad F = \frac{\sigma_{\max}\,I_{\mathrm{Z}}}{L_{\mathrm{BC}}\,H_{\mathrm{Y}}}, \tag{10}$$

where M is the moment about the neutral axis Z and H_{Y} is the perpendicular distance from the outermost layer of the beam to the neutral axis Z, which in this case corresponds to the half height of the section profile indicated in Table 2.

Equations (9) and (10) are used to estimate the values of w corresponding to $\sigma_{\max}$ with additional input of parameters specific for each experiment like I_{Z}, L_{AB} and L_{BC}. The resultant values of the displacement w, which is applied to the right end of bottom member as illustrated in Fig. 5, are reported in [4, 5].

The comprehensive geometrical models of welded cross-beam connections were created in CAD-software SolidWorks in solids and surfaces presentation including top and bottom members, top and bottom supports, and the weld seam. The dimensions of beam profiles are taken from Australian / New Zealand standard [18] and technical specification [19]. All variants of tubes (RHS/SHS), angles (CA) and channels (CC) are listed in detailed reports [4, 5]. It should be noted that the specimens geometry can be simplified to a half model (not a quarter) using the vertical symmetry plane along the bottom member (except channel-to-channel connection which requires full model), because of the unsymmetric loading and deformation. The example of the geometry for the connection of 75x75x4 CA (top member) to 50x50x3 SHS (bottom member) is shown in Fig. 7a using solids and in Fig. 7b using surfaces. The distance between centres of supports in this work is assumed to be 460 mm. In accordance to the 2D model of a simply supported beam in Fig. 6, the points A, B and C are denoted in Fig. 7. The legs of welds around the rounded corners are 9 mm (horizontal) and 7 mm (vertical) giving 8 mm in average in compliance with experiments [9]. The weld face length is 9.9 mm, and the weld throat

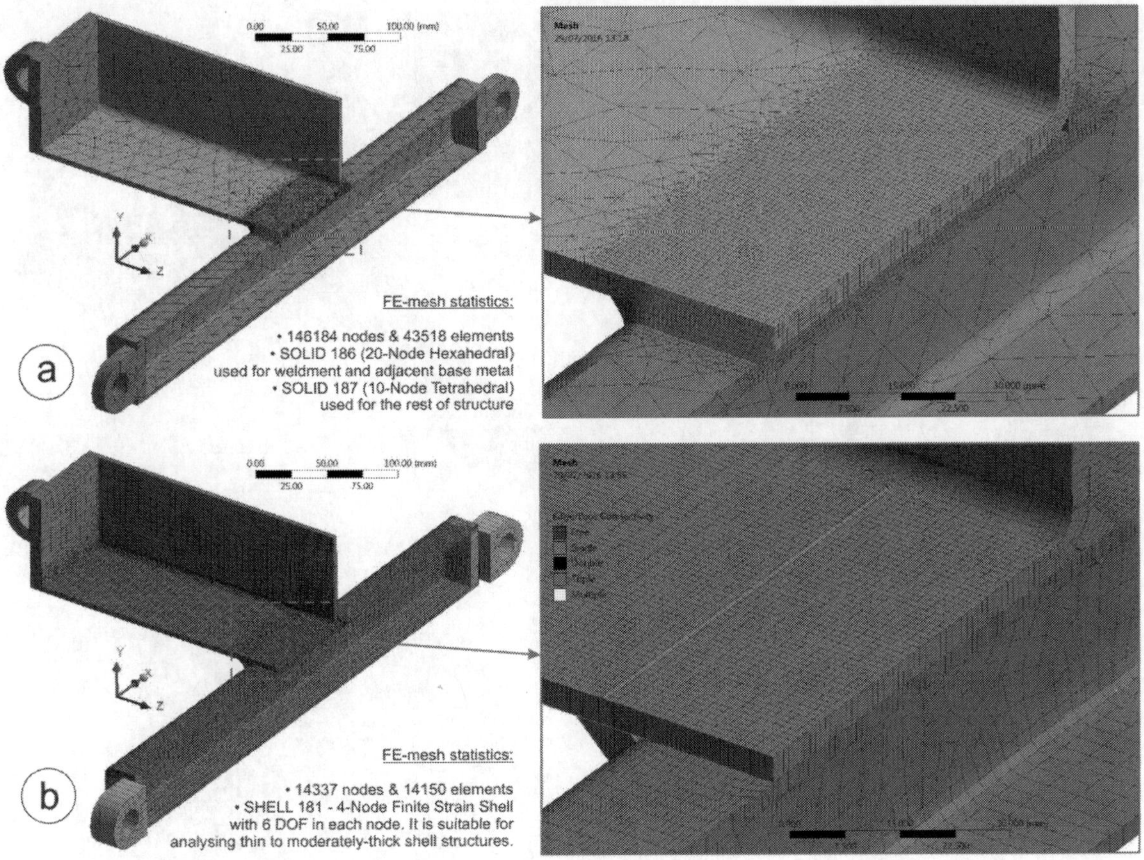

Figure 8. FE-mesh, statistics and blowup of the weldment area for 75x75x4 CA to 50x50x3 SHS connection using (a) solid FEs and (b) shell FEs.

considered in shell model formulation is 6 mm. The example of the FE-mesh corresponding to the configuration 75x75x4 CA to 50x50x3 SHS, together with mesh statistics and blowup of the location of contact between the beams are shown in Fig. 8a using solid FEs and Fig. 8b using shell FEs. It should be noted that the solution of the shell model is performed much quicker than the solution of the solid model since the number of solved equations expressed in the number of nodes is 10 times smaller for the shell model.

In solid model setup, the hinge type of constraint is applied to 3 constrained supports (points A and B in Fig. 7) as illustrated in Fig. 5, which assumes only 1 rotational degree of freedom (DOF) around the longitudinal axis and excludes any translation along the same axis. In shell model, the same boundary condition (BC) applied to same 3 supports is carried out by constraining: {1} 2 in-plain displacements (equivalent to constraining the radial displacement), {2} 2 out-of-plain rotations (which are constrained in cylindrical coupling automatically) and {3} axial translation in horizontal direction. This results in single rotational DOF in each of 3 supports and the vertical displacement applied to the centre of 4th support directed downwards. It should be noted that compared to previous study [4], where the cylindrical type of constraint (1 rotational DOF + 1 translational DOF) was applied to 3 constrained supports, this study uses the hinge type of constraint to agree more accurately with experimental setup [8, 9]. This modification of BCs results in slight increase of equivalent von Mises stress up to 10% and corresponding decrease in predicted fatigue life using shell FE-models when compared to the

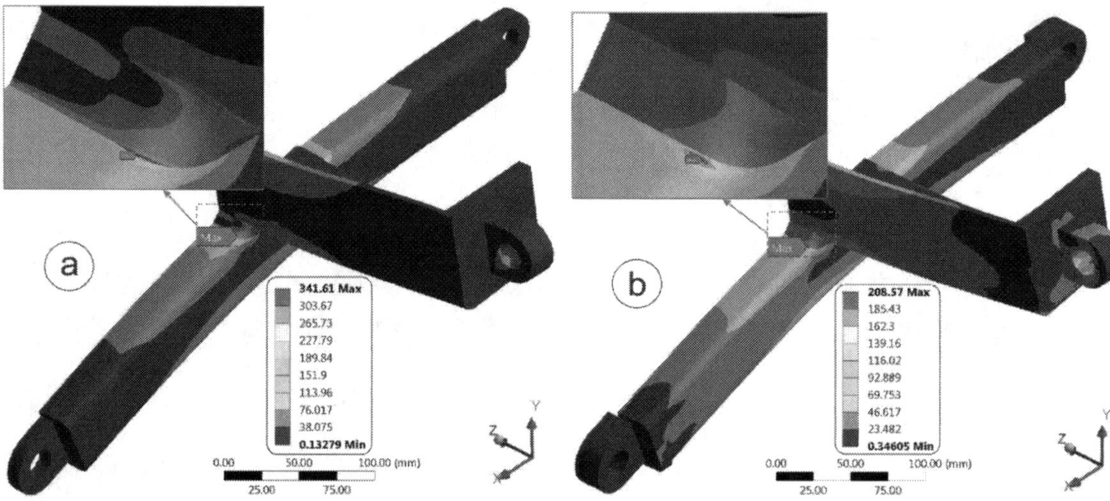

Figure 9. Equivalent von Mises stress (MPa) for 75x75x4 CA to 50x50x3 SHS connection for (a) solid FE-model and (b) shell FE-model.

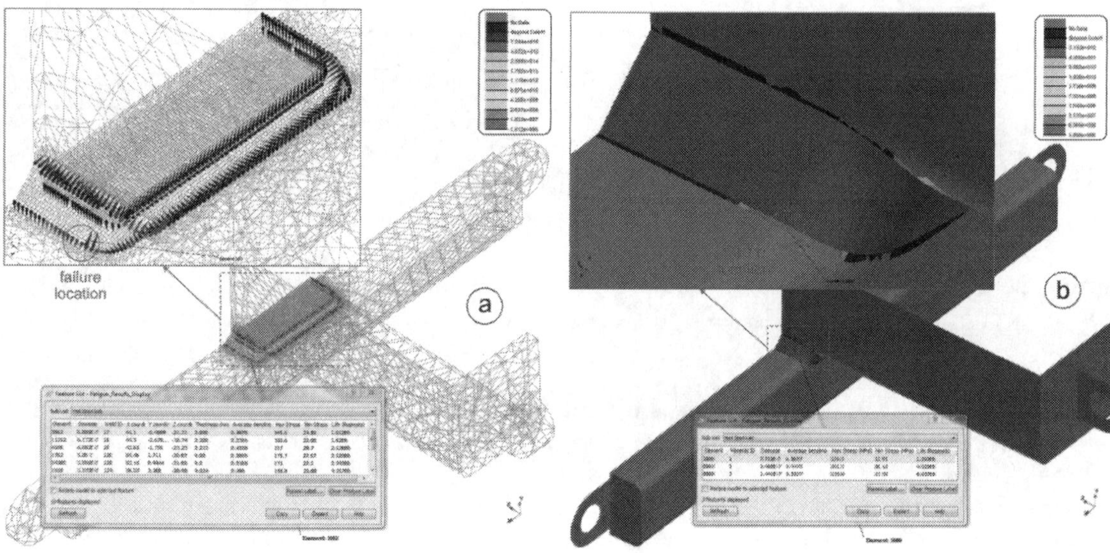

Figure 10. Result of fatigue life predictions (cycles) for 75x75x4 CA to 50x50x3 SHS connection for (a) solid FE-model and (b) shell FE-model.

results reported in previous study [4].

With equivalent BCs and loads applied to FE-models, the maximum values of equivalent von Mises stress obtained in result of structural analyses in solid models are up to two times bigger than in shell models. The examples of stress distribution together with blowup of the area of the highest stress are shown in Fig. 9a using solid FEs and in Fig. 9b using shell FEs for 75x75x4 CA to 50x50x3 SHS connection. Such a difference in maximum stress values is explained by the assumption that shell FE-model output a maximum hotspot stress in vicinity of the weld toe, while the maximum stress in solid FE-model is observed exactly on the sharp edge and mostly contributed by a non-linear component of structural stress caused by geometrical singularity.

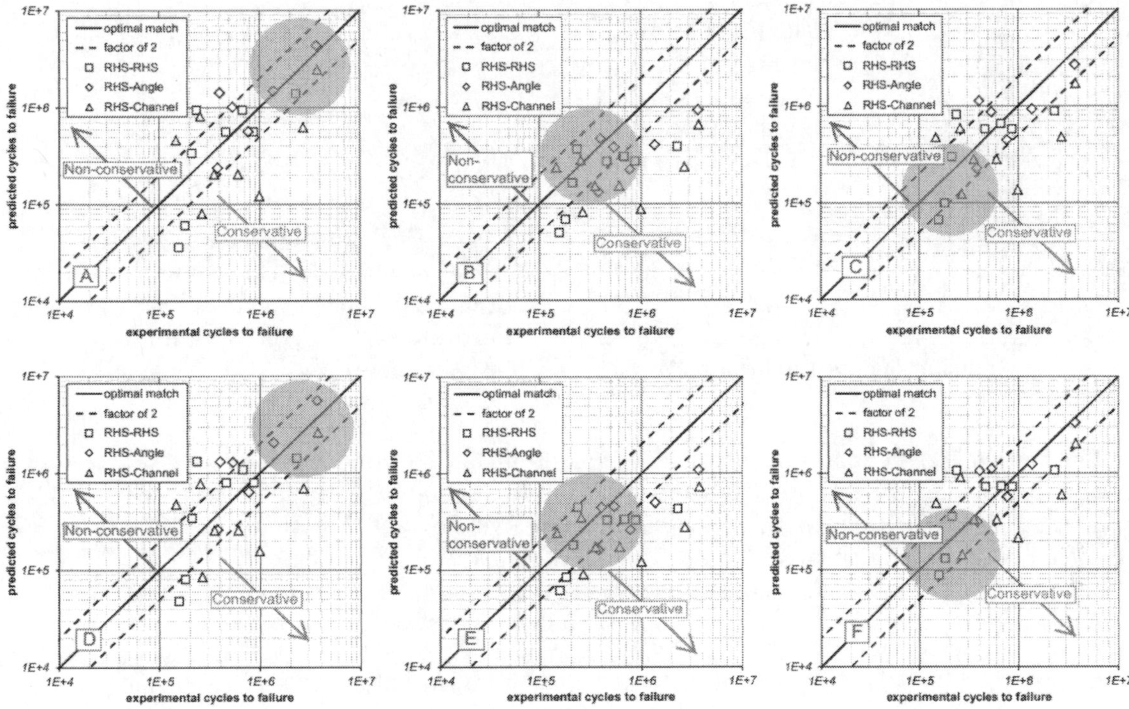

Figure 11. Comparison of the observed and predicted cycles to failure using solid (top) and shell (bottom) FE-models for: (A & D) S-N curves for SM490B steel welds {1}; (B & E) new generic S-N curves in nCode DesignLife {2}; (C & F) old generic S-N curves in nCode FE-Fatigue{3}.

4. Discussion

The results of fatigue life predictions are obtained for solid and shell formulation for 3 variants of cross-beam connections: {1} tube [RHS] to tube [RHS/SHS], {2} angle [CA] to tube [RHS/SHS], and {3} channel [CC] to tube [RHS/SHS]. Three variants of fatigue data input for CA and $t_{\mathrm{ref}} = 1$ mm from Table 1 are used in predictions: {1} S-N curves of SM490B steel welds (set no. 1b), and generic S-N curves from {2} nCode DesignLife (set no. 2b) and {3} nCode FE-Fatigue (set no. 3c). The examples of fatigue life predictions for the connection of 75x75x4 CA to 50x50x3 SHS beams using S-N curves {1} together with blowup of the crack location are shown in Fig. 10a using solid FEs and in Fig. 10b using shell FEs. The advantage of all performed numerical predictions is that the crack has been predicted exactly in the same location as in experiments [8, 9] – front part of the weld toe on the fillet of the bottom member. The numerical predictions N_*^{FE} are compared to the experimental fatigue life N_*^{exp} numerically [4, 5] and graphically in Fig 11 (top) for solid FE-models and Fig 11 (bottom) for shell FE-models using the following formula for discrepancy in percents %:

$$\Delta N_* = \frac{100\,(N_*^{exp} - N_*^{FE})}{\min(N_*^{exp}, N_*^{FE})}. \tag{11}$$

This value characterises not only the relative deviation, but also the amount of conservatism, which is defined by the sign of this value: positive – conservative and negative – non-conservative. The total value of discrepancies are calculated for three groups of experiments (RHS-RHS/SHS, CA-RHS/SHS, CC-RHS/SHS) – these and aggregate values for all considered experiments are reported in [4, 5].

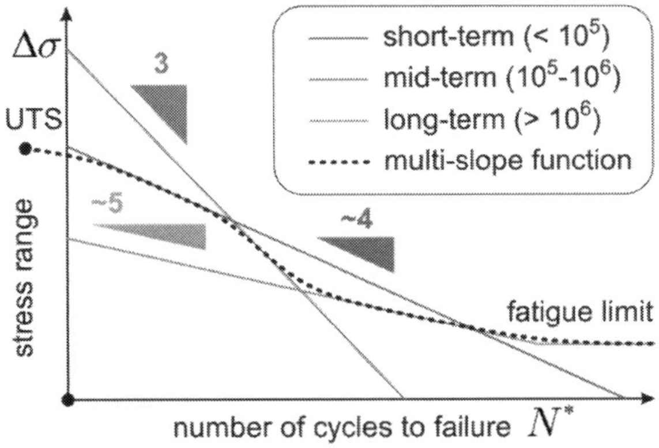

Figure 12. Multiple slopes of S-N curves are used for the formulation of a continuous function for the integral S-N curve covering all fatigue ranges.

The fatigue predictions for channel-to-tube group are the most conservative, for corner-to-tube group are less conservative, and for tube-to-tube group are non-conservative. When comparing the S-N curves input, it is clear from Figs 11b and 11e that the S-N curves {2} produce over-conservative predictions. The predictions using S-N curves {1} in Figs 11a and 11d and S-N curves {3} in Figs 11c and 11f have approximately equal small level of conservatism.

Assessment results of the failure prediction quality using Eq. (11) reported in [5] comply very well with a graphical representation of comparison of the observed and predicted cycles to failure in Fig. 11. Comparison of the total aggregate discrepancies for all considered experiments doesn't indicate the best S-N curves set, but it characterises the balance of fatigue predictions. The predictions using S-N curves {2} are unbalanced and shifted down into conservative domain. The predictions using S-N curves {1} and {3} are almost optimally balanced around the diagonal of optimal match. However, the predictions with S-N curves {3} can be considered as more accurate since all the points are closer to the diagonal with less conservative and non-conservative predictions being out of the domain for factor of 2, when compared to results obtained with S-N curves {1}. Closer look at Fig. 11 reveals that predictions with specific S-N curves {1} for SM490B steel welds have the best accuracy in long-term domain (see shaded area with $N_* > 10^6$ in Figs 11a and 11d). Predictions with generic S-N curves {2} from nCode DesignLife have the best accuracy in mid-term domain (see shaded area with $10^6 > N_* > 10^5$ in Figs 11b and 11e). Predictions with generic S-N curves {3} from nCode FE-Fatigue have the best accuracy in short-term domain (see shaded area with $N_* \sim 10^5$ in Figs 11c and 11f).

5. Conclusions

Therefore, all examined S-N curve inputs are equally good, but for different specific fatigue domains as illustrated in Fig. 12, since they all have different slopes: ~5 for S-N curves {1} for SM490B steel welds, 3 for S-N curves {2} from nCode DesignLife and ~4 for S-N curves {3} from nCode FE-Fatigue. So the most accurate fatigue predictions would be expected from a multi-slope S-N curve as suggested in Fig. 12, which is not yet available as a standard functionality for fatigue analysis of welds in nCode DesignLife. Possible candidates for the role of S-N curve with continuous range-dependent slope would be functions suggested by Bastenaire [21] or Lemaitre & Chaboche [22]

Comparing Figs 11a, 11b and 11c with Figs 11d, 11e and 11f, one can conclude that predictions using solid and shell FE-models are consistent. However, comparison of numbers N_* reported in [4, 5] indicates that predictions with solid FEs are more conservative compared to shell FEs with average differences of 20.3 %, 16.8 % and 24.9 % for each S-N curves input calculated using Eq. (11) resulting into 20.7% average difference, which is quite good. Based upon availability of ACT extension "nCode Weldline" [14] in ANSYS Workbench to facilitate the fatigue analysis in solid formulation and reasonable conservatism of corresponding predictions, the solid FE-model fatigue analysis would be recommended as more accurate, while the shell FE-model fatigue analysis might be used for quick assessment with minimum preprocessing.

The future modelling and analysis work will address the remaining 10 fatigue tests for welded Channel-to-Channel (CC-CC) cross-beam connections (all 100 x 50 x 4CC) reported in [9]. These experiments were not included in scope of current study, since their geometry couldn't be simplified using the symmetry condition and require the consideration of whole geometry in stress and fatigue analyses. The predictions for the remaining test group are bound to prove the suggested conclusions and explanations in this study.

Acknowledgements
The authors deeply appreciate the experts of HBM - nCode for the workshops, seminars and comprehensive technical support of their leading product – nCode DesignLife. Particular gratitude is expressed to Jeffrey Mentley for an essential assistance and consultations during the course of this work.

References

[1] Moore P and Booth G 2015 *The Welding Engineer's Guide to Fracture and Fatigue* (Oxford, UK: Woodhead Publishing)
[2] Heyes P (Mar 2013) Fatigue analysis of seam welded structures using nCode DesignLife Whitepaper HBM nCode Rotherham, UK
[3] Munse W H 1978 *Fatigue Testing of Weldments (ASTM STP 648)* ed Hoeppner D W (Philadelphia, PA, USA: ASTM) pp 89–112
[4] Gorash Y, Comlekci T and MacKenzie D 2015 *Procedia Engineering* **133** 420–432
[5] Gorash Y, Comlekci T and MacKenzie D 2017 *Thin-Walled Structures* 17 p., under review
[6] Chang K H 2013 *Product Performance Evaluation with CAD/CAE* (Boston, USA: Academic Press) Chapter 4, pp 205–273
[7] Newbold P 2013 *DesignLife Theory Guide* HBM nCode Rotherham, UK Version 9 ed
[8] Mashiri F R and Zhao X L 2010 *Thin-Walled Structures* **48** 159–168
[9] Mashiri F R, Zhao X L and Tong L W 2013 *Thin-Walled Structures* **63** 27–36
[10] Fatigue Testing Division (2003, 2004, 2006, 2009, 2011) Data sheet on fatigue properties of non-load-carrying cruciform welded joints of SM490B rolled steel for welded structure: NIMS Fatigue Data Sheets No. 91, 96, 99, 108, 114 National Institute for Materials Science Tsukuba, Japan
[11] EasySteel™ 2016 *The Steel Book* (Auckland, New Zealand: Fletcher Building Ltd.)
[12] Fermér M, Andréasson M and Frodin B 1998 *SAE Technical Papers No. 982311* 1–9
[13] ASME BPV Committee on Pressure Vessels 2010 *2010 ASME Boiler & Pressure Vessel Code: Section VIII Division 2 – Alternative Rules* 2010th ed (New York, USA: ASME)
[14] Mentley J 2015 nCode DesignLife Weldline ACT Extension PowerPoint Presentation, HBM - nCode Products (Southfield, MI, USA)
[15] Haibach E 2003 *FKM-Guideline: Analytical Strength Assessment of Components in Mechanical Engineering* 5th ed (FKM, Frankfurt am Main, Germany: VDMA Verlag)
[16] British Standard 2014 *Guide to fatigue design and assessment of steel products* BS 7608:2014 (London, UK: The British Standards Institution)
[17] Budynas R G and Nisbett J K 2010 *Shigley's Mechanical Engineering Design* 9th ed (New York, USA: McGraw-Hill)
[18] Australian / New Zealand Standard 2009 *Cold-formed structural steel hollow sections* AS/NZS 1163:2009 (Sydney / Wellington: Standards Australia / Standards New Zealand)
[19] AustubeMills 2013 *TS100 – angles, channels, flats* 7th ed DuraGal® Technical Specifications (Acacia Ridge, QLD, Australia: Australian Tube Mills Pty Ltd.)
[20] Zhao X L, Herion S, Packer J A, Puhtli R S, Sedlacek G, Wardenier J, Weynand K, van Wingerde A M and Yeomans N F 2001 *Design guide for circular and rectangular hollow section welded joints under fatigue loading* (Köln, Germany: TÜV-Verlag GmbH)
[21] Bastenaire F 1972 *Probabilistic Aspects of Fatigue* (*ASTM special technical publication* no STP 511) ed Heller R A (Philadelphia, USA: ASTM) chapter 1, pp 3 – 28
[22] Lemaitre J and Chaboche J-L 1994 *Mechanics of Solid Materials* (Cambridge, UK: Cambridge University Press)

IOP Conf. Series: Journal of Physics: Conf. Series **843** (2017) 012026 doi:10.1088/1742-6596/843/1/012026

Anisotropic degradation behavior of moduli of extruded pure magnesium during low cyclic-tension fatigue

H. Yamagishi

Institute for Manufacturing Research and Development, Toyama Industrial Technology Center, 150 Futagami, Takaoka 933-0981, Japan

E-mail: yamagisi@itc.pref.toyama.jp

Abstract. The mechanical properties of extruded pure magnesium during low cyclic-tension fatigue at room temperature were investigated in the extrusion and transverse loading directions (LDs) using ultrasonic reflection methods with longitudinal and shear waves (SWs), to clarify if there was an anisotropic degradation behavior between the extruded directions (EDs), which had not yet revealed. Regardless of the LD, the acoustic velocities and calculated Young's and shear moduli decreased significantly with an increasing number of cycles because of the growth of voids at the grain and twin boundaries. An anisotropic behavior was revealed: when the deflection surface of the SWs was aligned in the LD, the amount of decrease in the moduli was greater than when the alignment was in the transverse direction (TD). Additionally, when the stress amplitude was adjusted to provide the same number of cycles to failure, the decrease in the moduli was somewhat greater when the LD was parallel to the extruded direction than to the TD. Longitudinal and SW propagation characteristics, and investigations of grain and twin boundaries after fatigue using electron backscatter diffraction (EBSD) based on field-emission scanning electron microscopy (FE-SEM), revealed that the most of the void-gap width was less than several nanometers (almost closed), which corresponded to the longitudinal wave amplitude. Other damaged-phase data were obtained using X-ray diffraction (XRD), optical microscopy (OM), scanning electron microscopy (SEM) and Vickers-hardness test; the anisotropic degradation behaviors were attributed to the void morphology and the slight difference in orientation.

1. Introduction

Fatigue fracture often brings great social loss, and no evaluation method is sufficiently rigorous for field applications that could be applied to Stage I crack growth, in general which corresponds to most life of fatigue. Thus, a variety of approaches of nondestructive evaluation have been studied to detect slight changes of material during fatigue. For instance, eddy current [1], impedance-based monitoring [2], X-ray diffraction (XRD) [3], positron annihilation (PA) [4] and neutron diffraction (ND) [5] have been conducted. And for fretting fatigue, which occurs under a special condition of repeated relative surface motion and reduces the fatigue strength, the numerical simulations have been conducted [6-9].

On the other hand ultrasonic technique, which is relatively inexpensive and easy to use, have been also studied for long [10-12]. There are roughly two analyses methods in the technique, linear and non-linear. Linear method have been mostly applied for a concrete damage such as a crack. While, in recent years non-linear methods using harmonics have been becoming applied for fatigue evaluation due to its sensitiveness to dislocation motion [13-15]. There are also roughly two types of propagation modes in ultrasonic, longitudinal- and shear waves. In general, longitudinal wave is widely applied for both the linear- and non-linear methods due to its usability. Recently, the author of this article has applied shear wave to fatigue evaluation in nondestructive by use of linear- and non-linear methods, since shear wave is known to have highly sensitive interaction with a slight change of material, and indicated the characteristic behaviors accompanied with dislocation multiplication and residual stress during the fatigue [16-18].

Light metals, such as magnesium, are the most promising for structural applications in a low-carbon society. The strong texture of wrought magnesium makes it more difficult to form and the uncertainty in its fatigue lifetime limits its widespread use as a structural material.

Previous studies [19-21] used an ultrasonic pulse technique to non-destructively evaluate metal fatigue at room temperature. In that research, the elastic behavior of pure magnesium under low-, high- and giga-cycle tension fatigue was quantified when the loading direction (LD) was parallel to the extruded direction (ED). Significant reductions in the moduli were accompanied by fatigue damage. Most of the degradation was attributed to the presence of voids at the grain boundary. In this study, since the mechanical properties of expanded magnesium depend strongly on the ED, the anisotropy of the moduli behavior during low-cycle fatigue was evaluated by changing the LD : the validation was conducted using the already-known data of ED = LD, which was from Ref. [20], and newly acquired data of ED = TD. These results could contribute to a database allowing nondestructive inspection using this technique.

2. Experimental procedure

The material tested in this study was pure magnesium (Timminco Corp., Aurora, CO, USA). The chemical composition is shown in Table 1, and the mechanical properties of the material are listed in Table 2. The orientations obtained by X-ray diffraction (XRD) for the (0001) pole, using an incident slit width of 0.84 mm and a height of 3 mm, were 0.83 in the normal direction (ND), 0.07 in the ED and 0.09 in the transverse direction (TD). The configuration of the fatigue test sample is shown in Figure 1(a). The ED was oriented parallel to the LD (ED = LD specimen) and the TD (ED = TD specimen). The samples were subjected to cyclic tensile stress under controlled conditions. The stress ratio, R, was 0, and the frequency was 30 Hz, which was controlled by a hydraulic servo fatigue tester at room temperature. Figure 1(b) shows the cyclic stress as a function of the number of cycles to failure (S–N) for the material under the test conditions. The ultrasonic data of fatigue failure in the low cycles for ED = LD and ED = TD specimens were obtained under stress amplitudes of 28.4 MPa and 36.5 MPa, respectively. The specimens oriented as ED = TD were removed from the fatigue machine grip at arbitrary numbers of cycles to test the ultrasonic waveform. Under the ED = LD condition, the low-cycle samples' stress amplitude, which was greater than the proof strength, made the specimen surfaces too rough to conduct ultrasonic measurements accurately. Hence, those fatigued specimens were separately prepared using four different numbers of cycles; the 69,478-cycle specimen was fatigue-fractured such that macroscopic cracks were evident in the evaluation area. Ultrasonic measurements were conducted before and after the fatigue tests; the ED = LD-fatigued specimens were polished again with 1,000-grit SiC paper for the measurements. The rates of change in the ultrasonic parameters were calculated by the fatigue progress of each sample. The ultrasonic data were collected until a normalized fatigue ratio, N/N_f, of 1.0 was reached. The ultrasonic parameters of the as-extruded materials are listed in Table 3.

Table 1. Chemical composition of the magnesium material (mass%).

Material	Al	Zn	Mn	Fe	Ni	Cu	Si	Pb	Ca	Sn	Cd	Mg
Pure Mg	0.004	0.006	0.0074	0.0021	0.0007	0.004	0.003	0.001	0.001	<0.001	<0.0001	Bal.

Table 2. Mechanical properties of the material.

Temper	0.2% Proof stress (MPa)	Tensile strength (MPa)	Elongation (%)
as-extruded (ED = LD)	53	147	9.6
as-extruded (ED = TD)	131	187	2.8

Table 3. Acoustic properties of the as-extruded material.

Deflected surface of SVs	Longitudinal wave velocity (V_l/m·s^{-1})	Shear wave velocity (V_s/m·s^{-1})	Young's modulus (E/GPa)	Shear modulus (G/GPa)	Poisson's ratio (ν)
ED	5585	3132	43.38	17.07	0.271
TD	5585	3081	42.33	16.52	0.281

To evaluate variation in the elastic moduli as fatigue progressed, the longitudinal and shear wave (SW) velocities in the middle of the samples (Figure 1(a)) were measured separately using longitudinal wave and vertically polarized SW transducers, with a center frequency set to 5 MHz. The diameters of the longitudinal and SW probes were 4 and 10 mm, respectively. The transducers made contact with the sample under a load of 55 N using water-free naphthenic hydrocarbon oil at room temperature. The deflected surface of the SWs was aligned in the LD and TD. Ultrasonic wave pattern analysis was performed using an elastic modulus system (UMS-H; Toshiba Tungaloy, Kawasaki, Japan). The input voltage, sampling frequency and averaging number were 100 V, 100 MHz, and 16, respectively. The measuring accuracy of the system is within 0.1 % in velocity.

The detailed procedure is described in Ref. [19]. The wave velocities, i.e., the longitudinal V_l and shear V_s, gave the Young's modulus, E, and the shear modulus, G, according to the following relationships [22], where ρ is the density of the sample:

$$E = \frac{3\rho V_s^2\left(V_l^2 - \frac{4}{3}V_s^2\right)}{V_l^2 - V_s^2} \tag{1}$$

$$G = \rho V_s^2 \tag{2}$$

The structural variations in the material during the fatigue process were evaluated using the ultrasonic evaluation samples. After fatigue failure, the fracture surface and cross-sectional texture in the LD were examined using scanning electron microscopy (SEM), optical microscopy (OM) and an electron backscattering diffraction pattern (EBSD) method based on field-emission scanning electron microscopy (FE-SEM). In addition, using the specimens with different number of cycles the Vickers hardness were measured at the center of the samples in the LD. The numbers for averaging were 10 points.

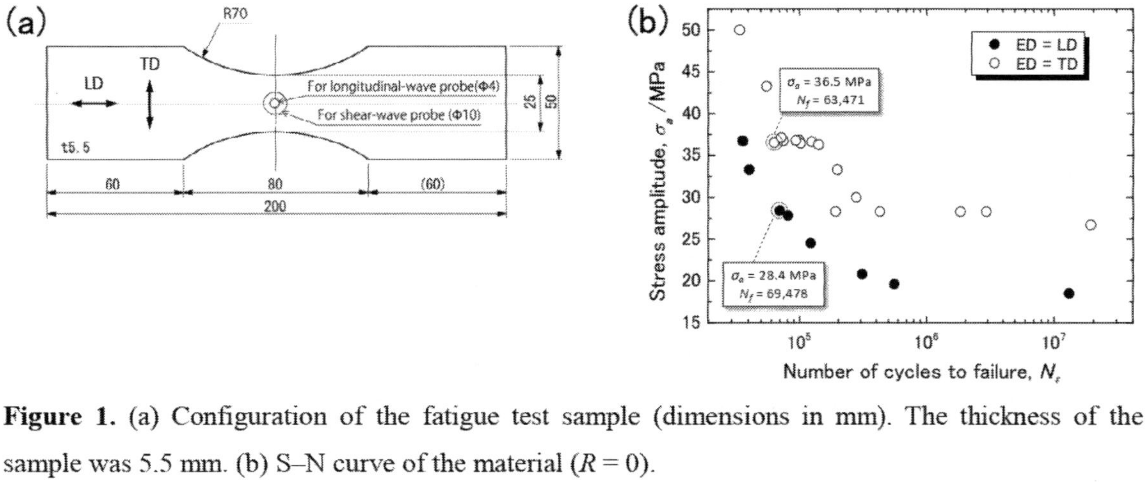

Figure 1. (a) Configuration of the fatigue test sample (dimensions in mm). The thickness of the sample was 5.5 mm. (b) S–N curve of the material ($R = 0$).

IOP Conf. Series: Journal of Physics: Conf. Series **843** (2017) 012026 doi:10.1088/1742-6596/843/1/012026

3. Results

The change ratios for the velocities of the longitudinal waves and SWs, as a function of N/N_f during the low-cycle fatigue, are shown in Figure 2(a) and (b), respectively. The thick and slim arrows in the graphs indicate the ED and the directions of the deflected surface of the SWs, respectively. The results reveal the following two tendencies that did not depend on the ED. First, for a given specimen, the SW velocity decreased to a greater extent than the longitudinal wave. Second, for the same specimen, the SW velocity decreased more when the deflected surface was aligned with the LD than with the TD. Furthermore, although the cycle to fracture was almost same, by comparing the two behaviors of the ED = LD and ED = TD specimens, the following tendency, which depended on the ED, was revealed. The velocities of the longitudinal and SWs decreased more when the LD was parallel to the ED (ED = LD specimen) than in the transverse setting (ED = TD specimen). Notably, the stress amplitude of the ED = LD specimen was 22 % less than that of the ED = TD specimen.

Figure 3(a) and (b) shows the change ratios for the Young's and shear moduli, respectively, resulting from the velocities as a function of the N/N_f during the low-cycle fatigue. The thick and slim arrows in the graphs indicate the ED and the directions of the deflected surfaces of the SWs, respectively. As well as the velocities, the results reveal the following two tendencies that did not depend on the ED. First, for a given specimen, the shear modulus decreased slightly more than the Young's modulus. Second, for the same specimen, the Young's and shear moduli decreased more when the deflected surface of the SWs was aligned in the LD rather than in the TD. Furthermore, by comparing the two behaviors of the ED = LD and ED = TD cases, the following tendency, which depended on the ED, was revealed. When the LD was parallel to the ED (ED = LD specimen), the Young's and shear moduli decreased more than with the transverse setting (ED = TD specimen).

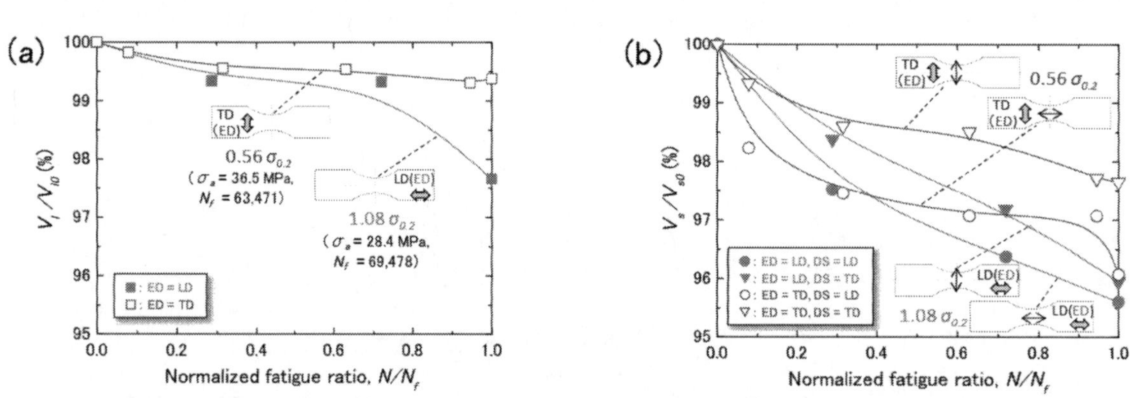

Figure 2. Change ratios during low-cycle fatigue for the velocities of the (a) longitudinal waves and (b) shear waves (SWs) as a function of the normalized fatigue ratio, N/N_f (DS: deflected surface).

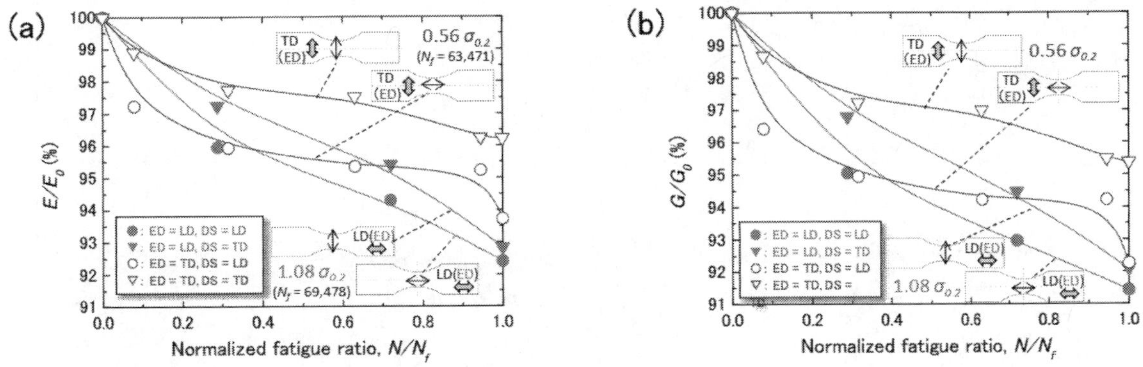

Figure 3. Change ratios during low-cycle fatigue for the (a) Young's modulus and (b) shear modulus as a function of N/N_f (DS: deflected surface).

The fracture surfaces near the crack initiation site of both loading modes, ED = LD and ED = TD, are shown in Figure 4. All of the surfaces exhibited crack propagation at the grain boundary, i.e., the grain boundary was considerably degraded during both fatigue modes. Optical micrographs of the cross-sections of the LD of the ED = LD specimen, and the EBSD analysis of the LD of the ED = TD specimen, are shown in Figure 5(a) and (b), respectively. Both materials showed interface voids at the grain or twin boundaries that corresponded to the LD, and which were mainly normal to the principal stress or oriented at 45° to it. However, recognizable voids with openings in the micrometer-range were few in total in a 10-mm-wide view. Figure 6 shows the variation of the Vickers hardness with the increasing number of fatigue cycles. The ED = LD specimen indicated a work-hardening behavior in the early cycles of fatigue. On the other hand the ED = TD specimen relatively stayed unchanged during the early period.

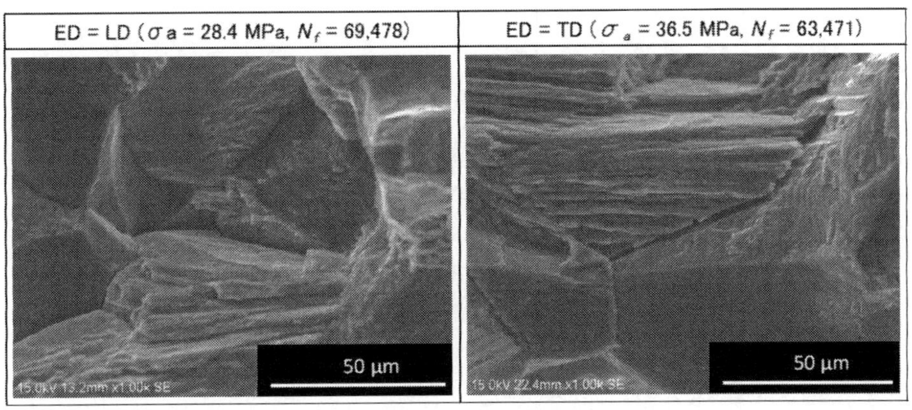

Figure 4. Scanning electron microscopy (SEM) images of the fracture surface near the crack initiation site.

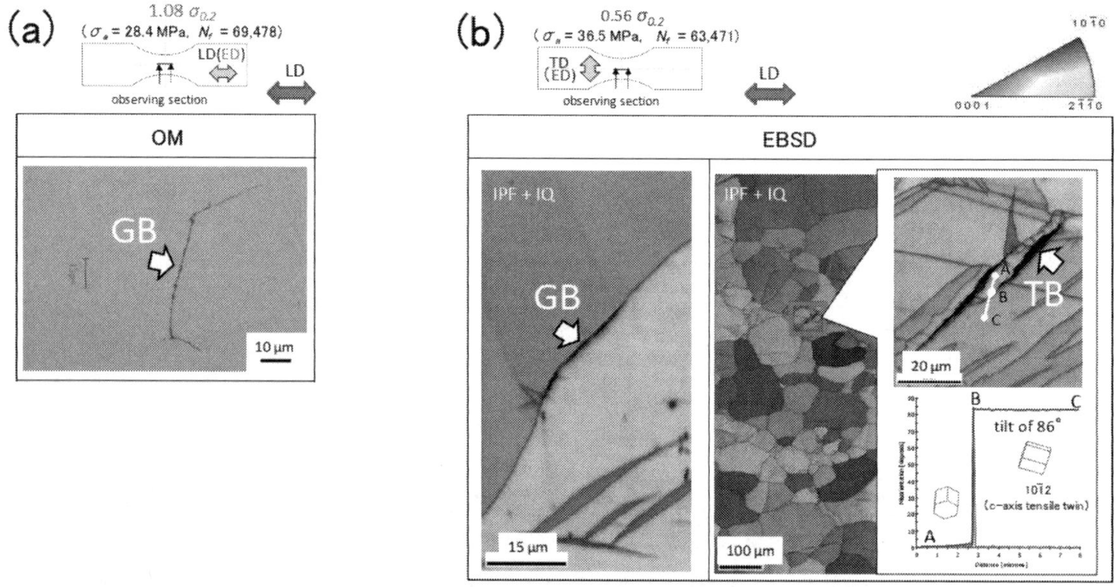

Figure 5. (a) Optical microscopy (OM) image and (b) electron backscatter diffraction (EBSD) analysis of the cross-sections in the loading direction (LD) for the extruded direction (ED) and transverse direction (TD) specimens, respectively (GB = grain boundary, TB = twin boundary).

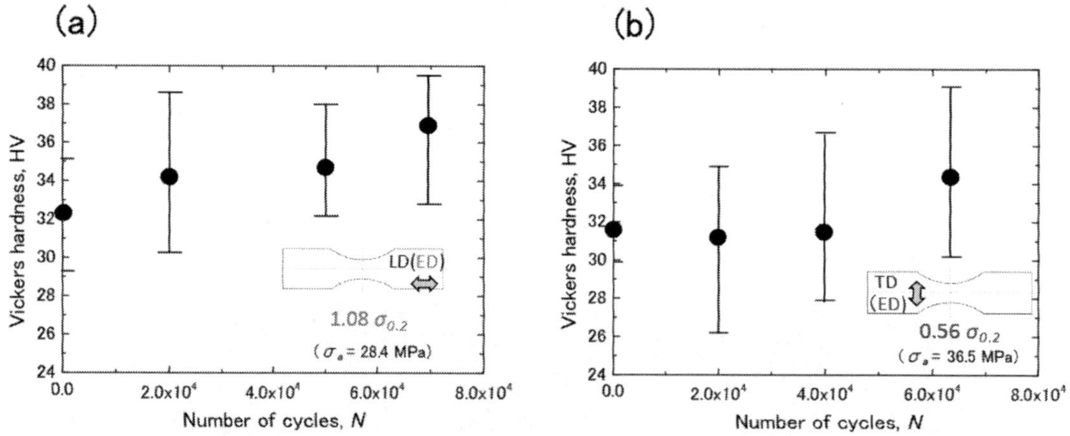

Figure 6. Variations of the Vickers hardness of the (a) ED = LD specimen and (b) ED = TD specimen.

4. Discussion

4.1. Anisotropy derived from defect morphology (anisotropy did not depend on the ED)

First, we consider the cause of the large decrease in the velocities with fatigue damage (Figure 2). This mechanical degradation, i.e., the several percentage-point decrease in velocities and resulting moduli, cannot be explained by common lattice defects, such as strain, twin, or dislocations; the influence of the velocity was estimated at less than 1 % [20]. Thus, voids at the grain and twin boundaries evident in Figure 5 were considered as the cause of the degradation. However, since the recognized voids were quite few, the defect that grew under the fatigue modes was thought to be mainly in a closed state [21]. In fact, regarding the grain-boundary damage, the deformation mechanism map of pure magnesium [23] indicates the possibility of grain-boundary creep under the fatigue test condition, even at room temperature. Furthermore, the two velocity anisotropies noted in Figure 2 support the boundary degradation concept resulting from the growth of a void. First, as shown in Figure 7(a), a longitudinal wave propagates a nanometer-sized gap in a state of compressive stress. In principle, a SW cannot propagate such a void, even if it is completely closed because of a lack of elastic restoration. Thus, under the development of voids, the defect is influenced largely by the SWs rather than by the longitudinal waves. Second, under uniaxial loading, a diffusional void can be dynamically induced in the normal plane against the LD. Thus, as shown in Figure 7(b), the SW propagates more easily, with a deflected surface aligned in the TD rather than in the LD (Figure 2(b)). The behaviors mentioned above are anisotropies derived from the defect morphology, which depended on the LD and not on the ED; this was recognized for both ED = LD and ED = TD specimens. The anisotropic behavior of the moduli calculated from the velocities was the same. Concerning the slight difference between the Young's and shear moduli, it is clear from Equations (1) and (2) that the shear modulus is more affected by an SW.

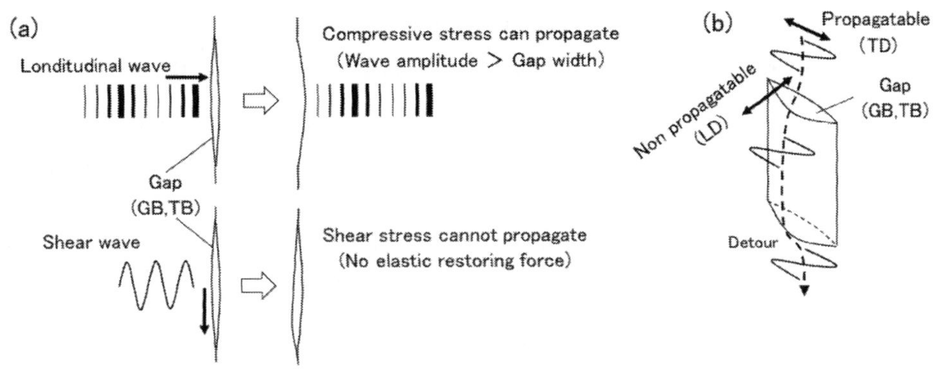

Figure 7. Propagation advantage between (a) longitudinal and SWs and (b) SWs for which the deflected surfaces are aligned in the LD and TD.

4.2. Anisotropy derived from texture (anisotropy depended on the ED)

The anisotropies discussed in the previous section derived from the void morphology that depended on the LD, not on the ED. On the other hand, the observed anisotropic behavior, whereby the velocities and moduli decreased more when the LD was parallel to the ED (ED = LD specimen) than in the transverse setting (ED = TD specimen), cannot be explained by the defect morphology. This anisotropy is considered to result from the texture, which was determined by the ED.

The development mechanism of a void during the fatigue test is now considered. In pure magnesium, for which the grain-boundary diffusion coefficient is relatively high [24], Coble creep, grain-boundary sliding (GBS) or cleavage crack originating from the two creeps could contribute to the interface degradation. However, there is a conflict, i.e., although the development of all of these boundary defects is dependent on the driving stress, the decrement of the mechanical properties was much larger in the ED = LD specimen, for which the stress amplitude was 22 % lower, than in the ED = TD specimen. The following is a plausible explanation. The value of the critical resolved shear stress (CRSS) of magnesium suggests that a basal slip is the dominant slip system at room temperature [25]. The specimen in this fatigue test had a strong texture; the *c*-axis was oriented in the through-thickness direction, yet the orientation was slanted slightly to the ED (ND: 0.83, ED: 0.09, TD: 0.07). Since the strong texture restricts the slip system considerably, the slight difference of *c*-axis orientation between the ED and TD led to the different mechanical behaviors. Considering the effect of shear stress on basal slip in this test, it was noted that the Schmidt factor of the ED = LD specimen was higher than that of the ED = TD specimen (Figure 8). The data of Table 2 suggest that the anisotropy of the mechanical properties could result from the orientation difference. In fact, as shown in Figure 6 the difference in behavior of the variations of Vickers-hardness supports this: the data of ED = LD specimen show explicit work-hardening due to a basal slip. Facile basal slip in the ED = LD specimen would result in the accumulation of dislocations at the grain or twin boundaries, leading to stress concentrations at the boundaries. Thus, a relatively small driving stress would lead to significant degradation at the interfaces. Nevertheless, the cause of the anisotropy that depended on the ED remains incompletely understood; with verification of quantitative repeatability the relationship between the morphology of lattice defects and fatigue progress, and an analysis of the Schmidt factor of the texture, are required.

IOP Conf. Series: Journal of Physics: Conf. Series **843** (2017) 012026 doi:10.1088/1742-6596/843/1/012026

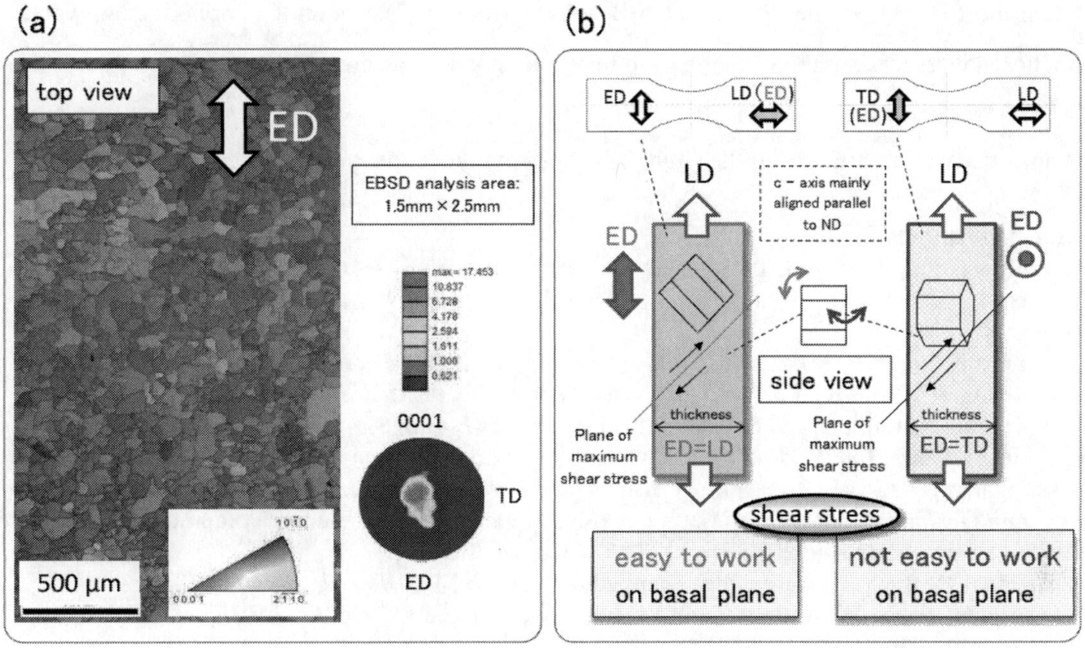

Figure 8. (a) Inverse pole figure (IPF) map of the pure magnesium. (b) Conceptual rendering of ease of basal-plane sliding by maximum shear stress between the ED = LD and ED = TD specimens.

5. Conclusions

The mechanical properties of extruded pure magnesium during low cyclic-tension fatigue were investigated at room temperature, in the extruded and transverse LDs, using ultrasonic reflection methods with longitudinal waves and SWs. The following anisotropies were found:

For a given specimen,

- Both longitudinal and SW velocities decreased significantly with fatigue progress. Comparing the two wave modes, the decrease was clearly greater for the SW than for the longitudinal wave. The resulting Young's and shear moduli, calculated from the velocities, also decreased significantly. The decrease in the shear modulus was slightly greater than that of the Young's modulus, because the shear modulus is more strongly affected by SWs.
- The SW velocity decreased more when the deflected surface was aligned with the LD than with the TD. The resulting moduli were similar.

These anisotropies were attributed to the defect morphology.

At nearly the same number of cycles to failure, the ED = LD and ED =TD orientations showed the following anisotropies:

- When the LD was parallel to the ED (ED = LD specimen), although the applied stress was 22 % smaller, the decrease in the velocities and resulting moduli was greater than in the transverse setting (ED = TD specimen).

This anisotropy was attributed to the slight orientation of the texture.

Acknowledgements
This work was supported by JSPS KAKENHI Grant Number JP 50416153.

References

[1] Yusa N, Perrin S, Mizuno K, Chen Z and Miya K. 2007 *Meas Sci Tech* **18** 3403
[2] Chang YZ, Liu MC, Pai CA, Lin CL and Yen KI. 2011 *Med Eng Phys* **33** 604
[3] Kamura N, Fujita T and Sasaki T. 2017 *Mater Sci Forum* **879** 891
[4] Hori F. 2005 *Appl Surf Sci* **242** 304
[5] Takahashi S, Sasaki T, Suzuki H, Kato T and Hirose Y. 2008 *Proc. 43th Symp on X-ray Studies on Mechanical Behavior of Materials* (Kyoto, Japan / The Society of Materials Science) p 184
[6] Ali Bhatti N and Abdel Wahab M 2017 *Tribol Int* **109** 552
[7] Kumar D, Biswas R, Hien Poh H and Abdel Wahab M 2017 *Tribol Int* **109** 124
[8] Ferjaoui A, Yue T, Abdel Wahab M and Hojjati Talemi R 2015 *Int J Fatigue* **73** 66
[9] Hojjati Talemi R and Abdel Wahab M 2013 *Tribol Int* **60** 176
[10] Bratina WJ and Mills D. 1962 *Can Metall* **1** 83
[11] Katsumura Y, Takatsu S and Kobori K. 1973 *J JSPM* **20** 23
[12] Liu JH, Li GL, Hao XY, Zeng DB and Sun ZhH. 2001 *Mater Lett* **50** 194
[13] Metya A, Parida N, Bhattacharya DK, Bandyopadhyay NR and Sagar SP. 2007 *Metall Mater Tranas A* **38A** 3087
[14] Shui G, Kim JY, Qu J, Wang YS and Jacobs LJ. 2008 *NDT & E Int* **41** 326
[15] Kumar A, Adharapurapu RR, Jones JW and Pollock TM. 2011 *Scripta Mater* **64** 65
[16] Yamagishi H, Fukuhara M and Chiba A. 2010 *Mater Trans* **51** 962
[17] Yamagishi H, Fukuhara M, Matsumoto H and Chiba A. 2010 *Metall Mater Trans A* **41** 2151
[18] Yamagishi H and Fukuhara M 2015 *Metall Mater Trans A* **46** 5114
[19] Yamagishi H, Fukuhara M and Chiaba A. 2010 *Mater Trans* **51** 1255
[20] Yamagishi H, Fukuhara M and Chiaba A. 2010 *Mater Trans* **51** 2025
[21] Yamagishi H and Fukuhara M 2012 *Acta Mater* **60** 4759
[22] Fukuhara M and Yamauchi I 1993 *J Mater Sci* **28** 4681
[23] Sato H 2000 *Handbook of advanced magnesium technology*, ed Kojima Y (Tokyo: Kallos) p125
[24] Mabuchi M 2004 *Magnesium processing technology* (Tokyo: The Japan Society for Technoloogy of Plasticity) p140
[25] Yoo MH 1981 *Metall Trans A* **12** 409

Fatigue in a heat treatable high silicon containing aluminium alloy

J.A. González[1], J. Talamantes-Silva[1], S. Valtierra[1] and Rafael Colás[2]

[1] Nemak México, S.A., Libramiento Arco Vial Km. 3.8, 66000 García, N.L., México.
[2] Facultad de Ingeniería Mecánica y Eléctrica, Universidad Autónoma de Nuevo León, 66455 San Nicolás, N.L., México.
Corresponding author: rafael.colas@uanl.edu.mx

Abstract. The use of cast aluminium alloys in automobiles contributes to reductions in weight and fuel consumption without impairing the safety for the occupants or the performance of the car. Most of the alloys used are heat treatable hypoeutectic Al-Si alloys, which have the drawback of exhibiting low wear resistance. So industry relies in wear resistant alloys, such as grey iron, for the liners of the combustion chambers in engine blocks, which increase the weight of the engine. Therefore, it is of interest to cast high silicon containing alloys into engine components that are able to resist wear while maintaining the mechanical properties required by the components. This work presents the result of the work carried out in a high silicon containing heat treatable aluminium alloy as it is subjected to high cycle fatigue. The alloy was prepared and cast in ingots designed to promote one dimensional solidification gradient to obtain samples to study the high cycle fatigue. The material was machined into hour-glass specimens that were tested at room temperature in a servohydraulic machine under load control following the stair case method. The results show that the resistance to fatigue depends on the microstructure of the sample, as the fatigue cracks originate in pores close to the surface of the sample and propagate through the eutectic aggregate. The results from this work are compared with those from previously obtained with hypoeutectic alloys.

1. Introduction

The engine block in automobiles converts chemical into mechanical energy by burning fuel within the combustion chambers. Most of the energy produced by the explosion of the air-fuel mixture is converted into heat, which has to be extracted through the walls of the chambers to the proper block and to the refrigerating fluids. Heat extraction has to be efficient otherwise the engine block may be subjected to excessive thermal loads that will modify the mechanical properties of the alloy from which the block is made, leading to its catastrophic failure [1,2]. Engine blocks in automobiles have been migrating from cast iron to aluminium alloys [3,4] to reduce their weight, improve fuel consumption and comply with environmental regulations imposed by governments in different countries. Unfortunately, conventional aluminium alloys do not exhibit the resisting capabilities required for sustaining the wear caused by the reciprocating movement of the pistons, so most engine blocks are manufactured with sleeves or liners made from wear resistant alloys, which will result in the increase of around 6 kg in a V8 engine when grey iron liners are used [2,5].

Hypereutectic Al-Si alloys are used in applications in which resistance to wear is adamant, due to the presence of primary silicon crystals, nevertheless these crystals pose a machinability problem for engine plants, and may not be able to comply with the rigorous static and dynamic mechanical properties demanded by the use in engine blocks [5-10].

The defects that are present in cast aluminium alloys affect their resistance to fatigue; the flaws that exert the higher influence are pores, either due to shrinkage or gas evolution, hard components of eutectic aggregates, intermetallic inclusions and precipitated particles, as cracks nucleate on the defects and grow following paths in which other smaller faults are concentrated [11-22]. The distribution and size of such defects are modified by augmenting the rate at which the casting solidifies; as smaller sizes and finer distribution are obtained by the increased nucleation and reduction in the time available for growing [3,4,23,24]. The most common method to assess the degree of microstructural refining is by means of the secondary dendrite arm spacing, DAS, which is related to the heat transfer rate occurring during solidification [4,23,24]. As fatigue cracks nucleate and grow from existing defects, it is of critical importance to determine the type of defects that exert the higher influence [11-22].

The aim of this work is to present the results of a series of studies carried out to evaluate the fatigue resistance of a cast aluminium alloy containing high amounts of silicon, as it is intended for the manufacture of automotive components. The study was conducted on heat treated samples that had different degrees of microstructural refining as they were obtained from ingots designed for this purpose.

2. Experimental procedure

The alloy (13.5 Si, 2.9 Cu, 1.4 Ni, 0.3 Mg, 0.15 Fe, 0.12 Ti, 0.07 Sr, wt.%) was prepared in a gas fired furnace of 250 kg of capacity. The liquid was kept at 760°C as it was degassed with nitrogen before adding titanium and strontium to refine the grain size and modify the eutectic aggregate; the melt was poured into sand moulds that contained a grey iron chilling plate at their bottom to promote directional solidification, Fig. 1a. The ingots were cut to obtain prismatic bars that were heat treated to obtain a T7 condition. Solution was carried out at 495°C for 5 h, followed by cooling in water at 90°C and aged at 240°C for 5 h. Three different levels of microstructural refining were considered. The bars that correspond to Level I are 1 to 4, those of Level II are 7 to 13 and for Level III are 18 to 25, Fig. 1b.

The material was machined into cylindrical bars of 15.0 ± 0.02 mm to produce hour-glass shape specimens that had a reduced diameter of 6.0 ± 0.05 mm at their centre, following a radius of curvature of 48.0 ± 0.25 mm. Tensile test were carried out the ASTM standards B-557 [25] and E-8 [26] in specimens of 12.7 mm in diameter by 62.7 mm in gauge length; the values of the yield (σ_y) and tensile (σ_u) strength, and total elongation (Δl) were recorded.

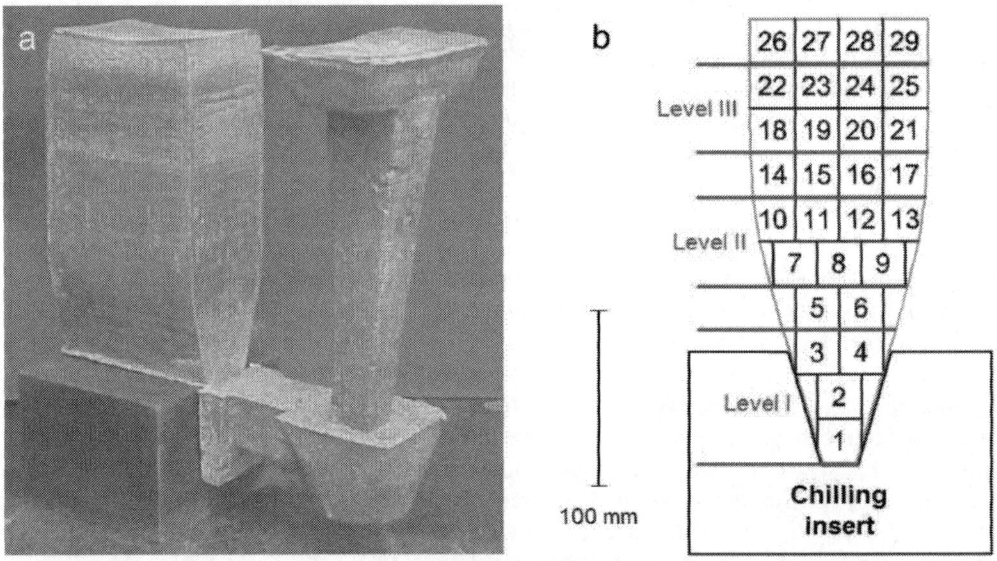

Figure 1. Photograph of one of the ingots into which the melt was cast, a; identification of the bars that were sectioned and treated to obtain the three levels of microstructural refining, b.

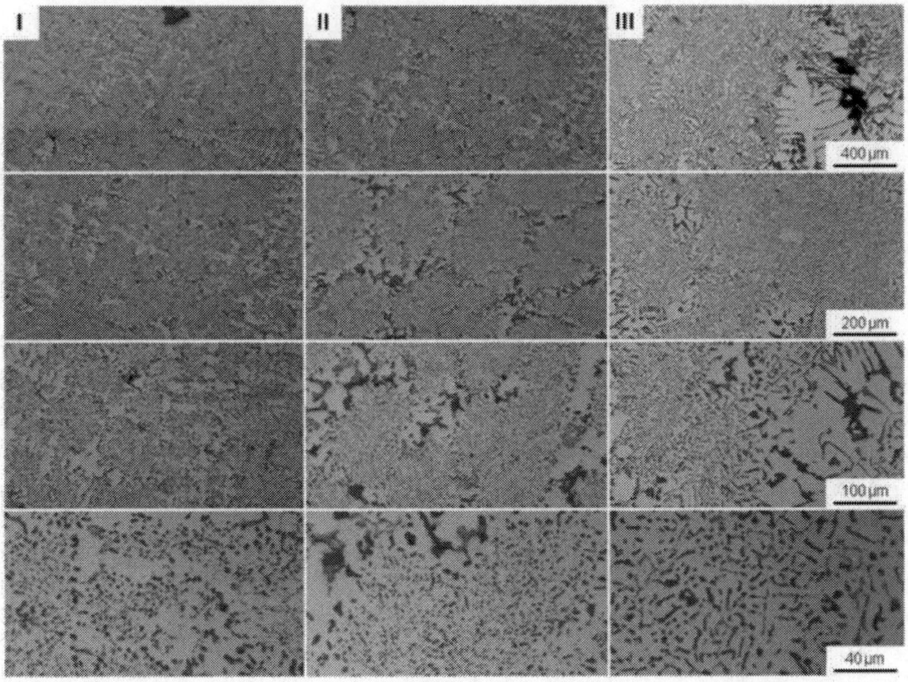

Figure 2. Microstructures of the samples at different magnification at the three different levels.

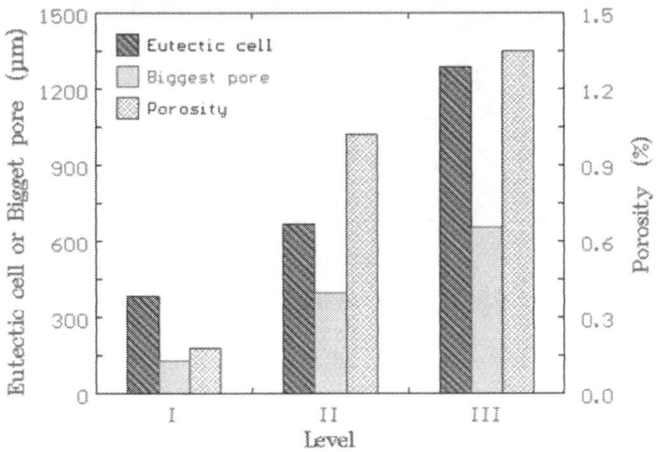

Figure 3. Sizes of the eutectic cell and biggest pore, and that of the porosity detected in the samples.

Fatigue tests were fully reversible (R = -1) and were conducted at 70 Hz at room temperature in a ±25 kN servohydraulic testing machine; chucks commonly in lathes for turning operations were used as grips to avoid the mechanical inertia caused by the hydraulic grips normally use in fatigue tests. These chucks had the limitation of not being able to sustain loads higher than 4.5 kN. Fatigue testing followed a reduced staircase schedule [27]; the specimens were tested up to failure or until a number of 10^7 cycles was achieved; the reversible stress was computed from the original cross-section of the specimen. The trials started at values of stress that did not surpassed the yield point and these values were then reduced by 10 MPa in the following test upon failure; this procedure was modified to reduce or increase the stress in 5 MPa steps once the sample resisted the 10^7 cycles. The samples that resisted 10^7 cycles were then subjected to fatigue at a stress level 10% higher to promote their failure and measure the size of the defect on which the crack was nucleated. All results are referred to the reversing cyclic stress or $\Delta\sigma/2$.

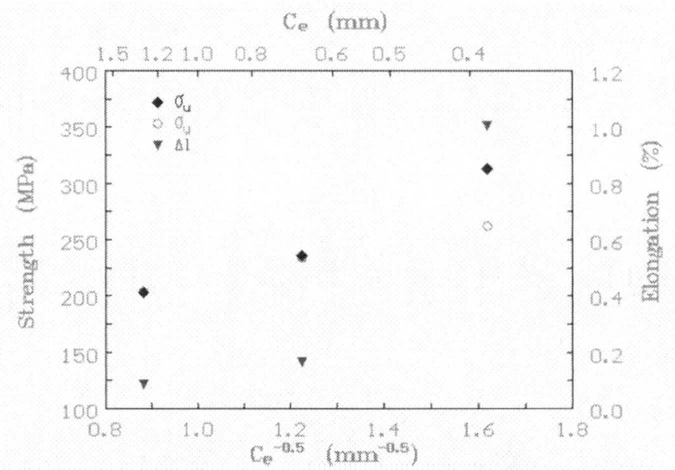

Figure 4. Dependence of the mechanical properties in tension as a function of the square root of the dendritic cell

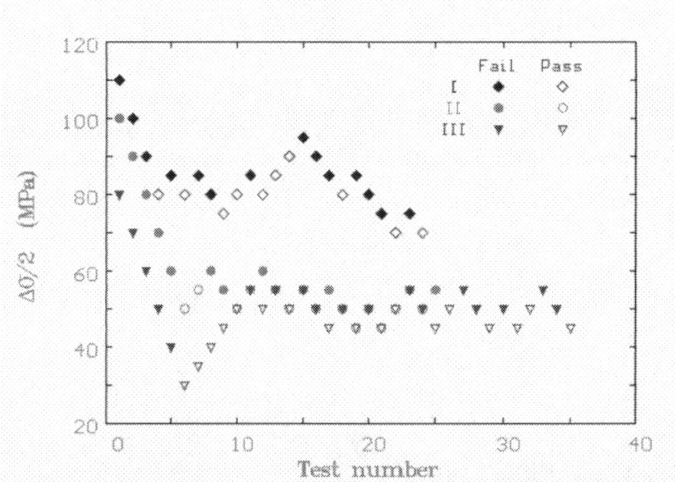

Figure 5. Staircases conducted on samples from the three different levels.

Failure and metallographic analyses were conducted on fractured specimens by scanning electron microscopy (SEM) and optical microscopy (OM). The size and type of defect that originated the failure by fatigue, as well as the distance and area of propagation of the crack was recorded by observation by SEM. This information was used to calculate design parameters that included fracture toughness (K_{IC}), maximum stress intensity factor (ΔK_{max}), as well as relationship between these parameters and the total number of cycles and maximum stress amplitude that a particular specimen was able to endure [28-30].

3. Results

The microstructures of the samples heat treated to the T7 condition are shown in Fig. 2, in which some dendrites can be distinguished, although the alloy does not solidify following the traditional dendritic growth characteristic of hypoeutectic Al-Si alloys. The dominant microstructural parameter in this alloy is the eutectic cell (C_e), which is affected by the solidification rate. Fig. 3 shows the values that this parameter, as well as the size of the biggest pore and the amount of porosity encountered at the different levels being considered. It was found that the mechanical properties depend on the inverse of the square root of the dendritic cell, Fig. 4. Table 1 shows the values of the microstructural features and those of the mechanical properties recorded in tensile tests and deduced from the fatigue tests.

Table 1. Properties and characteristics of the material at different levels.

Level	σ_y (MPa)	σ_u (MPa)	Δl (%)	σ_f (MPa)	K_{IC} (MPa·m$^{0.5}$)	C_e (μm)	BP* (μm)	Porosity (%)
I	262	313	1.01	81.4	32.7	387	128	0.18
II	234	236	0.17	52.5	29.2	635	396	1.02
III	203	203	0.09	48.2	22.3	1250	656	1.35

* Biggest pore

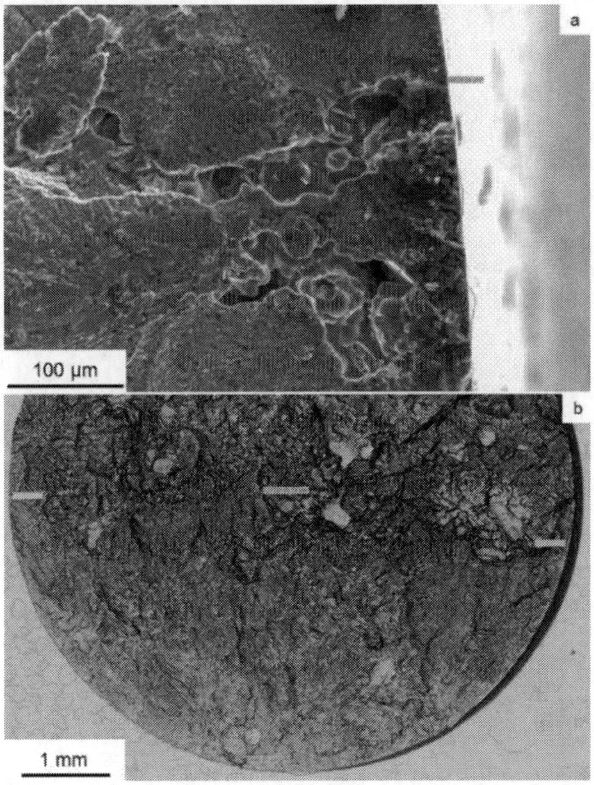

Figure 6. Pore that originated the crack, a, and distance and surface that the crack propagated in fatigue, b, before causing the failure of the sample.

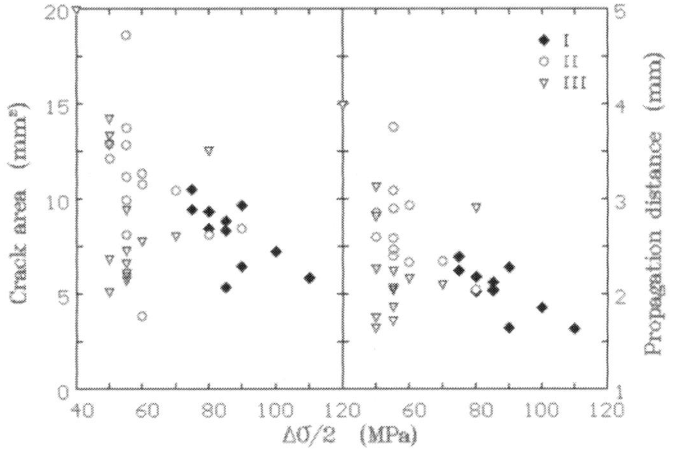

Figure 7. Area and distance that the crack propagated as a function of the nominal applied stress.

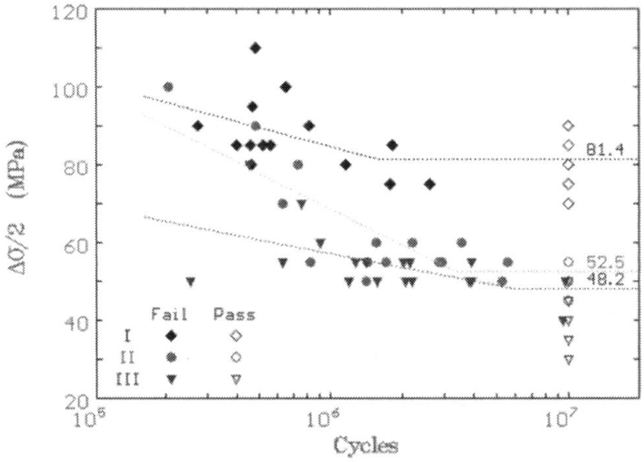

Figure 8. S-N plots for the three different levels.

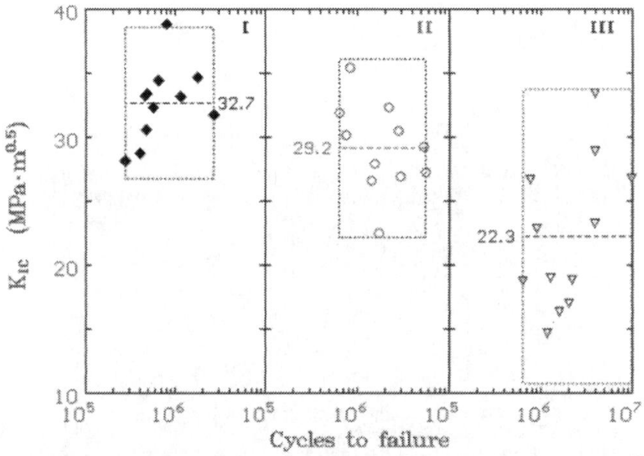

Figure 9. $\boldsymbol{K_{IC}}$ for the specimens that failed before reaching 10^7 cycles.

Fig. 5 shows the staircase schedules for the tests conducted, the specimens that failed before reaching the 10^7 cycles, or that resisted this number are identified by either full or empty symbols; the stress applied to a given specimen was reduced when the previous one failed, or increased, when the full number of cycles was achieved. The samples that survived the 10^7 cycles were then fatigued at a stress level 10% higher to promote their failure and measure the size of the defect on which the crack was nucleated.

The fracture surface of the samples was observed by SEM to detect the origin of the failure and measure its size and the area and distance that the fracture was able to propagate in fatigue before causing the failure of the sample. An example of the measurements carried out on every sample is shown in Fig. 6. It was found that the distance and area that the cracks propagated are related to the cycles to failure and to the nominal stress applied, Fig. 7 shows the relationship between these parameters and the nominal applied stress.

The fatigue stress limit (σ_f) was computed following the procedure for evaluating castings [27]:

$$\sigma_f = \sigma_o + \delta\sigma\left(\frac{\sum i n_i}{N} + \frac{\delta\sigma}{2}\right) \quad (1)$$

where σ_o corresponds to the lowest value of stress that was tested, $\delta\sigma$ is the step size in the final part of the staircase schedule (5 MPa), n_i is the number of pieces that resisted 10^7 cycles at each i stress interval, N is the total number of samples that resisted all the cycles.

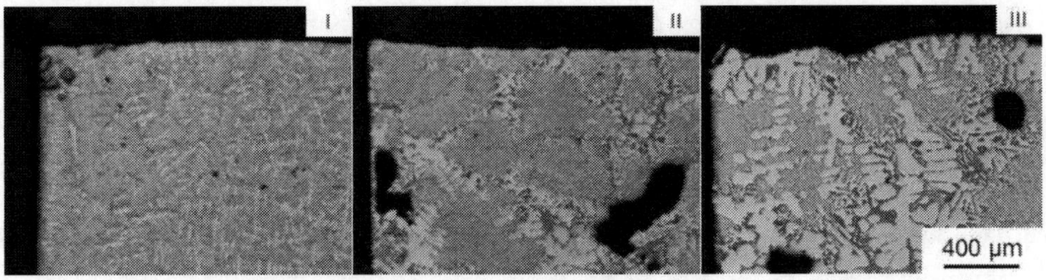

Figure 10. Metallographic observation of samples fatigued at different levels.

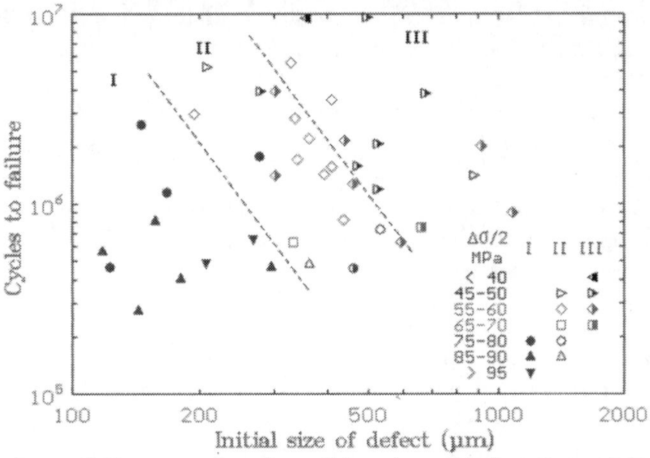

Figure 11. Cycles to failure as a function of the size of defect that originated the crack.

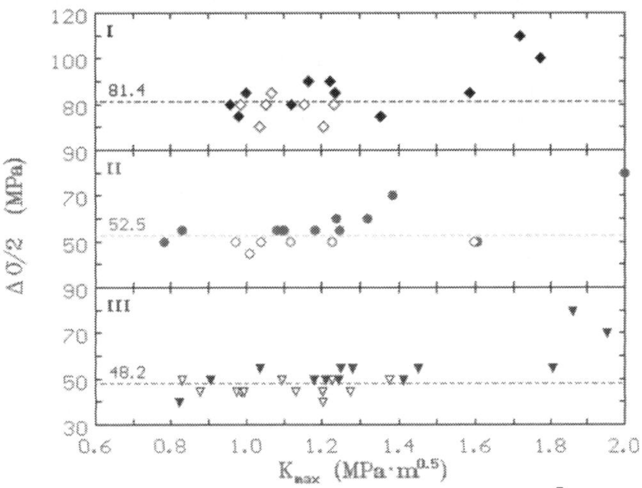

Figure 12. ΔK_{max} for the material tested. The specimens that resisted 10^7 cycles were then fatigued at a 10% stress to obtain the size of the pore that originated the crack.

4. Discussion

The S-N curves of the material are shown in Fig. 8. The fatigue limits shown in the figure and in Table 1 were calculated from the reduced staircases. The values of the mechanical properties and for the limit of fatigue for the samples from level I are comparable to those reported for a hypoeutectic alloy used in the manufacture of automotive blocks [21,22]. The toughness of the material obtained at different levels, K_{IC}, is calculated by [30]:

$$K_{IC} = K_T \alpha \sigma_{mx} \sqrt{\pi a_c} \qquad (2)$$

where α is a parameter that takes the value of 1.12 and represents the edge factor, K_T represents the concentration of stress due to the geometry of the defects and takes the value of 3 as it is considered that the crack follows a circular shape, σ_{mx} is the stress that promoted the fracture and a_c corresponds to the size of the crack, which corresponds to the propagated distance in Fig. 6. Fig. 9 shows the relationship between the value of K_{IC} and the number of cycles to fracture; the average values of K_{IC} for the specimens that did not resist the full 10^7 cycles are shown in Table 1.

Figure 10 shows a series of samples that were prepared for their metallographic observation after failing the fatigue test. The crack originated in all cases in a pore located at the upper left corner in each sample. The presence of massive silicon particles can be appreciated close to the origin of the fracture in the sample from level I which may be indicative that the pores are the predominant defect in which the cracks are formed. Such observation is confirmed in Fig. 11 in which the number of cycles to failure is plotted as a function of the size of the defect on which the crack initiated; it is worth noticing how the data can be segregated in regions for each level, as the size of the pores in level III samples are bigger than in the other two conditions. It is also clear from this figure that the samples from level I fail at a higher reversing stress.

Murakami et al. [28,29] showed that the maximum value for the stress intensity factor (ΔK_{max}) depends on the area of the defect and its location, rather than on its shape; with this assumption, the threshold for R = -1 can be calculated by:

$$\Delta K_{\max} = \beta \Delta\sigma \sqrt{\pi \sqrt{A_d}} \tag{3}$$

where β is a parameter that takes the value of 0.65 for cracks originated at pores close to the surface and 0.50 for cracks that originated in internal pores. The area of the defect that originated the crack, A_d, and its location was evaluated by **SEM**. For fully reversed fatigue cycling (R = -1) $\Delta\sigma = \sigma_{mx} - \sigma_{mn}$; since $\sigma_{mn} = -\sigma_{mx}$, then $\Delta\sigma = 2\sigma_{mx}$. The values for ΔK_{max} computed from the fractographic analyses and the experimental data are shown in Fig. 12. The full symbols represent samples that failed before the 10^7 cycles were achieved, the open symbols correspond to samples that resisted such level of cycles, but were broken at a 10% higher stress. A problem that was encountered in this type of alloy is the overlapping of the values corresponding for ΔK_{max} for samples that resisted the test and those that did not, contrary that is found in hypoeutectic alloys [21,22]. A reason for this behaviour may be the way in which the crack propagates. Low silicon containing alloys are made of a dendritic aluminium matrix with the eutectic aggregate around them; in contrary, the high silicon containing alloy is made of eutectic cells with aluminium dendrites or primary silicon particles, Figs. 2 and 10. In the hypoeutectic case, the aluminium dendrites are ductile enough and are able to resist the propagation of the crack, whereas in the present case, the crack is able to propagate rapidly following the silicon particles that form part of the eutectic cell, as can be seen by observing the contour of the cracks shown in Fig. 13.

5. Conclusions

An aluminium alloy containing high levels of silicon and other elements was tested to study its behaviour a high cycle fatigue. The alloy was cast in ingots designed to produce a solidification gradient to vary the microstructure of the samples intended to study.

It was found that the mechanical properties and the limit of fatigue depended on the degree of microstructural refining, as the cracks that cause the failure in fatigue are nucleated in pores generated during casting. The mechanical properties of the alloy were found to be comparable to those characteristic of a hypoeutectic alloy used in the manufacture of engine blocks.

The toughness of the material was calculated by means of the size of the defect that originated the crack and that of the area that the crack propagated in fatigue before failure. It was not possible to assess the stress intensity factor as the crack propagates following the boundaries of the eutectic cells.

6. Acknowledgements

The authors acknowledge the support provided by the National Council for Science and Technology (CONACYT), Mexico.

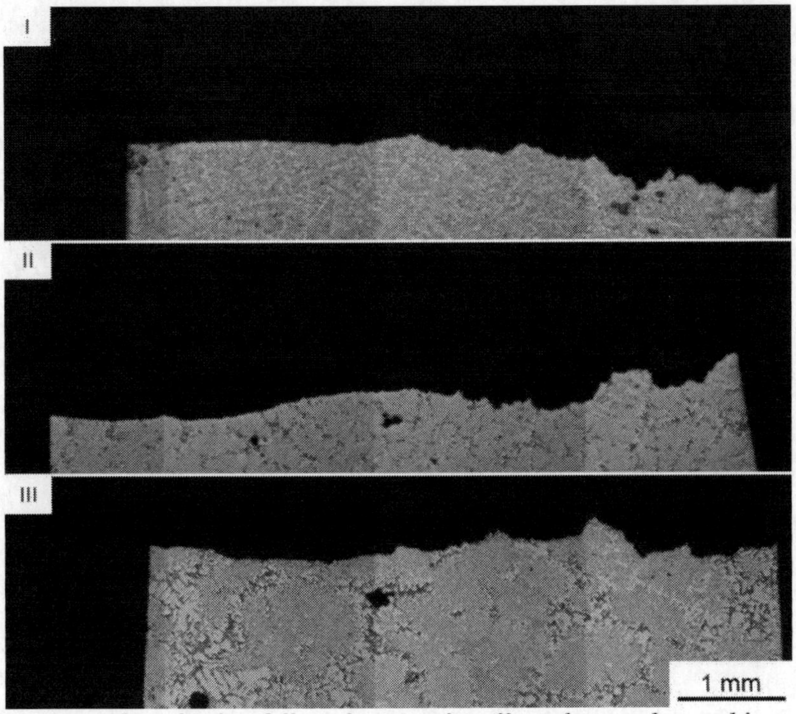

Figure 13. Contour of cracks that follow the eutectic cells and were observed in samples from different levels.

References

[1] Heywood, J.B., Internal Combustion Engine Fundamentals, McGraw-Hill, New York, 1989.

[2] Hoag, K., Vehicular engine design, Springer Science & Business Media, Wien, New York, 2007.

[3] Campbell, J., Castings, 2nd Ed., Butterworth-Heinemann, Oxford, 2003.

[4] Colás, R., Velasco, E. and Valtierra, S., Castings, Handbook of Aluminum, Vol. 1: Physical Metallurgy and Processes, G.E. Totten and D.S. MacKenzie (eds.), M. Dekker, New York, 2003, 591-641.

[5] Foss, J.E., Krug, P.L. and Kennedy, M., New aluminum alloys for cylinder liner applications, SAE Techn. Paper 2006-01-0983, SAE International, Warrendale, 2006.

[6] Tenekedjiev, N. and Gruzleski, J.E., 1990. Hypereutectic aluminium-silicon casting alloys-a review. *Cast Met.*, **3**, 96-105.

[7] Vijeesh, V. and Prabhu, K.N., 2014, Review of microstructure evolution in hypereutectic Al–Si alloys and its effect on wear properties. *Trans. Ind. Inst. Met.*, **67**, 1-18.

[8] Donahue, R. and Fabiyi, P.A., 2000. Manufacturing feasibility of all-aluminum automotive engines via application of high silicon aluminum alloy (No. 2000-01-0061). SAE Technical Paper.

[9] Doty, H.W., General Motors Corporation, 2005. Aluminum alloy for engine blocks. U.S. Patent 6,921,512.

[10] Kasprzak, W., Sokolowski, J.H., Yamagata, H., Aniolek, M. and Kurita, H., 2011. Energy efficient heat treatment for linerless hypereutectic Al-Si engine blocks made using vacuum HPDC process. *J. Mat. Eng. Perf.*, **20**, 120-132.

[11] Surappa, M.K., Blank, E. and Jaquet, J.C., 1986, Effect of macro-porosity on the strength and ductility of cast Al-7Si-0.3Mg alloy, *Scr. Metall.*, **20**, 1281-1286.

[12] Skallerud, B., Iveland ,T. and Harkegard, G., 1993, Fatigue life assessment of aluminum alloys with casting defects, *Eng. Fract. Mech.*, **44**, 857-874.

[13] Stanzl-Tschegg, S.E., Mayer, H.R., Tschegg, E.K. and Beste, A., 1993, In-service loading of AlSi11 aluminium cast alloy in the very high cycle regime, *Int. J. Fat.*, **15**, 311-316.
[14] Major, J.F., 1998, Porosity control and fatigue behavior in A356-T61 aluminum alloy, *Trans. AFS*, **105**, 901-906.
[15] Seniw, M.E., Conley, J.G. and Fine, M.E., 2000, The effect of microscopic inclusion locations and silicon segregation on fatigue lifetimes of aluminum alloy A356 castings, *Mat. Sc. Eng. A.*, **285**, 43-48.
[16] Gall, K., Yang, N., Horstemeyer, M., McDowell, D.L. and Fan, J., 2000, The influence of modified intermetallics and Si particles on fatigue crack paths in a cast A356 Al alloy, Fat. *Fract. Eng. Mat. Struct.*, **23**, 159-172.
[17] Chen, W., Zhang B. and Poirier, D.R., 2000, Effect of solidification cooling rate on the fatigue life of A356.2-T6 cast aluminium alloy, *Fat. Fract. Eng. Mat. Struct.*, **23**, 417-423.
[18] Boileau, J.M. and Allison, J.E., 2003, The effect of solidification time and heat treatment on the fatigue properties of a cast 319 aluminum alloy, *Metall. Mat. Trans. A*, **34A**, 1807-1820.
[19] Chan, K.S., Jones, P. and Wang, Q., 2003, Fatigue crack growth and fracture paths in sand cast B319 and A356 aluminum alloys, *Mat. Sc. Eng. A.*, **341**, 18-34.
[20] Wang, Q.G., Davidson, C.J., Griffith, J.R. and Crepeau, P.N., 2006, Oxide films, pores and fatigue lives of cast aluminum alloys, *Metall. Mat. Trans B*, **36B**, 887-895.
[21] González, R., Martínez, D.I., González, J.A., Talamantes, J., Valtierra, S. and Colás, R., 2011, Experimental investigation for fatigue strength of a cast aluminium alloy, *Int. J. Fat.*, **33**, 273-278.
[22] González, R., González, A., Talamantes-Silva, J., Valtierra, S., Mercado-Solís, R.D., Garza-Montes-de-Oca, N.F. and Colás, R., 2013, Fatigue of an aluminium cast alloy used in the manufacture of automotive engine blocks, *Int. J. Fat.*, **54**, 118-126.
[23] Jaquet, J.C. and Hotz, W., 1992, Quantitative description of the microstructures of aluminum foundry alloys, *Cast Met.*, **4**, 200-225.
[24] Biloni, H. and Boettinger, W.J., 1996, Solidification, Physical Metallurgy, 4th Ed., R.W. Cahn and P. Haasen (eds.), North Holland, Amsterdam, 669-842.
[25] ASTM International, 1994, ASTM B557M-94: Standard Test methods of tension testing wrought and cast aluminum and magnesium alloy products (metric), West Conshohocken.
[26] ASTM International, 1999, ASTM E8M-99: Standard Test methods for tension testing of metallic materials (metric), West Conshohocken.
[27] GM Engineering Standards, 2002, GMN7152: Specification and verification of tensile and fatigue properties in cast components, General Motors Corporation, Detroit.
[28] Murakami, Y. and Endo, M., 1994, Effect of defects, inclusions and inhomogeneities on fatigue strength, Int. J. Fatigue, 16, 163-181.
[29] Murakami Y. and Beretta, S., 1999, Small defects and inhomogeneities in fatigue strength: Experiments, models and statistical implications, Extremes, 2, 123-147.
[30] Stephens, R.I., Fatemi, A., Stephens, R.R. and Fuchs, H.O., 2001, Metal Fatigue in Engineering, 2nd Ed., John Wiley & Sons, Inc., NY.

IOP Conf. Series: Journal of Physics: Conf. Series **843** (2017) 012028 doi:10.1088/1742-6596/843/1/012028

Critical plane analysis of multiaxial fatigue experiments leading to White Etching Crack formation

S Averbeck[1], E Kerscher[1]

[1] Materials Testing, University of Kaiserslautern, Gottlieb-Daimler-Strasse, 67663 Kaiserslautern, Germany

stefan.averbeck@mv.uni-kl.de

Abstract. Various researchers have shown that rolling contact fatigue can be reproduced with cyclic compression-torsion experiments, with the load components either in-phase or out of phase. As reported previously, the authors used such experiments to reproduce the rolling contact fatigue phenomenon "White Etching Cracks" which can cause premature failures of rolling element bearings. It is characterized by subsurface crack initiation and propagation coupled with microstructural changes alongside the crack flanks. Surprisingly, only in-phase load superposition caused these microstructural changes to occur. This suggests that White Etching Crack formation is somehow linked to the multiaxial stress state in the specimens, as this was the only variable that changed between in-phase and out-of-phase testing. In this study, the multiaxial stress state in the two experiments is analysed and compared using different critical plane criteria. In contrast to common ways of characterizing the stress state, e.g. equivalent stress approaches, this class of criteria is explicitly designed for multiaxial stress states. Special attention is given to the Dang Van criterion, which has been used in a number of rolling contact fatigue studies.

1. Introduction

White Etching Cracks (WECs) are a premature damage and failure mode found in rolling element bearings across many different industries, bearing manufacturers, and bearing geometries. Due to the important function of bearings and the unpredictability of this damage phenomenon, White Etching Cracks can cause costly secondary damages and downtimes.

WECs are groups or networks of subsurface cracks which are accompanied by zones of microstructurally transformed material [1]. These zones consist of nanocrystalline ferrite with grain sizes of 10-100 nm. Additionally, the primary carbides which are dispersed in the original tempered martensite are dissolved in the transformed areas [2]. In a metallographic analysis, the transformed regions appear white when etched with 3% nitric acid (nital), which explains how WECs get their name.

WEC formation seems to be influenced by a great number of factors, which renders the search for the root causes and mechanisms difficult. The factors can be separated into three main sources: operating conditions (vibration, load peaks, slip), tribochemistry (lubricant composition, water contamination), and surface damage (corrosion, electrical current) [3]. The importance of sliding speed is highlighted by two relatively recent studies proposing characteristic values for WEC formation [4,5]. In terms of mechanisms, two basic directions of thought exist, which explain WEC formation either as hydrogen-induced or stress-induced (both hypotheses are discussed e.g. in [6]). It has also been proposed that both interpretations are correct, and that there are in fact two similar but separate phenomena that are only summarised under the term "White Etching Cracks" [7]. It should be remembered that while the two approaches emphasise different factors, both generally recognise that the respective other factor also can contribute to WEC formation, and a combined-cause mechanism was proposed by [8].

Investigation of WECs is usually based upon full-scale bearing tests and rolling contact fatigue tests. Several reviews [3,6,9] over the past few years on WEC studies are testimony to the great research activity in this field. In an earlier publication [10], the authors presented a novel experimental approach for reproducing WECs, using cyclic compression-torsion loads on cylindrical specimens. The experiments were based on the observations by other researchers that rolling contact fatigue can be approximated by superimposed compression and torsion, at which both in-phase [11] and out-of-phase loads [12] can be used. In our experiments, it appeared that cracks formed under both load variants; however, WEC-like microstructural transformation in the vicinity of the cracks only occurred during in-phase testing. At the time, we hypothesised that the different stress conditions of the two load cases caused this behaviour, as this was the only parameter that was changed. An exploratory critical plane analysis of the stress conditions suggested that cyclic plasticity processes should be more probable during in-phase testing [13]. In this paper, a more detailed and expanded analysis will be presented. In addition, the Dang Van criterion will be used to analyse the stress conditions in the experiment. As this criterion has been used a number of times in rolling contact fatigue analyses, a comparison between the present results and original RCF cases enables a transfer of the findings from compression-torsion experiments to rolling element bearing conditions.

Table 1. Material properties and testing conditions

Material	
Specification	100Cr6 (AISI 52100, SUJ-2) steel
Yield strength, R_e	1980MPa
Ultimate tensile strength, R_m	2446MPa
Young's modulus, E	208GPa
Hardness	730HV
Testing conditions	
Max. v. Mises equivalent stress	1080MPa
No. of cycles	10^6
Testing frequency	10Hz
Temperature	25°C

2. Methods

The full details of the compression-torsion testing setup can be found in [10] and are summarised in table 1. The most relevant information for this study – the stress values and variation in time of the two stress patterns – is shown in figure 1. In both cases, the compressional and torsional stresses are periodic in time and can be described as a sine wave:

$$[\sigma(t), \tau(t)] = [\sigma_m, \tau_m] + [\sigma_a, \tau_a]\sin(\omega t + \delta) \qquad (1)$$

The phase shift δ is zero for all cases except the out-of-phase torsional stress, where it is π. The two load patterns will be referred to as IP (for in-phase) and OP (for out-of-phase) in the following.

IOP Conf. Series: Journal of Physics: Conf. Series **843** (2017) 012028 doi:10.1088/1742-6596/843/1/012028

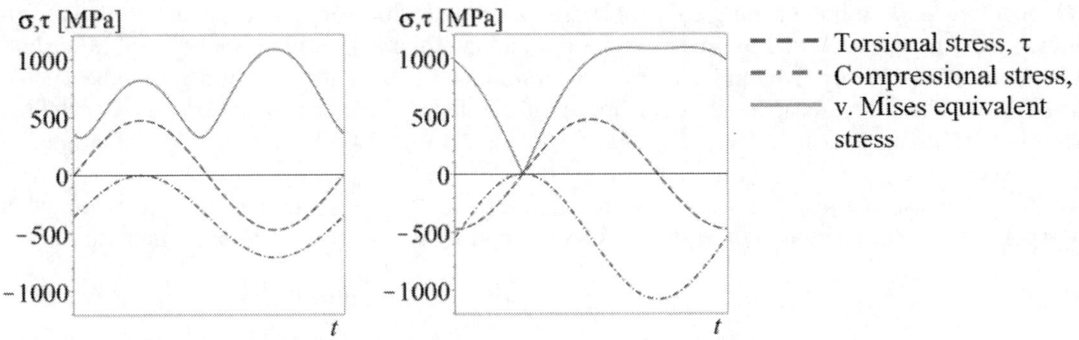

Figure 1. Load variants during testing. Left: in-phase loading, right: 90° out-of-phase loading.

As figure 1 shows, the v. Mises equivalent stress maxima are the same for both load cases, aside from the fact that the maximum equivalent stress acts nearly constantly over a longer portion of the stress cycle in the OP setup. Yet only in-phase loading resulted in WEC-like microstructural changes, which led to the conclusions that a) the stress conditions significantly influence WEC formation and b) that an equivalent stress approach is not suitable to reflect the different stress conditions between the experiments. This is especially true for the effects of phase shifted loads, which generally are more detrimental to the material than in-phase loads [14]. As an alternative means of characterizing the multiaxial stress states, the critical plane approach was chosen.

The common goal of critical plane approaches is to describe multiaxial fatigue better than conventional stress analysis (e.g. equivalent stresses) does, and to predict whether a given multiaxial load can be endured or will lead to failure (for an overview, see [14]). Multiaxial loading causes complex stresses which are highest on certain material planes in the material. These can be described by their orientation angles φ and θ relative to an arbitrary coordinate system xyz (see figure 2). For each plane Δ, a damage parameter can be calculated and compared against a fatigue limit. If the damage parameter exceeds the fatigue limit, fatigue damage will occur, ultimately leading to failure of the specimen or component. The damage parameter and stress limit can be in the form of stresses, strains, or energies.

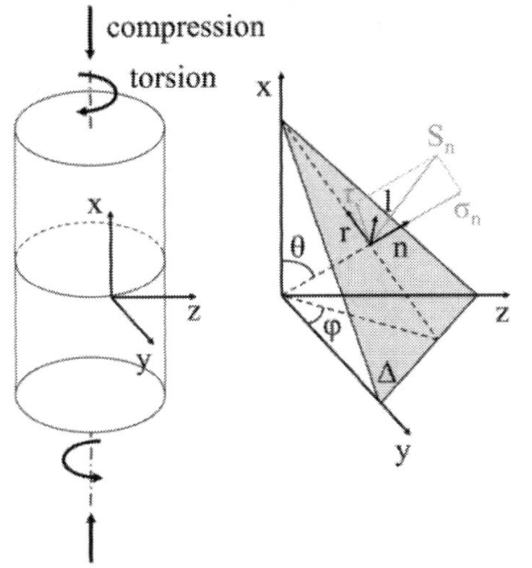

Figure 2. Directions and angles defining the critical plane. Adapted from [15].

For the present evaluation, different stress-based criteria were chosen for the following reasons: 1) The experiments were run in load-controlled mode, so that the macroscopic stresses are clearly defined and their variation over time is well known. 2) The fatigue stress limits of 100Cr6 steel, which is the steel most frequently used in rolling element bearings, are well documented. Especially relevant for the following are the uniaxial fatigue limit f_{-1}=650MPa and the torsional fatigue limit t_{-1}=380MPa (both at a stress ratio of R=-1) [16]. In contrast, energy- or strain-based fatigue limits would have to be calculated first-hand, without reference values to compare against. 3) Stress-based criteria were found to yield good results in the high cycle fatigue range (i.e., up to 10^7 load cycles) [14]. This is the relevant fatigue range for the experiments, which were run up to 10^6 cycles.

Within the class of stress-based criteria, the critical plane can be defined as the orientation where a linear combination of the normal and shear stresses amplitudes, N_a and C_a, reaches a maximum:

$$\{\theta^*, \varphi^*\}: \mathrm{Max}(a\, C_a(\theta,\varphi) + b\, N_a(\theta,\varphi)) \tag{2}$$

In many cases, a and b are one and zero or vice versa, so that the critical plane orientations θ^* and φ^* are only defined by the normal or shear stress amplitude. Care must be taken to choose an appropriate criterion for a given material, as some are only applicable to certain types of materials (e.g. depending on whether the material is ductile or brittle) [17].

The compression-torsion load case is not very well investigated, and consequently, reference values and even information on the applicability of different criteria are scarce. For this reason, we chose a set of criteria that includes different definitions of the critical plane and the damage parameter. At the same time, the criteria are still similar enough for comparison, so that contradictions and singularities in the results can be identified. The criteria will be presented briefly in the next section. All of them are based on searching for the critical plane(s) caused by the macroscopic stresses. This is in contrast to the Dang Van criterion presented in subsection 2.2, which uses a mesoscopic approach. All calculations were performed using the computer algebra system MAPLE.

2.1. *Macroscopic critical plane criteria*

It is relatively common for stress-based critical plane criteria to define the critical plane as the plane of maximum shear stress. This definition is based upon the usual failure mode of metallic materials, where shear-controlled failure is most important, except for some very brittle material conditions. As examples of this critical plane definition, the criteria of McDiarmid [18], Matake [19], and Susmel and Lazzarin [20] are used in this study. Differences exist in the respective damage criterion. Both McDiarmid and Matake use a linear combination of shear and normal stresses to determine the critical angles θ^* and φ^*:

$$\sigma_{\mathrm{lim}} \le C_a(\theta^*, \varphi^*) + \kappa\, N_{\mathrm{max}}(\theta^*, \varphi^*) \tag{3}$$

C_a is the shear stress amplitude on the critical plane, N_{max} the maximum normal stress perpendicular to the plane. The constant κ is defined as $t_{-1}/2R_e$ by McDiarmid and $2(t_{-1}/f_{-1})-1$ by Matake. Susmel and Lazzarin use a similar but modified criterion, where the maximum normal stress is put in relation to the shear stress amplitude:

$$\sigma_{\mathrm{lim}} \le C_a(\theta^*, \varphi^*) + \kappa \frac{N_{\mathrm{max}}(\theta^*, \varphi^*)}{C_a(\theta^*, \varphi^*)} \tag{4}$$

In order to keep the dimension MPa, κ in this case is $t_{-1} - 0.5 f_{-1}$. Consequently, all three criteria are evaluated against a stress limit, i.e. the fatigue strength of the material at a given number of cycles.

Liu and Mahadevan [17] proposed a criterion using the highest normal stress amplitude, $N_a(\theta^*,\varphi^*)$, as the defining parameter for the critical plane orientation. This criterion was developed specifically with regard to brittle and extremely brittle materials, where normal stresses are failure-critical instead of shear stresses. The damage criterion differs from the other criteria inasmuch as it is dimensionless. It also takes into account the hydrostatic stresses σ_{ha}. The criterion reads

$$\beta \leq \left(\left(\frac{N_a(\theta^*, \varphi^*)}{f_{-1}} \right)^2 + \left(\frac{C_a(\theta^*, \varphi^*)}{t_{-1}} \right)^2 + \kappa \left(\frac{\sigma_{ha}(\theta^*, \varphi^*)}{f_{-1}} \right)^2 \right)^{0.5} \tag{5}$$

The material parameters β and κ can be determined from established fatigue limits. For 100Cr6 steel, κ is zero and β is 0.99. This means that the hydrostatic stress on the critical plane, $\sigma_{ha}(\theta^*,\varphi^*)$, has no influence on the fatigue limit under torsion, which is in agreement with experimental data.

The critical plane definition in the criterion developed by Carpinteri et al. [21] is based on the weighted direction of principal stresses. The endurance limit is formulated as a stress:

$$\sigma_{lim} \leq \left(\left(N_a(\theta^*, \varphi^*) + f_{-1} \frac{N_m}{R_m} \right)^2 + \left(\frac{t_{-1}}{f_{-1}} \right)^2 C_a(\theta^*, \varphi^*)^2 \right)^{0.5} \tag{6}$$

The procedure to calculate the critical plane orientations and stresses is outlined by the authors of each criterion in the cited publications and, more generally, in [22] and [15].For the application of the general procedure to the specific case of our experiments, see [13]. The shear stress amplitude C_a in particular is not trivial to evaluate, as it is a vector function changing both its direction and value over time and the possible critical plane orientations.

2.2. *Dang Van criterion*

In addition to the aforementioned five criteria, an analysis with the Dang Van criterion [23] was performed. This criterion has been widely used for rolling contact fatigue analysis, providing reference values for our own results (e.g. [24,25]). One main difference to the criteria cited above is that the scale considered is not macroscopic but mesoscopic. First, the macroscopic stress tensor $\boldsymbol{\Sigma}$ is split between the hydrostatic part $PH(t)$ and the deviatoric part $\boldsymbol{S}(t)$. In the next step, $\boldsymbol{S}(t)$ is transferred onto the mesoscopic scale, where the mesoscopic stress deviator

$$\boldsymbol{s}(t) = \boldsymbol{S}(t) + \boldsymbol{\rho}^* \tag{7}$$

acts. $\boldsymbol{\rho}^*$ is the mesoscopic residual stress tensor in the stabilised state, which can be calculated as

$$\boldsymbol{\rho}^* = -z : \min_z \left[\max_t \sqrt{(\boldsymbol{S}(t) - z) : \boldsymbol{S}(t) - z} \right] \tag{8}$$

With $\boldsymbol{s}(t)$, the shear stress in Dang Van's criterion

$$\tau_{DV}(t) = \mathrm{Tresca}[\boldsymbol{s}(t)] = \sqrt{s_I(t) - s_{III}(t)} \tag{9}$$

can be calculated and finally, the damage parameter

$$d = \mathrm{Max}_t \left[\frac{\tau_{DV}(t)}{b} - a\, PH(t) \right] \tag{10}$$

is evaluated over t. If $d \geq 1$, fatigue damage will occur, eventually leading to failure. This general procedure and its theoretical foundations are outlined in more detail in [23] and [26]. For the present case, a simplified evaluation can be used, as all stresses vary periodically in time [24,26]. In contrast to the five criteria described above, searching for the critical plane is not necessary in this criterion. It is implicitly given by the orientation of the principal stresses which are used to calculate $\tau_{DV}(t)$. The endurance domain and load path of a given material and multiaxial load can be visualised in a diagram like figure 3.

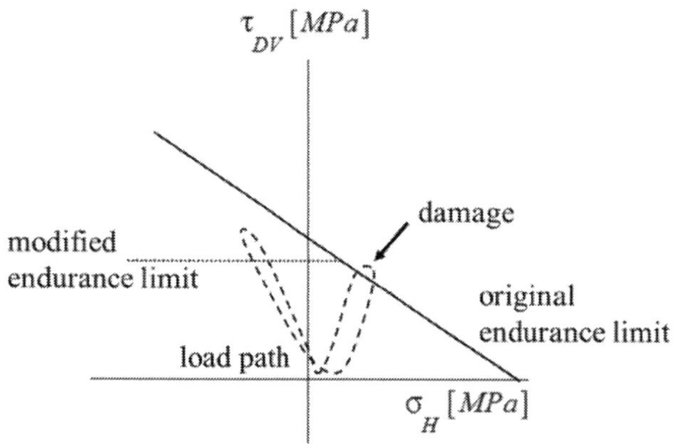

Figure 3. Endurance domain and an exemplary load path in the Dang Van criterion (adapted from [23])

Since its first publication, the criterion has been refined and extended to better cover the real behaviour of materials. In its first form, the criterion predicted an increasing shear fatigue limit when negative (i.e. compressive) hydrostatic stresses are applied [23]. Finding that this disagrees with experimental observations, Desimone et al. [25] proposed a modified fatigue limit composed of two segments with different slopes, as shown by the dotted line in figure 3. They noted that this leads to a conservative estimation, as the fatigue limit for pure torsion is outside the endurance domain; yet, if compressive hydrostatic stresses occur during the stress history, which is invariably the case in rolling contact fatigue, the modified curve better predicts fatigue behaviour. Further refinement of the endurance limit especially for compressive hydrostatic stresses may be necessary. For example, there is experimental evidence that the endurance limit actually decreases as the hydrostatic stresses increase [24]. As these results were obtained for rail steel, which is in many ways different from bearing steels, the constant endurance limit will be used in this study.

Another benefit of using the modified limit is that, for the present case, no exact knowledge of the residual stresses in the material is necessary. At the surface of both bearing rings and the specimens used in this study, compressive residual stresses can be expected from the final stages of the manufacturing process, namely grinding and polishing [27]. These stresses, however, have only the effect of moving the load path further into the region of compressive hydrostatic stresses (i.e., to the left in figure 3) [25]. This does not affect the endurance limit, so that macoscopic residual stresses can, as a first approximation, be ignored.

3. Results

3.1. Macroscopic critical plane criteria

The different definitions of the critical plane lead to different predictions regarding the orientations of critical planes. As the McDiarmid, Matake, and Susmel-Lazzarin criterion all define the critical plane as the plane of highest shear stress amplitude, their predicted critical orientations coincide with the global maxima of figure 4a, which shows the evaluation of $C_a(\theta,\varphi)$ for the in-phase load. The diagrams are periodic in θ and φ, as adding multiples of π to the angles result in the same planar orientation, only with alternating normal vectors. The location of the OP maxima is slightly offset by 0.06π along the φ axis. The Liu-Mahadevan criterion, using the normal stress amplitude to determine the critical planes, predicts very different angles of φ, as shown in figure 4b. Again, the OP case has its maxima slightly shifted along the φ axis, this time by 0.02π. The variation over θ is the same for both IP and OP, with maxima of equal height at 0.5π and 1.5π. It is therefore sufficient to conduct an analysis of the influence of φ on the damage parameters at constant $\theta=(j+0.5)\pi$, with $j=0,1,2,\ldots n$. Consequently, the following diagrams will be only two-dimensional.

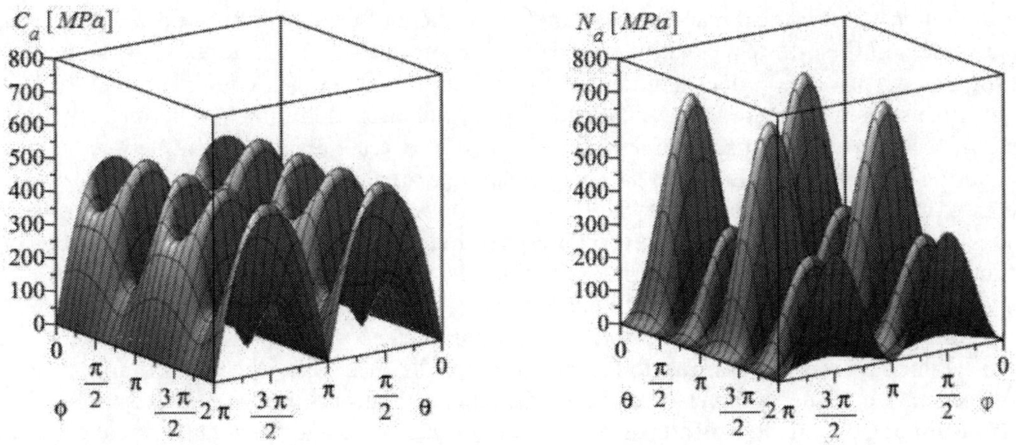

Figure 4. Shear stress amplitude (left) and normal stress amplitude (right) under IP load.

As figure 4 shows, the criteria using C_a to define the critical planes predict a higher number of critical orientations. The Carpinteri-Spagnoli criterion has a unique critical plane definition, using the weighted principal stresses. This leads to a single critical plane at $\theta=0.5\pi$, $\varphi=0.44\pi$.

When evaluating the damage parameter of the criteria by Matake, McDiarmid, and Susmel and Lazzarin, one finds that the predicted critical plane orientations do not always correspond to global maxima of the damage parameter. In the Matake criterion, this behaviour is less pronounced for IP load (figure 5a), where the values of the global and local maxima are very close to each other, than for OP load (figure 5b), where distinct global maxima exist. In both cases, all maxima stay below the uniaxial fatigue limit of 650MPa, which corresponds to none of the specimens having failed during 10^6 cycles. The torsional fatigue limit of 380MPa is exceeded on a number of orientations; however, this is not the limit that critical plane criteria are usually compared against.

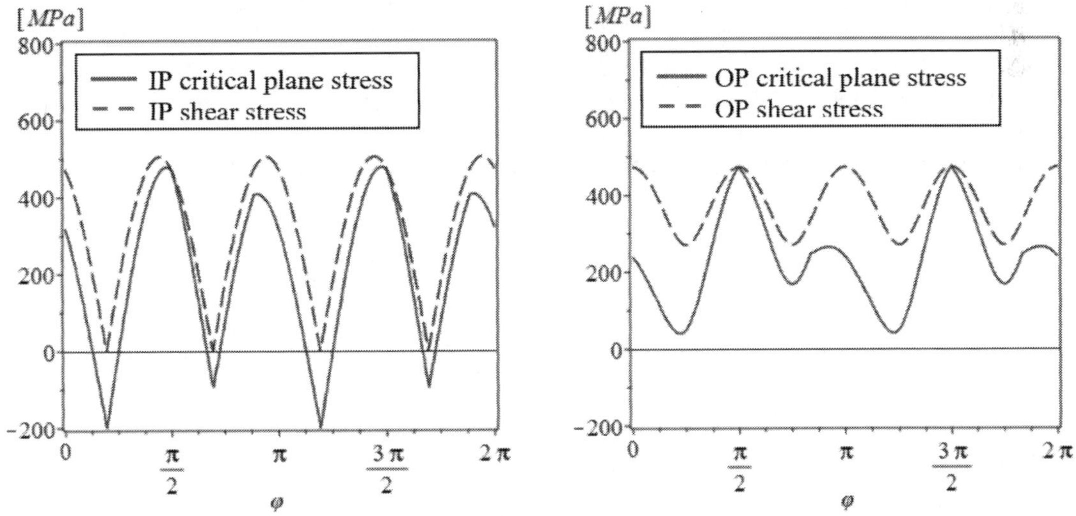

Figure 5. Damage parameter of the Matake criterion for IP (left) and OP load (right). The shear stress amplitude C_a for both cases is also shown.

In addition to the damage parameter curve, the evolution of the shear stress amplitude C_a is shown in figure 5. Comparing the location of the maxima of the damage parameter among the φ axis with the critical orientations predicted by C_a, a small but noticeable shift between the two can be observed. From the basic definition of the critical plane as the plane experiencing the highest stresses, one would expect the two to coincide. The difference can be traced to the contribution of N_{max} to the damage parameter. As the evolution of N_{max} over φ does not coincide with that of C_a, the linear combination $C_a+\kappa N_{max}$ will have its maxima shifted compared to C_a. The same behaviour is found in the McDiarmid criterion, the results of which are overall similar enough to figure 5 that a separate diagram is omitted.

The criterion by Susmel and Lazzarin has distinct singularities at $\varphi=[0.23\pi,0.77\pi,1.23\pi,1.77\pi]$ in the in-phase case, where very high negative values are found (see figure 6). The absolute values at these orientations exceed realistic stresses by several orders of magnitude. An explanation for this behaviour can be found in the second term of the criterion. When $N_a>N_m$ and $C_a<1$, the expression $(-N_a+N_m)/C_a$ assumes very high negative values and with the other summand C_a being negligibly small in this case, the singularities occur. The orientations and damage parameter values of the critical planes remain unaffected by this and are very similar to the previous results. Likewise, the prediction that in-phase loads will generate a higher number of critical orientations than out-of-phase loads stands as in the previous two criteria, and again a discrepancy between the predicted and the actual critical orientations is observed (figure 6).

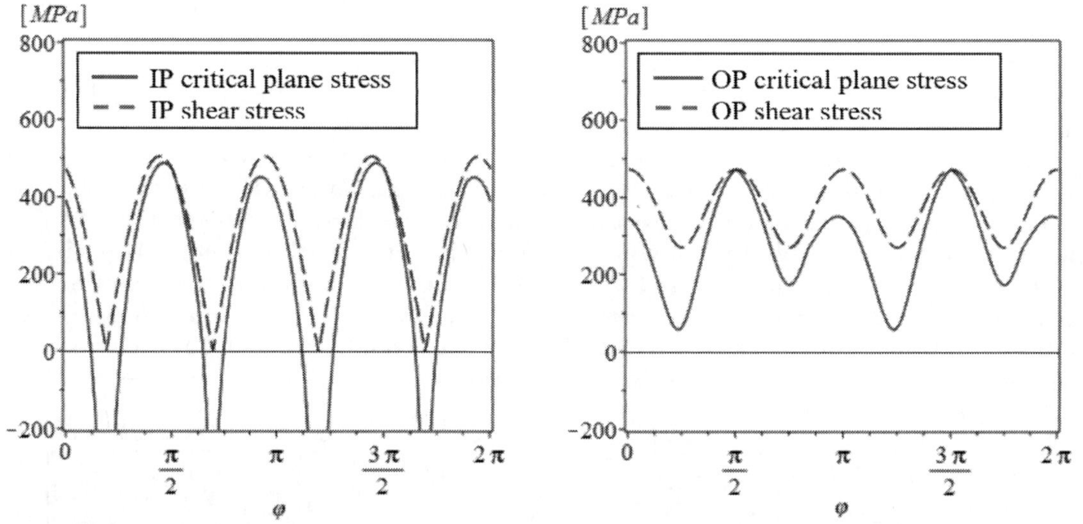

Figure 6. Damage parameter of the Susmel-Lazzarin criterion for IP (left) and OP load (right). The shear stress amplitude C_a for both cases is also shown.

Results for the Liu-Mahadevan criterion are presented in figure 7. As discussed above, determining the critical planes via the normal load amplitude leads to very different predictions regarding the orientation of the critical planes. Comparing these predicted values with the location of the actual maxima of the damage parameter, one finds a much more pronounced discrepancy between the two. This is especially true in the IP case, where the global maxima of N_a correspond to local minima of the damage parameter. The resulting difference between the predicted and the actual critical orientation is 0.25π. Moreover, the stress distribution is much more uniform than it would be expected from the N_a curve. Another major difference is that the damage parameter is above the endurance limit of β=0.99 for most orientations. This is not in agreement with the experimental data, as none of the specimens failed during testing.

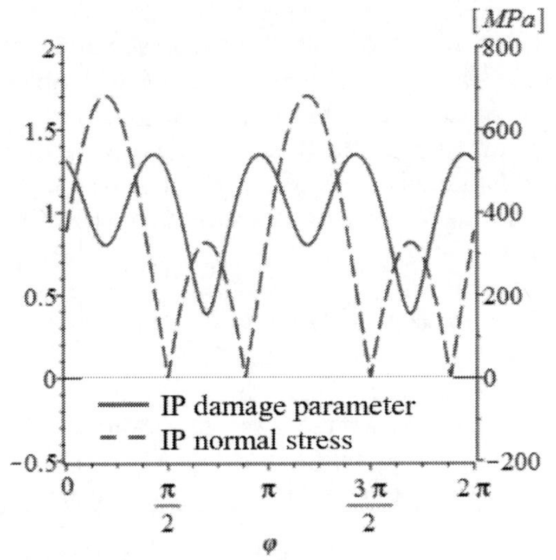

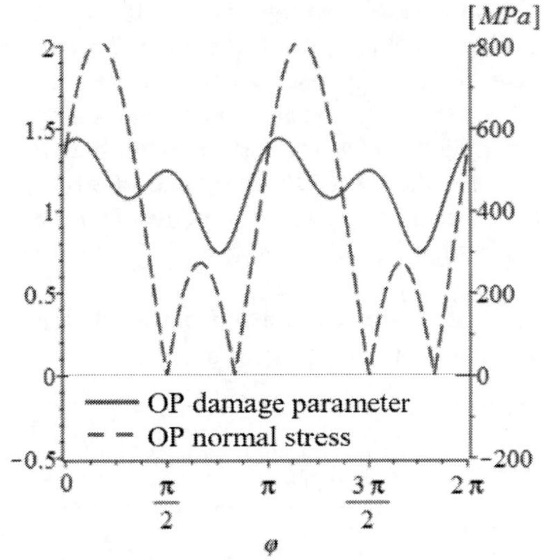

Figure 7. Damage parameter of the Liu-Mahadevan criterion for IP (left) and OP load (right). The normal stress amplitude N_a for both cases is also shown.

The Carpinteri-Spagnoli criterion, due to its critical plane definition via the principal stress directions, results in only one predicted critical orientation. Therefore, the diagram in figure 8 is different from figures 5-7, showing only the graphs of the damage parameter for IP and OP load. The critical angles θ^* remain at $(j+0.5)\pi$, $j=0,1,2,\ldots n$, which is the same as for all other criteria. The prediction for φ^* is close to the real value in the OP case, where only a difference of 0.04π is found. In the IP case, meanwhile, the difference is much greater at 0.11π. Secondary maxima are found at $\varphi=(j+0.96)\pi$ which reach almost the same damage parameter value as the global maxima. Generally, the stresses on the critical planes are significantly higher (+60% at the in-phase maxima, +100% at the out-of-phase maxima), exceeding the uniaxial fatigue limit on most orientations just like the Liu-Mahadevan criterion.

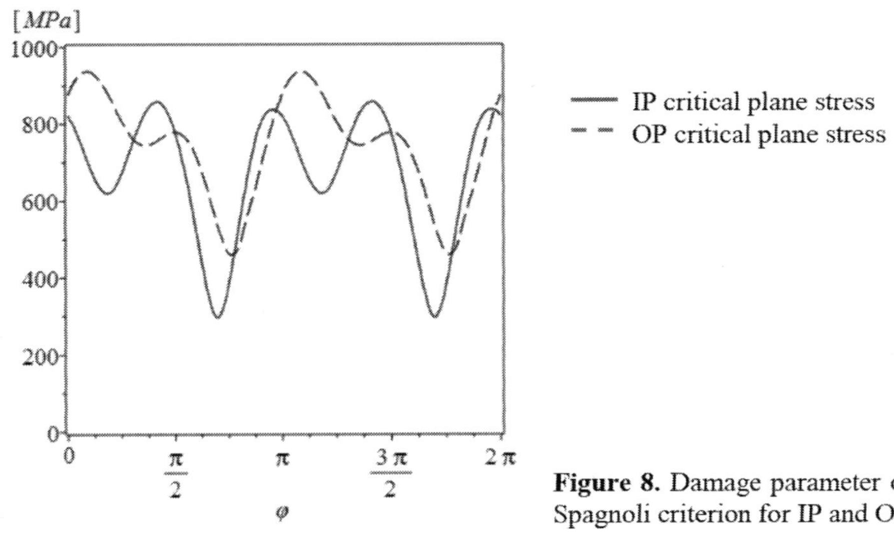

Figure 8. Damage parameter of the Carpinteri-Spagnoli criterion for IP and OP load.

Table 2 shows a comparison of the discrepancies between predicted and real critical plane orientation encountered in the different criteria. Only the plane of highest damage parameter value is shown. To all stated orientations, multiples of π can be added for equivalent orientations. For the shear stress-based critical plane definition, the shift is limited to less than 10%. Likewise, the underestimation of the stresses on the critical plane is of the same magnitude and is usually small enough to be negligible. The Liu-Mahadevan criterion is the notable exception, with significant differences between predicted at real value in both load cases. The orientation difference is up to 0.25π. In the IP case, the damage parameter on the predicted critical plane is almost 40% less than on the actual critical plane. Finally, the Carpinteri-Spagnoli criterion achieves good results in the OP case but underestimates the critical plane stresses in the IP case.

Table 2: Differences between predicted and real values for the orientation of the critical plane and the maximum damage parameter

Criterion	Critical plane angle φ		Maximum damage parameter value	
	As defined	Real	On definitional critical plane	On real critical plane
Matake				
IP	0.44π	0.48π	462.1MPa	478.0MPa
OP	0.50π	0.50π	473.0MPa	473.0MPa
McDiarmid				
IP	0.44π	0.46π	481.0MPa	486.7MPa
OP	0.50π	0.50π	473.0MPa	473.0MPa
Susmel and Lazzarin				
IP	0.44π	0.46π	483.5MPa	488.2MPa
OP	0.50π	0.50π	473.0MPa	473.0MPa
Liu and Mahadevan				
IP	0.19π	0.43π	0.83[a]	1.36[a]
OP	0.17π	0.09π	1.59[a]	1.67[a]
Carpinteri et al.				
IP	0.52π	0.41π	737.5MPa	858.2MPa
OP	0.05π	0.09π	925.2MPa	936.8MPa

[a] Dimensionless damage parameter

3.2. Dang Van criterion (mesoscopic scale)

The general stress tensor for the two load cases in this paper is (cf. [24])

$$\boldsymbol{\sigma} = \begin{bmatrix} \sigma_{\mathrm{m}} + \sigma_{\mathrm{a}} \sin(\omega t) & \tau_{\mathrm{a}} \sin(\omega t) & 0 \\ \tau_{\mathrm{a}} \sin(\omega t) & 0 & 0 \\ 0 & 0 & 0 \end{bmatrix} \tag{11}$$

The resulting residual stress tensor is

$$\boldsymbol{\rho}^* = \begin{bmatrix} -\frac{2}{3}\sigma_{\mathrm{m}} & 0 & 0 \\ 0 & \frac{\sigma_{\mathrm{m}}}{3} & 0 \\ 0 & 0 & \frac{\sigma_{\mathrm{m}}}{3} \end{bmatrix} \tag{12}$$

which leads to a mesoscopic stress deviator of the form

$$\hat{s}(t)=\begin{bmatrix} \frac{2}{3}\sigma_a \sin(\omega t) & \tau_a \sin(\omega t+\delta) & 0 \\ \tau_a \sin(\omega t+\delta) & -\frac{\sigma_a}{3}\sin(\omega t) & 0 \\ 0 & 0 & -\frac{\sigma_a}{3}\sin(\omega t) \end{bmatrix} \tag{13}$$

The hydrostatic stress is

$$PH(t)=\frac{\sigma_m}{3}+\frac{\sigma_a}{3}\sin(\omega t) \tag{14}$$

and the Dang Van shear stress is

$$\tau_{DV}(t)=0.5\sqrt{(\sigma_a \sin(\omega t))^2+4(\tau_a \sin(\omega t))^2} \tag{15}$$

With the last two equations, the load path at the critical orientation can be drawn in a diagram like figure 3 for both load variants, as depicted in figure 9. The modified endurance limit is also shown.

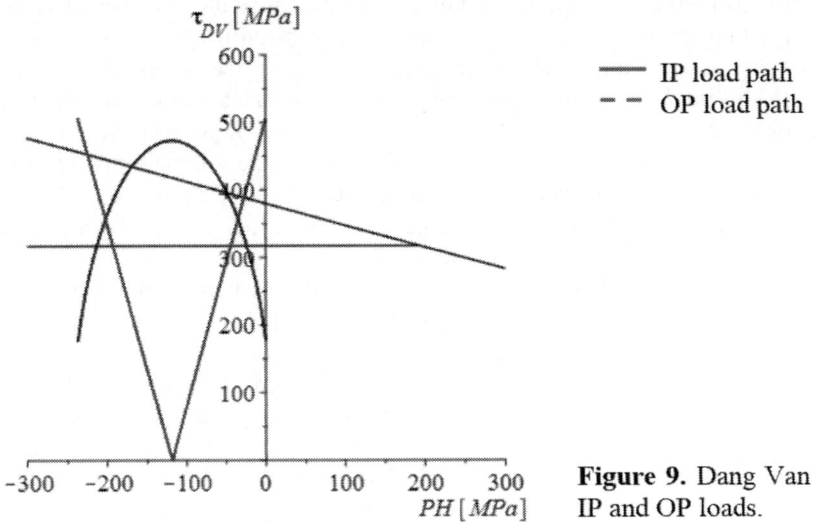

Figure 9. Dang Van criterion. Load paths for IP and OP loads.

The exact position of the load curves along the compressive stress axis cannot be determined without information on the residual stresses present in the material. However, as discussed in section 2.2, the compressive residual stresses that are likely to be present in the specimens would only effect a translation towards higher negative hydrostatic stresses, meaning that the portion of the curve beyond the endurance limit remains unaffected for the modified endurance domain.

This limitation notwithstanding, the Dang Van criterion correctly predicts that fatigue damage will occur in both cases, as parts of both load paths are beyond the endurance limit. It should be noted that surpassing the endurance limit of this criterion only means that crack initiation rather than failure will occur [23]. The load paths themselves are very different. The IP curve consists of two straight lines meeting at the x-axis. This point is the instant of zero torsional stress. Only the minor portion of one load cycle is spent beyond the endurance limit, but the shear stress maxima are fairly high. The OP curve is a parabola with its maximum at slightly lower shear stresses than in the IP case. The shear stresses are over the endurance limit during a greater part of the OP cycle and never reach the value zero. One major difference is that the IP shear stresses are highest at the extreme points of hydrostatic stress, while the shear stress maximum of the OP load is achieved in the presence of moderate compressive stresses.

For both load paths, comparable curves can be found in rolling contact studies by other authors. The in-phase curve resembles, for example, the ones shown in [23]. The out-of-phase curve is similar to the curves in [24].

It is important to notice one fundamental difference between the Dang Van criterion and the five multiaxial fatigue criteria presented above. In the latter, we were concerned with the variation of the damage parameter over different *orientations*, whereas figure 9 shows the variation in *time*. The critical orientation(s) of the Dang Van criterion can be determined from the principal stresses, as they are used in the calculation of $\tau_{DV}(t)$ in equation (9). For the two load cases discussed here, the critical orientations are $\theta^{*}=(j+0.5)\pi, j=0,1,2,\ldots,n$ for both load cases. In the OP case, $\varphi^{*}=j\pi$ and in the IP case, $\varphi^{*}=(j+0.5)\pi$.

4. Discussion

It might be appropriate to begin discussing the results by remembering that the present analysis is of a somewhat experimental nature. One could say that it repurposes critical plane criteria, which were originally developed to predict component lifetime under multiaxial fatigue loads. This aspect is of little concern for the present study, as the underlying experiments were run well below the failure limit. The actual goal is to understand the stress conditions and processes in the material leading to the microstructural changes (especially grain refinement) that are typical for White Etching Cracks. In this context, any circumstances influencing the mobility of dislocations are of particular interest. The reason for this is that we suppose WEC formation to be controlled by a mechanism similar to those responsible for nanograin formation in very high cycle fatigue [28] and severe plastic deformation [29]. In both of these cases, the formation of dislocation cells leads to sub-grain formation and ultimately to a grain refinement compared to the original structure. Naturally, drawing conclusions on grain-scale mechanisms based on macroscopic critical plane criteria risks a too high degree of simplification. However, there might still be some useful insights to be gained from a better characterization of the multiaxial stress state compared to other approaches such as equivalent stresses or stress intensity factors. Considering this background, the Dang Van criterion offers perhaps the greatest potential, as its mesoscopic approach is well suited for a grain-level interpretation.

From the macroscopic criteria, it seems that the shear stress amplitude is a better determinant of the critical planes than the normal stress amplitude. The criteria by Matake, McDiarmid, and Susmel-Lazzarin predict the local and global stress maxima with good accuracy. This is especially important in the IP load case, where the local maxima of the damage parameter are almost as high as the global maxima. For the original use of critical plane criteria, i.e. lifetime prediction, this is (at least in a first approximation) of no great relevance, as the highest load determines lifetime. For the present study, these secondary critical planes are important, as their presence indicates that dislocation movement should generally be easier in the IP than in the OP case. However, too many moving dislocations of similar orientation can block each other's movement. In this context, it is interesting to note that the three abovementioned criteria also predict steeper gradients of the damage parameter for IP. This could mean that the IP load case strikes a favourable balance between dislocation mobility on certain planes and inactive dislocations on others, while the OP load activates too few slide systems to cause significant rearrangement. However, due to the very high dislocation density in hardened bearing steels, conclusions on this subject are difficult. In high carbon steels like 100Cr6, the influence of carbon on dislocation mobility must also be considered.

In order to further validate the criteria, we plan to relate the critical plane orientations with the growth direction of the fatigue cracks found on the specimens. With the present results, this is not yet possible, as the critical plane and the actual crack plane are not necessarily the same [30]; however, several methods for calculating the fracture plane do exist [17].

The discrepancies between predicted and real critical orientation are rather high for the criterion by Liu and Mahadevan. This may seem surprising, as the criterion was developed especially for brittle materials, which hardened steel would usually assumed to be. Yet, bearing applications require sufficient toughness in addition to high hardness. Tempering treatment and residual austenite content of bearing steel could mean that this material is still brittle when compared to other metals but not brittle in the sense of the criterion (although the fatigue strength ratio of 100Cr6 is within the range for which the criterion is defined). The criterion also predicts failure of the specimens under both load variants, which is not in agreement with experimental evidence. It is conceivable that the high compressive, i.e. negative normal stresses pose a problem in the applicability of the criterion. Little information is available on this subject, as compression-torsion studies based on critical plane criteria are scarce.

The Carpinteri-Spagnoli criterion accurately predicts the orientation of the most critical plane for the OP case. However, like the Liu-Mahadevan criterion, it tends to give too conservative a lifetime estimation under the investigated loads. In the IP case, the critical plane angle φ^* and the critical plane stresses are not predicted correctly. For this study in particular, there is the added problem that only one critical plane orientation is found, which means that the secondary maxima can be overseen.

This last criticism can also be levelled at the mesoscopic Dang Van criterion. Its formulation with the Tresca equivalent stress, which in turn is based on the principal stresses, means that only one, namely the most critical, orientation will be identified. The criterion still is a valuable addition for studying the stress conditions induced by the two load patterns, as it provides information on the stress variation on the critical plane over time. Studying the markedly different load paths of one IP and OP cycle and comparing them with examples from other rolling contact fatigue studies opens opportunities for a better understanding of the differences between the load variants. At a first glance, it is surprising that two load patterns leading to markedly different multiaxial stress states can both be used to recreate rolling contact fatigue conditions. However, the stress field in a Hertzian contact is complex and depends both on time and the observed location. It is probable that the two load variants correspond to different points in the Hertzian contact. If this is the case, the region that is critical for WEC formation could be pinpointed. Unfortunately, the references used to compare the two Dang Van load paths seldom state the exact location leading to one or the other shape; further investigation is necessary on this subject.

5. Conclusion

The stress conditions of two cyclic compression-torsion load patterns used in White Etching Crack reproduction experiments were analysed using five stress-based macroscopic critical plane criteria and the mesoscopic Dang Van criterion. This approach repurposes critical plane criteria, which were developed for lifetime prediction, and continues earlier work to make critical plane criteria useful for the study of WECs. It was found that of the criteria studied, those which define the critical plane as the plane on which the highest shear stress amplitude occurs yielded the most useful results. The findings are interpreted as to their significance for dislocation mobility and dislocation cell formation.

The normal stress amplitude is found to be not a good indicator of critical plane orientations for both compression-torsion load cases. In addition, the lifetime estimation of the criterion is very conservative, predicting failure where none occurred during the experiments. This was also true for the Carpinteri-Spagnoli criterion, which, while very accurately predicting the critical plane in the OP case, underestimates lifetime and does not point to planes on which the damage parameter is close but not equal to the global maximum, especially for the IP load.

The Dang Van criterion is a valuable addition to the criteria based on shear stress amplitude. While it can only be used to search for the plane of the highest stresses, it shows the evolution of the stresses on this plane over time. A comparison of the load paths representing the experimental loads leads to the assumption that the load variants represent different locations and/or instants of a Hertzian contact. Further work should focus on correlating the predicted fracture orientations of different criteria to the crack angles found in tested specimens.

Acknowledgements

We gratefully acknowledge financial support by the German Research Foundation (Deutsche Forschungsgemeinschaft, DFG) under grant KE1426/6-1.

IOP Conf. Series: Journal of Physics: Conf. Series **843** (2017) 012028 doi:10.1088/1742-6596/843/1/012028

References

[1] Kotzalas M N and Doll G L 2010 *Phil. Trans. R. Soc.* A **368** 4829–50
[2] West O, Diederichs A M, Alimadadi H, Dahl K V and Somers M 2013 *Pract. Metallogr.* **50** 410–31
[3] Evans M-H 2012 *Tribol. Int.* **28** 3–22
[4] Ruellan Du Crehu, Arnaud 2014 *Tribological Analysis of White Etching Crack (WEC) Failures in Rolling Element Bearings* Ph.D. thesis (INSA Lyon: Lyon)
[5] Loos J 2014 Einfluss der Reibbeanspruchung auf die WEC-Bildung in Wälzlagern *55. GfT-Tagung* (Göttingen) chapter 23 pp 1-13
[6] Stadler K, Lai J and Vegter R H 2014 A review: The dilemma with premature white etching crack (WEC) bearing failures *Bearing Steel Technologies* ed Beswick J M (West Conshohocken: ASTM International) pp 487–508
[7] Alberdi A and Gomez R 2016 Typical bearing damage modes in wind turbines *Proc. Lubmat 2016 Conf.* (Bilbao, Spain)
[8] Gegner J 2011 Tribological aspects of rolling bearing failures *Tribology - Lubricants and Lubrication* ed Kuo C-H (Rijeka: InTech) pp 33–93
[9] Evans M-H 2016 *Mater. Sci. Technol.* **32** 1133–69
[10] Averbeck S and Kerscher E 2016 *Frattura ed Integrità Strutturale* **38** 12–8
[11] Burkart K, Bomas H, Schroeder R and Zoch H-W 2012 Rolling contact and compression-torsion fatigue of 52100 steel with special regard to carbide distribution *Bearing Steel Technologies* ed Beswick J M (West Conshohocken: ASTM International) pp 218–36.
[12] Beretta S and Foletti S 2012 Propagation of small cracks under RCF: a challenge to multiaxial fatigue criteria *Proc. of the 4th Int. Conf. on Crack Paths (CP 2012)* ed Carpinteri A et al. (Cassino: Gruppo Italiano Frattura) pp 15–28
[13] Averbeck S and Kerscher E 2016 *Eng. Fract. Mech.* (*Preprint* j.engfracmech.2016.12.011)
[14] Fatemi A and Shamsaei N 2011 *Int. J. Fatigue* **33** 948–58
[15] Papadopoulos I V 1998 *Fatigue Fract. Eng. M.* **21** 269–85
[16] Gabelli A, Lai J, Lund T, Rydén K, Strandell I and Morales-Espejel G E 2012 *Int. J. Fatigue* **38** 169–80
[17] Liu Y and Mahadavan S 2005 *Int. J. Fatigue* **27** 790–800
[18] McDiarmid D L 1991 *Fatigue Fract. Eng. M.* **14** 429–53
[19] Matake T 1977 *Bulletin of JSME* **20** 257–63
[20] Susmel L and Lazzarin P 2002 *Fatigue Fract. Eng. M.* **25** 63–78
[21] Carpinteri A, Spagnoli A and Vantadori S 2011 *Int. J. Fatigue* **33** 969–76
[22] Papadopoulos I V 1997 *Int. J. Fatigue* **19** 219–35
[23] Dang Van K, Griveau B and Message O 1989 On a new multiaxial fatigue limit criterion: theory and application *Biaxial and Multiaxial Fatigue* ed Brown M W and Miller K J (London: Mechanical Engineering Publications) pp 479–96
[24] Bernasconi A, Filippini M, Foletti S and Vaudo D 2006 *Int. J. Fatigue* **28** 663–72
[25] Desimone H, Bernasconi A and Beretta S 2006 *Wear* **260** 567–72
[26] Dang Van K, Cailletaud G, Flavenot J F, Le Douaron A and Lieurade H P 1989 Criterion for high cycle fatigue failure under multiaxial loading *Biaxial and Multiaxial Fatigue (EGF 3)* ed Brown M W and Miller K J (London: Mechanical Engineering Publications) pp 459–78
[27] Bhadeshia H 2012 *Prog. Mater. Sci.* **57** 268–435
[28] Grad P, Reuscher B, Brodyanski A, Kopnarski M and Kerscher E 2012 *Scr. Mater.* **67** 838–41
[29] Estrin Y and Vinogradov A 2013 *Acta Mater.* **61** 782–817
[30] Karolczuk A and Macha E 2005 *Int. J. Fract.* **134** 267–304

Fatigue fracture of fiber reinforced polymer honeycomb composite sandwich structures for gas turbine engines

Mikhail Nikhamkin, Nikolai Sazhenkov, Danil Samodurov

Perm National Research Polytechnic University, 29 Komsomolsky prospekt, Perm, Perm krai, Russia, 614990

samodurovdanil@gmail.com

Abstract. Fiber reinforced polymer honeycomb composite sandwich structures are commonly used in different industries. In particular, they are used in the manufacture of gas turbine engines. However, fiber reinforced polymer honeycomb composite sandwich structures often have a manufacturing flaw. In theory, such flaws due to their rapid propagation reduce the durability of fiber reinforced polymer honeycomb composite sandwich structures. In this paper, bending fatigue tests of fiber reinforced polymer honeycomb composite sandwich structures with manufacturing flaws were conducted. Comparative analysis of fatigue fracture of fiber reinforced polymer honeycomb composite sandwich specimens was conducted before and after their bending fatigue tests. The analysis was based on the internal damage X-ray observation of fiber reinforced polymer honeycomb composite sandwich specimens.

1. Introduction

Application of fiber reinforced polymer honeycomb composite sandwich (FRPHCS) structures is constantly growing. Currently FRPHCS structures are used in the aircraft, ships, cars, bridge construction and wind energy systems. The FRPHCS structures application in the aviation industry, specifically in gas turbine engine manufacture has the greatest interest. This is because that FRPHCS structures have a high fatigue resistance and stiffness, high strength-to-weight ratio and low acoustic conductivity [1, 2]. The last two features is most important due to the current trend of gas turbine engine weight reduce.

However, FRPHCS structures can have internal manufacturing flaws due to imperfect manufacturing technique. Such flaws can significantly reduce the static and fatigue life of FRPHCS structures.

Some authors [3,4] note that the durability of damaged FRPHCS structures under fatigue bending is lower than that of undamaged ones.

Theoretically, manufacturing flaws reduce the FRPHCS structures durability due to the fact that their propagation is rapid. Therefore, fracture pattern comparative analysis of FRPHCS specimens before and after their bending fatigue tests is the purpose of this paper. The analysis was based on X-ray observation of internal FRPHCS damage.

2. Method and materials

IOP Conf. Series: Journal of Physics: Conf. Series **843** (2017) 012029 doi:10.1088/1742-6596/843/1/012029

The cantilever bending fatigue tests are conducted on V-850-400 electrodynamic shaker with a power amplifier and APM-4M power regulator (Figure 1).

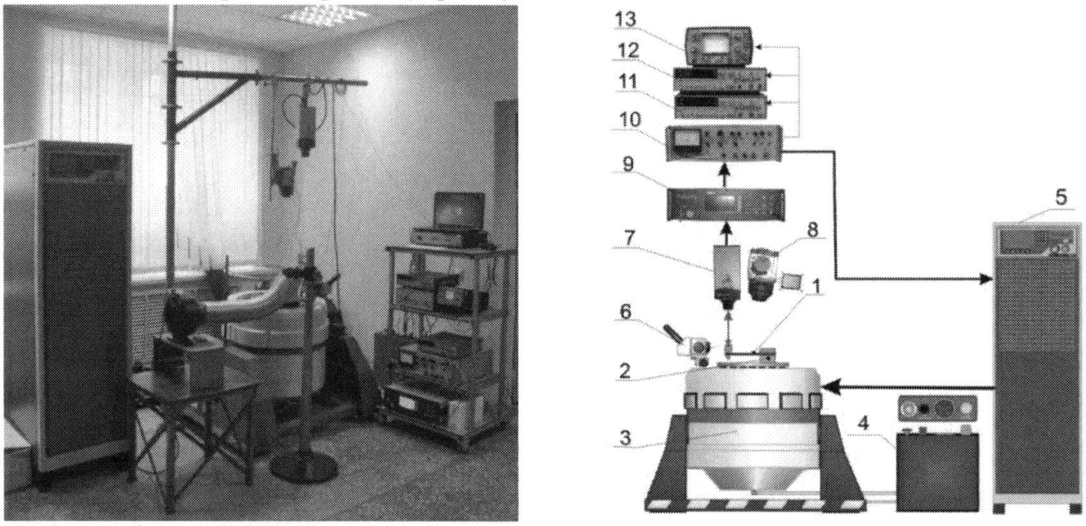

a) b)

Figure 1 – Experimental unit: external appearance (a) and scheme (b). 1 – Test specimen, 2 – Fixing spot, 3 – Electrodynamic shaker V850-400, 4 – Pneumatic control unit, 5 – Power amplifier, 6 – Microscope MBS-9, 7 – Sensor head of a laser vibrometer OFV-525 8 – Infrared camera, 9 – Laser vibration sensor OFV 5000, 10 – APM-4M power regulator , 11 – Frequency indicator, 12 – Cycle counter , 13 – Oscilloscope.

Fatigue tests are based on the ASTM C393/C393M standard [5]. This standard regulates the static bending tests of three-honeycomb-layer sandwich panel. In this study the cantilever bending fatigue tests, as opposed to static bending tests, involve cyclic load applying.

FRPHCS cantilever bending fatigue tests were conducted by fully reversed cycle on first flexural own shape. Initialization and maintenance of the cycling stress amplitude are carried out by the control point oscillation amplitude which is located at the free end of the specimen. The amplitude and frequency of bending is controlled by one-dimension laser vibration sensor OFV5000. Indirect monitoring of amplitude is carried out by microscope MBS-9. To reduce the testing frequency the weigh is fastened to the free end of the specimen.

In this study, the control point oscillation amplitude of the FRPHCS specimen is 1 mm. The tests are stopped after the 10^5 cycles.

The test specimens are subjected to X-ray detection based on the Nikon Metrology XT X 450 (Figure 2), which includes a microfocus X-ray tube with a minimum focal spot diameter of 30 μm (at 80 kW) and a plane-parallel detector Perkin Elmer XRD 1621 with a pixel size of 200 μm. X-ray photography of the specimen was carried out before and after the tests in the same plane and with an equal step.

As test specimens, two three-honeycomb-layer sandwich have been used (Figure 3a). Specimens have a carbon fiber reinforced skin and glass fiber reinforced skin. One of the ends of the panels is filled with foam. The test specimen was fixed

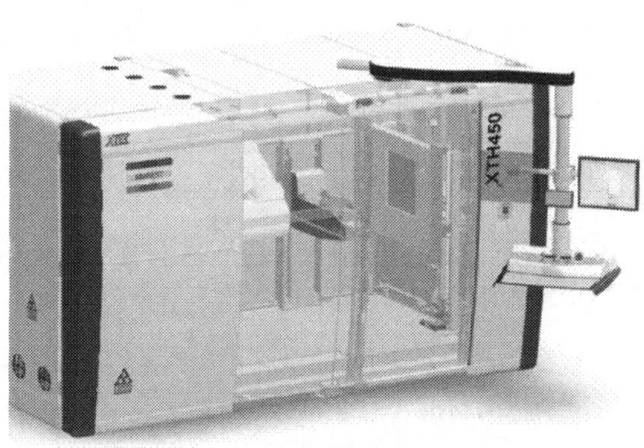

Figure 2 – Nikon Metrology XT X 450

IOP Conf. Series: Journal of Physics: Conf. Series **843** (2017) 012029 doi:10.1088/1742-6596/843/1/012029

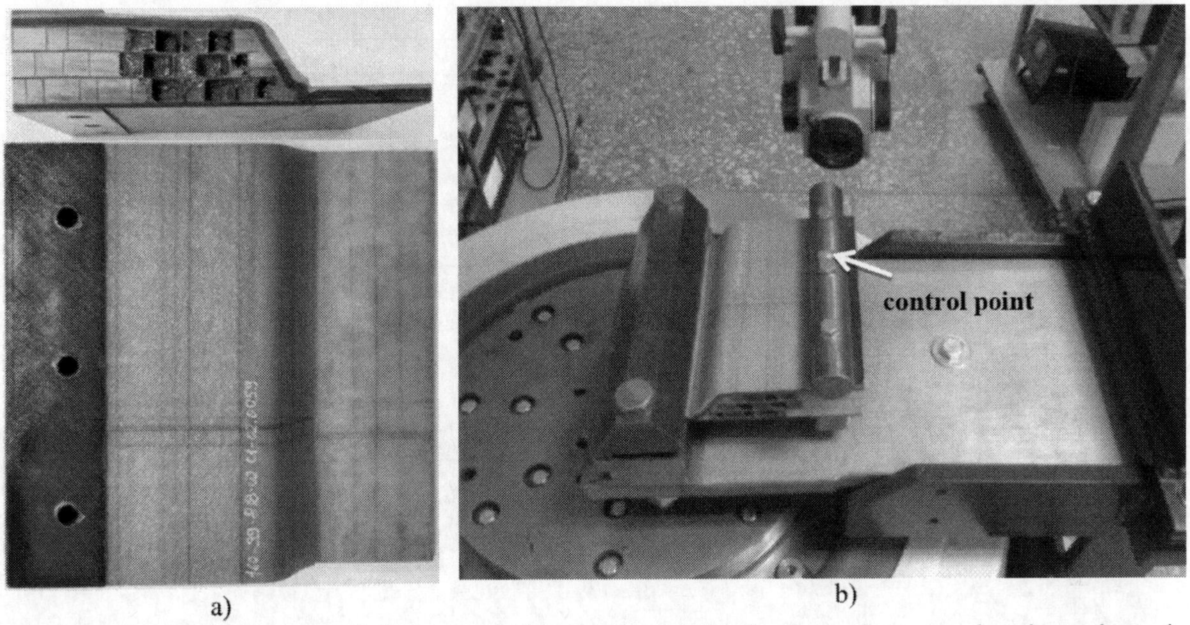

Figure 3 –Three-honeycomb-layer sandwich specimen (a); fixed specimen on the electrodynamic shaker (b)

cantilever on the electrodynamic shaker (Fig. 3b).

3. Results

Figure 4 shows the testing frequency relative change of the FRPHCS specimens during cantilever bending fatigue tests. The Table1 shows the FRPHCS specimens actual frequencies. The testing frequency, reducing of the first specimen was 3.8% and the second one is 8.65%.

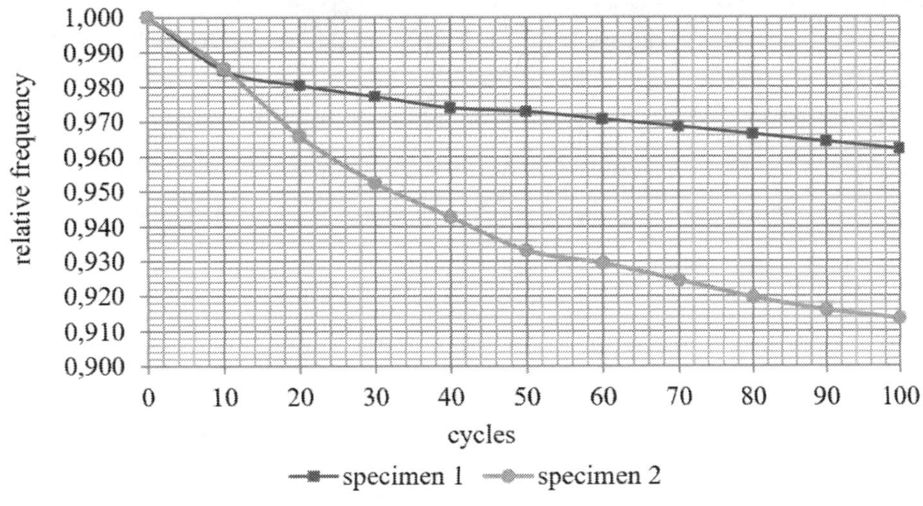

Figure 4 – Frequency relative change of the FRPHCS specimens during cantilever bending fatigue tests

IOP Conf. Series: Journal of Physics: Conf. Series **843** (2017) 012029 doi:10.1088/1742-6596/843/1/012029

Table 1 – Specimens frequencies, Hz

Cycles, 10^3		0	10	20	30	40	50	60	70	80	90	10
Specimen number	1	92,2	90,8	90,4	90,1	59,8	89,7	89,5	89,3	89,1	88,9	88,7
	2	82,1	80,9	79,3	78,2	77,4	76,6	76,3	75,9	75,5	75,2	75

Figures 5a and b show X-ray shadowgraphs of the first specimen before and after the test. It can be seen from the figures that there is no significant fatigue failure. The size of the flaws remained at about the same level. However, a new crack formation occurs in the glass fiber reinforced shell.

When X-ray shadowgraphs of the second specimen was examined (Fig. 6a and b), it can be noted that there is no significant fatigue failure. However, a flaw size slight increase is observed after 10^5 loading cycles.

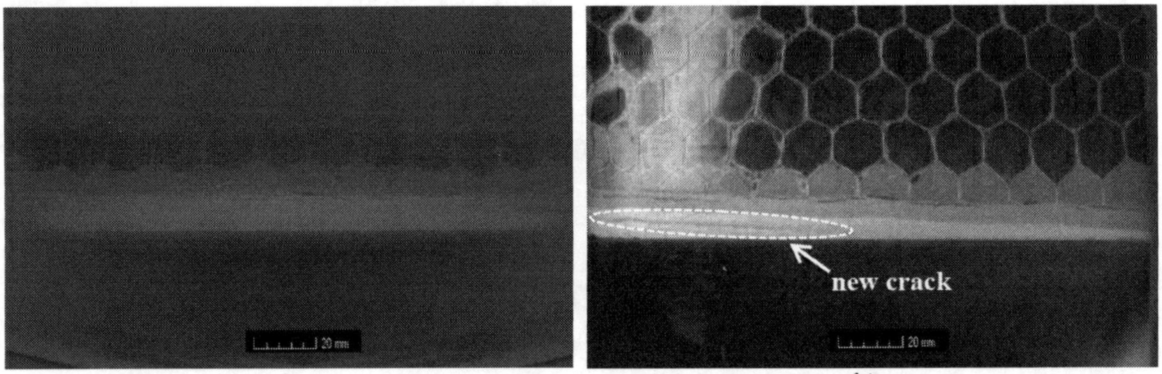

a) b)

Figure 5 – X-ray shadowgraphs of the first specimen before (a) and after (b) the test

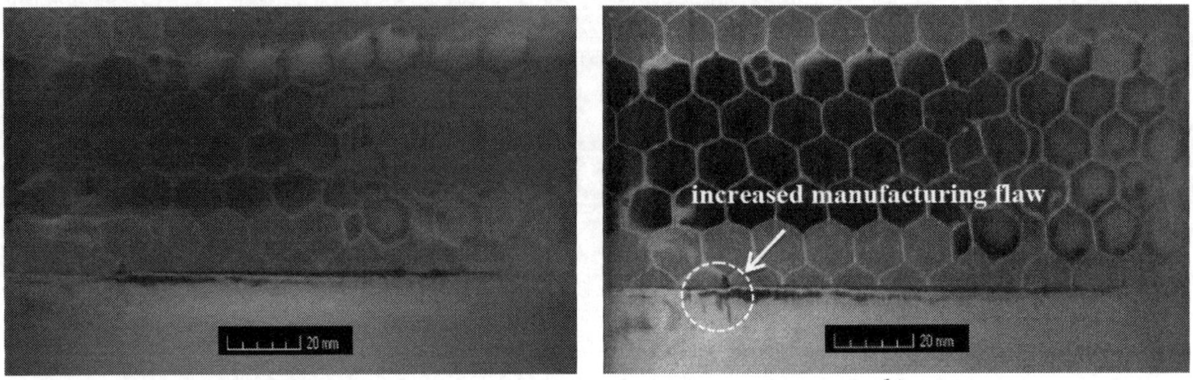

a) b)

Figure 6 – X-ray shadowgraphs of the second specimen before (a) and after (b) the test

4. Conclusion

The assumption that manufacturing flaws due to their rapid propagation reduce the fiber reinforced polymer honeycomb composite sandwich structures durability led to the following results. No significant fatigue failure has been observed after cantilever bending fatigue tests of polymeric honeycomb sandwich panels with manufacturing flaws. The flaws sizes light increase occurs on a par with newly crack. Therefore, one cannot reliably state that manufacturing flaws propagate rapidly.

IOP Conf. Series: Journal of Physics: Conf. Series **843** (2017) 012029 doi:10.1088/1742-6596/843/1/012029

However, in this paper, there was no comparison of the fatigue fracture of fiber reinforced polymer honeycomb composite sandwich specimen with manufacturing flaws and specimens without them.

In addition, the selected number of cycles of load may not be enough to cause significant fatigue damage.

Therefore, further investigation of this problem requires fatigue testing of fiber reinforced polymer honeycomb composite sandwich structures with manufacturing flaws and without them by more than 10^5 cycles.

5. References

[1] Vinson J. R. The behavior of sandwich structures of isotropic and composite materials. – CRC Press, 1999.

[2] S. Belouettar, A. Abbadi, Z. Azari, "Experimental Investigation of static and fatigue behaviour of composites honeycomb materials using four point bending tests", Composite Structures, Vol. 87, Issue 3, 265-273, 2009

[3] Burman M, Zenkert D. Fatigue crack initiation and propagation in sandwich structures. Doctoral thesis, Department of Aeronautics, Royal Institute of Technology, 1998

[4] M. Burman, D. Zenkert Fatigue of undamaged and damaged honeycomb sandwich beams J Sandwich Struct Mater, 2 (2000)

[5] ASTM C 393/C 393M Standard – Standard Test Method for Properties of Sandwich Constructions by Beam Flexure

Experimental and numerical investigation of crack initiation and propagation in silicon nitride ceramic under rolling and cyclic contact

Rahul Raga[1], Iyas Khader[1,2], Zdeněk Chlup[3], Andreas Kailer[1]
[1]Fraunhofer Institute for Mechanics of Material IWM, Wöhlerstr. 11, 79108 Freiburg, Germany
[2]Department of Industrial Engineering, German-Jordanian University, P.O. Box 35247, Amman 11180, Jordan
[3]Institute of Physics of Materials of the Academy of Sciences of the Czech Republic, v. v. I, Zizkova 22, 616 62 Brno, The Czech Republic

Email: rahul.raga@iwm.fraunhofer.de

Abstract. The focus of the work was to investigate crack initiation and propagation mechanisms in silicon nitride undergoing non-conforming hybrid contact under various tribological conditions. In order to understand the prevailing modes of damage in silicon nitride, two distinct model experiments were proposed, namely, rolling contact and cyclic contact experiments. The rolling contact experiment was designed in order to mimic the contact conditions appearing in hybrid bearings at contact pressures ranging from 3 to 6 GPa. On the other hand, cyclic contact experiments with stresses ranging from 4 to 15 GPa under different media were carried out to study damage under localised stresses. In addition, the experimentally observed cracks were implemented in a finite element model to study the stress redistribution and correlate the generated stresses with the corresponding mechanisms. Crack propagation under rolling contact was attributed to two different mechanisms, namely, fatigue induced fracture and lubricant driven crack propagation. The numerical simulations shed light on the tensile stress driven surface and subsurface crack propagation mechanisms. On the other hand, the cyclic contact experiments showed delayed crack formation for lubricated cyclic contact. Ceramographic cross-sectional analysis showed crack patterns similar to Hertzian crack propagation under cyclic contact load.

1. Introduction

Out of various structural ceramics such as zirconium oxide, aluminium oxide, SiAlONs, silicon nitride and silicon carbide; silicon nitride ceramics are considered to be the optimal choice of material for modern bearing applications due to their excellent properties [1]. The combination of properties such as, high hardness, adequate fracture toughness, high strength, superior corrosion resistance, thermal stability at elevated temperatures and longer service life makes silicon nitride attractive for hybrid bearing applications [2,3]. In addition, low adhesion affinity between steel and silicon nitride results in low friction and low heat generation, thus, extends the range of potential loading conditions even under marginal lubrication [4,5]. The emerging applications of hybrid bearings are seen in wind turbine generators where silicon nitride bearing balls of diameters reaching nearly 48 mm are used and in traction motors in the railway industry [6]. In spite of industrial deployment, the reliability of the rolling elements made from silicon nitride undergoing non-conforming contact is still a topic of research due to their brittle nature and scatter in strength. The most widely used grades for hybrid bearing application are hot isostatically-pressed silicon nitride (HIPSN) ceramics. This is mainly due to ability of achieving flaw-free dense material with superior mechanical properties. However, high manufacturing cost is still a concern in applying HIPSN [7]. Alternatively, silicon nitride balls produced from gas-pressure sintered techniques (GPSN) are slowly gaining importance in the bearing industry due to their cost effectiveness and reliable reproducibility in comparison to other manufacturing processes. Nonetheless, GPSN grades may suffer from intrinsic inhomogeneities known as snowflakes [8]. Snowflakes are undensified regions, in which only the grain skeleton is present and the surrounding glassy phase is missing, which may attribute to premature material failure under service. Therefore, understanding damage mechanisms under various tribological loading becomes inevitable[9–15].

The distinctiveness in this work arises from the combination of experimental (rolling and cyclic) work and numerical simulations, which enabled studying crack initiation and propagation in GPSN. The rolling contact experimental results have been extensively described in [16] and cyclic contact results have been addressed for the first time. For both cases most crucial parameters influencing crack initiation and evolution are described by a systematic study with different loads. Minute details in terms of crack initiation and propagation are illustrated by performing extensive post-experimental analysis with focused ion beam (FIB) as well as conventional cross-sectioning. Additionally, finite element (FE) simulations were employed to explain the prevailing complex stress state in both cases.

2. Experimental

2.1. Experimental procedures

The rolling contact experiments were carried out on a twin-disk tribometer as shown in the figure 1. The roll was made of silicon nitride and the disk was made up of hardened 100Cr6 steel. The samples were designed in such a way that when they come in contact they make an elliptical contact. Furthermore, various normal loads from 500 – 4150 N were applied to induced contact pressure ranging from 3 – 6 GPa, respectively. Once in contact, both samples were rotated at same speed making the system run under pure rolling (i.e., without induced slip). Experiments were performed ranging from 7500 revolutions to 50 million revolutions (Mrev) under continues lubrication. A non-additivated mineral oil (SKF-TT9) was used.

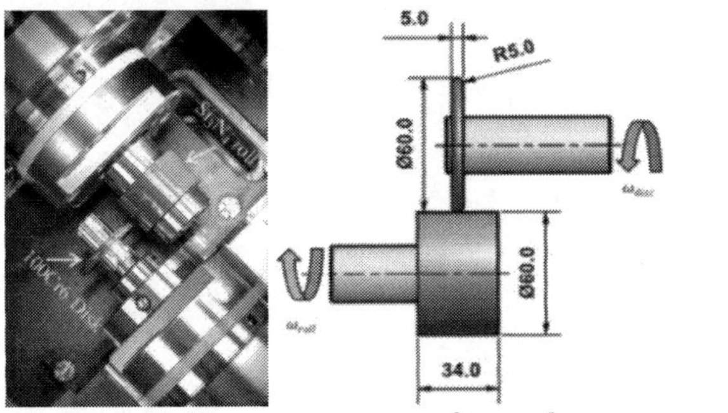

Figure 1. Rolling contact experimental set-up.

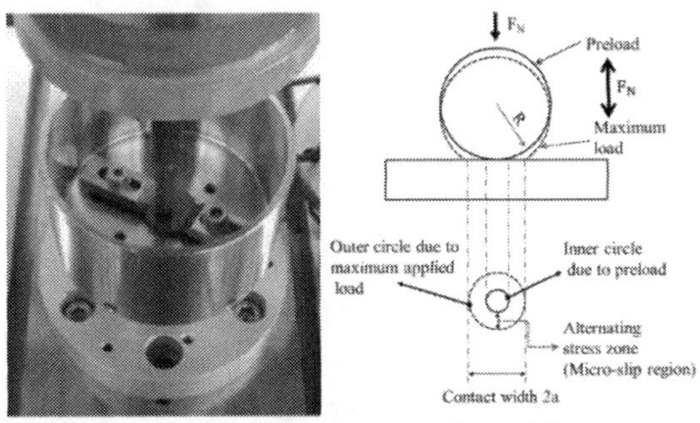

Figure 2. Cyclic contact experimental set-up.

The cyclic contact experiments were carried out on an electro-dynamic testing machine (ET10000). A silicon nitride disk was loaded in cyclic contact with a tungsten carbide indenter. Figure 2 shows the experimental setup and the circular contact area arising during loading. In these experiments, an initial preload was applied to hold the samples in contact, after which the load is ramped up until reaching its maximum designated value; the contact area increases accordingly as shown in figure 2. Initially, the experiments were carried out within application relevant contact pressure range (i.e., 4 – 6 GPa with a preload of 2.5 GPa). Subsequently, and in order to accelerate damage, high load experiment with contact pressure ranging from 10 – 15 GPa with a preload of 4.3 GPa were carried out. The tests were run for 1 to 40 million cycles (Mcycles) at a constant frequency of 80 Hz under dry and lubricated (SKF-TT9) conditions.

2.2. *Material specifications*

The ceramic rolls and disks were provided by FCT Ingenieurkeramik GmbH, Germany, the commercial designation for this silicon nitride is SN-GP black. The microstructure consists of β-Si_3N_4 grains with an aspect ratio of 7.4 and ca. 10 wt% glassy phase. The sintering additives consist of aluminium oxide (Al_2O_3) and yttrium oxide (Y_2O_3). The counter partners were hardened 100Cr6 steel (HV10>800) and tungsten carbide (WC) indenter for rolling and cyclic contact experiments respectively. The physical properties of the both materials are listed in table 1.

Table 1. Experimental sample material properties.

Physical property	Silicon nitride (Si_3N_4)	Steel (100Cr6)	Tungsten carbide (WC)
Density(g/cm^3)	3.24	7.28	15.63
Elastic modulus (GPa)	300	212	640
Poisson's ratio	0.26	0.3	0.24

3. Numerical simulations

The purpose of these simulations was to illustrate the tribological scenarios prevailing under rolling and cyclic contact loading. The physical geometry was discretised using a three-dimensional mesh for rolling contact as shown in figure 3a. A two-dimensional axisymmetric model was used for cyclic contact as shown in figure 3b.

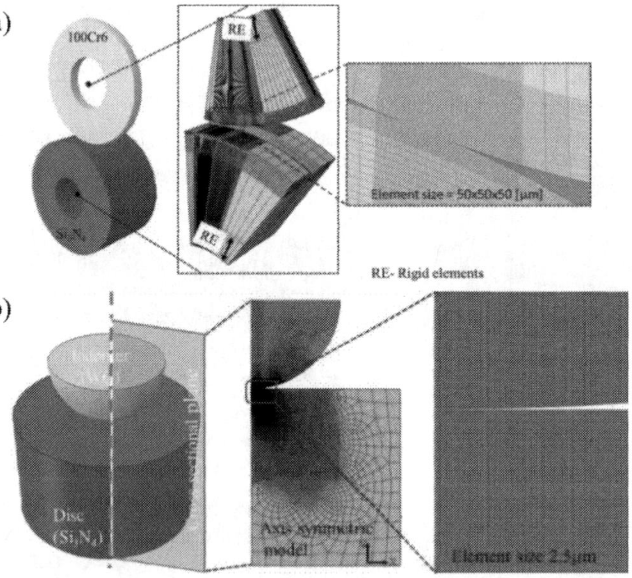

Figure 3. Finite element model to mimic the contact conditions as in (a) rolling contact, (b) cyclic contact experiments.

For rolling contact, the initial sets of simulations were performed considering both materials to be isotropic linear elastic. In order to study the influence of plastic deformation on the contact stresses, a second set of simulations was performed taking the plastic behaviour of steel into consideration. A uniaxial yield strength value of 1100 MPa was assigned to 100Cr6 steel. A quasi-static analysis procedure was employed to initiate contact and to obtain a steady-state contact stress distribution using Abaqus/Standard. For cyclic contact, the simulations were performed without cyclic behaviour. A single loading cycles was sufficient to capture the entire stress regime prevailing under cyclic loading due to the absence of accumulative plastic strain in both materials. A quasi-static analysis procedure in Abaqus/Standard was employed to run the simulations. Initially, frictionless simulations were performed in order to validate the model as well as to illustrate the stress state for lubricated contact condition with various loads. Additional simulations were performed with the Coulomb friction model used to introduce friction for unlubricated contact.

For both loading cases, additional simulations were performed with cracks of different lengths implemented in silicon nitride. Cracks were created using the partition function in Abaqus/CAE; the partitioned section was treated as a seam [17], so that during mesh regeneration it places overlapping nodes throughout the partitioned face, which act as crack during the analysis.

4. Results and Discussion

4.1. Rolling contact

Surface profilometry across the wear tracks resulted for all loads in barely quantifiable wear volumes and change in surface roughness in comparison to the virgin surface. Up to 2500 N and 50 Mrev, the damage was mainly in form of minute grain break-down from the edges of the machining striation and formation of micro-spalls as shown in figure 4a. These micro-spalls occurred only where snowflake were present. In other words, grain pulled out regions from snowflakes appeared as micro-spalls on the surface. Under certain condition (i.e., with excessive grain pull-out), the micro-spall formed craters as shown in figure 4b, which in turn trapped wear debris and lubricant, thus, allowing the lubricant to build up hydrostatic pressure within the crater during rollover. As a consequence, when sufficient hydrostatic pressure was reached within the crater, cracks would initiate and propagate in the depth as shown in the figure 4c.

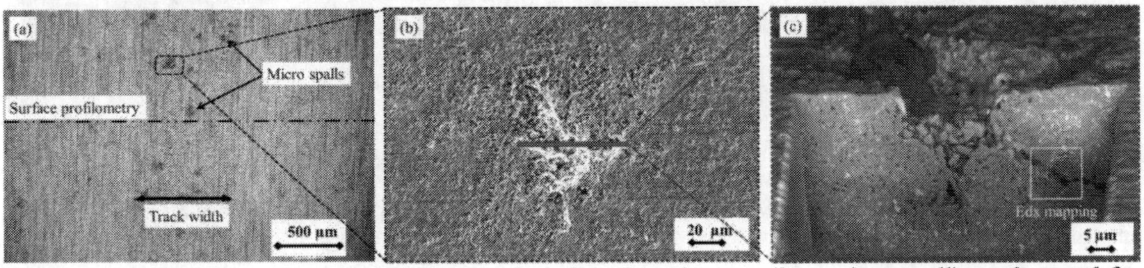

Figure 4. Optical image shows the wear track and surface damage (i.e., micro-spall) as observed for all loads and SEM micrographs shows the lubricant driven crack underneath a micro-spall.

For the highest applied load (4150 N), the damage appeared in form of macro-cracks and increased number of micro-spall regions on the surface. Figure 5a shows the top view of the wear track for 4150 N after 10 Mrev with the locations of cross-sections performed for subsurface analysis. The cross-sectional analysis performed at the crack free zones showed no signs of subsurface damage. However, the cross-sectional analyses performed on the surface crack showed that these cracks propagated in radial direction in the subsurface as shown in figure 5b.

The maximum crack length in the depth was measured and found to be 120±30 µm after 10 Mrev. Short experiments with a lower number of load cycles (7500 rev) suggested that crack formation was not spontaneous (i.e., not just after the initial contact or initial few rolling cycles) but rather due to fatigue [16,18]. No micro-spalls were observed for the short experiments and initial cracks on the surface were observed after 2 Mrev.

The FE simulations of pristine (uncracked) and cracked surfaces explained the fatigue crack propagation behaviour under rolling contact, as shown in figure 6. Figure 6a depicts the tensile stress (S_{max}) contours and the corresponding trajectories in the silicon nitride roll in a pristine surface. When

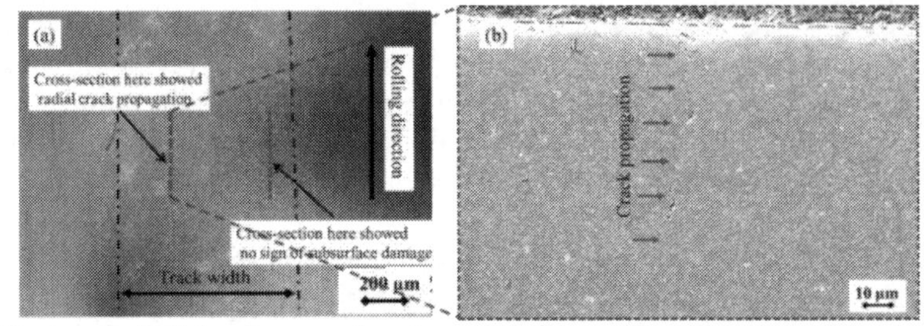

Figure 5. SEM micrographs show (a) the surface damage, (b) crack propagation in radial direction for highest load after 10 Mrev.

the contact ellipse approaches a crack (figure 6b), the tensile stress on leading edge drives the crack in the rolling direction; during rollover when the crack is engulfed by the contact ellipse (figure 6c) no tensile stress will be acting on it and compressive stresses retard crack propagation. However, when the crack is about to exit the compression dominated region, the trailing edge tensile stresses begin to dominate (figure 6d), thus, propagating the crack in the direction opposite to rolling. This continuous shifting of S_{max} trajectories at crack tip during rolling contact results in crack propagation in the radial direction (mode *I* propagation).

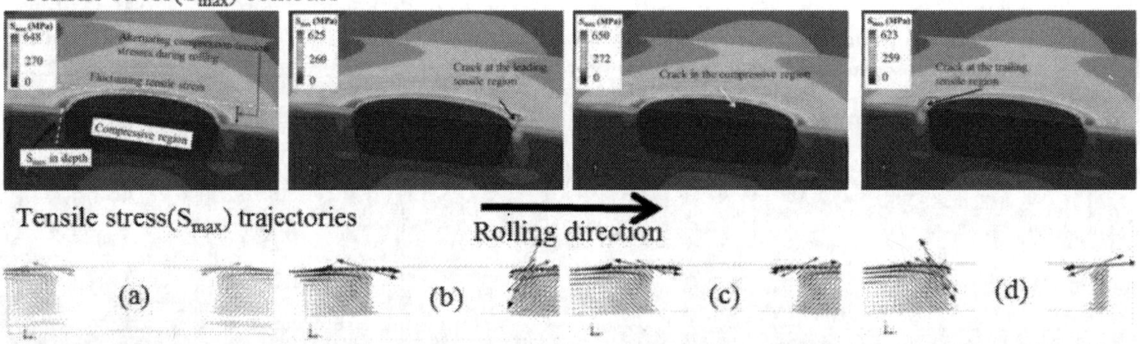

Figure 6. FEM-generated contours of S_{max} and the S_{max} trajectories (a) without crack, (b) with crack at leading edge S_{max}, (c) crack in compressive region and (d) crack at trailing edge S_{max}.

4.2. Cyclic contact

Table 2 shows an overview of the damage as observed with low load experiments under lubricated and unlubricated conditions.

Table 2. Experimental sample material properties.

Cycles (Millions)	Contact pressure (GPa)	Medium	Remarks
20 (2)* and 40 (1)	4.0	Lubricated	No damage
40 (2)	4.0	Unlubricated	No damage
20 (3)* and 40 (1)	5.0	Lubricated	Damage in form of surface staining
40 (2)	5.0	Unlubricated	Damage in form of surface fretting
20 (1)* and 40 (2)	6.0	Lubricated	Damage in form of surface fretting
20 (1)* and 40 (2)	6.0	Unlubricated	Damage in form of surface striations

()* The number in brackets shows the number of tests

Initially, a set of experiments was carried out at 4 GPa and a number of load cycles ranging from 20 – 40 Mcycles. The surface analysis indicated no signs of damage on silicon nitride for both unlubricated (dry) and lubricated tests. Starting from 5 GPa, three different patterns of damage were observed, namely, surface staining (figure 7a), surface fretting (figure 7b), and surface striation (figure 7c). The in-and-out squeezing of lubricant within the micro-slip zone triggered chemo-mechanical interactions between silicon nitride and the lubricant under the influence of alternating stresses which led to surface staining. Furthermore, the frictional stresses arising between the contact partners due to difference in elastic constants induced the surface fretting and striations in the microslip zone. Focused ion beam (FIB) cross-sections in the fretted zone suggest that damage is confined to a superficial layer.

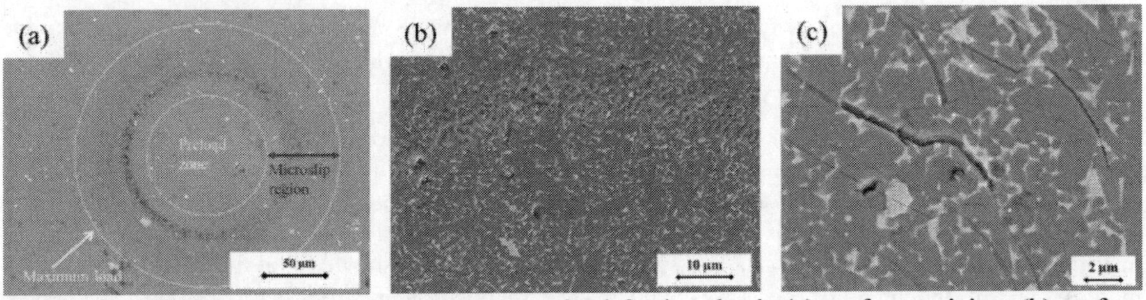

Figure 7. Different type of damage patterns noticed for low loads (a) surface staining, (b) surface fretting and (c) surface striations.

The tests with high loads (10 – 15 GPa) showed different behaviour under dry and lubricated conditions. Firstly, upon reaching a threshold number of cycles of 30 Mcycles, similar damage patterns were observed in lubricated tests up to 13 GPa. The damage was mainly in form of grain pull-out from the surface of both materials with an extended fatigue zone, figure 8. Secondly, delayed crack formation without any transfer layer was observed in lubricated tests in comparison to dry tests. In dry tests, Hertzian ring cracks were initially observed from 11.5 GPa, whereas crack formation was observed only at 15 GPa in the lubricated tests.

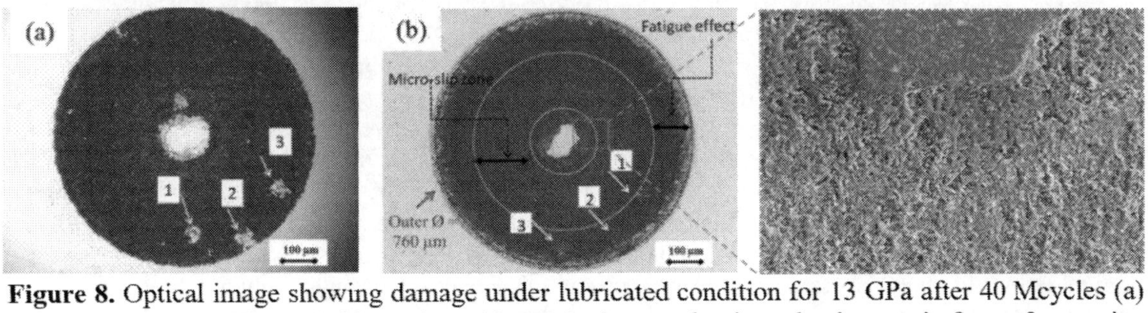

Figure 8. Optical image showing damage under lubricated condition for 13 GPa after 40 Mcycles (a) on indenter, (b) on silicon nitride surface. (c) SEM micrographs show the damage in form of excessive grain pull-out.

Figure 9a and b show a Hertzian ring crack and transfer layer on the surface of the ceramic and figure 9c shows the subsurface crack propagation in a dry test after 10 Mcycles at 13 GPa. The surface analysis (SEM and EDX) suggested that the transfer layer consisted of W, Co and O. Thus, suggests formation of tungsten trioxide (WO_3) and cobalt oxide (Co_2O_3) on the surface of silicon nitride due to initiation of damage in the WC indenter. A comparison between crack initiation under static and cyclic lubricated contact at 11.5 GPa showed no particular difference. The crack initiation was nearly at the same location. Thus, indicating that the experimentally observed crack under cyclic loading spontaneously forms upon initial contact and propagates with repeated loading on the surface and in the depth with severe sliding between crack faces.

Figure 9. Damage under unlubricated condition. (a) Optical image shows the surface ring cracks with transfer layer. SEM micrographs (b) transfer layer formation and location of FIB cut and (c) Subsurface crack propagation.

Figure 10 and figure 11 show distribution of contact pressure, tensile stresses and the location of crack initiation measured from the lubricated and dry test. Under lubricated condition the experimentally observed initial crack appears exactly at the peak tensile stress; however the experimental crack for dry case is shifted. This behaviour was explained by considering the influence of friction in numerical simulations. The simulations with friction showed that the peak tensile stress increases in the indenter and decreases with a shift in position in the silicon nitride disk, hence, explains the shifted crack initiation in dry contact.

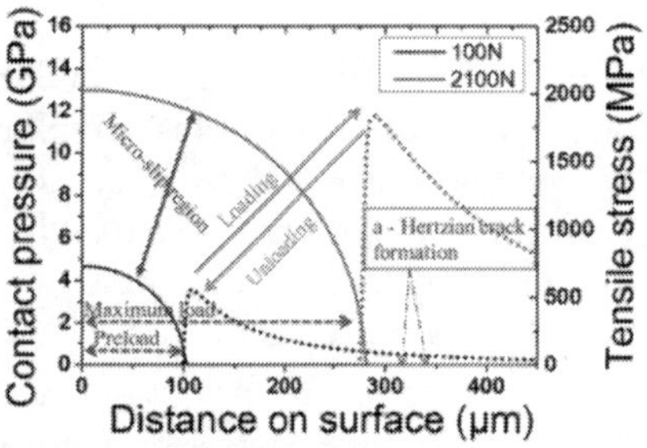

Figure 10.FEM generated contact pressure and tensile stresses for 2100 N.

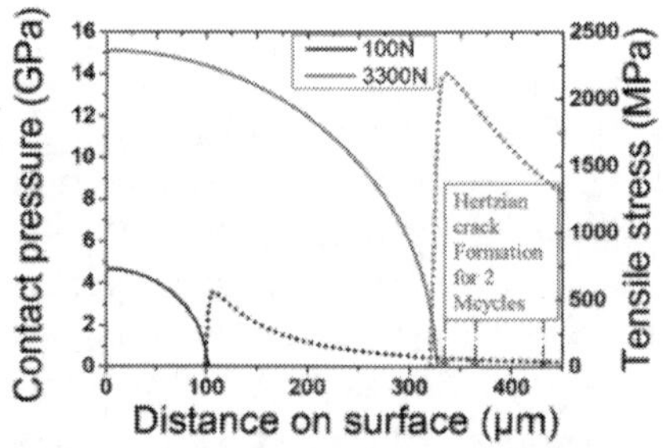

Figure 11. FEM generated contact pressure and tensile stresses for 3300 N.

In addition, the crack initiation location on the surface as well plays a decisive role in its propagation. Figure 12 shows the stress intensity factor (K_I and K_{II}) with loading time for 10 µm crack length with different initiation locations outside the contact zone. The dashed line in the graph represents the fracture toughness of this particular material. It can be clearly seen from the results that when crack initiates just outside the contact zone (i.e., at 290 µm for this case) both K_I and K_{II} dominates nonetheless, K_{II} stress intensity tends to diminish as the crack initiates away from the contact zone. Thus, pure mode *I* crack propagation dominates when the crack initiates at certain distance outside the contact zone.

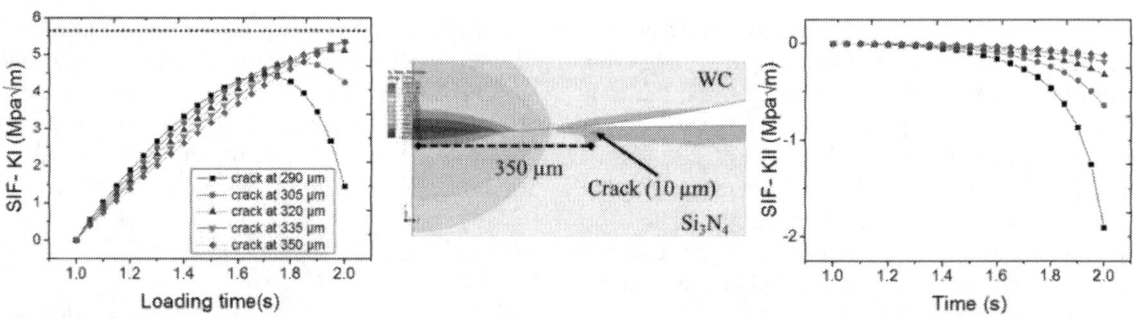

Figure 12. Stress intensity values (K_I and K_{II}) with loading time for 10 µm crack with different initiation locations (13 GPa).

5. Concluding remarks

5.1. Rolling contact

- Surface damage was dominated by the formation of micro-spalls up to a certain load (i.e., 2500 N), above which it appeared in form of macrocracks and micro-spalls
- Micro-spalls formed due to grain pull-outs from the snowflake structure undergoing rolling contact (i.e., alternating cyclic tension-compression fields).
- The crack initiation and propagation were attributed to two different mechanisms: (i) Fatigue induced crack propagation and (ii) lubricant driven crack propagation

5.2. Cyclic contact

- Different types of damage behaviour depending on loading (low and high) and lubrication conditions.
- No peculiar difference was noticed for the surface and the subsurface crack propagation mechanism under cyclic and static contact loading (Hertzian crack). However, excessive sliding occurs at crack face under cyclic loading conditions.
- Frictional forces play a vital role in crack initiation location under cyclic contact configuration.

Acknowledgement
The authors would like to thank the European Union's Seventh Framework Programme FP7/2007-2013 under Grant Agreement No. 263476 (RoLiCer Project) for the financial support.

References

[1] Popp M and Strenagel R 1999 Hybrid ceramic and all ceramic anti friction bearings *Space Mechanisms and Tribology, Proceedings of the 8th European Symposium* vol 438, ed D. Danesy. European Space Agency (Toulouse) p 105
[2] Katz R N 1995 *Ceramic materials for rolling element bearing applications (ARL-TR-751)* (Army Research laboratory, watertown, Massachusetts)
[3] Freitag D W and Richerson D W 1998 *Opportunities for Advanced Ceramics to Meet the Needs of the Industries of the Future (DOE/ORO 2076)* (U.S. Advanced Ceramics Association and Oak Ridge National Laboratory, Salt Lake City, UT 84117)
[4] Zhao X Z, Liu J J, Zhu B L, Miao H Z and Luo Z B 1997 Friction and wear of Si3N4 ceramic/stainless steel sliding contacts in dry and lubricated conditions *J. Mater. Eng. Perform.* **2** 203–8
[5] Reis P, Davim J P, Xu X and Ferreira J M F 2006 Friction and wear behaviour of beta-silicon nitride–steel couples under unlubricated conditions *Mater. Sci. Technol.* **22** 247–52
[6] Schöppl O 2012 Developments in ceramic bearing balls *Evol. Bus. Technol. Mag. skf* 1–5
[7] Wang L, Snidle R W and Gu L 2000 Rolling contact silicon nitride bearing technology: a review of recent research *Wear* **246** 159–73
[8] Herrmann M, Schulz I, Bales A, Sempf K and Hoehn S 2008 "Snowflake" structures in silicon nitride ceramics - Reasons for large scale optical inhomogeneities *J. Eur. Ceram. Soc.* **28** 1049–56
[9] Ferjaoui A, Yue T, Abdel Wahab M and Hojjati-Talemi R 2015 Prediction of fretting fatigue crack initiation in double lap bolted joint using Continuum Damage Mechanics *Int. J. Fatigue* **73** 66–76
[10] Khader I, Kürten D and Kailer A 2012 A study on the wear of silicon nitride in rolling-sliding contact *Wear* **296** 630–7
[11] Kumar D, Biswas R, Poh L H and Abdel Wahab M 2017 Fretting fatigue stress analysis in heterogeneous material using direct numerical simulations in solid mechanics *Tribol. Int.* **109** 124–32

IOP Conf. Series: Journal of Physics: Conf. Series **843** (2017) 012030 doi:10.1088/1742-6596/843/1/012030

[12] Pereira K, Yue T and Abdel Wahab M 2017 Multiscale analysis of the effect of roughness on fretting wear *Tribol. Int.* **in press**

[13] Yue T and Abdel Wahab M 2014 Finite element analysis of stress singularity in partial slip and gross sliding regimes in fretting wear *Wear* **321** 53–63

[14] Yue T and Abdel Wahab M 2016 A Numerical Study on the Effect of Debris Layer on Fretting Wear *Materials (Basel).* **9** 597

[15] Yue T and Abdel Wahab M 2017 Finite element analysis of fretting wear under variable coefficient of friction and different contact regimes *Tribiology Int.* **107** 274–82

[16] Raga R, Khader I, Chlup Z and Kailer A 2016 Damage initiation and evolution in silicon nitride under non-conforming lubricated hybrid rolling contact *Wear* **360** 147–59

[17] Dassault Systemes Simulia Crop 2013 *Abaqus Theory Manual, version 6.13*

[18] Khader I, Rasche S, Lube T, Raga R, Degenhardt U and Kailer A 2016 Lifetime prediction of ceramic components – A case study on hybrid rolling contact *Eng. Fract. Mech.* **169** 292–308

Monitoring microstructural evolution in-situ during cyclic deformation by high resolution reciprocal space mapping

Annika M. Diederichs[*1], **Felix Thiel**[1,2], **Torben Fischer**[3], **Ulrich Lienert**[4], **Wolfgang Pantleon**[1]

1: Department of Mechanical Engineering, Materials and Surface Engineering, Technical University of Denmark, 2800 Kgs. Lyngby, Denmark.

2: Institute for Metallic Materials, Leibniz Institute for Solid State and Materials Research, 01690 Dresden, Germany.

3: Helmholtz-Zentrum Geesthacht, Institute of Materials Research, 21502 Geesthacht, Germany.

4: DESY Photon Science, Deutsches Elektronen-Synchrotron, 22607 Hamburg, Germany.

*anmad@mek.dtu.dk

Abstract. The recently developed synchrotron technique High Resolution Reciprocal Space Mapping (HRRSM) is used to characterize the deformation structures evolving during cyclic deformation of commercially pure, polycrystalline aluminium AA1050. Insight into the structural reorganization within single grains is gained by in-situ monitoring of the microstructural evolution during cyclic deformation. By HRRSM, a large number of individual subgrains can be resolved within individual grains in the bulk of polycrystalline specimens and their fate, their individual orientation and elastic stresses, tracked during different loading regimes as tension and compression. With this technique, the evolution of dislocation structures in selected grains was followed during an individual load cycle.

1. Introduction

Fatigue-related damage due to repeatedly changing mechanical loading is one of the major failure reasons in structural materials. During mechanical loading of metals, plastic deformation occurs by motion of dislocations causing dislocations to be stored in the material [1]. For an understanding of the materials response to repeated mechanical loading, it is central to understand the interaction between these dislocations. While final failure occurs on the macroscale, changing mechanical loads cause on the micro- and nanoscale formation of characteristic dislocation structures, which play a decisive role for the materials life time.

During cyclic deformation of face-centered cubic metals with high stacking fault energy such as aluminium, dislocations self-organize into ordered structures, consisting of dislocation walls of high dislocation density separating almost dislocation-free subgrains with dimensions of 2-5 μm [2-7]. Their specific morphology is influenced by the deformation conditions such as the strain (or stress) amplitude, the strain rate and the number of cycles during cyclic deformation, where higher strain amplitudes causes an earlier onset of structure formation and larger misorientations [7-12]. The internal dislocation structure influences the lifetime of the material during cyclic deformation, because different mechanical properties are associated with dislocation walls and subgrains as rationalized in the composite model [13-14]. Subgrains yield more easily (at lower stresses) than dislocation walls and intergranular stresses arise to ensure compatible deformation. Information about internal strains can be gained from local changes of the lattice parameter, which can be determined by X-ray diffraction [13]. The characteristic dislocation structures are known to determine the plastic deformation behavior and thus the flow stress, but the mechanisms on the microscale are still insufficiently understood and their relation to macroscopic failure unsettled, because of limitations in the characterization techniques available.

Traditionally, dislocation structures were studied with classical X-ray line profile analysis using conventional X-ray diffractometers [15], but recently synchrotron-based techniques were developed as tools for non-destructive microscale characterization of bulk samples [16-18]. Due to their large penetration depth, high energy X-rays are suitable for non-destructive investigations on macroscopic specimens. While the signal in classical analysis originates from many different grains with different orientations, individual diffraction peaks of single grains can be selected by High Resolution Reciprocal Space Mapping (HRRSM) and the intensity distribution of a diffraction peak mapped in reciprocal space with high resolution ($\Delta q = 5 \cdot 10^{-4}$ Å^{-1} [16]). In this manner, structural features within individual grains can be identified due to their slightly different and unique orientations. By purposely designed load frames, metallic specimens can be deformed in-situ and, hence, the microstructural evolution in single grains followed during changing mechanical loads [16-17].

The feasibility of high resolution reciprocal space mapping has been demonstrated and successfully applied to in-situ tensile testing, loading/unloading sequences and strain path changes [14,16-17,19-24]. The aim of the present study is to gain insight into cyclic deformation by applying HRRSM to monitor individual grains in commercially pure, polycrystalline aluminium AA1050 bulk samples in-situ during a tension/compression sequence.

2. Test specimen

Tensile test specimens were manufactured by spark cutting from an AA1050 sheet cold-rolled to 90% thickness reduction to a final thickness of 1 mm. The geometry of the dog bone-shaped specimens (with a gauge section of 15 mm in length and 5 mm in width) is shown in Figure 1. The tensile specimens were annealed at 600 °C for 2 h to ensure homogeneous recrystallization. The microstructure after annealing was investigated metallographically to confirm homogeneous recrystallization throughout the entire gauge section of the specimen. Grain sizes were estimated to be between 30 µm and 100 µm using both, light optical microscopy and scanning electron microscopy in particular Electron Backscatter Diffraction and Electron Channeling Contrast Imaging.

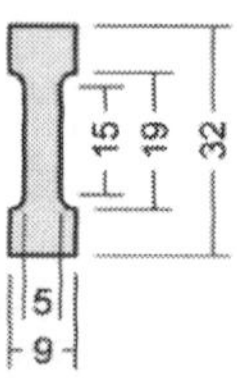

Figure 1. Geometry of dog bone-shaped tensile specimen manufactured from a cold-rolled AA1050 sheet (thickness 1 mm) for cyclic testing (lengths are in mm).

The tensile specimens were pre-fatigued using an MTS Acumen 3 kN Electrodynamic Test System equipped with Station Manager MTS FlexTest 40 and pneumatic grips. Cyclic pre-deformation was carried out in order to introduce a microstructure conform to cyclic deformation (and potentially giving rise to fatigue failure) in the specimen prior to the in-situ investigations by HRRSM. The sample was initially deformed by tension to a strain of 2% with a grip speed of 0.015 mm/s. Afterwards, displacement-controlled tension-compression cycling was performed with a rate of 0.5 Hz and a

IOP Conf. Series: Journal of Physics: Conf. Series **843** (2017) 012031 doi:10.1088/1742-6596/843/1/012031

displacement amplitude of 20 μm corresponding to a nominal strain amplitude of $\hat{\varepsilon} = 1.3 \cdot 10^{-3}$. Cyclic deformation was stopped after 44350 cycles at the end of the compression half cycle and unloaded. At this stage, no significant changes of the maximal tensile and compressive stresses were recorded during cyclic testing.

3. High Resolution Reciprocal Space Mapping (HRRSM)

The experiment was carried out at beam line P07 at PETRA III with a monochromatic beam of 53 keV. HRRSM was performed in-situ while loading the pre-fatigued sample using the custom-made load frame shown in Figure 2a. The setup was essentially similar to an earlier setup at APS [16-17] developed to obtain three-dimensional reciprocal space maps with high resolution to monitor individual bulk grains during deformation of polycrystalline samples. A sketch of the modified setup used in the experimental hutch EH3 at P07 is presented in Figure 2b.

An unloaded, pre-fatigued sample was equipped with two HBM LY4 standard strain gauges (glued on the opposite sides of the specimen, aligned with the tensile axis and centered with respect to gauge section) to monitor the strain in-situ and to reveal potential specimen bending. The sample was mounted in a transportable screw-driven load frame equipped with a 5 kN load cell. Both strain and axial force were monitored during mounting, deformation and dismounting of the sample. The load frame with the sample was placed with the load axis horizontally on a *xy* translation stage on top of a rotation stage allowing rotation of the entire load frame around the vertical *z* axis. Translation along *z* could be achieved by the heavy duty hexapod in EH3. The beam was focused in vertical direction to a Gaussian width of 10 μm and narrowed in horizontal direction to 50 μm allowing complete illumination of individual grains with sizes up to 30 μm. Grains are selected with the help of a Perkin Elmer XRD 1621 xN detector placed 70 cm behind the sample on a horizontal translation by finding isolated 400 diffraction peaks not overlapping with peaks of other grains. After identifying suitable grains, the near detector is moved out of the beam and the diffraction peaks are investigated by a MarCCD165 placed 3.9 m behind the sample on the location of a 400 diffraction peak with diffraction vector close to the tensile axis, i.e. in the horizontal diffraction plane at a diffraction angle $2\theta_{400}$ of 13.63° for aluminium.

An entire sequence of two-dimensional images of the 400 diffraction peak are acquired with the distant detector, while rocking the sample around the vertical axis perpendicular to the scattering plane in small intervals of the rocking angle ω. By stacking the images recorded for several adjacent ω intervals, three-dimensional distributions (two directions on the detector and the additional rocking) of the diffracted intensity are obtained representing three-dimensional reciprocal space maps. The desired high resolution is determined by the monochromaticity and divergence of the beam, the point spread function of the detector as well as by the chosen angular intervals of the rocking (for the presented data, the latter being 0.015° corresponding to $2 \cdot 10^{-3}$ Å^{-1}).

Three-dimensional reciprocal space maps can be assessed in two complementary projections: azimuthal projections and radial peak profiles, which represent the distributions of lattice plane inclinations and normal strains, respectively. As the crystalline lattice within individual grains becomes locally distorted by dislocation structures with subgrains separated by dislocation walls, the intensity distributions of individual diffraction peaks are not completely smooth. Two components can be distinguished [16,19], where sharp peaks of high intensity correspond to individual subgrains and a smooth cloud of lower intensity results from dislocations walls. By separating the two contributions numerically [22], intensity distributions corresponding to individual subgrains can be identified and analyzed individually for all load steps.

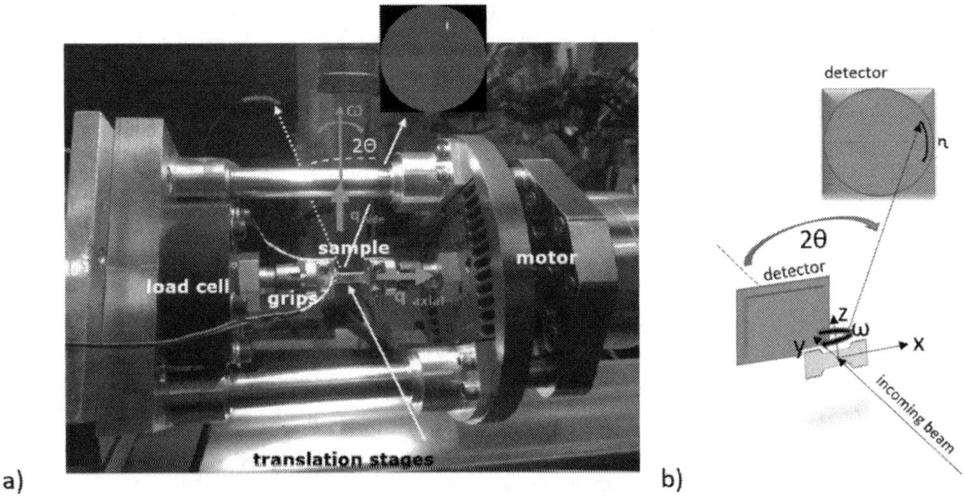

Figure 2. a) Load frame (from P21 at PETRA III) used for tensile and compressive deformation in-situ during High Resolution Reciprocal Space Mapping. A dog bone-shaped sample (with two strain gauges) is mounted in the load frame and loaded in horizontal direction by a screw-driven motor. The load frame is positioned on translation stages on top of a fast rotation stage with rotation axis perpendicular to the scattering plane (i.e. allowing rocking in ω). b) Sketch of the HRRSM set-up in EH3 at P07. The near detector is used for finding suitable grains and will be moved out of the beam path for reciprocal space mapping. The latter is achieved by acquisition of intensity distributions of the 400 diffraction peak from a selected grain with the distant detector during rocking of the sample using the ω rotation.

4. Experimental results

The aim of the experiment was to follow changes in the deformation structure of selected grains (having a ⟨400⟩ direction parallel to the tensile axis) during a tensile/compression sequence (or in terms of cyclic deformation: during an individual load cycle). By using the pre-fatigued specimen unloaded from maximum compression, an incomplete load cycle (cf. Figure 3) was achieved by following the hysteresis curve to maximal tensile load and continuing into compression to a relevant load. For obtaining reciprocal space maps, the deformation was paused at selected forces:

Since the pre-deformation was stopped at the highest compressive load in the hysteresis curve, it was aimed for characterizing the deformation structure after loading the sample with a load as small as permissible to hold the specimen tight (almost 0 N, L0). The tensile loading was interrupted and kept at the force for a first time at 70 N (L1) to analyze the behavior during elastic loading. For the following mapping (at 100 N, L2), the deformation was paused after a significant change in strain. Next, the tensile loading was stopped at the maximum force of 140 N (L3) corresponding to the highest tensile load the sample experienced during cyclic pre-deformation. During unloading along the hysteresis curve to 60 N (U1) only small changes in strain were observed, whereas at a compressive load of -100 N (U2), which is slightly below the highest compressive force during pre-deformation, a significant compressive strain was attained. For halting the deformation manually when reaching the desired loads, a slow grip speed of 0.003 mm/s was chosen. By utilizing two strain gauges at the opposite sides of the specimen, bending during compression can be detected. Such bending has been observed for similarly pre-fatigued specimens when exceeding compressive loads of -180 N corresponding to the theoretical buckling stress. Due to the low material strength, a slight bending occurred when mounting the specimen in the grips leading to internal stresses and strains in the specimen.

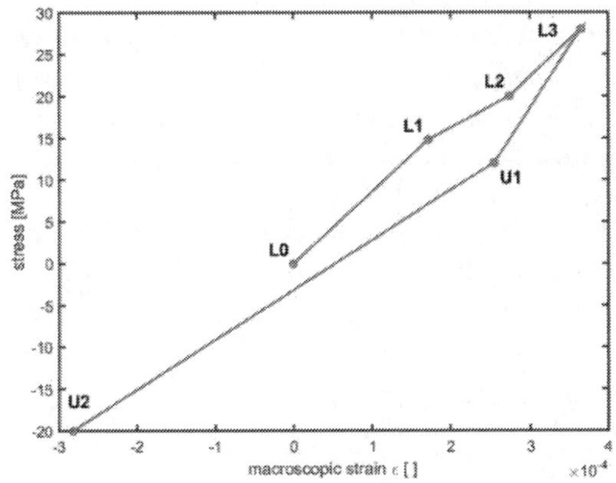

Figure 3. Hysteresis curve showing the incomplete load cycle performed in-situ. Reciprocal space maps were obtained at the designated points (L0-3, U1-2) while pausing the deformation for HRRSM. The maximal force of 140 N corresponds to the maximal tensile force during the cyclic pre-deformation of the sample (due to time limitations, no data were recorded at the maximal compressive force of -140 N).

At each of the load steps where the deformation was paused, reciprocal space maps of six different grains were collected; all having a ⟨400⟩ direction close to the tensile axis. The overall behavior of all grains is similar and only exemplary data are presented. Figure 4 shows the azimuthal projection, a projection of the intensity distribution in reciprocal space parallel to the diffraction vector, representing the different inclination of the (400) planes in grain A for the initial configuration (load step L0). Such maps form the basis for an identification of individual subgrains by their diffraction signature – the high-intensity peaks. Up to 100 high-intensity peaks corresponding to 100 individual subgrains were identified in each reciprocal space map. One high-intensity peak corresponding to a subgrain of grain A, which could be identified in the azimuthal projections of all load steps, is highlighted in Figure 4 and selected for exemplary analysis.

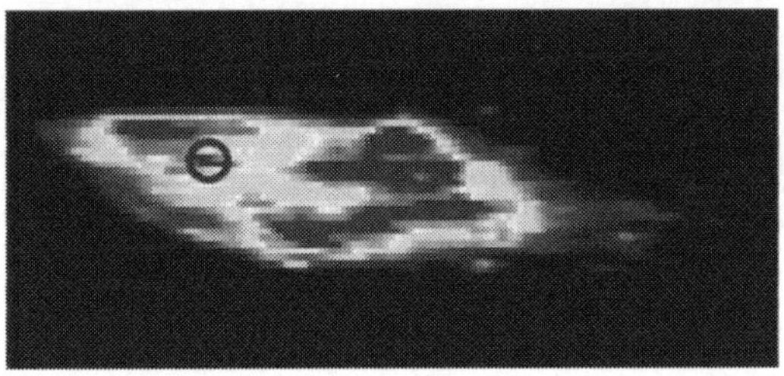

Figure 4. Azimuthal projection (with η horizontally covering a range of 0.3° and the rocking angle ω vertically covering a range of 1.2°) of the 400 diffraction peak from grain A before any significant tensile loading (L0). The intensity distribution is not smooth and one obvious high-intensity peak corresponding to a single subgrain is marked by a black circle.

Figure 5a shows radial peak profiles, which are projections of the reciprocal space maps on the direction of the diffraction vector, for grain A for all different loading steps. The radial position can be characterized either by the diffraction angle 2θ (from Bragg's equation $2d_{400}\sin\theta = \lambda$ with the wave length λ of the used X-rays and the lattice plane spacing d_{400}) or by the amount of the diffraction vector

$$q = \frac{4\pi}{\lambda}\sin\theta\,. \tag{1}$$

The obtained radial profiles can be further analyzed regarding their position, width and asymmetry and provide information on the lattice strain along the diffraction vector. The peak position can be quantified by the average position $q_{mean} = 2\pi/d_{400}$ which is inversely proportional to the lattice plane spacing. As seen from Figure 5a, the peak position changes to smaller values during tensile loading (L1-3), due to increasing elastic strain and increasing lattice plane spacing, and, vice versa, to higher values during unloading and compression (U1-2), due to decreasing or negative strain, respectively, and a smaller lattice plane spacing as a consequence of the elastic deformation caused by the applied load. No significant change in the peak shape is observed and the peak width seems to stay rather constant.

Figure 5b shows the mean peak positions q_{mean} of three different grains for all six different loading steps (i) in dependence of the total macroscopic strain (in colors) measured by the strain gauges and (ii) in dependence of the purely elastic strains (green)

$$\varepsilon = \frac{d - d_0}{d_0} = \frac{q_{mean,\mathrm{L0}}}{q_{mean}} - 1 \tag{2}$$

calculated from the deviation of the actual peak position from its initial value $q_{mean,L0}$. The peak position behaves similar for all three presented grains and reproduces in general the hysteresis curve of the in-situ experiment (Figure 3). The absolute values of the diffraction vector q_{mean} are surprisingly different from grain to grain and differ by 0.08 Å^{-1} between the three presented grains. Correspondingly, the elastic strain ε calculated from the mean peak position with respect to the presumed strain-free initial value not only differs from the total strain measured by the strain gauges, but also varies from grain to grain. This indicates the presence of large internal elastic stresses within the specimen, most likely caused by the bending through clamping when mounting the sample in the testing machine or in the load frame. Notably, the changes in the calculated elastic strains follow the applied load as expected from Hooke's law with the proper value of Young's modulus for aluminium.

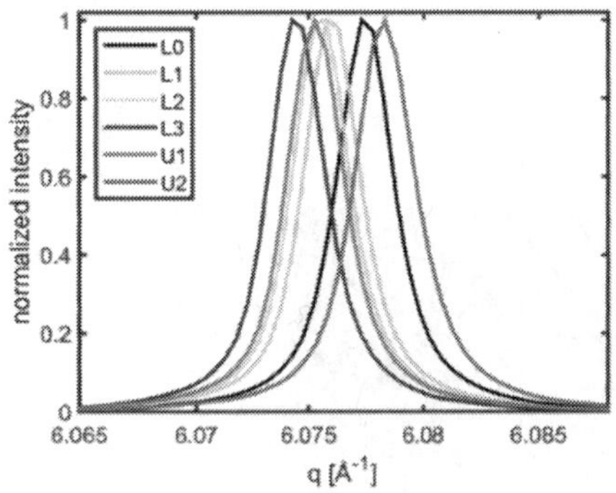

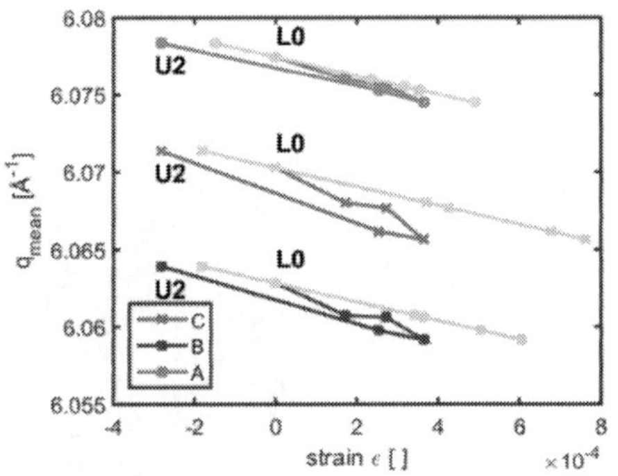

Figure 5. a) Radial peak profiles of grain A for the six different loading steps during tensile and subsequent compressive loading of the pre-fatigued sample. b) Mean peak position of three selected grains as function of the total strain measured by the strain gauges (magenta, red, blue) and the elastic strain (green) calculated from the mean peak position according to equation (2).

Radial peak profiles of subgrains are analyzed in the same way. Figure 6a shows the radial profiles for the single subgrain of grain A marked in Figure 4. This particular subgrain (among others) could be identified for all six load steps by its specific position in the azimuthal projections for grain A. The radial profiles of the subgrain behave similar to the radial profiles observed for the entire grain A. As shown in Figure 6b, the peak position of the particular subgrain shifts to lower values of q during tensile loading and to higher values of q during unloading and compression, while neither peak shape, nor peak width changes significantly.

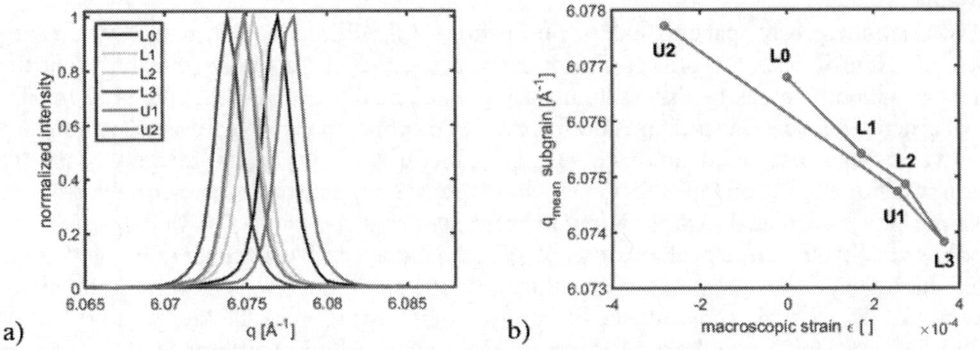

Figure 6. a) Radial peak profiles for the selected subgrain from grain A marked in Figure 4 for all load steps. The profiles shift similar to the profile of the entire grain A. b) Mean peak position of the selected subgrain as a function of the macroscopic total strain for all load steps.

Since up to 100 high-intensity peaks are identified in each reciprocal space map, a statistical analysis of their peak positions can be attempted in addition to the analysis of the fate of individual subgrains. Figure 7 presents a normal probability plot of the peak positions of the 80 largest subgrains. Essentially each subgrain has a different peak position (and hence experiences a different elastic strain); these peak positions are Gaussian distributed for each load step and shift in accordance with the applied load (cf. [14,24]).

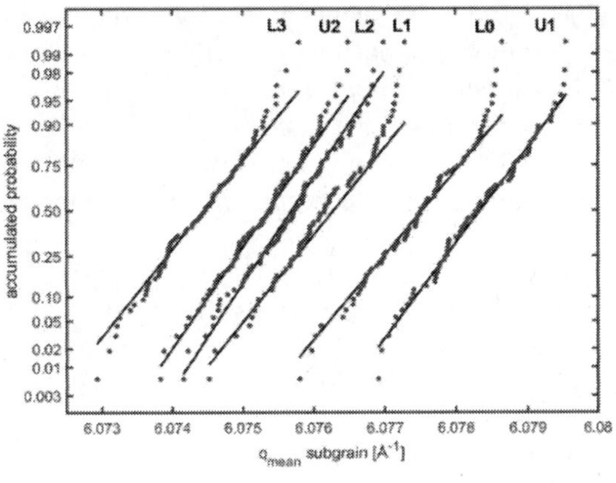

Figure 7. Normal probability plot of the peak positions of the 80 most intense high-intensity peaks corresponding to the 80 largest subgrains of grain A for all six load steps. Best linear fits to the data are shown as black lines indicating that the peak positions of the subgrains follow a Gaussian distribution at each load step.

5. Discussion

From the obtained reciprocal space maps certain features of the dislocation structure within the grains were revealed: High-intensity peaks corresponding to individual subgrains can be identified and separated from smooth intensity distributions using numerical methods [22]. In this manner, the existence of ordered dislocation structures during cyclic deformation is proven in-situ without any need for destructive post-mortem investigations by e.g. electron microscopy. Spatial imaging of the evolving deformation structure in-situ, on the other hand, would require further development of other synchrotron diffraction techniques such as dark field X-ray microscopy [18].

It was observed that the mean peak position q_{mean} of the radial profiles from an individual grain shifts as expected during tensile and compressive loading to lower and higher values of the diffraction vector (cf. Figure 5), respectively, as consequence of purely elastic distortion of the lattice [13-14,24]. While the radial profiles only shift about 0.05 $Å^{-1}$ during the load cycle, a higher difference of 0.08 $Å^{-1}$ is found between the peak positions of different grains. The observed difference between different grains (cf. Figure 5b) is attributed to different internal stresses caused mainly by bending during mounting of the pre-fatigued specimen before any loading took place. As the method is sensitive to local elastic strains, already minor specimen bends can cause the observed effect.

Analysis of the 80 largest subgrains in a selected grain has shown that their radial peak position follow a Gaussian distribution at each step in the load cycle (as revealed earlier for different loading conditions [14,24]). As only minor changes in the overall deformation structure occur during a single load cycle, individual subgrains could be followed in-situ during the (incomplete) load cycle. Their radial profiles show the same characteristics as the radial profile of the entire grain. Nevertheless, we expect to be able to reveal more subtle changes during an individual cycle, by analyzing the behavior of several subgrains individually in more detail and a thorough analysis of their statistical distributions. A hint on such slight changes (which have to be substantiated by findings on other grains) can be gained from the best linear fits in Figure 7 which do not follow exactly the same slope revealing a slight change in their distribution. In the end, there are these small changes which during repeated tension/compression cycles will evolve the deformation structures finally causing fatigue failure. For a complete description of the microstructural changes, the evolution of the microstructure should be followed over a number of load cycles by pausing the cyclic deformation after numerous load cycles.

6. Conclusions

High Resolution Reciprocal Space Mapping was successfully applied to follow the evolution of the deformation structure in several grains (which all had in common a ⟨400⟩ direction close to the loading direction) embedded in a pre-fatigued AA1050 sample in-situ during an individual load cycle, while monitoring stress and strain simultaneously.

The preliminary results show that High Resolution Reciprocal Space Mapping is a promising technique for in-situ investigations of cyclic deformation of metals providing detailed insight in the development of deformation structures during varying loads. In this manner, quantitative information about dislocation structures in individual bulk grains embedded in test specimens can be obtained during individual load cycles.

Acknowledgements

The authors gratefully acknowledge the financial support from MAX4ESSFUN of the European Regional Development Fund Interreg Öresund-Kattegat-Skagerrak (project DTU-007) and DANSCATT. The authors also acknowledge beamtime by DESY under proposal number I-20150204 EC. We would like to thank Dmytro Orlov for his participation and valuable discussions.

References

[1] Laird C, Charsley P and Mughrabi H 1986 Low energy dislocation structures produced by cyclic deformation *Materials Science and Engineering* **81** 433-450

[2] Li P, Li S X, Wang Z G and Zhang Z F 2011 Fundamental factors on formation mechanism of dislocation arrangements in cyclically deformed fcc single crystals *Progress Materials Science* **56**(3) 328-377

[3] Nellessen J, Sandlöbes S and Raabe D 2016 Low cycle fatigue in aluminum single and bi-crystals: On the influence of crystal orientation *Materials Science and Engineering A* **668** 166-179

[4] Wang J, Zhu Z G, Fang Q F and Liu G D 1999 The influence of the crystallographic orientation on the behavior of fatigue in Al single crystals *Materials research bulletin* **34**(3) 407-413

[5] Vorren O and Ryum N 1988 Cyclic deformation of Al single crystals: Effect of the crystallographic orientation *Acta Metallurgica* **36**(6) 1443-1453

[6] Xia Y B 1993 The effect of crystal orientation on mechanical behavior during fatigue in aluminium single crystals *Scripta metallurgica et materialia* **29**(7) 999-1003

[7] Snowden K U 1963 Dislocation arrangements during cyclic hardening and softening in A1 crystals *Acta metallurgica* **11**(7) 675-684

[8] Li P, Li S and Wang Z 2010 Cyclic deformation behaviors of [-5 7 9] oriented Al single crystals *Metall. Mater. Transactions A* **41** 2532-2537

[9] Fujii T, Sawatari N, Onaka S and Kato M 2004 Cyclic deformation of pure aluminum single crystals with double-slip orientations *Materials Science and Engineering: A* **387** 486-490

[10] El-Madhoun Y, Mohamed A and Bassim M N 2003 Cyclic stress–strain response and dislocation structures in polycrystalline aluminum *Materials Science and Engineering: A* **359**(1) 220-227

[11] Grosskreutz J C and Waldow P 1963 Substructure and fatigue fracture in aluminum *Acta metallurgica* **11**(7) 717-724

[12] Mitchell A B and Teer D G 1970 The analysis of dislocation structures in fatigued aluminium single crystals exhibiting striations *Philosophical Magazine* **22**(176) 399-417

[13] Mughrabi H 1983 Dislocation wall and cell structures and long-range internal stresses in deformed metal crystals *Acta metallurgica* **31**(9) 1367-1379

[14] Pantleon W, Wejdemann C, Jakobsen B, Lienert U and Poulsen H F 2010 Advances in characterization of deformation structures by high resolution reciprocal space mapping *Proceedings Risø International Symposium on Materials Science* DTU Risø **31** 79-100

[15] Ungár T 2001 Dislocation densities, arrangements and character from X-ray diffraction experiments *Materials Science and Engineering: A* **309** 14-22

[16] Jakobsen B, Poulsen H F, Lienert U, Almer J, Shastri S D, Sørensen H O and Pantleon W 2006 Formation and subdivision of deformation structures during plastic deformation *Science* **312**(5775) 889-892

[17] Pantleon W, Wejdemann C, Jakobsen B, Lienert U and Poulsen H F 2009 Evolution of deformation structures under varying loading conditions followed in situ by high angular resolution 3DXRD *Materials Science and Engineering: A* **524**(1) 55-63

[18] Simons H, King A, Ludwig W, Detlefs C, Pantleon W, Schmidt S, Stöhr F, Snigireva I, Snigirev A and Poulsen H F 2015 Dark-field X-ray microscopy for multiscale structural characterization *Nature communications* NC06(2015)6098

[19] Jakobsen B, Poulsen H F, Lienert U, and Pantleon W 2007 Direct determination of elastic strains and dislocation densities in individual subgrains in deformation structures *Acta materialia* **55** 3241-3430

[20] Pantleon W, Poulsen H F, Almer J and Lienert U 2004 In situ X-ray peak shape analysis of embedded individual grains during plastic deformation of metals *Materials Science and Engineering: A* **387** 339-342

[21] Jakobsen B, Poulsen H F, Lienert U, Bernier J, Gundlach C and Pantleon W 2009 Stability of dislocation structures in copper towards stress relaxation investigated by high angular resolution 3D X-ray diffraction *physica status solidi (a)* **206**(1) 21-30

[22] Wejdemann C, Lienert U, Nielsen H B and Pantleon W 2010 Identifying individual subgrains in evolving deformation structures by high angular resolution X-ray diffraction *Proceedings Risø International Symposium on Materials Science* DTU Risø **31** 477-487

[23] Wejdemann C, Lienert U and Pantleon W 2010 Reversal of asymmetry of X-ray peak profiles from individual grains during a strain path change *Scripta Materialia* **62**(10) 794-797

[24] Wejdemann C, Poulsen H F, Lienert U and Pantleon W 2013 In situ observation of the dislocation structure evolution during a strain path change in copper *JOM* **65**(1) 35-43

Monitoring fatigue damage of an adhesive joint using fiber optics sensors

C S Shin[1,3], Y J Yang[1] and S K Liaw[2]

[1]Mechanical Engineering, National Taiwan University, No.1, Sec.4, Roosevelt Road, Taipei, 10617, Taiwan, (R.O.C.)

[2] Elelctronic and computer engineering, National Taiwan Univ of Science and Technology, Taipei, Taiwan, No.43, Sec. 4, Keelung Rd., Taipei 106, Taiwan (R.O.C.)

[3] Author to whom any correspondence should be addressed (csshin@ntu.edu.tw).

Abstract. Adhesive joining has many merits over traditional joining techniques. However, adhesive joints are susceptible to damage and degradation caused by service loading. If such degradation went undetected, serious structural failures and catastrophic outcome might follow. It is difficult and economically unviable to carry out regular examination on the joint integrity using conventional non-destructive evaluation techniques, especially in the case of practical structures with large scale adhesive joining. In this work, optical fibers with Bragg grating (FBG) sensors embedded in single lap joints has been demonstrated to be able to detect internal damage caused by monotonic and cyclic loading. FBG sensor works by reflecting specific wavelengths from a broad spectrum incident light. The wavelengths reflected depend on the strain on the FBG. Internal damages will perturb the strain field in the joint and thus change the shape of the reflected FBG spectrum. With embedded FBG sensors, it is possible to achieve on-line monitoring of the joint integrity in an economical way.

1. Introduction

Traditional joining using bolts or rivets requires stress concentrating holes to be drilled in a structure. Joining by welding may degrade the materials in the heat affected zone. These shortcomings may be overcome by using adhesive joining [1-3]. The availability of adhesive joints also allows light alloys and fiber reinforced composites structures to overcome the problems of weak out-of-plane bending strength/stiffness, low joint toughness while improving their crash-worthiness associated with traditional joining. However, occasional overloading and fluctuating service loading may degrade these joints. If such degradation went undetected, serious structural failures and catastrophic outcome might follow.

Non-destsructive examination (NDE) techniques such as ultrasonic scan are commonly used to examine the integrity of adhesive joints [4-9] and neutron irradiation was found to be more effective than ultrasonic to detect internal damages in a joint [9]. The use of thermography [10] and shearography [11] has also been proposed. Current NDE techniques can detect debonding readily but cannot reliably detect all internal defects and degradation of an adhesive joint [12, 13]. For example, a kissing joint that has full interfacial contact but does not possess the required structural strength is difficult to detect using the current techniques [14]. Moreover, It is economically unviable to carry out regular examination of the joint integrity using traditional NDE techniques, especially in structures with large scale adhesive joining. On the other hand, strain gages had been employed for monitoring the integrity of the joint in an adhesive patch repair. The latter method not only allows continuous monitoring but also is much more economical than periodic examination of the whole bonded region using ultrasonic techniques [15]. However, as strain gages can only be applied to the outer surface, they are susceptible to environmental degradation and may not possess sufficient fatigue life for long term monitoring [15]. Optical fiber sensors are known to have much better fatigue strength. They can be embedded inside the bond and relatively free from environmental attack. The feasibility of using fiber Bragg grating (FBG) sensors for monitoring has been investigated in this work.

2. Material and methods

2.1. Fiber Bragg Grating sensor and its basic properties

When a broadband light is coupled into an optical fiber with a uniform Bragg grating, a single peak with wavelength λ satisfying the Bragg diffraction criterion will be reflected:

$$\lambda = 2n_e\Lambda \quad (1)$$

where n_e is the effective refractive index and Λ is the periodicity of the grating. When either or both of the n_e and Λ change, the center wavelength of the reflected spectrum shifts. Λ will be changed if the FBG is subjected to a deformation. Such deformation may be caused by mechanical or thermal strains. n_e will be affected by variation in temperature and the triaxial stress state acting on the fiber. If the FBG is subjected to a non-uniform strain field, uniformity of the grating period is perturbed, the single peak reflected spectrum will broaden or chirped. In general, the reflected wavelength will shift by ~1*pm* under a strain of 1 $\mu\varepsilon$ or a temperature change of 0.1°C. If temperature variation is negligible, the change in the spectrum basically reflects a change in the stress/strain status along the FBG.

FBGs used were fabricated in a Ge−B co-doped single mode fiber by side writing using a phase mask with a period of 1.05 μm. The sensing length of the FBGs was about 10 mm. The reflectivity of the resulting FBG was about 99%, and the peak wavelengths were between 1545 and 1547nm. The FWHM (full width half maximum) of the FBGs was about 0.175 nm. The reflected spectra from the FBGs were interrogated using an optical spectrum analyzer (Anritsu MS9710C OSA) .

2.2. Single lap joint specimens

6061-T6 aluminum strips were glued together with Loctite E-30CL structural epoxy adhesive to forms single lap joint specimens (Fig.1). Optical fibers with FBG are embedded at either 0° (parallel to loading direction), 45° or 90° in the lap joint as shown schematically in Fig.2. The sensor designation

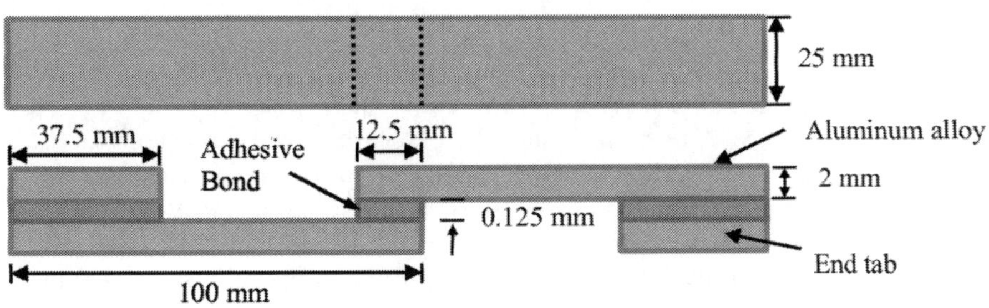

Fig.1 Dimensions of the single lap joint specimen.

is in the form DnPx, where n represents the orientation angle and x denotes the location. x may take the letter E (edge), L(left), M(middle) or R(right). The relative positions of these locations are indicated in Fig.2. The bond line is approximately as thick as the diameter of the optical fiber, i.e. 125 µm. The specimens were subjected to tensile or cyclic loading. Testings were interrupted periodically to allow the reflected light spectra from the FBGs to be recorded at the instantaneous loading and at 0 N. These spectra were compared with the respective original spectra before testing. The recorded spectra covered a span of 10 nm to ensure the characteristic FBG wavelength was totally included. For each set of testing conditions, the tests have been repeated at least three times to ensure the results are consistently reproducible.

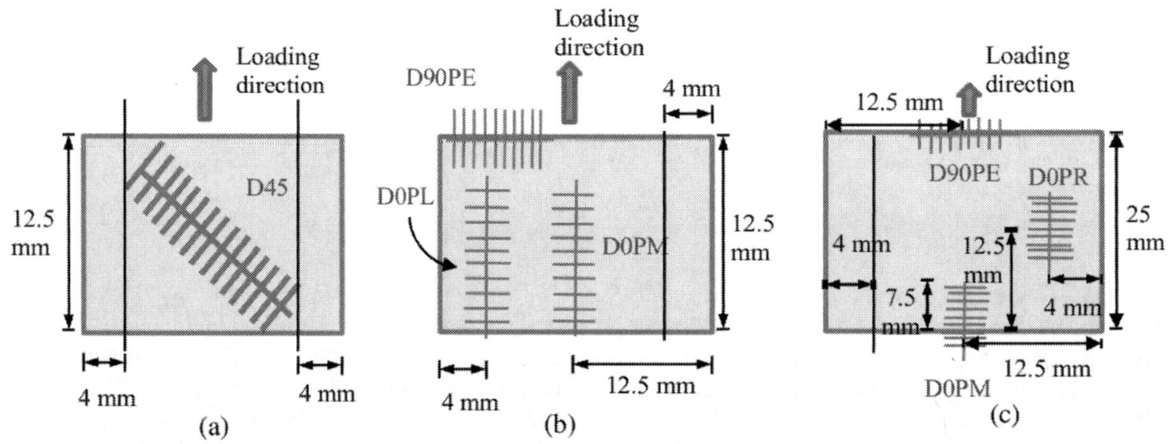

Fig.2 Different layouts of the FBG sensors in the adhesive bond.

3. Results and discussion

3.1. Damage monitoring during tensile tests

3.1.1. Sensitivities of different FBG orientations in response to damage under tensile tests. Fig.3 shows the spectrum from the 45° FBG configuration (Fig.2a) under different loading. As the applied tensile loading on the specimen increased, the FBG spectra initial shifted to the longer wavelength and eventually chirped and changed shape significantly. The reflected wavelengths from an FBG are affected by the strain acting on the FBG. The presence of internal damages will perturb the strain field in the joint and thus change the shape of the reflected FBG spectrum. On the other hand, loading the specimen will stress the joint and cause a wavelength shift of the spectrum. Non-uniform strain induced by stress concentration near the edges of the joint will chirp or even split the spectrum. To differentiate spectrum change caused by damage from the latter effects, spectra were recorded with the specimens fully unloaded after loading to different levels.

On unloading from 90% of failure load, the FBG spectrum started to show slight change in shape (Fig.3b), suggesting the existence of internal damages in the adhesive bond at this load level. Damage became more intensive on loading up to 95% of failure load and thus induced a more significant change in the FBG spectrum on unloading (Fig.3c).

Fig. 4 shows the spectrum from the 0° and 90° FBG configurations (Fig.2b) under different loading. For the 0° FBGs, the spectra changed shapes much more heavily (Figs. 4a & b) than the 45° FBG under similar loading. On the other hand, spectra from the 90° FBG changed very little shape even near failure (Fig. 4c). The same trends occur in the unloading spectra (Figs. 4d-f). As a result, it may be concluded that the 0° FBG is the most sensitive to damage inside the bond and the 90° FBG is the least sensitive while the sensitivity of the 45° FBG lies in between.

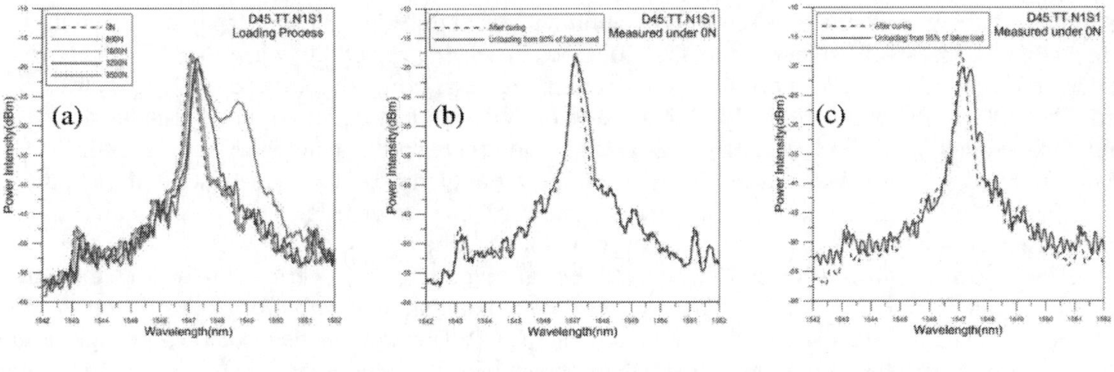

Fig.3 Spectrum measured from the 45° FBG under various loading.

The D0PM unloading spectrum from 87% of failure load (Fig. 4e) showed slightly more well-defined changes than that from D0PL. Results from two other specimens tested under the same conditons also exhibited the same phenomenon. This suggests that the 0° FBG located in the middle of the bond (D0PM) is slightly more sensitive than the one located on the left (D0PL). This may be due to damages started to occur earlier in the middle.

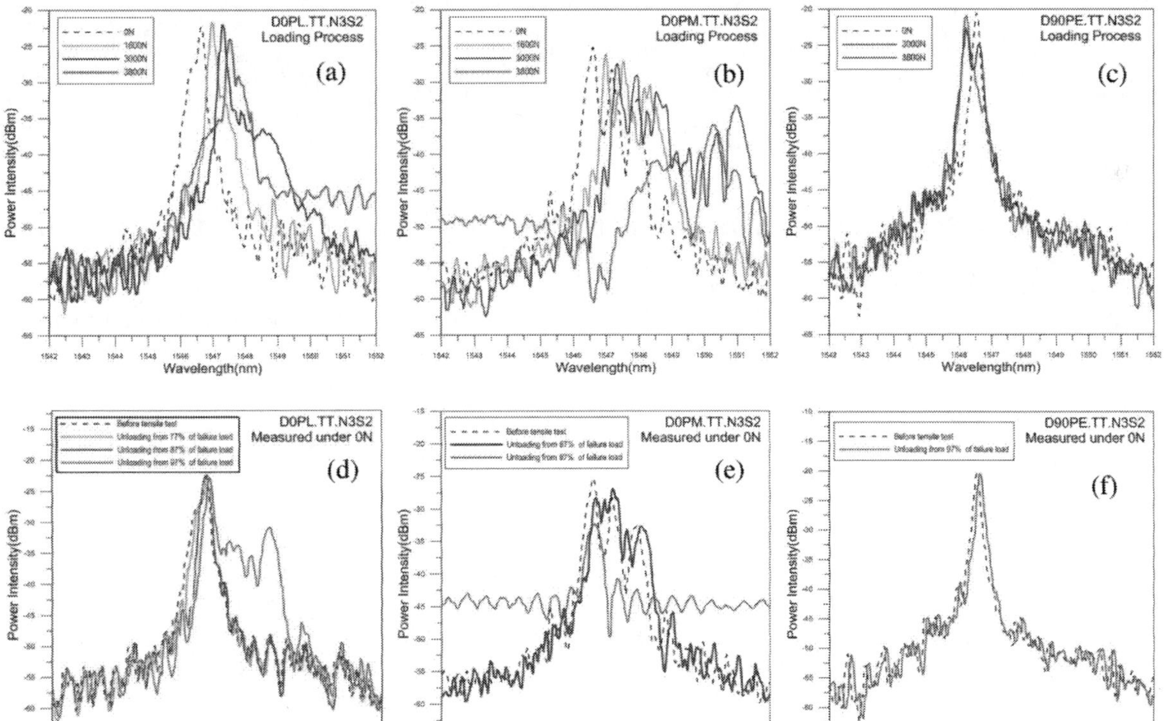

Fig.4 Spectra measured from the 0° and 90° FBGs under various loading.

3.1.2. Effect of of FBG locations on sensitivities in response to damage. It is known that stress concentration occurs near the edges where the adhesive bond lines end in the loading direction. Since the sensing length of the FBG employed was about 10mm and there was likely a ±1mm deviation in positioning it in the bond, the 0° FBGs in configuration depicted in Fig.2b may or may not touch the ending edges of the bond. To investigate the difference in sensing sensitivity when the 0° FBG covers the edge of the bond or well clear of it, a joint overlapping length of 25mm instead of 12.5mm was used. The D0PM FBG was deliberately placed across one edge so that 2.5mm of it was outside the bond. On the other hand, the D0PL FBG was placed in the center of the bond so that it was about 7.5mm from either bond edges. As results above suggest that damages may start earlier in the middle, a 90° FBG (D90PE) was also included and was placed very close to the edge and centered in the middle instead of on the left. The complete lay out is shown schematically in Fig.2c.

Fig.5 shows the unloading spectra from each FBG after loading up to various loads. The unloading spectra from the 90° FBG and the 0° FBG well inside the bond showed virtually no change up to 95% of the failure load. The one straddling one bond edge (D0PM) started to show marked change in the unloading spectra at 84% of the failure load. This shows clearly 0° FBGs in the vicinity of the ending edges of an adhesive bond has much greater sensitivity than FBG well clear of the edge.

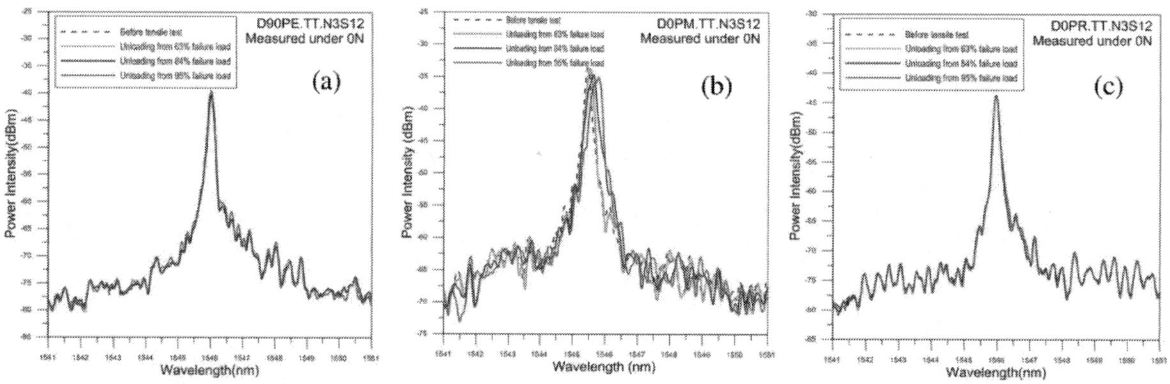

Fig.5 Unloading spectra measured from the 0° and 90° FBGs using the configuration in Fig.2(c).

3.2. Damage monitoring under fatigue tests

In these tests, the lap joints were 12.5mm in length. FBG layout was similar to Fig.2b except that the 90° FBG was changed to 0° and was placed 4mm from the right side. Also the center of the FBGs were positioned at the transverse center line of the bond as far as possible so that their ends were within 1.5mm from either ending edge of the bond. The specimens were cycled between 0.156kN and 1.56kN at a frequency of 5Hz. The minimum and maximum loads corresponds to 4% and 40% of the average tensile failure load. The resulting average fatigue life was 144860 cycles.

3.2.1. Virgin specimens fatigue testing. Fig.6 shows the unloading spectra from the three FBGs near the end of fatigue life. For the D0PL FBG, no marked change in the shape of the spectrum occurred up to 125000 cycles (95% life). At 130000 cycles (98% life), a prominent rise in background intensity was observed (Fig.6a). The D0PM spectra started to show discernible change at 115000 cycles, or 87% life (Fig.6b). The spectra continued to evolve and at 130000 cycles, it differed quite significantly from the initial spectrum (Fig.6c). The D0PR spectra started to show slight change at 95000 cycles (72% life). The change became quite marked at 115000 cycles (Fig.6d) and 130000 cycles (Fig.6e).

Fatigue damage is a highly localized and stochastic in nature. After damage initiated, it develops slowly under the small cyclic loading and may only affect the FBG in its vicinity. This explains why some FBG showed spectrum change much earlier and more significantly than the others.

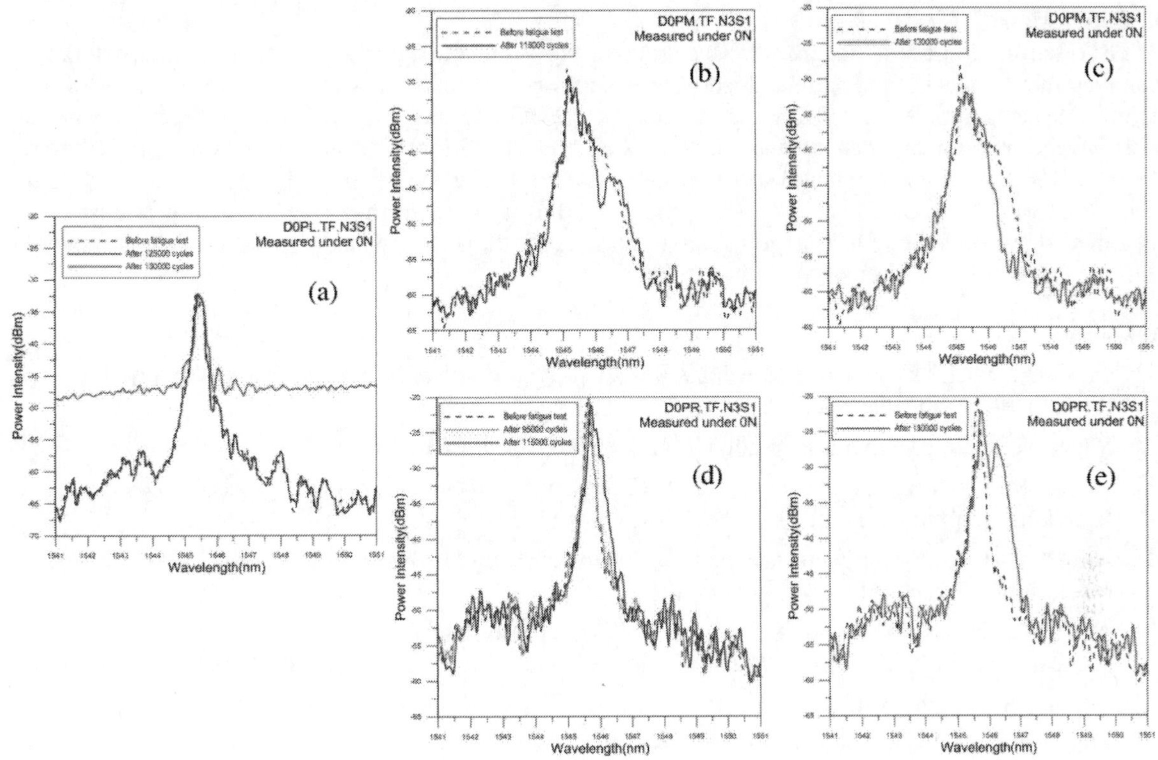

Fig.6 Unloading spectra from the three 0° FBGs near the end of fatigue life.

3.2.2. Pre-tensioned specimens fatigue testing. Tensile loading was applied to two batches of specimens. In the first batch, applied loading was increased stepwise until the unloading spectra of all three 0° FBG started to show discernible changes. In the second batch, stepwise increase of the applied loading was just stopped short of causing discernible change in the unloading spectra. Fig.7 shows the spectra of one specimen in the first batch unloaded from a maximum load of 2500 N, which corresponds to about 63% of the average tensile failure loading. On fatigue cycled between 0.156kN and 1.56kN, it got a life of 34600 cycles, significantly shorter than the average life of 144860 cycles.

A specimen in the second batch was loaded to 2000N, which corresponded to about 50% of the average tensile failure loading and was higher than the maximum cyclic loading of 1560N. The unloading spectra from all three 0° FBG showed no discernible changes. Follow up fatigue test gave a life of 137250 cycles, which is within the scatter of the virgin specimen fatigue life.

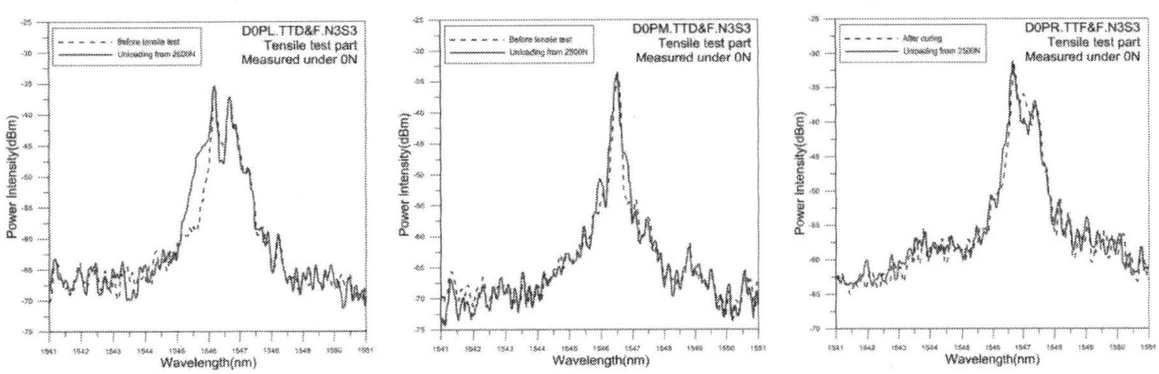

Fig.7 Unloading spectra from the three 0° FBGs which just started to show discernible changes.

4. Conclusions

FBG sensor embedded in an adhesive bond is able to detect the occurrence of internal bond damages due to tensile and fatigue loading and the signal is exhibited as a change in shape of its reflected spectrum under the load free condition. A 0° FBG aligning with the loading direction is more sensitive to bond damage detection than a 45° one. A 90° FBG is the least sensitive among the three. The 0° FBG detects bond damage sooner when it is straddling the ending edge of of the bond than when it is well clear of the edges. When the unloading spectra from a 0° FBG in a pre-tensioned specimen showed discernible change in shape, its fatigue life is significantly reduced when compared with the one that has not yet showed discernible change.

References

[1] da Silva L F M, Öchsner A. and Adams R D 2011 *Handbook of Adhesion Technology* (Berlin Heidelberg: Springer)

[2] Banea M D and da Silva L F M 2009 *J. Mater. Des. Appl.* **223** 1-18.

[3] Adams R D, Comyn J. and Wake W C 1997 Structural adhesive joints in engineering *2nd ed.* (London: Chapman & Hall)

[4] Allin J M 2002 *Disbond detection in adhesive joints using low frequency ultrasound* PhD thesis, Mech. Engng. Imperial College London

[5] Amami S, Santos S D, Savin A, Jeong W J, Bae J H, Joh K H and Ha M K 2010 Société Française d'Acoustique - SFA. *10ème Congrès Français d'Acoustique* (Lyon France)

[6] Hemantha L M and Mallika M S L R 2013 *Int. J. Engng. Research & Technology*, **2(1)**

[7] Zhang H 2013 *J. Appl. Sci.* 13 (**22**) 5298-5308

[8] Elgharib E E, Abdelhay A M and Etman M I 2013 *Int. J. Engng. Res. & Tech.* **2(12)**

[9] Michaloudaki M 2005 *An Approach to Quality Assurance of Structural Adhesive Joints* (Mu"nchen: Techn. Univ.) PhD Thesis

[10] Meola C, Bruzzone A, Giorleo L, Morace R E and Vitiello A. 2004 *J. Adhesion Sci. Tech.* **18** 635-654.

[11] Hung M Y Y, Chen Y S, Ng S P, Shepard S M, Hou Y and Lhota J L 2007 *Op. Engng.* **46 (5)** 051007

[12] Roach D, Rackow K and Duvall R 2010 *Proc. 10th Eur. Conf. on Non-Destructive Testing* (Moscow, Russia) **5** 3263

[13] Kalms M, Hellmers S, Huke P and Bergmann R B 2014 *Proc. SPIE* 9063, *Nondestructive Characterization for Composite Materials, Aerospace Engineering, Civil Infrastructure, and Homeland Security* (San Diego, California, USA) 906327

[14] Yan D, Drinkwater B W and Neild S A 2009 *NDT&E International* **42** 459-466

[15] Baker A A 2008 *Structural Health Monitoring of a Bonded Composite Patch Repair on a Fatigue-Cracked F-111C Wing* Australia DSTO-RR-0335

Acknowledgement

The authors would like to thank the Ministry of Science and Technology, ROC for funding this work through the project MOST 104-2221-E-002 -036 -MY3.

Fatigue of reinforcing bars during hydro-demolition

C W K Hyland[1,3], **A Ouwejan**[2]

[1]Hyland Consultants Ltd, Level 10, 300 Queen St, Auckland 1010, New Zealand
[2]Metlab Ltd, 1 Bushlands Place, Papakura, Auckland 2113, New Zealand

[3] Email: clark@hylandconsultants.com

Abstract. Reinforcing steel fractured during hydro-demolition of a reinforced concrete pier head due to low cycle flexural fatigue from vibration caused by impact of the high pressure water jet on the exposed length of the bars. Research into the fatigue performance of steel reinforcing steel tends to focus on the high cycle axial performance in reinforced concrete members and re-bending behaviour. However with the increasing use of hydro-demolition of concrete structures as part of remediation works care is required to ensure the steel reinforcement exposed to the high pressure jet of water is not going to suffer relatively low cycle flexural damage that may compromise the designed performance of the completed reinforced concrete structure. This paper describes the failure assessment, fatigue analysis, and metallographic examination that was undertaken. It was found that the rib to flank transition radius on the reinforcement steel was small enough to cause a significant stress concentration effect and was the location of fatigue crack growth. A relatively simple analysis using the maximum unrestrained cantilevered bar length and force exerted by the water jet was used to calculate the maximum expected bending moment. This was compared to the bending capacity at initiation of yielding at the rib flank transition accounting for stress concentration effects. This showed that the observed cyclic reversing ductile crack growth and fracture of the H25 bars was consistent with the loading applied. A method is proposed based on these observations to assess suitable limits for unrestrained bar lengths or maximum working offset of the water jet from the point of bar restraint when undertaking hydro-demolition work. The fatigue critical performance requirements of AS/NZS4671 500E bars are also therefore compared with those of BS4449:2005 and PN EN/ISO 15630-1:2011 for comparable 500C bars

1. Introduction

This paper describes an investigation into the causes of the bending, cracking and fracture of a number of reinforcing steel bars in a pier head that occurred during hydro-demolition of concrete around them.

Three H25 vertical bars with long unrestrained lengths on the same face of the pier head fractured and others developed fine flexural fatigue cracking at the rib flank transitions near their base during the hydro-demolition of the concrete.

Research into the fatigue performance of steel reinforcing steel bars tends to focus on their high cycle axial performance in reinforced concrete members and bending behaviour rather than low cycle flexural fatigue crack growth and fracture [1][2].

The site observations, metallographic examination undertaken and the engineering calculations prepared to assist with the assessment of the failure analysis are described.

A method is then presented for assessing appropriate maximum unrestrained bar lengths or working offsets for the jet relative to the point of bar restraint when undertaking hydro-demolition. The method accounts for the stress concentration effect at the rib flank transition.

The manufacturing requirements of AS/NZS4671 [3] with respect to controlling the rib flank transition relevant to fatigue performance are also compared to those of BS4449:2005 [4] and PN EN/ISO 15630-1:2011 [5] for similar 500C bars

IOP Conf. Series: Journal of Physics: Conf. Series **843** (2017) 012033 doi:10.1088/1742-6596/843/1/012033

2. Methodology

2.1 Site inspection
An inspection was made of the pier heads after the fractures occurred. The fractured H25 stubs of East Pier Head bars designated with identification numbers EN5 and EN7 were removed using a disc cutter for metallographic examination. The longer portion of bar EN7 on the other side of the fracture was also recovered

2.2. Determination of hydro-demolition method and sequence
A drawing was prepared by site staff identifying the location of each bar, which had fractured, or had been examined using MPI. Hydro-demolition sequence diagrams with dates and extent demolished were prepared from site records and matched with photographs taken during the works to identify what amount of bar restraint was in place during eth process and particularly at the time of the fractures.

2.3. Materials test reports and testing
Steel mill certificates were sourced for the bars supplied, along with bend, tensile and chemical analysis tests that had been conducted on specimens extracted from one of the fractured H25 bars.

2.4. Magnetic particle inspections
Magnetic particle inspections had been previously been undertaken by others using dual permanent magnets and Ardox 8901 white contrast with Ardox 800/3 black ink in accordance with AS1171:1998 [6] ("MPI"). This had found cracks in a number of bars that hadn't fractured.

2.5. Metallographic procedure
The fracture samples from EN5 and EN7 were prepared for metallographic examination at macro and micro level. The fracture faces were examined under a Meiji brand stereo microscope before the two faces were scrubbed with a hand-brush under running water. The faces were then sprayed in alcohol, dried and re-examined under the stereo microscope.

The longer portion of bar EN7 was cut 100 mm from the fracture to allow a smaller sample to handle. This length was wire-brushed around the diameter to remove dirt and scale and examined under the stereo microscope. Three sections were mounted from the area at and near the fracture face, and two sections were mounted at an area 1300-1400 mm away from the fracture. The sections were cold-cut on a metallographic cut-off saw flooded with coolant, and cold-mounted in resin so as not to induce any heat onto the samples.

The mounts were then ground and polished to a 1 micron diamond finish and examined on a metallurgical microscope in the as-polished condition. Etching to reveal the microstructure was carried out using 2% Nitric acid in alcohol.

Micro-hardness testing was carried out on a calibrated Matsuzawa brand micro-hardness tester at various locations at and around the ribs and also away from the ribs.

3. Results

3.1. Site inspection findings

Some of the H25 bars were found to have been bent up to 16 degrees either side of vertical but hadn't fractured. A number of the H25 bars had also been manually straightened since the fracturing of the bars in question (Figure 3). The larger diameter H40 bars however had not sustained any permanent deformation and appeared to have remained in the condition in which they had been originally cast into the concrete.

IOP Conf. Series: Journal of Physics: Conf. Series **843** (2017) 012033 doi:10.1088/1742-6596/843/1/012033

3.2. Hydro-demolition method and sequence

The hydro demolition had been conducted using a machine that had one jet operating at 1450 bar through a 2.3 mm diameter nozzle. Its movement and angle of attack to concrete surfaces could be adjusted by the operator. Restraining tie bars to the vertical bars had been removed progressively by oxy acetylene cutting as they were exposed by the hydro-demolition process. Each pier head had taken approximately 10 days to remove the concrete down to the original construction joint level

3.3. Findings from Steel Mill Certificates and Testing

The steel mill certificates and specific bend, tensile and chemical analysis tests of specimens extracted from one of the fractured H25 bars showed that they conformed with the manufacturing requirements for D500E 25mm micro-alloy bars to AS/NZS 4671.

3.4. Findings from magnetic particle inspections

The MPI reports showed that significant cracking occurred at the base of all the H25 bars on the north face of the east pier head at and up to 75 mm above the main fracture locations. This cracking was consistent with fully reversed cyclic displacement of the vertical bar acting as a cantilever being buffeted by the water jet.

3.5 Observations on the stereo microscope

The outer portions of the failure surfaces of the fracture in EN7 (Figure 4 and Figure 13) had a distinctive stepped profile (Figure 10 and Figure 15) consisting of relatively flat transverse fatigue cracks joined to the adjacent fatigue crack above and below by sloping failure surfaces (Figure 5). The stepped profiles on the failure surfaces were consistent with fatigue cracking initiating at the rib flank transition radii and stepping by means of localised shear failure at 45 degrees to the cracks above and below as crack growth progressed into the bar (
Figure 10 to 13).

The cracking and fracture followed the spiral of the ribs, and cracking occurred predominantly at the rib-flank transition (Figure 4 and Figure 5). The reinforcing bar ribs had a varying rib flank inclination pattern with differing rib flank inclination angles on each face of the rib (Figure 7). The rib flank inclination was measured perpendicular to a rib to be 58 degrees with a transition radius of 0.8 mm. The rib flank inclination angle on another was 42 degrees with a transition radius of 1.1 mm. This angle was slightly less than the rib flank inclination required of 45 degrees by cl 7.4.2.2 AS/NZS 4671.

3.6 Observations on the metallurgical microscope

On one side of the bar at the ribs there are a series of cracks and compression of the ferrite/pearlite microstructure. This is accompanied by some areas which appear to have 'bulged' and smaller micro-cracks have occurred in these regions, (Figure 7 to Figure 9). This is consistent with low cycle high stress range loading having occurred during crack development.

Micro-hardness testing in this compressed region shows a hardness of 310-325HV0.5. By comparison, at the outside diameter away from cracking or compressed microstructure the hardness was measured to be 250HV0.5. 8mm below the surface in an unaffected area the hardness was measured to be 235-245HV0.5.

At the opposite side of the diameter there is no such compression of the microstructure and no cracks evident. Micro-hardness testing in this region shows a hardness of 225-235HV0.5.At the fracture face of EN7 shown in Figure 13 there is cracking perpendicular to the main crack surface at the edge of the final crack process zone indicating volumetric plastic deformation (Figure 14).

Figure 1 West Pier Head

Figure 2 East Pier Head with H25 bars on north face bent inwards

Figure 3 Bar stub in East Pier Head

Figure 4 EN7 fracture surface and fatigue cracking at adjacent rib flank transition

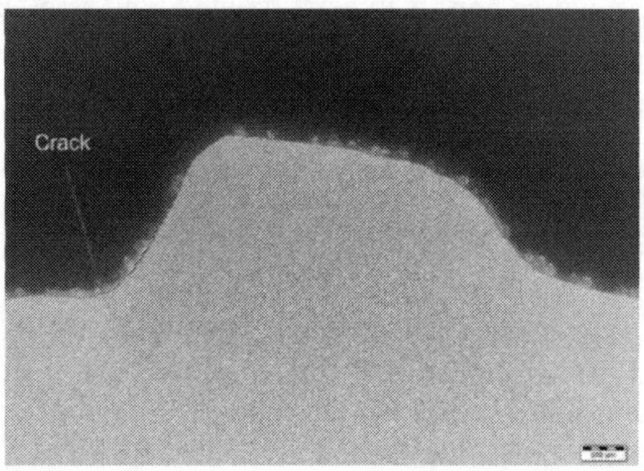

Figure 5 Crack at rib flank of EN7 with 0.8mm transition radius (90 degrees to rib)

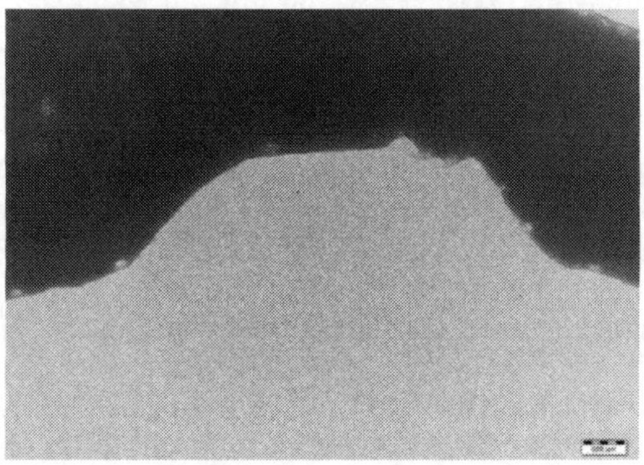

Figure 6 Varying rib profile on EN7 (90 degrees to rib)

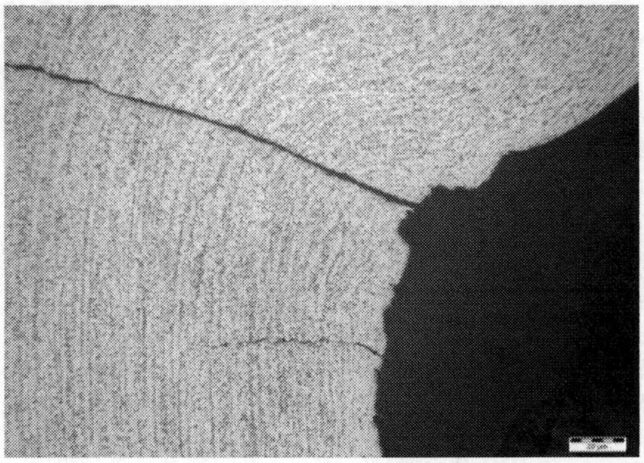

Figure 7 Etched cracks in EN7 rib flank transition radius above fracture surface (50x)

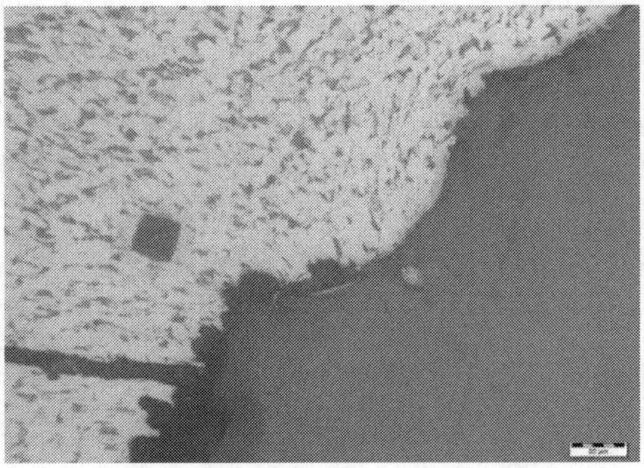

Figure 8 Etched localised bulging and compression deformation of grain structure near crack (200x)

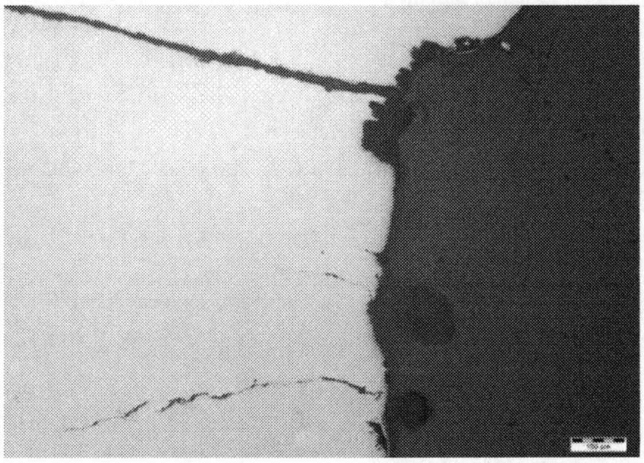

Figure 9 Bulging of surface at crack fronts indicating compressive yielding (100x)

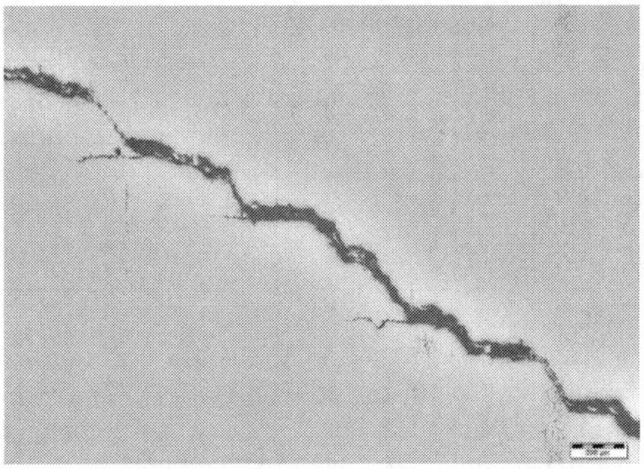

Figure 10 Stepping crack with flat fatigue cracks connected by sloping shear surfaces (50x)

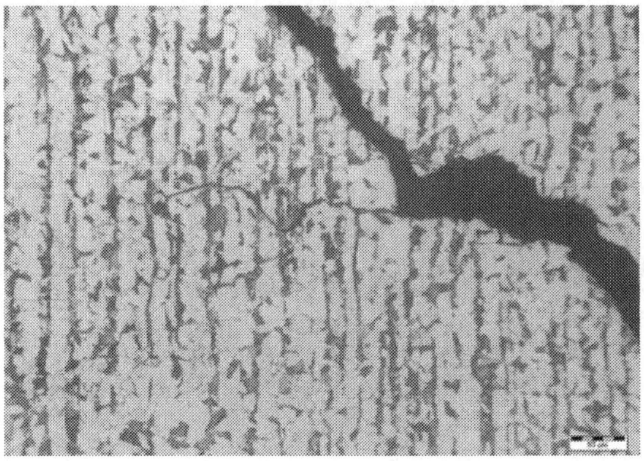

Figure 11 Etched crack steps in Figure 11 (200x)

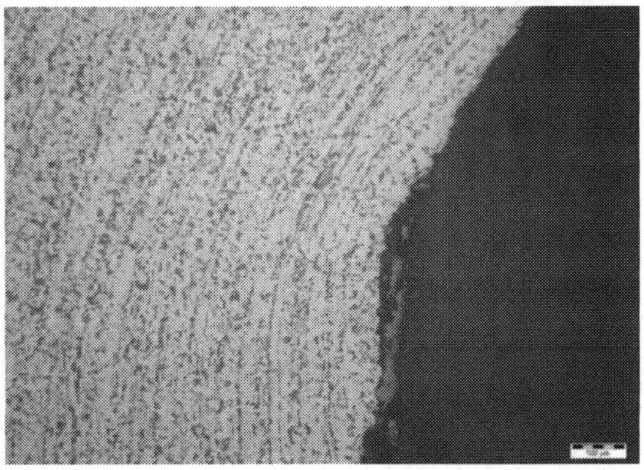

Figure 12 Etched rib flank transition 1300mm away from fracture in EN7 (100x)

Further along the fracture face there is stepping and smaller cracking similar to that observed in Figure 10. Examination of the profile of the ribs away from the fracture surface shows a crack in the tighter portion of the varying rib to flank transition angle in one of the ribs (Figure 5 compared with Figure 6). Fatigue cracking occurred at the base of a number of rib flank transitions up to two or three ribs or up to 75 mm away from the fracture surface. This is also consistent with the MPI site findings.

Cross-sections were also taken 1300-1400mm away from the fracture surfaces to provide a control. These sections showed the same rib profile as nearer the fracture. Examination of the microstructure around the ribs in these locations showed no deformation of the hot rolled ferrite/pearlite microstructure well distant from the fractures (Figure 12).

It is reasonable to conclude therefore that the compression bulging and cracking of the microstructure at the rib transition near the fracture was a result of localised cold working and not due to manufacture. Considering this compression bulging was near the fracture it is likely that this occurred during the failure process.

There is banding of the ferrite/pearlite phases in the microstructure and non-metallic inclusions running parallel with the rolling direction. However there is no microstructural evidence that these contributed to the failure. These would not be expected to have contributed significantly to fatigue crack initiation or growth as they would have been transverse to the crack propagation plane.

3.7 Discussion of Observations

The failure surface of bar EN7 shows that the development of the majority of the fatigue cracking occurred in a short number of high stress range cycles consistent with low cycle fatigue. The final crack tip process zone is approximately 2.5 mm deep (Figure 8). It appears that 4 or 5 high stress reversing cycles occurred before fracture. This is based on the crack process zone surface beach marking indicated by the slightly varied angles of the stepped failure surfaces of sample EN7.

The large crack tip process zone failure surface marks on the EN7 fracture surface are consistent with steel with high ductility and therefore large localised plastic zones ahead of the crack tip and ductile crack growth.

The final fracture appears to have occurred once the fatigue crack grew to approximately 8 mm depth from the bar surface (Figure 14). This is because the process zone on the right side is relatively bright and hasn't been dulled by a reversing compressive action, unlike the other stepped crack process zones either side of the bar centreline.

A photo transverse to the failure surface of EN7 also shows the final process zone slightly raised above the previous crack surface indicating it hadn't been compressed by a reversing compressive stress (Figure 15).

The longitudinal crack perpendicular to the fracture surface at the start of the final process zone appears to be due to both localised constant volume plastic volumetric strain and necking effects as the crack tip process zone elongated as the crack opened. Similar cracks at the prior crack tip process zones are apparent. Such cracking is consistent with very high bending stress cycles in the bar.

At the end of the final process zone the bar fracture then appears to have progressed into less constrained gross section fibrous fracture indicted by significant micro-void coalescence in the centre region of the bar. It seems that the fracture then finished with a small zone of cleavage fracture at the bottom left corner. This is characterised by the swirling leaf-like pattern possibly originating from localised non-metallic inclusions.

Bar EN5 had similar characteristics however the fracture plane shows greater distinct steps on each side of the fracture with a large vertical step the height of a rib and runs across the surface at 45 degrees to the longitudinal plane of the bar. The longitudinal plane runs along the bar between the two longitudinal ribs on each side (Figure 16).

4 Results of the failure analysis

The force applied by the hydro demolition jet was estimated based on the 1450 bar working pressure through its 2.3 mm nozzle, to be 602 N. This is likely to have been applied to the top of the vertical bars cantilevering 1.85 metres above the construction joint, as the hydro-demolition unit was seated on top of the pier head or adjacent to it with its jet angled downwards when exposing these bars.

Figure 13 The Fracture surface of EN7 showing low cycle fatigue crack development with large stepped crack growth marks at slightly varying angles each side of the dull fibrous fracture central area. The brighter band to the right of the central fibrous fracture zone is likely the final fatigue crack growth step prior to fracture progressing to gross section fibrous fracture. A portion of cleavage fracture with a swirling pattern is also apparent in the lower left of the failure surface.

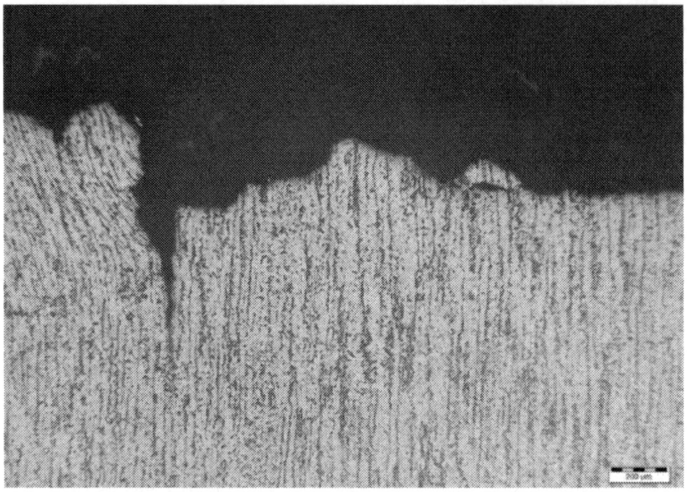

Figure 14 Cross section taken at the crack to the right of the final low cycle fatigue crack process zone in Figure 14. This shows the slightly uplifted nature of the final process zone to the left of the crack and the deformed grain structure indicating volumetric plastic deformation as the crack tip elongated on opening and the crack front progressed.

Figure 15 Fracture surface of EN5 showing vertically displaced steps

Even accounting for a 30 degree angle of attack of the jet giving a lateral component of thrust of 425 N, this would result in a bending moment of at least 0.79 kNm. This compares to a yield moment in the H25 of 0.83 kNm, calculated using the tensile test result of fy = 542 MPa. So the bars could have been loaded to close to gross section yield by the water jet. However this would not have been sufficient on its own to cause the fractures without the additional effect of stress concentration at the rib flank transition.

Based on the metallographic observations a static normal stress concentration factor of

$$K_t = A\left(\frac{r}{d}\right)^b = 2.16 \tag{1}$$

was calculated using the measured rib flank transition radius of r = 0.8 mm, bar diameter d = 25 mm, diameter at rib D = 29 mm, A = 0.97722 and b = -0.23093 [7]. The notch sensitivity was calculated to be q = 0.69 from the Kuhn-Hardrath formula [8].

This all resulted in a dynamic or fatigue stress concentration factor of

$$K_f = 1 + q(K_t - 1) = 1.8 \tag{2}$$

for steel with ultimate tensile strength UTS = 550 MPa .

The cantilevered H25 bars were therefore susceptible to localised plastic strain reversals at the rib flank transitions due to the stress concentration effect at bending moments as low as 0.46 kNm, which is well below the bending moment able to have been applied by the water jet. The H40 bars with similar unrestrained cantilever length of 1.85 m are likely to have been subjected to similar bending moments of 0.79 kNm during the hydro-demolition. However the average yield moment of the H40 based on the tensile properties listed on the mill certificates provided is 3.41 kNm. This is much greater than the demand on them even accounting for stress concentration effects at the rib flank transition. So no permanent damage due to low-cycle fatigue was expected to have occurred in the H40 bars from the hydro-demolition. However a further 10% of these were tested on site for cracks using the Magnetic Particle Inspection method, which confirmed that no cracking had occurred.

5 Method of assessment of unrestrained bar length during hydro-demolition

It is proposed that the maximum unrestrained length of any reinforcing steel bar can be calculated using equations (1) and (2) and any combination of end restraint conditions such as cantilevered, fixed or pin ended.

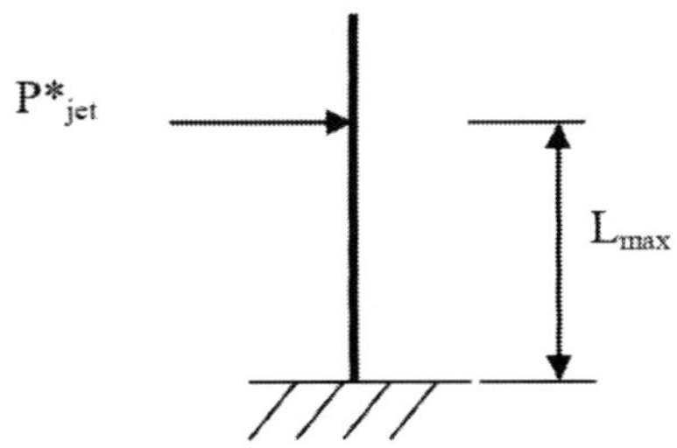

Figure 16 Unrestrained cantilever length of bar L_{max} relative to point of restraint and location of application of hydro-demolition jet

The limiting bending moment so that yield is prevented from occurring at the rib flank transition is

$$M^* \leq \varphi M_{cr} = \varphi\pi\frac{d^3}{32}\frac{f_y}{K_f} \tag{3}$$

for ϕ = 0.9.

For the H25 and H40 bars in this case with characteristic f_y = 500MPa, assuming the same rib profiles and rib flank transition radius of 0.8 mm with notch sensitivity q = 0.7, and horizontal water jet force of 602 N, the appropriate unrestrained cantilever lengths or offset of the jet from the point of bar fixity L_{max} (Figure 16) would have been as shown in Table 1.

Bar	D/d	A	b	Kt	Kf	φM_{cr} (kNm)	L_{max} (mm)
H25	1.16	0.97722	-0.23093	2.16	1.80	0.383	635
H40	1.10	1.01650	-021548	2.36	1.95	1.449	2410

Table 1 Example maximum unrestrained cantilever lengths or working offset of the jet from the point of bar fixity using Equation (3)

6 Comparison of fatigue critical provisions of AS/NZS 4671with BS and EN Standards

The steel mill certificates, the specific tensile and bend testing, and the chemical analyses showed that the bars complied with the tensile, bend and chemical analysis requirements of AS/NZS 4671:2001.

The rib height and rib flank inclination angles measured in a few locations were generally consistent with the requirements of AS/NZS 4671. The rib flank transition radii of 0.8 mm are a little larger but comparable to those of M24 PC8.8 bolts which the authors have observed to be in the order of 0.3 mm. The fatigue category of bolts in the Steel Structures Standard NZS 3404:1997[9] is 36 compared to that of typical hot rolled steel sections of 160. The Concrete Structures Standard NZS3101 [10] appears to assume that reinforcing bars would have a fatigue performance comparable to a Steel Structures Standard NZS 3404 [7] hot rolled section by designating theme to have fatigue category of 150.

The fatigue strength of steel reinforcing bars is generally recognised to be limited by the stress concentration effect of the rib flank transition radius. However there are no prescribed fatigue performance requirements for bar manufacturer's to comply with in AS/NZS 4671. It appears that AS/NZS4671 relies solely on the level of uniform elongation of the steel in a tensile test to provide reliable fatigue performance [11].

The British and European Standards BS4449:2005 and PN-EN ISO 15630-1:2011 specify minimum fatigue performance for bar manufacturer to ensure compliance with, for B500C bars which are comparable in many ways to AS/NZS 4671 500E bars.

BS4449:2005 and PN EN/ISO 15630-1:2011 require reversed bending for 25 and 40 mm diameter B500C bars. Whereas AS/NZS 4671 does not require reversed bend testing of 500E H25 or H40 bars which may have exposed the performance limiting stress concentration effect of the tight rib flank transition radius.

In summary the bars supplied appear to satisfy the requirements of AS/NZS 4671 but the Standard does not adequately specify critical fatigue performance requirements or rib flank radii limits for the bars to achieve the fatigue performance expected from comparable bars manufactured compliant with BS4449:2005 [12] and PN EN/ISO 15630-1:2011.

7 Conclusions

The force exerted by the hydro demolition jet were sufficient to develop gross section flexural yield stress in the H25 bars on the north face of the east pier head once the restraint of the concrete and tie bars were removed leaving the bars to cantilever 1.85 m above the lower concrete surface. However this would not have been sufficient on its own to cause the fractures without the additional effect of stress concentration at the rib flank transition.

A relatively simple analysis using the maximum unrestrained cantilevered bar length and force exerted by the water jet was used to calculate the maximum expected bending moment. This was compared to the bending capacity at initiation of yielding at the rib flank transition accounting for stress concentration effects. This showed that the cyclic reversing ductile crack growth and fracture of the H25 bars was consistent with the loading applied.

A method based on this finding is proposed to assess suitable limits for unrestrained bar lengths when planning hydro-demolition work.

The small rib flank transition radii measured on the H25 bars EN5 and EN7 extracted from the north face of the east pier head likely contributed to rapid reversing ductile crack growth or very low cycle fatigue as the vertically cantilevering H25 bars vibrated in response to the attack of the hydro demolition jet. The force exerted by the hydro demolition jet was however not sufficient to develop the same level of flexural stress in the H40 bars of the same length.

There was sufficient evidence to conclude that the H25 bars conformed to the requirements of AS/NZS 4671 in terms of tensile, bend and chemical analysis properties. Similarly the measurements of the ribs indicate the rib height inclination angles were generally consistent with requirements of AS/NZS 4671. However AS/NZS 4671 does not prescribe a minimum rib flank transition radius. Nor does it prescribe the fatigue performance criteria for steel manufacturers, or rebend test requirements for H25 and H40 bars. Whereas these are required by BS4449:2005 and PN EN/ISO 15630-1:2011 for comparable 500C bars in Europe.

The performance and detailed testing and examination of these bars raises concerns that bars compliant with AS/NZS 4671 may not give high and low cycle fatigue performance comparable to the equivalent British and European standards as perhaps expected by the New Zealand Concrete Structures Standard NZS3101.

References

[1] Tilly, G.P., April 1979, *Fatigue of steel reinforcement bars in concrete: A review*, Fatigue of Engineering Materials and Structures Vol. 2, pp. 251-268, Pergamon Press, Great Britain

[2] Krasnowski, K., 2015, *Concrete-reinforcing steel bars-applications and fatigue tests,* Vol 1/2015, Biuletyn Instytutu Spawalnictwa, Gliwice

[3] AS/NZS4671:2001 *Steel reinforcing materials*, Standards NZ, Wellington

[4] BS4449:2005 *Steel for the reinforcement of concrete. weldable reinforcing steel. bar, coil and decoiled product*, BSI

[5] PN EN/ISO 15630-1:2011, *Steel for the reinforcement and prestressing of concrete. Test methods. Part 1. Reinforcing bars, wire rod and wire*

[6] AS1171:1998, *Non-destructive testing - magnetic particle testing of ferromagnetic products, components and structures*, Standards Australia, Sydney

[7] Norton, R.L.N., 2011, *Machine Design an Integrated Approach*, 4th Edition, Pearson, Worcester, Massachusetts

[8] Kuhn,P., Hardrath, H.F., Oct 1952, *An engineering method for estimating notch-size effect in fatigue test on steel*, Technical Note 2805, NACA, Washington , D.C.,

[9] NZS 3404:1997, *Steel Structures Standard*, Standards NZ, Wellington

[10] NZS 3101:2006, *Concrete Structures Standard*, Standards NZ, Wellington

[11] Bull, D., Allington, C, 2002, *L, N and E Grade 500 reinforcing Steel,* Vol 15 No 1, SESOC Journal, Auckland

[12] UK CARES., 2011, *The CARES guide to reinforcing steels-Part 3,* P3/Rev1/2011/Aaron, Pembroke House, Kent

Study of fatigue crack propagation in Ti-1Al-1Mn based on the calculation of cold work evolution

O. A. Plekhov and A. A. Kostina

Institute of Continuous Media Mechanics of the Ural Branch of Russian Academy of Science, 1 Ac. Koroleva Street, Perm, 614013, Russia

E-mail: poa@icmm.ru

Abstract. The work proposes a numerical method for lifetime assessment for metallic materials based on consideration of energy balance at crack tip. This method is based on the evaluation of the stored energy value per loading cycle. To calculate the stored and dissipated parts of deformation energy an elasto-plastic phenomenological model of energy balance in metals under the deformation and failure processes was proposed. The key point of the model is strain-type internal variable describing the stored energy process. This parameter is introduced based of the statistical description of defect evolution in metals as a second-order tensor and has a meaning of an additional strain due to the initiation and growth of the defects. The fatigue crack rate was calculated in a framework of a stationary crack approach (several loading cycles for every crack length was considered to estimate the energy balance at crack tip). The application of the proposed algorithm is illustrated by the calculation of the lifetime of the Ti-1Al-1Mn compact tension specimen under cyclic loading.

1. Introduction

Nowadays the fatigue problem becomes especially acute due to the fact that many structures and machine parts are subjected to the alternating load. Since Paris' law [1] there were developed many fatigue crack growth rate models. Energy-based models allow one to propose a physical explanation of fatigue crack propagation process. Due to the fact that a plastic zone forms at the crack tip the most obvious parameter characterizing fatigue crack growth rate is plastic work. There are numerous works devoted to the description of the fatigue crack rate with the use of this parameter.

In [2] it has been shown that fatigue crack growth rate is proportional to the square of the plastic zone size in plane strain condition. N.W. Klingbeil [3] proposed a fatigue crack growth model according to which the fatigue crack growth rate is proportional to the energy dissipation per cycle. In [4] principles of irreversible thermodynamics were used for establishment of fatigue crack growth rate law. According to this law the crack growth rate is determined by the energy flux at the fatigue crack tip. Short [5] proposed a similar law on the base of the consideration of the energy balance at the crack tip. The crack rate was defined as function of the J-integral value and the stored energy value per cycle.

Works [6]-[9] devoted to the fretting fatigue life assessment on the base of the damage mechanics approach. The number of cycles to the macro-crack initiation is obtained by the integration of the evolution law for damage variable. The location of initial crack under such an approach coincides with the location of maximum value of equivalent multiaxial damage stress at the contact interface.

In this study the fatigue crack growth rate model proposed in [4, 5] is applied to the assessment of the lifetime of the metal specimen under cyclic loading. The key point of the algorithm is a calculation of the stored energy value in material under deformation process. The original thermomechanical model is proposed for this purpose on the base of the statistical model of the defect structure evolution developed in [10, 11]. The defects ensemble evolution in the material is modelled using the internal field variable representing a defect density and having a physical meaning of the additional deformation due to the nucleation growth and coalescence of defects. The model takes into account the evolution of all tensor components and allows us to calculate the energy balance in the material under inelastic deformation.

The paper is organized as follows. Section 2 is devoted to the energy balance model formulation. There is considered two approaches to the obtaining of the constitutive equations. The first one is based on the hypothesis of the local equilibrium state and the second one assumes the existence of two dissipative potentials. Section 3 describes an algorithm for fatigue life assessment. The efficiency of the proposed algorithm is illustrated by the simulation of the lifetime of Ti-1Al-1Mn compact tension specimen.

2. Mathematical model of energy balance

2.1. Elasto-plastic constitutive equations

In this paragraph we will consider elasto-plastic constitutive equations which are based on principles of linear nonequillibrium thermodynamics [12].

The key point in every thermomechanical model is the choice of independent internal variables describing evolution of a material structure in a deformation process. In this work the symmetrical defect density tensor $\boldsymbol{p}$ [10] was chosen as this variable:

$$\boldsymbol{p} = n < \boldsymbol{s} >, \tag{1}$$

where $\boldsymbol{s}$ - symmetrical tensor characterizing shear-type unit defect, n - volume concentration of defects, $<>$ - denotes averaging over the representative volume of defects. This variable has a physical meaning of an additional structural strain due to the initiation, growth and coalescence of the defects of the different type.

In [10] it has been shown that statistical function of defect distribution can be written in the form:

$$W = Z^{-1} \exp(-E/\theta), \tag{2}$$

where E - energy of the defect, Z - normalizing factor, θ - "effective" temperature factor.

The energy of the defect can be written up to a constant term E_0 as

$$E - E_0 = -\boldsymbol{H} : \boldsymbol{s} + \alpha s^2 . \tag{3}$$

The quadratic term is the internal energy of the defect, α characterizes susceptibility of the media to the initiation of the defect, $\boldsymbol{H}:\boldsymbol{s}$ describes interaction between defects and external stress field $\boldsymbol{\sigma}$:

$$\boldsymbol{H} = \gamma\boldsymbol{\sigma} + \lambda\boldsymbol{p}, \tag{4}$$

where γ, λ - constants of interaction defined by the properties of the considered media.

Self-consistency equation could be obtained by the averaging of $\boldsymbol{s}$ with the use of W:

$$\boldsymbol{p} = N \int \boldsymbol{s} W(\boldsymbol{s}) d\boldsymbol{s} . \tag{5}$$

Solution of (5) for the pure shear problem is given in [10]. Based on these results we can develop model for the description of the defect evolution and energy balance in the process of plastic deformation.

Let us obtain that strain increments are small on the every step of the loading process. Then full strain rate $\dot{\boldsymbol{\varepsilon}}$ can be decomposed into the sum of the elastic $\dot{\boldsymbol{\varepsilon}}^e$, plastic $\dot{\boldsymbol{\varepsilon}}^p$ and structural $\dot{\boldsymbol{p}}$ strain rates:

$$\dot{\boldsymbol{\varepsilon}} = \dot{\boldsymbol{\varepsilon}}^e + \dot{\boldsymbol{\varepsilon}}^p + \dot{\boldsymbol{p}} . \tag{6}$$

Thus, inelastic strain is separated on the "pure" plastic strain (dissipative) due to the movement of the defects and structural (non-dissipative) due to the initiation, growth and coalescence of the defects.

The internal variable approach allows defining of the free energy function F as

$$F = F(T, \boldsymbol{\varepsilon}^e, \boldsymbol{p}), \tag{7}$$

where T - temperature. Time derivative of (7) is

$$\dot{F} = \frac{\partial F}{\partial \boldsymbol{\varepsilon}^e} : \dot{\boldsymbol{\varepsilon}}^e + \frac{\partial F}{\partial \boldsymbol{p}} : \dot{\boldsymbol{p}} + \frac{\partial F}{\partial T} \cdot \dot{T}. \tag{8}$$

Substitution of (8) into the second law of thermodynamics gives the following inequality:

$$(\boldsymbol{\sigma} - \frac{\partial F}{\partial \boldsymbol{\varepsilon}^e}) : \dot{\boldsymbol{\varepsilon}}^e - (S + \frac{\partial F}{\partial T}) \cdot \dot{T} + \boldsymbol{\sigma} : \dot{\boldsymbol{\varepsilon}}^p + \boldsymbol{\sigma} : \dot{\boldsymbol{p}} - \frac{\partial F}{\partial \boldsymbol{p}} : \dot{\boldsymbol{p}} - \boldsymbol{q} \cdot \frac{\nabla T}{T} \geq 0, \tag{9}$$

where S - specific entropy, $\boldsymbol{q}$ - heat flux, ∇ - nabla-operator. Fulfillment of (9) for any temperature and elastic strain values let us obtain classical relations for stress and entropy:

$$\boldsymbol{\sigma} = \frac{\partial F}{\partial \boldsymbol{\varepsilon}^e}, \tag{10}$$

$$S = -\frac{\partial F}{\partial T}. \tag{11}$$

Using (10), (11) we can write (9) in the form:

$$\boldsymbol{\sigma} : \dot{\boldsymbol{\varepsilon}}^p + (\boldsymbol{\sigma} - \frac{\partial F}{\partial \boldsymbol{p}}) : \dot{\boldsymbol{p}} - \boldsymbol{q} \cdot \frac{\nabla T}{T} \geq 0. \tag{12}$$

For a thermomechanical process described by (12) $\dot{\boldsymbol{\varepsilon}}^p$, $\dot{\boldsymbol{p}}$ are thermodynamic fluxes and $\boldsymbol{\sigma}$, $(\boldsymbol{\sigma} - \frac{\partial F}{\partial \boldsymbol{p}})$ are thermodynamic forces associated with these fluxes. Relations between fluxes and forces can be represented in the following form under an assumption of a small deviation of equilibrium state:

$$\dot{\boldsymbol{\varepsilon}}^p = \Gamma_\sigma \boldsymbol{\sigma} + \Gamma_{p\sigma} (\boldsymbol{\sigma} - \frac{\partial F}{\partial \boldsymbol{p}}), \tag{13}$$

$$\dot{\boldsymbol{p}} = \Gamma_p (\boldsymbol{\sigma} - \frac{\partial F}{\partial \boldsymbol{p}}) + \Gamma_{p\sigma} \boldsymbol{\sigma}, \tag{14}$$

where Γ_σ, $\Gamma_{p\sigma}$, Γ_p - kinetic coefficients. Restrictions on the kinetic coefficients are followed from (12) and physical meaning of (13)-(14):

$$\Gamma_\sigma \geq 0,\ \Gamma_{p\sigma} \geq 0,\ \Gamma_p \geq 0,\ \Gamma_\sigma \Gamma_p - \Gamma^2_{p\sigma} \geq 0. \tag{15}$$

Solution to the statistical problem (5), the hypothesis of the coaxility between $\boldsymbol{\sigma}$ and $\boldsymbol{p}$ let us to propose approximation for $\boldsymbol{\sigma} - \frac{\partial F}{\partial \boldsymbol{p}}$ in the form:

$$\frac{1}{2\mathrm{G}}\left(\boldsymbol{\sigma} - \frac{\partial F}{\partial \boldsymbol{p}}\right) = \frac{1}{\delta}\left(\frac{1}{2\mathrm{G}}\boldsymbol{\sigma} + \boldsymbol{p}\right) - \left(\frac{f_1(|p|)}{|p|} + 1\right)\boldsymbol{p}, \tag{16}$$

where G - shear modulus, $|p| = \sqrt{p:p}$, δ - scaling parameter. Function $f_1(|p|)$ is defined as

$$f_1(|p|) = \mathrm{k}|p|^{\mathrm{n}} + \mathrm{m} \tag{17}$$

and characterizes nonlinear hardening of the material. In (16) m , k , n - material parameters.

To describe transition from linear elastic portion of the stress-strain diagram to the hardening part the following approximation of the kinetic coefficients were proposed:

$$\Gamma_\sigma = \frac{1}{2G\tau_\sigma} \frac{1}{1+\exp\left(-\frac{|\sigma|-\sigma_2}{a_2}\right)}, \tag{18}$$

$$\Gamma_p = \frac{1}{2G\tau_p} \frac{1}{1+\exp\left(-\frac{H(|\sigma|,|p|,\delta)-\sigma_1}{a_1}\right)}, \tag{19}$$

$$\Gamma_{p\sigma} = 0, \tag{20}$$

where τ_σ, τ_p - relaxation times, $|\sigma| = \sqrt{\sigma:\sigma}$, σ_1, σ_2 - material parameters, a_2, a_1 - normalizing constants. Condition (20) characterizes independence of a plastic deformation process on the structural deformation process. Solution to (5) which is given in [10] gives a condition of equilibrium concentration of defects for the one-dimensional case:

$$\sigma = \frac{\partial F}{\partial p}, \tag{21}$$

where $\sigma = \sigma_{xx}$, $p = p_{xx}$. This means that function $\mathrm{H} = \left(\sigma - \frac{\partial F}{\partial p}\right)$ can be considered as "degree of nonequillibrium of the system". Generalization of this function on the three-dimensional case gives

$$\mathrm{H}(|p|,|\sigma|,\delta) = |\sigma| - 2\mathrm{G}\left\{\delta\left[f_1(|p|)+|p|\right]-|p|\right\}. \tag{22}$$

Constitutive equations (13)-(14) together with (16)-(20), (22) defines kinetic of plastic and structural strains.

In this work it is assumed that elastic strain tensor is connected with stress tensor according to the Hook's law. Thereby, the volumetric stress tensor σ_0 and stress deviator tensor $\boldsymbol{\sigma}_d$ are defined as

$$\dot{\sigma}_0 = \mathrm{K}\dot{\varepsilon}_0^e, \tag{23}$$

$$\dot{\boldsymbol{\sigma}}_\mathrm{d} = 2\mathrm{G}\dot{\boldsymbol{\varepsilon}}_\mathrm{d}^e, \tag{24}$$

where $\varepsilon_0^e = \frac{1}{3}\boldsymbol{\varepsilon}^e:\mathbf{E}$, $\boldsymbol{\varepsilon}_\mathrm{d}^e = \boldsymbol{\varepsilon}^e - \frac{1}{3}(\boldsymbol{\varepsilon}^e:\mathbf{E})\mathbf{E}$, $\mathbf{E}$ - unit tensor, K - bulk modulus.

To define temperature changes during the process of inelastic deformation we will express the first law of thermodynamics in the form:

$$\dot{e} = \boldsymbol{\sigma}:\dot{\boldsymbol{\varepsilon}} + \nabla\cdot\boldsymbol{q} + r', \tag{25}$$

where e - internal energy, r' - heat source power. The energy balance equation (25) and relations (6), (8), (10)-(11) give

$$\nabla\cdot\boldsymbol{q} + r' + \dot{Q}_e + \dot{Q}_p = -c\dot{T}, \tag{26}$$

where

$$c = T\frac{\partial^2 F}{\partial T^2}, \tag{27}$$

$$\dot{Q}_e = T\frac{\partial \boldsymbol{\sigma}}{\partial T}:\dot{\boldsymbol{\varepsilon}}^e, \tag{28}$$

$$\dot{Q}_p = \boldsymbol{\sigma}:(\dot{\boldsymbol{\varepsilon}}^p + \dot{\boldsymbol{p}}) - \frac{\partial F}{\partial \boldsymbol{p}}:\dot{\boldsymbol{p}} = \dot{W}^p - \dot{E}^s. \tag{29}$$

Here c - specific heat, T - temperature, $\dot{Q}_e$ - the heat source associated with thermoelastic effect, $\dot{Q}_p$ - inelastic contribution to the heat generation, $\dot{W}$ - plastic work rate, $\dot{E}^s$ - the stored energy rate.

Constitutive equations (13)-(14) describes evolution of plastic and structural strains, equation (26) will be used for the energy balance calculation in the process of inelastic deformation.

2.2. Modification of elasto-plastic constitutive equations

Equation (13) describes rate dependent material behavior. However, metals are rate independent under quasistatic loading conditions.

Following [13] let us suppose that considered process of inelastic deformation is described in terms of two internal variables - $\boldsymbol{p}$ and r. A tensorial variable $\boldsymbol{p}$ has been introduced in section 2.1, a scalar parameter r characterizes isotropic hardening of the material:

$$V_k = \{\boldsymbol{p}, r\}. \tag{30}$$

The second law of thermodynamics under isothermal conditions has the form:

$$\boldsymbol{\sigma} : \dot{\boldsymbol{\varepsilon}}^p + (\boldsymbol{\sigma} - \frac{\partial F}{\partial \boldsymbol{p}}) : \dot{\boldsymbol{p}} - \frac{\partial F}{\partial r}\dot{r} \geq 0. \tag{31}$$

We introduce the following notations:

$$R \equiv \frac{\partial F}{\partial r}, \tag{32}$$

$$\boldsymbol{P} \equiv \boldsymbol{\sigma} - \frac{\partial F}{\partial \boldsymbol{p}}. \tag{33}$$

Vectors of thermodynamic forces $\boldsymbol{X}$ and fluxes $\boldsymbol{J}$ can be defined as

$$\boldsymbol{X} = \{\boldsymbol{\sigma}, \boldsymbol{P}, -R\}, \tag{34}$$

$$\boldsymbol{J} = \{\dot{\boldsymbol{\varepsilon}}^p, \dot{\boldsymbol{p}}, \dot{r}\}. \tag{35}$$

We will postulate the existence of a convex function Ψ which defines thermodynamic fluxes $\boldsymbol{J}$ by the relation:

$$\boldsymbol{J} = \dot{\Lambda}\frac{\partial \Psi}{\partial \boldsymbol{X}}, \tag{36}$$

where $\dot{\Lambda}$ - undetermined multiplier. Assumption of independence of processes of plastic and structural deformation let us decompose Ψ into the sum:

$$\Psi = \Psi(\boldsymbol{\sigma}, \boldsymbol{P}, R) = \Psi^{\varepsilon^p} + \Psi^p, \tag{37}$$

where Ψ^{ε^p} defines plastic deformation process and Ψ^p defines structural deformation process.

Substitution of thermodynamic forces $\boldsymbol{X}$ (34), fluxes $\boldsymbol{J}$ (35) as well as equation (37) into (36) allows us to obtain constitutive relations for the evolution of isotropic hardening parameter, plastic strain and structural strain in the form:

$$\dot{\boldsymbol{\varepsilon}}^p = \dot{\Lambda}^{\varepsilon^p}\frac{\partial \Psi^{\varepsilon^p}}{\partial \boldsymbol{\sigma}}, \tag{38}$$

$$\dot{r} = -\dot{\Lambda}^{\varepsilon^p}\frac{\partial \Psi^{\varepsilon^p}}{\partial R}, \tag{39}$$

$$\dot{\boldsymbol{p}} = \dot{\Lambda}^p\frac{\partial \Psi^p}{\partial \boldsymbol{P}}. \tag{40}$$

Equations (38)-(40) characterize the fact that variables $\dot{\varepsilon}^p$, $\dot{r}$, $\dot{\boldsymbol{p}}$ are defined by the outer normal to the surface $\Psi = 0$, i.e. by the associated flow rule.

We will postulate function Ψ^{ε^p} in the form:

$$\Psi^{\varepsilon^p} = \frac{2a_2 b e^{-0.5x}\sqrt{e^y + e^x \ln(e^{0.5x} + \sqrt{e^y + e^x})}}{\sqrt{1+e^{y-x}}} + R - 2\sqrt{\frac{b}{c}}a_2 \ln\left(1+\sqrt{e^y+1}\right), \tag{41}$$

where b, c - material parameters, $x = |\sigma| / a_2$, $y = \sigma_2 / a_2$.

Unknown parameter $\dot{\Lambda}^{\varepsilon^p}$ is defined according to Prager consistency condition:

$$\Psi^{\varepsilon^p} = \frac{\partial \Psi^{\varepsilon^p}}{\partial \boldsymbol{\sigma}} : \dot{\boldsymbol{\sigma}} + \frac{\partial \Psi^{\varepsilon^p}}{\partial R} \dot{R} = 0 . \tag{42}$$

Substitution of $\dot{\Lambda}^{\varepsilon^p}$ into (38) and (39) let us obtain the final form for evolution equation:

$$\dot{\boldsymbol{\varepsilon}}^p = \frac{\boldsymbol{\sigma} : \dot{\boldsymbol{\sigma}}}{\tau_\sigma \left[1 + \exp\left(-\dfrac{|\sigma| - \sigma_2}{a_2} \right) \right] |\sigma|} \cdot \frac{\boldsymbol{\sigma}}{|\sigma|} \frac{1}{2\mathrm{G}} , \tag{43}$$

$$\dot{r} = -\frac{\boldsymbol{\sigma} : \dot{\boldsymbol{\sigma}}}{\sqrt{c\tau_\sigma} \left[1 + \exp\left(-\dfrac{|\sigma| - \sigma_2}{a_2} \right) \right]^{1/2}} \frac{1}{2\mathrm{G}} , \tag{44}$$

where $\tau_\sigma = c / b$.

In order to derive $\dot{\Lambda}^p$ we will use quasi-standard thermodynamic approach for rate-dependent materials which is described in [13]. In this case $\dot{\Lambda}^p$ is defined by the relation:

$$\dot{\Lambda}^p = \left\langle \frac{\Psi^p}{k'} \right\rangle^v . \tag{45}$$

Here, $\langle \ \rangle$ denotes Macaulay brackets, $\Psi^p = |P| = \sqrt{\boldsymbol{P} : \boldsymbol{P}}$ - part of the Ψ connected with the change in the material structure. Equation (40) allows obtaining of equation for structural strain:

$$\dot{\boldsymbol{p}} = \left\langle \frac{\Psi^p}{k'} \right\rangle^v \frac{\partial \Psi^p}{\partial \boldsymbol{P}} . \tag{46}$$

Taking $v = 1$, $k'(|\sigma|, |p|, \delta) = \tau_p \left[1 + \exp\left(-\dfrac{H(|\sigma|, |p|, \delta) - \sigma_1}{a_1} \right) \right]$ and substituting expressions for $\dfrac{\partial \Psi^p}{\partial \boldsymbol{P}}$ and Ψ^p into (46) we will get equation for the structural strain obtained in the section 2.1:

$$\dot{\boldsymbol{p}} = \frac{1}{2\mathrm{G}\tau_p} \frac{1}{1 + \exp\left(-\dfrac{H(|\sigma|, |p|, \delta) - \sigma_1}{a_1} \right)} \left(\boldsymbol{\sigma} - \frac{\partial F}{\partial \boldsymbol{p}} \right) . \tag{47}$$

In following calculation we will suppose that contribution of r into the stored energy is small compared with contribution of $\boldsymbol{p}$. Thereby, the energy balance is still determined by (26)-(29).

The full thermomechanical model of inelastic deformation of metal consists of equations (6), (14), (16)-(17), (19)-(20), (22), (23)-(24), (43) describing stress-strain state of the material and equations (26)-(29) for energy balance calculation.

3. Numerical simulation of a fatigue life in Ti-1Al-1Mn alloy

One of the important applications of the energy balance evaluation is possibility of a calculation of a fatigue crack growth rate on the base of the energy equations. This information can be used for an assessment of a lifetime (number of cycles to specimen failure) of a specimen. In this work we propose a method for a lifetime assessment on the base of the energy equation derived in [4, 5]:

$$\frac{da}{dN} = \frac{1}{h(J_c - J)} \frac{\mathrm{d}E_s}{\mathrm{d}N} , \tag{48}$$

where $\dfrac{da}{dN}$ - fatigue crack rate, $\dfrac{\mathrm{d}E_s}{\mathrm{d}N}$ - the stored energy value per cycle, J - J-integral, J_c - critical value of J-integral, h - specimen thickness.

The algorithm for a lifetime assessment includes a several steps.

First, we chose several stress amplitudes for numerical simulation. For every force the critical crack length is calculated as [14]:

$$K_c = \frac{F}{hw^{1/2}} f(a_c / w), \tag{49}$$

where K_c - fracture toughness, F - force magnitude, w - width of the specimen, a_c - critical length, f - polynomial function depending on the crack size and geometric parameters of the considered sample.

Second, for every chosen force, the fatigue crack rate was determined. In order to define crack rate equation (48) well be used. To apply (48) we calculated the value of a stored energy per cycle and the J-integral value. The stored energy value is evaluated according to equation (29). A stationary crack approach for a fatigue crack growth rate calculation proposed in [3] is used. According to this approach, a few loading cycles for every given crack length a are considered and stabilized value of the stored energy per cycle is evaluated. The J-integral values are calculated in the finite-element package Simulia Abaqus which is run under academic license. Thus, on the second step of the algorithm the function $l(a) = \frac{1}{h(J_c - J(a))} \frac{dE_s(a)}{dN}$ is determined.

Third, the number of cycles to fracture N_* is calculated as

$$N_* = \int_{a_0}^{a_c} \frac{dl}{l(a)}, \tag{50}$$

where a_0 - initial crack length.

We will illustrate the abovementioned algorithm by the simulation of the lifetime of the Ti-1Al-1Mn specimen subjected to the cyclic loading.

For calibration of the material parameters of the proposed mathematical model, a quasistatic tensile experiment of a Ti-1Al-1Mn dogbone specimen with a strain rate of 1.5 s^{-1} was simulated. Equations (6), (14), (16)-(17), (19)-(20), (22), (23)-(24), (43), (29) were implemented in the finite-element package Simulia Abaqus with the use of UMAT procedure. Arrays of material constants, strain, strain increments and the time step were passed as the input data to the procedure. Values of plastic strain tensor components, structural strain tensor components and stress tensor components were needed to be defined. For this purpose, equations (6), (14), (16)-(17), (19)-(20), (22), (23)-(24), (43) were integrated by the midpoint method and increments of the unknown variables were defined. Values of the unknown variables at the next time step were defined as the sum of the values on the previous time step and the appropriate increments. Following values of the material parameters were obtained as a result of the stress-strain state and energy balance simulation: $\tau_p = 0.003$, $\tau_\sigma = 0.003$, $\sigma_1 = 3\cdot 10^8$, $a_1 = 10^8$, $\sigma_2 = 625\cdot 10^6$, $a_2 = 10^5$, $n = 0.85$, $k = 0.007$, $m = 0.004$, $\delta = 1$. More detailed information about this procedure can be found in [15].

To illustrate the algorithm we estimate a lifetime of Ti-1Al-1Mn compact tension specimen (CT-specimen) subjected to the cyclic loading with stress the ratio $R = 0.1$. Geometry of the specimen is presented in figure 1. Figure 2 shows finite-element mesh of the specimen. Symmetry of the specimen allows us to simulate only half of the specimen. In the vicinity of the crack tip the more refined mesh was used. Convergence of the solution was verified by doubling of a mesh size in each direction which produced a negligible change.

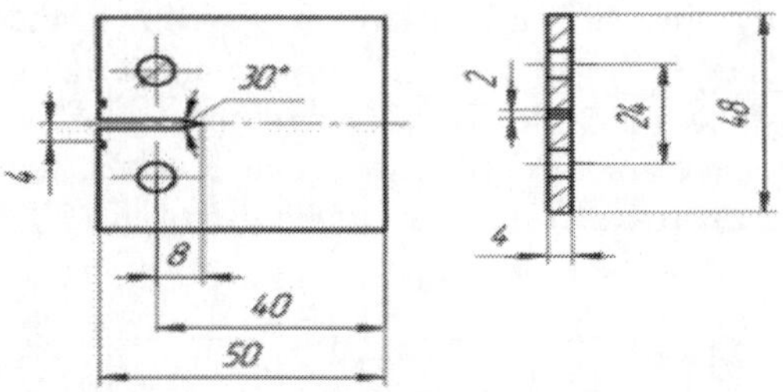

Figure 1. Geometry of the CT-specimen. All dimensions are in millimeters.

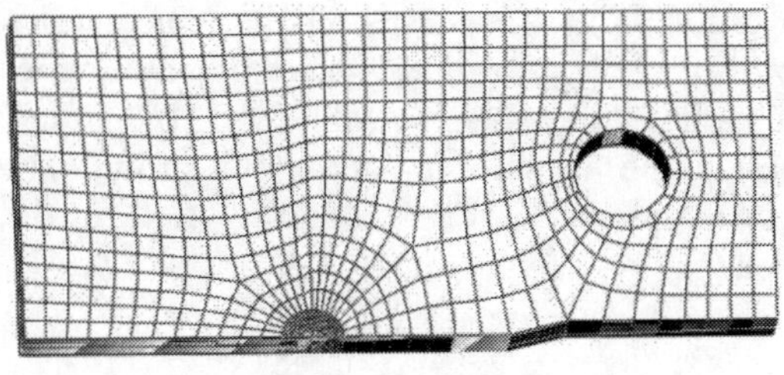

Figure 2. Finite-element model of the CT-specimen.

Four values of a force magnitude has been chosen: $F_1 = 2500N$, $F_2 = 3500N$, $F_3 = 4300N$, $F_4 = 5000N$. For every considered force magnitude a critical crack length was obtained with the use of (49). Polynomial function f for the considered specimen has the form [14]:

$$f = \frac{(2+f')(0.886+4.64f'-13.32(f')^2+14.72(f')^3-5.6(f')^4)}{(1-f')^{1.5}}, \tag{51}$$

where $f' = \dfrac{8\cdot 10^{-3} + a_c}{4\cdot 10^{-2}}$. Substitution of (51) into (49) for every considered force magnitude allows us to obtain the following critical crack lengths: $a_{c1} = 20$ mm, $a_{c2} = 17$ mm, $a_{c3} = 15$ mm, $a_{c4} = 13$ mm. The value of 3 mm was chosen as the first approximation for the initial crack length for every considered force magnitudes.

To calculate the functions $l(a)$ for every force magnitude we estimate J-integral values and values of the stored energy per cycle should be obtained. Figure 3 shows dependence of the J-integral values upon the crack length for the considered force magnitudes. We can note that J-integral values increase with the rise of the force magnitude. Figure 4 depicts stored energy values per cycle in semi-log scale. In order to obtain stored energy values per cycle the proposed in section 2 model was applied. This model allows us to describe an upward trend in the stored energy value which can be observed under the increase of the applied force.

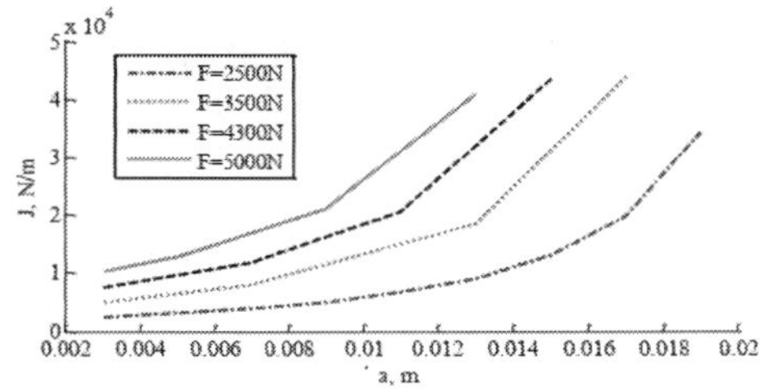

Figure 3. J-integral values versus crack length for the applied force magnitudes.

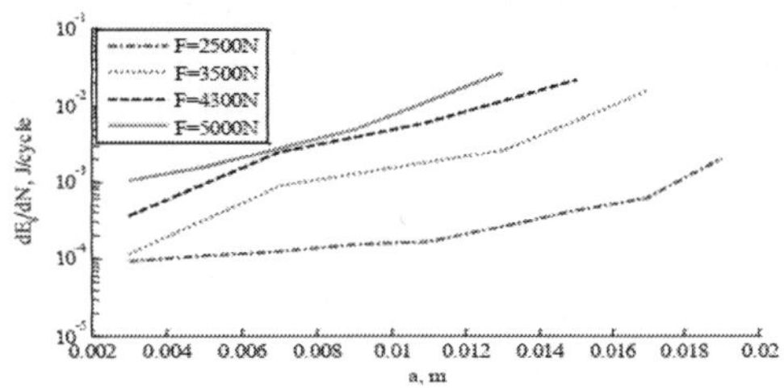

Figure 4. The stored energy values per cycle versus crack length for the applied force magnitudes.

Figure 5 shows crack growth rate obtained by equation (48) in semi-log scale. It can be noted that the increase of the applied load magnitude leads to the rise in the fatigue crack growth rate. The chosen force magnitudes span three orders of magnitudes of the fatigue crack growth rate and give us values for $l(a)$ functions. Substitution of $l(a)$ and values of critical crack lengths which have been obtained on the first step of the algorithm into (50) give number of cycles to fracture for the CT-specimen with the given crack length of 3 mm. These results are plotted in figure 6. As can be seen figure 6 shows continuous decreasing curve without pronounced endurance limit. This can be explained by the two reasons. The first reason is insufficient number of chosen points. So, one of the key moment of the proposed method is to provide sufficient number of the applied forces to cover all curve characterizing fatigue life of the specimen. The second reason is the use of the fixed initial crack length instead of the crack length at which fatigue crack growth starts to propagate for the given force range. Thus, the proposed method in this study is applied for the first-order approximation of a fatigue life of the CT-specimen.

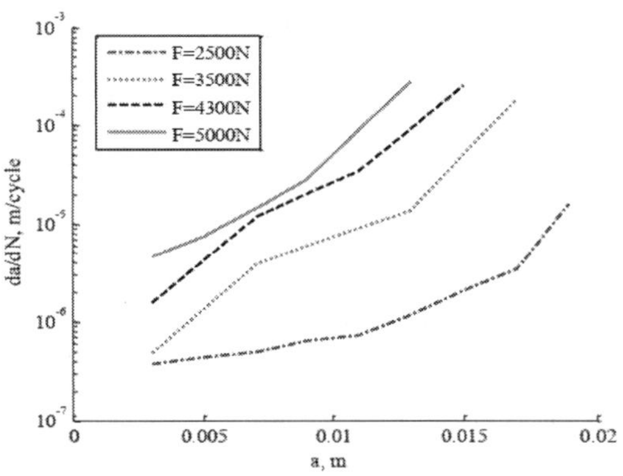

Figure 5. Fatigue crack growth rate versus crack length for the applied force magnitudes.

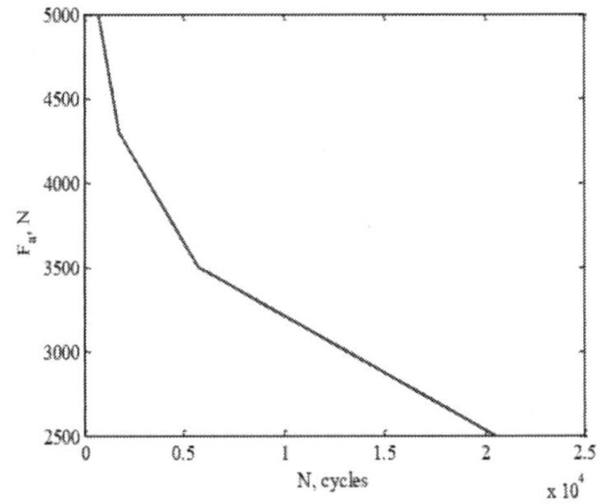

Figure 6. Applied force magnitudes versus number of cycles to fracture.

4. Conclusions

In this study a numerical method for the fatigue life assessment was proposed. This method is based on the calculation of a stored energy of a cold work. To calculate the energy balance in the material, original thermomechanical model was proposed. According to this model, the stored energy value is defined by the structural parameter which has a physical meaning of an additional strain induced by initiation, growth and coalescence of defects. Thereby, plastic strain is responsible to the energy dissipation and structural strain is responsible for the energy storage during the process of inelastic deformation. Two types of constitutive equations for plastic strain and structural strain were derived under an assumption of a local equilibrium and with the use of the flow plasticity theory.

The possibility of energy balance calculation gives us opportunity to evaluate fatigue crack growth rate on the base of the energy law and propose an algorithm for fatigue life assessment. The application of the algorithm is illustrated by the numerical simulation of the fatigue life of Ti-1Al-1Mn compact tension specimen. The material parameters were defined by the simulation of a mechanical and thermodynamic behavior of a dogbone specimen subjected to quasistatic tensile loading. Results of the residual life simulation according to the proposed algorithm show continuous decreasing trend of the obtained curve and can be considered as first-order approximation of a fatigue life of the CT-specimen.

Acknowledgments
The reported study was funded by the Russian Foundation for Basic Research in the framework of research projects No. 16-31-00156 and No. 16-48-590148

References

[1] Paris P and Erdogan F 1963 *J. Basic Eng-T ASME* **85** 528-34
[2] Iino Y 1979 *Eng. Fract. Mech.* **12** 279-99
[3] Klingbeil N W 2003 *Int. J. Fatigue* **25** 117-28
[4] Chudnovsky A and Moet A 1985 *J. Mater. Sci.* **20** 630-35
[5] Short J S and Hoeppner D W 1989 *Eng. Fract. Mech.* **33** 175-84

IOP Conf. Series: Journal of Physics: Conf. Series **843** (2017) 012034 doi:10.1088/1742-6596/843/1/012034

[6] Ferjaoui A, Yue T, Abdel Wahab M and Hojjati-Talemi R 2015 *Int. J. Fatigue* **73** 66-76
[7] Kumar D, Biswas R, Poh LH and Abdel Wahab M 2017 *Tribology International* **109** 124-132
[8] Resende Pereira KdF, Bordas S, Tomar S, Trobec R, Depolli M, Kosec G and Abdel Wahab M 2016 *Materials* **9** 639
[9] Hojjati Talemi R and Abdel Wahab M 2013 Tribology International 60 176-186
[10] Naimark O B 2003 *Phys. Mesomech.* **6** 39-63
[11] Plekhov O A and Naimark O B 2009 *J. Appl. Mech. Tech. Phys.* **50** 127-36
[12] Glansdorff P and Prigogine I 1971 *Thermodynamic Theory of Structure, Stability and Fluctuations* (London: Wiley-Interscience)
[13] Murakami S 2012 *Continuum Damage Mechanics – A Continuum Mechanics Approach to the Analysis of Damage and Fracture* (Dordrecht, Heidelberg, London, New York: Springer)
[14] ASTM E 1820-01 2001 *Standard Test Method for Measurement of Fracture Toughness* (Philadelphia: ASTM)
[15] Iziumova A, Vshivkov A, Prokhorov A, Kostina A and Plekhov O 2016 *Quant. Infrared Thermogr. J.* **13** 242-56

Comparison of fatigue crack growth of riveted and bonded aircraft lap joints made of Aluminium alloy 2024-T3 substrates – A numerical study

S Pitta[1], **J I Rojas**[1] **and D Crespo**[2]

[1] Dept. of Physics-Division of Aerospace Engineering, C/Esteve Terradas 5, Castelldefels, 08660 Spain

[2] Dept. of Physics, C/Esteve Terradas 5, Castelldefels, 08860 Spain

E-mail: siddharth.pitta@estudiant.upc.edu, josep.ignasi.rojas@upc.edu, daniel.crespo@upc.edu

Abstract. Aircraft lap joints play an important role in minimizing the operational cost of airlines. Hence, airlines pay more attention to these technologies to improve efficiency. Namely, a major time consuming and costly process is maintenance of aircraft between the flights, for instance, to detect early formation of cracks, monitoring crack growth, and fixing the corresponding parts with joints, if necessary. This work is focused on the study of repairs of cracked aluminium alloy (AA) 2024-T3 plates to regain their original strength; particularly, cracked AA 2024-T3 substrate plates repaired with doublers of AA 2024-T3 with two configurations (riveted and with adhesive bonding) are analysed. The fatigue life of the substrate plates with cracks of 1, 2, 5, 10 and 12.7mm is computed using Fracture Analysis 3D (FRANC3D) tool. The stress intensity factors for the repaired AA 2024-T3 plates are computed for different crack lengths and compared using commercial FEA tool ABAQUS. The results for the bonded repairs showed significantly lower stress intensity factors compared with the riveted repairs. This improves the overall fatigue life of the bonded joint.

Keywords: Fatigue, stress intensity factor, finite element analysis, crack, joint, repair

1. Introduction

Aircraft structures are made of many small parts, assembled to form major structures. Most of the aircraft skin is made of aluminium alloys and composites. During the operation of the aircraft, structures are subjected to static and dynamic loading. A safety factor of 1.5 is incorporated in the design of aircraft structures, so to prevent the structure from failing under static loading. On the other side, metals have tendency to form fatigue cracks due to repetitive loading that can be of tensile, torsion or vibrational nature. Hence, fatigue of aircraft structures must be well-understood to prevent accidents because the majority of catastrophic failures of aerospace parts occur due to cyclic loading during the aircraft service life. Early detection of cracks in the structure allows calculating the expected life of the part before failure and hence it can be repaired at critical size to prevent final failure.

The common repair options for cracked metal plates are patches or doublers joined with mechanical fasteners and adhesive bonding. Rivets are also widely used in aircrafts to join the skin panels. The fastened joint strength depends on various parameters such as rivet diameter, material properties, fastening pattern and clamping force [1-3]. Riveting of the joints induces residual stresses in the substrate plates, and the residual stresses are directly proportional to rivet squeeze force. A major disadvantage of riveted joints are the holes created in the plate as they cause high stress concentrations. Cold expansion [4], clamping force [5] and interference fit [6] can reduce the stress concentrations around the fastened holes, improving the fatigue life. Iyer *et al.* [7] investigated the fatigue behaviour of single and double self-piercing rivets with experiments and finite element methods. Rijck *et al.* [8] investigated the implications of rivet driven head dimensions on the fatigue performance of aircraft lap joints. Skorupa *et al.* [9] investigated fatigue life and fatigue crack location in riveted aircraft lap joints in fuselage. Esmaeili *et al.* [10] investigated the effects of torque tightening on fatigue strength of aluminium alloy (AA) 2024-T3 double-lap bolted joints. Carter *et al.* [11] investigated on production quality of aircraft fastener holes on fatigue life. Their investigation showed that machining of fastener holes has a more important effect on fatigue life than residual stresses. Fretting fatigue damage occurs when contact regions experience cyclic stresses, initiating micro crack formation and leads to failure. Fretting fatigue crack initiation in double lap bolted joints was investigated by Ferjaoui *et al.* [12]. Hojjati *et al.* [13] developed fretting fatigue crack initiation lifetime predictor tool using damage mechanics approach. Corrosion also plays an important role in fatigue life. Kermanidis *et al.* [14] investigated fatigue and damage tolerance behaviour of corroded AA 2024-T351 for different stress ratios. Corrosion of aluminium alloys is classically attributed to complex processes of oxidation [15, 16]. Investigations performed on aircraft alloys showed that corrosion is a damage process not only limited to the surface (affecting the yield strength (YS) and fatigue life through occurrences of corrosion notches or pits), but which also causes diffusion-controlled hydrogen embrittlement of the materials [17-21].

Adhesive bonding of structures is another most used technique in joining parts. In bonded joints, the stresses are distributed uniformly along the bond area, hence reducing the stress concentrations in the joint. Most of the bonded repairs on cracked metallic plates consist of bonded composite patches [22-24]. The strength of the adhesive and surface preparation play a key role on the strength of the joint and fatigue resistance [25]. Bonded composite repairs of cracked primary aircraft structures under fatigue loading were investigated by Baker [26]. Uncertainties of the material and their effect on aircraft fuselage fatigue life was investigated by Koutsourelakis *et al.* [27]. They proposed a cohesive element model to calculate the fatigue life of aircraft fuselages. Fatigue durability of aircraft repairs with clad AA 7075-T6 bonded joints was investigated experimentally and numerically by Harman *et al.* [28].

In this paper, the fatigue life of AA 2024-T3 is computed for different crack lengths using FRANC3D. Second, the performance of AA 2024-T3 plates repaired with mechanically fastened joints and adhesively bonded joints is compared. A three dimensional (3D) finite element model is developed in ABAQUS and imported to FRANC3D to obtain the fatigue life. Later, the repair configurations are modelled and analysed in ABAQUS to compute the stress intensity factors.

2. Materials and Methods

2.1 Materials

AA 2024-T3 is widely used in manufacturing aircraft skin panels due to its high strength and fatigue resistance properties. Table 1 and Table 2 show, respectively, the mechanical properties and chemical composition of AA 2024-T3, used for the substrate and doublers. The thickness of aircraft skin at high load areas is 3.175mm, while the doublers have thickness of

1.5875mm. The length/width of the substrate plates used in this investigation is 342.8mm/25.4mm with an aspect ratio of 13.5. The doublers have a length of 203.2mm and 152.4mm, width of 25.4mm with aspect ratios of 8 and 6. A two-part adhesive HYSOL 9480 is used for the adhesive bonding of the doublers to the cracked plates. The Young's modulus of Hysol 9480 is 4.5 GPa (ISO 527-3) and the tensile strength (TS) is 47 MPa (ISO 527-3).

Table 1 Mechanical properties of the as-received, commercial aluminium alloys (AA) 2024-T3.

Aluminium alloy	YS (MPa)	TS (MPa)	% Area Reduction	Brinell hardness
AA 2024-T3	316	464	20.2%	HB 123

Table 2 Chemical composition of the as-received, commercial aluminium alloys (AA) 2024-T3 [27].

Aluminium alloy	Units	Si	Fe	Cu	Mn	Mg	Zn	Ti	Cr	Al
AA 2024-T3	Wt. %	0.50	0.50	4.9	0.9	1.8	0.25	0.15	0.10	90.9
ASTM E8/B557	At. %	0.07	0.20	4.4	0.59	1.4	0.09	0.02	0.01	93.22

2.2 Modelling in FRANC3D

A simple 3D finite element model is modelled using FEA tool ABAQUS with material properties and boundary conditions. Pressure is applied on the ends of the substrate plates with different magnitudes of 93.6 MPa (design limit load) and 140 MPa (ultimate load) in the tensile direction. The ultimate load the joint can carry was calculated from the strength of the rivets (rivet value times number of rivets). A safety factor of 1.5 applied to the ultimate load is used to calculate the design limit load of the joints. The aluminium alloy substrate plates are modelled with solid elements of 8-node linear brick elements with reduced integration. These plates are meshed with an element size of 2mm and static general step is considered to generate the input file in ABAQUS. Then, the input file is imported to FRANC3D where cracks of 1, 2, 5, 10 and 12.7mm are induced in the centre of the model, as shown in Fig. 1(a). FRANC3D has an inbuilt meshing algorithm to mesh the crack fronts with an option to refine depending on the complexity of model. The model meshed in FRANC3D is shown in Fig. 1(b). Static analysis is performed on the cracked plate to obtain stress intensity factors at the crack fronts. The cracks are allowed to grow under quasi-static power law criterion. Based on the fatigue crack growth (FCG) for the given loads, the fatigue life of the plate is computed in FRANC3D. For this purpose, the material properties shown in Table 3 were used, together with the NASGRO version 3 equation:

$$\frac{da}{dN} = C\left[\frac{1-f}{1-R}\Delta K\right]^n \frac{\left(1-\frac{\Delta K_{th}}{\Delta K}\right)^p}{\left(1-\frac{K_{max}}{K_c}\right)^q} \tag{1}$$

where $\frac{da}{dN}$ is the crack growth rate, ΔK is the applied stress-intensity factor, R is the stress ratio, ΔK_{th} is the fatigue threshold, K_{max} is the stress-intensity factor corresponding to peak applied load, K_c is the critical stress intensity, f is Newman's crack opening function, C, n, p and q are curve fitting constants.

Table 3 Material properties as used in the NASGRO version 3 equation (Eq. 1) for AA 2024-T3.

Material	TS (MPa)	YS (MPa)	K_c (MPa.mm$^{0.5}$)	C (mm/cycle)	n	p	q
AA 2024-T3	455.07	365.44	36.26	1.832e-12	3.284	0.5	1

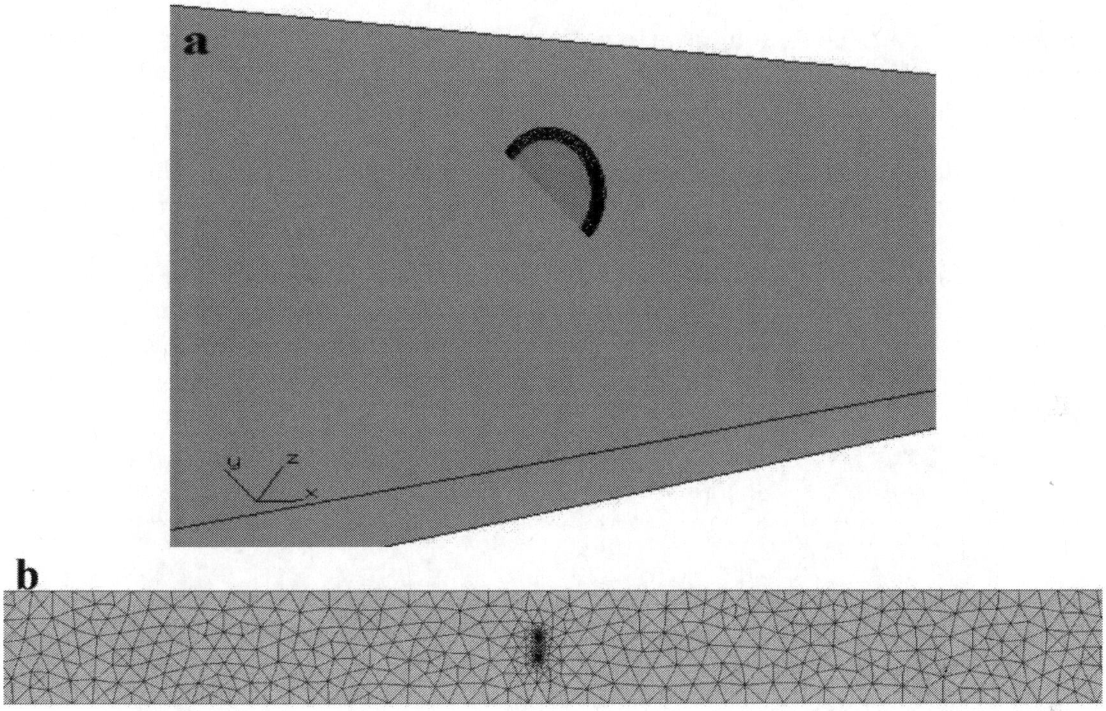

Fig. 1 (a) Isometric view of insertion of crack in the substrate plate in the graphical user interface of FRANC3D, and (b) lay out view of the mesh in the plate as generated by the FRANC3D mesh algorithm.

2.3 Modelling in ABAQUS

Cracked aluminium substrate plates are repaired with two doublers with riveted and bonded joint configuration. Models of these repairs are analysed in ABAQUS to obtain the stress intensity factors and the results study is performed for various crack lengths are compared. The aluminium alloy substrates and doublers are modelled as 3D shell elements to increase the computational efficiency. Assumption of shell elements considered in this study provides reasonable results compared with solid elements. If two dimensions of the specimen are much greater than the third dimension, then shell elements provide good results. Table 1 shows the properties as used for substrates and doublers. The rivets are modelled as point-based fasteners, as shown in Fig. 2(a). Bushing elements are considered to simulate the stiffness of the rivets in six directions. These stiffness values are calculated using the following formulae [29]:

$$K_1 = \left\{\frac{k}{m}\left[\left(\frac{t_{shell1}+t_{shell2}}{2d}\right)^{\lambda}\left(\frac{1}{E_{11,shell1}t_{shell1}}+\frac{1}{m.E_{11,shell2}t_{shell2}}+\frac{1}{2.E_{fast}t_{shell1}}+\frac{1}{2m.E_{fast}t_{shell2}}\right)\right]\right\}^{-1} \quad (2)$$

$$K_2 = \left\{\frac{k}{m}\left[\left(\frac{t_{shell1}+t_{shell2}}{2d}\right)^{\lambda}\left(\frac{1}{E_{22,shell1}t_{shell1}}+\frac{1}{m.E_{22,shell2}t_{shell2}}+\frac{1}{2.E_{fast}t_{shell1}}+\frac{1}{2m.E_{fast}t_{shell2}}\right)\right]\right\}^{-1} \quad (3)$$

$$K_3 = \frac{\pi d^2 E_{fast}}{4(t_{shell1} + t_{shell2})} \tag{4}$$

$$K_4 = K_5 = \frac{\pi d^2}{16}\left(\frac{E_{fast} d^2}{4L} + G_{fast} \cdot L\right) \tag{5}$$

$$K_6 = \frac{G\pi d^4}{32L} \tag{6}$$

Where k is 2.2 and λ is 0.4 for solid rivet, m is 1 for single shear, $t_{shell1}$and t_{shell2} are thickness of shell 1 and shell 2, d is diameter of rivet hole, E_{11},E_{22},E_{fast} are elastic modulus of plates and rivet, $G_{fast}$is shear modulus of rivet and L is length of the fastener.

The adhesive bonding of the substrate and doublers is modelled as cohesive contact between the surfaces. The adhesive properties are mentioned Section 2.1. Traction-separation law criteria is used in modelling cohesive with parameters (*E/Enn*, *G1/Ess* and *G2/Ett*). These values are calculated based on thickness of the layer which is 0.5mm. Epoxy properties are assumed to be isotropic due to lack of data. Contour integral cracks were inserted in the substrates with the seam length equal to the crack length, as shown in Fig. 2(b). Stress intensity factors at the crack fronts were obtained from history outputs at maximum energy release rate criterion for FCG.

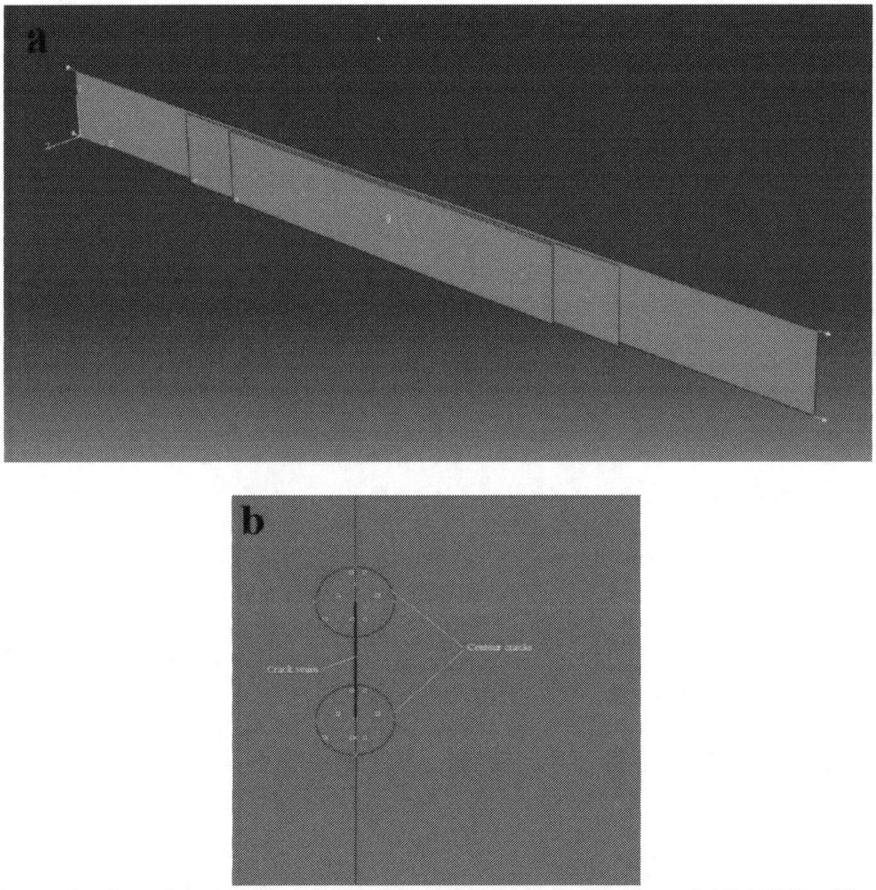

Fig. 2 (a) Isometric view of riveted joint model in the graphical user interface of ABAQUS with point based fasteners, and (b) detail of the insertion of a contour integral crack in the joint model.

IOP Conf. Series: Journal of Physics: Conf. Series **843** (2017) 012035 doi:10.1088/1742-6596/843/1/012035

3. Results and Discussions

Table 4 shows the fatigue life of the substrate plate with various crack lengths, as obtained from FRANC3D. A null stress ratio is applied for all crack sizes. Fig. 3 shows the stress intensity factors as obtained with FRANC3D for three contours of FCG and various crack lengths at applied stress of 93.6 and 140 MPa. The zeroth contour corresponds to the initial stress intensity factor without growth and the later three contours are the stress intensity factors after initiation of FCG. Fig. 4 shows the crack propagation of 5mm initial crack at applied stress of 93.6 MPa. The latter values for the substrate plates are below the fracture toughness K_C of AA 2024-T3 (see Table 3).

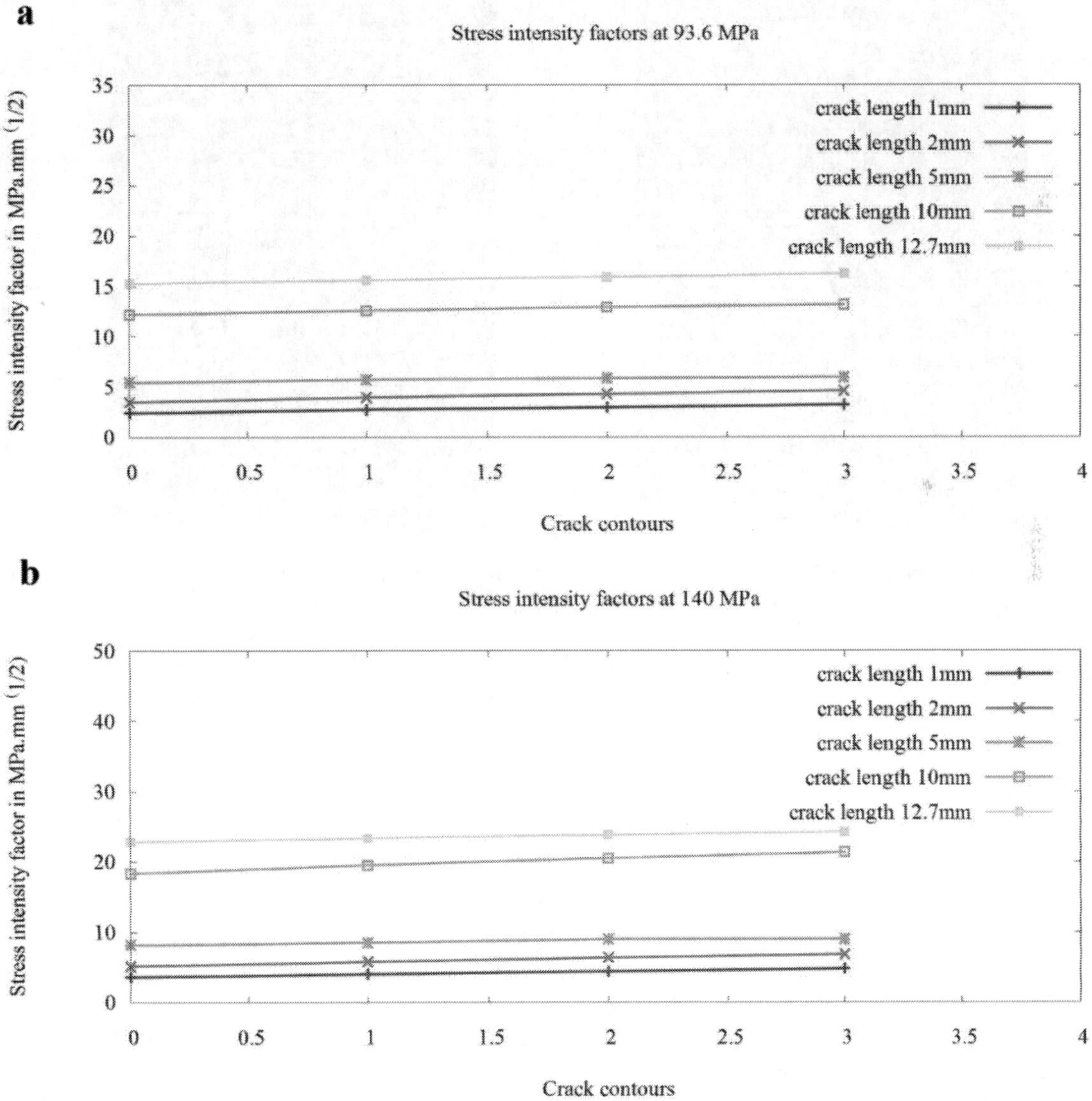

Fig. 3 Stress intensity factors for different crack lengths and contours for a tension of (a) 93.6 MPa and (b) 140 MPa.

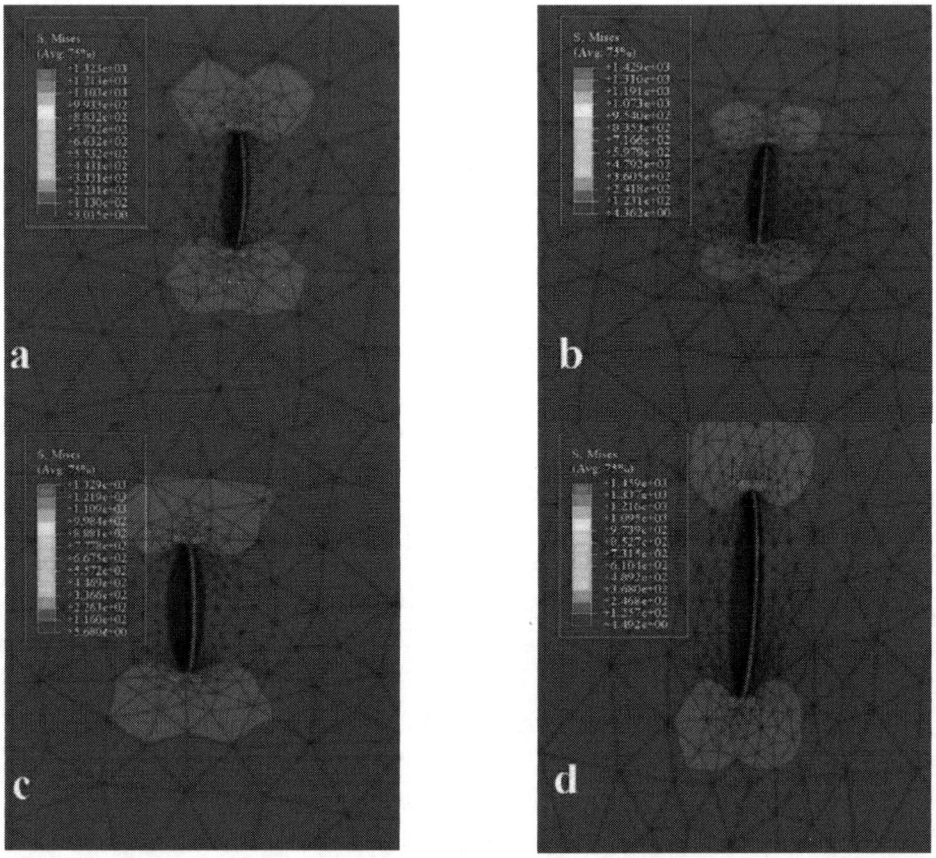

Fig. 4 Crack propagation for initial crack size 5mm for a tension of 93.6 MPa (a) static analysis (b) contour 1 (c) contour 2 and (d) contour 3

Table 4. Fatigue life for AA 2024-T3 substrate plate for different crack length and stresses

Crack length (mm)	Fatigue Life (N_f) for 93.6 MPa	Fatigue life (N_f) for 140 MPa
1	314362	48193
2	178451	30024
5	14447	3670
10	1206	616
12.7	576	126

Average aircraft structures are designed for 50000-100000 flights. Based on the crack size, the number of cycles to failure can be calculated. The repairs with doubler plates should arrest further FCG in the substrate plate. FRANC3D is used to obtain the number of cycles to failure and ABAQUS is used to obtain the stress intensity factors for the studied joint configurations. Repaired substrates are further modelled and analysed in FEA tool ABAQUS. Fig. 5 shows the obtained stress intensity factors for the riveted and bonded joints, stressed at 93.6 and 140 MPa.

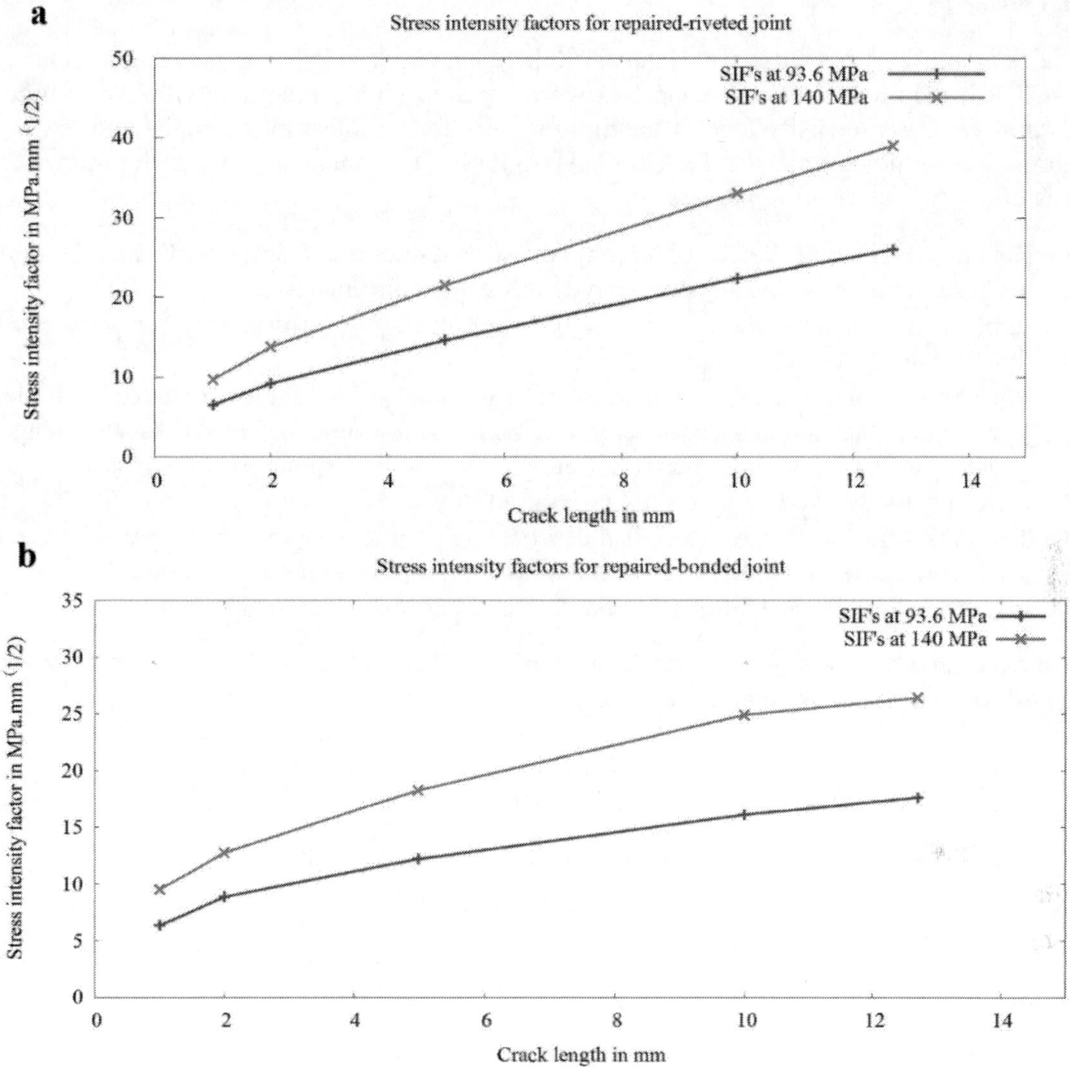

Fig. 5 Stress intensity factors for (a) riveted joint of aluminium alloy (AA) doubler with AA substrate plate, and (b) bonded joint of AA doubler with AA substrate plate.

The results from ABAQUS show that bonded joints result in lower stress intensity factors than riveted joints for crack lengths higher than 2mm. For the cracks of 1 and 2mm, the stress intensity factors at the crack fronts for the riveted and bonded joints are nearly equal. At a stress of 93.6 MPa, the bonded joints have stress intensity factor of maximum 17.63 MPa.mm$^{0.5}$, whereas for the riveted joints the maximum was 26 MPa.mm$^{0.5}$. It is remarkable the difference between the stress intensity factors given by FRANC3D and ABAQUS. This is because they use different FCG criterion; namely, FRANC3D uses a quasi-static power law to compute FCG, while ABAQUS uses a maximum energy release rate criterion.

4. Conclusions

The fatigue life of AA 2024-T3 plates with different crack lengths was investigated using FRANC3D. In addition, cracked AA plates were repaired with riveted and adhesive bonded joints, and the stress intensity factors for these repairs for various crack lengths and loading conditions were studied with the FEA tool ABAQUS. The conclusions from the numerical analysis of the studied repair joints are:

- Repairs made with riveted joints may arrest or reduce the crack growth in substrates but fastener holes in the substrate have high stress concentrations.
- Repairs made with bonded joints show lower stress intensity factors compared with those made with riveted joints.
- The stress intensity factors at the crack front are nearly equal for the riveted and bonded joint configurations for a crack length of 2mm. This suggests that cracks with length under 2mm cause no significant differences in the behaviour of the stress intensity factor for riveted and bonded joint configurations.
- The stress-intensity factor exceeds the critical stress intensity value for repaired-riveted joint with crack length of 12.7mm at 140 MPa, implies the failure is static.
- Load transfer in bonded joints is smoother and has lower secondary bending.

Further experimental investigations on repairs made with riveted and bonded joints are on-going with the purpose of validating the numerical results.

References

[1] Yan Y, Wen W D, Chang F K and Shyprykevich P 1999 *Compos. Part. A* **30** pp 1215-29

[2] Minguez J M and Vogwell J 2006 *Eng. Fail. Anal.* **13** pp 1410-21

[3] Oskouei R H, Keikhosravy M and Soutis C 2009 *Proc. Inst. Mech. Eng. Part. G J. Aerosp. Eng.* **223** p 863

[4] Chakherlou T N and Vogwell J 2003 *Eng. Fail. Anal.* ***10*** 2003 pp 13-24

[5] Chakherlou T N, Oskouei R H and Vogwell J *Eng. Fail. Anal.* ***15*** 2008 pp 563-74

[6] Chakherlou T N, Mirzajanzadeh M, Abazadeh B and Saeedi k *Eur. J. Mech. A Solid* ***29*** 2010 pp 675-82

[7] Iyer k, Hu S J, Brittman F L, Wang P C, Hayden D B and Marin S P *Fatigue Fract. Eng. Mater. Struct.* **28** 2010 pp 997-1007

[8] Rijck J J M, Homan J J, Schijve J and Benedictus R *Int. J. Fatigue* **29** 2007 pp 2208-18

[9] Skorupa M, Skorupa A, Machneiwicz T and Korbel A *Int. J. Fatigue* **58** 2014 pp 209-17

[10] Esmaeili F, Chakherlou T N and Zehsaz M *Mat. And Design* **59** 2014 pp 430-38

[11] Carter Ralph W, Steven Johnson W, Andrew M and James C *Int. J. Fatigue* **29** 2007 pp 1319-27

[12] Ferjaoui A, Yue T, Abdel Wahab M and Hojjati-Talemi R *Int. J. Fatigue* **73** pp 66-76

[13] Hojjati Talemi R and Abdel Wahab M *Tribology Int.* **60** pp 176-186

[14] Inman M E, Kelly R G and Williard R S *Proc. FAA-NASA symposium on the continued Airworthiness of Aircraft Structures (Virginia, USA)* p 129

IOP Conf. Series: Journal of Physics: Conf. Series **843** (2017) 012035 doi:10.1088/1742-6596/843/1/012035

[15] Wallace W, Hoeppner D W and Kandachar P V *AGARD Corrosion handbook I* 1985 p 278

[16] Pantelakis Sp G, Vassilas N I and Daglaras *Metall* **47** 1993 pp 135-41

[17] Pantelakis Sp G, Daglaras P G and Apostolopoulos Ch Alk *J. Theor. Appl. Fract. Mech* **33** 2000 p 117

[18] Papnikos P and Kermanidis Al Th *Proc. 4th Int. Conf. on new challenges in Mesomechanics (Denmark)* 2002

[19] Smiyan O D, Coval M V and Melekhov R K *Soviet Mater. Sci* **19** 1983 pp 422-30

[20] Charitidou E, Papapolymerou G, Haidemenopooulos G and Hassiotis N *Script. Mater.* **41** 1999 p 41

[21] Petroyiannis P V, Kermandis Al Th, Kamoutsi E, Pantelakis G Sp, Bontizoglou and Haidemenopoulos G V*Fract. Eng. Mater. Struct.* **28** pp 565-574

[22] Baker A A and Jones R *Bonded Repair of Aircraft Structures* 1988 chapter 4-5

[23] Rose L R F *Int. J. Solids Structures* **17** 1981 pp 827-38

[24] Ratwani M M *J. Eng. Mat. And Tech.* **100** 1978 pp 46-51

[25] Banea M D and da Silva L F M *Proc. IMECHE Vol. 223 Part I, J. Mater. Design and appl.* 2009

[26] Baker A *Compos. Struct.* **47** 1999 pp 431-43

[27] Koutsourelakis P S, Kuntiyawichai K and Schuëller G I *Eng. Fract. Mech.* **73** 2006 pp 1202-19

[28] Harman A B and Rider A N *Int. J. Adh. & Adh.* **44** 2013 pp 144-56

[29] Kaiser Aluminium Frabricated Products, USA, *Material properties AA 2024-T3* 2016

[30] Swift T *ASTM STP 842* 1984 pp 69-107

Repeated Impact Method and Devices to Simulate the Impact Fatigue Property of Drillstring

Y H LIN[1], B LI[2], J PAN[2], Q LI,[2] W Y LIU[2], Y PAN[2]

1 Professor,State Key Laboratory of Oil & Gas Reservoir Geology and Exploitation,Southwest Petroleum University,chengdu,China.

2 Lab Assistant, School of materials science and Engineering,Southwest Petroleum University, chengdu, China.

E-mail: yhlin28@163.com; 820634062qq.com

Abstract. It is well known that drillstring failures are a pendent problem in drilling engineering, because of the fatigue accumulation caused by the low amplitude-repeated impact. In order to reveal the effect of low amplitude-repeated impact on the failure mechanism of the drillstring, a repeated impact method and instrument have been developed based on the Charpy impact method, by which a series of tests have been performed in the condition of non-corrosive medium and with H_2S environment respective. Test results of non-corrosive medium environment indicates that, with the increase of single impact energy, the low amplitude-repeated impact resistance of drillstring decreases significantly; For H_2S corrosion environment, the low amplitude-repeated impact resistances with H_2S is much lower than that without H_2S corrosion, and high strength material such as V-150 drillstring is more sensitive to H_2S corrosion media. Furthermore, based on the experiment data, the accumulation fatigue model to predict the service life of the drillstring is developed, which could be used to predict the fatigue life. Research fruits are very vital to select a suitable rotational speed for drilling job and drillstring design.

1. Introduction

Drilling plays an important role in the exploration and exploitation of oil and gas fields all over the world. And drillstring is an important downhole tool in oil drilling operation. Because of the change of formation characteristics, the rotating drillstring will bend in the drilling process alternative, bearing a periodic rotational bending stress[1]. The cyclic load will cause partial damage on the surface of the drillstring, which is not easy to be found. Charpy impact property, which is regarded as the evaluation index of impact load capacity of materials, is commonly used in engineering industry[2]. But in the process of drilling, the material of drillstring will always collide with the borehole wall, which will produce low amplitude-repeated impact load[3]. This kind of load can cause local microcracks, which mainly occur in the parts of threaded connection of drillstring and the parts of local superficial damage caused by collision. The local microcracks will expand and cause fracture failure of drillstring. Therefore, even though the drillstring material conforms to the API 5D standard, it still fails constantly in the oilfield, which brings about huge economic losses to the drilling industry.

Repeated impact resistance is different from a shock resistance, and it also differs from fatigue resistance. It reflects most of the works under impact load in the process of actual service situation, and it is the bridge between them[4-5]. Traditional mechanical testing methods can't describe fatigue failure of drillstring material under repeated impact load. As for the petroleum industry, no mechanical testing method can test the failure of drillstring material caused by repeated impact load in API 5D standard. But for the other area, since the beginning of the 20th century, many scholars have done the repeated impact experiments[6]. In recent years, the theoretical study of low amplitude-repeated impact is expanded gradually by scholars[7]. Although the repeated impact of different materials have been studied, but further systematic and comprehensive experiment has to be developed, especially drillstring material of low amplitude-repeated impact.

As the world's increasing demand for oil and gas resources continues, the number of deep wells and ultra-deep wells grows rapidly, and drilling tool service conditions are more and more severe[8]. Furthermore, hydrogen sulfide(H_2S) often presents in deep wells and ultra-deep wells with complex drilling conditions[9]. It's extremely difficult to develop sour gas reservoir with conventional technology and drilling materials because of the corrosion of H_2S. When the partial pressure of hydrogen sulfide is high, the mechanical properties of the drillstring are obviously decreased. Especially the drillstring material can occur stress-corrosion-cracking and corrosion fatigue fracture with the underground alternating stress[10-12]. The corrosion time in hydrogen sulfide situation is an important factor to destruct drillstring material[13]. Different materials have different sensitivity to H_2S and different rate of crack propagation under external force[14-15]. Loss of drill pipe material often causes serious casualties and property damage. From the existing research situation, people have a rich understanding of the corrosion mechanism of hydrogen sulfide and other media, and put forward valuable conclusions and opinions on corrosion prevention[16-18]. However, the variation law of low amplitude-repeated impact fatigue behavior of drillstring material is not studied deeply under the influence of corrosive medium. Therefore, it is very important for the improvement of the drilling efficiency and drilling safety by studying the variation law of a low amplitude-repeated impact fatigue behavior of the drillstring material under the hydrogen sulfide environment. In this paper, three kinds of commonly used drillstring materials are measured under the low amplitude-repeated impact experiment, and the variation law for low amplitude-repeated impact resistance of typical drillstring material under normal and hydrogen sulfide environment is obtained.

2. Experimental Methods

2.1. Presentation of low amplitude-repeated impact experiment method

Charpy impact energy (a_{kv}) is often seen as the mechanical properties index of resistance to impact load for the material. Charpy impact energy is obtained by the single impact of the sample under the large enough impact load, but downhole drillstring crack propagation is caused by the low amplitude-repeated impact. So the common Charpy impact experiment can not reflect the actual working conditions underground for the drillstring, and cannot be used to evaluate the characteristics of materials under the low amplitude-repeated impact load. Hence, in order to accurately evaluate the damage mechanism of material under the repeated impact load, a testing device to simulate drillstring for actual downhole conditionis developed, and it is based on principles of Charpy impact experiment and depends on the the characteristics of rotary body. The device has the advantages of simple structure, easy maintenance and low cost. The picture of device as shown in Figure 2-1. Furthermore, an experimental specimens with prefabricated notch is designed for the experiment. The notch which has an angle of 120° on the specimen is a fixed gap of the fastening bolt. We can derive the single impact energy (*E*), and record the number of the impacting (*N*). The product of *N* and *E* is the cumulative impact energy (*A*) which can characterize the property at the low amplitude-repeated impact. There are the calculation formulas as type (2-1) and (2-2). The pictures of counter and sensor schematic are shown in figure 2-2 and figure 2-3.

$$E=m(\omega r)^2/2 \quad (2\text{-}1)$$

$$A=NE \quad (2\text{-}2)$$

Fig.2-1 Experimental apparatus for low amplitude-repeated impact fatigue fracture

Fig.2-2 Counter schematic

Fig.2-3 Sensor schematic

2.2. Experimental material

This study used G-105、S-135 drillstring materials which commonly used in deep or ultra deep well, and high strength V-150 drillstring materials, after mechanical processing, wire cutting with prefabricated crack and specifications for specimens of 50mm×10mm×2.5mm, and these samples conform to the requirements of GB and ASTM. The real impact specimen photos as shown in Figure 2-4, which has a 45 degree prefabricated breaking gap and a 120 degree fastening bolt fixed gap, and notch root radius is 0.25mm. In order to record the crack length, the samples surface was polished.

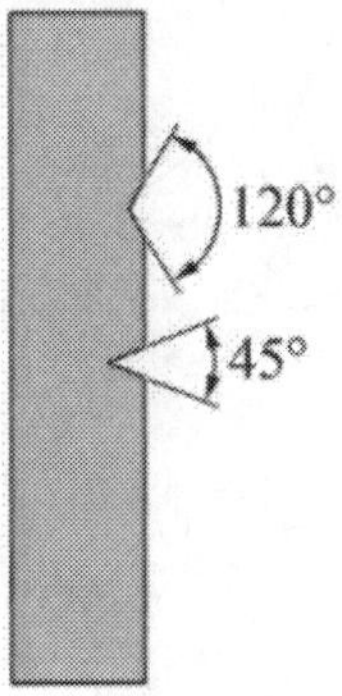

Fig.2-4 Profile and real picture of test blocks

We formulate the corrosion solution based on the standard NACE TM177-96, and this solution will be used in the corrosion immersion test. The formula is 5% NaCl and 0.5% acetic acid dissolved in distilled water, and with saturation of H_2S , whose concentration is about 2219~2337mg/L. Corrosion immersion test was performed in a transparent glass vial. Before the experiment, samples need to be cleaned and surface corrosion products need to be removaled and dried.

3. Rusults and Discussion

3.1. Relation of single impact energy and low amplitude repeated impact resistance of drillstring material under non-corrosive medium environment

In order to deeply study the relationship between single impact energy and dynamic fatigue performance of drillstring materials, the testing with different single energy is carried out. The testing samples are from G-105, S-135 and V-150 three kinds of drillstring steel body. The cumulative impact energy A presents low amplitude-repeated impact resistance. The experimental results are shown in table 3-1.

Table 3-1 The cumulative energy of G-105, S-135andV-150 under different single impact energy

Single impact energy （J）	**2.21**	**2.50**	**3.00**	**3.58**	**3.90**	**4.39**	**4.90**
Cumulative energy of G-105 （J）	475.2	450.2	354.3	225.5	195.0	175.6	166.6
Cumulative energy of S-135 （J）	640.9	627.5	483.1	358.2	315.9	219.5	186.2
Cumulative energy of V-150 （J）	897.3	562.5	274.2	199.4	219.1	109.6	119.5

We mapping the curve of low amplitude-repeated impact resistance under different single impact energy according to the above data, and the curve as shown in figure 3-1,3-2,3-3.

IOP Conf. Series: Journal of Physics: Conf. Series **843** (2017) 012036 doi:10.1088/1742-6596/843/1/012036

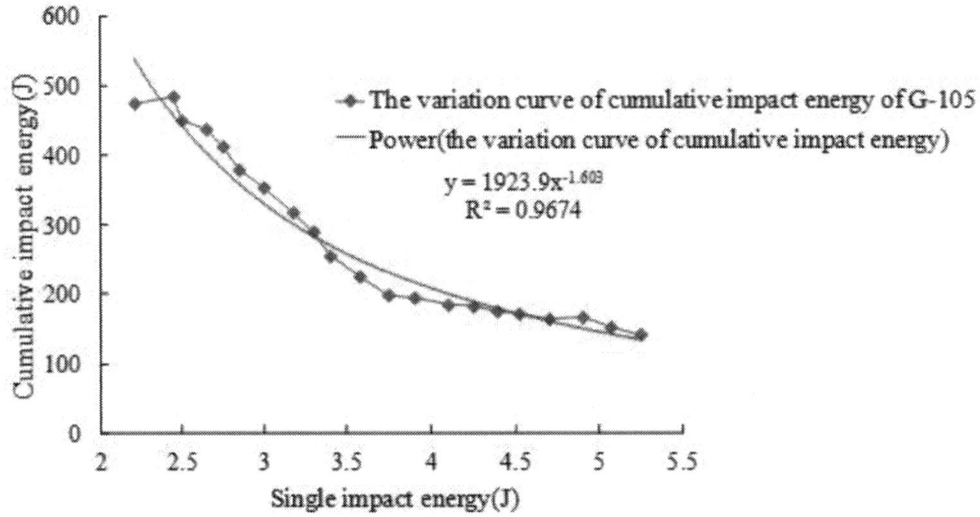

Fig.3-1 Relation of single impact energy and cumulative thrust energy of G-105

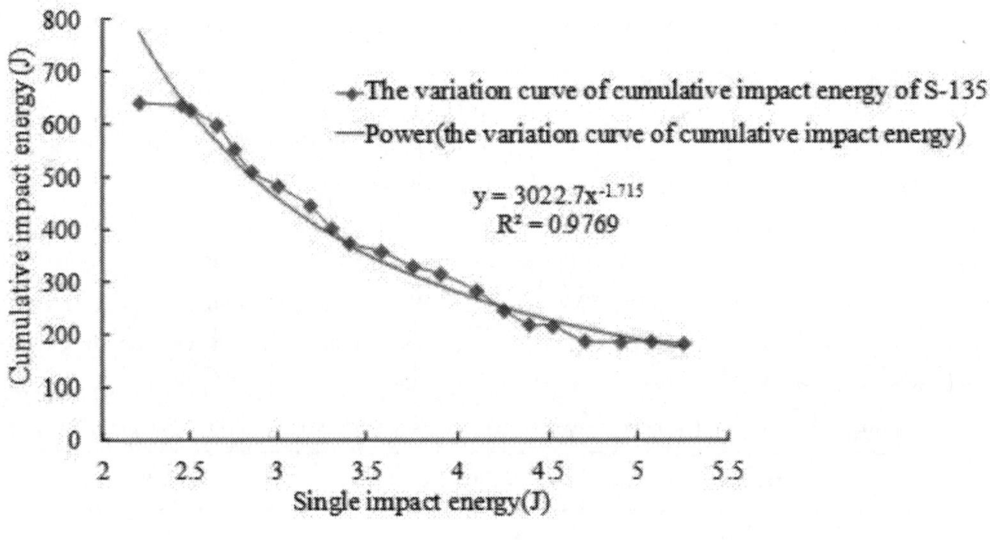

Fig.3-2 Relation of single impact energy and cumulative thrust energy of S-135

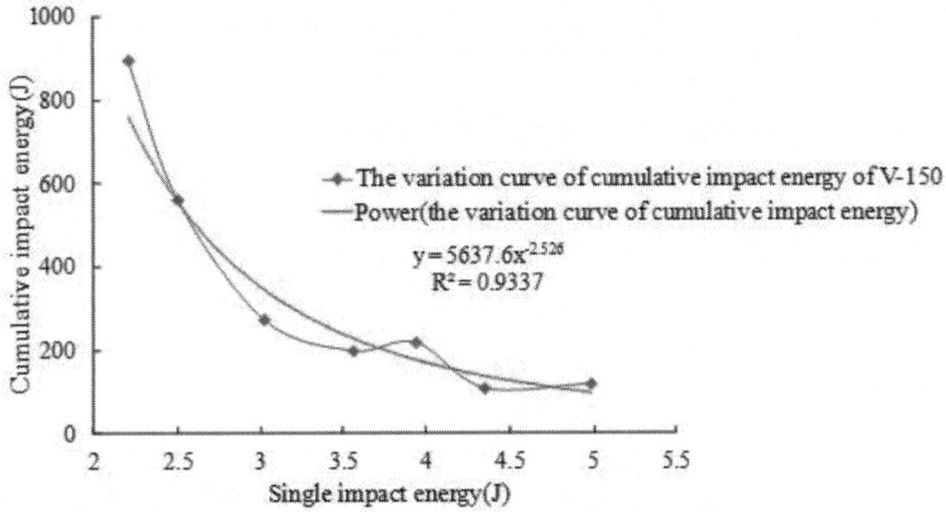

Fig.3-3 Relation of single impact energy and cumulative thrust energy of V-150

It can be seen from the figures, G-105, S-135 and V-150 three kinds of drillstring materials' cumulative energy decreases sharply with the increase of single impact energy. And the cumulative impact energy has a exponential relationship with single impact energy, which is similar to the fatigue "S-N" curve. The results show that the low amplitude-repeated impact resistance of drillstring material decreases obviously with the increase of single impact energy. Therefore, the single impact energy, that is, the rotation speed of the drillstring in the working process has a great influence on the low amplitude-repeated impact resistance. In the actual drilling operation, reasonable drillstring rotation speed should be selected to avoid the early failure of drill string.

The relationship between the low amplitude-repeated impact and the material strength index is established as type (3-1).

$$A = 1092.2e^{0.5376\sigma} E^{(-0.3495\sigma^2 + 0.9365\sigma - 2.19)} \quad (3\text{-}1)$$

3.2. Presentation of low amplitude-repeated impact energy conversion efficiency

Under the condition of low amplitude-repeated impact, the initial damage of the structure will increase slowly. The cumulative impact energy is different under the different single impact energy.

By comparing the Charpy impact energy and the repeated impact cumulative energy, we use the Charpy impact energy to divide by the repeated impact cumulative energy, and the percentage of the ratio is obtained under different single impact energy. We define the energy conversion efficiency is above percentage, which as shown in table 3-2.

Table 3-2 The percentage of Charpy impact energy divided by the repeated impact cumulative energy

Single impact energy *E* (J)		**2.21**	**2.50**	**3.02**	**3.56**	**3.94**	**4.35**	**4.98**
	The Charpy impact energy of G-105 is 21.66J							
G-105	Repeated impact cumulative energy (J)	475.1	450.2	354.3	225.5	195.2	175.6	166.6
	Percentage (%)	4.56	4.81	6.11	9.60	11.11	12.33	13.00

	The Charpy impact energy of G-105 is 20.25J							
S-135	Repeated impact cumulative energy（J）	640.9	627.5	483.1	358.2	315.9	219.5	186.2
	Percentage（%）	3.16	3.23	4.19	5.65	6.41	9.23	10.88
	The Charpy impact energy of G-105 is 18.86J							
V-150	Repeated impact cumulative energy（J）	897.3	562.5	274.2	239.4	219.1	139.6	119.5
	Percentage（%）	2.10	3.35	6.88	7.88	8.61	13.51	15.78

It can be seen from the chart, the Charpy impact energy of drillstring material is less than the low amplitude-repeated impact cumulative energy. It is obvious that just some of the energy of repeated impact load on the specimen do good to crack initiation and propagation process. With the increase of single impact energy, the energy for the crack initiation and crack propagation is higher. Moreover, the impact sensitivity of high strength V-150 drillstring material is obviously improved with the increase of single impact energy. However, in the case of lower impact energy(such as 2.21J), the sensitivity of high grade steel V-150 is significantly lower than that of low grade steel G-105. The sequences of sensitivity is V-150 > S-135 > G-105. Therefore, in the condition of low speed drilling operation, the selection for high grade steel drillstring could have longer service life.

3.3. Effect trend of H_2S on repeated impact resistances of drillstring

Set the rotation speed to 150 r/min, and the low amplitude-repeated impact experiments for different immersion time of G-105, S-135 and V-150 respectively are made, then the accumulated impact energy is obtained. Based on the experimental data, the variation curves of the cumulative impact energy of the drillstring materials with different corrosion time are obtained, and the curves as shown in figure 3-4,3-5,3-6.

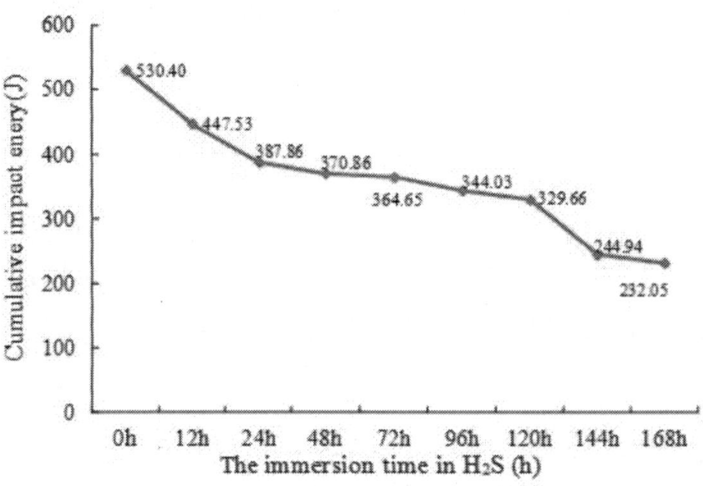

Fig.3-4 Effect of H_2S immersion time on cumulative thrust energy of G-105

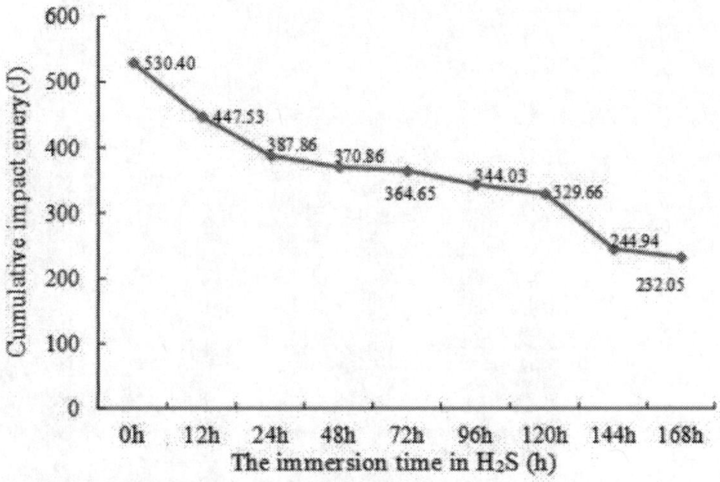

Fig.3-5 Effect of H_2S immersion time on cumulative thrust energy of S-135

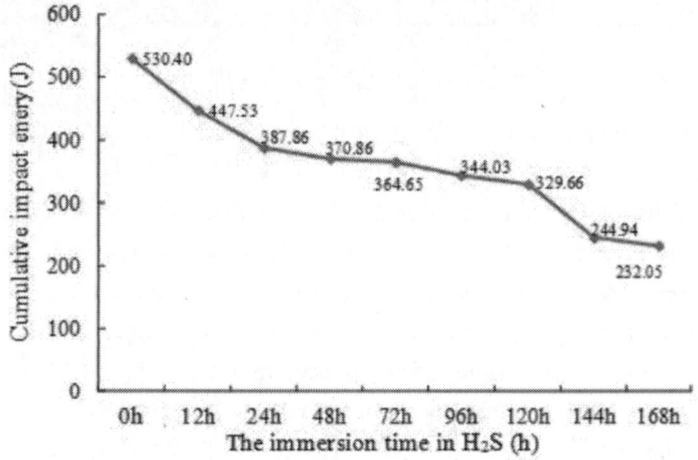

Fig.3-6 Effect of H_2S immersion time on cumulative thrust energy of V-150

It can be seen from the above three diagrams that the low amplitude-repeated impact resistance of the three kinds of drillstring materials with H_2S environment is much lower than that withou H_2S, and the longer corrosion time causes the low amplitude-repeated impact resistance to decrease rapidly.

In order to compare the influence of different corrosion time on the impact resistance of different drillstring materials, the percentage of the low amplitude-repeated impact resistance with different corrosion time of three kinds of materials was calculated, and the chart of comparative analysis as shown in Figure 3-7.

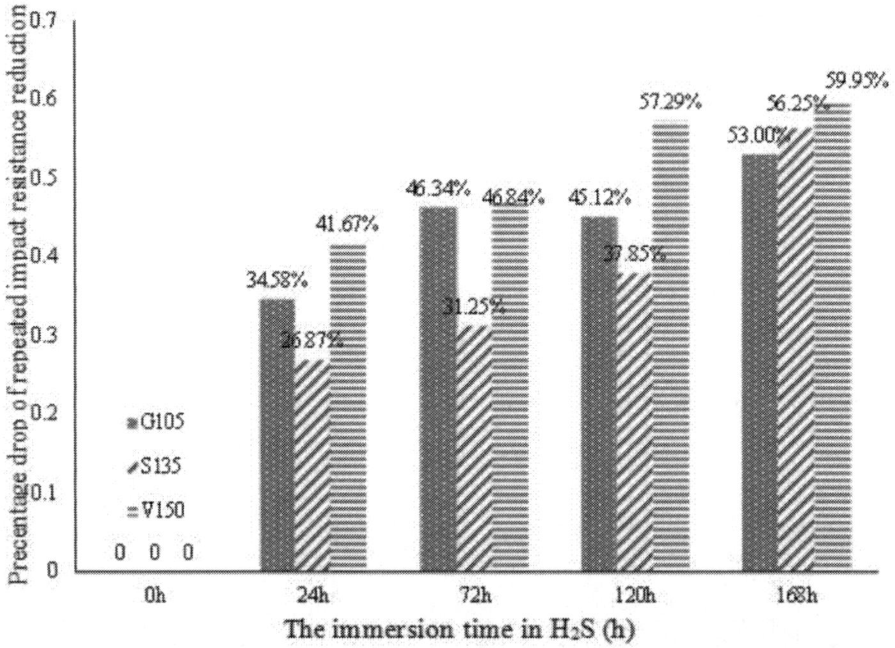

Fig.3-7 Three kinds of drill pipe materials in different hydrogen sulfide corrosion time the percentage of resistance reduction

It is obtained from the experimental data that with the immersion time less than 120h, the decrease in cumulative impact average energy of S-135 is lowest, and the cumulative impact average energy of V-150 drops at the most in the all stage of corrosion. But the low amplitude-repeated impact resistance of three kinds of drillstring material decreased constantly with longer immersion time. When the corrosion time is 168h, the average value of cumulative impact energy of G-105 decreased by 53%, and the average value of cumulative impact energy of S-135 decreased by 56.25%, while the average value of cumulative impact energy of V-150 decreased by 59.95%. It can be seen that the low amplitude-repeated impact resistance of high strength material such as V-150 drillstring decreased significantly, which indicates that it is more sensitive to H_2S corrosion environment. Hence, this type drillstring should avoid contacting with hydrogen sulfide corrosion environment for long time. This low amplitude-repeated impact method is better to reflects the influence of H_2S corrosion environment on drillstring material, because it can describe the failure of drillstring in the underground actual service situation.

3.4. Analysis of fractures

The EDS energy spectrum analysis of V-150 fracture surface after the corrosion time of 12h as shown in Figure 3-8 below.

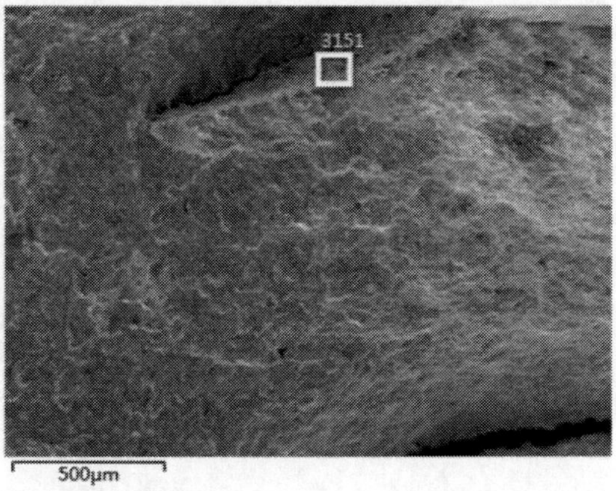

(a) SEM fracture

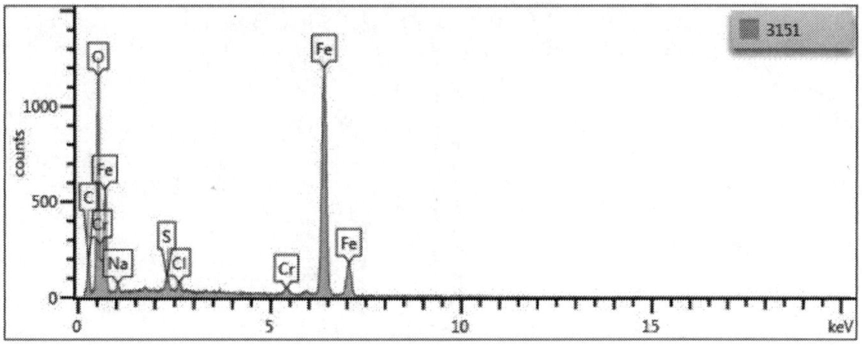

(b) EDS map

Fig.3-8 SEM and EDS diagram of multi punch fracture of V-150 drill pipe sample after hydrogen sulfide soaking 12h.

EDS has detected the Na and Cl elements which have been permeated into the fracture surface. It is indicated that H_2S corrosion environment causes the crack of the metal surface, and the Cl^- can penetrate into the surface of metal material. Then the passive film is ruptured and will develop pitting corrosion, so it will reduce the mechanical properties of metal materials greatly. The results verify the above conclusion which is mentioned, it means that the low amplitude-repeated impact resistance of drillstring material in H_2S environment is far less than that in non-corrosive medium condition.

4. Establishment of cumulative fatigue model

The low amplitude-repeated impact is different from from the Charpy impact. The low amplitude-repeated impact load is a kind of dynamic load. In this paper, the rupture process is divided into two aspects from the angle of damage accumulation and energy. The impact process of specimen is also divided into two parts. It means that the energy consumption is also divided into two parts. One is that the repeated impact energy causes the specimen to accumulate damage and to initiate crack in prefabricated gap; The other one is that the stress concentration causes the crack propagation at the notch root.

Make the model assumptions as follows:

1)The sample is continuous and uniform, and the density is unchanged after deformation;

2)The chopping block is considered as a rigid body, and the plastic deformation of the specimen is negligible afte the collision; The specimen is regarded as an elastic plastic body, which can produce elastic plastic deformation;

3) The impact energy is absorbed completely and the energy loss is negligible;

4) It is linear elastic relation before crack formation at notch root.

The energy of each impact is equivalent to a small rectangular area, the sum of a several small rectangles area is the area of entire curve around, that is the consumed energy (A_n) of the crack initiation and expansion, as shown in figure 4-1. The energy absorption value before crack initiation is A_{kn}, and the energy absorption value after crack initiation is Q, so that the energy calculation method of the whole repeated break process is shown (4-1):

$$A = A_{kn} + Q \quad (4\text{-}1)$$

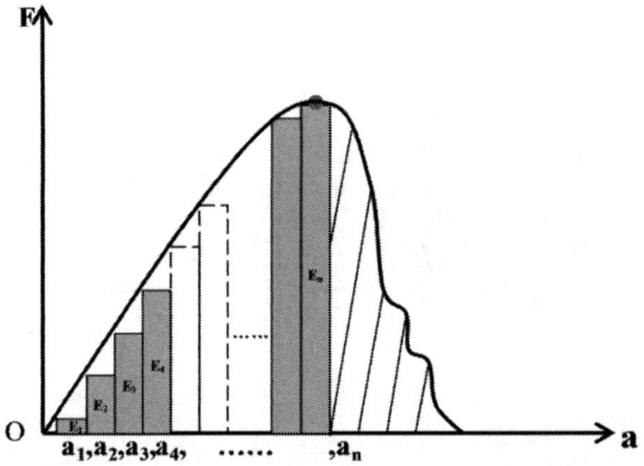

Fig. 4-1 Schematic diagram of energy accumulation before crack initiation

The detailed analysis of the two phase energy equivalence method is as follows:

(1)Under the condition of low amplitude-repeated impact, the crack initiation at the notch requires the stress concentration to cause the root of the notch to yield and to produce a certain plastic deformation. The specimen prefabricated notch root absorbs a certain amount of energy when the specimen fixed on the specimen holder impacts the chopping block each time. The picture of specimen holder as shown in Figure 4-2. When the edge of the specimen impacts the chopping block, it can produce a certain elastic deformation. The specimen restores original shape after knocking, but the gap has been damaged. After several times, the damage has been accumulated to a certain extent, and the specimen was no longer return to its original shape, namely it has develped the plastic deformation. At the same time, the crack begins to start from the notch root initiation and begins to expand, the absorbed energy is the energy A_{kn} in type (4-1), which corresponds to blue small rectangular area in Figure 4-1.

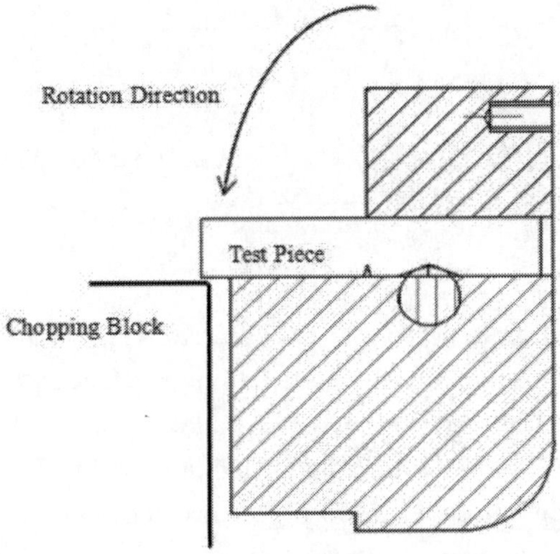

Fig. 4-2 The sample block diagram of percussion

Based on the crack nucleation criterion of Stroh[19,20], the method of crack nucleation and crack initiation is derived:

$$A_{kn} = k\frac{(1-\nu^2)\pi^2 S_c B}{12a_0 E}\rho d\,\sigma_y^2 \tag{4-2}$$

The K value in the formula (4-2) is a constant with respect to the material.

(2)When the tip of the prefabricated gap is yielding, the stress intensity factor at the crack tip reaches the fracture toughness K_{IC}. And the crack at the notch is produced and expand rapidly. K_{IC} can be obtained by formula. The energy causes stress concentration and causes the plastic zone. When the notch plastic zone reaches the yield strength of the material and the cumulative damage satisfy the crack initiation condition, the crack starts initiate from the notch root. To define the initial crack opening angle is θ_0 (it is 45° of prefabricated gap in this paper), the notch crack opening angle change into θ_0, θ_1, θ_2, θ_3, ……, θ_n after the impact load acting on the specimen. The initial crack length is also defined as a_0. With the expansion of the crack, the crack length becomes into a_1, a_2, a_3, ……, a_n.

When the specimen impacts the chopping block, the crack length increses Δa, then the crack length becomes into a_n. The a_n is the initial crack of the next crack expansion. The crack propagation stress can be calculated based on the value of K_{IC}. Also the the total power consumption (Q) of the crack propagation process can be caculated by using the geometric integration principle of the stress and displacement of each impact.

The extended function of K_{IC} calculation method is derived based on the crack, crack under plane strain condition for the total power consumption growth expression:

$$Q=\frac{1}{2}\sum_{1}^{n}\frac{\sqrt{0.17F_m^{\ 2}}}{0.35g_1\left(\frac{a_n}{W}\right)}\cdot\Delta a_n \tag{4-3}$$

To put the formula (4-2) and (4-3) into (4-1), then we can get a complete multi break cumulative energy model.

5. Conlusions

Based on the above mentioned research, the following conclusions can be drawn:

(1) The method and a device to measure the material property under the low amplitude-repeated impact has been perfected, and the three types of material G-105, S-135 and V-150 drillstring materials have been done under non-corrosive medium conditions and H_2S environments. This method is better than the traditional Charpy impact experiment, because it can describe the failure of drillstring in the underground actual service situation.

(2) Under non-corrosive medium conditions, change the single impact energy, with the increase of single impact energy, the low amplitude-repeated impact resistance of three types of drillstring materials decreased significantly. According to the energy conversion efficiency curve, it is concluded that G-105 is more prone to failure than the S-135 and V-150 in low amplitude-repeated impact.

(3) In the constant single impact energy, the experimental results show that the hydrogen sulfide corrosion has a great influence on the G-105, S-135 and V-150, and the influence of V-150 drillstring steel is the highest obviously. Through the SEM observation of the typical fracture surface, the brittle phase caused by hydrogen sulfide immersion corrosion was found at the fracture surface.

(4) The mechanical properties of drillstring material can rely on the traditional analysis testing methods, but the new method to consider the the actual load of the drillstring in the drilling process is the best. Because high tensile strength steel is more sensitive to hydrogen sulfide, low level and wall thickness steel in H_2S environment is suggested to use. Moreover, long time working in the corrosive environment is not permitted to avoid drillstring failure.

References

[1] Hansford J E and Lubinski A 1966 Cumulative fatigue damage of drill pipe in dog-legs *Journal of Petroleum Technology* **18** pp 359-363.

[2] Lin Y H *et al* 2013 A repeated impact method and instrument to evaluate the impact fatigue property of drill-pipe *Journal of Materials Engineering and Performance* **22** pp 1064-1071.

[3] American Petroleum Institute 2001 Drill-pipe standard, Chapter 6 Mechanical performance requirements: API Spec 5D.

[4] Zhou H J and Huang M Z 1962 Study on the steel fracture resistance under repeated impact loads *Journal of Xi'an Jiaotong University* **2** pp 1-19.

[5] Zhou H J, Huang M Z and Zhu J H 1965 The crack development process of steel under repeated impact loads *Journal of Xi'an Jiaotong University* **5** pp 25-35.

[6] Stanton T E and Bairstow L 1908 The resistance of materials to impact *Proc. Inst. Mech. Engrs*. p 889.

[7] Zhou H J, Huang M Z and Zhu J H 1965 The crack development process of steel under repeated impact loads *Journal of Xi'an Jiaotong University* **5** pp 37-40.

IOP Conf. Series: Journal of Physics: Conf. Series **843** (2017) 012036 doi:10.1088/1742-6596/843/1/012036

[8] Zeng Y J and Liu J L 2005 Technical status and developmental trend of drilling techniques in deep and ultra-deep wells *Petroleum drilling techniques* **33** pp 1-5.

[9] Rordrigues R et al 2014 New generation of high strength sour service drill pipe: a breakthrough innovation to address ultradeep and extended-reach drilling challenges combined with H_2S environments// paper SPE-167939-MS presented at the IADC/SPE Drilling Conference and Exhibition. http://dx.doi.org/10.2118/167939-MS.

[10] Qi Y M *et al* 2014 Comparison of tensile and impact behavior of carbon steel in H_2S environments *Materials and Design* **58** pp 234-241.

[11] Wang Xia *et al* 2006 Study on corrosion factors and related prevwntive measures for drilling H_2S-containing gas wells *Natural Gas Industry* **26** pp 80-84.

[12] Huang B S, Lu X and Liu Q Y 2011 Research progress of H_2S-induced corrosion and relevant measure for corrosion prevention of oil drill pipes *Corrosion science and protection technology* **23** pp 205-208.

[13] QI Y M, Luo H Y and Zheng S Q 2013 Effect of immersion time on the hydrogen content and tensile properties of A350LF2 steel exposed to hydrogen sulphide environments *Corrosion Science* **69** pp 164-174.

[14] Huang F, Liu J and Deng Z J 2010 Effect of microstructure and inclusions on hydrogen induced cracking susceptibility and hydrogen trapping efficiency of X120 pipeline steel *Materials Science and Engineering* **527** pp 6997-7001.

[15] Arafina M A and Szpunar J A 2011 Effect of bainitic microstructure on the susceptibility of pipeline steels to hydrogen induced cracking *Materials Science and Engineering* **528** pp 4927-4940.

[16] Xiong J X 2009 Research on H_2S corrosive operation condition for drill pipe in G105 Steel *Steel Pipe* **38** pp 28-32.

[17] Nie C J, Lu S L and Yuan P B 2010 Analysis of fracture cause of S135 drill pipe *Corrosion & Protection* pp 820-822.

[18] Liu Y G *et al* 2009 Research on drill tools failure in hydrogen sulfide environment *Oil Field Equipmet* **38** pp 62-65.

[19] Stroh A N 1954 The formation of cracks as a result of plastic flow *Royal Society of London Proceedings* **223** pp 404-414.

[20] Zhang Y X 1993 A calculation method of crack nucleation energy for Charpy specimen *Physics Examination and Testing* pp 23-26.

Correlation between cavitation erosion resistance and cyclic mechanical properties of different metallic materials

Corinna Kaufhold[1], Fabian Pöhl and Werner Theisen

Chair of Materials Technology, Ruhr University Bochum, 44801 Bochum, Germany

[1]Corresponding author: hardes@wtech.rub.de

Abstract. Machine components in contact with flowing fluids are especially prone to cavitation erosion, where plastic deformation and material loss occur due to the repeated implosion of cavitation bubbles in the vicinity of a solid surface. Identifying a correlation between experimentally derivable material properties and resistance against cavitation erosion could help improve the lifetime of cavitation-affected components. Cavitation erosion is a predominantly fatigue-driven phenomenon. In this investigation, we conducted nanoindentation experiments to examine cyclic micromechanical material properties in response to an increasing number of cycles. The experiments were performed on pure iron and different steel grades, i.e., austenitic stainless CrMnCN steels, interstitially alloyed with carbon and nitrogen. We confirmed the view, also proposed in literature, that indentation hardness is inappropriate for ordering the investigated materials by incubation period or maximum erosion rate. We found that the percentage increase of nanoindentation contact stiffness, after an increasing number of cycles, is a promising indicator in terms of the overall ranking of cavitation erosion resistance among the considered materials. Although a single cavitation impact is associated with a significantly higher strain rate than nanoindentation experiments, it is shown that the plastically deformed area around each indent exhibits indications of deformation, such as the formation of slip lines that are also observable after cavitation-induced impacts.

1. Introduction

Cavitation erosion is a specific form of wear that reduces the quality and lifespan of engineering units and components employed in streaming liquids, e.g., marine propellers or hydraulic machinery. The implosion of a cavitation bubble near a solid structure triggers a succession of hydrodynamic events, including a wall-directed 'liquid-jet' and the subsequent emergence of shockwaves, mechanically impacting the solid surface [1]. The repeated implosion of cavitation bubbles leads to surface spallation and material fatigue [2]. While ductile materials deform plastically before material is removed, brittle materials show less plastic deformation but form fatigue cracks instead [2]. Feller and Kharazzi [3] relate a material's ability to resist cavitation erosion to a number of influencing factors: binding and stacking fault energy, crystal structure and corrosion resistance. One decisive factor is a material's deformation and transformation ability [3], elsewhere described as the materials' capacity to absorb energy [4].

Various prior investigations have discussed the correlation between material properties and resistance against cavitation erosion [3, 5–10]. Important indicators of cavitation erosion resistance include the incubation period (IP) and the maximum erosion rate (MER). Material hardness is also a widely discussed material property in this context [5–8, 10]. Richman and McNaughton [5], working

with cavitation erosion data from various researchers, found no correlation between incubation time and Vickers hardness. They attribute the lack of correlation between statically measured material properties and cavitation erosion resistance to cyclic strain hardening [5]. They identify convincing correlations between material removal rates and cyclically obtained material properties based on cyclic stress strain-curves from different authors [5].
Hattori et al. [11] developed a material database and identified a correlation between hardness and cavitation erosion resistance for carbon steels. Some stainless steels undergo considerable work-hardening, making them an exception to this tendency [11]. They ultimately found a good correlation between the cavitation erosion resistance and material hardness of stainless steels by correcting the hardness values using a material factor, taking work-hardening into account [11].

The current investigation focuses on CrMnCN steels. In these steels, the expensive alloying element nickel is replaced by manganese, which additionally enhances nitrogen solubility and thus allows production under ambient pressure [12]. The combined alloying of C and N improves yield strength, fracture strength and tensile strength [12]. The hardness and degree of work-hardening are likewise increased compared to CrNi austenites [12]. Niederhofer et al. [8] showed that the cavitation erosion resistance of CrMnCN steels is greatly enhanced compared to conventionally alloyed austenitic steel grades. Among their investigated materials, they were unable to identify a correlation between statically determined indentation hardness, H_i, and cavitation erosion resistance, determined by means of an ultrasonic horn according to ASTM G-32. However, nanoindentation was determined to be a more promising approach to correlate the mechanical properties of CN steels with cavitation erosion resistance than tensile experiments on the macroscale [8].

In this study, cyclic nanoindentation experiments were performed to find a possible correlation between mechanical properties of different metallic materials, cyclically determined on the microscale, and the ranking of these materials in terms of cavitation erosion resistance. The following questions are addressed:

- Is cyclic contact stiffness a suitable indicator for ranking the investigated materials in terms of cavitation erosion resistance?
- Does the surface pile-up in the periphery of the residual indentation imprint, after exposure to an increasing number of loading cycles, provide information about resistance against cavitation erosion?
- Does cyclic nanoindentation induce microstructural changes that are comparable to the surface indications occurring in the first stages of cavitation erosion?

2. Materials and Methods

2.1. *Investigated materials and sample preparation*

Pure iron, austenitic stainless steel X2CrNiMo17-12-2 (1.4404), hardened carbon steel C45, and two different grades of austenitic stainless CrMnCN steels are examined. The steels CN0.96 and CN1.07 are named after their total carbon and nitrogen content. Table 1 details the chemical composition of the investigated materials, except for pure iron, measured via spark spectroscopy. The relevant heat treatment parameters are given in Table 2.

Table 1. Chemical composition of the investigated steels in wt%.

Alloy	C	N	Cr	Mn	Mo	Ni	Si	P+S
C45	0.437		0.052	0.761	0.014	0.068	0.245	0.084
1.4404	0.011		16.59	1.592	1.913	9.940	0.381	0.064
CN0.96	0.340	0.610	18.20	18.90	0.060	0.340	0.300	
CN1.07	0.490	0.580	18.80	18.80	0.070	0.400	0.430	

Table 2. Heat treatment parameters of the investigated materials.

Alloy	Condition	Heat treatment	Temperature in °C	Time in min.
C45	As cast	Quench hardened	860	20
1.4404	As cast	Solution annealed	1070	30
CN0.96	Hot worked	Solution annealed	1100	30
CN1.07	Hot worked	Solution annealed	1175	30

In both tables, the values of CN0.96 and CN1.07 are adapted from a previous publication [8]. All austenitic steel grades were solution annealed; however, 1.4404 was quenched in oil, while the CN steels were quenched in water, due to their increased tendency to form carbide or nitride precipitates at lower cooling rates [8, 12]. All samples were mechanically grinded, polished, and finally treated with oxide polishing suspension.

2.2. *Cyclic indentation testing*

Micromechanical tests were performed with a nanoindentation test device (Nanomechanics, iMicro) using a Berkovich diamond tip. One approach to cyclic indentation testing is a multistep method in which multiple test cycles are applied under an increasing load, i.e., to measure the mechanical properties at different depths [13]. Alternatively, cyclic unloading and loading is performed by keeping the maximum load at the same level for each loading cycle [14]. Figure 1 shows the test method used in this study. Force-controlled experiments with a cyclically applied maximum load of 1000 mN are performed. The maximum load was kept in place for one second and fully removed after three seconds at exactly the same surface position. The corresponding load-time evolution of the applied method is schematically outlined in Figure 1.

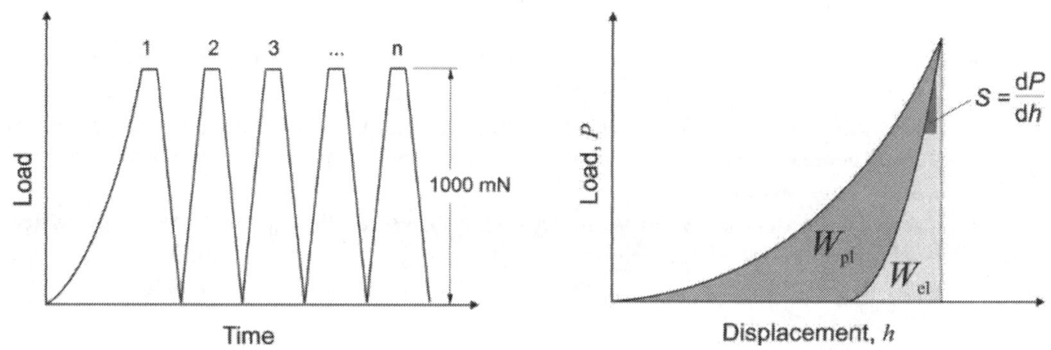

Figure 1. Schematic overview of the cyclic load-time development of the nanoindentation method (left). Simplified illustration (cf. [15]) of a load-displacement curve including the indentation stiffness, S_i, elevated using the method by Oliver and Pharr [15, 16] (right).

In each cycle, the indentation tip penetrates the previously induced indentation imprint and thus reloads the predeformed material surface. The stiffness of the indentation contact, referred to as 'indentation stiffness,' S_i, such as the indentation hardness H_i, is automatically evaluated based on the measured load-displacement curves using the method by Oliver and Pharr [16]. As illustrated in Figure 1, S_i is defined as the slope of the upper part of the unloading curve during the unloading stage under the assumption that only elastic displacements are recovered during unloading. To determine the average percentage increase in indentation stiffness after a certain number of cycles, five indentation experiments up to 250 cycles were performed. The indentation imprints were located at different positions on the sample to average the mechanical properties from differently oriented grains. The static indentation hardness and stiffness were averaged from nine different indentation locations. The induced strain rate was set at 0.2 %/s. Poisson's ratio was defined as 0.3 for all investigated materials.

2.3. Scanning electron microscopy and atomic force microscopy

The nanoindentation experiments were performed *ex situ* and the remaining indetation imprints were subsequently analyzed with a scanning electron microscope (SEM, Tescan Mira) using a working distance of 11.09 mm, an acceleration current of 20 kV, and a magnification of 3.5 kx. To analyze the resulting surface deformation depending on an increasing number of indentation cycles, the indentation experiments were stopped after a certain number of cycles to remove the sample and to install it within the optical devices. Due to experimental restrictions, the subsequent indentation experiments were restarted from a different sample position. This only affected the indentation imprints used for the optical analysis.

The three-dimensional measurement of the indent, including the adjacent surface area, was achieved using atomic force microscopy (AFM, Bruker Nanos). The contact operation mode was used, with a scanning velocity of approx. 40 µm/s. The AFM images were analyzed using the software Gwyddion. The surface topography of the peripheral indentation area was examined along three paths. As shown in Figure 2, paths 1, 2, and 3 traverse the indent from each triangle corner to the middle of the opposite side. The resulting height profile is schematically indicated with the help of the additional image in the upper right corner of the figure. The resulting graphs are mathematically averaged and subsequently shifted along the height-axis, so that the first recorded point of each graph is set as zero level. The graphs are finally transferred along the abscissa, so the height maxima overlap at the same x-position.

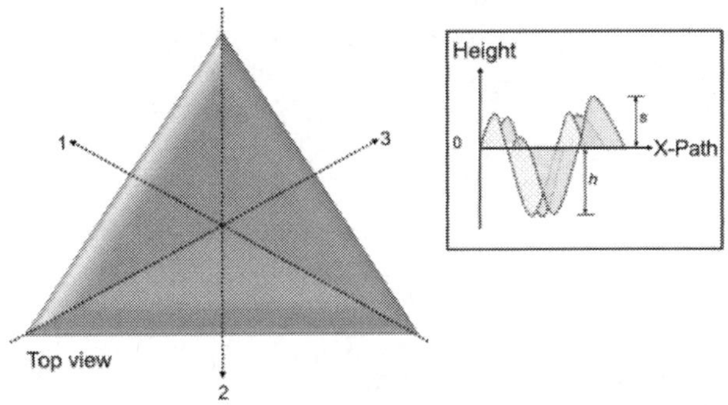

Figure 2. Schematic overview of the positions of the height profile paths traversing the surface indent, accounting for the evaluation of the average pile-up behavior of the considered materials.

To evaluate the quantitative amount of plastic deformation in response to cyclic nanoindentation, we calculated the ratio s/h. The height of the piled-up material around the residual indentation imprint is thereby described by s, while h represents the indentation depth (Figure 2).

3. Results and discussion

Based on cavitation erosion data adapted both from a previous investigation [8] and from literature [3, 5, 11], Table 3 shows a ranking of cavitation erosion resistance between the investigated materials.

It is known from literature that pure iron is less resistant to cavitation erosion than austenitic stainless steels in terms of incubation time and cavitation erosion rate [3]. The cavitation erosion rate is much more pronounced for iron than for all other steel grades investigated in [3]. Due to its high erosion rate and its incubation period of about 100 minutes [3], iron has the lowest cavitation erosion resistance of the investigated materials (see Table 3).

As shown in [8], CN steels exhibit enhanced cavitation erosion resistance compared to conventionally alloyed austenitic steels. Whereas the stainless austenitic steel 1.4301, investigated in [4], has an incubation period of about 152 minutes, CN steels exhibit incubation periods of 30 to 35 hours and 15 times lower erosion rates [4, 8]. CN steels are therefore the most resistant against cavitation erosion of the materials in Table 3.

In spite of their high hardness, very brittle materials show a low cavitation erosion resistance [17]. Quenched C45, in the not-annealed state, was chosen for the cyclic nanoindentation experiments, due to its brittle martensitic microstructure. As the cavitation erosion experiments of a non-stainless martensitic steel might be overshadowed by corrosion influences, stainless martensitic steels are taken as reference material in Table 3. Hattori and Ishikura [11], examining the correlation between cavitation erosion resistance and hardness, showed that austenitic stainless steels exhibit up to twice as much cavitation erosion resistance as carbon steels of the same hardness. All other investigated stainless steels, with no pearlitic content, exhibit lower cavitation erosion resistance in this case. In Table 3 stainless martensitic steel and hence C45 is ranked below the austenitic steels. Further details are discussed later.

Table 3. Investigated materials in decreasing order of cavitation erosion resistance.

Material
CrMnCN-steels
1.4404
Stl. Martensitic steel
Fe

3.1. Mechanical properties derived from cyclic nanoindentation

The indentation hardness, H_i, was derived from static nanoindentation experiments. Figure 3 shows the averaged indentation hardness, measured on various sample positions and differently orientated grains. The martensitically hardened steel C45 is the hardest of the investigated materials, whereas pure iron is the softest. CN0.96 has a higher indentation hardness than CN1.07, which was also investigated in [8]. Figure 3 shows that the statically determined hardness values do not correspond with the ranking of the investigated materials by cavitation erosion resistance. In addition to static hardness, the static indentation stiffness, S_i, and percentage stiffness increase were examined as possible indicators to estimate the ranking of this group of materials in terms of the resistance against cavitation erosion.

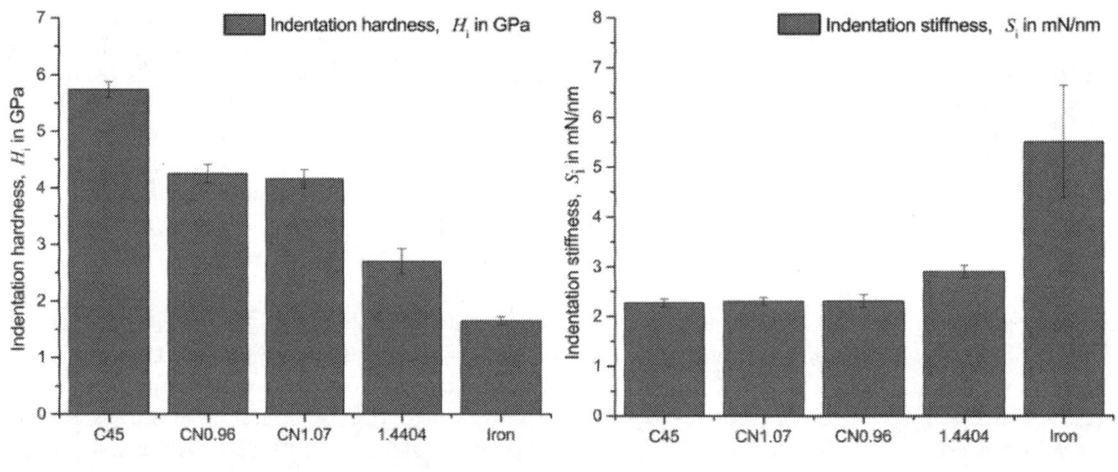

Figure 3. Indentation hardness, H_i, in GPa (left) and indentation stiffness, S_i, in mN/nm (right) averaged from nine indentations at different sample positions.

As described in Section 2.2., S_i represents the slope of the upper unloading part of the load displacement curve. A large S_i value represents a steeply sloping unloading curve and hence a large amount of plastic

deformation. According to Figure 3, CN1.07 and CN0.96 only differ slightly from each other, but differ considerably from 1.4404. Iron shows the largest static indentation stiffness, whereas C45 has the lowest. Within the group of the investigated materials, statically derived indentation stiffness seems to be just as unsuitable for determining cavitation erosion resistance as indentation hardness.
Figure 4 depicts the increase in indentation stiffness, S_i, averaged from five indentation experiments over 250 loading-unloading cycles at the same indentation imprint. An increase in indentation stiffness with an increasing number of cycles means a greater plastic share of the total deformation. CN0.96 has shown a higher hardness and a lower cavitation erosion resistance in a prior investigation [8]. It can be seen from Figure 4 that CN1.07 shows a lower S_i increase than the less cavitation-erosion-resistant CN0.96. The increased amount of work-hardening for CN1.07 is a possible explanation for the smaller S_i increase and the improved CE resistance.

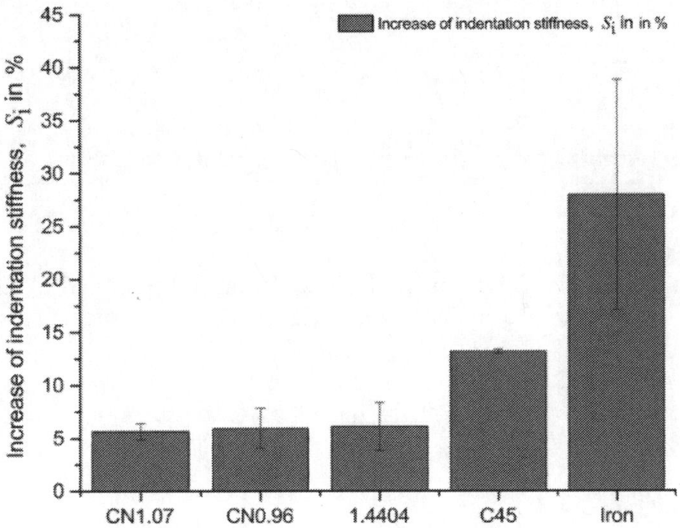

Figure 4. Increase in indentation stiffness, S_i (in %), averaged from five nanoindentation experiments over 250 loading-unloading cycles, located at different sample positions.

Compared to the other investigated austenitic steel grades, the increase in indentation stiffness exhibited by C45 and pure iron is in a completely different range. Hence, the overall ranking of the investigated materials is correct in terms of cavitation erosion resistance. However, the percentage stiffness increase shown by 1.4404 is only slightly greater compared to the CN steels, whereas its cavitation erosion resistance is many times greater, as shown in [8].
To analyze these large differences in cavitation erosion resistance in terms of mechanical properties, the indentation hardness H_i, (see Figure 3) was also taken into account. It is clear that even though the stiffness increases of the three stainless austenitic steel grades fall within a similar range, the initial hardness of the CN steels is about thirty percent greater than that of 1.4404. In this case, the stiffness increase alone is less useful for estimating the ranking of cavitation erosion resistance than the stiffness increase with respect to the indentation hardness, H_i.
Among the considered materials, iron exhibits the largest increase in indentation stiffness. This accords with its significantly lower cavitation erosion resistance compared to the other investigated materials.

3.2 Analysis of surface deformations after cyclic nanoindentation
The amount of piled-up material was examined via atomic force microscopy (AFM). Iron exhibited such extensive depth that the AFM recording was experimentally limited. Hence, iron is excluded from the following results. Figure 5 shows the averaged depth, h, of the remaining indentation imprints and the

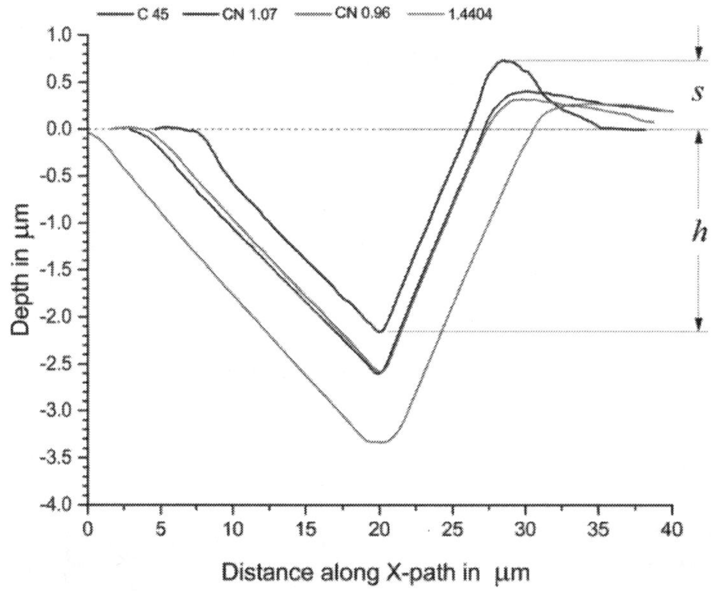

Figure 5. Averaged depth of the remaining indentation imprints and height of the piled-up material in μm, relative to the undeformed surface of the investigated materials, after 250 loading-unloading cycles.

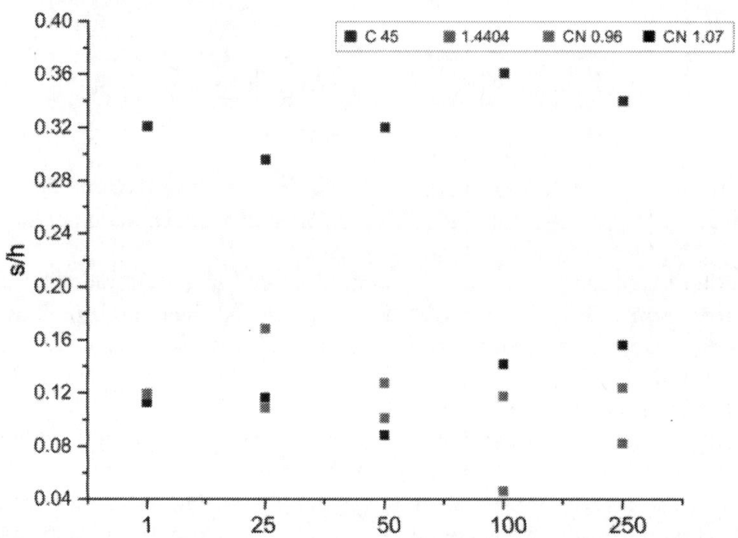

Figure 6. Ratio of pile-up, *s*, and maximum indentation depth, *h*, measured via atomic force microscopy.

amount of piled-up material, *s*, in the periphery of the penetrated surface after 250 cycles. C45 has the smallest indentation depth by a significant margin, and also the highest amount of pile-up.
Figure 6 provides an overview of the calculated *s*/*h* ratios determined from the resulting indentation imprints. The results are shown after 1, 25, 50, 100, and 250 cycles, after which the experiment was stopped to perform the optical experiments, as explained above. The investigated austenitic steels exhibit

a significantly lower *s/h* ratio overall compared to the martensitically hardened C45. Pile-up or sink-in depends on several influencing factors, including work-hardening behavior [18–20].
Yuan et al. determined, based on data by different authors (e.g. [18]), that materials with very low strain hardening coefficients show considerably larger *s/h* ratios than materials with large strain hardening coefficients. This is in approximate accordance with the differences between martensitic C45 and the austenitic steel grades depicted in Figure 5. The investigated austenitic steels reach significantly smaller *s/h* values than martensitic C45, presumably due to their greater work-hardening. No clear difference among the austenitic steel grades is distinguishable based on Figure 6. In addition to work-hardening behavior, the *s/h* ratio also depends on the grain orientation [21]. The strong fluctuations between the *s/h* values of the same material for the different cycles, plotted against the abscissa, clearly shown in Figure 6, might be due to differently oriented grains being penetrated in each cyclic indentation experiment, performed between the optical analyses.
Figure 7 shows the three-dimensional images of the indentation imprints of 1.4404 and C45 after 250 cycles. It is clear from the figure that the resulting imprint of 1.4404 occupies more of the evaluated surface area than C45. The outer edges of the originally triangular imprint are strongly curved for 1.4404, representative of the group of investigated austenitic steels. For 1.4404, the image indicates no clear amount of piled-up material. In contrast, C45 shows a significant amount of piled-up material on each side of the triangular imprint, whereas almost no material has piled-up in the corners. This behavior is also mentioned in [20] for investigations on quenched C45, using a sharp indenter. Pile-up occasionally occurs during nanoindentation experiments. If pile-up emerges, the true contact area is larger than predicted by the method of Oliver and Pharr [16], leading to an overestimate of the indentation hardness [15]. Hence, the hardness results for C45 are slightly overrated, due to the formation of pile-up. However, the overall tendencies described in this study are judged to be unaffected by this.

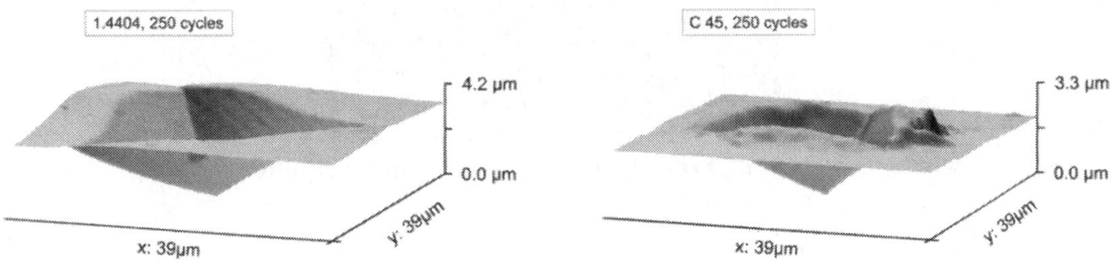

Figure 7. Three-dimensional view of the indentation imprints of 1.4404 and C45 after 250 indentation cycles.

3.3 Microstructural surface changes after cyclic nanoindentation compared to cavitation erosion

Figure 8 shows the nanoindentation imprints resulting from each 1 and 50 loading-unloading cycles. It is clear from the figure that the size of the resulting imprints increases with the number of cycles. Pronounced slip lines occur in response to the penetration of the surface with the sharp indenter. The presence of slip lines is only visible along the sides of the indentation imprint, not along the three edges.
Figure 9 shows the surface of CN1.07 after exposure to cavitation erosion in an ultrasonic horn, adapted from [8]. The strong formation of slip lines is revealed. In addition, pronounced grain boundaries are apparent [8], a typical reaction of homogeneous polycrystalline materials to cavitation impacts [22].
While the formation of slip lines is detectable after both cyclic nanoindentation (Fig. 8) and cavitation erosion (Fig. 9), the material accumulation along the grain boundaries, which is relatable to a locally increased number of dislocations [22], was not apparent within the cyclic nanoindentation experiments.

Hence, the deformation mechanisms after cyclic nanoindentation experiments on up to 250 cycles can only be compared to a certain extent.

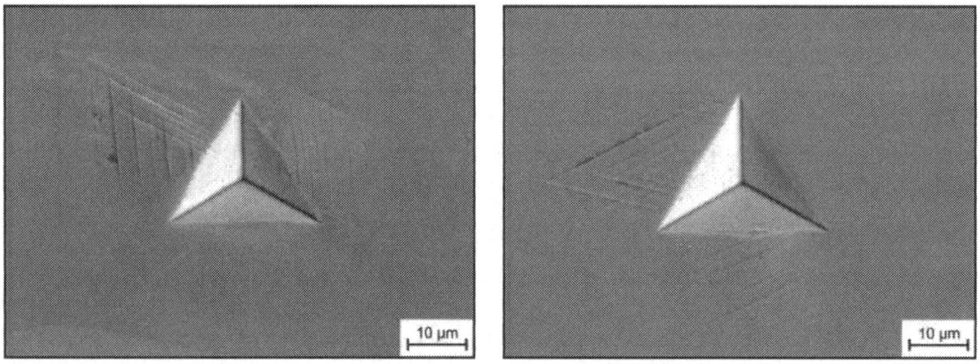

Figure 8. Resulting nanoindentation imprints in CN1.07 after 1 cycle (left) and 50 cycles (right).

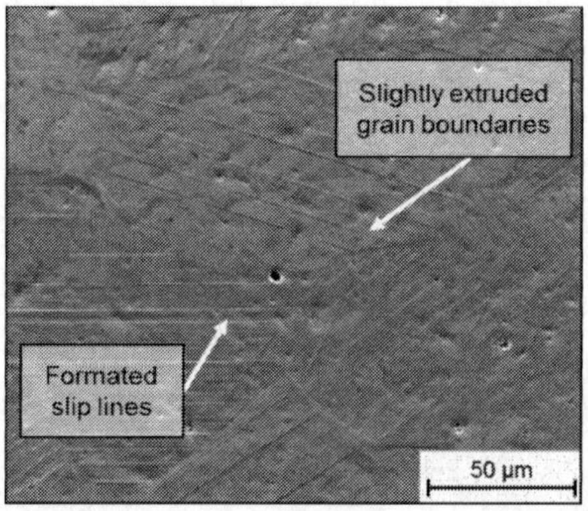

Figure 9. Electropolished surface of a CN1.07 specimen, after exposure to 2000 s of cavitation erosion in an ultrasonig horn, adapted from [8].

The deformation induced by cyclic nanoindentation experiments might vary from cavitation-induced deformations for different reasons. The material deformation resulting from cavitation-induced impacts (e.g. numerically investigated in [23]), depends on several influence factors, including the distance from the bubble center to the solid wall or the ambient pressure of the surrounding fluid [24].
In addition, the impact of cavitation bubbles is often accompanied by high strain rates in an average range of $5 * 10^3\ s^{-1}$ [6], which cannot be achieved during nanoindentation experiments.

4. Conclusions

The present study has shown that:

- the overall ranking of cavitation erosion resistance among the investigated materials generally agrees with the inverse increase of S_i
- the percentage differences between the S_i increases among the investigated materials do not match the differences in cavitation erosion resistance
- for CN1.07 and CN0.96, the percentage increase in S_i is a more appropriate indicator than indentation hardness
- when comparing the CN grades and 1.4404, the percentage increase in S_i seems to be more suitable as a co-factor to indentation hardness than as a standalone indicator

- even if clear differences are visible between C45 and the group of stainless austenitic steels, the *s/h* values of the three investigated austenites are out of order and do not correlate with the order of cavitation erosion resistance. The influence of the grain orientation will be further investigated
- cyclic nanoindentation leads to the formation of slip lines in the CN steels, which is also observable for cavitation erosion

5. Acknowledgements

The authors acknowledge the financial support provided by the Competence Center for Hydraulic Fluid Flow Systems at Ruhr University, Bochum (No: IV B 3-43-02/2005-WFBO-01 1Z) and especially thank U. Föckeler and O. Yilmaz for their involvement in part of the experiments.

References

[1] Philipp A and Lauterborn W 1998 Cavitation erosion by single laser-produced bubbles *J. Fluid Mech* **361** 75–116
[2] Pohl M 1990 Mikroverschleißmechanismen bei der Kavitationserosion *14. Vortragsveranstaltung des Arbeitskreises Rastermikroskopie in der Materialprüfung* ed DVM 25–36
[3] Feller H G and Kharrazi Y 1984 Cavitation erosion of metals and alloys *Wear* **93** 249–260
[4] Stella J 2004 Metallkundliche Vorgänge in der Inkubationsphase der Kavitationserosion *Ph.D. Thesis* Ruhr Universität Bochum
[5] Richman R and McNaughton W 1990 Correlation of cavitation erosion behaviour with mechanical properties of metals *Wear* **140** 63–82
[6] Kim K H, Chahine G, Franc J P and Karimi A (eds) 2014 *Advanced experimental and numerical techniques for cavitation erosion prediction* (Springer)
[7] Pohl M, Stella J and Hessing C 2003 Comparative study on CuZnAl and CuMnZnAlNiFe shape memory alloys subjected to cavitation erosion *Adv. Eng. Mater* **5** 251–256
[8] Niederhofer P and Huth S 2013 Cavitation erosion resistance of high interstitial CrMnCN austenitic stainless steels *Wear* **301** 457–466
[9] Stella J, Pohl M, Bock C and Kunze U 2014 Influence of grain orientation on the local deformation mode induced by cavitation erosion in a CuSnNi alloy *Wear* **316** 1–5
[10] Duraiselvam M, Galun R, Wesling V, Mordike B L, Reiter R and Oligmüller J 2006 Cavitation erosion resistance of AISI 420 martensitic stainless steel laser-clad with nickel aluminide intermetallic composites and matrix composites with TiC reinforcement *Surf Coat Tech* **201** 1289–1295
[11] Hattori S and Ishikura R 2010 Revision of cavitation erosion database and analysis of stainless steel data *Wear* **269** 109–116
[12] Berns H, Riedner S and Tyshchenko A I Stainless Austenitic CrMnCN Steels of Superior Strength Part I. Alloy Design and Properties *Materials Science Forum* **539-543** 4956–4961
[13] Shuman D, Costa A L M and Andrade M S 2007 Calculating the elastic modulus from nanoindentation and microindentation reload curves *Materials Characterization* **58** 380–389
[14] Nêmeček J 2009 Creep effects in nanoindentation of hydrated phases of cement pastes *Materials Characterization* **60** 1028–1034
[15] Oliver W C and Pharr G M 2004 Measurement of hardness and elastic modulus by instrumented indentation: Advances in understanding and refinements to methodology *J Mater Res* **19** 3–20
[16] Oliver W C and Pharr G M 1992 An improved technique for determing hardness and elastic modulus using load and displacement sensing indentation experiments *J Mater Res* **7** 1564–1583

[17] Gülich J F 2014 *Centrifugal Pumps* 3rd Edition (Springer Berlin Heidelberg)
[18] Taljat B and Pharr G M 2004 Development of pile-up during spherical indentation of elastic-plastic solids *Int J Solids Struct* **41** 3891–3904
[19] Yuan F, Jiang P., Xie J and Wu X 2012 Analysis of spherical indentation of materials with plastically graded surface layer *International Journal of Solids and Structures* **49** 527–536
[20] Zhu L N, Xu B S, Wang H D and Wang C B 2013 Pileup behavior in sharp nanoindentation of AISI 1045 steel *Phys Procedia* **50** 214–218
[21] Renner E, Gaillard Y, Richard F, Amiot F and Delobelle P 2016 Sensitivity of the residual topography to single crystal plasticity parameters in Berkovich nanoindentation on FCC nickel *Int J Plasticity* **77** 118–140
[22] Pohl M 1990 *Werkstoffverhalten bei Kavitation* Tagungsband der Tribologie Fachtagung der Gesellschaft für Tribologie e.V. "Verschleiß-Technik" 7/1
[23] Pöhl F, Mottyll S, Skoda R and Huth S 2015 Evaluation of cavitation-induced pressure loads applied to material surfaces by finite-element-assisted pit analysis and numerical investigation of the elasto-plastic deformation of metallic materials *Wear* **330-331** 618–628
[24] Lauer E, Hu X Y, Hickel S and Adams N A 2012 Numerical modelling and investigation of symmetric and asymmetric cavitation bubble dynamics *Comput Fluids* **69** 1–19

IOP Conf. Series: Journal of Physics: Conf. Series **843** (2017) 012038 doi:10.1088/1742-6596/843/1/012038

Fatigue assessment of an existing steel bridge by finite element modelling and field measurements

J Kwad[1,2(*)], G Alencar[3], J Correia[3], A Jesus[3], R Calçada[3], P Kripakaran[1]

[1]University of Exeter, College of Engineering, Mathematics and Physical Sciences Vibration Engineering Section, North Park Road, EX4 4QF,+44 (0)1392 72 6421, Exeter, United Kingdom

[2]Civil Engineering Department, University of Anbar, Ramadi, Anbar, Iraq, 31001.

[3]CONSTRUCT-LESE, Faculty of Engineering (FEUP), University of Porto, Rua Dr. Roberto Frias, s/n, 4200-465, Porto, Portugal.

[(*)]E-mail: jeek201@exeter.ac.uk.

Abstract. The evaluation of fatigue life of structural details in metallic bridges is a major challenge for bridge engineers. A reliable and cost-effective approach is essential to ensure appropriate maintenance and management of these structures. Typically, local stresses predicted by a finite element model of the bridge are employed to assess the fatigue life of fatigue-prone details. This paper illustrates an approach for fatigue assessment based on measured data for a connection in an old bascule steel bridge located in Exeter (UK). A finite element model is first developed from the design information. The finite element model of the bridge is calibrated using measured responses from an ambient vibration test. The stress time histories are calculated through dynamic analysis of the updated finite element model. Stress cycles are computed through the rainflow counting algorithm, and the fatigue prone details are evaluated using the standard SN curves approach and the Miner's rule. Results show that the proposed approach can estimate the fatigue damage of a fatigue prone detail in a structure using measured strain data.

1. INTRODUCTION

During the lifetime of metallic bridges, traffic loads can induce fatigue issues in its structural details. For these structures there are strong motivations to evaluate not only the remaining fatigue life, but also to develop a reliable and cost-effective approach to ensure appropriate maintenance and management. Relating the evaluated fatigue life to stress ranges is a challenging task. The stress range spectra for bridges are strongly site-specific depending upon vehicle types, range of vehicle speeds, road roughness conditions, ambient environment and bridge type [1]. Field measurements are normally used to consider some of these uncertainties. However, in situ monitoring of all the details of a structure is not feasible in practice. An effective Finite Element model (FEM) is therefore normally deployed in the fatigue analysis. Nevertheless, for a numerical model to accurately reproduce the real response of a bridge, it must be calibrated and updated based on experimental data, gathered in in situ experimental campaigns. [2] asserted that an effective FEM is usually calibrated using field measurements such as static and dynamic load tests to update parameters of bridge models. According to Friswell & Mottershead as cited in [3] a Finite Element Model calibrated or updated using modal properties identified through

IOP Conf. Series: Journal of Physics: Conf. Series **843** (2017) 012038 doi:10.1088/1742-6596/843/1/012038

system identification provides are the best tool for understanding and physical interpretation of the causes and effects of the various loading regimes.

This paper, presents the dynamic behavior of numerical model for the case of a bridge of for the passage of roadways traffic, the Bascule Bridge, which is located at the A379 across the Exeter Canal of the UK roadways. The results of a preliminary experimental campaign are also presented, which considered an ambient vibration data obtained from real traffic. The results of the experimental work were used for the updating and validation of the numerical model of the bridge. Finally, some of the results of the fatigue damage analysis performed on the structure were presented, according to the BS NA EN 1993 1-9 [4].

2. Case study

A roadway bridge, the Bascule Exeter Bridge, is used as a case study for fatigue evaluation. The fatigue critical connections examined in this study were identified by a preliminary fatigue assessment based on the nominal stress method.

2.1. Bascule Bridge

The Bascule Bridge was built in 1972. It takes the A379 across the Exeter Canal. It is designed to be opened to allow shipping to pass up and down the canal. The bridge may not often open for water-borne, but they are kept busy with vehicles passing over it along the A379. Figure 1 shows overview of the bridge (right) and bridge location.

(a) (b)

Figure 1. (a) Bascule Bridge Location and (b) Bridge overview

2.2. Geometry

The Bascule Bridge is a single span structure with a light-weight aluminium deck supported on cross beams which are in turn supported by two steel longitudinal girders (W36x12 sections) that are simply supported as illustrated in Figure 2.

IOP Conf. Series: Journal of Physics: Conf. Series **843** (2017) 012038 doi:10.1088/1742-6596/843/1/012038

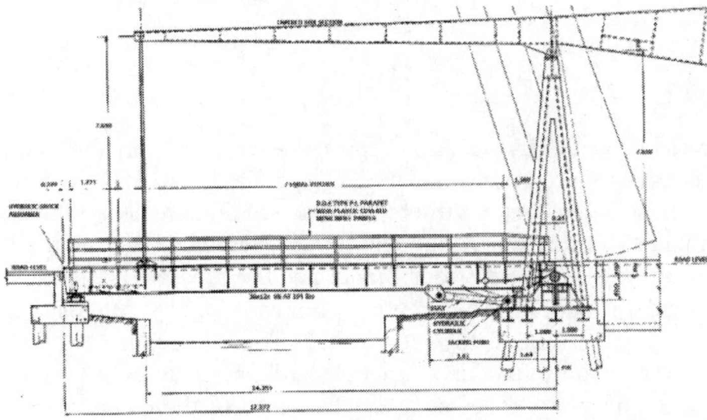

Figure 2. Side view of the bridge "Devon County Council"

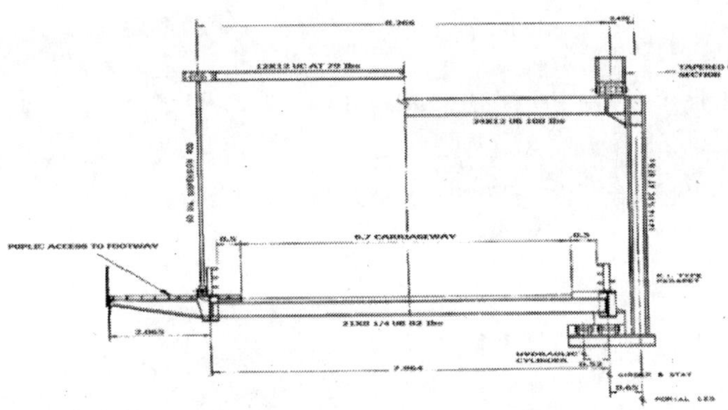

Figure 3: Typical cross section of the bridge "Devon County Council"

The lifting section is 17.28m long with a carriageway width of 6.7m and a footway on the outside edge of the parapet of 2m. The width of the bridge is 8.116 m, a distance spanned by a series of transverse cross beams (W21x8.25 sections), that are spaced 0.97 m apart as showed in Figure 3.The deck of this bridge consists of two universal beams with cross beam supporting an aluminium deck planks Figure 3. The aluminium planks are held down onto steel cross beams with bolts Fig. 4. The cross beams are bolted to the web stiffeners on the main beam.

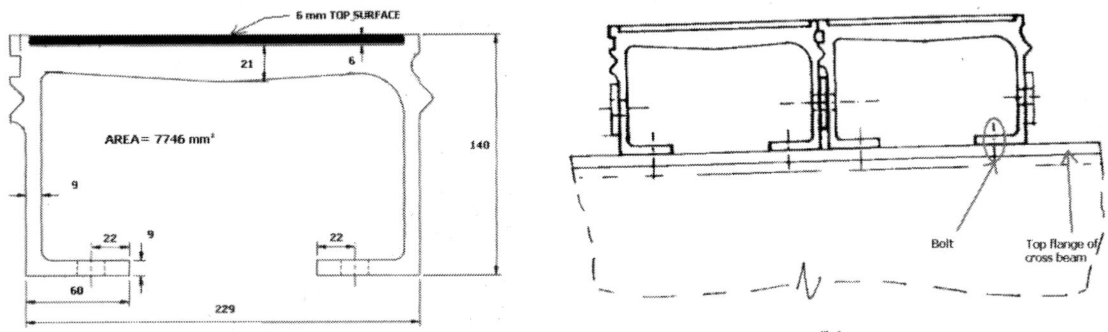

(a) (b).

Figure 4. (a) Deck plank cross-section and (b) Deck plank and beam top flange connection "Devon County Council"

IOP Conf. Series: Journal of Physics: Conf. Series **843** (2017) 012038 doi:10.1088/1742-6596/843/1/012038

3. The numerical model

In this section, the finite element (FE) which is representative of a common number of short-span, riveted and welded, roadway bridges around the UK. Three-dimensional 3D 8-noded shell elements and were used to model the cross beams, main girders, stiffeners. Solid element 3D 8-noded were used to model equivalent aluminium deck of the bridge and MASS21 to simulate the mass of non-structural elements. Figure 5a,b show the complete bridge model and the model without the deck, respectively. Beam elements were used to model the hydraulic cylinders (Figure 5c), that together with other modelling strategies revealed essential to get right calibrated frequencies. The welded-riveted connections were assumed to be fully fixed in a finite element perspective, i.e. within the FE model this is achieved by tying all the members together at the locations of the connections fixed as shown in Figure 5d.

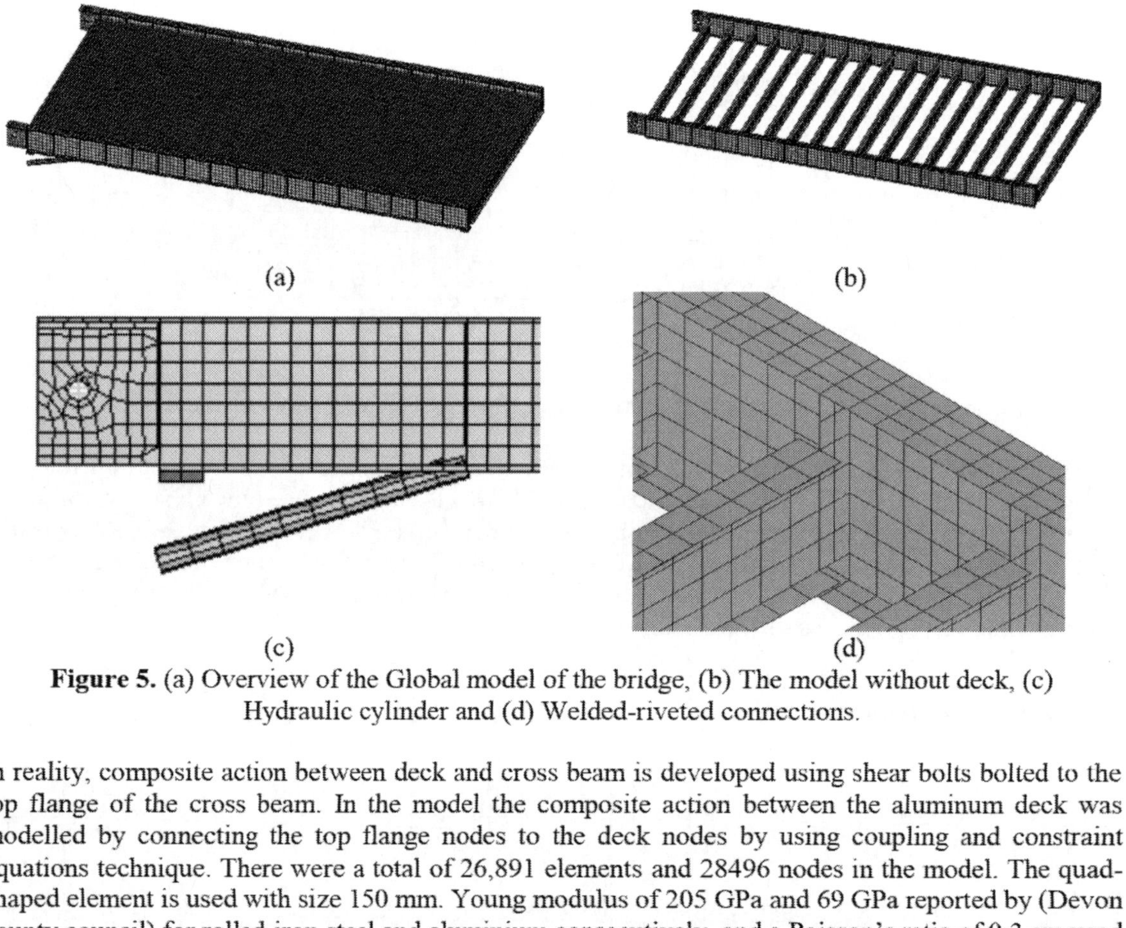

(a) (b) (c) (d)

Figure 5. (a) Overview of the Global model of the bridge, (b) The model without deck, (c) Hydraulic cylinder and (d) Welded-riveted connections.

In reality, composite action between deck and cross beam is developed using shear bolts bolted to the top flange of the cross beam. In the model the composite action between the aluminum deck was modelled by connecting the top flange nodes to the deck nodes by using coupling and constraint equations technique. There were a total of 26,891 elements and 28496 nodes in the model. The quad-shaped element is used with size 150 mm. Young modulus of 205 GPa and 69 GPa reported by (Devon county council) for rolled-iron steel and aluminium consecutively, and a Poisson's ratio of 0.3 are used for the FE analyses. The commercial finite element code ANSYS® was used to perform numerical analyses. The ANSYS® parametric design language (APDL) was used to build the model of the bridge.

IOP Conf. Series: Journal of Physics: Conf. Series **843** (2017) 012038 doi:10.1088/1742-6596/843/1/012038

4. Experimental characterization of the structure

4.1. Modal identification and numerical modal analysis

In order to update the numerical model of the bridge, a preliminary experimental modal identification was performed by two different accelerometers setups which involved an ambient vibration test and a dynamic test for the passage of a truck used in load test. Regarding the first setup, the ambient structural response was measured during period of 30 minutes at 4 points along the one of the main beam of the bridge. ARTeMIS Modal 5.1 with time domain modal analysis from accelerations data using these techniques: Stochastic Subspace Identification (SSI) with Unweighted Principal Component (SSI-UPC) was used for modal identification.

A global mode was identified from ambient vibration test that was a first global bending mode only due to a limitation with ambient test was just for one side of the bridge. The second accelerometers setup which has 4 points were distributed on both side of the bridge at quarter and mid span during period of 20 minutes. A first global bending and torsion modes were identified. Figure 7 shows stabilization diagrams associated to the application of these two accelerometers setup. The red line is "stable mode", the other (yellow) is a "noise mode", and both have been identified on accelerometers setups.

On the other hand, a modal analysis for the calibrated bridge model was performed using ANSYS. Only 100 modes were considered in dynamic analysis to predict the bridge performance. A first global bending and torsion modes were also identified from the numerical model. The numerical analysis has showed that both global bending and torsion modes are responsible for almost all the stress range in the main girders of the bridge. The asymmetric behaviour observed in the first bending mode is due to the difference of mass included more next to this lane due to existence of a footway. Figure 7 show a comparison between the experimental and numerical identified modes with 2% of error.

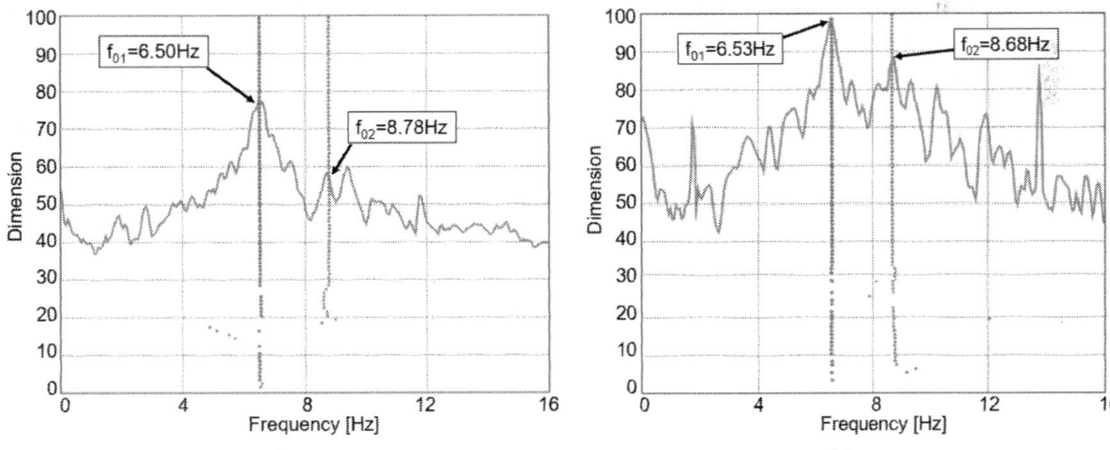

a) b)

Figure 6. (a) Stabilization diagram of the accelerometers setup one. The red line is "stable mode" and the other (yellow) is a "noise mode and (b) Accelerometers setup two. The red line is "stable mode" and the other (yellow) is a "noise mode"

$f_{01exp} = 6.53$ Hz $f_{01num} = 6.50$ Hz

(a)

$f_{02exp} = 8.68$ Hz $f_{02num} = 8.87$ Hz

(b)

Figure 7. (a) Comparison between 1st numerical and experimental vibration mode shape and (b) Comparison between 2nd numerical and experimental vibration mode shape.

4.2. Strain analysis

Weldable strain gauge installed within 7 cm of the exterior edge of the bottom flange as shown in Figure 8a,b and Figure 9a. To avoid any notch effect induced by the welded connection near to the weld toe, the sensor was offset a distance of 40 mm to the left of the cross beam. A controlled load test was conducted on the bridge for collecting data to calibrate the finite element models of the bridge. A truck used in the load test, with weighed axle, described in next section and shown in Figure . During the test the speed of the truck was measured equal to almost 17 km/h. The bridge was closed during the test for local traffic. Bending strains were measured from the live-load response at 2000 Hz sample rate. Using the FE models, the field test run using the truck was simulated. To simulate the truck pass, the two wheel loads for truck were applied as nodal forces time steps on the deck nodes at the dimensions shown in next section. At Figure b strains were multiplied by $E = 2.05\times10^{11}$ to get the stress-history response in terms of Megapascals for the truck crossing. The results would indicate agreement between the FEM and the field test results, presenting a good agreement between numerical and experimental. Comparable maximum stresses would indicate agreement between the FE model and the field test results.

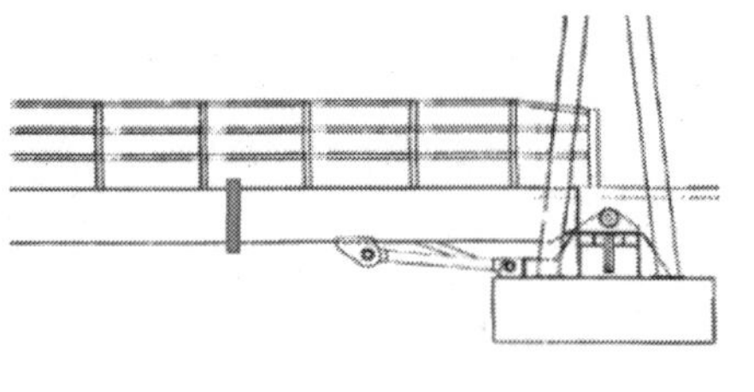

(a)

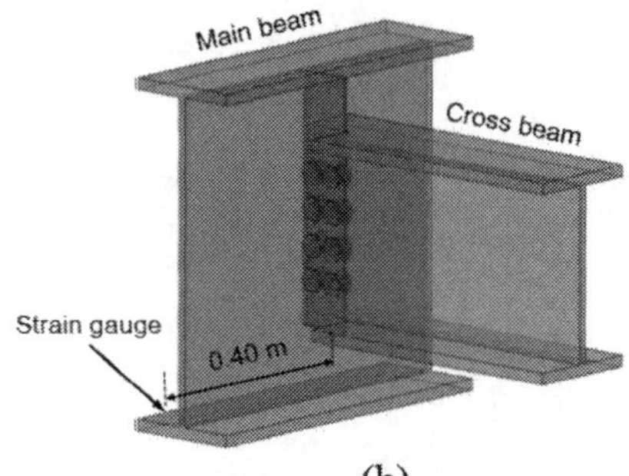

(b)

Figure 8. (a) Schematic of strain gauge Section and (b) Strain gauge at main beam location.

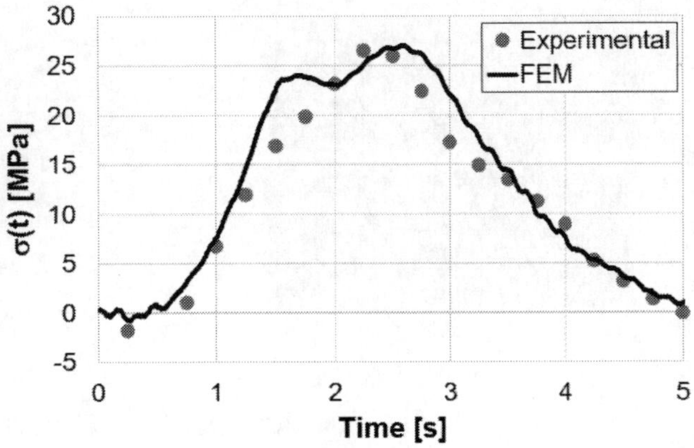

(a) (b)

Figure 9. (a) Weldable strain gauge installed to bottom flanges and (b) FEM vs Experimental stress comparison.

5. Fatigue life evaluation

5.1. Identification and classification of fatigue prone details

An identification and classification of bridge's fatigue prone detail connections is presented first to calculate fatigue damage using SN method. The design information based on the Eurocode 3 [5] was used for this classification. Among the details, where are not considered in the present study, the most vulnerable position with respect to fatigue would be the fillet weld between the vertical stiffener welded to the web of the left main girder of the bridge. Figure shows the detail which is classified as class "80". In the current paper, the fatigue damage assessment for the Bascule Bridge was limited only to this detail at an initial stage of this research work.

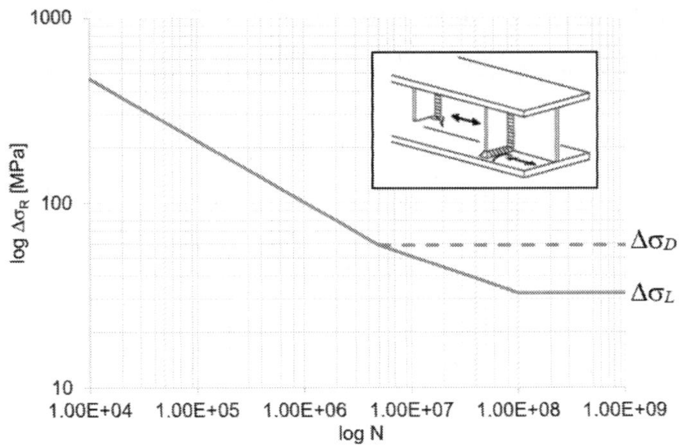

(a) (b)

Figure 9: (a) SN characteristic curve Detail 80 ($\Delta\sigma_c = 80$ MPa) and (b) Potential prone fatigue joint.

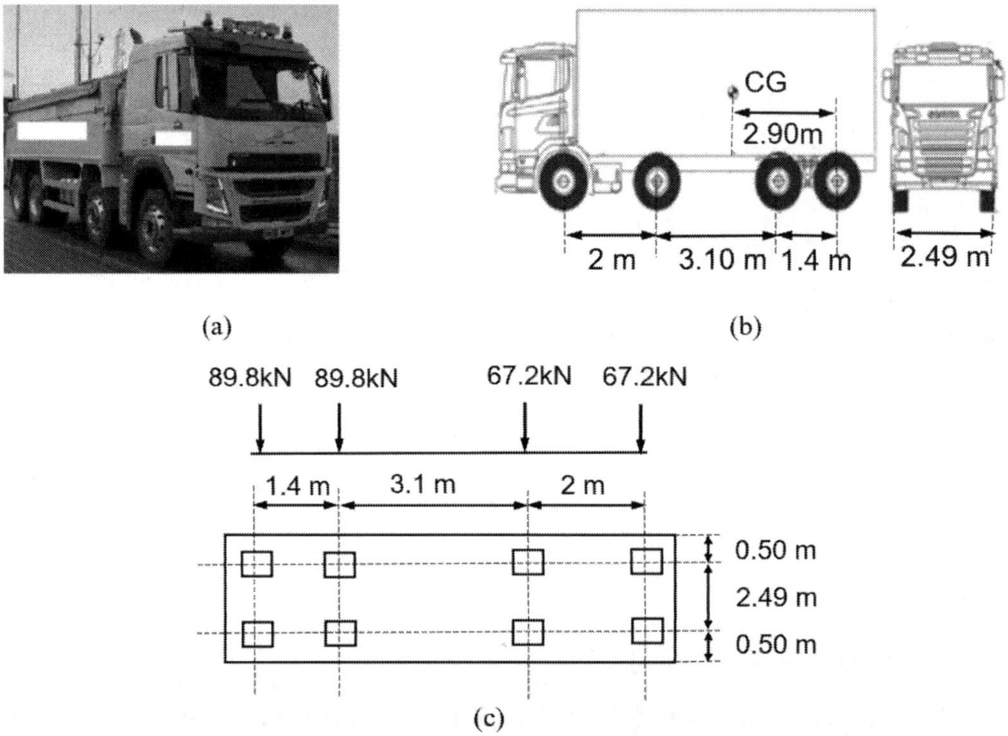

Figure 10. (a) A picture of the truck used in the test load, (b) Truck scheme dimensions and (c) Axle configurations and axle loads for trucks.

5.2. Fatigue damage calculation based on design standard

In this section, an initial estimation of the fatigue damage of the aforementioned detail is presented and the research will continue studying and investigating more details on the bridge. The damage linear accumulation method, described in Eurocode 3 [5] was used to analysis the critical detail. In order to extract stress history responses of the critical detail of the bridge, a design loads based on NA to BS EN 1991-2-3 [6] were used to represent the service period from 1973 up to the present.

The most common practices for the assessment of fatigue strength of Civil Engineering structures are currently available in the Eurocodes. In general, Eurocode 3 [5] allows 3 different methods for fatigue analysis of steel structures: (a) the simplified method of equivalent constant amplitude stress range, (b) the linear damage accumulation method (Annex A) and (c) the hot-spot stress approach (Annex B). At a preliminary stage, the damage accumulation method is used in this work, and its application in the context of steel and steel-concrete roadway bridges involves the following steps Nussbaumer et al. as cited [7].

1. definition of the traffic scenario (vehicle types, annual traffic volume);
2. calculation of the stress history in potential critical structural details, including the dynamic effects on structural response;
3. calculation of stress histograms, representing the number of cycles versus the corresponding distribution of stress ranges, by using a cycle counting algorithm, e.g. the rainflow algorithm [8];

IOP Conf. Series: Journal of Physics: Conf. Series **843** (2017) 012038 doi:10.1088/1742-6596/843/1/012038

4. adoption of suitable SN curves to describe the fatigue resistance of the detail; the curve proposed by the Eurocode 3 [5] are identified; moreover, as stated by NA BS EN 1993-1-9 [4], the fatigue strength reduction factor, γ_{Mf}, should be taken as 1.1;
5. computation of the fatigue damage, by using a linear damage accumulation model, as proposed by Miner[8], according to which the damage factor, D, is given by Equation (1), where n_i is the number of applied load cycles for a given stress range and N_i denotes the number of resisting load cycles for a given stress range. Fatigue failure is reached for D = 1.

$$D = \frac{n_1}{N_1} + \frac{n_2}{N_2} + \frac{n_3}{N_3} + ... = \sum_{i=1}^{k} \frac{n_i}{N_i} \leq 1.0 \tag{1}$$

5.2.1. *Stress ranges calculation*

Although the estimation of the fatigue damage obtained based Eurocode 3 [5], there are assumptions adopted based on the UK specific National Annex NA BS EN 1991-2-3 [6]. The bridge was loaded in both lanes with the Fatigue Load Model 4 Lorries as described in NA BS EN 1991-2-3 [6] and shown in Table 1. These lanes are the two notional lanes that individually cause the most theoretical fatigue damage in the component under consideration. Vehicle numbers in these lanes were obtained from Table NA4 in NA BS EN 1991-2-3 [6]. An equal annual frequency of the vehicles was assumed to represent the period between 1973, opening time, to the 2017. Fatigue damaging stress cycles in the investigated detail were calculated from two traffic lanes based on NA BS EN 1991-2-3 [6] and counted using the rainflow algorithm [9]. The calibrated FE model of the bridge was used to extract the fatigue stress time history.

Table 1. Set of equivalent lorries for Fatigue Load Model 4 NA to BS EN 1991-2-3 [6].

Chassis type	Average Spacing (m)	Loading Group	Total Weight kN	Axle Loads kN
Girder trailer and 2 tractors	1.5 4.5 4 1.5 [No.5] 13.5 1.5 [No.5] 4 1.54.5	H	3680	80 160 160 240(6no.) 240(6no.) 80 160 160
		M	1520	80 160 160 60(6no.) 6(6no.) 80 160 160
Girder trailer and tractor	4.5 1.5 4.0 1.5 [No.5]	H	1610	70 140 140 210 210 210 210 210 210
		M	750	50 110 110 80 80 80 80 80 80
Girder trailer and 2 tractors	4.5 1.5 4.0 2.0 10.0 2.0	H	1310	70 140 140 240 240 240 240
		M	680	60 130 130 90 90 90 90
Articulated	3.0 1.5 9.5 1.51.51.5	H	790	70 100 100 130 130 130 130
Articulated	3.0 1.5 9.5 1.5	H2	630	70 130 130 150 150
Articulated	3.5 5.5 1.5 1.5	H	380	70 100 70 70 70
		M	300	50 70 60 60 60
		L	190	40 60 30 30 30
Articulated	3.0 6.5 1.5	H	240	40 80 60 60
		M	175	40 55 40 40
		L	145	35 50 30 30
Rigid	1.5 3.5 1.5	H	280	50 50 90 90
		M	240	40 40 80 80
		L	120	20 20 40 40
Rigid	4.0 1.5	H	240	60 90 90
		M	195	45 75 75
		L	120	60 45 45
Rigid	4.0	H	135	50 85
		M	65	30 35
		L	30	15 15

Figure 11 illustrates the different stress history responses of the investigated detail of the bridge to the passage of the 10 heaviest standard fatigue lorries. Dynamic analyses using the Fatigue Vehicles with a crowd speed of 17 km/h were performed on the basis of the dynamic superposition method using ANSYS and Matlab, after a parametric analysis varying speed between 10 and 120 km/h had shown relatively low influence over the stress ranges. In this work, dynamic analyses were performed using a damping ratio value of 2.45% for all vibration modes, due to the fact that this was the damping ratio measured from the ambient test for the 1st vibration mode.

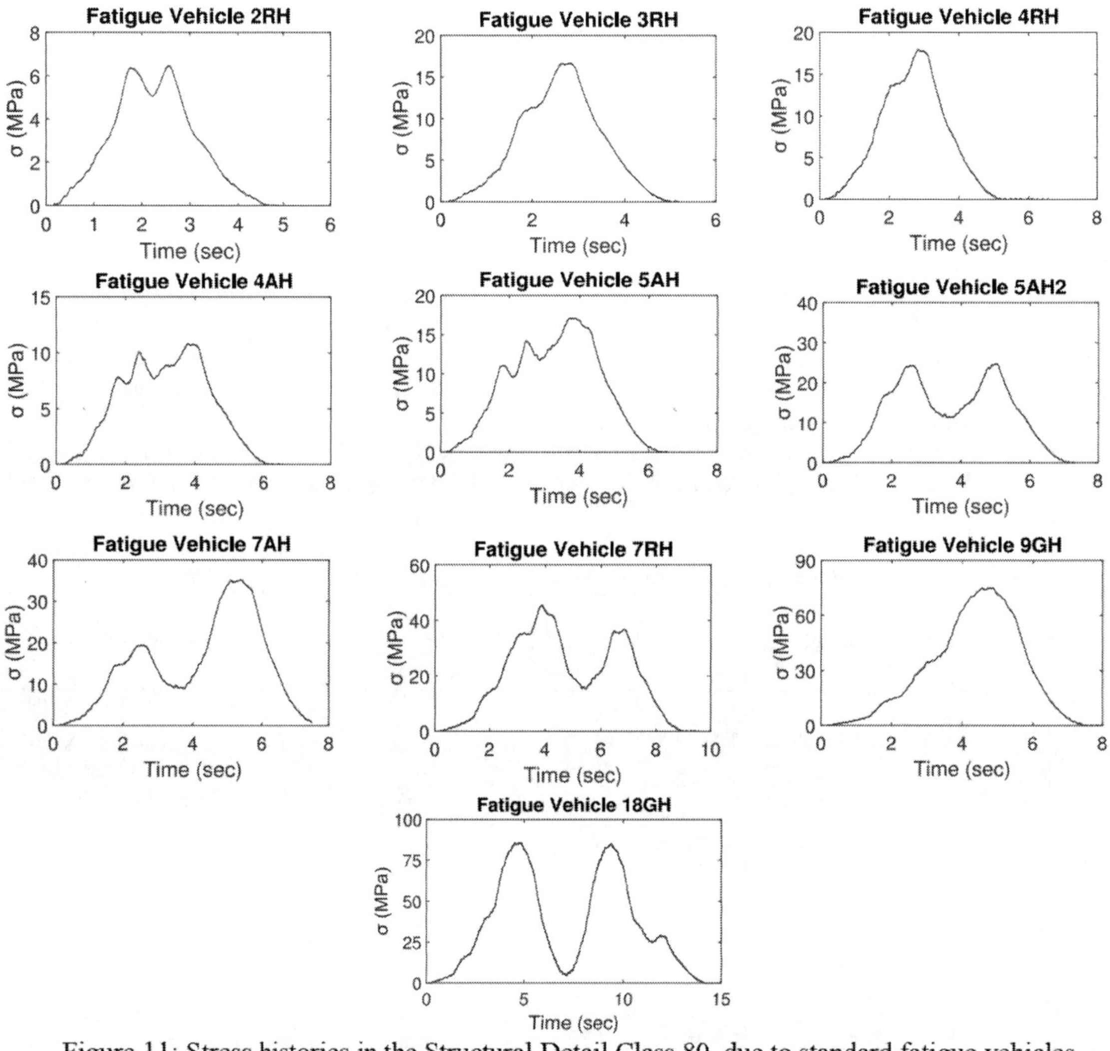

Figure 11: Stress histories in the Structural Detail Class 80, due to standard fatigue vehicles.

It can be seen that the behaviour was axle-dominated and that the stresses vary between 6.5 and 86 MPa, In this case the damage was vary and the most conditioning vehicle was the vehicle 15. Total damage summation, D_d, is obtained by adding contributions from the Lane 1 and Lane 2 traffic as given in Equation (2).

$$D_d = D_1 + D_2 \quad (2)$$

IOP Conf. Series: Journal of Physics: Conf. Series **843** (2017) 012038 doi:10.1088/1742-6596/843/1/012038

where D_1, D_2 ate the damages from Lane 1 and Lane 2 respectively. The effect of side-by-side running also was allowed as recommended in NA to BS EN 1991-2-3 [6] and given by Equation (3).

$$D_d = K_b \cdot Z \tag{3}$$

where: K_b = ratio of the maximum stress range caused by single vehicles in lane 2 to the maximum stress range caused by single vehicles in Lane 1, and Z varies linearly in proportion to the logarithm of the loaded length from 1.0 to 1.5 for loaded lengths between 3.0 m and 20.0 m. In this case study the *Kb* and Z were calculated as 1 and 1.448, respectively. Figure 12 shows the total damage for each fatigue vehicle types that were considered in this study and the total annual damage for all vehicles was 2.428×10^{-4}, considering 1 million vehicles per year per lane.

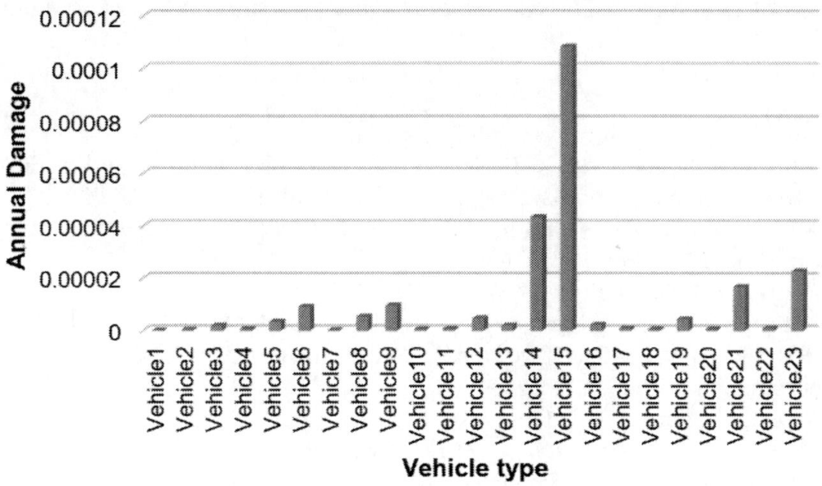

Figure 12: Fatigue damage in the critical detail, due to standard fatigue vehicles.

However, this damage was for 2 million vehicles (1 million per lane per day). As the bridge is located at the A379 roadway, classified as all-purpose type, has a single traffic categories and two lanes only. From Table NA4 BS NA EN 1991-2 [6] number of heavy goods vehicles expected per year was used 0.5 million per lane. Therefore, the total annual damage become 1.214×10^{-4}. Finally, the total damage for the critical for a 42-year period is 5.1×10^{-3}.

6. Conclusions

The steel Bascule Exeter Bridge is under study concerning its fatigue behaviour. In this kind of analysis, the finite element model of the structure is a very important tool. In the present work a finite element mode of the bridge was calibrated through measured responses for ambient vibration test and updated using output-only modal techniques. Strain measurement recorded from dynamic testing of the bridge was used to crosscheck results. A good agreement between experimental strains and numerical were achieved, this was possible due to a calibration of the first two global modes with less than 2%, that for strains measured in the main girders were responsible for great part of the numerical response computed using the dynamic mode superposition method. Analysis of the fatigue damage was conducted using code-specified loadings. Among all the stress range from the Fatigue Vehicles the higher stress range is obtained for the Standard Fatigue Vehicle 23. Moreover, in this case study, the damage in the structural detail 80 was vary and the most conditioning the Vehicle 15. In general, low damage values were found.

This study is currently under way in order to further validate these results and obtain better estimates of fatigue life. Future work will also focus on assessing other potential fatigue prone details combining both methodologies to update and calibrate a model: (1) static analysis versus strain measurements from static load test (2) dynamic analysis (modal and transient) versus acceleration measurements. In addition, study using traffic survey is under way to predict fatigue damage at fatigue prone details of the bridge. Finally, a local model, which will allow better assessment of local stresses, is currently being developed for investigating model uncertainties such as a mesh sensitivity and lifting rods.

Acknowledgements
The first author would like to acknowledge the financial support of The Higher Committee for Education Development in Iraq (HCED IRAQ) scholarship reference D11000218, and the second author would like to acknowledge the final support of the National Council for Scientific and Technological Development, CNPq, Brazil, scholarship reference 203662/2014-8. The authors would also like to thank the Bridges and Structures team at Devon County Council for sharing data and providing access to the Bascule Bridge.

References

[1] Laman B J A and Members A S N 1996 FATIGUE-LOAD MODELS FOR GIRDER BRIDGES By Jeffrey A. Laman 1 and Andrzej S. Nowak/ Members, ASeE **122** 726–33
[2] Goulet J-A, Kripakaran P and Smith I F C 2010 Multimodel structural performance monitoring *J. Struct. Eng.* **136** 1309–18
[3] Rahbari R, Niu J, Brownjohn J M W and Koo K Y 2015 Structural identification of Humber Bridge for performance prognosis *Smart Struct. Syst.* **15** 665–82
[4] BS NA EN 1993-1-9 2005 UK National Annex to Eurocode 3: Design of Steel Structures: Fatigue strength of steel structures.
[5] EN 1993-1-9 2005 Eurocode 3: Design of steel structures - Part 1-9: Fatigue Fatigue strength of steel structures, *European Comitte for Standardization,* Brussels.
[6] BS NA EN 1991-2 2003 (English) UK National Annex to Eurocode 1. Actions on structures.Traffic loads on bridges
[7] Alencar G 2016 Fatigue Assessment and Dynamic Monitoring of Railway Bridges
[8] Miner M 1945 Miner_Cumulative Damage in Fatigue.pdf *J. Appl. Mech.* **12** A159–64
[9] ASTM E1049 - 85 2005 Standard Practices for Cycle Counting in Fatigue Analysis 1 *ASTM Int. West Conshohocken, PA.* **85** 1–10

Fatigue crack propagation behavior and debris formation in Ti-6Al-4V alloys with different grain size

H J Kim[1, 2], **J Nakahigashi**[3], **R Ebara**[2] **and M Endo**[1, 2]

[1] Department of Mechanical Engineering, Fukuoka University
8-19-1, Nanakuma, Jonan-ku, Fukuoka, 814-0180, Japan
[2] Institute of Materials Science and Technology, Fukuoka University
8-19-1, Nanakuma Jonan-ku, Fukuoka, 814-0180, Japan
[3] Department of Mechanical System Engineering, Fukuyama University
1-3, Gakuen, Fukuyama, 729-0292, Japan

E-mail: hjkim@fukuoka-u.ac.jp

Abstract. Titanium alloy is widely used in applications where high specific strength as well as good heat and corrosion resistance is required. Consequently, there are a number of studies on the fatigue characteristics of titanium alloys. In recent years, grain refinement for metallic materials processed by several methods, such as severe plastic deformation, has been studied to improve the mechanical properties. Grain refinement of titanium alloy by the protium treatment is a new technology, and the fatigue properties of this material have yet to be sufficiently studied. Therefore in this study, tension-compression fatigue tests were conducted for a protium treated Ti-6Al-4V alloy with ultra-fine grains of 0.5 μm in average size as well as for an untreated alloy with conventional grains of 6 μm. Specimens had shallow, sharp notches with the depth of 50 μm and the root radius of 10 μm, which enabled successive observation of the initiation and early propagation behaviors of small fatigue cracks. Substantial amount of oxide debris was formed along the crack during crack propagation. The role of debris was discussed in association with propagation resistance.

1. Introduction

Use of titanium alloy as a structural material spreads in wide area of industry, having attractive properties such as high specific strength and good heat and chemical resistances. The fatigue properties of titanium alloys have thus been studied with great effort. Structural refinement of materials including titanium alloy has attracted attention since improvement of mechanical property is attained in many cases. To produce the fine-grained materials, the severe plastic deformation (SPD) processes such as the equal channel angular pressing (ECAP), the hot isostatic pressure (HIP) and the high pressure torsion (HPT) are commonly used. It is reported that fine-grained materials obtained by the SPD for Al alloy [1, 2], Mg alloy [3, 4] and Ti alloy [5] show an increase in monotonic and fatigue strength compared to the unprocessed materials with conventional-size grains. Another process with the hydrogen storage material for refinement is recently developed, which has mainly been applied to Mg-based and Ti-based alloys. One of the authors [6-7] developed a new method for refinement of Ti-6Al-4V alloy for medical use by means of the protium treatment. They reported that the ultra-fine grained Ti-6Al-4V alloy produced by this method, which had grains with an average grain size of 0.5 μm, had an enhanced 0.2% yield strength and exhibited a superplastic elongation of over 9000% at 1123K [6]. However, the grain refinement of titanium alloy by means of the protium treatment is a new technology, and the fatigue properties of this material are not well understood.

It is well known that oxide debris is formed due to the relative sliding of two material surfaces. There are thus many researches on debris formation in wear and fretting fatigue. For titanium alloy, there are reports that oxide debris was observed in sliding wear [8] and fretting fatigue tests [9]. Moreover, it was reported that oxide debris was also observed at a fatigue crack that propagated in shallow-notched Ti-6Al-4V alloy specimens subjected to tension-compression ($R = -1$) fatigue loading [10]. However, the process of oxide debris formation and its role in determining the fatigue properties have yet to be discussed in detail.

The objective of this study is to understand the fatigue crack propagation behavior of a titanium alloy, Ti-6Al-4V, refined to an ultra-fine grain level by the newly-developed protium treatment. The tension-compression ($R = -1$) fatigue tests of specimens with sharp, shallow notches were conducted for an ultra-fine grained alloy (0.5 μm in average grain size) as well as a conventional alloy (6 μm in average grain size). The behaviors of initiation and early propagation of small fatigue cracks at notch root were successively observed by the plastic replica technique. The focus of this study was also on the effect of oxide debris on the crack propagation properties.

2. Experimental procedures

The base material investigated was an $\alpha + \beta$ type Ti-6Al-4V alloy. The chemical composition is listed in Table 1. Ultra-fine grained Ti-6Al-4V alloy was obtained by the protium treatment described in the literature [6-7, 11]. The Ti-6Al-4V alloy with grains of conventional, regular size was obtained by the same process by using argon gas, instead of hydrogen gas. Figure 1 shows the microstructures of these Ti-6Al-4V alloys. The average grain sizes of the conventional and the ultra-fine grained Ti-6Al-4V alloys were 6 and 0.5 μm, respectively. For alloys with regular-sized and ultra-fine grains, the tensile strengths: 1026 MPa and 1266 MPa; the yield strengths: 955 MPa and 1125 MPa; the percent elongations: 21 % and 9.0 %; and the Vickers hardness values measured under a 9.8 N load: 314 and 351, respectively.

Table 1. Chemical composition (mass %)

Al	V	Fe	O	N	C	H
6.15	3.18	0.16	0.14	0.01	0.01	0.001

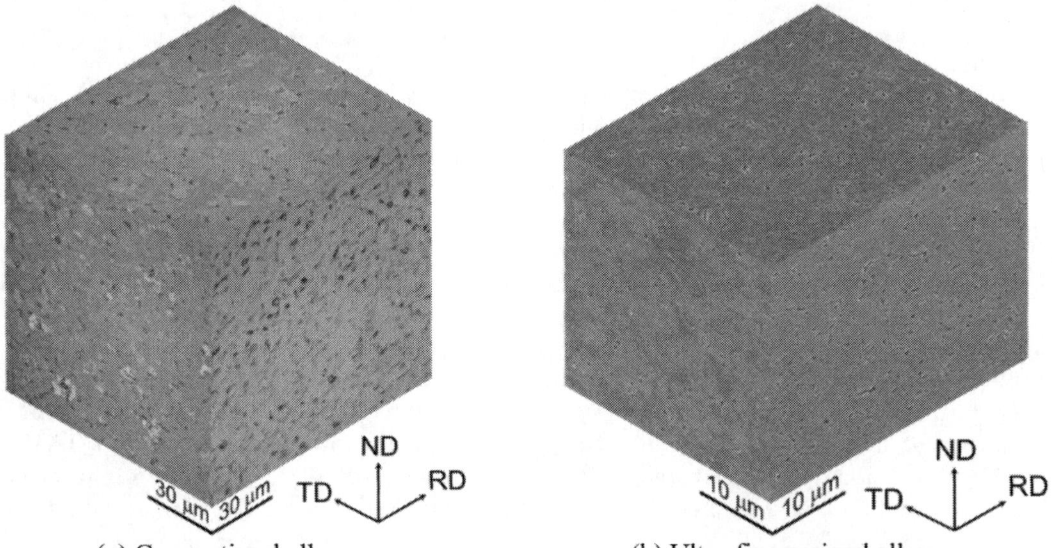

(a) Conventional alloy (b) Ultra-fine grained alloy

Figure 1. Microstructure.

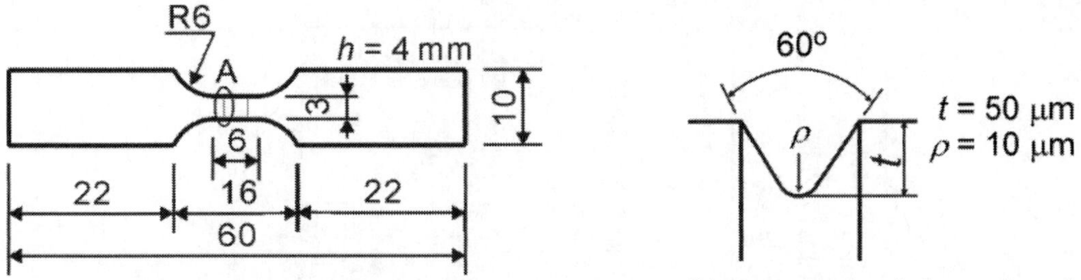

(a) Tension-compression plate specimen (b) Cross sectional shape of notch

Figure 2. Shapes and dimensions of specimen and notch.

The shapes and dimensions of 4 mm thick plate specimens are shown in Figure 2(a). After surface finish with an emery paper of #2000, the stress relief annealing was performed at 600°C for 4 h [12], and about a 50 μm of surface layer was removed by electro-polishing. Two notches normal to the loading direction were introduced at both sides of a specimen, i.e. a specimen had four notches in total.

The cross-sectional shape and dimensions of notch is shown in Figure 2(b). The root radius, ρ, was 10 μm and the depth, t, was 50 μm, giving a stress concentration factor of about 5.5 under tension. After introducing the notches, about a 10 μm deep surface layer was removed by electro-polishing. In practice, by observing these notches, the crack initiation and propagation processes were studied.

The tension-compression fatigue tests were carried out using a uniaxial hydraulic fatigue testing machine at a frequency of 30 Hz and at a stress ratio, R, of -1. The fatigue limit was defined by the maximum stress amplitude under which a specimen did not break for $N = 2\times10^7$ cycles. To avoid the effect of bending, for every fatigue test, the strain at smooth surfaces of both sides were monitored with four strain gages, and then the alignment was adjusted using an alignment fixture until bending stresses were in the range of ±10 MPa. The plastic replicas were taken at appointed numbers of cycles by interrupting fatigue test. Gold was evaporated on the surface of these replicas before microscopic observation.

3. Results and discussion

3.1. S-N characteristics of notched specimens and fatigue process at notch root

In the case of $\alpha + \beta$ type Ti-6Al-4V alloy, the α-phase that plays a dominant role for fatigue crack initiation has a HCP crystal structure with very few slip systems. In the case of smooth specimen, the number of cracks that can initiate at the surface is quite few because of the small number of α-grains with a preferred crystallographic orientation with respect to loading direction. Therefore, it is hard to systematically investigate a typical fatigue process from crack initiation to failure by using the smooth specimens. In this study, thus, four notches with the notch root radius of 10 μm and the notch depth of 50 μm were introduced into the surface of a specimen to make a successive observation of fatigue process easier.

Figure 3 shows the *S-N* data for notched specimens of ultra-fine grained and conventional alloys. The fatigue limit of ultra-fine grained Ti-6Al-4V alloy was 270 MPa and that of conventional alloy was 240 MPa. Many researchers reported that the smooth specimen of the grain refined materials obtained by the SPD process for carbon steel [13, 14], stainless steel [15] and etc. [16-17] showed a significantly enhanced fatigue limit due to the grain size refinement. The smooth specimens of grain-refined pure titanium [18-19] and titanium alloy [5, 11, 20-21] showed similarly that the fatigue limit increased with the grain size refinement. In this study, however, the fatigue limit of ultra-fine grained Ti-6Al-4V alloy specimen with sharp, shallow notches did not show such a significant enhancement compared to that of conventional Ti-6Al-4V alloy specimen with identical notches. The ratio of fatigue of these alloys is calculated to be 270 MPa/240 MPa = 1.13, which would rather be a reasonable value, given the ratio of Vickers hardness of 351/314 = 1.12 is taken into account.

IOP Conf. Series: Journal of Physics: Conf. Series **843** (2017) 012039 doi:10.1088/1742-6596/843/1/012039

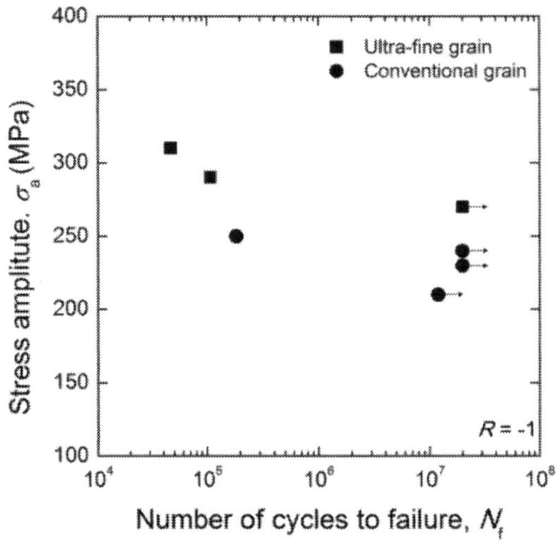

Figure 3. *S-N* data.

Figures 4 shows the fatigue crack initiation and propagation processes for conventional alloy (Figure 4(a)) and ultra-fine grained alloy (Figure 4(b)), which were successively observed at notch root at fatigue limit by the plastic replica technique. Only the micrographs taken at $N = 2.0\times10^7$ cycles show the real feature of notch roots, which were directly observed for specimens dismounted from the testing machine after the test. For both the conventional and ultra-fine grained alloys, numerous hairline cracks were observed at notch root, but slip bands could not be detected at crack initiation sites. Many cracks initiated before $N = 10^5$ cycles at notch root in the case of conventional-sized grain (Figure 4(a)), while in the case of ultra-fine grain (Figure 4(b)), the crack initiation was observed between $N = 3.0\times10^6$ and 6.0×10^6 cycles. Namely, more than 30 times of load cycles were required to give rise to crack initiation for ultra-fine grained alloy specimen subjected to an even higher stress of $\sigma_a = 270$ MPa. This result suggests that grain refinement would work to enhance the crack initiation threshold. This is presumably because of a decrease in the amount of cyclic slip resulting from an increase in yield strength associated with an increase in grain boundary density. Conversely, no marked difference of total number of cycles to failure in the finite life region (cf. Figure 3) was observed between two alloys though the number of data is not enough.

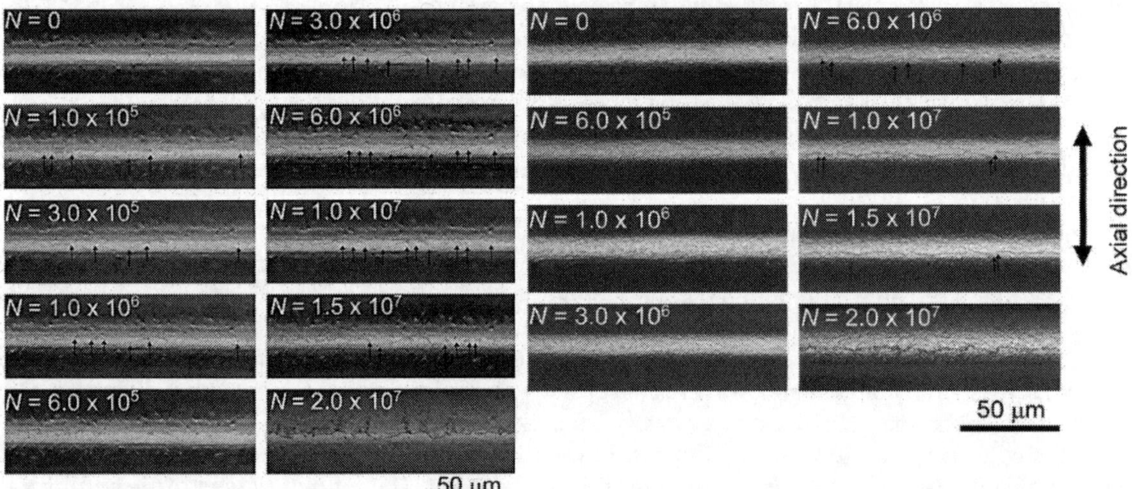

(a) Conventional alloy (σ_a = 240 MPa) (b) Ultra-fine grained alloy (σ_a = 270 MPa)

Figure 4. Fatigue process observed at notch root of Ti-6Al-4V alloy specimens.

3.2. Roughness of fracture surface and oxide debris formation

Owing to the notch, several fatigue cracks initiated simultaneously at the notch root, and they coalesced to each other. At fatigue limit, the crack path at notch root was more tortuous for conventional alloy (Figure 4(a)) than for ultra-fine grain alloy (Figure 4(b)) but the difference was small. The fracture surfaces originated from notches of specimens under the stress above fatigue limit are shown for conventional alloy in Figure 5 and for fine-grained alloys in Figure 6. The fracture surface very close to the notch root was relatively smooth for both alloys but at the place apart from the notch root, it was much rougher for conventional alloy than for ultra-fine grain alloy.

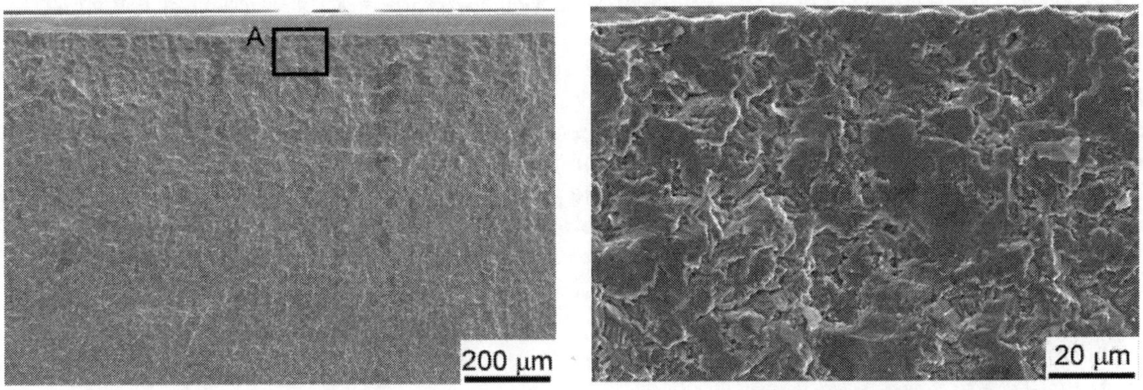

(a) Low-power field (b) Magnified image at A

Figure 5. Fracture surface of conventional alloy ($\sigma_a = 250$ MPa, $N_f = 1.9\times10^5$ cycles).

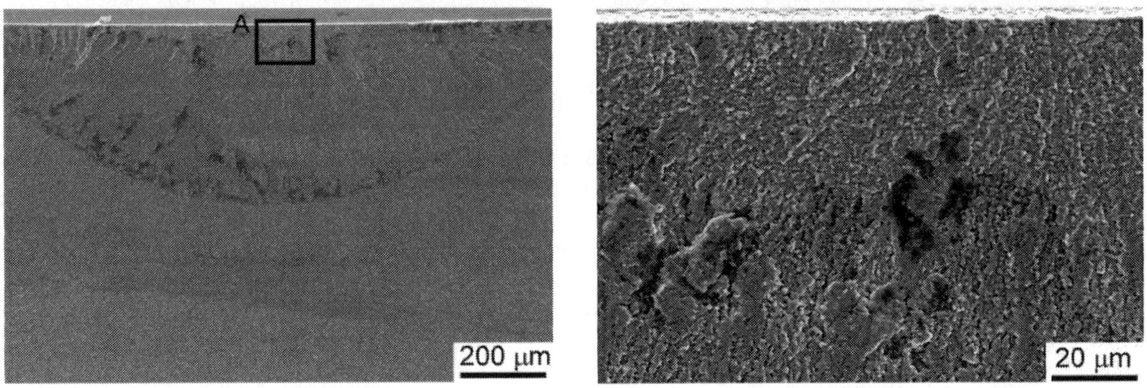

(a) Low-powered field (b) Magnified image at A

Figure 6. Fracture surface of ultra-fine grained alloy ($\sigma_a = 290$ MPa, $N_f = 1.1\times10^5$ cycles)

Figure 7 shows the debris (the color is reddish brown) observed along the crack path at notch root, which is considered to be oxide particles emerged out from the inside of cracks as a result of repeated contact of crack face asperities. The occurrence of oxide debris is an evidence that the crack surfaces interfere to each other. The amount of debris was larger for conventional alloy than fine-grained alloy as seen in Figure 7. This may be because more debris is likely to be produced by contact of rougher surfaces. It is reported that in titanium alloys, the crack growth rate and the threshold stress intensity factor range, ΔK_{th}, is primarily influenced by the roughness-induced crack closure especially for low stress ratio [22] but in this literature, no report about debris is provided. The debris produced within the crack would possibly develop the oxide-induced crack closure. It is thus expected that the effect of crack

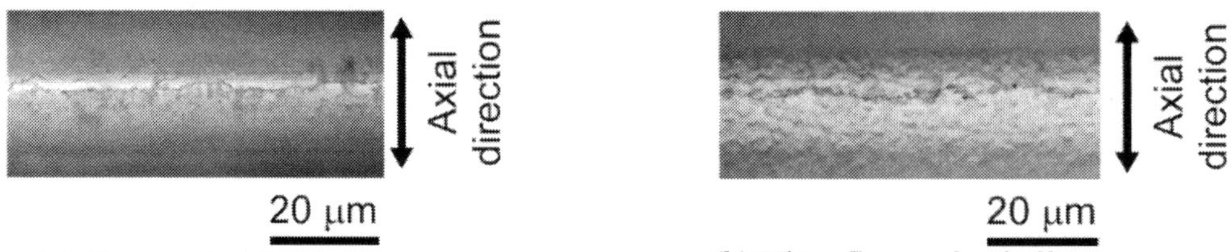

(a) Conventional alloy (b) Ultra-fine grained alloy

Figure 7. Oxide debris observed along crack path at notch root at fatigue limit.

closure is relatively larger for coarser-grained alloy that tends to form a rougher crack plane and thereby produces more amount of debris.

The fatigue life is the sum of crack initiation life and crack propagation life. As mentioned, with regard to crack initiation life, for the same stress level, more number of loading cycles was necessary for ultra-fine grain alloy than for conventional grain alloy. Conversely, with regard to crack propagation life, less contribution of crack closure is expected for ultra-fine grain alloy than for conventional grain alloy. As a result, it is considered that the total fatigue life had no marked difference because of these opposing effects. For the threshold condition, the effect of crack closure may be smaller since the crack surface roughness is smaller in the vicinity of notch root and the debris may possibly be easily ejected from shallow cracks. Consequently, the threshold levels are supposed to be mainly determined by the material factor, i.e. the Vickers hardness.

4. Conclusions

The influences of oxide debris and crack surface roughness on the fatigue crack propagation behavior were investigated by conducting high-cycle tension-compression fatigue tests using both conventional grain and ultra-fine grain Ti-6Al-4V alloy specimens containing sharp, shallow notches. The results of the present study are summarized as follows:

(1) The fatigue limit of notched specimen for ultra-fine grained alloy was increased by 12% compared to that of the conventional alloy. This is mainly because of the difference of material hardness.
(2) The fatigue crack initiation was delayed for ultra-fine grain alloy compared to conventional alloy. Inversely, the difference in total fatigue life was small.
(3) The oxide debris was observed at notch root of both conventional and the ultra-fine grained alloy specimens. The amount of the oxide debris was smaller for ultra-fine grained alloy than for conventional alloy.
(4) The crack closure effect was larger for conventional grain alloy than ultra-fine grain.

References

[1] Valiev R Z, Enikeev N A, Murashkin M Y, Kazykhanov V U and Sauvage X 2010 *Scr. Mater.* **63** 949
[2] Höppel H W, May L, Prell M and Göken M 2011 *Int. J. Fatigue* **33** 10
[3] Zhu R, Wu Y J, Ji W Q and Wang J T 2011 *Mater. Lett.* **65** 3593
[4] Jahadi R, Sedighi M and Jahed H, 2014 *Mater. Sci. Eng.* A **593** 178
[5] Saitova L R, Höppel H W, Göken M, Semenova I P, Raab G I and Valiev R Z 2009 *Mater. Sci. Eng.* A **503** 145
[6] Nakahigashi J and Yoshimura H 2002 *Mater. Trans.* **43** 2768
[7] Yoshimura H and Nakahigashi J 2002 *Int. J. Hydrog. Energy* **27** 769
[8] Rastkar A R Bloyce A and Bell T 2000 *Wear* **204** 19
[9] Antoniou R A and Radtke T C 1997 *Mater. Sci. Eng.* A **237** 229
[10] Shojima K, Weldle S, Okazaki S, Endo M, Eifler D and Balle F 2013 *Mater. Sci. Forum* **750** 232
[11] Ebara R Endo M Kim H J and Nakahigashi J 2014 *Fatigue of Materials III: Advances and*

Emergences in Understanding (Edited by Srivatsan T S Ashraf Imam M and Srinivasan R) TMS (The Mineral Metals & Material Society) 33
[12] Donachie M J 2000 *Titanium: A Technical Guide* (ASM International)
[13] Hirukawa H and Furuya Y 2009 *Tetsu to Hagane* **95** 79
[14] Chapetti M D Miyata H Tagawa T Miyata T and Fujioka M 2004 *Mater. Sci. Eng.* A **381** 331
[15] Matsumura Y Kishi K Ishibashi Y and Ebara R 2008 *J. Soc. Mater. Sci., Japan* **57** 134
[16] Vinogradov A Patlan V Suzuki Y Kitagawa K and Kopylov V I 2002 *Acta Mater.* **50** 1639
[17] Vinogradov A Hashimoto S and Kopylov V I 2003 *Mater. Sci. Eng.* A **355** 277
[18] Vinogradov A Y Stolyarov V V Hasimoto S and Vailev R Z 2001 *Mater. Sci. Eng.* A **318** 163
[19] Kim W J Hyun C Y and Kim H K 2006 *Scr. Mater.* **54** 1745
[20] Saitova L R Höppel H W Göken M Semenova I P and Vailev R Z 2009 *Int. J. Fatigue* **31** 322
[21] Nakaiso T 2005 Master thesis of Kagawa University
[22] Halliday M D Beevers C J 1981 *J. Test. Eval.* **9** 195

Fatigue Properties of Butt Welded Aluminum Alloy and Carbon Steel Joints by Friction Stirring

M Okane[1], T Shitaka[2], M Ishida[3], T Chaki[3], T Yasui[4] and M Fukumoto[4]

[1]Department of Mechanical Engineering, National Institute of Technology, Toyama College, Japan.
[2]Advanced course, National Institute of Technology, Toyama College, Japan.
[3]National Institute of Technology, Toyama College, Japan.
[4]Department of Mechanical Engineering, Toyohashi University of Technology, Japan

E-mail: okane@nc-toyama.ac.jp

Abstract. The butt dissimilar joints of Al-Mg-Si alloy JIS A6063 and carbon steel JIS S45C by means of friction stir welding were prepared for investigating fatigue properties of the joints. The joining tool used has cemented carbide thread probe and a shoulder made of alloy tool steel. All the fatigue tests were carried out under a load-controlled condition with a load ratio R=0.1 in air at room temperature. From the experimental results, it was found that hardness near the interface in A6063 was lower than that of base material. Three types of fatigue fracture occurred even in case of same welding condition. The first one was fracture at boundary between the lower hardness region and base material in A6063, the second type was initiated in the stir zone by FSW process and the last one was fracture at interface. Fatigue strength in case of the second one was lower than others. Furthermore, to investigate the effect of heat treatment on fatigue properties of the dissimilar joints, fatigue tests were also carried out with using the specimens which were heat treated under the same condition to aging process in T6 treatment. Fatigue fracture was initiated at interface between A6063 and S45C in case of the heat treated specimen, but fatigue strength was improved approximately 25% as compared with that of the non-heat treated specimen.

1. Introduction

It is a common knowledge that for reducing weight of mechanical structures, using lightweight metal such as aluminum alloy or magnesium alloy instead of steel is very much effective. However, since lightweight metal has generally lower strength as compared with steel, constructing all structures with lightweight metal is hardly possible. Friction stir welding (FSW) is one of the latest joining techniques which can weld the materials in state of solid phase. Therefore, it is expected for joining of dissimilar materials. The authors have previously reported characteristics and mechanical properties of Al alloy and steel joint by FSW [1]-[5] or fundamental fatigue properties of friction stir spot welding [6]-[8].

It is important to clarify the mechanical properties especially fatigue properties of the dissimilar joint before it will be used in the actual components and structures, however there are few reports regarding fatigue properties of the joints. In this study, the butt dissimilar joints of Al-Mg-Si alloy JIS A6063 and carbon steel JIS S45C by means of friction stir welding were prepared for investigating the basic fatigue properties of the joints. On the other hand, there are some reports regarding the aging effect for improving mechanical properties on friction stir welds of precipitation hardening aluminum alloy [9]-[11]. However, there are very few reports about the effect of aging on the dissimilar joints. In this study, effect of heat treatment on fatigue properties of the A6063/S45C dissimilar joint was also investigated.

2. Experimental procedures

2.1. Materials and FSW conditions

In this study, carbon steel JIS S45C and Al-Mg-Si alloy JIS A6063-T5 which were both 60 mm in width, 300 mm in length and 6 mm in thickness, were used for the butt dissimilar joints by means of friction stir welding (FSW). The chemical compositions and basic mechanical properties of both materials are summarized in Table 1 and Table 2, respectively. The welding tool used in this study has cemented carbide M5 thread probe and a shoulder made of alloy tool steel with a diameter of 20 mm. Figure 1 shows schematic illustration of FSW. The probe was plunged into A6063 and moved along the weld line. The offset displacement which means deviation value between weld interface of A6063 and S45C and side surface of the tool was 0.05 mm in S45C as illustrated in Fig. 1. The rotating direction and welding direction were identical in S45C side, i.e. the advancing side (AS) was S45C and retreating side (RS) was A6063 as also illustrated in Fig. 1. Rotating speed of the tool was 2000 rpm and welding speed was 500 mm/min in this study.

2.2. Heat treatment of the dissimilar joints

In this study, some butt welded joints by FSW were heat treated. The heat treatment condition was as follows, the joints were heated for eight hours by using air furnace under the condition of 523 K (250 ℃), after that the joints were air cooled. This condition was in compliance with the aging process of T6 treatment for A6063. After the aging process, fatigue specimens were machined from the joints as described later in detail.

2.3. Hardness distribution

To clarify hardness distribution around the welding interface, Vickers hardness test was carried out on the cross section of the joint. Testing load were 300 gf for A6063 and 1000 gf for S45C

Table1 Chemical compositions of the materials used. (mass %)

	Si	Fe	Cu	Mn	Mg	Cr	Zn	Ti	Al	C	P	S	Ni
A6063-T5	0.43	0.17	0.00	0.03	0.48	0.00	0.00	0.01	Re	-	-	-	-
S45C	0.18	Re	0.12	0.64	-	0.07	-	-	-	0.46	0.14	0.03	0.06

Table2 Mechanical properties of the materials used.

	Tensile strength σ_B, MPa	0.2%Proof stress, MPa	Yield stress, MPa	Young's Modulus, GPa
A6063-T5	205	145	-	68.6
S45C	605	-	345	205

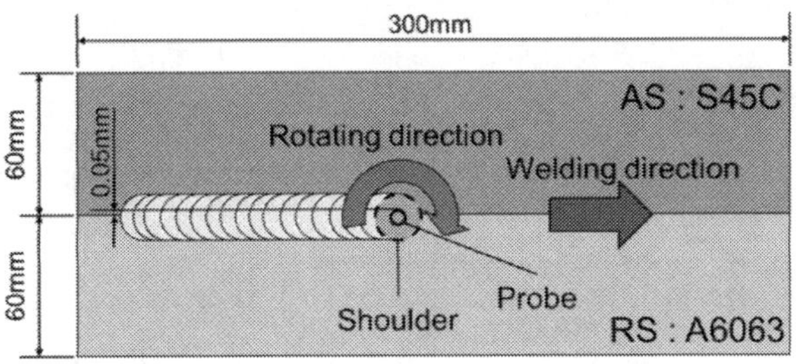

Fig. 1 Schematic illustration of FSW.

IOP Conf. Series: Journal of Physics: Conf. Series **843** (2017) 012040 doi:10.1088/1742-6596/843/1/012040

respectively, and holding time was 10 seconds. The relationship between distance from the interface and Vickers hardness is shown in Fig. 2. Data plots located in right side of the interface indicate the hardness of A6063, and data plots in left side indicate the hardness of S45C. It can be seen from the figure that there is lower hardness region near the interface in case of non-heat treated A6063 specimen (○). The average hardness of this region where distance from the interface is less than approximately 9mm was 62 Hv, which was significantly lower than that of A6062 base metal (77 Hv). This region seems to be the stir zone (SZ) and the thermo-mechanically affected zone (TMAZ) by FSW process [9]. On the other hand, in case of heat treated specimen (▲), the average hardness of this region recovered to approximately 89 Hv. And the hardness distribution in the region where distance from the interface is larger than 10mm, was almost similar to that in case of non-heat treated specimen. This characteristic is similar to the aging effect of friction stir welds between the precipitation hardening aluminum alloys [9]-[11], it had been clarified that in case of the dissimilar joints same phenomenon occurs.

2.4. Specimen and tests conditions

In this study two kinds of fatigue specimen were used. The shapes and dimensions of the fatigue specimens are shown in Fig. 3, respectively. Type-A specimen where the welding interface aligned with the center of specimen was used for both tensile and fatigue tests, and Type-B specimen where the distance between welding interface and specimen center was 5 mm, was used for fatigue tests. The gauge parts of both specimens were polished with emery paper up to grade of #2000 longitudinally and final polishing was conducted by using diamond paste with grain size of 10 μm.

Tensile tests were carried out by using universal testing machine with crosshead speed of 0.5 mm/min. A fatigue test machine with 19.6 kN capacities was used for fatigue tests. All the fatigue tests were carried out under a load-controlled sinusoidal wave form condition with a stress ratio R=0.1 and frequency f=20Hz in air at room temperature. After the fatigue tests, fracture surfaces were observed in detail by using a scanning electron microscope (SEM).

3. Results and discussions

3.1. Tensile tests

From the results of tensile tests, it was found that the average tensile strength σ_B in case of non-heat treated specimen was approximately 140 MPa. On the other hand, in case of heat treated specimen, the average tensile strength σ_B was approximately 195 MPa which was about 40% improvement compared with that of the non-heat treated specimen. In both cases, tensile fracture

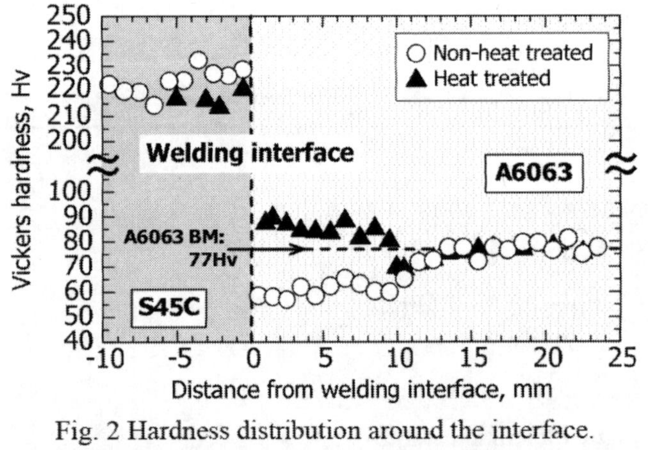

Fig. 2 Hardness distribution around the interface.

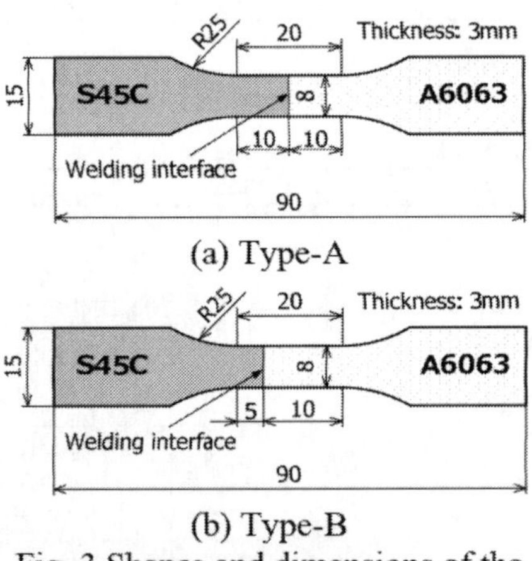

Fig. 3 Shapes and dimensions of the specimen used.

occurred at the site where distance from the interface is approximately 10mm in A6063. This site almost correspond to a location of boundary between the lower hardness region in case of non-heat treated specimen or recovered hardness region in case of heat treated specimen, and base metal respectively, as described in Fig. 2.

3.2. Fatigue fracture modes

In this study, three types of fatigue fracture modes occurred in case of non-heat treated specimen. Overviews of the fractured specimens are shown in Fig. 4. Figure 4(a) is a typical observation of the specimen fractured at a boundary between the lower hardness region and base metal in A6063. It is clearly found that significant plastic deformation occurred around the fracture surface. Figure 4(b) shows the specimen fractured at around the location where distance from the interface is approximately 2-3 mm in A6063. This location seems to be a center of the tool path in FSW process, i.e. a center of the stir zone (SZ). Figure 4(c) shows the specimen fractured at interface, and significant plastic deformation was observed clearly where the fatigue fracture occurred in the case of Fig. 4(a).

On the other hand, in the case of heat treated specimen, fatigue fracture occurred at welding interface in all specimens. Figure 5 is a typical observation of the fractured specimen. It is found that there are other cracks near fracture surface. Since striation was observed on the crack surface, this

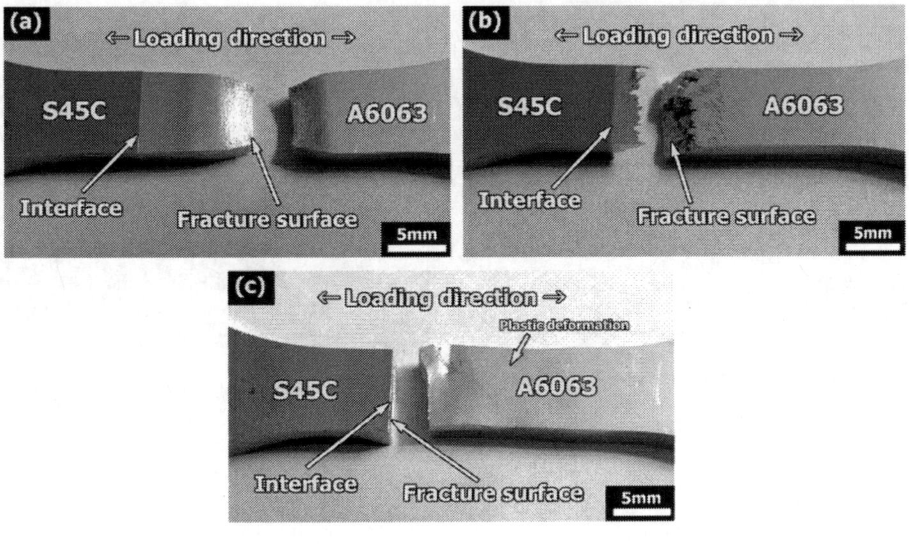

(a) σ_a: 55MPa, N_f: 1.64×10^5, (b) σ_a: 40MPa, N_f: 3.41×10^4, (c) σ_a: 50MPa, N_f: 2.02×10^5
Fig. 4 Overviews of fatigue fractured non-heat treated specimens.

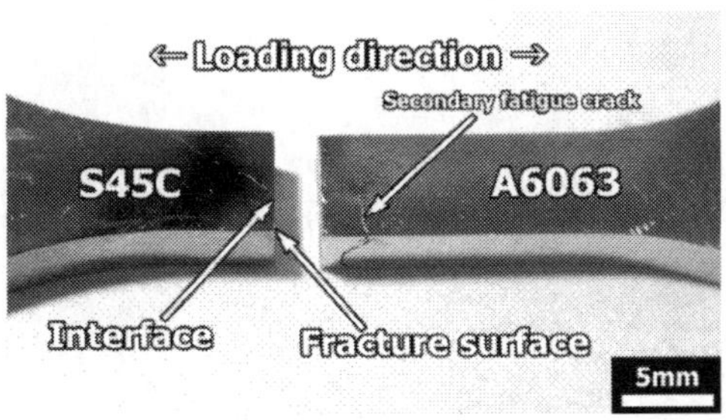

Fig. 5 Overview of fatigue fractured heat treated specimen.
(σ_a: 60MPa, N_f: 4.06×10^4)

kind of crack seems to be a secondary fatigue crack, as described later in detail. And no significant plastic deformation was observed where the fatigue fracture occurred in the case of Fig. 4(a).

3.3. S-N curves

The relationship between applied stress amplitude and number of cycles to failure (S-N curve) is shown in Fig. 6. In the figure, data points with arrow show that no failure occurred up to the testing cycles of 1×10^7. Circle plots indicate the results of non-heat treated specimen and triangle plots indicate the results of heat treated specimen. The difference in painting on data plots indicates the difference of fatigue fracture mode. Further, data points with * or ** show the results of using Type-A specimen, and the other plots show results of using Type-B specimen, as described later in detail.

It can be seen from the figure that for non-heat treated specimen, fatigue strength in case of fracture in stir zone (○)was lowest and fatigue strength in case of fracture at boundary between the lower hardness region (TMAZ) and base metal (BM) in A6063 (◑) was highest, although there is a large dispersion. The fatigue strength at 1×10^7 cycles was approximately 40 MPa (maximum stress: 89 MPa). On the other hand, for heat treated specimen, fatigue strength was higher than those of non-heat treated specimen. The fatigue strength at 1×10^7 cycles was around 50 MPa (maximum stress: 111 MPa) which was about 25% improvement compared with that of non-heat treated specimen.

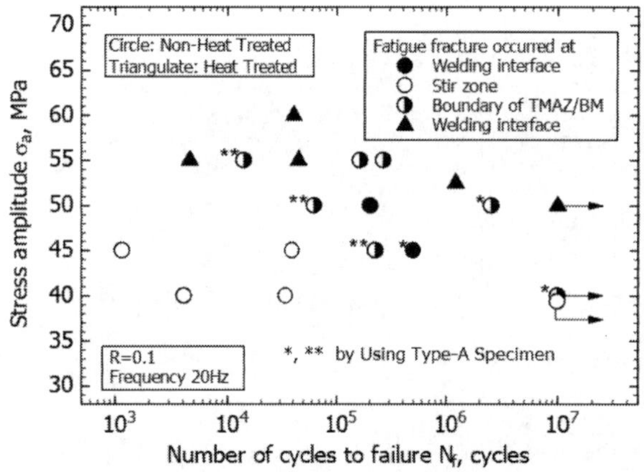

Fig. 6 S-N curves.

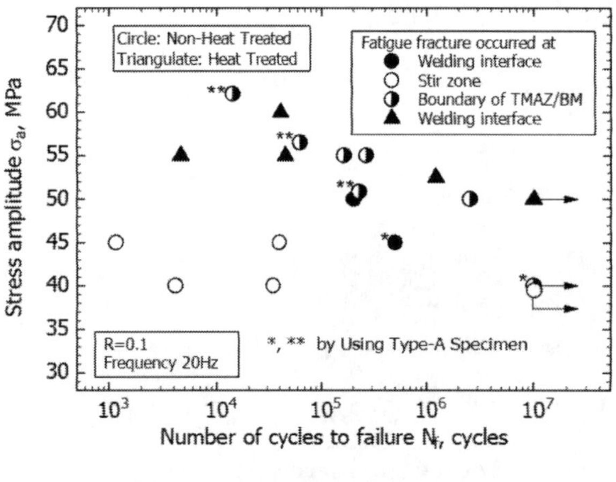

Fig. 7 Revised S-N curves.

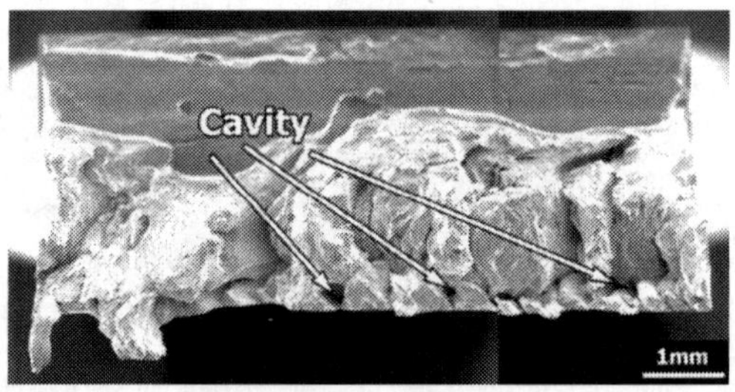

Fig. 8 Fracture surface of the non-heat treated specimen (σ_a: 40MPa, N_f: 3.41×10^4).

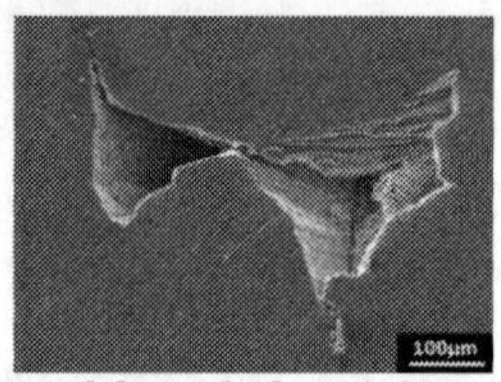

Fig. 9 Cavities (defects) in stir zone.

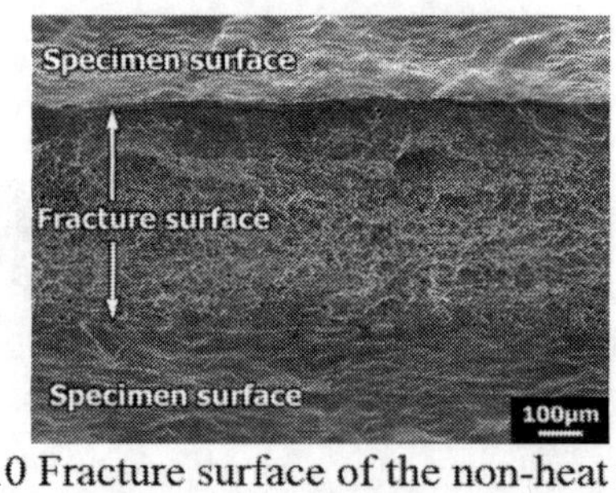

Fig. 10 Fracture surface of the non-heat treated specimen (σ_a: 55MPa, N_f: 1.64×10^5).

According to the data base of The Japan Aluminium Association [12], fatigue strength of A6063-T5 at 1×10^7 cycles under the condition of stress ratio R=-1 is 70 MPa. By using the modified Goodman's line, fatigue strength of A6063-T5 under the condition of R=0.1 had been estimated approximately 49.4 MPa in stress amplitude, which is almost similar level to that of the heat treated specimen. Therefore, the joint efficiency in fatigue strength of the heat treated dissimilar joint used in this study is thought to be almost 100 %.

3.4. Considering the shape effect of fatigue specimen

In Fig. 6, data points with * or ** show the results of using Type-A specimen that the welding interface and center of specimen were aligned. Furthermore, the plots with ** indicate that fracture occurred at boundary between the lower hardness region and base metal, that was approximately 10 mm from the interface in A6063. In this region, the stress seemed to concentrate by influencing of the fillet part (R=25) in Type-A specimen. The stress cocentration factor α is 1.129 for this shape [13]. Therefore, although it is simplified method, applied stress in these cases must be correct conveniently by multiplying α. The revised S-N curve is shown in Fig. 7. From the figure, it is clarified that the dispersion in fatigue lives become smaller in case of the specimen fractured at boundary between TMAZ and BM.

3.5. Observation of the fracture surface and specimen surface

SEM micrograph of the fracture surface on the non-heat treated specimen which fractured in stir zone is shown in Fig. 8. There are many large grains which are millimeter size and a lot of cavities are also found. Cross sectional observation of stir zone in the non-heat treated specimen unused is shown in Fig. 9. Some cavities (defects) which are similar to the cavities observed on the fracture surface are found, and these cavities were thought to generate during FSW process. Since fatigue strength of the specimens which fractured in stir zone was lower than others, fatigue fracture was thought to initiate from the cavities in this case. Figure 10 shows SEM micrograph of the fracture surface on the non-heat treated specimen fractured at boundary between TMAZ and BM as shown in Fig.4(a). Dimples are dominant on the fracture surface.

On the other hand, SEM micrograph of the heat treated specimen near the fracture surface in A6063 is shown in Fig. 11. Figure 11(a) is a macroscopic view, and a lots of small chips and some

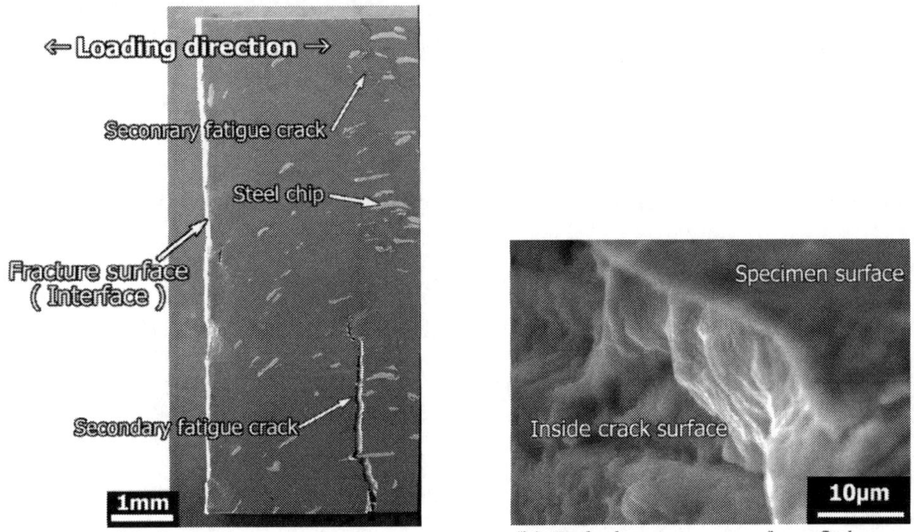

(a) Overview (b) Striations on secondary fatigue crack

Fig. 11 SEM micrographs around fracture surface of the heat treated specimen.
(σ_a: 60MPa, N_f: 4.06×10^4)

cracks are found. In the FSW process performed in this study, S45C on the welding interface was slightly removed by probe of FSW tool. From the previous studies [4][5], it was clarified that removed steel was distributed in Al alloy as small chips but the strength of the joint was not affected by steel chips significantly. Figure 11(b) is a close up of the crack observed in Fig. 11(a). Since some striations are found on crack surface, these kind of cracks are thought to be the secondary fatigue cracks.

In this study, FSW condition was same for all specimens, however fatigue fracture modes and strengths were not necessarily similar or same level. Some environmental factors, for example temperature, humidity and so on during FSW process may affect the quality or properties of the joints, and this should be investigated in detail. Furthermore, there is a report that heat treatment of the dissimilar joint make the compound layer on the welding interface thicker and strength of the interface may become weak [14]. Since the thickness of compound layer, i.e. strength of the interface depends on heat treatment condition, further investigation will be need.

4. Conclusions

The butt dissimilar joints of Al-Mg-Si alloy JIS A6063 and carbon steel JIS S45C by means of friction stir welding were prepared for investigating fatigue properties of the joints. And the effect of heat treatment on fatigue properties of the dissimilar joints was also investigated. The main results obtained are summarized as follows:

1. There was lower hardness region near the interface in A6063, and the hardness of this region was recovered by the heat treatment which is similar to the condition of aging process in T6 treatment for A6063.
2. The average tensile strength of the non-heat treated specimen was approximately 140 MPa and that of the heat treated specimen was 195 MPa which was about 40% improvement compared with that of the non-heat treated specimen.
3. Three types of fatigue fracture modes occurred in case of the non-heat treated specimen. The first one was fracture at boundary between lower hardness region and base material in A6063, the second type was fracture initiated in stir zone and the last one was fracture at welding interface. Since the fatigue fracture initiated at cavities in the stir zone, fatigue strength in case of the second one was lower than others.
4. Fatigue fracture was initiated at welding interface in all case of the heat treated specimen, and fatigue strength was improved approximately 25% as compared with that of the non-heat treated specimen.

Acknowledgements

This study was supported by JSPS KAKENHI Grant Number 15K05704. The authors thank Mr. M. Tsubaki who is a technical staff of Toyohashi University of Technology, for his support in FSW.

References

[1] M Fukumoto, M Tsubaki, Y Shimoda and T Yasui, *Welding between ADC12 and SS400 by Means of Friction Stirring*, J. of the Japan Welding Society, 22-2, pp. 309-314, 2004.

[2] T Yasui, Y Shimoda, M Tsubaki, T Ishii and M Fukumoto, *Characteristics of High Speed Welding between 6063 and S45C by Means of Friction Stirring - Study on Welding in Dissimilar Metals by Means of Friction Stirring (1st Report)-*, J. of the Japan Welding Society, 23-3, pp. 469-475, 2005.

[3] Y Shimoda, M Tsubaki, T Yasui, M Fukumoto, T Fujita and J Osawa, *Effect of Tool Shape on Material Flow in Welding between Aluminum and Steel by Friction Stirring*, Steel Research International, 81-9, pp. 1108-1111, 2010.

[4] T Yasui, S Ishida and M Fukumoto, *Material flow observation for dissimilar friction stir welding of aluminum alloy and steel by X-ray CT method*, J. JILM, 64, pp. 604-610, 2014.

[5] T Shitaka, M Okane, T Yasui, A Ito and M Fukumoto, Strength Properties of Butt Welded A6063/S45C Joints by Friction Stirring, Proceedings of MECJ-16, 2016.

[6] M Okane, Y Takami, K Miyagawa, T Yasui and M Fukumoto, *Fatigue Behavior of Aluminum Alloy / Steel Joints by Spot Friction Stirring*, J. of the Japan Welding Society, 28-4, pp. 395-401, 2010.

[7] M Okane, Y Takami, K Miyagawa, T Yasui and M Fukumoto, *Fatigue Behaviors of Aluminum Alloy / Steel Dissimilar Joint by Friction Stirring*, Steel Research International, 81-9, pp. 1120-1123, 2010.

[8] M Okane. T Shimizu, T Yasui and M Fukumoto, *Mechanical Factors Influencing Fatigue Behavior of A5052/SPC270C Joints by Friction Stir Spot Welding*, APCFS2016, JSME-MMD, pp. 385-386, 2016.

[9] Y S Sato, H Kogawa, M Enomoto, S Jogan and T Hashimoto, *Precipitation sequence in friction stir weld of 6063 aluminum during aging*, Metallurgical and Materials Transactions A, 30-12, pp. 3125–3130, 1999.

[10] K V Jata, K K Sankaran and J J Ruschau, *Friction-stir welding effects on microstructure and fatigue of aluminum alloy 7050-T7451*, Metallurgical and Materials Transactions A, 31-9, pp. 2181–2192, 2000.

[11] B Heinz and B Skrotzki, *Characterization of a friction-stir-welded aluminum alloy 6013*, Metallurgical and Materials Transactions B, 33-3, pp. 489–498, 2002.

[12] Data base of The Japan Aluminium Association, http://metal.matdb.jp/JAA-DB/

[13] M Nishida, *Stress Cocentration Factor Handbook*, ISBN 4-627-94029-7, Morikita Publishing Co., Ltd., 2001.

[14] M Kuwahara, M Tsubaki, T Yasui and M Fukumoto, *Separation by heating for weld joint between aluminum and steel by friction stirring*, Preprints of The National Meeting of J.W.S. 2007f, pp. 172-173, 2007.

Compressive residual stress relaxation in hardened steel during cyclic and static load

D Cseh[1], V Mertinger[1]

[1]Institute of Physical Metallurgy, Metalforming and Nanotechnology, University of Miskolc, H 3515 Miskolc-Egyetemvaros, Hungary; corresponding author e-mail address: cseh88david@gmail.com

Abstract. The benefits of applied compressive residual stress on fatigue properties of materials is a well-known phenomenon, but not well described in all respects. The fatigue life and the fatigue limit could be improved by targeted created compressive residual stress in the surface layers therefore, diversified surface compressing methods are developed and used in the engineering industry. The relaxation of the compressive residual stress state during a cyclic and static load is determinative for the life time of a componenet. Compressive stress relaxation was experimentally determined during the cyclic and static load. The compressive residual stress was induced by shot penning on the surface of stainless steel, micro alloyed high strength steel and hardened steel specimens. The residual stress state was investigated nondestructively by X-ray diffraction method then these specimens were load. After a certain number of cycles the fatigue load was stopped and the residual stress state was recorded again and again until fracture. To investigate the relaxation process during static load a four-point bending bench was used. The compressive residual stress relaxation was correlated to the applied fatigue stress level, the cycle number the quality of alloys.

1. Introduction

Various procedures exist, which result compressive residual stress at the surface and the shot peening is the most wides spread surface compressing method [1], [2]. The applied compressive residual stress has a beneficial effect on the fatigue property of metals, especially on fatigue strength [3], [4], [5]. For the life time estimations, the behaviour of this directly induced compressive residual stress field during the operation of the with enhanced machine element must be well known. During the operation, which is a mixture of different kind of external loads the macro and micro residual stress states interact with the cyclic, quasi-static or dynamic load stresses and the work hardening and softening processes.

The relaxation susceptibility of the surface compressive residual stress during the operation has a determinative importance. The constancy of the compressive residual stress state during fatigue load in annealed micro alloyed, medium carbon, perlitic and austenitic steels has been already investigated [6], [7], [8], [9]. It has showed that the residual stress relaxation is not monotonic in these alloys. Nevertheless, a significant portion of machine elements, which are typically exposed to fatigue loads, are mainly made of high strength hardened steels. One of our aim to investigate the stability of residual stress in hardened steels. The relaxation susceptibility of hardened, shot peened CM45MV specimens during cyclic load was investigated in one of earlier project of us [10]. The relaxation of the compressive residual stress was not observable in that steel, because of microstructure defects and specimen geometry.

In this paper the stress relaxation susceptibility of another kind of widely used hardened steel 42CrMo4 is presented. Fatigue specimens were worked out, shoot peened and exposed to cyclic load. The residual stress state was investigated non-destructively after the shot peening and recorded repeatedly during the fatigue load. The effect of static load on the residual stress was also observed. Comparative experiments were carried out on HSLA, 42CrMo4 and austenitic stainless steel specimens. The compressive residual stress field in static loaded specimens was created by used shot peening indirectly. The static load was implemented by a four-point bending bench. The residual stress was measured by X-ray diffraction in both experiment.

2. Materials

Fatigue specimens were cut from a hot rolled bar made of 42CrMo4 with 20 mm diameter (figure 1). Some specimens with 5x5 cross section on the middle quadratic part were also used, for check the size effect. The yield strength (1200 MPa) and tensile strength (1250 MPa) were measured by three parallel tensile tests on hardened test specimens made of the same piece of bar.

The faults of the heat treatments could strongly influence the residual stress state [11], [12], [13] Our specimen geometry (long, thin) is especially sensitive on distortions, so the heat treatment was carried out with care.

The specimens were induction hardened and oil quenched, immersed by hand. The austenitizing temperature of the specimens were controlled by Ni-NiCr thermocouple, soldered to the surface. First, the specimens were rapidly heated to 1000°C during a few seconds then the temperature was held for 2 seconds When the sample reached 850°C, the thermocouple was cut from the specimens and the specimens were immersed in quenching oil media, During the immersion the longitudinal axis was perpendicular to the surface of the media. The tempering was carried out at 600°C in an inductive furnace for half an hour, and specimens were subsequently cooled in air. The figure 2 shows the homogeneous martensitic microstructure of the cross section of the quenched and tempered sample. The HV10 (subscript) hardness data on the cross section (332, 337, 339, 361, 341, 333) confirms that the quenching was successful.

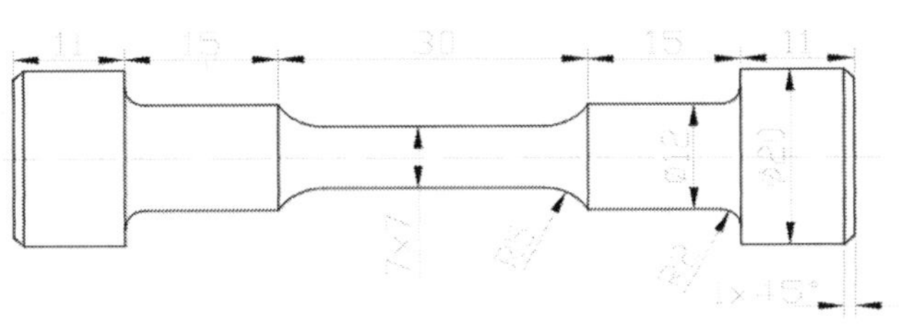

Figure 1. Size of the shot peened specimens

Figure 2. Light microscopy image of the hardened sample, cross section.

The effect of static load on compressive residual stress state were observed in three alloys. Hardened EN 42CrMo4, MIL–A 46100 HSLA as well as AISI 316Ti, austenitic stainless steel specimens were used in an elongated plate geometry (l:105, w:20, h:3mm). The geometry was given by the four-point bending device. The 42CrMo4 and the HSLA specimens were machined out from bulk material, after the cutting the 42CrMo4 were hardened, the HSLA specimen were used as received. The austenitic stainless steel was available in 20*3 cross section hot rolled form and after cutting the length used as received. The figure 3 and figure 4 shows homogeneous tempered martensitic microstructure in the cross section of the hardened and the HSLA specimens. The stainless-steel specimen bears a recrystallized cold formed characteristic figure 5.

IOP Conf. Series: Journal of Physics: Conf. Series **843** (2017) 012041 doi:10.1088/1742-6596/843/1/012041

After the cutting and heat-treating the subsequent step was the shot peening. It was performed in the Raba Automotive Holding Plc. The specimens were treated together with Raba manufactured shafts in the same time. The parameters of the shot peening were given by the requirements of the shaft. This means that the parameters of the applied peening method were industrial parameters. The shot peening was carried out in a WMKD 3 type shot peening equipment. The peening time was 9 minutes. Two Almen C strips were fixed on the sample holders to control the peening. The values of the Almen C probes were 0.26 and 0.24. The used peening balls were made of 1.8 mm diameter steel wire (430 HV), cut in 1.8 mm length pieces and conditioned for 4 hours. The minimal peening coverage was 150%. The average speed of the balls, which left the peening wheels was 55 m/s. As an average 117 J/mm2 energy was imparted with the surface. The performance of the peening wheels was 252 kg/min.

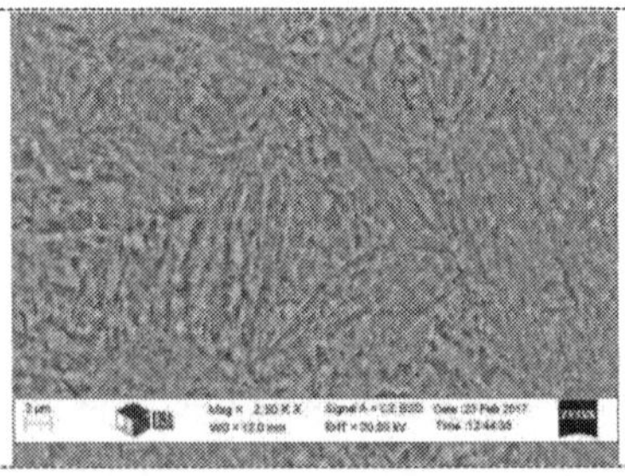

Figure 3. SEM microscopy image of the hardened specimen cross section. (Mg:2500x)

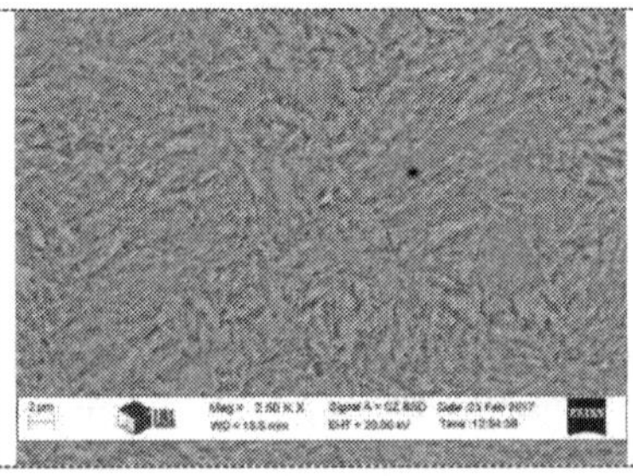

Figure 4. SEM microscopy image of the HSLA specimen cross section. (Mg:2500x)

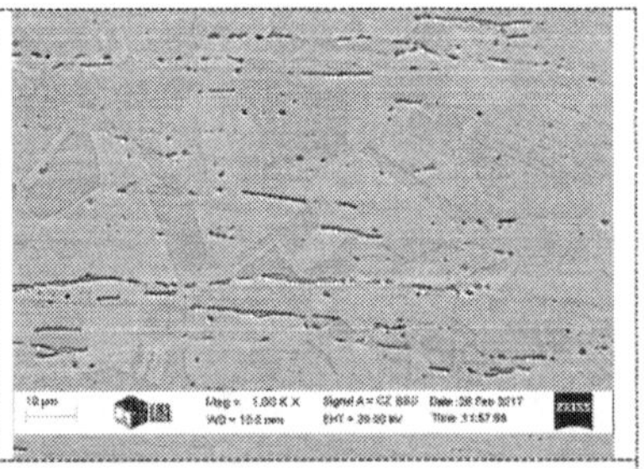

Figure 5. SEM microscopy image of the austenitic stainless steel specimen cross section. (Mg: 1000x)

After the cutting and heat-treating the subsequent step was the shot peening. It was performed in the Raba Automotive Holding Plc. The specimens were treated together with Raba manufactured shafts in the same time. The parameters of the shot peening were given by the requirements of the shaft. This means that the parameters of the applied peening method were industrial parameters. The shot peening was carried out in a WMKD 3 type shot peening equipment. The peening time was 9 minutes. Two Almen C strips were fixed on the sample holders to control the peening. The values of the Almen C probes were 0.26 and 0.24. The used peening balls were made of 1.8 mm diameter steel wire (430 HV), cut in 1.8 mm length pieces and conditioned for 4 hours. The minimal peening coverage was 150%. The average speed of the balls, which left the peening wheels was 55 m/s. As an average 117 J/mm2 energy was imparted with the surface. The performance of the peening wheels was 252 kg/min.

Figure 6. Fatigue specimens, after shot peening in socket

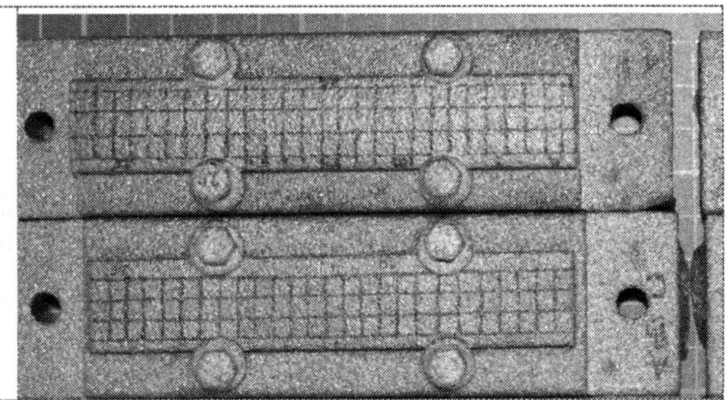

Figure 7. Static specimens after shot peening in Almen analogous clamping tool

The fatigue specimens were peened in socket, such way the heads of the fatigue specimens were shadowed. The non-peened machined surface in the heads required for the clamping during the fatigue load (figure 6). The specimens, used for static measurements, were also treated by peening in lengthened, Almen analogous clamping tools (figure 7).

3. Experiments

The residual stress was measured non-destructively by a Stresstech Xstress 3000 G3R X-ray diffractometer, resident at the Institute of Physical Metallurgy, Metalforming and Nanotechnology at the University of Miskolc. Our surface results regard to residual stress values from approximately 10μm depth of the surface layer. All our residual stress values refer to this layer. In case of hardened and HSLA steels Cr K_α radiation was used and the stress was calculated from the shift of the {211} reflection of ferrite according to the sin2ψ method using Young's modulus (E) of 210000 MPa and Poisson's ratio of 0.29. The austenitic specimens were irradiated with Mn K_α radiation and the stress was calculated from the shift of the {311} reflection of austenite. Young's modulus (E) of 196000 MPa and Poisson's ratio of 0.28 were used for the calculations. ψ equipment geometry and 3 mm spot size was used in all measurements
In case of the fatigue specimens Reflections were obtained in 3 tilting positions in the -45°/+45° range according to figure 8. The stress was measured with less error (scatter) than ±10 MPa in every point All four sides of the specimens were monitored. The axial residual stress state was recorded at 3 spots with 3 mm spot size areas, on each side of each specimen in the initial shot peened state and during the fatigue loads. The same areas were investigated in each step of the monitoring. The represented stress value is always an average of twelve data after each step. First, the residual stress field of the shot peened state was recorded then, the specimens were fatigued with a certain number of cycles, finally the residual stress state was measured. This method repeated until the fracture of the specimen was reached. Several specimens were fatigued until fracture without stop for expected lifetime estimations. The fatigue load was tensile-compressive, the stress ratio (Rσ) was -1 (Rσ = σmin/σmax; σmax = 350; 400; 450; 500 MPa); so the absolute values of the tensile and compressive stresses were equal. The cyclic load was implemented by an MTS type universal electro-hydraulic testing machine of the Institute of Materials Science and Technology at the University of Miskolc.

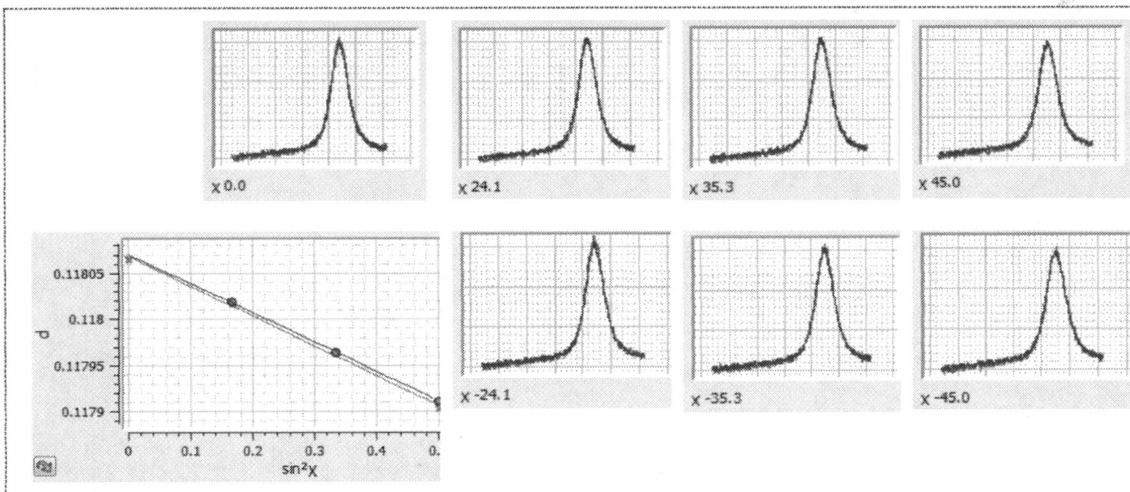

Figure 8. The result of one measured data of sample No. 1, χ: the tilting angle [°] , d: lattice distance [nm]

The statically loaded specimens residual stress measurements were recorded in 5-5 tilting positions in the (-45°/+45°) range. Two specimens were used from all type of steels. The residual stress state of the

shot-peened and from the clamping tool discharged specimens were recorded along a longitudinal stripe in 3*3 mm grid. The back side of the sample were also investigated and the longitudinal as well as the transverse residual stress distribution were docketed. All measured stresses lay in longitudinal direction, aside from a transverse travers on the back side.
In the possession, of residual stress data on the peened front and the back side, finally the back side were sorted out for the static load measurements, but previously the surface of the specimens were 3D scanned.
The applied bending device is a BC500 modulus measurement accessory of the X-Ray diffractometer. In this work used as a force controlled in situ external static loader machine. The specimens were fixed in the four-point bending bench under the diffractometer and the residual stress state was recorded in middle point of the sample surface bounded by the bending points. The residual stress was recorded with increasing and backward, decreasing external loads. The external loads were consecutively increased with 50MPa from zero. The needed bending force for the discrete bending stresses was calculated by the recombination of the

$$\sigma=3Fa/(bh^2) \quad (1)$$

equation, were σ: generated stress on the middle point of the surface bounded by the bending points (MPa), F: the applied bending force (N), a: the distance of the bending points (mm), b: the width of the strips (mm), h: the thickness of the strips (mm).

4. Results

The effect of the fatigue load on the compressive residual stress field in the surface layer during the cyclic load is illustrated by figure 9. The results of other stress levels were negotiated in another paper of us [14]. The most conspicuous common characteristic of the fatigues specimens residual stress relaxation is the strong initial fall. The approximately 25% residual stress relaxation happen in the beginning of the cyclic load, before the first residual stress monitoring step.

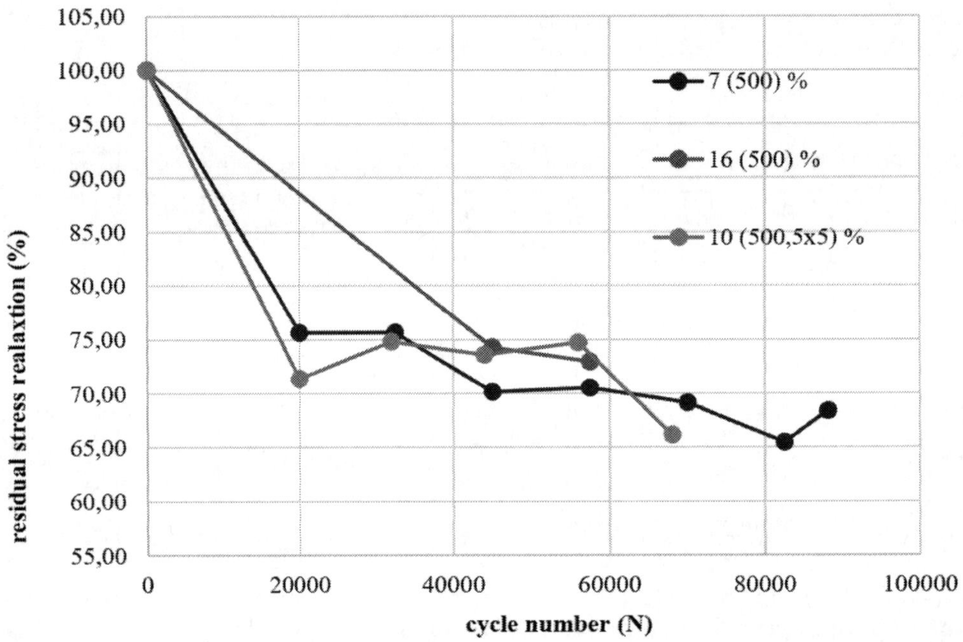

Figure 9. The residual stress relaxation at 500 MPa stress level, during fatigue loads

Figure 10.; figure 11.; figure 12.; figure 13.; figure 14.; figure 15. shows the residual stress distribution of the shot peened surface of the specimens made of different alloys in clamping tool, and after the discharging from it along 3*3 mm grid. The grid shown on figure 7. However, two piece of specimens were used made from all of the investigated alloys. Each pair show identical results so only one result presented here in case of each alloys.

All the investigated specimens were shot-peened simultaneously. The residual stress surface distribution results have common and dissimilar specifics. The HSLA specimen retains the highest residual stress values, nevertheless the same surface compressing treatment eventuate a significantly lower average residual stress extent in the hardened steel and in the austenitic stainless steel specimens. The discharging of the specimens from the clamping tool cause similar changes in case of all investigated alloys.

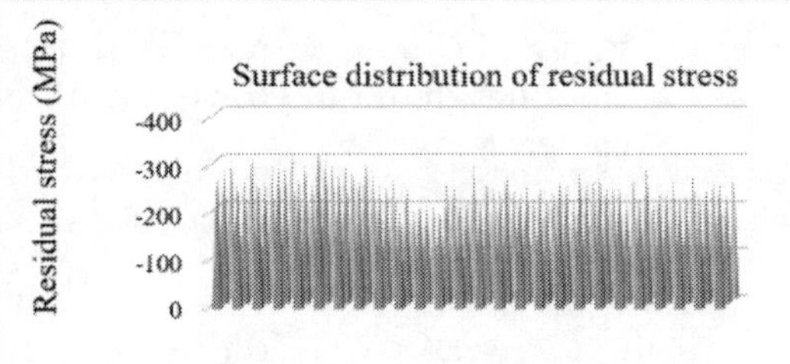

Figure 10. residual stress distribution of the peened, clamped, hardened sample surface

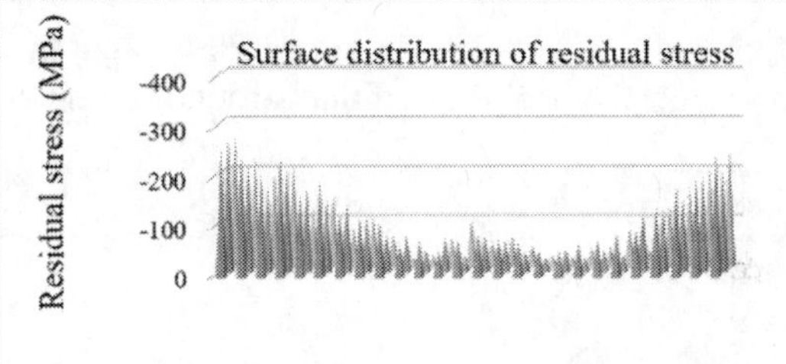

Figure 11. residual stress distribution of the peened, discharged hardened sample surface

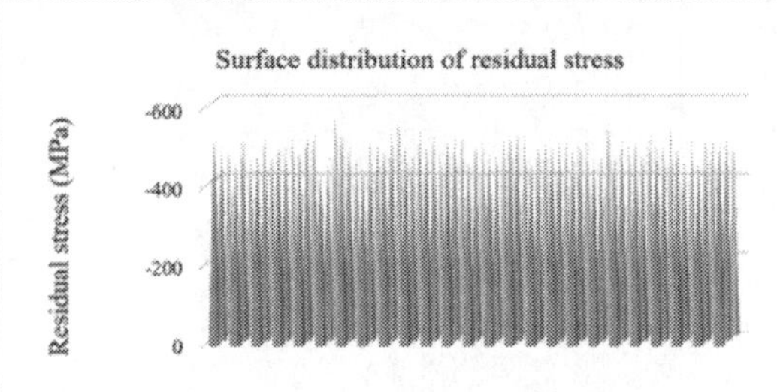

Figure 12. residual stress distribution of the peened, clamped HSLA sample surface

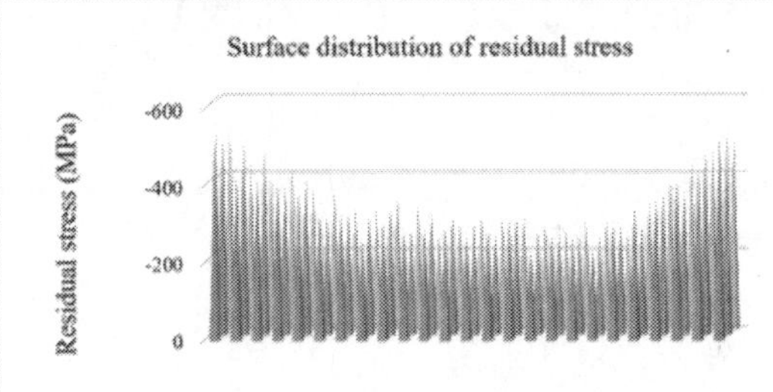

Figure 13. residual stress distribution of the peened, discharged HSLA sample surface

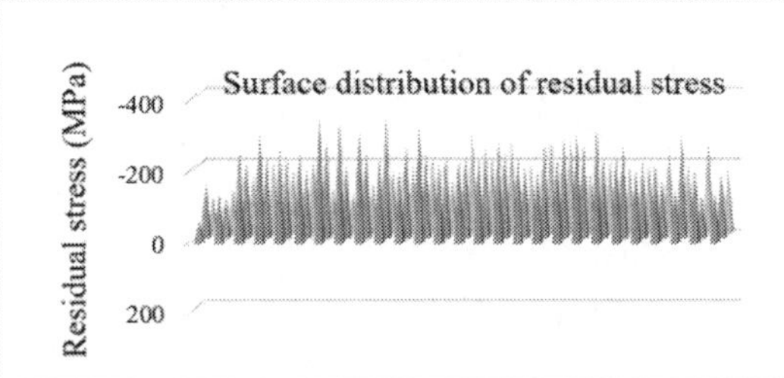

Figure 14. residual stress distribution of the peened, clamped stainless sample surface

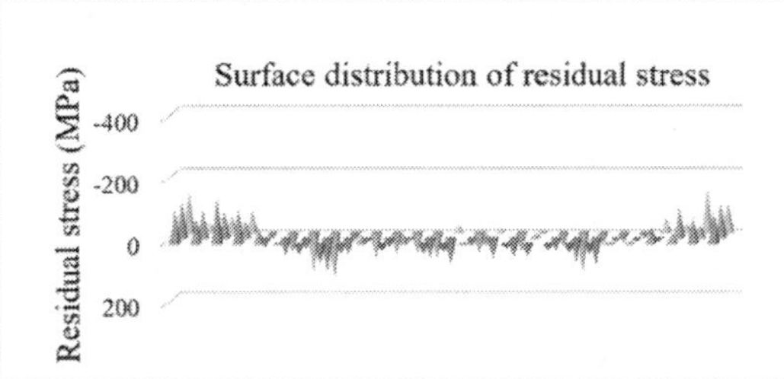

Figure 15. residual stress distribution of the peened, discharged stainless sample surface

On figure 16. colored squares show the investigated areas of the backside of a specimen. The longitudinal directional residual stress travers recorded in longitudinal and in transverse direction also. The transverse directional residual stress recorded along the transverse travers. Figure 16.; figure 17.; figure 18.; figure 19.; figure 20.; figure 21.; shows the typical residual stress travers of the back side of the specimens. The longitudinal residual stress value of the shot-peened side are substantially smaller than the back side.

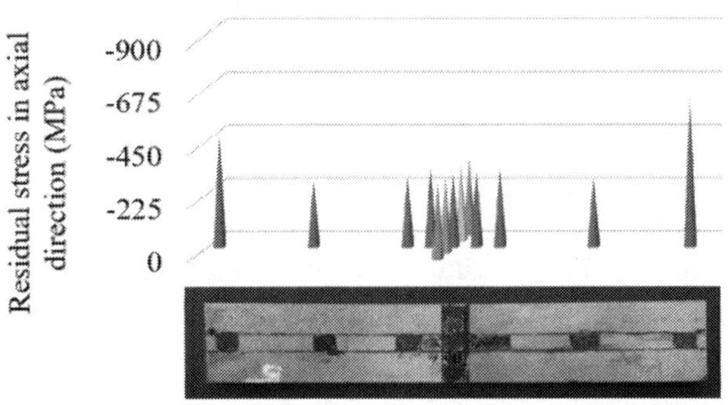

Figure 16. longitudinal residual stress length and cross travers on the back side of the hardened sample

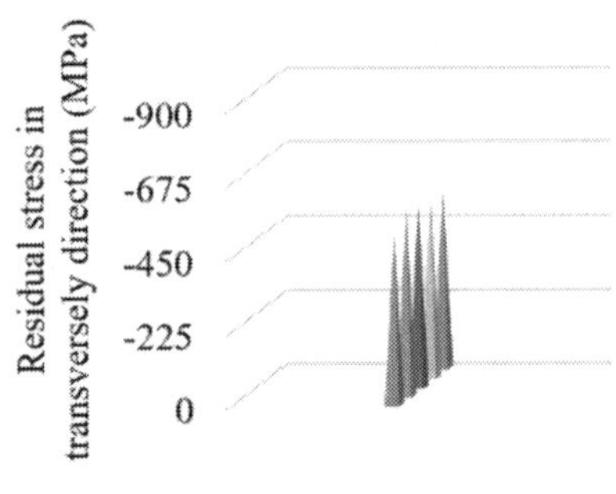

Figure 17. transverse residual stress cross travers on the back side of the hardened sample

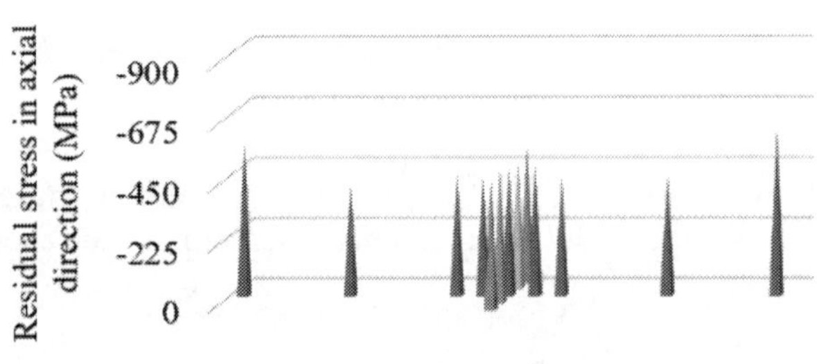

Figure 18. longitudinal residual stress length and cross travers on the back side of the HSLA sample

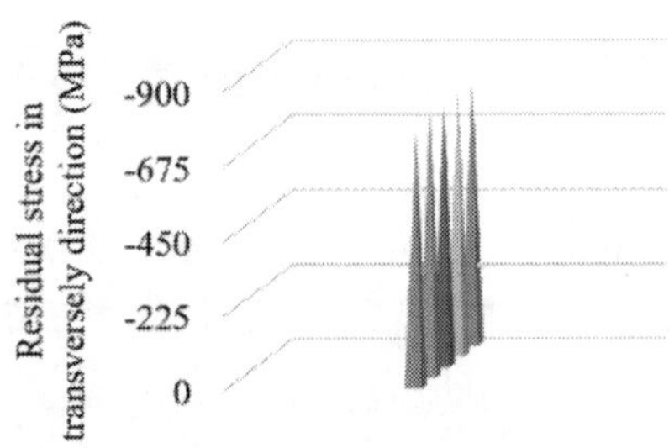

Figure 19. transverse residual stress cross travers on the back side of the HSLA sample

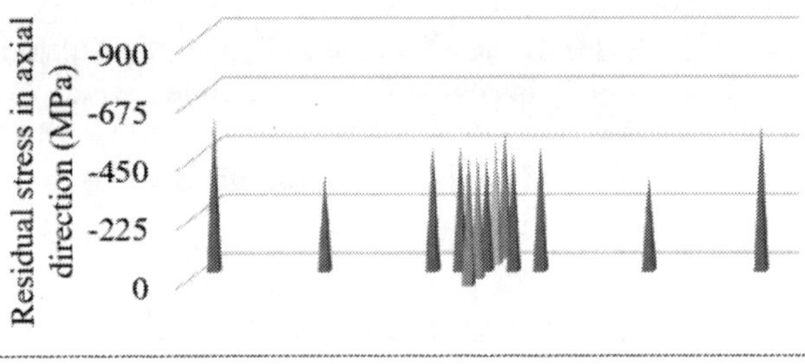

Figure 20. longitudinal residual stress length and cross travers on the back side of the stainless sample

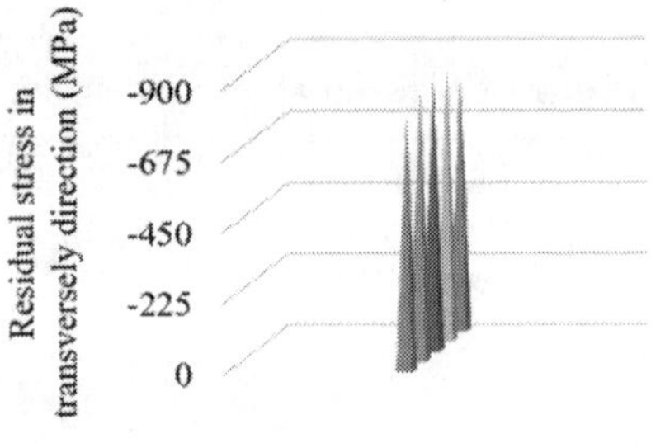

Figure 21. transverse residual stress cross travers on the back side of the stainless sample

The asymmetric cold forming (the shot-peening stretched only the upper side of the stripes) caused residual stresses were rearranged and generated a remarkable, at least 300MPa magnitude longitudinal compressive residual stress on the back side. The transverse directional residual stress traverses are interesting. The transverse residual stress values largely exceed the longitudinal ones, which is an unexpected phenomenon.

All specimens exposed to increasing static load in discrete steps ant he residual stress state was recorded in every step. The applied bending force were controlled, and fixed on calculated values during the measurements. The applied force calculated form equation (1) so the generated stress was 50MPa or the multiple of it. The external load was increased until the measurable stress on the investigated point of the sample surface shown approximately stress less state. The unloading was done by the same schedule, and the residual stress recorded during the unload also at the same bending forces. Figure 22. present a snapshot about the residual stress measurement, while the sample loaded by the four-point bending bench. The white arrows show the impact lien of the force, black ones show the lines hold against. The side line of the specimen highlighted with white. The yellow arrow denotes the primary beam and shows on the measured point what the calculated stress values concern. Orange arrows denotes the reflected radiation and show on the detectors.

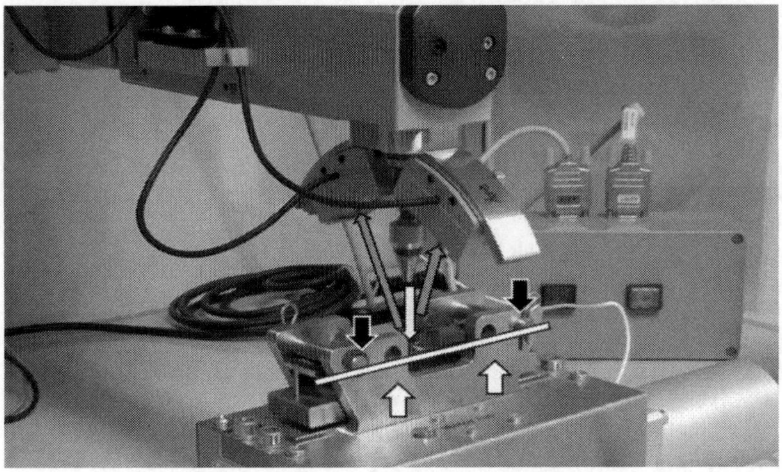

Figure 22. the bending device with a specimen under the X-ray diffractometer

Figure 23. present an example of the static loaded specimens measurement result. The measured stress values are illustrated in the function of the from force calculated external loads. Trendline was fitted on the results. The fitting and the slope (linear equation) was calculated.

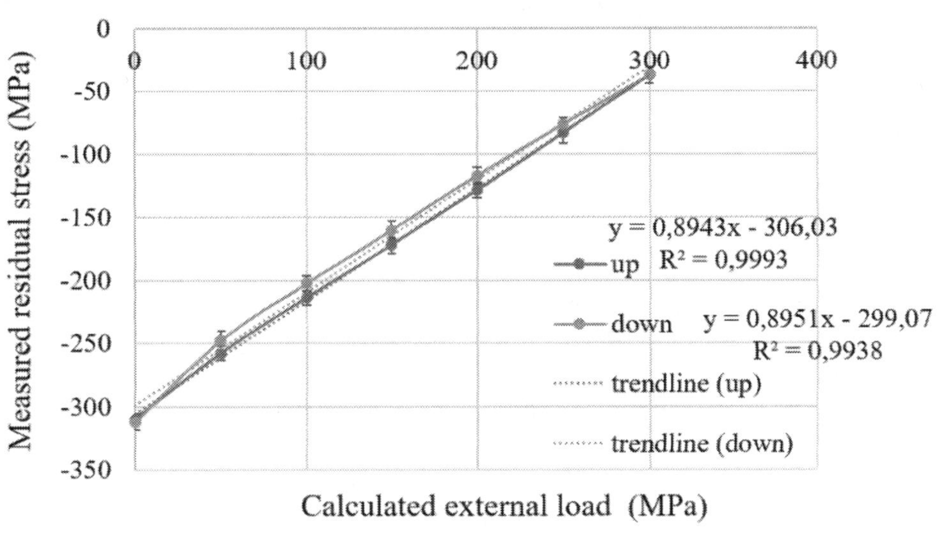

Figure 23. the residual stress behavior during static load in a hardened specimen

IOP Conf. Series: Journal of Physics: Conf. Series **843** (2017) 012041 doi:10.1088/1742-6596/843/1/012041

Six specimens were investigated by the same method as it is presented in figure 23. As a result of the bending tests linear equations and fitting (R^2) data were gained, represented in table 1.

Table 1. Linear equation and fitting datas of bending tests

line identification		linear equation	fitting (R^2)	average slope
Hardened 1	up	y = 0,8943x - 306,03	0.9993	0.878
	down	y = 0,8951x - 299,07	0.9938	
Hardened 2	up	y = 0,8492x - 331,24	0.9997	
	down	y = 0,8736x - 334,36	0.9983	
HSLA 1	up	y = 0,8887x - 431,96	0.9993	0.897
	down	y = 0,8948x - 427,66	0.9978	
HSLA 2	up	y = 0,889x - 452,39	0.998	
	down	y = 0,9147x - 451,36	0.999	
Austenitic stainless 1	up	y = 0,6442x - 346,36	0.9976	0.653
	down	y = 0,6659x - 337,77	0.9882	
Austenitic stainless 2	up	y = 0,6351x - 398,2	0.9979	
	down	y = 0,665x - 399,9	0.9951	
As received, not peened austenitic stainless	up	y = 0,968x + 62,959	0.9988	0.943
	down	y = 0.917x + 121.56	0.998	

Except the austenitic stainless 1 down, which fit with 0.988 R^2, all of the trendlines fit at least 0.99. The fitting is well generally. The different materials have different average slopes. The 3D scanning results of the surfaces of the static specimen help understand this phenomenon.

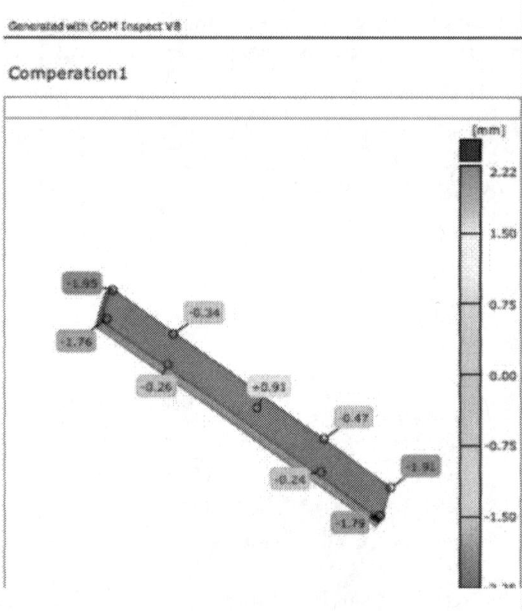

Figure 24. the 3D scanned image of the shot peened side of the stain less steel specimen 1, with 9 height data in the indicated points

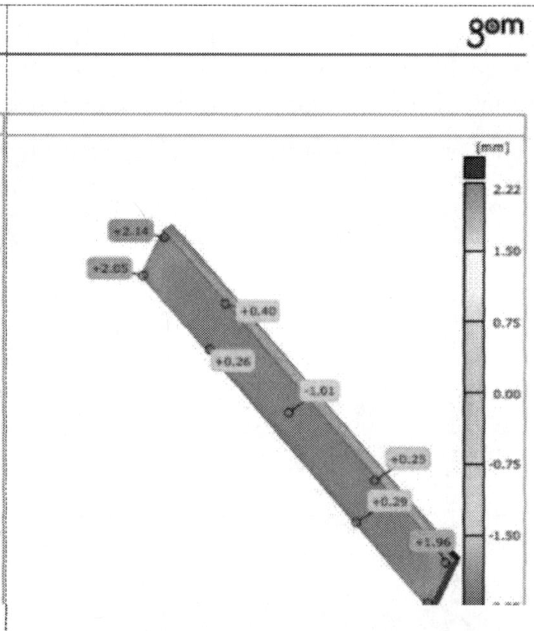

Figure 25. the 3D scanned image of the back side of the stain less steel specimen 1, with 9 height data in the indicated points

Table 2. height data in the 9 indicated points

scanned surface		**height data (mm)**				
Hardened 1	top	-1.03	-0.24	0.46	-0.01	-1.12
		-1.04	-0.04		-0.02	-1.15
	back	0.91	-0.17	-0.55	-0.11	1
		0.97	-0.23		-0.05	0.98
Hardened 2	top	-0.99	0.3	0.43	0.9	-1.09
		-1	0.16		-0.13	-1.12
	back	0.79	-0.08	-0.56	-0.3	0.9
		0.87	-0.24		-0.15	0.9
HSLA 1	top	-1.16	-0.20	0.30	-0.40	-1.07
		-1.18	-0.10		-0.34	-1.15
	back	0.98	0.12	-0.35	0.02	1.03
		0.97	0.16		0.24	1.07
HSLA 2	top	-1.06	-0.08	0.39	0.09	-1.3
		-1.01	0.04		-0.04	-1.07
	back	0.94	-0.13	-0.57	-0.2	1.09
		1.01	-0.24		-0.08	1.16
Austenitic stainless 1	top	-1.95	-0.34	0.91	-0.47	-1.91
		-1.76	-0.26		-0.24	-1.79
	back	2.14	0.4	-1.01	0.25	1.96
		2.05	0.26		0.29	1.91
Austenitic stainless 2	top	-1.9	-0.6	0.88	-0.55	-1.84
		-1.91	-0.31		-0.37	-1.93
	back	1.91	0.37	-0.97	0.2	1.79
		1.98	0.38		0.31	1.94

After the discharging from the clamping tools the specimens were warped to trough form as it shown on the 3D scanned results on figure 24. and figure 25. The shot peened (top) surface of the specimens is convex in every case. Table 2. contains the 3D scanned results of all the specimens. The curvature intensity is similar the height differences on the surface. The intensity of the curvature shows material dependent characteristic. The different average slopes in table 1. correlate to the different intensity of the curvature in table 2. A not shot peened, perfectly flat specimen were also submitted to static load measurements. The slope of the not peened specimens is approximately 1, this verify the correlation between the curvature and the slope of the bending test trend lines. The higher curvature cause the lower slope, and the perfectly flat specimen has a slope of almost 1.

No relaxation was observable after the static load. Figure 23. shows result of only one hardened specimens. None residual stress values of any specimens show relaxation after the bending tests.

5. Summary

The residual stress holding stability of an untested kind of steel 42CrMo4 was investigated during fatigue load. The alloy shows 25% residual stress relaxation on the applied fatigue stress level. The most of the total stress relaxation extent happen until the first several thousands of cycles.

Three alloy, hardened, HSLA and austenitic stainless steel were examined in the aspect of residual stress holding capability during static load. The compressive residual stress of the samples has proven stable at the applied static load. A detailed description of the surface residual stress distribution evolution of the static samples was presented. The measurable stress during the bending test were fallen short of the from external force calculated stress. Correlation has found between this phenomenon and the extent of the warping (curvature) of the specimens. The warping was caused by the asymmetric cold forming of the surface, only one side of the specimen plates were treated by shot-peening. The correlation was verified by bend testing of a not peened, perfectly flat specimens. The extent of the warping shows material dependence. The lower strength allows higher plastic deformation, and also play important rule during the stress relaxation accompanied by permanent deformation.

6. Acknowledgement

The authors say thanks to Szabolcs Nyitrai for the 3D scanning.
The described article was carried out as part of the EFOP-3.6.1-16-00011 "Younger and Renewing University – Innovative Knowledge City – institutional development of the University of Miskolc aiming at intelligent specialisation" project implemented in the framework of the Szechenyi 2020 program. The realization of this project is supported by the European Union, co-financed by the European Social Fund.

SUPPORTED THROUGH THE NEW NATIONAL EXCELLENCE PROGRAM OF THE MINISTRY OF HUMAN CAPACITIES

7. References

[1] Davis, J., 2002. *Surface hardening of steels.* USA: ASM International.
[2] Timoshenko, S., 1983. *History of strength of materials.* Dover: Dover Publications.
[3] ASM International, 2002. *Handbook of Residual Stress and Deformation of Steel.* USA: ASM International.
[4] Stephens, R. I., Fatemi, A., Stephens, R. R. & Fuchs, H. O., 2001. *Metal fatigue in Engineering 2nd Edition.* USA: John Wiley & Sons.
[5] Macherauch , E. & Hauk, V., 1987. *Residual stess in Scinence Technology Vol 2.* Németország: I.R. Publications.
[6] Nikitin, I. & Besel, M., 2008. Correlation between residual stress and plastic strain amplitude during low cycle fatigue of mechanically surface treated austenitic stainless steel AISI 304 and ferritic–pearlitic steel SAE 1045. *Materials Science and Engineering A,* pp. 297-303.
[7] Dalaeia, K., Karlssona, B. & Svenssona, L.-E., 2010. Stability of shot peening induced residual stresses and their influence on fatigue lifetime. *Materials Science and Engineering A,* pp. 1008-1015.
[8] Kim, J.-C., Cheong, S.-K. & Nog, H., 2013. Residual stress relaxation and low- and high-cycle fatigue behavior of shot-peened medium-carbon steel. *International Journal of Fatigue,* pp. 114-122.
[9] Kim, J.-C., Cheong, S.-K. & Nog, H., 2014. Proposed fatigue damage measurement parameter for shot-peened carbon steel based on fatigue crack growth behavior. *International Journal of Fatigue,* pp. 97-106.
[10] Cseh, D., Mertinger, V. & Lukacs, J., 2013. Residual Stress Evolution During Fatigue Test of a Shoot Peened Steel Sample. *Materials Science Forum,* Vol. 752. kötet, pp. 95-104.
[11] ASM International, 1991. *Heat treating ASM Handbook Vol. 4,.* USA: ASM International.
[12] ASM International, 1996. *Cast Irons ASM Specialty Handbbok.* USA: ASM International.
[13] Haimbaugh, R., 2001. *Induction Heat Treating.* USA: ASM International.
[14] Cseh , D., Mertinger , V. & Lukacs, J., 2016. *Residual Stress Behavior in Hardened Shot Peened 42CrMo4 specimens during fatigue load.* Sydney, pp. 497-501.

IOP Conf. Series: Journal of Physics: Conf. Series **843** (2017) 012042 doi:10.1088/1742-6596/843/1/012042

Numerical determination of Paris law constants for carbon steel using a two-scale model

M Mlikota[1], S Staib[1], S Schmauder[1] and Ž Božić[2]

[1] Institute for Materials Testing, Materials Science and Strength of Materials (IMWF), University of Stuttgart, Pfaffenwaldring 32, 70569 Stuttgart, Germany
[2] Faculty of Mechanical Engineering and Naval Architecture, University of Zagreb, I. Lučića 5, 10000 Zagreb, Croatia

marijo.mlikota@imwf.uni-stuttgart.de (corresponding author)

Abstract. For most engineering alloys, the long fatigue crack growth under a certain stress level can be described by the Paris law. The law provides a correlation between the fatigue crack growth rate (FCGR or da/dN), the range of stress intensity factor (ΔK), and the material constants C and m. A well-established test procedure is typically used to determine the Paris law constants C and m, considering standard specimens, notched and pre-cracked. Definition of all the details necessary to obtain feasible and comparable Paris law constants are covered by standards. However, these cost-expensive tests can be replaced by appropriate numerical calculations. In this respect, this paper deals with the numerical determination of Paris law constants for carbon steel using a two-scale model. A micro-model containing the microstructure of a material is generated using the Finite Element Method (FEM) to calculate the fatigue crack growth rate at a crack tip. The model is based on the Tanaka-Mura equation. On the other side, a macro-model serves for the calculation of the stress intensity factor. The analysis yields a relationship between the crack growth rates and the stress intensity factors for defined crack lengths which is then used to determine the Paris law constants.

1. Introduction

Structural health monitoring and detection of failures are highly important for fitness and service assessment of structures. Failure of structures under fatigue loading can occur at load levels below the yield stress of the used material. Therefore, it is of special importance to be able to make predictions of life cycles until catastrophic fracture occurs. A well-known example of such catastrophic failure is the huge train accident in Eschede where one wheel of the train broke due to cyclic loading (figure 1) that has not been considered during construction [1]. On the other hand not every single crack has to be critical immediately. Often structures can withstand cracks up to a certain crack length until unstable fracture occurs. In order to determine the point where crack growth becomes unstable, it is necessary to develop methods to describe crack growth mathematically.

Figure 1. Fracture surface of the broken train wheel [1].

A very common and often used method for the characterization of the long crack growth is the Paris law which gives a relationship between the fatigue crack growth rate (FCGR or da/dN) and the stress intensity factor (ΔK) at the crack tip during the stable crack growth [2], equation (1). A typical fatigue growth rate curve – also known as da/dN versus ΔK curve – is shown in figure 2. The curve is defined by regions I, II and III. The Paris law relationship can be visualized as a straight line for the region of stable crack growth – region II [3]. The accompanying mathematical equation contains two material parameters C and m, where m represents the slope of the line in figure 2 and C the y-axis-intercept [4]:

$$\frac{\mathrm{d}a}{\mathrm{d}N} = C(\Delta K)^m \tag{1}$$

The crack growth rates in the region II are typically in the order of 10^{-9} to 10^{-6} m/cycle and correspond to stable crack growth. The constants C and m are usually determined in experiments [5-9] and depend on the material and various influencing factors such as temperature, environmental medium and loading ratio [5, 10]. The last is probably the most significant and usually results in closely spaced lines parallel to each other. For metals the exponent is typically of the order m = 2…4 [10].

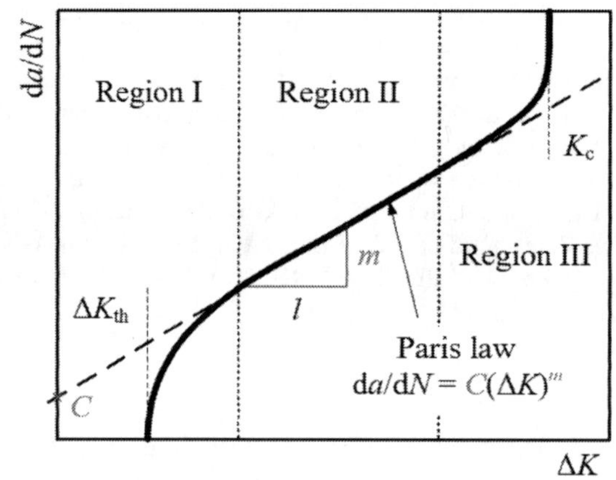

Figure 2. Geometry of the used specimen with a central crack [3].

Since the experimental determination of the Paris law constants is typically tedious and time consuming, the objective of this paper is to determine them numerically considering the influence of microstructure on the crack growth rate.

2. Modelling approach and details

The simulations for determination of the Paris law constants are done with the Finite Element (FE) software ABAQUS and by applying a two-scale model. The model contains the microstructural submodel of a material that is used to calculate the fatigue crack growth rate d*a*/d*N* at a crack tip and the macro-model that serves for the calculation of the stress intensity factor ΔK. In order to get the typical d*a*/d*N* versus ΔK curve, the crack growth rate and stress intensity factors for six different crack lengths are determined. The constants are then determined from the interpolated straight line, representing region II in figure 2, as the slope *m* and the y-axis-intercept *C*.

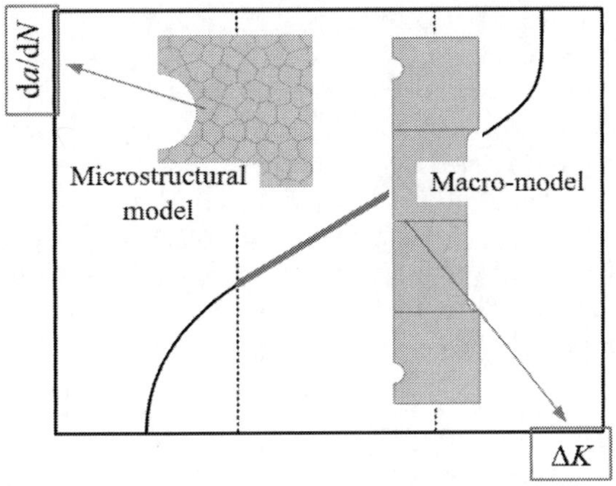

Figure 3. Schematic illustration of the two different approaches.

The input data for the common diagram is calculated using the two aforementioned approaches, as shown in figure 3. On the one hand the stress intensity factor is determined using the full-scale macro-model with a center crack and on the basis of Linear Elastic Facture Mechanics (LEFM). The crack growth rate on the other hand is determined using the submodel based on the Tanaka-Mura equation [11, 12]; the submodel is placed in the area behind the crack tip. The grain structure is assigned to the submodel to consider micro-crack initiation processes in the vicinity of the crack tip.

The Tanaka-Mura equation is typically used to estimate the number of cycles that are attributed to the crack nucleation in a single grain [11, 12]. Jezernik *et al.* [13, 14] introduced a modification of the original equation in the sense that the crack does not form instantaneously through the whole grain but it forms in segmental manner:

$$N_s = \frac{8GW_c}{\pi(1-\nu)d_s(\Delta\bar{\tau} - 2CRSS)^2} \tag{2}$$

where N_s is the number of cycles required to form a crack segment in a single grain, i.e. crystal. Furthermore, equation (2) contains two microstructure-related parameters, namely the length of slip line segment d_s and the average shear stress range $\Delta\bar{\tau}$ on it. Other material constants (such as shear modulus G, crack initiation energy W_c and Poisson's ratio ν) can be either found in literature for most materials or can alternatively be obtained experimentally. The critical resolved shear stress (*CRSS*) is particularly interesting since it can be obtained by means of Molecular Dynamics (MD) simulations [15]. The *CRSS* represents a critical value of the shear stress along the slip direction that must be overcome for the dislocation to move, i.e. if the magnitude of the resolved shear stress is below the value of the *CRSS*, no dislocation movement is allowed and consequently no pile up at the grain boundary takes place [16].

2.1. Geometry of the macro-models

In order to accomplish the goal of the present paper it was necessary to create two slightly different full-scale macro-models. Both models represent the same geometry, however, the ways of modelling the crack for determination of stress intensity factor ΔK on the one side and fatigue crack growth rate da/dN on the other require certain adjustments at the regions of interest, i.e. on the crack-affected path. More specifically, the macro-model for the calculation of ΔK requires the usage of some special ABAQUS techniques to represent the crack – in this case the seam crack is used – while the macro-model (further: global model) for determination of da/dN needs to be geometrically adjusted to the submodel that is placed at the tip of the structural crack. A seam defines an edge or a face in a model that is originally closed but can open as a crack during an analysis [17].

The models were built on the basis of a specimen with central pre-crack prepared for fatigue testing, shown in figure 4, considered in the paper of Božić *et al.* [3]. At the initial state the specimen has a notch and a pre-crack of 2a = 8 mm (figure 4, A – Crack detail). Only half of the specimen has to be modeled due to symmetry with respect to the vertical y-axis. The symmetry with respect to the x-axis is aligned with the crack and, therefore, cannot be used in either case; for ΔK determination the applied seam cannot extend along the boundaries of a part and must be embedded within a face of a two-dimensional (2D) part or within a cell of a solid part [17]. In the other case, the used submodel (microstructural model) at the crack tip area requires the transfer of boundary conditions – in this case displacements – from both parts of the global model, the upper and the lower one.

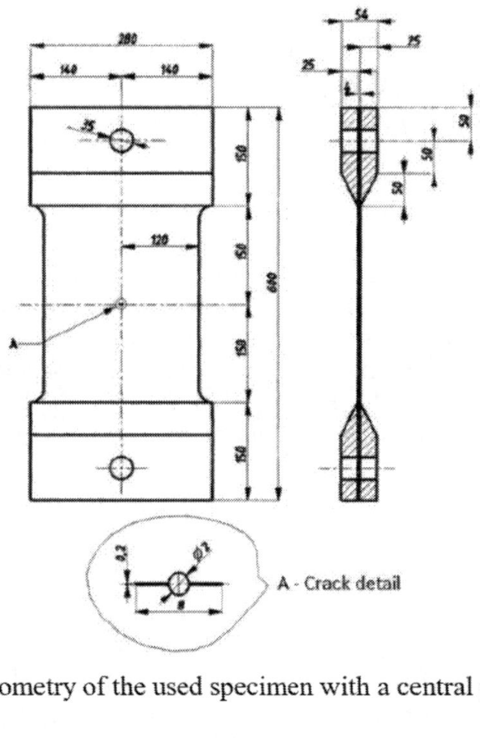

Figure 4. Geometry of the used specimen with a central pre-crack [3].

The structural crack lengths which were used for this study were taken from [3], where ΔK values were determined using the FE software ANSYS. Nevertheless, ΔK values were calculated again – this time with ABAQUS – first to compare the results with previous simulations and second to ensure that the models are built properly. The considered crack lengths for the simulations are listed in table 1.

Table 1. Considered structural crack lengths.

Crack designation	Crack length (mm)
a_1	9.9
a_2	20.1
a_3	29.9
a_4	39.1
a_5	49.3
a_6	59.3

All models which are used here are built as 2D, in accordance to the geometry of the specimen and the fact that the main central part has a thickness of just 4 mm along the z-direction. The areas in the upper and lower part of the models with relatively higher thickness (54 mm) were also modelled as 2D. Those two parts are used to apply loading conditions and movement constraints, respectively, as shown in figure 5. The stress state in the middle part of the cracked plate is assumed to be plane stress. The thicknesses of the specimen in both regions, the central and outer, were considered by assigning plane stress thickness to their belonging sections.

As already mentioned only one half of the specimen has to be modelled. The symmetry is realized by using boundary conditions on the vertical y-axis which fix displacements in x-direction as well as rotations of any kind (figure 5).

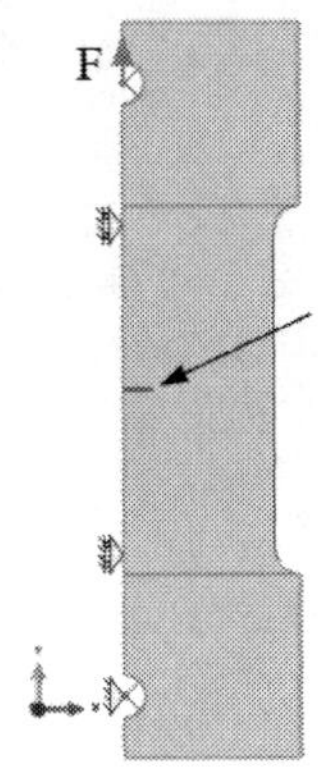

Figure 5. The half of full-scale model with applied boundary and loading conditions, and with the seam crack in the ΔK case.

The material behaviour is assumed to be purely linear elastic; only a small plastic zone at the crack tip is expected (figure 7), therefore, no plastic material data are necessary to be used. The isotropic material data for the specimen made of conventional mild carbon steel were adopted from the study of Božić *et al.* [3]. The material parameters applied to both models are: Young modulus E = 206 GPa, Poisson's ratio ν = 0.3, yield stress R_e = 235 MPa and shear modulus G = 80 GPa. It is opportune to introduce the experimentally obtained Paris law constants for the selected material in order to provide validation data for the numerical results that follow later; m = 2.75 and C = 1.43x10^{-11} [3].

As the material data, loading and boundary conditions were taken from [3]. The testing specimen was exposed to constant amplitude cyclic tension load in a hydraulic fatigue testing machine. The load was applied to the pin which was placed in the hole in the upper part of the model while the pin in the lower hole was fixed (figure 5). The force range and the stress ratio applied to the half-model are denoted by $F = F_{max} - F_{min}$ and $R = F_{min} / F_{max}$, and were $F = 76\ 800$ N and $R = 0.0253$ [3]. In contrast to the experiments, simulations on the macro-models were performed with static loading conditions, however, in combination with LEFM for the ΔKs determination and with the Tanaka-Mura equation for the determination of da/dN.

For both models continuum plane stress 4-node bilinear elements with reduced integration and hourglass control (CPS4R) were used. In order to deal with the stress singularity at the crack tip in the case of ΔKs calculation, special elements have to be used that are able to indicate the infinite stresses properly. In ABAQUS this is done by collapsing one side of an 8-node isoparametric element connected to the crack tip [17], as can be seen in figure 7.

2.2. Submodel (microstructural model) details

The submodel or microstructural model is placed right at the tip of the global model structural cracks (table 1) where their extension is expected. As already mentioned, the Tanaka-Mura equation is typically used to estimate the duration of the short crack initiation stage. In this paper, however, the equation is applied within the microstructural model (figure 6) with the aim to estimate the rate of the infinitesimal crack extension. The geometry of the submodel can be seen in figure 6, as well as the location where it is placed with respect to the global model. Its size is selected to 0.4 mm x 0.4 mm including the tip of the structural crack of 0.1 mm radius, which can be identified as a notch. As mentioned before, displacements of the global model are applied to the coinciding edges of the submodel, marked bold in figure 6.

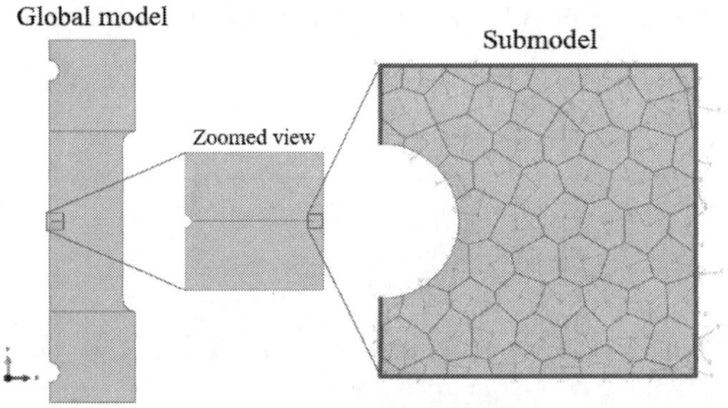

Figure 6. Geometry of the submodel containing a microstructure of the material and its location with respect to the global model.

Moreover, the submodel takes into consideration the microstructure of the investigated material and its influence on the crack growth rate in the initial extension phase. Accordingly, the submodel is partitioned into individual grains using the random Voronoi tesselation. Each individual grain possesses a local coordinate system (figure 6) and individual slip band orientation. The resulting average grain size, i.e. average slip band length is 50 μm, what is appropriate for the used material.

Concerning the material description, the microstructural model requires orthotropic material data, i.e. elastic constants of cubic crystals. These constants reflect cubic symmetry where $C_{11} = C_{22} = C_{33}$, $C_{12} = C_{13} = C_{23}$ and $C_{44} = C_{55} = C_{66}$, and are calculated using the following equations: $C_{11} = E(1 - \nu)/(1 - \nu - \nu^2)$; $C_{12} = E\nu/(1 - \nu - 2\nu^2)$; $C_{44} = G$. With the isotropic material data from above this gives $C_{11} = 277\ 307$ MPa, $C_{12} = 118\ 846$ MPa and $C_{44} = 80\ 000$ MPa. Furthermore, the microstructural model enriched with the Tanaka-Mura equation requires two additional material properties, namely the critical resolved shear stress ($CRSS$) and the crack initiation energy (W_c). With respect to this, the following values were applied: $CRSS = 117$ MPa [15] and $W_c = 69$ N/mm [18].

IOP Conf. Series: Journal of Physics: Conf. Series **843** (2017) 012042 doi:10.1088/1742-6596/843/1/012042

The submodel was very fine meshed with the same elements as the global model (CPS4R); what in the end gives smooth stress distribution in the undamaged as well as in the damaged submodel (figures 9-12). To depict the mesh fineness, a single grain has 120 elements in average.

3. Results

The stress intensity factor ΔK as well as the *J*-Integral and other fracture mechanics characteristics, can be requested in ABAQUS as history output data. In order to get ΔK for each individual crack of six in total from table 1, the same number of variations of the macro-model were modelled. Figure 7 gives the von Mises stress distribution at the crack tip of a macro-model. As already mentioned in Chapter 2, the geometry itself with the structural pre-crack stays the same for all variations, however, the used seam length varies. By evaluating ΔK for each crack length and putting the results into a common diagram gives the expected linear relationship (see region II in figure 2) between ΔK and the crack length a (figure 8).

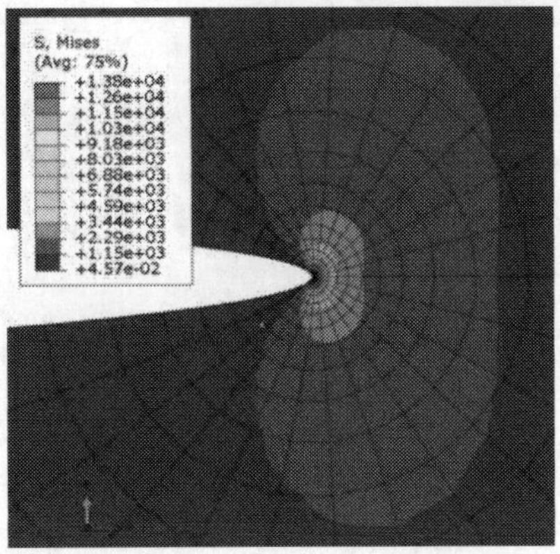

Figure 7. Stress field at the crack tip of the macro-model that is used to calculate the ΔK values.

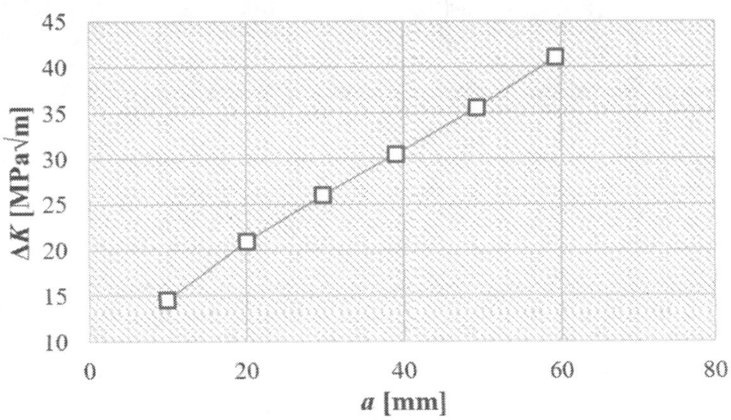

Figure 8. ΔK in dependence of the crack length.

In the submodel, the crack can form in a grain on different slip bands depending on the grain orientation, i.e. on the stress field inside and in the vicinity of the grain – generally, the stress field in the microstructural model (figures 9-12) is influenced by boundary conditions (displacements from the global model), microstructural configuration (i.e. orientation and shape of crystals), elastic constants and geometrical factors (notch, formed cracks etc.). The slip bands of a certain grain have the same orientation and are distanced one from each other with an offset. Furthermore, each slip band is divided in four segments meaning that a crack does not form instantaneously through the whole grains but it forms in a segmental manner. The condition for the formation of crack segments is contained in the Tanaka-Mura equation (2) and it reads as follows: the average shear stress $\Delta\bar{\tau}$ on a slip band segment needs to exceed two times the *CRSS*. Additionally, the equation (2) gives the number of cycles dN (i.e. N_s) that are spent for every formed crack segment. Furthermore, after the crack condition has been satisfied in one step, the model gets updated with the latest crack and remeshed in the following step where the condition is applied again and a new weakest crack segment is traced.

The von Mises stress state before the first formed crack segment is shown in figure 9. Figures from 10 to 12 show the stress states after 5, 15 and 25 broken crack segments. Those four figures show the simulation results for the crack length of 20.1 mm. Results for other considered structural cracks from table 1 are principally similar and, therefore, not shown in the paper.

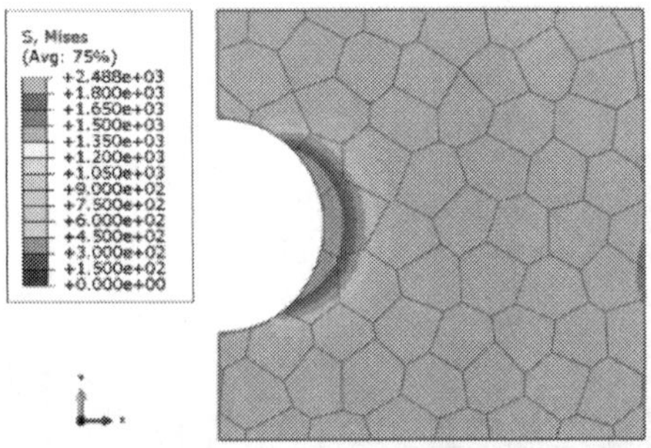

Figure 9. Von Mises stress distribution in the undamaged microstructural model (a = 20.1 mm).

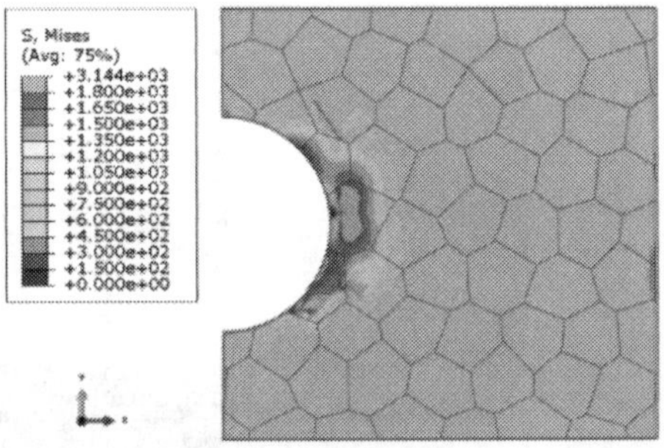

Figure 10. Microstructural model with 5 broken crack segments.

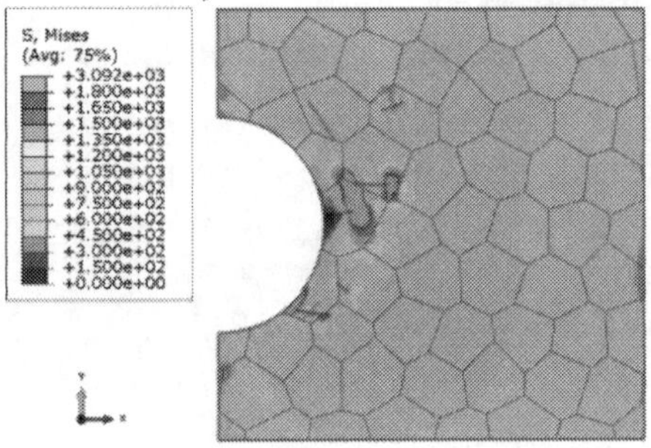

Figure 11. Microstructural model with 15 broken crack segments.

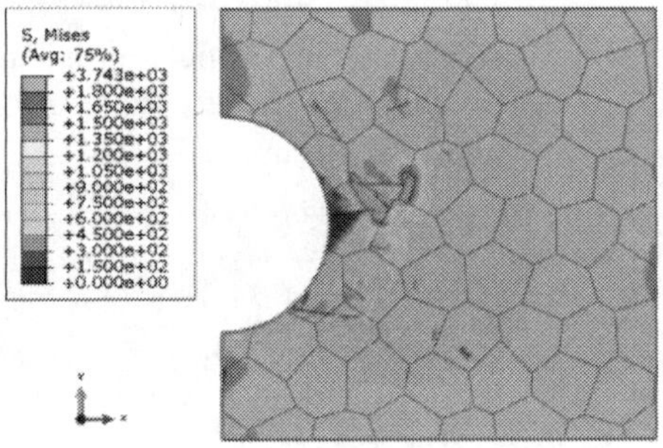

Figure 12. Microstructural model with 25 broken crack segments.

Generally what happens for all considered structural cracks is that the formation of cracks in the submodel ceases after a certain number of steps – the number varies from one model setup to another. The reason for this is that grains that are favourable for cracking on the basis of the aforementioned condition fade out.

In order to determine the fatigue crack growth rate da/dN, both the cycles dN and the accompanying length da of each individual crack segment that formed in the microstructural model have to be determined. The crack length da can easily be quantified using the ABAQUS graphical interface or can be rather gathered from output data.

By taking the measured da and the correlated dN one can evaluate the growth rate da/dN for each step. It is necessary to indicate that the common growth rate da/dN is not a constant value, but it fluctuates as it can be seen in figure 13 for the case of the 20.1 mm long structural crack. It can be noticed that figure 12 contains 25 broken segments while only 11 are considered in figure 13 where

da/dN is plotted. An explanation for this is that starting from step 12 in figure 13 the number of required cycles for further formed cracks becomes significantly higher, meaning directly that the da/dN drops down to a negligible value. This leads to the assumption and consequence that the crack growth rate should be neglected in this relatively advanced stage.

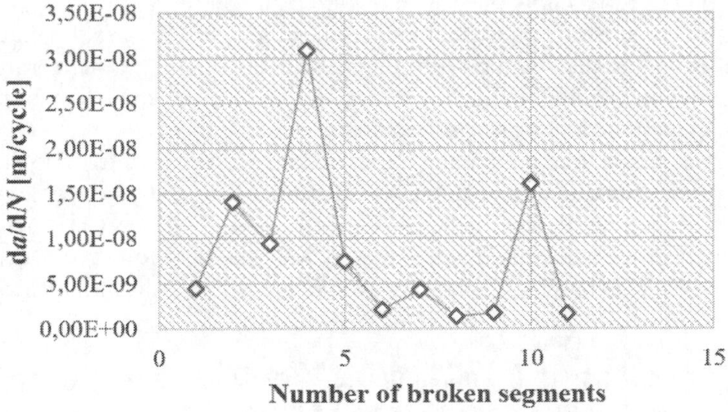

Figure 13. Crack growth rate in relation to the number of broken segments for the extension of the 20.1 mm long structural crack.

In order to get a typical da/dN versus ΔK plot as in figure 2, the fluctuating crack growth rate from, e.g. figure 13, needs to be averaged. For the structural crack length of 20.1 mm this results in $da/dN = 8.49421\text{x}10^{-9}$, which seems to be appropriate according to literature [3, 19]. This averaged growth rate can be considered as extensional growth rate for the structural crack. The same procedure was applied for remaining structural crack lengths. Finally, the results for da/dN and for ΔKs (figure 8) are put into a common log da/dN versus log ΔK diagram, figure 14. The resulting six single points were interpolated with a straight line from which the material constants of Paris equation (1) were determined, the slope of the line is $m = 2.75$ and the y-axis intercept $C = 3.8\text{x}10^{-12}$, which agree quite well with the experimentally determined values of $m = 2.75$ and $C = 1.43\text{x}10^{-11}$ [3].

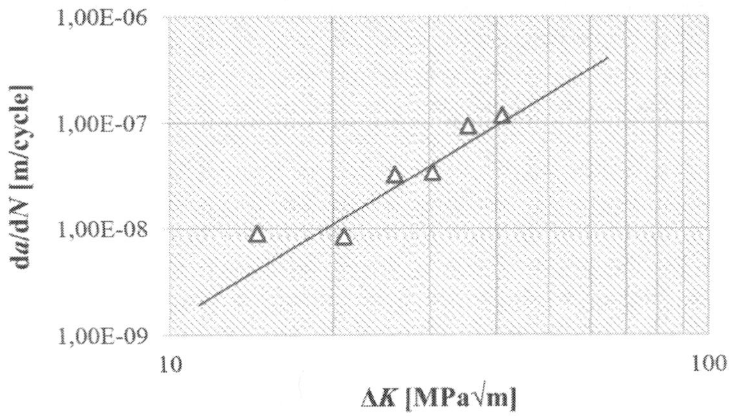

Figure 14. Extensional crack growth rate da/dN of different structural cracks versus ΔK of the same cracks. Data are shown in log-log scale.

4. Discussion and conclusions

The aim of this paper was to numerically determine the Paris law material constants *C* and *m* and this was successfully accomplished by applying a two-scale model. Namely, stress intensity factors for six different structural cracks were calculated using the macro-model based on classical LEFM. In the second part, the fatigue crack growth rates at the tips of same structural cracks were determined by using the microstructural model enriched by the Tanaka-Mura equation.

Having determined the stress intensity factors and the crack growth rates separately, the values have been visualized afterwards in the common log-log diagram (figure 14). Considering that the Paris law can be approximated by a straight line in the diagram for the case of stable crack growth, the material constants have been determined then as $m = 2.75$ and $C = 3.8\text{x}10^{-12}$.

Since this study was the first, to the authors' knowledge, attempt to determine the Paris constants in a numerical way as well as taking into account the microstructure of the material, and considering that the literature values of the investigated material ($m = 2.75$ and $C = 1.43\text{x}10^{-11}$) match the numerically obtained ones quite good, the study and its working hypothesis seem to prove their worthiness. It is to be supposed that the results determined here may even become better during further research.

References

[1] Esslinger V, Kieselbach R, Koller R and Weisse, B 2004 *Eng. Fail. Anal.* **11**(4) 515-35
[2] Paris P and Erdogan F 1963 *J. Basic Eng.* **85**(4) 528-33
[3] Božić Ž, Mlikota M and Schmauder S 2011 *The. Vjesn.* **18**(3) 459-66
[4] Broek D 1988 *The Pratical Use of Fracture Mechanics* (Dordrecht: Kluwer Academic Publishers)
[5] Branco R, Antunes F, Ferreira JM and Silva M 2009 *Eng. Fail. Anal.* **16** 631-38
[6] Branco R, Antunes F, Costa D, Yang FP and Kuang ZB 2012 *Eng. Fract. Mech.* **96** 96-106
[7] Carrascal I, Casado J, Diego S, Lacalle R, Cicero S and Alvarez J 2014 *Polym. Test.* **40** 39-45
[8] Chauhan S, Pawar AK, Chattopadhyay J and Dutta BK 2016 *T. Indian I. Metals* **69**(2) 609-15
[9] Ancona F, Palumbo D, Finis RD, Demelio G and Galietti U 2016 *Eng. Fract. Mech.* **163** 206-19
[10] Szata M and Lesiuk G 2009 *Arch. Civ. Mech. Eng.* **9**(1) 119-34
[11] Tanaka K and Mura T 1981 *J. Appl. Mech.* **48**(1) 97-103
[12] Tanaka K and Mura T 1982 *Metall. Trans. A* **13**(1) 117-23
[13] Jezernik N, Kramberger J, Lassen T and Glodez S 2010 *Fatigue Fract. Eng. M.* **33**(11) 714-23
[14] Glodez S, Jezernik N, Kramberger J and Lassen T 2010 *Adv. Eng. Softw.* **41**(5) 823-29
[15] Božić Ž, Schmauder S, Mlikota M and Hummel M 2014 *Fatigue Fract. Eng. M.* **37**(9) 1043-54
[16] Mlikota M, Schmauder S, Božić Ž and Hummel M 2017 *Fatigue Fract. Eng. M.* (accepted)
[17] SIMULIA *ABAQUS Documentation*
[18] Roos E and Eisele U 1988 *J. Test. Eval.* **16**(1) 1-11
[19] Božić Ž, Schmauder S and Mlikota M 2011 *Key Eng. Mater.* **488-489** 525-28

Comparative Study on Prediction Effects of Short Fatigue Crack Propagation Rate by Two Different Calculation Methods

Bing Yang, Zhen Liao, Yahang Qin, Yayun Wu, Sai Liang, Shoune Xiao, Guangwu Yang, Tao Zhu

State Key Laboratory of Traction Power, Southwest Jiaotong University, Chengdu, 610031, China

Email: yb@swjtu.cn

Abstract. To describe the complicated nonlinear process of the fatigue short crack evolution behavior, especially the change of the crack propagation rate, two different calculation methods are applied. The dominant effective short fatigue crack propagation rates are calculated based on the replica fatigue short crack test with nine smooth funnel-shaped specimens and the observation of the replica films according to the effective short fatigue cracks principle. Due to the fast decay and the nonlinear approximation ability of wavelet analysis, the self-learning ability of neural network, and the macroscopic searching and global optimization of genetic algorithm, the genetic wavelet neural network can reflect the implicit complex nonlinear relationship when considering multi-influencing factors synthetically. The effective short fatigue cracks and the dominant effective short fatigue crack are simulated and compared by the Genetic Wavelet Neural Network. The simulation results show that Genetic Wavelet Neural Network is a rational and available method for studying the evolution behavior of fatigue short crack propagation rate. Meanwhile, a traditional data fitting method for a short crack growth model is also utilized for fitting the test data. It is reasonable and applicable for predicting the growth rate. Finally, the reason for the difference between the prediction effects by these two methods is interpreted.

Keywords: short fatigue crack; replica test; crack propagation rate; genetic algorithm; date fitting

1. Introduction

Fatigue fracture is one of the main failure modes for metal structures and components in engineering field. The entire fatigue life is spent in the crack initiation and the crack propagation, in which the latter can be divided into two stages, i.e. the long crack propagation and the short crack propagation. The initiation and propagation of short cracks can account for a large portion of fatigue life, so it's important of achieving a more precise understanding of short crack initiation and propagation process in fatigue-related components [1]. Propagation law for long crack is based on the linear elastic fracture mechanics (LEFM), while propagation law for short crack which usually will be affected by many uncertain factors is different from that for long crack. Therefore, the evolution behavior of short crack, including the fretting fatigue crack, is a complex process [1-4]. The actual propagation rate of short crack is greater than that calculated by the theory of LEFM. Dangerous conclusions may be obtained when the LEFM is directly applied to describe the behavior of short crack.

Neural network, which is provided with the ability of strong non-linear processing, can provide a new way to solve various nonlinear dynamics problems. In recent years, neural network has been widely used in solving the complex nonlinear problems in materials science [5]. Zhang put forward the Wavelet Neutral Network (WNN) on the basis of the fast decay and the nonlinear approximation ability of wavelet analysis and the self-learning ability of Neural Network [6]. Meanwhile, the Wavelet Neural Network can be trained by various algorithms, in which the gradient method is the commonly used one. However, the computing speed is slow and easy to fall into local extreme value while using this method. Instead, genetic algorithm is a common global optimization method which can effectively avoid the influence of local extreme value [7]. Therefore, the genetic wavelet neural network combined with the advantages of genetic algorithm and wavelet neural network can obviously improve the performance of the wavelet neural network [8]. At present, the method has been successfully applied in many aspects, such as speech recognition [9], fault diagnosis [10], GPS positioning [11], Multi-Sensor Information Fusion Technique [12] and so on.

In present study, the short fatigue crack propagation rate model based on genetic wavelet neural network is established. The predicting results of this model are compared with that of a multi-microstructural barriers model. The comparison results show that these two different models can both reflect the trend of the short fatigue crack propagation rate reasonably and effectively. However, the prediction accuracy of the model based on genetic wavelet neural network is higher while the multi-microstructural barriers model can indicate the periodic deceleration behavior of growth rate during the whole short fatigue crack propagation process.

2. Experimental details

Test material applied in present work is LZ50 axle steel, which is originated from an RE2B type railway axle. Test specimens were machined into a smooth axial funnel shape, while the axial of specimens were consistent with the axial of the RE2B axle. The minimum diameter of the specimen was 8mm while the arc segment of the specimen was polished to a mirror effect as shown in Figure 1. Test results of mechanical properties and fatigue limit at room temperature are listed in Table 1. Metallographic test of LZ50 steel showed that the material has the typical banded structure (Figure 2a) due to the forging process in the manufacture of the axle and "ferrite and lamellar pearlite" structure (Figure 2b).

Before the test, the central arc surface of the specimen was etched with 4% nitric acid alcohol solution so as to expose the metallographic structures. It has been proved that the short crack behavior was not affected by surface etching in previous research because the etching depth is smaller than the roughness of finest engineering surface [13]. All replica tests were carried out on MTS 809 type axial tension and torsion fatigue testing machine following ASTM E647-11 Standard Test Method for Measurement of Fatigue Crack Propagation Rates. The tests were performed under sinusoidal loading wave with a frequency of 15 Hz in air and at room temperature. The stress amplitude was controlled to be 225 MPa and the stress ratio was 0.1. In the loading process, the test was interrupted at predetermined cycle number. Then, cellulose acetate film softened by acetone was pasted on the specimen arc surface. Once the film was dry, it was peeled off and saved by two glass slides. Above steps were repeated until the specimen was finally fractured. To obtain the evolution information of the short fatigue crack as much as possible, the given cyclic interval was relatively small to ensure that the effective replicating times for each specimen was no less than 10 times. In the meantime, a tensile stress of 10 MPa was retained when the fatigue machine was stopped temporarily to keep the crack in open state during surface replicating process [14].

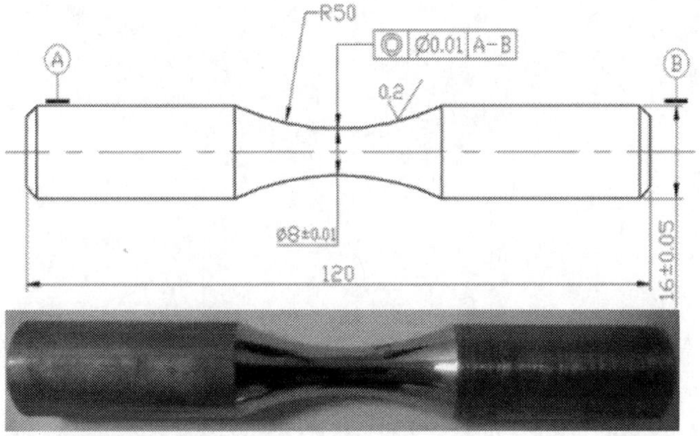

Figure 1. Schematic of shape and dimension of the specimen for fatigue test (unit: mm)

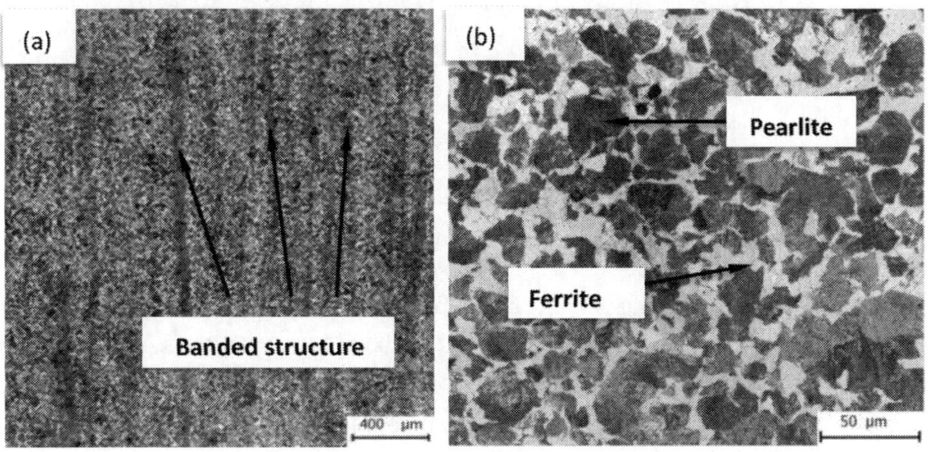

Figure 2. Metallographic test photograph of LZ50 axle steel

Table 1. Test results of mechanical properties and fatigue limit

Tensile Strength/MPa	Yield Strength/MPa	Elongation	Reduction of area	Fatigue limit/MPa
674	342	20.79%	40.38%	264.88

The dried replica films were examined step by step according to the "reverse observation method" [15] from failure to crack initiation using Olympus OLS4100 laser microscope. Ultimately, cracks information, i.e., the crack length, crack angle and number of cracks, of each specimen at every replicating step were obtained. The "effective short fatigue crack criterion" proposed by Zhao [15] has established the basic frame for the study of short fatigue crack behavior. Based on this criterion, the original data of dominant effective short fatigue crack (DESFC) size at each replica were obtained and the DESFC growth rates with corresponding DESFC size were calculated and illustrated in Figure 3.

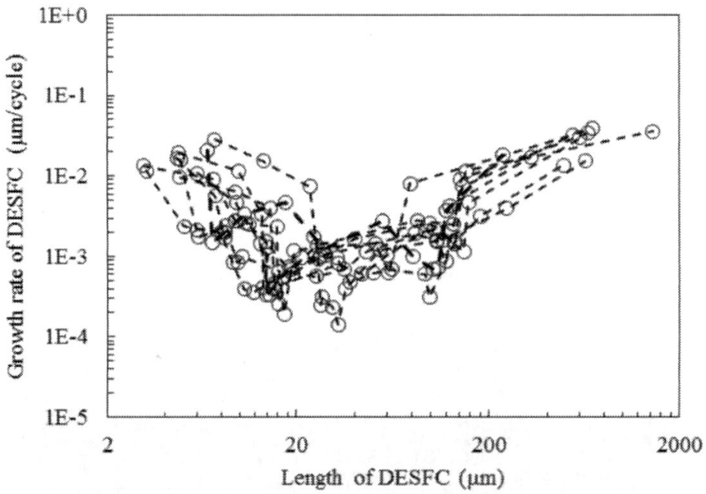

Figure 3. DESFC growth rate of LZ50 axle steel with corresponding DESFC size

It can be seen from Figure 3 that the DESFC growth rates of all specimens are different to a certain extent, but the growth of DESFC decreases twice during the propagation process. Combined with the work in reference [16], conclusions can be drawn that the reason for these two significant decreases of DESFC growth rate lies in the restraints of ferrite grain boundary and the rich pearlite banded structure.

3. Genetic Wavelet neural network model

3.1 Wavelet Neural Network

Wavelet neural network is a feed-forward neural network combined with wavelet analysis and neural network, of which the essence is using wavelet basis function instead of the neural network hidden layer nodes as incentive function, and expressing the signal of the hidden layer through linear superposition of wavelet [11]. In this study, the wavelet neural network with three layers is shown in Figure 4.

The wavelet neural network output can be expressed as,

$$f(x) = \sum_{j=1}^{m} \omega_j \, \psi \left[\frac{\sum_{i=1}^{n} v_{ij} x_i - b_j}{a_j} \right] \tag{1}$$

Where x_i is the node of i in the input layer, v_{ij} is the connection weight value of the node of i in the input layer and the node of j in the hidden layer, ω_j is the connection weight value of the node of j in the hidden layer and the node in the output layer, a_j and b_j are the dilation factor and translation factor of wavelet basis function respectively.

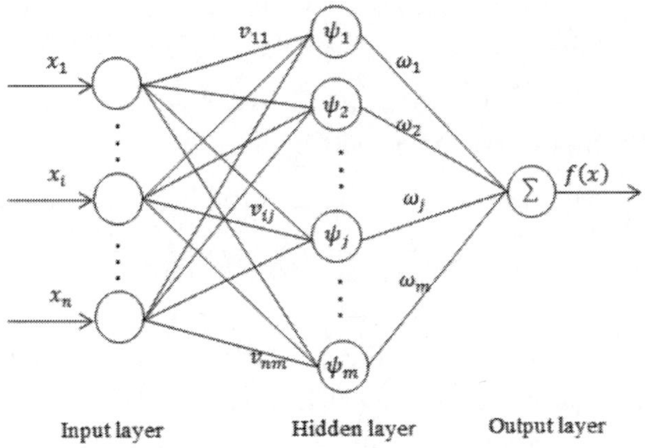

Figure 4. Wavelet neural network structure

The hidden layer is Morlet wavelet basis function,

$$\psi(x) = cos(1.75x) \cdot exp\left(-\frac{x^2}{2}\right) \tag{2}$$

The output layer is Sigmoid function,

$$f(x) = \frac{1}{1+exp\left(-\frac{x^2}{2}\right)} \tag{3}$$

Network learning objective function can be defined as selecting a set of training parameters in the sample set so as to ensure the minimum sample variance.

$$E = \frac{1}{N}\sum_{t=1}^{N}\left(f(X_t) - f^*(X_t)\right)^2 \tag{4}$$

Where X_t is the training samples for group of t, $f(X_t)$ and $f^*(X_t)$ are the corresponding actual and expected output respectively when the training samples of t as the input.

3.2 Wavelet neural network optimized by genetic algorithm

Genetic algorithm (GA) has the typical characteristics of random search and global optimization, and its application object is not limited only if the questions are solvable. The WNN structure is not easy to determine because the node number of hidden layer is not easy to determine. In the present study, the total number of WNN hidden layer nodes is initially determined by trial and error method firstly, but the convergence speed is slow, and it is difficult to obtain the optimal parameters quickly. Therefore, GA is applied to optimize wavelet neural network, which can effectively overcome the characteristic of the uncertainty, avoid the feedback neural network falling into local optimum, and simplify the network training process to a great extent.

GA optimization of WNN includes population initialization, fitness function, crossover operation and mutation operation. Therefore, GA is used to encode the chromosome parameters of WNN, which is composed of wavelet network structure, weight factor, dilation factor and translation factor. The coding method of the specific chromosome is shown in Figure 5, and the j hidden of layer wavelet neural network is taken as an example to display coding method. v_{1j}, v_{2j}, …, v_{nj} is the connection weight value of the node in the input layer to the hidden layer of j, ω_j is the connection weight value of the hidden layer of j to the input layer, a_j and b_j are the wavelet dilation factor and translation factor of j respectively. $C_j \in [0, 1]$ is the hidden layer node characteristic coefficient. The structure of WNN can be determined based on the value of C_j. According to the experience, when $C_j \geq 0.5$ the wavelet neurons exist, otherwise, it does not exist [17].

…	C_j	v_{1j}	…	v_{nj}	ω_j	a_j	b_j	…

Figure 5. GAWNN real coding

The parameters of the GA chromosome were assigned to the corresponding WNN structure, and then the fitness of each individual in the population was calculated, which could reach or approach the optimal solution. The fitness function can be expressed,

$$F = \frac{1}{E} \tag{5}$$

When the value of E is minimum, that is, when the value of F is the maximum, the WNN structure determined by the parameters of the GA chromosome is the best.

Because the nature of the genetic evolution is dynamic and random, it is necessary to carry out the crossover and mutation operation to ensure the diversity of the population. In order to avoid falling into the local optimal solution of WNN, the value of P_C and P_m is changed according to the fitness function value in the process of operation. This adaptive method is to make the value of the P_C and P_m self-adaption change relative to the variety of the maximum and minimum value of F_{cmax}, F_{mmax}, F_{cmin} and F_{mnin} and the mean fitness of population F_{avg}.

$$P_c = \begin{cases} \dfrac{P_C - P_{Cmin}}{1+exp\left\{\beta\left[\frac{2(F_c - F_{avg})}{F_{cmax}-F_{avg}} - 1\right]\right\}} + P_{Cmin} & F_c \geq F_{avg} \\ P_{Cmax} & F_c < F_{avg} \end{cases} \tag{6}$$

$$P_m = \begin{cases} \dfrac{P_m - P_{mmin}}{1+exp\left\{\beta\left[\frac{2(F_m - F_{avg})}{F_{mmax}-F_{avg}} - 1\right]\right\}} + P_{mmin} & F_m \geq F_{avg} \\ P_{mmax} & F_m < F_{avg} \end{cases} \tag{7}$$

Where F_c and F_m are the fitness values of the crossed chromosomes and the variant chromosomes respectively. Considering the influence of GA on the global convergence of GAWNN, the value of adjustment factor β can be determined by several experiments.

3.3 Genetic Wavelet neural network model

In this study, the number of wavelet neural network input nodes is set to be n while the output node number is 1. The initial number of nodes in the hidden layer is determined by empirical formula [18]. The final number of nodes in the hidden layer needs to be determined by the characteristic coefficients obtained by GAWNN training. The flow chart is shown in Figure 6.

4. Multi-microstructural barriers short crack growth rate model

In early research on the short fatigue crack behavior of LZ50 axle steel, it has been pointed out that the ferrite grain boundaries and the pearlite banded structure are the main micro-structures which constrain the propagation of short crack [16]. A short fatigue crack growth rate model including multi-microstructural barriers (MMB) has been proposed by introducing a resistance coefficient function to comprehensively consider the effects of ferrite grain boundary and pearlite banded structure on short fatigue crack behavior [16],

$$\frac{da}{dN} = G_0 + A[\Delta W_t a - \Delta W_t \sum_{i=1}^{n} f_i(\Delta d_i) d_i]^m \tag{8}$$

Where a is the size of DESFC, N is the cyclic number, G_0 is the minimum growth rate in the first cycle of microscopic scale, ΔW_t is the total energy density of remote fields, di is the characteristic microstructural barrier size, i is the subscript to specify the type of barriers, i.e., d_1 for the average equivalent diameter of ferrite grains and d_2 for the mean value of intervals between two rich pearlite bands; A and m are corresponding material constants; $f_i(\Delta d_i)$ is the resistance coefficient function, which can reflect the behavior that the closer the crack tip is to the microstructural barrier, the stronger the constraint effect is. Particularly, while the total energy density of remote fields is constant, the model for LZ50 axle steel can be simplified as,

$$\frac{da}{dN} = G_{oi} + A_i[a - \sum_{i=1}^{2} f_i(\Delta d_i) d_i]^{m_i} \tag{9}$$

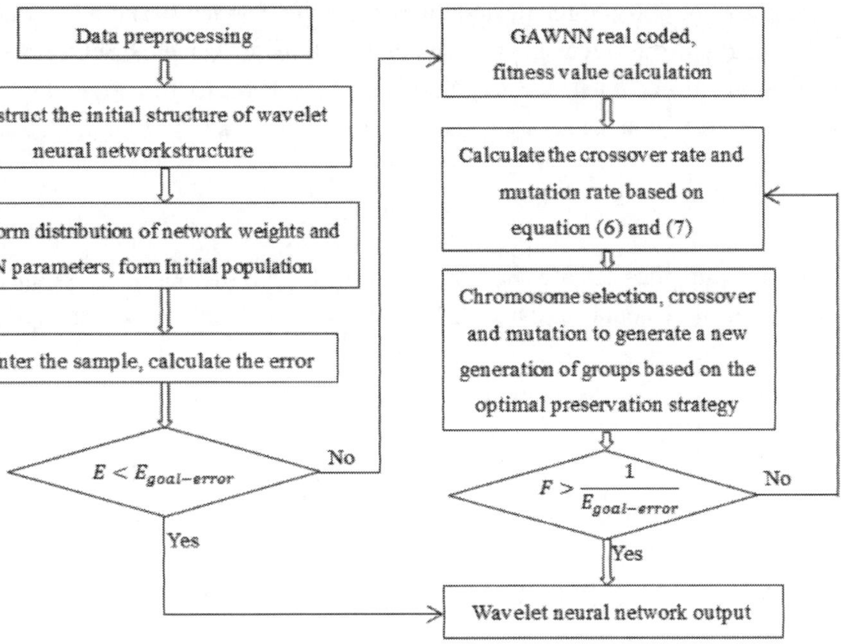

Figure 6. Optimization of wavelet neural network with genetic algorithm

5. Predicting results

Genetic neural network model is used to simulate and predict. 90 groups of data were selected from the test results of all the 9 specimens, and a total of 80 groups of data were randomly selected to train the neural network as the learning samples, while the other 10 groups of data were used as the test samples to verify the performance of the network. According to the relationship between the DESFC scale and the DESFC expansion rate, the network is trained. The crack length, the number of cracks, the applied stress amplitude, the stress ratio, the material fracture toughness and the life fraction are selected as the input node variables. The ESFCs density $n(N)$ is used as the output node variable, that is to say, the number of neurons in the input layer is 6, while the number of neurons in the output layer is up to 1. The optimization result of the number of neurons in the hidden layer is 12. Error limits $E_{goal\text{-}error}$ are all 10^{-4}. Meanwhile, the fitting parameters of the fatigue short crack growth rate model as shown in equation (9) based on above 10 groups of data which were applied as the test samples are also calculated and listed in table 2. The fitting curves derived from two different calculation methods are compared with the experimental data as shown in table 3 and figure 7.

It can be seen that both the GAWNN model and the MMB model can reflect the growth trend of short fatigue crack well. In general, the predicted growth rates of the GAWNN method are closer to the test data due to the consideration of synergetic effects of multi-factors. The GAWNN method has smaller absolute values of prediction error at seven of all the ten given DESFC sizes according to test data. However, when the DESFC was initially observed with a crack length of 8.0 μm and propagated at 4.83×10^{-4} μm/cycle, and then when it grew to 10.6 μm while the first significant deceleration occurred, and finally when it extended to 131.1 μm according to the second deceleration, the fitting precision of MMB model are much better than or at least the same as that of GAWNN model. To some extent, this is because the MMB model has more clear physical meaning. It takes the influence of microstructural barriers, i.e. the grain boundary of ferrite and the rich pearlite banded structure, into account, so that it can indicate the two significant decreases of crack propagation rate when DESFC propagates to the sizes which are close to the characteristic sizes of microstructural barriers. Moreover, the MMB model can reveal the periodic deceleration behavior of growth rate during the whole short fatigue crack propagation process.

6. Conclusions

Based on the results of replica fatigue short crack test with nine smooth funnel-shaped specimens, the DESFC growth rates are calculated. The DESFC growth rate decreases twice significantly with the increase of DESFC size during the whole propagation process.

Both the short fatigue crack propagation rate model based on genetic wavelet neural network and the multi-microstructural barriers model can describe the change of the short fatigue crack propagation rate reasonably and effectively. The prediction accuracy of the model based on genetic wavelet neural network is relatively higher due to the consideration of synergetic effects of multi-factors while the multi-microstructural barriers model can indicate the periodic deceleration behavior of growth rate during the whole short fatigue crack propagation process due to the physical meaning of its construction.

Table 2. The model parameters of short crack growth rate of specimens

d_1 (μm)	G_{o1} (μm/cycle)	A_1	m_1	d_2 (μm)	G_{o2} (μm/cycle)	A_2	m_2
10.60	5.20×10^{-8}	2.71×10^{-10}	2.88	131.12	5.81×10^{-7}	6.10×10^{-11}	1.38

Table 3. Growth rate comparison between test data and predicted value by two calculation methods

DESFC length (μm)	Test data (μm/cycle)	GAWNN model		MMB model	
		Predicted value (μm/cycle)	Difference ration (%)	Predicted value (μm/cycle)	Difference ration (%)
8.0	4.83×10^{-4}	5.10×10^{-4}	5.59	4.74×10^{-4}	-1.86
9.2	2.33×10^{-4}	2.2×10^{-4}	-5.58	2.74×10^{-4}	17.60
10.6	1.05×10^{-4}	1.06×10^{-4}	0.95	1.06×10^{-4}	0.95
11.9	1.27×10^{-4}	1.39×10^{-4}	9.45	6.06×10^{-4}	377.17
18.7	1.87×10^{-4}	2.43×10^{-4}	29.95	6.24×10^{-4}	233.69
25.7	5.08×10^{-4}	4.81×10^{-4}	-5.31	6.47×10^{-4}	27.36
85.4	7.35×10^{-4}	8.32×10^{-4}	13.20	8.47×10^{-4}	15.24
131.1	5.81×10^{-4}	6.32×10^{-4}	8.78	5.88×10^{-4}	1.20
142.7	3.84×10^{-3}	3.87×10^{-3}	0.78	2.26×10^{-3}	-41.15
654.6	1.06×10^{-2}	1.28×10^{-2}	20.75	1.52×10^{-2}	43.40

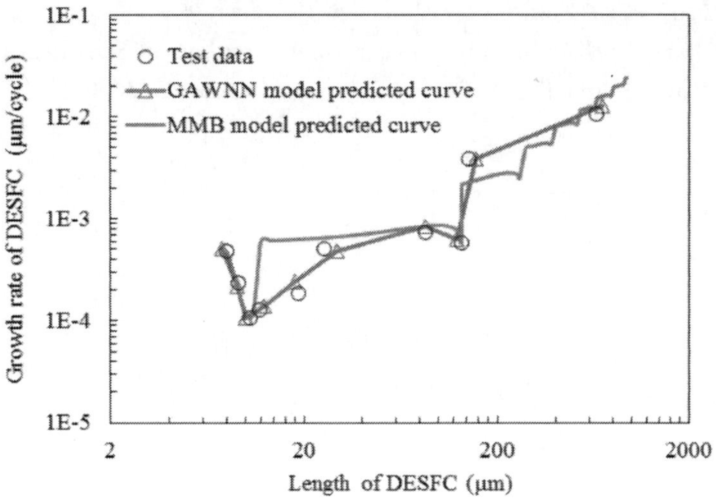

Figure 7. Fitting effects of two different models to test data

Acknowledgements

Present work is supported by the National Natural Science Foundation of China (51675446 and U1534209), the National Key Research and Development Program of China (2016YFB1200403), and the Opening Project of State Key Laboratory of Traction Power (2015TPL_T13).

Reference

[1] Ferjaoui A, Yue T, Abdel Wahab M and Hojjati-Talemi R 2015 *International Journal of Fatigue* **73** 66

[2] Kumar D, Biswas R, Poh LH and Abdel Wahab M 2017 *Tribology International* **109** 124

[3] Resende Pereira KdF, Bordas S, Tomar S, Trobec R, Depolli M, Kosec G and Abdel Wahab M 2016

IOP Conf. Series: Journal of Physics: Conf. Series **843** (2017) 012043 doi:10.1088/1742-6596/843/1/012043

Materials **9** 639
[4] Hojjati Talemi R and Abdel Wahab M 2013 *Tribology International* **60** 176
[5] Xu Q, Zhang X H, Han J C, He X D and Pan W 2005 *Materials Science &Technology* **13** 352
[6] Zhang Q H and Benveniste A 1992 *IEEE Transactions on Neural Networks* **3** 889
[7] Lin X Y 2011 *Ordnance Industry Automation* **30** 42
[8] Zhang X L, Gao M J and Gu F C 2003 *Journal of System Engineering* **18** 147
[9] Han Z Y, Wang J and Lun S X 2010 *Computer Science* **37** 243
[10] Liu M R 2009 *The Research on Method of Fault Diagnosis for Analog Circuits Based on Genetic Algorithm, Wavelet Analysis and Neural Networks* (Changsha: Hunan University) 25
[11] Li J P 1997 *Wavelet Analysis and Signal Processing* (Chongqing: Chongqing Publishing Group) 259
[12] Gao M J, Zhao Y and TAN A 2007 *Chinese Journal of Scientific Instrument* **28** 2103
[13] Zhao Y X, Gao Q and Wang J N 2000 *Acta Metallurgica Sinica* **36** 931
[14] Zhao Y X, Gao Q and Wang J N 2000 *Acta Metallurgica Sinica* **36** 937
[15] Zhao Y X, Gao Q and Wang J N 1999 *Fatigue & Fracture of Engineering Materials & Structures* **22** 459
[16] Yang B and Zhao Y X 2012 *International Journal of Fatigue* **35** 71
[17] Hu X H and Chen L 2009 *Journal of Northwest University: Natural Science Edition* **39** 203
[18] Luo Y C, Du H J amd Cui F F 2007 *Modern Electronics Technique* **30** 88

Effect of Fillet Rolling Load on the Fatigue Performance of a Micro-Alloy Steel Diesel Engine Crankshaft

Gül Çevik[1] and Rıza Gürbüz
[1]Ford Otomotiv San. A.Ş. Akpinar M. Hasan Basri C. No: 2 Sancaktepe/İstanbul TURKEY
gcevik@ford.com

Abstract. Fillet rolling process is an effective method used to improve the fatigue performance of crankshafts by hardening the fillet region and inducing compressive residual stresses. This paper summarizes the work conducted to investigate the effect of rolling load on fatigue behaviour of a micro-alloy steel crankshaft used in diesel engine applications. Based on the staircase test methodology, component-scale resonant bending fatigue tests were conducted to obtain stress versus number of cycles curves and to evaluate the fatigue endurance limits of the crankshaft at un-rolled condition and fillet-rolled conditions at three different loads. Test data was analysed by Dixon-Mood method to calculate the endurance limits. Results showed that the endurance limit increased significantly with fillet rolling process in comparison to un-rolled condition. Endurance limit further increased with the increasing rolling load however with a limited extent above which excessive hardening deteriorates the fillet region; that is the workability limit. The outcomes of this study has shed light on the fillet rolling process to select the optimum rolling load for the used design and material conditions.

1. Introduction

Crankshaft is one of the most critical parts of an internal combustion engine and drives the engine system parts by converting the reciprocating motion of the piston into rotational motion. During service, crankshafts are subjected to a significant number of cyclic loadings; such as inertia forces due to rotating and reciprocating parts and gas explosion forces. Main stresses created on the crankshaft due to these forces are the cyclic bending and torsion.

Fatigue failures on crankshafts are mostly observed at fillet regions since they are dynamically and geometrically the most critical regions and stress concentrations are eventually present. Other critical regions for crack initiation are the pin surface and oil hole ends on the crankshaft surface.

In order to improve the fatigue and wear performance of the crankshafts a variety of surface treatment processes are applied such as; nitriding, induction hardening and fillet rolling. Among these, fillet rolling process is a widely used process in automotive industry which improves fatigue strength of the crankshaft fillet region [1].

Fillet rolling process aims to develop compressive residual stresses and increase the hardness at the fillet region by for local mechanical loading. Developed compressive stresses increase the fatigue life by compensating the tensile stress components created during working conditions in addition to the increased surface hardness which also improves the fatigue strength. In order to obtain a suitable geometry for the roller to fit onto, an undercut having a radius of curvature is formed at this region, which is called the undercut region.

As any failure on the crankshaft will result in catastrophic failures on the engine, studies on the crankshaft durability optimization by fillet rolling process have been in the scope of a large number of studies which are based on experimental and analytical approaches.

Effect of process parameters have been studied by numerous researchers. By crankshaft rig tests of crankshafts with different rolling conditions and by the X-Ray measurement of residual stresses, Ko et.al. [2] have developed a correlation between rolling load, surface hardness and residual stresses with the applied bending moment. By using the experimental data, optimum conditions for rolling load, fillet geometry and material were identified by CAE analysis. Effect of fillet rolling process was on a ductile cast iron crankshaft was evaluated by Çevik et.al. [3] by resonant bending fatigue tests of crankshafts at the fillet- rolled and un-rolled conditions. Regul'skii et al [4], have studied the effect of rolling by the use of fatigue tests of non-hardened crankshafts with fillet and hardened and non-hardened crankshafts without a fillet. By analysing of the test results a procedure of fatigue testing was proposed to obtain the in-service fracture patterns of motorcycle crankshafts.

Various studied have been conducted on component scale testing of the crankshafts to determine the stress versus numbers of cycles (S-N) curves. The possibility of the damage formation on the test rig in case of two-piece failure of the crankshaft has been the starting point of the studies concerning proposal of failure criteria of laboratory tests for many years. Later on, resonance shift failure criterion has been one of the widely used methods which is based on the resonance shifts induced by crack formation and subsequent stiffness drop in resonant bending fatigue rig tests. Feng and Li [5] developed an electrodynamic test machine and automotive component testing procedure and proposed the relationship between crack formation, stiffness and resonance shift. Surface failure criterion which is based on the crack initiation has also been a subject of many studies. According to this criterion, any crack that can be identified visually can be accepted as failure [1, 6]. Chien et al [7] studied the crack arrest phenomenon on the subsurface due to the presence of compressive residual stresses by finite element mapping of stress intensity factors against the fillet depth. Spiteri et al [8] conducted an experimental work to qualify and validate the crack arrest theory proposed by Chien et al (8) followingly.

This study aimed to evaluate the effect fillet rolling process and the applied rolling loads on the fatigue performance of a micro-alloy steel crankshaft. For this purpose, component scale resonant bending fatigue tests were conducted with the crankshaft samples at 4 different fillet rolling conditions; un-rolled at fillet-rolled at 3 different loads. Moment amplitude versus number of cycles to failure curves were obtained at these 4 conditions. The endurance limits were calculated by Dixon-Mood analytical method and compared with respect to the fillet-rolling condition. The correlation of endurance limit improvement with the rolling load and limits of improvement were evaluated.

2. Experimental Procedure

2.1 Material

A micro-alloyed steel of 38MnVS6 grade crankshaft, produced by forging, was used in this study. After forging process, crankshafts were controlly cooled down and no sub-sequent heat treatment is applied after cooling. Spectrometric analysis were carried out with the specimens from the steel crankshafts and chemical composition is shown on Table 1.

Table 1. Chemical Composition of 38MnVS6 Forged Steel.

	Chemical Composition, wt%												
Material	**C**	**Si**	**Mn**	**V**	**P**	**S**	**Cr**	**Mo**	**Ni**	**Cu**	**N**	**Al**	**Fe**
38MnVS6	0.38	0.56	1.42	0.10	0.01	0.01	0.14	0.02	0.09	0.16	0.01	0.01	Rest

2.2 Fillet Rolling Process

In the frame of this study, crankshafts with 4 different fillet rolling conditions were studied in order to correlate the fatigue strength with the amount of rolling load. Namely, crankshafts with the unrolled condition and fillet-rolled at 15kN, 20kN and 24kN were used at the pin fillet region.

In this study, an undercut depth of 1.55mm was applied on the fillet regions. An undercut is machined at the journal corners during machining process of the crankshaft to create a suitable form for the roller to fit on. Figure 1 shows the 2D projection view of the undercut region, describing the radius and depth profiles.

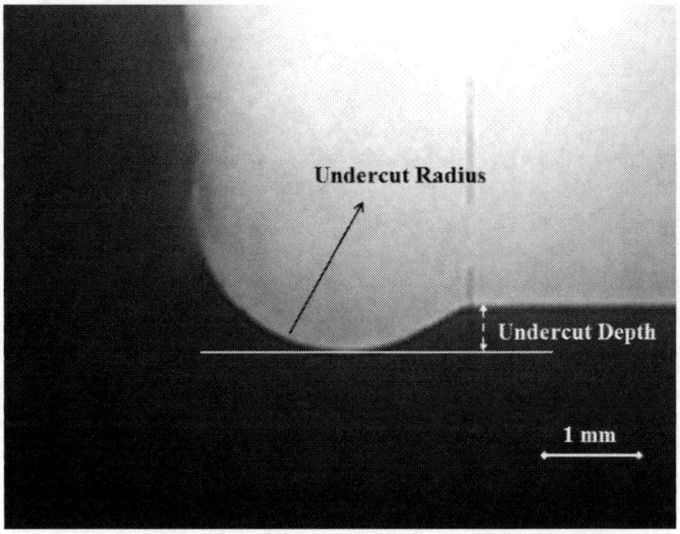

Figure 1. 2D view of the undercut region.

Fillet rolling is applied to the undercut region of main and pin journals of the crankshaft hydraulically by the use of a rigid rolling apparatus. Rolling apparatus is composed of a primary roller associated with two secondary rollers that deform both corners of the crankshaft journal simultaneously during rolling operation. Rollers apply to the crankshaft journal at an angle of 55° to crank axis. Figure 2 demonstrates the main components of the crankshaft; undercut region and rolling operation schematically.

(a) Schematic representation of the crankshaft and its main features

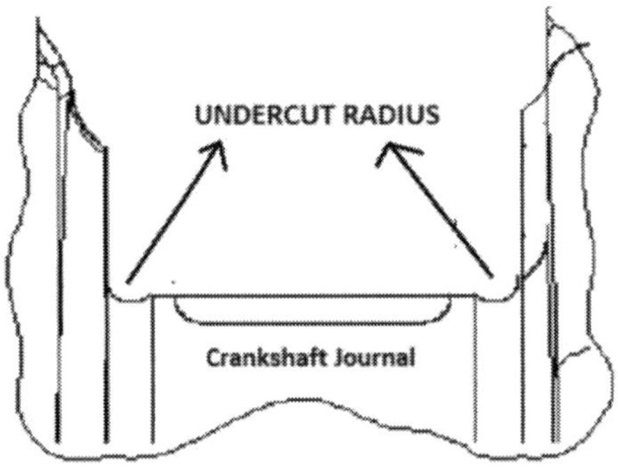

(b) Undercut region

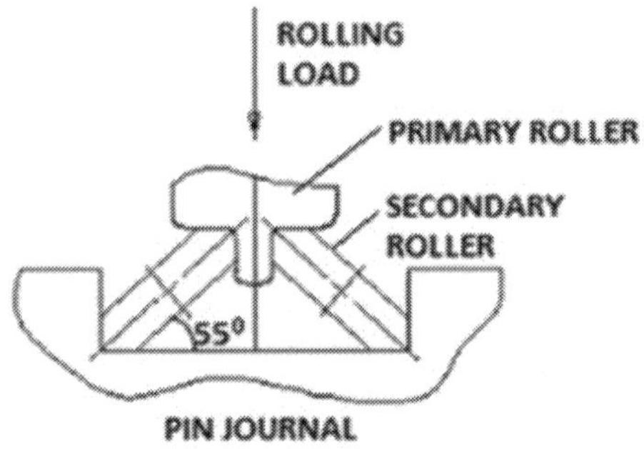

(c) Crankshaft rolling operation

Figure 2. Schematic representations of the crankshaft and rolling operation

Rolling operation was conducted at the pin fillet region with the given loads (15kN, 20kN and 24kN) on the primary roller at a frequency of 80 rpm with 12 crank revolutions. The secondary rollers used in this study are of disc shape with a diameter of 15 mm and a thickness of 5 mm. Radius of curvature of the roller at the rolling contact area is 1.45 mm.

2.3 Resonant Bending Fatigue Tests

Resonance fatigue testing method was used to evaluate the moment versus cycles to failure curve data of the crankshafts at 4 different fillet-rolled conditions. The resonant fatigue testing method is based on the mechanical stiffness theory. Under testing, the frequency of the test is stabilized at the resonance frequency of the component on the test rig, depending on the material's Elastic Modulus and geometry of the system. With the formation and propagation of a crack on the specimen, the resonant frequency of the test rig decreases due to the decreasing cross-section and thus the decreasing mechanical stiffness of the system [9].

Specimens from crankshafts were cut for fatigue rig tests and tested under cyclic bending conditions on the resonant type fatigue test machine. Cyclic bending moments were applied on the fillet region of 5th pin journal, which is the most critical region of the crankshaft under service. Tests were conducted at completely reversed constant amplitude cyclic loads.

During the tests, the resonance test frequencies observed to be approximately 40 Hertz. In order to cover both crack initiation and propagation stages on the test specimens, a relatively large value of frequency limit of ±4 Hertz, 10% of the resonance frequency, was used.

For the applied bending moments, number of cycles to failure were recorded to construct the moment amplitude versus number of cycles to failure (M-N) curves of the crankshafts at 4 different fillet rolling conditions. Test run-out criterion was used as 10 million cycles. Thus, the fatigue strength values used in this study define the fatigue strengths at 10 million cycles.

Predefined bending moments, at a span length of 66.1mm, were applied to the pin fillet region next to main journal side throughout the tests. Figure 3 shows the schematic and actual views of the test specimen, test set-up and demonstrates the test failure region.

IOP Conf. Series: Journal of Physics: Conf. Series **843** (2017) 012044 doi:10.1088/1742-6596/843/1/012044

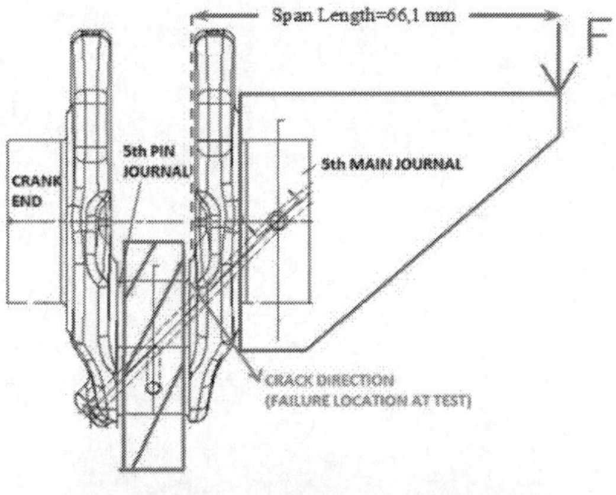

(a) Schematic view of the test specimen and test set-up

(b) Photograph of the test set-up

Figure 3. Schematic and actual views of the test set-up.

Staircase test methodology [10] was used for load amplitude selection for the fatigue tests. By the use of this method, first specimen was tested at a predefined load amplitude level based on experience. If the specimen failed, load amplitude level was decreased one step for the next test. If the specimen did not fail, a one step higher stress amplitude was used for the following test. This procedure was repeated until a valuable number of data is obtained to construct the M-N curve and for the endurance limit calculations.

3. Test Results

The number of cycles to failures were recorded against applied bending moment amplitude throughout the tests. The test data are demonstrated as the applied moment in in Nm versus number of cycles to failure curves for each rolling condition on Figures 4, 6, 8 and 10 on a semi-logarithmic scale. The number of run-outs for each moment amplitude value were also shown on the figures. Figures 5, 7, 9 and 11 show the staircase moment steps and failure/run-out history around the endurance limit for each case.

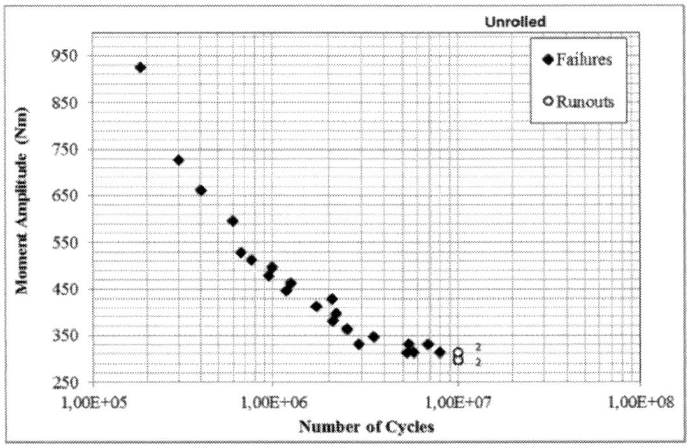

Figure 4. M-N curve for the un-rolled condition.

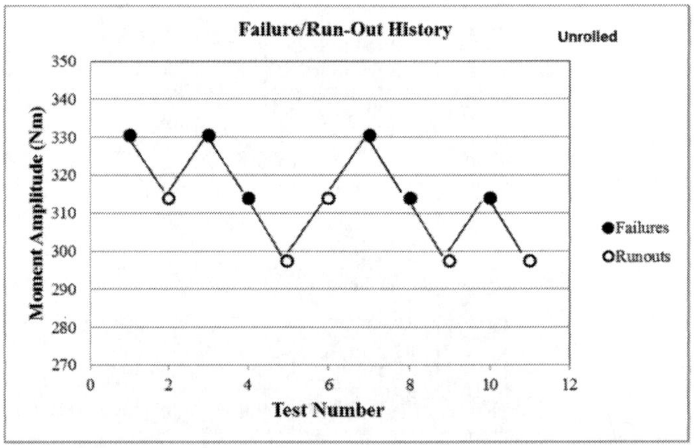

Figure 5. Failure/Run-out history for the un-rolled condition.

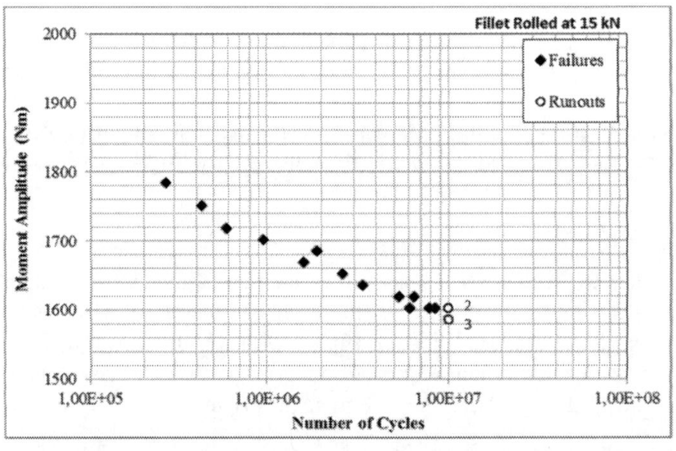

Figure 6. M-N curve for the crankshaft fillet-rolled at 15kN.

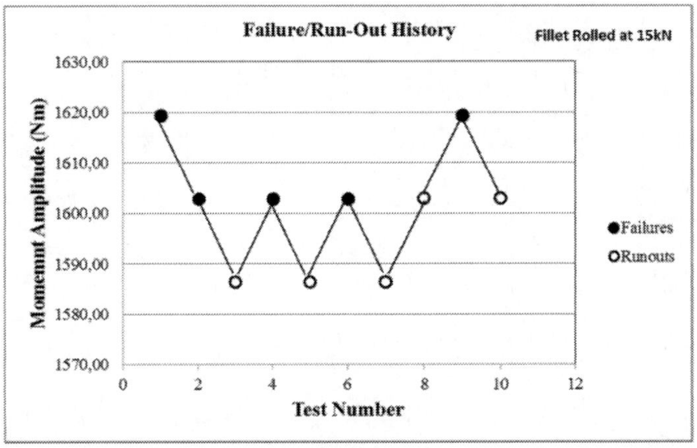

Figure 7. Failure/Run-out history for the crankshaft fillet-rolled at 15kN.

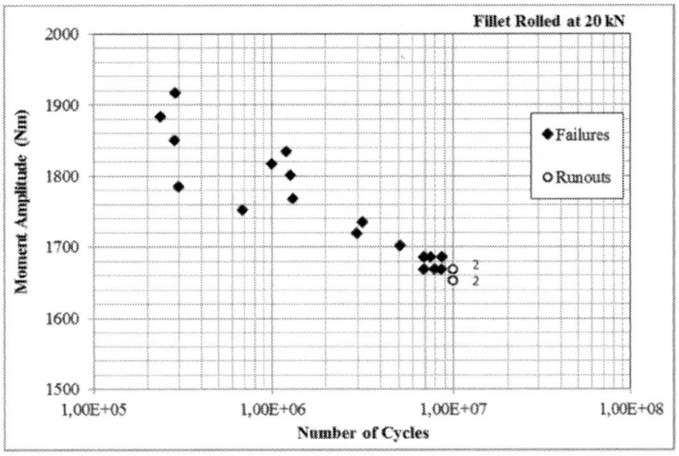

Figure 8. M-N curve for the crankshaft fillet-rolled at 20kN.

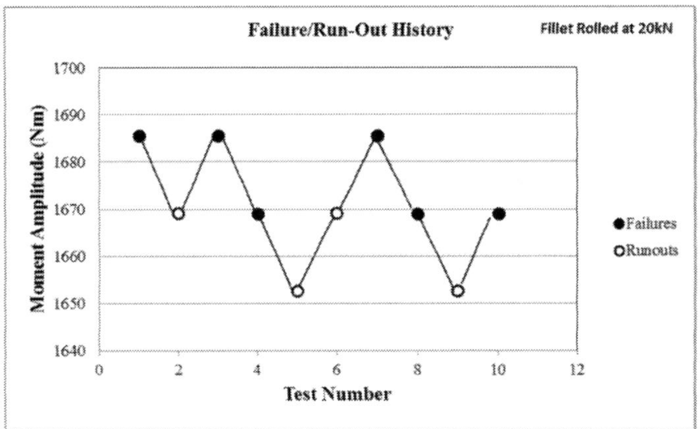

Figure 9. Failure/Run-out history for the crankshaft fillet-rolled at 20kN.

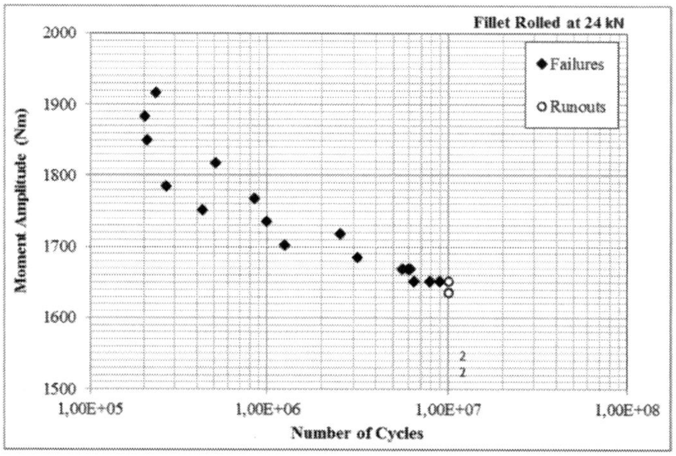

Figure 10. M-N curve for the crankshaft fillet-rolled at 24kN.

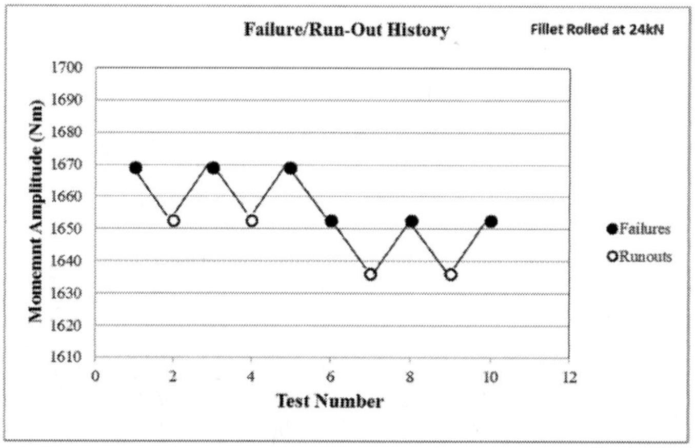

Figure 11. Failure/Run-out history for the crankshaft fillet-rolled at 24kN.

4. Endurance Limit Calculations

Dixon-Mood [11] method was used to calculate the endurance limits from the failure/run-out data around the transition region. This method is based on Maximum Likelihood Estimation which assumes a normal distribution of fatigue limit to calculate the mean (μ) and standard deviation (σ). The equally spaced stress (S) levels are sorted and numbered starting from the lowest stress level, S_0. Number for S_0 is denoted by i=0. Stress increment or stress step is denoted by S_d. The number of less frequent event at a stress level is defined by n_i. For statistical analysis, three parameters, A, B and C, are calculated where

$$A = \sum_{i=0}^{i_{max}} n_i \tag{1}$$

$$B = \sum_{i=0}^{i_{max}} i x n_i \tag{2}$$

$$C = \sum_{i=0}^{i_{max}} i^2 x n_i \tag{3}$$

Mean fatigue limit is calculated by the equation

$$\mu = S_0 + S_d x(\frac{B}{A} \pm 0.5) \tag{4}$$

IOP Conf. Series: Journal of Physics: Conf. Series **843** (2017) 012044 doi:10.1088/1742-6596/843/1/012044

If the more frequent event is failure, the plus sign is used in the above equation and if the more frequent event is survival the negative sign is used.

Standard deviation is calculated by the below formula

$$\sigma = 1.62xS_dx(\frac{AxC-B^2}{A^2} + 0.029) \text{ if } \frac{AxC-B^2}{A^2} \geq 0.3 \quad (5)$$

or

$$\sigma = 0.53xS_d \text{ if } \frac{AxC-B^2}{A^2} < 0.3 \quad (6)$$

Standard deviation equations proposed by Dixon and Mood are based on the assumption that stress increment value is on the order o 0.5σ to 2.0σ.

Within this study, Moment values were studied instead of Stress values; and denoted by M.

Dixon-Mood calculations of endurance limits in terms of moment amplitudes are summarized on Table 2. Calculated mean endurance limit, standard deviation and lower and upper endurance limits for 95% confidence are listed. Results are discussed on the following section.

Table 2. Endurance Limit Calculation Results

		Fatigue Limit at 95% Confidence				
	Mean Fatigue Limit (Nm)	**Lower (Nm)**	**Upper (Nm)**	**Standard Deviation (σ) (Nm)**	**Moment Increment (M_d) (Nm)**	**M_d/σ**
Un-Rolled	322.24	311.01	333.46	8.77	16.53	1.89
Rolled at 15kN	1602.93	1591.70	1614.15	8.77	16.53	1.89
Rolled at 20kN	1677.29	1666.06	1688.51	8.77	16.53	1.89
Rolled at 24kN	1660.76	1649.54	1671.99	8.77	16.53	1.89

5. Discussions and Conclusions

In order to evaluate the success of Dixon-Mood method calculations of staircase test data, moment increments to standard deviation ratio for un-rolled and rolled crankshafts are calculated for each case which is found to be 1.89 for all cases; as same moment increments were used and a 8.77 Nm of standard deviation was obtained for all cases. For Dixon-Mood results to be evaluated as successful, this ratio should be within which is within the range of 0.5 to 2.0 where Dixon-Mood equations are based on. Thus it can be concluded that the calculated endurance limits can be regarded as successful evaluations. When mean fatigue endurance limits of un-rolled and rolled crankshafts are compared mathematically, it can be derived that fatigue strength of rolled crankshafts are at least 5 times of the unrolled crankshaft. This fact describes that the fillet rolling is an effective process to improve the fatigue strength of crankshafts significantly by a convenient local deformation and hardening process. The mechanism of strengthening by fillet rolling process is mainly the compressive residual stresses developed and local hardening as a result of local plastic deformation. Compressive residual stresses compensate the tensile components of the service loads, and local hardness increase can retard the crack initiation which both improve the fatigue strength.

An improvement factor of 4.97 was calculated when the endurance limits of un-rolled and rolled crankshafts were compared. As seen from Table 2, increasing the fillet rolling load to 20 kN has resulted in a further increase in fatigue strength. However when a fillet rolling load was increased to 24 kN, a slight decrease of endurance limit was observed. This result shows that the maximum achievable fatigue strength can be obtained within the rolling load range of 20 kN and 24 kN. Depending on this data, the surface qualities of crankshafts rolled with 20 kN and 24 kN were examined.

Figure A on the appendix section shows the fillet surfaces of crankshafts rolled at 20kN and 24kN. An apparent difference in surface qualities of the crankshafts were observed as demonstrated on this figure. On the crankshaft where a rolling load of 24kN was applied, surface irregularities were observed which can act as crack initiation points and decrease the duration of fatigue crack initiation stage under cyclic loading. These finding explains the lower fatigue strength observed at 24kN rolling load condition with respect to 20kN.

Thus, it can be derived that the optimum rolling load for the crankshaft design and material combination used in the frame of this study is between 20 and 24kN and the workability limit of the design and material combination was reached within this fillet rolling load range. This finding has accepted to be an important guideline for selecting the correct fillet rolling load with this design optimization study.

6. Appendix

(a) 24kN (b) 20kN

Figure A. Fillet surface qualities of 38MnVS6 crankshafts at 24 and 20kN rolling conditions.

7. References

[1] Love R J, and Waistall D N, 1954, The Improvement in the Bending Fatigue Strength of Production Crankshafts by Cold Rolling, M.I.R.A., Report No. 1954/1, pp 1-8.

[2] Ko YS, Park J W, Bhan H O, Park H, and Lim J D, 2005, Fatigue Strength and Residual Stress Analysis of Deep Rolled Crankshafts, SAE World Congress, 2005-01-0988, Detroit, Michigan, USA.

[3] Çevik G, Gürbüz R, 2013, Evaluation of Fatigue Performance of a Fillet Rolled Diesel Engine Crankshaft, Engineering Failure Analysis, V 27, pp 250–261

[4] Regul'skii M N, Pogrebnyak A D and Balakowski O B, 2002, Procedure and Results of Investigation into Fatigue Strength Characteristics of Motorcycle Engine Crankshafts, Strength of Materials, 34(6), pp 629-635.

[5] Feng M, Li M, 2003, Development of a Computerized Electrodynamic Resonant Fatigue Test Machine and Its Applications to Automotive Components, SAE World Congress, 2003-01-0951.

[6] Watmough T, Malatesta M J, 1984, Strengthening of Ductile Iron for Crankshaft Applications, Proceedings of Expendable Pattern Casting: Managing the Technology, AFS Trans., 92 (83).
[7] Chien W Y, Pan J, Close D, Ho S, 2005, Fatigue Analysis of Crankshaft Sections under Bending with Consideration of Residual Stresses, Int. J. of Fatigue, 27 (1), pp 1-19.
[8] Spiteri P, Ho S, Lee Y L, 2007, Assessment of Bending Fatigue Limit for Crankshaft Sections with Inclusion of Residual Stresses, Int.J. of Fatigue, 29 (2), pp. 318-329.
[9] Zhou X, Yu X, 2008, Fatigue Crack Growth Rate Testing Using a Frequency Sweep Method, J. of Zhejang University, Science A, 9 (3).
[10] Ransom J T, and Mehl R F, 1949, The Statistical Nature of the Endurance Limit, Metals Trans,185, pp 364-365.
[11] Dixon W J, and Mood A M, 1948, A Method for Obtaining and Analyzing Sensitivity Data, J. of Amer. Statist. Assoc., 43, pp. 109-126.

IOP Conf. Series: Journal of Physics: Conf. Series **843** (2017) 012045 doi:10.1088/1742-6596/843/1/012045

Crack propagation modelling for high strength steel welded structural details

B J Mecséri[1], **B Kövesdi**[1]

[1]Department of Structural Engineering, Budapest University of Technology and Economics, Műegyetem rkp. 3, Budapest, H-1111, Hungary

mecseri.balazs@epito.bme.hu

Abstract. Nowadays the barrier of applying HSS (High Strength Steel) material in bridge structures is their low fatigue strength related to yield strength. This paper focuses on the fatigue behaviour of a structural details (a gusset plate connection) made from NSS and HSS material, which is frequently used in bridges in Hungary. An experimental research program is carried out at the Budapest University of Technology and Economics to investigate the fatigue lifetime of this structural detail type through the same test specimens made from S235 and S420 steel grades. The main aim of the experimental research program is to study the differences in the crack propagation and the fatigue lifetime between normal and high strength steel structures. Based on the observed fatigue crack pattern the main direction and velocity of the crack propagation is determined. In parallel to the tests finite element model (FEM) are also developed, which model can handle the crack propagation. Using the measured strain data in the tests and the calculated values from the FE model, the approximation of the material parameters of the Paris law are calculated step-by-step, and their calculated values are evaluated. The same material properties are determined for NSS and also for HSS specimens as well, and the differences are discussed. In the current paper, the results of the experiments, the calculation method of the material parameters and the calculated values are introduced.

1. Introduction

To determine the fatigue lifetime of welded structural details modelling the fatigue crack propagation can be a more precise method than using the nominal stress based design approach. However, to apply this method specific material constants and loading parameters are required. The tests, which are necessary to determine the required material constants are usually expensive and cumbersome. The most of these experiments needs special equipment, and specimens with specific geometry (C(T) specimen). Another difficulty is, that the most of the previous tests were carried out on C(T) specimens and the main part of the applied structural details in bridges were not tested to determine these material constants. In the current paper a favourable method is introduced, which is able to determine the material constants of the Paris law of fatigue crack propagation for a selected structural detail. To calculate the constant parameter, strain gauge measurements are necessary during the laboratory tests. Using the measured data and the numerical model of the investigated detail the material constants can be determined. The fatigue behaviour of a gusset plate connection was investigated to present the applied calculation method. Fatigue laboratory tests are performed and the numerical model of the structural detail is developed. Applying the measured data, the C and m parameters of the Paris law (equation (1)) are determined, and the calculated values of specimens from S235 and S420 structural steels were compared.

2. Literature review

2.1. General remarks

The previous results show, that the dominant regime of the fatigue lifetime of a welded structural detail belongs to the crack propagation. The fatigue cracks readily initiate due to the presence of the welding, and the crack initiation part of the fatigue lifetime can be neglected [1]. Therefore, the fatigue lifetime of a welded structural detail can be determined using fracture mechanics. One of the most common and simplest form of the mathematical term of the crack propagation is the Paris law, shown by equation (1).

$$\frac{\mathrm{d}a}{\mathrm{d}N} = C \cdot (\Delta K)^m \tag{1}$$

where C and m are material parameters, ΔK is the stress intensity factor range, a is the length of the fatigue crack and N is the number of cycles. To apply the Paris law, the determination of the material constants is necessary. These parameters can be calculated from test results. The length of the crack, the number of cycles and the nominal stress range ($\Delta\sigma$) are necessary to be measure during the tests. From the nominal stress range and the length of the crack, the stress intensity factor range can be calculated by equation (2).

$$\Delta K = \Delta\sigma\sqrt{\pi a} \tag{2}$$

The number of cycles and the nominal stress range is easy to measure, because they depend on the loading method and the geometry of the specimen, which is defined by the researcher. The measuring of the length of the fatigue crack is the most difficult part of the experiments. According to previous studies and standards there are two main methods to determine the length of the cracks. These methods are the visual method and the indirect methods [2].

2.2. Visual method of the measuring of the length of the fatigue crack

The initial and the final crack size can be determined from the crack surface after the failure of the specimen. The length of the initial or the final crack has to be measured several individual local points, then their lengths can be calculated by averaging the measurements. To measure the change in the crack size (Δa), the front of the crack has to be marked by some treatment. One of the crack-front marking methods is cooling down the specimen to low temperatures, for example by immersing in liquid nitrogen. The fracture mechanics analyses are usually based on two-dimensional methods, the crack must be defined with a single length, but the crack exhibits a curved front, thus the averaging of the measured lengths is necessary. Visual crack-length measurements are usually done with a microscope on a calibrated traveling stage [2]. Another way to determine the length of a fatigue crack is measuring the surface crack length by an optical (Figure 1 a)) or scanning electron microscope. In this case the surface crack length has to be measured on the both sides of the specimen, then the crack length can be determined by statistical methods [3].

2.3. Indirect methods of the measuring of the length of the fatigue crack

Two methods will be presented in the following. On one hand these procedures do not measure the crack length directly, they use the deformation properties of the specimen or electric potential techniques to determine the length of the fatigue crack. On the other hand, it is a big advantage of these ways, that they average the crack length over the cross-section of the specimen.

Elastic compliance method

This method uses the changing of the elastic compliance of the specimen to calculate the length of crack. When the crack length increases, the slope of the force-deformation diagram of the specimen varies, thus the tangential elastic compliance of the specimen varies as well. For a standard C(T) (compact tension) specimen the normalized crack length-plane stress elastic compliance function is given in the standards [3], from which the actual length of the crack can be determined.

Electrical potential difference (EPD) method
This method is based on the principle, that the electrical field in a cracked specimen with a current flowing through it is a function of the crack size. For a constant current flaw, the voltage drop across the crack plane will increase with increasing the length of the crack due to modification of the electric field. The alteration in voltage can be related to crack length through analytical or experimental calibration relationships (Figure 1 b)).

a)

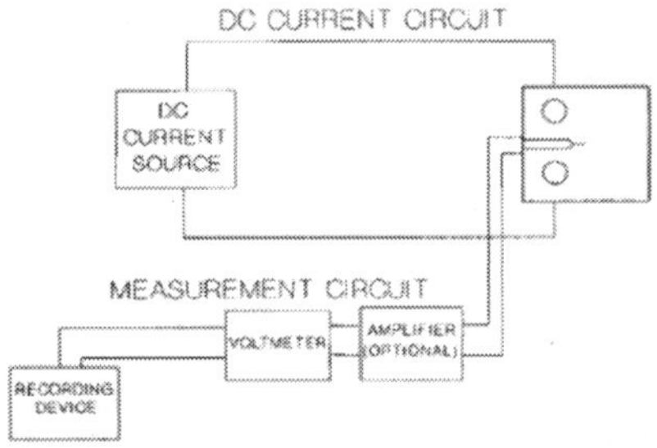

b)

Figure 1 a) Visual crack measuring; b) EPD test set up [ ]

3. Experimental research program

3.1. Test specimens and test set up
A total of 6 fatigue tests are conducted on flange gusset joint specimens. Two types of geometries are investigated within the current research program. The geometries and the dimensions of the specimens (II-10-norm and II-18-norm) are shown in Figure 2.

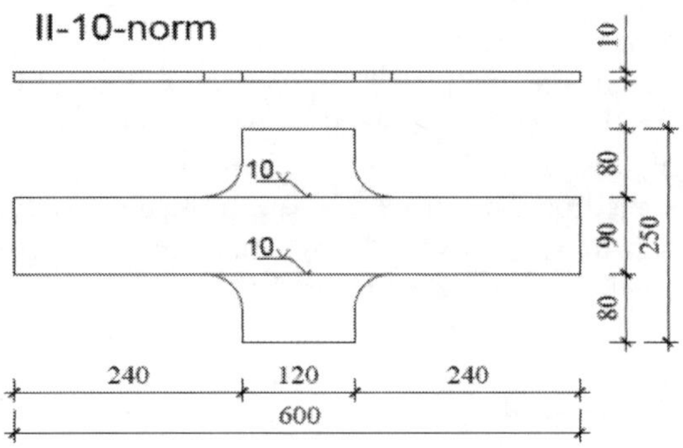

a)

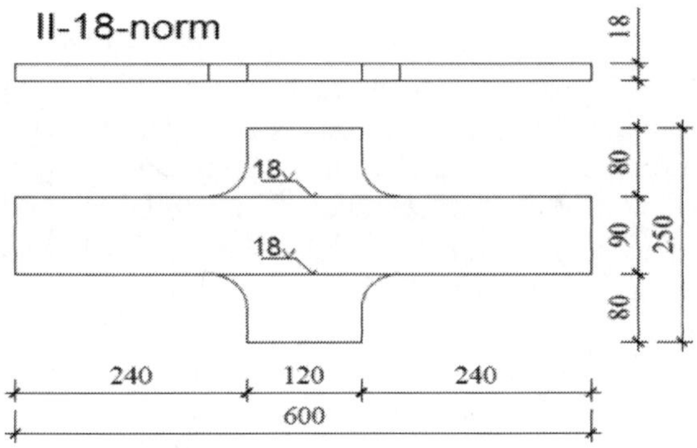

b)

Figure 2 a) Geometry of II-10-norm specimens; b) Geometry of II-18-norm specimens

These test specimens are produced from NSS (S235) and from HSS (S420) material to be able to compare the fatigue behaviour of NSS and HSS structures. The another type of geometry and the dimensions of the specimens (X-18-norm) are also shown in Figure 2 b). These test specimens are produced only from NSS (S235) material to investigate the size effect on the fatigue behaviour. The post-weld treatment methods are also investigated on this structural detail. Several specimens are treated by disc grinding, and these specimens are also tested to determine the effect of the post-weld treatment method on the fatigue lifetime. The loading method of these specimens is the same as the loading method of the cruciform joint specimens.

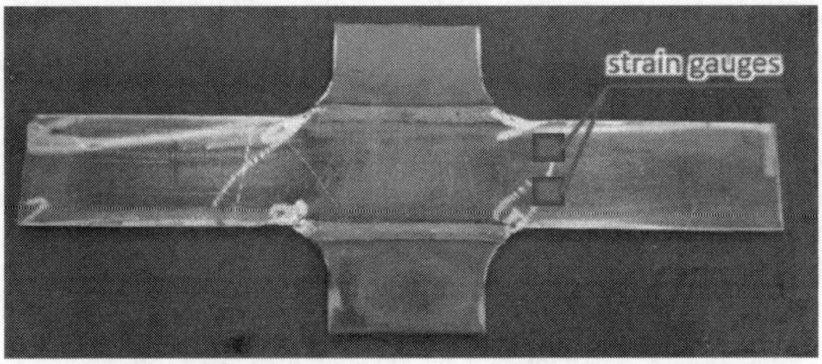

Figure 3 Position of strain gauges

3.2. Results of experiments

The aim of the research program is to recommend a fatigue crack propagation method. Therefore the main results of the experiments are the cyclic sample – stress range diagrams of the test specimens (Figure 4 a)), from which the material constants of the Paris law are recalculated. The other important information from the experiments are the crack patterns. One typical crack pattern observed for this specimen type is presented in (Figure 4 b)). From the pictures the point of the crack initiation and the path of the crack propagation can be clearly determined.

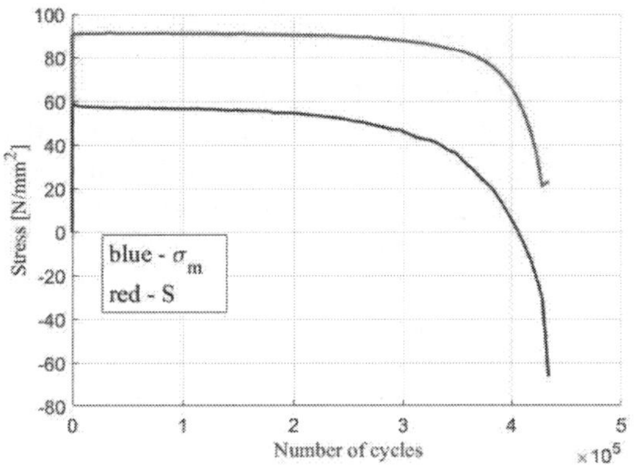

a)

b)

Figure 4 a) Data measured by a strain gauge at the crack; b) crack pattern of the specimen

4. Processing of the test results

The experimental and numerical calculation methods to determine the material properties of the Paris law require special test specimens and special facilities, thus the tests become more expensive and more complicated. In this research program a different method is applied to determine the crack propagation constants of a structural detail. This method is based on the measured data of the strain gauges. When the length of the fatigue crack varies, the measured strains and therefore the calculated stresses are changing in the surrounding area of the fatigue crack. Investigating the measured data of the strain gauges it can be observed, as the length of the crack increases, the stresses next to the crack decrease, but the stresses on the opposite side increase. The reason of this phenomenon is that the path of the stress lines vary close to the fatigue crack, as presented in Figure 5. To calculate the material constants (m and C), the crack propagation rates and the length of the cracks have to be determined from the measured data [4]. In this method an equivalent crack length (ECL) is used which differs from the real crack in its size and shape. The equivalent crack is a through-thickness crack and it has a crack front which is always perpendicular to the surface of the base plate. The direction of the crack propagation is always perpendicular to the edge of the base plate. This neglecting is in good agreement with the analysis of crack pattern showed in Figure 4 b).

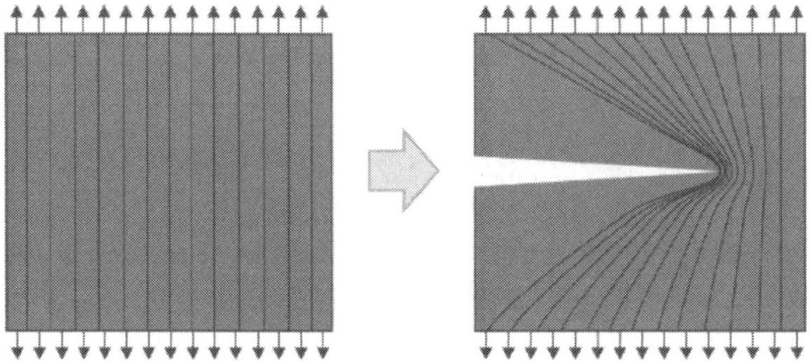

Figure 5 Alteration of the stress field due to crack propagation

Another approximation is the simplification of the equation (1). The original differential equation was reduced to a simplified version as shown by equation (3).

$$\frac{\Delta a}{\Delta N} = C \cdot (\Delta K)^m \quad (3)$$

4.1. Determination of the ECL

The numerical model of the investigated specimen is developed in ANSYS 17.2 [5]. The model is based on a full solid model using eight-node structural solid elements. To analysing the tip of the equivalent crack, a concentration keypoint is defined in the crack tip, thus the stress intensity factor could be determined in the crack tip using the FE model (Figure 6 a)). The geometry of the models is the same as the specimens' (Figure 6 b)). The ECL is a variable parameter of the numerical model from 0 to 40 mm. The stresses at the position of the strain gauges are calculated, hence the ECL can be plotted as a function of the measured stresses or the function of the stress ranges (Figure 7). Thus for an optional stress range ($\Delta\sigma_i$) the ECL (a_i) can be determined.

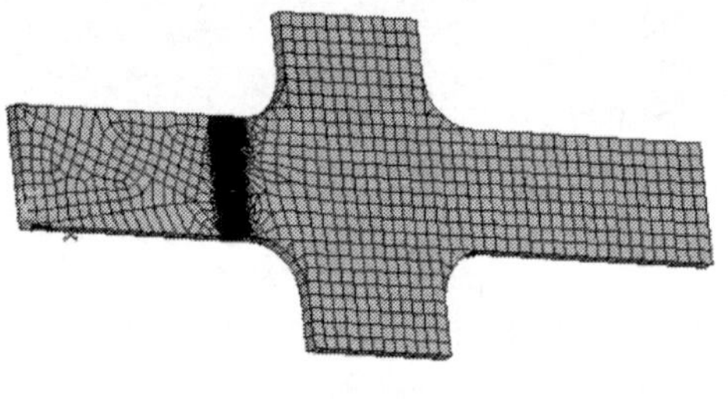

a)

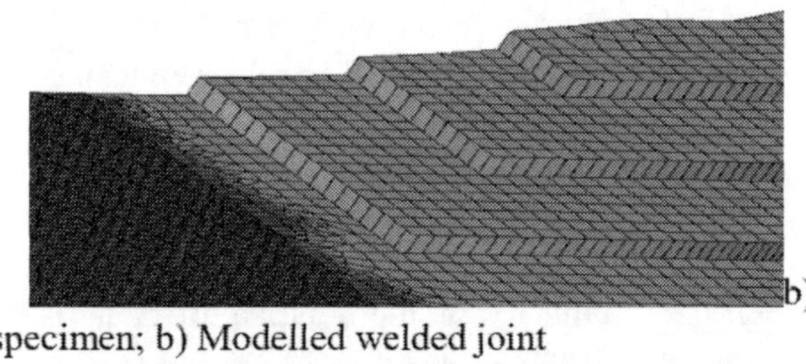

b)

Figure 6 a) FE model of the specimen; b) Modelled welded joint

4.2. Determination of the number of cycles

According to the recorded data the length of a measurement (10 s) and the length of the break (900 s) between the measurements and the time of failure could be localized. By the frequency of the loading and the calculated time of a cycle the number of this cycle are determined. Using this calculation method for all measured parameter, a number of cycle – stress range diagram could be plotted. From this diagram the stress range ($\Delta\sigma_i$) are determined for every number of cycles (N_i).

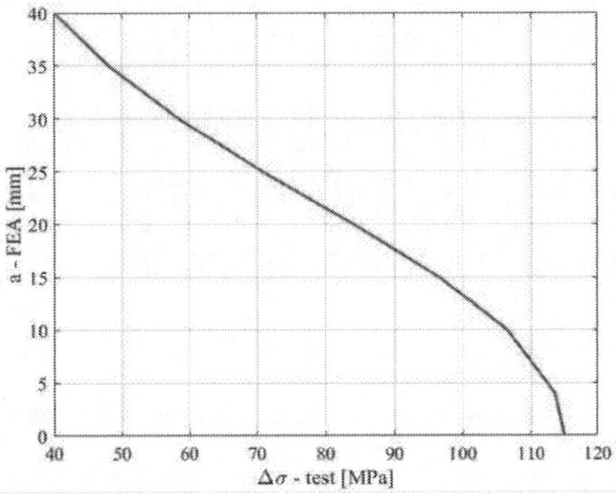

Figure 7 The ECL (a) values as a function of the measured stress ranges ($\Delta\sigma$)

4.3. Determination of the crack growth rate

According to the simplified equation (3) the crack growth rate is calculated by the ratio of Δa and ΔN, where Δa is the difference of two ECLs and ΔN is the difference of two numbers of cycles (Figure 8), which are calculated from two next test data. The crack propagation rate is calculated by equation (4).

$$\frac{\Delta a_i}{\Delta N_i} = \frac{a_{i+1} - a_i}{N_{i+1} - N_i} \tag{4}$$

4.4. Determination of the stress intensity factor range (SIFR)

The method, introduced in Section 4.1. can be applied to determine the stress intensity factor range (ΔK^*_i). Using the numerical model of the specimen, the stress range – SIFR diagram is plotted in Figure 9. However, there is a difference between the two cases. When the Δa is determined, two ECLs were used, thus two different stress ranges ($\Delta\sigma_i$ and $\Delta\sigma_{i+1}$) are applied for the calculation. Determining the stress intensity factor range is needed only one stress range. In this method the average of the two stress range is calculated (equation (5)), and it is used to get the SIFR from the stress range – SIFR diagram.

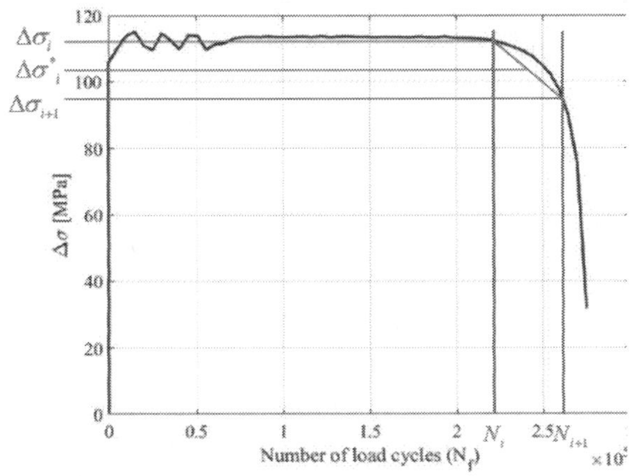

Figure 8 Determination of SIFR (ΔK) and ΔN

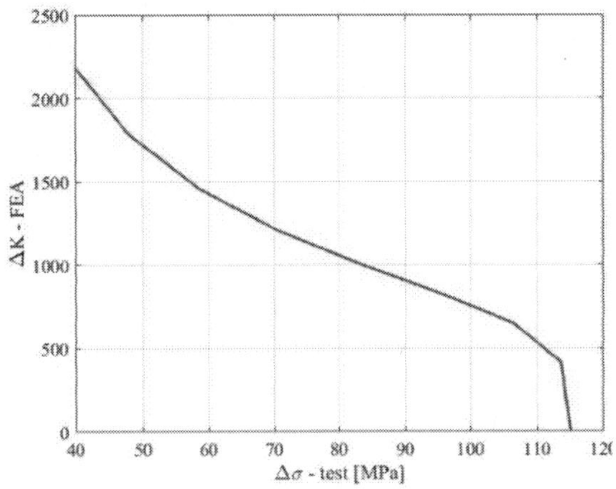

Figure 9 The SIFR (ΔK) values as a function of the measured stress ranges ($\Delta\sigma$)

IOP Conf. Series: Journal of Physics: Conf. Series **843** (2017) 012045 doi:10.1088/1742-6596/843/1/012045

$$\Delta\sigma^*_i = \frac{\Delta\sigma_i + \Delta\sigma_{i+1}}{2} \tag{5}$$

4.5. Determination of the material constants (m and C) of the Paris law

When all required variables are calculated for all measurements, the crack propagation rates and the SIFRs can be write in the equation (1). This equation is an exponential equation, but the solution can be simplified if some reforming is applied, as shown by equation (6).

$$\log\frac{da}{dN} = \log C + m\cdot\log(\Delta K) \tag{6}$$

After the reformulation of the equation, the method of linear least squares is applied to calculate m and log(C) from equation (6) using the previously determined parameters. The calculation results are summarized in
Table **1**.

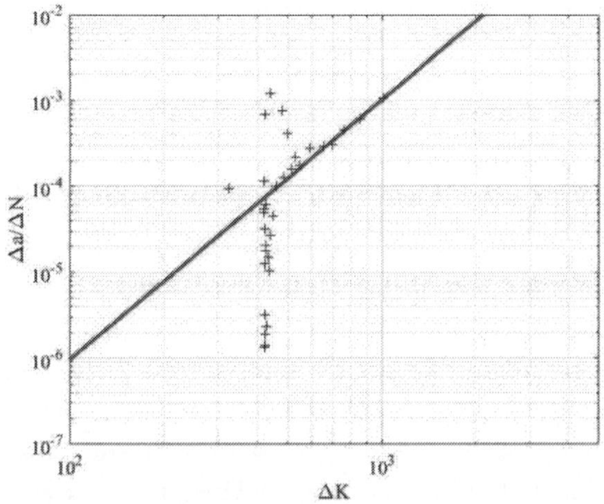

Figure 10 Fatigue crack propagation diagram of specimen No. 11

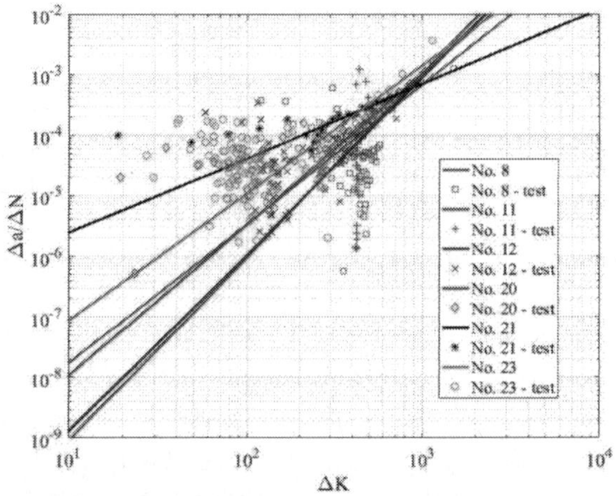

Figure 11 Fatigue crack propagation diagram of all specimens

Moreover, the final ECL is determined from the final measured data of the experiment. This ECL belongs the measured stress range next to the first value, which drops off to zero.

Table 1 Parameters and results of the experiments

No. of the specimens	Material	Thickness [mm]	Post weld treatment	$\Delta\sigma$ [MPa]	N_f	m	C	a_f [mm]
8	S235	10	disc grinding	100	672400	2.30	8.51E-11	33.533
11	S420	10	none	100	275200	3.03	8.31E-13	23.152
12	S420	10	disc grinding	78	530800	2.92	1.50E-12	33.855
20	S235	18	none	86	442200	2.50	3.34E-11	36.056
21	S235	18	none	74	519200	1.23	1.44E-07	35.654
23	S235	18	disc grinding	136	1310000	2.11	2.11E-10	32.641

$\Delta\sigma$: stress range; N_f : fatigue lifetime; m and C: material constants of Paris law; a_f : final/critical equivalent crack size

5. Evaluation of the test results

Investigating the results and the calculated material constants the following conclusion are drawn. The most conspicuous one is that the slope of the crack propagation diagram in stage II is the commonly used value [6] ($m = 3$) for S420, but it is smaller in the case of the S235 material. According to this conclusion if this type of structural details is analysed, the "$m = 3$ constant" can be used for the crack propagation calculations, however using S235 steel, a more favourable behaviour is observed. The main part of the fatigue lifetime of the welded details comes from the crack propagation, thus applying the Paris law of crack propagation the fatigue lifetime can be calculated for the welded joint. To use this method, the material constants and the initial and final crack length (or ECL) is necessary. As the results of

Table **1** shows, the final ECL is approximately the same in all the analysed cases. There are different recommendations to the initial crack length for different structural details in [4] and [7]. Using the material constants of

Table **1** the fatigue lifetime of the investigated structural detail can be calculated. Investigating the effect of the thickness of the material it can be stated that the plate thickness has no significant effect on the slope of the crack propagation diagram. The same phenomenon is determined when the effect of the post weld treatment is analysed. The material constant (m), which represents the slope of the crack propagation diagram in stage II, is the same to both as-welded and disc grinded specimens.

6. Summary

In this paper an alternative method is introduced to determine the material constants of Paris law of crack propagation. Using the measured data from strain gauges and the calculated results from the FEM, an equivalent crack length (ECL) is determined for every measurement regime. According to the results of the FEM, the stress intensity factor ranges (SIFR) are calculated for every ECL. Knowing the number of load cycles, the crack propagation rates are determined for every calculated SIFR, thus the crack propagation diagram are plotted for each specimen. If these diagrams are given, the material constants (C and m) and the final (critical) ECL could be calculated. These parameters were compered in the cases of different plate thicknesses (10 and 18 mm) and different types of material (S235 and S420). It can be observed, that the crack propagation behaviour of the S235 material is more favourable in the cases examined. On the other hand, there is no significant effect of the disc grinding on the crack propagation behaviour. Thus it can be declared, that the higher fatigue lifetime of the specimen with post weld treatment comes from the crack development phase and it has no impact on the crack propagation phase.

References

[1] Maddox S J 1997 Fatigue design and assessment of welded joints *Int. Conf. Fracture 9 (Sydney)* pp 2287-2298
[2] Schwalbe K-H, Landes J D and Heerens J 2007 Classical Fracture Mechanics Methods *Comprehensive Structural Integrity, Online Update* Milne R O, Ritchie B, Karihaloo (Geesthacht: Helmholtz-Zentrum) **11** pp 3-42
[3] ASTM E 647 – 00 Standard Test Method for Measurement of Fatigue Crack Growth Rates: 2004
[4] Maddox S J 1974 Assessing the significance of flaws in welds subject to fatigue *Welding J, Res Suppl* **53** pp 401-409
[5] ANSYS® v17.2, Canonsburg, Pennsylvania, USA
[6] Hobbacher A 2008 Recommendations for fatigue design of welded joints and components *International Institute of Welding*
[7] Nykänen T, Marquis G and Björk T 2006 Fatigue analysis of non-load-carrying fillet welded cruciform joints *Engineering Fracture Mechanics* **74** pp 399-415

IOP Conf. Series: Journal of Physics: Conf. Series **843** (2017) 012046 doi:10.1088/1742-6596/843/1/012046

Effect of Si content on fatigue fracture behavior of hot-rolled high-silicon steels

Osamu Umezawa[1], **Jyunichi Kanda**[2], **Takao Yamazaki**[2]

[1]Faculty of Engineering, Yokohama National University, 79-5 Tokiwadai, Hodogaya, Yokohama, 240-8501, Japan
[2]Graduate School of Engineering, Yokohama National University, 79-5 Tokiwadai, Hodogaya, Yokohama, 240-8501, Japan

E-mail: umezawa-osamu-fv@ynu.ac.jp

Abstract. As the Si content was increased from 1.5 to 5 mass%, both the yield stress and ultimate tensile strength were increased, respectively. The work hardening rate was also increased as the increase of Si content. On the contrary, the elongation was decreased as the increase of Si content, and the fracture manner was shifted from ductile to brittle. The 10^7 cycles fatigue strength was higher as the increase of Si content. The small misorientation distribution as ladder-like was detected in the grains of 1.5 mass%Si steel. Around the grain boundary, the strain incompatibility was detected in the steels containing over 3 mass%Si. The lattice rotation was locally detected in the vicinity of grain boundaries.

1. Introduction

The addition of Si in α-iron (bcc) presumably lowers its stacking fault energy so that cross slip is strongly suppressed.[1] Such slip manner provides planar dislocation structures under fatigue deformation. The subsurface fatigue crack generation in titanium alloys (hcp) and austenitic steels (fcc), which has been commonly reported in high-cycle regime and at low temperature,[2,3] is closely related with heterogeneous deformation due to the restriction of slip systems.[4,5] Si addition in steel causes the restriction to {110} <111> primary slips, especially over 3.8 mass% Si content at room temperature.[6] The deformation and fatigue fracture behavior in Si steels depend on their Si content and are highly reflected on their crystal orientation and microstructure. Ushioda *et al*.[7] investigated the evolution of dislocation structures in Fe–Si alloys with 0, 0.5 and 1 mass%Si during a cyclic bending test. In the 1 mass%Si steel, the vein structure developed, and many intergranular cracks were detected. The mixed cleavage mode and crack front branching on the fatigue crack growth were detected in 2 mass%Si steel.[8] In 3 mass%Si steel, the fatigue crack initiation sites were mostly located at the grain boundaries and that fatigue crack growth rate was lowered at some grain boundaries.[9] The <111> slip bands and regular arrays of dislocation wall structure like "labyrinth" were seen around fatigue crack tip region.[10] Therefore, the deformation behavior in the vicinity of grain boundaries should make an important role on fatigue crack generation and crack growth in high Si steels. Few studies, however, have been done on the cyclic deformation structure and crack generation in polycrystalline high Si steels. In the present study, the effects of Si content on high-cyclic fatigue fracture and deformation structure in the polycrystalline steels containing Si as 1.5, 3, 4 and 5 mass% have been studied.

2. Experimental

2.1. Test materials

The Fe-Si alloys containing 1.56, 3.00, 3.95 or 4.86 mass%Si were hot-rolled either by flat rolling (FR: heating at 1373 K for 60 min + heating at 1373 K for 60 min) or grooved-rolling (GR: heating at 1373 K for 30 min + 1273 K for 30 min). The materials were named as FR1.5, FR3, FR4, FR5, GR1.5, GR3, GR4 and GR5. Their average grain sizes were from 535 μm to 570 μm in diameter for the FR and from 263 μm to 298 μm in diameter for the GR. Although almost no dislocations were detected in the FR materials, some mobile dislocations were dispersed in the GR materials.

2.2. Testing

The round bar type specimens with a diameter of 4 mm for both tensile and fatigue tests were machined from the plates or rods as parallel to the rolling direction. Tensile test (displacement-controlled) was carried out at an initial strain rate of about 4×10^{-4} sec^{-1} using a screw-driven machine at room temperature. Yield strength was evaluated by the off-set value of 0.2% proof stress using an extensometer. Load-controlled fatigue test was carried out in the air using a servo-hydraulic machine. The sinusoidal waveform loading was uniaxial with a minimum-to-maximum stress ratio, R ($\sigma_{min}/\sigma_{max}$), of 0.01 and the test frequency was 20 Hz.

2.3. Microscopy and characterization

The microstructure and fatigue-crack-initiation sites and fracture surfaces were studied by scanning electron microscopy (SEM). The electron backscatter diffraction (EBSD) analysis with SEM was performed to characterize deformation microstructure and orientation. The EBSD-Wilkinson measurements [11] was adopted to elastic strain field analysis. Transmission electron microscopy (TEM) was also applied to determine the dislocation structure. For fatigued specimens, TEM disks about 600 μm thick were sectioned from beneath the fracture surface. They were cut perpendicular to the principal stress axis and were reduced to less than 100 μm in thickness by abrasion with No. 600 emery paper. Both the TEM foils and EBSD samples were finally prepared by a conventional electrical-jet polishing technique in a stirred solution of 20 pct perchloric acid, 70 pct ethanol and 10 glycerol.

3. Results and discussion

3.1. Tensile properties

Figure 1 represents the stress – strain curves of the GR materials, and Table 1 summarized the tensile properties of tested materials. As the Si content was increased from 1.5 to 5 mass%, both the yield stress and ultimate tensile strength were increased from 239 MPa to 471 MPa and from 365 MPa to 713 MPa, respectively. The FR materials showed similar stress – strain curves as the GR ones, but their tensile strength was lower as from 322 MPa to 682 MPa due to larger grain sizes. The work hardening rates were increased as the increase of Si content, too. On the contrary, the total elongation and reduction of area were decreased from 25.5% to 6.9% and from 88% to 5% in the GR as the increase of Si content. The FR1.5 and GR1.5 appeared a ductile fracture surface covered with dimples, and the fracture manner was shifted from ductile to brittle as the increase of Si content.

3.2. Fatigue properties

The 10^7 cycles fatigue strength was higher as the increase of Si content in the FR and GR materials. In every alloy, the 10^7 cycles fatigue strength in the GR material was higher than that in the FR one. It may be resulted from the smaller grain size in the GR materials. On the contrary, the ratio of 10^7 cycles fatigue strength by ultimate tensile strength was lowered as the increase of Si content. Their ratios were less than 0.5 in the FR materials. The GR1.5 was 0.84, and the others were about 0.75.

The FR1.5 showed a ductile fracture manner in the stage III as shown in Figure 3(a). The GR1.5 and GR3 fractured with low cycles also showed a ductile fracture manner in the stage III. The others appeared brittle manner. Fatigue crack initiation sites in FR5.0 were detected at the specimen interior as shown in Figure 3(b).

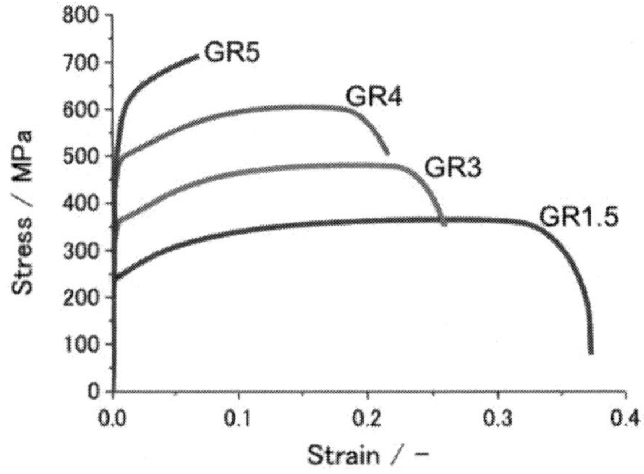

Figure 1. Nominal stress – nominal strain curves of the GR materials.

Table 1. Tensile properties of test materials.

Materials	Average grain size (μm)	Ultimate tensile strength (MPa)	0.2% proof stress (MPa)	Elongation (%)	Reduction of area (%)
FR1.5	570	322	207	21	90
FR3	547	432	338	14	73
FR4	561	566	472	15	48
FR5	535	682	560	11	8
GR1.5	263	365	239	37	88
GR3	298	481	312	26	62
GR4	294	605	428	22	55
GR5	286	713	471	7	5

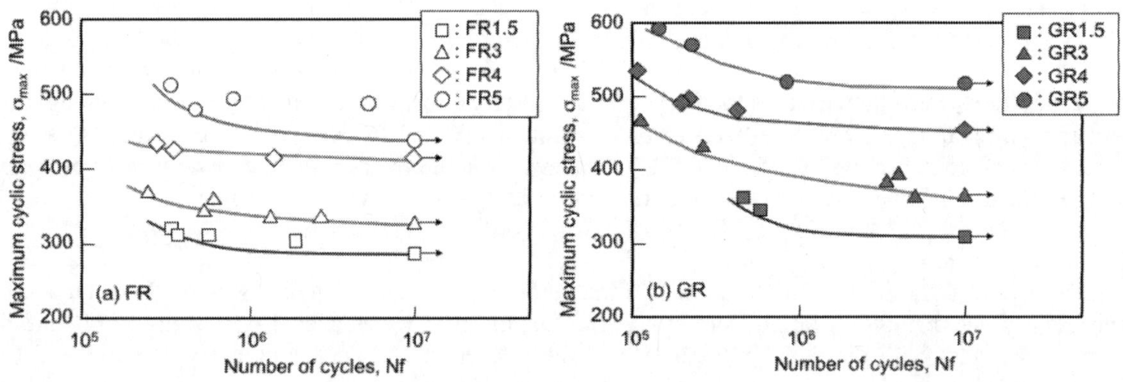

Figure 2. S-N data of the FR (a) and GR (b).

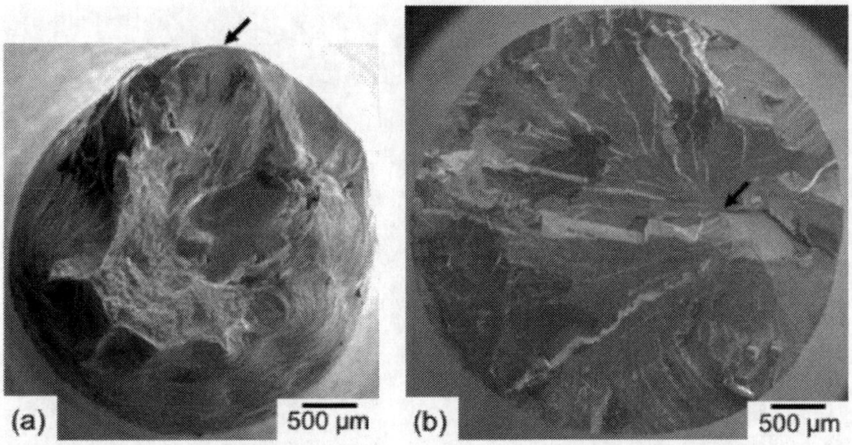

Figure 3. Fracture surfaces of (a) FR1.5 (σ_{max}=306 MPa) and (b) FR5 (σ_{max}=450 MPa). Arrows indicate the crack initiation site.

3.3. Strain incompatibility at grain boundaries

Both Si content and cycle stress were significantly influenced the evolution of the dislocation structure. The dislocation structure was more rectilinear as the increase of Si content or the decrease of cycle stress. In the GR1.5, the Vein structure of saturated dislocation arrangement was well developed. In the FR1.5, the cellular structure was also developed by cyclic deformation as shown in Figure 4(a). As Si content was increased, dislocation arrangement became planar and its density was lowered as the sub-boundary structure shown in Figure 4(b).

The strain and misorientations were detected in both grain boundaries and cell boundaries. The small misorientation distribution as ladder-like was detected in the grains of the FR1.5 and GR1.5 as gown in Figure 5(c) and Figure 6(a). It may reflect on the cellular structure or Vein structure. In the FR5 and GR5, on the other hand, no strain distribution in grains but sub-grains along the grain boundary were detected as shown in Figure 5(d) and Figure 6(c).

According to the EBSD-Wilkinson measurements, the elastic strain was partitioned among the grains in the FR1.5 as shown in Figure 6(b). On the other hand, the lattice rotation was detected in the vicinity of grain boundaries in the FR5 as shown in Figure 6(d). It may give a local stress concentration and a site of cracking.

Therefore, Si addition enhanced the strain gradient at grain boundaries, which may strongly influence on the fatigue crack generation and fatigue strength of high Si steels.

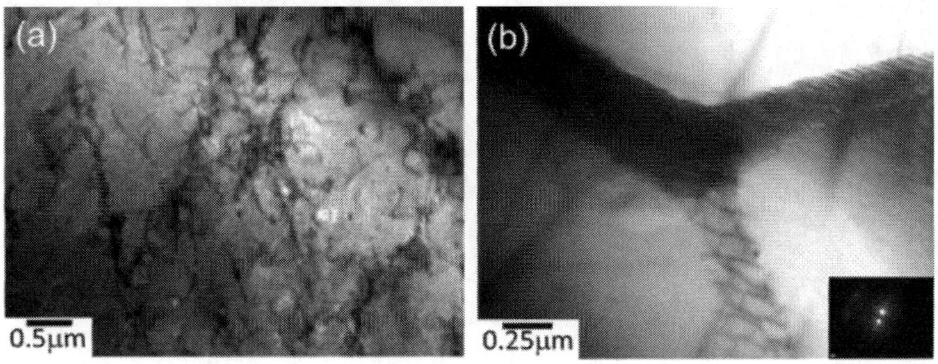

Figure 4. A ladder-like dislocation structure in the FR1.5 (σ_{max}=266 MPa) (a) and sub-boundary structure in the FR5 (σ_{max}=433 MPa) (b).

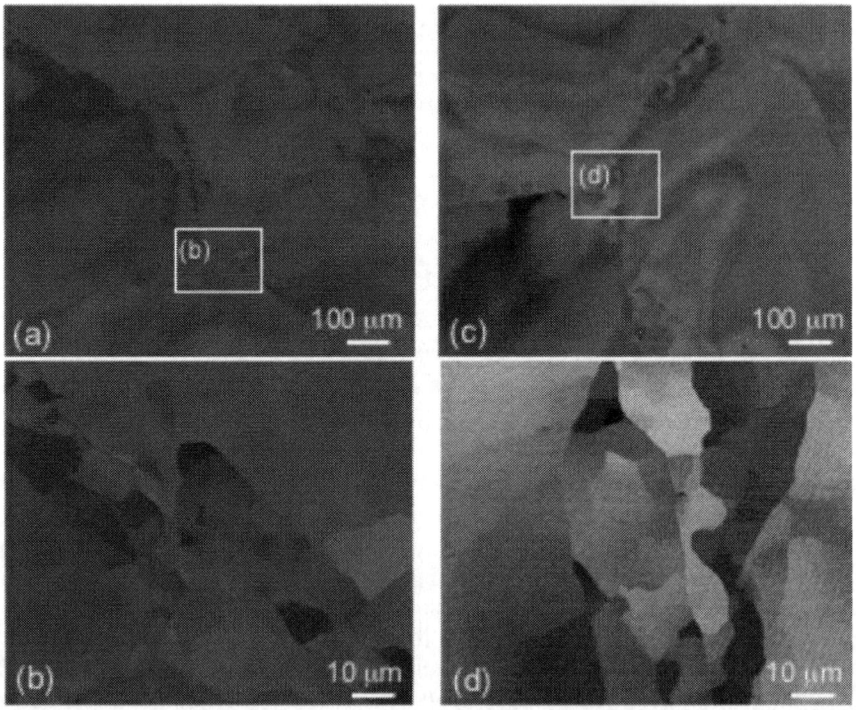

Figure 5. Back scatter electron images of microstructure after 10^7 cycles: (a),(b) FR1.5 (σ_{max}=258 MPa) and (c),(d) FR5 (σ_{max}=362 MPa).

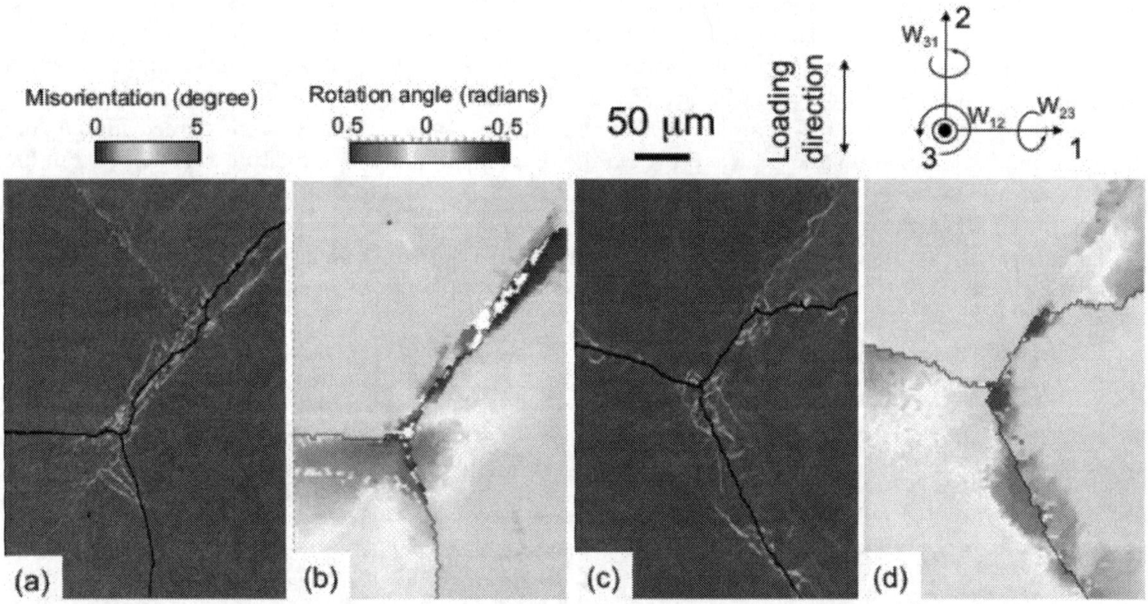

Figure 6. Kernel average misorientation and rotation angle (W_{12}) of near the boundaries after 10^7 cycles: (a),(b) FR1.5 (σ_{max}=258 MPa) and (c),(d) FR5 (σ_{max}=362 MPa).

4. Conclusions

The effect of Si content on high-cyclic fatigue fracture behaviour in the steels containing Si as 1.5, 3, 4 and 5 mass% were studied. Major results were summarized as follows:

(1) Both tensile strength and yield strength were increased as the increase of Si content, although the fracture manner was shifted from ductile to brittle. The work hardening rate was increased as the increase of Si content, too.
(2) The 10^7 cycles fatigue strength was higher as the increase of Si content.
(3) The small misorientation distribution as ladder-like was detected in the grains of 1.5 mass%Si steel. It may be reflected on the cellular structure or Vein structure of saturated dislocation arrangement developed by cyclic deformation. The elastic strain was partitioned among the grains.
(4) Around the grain boundary, the strain incompatibility was detected in high Si steels, e.g. 5 mass%Si steel. The lattice rotation was detected in the vicinity of grain boundaries. It may give a local stress concentration and a site of cracking.

Acknowledgments

The authors wish to thank Dr. S. Morooka of Yokohama National University (now Japan Atomic Energy Agency) for his help on the experiments. This work was supported by KAKENHI 16K06769.

References

[1] Griffiths D and Rilley J N 1966 *Acta Metallurgica* **14** 755
[2] Umezawa O, Morita M, Yuasa T, Morooka S, Ono Y, Yuri T and Ogata T 2014 *Advances in Cryogenic Engineering Materials* vol 60 AIP Conf. Proc. 1574 p 34
[3] Umezawa O and Nagai K 1997 *ISIJ International* **37** 1170
[4] Morita M and Umezawa O 2011 *Materials Transactions* **52** 1595
[5] Morita M and Umezawa O 2012 *ISIJ International* **52** 1153
[6] Barrett C S, Ansel G and Mehl R F 1937 *Trans. Amer. Soc. Metals* **25** 702
[7] Ushioda K, Goto S, Komatsu Y, Hoshino A and Takebayashi S 2009 *ISIJ International* **49** 312
[8] Chakravarthula S S and Qiao Y 2005 *International J. Fatigue* **27** 1210
[9] Hojo M, Tanaka K and Nakai Z 1983 *J. Soc. Materials Science Japan* **32-358** 65
[10] Takahashi Y, Tanaka M, Higashida K, Yamaguchi K and Noguchi H 2010 *Acta Materialia* **58** 1972
[11] Wilkinson A J, Dingley D J and Meaden G 2009 *Electron Backscatter Diffraction in Materials Science* eds A J Schwartz, M Kumar, B L Adams and D P Field (Dordrecht: Kluwer Academic Pub.) chapter 19 pp 231-246

Experimental characterization of fatigue strength in butt welded joint considering the geometry and the effect of cooling rate of the weld

Nelson Arzola[1] and Edgar Hernández[2]

Research Group in Multidisciplinary Optimal Design, Department of Mechanical and Mechatronics Engineering, National University of Colombia, Cra. 30 No. 45-03 Bogotá D.C., Colombia, Phone (571) 316 5000 Ext. 14 062, Fax (571) 316 5333.

Email: [1] narzola@unal.edu.co; [2] edhernandezl@unal.edu.co

Abstract. In this work the experimental characterization of fatigue strength in butt welded joints considering the geometry and the post-weld cooling cycle was performed. ASTM A-36 structural steel was used as the base metal for the shielded metal arc welding process, with welding electrode E6013. Two experimental factors were established: weld bead geometry and the post-weld cooling rate. Two levels for each factor, the welding reinforcement (1 and 3 mm), and the rate of cooling, slow (quiet air) and fast (immersion in water) are evaluated respectively. For the uniaxial fatigue tests, 8 samples were selected for each treatment for a total of 32 specimens. The mechanical and fractomechanical properties of fusion zone, heat affected zone and base metal in relation to the analysis of failure mechanisms were analysed. The fatigue crack growth rates were estimated based on the counting of microstrations. Furthermore, experimental tests, such as uniaxial tension, microindentation hardness, Charpy impact and metallographic analysis, were made to know the influence of the experimental factors in the fatigue strength. On this research, about the 78.13% of the samples obtained a resistance higher than the recommended one by class FAT 100. The results showed that the geometry of the joint is the factor of greatest influence on fatigue strength for butt welded joints; the greater the weld reinforcement the lower the fatigue strength of the joint. Although it is also important to consider other geometric factors of less impact as it is the weld toe radius and the welding chord width.

Keywords: Butt weld joint; fatigue; crack growth rate; weld reinforcement; rate of cooling.

1. Introduction

The origin of fatigue failure in butt welded joints is caused by different factors, it could be human factor or factors inhered to the process of fabrication employed. The most common imperfections are discontinuities like cracks, undercuts, lack of fusion, porosities and inclusions among others. The fatigue strength in the joint is related to the geometric profile and the coalition between the filler material and the base material. As a result of the thermal cycle experienced during the joint process, there is produced three zones with microstructural differences: the fusion zone (FZ), the heat affected zone (HAZ) and the base metal (BM).

A main evaluation of fatigue strength in the welded joints is thought the initiation process and crack propagation, which depends on the local stress notch in the weld toe and weld root, and the stress intensity factor (SIF) respectively [1]. To predict the fatigue strength from the structures or welded joints it is common to use the S-N curve method, where S is the nominal stress amplitude as a function of N that is the failure number of cycles. Besides, the S-N curve characterizes the limit stress from the notch or the class of detail establish by the British Standard Institution (BSI) [2]. In butt welding joints with a transversal load applied to the weld, the class is between 80-90 according to [3] and C class according to [4], as it is shown in figure 1. Those curves are developed up to two million cycles and are called class FAT.

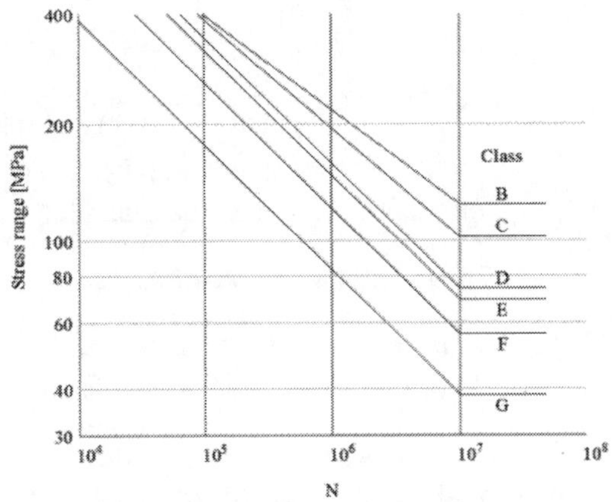

Figure 1. The S-N Curves for the various classes of welded joints [4].

Because the S-N curve is not intended to identify the crack propagation stage, it is necessary to use the empiric relation given by Paris and Erdogan [5]. This expression allows the evaluation of the crack propagation rate (da/dN) as function of the SIF range (ΔK), where C and m are material constant values:

$$\frac{da}{dN} = C(\Delta K)^m \qquad (1)$$

In a double logarithm graphic from de log (da/dN) in function of log (ΔK) it is possible to obtain three characteristic regions shown in figure 2. The region I, known as the fatigue crack initiation stage, is used to determine the threshold stress intensity range ΔK_{th} for large cracks, although the classification between large and shorts cracks is not clearly established. If a crack is completely in the plastic zone and/or its size related to the propagation direction it's than lower to 0.5 mm it is possible to say that it is a short crack [6]. The region II could be represented by equation (1), which occupies the greater fatigue life percentage, related to regions I and III. In the same way it describes a stable crack propagation stage. Finally, region III presents a sudden increase of the crack propagation rate, giving as a result the final fracture when the fracture toughness of the material is reached.

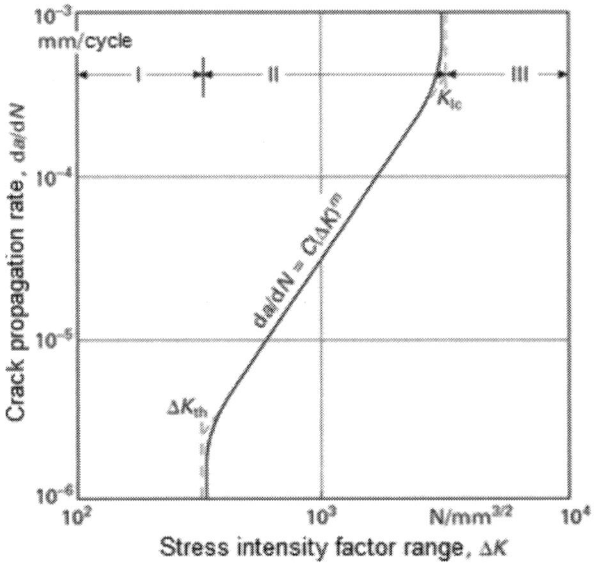

Figure 2. Fatigue crack growth diagram [6].

One of the current methods used to evaluate the fatigue in welding joints is the Fracture Mechanics. Its principal application is the element analysis that contains any type of imperfection, especially elliptic cracks as it is shown in figures 3 and 4. The Fracture Mechanics considers every single stress and its distribution through the thickness of the plate, where the calculation of the crack propagation is governed by the maximum principal stress, because the maximum energy released is presented in the crack propagation perpendicular to the maximum principal stress direction [6].

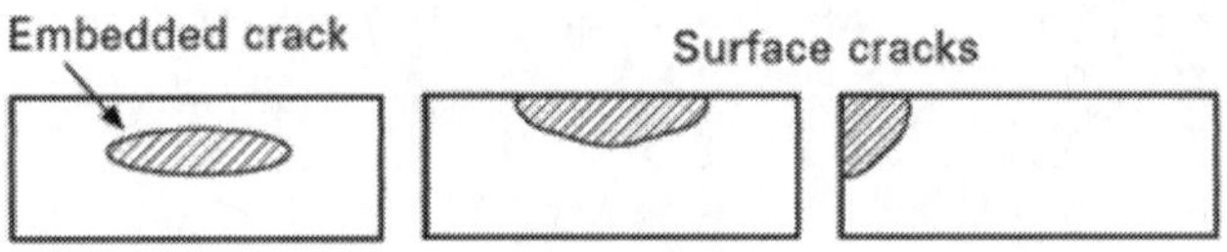

Figure 3. Types of semielliptical cracks [6].

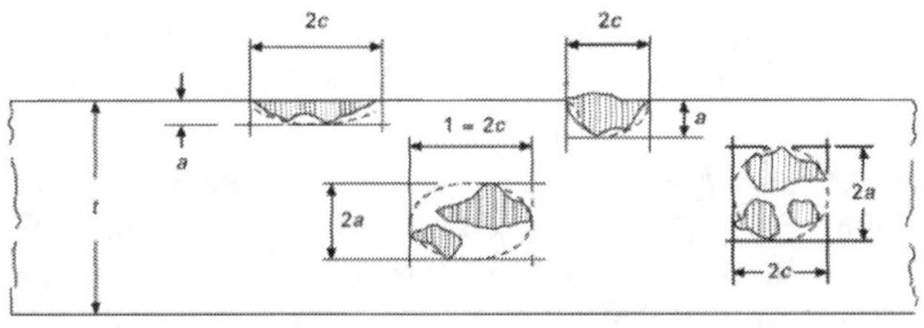

Figure 4. Transfer to NDT into elliptical cracks [6].

In the fracture surface it is common to observe different characteristic scratch that depends on the load type and the environment, among other factors [7]. Two characteristic zones are observed by the uniaxial fatigue on the fracture surface. The first of them is the stable crack propagation which has a smooth or granular texture that can be seen by the human eye. The second zone is called the final fracture because in this zone the mechanical element is not able to resist the cyclic peak load, and generates the final fracture. The common fractures surfaces marks are shown in figure 5 those marks are related to the mechanical uniaxial fatigue like: river marks, beach marks, ratchet marks, the crack origin and in some cases the striations can be observed through a scanning electron microscope (SEM).

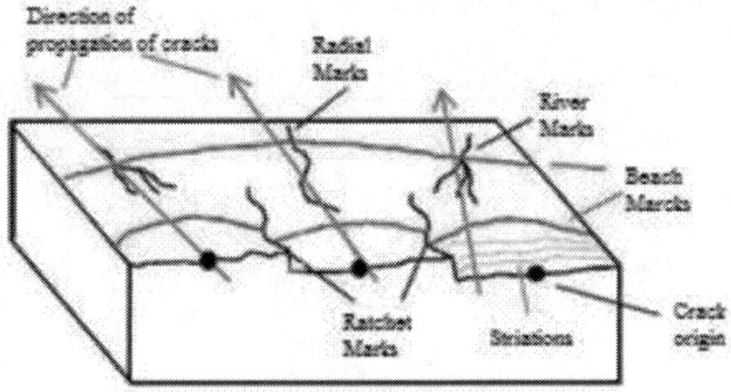

Figure 5. Common marking on fatigue surfaces [6].

The beach marks location determines the fatigue stable crack propagation zone, although it is possible to find them in the fracture surface by stress-corrosion. In the weld joint failure, it is possible to find the crack origin by the beach marks. In that way the striations are fracture surface marks that indicate the crack front shape and size after the fatigue cycle load, however in order to observe them it is necessary to use a SEM. Moreover, the ratchet marks are generated by the interaction of different crack origins, where each one of them produces its own fatigue crack stable propagation zone. Despite this, the different crack fronts can coalesce in a single progressive main crack front as it moves [8].

Different mechanical properties are not constant in weld joints such as yield strength, tensile strength, hardness, fracture toughness, among others. This is caused because of the thermal gradient generated in the fabrication process and the cooling process which has been subjected the welded joint. For that reason, it is necessary to evaluate those properties for the three welded joint zones, the FZ, HAZ and BM, in order to obtain accurate results in terms of welded joint strength [9]. The more common tests used in mechanical characterization for the welded joints are: microindentation hardness [10], uniaxial mechanical stress [11] and Charpy Impact test [12-15], among others. Besides the mentioned tests, the metallographic characterization plays an important role in the welded joint mechanical properties; it allows linking the different microstructural mechanic phases on the joint with its mechanical behavior.

This article describes an experimental study of the fatigue failures on butt welded joints for structural carbon steel ASTM A-36 used in the electric arc welding process with coated electrode E6013. The influence in the geometric joint was evaluated as well as the post weld cooling environment by the uniaxial fatigue tests and the failure mechanism analysis present in the fracture surface. In collateral tests, the mechanical and fractomechanical properties from the welded joints were analyzed in relation with the crack origin and propagation caused by fatigue.

2. Materials and methods

2.1. Experimental design

For the uniaxial fatigue tests, 8 replicas were selected for each treatment accounting for a total of 32 test specimen. Two experimental factors were established: weld bead geometry and the post-weld heat treatment. Two levels for each factor the welding reinforcement (1 and 3 mm), and the rate of cooling, slow (quiet air) and fast (immersion in water) are evaluated, respectively. For the uniaxial tension test three samples were evaluated, as total sampling amount for each treatment, accounting a total of 12 test specimens, and for the Charpy impact test six replicas for each treatment were done, 3 in the (FZ) and

IOP Conf. Series: Journal of Physics: Conf. Series **843** (2017) 012047 doi:10.1088/1742-6596/843/1/012047

3 in the (HAZ), respectively; additionally 3 more replicas were analyzed for the base metal (BM), thus 27 test specimens were obtained for the late type of test. In order to make easier the data management a codification system was assigned for each test type and cooling medium correspondingly, which is found in table 1 for the uniaxial fatigue test specimens, table 2 for the uniaxial tension test and finally in table 3 for the Charpy Impact samples, respectively.

Table 1. Codification system for the uniaxial fatigue test specimens.

1th digit	Weld Reinforcement	~1 mm	1
		~3 mm	3
2nd digit	Cooling Medium	Air	A –
		Water	W –
3rd digit	Consecutive numbering for replicas		1 – 8

Table 2. Codification system for the uniaxial tension test specimens.

1th digit	Weld Reinforcement	~1 mm	1
		~3 mm	3
2nd digit	Cooling Medium	Air	A
		Water	W
3rd digit	Uniaxial tension	– T–	
4th digit	Consecutive numbering for replicas		1 – 3

Table 3. Codification system for the Charpy-V notch test specimens.

1th digit	Weld Reinforcement	~1 mm	1
		~3 mm	3
2nd digit	Cooling Medium	Air	A
		Water	W
3rd digit	Charpy Impact		– CVN
4th digit	Impact Zone	Fusion Zone	FZ –
		Heat Affected zone	HAZ –
		Metal Base	MB –
5th digit	Consecutive numbering for replicas		1 – 3

2.2. Materials

The metal base (MB) employed on the welded samples fabrication for the uniaxial fatigue and uniaxial tension, was the rolled carbon structural steel ASTM A-36 with a thickness of 6 mm, while for the impact test a thickness of 12 mm was used. The chemical composition of the material was determined by spark spectrometry test UV-VIS, results are presented in table 4 and the mechanical properties are shown in table 5.

Table 4. Chemical composition for ASTM A-36 steel, (Wt. %).

C	Si	Mn	P	S
0,202	0,030	0,532	0,028	0,011

Table 5. Mechanical properties for ASTM A-36 steel.

Yield Stress [MPa]	Ultimate Tensile Strength [MPa]	Elongation [%]
262	434	38

2.3. Welding procedure

On the welded joints fabrication as providing material the welding electrode E6013 was employed. The welding direction was performed perpendicular to laminar direction of MB. For welding procedure the plane to top position was utilized with a square joint design following [12], voltage ranging between 15 – 42 V and current between 85 – 140 A (direct current). Approach speed between 2 and 3 mm/s. The applied passes were varied according to the over-thickness width s following: single pass for $s = 1$ mm and two passes for $s = 2\ y\ 3\ mm$ respectively.

2.4. Uniaxial fatigue test

The test specimen dimensions for the fatigue and uniaxial tension test are shown in figure 6. The same geometry was used for the samples on both tests in order to facilitate their fabrication. However, the AWS B4.0:2007 and ASTM E606 standard were taken into account respectively. Subsequently to post-welding procedure, the final geometry of test specimens was achieved with a waterjet cutting machine to avoid modifications on post-welding treatment. For uniaxial fatigue experimental test a non-standard installation was used. Holding and measuring machine mechanism is supported on two fixed plates where the assembly, for load-cell and sample to be tested, is placed as it is shown in figure 7. Frequency used on uniaxial fatigue test was 4 Hz, peak applied load was 12 kN with a load cycle ratio. $R = 0.05$. A load cycle ratio $R > 0$ in order to guarantee non compressive contact between the crack sides and the information lost concerning its propagation rate is selected.

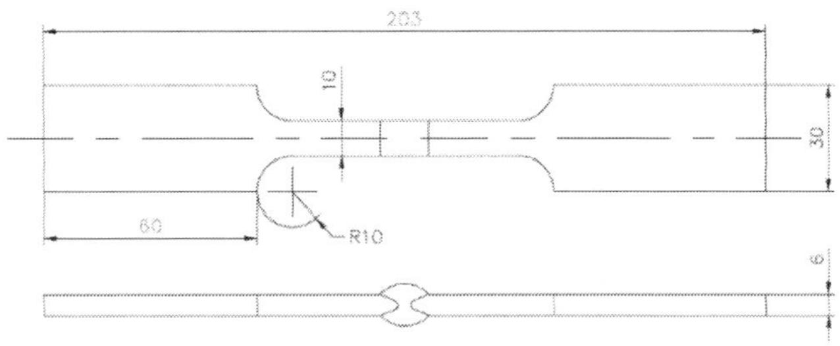

Figure 6. Dimensions for tensile and fatigue test specimens (mm).

Figure 7. Butt welded specimen in the axial fatigue equipment.

2.5. Uniaxial tension test, microindentation hardness test and Charpy V-notch test

The uniaxial tension test were performed following the standard ASTM E8 in a universal test machine Shimadzu UH-500 kNI aiming to obtain the mechanical tension characterization of the studied samples. The hardness test by microindentation were performed in relation with sample thickness (6mm) on three equidistant lines and a third of the sample thickness, employing a durometer LECO M-400-G2 Hardness Tester with a load of 500 gr during 10 s ($HV_{0,5}$). Fabrication of the sample for the energy of impact characterization test, its geometry and percentage of fracture due to shearing were done accordingly the standard ASTM A370 as it is shown in figure 8. The tests were performed at environmental temperature ranging from 19.06 – 19.10°C. A chemical macro attached was performed with Nital at 3% over the welding profile in order to place the notch in the center of zones HAZ and FZ as it is shown in figure 9 following standard ISO 15653.

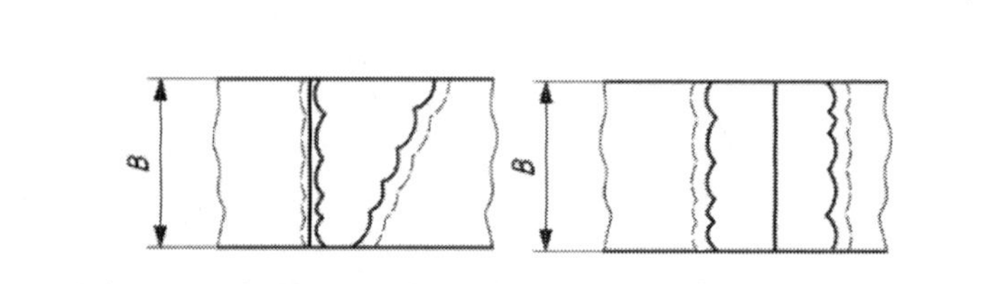

Figure 8. Specimen for impact Charpy V-notch test (mm) following standard ASTM A370.

Figure 9. Notch location in HAZ (left) and FZ (right), following standard ISO 15653.

2.6. *Microstructure and fractography*

Samples used for metallographic test were polished and subsequently attached with Nital at 3%. The microstructural characterization of zones FZ, HAZ and BM was performed by an optical microscope and SEM. Alike the analysis of failures over crack surface was performed on the welded joints in order to identify the crack origin, stable propagation crack zone and its limit. Through observation with stereoscope the stable propagation crack zone to readily use the SEM, which was used for the counting procedure of striations close to initiation and near the end of stable propagation crack zone, as it is described on the study carried out by DeVries et al. [16].

3. Results and discussion

Using as reference FAT 80 and FAT 100 classes in figure 10 the comparison between results for uniaxial fatigue of butt welded joints for the four studied treatments through curve S-N is presented. In this case treatment 3A shows 50% of its results under limit of class FAT 100. Further it is seen that only one sample of treatments 3A and 3W are under limit of class FAT 80; while in majority are the results for samples from treatments 1W and 1A those which represent a greater deviation above limit of class FAT 100. Accordingly with results from figure 10, it is observable the stochastic nature on the fatigue behaviour of butt welded joints. Owing to this, it is necessary to perform a statistic evaluation of the results. Based on [3] a statistic procedure to evaluate surviving probability as function of fatigue capacity (log C), which is obtained through equation (2). Where it is set $m = 3$ for the steels was employed.

$$logN = logC - m \cdot log\Delta\sigma \qquad (2)$$

Subsequently the surviving probability in relation to values of log(C) for 32 analysis samples is obtained, in a normal probability graphic as it is presented in figure 11.

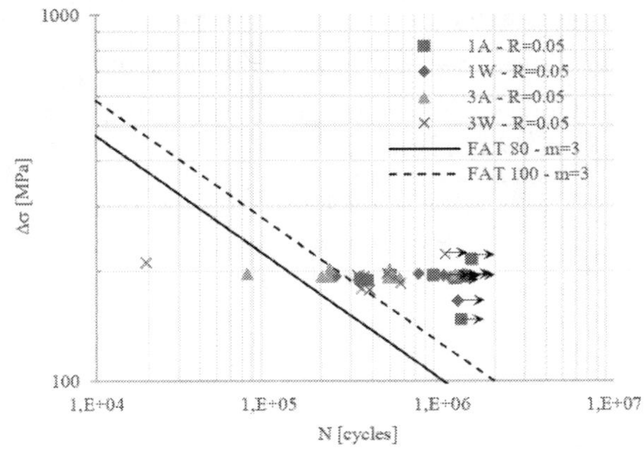

Figure 10. The S-N Curves for treatments of study accord with the classes FAT 80 and FAT 100.

In figure 11 it is possible to identify the behavior of welded samples in comparison with figure 10, even though initially it is necessary to obtain curve S-N to determine the dispersion among analysis information. In figure 11 it is observable that in treatment 1A approximately 63% of the samples are above the 70% of surviving probability, remaining are between 30 and 50% of surviving. In the treatment 1W the 87.5% of the samples are between the 60 and 90% of surviving and remaining in 19%. In treatment 3A the 50% of the samples are between 40 and 80% of surviving and the other 50% of the samples are between 2 and 20% of surviving. Finally in the samples of treatment 3W, the 87.5% of the samples are between 30 and 70% of surviving and remaining in 0.025%. Thus it is possible to affirm that treatment 1W is the one which has the highest surviving probability in comparison with the other treatments. On the contrary the treatment 3A shows the lowest probability of surviving.

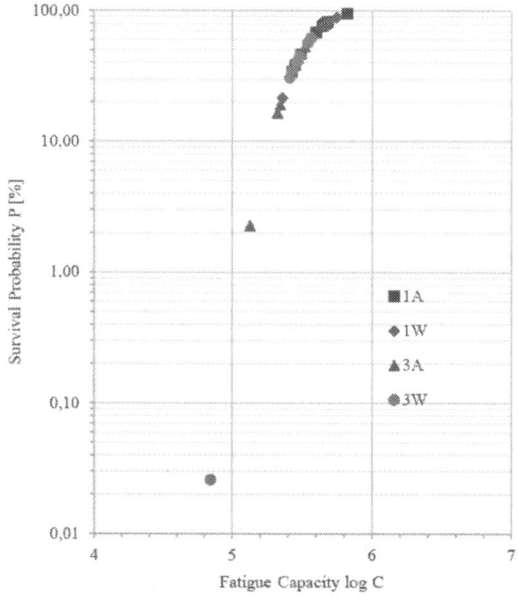

Figure 11. Survival probability for the specimens of fatigue butt welded joint test.

In figure 12 the counting of striations performed on sample 3W-8 is shown. Obtaining two crack propagation rates at beginning of stable zone $\left(\frac{da}{dN}\right)_1$ and close to its end $\left(\frac{da}{dN}\right)_2$. The propagation rates for both locations were estimated by counting and measurement of separation among microstrations, according with [16]. Where C y m are obtained by:

$$C = \frac{\left(\frac{da}{dN}\right)_1}{\left(\Delta K_{I,1}\right)^m} \quad (3)$$

$$m = \frac{\log\left(\frac{da}{dN}\right)_1 - \log\left(\frac{da}{dN}\right)_2}{\log \Delta K_{I,1} - \log \Delta K_{I,2}} \quad (4)$$

Nevertheless, there are eight replicas for each treatment, in table 6 are shown only the results more representative and with better quality for the constant values C and m. In some of the samples it is not possible to establish the distance among striations, because these are not visible or are confuse among other superficial details.

The propagation curves of cracks for fatigue are shown in figure 13 accordingly with the results obtained in the striation counting from table 6. The curve of sample 1W-8 highlights for the right displacement, which does not mean it has a higher fatigue resistance instead the striation counting was performed for an initial crack size a_I much higher than used for the other samples. Furthermore it is observed that maximum value for the propagation rate of the cracks is for sample 1A-7 and it is 1.34e-6 [m/cycle], while the sample 1W-8 presents a maximum value of 7.97e-7 [m/cycle] much lower than the previous. This is consistent with experimental results where the life to fatigue for the sample 1A-7 and 1W-8 are $8.98 \cdot 10^5$ and $2.45 \cdot 10^5$ cycles respectively.

The results of uniaxial tension test for the analysis samples from table 2 are shown in figure 14 and the average of the results for each treatment is presented in table 7. The differences among treatments are not too significant in means of resistance to tension and elastic limit, however. It is important to note that area under the curve for treatment 1W in contrast with treatment 3A present a well-defined difference. This indication allows establishing that treatments 1A and 1W will have a higher resistance to crack propagation than treatments 3A and 3W, as it is observed in results from S-N curve and the surviving probability from figures 10 and 11 respectively. Broadly it is observed that value of tension resistance and elastic limit in butt welded joints increase in relation with properties of BM from table 5. However, the elongation percentage decreases considerably, which leads to a reduction of material ductility of HAZ and FZ.

Table 6. Summary of fatigue striation counting for each treatment.

Specimen	$\left(\frac{da}{dN}\right)_1$ [m/cycle]	$\left(\frac{da}{dN}\right)_2$ [m/cycle]	a_1 [m]	a_2 [m]	$\Delta\sigma$ [MPa]	$\Delta K_{I,1}$ [MPa √m]	$\Delta K_{I,2}$ [MPa √m]	C	m
1A-7	6.28E-7	1.34E-6	9.00E-4	2.45E-3	192	10.23	16.87	6.14E-8	1.51
1W-8	2.34E-7	7.97E-7	2.40E-3	5.29E-3		16.70	24.79	1.40E-8	3.10
3A-4	5.41E-7	1.28E-6	1.09E-3	5.41E-3		11.25	25.07	4.81E-8	1.08
3W-8	3.48E-7	1.14E-6	6.50E-4	4.42E-3		8.69	22.66	4.01E-8	1.24

0 – 0.001 m: 3.48x10-7 m/cycle (9000x)

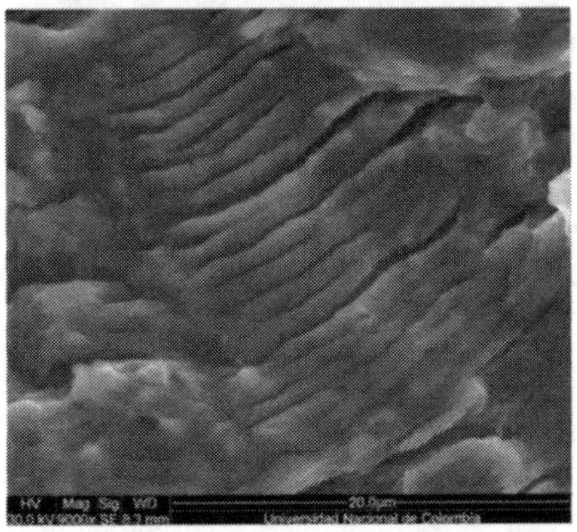

0.004 – 0.0054 m: 1.17x10-6 m/cycle (9000x)

Figure 12. Fatigue striation counting close to the crack origin and close to the end of propagation stable stage (stage II), specimen 3W-8.

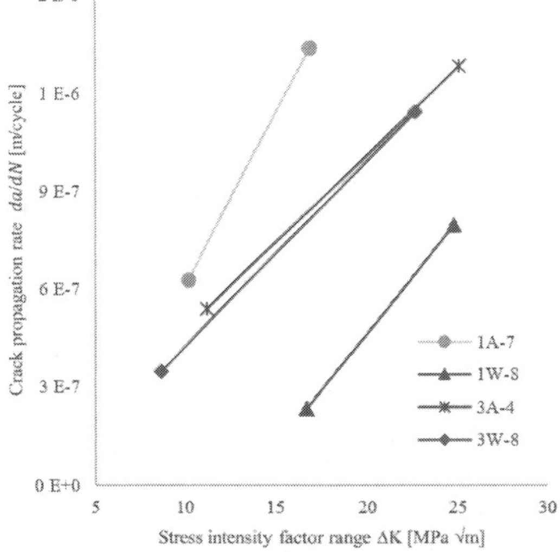

Figure 13. Fatigue crack growth rate in butt joint welds.

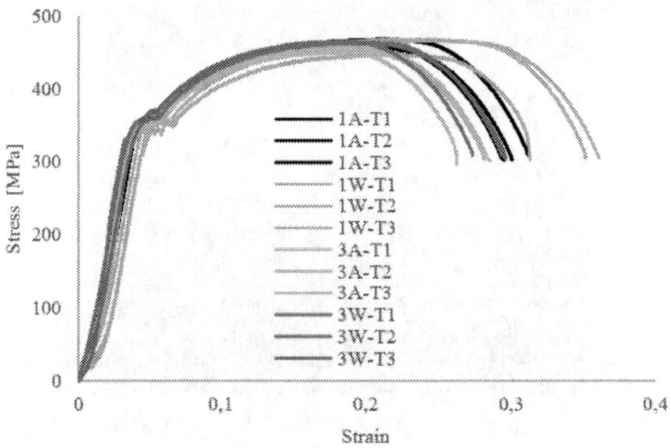

Figure 14. Stress-strain curves of butt welded joints.

Table 7. Results for the mechanical properties obtained by uniaxial tensile test (average values).

Treatment	Ultimate tensile strength [MPa]	Yield stress 0.20% [MPa]	Percent reduction of area [%]	Elongation [%]
1A	462.2	331.6	31.4	19.4
1W	461.3	332.3	32.1	23.3
3A	454.8	321.6	32.1	18.3
3W	465.0	322.2	28.9	19.6

The mean of the results for the Charpy V-notch impact test is shown in table 8. The fracture toughness of the material was estimated indirectly with the Rolfe-Barsom correlation [17]. In figure 15 it is presented a comparison among the percent shear fracture for the highest and lowest percentages obtained following the standard ASTM A370-14 [13]. Which suggests that percent shear is proportional to material ductility.

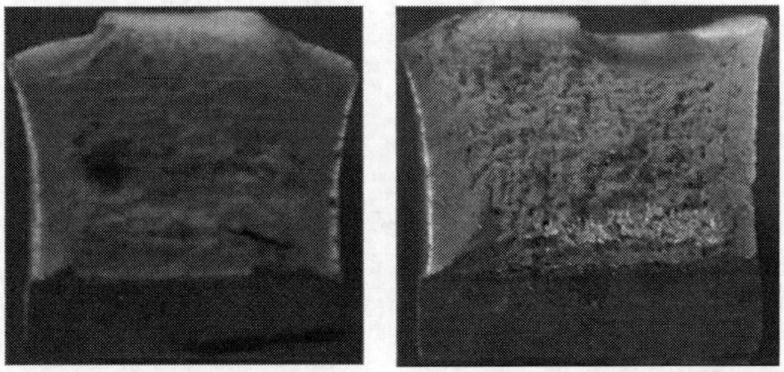

Figure 15. Percent shear fracture 1A-CVN-ZAT (56%) (left); 3W-CVN-ZF (21%) (right).

In relation with cooling cycle it is possible to affirm that a fast cooling rate produces a reduction in the final grain size, thus increasing the material hardness. Moreover, when the welded joint is fast cooled in water immediately after applying the welding chord, a quench is produced due to the fast cooling rate of the welding chord; subsequently if a new welding pass is applied, a tempering and a reduction of the grain size in nears of the late welding chord are produced. In figure 16 it is shown a scheme about grain refinement as a result of multiple passes of welding in a butt welded joint with base on [18]. The columnar microstructure domains in the zone FZ, in interface FZ-HAZ a coarse grain is present, in central zone of HAZ there is a fine grain size; and finally in the interface HAZ-BM it is present a partial refinement.

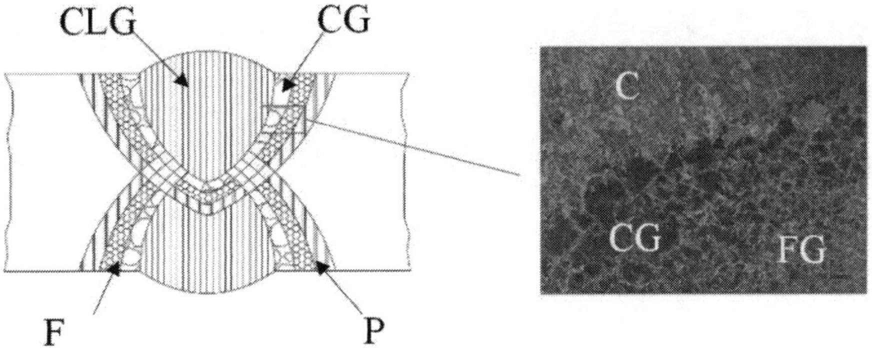

Figure 16. Scheme of grain refinement product of multi-pass welds in a butt welded joint. CLG (columnar grain); CG (coarse grain); FG (fine grain) and RP (partial refinement) [18].

The results obtained from microindentation hardness testing are shown in figure 17, for the 4 studied treatments. The presented profiles are the real ones. These were obtained by means of digitalization of macro structure image and later processing on Academic suite of AutoCAD. A strong tendency to present softness for zone HAZ is presented in these results, owing to the fine grain size as it is shown in figure 16. Though, in samples 1W and 3W this behavior is present in a more pronounced way. Although in figure 17 (superior and inferior image respectively) the hardness increase is observed on rows 1 and 3 for the samples 1W and 3W respectively, the hardness value increases for sample 1W on the interfaces FZ-HAZ and HAZ-BM, while in sample 3W that happens on the interfaces TAZ-BM. The possible reasons for the previous behavior could be attributed to fast cooling rate and number of welding applied passes, thus generating microstructural variations in size and shape of the grain as it is appreciable in figure 16.

Table 8. Results of impact energy tests CVN (average values).

Treatment	CVN [J]	K_{Ic} [MPa √m]	Shear fracture [%]	Treatment	CVN [J]	K_{Ic} [MPa √m]	Shear fracture [%]
1A-CVN-ZF	100.29	163.33	33	1A-CVN-ZAT	157.04	157.42	56
1W-CVN-ZF	81.17	146.17	33	1W-CVN-ZAT	138.67	147.77	40
3A-CVN-ZF	75.97	141.15	21	3A-CVN-ZAT	125.98	140.73	45
3W-CVN-ZF	66.91	131.94	21	3W-CVN-ZAT	187.01	172.01	31
MB	204.63	180.03	27				

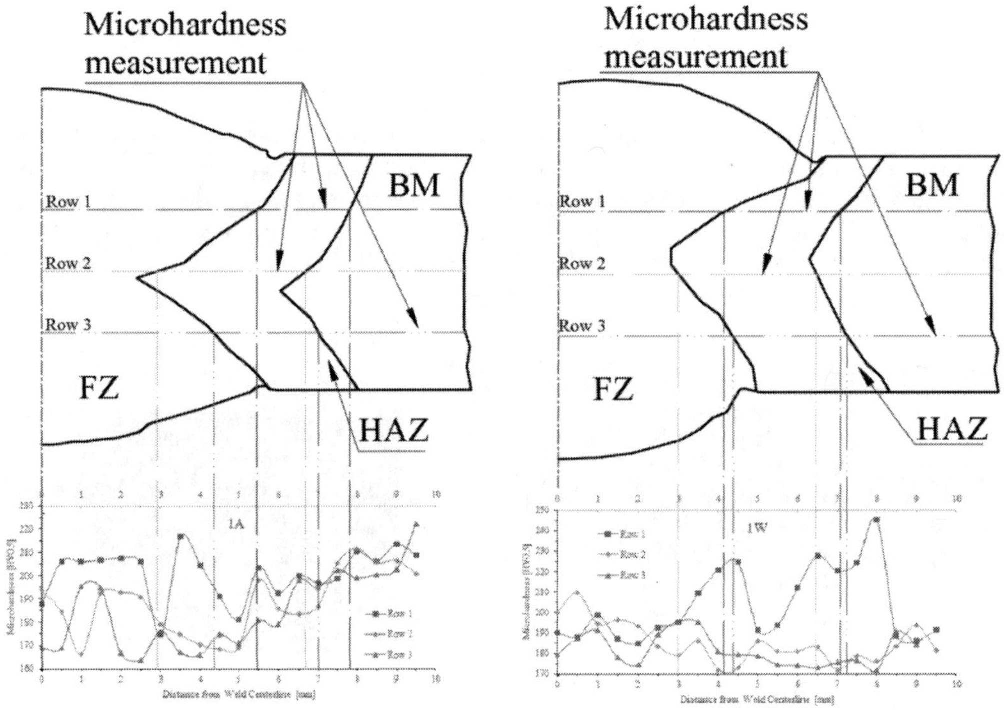

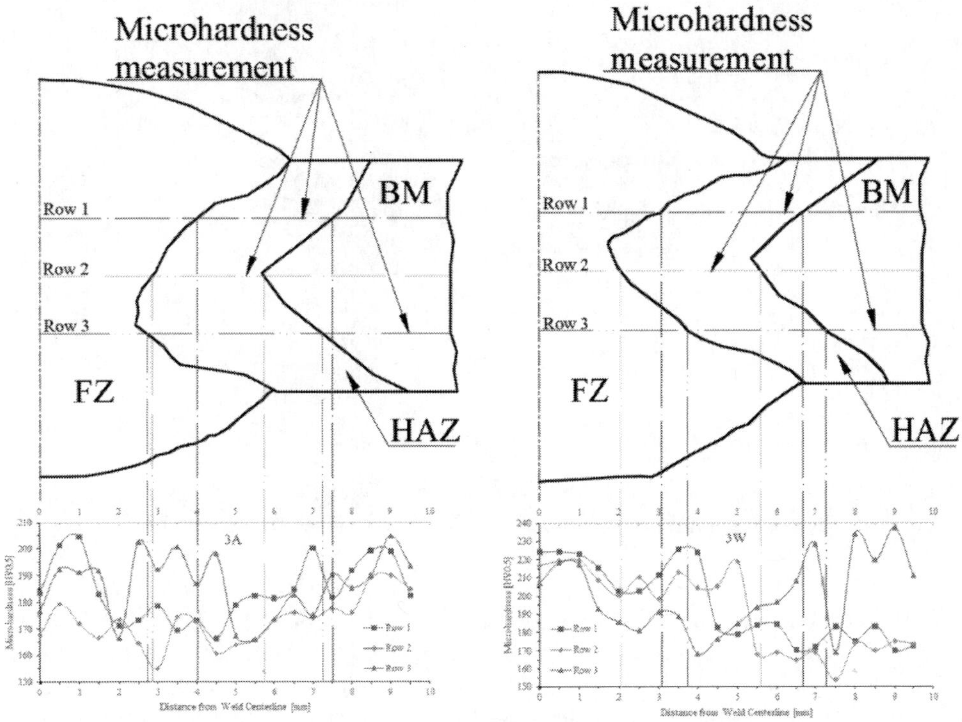

Figure 17. Microindentation hardness profile, specimen 1A (upper left); specimen 1W (upper right); specimen 3A (lower left); specimen 3W (lower right).

For microstructural evaluation of zones FZ, HAZ and BM, the phase diagram Fe-Fe3 C was used as it is shown in figure 18. This method allowed to establish a first correlation between cooling cycle, carbon percentage in the material from table 4 and the microstructural phases present in the different zones of welded joints [19]. Due to the maximum reached temperature and the size of welding pool on process SMAW, it is common to find a columnar ferritic microstructure or an epitaxial growth of the grains in the zone FZ (see figure 18 upper right image). The last is generated by the atomic disposition of liquid metal in relation with the crystallographic orientation of BM [20]. Although when several welding passes are made on zone FZ a grain refinement is produced on the interface of each welding pass [21]. At the limit between zones FZ and HAZ a partial melted zones might be formed as a result of thermic gradient on the material. On this zone a coarse grain size predominates (see figure 18 central left image), where the increase of the grain size decrease the material ductility besides increases the cracking susceptibility due to liquefaction. In figure 18 central right image, the microstructure of zone HAZ is shown. That region clearly contains the microstructural phase ferrite and perlite, bright and dark regions respectively. Where a microstructure with grain size particularly refined, which implies as a consequence a higher ductility. In some cases it is possible to find austenite due to perlite transformation when exceeding 727 °C. However, the austenite in majority decomposes into fine grains of perlite and ferrite. The BM zone figure 18 bottom right image has a microstructure with homogenous ferrite grains and perlite colonies at the ferrite grain limits.

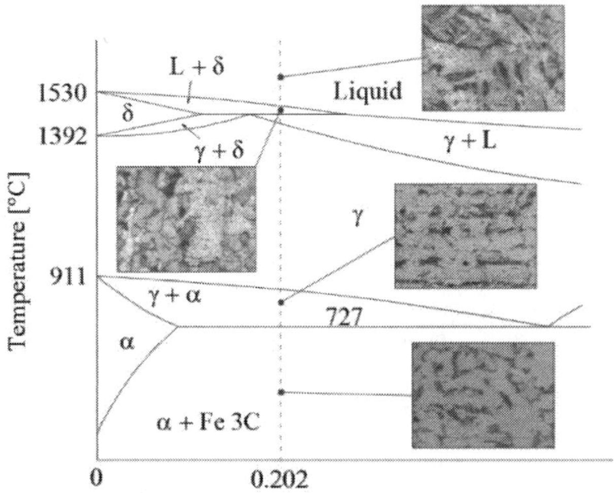

Figure 18. Phase diagram Fe – Fe_3C based on the treatment 1A.

Furthermore of considering the chemical composition of material, it is necessary to evaluate the quenching rate to establish the result of transformations on the zone HAZ. In figure 19 it is shown a scheme of transformations by continuous quenching of steels [19]. Dotted line with double dot (left) defines a fast cooling and the obtainment of a martensitic structure (treatments 1W and 3W) and the dotted line with single dot (right) defines a slow cooling with possible austenitic, ferritic and/or perlitic structures (treatments 1A and 3A).

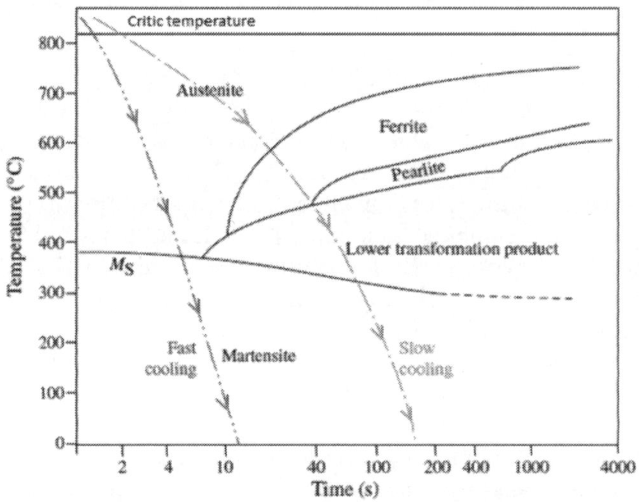

Figure 19. Continuous cooling transformation diagram for steels [19].

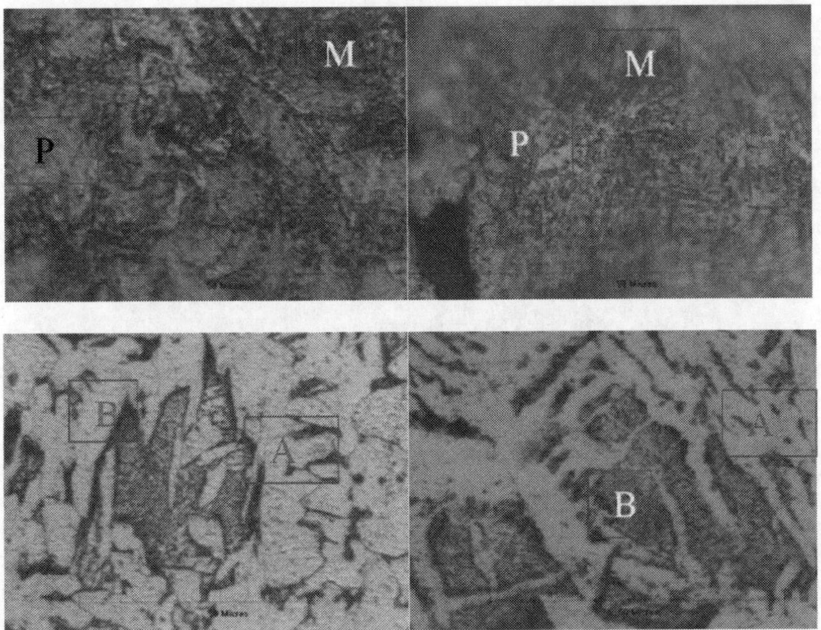

Figure 20. Microstructural phases of ZF-HAZ interface at the weld toe (2000x).

In consistency with the obtained results from uniaxial fatigue test for the failed samples, it was possible to observe that at the crack initiation occurred at the weld toe, mainly on the interface FZ-HAZ, consequently it was decided to analyze the microstructure present in that zone in order to relate the fatigue crack initiation and the microstructure influence. In figure 20 the results of interface analysis are shown for interface FZ-HAZ. Where it is observed the presence of martensitic microstructure (M) and polygonal ferrite (PF) of fine grain in samples 1W and 3W (figure 20 upper left and right images). This is consistent with a fast cooling rate, where pre-eutectoid ferrite disappears and the austenite grains are transformed into acicular structures as: martensite or inferior bainite or a mixture of those [22]. These micro structures have an elevated hardness although a low ductility too, thus favoring the crack propagation due to fatigue and a lower critical crack size, owing to a low fracture toughness to material as it is observed in table 8 when comparing values of the zone FZ and zone HAZ. This agrees with the results shown in figure 10 for the treatment 3W where this treatment exposes the lowest fatigue resistance compared with others. Nevertheless, the treatment 1W, according with results from figure 10 is among the highest resistance values. Which allows concluding that geometric factor of over-thickness has a greater influence on the fatigue resistance than microstructure resulting from cooling cycle, for each of the studied treatments. However, it is important too to consider the amount of welding passes applied in the joint owing to the microstructural variations that might produce. On the other side, for treatments 1A and 3A secondary structures were obtained such as Bainite (B) and acicular Ferrite (AF) (see figure 20 bottom images), as a result of slow cooling rates. Likewise the grain size is much smaller than that obtained in samples 1W and 3W.

Finally, in figure 21, comparison between samples 1A-7 and 3A-5 is carried out in terms of appearance of the failure zone due to fatigue. It is observed that sample 1A-7 presents a single visible crack origin owing to the radial marks point toward a single direction, in opposition the sample 3A-5 contains three dots which indicate the crack origin of superficial semielliptical kind, based on figure 4. In sample 3A-5 (right) it is observed clearly how the three crack fronts, as crack propagation moves along, tend to convert into a single crack front, while in sample 1A-7 (left) the only crack front present propagates more or less in a concentric way around its origin. The bigger the size of stable growth zone of cracks on the surface of fracture for a same load cycle, it means a higher tolerance to damage of welded joint.

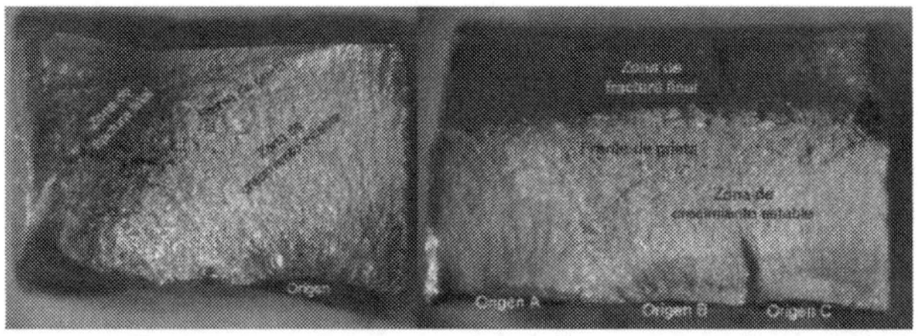

Figure 21. Fatigue fracture surfaces. Specimen 1A-7, Nf = $8.98 \cdot 10^5$ cycles (left) and specimen 3A-5, Nf = $0.75 \cdot 10^5$ cycles (right).

4. Conclusions

According with the results obtained experimentally on this research, it is possible affirm that:

The fatigue resistance in butt welded joints is strongly influenced by the weld joint geometry, specifically the height of reinforcement weld (*s*), although it is also important to consider other geometric factors of less impact as it is the weld toe radius to plate, the welding chord width, and material thickness among others.

On this research 93.75% of the samples presented a fatigue resistance superior to the one of class FAT 90, while about the 78.13% of the samples obtained a resistance higher than the recommended one by class FAT 100. If a more conservative approach is preferred, with class FAT 100 a higher security factor will be obtained that counteracts the variability typical of the fatigue behaviour in the welded joints.

The fatigue striation counting allows estimating and validating the fatigue crack growth rate. However, in some cases owing to the post-welding procedure it was not possible, because the zone of crack propagation for fatigue was not ductile enough to generate characteristic marks of striations.

In consistency with welding procedure and post-welding procedure the experimental test as the uniaxial tension, Charpy impact, microindentation and the metallographic observation. Allowed to characterize the mechanical properties of butt welded joints in relation with base metal, where it was clearly observed that the cooling cycle and the amount of passes influence the material ductility and thus the fracture toughness of the material too

In agreement with experimental results, it was found that crack front on the failure surface for fatigue presented an approximated semielliptical shape. This allows establishing in an easier way the numeric modelling for butt welded joints, in order to establish correlations between experimental results and theoretical calculations. On the other hand, it is highlighted the importance of performing additional studies where the nucleation stage and initiation of cracks for fatigue are examined, for that the microstructural phases and the size of the microstructures are more relevant in comparison with the stage of stable propagation of cracks.

References

[1] Radaj D, 1990 Design and analysis of fatigue resistant welded structures. Abington Publishing.
[2] BS7608, 2014 Guide to fatigue design and assessment of steel products, BCI Standards Publication.
[3] Hobbacher A F, 2016 Recommendations for fatigue design of welded joints and components. Cham: Springer International Publishing.
[4] Schijve J, 2001 Fatigue of structures and materials. Dordrecht; Boston MA: Kluwer Academic.
[5] Paris P and Erdogan F, 1963 A critical analysis of crack propagation laws, *J. Basic Eng.*, **85**, no. 4, p. 528.
[6] Hobbacher A, 2011 The use of fracture mechanics in the fatigue analysis of welded joints,

Fracture and Fatigue of Welded Joints and Structures, pp 91–112.
[7] Espejo E and Hernandez E, 2002 Mecánica de fractura y análisis de falla, Colección Universidad Nacional de Colombia
[8] Davis J R, 1998 *Metals handbook Desk Edition*. ASM International.
[9] Belen'kii D M, Beskopyl'nyi A N, Vernezi N L, and Shamraev L G, 1997 Determination of the strength of butt welded joints, *Weld. Int. J.*, **11**, no. 8, pp 642–645.
[10] ASTM E384-16, 2016 Standard test method for microindentation hardness of materials.
[11] Standard test methods for tension testing of metallic materials. 2012 West Conshohocken PA: ASTM International.
[12] Structural Welding Code Steel 2015 Steel 23rd Edition Supersedes AWS D1.1/D1.1M:2010.
[13] ASTM A370-16 2016, Standard test methods and definitions for mechanical testing of steel products.
[14] ISO 15653 2010 Metallic materials–Method of test for the determination of quasistatic fracture toughness of welds, *International Standards Organisation*.
[15] Potter J and McHenry H 1990 Fatigue and Fracture Testing of Weldments. 100 Barr Harbor Drive, PO Box C700, West Conshohocken, PA 19428-2959: ASTM International.
[16] DeVries P, 2010 Counting on Fatigue: Striations and Their Measure, *J. Fail. Anal. Prev.*, **10**, no. 2, pp 120–137.
[17] Barsom J and Rolfe S, 1970 Correlations between K_{ic} and charpy v-notch test results in the transition-temperature range, 100 Barr Harbor Drive, PO Box C700, West Conshohocken, PA 19428-2959: ASTM International, pp 281-281–22.
[18] Boukharouba T, Elboujdaini M, and Pluvinage G 2009 Damage and Fracture Mechanics. Dordrecht: Springer Netherlands.
[19] Lippold J C, 2015 Welding metallurgy and weldability, *John Wiley & Sons, Inc.* DOI: 10.1002/9781118960332.ch1
[20] Kou S, 2002 Welding metallurgy, *John Wiley & Sons, Inc.* DOI: 10.1002/0471434027
[21] Kerr H, 1976 A review of factors affecting toughness in welded steels," *Int. J. Press. Vessel. Pip.*, **4**, no. 2, pp 119–141.
[22] Lancaster J F, 1980 Carbon and ferritic-alloy steels, in Metallurgy of Welding, Dordrecht: Springer Netherlands, pp 110–177.

Crack Nucleation in β Titanium Alloys under High Cycle Fatigue Conditions - A Review

Benjamin Rohit[1], Nageswara Rao M[2]

[1]Assistant Professor, Department of Aerospace Engineering, RVCE, Bengaluru, Karnataka, India +919986735896

[2]Senior Professor, School of Mechanical Engineering, VIT University, Vellore, Tamil Nadu, India, +919629074177

E-mail: [1]benjaminr@rvce.edu.in, [2]m.nageswararao@vit.ac.in

Abstract. Beta titanium (β-Ti) alloys have emerged over the last 3 to 4 decades as an important class of titanium alloys. Many of the applications that they found, particularly in aerospace sector, are such that their high cycle fatigue (HCF) behavior becomes critical. In HCF regime, crack nucleation accounts for major part of the life. Consequently it becomes important to understand the mechanisms underlying the nucleation of cracks under HCF type loading conditions. The purpose of this review is to document the best understanding we have on date on crack nucleation in β-Ti alloys under HCF conditions. Role of various microstructural features encountered in β-Ti alloys in influencing the crack nucleation under HCF conditions has been reviewed. It has been brought out that changes in processing can result in changes in microstructure which in turn influence the time for crack nucleation/fatigue life and fatigue limit. While majority of fatigue failures originate at the surface, subsurface cracking is not uncommon with β-Ti alloys and the factors leading to subsurface cracking have been discussed in this review.

Keywords: β **Titanium alloys, Fatigue crack nucleation, Super-transus and sub-transus processing, Grain boundary alpha, Subsurface crack nucleation, Duplex aging, Precipitate free zones**

1. Introduction

1.1. Importance of β Titanium Alloys

Beta titanium (β -Ti) alloys constitute an important class of titanium alloys; they offer the highest strength to weight ratio and very attractive combinations of strength, toughness and fatigue resistance [1]. Some grades are produced as forged components, while some others are produced as strips. Through the aging process a wide range of yield strengths can be obtained in β-Ti alloys. An important benefit provided by a beta structure is the increased formability of β-Ti alloys, relative to α and α+β alloys [2]. β-Ti alloys have been widely used in the aerospace industry - load-bearing fuselage components and high-lift devices of Russian aircraft, forgings for airframe and landing gear applications, such as for Boeing 787, springs and fasteners [3]. Table 1 gives details of β-Ti alloys which have been the subjects of detailed investigation for high cycle fatigue (HCF) behaviour.

1.2. Importance of the HCF Strength of β titanium alloys

High cycle fatigue properties and the fatigue limit are important considerations in design of structural parts in many aerospace applications [4, 5]. The majority (>50%) of fatigue fractures in all manufactured components are caused by HCF failures, particularly in the aerospace industries [6]. Accordingly much of the research on β-Ti alloys has been dedicated to develop new grades with high HCF life or improve the HCF life of the existing grades.

1.3. Importance of Crack nucleation phase in determining the HCF life

It is well known that fatigue crack initiation process occupies the main portion of the total fatigue life of the material in the HCF regime (fatigue life larger than $10^4 - 10^5$ cycles) [7]. Thus the HCF strength is a good measure of the resistance of the material to crack nucleation under HCF loading conditions. Lower HCF strength indicates that crack nucleation is easy; materials offering high resistance to fatigue crack nucleation under HCF loading conditions have a high HCF strength [8]. The present review of fatigue crack initiation in β-Ti alloys under HCF conditions is thus directly relevant to the HCF life of these alloys.

1.4. Solution Treated and Aged Condition vs Solution Treated Condition

The standard heat treatment of β-Ti alloys comprises of solution treatment and aging (STA) [2, 3]. Solution treatment may be super-transus (at a temperature higher than β-transus) or sub-transus (at a temperature below β-transus) [9]. Most of the applications for β-Ti alloys involve service in STA condition, because of the high strength they have in this condition. After solution treatment (ST) but before aging, they do not possess high strength. Hence the ST condition is not of much importance in practice. Hence this review gives more attention to the STA condition. Brief reference is however made to fatigue behavior in ST condition (Section III). Section II is a detailed treatment of fatigue behavior of β-Ti alloys in STA condition with super-transus solution treatment. Section IV deals with β-Ti alloys processed based on sub-transus solution treatment.

2. Crack nucleation in super-transus solution treated and aged condition

2.1. Microstructural features influencing the crack nucleation

Studies have shown that fatigue crack nucleation is controlled by different microstructural features, importantly grain boundary α, precipitate free zones (PFZs) and beta grain size.

2.1.1. Grain boundary alpha It is well known that soft α layers located at prior β grain boundaries are the most typical feature in metastable β-Ti alloys, which can dramatically decrease HCF properties of these alloys [10, 11]. It is widely believed that α phase is softer than the β matrix. As a result HCF damage is importantly governed by grain boundary α layers in metastable β-Ti alloys. Schmidt et al [5] showed that formation of grain boundary α in Ti 38-644 lowers the fatigue limit. Santosh et al [12] concluded that grain boundary α was facilitating easy and early crack initiation during fatigue loading of Ti 15-3 alloy. Kim et al [13], based on their studies on Ti 15- 3 alloy, concluded that concentrated stress in grain boundary region first causes the debonding between the favorably oriented (i.e. 45° to loading axis) alpha and beta phases in the close vicinity of grain boundary. Studies by Kokuoz et al on TIMETAL-LCB [14] showed that damage accumulation within grain boundary α and fatigue crack nucleation are related. The authors also concluded that HCF behavior is controlled by the contiguity of the grain boundary α and that a higher degree of contiguity leads to higher chances of fatigue crack nucleation. In case of lamellar microstructures, the fatigue cracks usually nucleate at grain boundary α layers. Even in case of bimodal microstructure, grain boundary α layers serve as nucleation sites [11].

IOP Conf. Series: Journal of Physics: Conf. Series **843** (2017) 012048 doi:10.1088/1742-6596/843/1/012048

2.1.2. Precipitate free zones Precipitate free zones are important microstructural features in β-Ti alloys. PFZs are soft zones in the microstructure and hence preferentially deform on application of load. They exert a strong effect on the fatigue and fracture behavior of Ti and Al alloys [15]. HCF life of β-Ti alloys is very sensitive to the presence of PFZs. In single aged condition, the PFZs were found to be the cause for early crack nucleation in Ti 38-644 under HCF conditions, resulting in a lower fatigue limit [5, 16]. The amount and size of PFZs in Ti 38-644 increased with aging temperature. Reduction in aging time at a constant temperature lead to increase in the volume fraction of PFZs [5]. Thus the aging temperature and time come importantly into play in determining the HCF life of the alloy. The micro-PFZs present in Ti15-3 were found to contribute to reduced resistance to fatigue crack nucleation [12].

2.1.3. Grain size An increase in solution treating temperature results in an increase of grain size in β-Ti alloys. A grain size reduction leads to increase in the fatigue strength of β titanium alloys as reported in [1]. Tokaji et al [17] reported that for β-Ti alloys in STA condition largest β grain size value is associated with considerably lower fatigue strength. Schmidt et al [5] reported that β grain size is the primary factor controlling the fatigue life of Ti 38-644. In the case of coarse grained beta structure, fatigue cracks always nucleated on the continuous alpha layers at beta grain boundaries [11] which was also confirmed by [14] and [5]. Reduction in grain size also resulted in increase in the HCF strength to Yield Stress ratio [11]. The operative slip length in the crack nucleation phase is proportional to the grain size; the resistance to fatigue crack nucleation is increased by a reduction of the slip length, i.e. by a reduction in β grain size [11]. The dependence of fatigue life on β grain size can thus be explained.

2.2. Effect of aging on crack nucleation - Single aging vs. Duplex aging
There is a drastic improvement in HCF life of Ti15-3 alloy on switching over from single aging to double aging [12, 18]. Similar improvement was reported by Schmidt et al for Ti 38-644 alloy[5]. After double aging Ti15-3 alloy, it was found that there were no PFZs in the microstructure and grain boundary α was present to a very small extent; in contrast, after single aging PFZs were present and there was a higher amount of grain boundary α [12]. Schmidt et al [5] also reported that duplex aging increases the ductility, suppresses the occurrence of PFZs and eliminates grain boundary α in Ti38-644. Duplex aging reduces the impact of PFZs and grain boundary α on fatigue crack nucleation. Consequently the crack nucleation time increases, leading to an increase in HCF life.

3. Crack nucleation in solution treated condition
HCF studies were carried out on three β-Ti alloys in solution treated condition [19]. The authors concluded that initial damage accumulation during HCF is associated with the formation of coarse, planar slip bands. Fatigue crack nucleation was associated with crack initiation at intersecting planar slip bands at or near the free surface. The increase in operative slip length occasioned by the presence of low-angle grain boundaries leads to accelerated crack nucleation. The authors concluded that the resistance to fatigue crack nucleation is inversely proportional to the square root of the operative slip length. Hu et al [20] carried out studies on fatigue crack nucleation in TIMETAL LCB in ST condition. They concluded that both slip bands and grain boundaries are the preferred sites for fatigue crack nucleation. They also found that high angle grain boundaries were preferred sites for intergranular fatigue crack initiation. Low angle grain boundaries were never found to be associated with formation of fatigue cracks.

4. Crack nucleation in materials produced by sub-transus processing
Huang et al [21] carried out HCF studies on Ti55531 alloy with a bimodal microstructure (BM) having 15% near globular α_p phase with an average grain size of 3 ± 1 mm, approximately 3%

IOP Conf. Series: Journal of Physics: Conf. Series **843** (2017) 012048 doi:10.1088/1742-6596/843/1/012048

continuous Grain Boundary Alpha phase with 200 ± 10 nm in width and a large volume fraction of secondary α (α_s) phase within the residual beta matrix. The authors demonstrated that the fatigue crack initiation occurs at severely deformed α_p particles below the surface and the direction of the initiation facet is inclined at about 43.8^o to the stress axis. Shear deformed bands or slip bands formed in the globular α_p phase promotes the nucleation of micro-voids, which would in turn play a significant role in the initiation of microcracks. Lenain and Jacques [22] studied the HCF behavior of TIMETAL LCB with a bimodal microstructure. They found that cracks initiate at the interface between α and β phases. Huang et al [23] also compared the HCF behavior of the alloy with bimodal microstructure (BM) and with lamellar microstructure (LM). Grain boundary α layers have more effect on promoting crack initiation in LM than in BM. As a result fatigue microcracks mainly initiate at the interface between grain boundary α films and prior β grains or at the secondary α/residual β interface for LM. However microcracks primarily nucleate at the α_p / transformed β interface or at α_p particles for BM. Shi et al [6] studied the crack initiation behavior and fatigue limit of Ti55511 alloy with different basket-weave features. They brought that difference in microstructure cause differences in crack initiation behavior, microstructure having primary α phase promoting subsurface nucleation. The differences in microstructure also cause differences in fatigue limit. Table 2 summarizes the microstructural features in β-Ti alloys which, as per published literature, serve as sites for crack nucleation during fatigue loading.

5. Factors influencing surface vs subsurface crack nucleation

There are different factors which can contribute to subsurface crack initiation in β-Ti alloys.

5.1. Effect of Grain Boundary Alpha and Precipitate Free Zones

The contribution of grain boundary alpha to subsurface crack initiation in single aged condition was reported by Santhosh et al in Ti 15-3 alloy [12]. Subsurface crack initiation was also reported by Schmidt et al in Ti38-644 and was attributed to extensive grain boundary α and the PFZs existing within the β grains [5]. Studies by Tokaji et al [17] also suggest that subsurface crack initiation in solution treated and aged β-Ti alloys occurs due to a slip-related fracture process on the grain boundary α phase.

5.2. Effect of primary α in sub-transus processed β-Ti alloys

Studies by Shi et al [6] showed that presence of coarse α phase in basket weave type microstructures in Ti55511 promotes subsurface crack nucleation. For microstructures without blocky α, crack nucleation is essentially at the surface. Subsurface crack initiation is also dominant in β-Ti alloys with bimodal microstructures, the primary α particles and interfaces of primary α particles and β matrix serving as preferred nucleation sites [21, 23].

5.3. Effect of aging

Santhosh et al [12] found instances of subsurface initiation on fatigue loading of single aged Ti15-3 material. On the other hand, crack nucleation in the material was found to have occurred at the surface in double aged specimens. Studies by Schmidt et al [5] concluded that fatigue crack initiation in Ti 38-644 alloy samples was subsurface in single aged materials; in contrast, initiation was associated with the surface in double aged condition. With the non-availability of potential crack nucleation sites beneath the surface in double aged condition, crack nucleation occurs at the surface. The aging temperature rather than aging time was found to be important in causing internal fatigue crack initiation in Ti15-3 alloy [13]. In unaged (solution treated) β-Ti alloys, no subsurface crack initiation was observed on fatigue loading [19]. This is believed to be a consequence of microstructural features which may cause subsurface crack nucleation such as grain boundary α, PFZs being absent in the microstructure.

5.4. Effect of shot peening

Shot peening is done in β-Ti alloys to harden the surface [2]. Shot peening induces changes in the surface layer properties [24]. Shot peening improves fatigue strength considerably at room temperature; on the other hand, at elevated temperatures, shot peening reduces fatigue strength [1]. Shot peening also resulted in an increase in surface roughness [25]. Fatigue crack nucleation was found below the surface after shot peening TIMETAL LCB alloy specimens [25]. This was interpreted in terms of the occurrence of high residual tensile stress regions below the surface. Studies by Dorr and Wagner [26] showed carried out shot peening experiments on two β-Ti alloys - Ti 10-2-3 in STA condition and Ti3Al8V6Cr4Mo4Zr in ST condition. The authors observed that fatigue cracks nucleated at the surface. Influence of shot peening in terms of shifting the nucleation site from surface to subsurface appears to depend on the process parameters used for shot peening and the condition of the starting material.

6. Conclusions

(i) Grain boundary α facilitates crack nucleation in β-Ti alloy microstructures obtained with both super-transus and sub-transus solution treatment. Triple points with continuous grain boundary α favor early fatigue crack initiation.

(ii) Presence of soft precipitate free zones (PFZs) in the microstructure leads to reduced resistance to fatigue crack initiation.

(iii) High angle grain boundaries are preferred location for fatigue crack initiation.

(iv) Reduction in β grain size reduces the slip length leading to an increase in resistance to fatigue crack initiation.

(v) Duplex aging delays fatigue crack nucleation by way of suppressing/ eliminating the potential crack nucleation sites - grain boundary α, PFZs

(vi) Crack nucleation in β-Ti alloys under fatigue loading can be surface or subsurface in nature. Microstructural features such as grain boundary α, PFZs, coarse primary α and surface treatments such as shot peening can lead to occurrence of subsurface crack nucleation.

7. Acknowledgments

The first author (Benjamin Rohit) would like to thank the Management and Principal of RVCE, Bengaluru for their support and encouragement to publish this paper. The second author (Nageswara Rao M) is indebted to the Management of VIT University for its kind consent to publish this paper.

References

[1] C. Leyens and M. Peters. *Titanium and Titanium Alloys*. Weinheim: WILEY-VCH Verlag GmbH & Co. KGaA, 2003.

[2] Matthew J. and Donachie Jr. *Titanium: A Technical Guide*. ASM International, 2 edition, 2001.

[3] James D. Cotton, Robert D. Briggs, Rodney R. Boyer, Sesh Tamirisakandala, Patrick Russo, Nikolay Shchetnikov, and John C Fanning. State of the art in beta titanium alloys for airframe applications. *Journal of Materials*, **67**:1281–1303, 2015.

[4] G.Q. Wu, C.L. Shi, W. Sha, A.X. Sha, and H.R. Jiang. Microstructure and high cycle fatigue fracture surface of a Ti-5Al-5Mo-5V-1Cr-1Fe titanium alloy. *Materials Science and Engineering*, **575**:111–118, 2013.

[5] P. Schmidt, El-Chaikh H, and J. Christ. Effect of duplex aging on the initiation and propagation of fatigue cracks in the solute-rich metastable beta titanium alloy Ti 38-644. *Mettalurgical and Materials Transactions*, **42**:2652–2667, 2011.

[6] Xiaohui Shi, Weidong Zeng, Shikun Xue, and Zhiqiang Jia. The crack initiation behavior and the fatigue limit of Ti-5Al-5Mo-5V-1Cr-1Fe titanium alloy with basket-weave microstructure. *Journal of Alloys and Compounds*, **631**:340–349, 2015.

[7] J A Hall. Fatigue crack initiation in alpha-beta titanium alloys. *International Journal of Fatigue*, **19**:23–37, 1997.

[8] J. O. Peters, G. Luetjering, R. K. Nalla, I. Altenberger, and R. O. Ritchie. High cycle fatigue of beta titanium alloys. *Fatigue*, 2002.

[9] R.Santhosh, M.Geetha, and M.Nageswara Rao. Recent developments in heat treatment of beta titanium alloys for aerospace applications. *Transactions of the Indian Institute of Metals*,1–8, 2016.

[10] Foltz JW, Welk B, Collins PC, Williams JC, and Fraser HL. Formation of grain boundary alpha in beta Ti alloys: its role in deformation and fracture behavior of these alloys. *Metallurgical and Materials Transactions A*, **42**: 645–650, 2011.

[11] G. Luetjering and J.C. Williams. *Titanium*. SPRINGER, 2 edition, 2007.

[12] R. Santhosh, M. Geetha, V.K. Saxena, and M. Nageswara Rao. Effect of duplex aging on microstructure and mechanical behavior of beta titanium alloy Ti-15V-3Cr-3Al-3Sn under unidirectional and cyclic loading conditions. *International Journal of Fatigue*, **73**:88–97, 2015.

[13] S.J. Kim, M. Hagiwara, Y. Kawabe, and S.S. Kim. Internal crack initiation in high cycle fatigued Ti-15V-3Cr-3Al-3Sn alloys. *Materials Science and Engineering*, **334**:73–78, 2002.

[14] Basak Yazgan Kokuoz, Yoji Kosaka, and H.J. Rack. High-cycle fatigue crack initiation and growth in timetal lcb. *JMEPEG*, **14**:773–777, 2005.

[15] G. Ltjering, J. Albrecht, C. Sauer, and T. Krull. The influence of soft, precipitate-free zones at grain boundaries in Ti and Al alloys on their fatigue and fracture behavior. *Materials Science and Engineering*, **468**:201–209, 2007.

[16] P. Schmidt, El-Chaikh H, and J. Christ. Fatigue properties of duplex-aged Ti 38-644 metastable beta titanium alloy. *Procedia Engineering*, **42**:1973–1982, 2010.

[17] K. Tokaji, K. Ohya, and H.Kariya. Subsurface fatigue crack initiation in beta titanium alloys. *Fracture of Engineering Materials and Structures*, **23**:759–766, 2000.

[18] R. Santhosh, M. Geetha, V.K. Saxena, and M. Nageswara Rao. Studies on single and duplex aging of metastable beta titanium alloy Ti-15V-3Cr-3Al-3Sn. *Journal of Alloys and Compounds*, **47**: 222–229, 2014.

[19] M. Long, R. Crooks, and H. J. Racks. High-cycle fatigue performance of solution- treated metastable-beta titanium alloys. *Acta Mater*, **47**:661–669, 1999.

[20] Y.M. Hu, W. Floer, U. Krupp, and H.-J. Christ. Microstructurally short fatigue crack initiation and growth in Ti-6.8Mo-4.5Fe-1.5Al. *Materials Science and Engineering*, **278**:170–180, 2000.

[21] Chaowen Huang, Yongqing Zhao, Shewei Xin, Wei Zhou, Qian Li, Weidong Zeng, and Changsheng Tan. High cycle fatigue behavior of Ti-5Al-5Mo-5V-3Cr-1Zr titanium alloy with bimodal microstructure. *Journal of Alloys and Compounds*, **695**:1966-1975, 2016.

[22] A. Lenain and P. J. Jacques. A study of fatigue crack initiation in a two phase beta metastable titanium alloy. *Fracture of Nano and Engineering Materials and Structures*, 165–166, 2006.

[23] Chaowen Huang, Yongqing Zhao, Shewei Xin, Changsheng Tan, Wei Zhou, Qian Li, and Weidong Zeng. Effect of microstructure on high cycle fatigue behavior of Ti5Al5Mo5V3Cr1Zr titanium alloy. *International Journal of Fatigue*, **94**:30–40, 2017.

[24] Juergen kiese, Jiulai zhang, Oliver Schauerte, and Lothar wagner. Shot peening to enhance fatigue strength of timetal application as suspension springs. *Journal of Materials*, 2006.

[25] Marcin Kocan, H.J. Rack, and Lothar Wagner. Fatigue performance of metastable β titanium alloys: Effects of microstructure and surface finish. *JMEPEG*, **14**:765–772, 2005.

[26] T. Dorr and L. Wagner. Fatigue response of various titanium alloys to shot peening. *Transactions on Engineering Science*, **25**:349–357, 1999.

Table 1. Details of β-Ti alloys studied for their HCF behavior

Sl.No	Chemical Composition	Designation	Comments/Applications
1	Ti10V2Fe3Al	10-2-3	High strength forgings
2	Ti15V3Cr3Sn3Al	Ti15-3	Sheet, plate, airframe castings
3	Ti5V5Mo5Al1Cr1Fe	VT22/55511	High strength forgings
4	Ti5Al2Sn2Cr4Mo4Zr1Fe	β-CEZ	High strength, medium temperature applications
5	Ti4.5Fe6.8Mo1.5Al	TIMETAL LCB	Spring suspension applications in automotive sector
6	Ti3Al8V6Cr4Mo4Zr	38-644/Beta C	Oil fields, springs, fasteners

IOP Conf. Series: Journal of Physics: Conf. Series **843** (2017) 012048 doi:10.1088/1742-6596/843/1/012048

Table 2. HCF Crack nucleation sites in β-Ti alloys

Sl.No	Site designation	β-Ti alloys where the site is active as per published literature
1	Grain boundary α	Different types of β-Ti alloys
2	Precipitate free zones	Ti15-3, 38-644 in super transus solution treated and single aged condition
3	Primary α	Ti 55511 subtransus heat treated with bimodal microstructure
4	Secondary α	Ti 55511 heat treated to a lamellar microstructure

Influence of mean stress on fatigue strength of ferritic-pearlite ductile cast iron with small defects

T Deguchi[1], **H J Kim**[2,3], **T Ikeda**[4] **and K Yanase**[2,3]

[1] Graduate School of Engineering, Fukuoka University, 8-19-1, Nanakuma, Jonan-ku, 814-0180, Japan
[2] Department of Mechanical Engineering, Fukuoka University, 8-19-1, Nanakuma, Jonan-ku, 814-0180, Japan
[3] Institute of Materials Science and Technology, Fukuoka University, 8-19-1, Nanakuma, Jonan-ku, 814-0180, Japan
[4] R&D Center HINODE, Ltd., Iwasaki, Harakoga, Miyaki-cho, Miyaki-gun, Saga, 849-0101, Japan

E-mail: td160004@cis.fukuoka-u.ac.jp

Abstract. Because of their excellent mechanical properties, low cost and good workability, the application of ductile cast iron has been increased in various industries such as the automotive, construction and rail industries. For safety designing of the ductile cast iron component, it is necessary to understand the effect of stress ratio, *R*, on fatigue limit of ductile cast iron in the presence of small defects.
Correspondingly in this study, rotating bending fatigue tests at $R = -1$ and tension-compression fatigue tests at $R = -1$ and 0.1 were performed by using a ferritic-pearlitic ductile cast iron. To study the effects of small defects, we introduced a small drilled hole at surface of a specimen. The diameter and depth of a drilled hole were 50, 200 and 500 μm, respectively.
The non-propagating cracks emanating from graphite particles and holes edge were observed at fatigue limit, irrespective of the value of stress ratio. From the microscopic observation of crack propagation behavior, it can be concluded that the fatigue limit is determined by the threshold condition for propagation of a small crack. It was found that the effect of stress ratio on the fatigue limit of ductile cast iron with small defects can be successfully predicted based on $\sqrt{area}$ parameter model. Furthermore, a use of the tensile strength, σ_B, instead of the Vickers hardness, *HV*, is effective for fatigue limit prediction.

1. Introduction

Because of their excellent workability, low cost as well as good mechanical properties, ductile cast irons are widely used for industrial application such as crankshafts, railways and gas pipes. Correspondingly, a number of researches have been conducted on the fatigue limit prediction to employ the ductile cast irons in the structural components. However, in essence, the fatigue limit prediction for ductile cast irons is difficult because it has intrinsic defects such as graphite and casting defects in the structure.

The fatigue strength of ductile cast iron is affected by the size and distribution of casting defects and graphite particles and by the phase distribution of matrix materials. It is known that the fatigue limit of many metallic materials with small defects can successfully be predicted with the $\sqrt{area}$ parameter model by using $\sqrt{area}$ as a geometrical parameter of defect and the Vickers hardness, *HV*, as a material parameter [1]. Several researchers [2-6] proposed the fatigue limit predictions using *HV* for ductile cast iron with single-phase matrix. The prediction using *HV* of the matrix showed a good agreement with the experimental data. However, ductile cast irons have a vast number of graphite particles in the complex matrix structure. The measurement of *HV* of ductile cast iron is affected by soft graphite particles. Furthermore, when a ductile cast iron contains two-phase matrix, the measurement of *HV* of the matrix is rather impossible. This is because ferrite and pearlite phases are evenly distributed within the matrix structure, in addition to the influence of interference by graphite particles.

In our previous study [7-9], we proposed a fatigue limit prediction for ferritic-pearlitic ductile cast iron with small and large defects based on the $\sqrt{area}$ parameter model by using the tensile strength, σ_B, as material parameter, instead of *HV*, for stress ratio $R = -1$ [7, 8]. The fatigue limit of ferritic-pearlitic ductile cast iron specimens with small defects could successfully be predicted by using σ_B. In addition, the fatigue limit for larger defects could be predicted based on the conventional fracture mechanics approach. However, ductile cast iron is frequently used under the condition where the mean stress is applied (i.e., $R \neq -1$). Accordingly, for safety designing of the products with ductile cast iron, it is necessary to understand the effects of small defects and stress ratio, *R*, on the fatigue limit.

In this study, we investigated the fatigue limit of a ductile cast iron with a two-phase matrix of almost evenly distributed ferrite and pearlite phases by using the specimens containing small defects at $R = 0.1$ in addition to $R = -1$. The purpose of this study is to present a simple yet useful method for prediction of the fatigue limit as a function of *R*.

2. Experimental procedure

The material investigated was an as-cast ductile cast iron. The chemical composition is listed in Table 1. The microstructure is shown in Figure 1. The area fractions in the microstructure were 10.5% for graphite, 45.3% for ferrite and 44.2% for pearlite. The ultimate tensile strength, σ_B, and 0.2% yield strength, σ_Y, were 552 and 335 MPa, respectively. The shapes and dimensions of test specimens are shown in Figure 2(a) and 2(b). After lathe turning of specimens, the surface was finished with an emery paper up to #1000 and then by buffing with an alumina paste. Thereafter, a small hole was introduced at the specimen surface, as shown in Figure 2(c). Before fatigue test, the surface layer of about 10 μm in thickness was removed from the specimens by electro-polishing.

Table 1. Chemical compositions (wt.%.).

C	Si	Mn	P	S	Cu	Mg
3.84	2.5	0.66	0.017	0.009	0.21	0.043

The rotating bending fatigue tests were carried out at $R = -1$ using a machine of uniform moment type, and its capacity was 100 Nm and the operating frequency was 50-67 Hz. The tension-compression

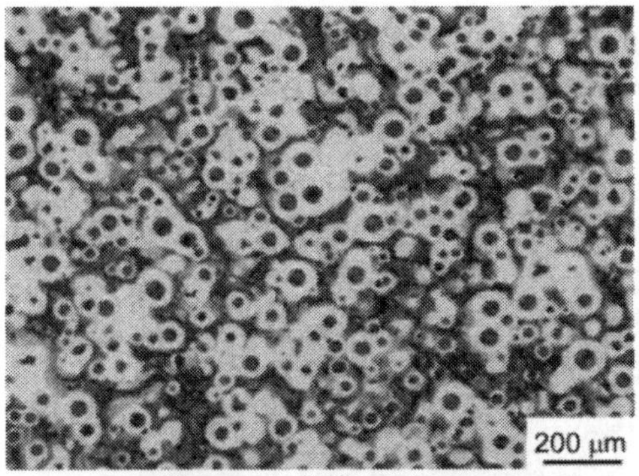

Figure 1. Microstructure.

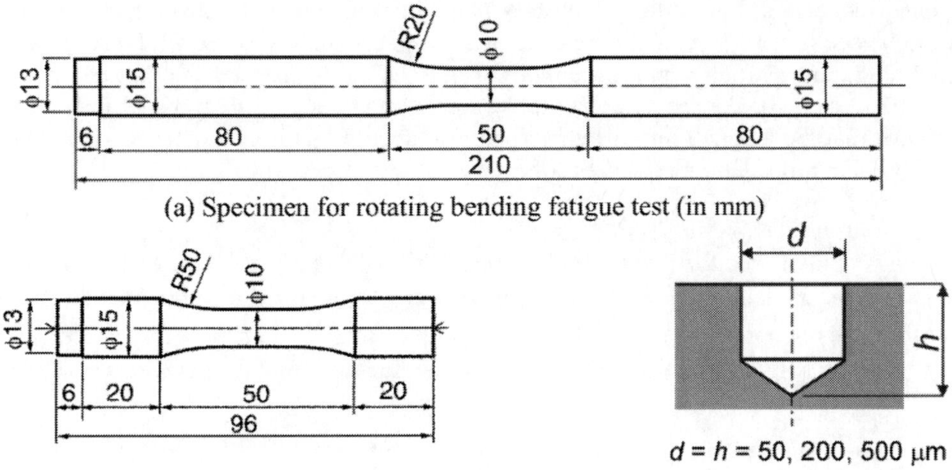

(a) Specimen for rotating bending fatigue test (in mm)

(b) Specimen for tension-compression fatigue test (in mm)

(c) Drilled holes

Figure 2. Shapes and dimensions of test specimens and artificial defects.

fatigue tests were carried out at $R = 0.1$ using a uniaxial hydraulic fatigue test machine and the operating frequency was 35 Hz. The fatigue limit was defined by the maximum stress amplitude, σ_a, for which a specimen endured $N = 10^7$ cycles without failure.

3. Results and discussion

3.1. S-N diagram and non-propagation behavior

The *S-N* diagram of smooth and holed specimens at the stress ratios of $R = -1$ and 0.1 are shown in Figure 3. The fatigue strength decreases with increase in the defect size at $R = -1$ as well as at $R = 0.1$. However, the fatigue strength of the holed specimen with $d = h = 50$ μm was not decreased. It means that the critical size of defect that does not affect the fatigue strength of ferritic-pearlitic ductile cast iron exists.

Figures 4 and 5 show the non-propagating cracks observed at the hole edge of holed specimens at fatigue limit. Figures 4(c) and 5(c) show the non-propagating crack observed at $N = 10^7$ cycles. In the

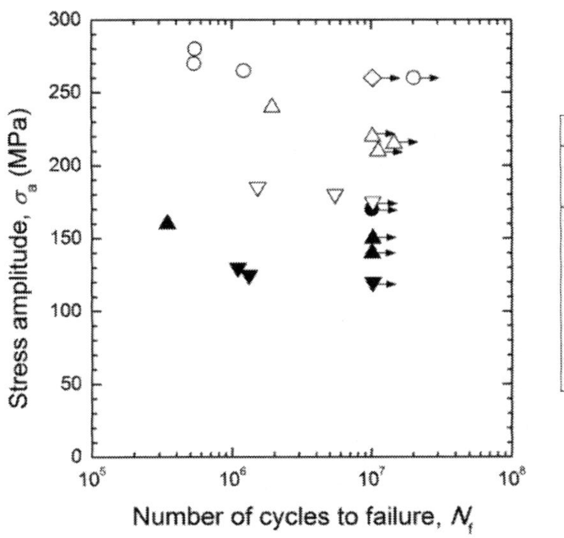

Specimen		Fatigue limit (MPa)		Loading Type	Symbol
Smooth	R = -1	-	260	RB	○
	R = 0.1	-	170	TC	●
Notched		$d = h$ (μm)			
	R = -1	50	260	RB	◇
		200	220	RB	△
		500	175	RB	▽
	R = 0.1	200	150	TC	▲
		500	120	TC	▼

RB : Rotating bending, TC : Tension-compression

Figure 3. *S-N* curve.

case of holed specimens, the fatigue cracks emanated from the hole edge, and they stopped the propagation before $N = 10^7$ cycles. In the case of smooth specimens, several fatigue cracks emanated from different graphite particles at the specimen surface and they coalesced, but they ceased propagation before $N = 10^7$ cycles [8]. Consequently, all those cracks can be regarded as non-propagating crack and it is concluded that the fatigue limit of ferritic-pearlitic ductile cast iron is determined by the threshold condition for crack propagation, regardless of hole sizes and stress ratios.

3.2. Fatigue limit prediction for ductile cast iron

In our previous study [7-9], the fatigue limit prediction for ferritic-pearlitic ductile cast iron based on the $\sqrt{area}$ parameter model by using ultimate tensile strength, σ_B, instead of the Vickers hardness, *HV*, was proposed. The relationship between the fatigue limit of smooth specimen, σ_{w0}, and σ_B was obtained by linear regression [7]. A predictive equation of fatigue limit for smooth specimens, σ_{w0}, was rendered as:

$$\sigma_{w0} = 0.25\sigma_B + 110 \tag{1}$$

where σ_{w0} and σ_B are in MPa. Further, the following form of a predictive equation for the fatigue limit in the presence of small defect, σ_w, for ductile cast irons was proposed by taking advantage of the $\sqrt{area}$ parameter model:

$$\sigma_w = \frac{F_{loc}(0.34\sigma_B + 170)}{(\sqrt{area})^{1/6}} \tag{2}$$

where F_{loc} is the correction coefficient for the location of small defect being 1.43 for surface defects, 1.56 for internal defects and 1.41 for defects just in contact with the surface [1]. Here, σ_w is in MPa and $\sqrt{area}$ is in μm. Further, based on the $\sqrt{area}$ parameter model, it is expected that the effect of stress ratio,

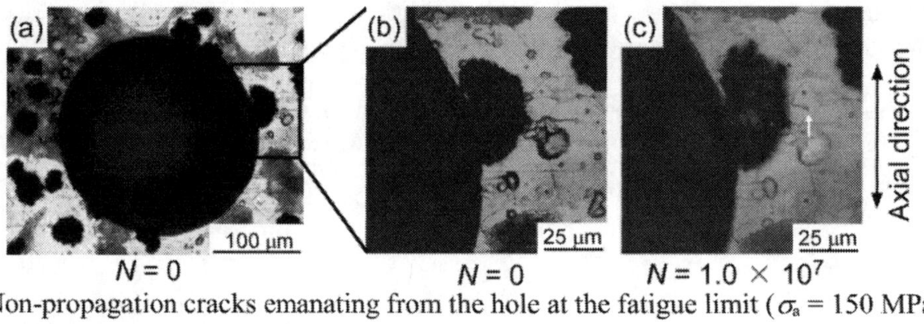

Figure 4. Non-propagation cracks emanating from the hole at the fatigue limit (σ_a = 150 MPa) at R = 0.1

R, can be accounted for by using a term of $\{(1-R)/2\}^{\alpha}$ as follows:

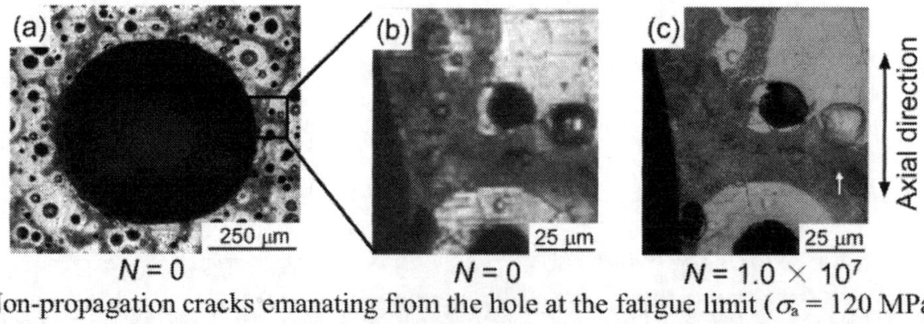

Figure 5. Non-propagation cracks emanating from the hole at the fatigue limit (σ_a = 120 MPa) at R = 0.1

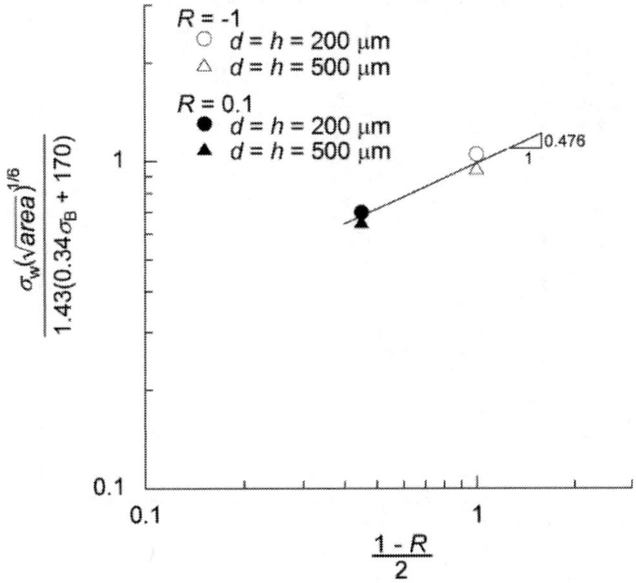

Figure 6. Influence of stress ratio on the fatigue limit.

$$\sigma_{\rm w} = \frac{F_{\rm loc}(0.34\sigma_{\rm B}+170)}{(\sqrt{area})^{1/6}}\cdot\left(\frac{1-R}{2}\right)^{\alpha} \qquad (3)$$

where α is a correction coefficient that depends on the material. Figure 6 shows the values of $\sigma_{\rm w}(\sqrt{area})^{1/6}/\{F_{\rm loc}(0.34\sigma_{\rm B}+170)\}$ as function of $(1–R)/2$ on a logarithmic graph. The slope of line in Figure 6 represents the value of α. In Figure 6, the results for the hole specimens with $d = h = 200$ μm and $d = h = 500$ μm were plotted. The slope indicated the values of $\alpha = 0.476$ and this value was adopted for the prediction of fatigue limit.

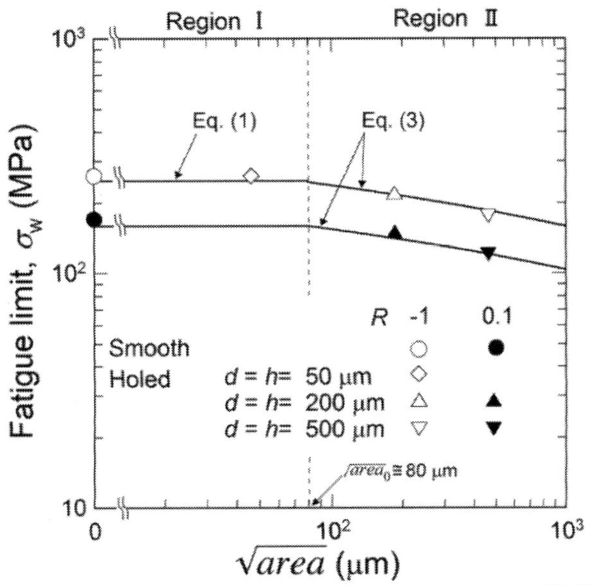

Figure 7. Relationship between fatigue limit and $\sqrt{area}$.

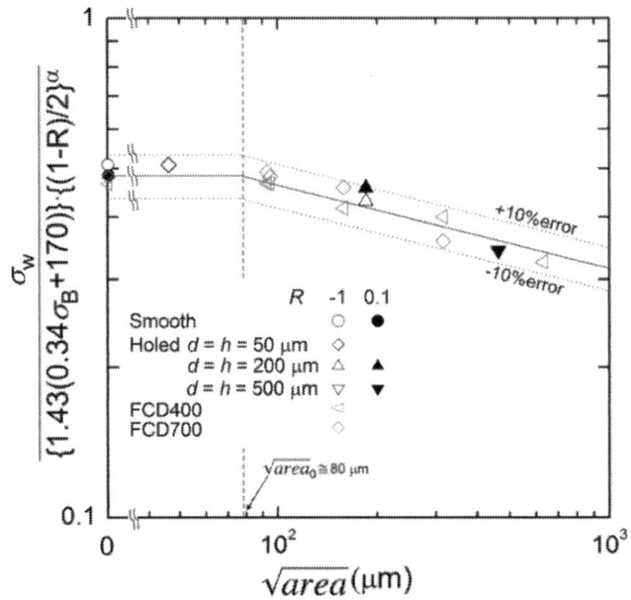

Figure 8. Relationship between normalized fatigue limit by $\{(1\text{-R})/2\}^\alpha$ and $\sqrt{area}$.

Figure 7 shows a comparison of prediction by Eqs. (1) and (3) with the experimental results for smooth specimens as well as holed specimens. The prediction of fatigue limit is in reasonable agreement with the experimental data. Irrespective of the stress ratio, there is a critical size of defect that has no effect on the fatigue limit of ferritic-pearlitic ductile cast iron. Defects with the $\sqrt{area}$ smaller than about 80 μm (region I), designated herein as $\sqrt{area_0}$, are not harmful to the fatigue limit. Prediction with Eq. (3) shows a good agreement with the experimental results for the artificial defects with 80 μm$\leq \sqrt{area} \leq$ 1000 μm (region II).

Figure 8 shows the relationship between the fatigue limit normalized by the term of $\{(1-R)/2\}^\alpha$ and $\sqrt{area}$. The data of FCD400 and FCD700 were taken from the previous study [5, 7]. The tensile strength

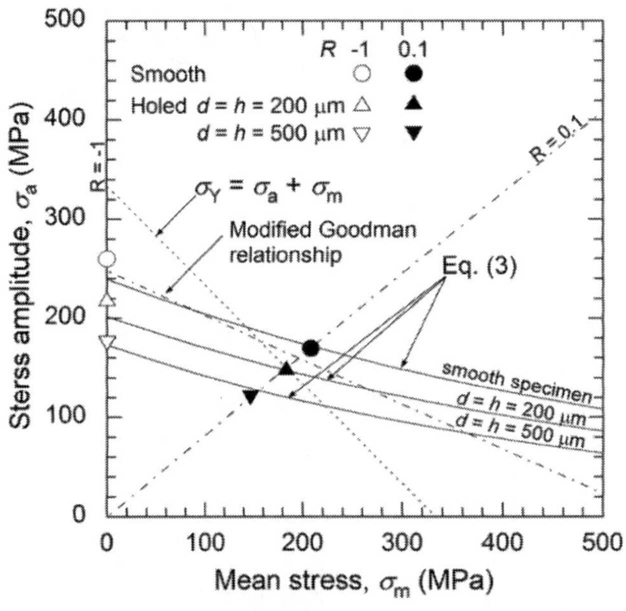

Figure 9. Effects of mean stress on fatigue limit.

of FCD400 and FCD700 were 420 MPa and 734 MPa, respectively. The prediction of fatigue limit is in reasonable agreement with the experimental data.

Figure 9 shows the effect of mean stress on fatigue limit. The solid lines calculated by Eq. (3) show the fatigue limit prediction. It is noted that to make use of Eq. (3) even for smooth specimen, we substituted $\sqrt{area} = 80$ μm (cf. Figure 8) in Eq. (3). The fatigue limit of smooth specimen is slightly higher than the modified Goodman relationship and the yield strength at $R = 0.1$. Near the fatigue limit at $R = 0.1$, the effect of plastic yielding would be negligible because of the possible activation of elastic shake-down. As a whole, the predictions with Eq. (3) show a reasonably good agreement with the experimental results both for smooth and holed specimens.

4. Conclusions

The effects of small defects and stress ratios on the fatigue limit of ferrite-pearlite ductile cast iron were investigated. The obtained conclusions are as follows:

(1) From the microscopic observation, it is revealed that the fatigue limit is determined by the threshold condition for propagation of a crack emanating from the hole edge, regardless of defect sizes and stress ratios.

(2) In the region I where $\sqrt{area}$ is smaller than 80 μm, the defect can be regarded as a non-detrimental defect and the fatigue limit is the same as that of smooth specimen.

(3) In the region II where 80 μm $\leq \sqrt{area} \leq$ 1000 μm, the fatigue limit can be quantitatively predicted based on the $\sqrt{area}$ parameter model in terms of a geometrical parameter of defects, $\sqrt{area}$, and a material parameter, σ_B. Furthermore, the stress ratio dependency of fatigue limit can be described by the term of $\{(1-R)/2\}^{\alpha}$. The value of $\alpha = 0.476$ was obtained for the investigated ferritic-pearlitic ductile cast iron.

References

[1] Murakami Y 2002 *Metal Fatigue: Effects of Small Defects and Nonmetallic Inclusions*, (Elsevier)
[2] Yamabe J, Kobayashi M and Nakajima N 2005 *Trans. Jpn. Soc. Mech. Eng.* **71** 116
[3] Endo M 1989 *Trans. Jpn. Soc. Mech. Eng.* **38** 1139
[4] Tanaka Y, Yang Z and Miyamoto K 1995 *Mater. Trans.* **36** 749
[5] Endo M and Yanase K 2014 *Theor. Appl. Fract. Mec.* **69** 34
[6] Costa N, Machado N and Silva F S 2010 *Int. J. Fatigue* **32** 988
[7] Deguchi T, Matsuo T, Takemoto S, Ikeda T and Endo M 2016 *International Journal of Fracture Fatigue and Wear* **4** 93
[8] Deguchi T, Matsuo T, Kim H J, Ikeda T and Endo M 2016 *11th Int. Symp. on Advanced Scienc and Technology in Experimental Mechanics* Ho Chi Minh
[9] Deguchi T, Kim H J and Ikeda T 2017 *12th Int. Conf. on Damage Assessment of Structures, DAMAS2017* Kitakyushu Japan (This thesis will be presented)

IOP Conf. Series: Journal of Physics: Conf. Series **843** (2017) 012050 doi:10.1088/1742-6596/843/1/012050

An investigation into the frequency dependence upon the fatigue crack growth rate conducted by a novel fatigue testing method with in-situ hydrogen-charging

T Yoshimoto[1] and T Matsuo[2, 3]
[1]Graduate School of Engineering, Fukuoka University
8-19-1 Nanakuma, Jonan-ku, Fukuoka 814-0180, Japan
[2]Department of Mechanical Engineering, Fukuoka University
8-19-1 Nanakuma, Jonan-ku, Fukuoka 814-0180, Japan
[3]Institute of Materials Science and Technology, Fukuoka University
8-19-1 Nanakuma, Jonan-ku, Fukuoka 814-0180, Japan

E-mail: tmatsuo@fukuoka-u.ac.jp

Abstract. The most widely used testing methods for evaluating the hydrogen embrittlement resistance are classified into the following two types: (1) testing in high-pressure hydrogen gas and (2) testing in air using hydrogen pre-charged specimens. Testing in hydrogen gas is technically difficult and expensive, because high-pressure gas equipments composed of such as a high-pressure vessel and piping system has to be introduced. On the other hand, in the case of pre-charging method, the outgassing of hydrogen from the specimen occurs during the test. Therefore, hydrogen embrittlement could not be evaluated accurately by the pre-charging method in long-term tests such as a fatigue test. In a previous study, to evaluate hydrogen embrittlement effectively, a novel experimental method was developed, in which four-point bending fatigue test system was performed with continuous circulation of a hydrogen-charging solution into a pipe specimen. This new testing system using hydrogen-charging solution enables an easy yet reasonable evaluation of hydrogen embrittlement for a long-term fatigue test. In this study, the frequency effect on the crack growth rate due to hydrogen was investigated by this new testing method. Fatigue crack growth tests at a test frequency ranging from 1 to 0.0002 Hz demonstrated that the fatigue crack growth rate was faster in the presence of hydrogen than in the absence of hydrogen at all test frequencies. Further, the increase in crack growth rate became more pronounced with decrease in test frequency. This frequency dependence upon crack growth rate was discussed by considering the difference between the effects of external hydrogen and internal hydrogen.

1. Introduction

Development of practical use of hydrogen as an environment-friendly energy is highly required to solve global warming and resource problem, and consequently the establishment of hydrogen energy society is being proceeded in USA, Europe, Japan, etc. However, in order to ensure the integrity of hydrogen equipments (e.g. hydrogen fuel cell vehicle and hydrogen station), problems of hydrogen embrittlement (i.e., degradation of strength properties of metallic material due to hydrogen penetration into the material) need to be well understood and properly solved.

The strain rate dependence is a well known feature of hydrogen embrittlement [1-7]; i.e., slow strain rate for monotonic loading or low frequency for fatigue loading intensifies the hydrogen effect. Regarding the hydrogen equipments mentioned above, many kinds of components are used and some of them (e.g., pressure vessel) are subjected to a cyclic load with extremely low frequency. Therefore, the frequency dependence of fatigue properties in the presence of hydrogen is a very important issue.

Recently, Matsuo et al. [8] reported a peculiar frequency dependence of fatigue crack growth in hydrogen gas, which is summarized as follows. The crack growth rate in hydrogen gas was increased with decrease in test frequency and it reached a peak at 0.1 Hz. This can be explained by the hydrogen enhanced successive fatigue crack growth (HESFCG) [9, 10] mechanism in which the increase in crack growth rate is caused by stress-induced hydrogen diffusion and accumulation in the limited area in the vicinity of the crack tip. Surprisingly, the further decrease in frequency inversely reduced the crack growth rate to the level of that in air. The HESFCG mechanism cannot describe this phenomenon. Matsuo et al. [8] pointed out that hydrogen penetrating into the material through the crack surface cannot remain within the small area in the vicinity of crack tip (i.e. hydrogen will diffuse to comparatively large area around crack tip) owing to extremely long loading time in a cycle. Someday et al. [11] also reported a similar frequency dependence of crack growth in hydrogen gas. They interpreted that the inhibitor effect against hydrogen penetration due to impurity such as oxygen contained in hydrogen gas became pronounced with decrease of test frequency. Both groups conducted the fatigue crack growth tests in hydrogen gas and considered the effect of the hydrogen which came from the outside of a specimen during the tests (so-called *external hydrogen*). In the case of the external hydrogen, both hydrogen penetration and subsequent diffusion need to be considered. This may be a cause that complicates the interpretation of the mechanism of peculiar frequency dependence. In the case of hydrogen that exists inside a specimen prior to the test (so-called *internal hydrogen*), it is possible to discuss about the frequency dependence by focussing only on the effect of hydrogen diffusion and accumulation at crack tip. However, in conventional fatigue test for investigating the effect of internal hydrogen (i.e. the test using pre-charged specimen), the crack growth property at low frequency cannot be evaluated successfully because of considerable hydrogen outgassing from a specimen during the early stage of long-term testing.

In this study, we investigated the frequency dependence of crack growth rate in the presence of internal hydrogen using the method newly developed in the previous study [12], which is a four-point bending test system with internal circulation of hydrogen-charging solution in a pipe specimen.

2. Testing method and specimen

2.1 Testing method

Figure 1 shows the concept of the testing method developed in the previous study [12]. The fatigue crack growth test was conducted by the four-point bending fatigue test system with internal circulation of hydrogen-charging solution in a pipe specimen. The hydrogen-charging solution (20 mass% NH_4SCN) was pumped into the pipe specimen through a flexible tube. Two screw holes were introduced on both ends of the specimen, which were connected to a flexible tube by a screw type fitting. This specimen with internal circulation of hydrogen-charging solution is hereafter called the *H-charging specimen*, while the specimen tested without hydrogen-charging is called the *non-charging specimen*.

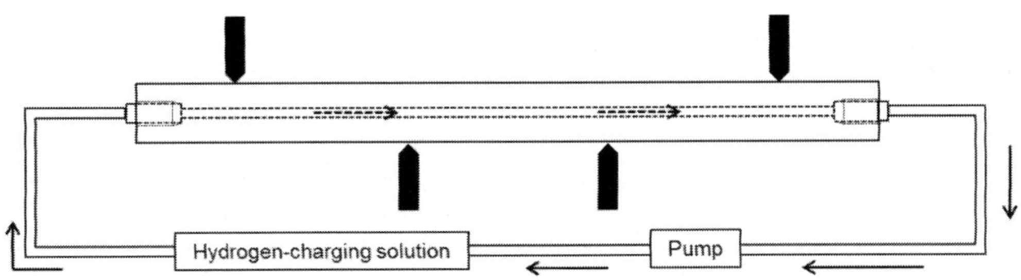

Figure 1. Concept of the testing method.
(Four-point bending fatigue test with internal circulation of hydrogen-charging solution).

2.2 Specimen

An annealed carbon steel (JIS-S35C) was used in this study. The Vickers hardness of this material is approximately 156. The microstructure is composed of ferrite-pearlite with an elongated texture in the rolling direction, as shown in Fig. 2.

Figure 3 shows the geometry of pipe specimen. The specimen surface was polished with a #2000 emery paper and subsequently buff-polished using alumina paste with a particle diameter of 1 μm. After polishing, three drilled holes shown in Fig. 4 were introduced on the specimen surface as a crack starter. Then, stress relief annealing in vacuum at 600 ℃ for 1h was conducted.

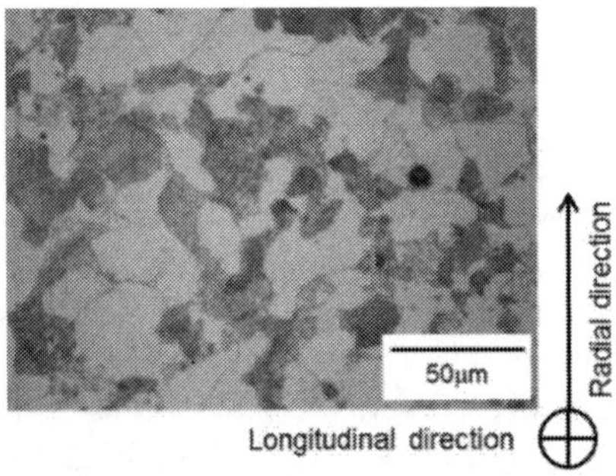

(a) Transverse sectional view

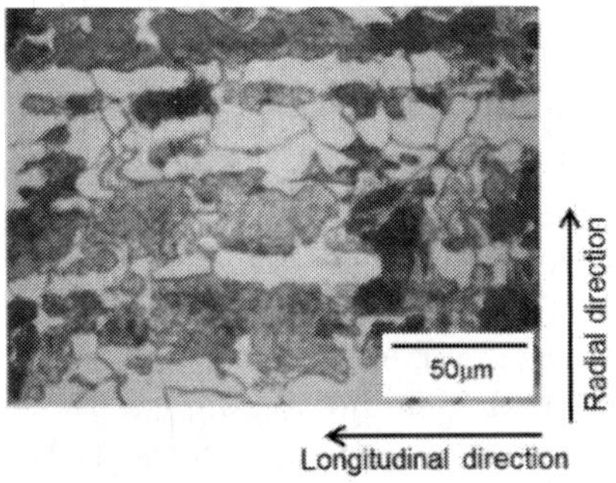

(b) Longitudinal sectional view

Figure 2. Microstructure of S35C.

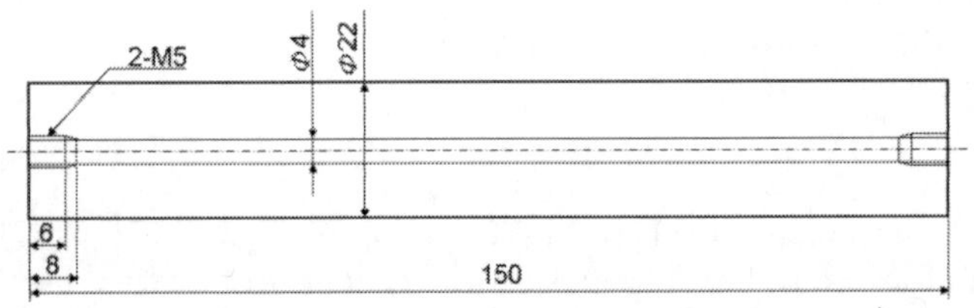

Figure 3. Geometry of pipe specimen (in mm).

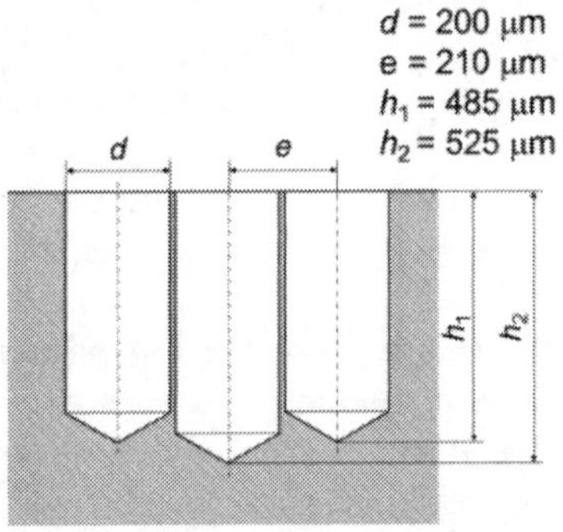

Figure 4. Shape and dimensions of drilled holes.

2.3 Fatigue crack growth test

Fatigue crack growth behavior was investigated by the four-point bending fatigue of a H-charging specimen and a non-charging specimen in air at room temperature. The stress amplitude was constant, and the stress intensity factor (SIF) range, ΔK, increased from 5.5 MPa·m$^{1/2}$ to 15 MPa·m$^{1/2}$ with crack growth (ΔK-increasing test). The stress ratio, R, was 0.1 and the test frequency, f, was 1 Hz. Then, the frequency dependence of fatigue crack growth rate was investigated at test frequencies of 0.0002 Hz, 0.002 Hz, 0.05 Hz, 0.2 Hz and 1 Hz. The stress amplitude was decreased every time the frequency was changed so that the SIF range could be kept around 15 MPa·m$^{1/2}$ (ΔK-constant test). The crack growth behavior was observed by an optical microscope at the appointed numbers of cycles to measure the crack length. The fracture surface was observed by a scanning electron microscope. Hydrogen-charging solution was changed every three days in order to maintain the enough amount of hydrogen to supply from the charging solution to the specimen.

3. Results and discussion

3.1 *Fatigue crack growth rate obtained by ΔK-increasing test*

ΔK-increasing tests were conducted at a stress ratio of $R = 0.1$ and at a test frequency of $f = 1$ Hz under the conditions of continuous hydrogen-charging as well as no hydrogen-charging. The SIF range was initially about 5.5 MPa·m$^{1/2}$ and gradually increased up to about 15 MPa·m$^{1/2}$ with crack growth. Figure 5 shows the relationship between the crack growth rate, d*a*/d*N* and the SIF range, *ΔK*. As reported in our previous study [12], the crack growth rate was enhanced in the H-charging specimen in the relatively high *ΔK* regime ($\Delta K > 7$ MPa·m$^{1/2}$). For the present H-charging specimen, after the SIF range reached 15 MPa·m$^{1/2}$ in the process of *ΔK*-increasing test, the test frequency was decreased in a stepwise fashion from 1 Hz to 0.0002 Hz, maintaining a constant SIF range of around 15 MPa·m$^{1/2}$ by properly decreasing the stress amplitude. As shown in Fig. 5, the enhanced crack growth rate with frequency effect was observed at $\Delta K = 15$ MPa·m$^{1/2}$. A further discussion on frequency effect is provided in *Section 3.2*. Many flat fracture facets of the order of a few grains in size were observed on the fracture surface of the H-charged specimen.

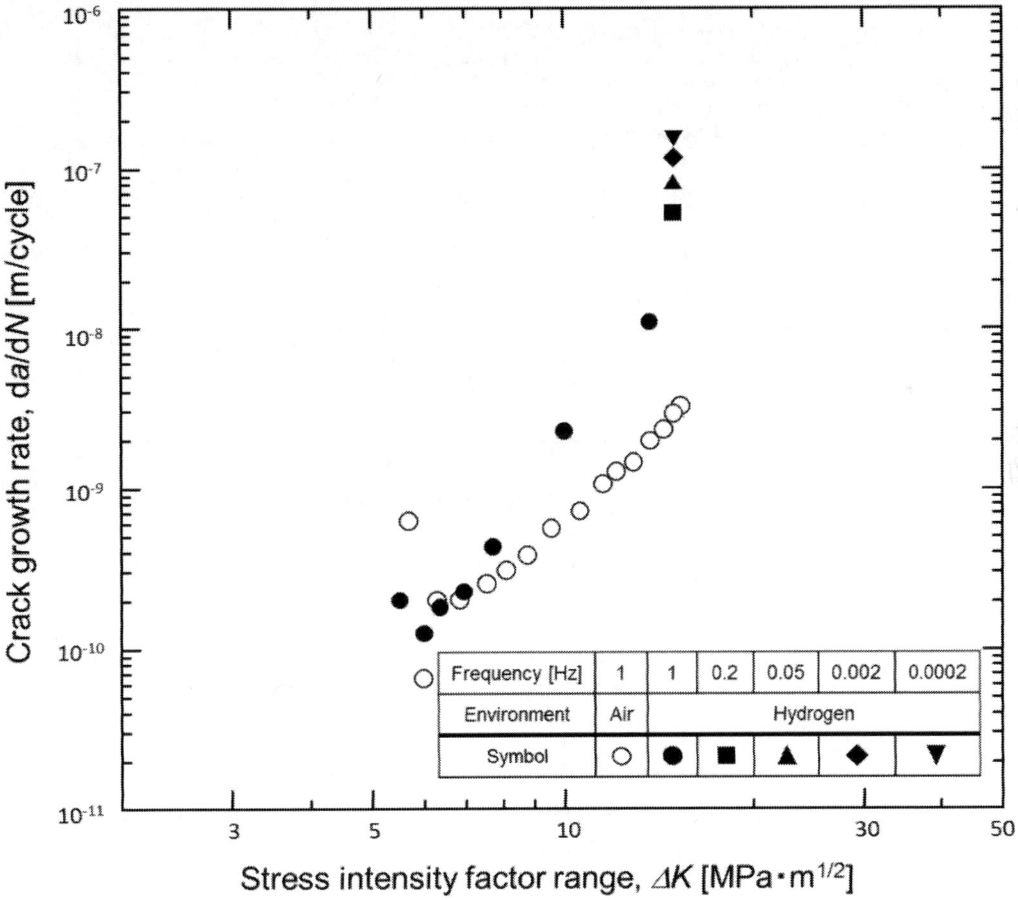

Figure 5. Relationship between crack growth rate and *ΔK*.

3.2 *Frequency dependence of fatigue crack growth rate under continuous hydrogen-charging*

The ratio of the crack growth rate of H-charging specimen $(da/dN)_H$ to that of non-charging specimen $(da/dN)_{H\text{-free}}$, is shown as a function of test frequency, *f*, in Fig. 6. In this study at all test frequency, the crack growth rate is higher for the H-charging specimen than for the non-charging specimen. The rate of $(da/dN)_H/(da/dN)_{H\text{-free}}$ increased as the test frequency decreased. The increase in crack growth rate

IOP Conf. Series: Journal of Physics: Conf. Series **843** (2017) 012050 doi:10.1088/1742-6596/843/1/012050

was noticeable when the frequency decreased from 1 Hz to 0.05 Hz, whereas the ratio moderately increased with decreasing from f = 0.05 Hz to 0.0002 Hz. The ratio reached 52 at 0.0002 Hz (the lowest frequency in this study). This frequency dependence of crack growth rate due to hydrogen is well known as a feature of hydrogen-related fracture [1-8], which has been explained by considering the increase in hydrogen accumulation at the crack tip with decrease in test frequency. As mentioned in the *Introduction*, a peculiar frequency dependence of hydrogen-induced crack growth rate in hydrogen gas was reported by Matsuo et al. [8]; that is, although the increase in crack growth rate became pronounced with decrease to f = 0.1 Hz, the crack growth rate did not increase at a frequency of 0.01 Hz anymore and it decreased with further decrease in f (cf. Fig 6). In this study, however, such the peculiar frequency dependence was not observed even at 0.0002 Hz, showing a monotonic increase with decrease in f. In the next section, this different characteristics for frequency dependence will be discussed by focusing on the possible different actions of internal hydrogen and external hydrogen.

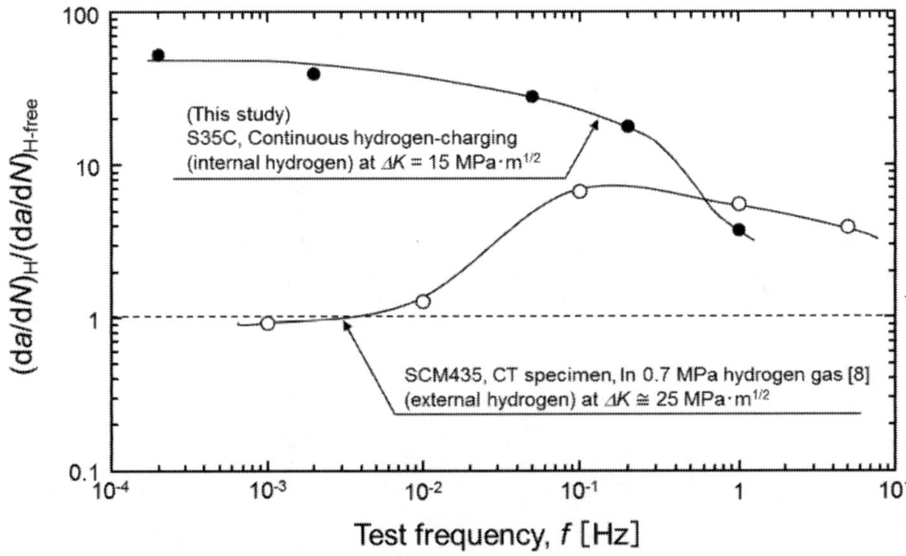

Figure 6. Frequency dependence of $(da/dN)_H/(da/dN)_{H\text{-free}}$ under ΔK-constant condition.

3.3 *Effect of internal and external hydrogens on the frequency dependence*

As shown in Fig. 6, the increase in crack growth rate due to internal hydrogen tends to be saturated at an extremely low frequency, approaching to a value of about 60 in the ordinate. In the case of external hydrogen, the increase rate reached a peak of about 7 at f = 0.1 Hz. This different actions of internal and external hydrogens can be explained from the viewpoint of the mechanism of hydrogen accumulation at the crack tip. Figure 7 shows the schematic illustrations of hydrogen accumulation at the crack tip. External hydrogen penetrates easily into the material directly through the newly formed crack surface and thereby more hydrogen accumulates at the crack tip. On the other hand, internal hydrogen needs to diffuse inside the material to accumulate at the crack tip. For this case, internal hydrogen requires much longer time for accumulation than external hydrogen. As a result, the frequency corresponding to the maximum influence on the crack growth is supposed to become lower for internal hydrogen than for external hydrogen. The peak of increase rate for internal hydrogen may exist at a further low frequency less than 0.0001 Hz. Unfortunately, this could not be verified in this study because of time-consuming fatigue test. However, the following two points are worthy to note: (i) the different accumulation mechanisms of internal and external hydrogens at the crack tip and (ii) the increase rate for internal hydrogen shows a tendency of saturation around f = 0.0001 Hz.

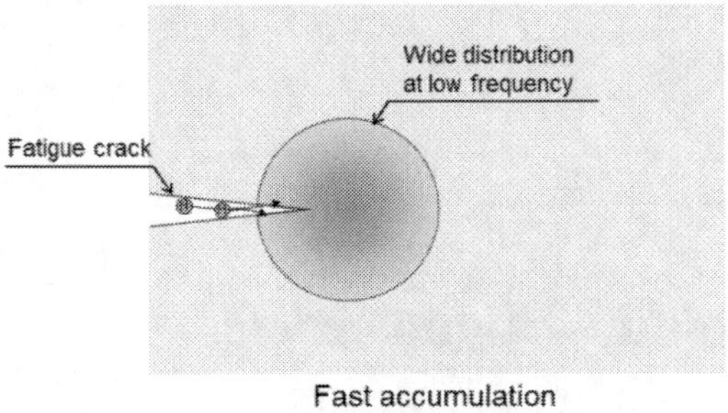

(a) External hydrogen

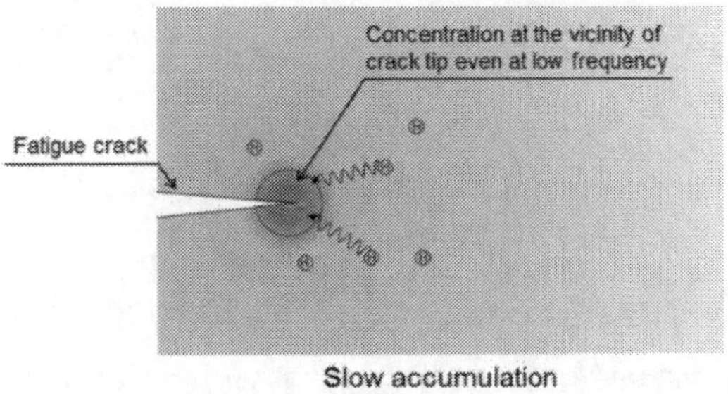

(b) Internal hydrogen.

Figure 7. Different mechanisms of hydrogen accumulation at crack tip between internal and external hydrogens.

4. Conclusions

The frequency dependence of the crack growth rate due to hydrogen was investigated by conducting four-point bending fatigue tests with internal circulation of hydrogen-charging solution in a pipe specimen. The conclusions are summarized as follows:

1. The crack growth rate was increased by continuous hydrogen-charging in the range of a frequency from 0.0002 Hz to 1 Hz, which became pronounced with decrease in frequency.
2. The frequency at which the degree of increase showed a peak was much lower for internal hydrogen than for external hydrogen.
3. The different effects of internal and external hydrogen on frequency dependence of crack growth rate could be explained by considering the different mechanisms of hydrogen accumulation at the crack tip.

References

[1] Matsunaga H, Yoshikawa M, Kondo R, Yamabe J and Matsuoka S 2015 *Int. J. Hydrogen Energy* **40** 5739
[2] Tanaka H, Homma N, Matsuoka S and Murakami Y 2007 *Trans. Jap. Soc. Mech. Eng.* **73** 1358
[3] Salival G C, Creighton D L and Hoeppner D W 1981 *Eng. Frac. Mech.* **14** 337
[4] Matsuoka S, Tanaka H, Homma N and Murakami Y 2011 *int. J. Fracture* **168** 101
[5] Yamabe J, Itoga H, Awane T, Matsuo T, Matsunaga H and Matsuoka S 2016 *J. Pressure Vessel Technology* **138** 011401-1
[6] Itoga H, Matsuo T, Orita A, Matsunaga H, Matsuoka S and Hirotani R 2014 *Pressure Vessels and Piping Conf.*
[7] Orita S, Matsuo T, Matsuoka S and Murakami Y 2013 *Proc. 19th European Conf. Frac. (ECF19)*
[8] Matsuo T, Matsuoka S and Murakami Y 2010 *Proc. 18th European Conf. Frac. (ECF18)*
[9] Matsuoka S, Tsutsumi N and Murakami Y 2008 *Trans. Jap. Soc. Mech. Eng.* **74** 1327
[10] Murakami Y, Kanezaki T, Mine Y and Matsuoka S 2008 *Metal. Mater. Trans.* A **39** 1327
[11] Somerday B P, Sofronis P, Nibur K A, San Marchi C and Kirchheim R 2013 *Acta Mater.* **61** 6153
[12] Yoshimoto T and Matsuo T, Development of fatigue testing method using continuously hydrogen-charging pipe specimen, *12th Int. Conf. Damage Assessment of Structures, DAMAS 2017*, 2017.07.10.

Microscopic observation of shear-mode fatigue crack growth behavior under the condition of continuous hydrogen-charging

Y Akaki[1], T Matsuo[2, 3], Y Nishimura[4], S Miyakawa[4] and M Endo[2, 3]

[1]Graduate School of Engineering, Fukuoka University, 8-19-1 Nanakuma, Jonan-ku, Fukuoka 814-0180, Japan
[2]Development of Mechanical Engineering, Fukuoka University, 8-19-1 Nanakuma, Jonan-ku, Fukuoka 814-0180, Japan
[3]Institute of Materials Science and Technology, Fukuoka University, 8-19-1 Nanakuma, Jonan-ku, Fukuoka 814-0180, Japan
[4]Materials Engineering R&D Division, DENSO CORPORATION, Japan, 1-1 Syowa-cho, Kariya, Aichi 448-8661, Japan

E-mail: tmatsuo@fukuoka-u.ac.jp

Abstract. Rolling bearings facilitate the smooth motion of moving mechanical transmission devices substantially reducing the energy loss. However, flaking failure caused in bearings by rolling contact fatigue can crucially deteriorate the integrity of an overall mechanical system composed using bearings. It is understood nowadays that this phenomenon is intimately associated with shear-mode (Mode II and III) fatigue crack growth caused by cyclic shear stress in the presence of large compression. Further, a problem has become of importance that the premature flaking in ball bearings could be caused by the assistance of the penetration of hydrogen into material during the operation. In this study, torsional fatigue test of a bearing steel (JIS SUJ2) was performed by using a newly-developed method of continuous hydrogen-charging. Based on the microscopic observation, it was found that continuous hydrogen-charging had an influence on the shear-mode fatigue crack growth behavior in the high-cycle fatigue regime and reduced the threshold level.

1. Introduction

White structure flaking (WSF) is a typical failure frequently observed in rolling bearings used in automotive components, wind turbine gearboxes, etc, which is attributed to the crack growth from a surface or subsurface inclusion under the combination of cyclic shear stress and heavy compressive stress. Some previous studies [1-10] reported that WSF could be caused by invasion of the hydrogen derived from decomposition of lubricant in operation. In order to understand the essential mechanism of WSF, it is necessary to clarify the effect of hydrogen on the shear-mode fatigue crack growth behavior and the threshold condition, in particular for a small crack of a few millimeters or smaller in size that determines the rolling contact fatigue strength. However, there is only a few researches [10] that quantitatively investigate the growth behavior of a small sized shear mode fatigue crack. Especially, there is no study on the *high-cycle* fatigue in hydrogen environment because a problem had been left that hydrogen will emit from the specimen during fatigue test.

In our previous study [11], an easy-to-use as well as effective method was newly developed to investigate the shear-mode fatigue crack growth behavior in the presence of hydrogen. The developed method demonstrated that hydrogen had an influence on the shear mode fatigue crack growth behavior in the *low-cycle* fatigue regime. However, in order to ensure the safety in long-term use of rolling bearings, the *high-cycle* fatigue property in the presence of hydrogen is also an important issue to be addressed. In this study, the effect of hydrogen on shear mode fatigue crack growth behavior in the regime of high-cycle fatigue was investigated by utilizing the continuous hydrogen-charging (H-charging) method developed in the previous study [11].

2. Experimental set-up

2.1 Material and specimen

The material used in this study was a bearing steel, JIS-SUJ2, which is substantially the same material used for the ball in a ball bearing. The shape and dimensions of specimen are shown in Figure 1.To circulate an H-charging solution, a through-hole of 2 mm in diameter was introduced coaxially with the specimen axis. Also, a blunt notch with the gage length of 2 mm was introduced to localize the crack initiation site and to facilitate the observation of fatigue crack growth. The surface of specimen was finished by alumina-buffing following polishing with an emery paper up to #2000. Then, electro polishing was conducted to remove a thin worked layer with residual stress. The residual stress measured by the X-ray diffraction method was within ± 30 MPa.

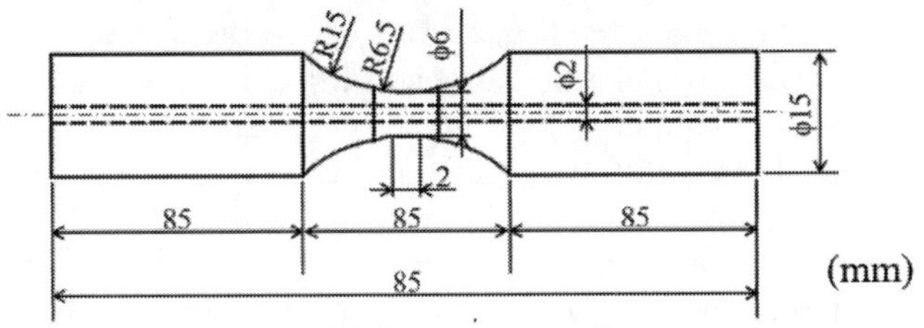

Figure 1. Fatigue test specimen.

2.2 Continuous hydrogen-charging method [11]

Continuous H-charging was performed by circulation of an aqueous solution of 20 mass% NH_4SCN in a specimen. The mechanism of hydrogen circulation is shown in Figures 2 and 3. This procedure was started 48 hours prior to the fatigue test under the H-charging condition. The hydrogen-charging solution was exchanged within 3 days monitoring the pH value to maintain a sufficient hydrogen supply to the specimen during fatigue test.

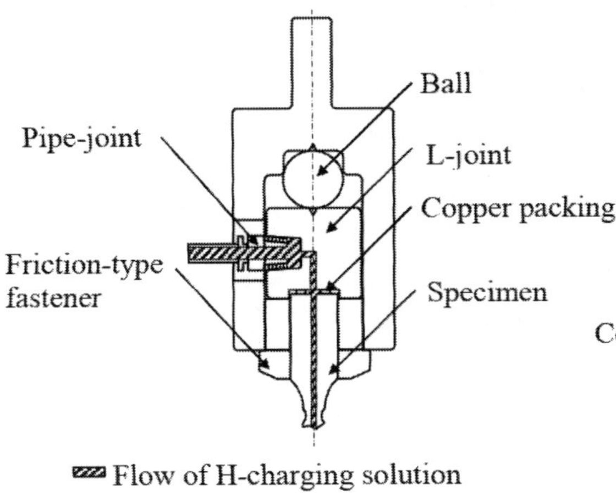

Figure 2. Specimen and fixture with channel of H-charging solution [11].

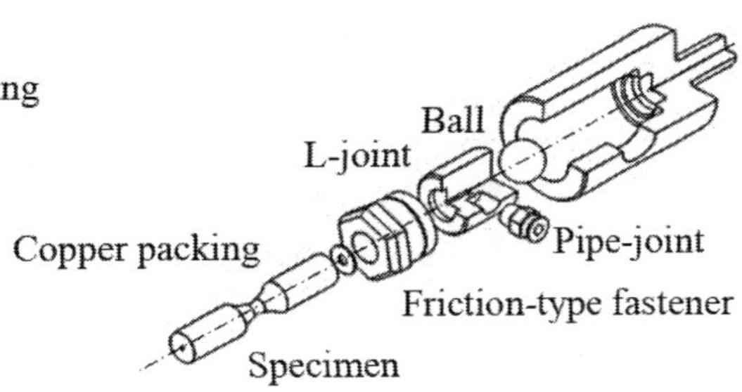

Figure 3. 3D illustration of components [11].

2.3 Fatigue test procedure

Fatigue test was carried out in air at room temperature by a servo-hydraulic combined axial/torsional loading fatigue testing machine. The frequencies were f = 20 Hz for testing under the non-charging condition and f = 2 Hz for testing under the H-charging condition. The stress ratio was $R = -1$. A statically-applied compressive axial stress of $\sigma_s = -1200$ MPa was superposed on the cyclic torsion to suppress the Mode I crack branching. The shear stress with a nominal amplitude of τ_a = 900 MPa was initially applied to the specimen under the non-charging condition. The shear stress amplitude, τ_a, was thereafter decreased in a stepwise manner with increase in the crack length. In this study, when the fatigue crack growth rate, da/dN, became less than 10^{-11} m/cycle over an additional 10^6 cycles after the nominal shear stress amplitude had been decreased, the crack was regarded as a non-propagating crack. The critical value of τ_a was defined to be the threshold stress, τ_{th}, for crack propagation. After the non-propagation condition was satisfied under the non-charging condition, the testing condition was switched from non-charging to H-charging condition; namely, internal circulation of hydrogen-charging solution in the specimen was undertaken. The fatigue test under the H-charging condition was conducted at the same stress amplitude at which the non-propagation of a crack was observed under the non-charging condition. Surface cracks were observed by the replica method at every scheduled number of cycles to measure the crack length.

3. Results and discussion

The fatigue test was conducted using a single specimen according to the following sequence:
(1) The test was initially performed under the non-charging condition at a high stress amplitude of τ_a = 900 MPa to generate shear-mode fatigue cracks on the specimen surface, cf. Figure 4.
(2) The stress amplitude was changed toward the threshold level in a stepwise manner, cf. Figures 5 and 6.
(3) When the threshold level was found, i.e. the crack can be regarded as virtually stopping propagation, then the environmental condition was switched from non-charging to H-charging to investigate the effect of hydrogen-charging on the near-threshold growth behavior of a shear-mode fatigue crack, cf. Figure 7.

At step (1), the fatigue test at τ_a = 900 MPa is in the regime of high-stress, low-cycle fatigue. A number of small fatigue cracks initiated on the specimen surface. The origin of crack initiation was mostly at the inclusions elongated in the axial direction. Figure 4 shows a typical process of the

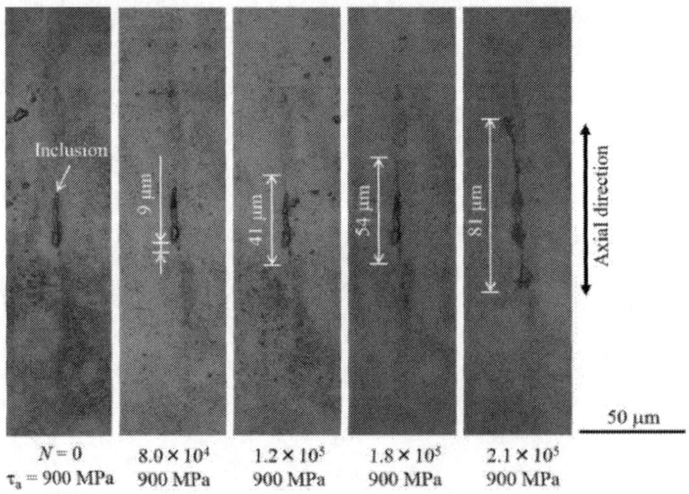

Figure 4. Shear-mode crack initiation and early growth behavior under non-charging condition. ($\sigma_s = -1200$ MPa τ_a = 900 MPa, f = 20Hz)

initiation and early growth of a shear-mode fatigue crack. The crack, initiated at a long shallow notch formed by dropping-off of inclusion, grew as a shear-mode crack in the axial direction. This crack was a leading crack with the maximum length throughout the investigation and therefore, its behavior was continuously observed afterward. In the previous study [11] in which a fatigue test was conducted under the identical test condition except for applying a little greater constant stress amplitude of τ_a = 1000 MPa, the similarly-initiated crack with the maximum length grew maintaining a clear shear-mode feature up to a length of 256 μm, then caused a typical Mode-I crack branching oriented at an angle of about 45° with respect to specimen axis.

At step (2), after the stage of the far-right figure in Figure 4, the stress was continuously changed to smaller values step by step to calibrate the level of fatigue threshold for propagation. This process is partly shown in Figure 5, in which the numerical numbers indicate the applied shear stress amplitudes, τ_a. The central part of crack that had experienced a relatively high stress of τ_a = 800-900 MPa exhibited a distinct feature of shear-mode crack growth (cf. B in Figure 5), while the outside that had been subjected to a lower stress of τ_a = 640-760 MPa showed an irregular crack path accompanied by small bends (cf. A and C in Figure 5), though the crack propagated approximately in the direction of specimen axis without causing a clear branching followed by macroscopic Mode I crack propagation. A large amount of fine oxide-particles was observed, which is considered to be produced by rubbing of opposing crack faces. They are not visible in Figure 5 since most of them had been removed in collecting replicas. In addition, the crack is thicker for the central part shown in B than for the crack tip shown in C probably because the crack faces were worn more by cyclic sliding contact with larger displacement for the central part than for the part near the tip. These results imply that the interference

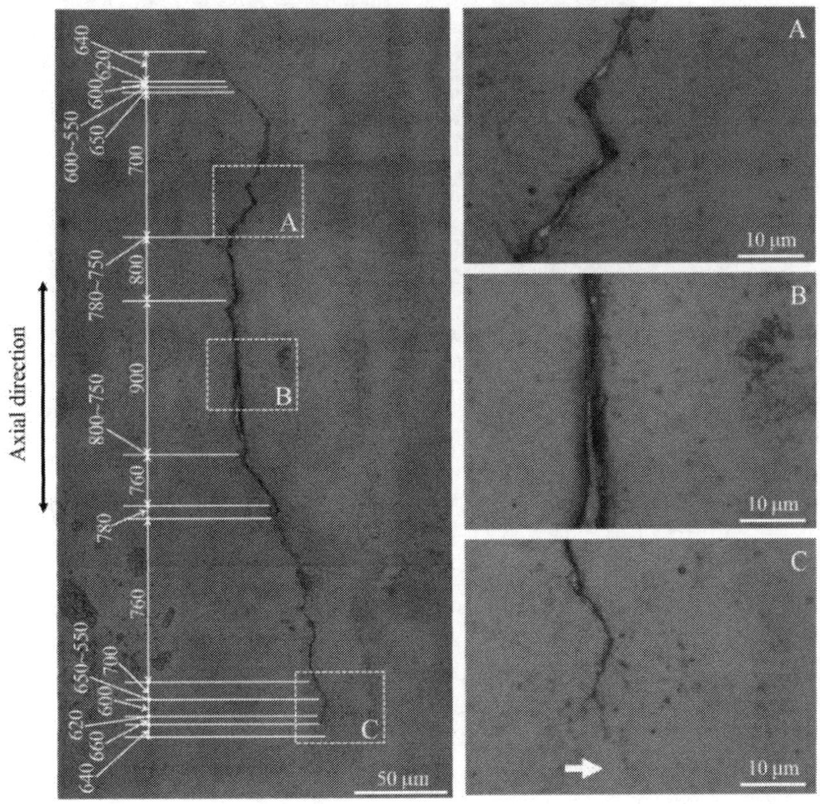

Figure 5. Process of a shear-mode crack growth in τ_a-decreasing test under non-charging condition (σ_s = −1200 MPa, f = 20 Hz). Numbers indicate the applied shear stress amplitude, τ_a, in MPa.

of crack faces would largely affect the shear-mode crack growth behavior. Figure 6 shows the final process approaching the threshold level under the non-charging condition. A slight crack extension is seen but the average crack growth rate, da/dN, is calculated to be 6.4×10^{-12} m/cycle. At this rate, the average increment of crack length per cycle is smaller than the interatomic distance and this means that the crack no longer propagated in a cycle-by-cycle manner. In the present study, therefore, the crack being at this level of growth rate was regarded as a non-propagating crack and accordingly, the applied stress, $\tau_a = 440$ MPa, was defined as the threshold stress, i.e. $\tau_{th} = 440$ MPa.

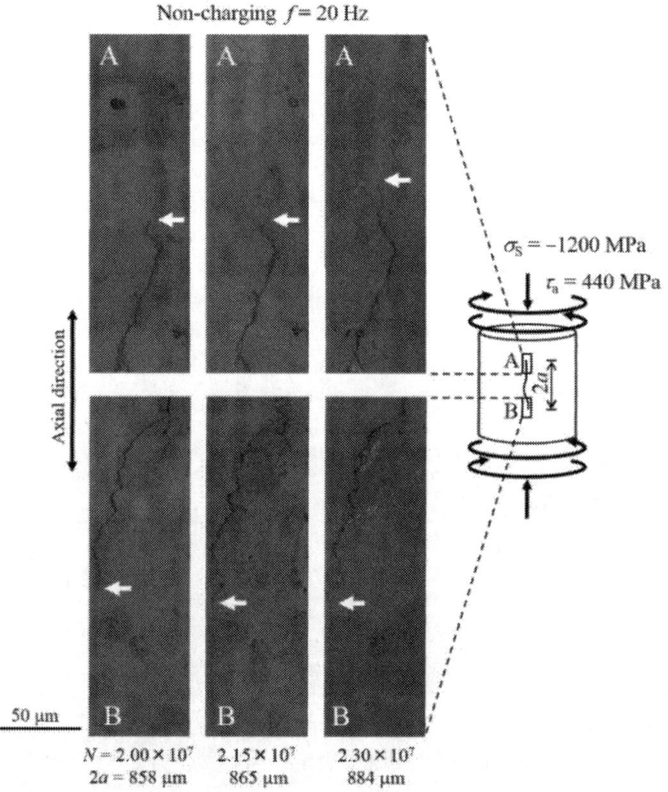

Figure 6. Shear-mode crack growth behavior at threshold level under non-charging condition. ($\sigma_s = -1200$ MPa, $\tau_a = \tau_{th} = 440$ MPa, f = 20Hz)

At step (3), to investigate the influence of hydrogen-charging, the specimen that had been at the threshold level at step (2) under the non-charging condition was next put under the H-charging condition. The mechanical conditions of shear stress amplitude ($\tau_a = 440$ MPa) and statically-applied compressive stress ($\sigma_s = -1200$ MPa) were kept unchanged and only the cyclic frequency was changed from 20 Hz to 2 Hz to facilitate the action of hydrogen. Figure 7 shows the growth behavior of a shear-mode crack under the H-charging condition. It is clearly shown that by hydrogen-charging, the crack growth was reactivated and the growth rate increased to the level of da/dN = 1.1×10^{-10} m/cycle, which is about 170 times greater than the previous growth rate under non-charging condition. The crack path seen in Figure 7 seems to be less irregular compared with that under the non-charging condition at the same stress amplitude level. This may be an indication that hydrogen-charging would facilitate more for shear mode growth than opening mode growth. In our previous [11] and Fujita et al.'s [10] experiments, fatigue tests for almost the same material and testing condition were performed under the high-stress, low-cycle fatigue testing condition, and early initiation of many cracks and acceleration of crack growth were observed. The present study provides a new result that the hydrogen-charging has also an influence on the low-stress, high-cycle growth behavior of a shear-mode fatigue crack including the threshold condition also in the regime of low stress, high-cycle fatigue. The knowledge of the near-threshold fatigue properties of metallic materials in the presence of hydrogen is considered to become more important in the future. It has previously been uneasy to perform a long-term test for investigation of the effect of hydrogen. However, it is expected that the newly-proposed method of continuous hydrogen-charging would play an important role to the advance of study in this field.

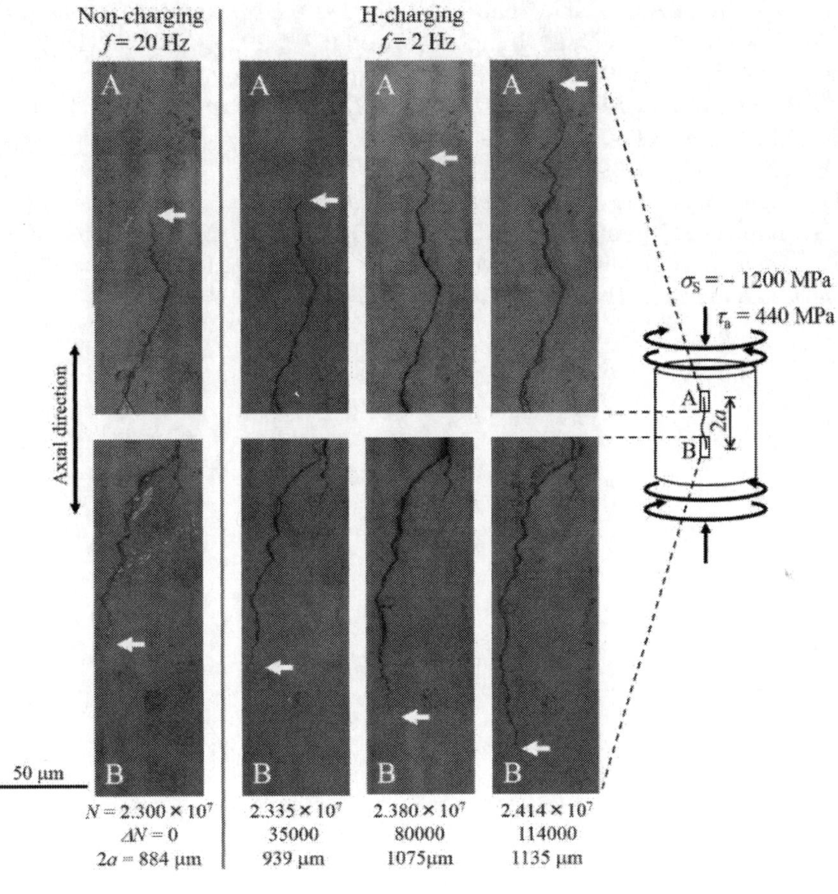

Figure 7. Shear-mode crack growth behavior successively observed after test condition was changed from non-charging to H-charging. ($\sigma_s = -1200$ MPa, $\tau_a = \tau_{th} = 440$ MPa)

4. Conclusions

The shear-mode fatigue crack growth behavior of a bearing steel was observed near the threshold level under the conditions of non-charging and continuous hydrogen-charging. The following results were found:

1. At a high stress of τ_a = 900 MPa under the non-charging condition, a number of shear-mode fatigue cracks initiated at inclusions on the specimens. The stress level was then decreased and finally a threshold stress level of τ_{th} = 440 MPa was obtained.

2. The crack that had been a non-propagating crack under the non-charging condition restarted propagation under the continuous hydrogen charging at the same stress level (i.e. τ_a = 440 MPa). Consequently, the continuous hydrogen-charging affected the growth behavior and reduced the threshold level of a shear-mode fatigue crack.

References

[1] Tamada K and Tanaka H 1996 *Wear* **15** 245
[2] Evans M-H. 2012 *Mater. Sci. Technol.* **28** 22
[3] Matsumoto Y, Murakami Y and Oohori M 2002 *ASTM Bearing Steel Technology* **7** 226
[4] Evans M-H, Richardson A.D, L. Wang and Wood R.J.K 2013 *Wear* **302** 1573
[5] Shravan J, Ole W, Peder K and Niels S.J 2015 *Inter. J. Fatigue* **77** 128
[6] Iso K, Yokouhci A, and Takemura H 2005 *SAE Technical Paper* **114** 1868
[7] Kino N and Otani K 2003 *JSAE Review* **24** 289
[8] Matsubara Y and Hamada, H 2006 *J. ASTM. Inter.* **3** 1
[9] Hamada H and Matsubara Y *NTN Technical Review* **74** 54
[10] Fujita S, Matuoka S, Murakami Y and Marquis G 2010 *Inter. J. Fatigue* **32** 943
[11] Akaki Y, Matsuo T, Nishimura Y, Miyakawa S and Endo M 2017 Proceedings of *Damage of Assesment of Stractures* To be presented.

Effect of cyclic fatigue on the fracture toughness of Polyoxymethylene

B. Ramoa[1], M. Berer[1], M. Schwaiger[1], G. Pinter[2]

[1] Polymer Competence Center Leoben GmbH, Leoben 8700, Austria
[2] Materials Science and Testing of Polymers, Montanuniversität Leoben, Leoben 8700, Austria

Correspondance to: bruno.ramoa@pccl.at

Abstract. Polymers are used in a wide range of applications and their properties are dependent upon the morphological development during processing and the specimen configuration which in turn define the mechanical properties. In this context fatigue and monotonic testing are part of the standard procedure to assess relevant mechanical and material parameters to ensure a better part design. The present work addresses the performance issues of a real component made of Polyoxymethylene (POM) which is subjected to cyclic loads from intermediate levels to high peak values inside a damping mechanism. For this linear elastic fracture mechanics concepts were used to characterize the behavior of a POM homopolymer resin used in this application. Injection molded compact tension specimens, with sharp and blunt notches, were tested under a combination of cyclic and monotonic loads and the fracture surfaces were examined. The critical stress intensity factor obtained by monotonic tests was evaluated as a function of the cycle number, where an increase after the first 1000 cycles followed by a continuous decrease with higher numbers of cycles was observed. A variation of approximately 50% and 70% were obtained along the duration of the tests for the sharp and blunt notch specimens, respectively. In light of the obtained results, a discussion is presented considering the dynamic specimen compliance and the structural features observed on the fracture surfaces in combination with the fracture mechanical response.

1. Introduction

Plastics are one of the most versatile materials in recent human history. They are used in a wide variety of applications, from agriculture to medicine, replacing in some cases ceramics and metals. Polyoxymethylene (POM) is a linear polymer composed of repeating –CH2-O- units. This material is used in many structural applications such as gears, sliding components and bearings due to its tribological properties, stiffness, dimensional stability and fatigue resistance [1–5].

In service application polymer components are frequently exposed to several combinations of loads during their life. Under cyclic loadings, polymeric structures can fail at load levels lower than they can withstand under monotonic situations [6]. Hence, there is a need to understand the effect of fatigue on the mechanical properties of the materials to ensure a better part design. Nowadays, fatigue and monotonic fracture testing are common procedures to assess relevant mechanical and material parameters. For their analysis different concepts of fracture mechanics can be applied to polymers depending on the test temperature, strain rate, specimen configuration and instrumental setup which in turn dictate the fracture behavior of a material. For POM, the concept of crack tip opening displacement

[7], essential work of fracture [8] and linear elastic fracture mechanics (LEFM) [9–11] have already been applied. There are some studies regarding the fatigue and monotonic fracture behavior of POM [9, 10, 12–14]. However, the effect of cyclic fatigue on the critical stress intensity factor for the chosen material and specimen configuration, to the best of authors' knowledge, has not yet been studied. There is one article which studies this effect for polycarbonate [15] but usually the influence of fatigue on the tensile properties rather than on the fracture toughness is addressed in the literature [12, 16, 17].

This work explores the fracture toughness of a commercial Polyoxymethylene homopolymer used to manufacture a thin-walled component. The component and the issue observed in service application are depicted in Figure 1. The component is used as part of a damping system which is composed of a piston and a cylinder with fluid inside. In application, this device is loaded monotonically with the compression of the fluid and it suffers fatigue with multiple usage. It was noticed that, after some cycles at medium to large load levels, the strength of the cylinder was noticeably reduced which caused a crack to be developed near the notch of the cylinder. The crack development process was rather spontaneous and hence was not considered as fatigue crack growth. To study this effect on specimen level, injection molded compact tension (CT) specimens were used and submitted to cyclic fatigue followed by monotonic quasi-static tensile testing.

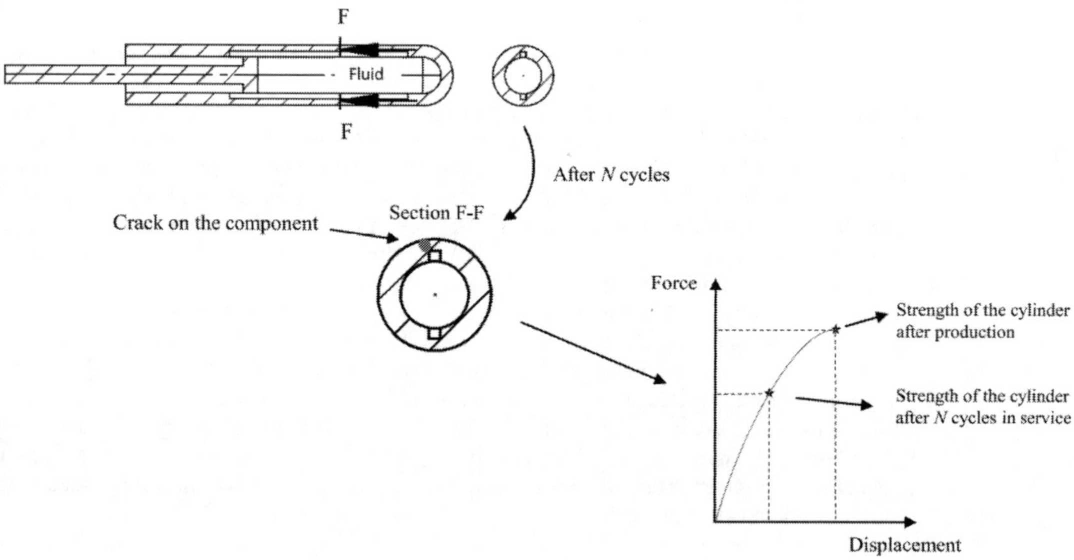

Figure 1: Schematic representation of the damping system and definition of the problem observed in service application

2. Experimental

2.1. Material and Specimen

The material used during this research was a commercially available Polyoxymethylene homopolymer resin produced by DuPont (DuPont, Wilmington, DE). The resin represents a modified version of "Delrin 300". From the manufacturer's datasheet some relevant properties are summarized in Table 1.

Table 1: Relevant material properties from the manufacturer's datasheet.

Property	**Value**	**Unit**
Density (ISO 1183)	1.42	g/cm^3
Melt Flow Rate (ISO 1133)	6	cm^3/10 min
Melting Temperature (ISO 11357)	178	°C
Yield Stress (ISO 527)	71	MPa
Tensile Modulus (ISO 527)	3.1	GPa

The fracture mechanics tests were performed on injection molded compact tension specimens with a width (*W*) of 40 mm and a thickness of 4 mm. Further details of the geometry are given in Figure 2. All the specimens were molded with the same processing conditions and in one production process.

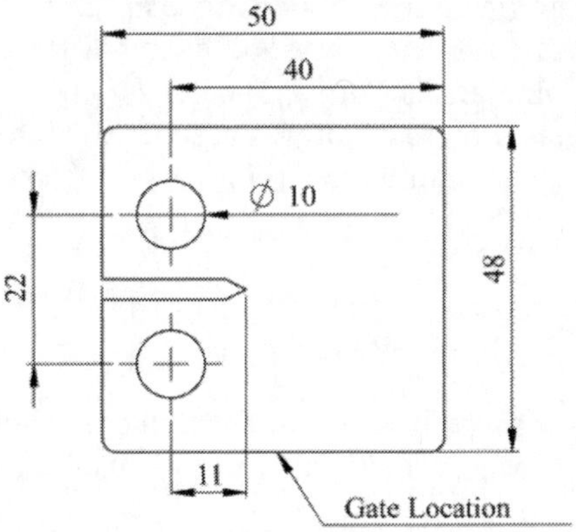

Figure 2: Compact tension test specimen following [18]

In order to assure plane strain conditions the size criteria given in Equation 1 should be met [18]. However, with the yield stress in Table 1 and the fracture toughness results in Table 2 this leads to a minimum required thickness of approximately 9 mm which is much higher than the 4 mm thick CT specimens used in this study. Hence, the test configuration does not ensure that the determined critical stress intensity factors are material properties and the denomination "K_C" is used instead of "K_{IC}". Nevertheless, the determined K_C values should suffice for comparing samples with the same geometry.

$$B, a, (W - a) \geq 2.5\left(\frac{K_C}{\sigma_Y}\right)^2 \quad (1)$$

Where σ_Y is the yield stress of the material, *B* the thickness and *a* the crack length

2.2. Fracture mechanics tests

The CT specimens were analyzed in two ways. One set of samples was notched prior to testing using a fresh portion of a thin razor blade. This was done to guarantee a uniform crack geometry with approximately 1-2 mm in depth. The second set was used as obtained from the injection molding process. The latter had a blunt notch that better represented the notch geometry of the component where the cracks were observed in the service application. For purposes of comparison, the first set of samples will be termed "Blade Notch" and the second set will be termed "Injection Molded Notch" or "IM Notch" for short.

The fatigue and monotonic fracture tests were performed on a servo-hydraulic testing machine of the type "MTS 858 horizontal" (MTS, Eden Prairie, MN, USA) in load controlled mode. The fatigue measurements were performed in tension with a sinusoidal signal of 10 Hz and a load ratio, R (F_{min}/F_{max}) of 0.1. To observe the actual crack length of the specimens a traveling microscope with a magnification of 20 (lens: Marcel Aubert SA, Bienne, Switzerland; adjustable desk: Mitutoyo, Tokyo, Japan) was mounted on the testing device. It is known that in the mechanically dominated fatigue region the fatigue life is highly dependent on the amplitude and frequency of loading. During this study, the experiments were conducted assuming plausible service conditions, i.e., maximum and minimum force of 450 and 45 N, respectively. The peak/valley pairs of force and displacement were recorded every 100 cycles, and every 1000 cycles complete hysteresis curves were recorded. The selected load level shows a rather limited life time in fatigue fracture tests (ca. 3x10^5 cycles). Additionally it is expected that this load level causes comparably big zones of irreversible deformation in the fatigue tests which was intended because it is expected to better represent the conditions in the real component. As mentioned above the formation of the cracks in the real component (Figure 1) happened rather spontaneous. Thus, it was expected that irreversible deformation played a significant role in the fatigue process of the cylinder.

The force and displacement pairs recorded every 100 cycles were used to calculate the dynamic specimen compliance, following Equation 2. This parameter has been found to be quite useful for mechanical studies during fatigue tests. Previously it was used to generate fatigue crack growth kinetics curves of POM and PEEK [9, 10] and to minimize the influence of creep on the dynamic mechanical analyses of fatigue tests [17, 19].

$$C_{Dym} = |C^*| = \frac{d_{max} - d_{min}}{F_{max} - F_{min}} \quad (2)$$

Where C_{Dyn} is the dynamic compliance which is equivalent to the magnitude of the complex specimen compliance. d_{max} and d_{min} are the peak/valley data of the displacement and F_{max} and F_{min} the corresponding force data.

After fatiguing for 10^3, 10^4 and 10^5 cycles, the tests with fatigue pre-load were stopped for one minute followed by a monotonic tensile test with a piston velocity of 10 mm/min. From the force – displacement curves, the maximum force was used to estimate the stress intensity factor using Equation 3 [18]. Additionally, monotonic fracture tests on unfatigued samples were conducted. For each testing condition at least 5 specimens were used.

$$K_C = \frac{F}{B\sqrt{W}} f\left(\frac{a}{W}\right) \quad (3)$$

$$f\left(\frac{a}{W}\right) = \frac{2 + \frac{a}{W}}{\left(1 - \frac{a}{W}\right)^{\frac{3}{2}}} \left[0.886 + 4.64\left(\frac{a}{W}\right) - 13.32\left(\frac{a}{W}\right)^2 + 14.72\left(\frac{a}{W}\right)^3 - 5.60\left(\frac{a}{W}\right)^4\right]$$

Where K_C is the critical stress intensity factor in mode I, F the maximum load and W the width

2.3. Light Microscopy

The cross section of the post-mortem representative specimens were studied under light microscopy in reflection mode using an "Olympus SZX12" microscope (Olympus Corporation, Tokyo, Japan) with a magnification of 25. The dimensions of the plastic zones and of discontinuous crack growth bands (DCG) were measured using the measuring tools of the integrated image acquisition software.

3. Results and Discussion

The effect of cyclic pre-loading on the measured critical stress intensity factor is shown in Figure 3. The corresponding values are given in Table 2. For the measurements with the highest number of pre-cycles (10^5 cycles) it was observed that crack growth initiation had already started. In this case K_C was corrected for the higher crack length using post-mortem micrographs and assuming that the increment in crack length was equal to the width of the DCG band. In order to assess the effect of the relaxation time, one additional test was carried out for each fatigue condition on the Blade Notch CT specimens. This time the test was stopped for 15 minutes between the cyclic pre-loading and the monotonic fracture test. The corresponding results are also included in Figure 3 and Table 2.

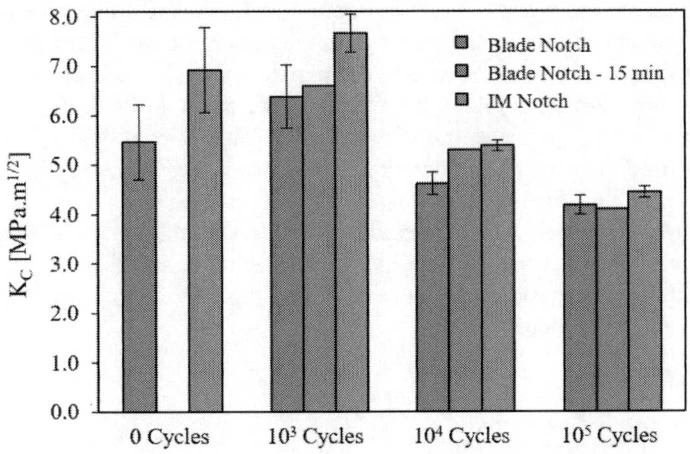

Figure 3: Evolution of K_C with increasing number of cycles in the fatigue pre-loading step

Analysis of Figure 3 indicates that both specimen types examined (Blade Notch and IM Notch) have the same trend which is important concerning the conclusions for the real component. The measured critical stress intensity factor increases for the first 10^3 cycles of fatigue followed by a continuous decrease afterwards. Considering the maximum and minimum values obtained, there is a relative variation in the critical stress intensity factor $((K_{C\ Max} - K_{C\ Min})/\ K_{C\ Min})$ of approximately 50% for the Blade Notch and of approximately 70% for the IM Notch specimens. For the IM Notch specimens, the average K_C values determined are always higher than the ones obtained for the Blade Notch ones. The specimens of the latter type have a sharper crack radius therefore the stress is more concentrated at the crack tip and less energy is needed to initiate crack growth. Interestingly, the difference decreases with increasing number of fatigue cycles which is attributed to the development of well-defined craze zones with similar geometric characteristics.

The samples tested with a relaxation time of 15 minutes display the same trend and similar values as the ones with 1 minute relaxation time thus one can assume that, for the chosen setup, relaxation has a low effect on K_C. Moreover, because of the long relaxation time of 15 min it can also be excluded that the changes in the K_C values are a thermal effect (due to hysteretic heating during fatigue pre-loading).

Table 2: Average values of the critical stress intensity factor measured for each testing condition

Number of fatigue cycles (N)	0	10^3	10^4	10^5
K_C(Blade Notch) [MPa.m$^{1/2}$]	5.5 ± 0.76	6.4 ± 0.64	4.6 ± 0.23	4.2 ± 0.19
K_C(Blade Notch -15min) [MPa.m$^{1/2}$]	-	6.6	5.3	4.1
K_C(IM Notch) [MPa.m$^{1/2}$]	6.9 ± 0.86	7.7 ± 0.38	5.4 ± 0.11	4.4 ± 0.11

A representative force-displacement curve for each testing condition is presented in Figure 4. All samples reveal brittle behavior. It can also be seen that both force and displacement values decrease with increasing number of fatigue cycles as was observed in service application for the component. The effect is more pronounced for the blunt specimens than for the sharp notched ones.

Table 3: Estimated and measured dimension of plastic zones of specimens without fatigue pre-load

Blade Notch			IM Notch		
$d_{plane\ strain}$ (mm)	$d_{plane\ stress}$ (mm)	$d_{observed}$ (mm)	$d_{plane\ strain}$ (mm)	$d_{plane\ stress}$ (mm)	$d_{observed}$ (mm)
0.63	1.88	0.16	1.01	3.02	0.34

Irreversible material deformation, such as crazing in polymers, leads to stress relaxation at the crack tip. The LEFM approach becomes more inaccurate as this plastic region at the crack tip grows. For the monotonous tests without fatigue pre-load the size of the plastic zone (*d*) was estimated using Equation 4 and 6, for plane strain conditions, and Equation 5 and 6, for plane stress conditions [20]. For the calculations, the yield stress in Table 1 and the K_C values in Table 2 were used. Micrographs of the corresponding unfatigued specimens of both notch conditions were used to assess the size of the plastic zone. For these specimens the measured *d* vales are reported in Table 3. They indicate the presence of a lower plastic zone than the ones estimated with Equation 6. Hence, at least for the monotonous tests without fatigue pre-load pure plane strain conditions can be assumed. For the specimens with fatigue pre-load this analysis could not be conducted because in this case the plastic zone dimensions were strongly influenced by the fatigue process.

$$r_{plane\ strain} = \frac{1}{6\pi}\left(\frac{K_C}{\sigma_Y}\right)^2 \tag{4}$$

$$r_{plane\ stress} = \frac{1}{2\pi}\left(\frac{K_C}{\sigma_Y}\right)^2 \tag{5}$$

$$d = 2r \tag{6}$$

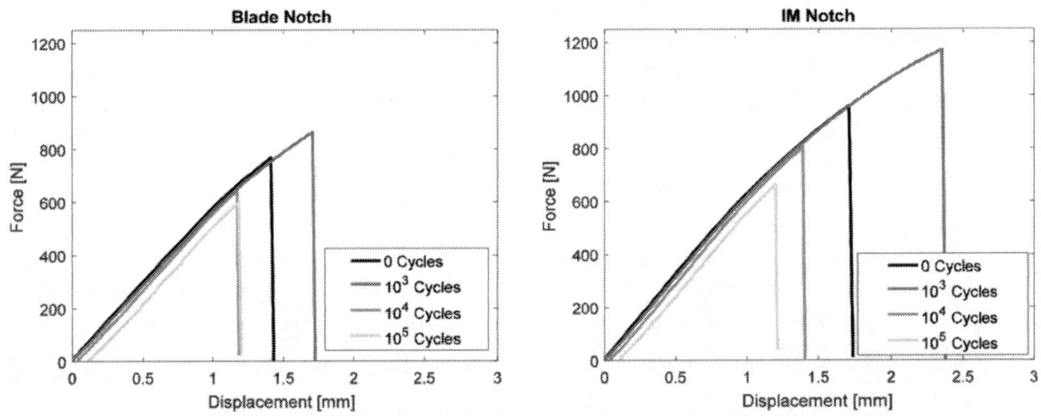

Figure 4: Representative force-displacement curves of Blade Notch (left) and IM Notch (right) specimens for each testing condition

For the sake of completeness, the dynamic compliance and hysteresis curves of a representative specimen fatigued for 10^5 cycles are presented in Figure 5 and 6 for each notch type used. For the Blade Notch specimen C_{Dyn} decreases slightly until right after $2x10^4$ cycles followed by a plateau and a sudden jump. Regarding the IM Notch specimen, a slight decrease in the compliance is also visible followed by several stair-like jumps. Berer et al., [9] studied fatigue crack growth of POM using the dynamic compliance method. They noticed a decrease in the initial portion of the dynamic compliance followed by a small jump for blade notched CT specimens. The authors attributed this to the formation of a craze zone and the subsequent drawing of fibrils which made the material stiffer thus reducing the compliance of the specimen. As for the following jump, it was attributed to the fracture of fatigued fibrils and to the initiation of crack growth. It is worth noting that in the present study cyclic fatigue was performed until a maximum of $1x10^5$ cycles, which, for the experimental setup used, represents a region slightly after crack growth initiation. Thus only slight changes are visible in the specimens' compliances shown in Figure 5. For comparison purposes the complete dynamic compliance curve of a cyclic fatigue test conducted until fracture is included in Figure 5 for a Blade Notch sample. From the insert in Figure 5 it can be seen that the maximum number of cycles (10^5) used for the fatigue pre-loading represents about one-third of the fatigue life time for the chosen setup and load level (F_{max} 450 N / F_{min} 45 N).

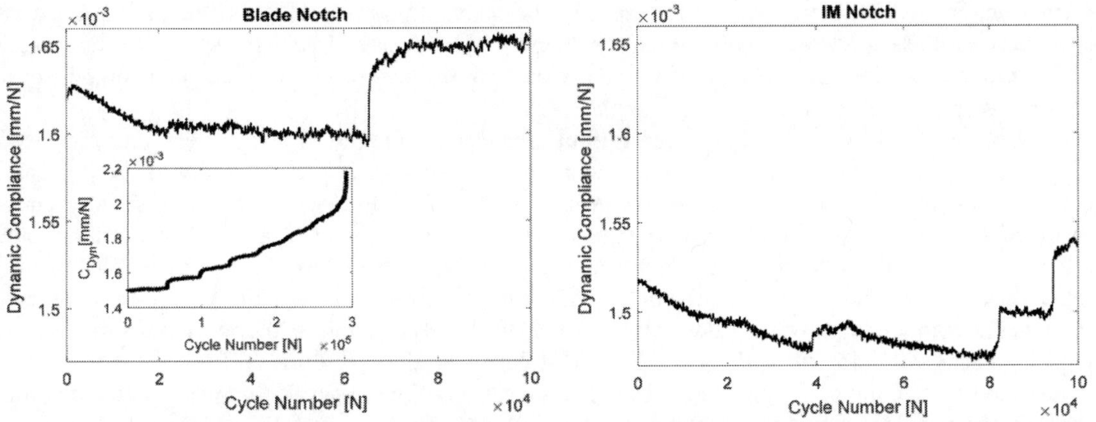

Figure 5: Dynamic compliance curve for the Blade Notch (left) and IM Notch (right) specimens with 10^5 cycles of fatigue pre-load (the insert in the plot to the left shows a complete fatigue fracture test with identical setup and load level)

The hysteresis curves shown in Figure 6 are very similar in shape but shifted rightwards with increasing number of fatigue cycles. The similarity in shape indicates that the dynamic mechanical behavior of the specimen is quite similar for the three different numbers of cycles used for the fatigue pre-load. The shift in displacement with cycle number can be attributed to the viscoelastic deformation (creep) coupled with crack growth for the highest cycle number shown. The latter is also indicated by a small decrease in the slope of the corresponding hysteresis curves. The hysteresis data were also used to split the dynamic compliance into its storage and loss parts as well as to calculate the loss factor. However, no significant differences could be detected and thus the results are not shown here.

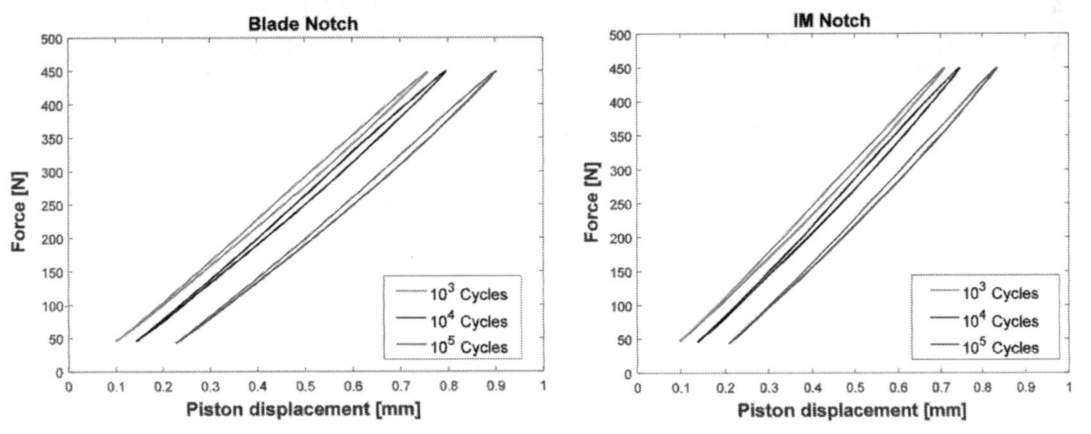

Figure 6: Hysteresis loops of Blade Notch (left) and IM Notch (right) specimens during the fatigue pre-load up to 10^5 cycles

Images of the fracture surfaces obtained by light microscopy are presented in Figure 7. For each testing condition a representative specimen was selected. It can be seen that all micrographs, depending on the amount of cyclic fatigue, present some level of stress whitening. From 0 to 10^4 cycles of fatigue pre-load all images show the same basic features. On the micrographs four distinct regions can be observed for the Blade Notch specimens and three for the IM Notch specimens (numbered from left to right). For the Blade Notch specimens, region one and two represent the notch from the manufacturing process and the one introduced by the blade, respectively. Region three will be termed as "plastic zone"

IOP Conf. Series: Journal of Physics: Conf. Series **843** (2017) 012052 doi:10.1088/1742-6596/843/1/012052

in the following discussion. It is formed by the coupled effect of cyclic fatigue and monotonic loading. Finally region four corresponds to spontaneous crack growth at the end of the monotonic tests. The surface of this region indicates that the specimen broke in a brittle manner without further plastic deformation. For the IM Notch specimens, region one represents the notch from the molding process. Region two and three are homologous to regions 3 and 4 of the Blade Notch specimens. Comparing the samples with and without fatigue pre-load it is obvious that the size of the plastic zone is bigger for the specimens with fatigue pre-load.

For the specimens with 10^5 cycles of fatigue pre-load, four and five different regions can be distinguished for the blunt and sharp notched specimens, respectively. Additionally to the previously discussed ones, the plastic zone is divided into two parts. The boundary between these two parts is interpreted as a discontinuous crack growth band. These are formed when a fatigue crack remains static for some period of cyclic loading in which a craze at the crack tip grows and matures. With further increase in cycle number, a sudden advance in the crack length occurs and the procedure is repeated until either unstable crack growth takes place or the mechanism changes to continuous crack growth [21, 22]. Images very similar to the ones presented here can also be found in [9], where the authors attributed the white color of the plastic zones to the formation of micro-voids that were generated during craze formation. This is supported by Plummer et al [23, 24] who observed that the yielding behavior in bulk POM was accompanied by stress whitening and cavitation.

Based on the compliance curves and the light microscope images it is interpreted that the increase in K_C during the first 1000 cycles is caused by the formation of a distinct craze zone with fibrils, which draws material from the polymer matrix resulting in a localized strain hardening. After a certain number of cycles the strain hardening effect produced by the fibril formation and orientation is counterpoised by the coalescence of voids (which were nucleated during the fibril formation) and by disentanglement and / or fracture of highly stressed fibrils [22, 23].

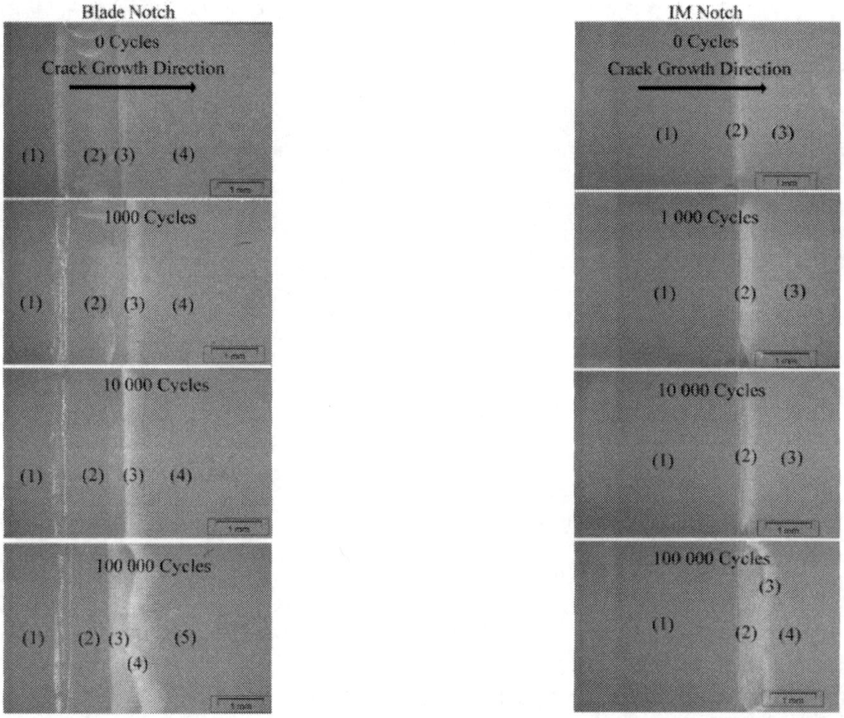

Figure 7: Sequence of fracture surfaces for both Blade Notch and IM-Notch specimens obtained for different levels of fatigue pre-load conditions

4. Conclusions

The effect of cyclic fatigue on the fracture toughness of Polyoxymethylene was investigated. The results were discussed in light of previous findings available in the literature. It was observed, within the experimental conditions used, that the critical stress intensity factor increases at low fatigue cycle numbers but strongly decreases with further fatigue loading. Supported by micrographs and plots of the dynamic specimen compliance, this was attributed to the formation and fatigue of a craze zone ahead of the crack tip. For low numbers of fatigue cycles the dominant effect was local strain hardening produced by drawing and orientation of fibrils. With further increasing fatigue the coalescence of voids and the breakdown of fibrils in the craze zone caused the critical stress intensity factor and thus the fracture toughness of the material to significantly decrease. Along the performed measurements a variation in the stress intensity factor of nearly 50% for the blade notched specimens and of 70% for the blunt notched specimens was observed.

The conclusion for the component is that the design of notched areas has to take into account the reduction of the fracture toughness in service applications when fatigue loading is present. Currently this mainly affects thin walled parts in combination with the material (Polyoxymethylene homopolymer) examined in this study. However, the formation and fatigue of craze zones is not an effect which was only observed for Polyoxymethylene. This effect is present in all polymers which show discontinuous crack growth bands in fracture mechanical fatigue tests [22].

Acknowledgements

The research work of this article was performed at the Polymer Competence Center Leoben GmbH (PCCL, Austria) within the framework of the COMET-programme of the Austrian Ministry of Traffic, Innovation and Technology with contributions by the Montanuniversität Leoben (Materials Science and Testing of Plastics). The PCCL is funded by the Austrian Government and the State Governments of Styria and Upper Austria.

References

[1] S. Lüftl, P. M. Visakh, and S. Chandran, 2014, Polyoxymethylene Handbook : Structure, Properties, Applications and their Nanocomposites: Wiley pp.153-163.
[2] M. Berer and Z. Major, 2010, "Characterization of the global deformation behaviour of engineering plastics rolls," *Int. J. Mech. Mater. Des.*, **6**, pp. 1–9.
[3] M. Berer and Z. Major, 2012, "Characterisation of the Local Deformation Behaviour of Engineering Plastics Rolls," *Strain*, **48**, pp. 225–234.
[4] Heym B. and Beitz W., 1995, "Zur Belastbarkeit von Stirnzahnrädern aus dem Hochtemperatur-Thermoplast PEEK," *Konstruktion*, **47**, pp. 351–357.
[5] J. Rösler, 2005, "Zur Tragfähigkeitssteigerung thermoplastischer Zahnräder mit Füllstoffen," Dissertation, Fakultät V der Technische Universität Berlin, Berlin, Germany.
[6] J. A. Sauer and G. C. Richardson, 1980, "Fatigue of polymers," *Int. Journ. of Fracture*, **16**, pp. 499–532.
[7] C. J. G. Plummer, P. Scaramuzzino, H.-H. Kausch, and J.-M. Philippoz, 2000, "High temperature slow crack growth in polyoxymethylene," *Polym. Eng. Sci.*, **40**, pp. 1306–1317.
[8] C. J. G. Plummer, P. Scaramuzzino, R. Steinberger, R. W. Lang, and H.-H. Kausch, 2000, "Application of the essential work of fracture concept to high temperature deformation in polyoxymethylene," *Polym. Eng. Sci.*, **40**, pp. 985–991.
[9] M. Berer, G. Pinter, and M. Feuchter, 2014, "Fracture mechanical analysis of two commercial polyoxymethylene homopolymer resins," *J. Appl. Polym. Sci*, **131**, 40831.
[10] M. Berer and G. Pinter, 2013, "Determination of crack growth kinetics in non-reinforced semi-crystalline thermoplastics using the linear elastic fracture mechanics (LEFM) approach," *Polym. Test.*, **32**, pp. 870–879.

[11] F. Arbeiter, B. Schrittesser, A. Frank, M. Berer, and G. Pinter, 2015, "Cyclic tests on cracked round bars as a quick tool to assess the long term behaviour of thermoplastics and elastomers," *Polym. Test.*, **45**, pp. 83–92.
[12] A. Lazzeri, A. Marchetti, and G. Levita, 1997, "Fatigue and Fracture in Polyacetal Resins," *Fatigue Fract. Eng. Mater. Struct*, **20**, pp. 1207–1216.
[13] A. J. Lesser, 1995, "Changes in Mechanical Behavior During Fatigue of Semicrystalline Thermoplastics," *J. Appl. Polym. Sci*, **58**, pp. 869–879.
[14] S. Bandyopadhyay, J. R. Roseblade, and R. Muscat, 1993, "Fracture study of a polyacetal resin," *J. Mater. Sci.*, **28**, pp. 3777–3782.
[15] X. Li, H. A. Hristov, A. F. Yee, and D. W. Gidley, 1995, "Influence of cyclic fatigue on the mechanical properties of amorphous polycarbonate," *Polymer*, **36**, pp. 759–765.
[16] Q.-Z. Fang, T. J. Wang, H. G. Beom, and H. M. Li, 2008, "Effect of cyclic loading on tensile properties of PC and PC/ABS," *Polym. Degrad. Stab.*, **93**, pp. 1422–1432.
[17] M. Berer, Z. Major, G. Pinter, D. M. Constantinescu, and L. Marsavina, 2014, "Investigation of the dynamic mechanical behavior of polyetheretherketone (PEEK) in the high stress tensile regime," *Mech Time-Depend Mater*, **18**, pp. 663–684.
[18] ISO 13586, 2000, Plastics — Determination of fracture toughness (G_{IC} and K_{IC}) — Linear elastic fracture mechanics (LEFM) approach.
[19] M. Berer, D. Tscharnuter, and G. Pinter, 2015, "Dynamic mechanical response of polyetheretherketone (PEEK) exposed to cyclic loads in the high stress tensile regime," *Int. J. Fatigue*, **80**, pp. 397–405.
[20] H. Tada, P. C. Paris, and G. R. Irwin, 2000, The stress analysis of cracks handbook, 3rd ed. New York: ASME Press; London : Professional Engineering Publishing pp.16.
[21] P. E. Bretz, R. W. Hertzberg, and J. Manson, 1982, "The Effect of Molecular Weight on Fatigue Crack Propagation in Nylon 66 and Polyacetal," *J. Appl. Polym. Sci*, **27**, pp. 1707–1717.
[22] R. W. Hertzberg and J. A. Manson, 1980, Fatigue of Engineering Plastics: Academic Press pp.146-183.
[23] C. J. G. Plummer, N. Cudré-Mauroux, and H.-H. Kausch, 1994, "Deformation and Entanglement in Semicrystalline Polymers," *Polym. Eng. Sci.*, **34**, pp. 318–329.
[24] C. J. G. Plummer, P. Menu, N. Cudré-Mauroux, and H.-H. Kausch, 1995, "The Effect of Crystallization Conditions on the Properties of Polyoxymethylene," *J. Appl. Polym. Sci*, **55**, pp. 489–500.

IOP Conf. Series: Journal of Physics: Conf. Series **843** (2017) 012053 doi:10.1088/1742-6596/843/1/012053

Minimum shear stress range: a criterion for crack path determination

K. Pereira, M. Abdel Wahab

Department of Electrical energy, Systems and Automation, Ghent University, Belgium

E-mail: Magd.AbdelWahab@UGent.be

Abstract. For problems under proportional mixed-mode conditions, various criteria are used to predict fatigue crack growth directions, most achieving reasonable accuracy. The crack propagation angle is often obtained by maximizing a quantity (for instance, energy or stresses) as function of the stress intensity factors K_I and K_{II}. This maximization is generally performed at the instant of maximum fatigue loading and a stress analysis at this instant is sufficient to predict the crack propagation angle and thus the fatigue crack growth direction. However, under non-proportional loading, the maximum values of K_I and K_{II} may occur at different instants of the fatigue cycle and so a simple analysis at the maximum loading instant is not appropriate; it is necessary to consider the entire loading cycle history. One possible criterion to treat problems under these circumstances is the minimum shear stress range criterion (MSSR). This paper presents a brief discussion of the most common criteria used for determination of crack propagation direction, focusing on an implementation of MSSR. Its performance is assessed in different conditions and the results are compared to literature data.

1. Introduction

Fracture mechanics has been widely used to predict crack growth of problems under mode I loading conditions, implying that propagation is co-planar and perpendicular to the loading direction [1]. However, most engineering applications are subjected to a combination of normal and shear loading (mixed-mode I and II conditions) and a definition of a propagation criterion that takes into account this loading scheme is necessary. There are several crack propagation direction criteria available in the literature that are used for proportional mixed-mode fatigue. A summary on the subject can be found in Rozumek and Mancha [2]. Here, we focus on the most widely used criteria: the maximum tangential stress (MTS) criterion and the maximum energy release rate (MERR) criterion.

The MTS was originally proposed by Erdogan and Sih [3] in the early 60s. It states that crack growth is in the radial direction θ_p of greatest tension, i. e., in the direction along which the tangential stress $\sigma_{\theta\theta}$ is maximized and exceeds a critical value σ_c (a material property). Considering the analytical expressions for the stress field near the crack tip, the tangential stress can be written as function of the stress intensity factors K_I and K_{II} as

$$\sigma_{\theta\theta} = \frac{1}{\sqrt{2\pi r}}\cos\frac{\theta}{2}\left(K_I\cos^2\frac{\theta}{2} - \frac{3}{2}K_{II}\sin^2\theta\right) \tag{1}$$

where r and θ are cylindrical coordinates as represented in **Error! Reference source not found.**. Imposing the conditions $\frac{\partial\sigma_{\theta\theta}}{\partial\theta} = 0$ and $\frac{\partial^2\sigma_{\theta\theta}}{\partial\theta^2} = 0$, the direction of propagation θ_p can be obtained as function of K_I and K_{II} as:

$$\theta_P = \cos^{-1}\left(\frac{3K_{II}^2 + \sqrt{K_I^4 + 8K_I^2 K_{II}^2}}{K_I^2 + 9K_{II}^2}\right) \tag{2}$$

The MERR was introduced independently by Hussain et al. [4] and Palaniswamy et al. [5]. It is based on the following hypotheses: crack will propagate at the crack tip in a radial direction θ_p along which the energy release rate $G(\theta)$ is maximized and exceeds the critical level G_c (a material property). For a co-planar crack growth under mixed mode, $G(\theta)$ can be written as

$$G(\theta) = \frac{1+\kappa}{8\mu}\left(K_I^2(\theta) + K_{II}^2(\theta)\right) \tag{3}$$

where κ is a function of the Poisson's ratio and the stress state (plane stress or plane strain), μ is the shear modulus and $K_I(\theta)$ and $K_{II}(\theta)$ are the stress intensity factors associated to the new crack tip at a branched crack from the original crack.

Based on the hypothesis of this criterion, the propagation direction can be predicted by the solution:

$$\left.\frac{\partial G(\theta)}{\partial \theta}\right|_{\theta=\theta_P} = 0 \tag{4}$$

$$\left.\frac{\partial^2 G(\theta)}{\partial \theta^2}\right|_{\theta=\theta_P} \leq 0 \tag{5}$$

Both criteria have been widely used for prediction of propagation direction with reasonable accuracy [6-9]. However, they fail to predict correct paths under non-proportional loading [10-12]. This inconsistency is that, for the case of fatigue under constant amplitude proportional loading conditions, the size of Mohr's circle for a two-dimensional stress state varies in time, as the amplitude of the applied stress also varies in time, but the principal stresses remain proportional to each other and, consequently, the angle of the principal axes remains constant throughout the fatigue cycle. This is not the case for non-proportional loading. Here, the directions of principal stresses rotate and, generally, the ratio between the principal stresses also vary within the cycle [13]. As a consequence, the maximum values of stress intensity factors K_I and K_{II} may occur at different points during the cycle [14], so it is not obvious a priori which points in a loading cycle will maximize the orientation criteria, increasing the difficulty to define crack propagation direction. This non-proportional loading causes extra complexity when modelling crack propagation phase as it invalidates the application of conventional linear elastic fracture mechanics orientation criteria.

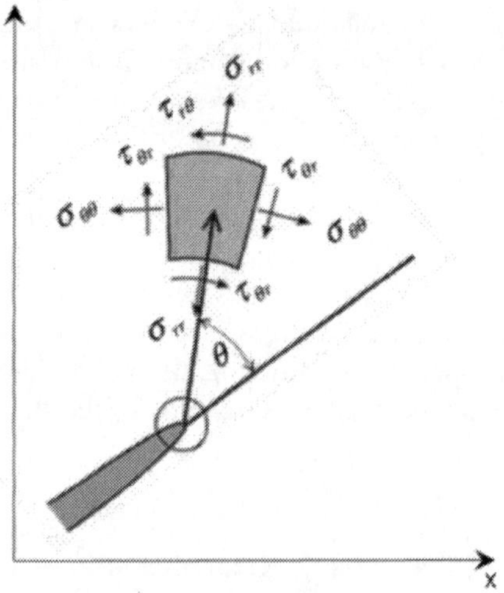

Figure 1: Stresses at a fixed position ahead of crack tip in cylindrical coordinates (based on [15]).

The focus of this paper is to discuss one possibility to deal with problems under non-proportional loading: the minimum shear stress range criteria (MSSR) [11]. Its implementation and performance under different fatigue problems is the main goal of this paper. To this end, we first present a discussion on the minimum shear stress range criteria and its implementation in section **Error! Bookmark not defined.**. Then, the criterion performance is accessed in different applications and the results will be compared with literature data.

2. A minimum shear stress range criterion

The minimum shear stress range criterion was proposed by Giner et al. [11]. As discussed in the paper, it can be seen as a generalization, for non-proportional loading conditions, of the "criterion of local symmetry", which states that crack will propagate in the direction that causes K_{II} to vanish. In general, for non-proportional conditions, K_{II} will not be zero at any direction and, therefore, it is logical to search for a direction that minimizes the range ΔK_{II} over the loading cycle.

Giner et al. [11] also mention that, in some conditions, obtaining the stress intensity factor K_{II} can be computationally expensive and also not very accurate. In order to circumvent this problem, they propose to search for the direction that minimizes the shear stress range $\Delta\tau_{r\theta}$ at the crack tip. From the two orthogonal planes on which $\Delta\tau_{r\theta}$ is minimized, the propagation direction is chosen as the plane with the maximum variation of the range $\Delta\sigma_{\theta\theta}$. This is justified by the fact that, under this plane, less frictional energy will be dissipated and more energy would be available for propagating the crack.

The authors further implemented the MSSRC in a XFEM framework. Here, we adapted their criterion for a conventional finite element (FE) analysis in the following way. Firstly, a local cylindrical coordinate system (r,θ) was defined at the crack tip and the stress results from the FE analysis were transformed from the Cartesian (x,y) to this local coordinate system. Then, as showed in **Error! Reference source not found.**, a circular path ahead of crack tip of radius R and centered at the crack tip was defined, with θ varying between -90° and +90°. The stresses at this fixed path were stored for each time increment in the loading cycle. These results were later used to obtain maximum and minimum envelopes and also the range of shear stress $\tau_{r\theta}$ and normal stress $\sigma_{\theta\theta}$ as function of θ. The propagation angle θ_p was then defined as the direction with the minimum $\Delta\tau_{r\theta}$ and with the highest value of $\Delta\sigma_{\theta\theta}$.

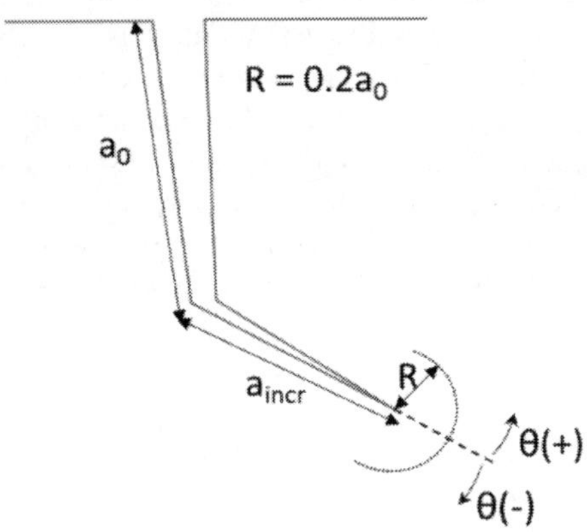

Figure 2: Circular path ahead of crack tip

A python script was created to post-process the results from the analysis and to define the propagation angle as discussed above. Figure 3 shows a flow chart summarizing the procedure adopted in this paper.

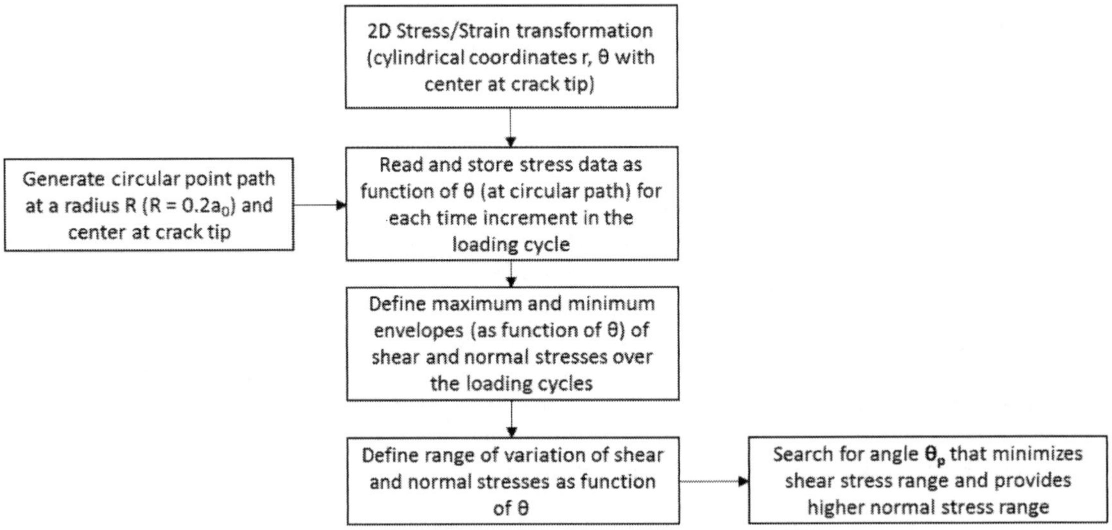

Figure 3: Flow chart of the implementation of minimum shear stress range criterion

3. Results

The performance of the minimum shear stress range criterion was checked under different conditions: proportional and non-proportional mixed-mode fatigue. For each situation sensibility checks were performed to analyse how different parameters, such as the path radius R, mesh size at crack tip and propagation increment length a_{inc}, influenced the propagation path.

The crack propagation was modelled using the conventional Finite Element Method (FEM) with remeshing technique. FEM has been widely used in the literature for many engineering applications that have been recently published [16-29]. The crack was simulated using a "seam" in ABAQUS®, a region in of the model that can open during analysis. The crack faces interaction may impact the results, especially in case under negative cyclic loading ratio. As the contact algorithm in ABAQUS® provides more accurate results with linear elements, all models were meshed using a 2D quadrilateral, 4-node (bilinear) elements and the general contact algorithm was used to define the contact interaction between the crack faces. Lagrange multiplier formulation was used to define the tangential behavior and a hard contact approach was used to define the normal behavior of the contact pair. In order to capture singularity, the crack tip was meshed using a ring of collapsed linear quadrilateral elements. A stepwise analysis was done and the whole model was remeshed after advancing the crack in each increment of crack propagation.

3.1. Proportional mixed-mode fatigue

A numerical prediction of a mixed-mode fatigue crack growth in a plane elastic plate was performed in ABAQUS®. The model consisted in a plate with a centre slant crack, as shown in **Figure 4**. The dimension and material properties are based on the model proposed by Yan [30] and are summarized in **Table 1**.

Each step of the propagation simulation consisted in the analysis of a full loading cycle, with stress σ applied in three loading steps varying from maximum, mean and minimum values (obtained from R_{cycle} and σ_m in **Table 1**). For the static analysis, each loading step was divided into 10 time increments and the solution for each of those increments was later used in the calculation of the propagation direction, as explained in section **Error! Reference source not found.**. The propagation path obtained using the minimum shear stress range criterion was compared with the predicted path using MTS criterion [30] and it is presented in **Figure 5**(a). Both criteria predict the same path, implying that MSSR criterion can also be a good choice under proportional mixed-mode conditions.

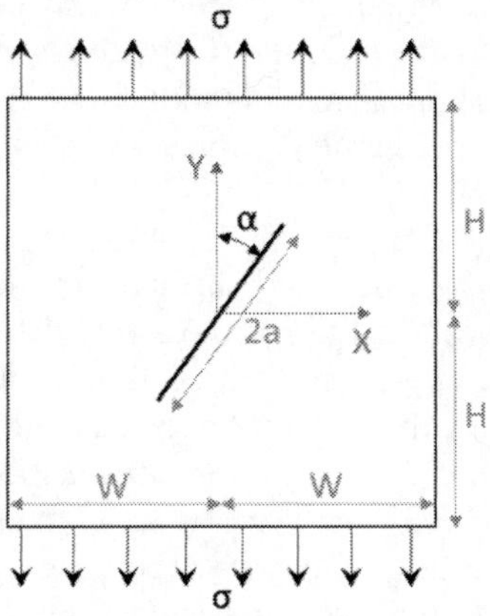

Figure 4: Slant centre crack model

Table 1: Model dimension and material properties (data from [30]).

Parameter	Value
Crack length, a	7mm
Plate half-height, H	17.5mm
Plate half-width, W	17.5mm
Slant crack angle, α	30°
Shear modulus, G	2744 kgf/mm^2
Poisson's ratio, ν	0.321
Mean stress, σ_m	15.33 kgf/mm^2
Characteristic of cyclic loading ratio, R_{cycle}	0.048

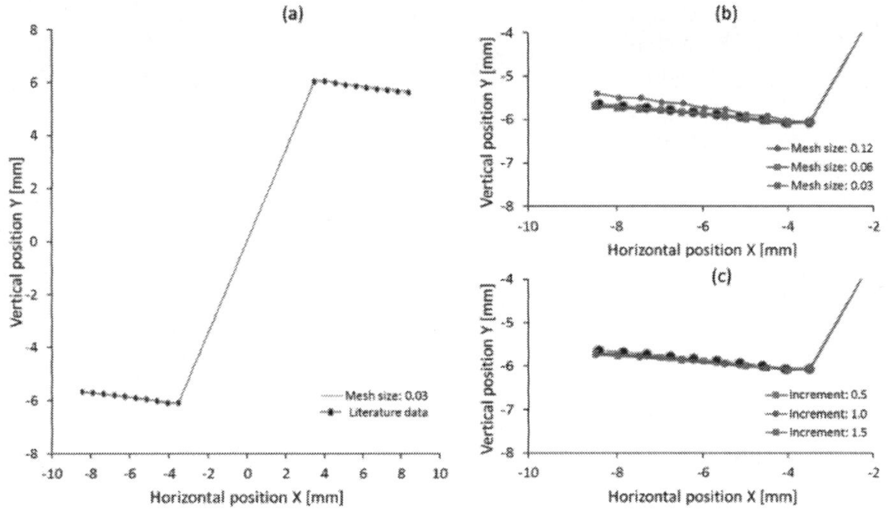

Figure 5: Results obtained using minimum shear stress range criterion: (a) Comparison of predicted crack propagation path with literature data from [30], (b) Sensitivity check: crack tip mesh size impact on predicted propagation paths, (c) Sensitivity check: impact of propagation increment length a_{inc} on predicted propagation paths

A sensibility check was also performed and the results are presented in **Figure 5**(b) and **Figure 5**(c). The mesh size has an influence on the results. However, as the size reduces, the analysis can capture more accurately the stress singularity at crack tip and the results converge to a unique path. In addition, the propagation increment length a_{inc} had insignificant influence on the prediction of path.

3.1.1. Non-proportional mixed-mode fatigue.
Fretting is a surface damage due to a very small relative slip between two surfaces and can lead to fretting wear [31-34] or fretting fatigue [35-39]. Under fretting conditions, the stress field near the contact region is non-proportional, even if the external loads are applied in a proportional way [40, 41]. This is caused by the non-linear characteristic of friction at contact interface between pad and specimen.

In order to study the performance of the MSSR criterion under non-proportional mixed-mode fatigue, the same elastic fretting model presented by Giner et al. [11] was analysed using conventional FE framework. The model details such as geometry, material properties, boundary conditions and loading history are presented in **Figure 6**. The model was composed of only two parts: a pad and a specimen, which represents a quarter of the experimental set-up, due to its symmetry. In this set-up, two flat pads are maintained in contact with a flat specimen through the application of a constant clamping or normal force F. The specimen is fixed at one end and the other end is subjected to an oscillatory bulk stress σ_{axial}. The master-slave algorithm in ABAQUS® was used to describe the contact and the Lagrange multiplier formulation was used to define the tangential behaviour of the contact interface between pad and specimen, with a coefficient of friction of 0.8. The surface to surface and finite sliding options were used to define the contact interaction. A 2D quadrilateral, 4-node (bilinear), plane strain, reduced integration element (CPE4R) was used to mesh the model with a ring of collapsed elements at crack tip.

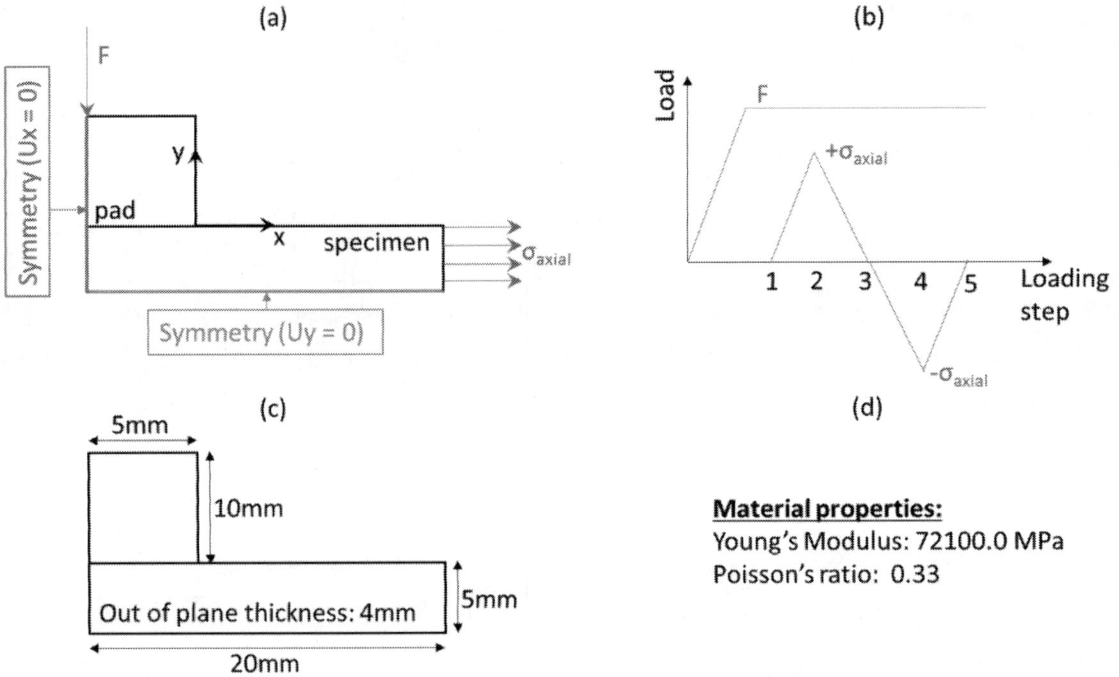

Figure 6: Fretting fatigue model details, based on the model from Giner et al. [11]: (a) Boundary conditions, (b) Cyclic loading steps of one full fretting cycle, (c) Model dimensions and (d) material properties

The initial crack was inserted at the contact edge in the same way as described by Giner et al. [11] (at an angle of -120° with horizontal x direction and with and initial length of 50μm). An analysis with 15 crack increments with fixed length a_{inc} equal to 50μm was performed. Each step of the propagation simulation consisted in the analysis of a full fretting loading cycle. This fretting cycle was divided in five loading steps (see Figure 6(b)). In the first loading step, the top pad was pressed against the specimen surface by a normal load F = 100N/mm and this compressed condition was held constant until the end of the cycle. Then, an oscillatory axial stress σ_{axial}=±110MPa was applied to the side of the specimen. The contact between the crack faces were also modelled using the general contact algorithm in ABAQUS® and a coefficient of friction of 0.8 was used at this interface.

Figure 7(a) shows experimental data from Giner et al. [11] that was then, in Figure 7(b), compared with the predicted path obtained when using MSSR and MTS criteria (in order to extract experimental data from the picture, the software Web Plot Digitizer [42] was used). One can conclude that the predicted path obtained by MSSR is in good agreement with the experimental observations, but that is not the case for MTS criterion. As also mentioned by Giner et al. [11], it is important to notice that the use of MTS criterion simplifies the problem because, for its computation, it is only necessary to consider the finite element results at the instant of maximum σ_{axial}, while computing the MSSR the entire loading cycle is considered. Therefore, this criterion does not provide correct crack propagation as it neglects the effect that the rest of the fretting cycle may have on the crack propagation direction.

A sensibility check was performed in order to check the influence of the path radius R, mesh size at crack tip and propagation increment length a_{inc} on the crack path prediction. The results are presented in Figure 8. It can be noticed that the path radius and the propagation increment length had no significant impact on the crack path prediction. However, as expected, the mesh size at crack tip elements can affect considerably the propagation path. As mentioned in section **Error! Reference source not found.**, in order to accurately predict the stresses field ahead of the crack tip, it is required a very fine mesh size at this region. Therefore, only for very small mesh sizes, the criterion can correctly predict the crack propagation behaviour.

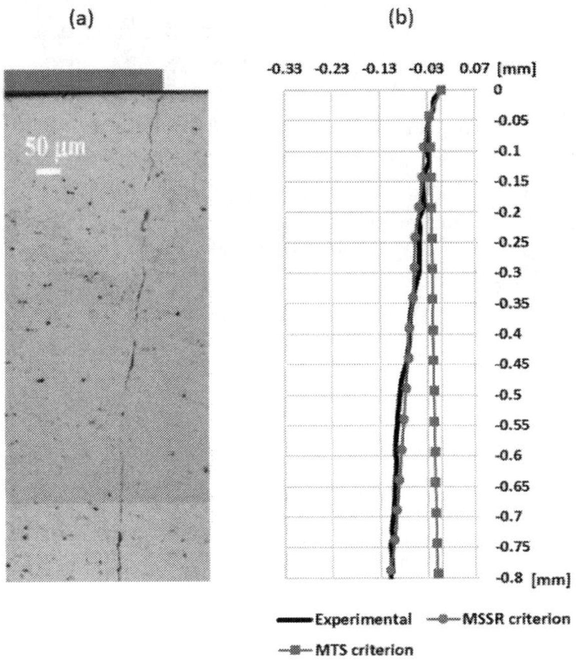

Figure 7: Fretting fatigue crack propagation path: (a) Experimental data from Giner et al. [11], (b) Comparison of predicted path using different criteria and experimental data

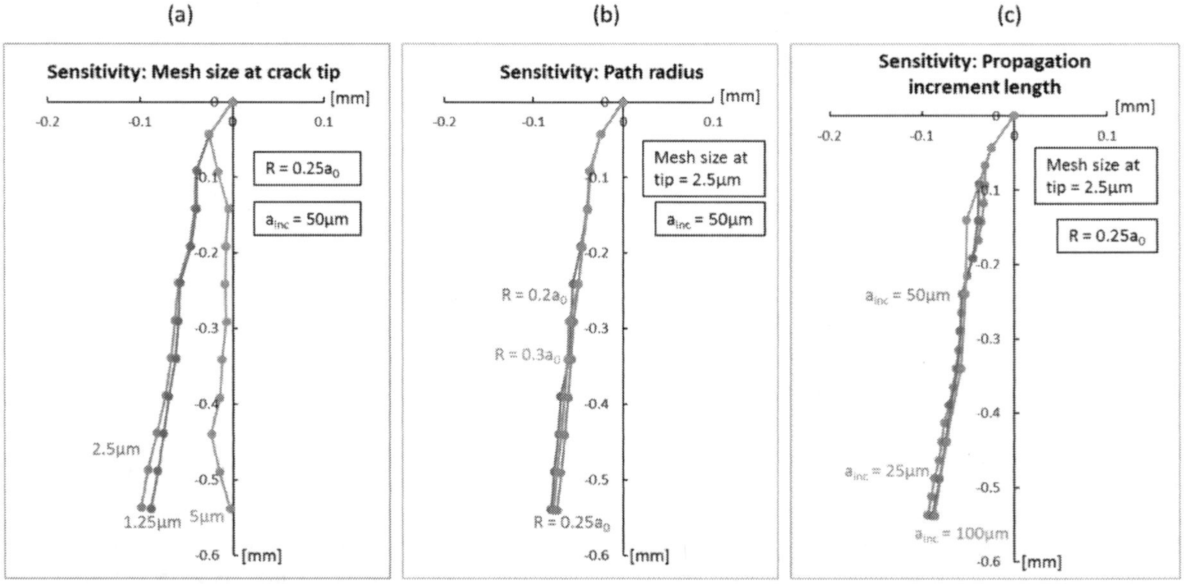

Figure 8: Sensitivity check: (a) crack tip mesh size impact on predicted propagation paths, (b) influence of different path radius and (c) impact of propagation increment length a_{inc} on predicted propagation paths

4. Conclusions

This paper focused on the discussion of the minimum shear stress range criterion and its applicability to different fatigue problems. Under proportional loading conditions, the MSSR provided the same results as MTS and both of them seem to correlate well with experimental data. In addition, the MSSR seems to be a good and simple alternative to deal with fatigue problems, not only under proportional, but also under non-proportional conditions. The performance of MSSR was also verified for a non-proportional loading scenario, under fretting fatigue condition. As showed by our results, this type of problem invalidates the application of conventional orientation criteria, such as MTS, but the MSSR criterion is capable of capturing the main characteristics of the crack path, providing a prediction that correlates well with experimental data. As future work, we intend to apply the discussed methodology in fretting simulations for different pad geometries and validate the results with experimental data.

It is also important to mention that the final results depend on the mesh size at crack tip. It requires a very fine mesh in order to accurately predict paths using MSSR criteria and a mesh refinement study is therefore also recommended.

Acknowledgements: The authors would like to acknowledge the financial support of the Research Foundation-Flanders (FWO), The Luxembourg National Research Fund (FNR) and Slovenian Research Agency (ARRS) in the framework of the FWO Lead Agency project G018916N ‘Multi-analysis of fretting fatigue using physical and virtual experiments’.

References

[1] Plank R and Kuhn G 1999 *Engineering Fracture Mechanics* **62**(2-3) 203-229
[2] Rozumek D and Macha E 2009 *Materials science* **45**(2) 190-210
[3] Erdogan F and Sih G 1963 *Journal of basic engineering* **85**(4) 519-527
[4] Hussain M, Pu S and Underwood J 1974 Fracture Analysis: Proceedings of the 1973 National Symposium on Fracture Mechanics, Part II ed editors) ASTM International
[5] Palaniswamy K and Knauss W 1978 *Mechanics today* **4**(30) 87-148
[6] Boljanović S and Maksimović S 2011 *Engineering Fracture Mechanics* **78**(8) 1565-1576

[7] Mageed AMA and Pandey RK 1992 *International Journal of Fatigue* **14**(1) 21-29
[8] Richard H, Fulland M and Sander M 2005 *Fatigue & fracture of engineering materials & structures* **28**(1-2) 3-12
[9] Miranda ACO, Meggiolaro MA, Castro JTP, Martha LF and Bittencourt TN 2003 *Engineering Fracture Mechanics* **70**(10) 1259-1279
[10] Hojjati-Talemi R. Numerical Modelling Techniques for Fretting Fatigue Crack Initiation and Propagation: Ghent University.
[11] Giner E, Sabsabi M, Rodenas JJ and Fuenmayor FJ 2014 *International Journal of Fatigue* **58** 172-180
[12] Giner EaSNaDFDaFFJ 2008 *International Journal of Solids and Structures* **45**(22) 5675--5687
[13] Dahlin P. Growth of fatigue cracks subjected to non-proportional mode I and II: KTH; 2005.
[14] Madge JJ. Numerical modelling of the effect of fretting wear on fretting fatigue: University of Nottingham; 2009.
[15] Kuna M 2013 Finite elements in fracture mechanics ed editors. Dordrecht: Springer
[16] X. Nguyen H, N. Nguyen T, Abdel Wahab M, Bordas SPA, Nguyen-Xuan H and P. Voa T 2017 *Computer Methods in Applied Mechanics and Engineering* **313** 904-940
[17] Tran Vinh L, Lee J, Nguyen-Van H, Nguyen-Xuan H and Abdel Wahab M 2015 *International Journal of Non-Linear Mechanics* **72** 42-52
[18] Tran LV, Phung-Van P, Lee J, Wahab MA and Nguyen-Xuan H 2016 *Composite Structures* **140** 655-667
[19] Thai CH, Ferreira AJM, Abdel Wahab M and Nguyen-Xuan H 2016 *Acta Mechanica* **227**(5) 1225-1250
[20] Thai C, Zenkour AM, Abdel Wahab M and Nguyen-Xuan H 2016 *Composite Structures* **139** 77-95
[21] Phung-Van P, Tran LV, Ferreira AJM, Nguyen-Xuan H and Abdel-Wahab M 2016 *Nonlinear Dynamics* 1-16; doi:10.1007/s11071-11016-13085-11076
[22] Phung Van P, Nguyen LB, Tran Vinh L, Dinh TD, Thai CH, Bordas SPA, Abdel Wahab M and Nguyen-Xuan H 2015 *International Journal of Non-Linear Mechanics* **76** 190-202
[23] Phung Van P, De Lorenzis L, Thai CH, Abdel Wahab M and Nguyen-Xuan H 2015 *Computational Materials Science* **96** 495-505
[24] Phung Van P, Abdel Wahab M, Liew KM, Bordas SPA and Nguyen-Xuan H 2015 *Composite Structures* **123** 137-149
[25] Noda N-A, Chen X, Sano Y, Wahab MA, Maruyama H, Fujisawa R and Takase Y 2016 *MATERIALS & DESIGN* **96** 476-489
[26] Tran Vinh L, Lee J, Ly HA, Abdel Wahab M and Nguyen-Xuan H 2015 *International Journal of Mechanical Sciences* **96-97** 65-78
[27] Junyan Ni and Wahab MA 2017 *Computers & Structures* **186** 35–49
[28] Phung-Van P, Qui LX, Nguyen-Xuan H and Wahab MA 2017 *Composite Structures* **166** 120–135
[29] Phung-Van P, Ferreira AJM, Nguyen-Xuan H and Abdel Wahab M 2017 *Composites Part B: Engineering* **118** 125-134
[30] Yan X 2010 *Fatigue & fracture of engineering materials & structures* **33**(6) 365-377
[31] Yue T and Abdel Wahab M 2017 *Tribology International* **107** 274-282
[32] Pereira K, Yue T and Abdel Wahab M 2017 *Tribology International* **110** 222-231
[33] Yue T and Abdel Wahab M 2016 *Materials* **9** 597; doi:510.3390/ma9070597
[34] Yue T and Abdel Wahab M 2014 *Wear* **321** 53-63
[35] Martínez JC, Vanegas Useche LV and Wahab MA 2017 *International Journal of Fatigue* **100, Part 1** 32-49
[36] Kumar D, Biswas R, Poh LH and Abdel Wahab M 2017 *Tribology International* **109** 124-132
[37] Bhatti NA and Abdel Wahab M 2017 *Tribology International* **109** 552-562

[38] Resende Pereira KdF, Bordas S, Tomar S, Trobec R, Depolli M, Kosec G and Abdel Wahab M 2016 *Materials* **9** 639; doi:610.3390/ma9080639
[39] Ferjaoui A, Yue T, Abdel Wahab M and Hojjati-Talemi R 2015 *International Journal of Fatigue* **73** 66-76
[40] Nowell D, Dini D and Hills DA 2006 *Engineering Fracture Mechanics* **73**(2) 207-222
[41] Tur M, Fuenmayor J, Rodenas J and Giner E 2003 *Finite Elements in Analysis and Design* **39**(10) 933-949
[42] Rohatgi A 2011 *URL http://arohatgi info/WebPlotDigitizer/app*

Numerical investigation of contact stresses for fretting fatigue damage initiation

N A Bhatti and M Abdel Wahab

Soete Laboratory, Faculty of Engineering and Architecture, Ghent University, Technologiepark Zwijnaarde 903, Zwijnaarde B-9052, Belgium

E-mail: magd.abdelwahab@ugent.be

Abstract. Fretting fatigue phenomena occurs due to interaction between contacting bodies under application of cyclic and normal loads. In addition to environmental conditions and material properties, the response at the contact interface highly depends on the combination of applied loads. High stress concentration is present at the contact interface, which can start the damage nucleation process. At the culmination of nucleation process several micro cracks are initiated, ultimately leading to the structural failure. In this study, effect of ratio of tangential to normal load on contact stresses, slip amplitude and damage initiation is studied using finite element analysis. The results are evaluated for Ruiz parameter as it involves the slip amplitude which in an important factor in fretting fatigue conditions. It is observed that tangential to normal load ratio influences the stick zone size and damage initiation life. Furthermore, it is observed that tensile stress is the most important factor that drives the damage initiation to failure for the cases where failure occurs predominantly in mode I manner.

1. Introduction

Fretting is a surface damage due to a very small relative slip between two surfaces and can lead to fretting wear [1-4] or fretting fatigue [5-9]. The failure process in fretting fatigue is usually characterized by the initiation phase and propagation phase. Latter it has more sophisticated theories to study the process, whereas the intricacies involve in modelling of initiation phase still hampers to develop the consent among the researchers on best damage model. The problem perhaps stays in defining the point where initiation phase has been ended. In this paper, the term damage nucleation is used as the process of damage accumulation and the term damage initiation is used for a particular instant where micro cracks have started to develop [10].

Many researchers have proposed various methods and damage models to find damage initiation location and initiation life. Based on experimental study, Ruiz et al. [11] proposed two parameters to define damage initiation. The first parameter is termed as damage parameter $F1$ which is obtained by multiplying relative slip and shear stress. The second parameter is termed as initiation parameter $F2$, which is obtained by multiplying tensile stress with $F1$ parameter. Later, other researchers also used Ruiz parameters to model initiation behaviour in fretting fatigue. Kuno et al. [12] related the initiation parameter to fretting fatigue life. Nowell and Hills [13] explained the contact size effect on fretting fatigue lives. According to them the fatigue lives are inversely proportional to the contact width, even by keeping the stress level constant. As slip amplitude in partial slip is proportional to contact width, therefore they concluded that fretting fatigue life is dependent on slip amplitude.

Szolwinski and Farris [14] proposed a method by combining contact mechanics approach with multiaxial fatigue life parameters. With this approach they have observed crack initiation location at the trailing edge of the contact. Later, they showed comparison of experimental nucleation lives correlating well with the predicted method for aluminium alloy structural components [15]. Iyer and Mall [16] analysed the effect of contact pressure and stress amplitude on fretting fatigue lives. Their study revealed that amplification in local stress range is partly due to the normal load and partly due to compressive stresses during unloading. Furthermore, they showed fretting fatigue life decreases with increasing contact pressure alone. Lykins et al. [17] studied fretting fatigue crack initiation life for titanium alloy. Crack initiation location and lives are evaluated using strain based parameters, critical plane models and Ruiz parameter. They observed the damage initiation location is near the trailing edge of the contact, which is in agreement with the experimental results, however the results showed no specific correlation between $F2$ parameter and cycles to damage initiation.

This paper mainly focuses on application of finite element methods to explore damage initiation in fretting fatigue condition. The first portion presents the effect of tangential to normal load ratio (Q/fP) on damage initiation location. In the second section the numerical results are combined with experimental results presented by Szolwinski and Farris [15], the relationship between initiation parameter and initiation life is studied. Also, the order of severity for internal stresses influencing the initiation life is analysed for the case where cracks grow under tensile stress or strain.

2. Experimental data and materials

Szolwinski and Farris [15] statistically designed experimental scheme to analyse the effect of different loads and pad radii to study fretting fatigue damage initiation. They studied contact under partial slip condition for the cylindrical pad and flat specimen. The normal load P is applied to maintain constant pressure. The cyclic axial load σ_A is applied to specimen and tangential load Q is generated through attached springs on the fretting pads. The configuration of experimental setup is shown in figure 1. Initiation life is determined from total number of cycles to failures and subtracting the propagation life. It is also pertinent to mention here that they computed the propagation life for 1mm crack length to the mid surface of the specimen.

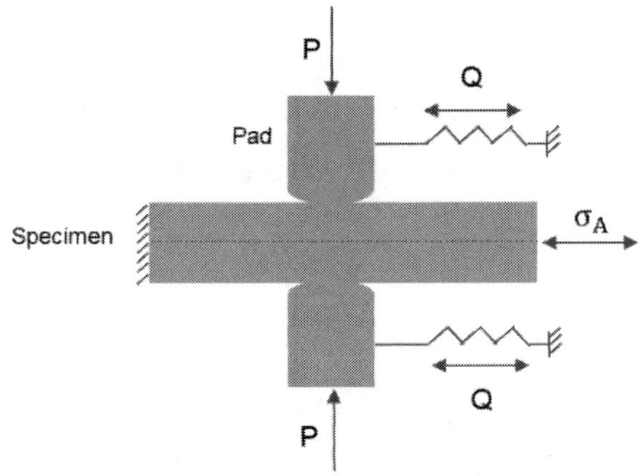

Figure1. Fretting fatigue test configuration

The data set used in this study is taken from the experimental data presented in [15]. Four selected experiments, shown in table 1, are used based on different Q/fP ratio (f is coefficient of friction). The pad radius selected is 178mm. The Aluminium 2024-T351 is used for both specimen and fretting pads. Poisson's ratio ν is 0.33 and modulus of elasticity E is 74.1 GPa.

Table 1. Experimental data from fretting fatigue test.

Experiment Number	P (N)	R (mm)	Q/fP (-)	σ_A (MPa)	$N_{i,exp}$ (cycles)	$N_{f,exp}$ (cycles)
1	7085	178	0.32	85.2	600638	665073
2	6176	178	0.42	84.7	555830	621442
3	6460	178	0.52	106.4	195200	225535
4	5319	178	0.55	97.4	418911	459882

3. Finite element model

Finite element analysis has been widely used in the literature for many engineering applications that have been recently published [18-31]. In this paper, the finite element model is generated using commercial software ABAQUS. The geometrical model including loads and boundary conditions is presented in figure 2. A constant normal load P is applied in the vertical direction at the top surface of the pad. Cyclic axial load σ_A is applied on one side of the specimen while the other end is retrained to move in horizontal direction. The bottom of the specimen is restricted to displace in vertical direction. To simulate the tangential load generated by the spring, Q is applied directly on one side of the pad. The value of coefficient of friction specified is 0.65. The master slave interaction algorithm is used to establish the contact between the pad and the specimen. The heights of 25mm and 6.35mm are used to model pad and specimen respectively. Thickness of both the fretting pad and the specimen is 12.7 mm. The lengths of fretting pad and specimen are 25mm and 40mm respectively.

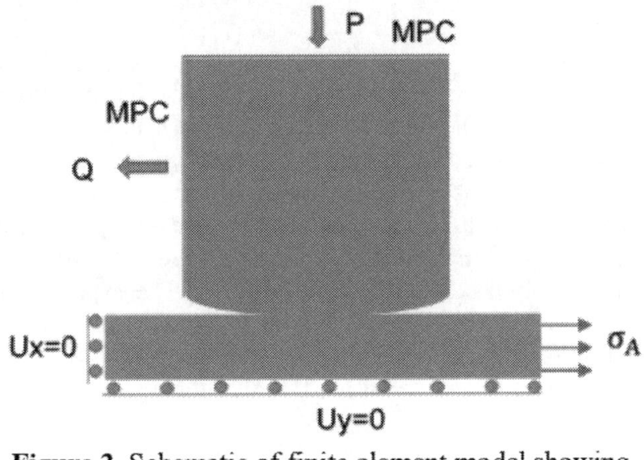

Figure 2. Schematic of finite element model showing loads and boundary conditions.

To accurately model the high stress gradient effects at contact surface, mesh size of 5µm by 5µm is used at and near the contact region. The mesh size of 5 µm x 5 µm is used, as some researchers [32] have modelled the fretting fatigue problem with reasonably accuracy. 4-noded bilinear elements CPE4R are used to mesh both parts using Plane Strain formulation. All the loads are applied in six steps for one complete loading cycle. The cyclic axial load σ_A and tangential load Q are applied in first step simultaneously, however the normal load is kept constant for all steps. Both the stress ratio R_σ and tangential load ratio R_Q are equal to -1. The finite element model is verified using half-space model for the data set of experiment 2. To satisfy the condition of half-space, the height of specimen modelled is b=8.35mm. The figure 3 shows the comparison of analytical and numerical profile for shear traction at the contact interface. For analytical solution readers are referred to [10]. The computed semi contact widths are a_{anl}=1.67mm and a_{FEA}=1.70mm. It is observed that finite element solution matches well with the analytical.

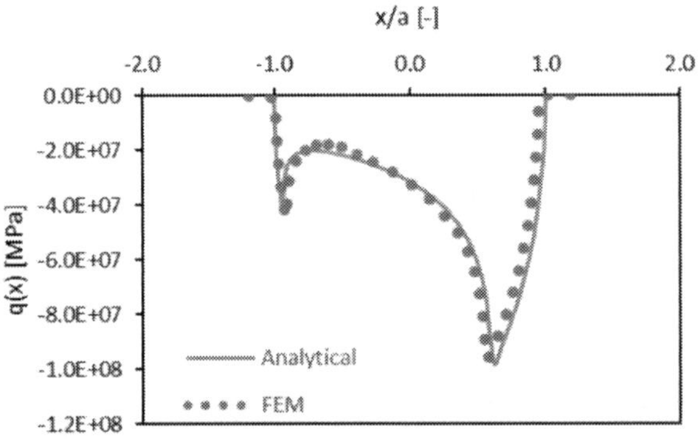

Figure 3. Comparison between analytical and FEM solution of shear traction *q(x)* at the contact interface for Exp. 2 using half space model.

4. Results and discussion

The important factor which directly effects relative slip between contacting bodies is load ratio *Q/fP*. In the first section of the present study, the effect of variation of this parameter on damage initiation location is analysed using Ruiz parameters *F*1 and *F*2. Later variation of *F*2 with initiation life is presented.

4.1. Effect of Q/fP ratio

To analyse the effect of tangential to normal load ratio, data set of four experiments is used as stated in table 1. The coefficient of friction is kept constant at 0.65. Figure 4 shows the effect of variable *Q/fP* on stresses and damage parameters. From figure 4(a) it can be seen that the stick and slip zone widths change with changing load ratios. The magnitude of maximum shear traction and locations for maximum shear tractions are also different for all cases. For lower values of *Q/fP*, larger stick zone width is observed and vice versa. This implied that stick zone width is inversely proportional to *Q/fP*. However, peak of the shear traction is directly linked with the axial stress. The experiment with the highest value of applied axial stress has shown the highest peak for shear traction. Same is true for the tensile stress as depicted in figure 4(c). The normal stress directly depends on magnitude of normal load as shown in figure 4(b). The higher the magnitude of normal load, higher the normal stress is, whereas no uniform pattern is observed with *Q/fP*.

The relative slip is computed for each of the case and it is observed, in figure 4(e), that outside the stick zones the relative slip's amplitude increases with increase of *Q/fP*. It is also intuitive because higher tangential load will produce larger slip between the contacting bodies. The values of Ruiz parameters calculated are shown in figures 4(d) and 4(f) and for all cases peaks are observed at the trailing edge of the contact. Since the damage parameter is dependent on shear traction and relative slip, higher values of *F*1 and *F*2 are found for the experiment with larger shear tractions.

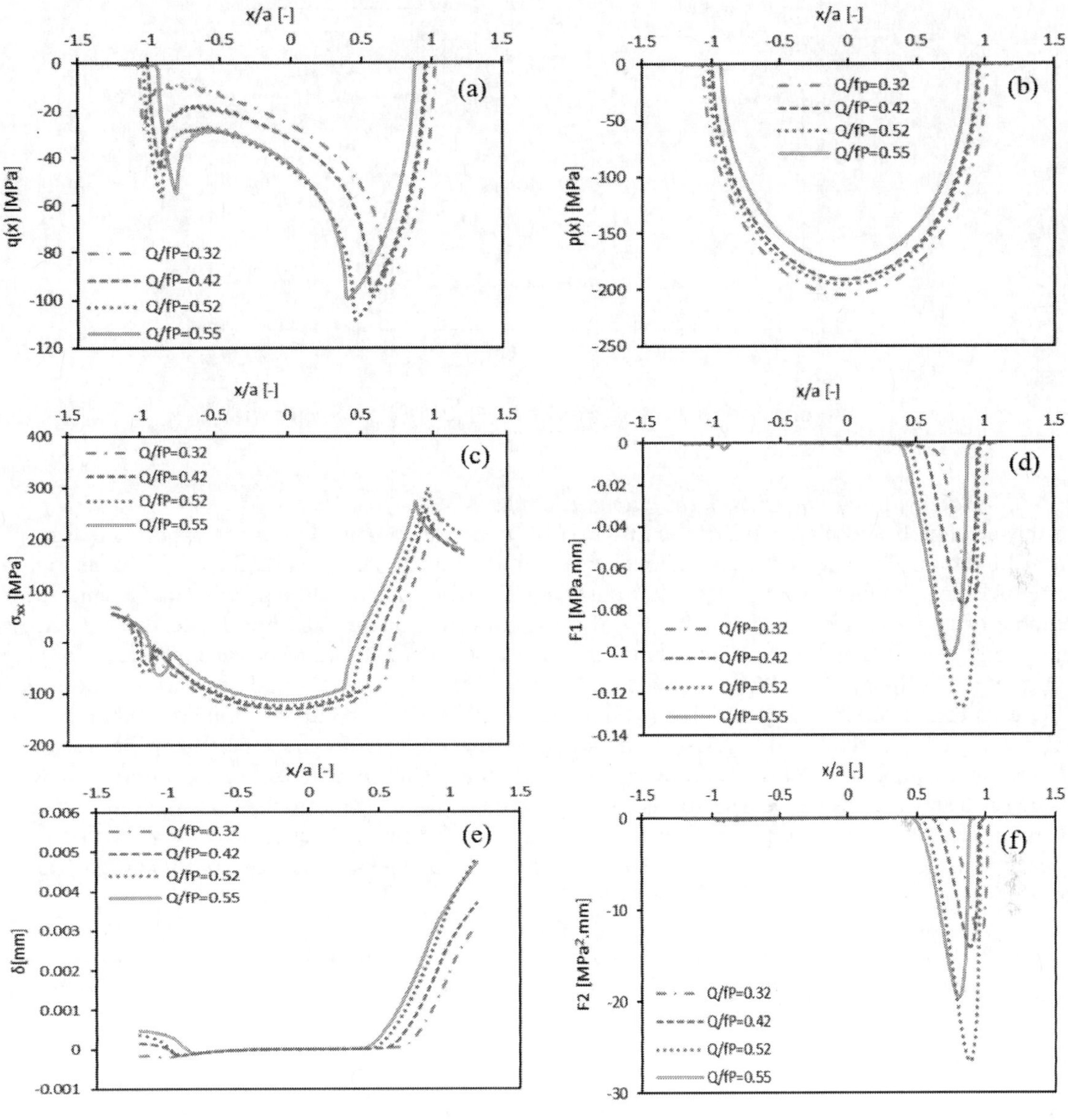

Figure 4. Effect of *Q/fP* on (a) shear traction *q(x)*, (b) normal stress *p(x)*, (c) tensile stress σ_{xx}, (d) Ruiz parameter *F*1, (e) Relative slip *δ* (f) Ruiz parameter *F*2.

Moreover, it is observed that for lower values of *Q/fP*, the initiation parameter is predicted more towards the trailing edge. Figure 5 shows the trend for variation of locations for maximum values of *F*2 parameter, normal stress, tensile stress and shear traction against *Q/fP*. It can also be interpreted that the magnitude of damage and initiation parameters are dependent on individual loads whereas the location is dependent more on the ratio *Q/fP*.

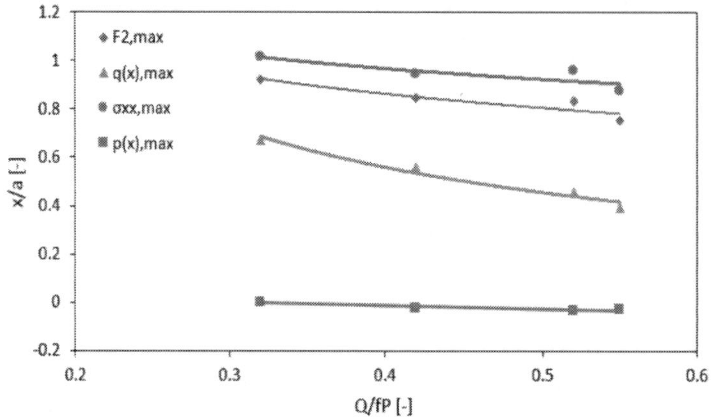

Figure 5. Variation of peak stresses and $F2$ parameter with Q/fP.

4.2. Variation of initiation life with initiation parameter

In this section, the results of the current numerical work are combined with the experimental results from literature. The stresses shown above are computed at the contact interface only, as in fretting fatigue damage initiates at the surface. It is observed that when the initiation parameter $F2$ is higher, the number of cycles to initiation is lower. It is also logical and intuitive to expect the decrease in damage initiation life on increase of damage parameter. The variation of $F2$ parameter versus cycle to initiation is presented in figure 6(a). However, this trend may be dependent on material properties. Lykins et al. [17] used this parameter for titanium alloy Ti-6Al-4V and they found no specific correlation between $F2$ parameter and number of cycles to initiation. Figure 6(b) shows the relationship between peak stresses and number of cycles to initiation. For cases with higher number of cycles to initiation, lower value of shear tractions and tensile stresses are observed while no specific trend is observed for normal stresses. For comparison purposes absolute values are plotted for each case. It is observed that for all experiments normal stresses are higher than shear tractions and tensile stresses are higher than normal stress.

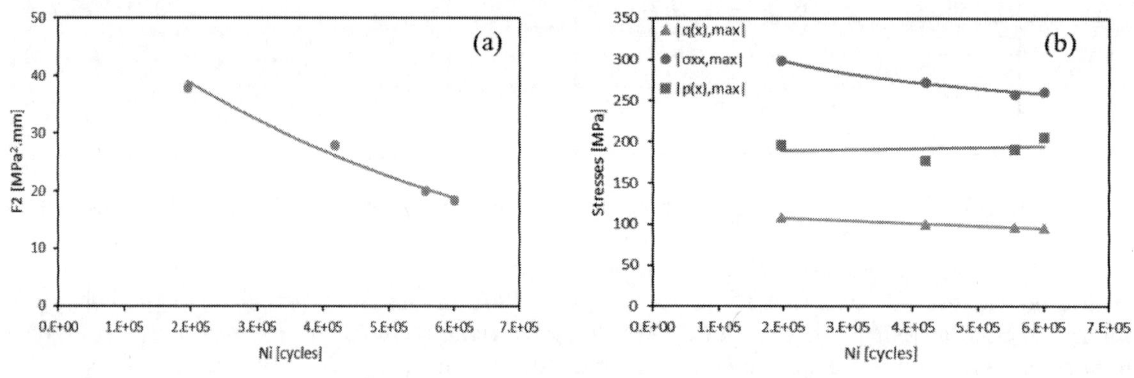

Figure 6. Variation of number of cycles to failure with (a) $F2$ parameter (b) absolute value of maximum tensile stress σ_{xx}, normal stress $p(x)$ and shear traction $q(x)$.

Considering the above discussion and the experimental results presented in [15], where it is shown that the failure occurred in the direction perpendicular to the specimen axis, it can be concluded that the peak in the tensile stress at the trailing edge of the contact is the primary reason that energies damage initiation to failure for the cases where failure occur in classical mode I manner.

5. Conclusion

In this paper various factors have been analysed, affecting the fretting fatigue damage initiation location. Based on the results presented above, following points can be concluded.

- In fretting fatigue configuration, damage initiation is dependent on *Q/fP* ratio. When *Q/fP* ratio decreases, the stick zone size increases and the initiation location shifts more towards the trailing edge of the contact.
- The peak of the shear traction and tensile stress is directly linked with applied axial stress. The higher the values of applied axial stress, the higher the values of shear traction and tensile stress.
- Results suggest that, higher the value of fretting fatigue damage initiation parameter *F*2, the lower number of cycles are required for damage initiation and vice versa.
- The magnitude for tensile stresses is much higher than shear and normal stresses for any particular case. Therefore, the most damaging stress for failure is tensile stress, then normal stress and the least effect is expected for shear traction.

Acknowledgements: The second author would like to acknowledge the financial support of the Research Foundation-Flanders (FWO), The Luxembourg National Research Fund (FNR) and Slovenian Research Agency (ARRS) in the framework of the FWO Lead Agency project G018916N 'Multi-analysis of fretting fatigue using physical and virtual experiments'.

References

[1] Yue T and Abdel Wahab M 2017 *Tribology International* **107** 274-282
[2] Pereira K, Yue T and Abdel Wahab M 2017 *Tribology International* **110** 222-231
[3] Yue T and Abdel Wahab M 2016 *Materials* **9** 597; doi:510.3390/ma9070597
[4] Yue T and Abdel Wahab M 2014 *Wear* **321** 53-63
[5] Martínez JC, Vanegas Useche LV and Wahab MA 2017 *International Journal of Fatigue* **100, Part 1** 32-49
[6] Kumar D, Biswas R, Poh LH and Abdel Wahab M 2017 *Tribology International* **109** 124-132
[7] Bhatti NA and Abdel Wahab M 2017 *Tribology International* **109** 552-562
[8] Resende Pereira KdF, Bordas S, Tomar S, Trobec R, Depolli M, Kosec G and Abdel Wahab M 2016 *Materials* **9** 639; doi:610.3390/ma9080639
[9] Ferjaoui A, Yue T, Abdel Wahab M and Hojjati-Talemi R 2015 *International Journal of Fatigue* **73** 66-76
[10] Hills DA and Nowell D 1994 Mechanics of Fretting Fatigue ed editors. Springer Netherlands
[11] Ruiz C, Boddington PHB and Chen KC 1984 *Experimental Mechanics* **24**(3) 208-217
[12] Kuno M, Waterhouse RB, Nowell D and Hills DA 1989 *Fatigue & Fracture of Engineering Materials & Structures* **12**(5) 387-398
[13] Nowell D and Hills DA 1990 *Wear* **136**(2) 329-343
[14] Szolwinski MP and Farris TN 1996 *Wear* **198**(1-2) 93-107
[15] Szolwinski MP and Farris TN 1998 *Wear* **221**(1) 24-36
[16] Iyer K and Mall S 2001 *Journal of Engineering Materials and Technology-Transactions of the Asme* **123**(1) 85-93
[17] Lykins CD, Mall S and Jain V 2000 *International Journal of Fatigue* **22**(8) 703-716
[18] X. Nguyen H, N. Nguyen T, Abdel Wahab M, Bordas SPA, Nguyen-Xuan H and P. Voa T 2017 *Computer Methods in Applied Mechanics and Engineering* **313** 904-940
[19] Tran Vinh L, Lee J, Nguyen-Van H, Nguyen-Xuan H and Abdel Wahab M 2015 *International Journal of Non-Linear Mechanics* **72** 42-52
[20] Tran LV, Phung-Van P, Lee J, Wahab MA and Nguyen-Xuan H 2016 *Composite Structures* **140** 655-667
[21] Thai CH, Ferreira AJM, Abdel Wahab M and Nguyen-Xuan H 2016 *Acta Mechanica* **227**(5) 1225-1250

[22] Thai C, Zenkour AM, Abdel Wahab M and Nguyen-Xuan H 2016 *Composite Structures* **139** 77-95

[23] Phung-Van P, Tran LV, Ferreira AJM, Nguyen-Xuan H and Abdel-Wahab M 2016 *Nonlinear Dynamics* 1-16; doi:10.1007/s11071-11016-13085-11076

[24] Phung Van P, Nguyen LB, Tran Vinh L, Dinh TD, Thai CH, Bordas SPA, Abdel Wahab M and Nguyen-Xuan H 2015 *International Journal of Non-Linear Mechanics* **76** 190-202

[25] Phung Van P, De Lorenzis L, Thai CH, Abdel Wahab M and Nguyen-Xuan H 2015 *Computational Materials Science* **96** 495-505

[26] Phung Van P, Abdel Wahab M, Liew KM, Bordas SPA and Nguyen-Xuan H 2015 *Composite Structures* **123** 137-149

[27] Noda N-A, Chen X, Sano Y, Wahab MA, Maruyama H, Fujisawa R and Takase Y 2016 *MATERIALS & DESIGN* **96** 476-489

[28] Tran Vinh L, Lee J, Ly HA, Abdel Wahab M and Nguyen-Xuan H 2015 *International Journal of Mechanical Sciences* **96-97** 65-78

[29] Junyan Ni and Wahab MA 2017 *Computers & Structures* **186** 35–49

[30] Phung-Van P, Qui LX, Nguyen-Xuan H and Wahab MA 2017 *Composite Structures* **166** 120–135

[31] Phung-Van P, Ferreira AJM, Nguyen-Xuan H and Abdel Wahab M 2017 *Composites Part B: Engineering* **118** 125-134

[32] Hojjati-Talemi R and Wahab MA 2013 *Tribology International* **60** 176-186

Frequency-Dependent Cohesive Zone Models for Fatigue

S Salih, K Davey, Z Zou

School of Mechanical, Aerospace and Civil Engineering, The University of Manchester
eng.sarmed@gmail.com

Abstract: This paper is concerned with a new cohesive zone model (CZM) to better describe the effects of rate and cyclic loading. Rate is known to affect the manner in which cracks propagate in materials, yet there presently exists no rate-dependent cohesive model for fatigue simulation. The frequency of the applied cyclic load is recognised to influence crack growth rates with crack growth significantly different at lower frequencies due to microstructural effects or other damage mechanisms such as creep or corrosion. A rate-dependent trapezoidal cohesive model is presented that has the ability to capture this behaviour and shows slower rates of crack propagation with higher loading frequencies. This is achieved by allowing the cohesive fracture energy to increase with frequency up to a specified limit. On unloading the cohesive model retains material separation, which accumulates with the number of loading cycles, leading to final failure. An experimental fatigue investigation is currently underway to validate the new cohesive model, which has been coded in a UMAT subroutine and implemented in ABAQUS.

Keyworks: Cohesive zone model; Rate effect; Fatigue; Frequency effect; Rate-dependent CZM

1. Introduction

The process of fatigue failure can be divided into three parts: crack initiation, crack propagation, and then fast fracture, which leads ultimately to the final failure [1], [2]. Fatigue can be classified as a subcritical failure mechanism, occurring when the applied stress is significantly lower than the yield stress of the material. Mechanical, microstructural and environmental factors can significantly influence the fatigue behaviour of a part. Consequently fatigue analysis is primarily an empirical study founded on extensive experimental investigations typically needed to evaluate and understand material behaviour and estimate the life of mechanical parts. These can be costly and extremely time-consuming giving rise to an urgent requirement for fatigue models possessing an ability to predict crack growth rates and fatigue life [2].

The limitation of conventional fatigue models has heightened the need for a model that is applicable at crack initiation, as well as for crack growth. Hence, the recent focus on cohesive zone models (CZMs), which were initially advanced by Dugdale [3] and Barenblatt [4] in order to overcome the unrealistic mathematical-stress singularity at a crack tip [5]. In Dugdale's model the stress in the small plastic zone ahead of the crack tip is finite and equal to the yield strength, while in Barenblatt's model the stress in the region in front of the crack is a function of the ligament length. The CZM became more popular following the work of Hillerborg et al. [6], which demonstrated that numerical analysis was possible using the finite element method and a bilinear cohesive zone model (BCZM).

The CZM is founded on a traction-separation law (TSL), where traction is related to material separation through a constitutive equation. On traction reaching a critical value σ_c, material damage initiates and material softening occurs. Following damage initiation, traction decreases until it finally reaches zero which is the point of material separation. This point is also identified by the separation between the surfaces of the cracked material attaining a critical value δ_c; at this point all cohesive energy Γ_0 has been dissipated and consequently the crack propagates. The advantage of using a cohesive model over other models is its ability to predict the initiation and the propagation of cracks along with its applicability to monotonic and cyclic loading.

It is well appreciated that cyclic loading presents added complication arising out of the unloading part of the loading cycle and the history dependence of the fatigue process. Most conventional CZMs are limited to monotonic crack growth prediction; see for example the models of: Needleman [7], Needleman [8], Scheider [5] and Tvergaard & Hutchinson [9], which are reversible and history-independent. There exists irreversible models such as those presented in references [10]–[13], but these are history-independent and do not include a mechanism to cater for cyclic-damage accumulation. These are consequently limited to the simulation of monotonic-crack propagation. A CZM for fatigue requires an irreversible and history-dependent cohesive law, which is needed to capture the damage accumulation associated with loading-unloading hysteresis. This is a feature of the model introduced by De-Andrés et al. [14], which involves a damage factor D that represents the value of dissipated energy in the fracture process divided by the critical fracture energy. Dahlan [2] adopted a similar damage concept involving a variable that cumulatively recorded the value of the separation at the end of each loading cycle with failure identified by the accumulated value reaching δ_c. A different damage accumulation mechanism has used in [15], involving a degrading cohesive stiffness with damage accumulation. Two stiffness parameters (K^+ and K^-) are used for loading and unloading, where the value of K^- is constant for each unloading, but K^+ is allowed to evolve with the number of cycles through a decay factor, until the complete separation of the cohesive element. An adapted Paris-like equation is used, where fatigue crack growth rate da/dN is related to the change in crack tip opening displacement range per cycle $\Delta\delta$ as in the equation $da/dN = C(\Delta\delta)^n$. However, the use of $\Delta\delta$ as a material property is questionable as it depends to some extent on the degree of mechanical constraint [16]. A similar but slightly different approach was adopted by Yang et al. [17], with a model also involving loading and the unloading stiffness parameters but in this case both are functions of a damage parameter that evolves during loading and unloading cycles. A recent study by Roe & Siegmund [16] investigated a CZM with a degrading cohesive strength as a function of the damage; the model involves two additional parameters, i.e. σ_f, which is the endurance limit and δ_Σ, which is the accumulated cohesive length. In this model, if the stress on the element ahead of the crack tip is less than the endurance limit, then the model will present infinite life (i.e. no crack propagation). If the stress is larger than the endurance limit, then material separation accumulates until it reaches the cohesive length δ_0, which represents separation at the critical cohesive stress σ_c. Subsequently on reaching δ_0, damage accumulates and the critical cohesive stress reduces as a function of the damage; the crack extends and the cohesive element fails when the maximum value of the stress in the loading cycle reaches the current critical cohesive stress. Other studies [18], [19] have considered relatively similar approaches, in which the damage will not accumulate if the stress is less than a predefined value (i.e. an endurance limit). All the methods mentioned above are rate-independent models and do not capture the frequency effect. In this study a trapezoidal rate-dependent CZM introduced by Salih et al. [13] is further developed to be used for the simulation of the fatigue crack growth.

2. Cohesive zone model for fatigue

Different types of TSL are presented in the literature, which include the: polynomial, exponential, bilinear, and trapezoidal as shown in Figure 1. The CZM is recognised not to be a physical material model but a phenomenological law that is designed to capture the essence of the fracture process. Therefore, there is no underpinning physical evidence to support a particular shape of the curve that describes the relationship between traction and separation. In this study the Trapezoidal TSL has been used because of its ability to capture plasticity locally.

The behaviour of the existing model introduced by Salih et al. [13] is best understood and described by an arrangement of 1-D spring, slider, dashpots, and linear cohesive elements as depicted in Figure 2. Rate dependency is recognised to arise from the behaviour of the dashpots. In this work a frequency-dependent cohesive zone model for fatigue is produced in a similar manner, but in this model the critical stress increases as a function of the loading frequency and the increase in the frequency-dependent critical cohesive stress is automatically limited by the rate equation. The traction-separation curve for this model is shown in Figure 3.

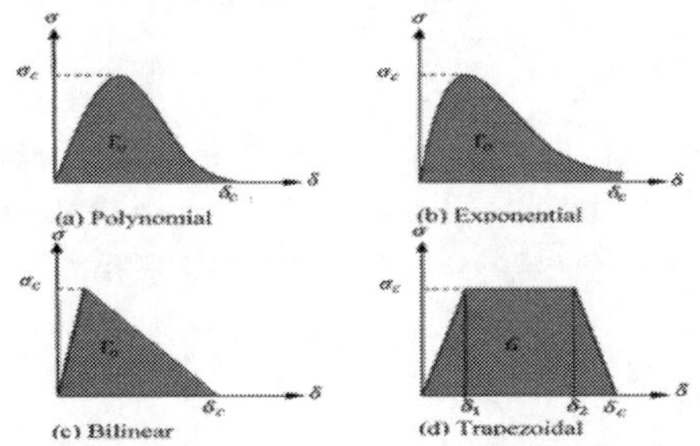

Figure 1. Common traction-separation

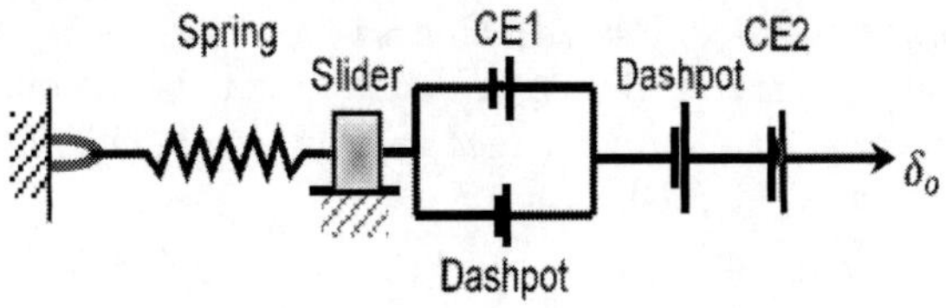

Figure 2. Mode I Cohesive Zone model

The area under the traction separation curve represents the fracture energy G_c, which is defined mathematically to be

$$G_c = \int_0^{\delta_c} \sigma_{(\delta)}\, d\delta \tag{1}$$

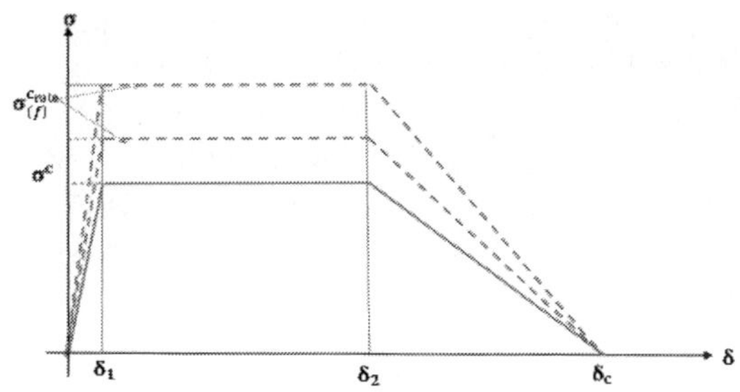

Figure 3. Mode I Cohesive Zone model

In the case of trapezoidal CZM the fracture energy can be considered to be formed in two parts, i.e. plastic G_p and cohesive Γ_o components to give

$$G_c = G_p + \Gamma_o \tag{2}$$

The area under the curve(s) in Figure 3 is equal to

$$G_c = \frac{1}{2}\sigma_{(f)}^{Crate}\left((\delta_2 - \delta_1) + \delta_c\right) \tag{3}$$

$$\sigma_{(f)}^{Crate} = \sigma_c \begin{cases} 1 & \text{if } f \leq 0.05 \\ \left(0.52 + \dfrac{1}{a+b*exp\left(-\left(\left(\frac{d*(f-f_{min})}{f_{min}}\right)^n\right)\right)}\right) & \text{if } f > 0.05 \end{cases} \tag{4}$$

The meanings the symbols relay are: G_c is the total dissipated energy per unit area, G_p is the local plastic dissipated energy, Γ_o is the energy dissipated in the formation of new surfaces, $\sigma_{(f)}^{Crate}$ is the frequency-dependent critical cohesive traction, σ_c is the rate-independent critical cohesive traction, δ_1 is the displacement when the traction first reach its critical value, δ_2 is the displacement at which the element deterioration is assumed to start, and δ_c is the rate-independent critical cohesive separation. a, b, d and n are fixed and are determined with experimental data; a first estimation of these constants is: 1.2, 0.85, 3 and 0.25. The lowest frequency f_{min} is equal to 0.05Hz, f is the applied frequency and

$$\delta_2 \leq 0.75 \times \delta_c \tag{5}$$

The monotonic trapezoidal TSL depicted in Figure 3 is represented mathematically as

$$\sigma(\delta) = \sigma_{(f)}^{Crate} \begin{cases} \frac{\delta}{\delta_1} & \text{if } 0 < \delta < \delta_1 \\ 1 & \text{if } \delta_1 \leq \delta \leq \delta_2 \\ (1 - D) & \text{if } \delta_2 < \delta < \delta_c \end{cases} \tag{6}$$

In the cohesive model the mechanism of crack propagation is void nucleation, growth and coalescence. The damage variable (D) can be understood to be the ratio between damaged area (A_d) (the accumulated areas of the voids) to undamaged reference area (A_e) [20]. In the present configuration damage can be represented by the relationship

$$D = \frac{A_d}{A_e} = \frac{\delta - \delta_2}{\delta_c - \delta_2} \tag{7}$$

Unloading and subsequent reloading occurs at a constant stiffness $K = \sigma_{max}/\delta_{max}$, where σ_{max} and δ_{max} are the stress and the separation at the point where unloading starts, respectively. In this case traction is evaluated by

$$\sigma(\delta) = K \times \delta \tag{8}$$

The cohesive model described thus far, will result in an infinite life when cyclic load is applied. So in order to capture finite life, it is necessary to use an irreversible and history dependent cohesive zone model. This can be done by identifying a damage variable that accumulates with the number of cycles. The damage mechanism in this model consists of two parts: first is the remnant plastic separation (δ^p), which accumulates as a result of cyclic plasticity, and second the damage variable D that represents the material deterioration in the CZM. The proposed damage accumulation mechanism is designed to provide the trapezoidal model with an ability to capture fatigue crack growth. A full mathematical description of the proposed new model is:

$$\sigma_{(\delta)} = \sigma_{max} \begin{cases} \left(\frac{\delta_{eff}}{\delta_{max} - \delta^p}\right) & \text{if } \delta_{eff} < 0 \\ \left(\frac{\delta_{eff}}{\delta_{max} - \delta^p}\right) H_{(\delta_{eff} - \delta^p)} & \text{if } 0 \leq \delta_{eff} \leq \delta_{max} \\ 1 & \text{if } \delta_{max} < \delta_{eff} \leq \delta_2 \\ (1 - D) \times \frac{\sigma_{(f)}^{Crate}}{\sigma_{max}} & \text{if } \delta_2 < \delta_{max} < \delta_{eff} < \delta_c \\ 0 & \text{if } \delta_{eff} \geq \delta_c \end{cases} \tag{9}$$

where σ_{max} and δ_{max} are the stress and the separation at the end of the loading cycle respectively, but at the beginning when there is no unloading these values set to be equal to $\sigma_{(f)}^{Crate}$ and δ_1, respectively. The parameter δ^p records the value of the remnant separation at the end of the loading cycle. This is required because at unloading, separation does not return to zero. The Heaviside function $H_{(\delta_{eff} - \delta^p)}$ equals zero if δ_{eff} is smaller than δ^p and one in other cases and finally D is a damage variable of equation (9) with δ_{eff} replacing δ. In addition

$$\delta_{eff} = \delta^{cyc} + \delta^p \tag{10}$$

$$\delta_{(i+1)}^p = \delta_{(i)}^p + \frac{\delta_{max}^{cyc}}{C} \tag{11}$$

where δ^{cyc} and δ^p are the cyclic applied displacement and the stored plastic displacement respectively. The material parameter C dictates the level of cyclic plasticity and its value is determined on tuning the analysis results to fit with the experimental data.

In this model, the value of the stored separation (δ^p) accumulates with increasing number of loading cycles until it reaches δ_2 at which point material softening begins and the cohesive element fails when the effective separation (the accumulated plus the cyclic) reaches the critical cohesive separation δ_c. Figure 4 shows the behaviour of the cohesive model under monotonic and cyclic loads.

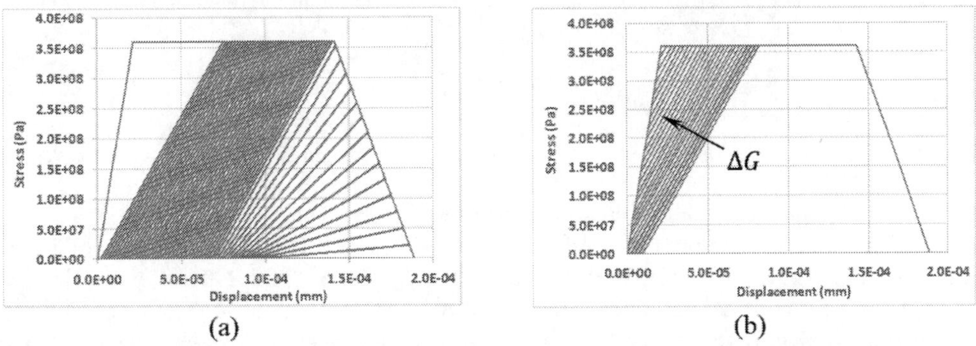

Figure 4. Behaviour of the cohesive model: (a) cyclic loads, (b) cyclic-dissipated energy

The accumulated dissipated energy (ΔG) as shown in Figure 4b can be calculated by:

$$\Delta G = \begin{cases} \frac{1}{2}\left[\sigma_{(f)}^{Crate}(\delta_{max} + \delta^p - \delta_1)\right] & \text{if } \delta_{max} \le \delta_2 \\ \frac{1}{2}\left[\sigma_{(f)}^{Crate}(\delta_{max} + \delta_2 - \delta_1) + \sigma_{max}(\delta^p - \delta_2)\right] & \text{if } \delta_{max} > \delta_2 \end{cases} \quad (12)$$

From the analysis is found that a minimum number of 20 elements in the cohesive zone should be used to obtain satisfactory results, although in this study higher numbers were used in order to have superior control of the crack growth. The length (l_{coh}) can be estimated through a similar formula used for the plastic zone evaluated by von Mises yield criterion, i.e.

$$l_{coh} = \frac{E}{2\pi}\frac{G_{Ic}}{\sigma_c^2} \quad (13)$$

3. Implementation of the cohesive zone model in abaqus

Numerical analysis is performed using the commercial finite element solver ABAQUS. Although a cohesive element exists in ABAQUS, the traction separation laws associated with this element are history independent and cannot be used for the simulation of the fatigue problem. The CZM described in Section 2 has been incorporated into ABAQUS though a user defined material subroutine UMAT. This subroutine provides the user with a facility to identify a specific constitutive behaviour and link this behaviour to any element.

3.1. *Implementing and testing the UMAT*

To test the UMAT, a model that consists of two material elements and one cohesive element connecting them was used as shown in Figure 5. The material properties were set to be elastic in the bulk material elements with 192 GPa for the elastic modulus and 0.29 Poisson's ratio, while the cohesive element set to have a critical stress of 360 MPa, critical separation of 0.189 mm and 17000 GPa elastic stiffness. The value of the parameter C is set to a small value (20) in order to artificially reduce the number of cycles required for the cohesive element to fail. The cyclic stress is shown in Figure 4a with respect to the separation and in Figure 6 with respect to the time. The figures show how the plastic separation is accumulating with the number of cycles until the accumulated plus the applied separation reaches δ_2. Subsequently the cohesive stiffness decreases with the number of cycles as a consequence of material deterioration and the cohesive element fails when the stress reaches zero.

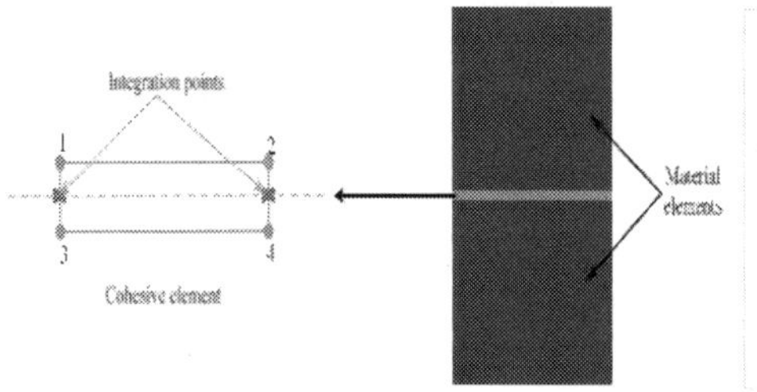

Figure 5. Implementation of the CE in the FE model

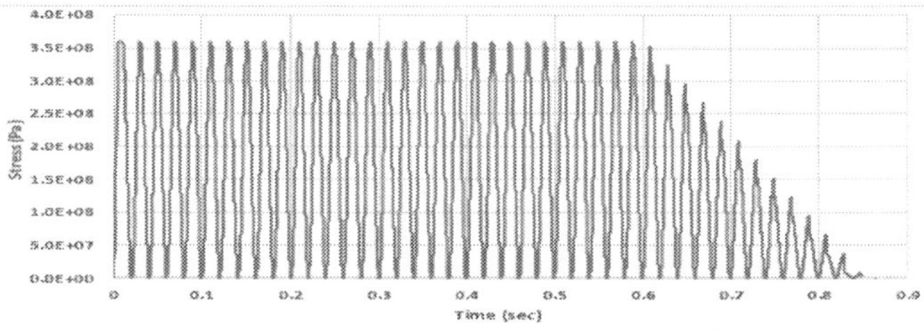

Figure 6. Stress levels in the cohesive element

3.2. *Analysis model: geometry and boundary conditions*

The finite element model used in this work is shown in Figure 7. The model consists of 23621 plane-stress elements of which 22991 (type CPS4) and 390 (type CPS3), and 240 cohesive elements (type COH2D4) [21]. In the commercial finite element solver ABAQUS, the cohesive behaviour can be identified either as a cohesive surface or by implementing cohesive elements along the crack path; these cohesive elements adjoin with the bulk material elements. In this study, the second method has been used, since in this method the cohesive behaviour is defined through the material behaviour attributed to those elements. A mesh sensitivity analysis has been performed which confirms that converged results are attained. Increasing the number of elements in the bulk material or the cohesive zone, has little impact on the simulation results. The material properties for the bulk material elements and the cohesive element are shown in Table 1.

Table 1. Material properties

σ_y (Pa)	E (GPa)	ϑ	Γ_o (N/m)	E_{coh} (GPa)	σ_c (Pa)	δ_c (m)	δ_1 (m)	δ_2 (m)	C
3.4e8	193	0.29	49320	17 e3	3.6e8	0.168e-3	0.02 e-3	0.126 e-3	20

The boundary conditions applied to the model are shown in Figure 7, with nodes at the lower edge fixed in all directions. The top edge is subjected to a uniform displacement in the y-direction. The load is applied in seven steps: first a ramp load that increases from zero to 5.8E-5m and then the following steps are set to have a sinusoidal load with 5.8E-5m mean displacement and amplitude of (5.8E-5, 6.2E-5, 6.5E-5, 8.6E-5, 11.2E-5, 12.6E-5, 20.6E-5m), respectively. An example of load incremental steps is shown in Figure 8.

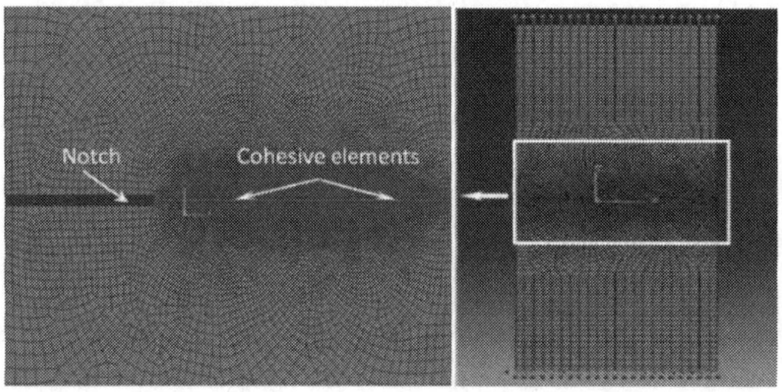

Figure 7. Boundary conditions and loading for FE model

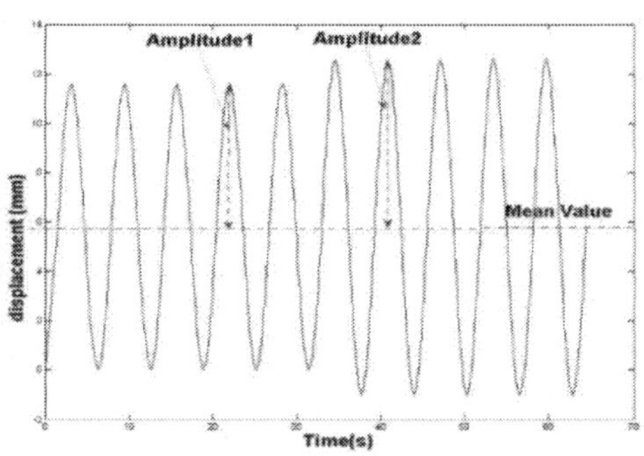

Figure 8. Load as function of time

The pre-cracked tension specimen is loaded in ABAQUS at four different applied frequencies (0.05, 0.1, 0.5 and 30Hz). The curves for crack length vs number of cycles for the four frequencies are shown in Figure 9. The predicted results match expectations since the model correctly predicts slower crack propagation at higher frequency. Figure 10a and Figure 10b shows contour graphs of the finite element simulation at 0.05Hz and 30Hz respectively after 80 loading cycles.

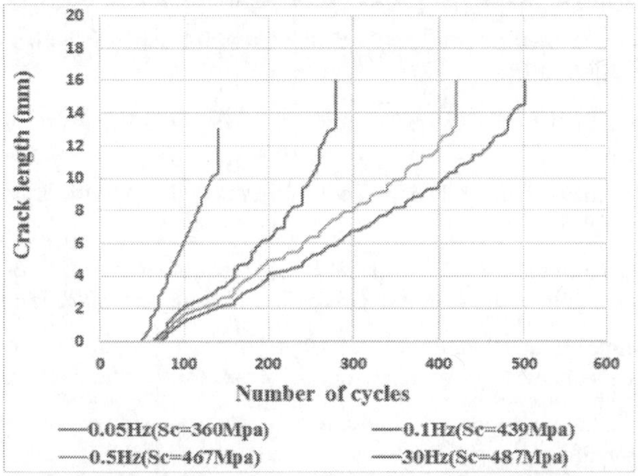

Figure 9. a-N curves at different frequencies

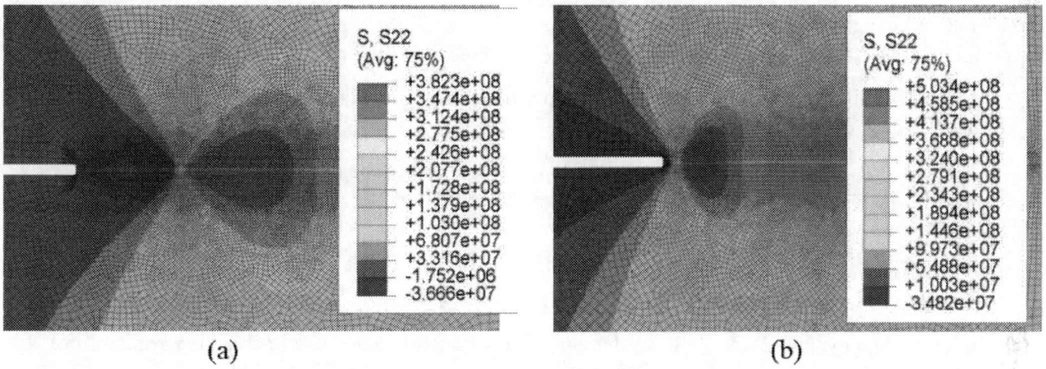

Figure 10. (a)Crack length after 80 cycles at 0.05Hz, (b) Crack length after 80 cycles at 30Hz

4. Conclusions and future work

Based on the finding of this study the following can be concluded:

- A new CZM that combines cyclic damage with the monotonic damage has been established and applied to simulate fatigue crack propagation.
- The new CZM is founded on a rate-dependent methodology previously established by the authors, and has been shown to capture frequency effects on fatigue crack propagation in stainless steel 304.
- The new rate-dependent CZM has been implemented in the finite element solver ABAQUS (via a bespoke UMAT routine) and has been tested at different applied frequencies (0.05, 0.1, 0.5 and 30Hz).
- An experimental fatigue investigation is currently underway to validate the new cohesive model.

Acknowledgments

The authors would like to acknowledge the Higher Committee for Education Development in Iraq and the College of Engineering at Babylon University for providing support for Sarmed Salih to facilitate his doctoral research at the University of Manchester.

References

[1] J. Schijve, *Fatigue of structures and materials*. Dordrecht: Kluwer Academic Publishers, 2001.

[2] H. Dahlan, "A Fast-Track Method for Fatigue Crack Growth Prediction with a Cohesive Zone Model," University of Manchester, 2013.

[3] D. Dugdale, "Yielding of steel sheets containing slits," *J. Mech. Phys. Solids*, vol. 8, no. 2, pp. 100–104, 1960.

[4] G. I. Barenblatt, "The mathematical theory of equilibrium cracks in brittle fracture," *Adv. Appl. Mech.*, vol. 7, no. 1, pp. 55–129, 1962.

[5] I. Scheider, "Cohesive model for crack propagation analyses of structures with elastic–plastic material behavior Foundations and implementation," *GKSS Res. Cent. Geesthacht, Dept. WMS*, 2001.

[6] A. Hillerborg, M. Modéer, and P. Petersson, "Analysis of crack formation and crack growth in concrete by means of fracture mechanics and finite elements," *Cem. Concr. Res.*, vol. 6, pp. 773–782, 1976.

[7] A. Needleman, "A continuum model for void nucleation by inclusion debonding," *J. Appl. Mech.*, vol. 54, no. September, pp. 525–531, 1987.

[8] A. Needleman, "An analysis of tensile decohesion along an interface," *J. Mech. Phys. Solids*, vol. 38, no. 3, pp. 289–324, Jan. 1990.

[9] V. Tvergaard and J. W. Hutchinson, "The relation between crack growth resistance and fracture process parameters in elastic-plastic solids," *J. Mech. Phys. Solids*, vol. 40, no. 6, pp. 1377–1397, Aug. 1992.

[10] G. Camacho and M. Ortiz, "Computational modelling of impact damage in brittle materials," *Int. J. Solids Struct.*, vol. 33, no. 20–22, pp. 2899–2938, 1996.

[11] G. Alfano, S. De Barros, L. Champaney, and N. Valoroso, "Comparison Between Two Cohesive-Zone Models for the Analysis of Interface Debonding," *Eur. Congr. Comput. Methods Appl. Sci. Eng.*, pp. 1–18, 2004.

[12] P. P. Camanho, C. G. Davla, and M. F. De Moura, "Numerical simulation of mixed-mode progressive delamination in composite materials," *J. Compos. ...*, vol. 37, no. 16, 2003.

[13] S. Salih, K. Davey, and Z. Zou, "Rate-dependent elastic and elasto-plastic cohesive zone models for dynamic crack propagation," *Int. J. Solids Struct.*, vol. 0, pp. 1–21, 2016.

[14] A. De-Andrés, J. Pérez, and M. Ortiz, "Elastoplastic finite element analysis of three-dimensional fatigue crack growth in aluminum shafts subjected to axial loading," *Int. J. Solids Struct.*, vol. 36, no. 15, pp. 2231–2258, 1999.

[15] O. Nguyen and E. Repetto, "A cohesive model of fatigue crack growth," *Int. J. Fract.*, vol. 110, pp. 351–369, 2001.

[16] K. L. Roe and T. Siegmund, "An irreversible cohesive zone model for interface fatigue crack growth simulation," *Eng. Fract. Mech.*, vol. 70, no. 2, pp. 209–232, Jan. 2003.

[17] B. Yang, S. Mall, and K. Ravi-Chandar, "A cohesive zone model for fatigue crack growth in quasibrittle materials," *Int. J. Solids Struct.*, vol. 38, no. 22–23, pp. 3927–3944, May 2001.

[18] P. Beaurepaire and G. I. Schuëller, "Modeling of the variability of fatigue crack growth using cohesive zone elements.," *Eng. Fract. Mech.*, vol. 78, no. 12, pp. 2399–2413, Aug. 2011.

[19] A. Ural, V. R. Krishnan, and K. D. Papoulia, "A cohesive zone model for fatigue crack growth allowing for crack retardation," *Int. J. Solids Struct.*, vol. 46, no. 11–12, pp. 2453–2462, Jun. 2009.

[20] A. Turon, J. Costa, P. P. Camanho, and C. G. Dávila, "Simulation of delamination in composites under high-cycle fatigue," *Compos. Part A Appl. Sci. Manuf.*, vol. 38, no. 11, pp. 2270–2282, Nov. 2007.

[21] ABAQUS, *ABAQUS 6.13 User Guide*. USA: Dassault Systèmes Simulia Corp, 2013.

Roughness Effects on Fretting Fatigue

Tongyan Yue, Magd Abdel Wahab

Ghent University, Laboratory Soete, Belgium

Magd.AbdelWahab@UGent.be

Abstract. Fretting is a small oscillatory relative motion between two normal loaded contact surfaces. It may cause fretting fatigue, fretting wear and/or fretting corrosion damage depending on various fretting couples and working conditions. Fretting fatigue usually occurs at partial slip condition, and results in catastrophic failure at the stress levels below the fatigue limit of the material. Many parameters may affect fretting behaviour, including the applied normal load and displacement, material properties, roughness of the contact surfaces, frequency, etc. Since fretting damage is undesirable due to contacting, the effect of rough contact surfaces on fretting damage has been studied by many researchers. Experimental method on this topic is usually focusing on rough surface effects by finishing treatment and random rough surface effects in order to increase fretting fatigue life. However, most of numerical models on roughness are based on random surface. This paper reviewed both experimental and numerical methodology on the rough surface effects on fretting fatigue.

1. Introduction

Fretting is a small oscillatory motion between two contact surfaces by normal load, which may cause fretting fatigue [1-5] and/or fretting wear [6-9] and/or fretting corrosion involving a large number of factors including both material properties and the working environment of fretting couples. Fretting corrosion is the deterioration at the contact interface as a result of corrosion and the micro oscillatory slip. This is a case that chemical reaction rules the extent of the damage during the fretting process, which is not discussed in the paper.

Besides fretting corrosion, fretting wear and fretting fatigue are two main fretting problems. Fretting wear is material removal between two contact surfaces, with oscillated micro relative slip under contact pressure. Given an external alternating load is superimposed on the contact load, fretting fatigue may occur [10]. It results in catastrophic failure at the stress levels below the fatigue limit of the material. Both fretting wear and fretting fatigue often coexist at the same time [11] and come out as competing process. However, generally, fretting wear and fretting fatigue are predominant in gross sliding and partial slip, respectively. The main reason of this is that the micro superficial crack generated in gross sliding condition could be eliminated by the material removal. While due to less wear and high velocity of cracking in partial slip condition, cracks initiated on the contact surface have more opportunity to propagate to the inner of the specimen.

Unlike rolling or reciprocating, usually, fretting happens where the contact surfaces are not supposed to move relatively to each other [12]. This movement is attributed to the deflection of machine components with clamped joints or press fits. Therefore, fretting problems could be found widely in almost all contact conditions due to this small displacement. The nature of fretting involves a large number of factors including both material properties and the working environment of fretting couples, such as the applied normal load and displacement, material properties, roughness of the contact surfaces[13].

As increasing surface roughness is one of key factors enhancing fretting fatigue performance [14], researchers have studied both regular rough surface and random rough surface influence on fretting fatigue by experimental method. Meanwhile, the numerical analysis on random rough surface effects has been conducted. The motivation of present work is to review the research of fretting fatigue related to the roughness effects in both experimental and numerical modelling aspects. This paper is divided 4 parts; after the introduction section, the experimental research of it is reviewed. Then, the numerical methods employed for roughness modelling is presented in section 3. Finally, a conclusion is presented.

2. Roughness effects on fretting fatigue

Surface roughness plays an important role on working performance of fretting couples. By different techniques, the surface presents various topography. By shot peening or laser peening, compressive residual stress and surface roughening are induced. Most of finishing methods, such as turning, grinding, lapping, honing result in irregular roughness pattern, which is described as random distribution. However, using vibration rolling process or laser surface texturing technology, it is possible to achieve regular rough contact surfaces. Therefore, it is worth to study the effects of rough surface by different surface finishing methods.

2.1. Rough surface by shot peening and shot blasting

Shot peening is a mechanical surface treatment used to produce a localised plastic flow at component surfaces [15, 16]. This plastic flow not only results in a work hardening and general roughening of surfaces, but also causes a compressive residual stress layer. All these factors may affect the resistance to plain fatigue and fretting fatigue of components.

By studying the experimental work of aluminium alloy with/without shot peening on plain fatigue and fretting fatigue reported in 1984, researchers found that the main reason for improving the fretting fatigue strength is the compressive residual stress induced by shot peening. After reducing the roughness produced by the shot peening, the strength of fretting fatigue decreased [7]. Later on, researches of the same group investigated the effect of the roughening produced by different levels of shot peening on fretting fatigue of an aluminium alloy. After removing the compressive residual stress by heat treatment, the increased peening roughness is progressively less damaging in fretting fatigue [9], as listed in Table 1. This tendency could be explained by Bramhall that the initiation of crack under fretting fatigue area needs a critical volume of material. On rough surfaces the volume of contact regions does not reach this critical volume, hence the fretting fatigue life is increased. In the study of fretting fatigue on aluminium alloy treated by shot peening and laser peening [10], experimental results show that the fatigue life increases with increasing the roughness of contact surfaces due to decreasing the stress intensive factor in mode II.

Table 1. Fatigue strengths (in MPa) after removal of residual compressive stress and work hardening[17]

Peening intensity	Roughness peak to valley (μm)	Number of cycles: 10^6		Number of cycles: 10^7	
		Fretting fatigue strength	Reduction (%)	Fretting fatigue strength	Reduction (%)
Unpeened		150		100	
12-16 A	32	75	25	93	46
16-20 A	62	77	23	83	34
8-10 C	98	82	18	50	20

Effects of shot peening on fretting fatigue of Titanium alloy were also studied under both room temperature [11] and elevated temperature (350 ℃ and 500 ℃) [12] conditions. Under the room temperature condition, the compressive residual stress and surface roughness are predominant factors in enhancing fretting fatigue resistance, presented in Figure 1 (a). In contrast to the case of room temperature, the surface roughness plays negative influence on fretting fatigue resistance of Titanium alloy at 350 ℃, shown in Figure 1 (b). This is related to the correlation between compressive residual stress and shot peening-induced surface morphology changes. They are different at elevated temperature compared to room temperature.

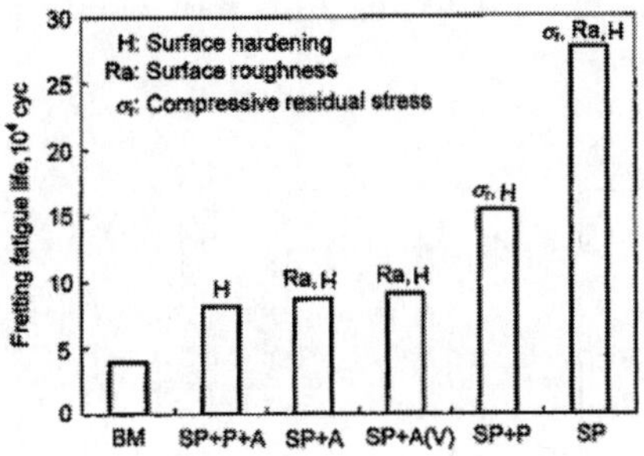

(a)

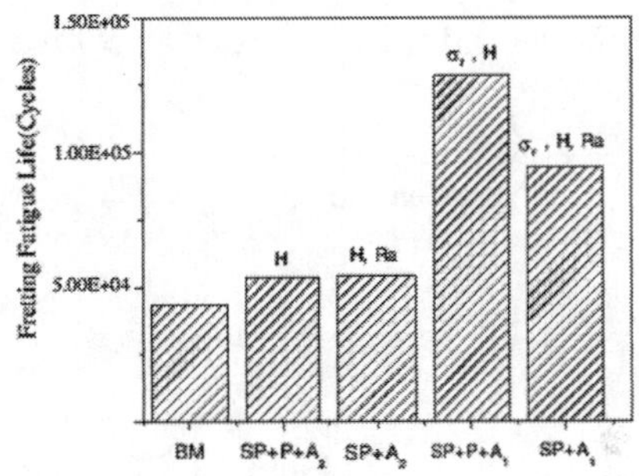

(b)

Figure 1. Fretting fatigue lives with different surface treatment,(a) under room temperature [18], (b) under 350 °C [19]

Shot blasting is a treatment for preparing the surface to be coated. By shot blasting, this surface is roughened to provide a strong mechanical bond between coating and the substrate. The objectives of shot peening and shot blasting are different. Shot peening is intentionally applied for improving the fatigue resistance. However, the shot blasting is a technical treatment before coating. In the experimental work on Al–Mg–Si alloy AA6061 as presented in [20], it is found that shot blasting significantly increased fretting fatigue life by a factor 2.4 at a maximum cyclic stress of 169 MPa, but at higher stress levels shot blasting slightly reduced the fretting fatigue life.

2.2. *Regular rough surface*

The vibration rolling process is a finishing technique that produces a regular surface microgrooves pattern. As shown in Figure 2, turning and polishing, which are conventional finishing methods, generate irregular pattern on the surface while the regular patterns are produced by vibrorolling, though the groove depths are equal by these three different methods. Researchers [21] of the St Petersburg Precision Mechanics and Optics University found that regular grooves help remove wear debris from the contact surface and distribute stress and plastic deformation, effectively reducing the fretting damage.

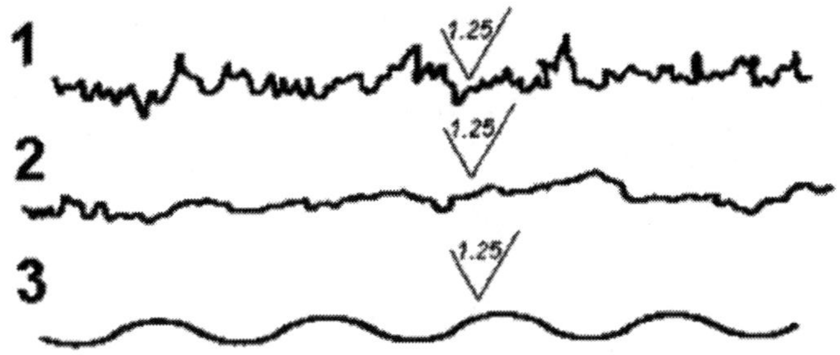

Figure 2. Topography of surfaces, finished by different methods: (1) turning, (2) polishing, (3) vibrorolling.[21]

The regular micropore is another type of regular surface roughness pattern, as shown in Figure 3. During fretting, the wear debris may escape into these micropores. In the study of fretting fatigue with cylinder on beam subjected the fatigue load contact [10], micropores of the contact surface are produced by the laser surface texturing. It is found that effects of micropores of component surface play different roles on fretting fatigue depending on if this component is subjected the fatigue load. As listed in Table 2, the fretting fatigue life of the beam with deeper regular micropores is reduced due to the generation of the stress concentration induced by the micropores on the beam surface. In contrast, the appearance of micropores on the cylinder surface has beneficial influence on the fretting fatigue life, which is twice compared to the case without textured surface. This is because that the wear debris moves into these micropores and thus reduces the relative slip between contact surfaces.

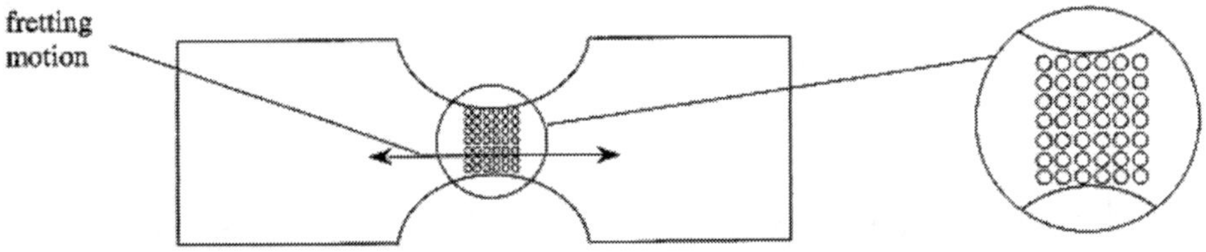

Figure 3. Micropores on the beam for fretting experiments [10]

Table 2. Mean life in the fretting fatigue experiments [10]

Experiment type	Pore depth (μm)	Mean fretting fatigue life ($\times 10^5$ cycles)
No textured	0.0	5.18
Textured beam	5.3	4.67
Textured beam	2.6	7.82
Textured cylinder	5.9	9.34
Textured cylinder	31.9	9.72

2.3. Random rough surface

In industry applications, most surfaces of parts are finished by turning, grinding, lapping, honing, etc. The surface microgrooves distribute vary in shape, size and relative location due to the asperities and waviness. In this case, many roughness parameters are created for describing the roughness of the surface. In the study of fretting fatigue, usually the average roughness *Ra* and the maximum height of the profile *Rt* are used to describe the roughness of the contact surface.

In the investigation of fretting fatigue on Titanium ally in low and high cycle fatigue life regions [22], researcher found that the fretting fatigue life with smoother surface (Ra = 13.2~32.2 nm) is longer than the case with rougher surface (Ra = 120~140 nm) in both low cycle and high cycle fatigue life conditions, displayed in Figure 4.

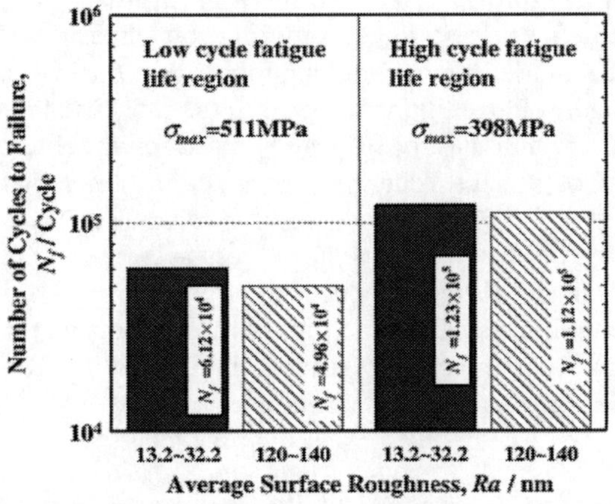

Figure 4. Fretting fatigue lives of fretting couples with different rough surfaces [22]

The roughness effects on fretting fatigue of Aluminium alloy also have been studied [23], especially on the process of the crack nucleation. By investigating the fretting fatigue results of three different rough surfaces shown in Table 3, researchers found that a smoother surface leads to a higher tangential load for the crack initiation presented as Figure 5 (a). By introducing effective linear loads, which is calculated by the effective contact length, it reveals that the crack initiation of 2024 Aluminium alloy depends on the local stress status determined by the real contact area between rough surfaces, shown as Figure 5 (b).

Table 3. Ra and Rt of different surfaces [23]

	R1	R2	R3
Ra (μm)	0.11	0.6	0.75
Rt (μm)	1.05	3.1	3.15

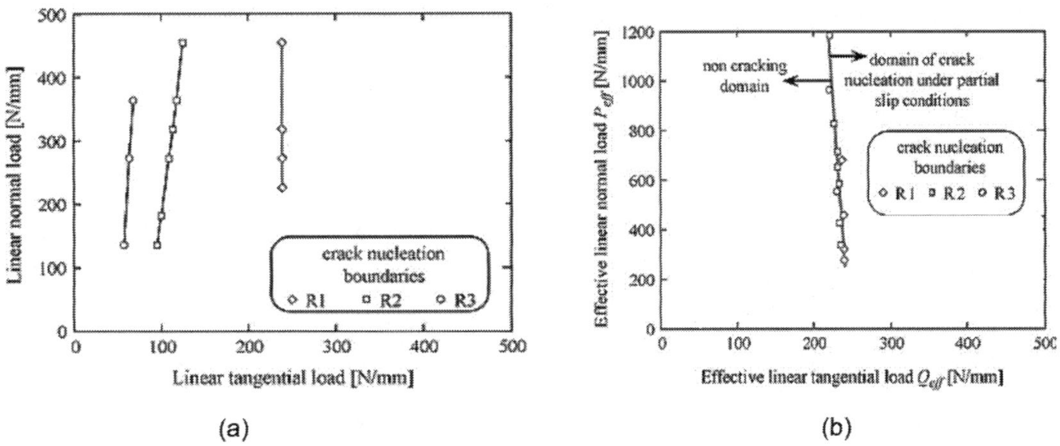

Figure 5. The relation between the crack nucleation and the linear loads/ effective linear loads with three different roughness [23], (a) linear loads, (b) effective linear loads

3. Numerical modelling on roughness effects on fretting fatigue

The research on modelling of fretting fatigue considering roughness effects is very limited. Kasarekar [24, 25] proposed a numerical modelling method that investigate the evolution of fretting fatigue life of Hertzian rough contacting bodies in fretting wear. One of the most used numerical modelling technique is Finite Element Analysis, which has been widely used in the literature for many engineering applications [26-39]. Effect of surface roughness parameters, i.e. root mean square, skewness and orientation is evaluated. The results show that the location of crack initiation changes from trailing edge to the stick-slip zone interface during fretting wear process according to the crack initiation parameter. Increasing the roughness of contact surfaces reduces the fretting fatigue life. Beheshti [40] introduced a robust approach for prediction of the surface tractions in rough contact under partial slip condition. This numerical model is applied to study the fretting fatigue crack initiation. Results also indicate that higher surface roughness leads to higher risk of crack initiation.

4. CONCLUSION

The research on effects of roughness of contact surfaces on fretting fatigue has been reviewed. Shot peening and laser peening are two finishing treatment used for improving the fretting fatigue resistance. This is because that they could generate compressive residual stress approaching the contact surface and roughen surface. Effects of roughness on fretting fatigue are contradictory. For the shot peening or laser peening experiments, rougher surface induces higher fretting fatigue resistance when removing the compressive residual stress. Introducing regular rough surface also improve the anti-fretting fatigue performance. However, on the random rough surface research, it is found that rougher surface has negative effect on the fretting fatigue life. The numerical study on the roughness effects on fretting fatigue is very limited. The reported results show that the higher roughness of surface reduce the fretting fatigue life.

References

[1] Martínez JC, Vanegas Useche LV and Wahab MA 2017 *International Journal of Fatigue* **100, Part 1** 32-49
[2] Kumar D, Biswas R, Poh LH and Abdel Wahab M 2017 *Tribology International* **109** 124-132
[3] Bhatti NA and Abdel Wahab M 2017 *Tribology International* **109** 552-562
[4] Resende Pereira KdF, Bordas S, Tomar S, Trobec R, Depolli M, Kosec G and Abdel Wahab M 2016 *Materials* **9** 639; doi:610.3390/ma9080639
[5] Ferjaoui A, Yue T, Abdel Wahab M and Hojjati-Talemi R 2015 *International Journal of Fatigue* **73** 66-76
[6] Yue T and Abdel Wahab M 2017 *Tribology International* **107** 274-282
[7] Pereira K, Yue T and Abdel Wahab M 2017 *Tribology International* **110** 222-231
[8] Yue T and Abdel Wahab M 2016 *Materials* **9** 597; doi:510.3390/ma9070597
[9] Yue T and Abdel Wahab M 2014 *Wear* **321** 53-63
[10] Volchok A, Halperin G and Etsion I 2002 *Wear* **253**(3–4) 509-515
[11] Vincent L, Berthier Y, Dubourg MC and Godet M 1992 *Wear* **153**(1) 135-148
[12] Neale MJ and Gee M. Chapter 2 - Industrial wear problems. A Guide to Wear Problems and Testing for Industry. Suffolk, UK: William Andrew Publishing; 2001. p. 3-III.
[13] Braunovic M 2009 *IEICE Transactions on Electronics* **92**(8) 982-991
[14] Fu Y, Wei J and Batchelor AW 2000 *Journal of Materials Processing Technology* **99**(1–3) 231-245
[15] Leadbeater G, Noble B and Waterhouse R 1984 Proc. Int. Conf. on Fracture ed editors (New Delhi) p. 2125–2132
[16] Liu KK and Hill MR 2009 *Tribology International* **42**(9) 1250-1262
[17] Waterhouse RB and Trowsdale AJ 1992 *Journal of Physics D: Applied Physics* **25**(1A) A236
[18] He DLJ 2001 *Acta Metallrugica Sinica* **37**(2) 5

[19] Zhang X and Liu D 2009 *International Journal of Fatigue* **31**(5) 889-893
[20] Naidu NKR and Raman SGS 2005 *International Journal of Fatigue* **27**(3) 323-331
[21] Bulatov VP, Krasny VA and Schneider YG 1997 *Wear* **208**(1–2) 132-137
[22] Takeda J, Niinomi M, Akahori T and Gunawarman 2004 *MATERIALS TRANSACTIONS* **45**(5) 1586-1593
[23] Proudhon H, Fouvry S and Buffière JY 2005 *International Journal of Fatigue* **27**(5) 569-579
[24] Kasarekar AT, Bolander NW, Sadeghi F and Tseregounis S 2007 *International Journal of Mechanical Sciences* **49**(6) 690-703
[25] Kasarekar AT, Sadeghi F and Tseregounis S 2008 *Wear* **264**(7–8) 719-730
[26] X. Nguyen H, N. Nguyen T, Abdel Wahab M, Bordas SPA, Nguyen-Xuan H and P. Voa T 2017 *Computer Methods in Applied Mechanics and Engineering* **313** 904-940
[27] Tran Vinh L, Lee J, Nguyen-Van H, Nguyen-Xuan H and Abdel Wahab M 2015 *International Journal of Non-Linear Mechanics* **72** 42-52
[28] Tran LV, Phung-Van P, Lee J, Wahab MA and Nguyen-Xuan H 2016 *Composite Structures* **140** 655-667
[29] Thai CH, Ferreira AJM, Abdel Wahab M and Nguyen-Xuan H 2016 *Acta Mechanica* **227**(5) 1225-1250
[30] Thai C, Zenkour AM, Abdel Wahab M and Nguyen-Xuan H 2016 *Composite Structures* **139** 77-95
[31] Phung-Van P, Tran LV, Ferreira AJM, Nguyen-Xuan H and Abdel-Wahab M 2016 *Nonlinear Dynamics* 1-16; doi:10.1007/s11071-11016-13085-11076
[32] Phung Van P, Nguyen LB, Tran Vinh L, Dinh TD, Thai CH, Bordas SPA, Abdel Wahab M and Nguyen-Xuan H 2015 *International Journal of Non-Linear Mechanics* **76** 190-202
[33] Phung Van P, De Lorenzis L, Thai CH, Abdel Wahab M and Nguyen-Xuan H 2015 *Computational Materials Science* **96** 495-505
[34] Phung Van P, Abdel Wahab M, Liew KM, Bordas SPA and Nguyen-Xuan H 2015 *Composite Structures* **123** 137-149
[35] Noda N-A, Chen X, Sano Y, Wahab MA, Maruyama H, Fujisawa R and Takase Y 2016 *MATERIALS & DESIGN* **96** 476-489
[36] Tran Vinh L, Lee J, Ly HA, Abdel Wahab M and Nguyen-Xuan H 2015 *International Journal of Mechanical Sciences* **96-97** 65-78
[37] Junyan Ni and Wahab MA 2017 *Computers & Structures* **186** 35–49
[38] Phung-Van P, Qui LX, Nguyen-Xuan H and Wahab MA 2017 *Composite Structures* **166** 120–135
[39] Phung-Van P, Ferreira AJM, Nguyen-Xuan H and Abdel Wahab M 2017 *Composites Part B: Engineering* **118** 125-134
[40] Beheshti A, Aghdam AB and Khonsari MM 2013 *International Journal of Fatigue* **56** 75-85

IOP Conf. Series: Journal of Physics: Conf. Series **843** (2017) 012057 doi:10.1088/1742-6596/843/1/012057

Tribologic analyses of a self-mated aluminium contact used for overhead transmission lines

V Franco Steier[1]

[1] Grupo de Mecânica Experimental e Computacional (GMEC), University of Brasília (UnB), 72444-240, Gama – DF- Brazil.
E-mail: vfs@unb.br

Abstract. The lifetime of aluminium components is often limited to their poor wear resistance. One example for such aluminium applications are overhead transmission lines. The sore points of these lines are the segments where the aluminium conductors are fixed to the line supports. The fixation is commonly realized via aluminium suspension clamps. Here, a superposition of different loads like traction and bending stresses, clamping forces and different types of wear occurs. To investigate the wear behaviour in these peculiar points, tribologic model tests were carried out. Within the tests, overhead conductor wires and aluminium plates, extracted from suspension clamps were reciprocally slid against aluminium plates (cylinder-on-plate test). The COF and a wear related parameter were recorded constantly. Subsequently, the loaded surfaces were analysed using confocal laser and electron scanning microscopy as well as energy dispersive X-ray spectroscopy. The investigation detected the formation of an oxidized tribologic layer between both components. The tribolayer, which mayor part adhered on the suspension clamps, was mostly formed from material removed from the conductor wires.

Introduction

The main damage of overhead transmission lines occurs at points where the conductors are connected to the line supports [1]. The fixation of the transmission line to the line support is realized with aluminium suspension clamps. The anchoring of the conductor is realized with two rounded half-shells that are bolted together with four screws (fig. 1b) [2]. Although the used devices prevent conductor movements directly under the clamping jaws (keeper/clamp), relative movements are unavoidable in the region between the keeper edge and the edge of the clamp (fig. 1b). The farther the distance to the keeper edge, the larger the movements between the components. These movements are the main reason for the damage of overhead conductors, as they are the cause for wear and fatigue of the conductor wires. The damage mechanism fretting fatigue was identified as the most server damage mechanism and the main reason for catastrophic transmission line breakdowns [3]. Fretting fatigue takes place in applications, which are exposed to a specific combination of loads: (a) a relative movement. In the analysed case, these movements are generated by wind induced oscillations of the conductor [4]; (b) a normal force, which is present due to the applied clamping force and (c) a tensile stress, which is generated by the dead weight of the conductor and the applied pre-load [5]. Further performed research found out that the conductor damage in the suspension clamps can be narrowed to an even smaller region (fig. 1a) [6].

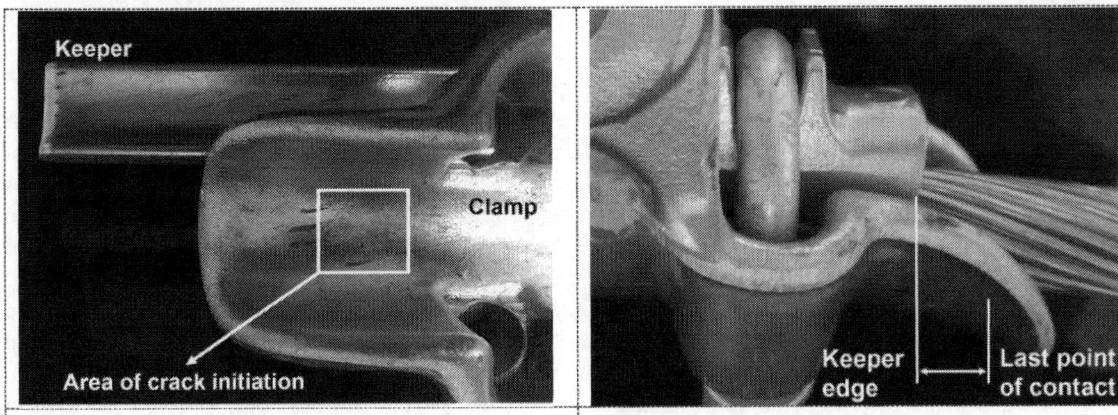

Figure 1a: Wear mark in a suspension clamp **Figure 1b:** Mounted suspension clamp

In fig. 1a, transverse lines are visible at the bottom of the suspension clamp and on the keeper surface. The visibility of these lines decreases distinctly with an increasing distance to the clamp edge. Whereas the last mm of these lines are coloured in deep grab and are therefore easy to distinguish, the rest of these marks did not undergo a colour change. The grey coloration of these marks is (a) related to a higher degree of oxidation and (b) to a topographic change on the clamp surface. Already published investigations revealed a material transference from the conductor wires to the clamp surface [6]. In contrast, the harder to visualize lines are generated by plastic deformation of the clamp material. The applied clamping force is strong enough that the conductor as well as the suspension clamps/keeper surfaces deforms plastically. The oxidized parts of the wear marks are indicating a gross slide wear regime [6]. In other word, huge relative movements between conductor and suspension clamps occur. The end of these wear marks are defined as the "last point of contact" between conductor and suspension clamp (fig. 1b). The area accentuated with a square (fig. 1a) indicates the region where stick and slip occurs (also called mixed mode). Such movements are known to be the most server mode for fretting fatigue crack initiation [5, 7]. On the right side of the square, no relative movements occur [7, 8].

As shown in fig. 1b, conductors for overhead transmission lines are built-up by many aluminium wires, which are organized in layers. Independent from the conductor type (e.g. pure aluminium conductors, steel reinforces conductors or aluminium alloy conductors), the outer layer of the conductors are made of aluminium, e.g. aluminium alloys. Even though wire breaks occur in the external and the internal layers, evidences for preferred first wire breaks in the external layer were found [8]. The reason for a preferred crack initiation in the outer layer is seen in the contact with the suspension clamp.

So far, no detailed tribological analysis of the conductor wire/clamp contact was carried out. As shown in fig. 1 and described elsewhere [6], a material transfer from the conductor wire to the suspension clamp occurs. This means, that the mayor damage occurs in the conductor and not in the clamp. An opposite behaviour would be more desirable, as suspension clamps could be renewed easier that the large conductors. Considering that tests with real conductors are very costly, the aim of the present investigation is to verify, if tribological model tests with dry sliding conditions can reproduce the damage type observed in the conductor/clamp contact. It needs to be mentioned that the used model

test was chosen to reproduce the tribological aspect of the contact. The fretting fatigue behaviour of the conductor wires cannot be investigated with the chosen test method, as no tensile stresses have been applied on the tested wire.
It is known that a reduced coefficient of friction (COF) has strong effects on the wear behaviour and the generation of fretting fatigue cracks [8, 9]. Therefore, it is expected that the lifetime of overhead transmission lines can be extended by modifying the surface of suspension clamps. As described above, this work focuses on the evaluation of the tribologic contact between conductor and clamp. The impact of the COF on the wear and fretting fatigue behaviour will be part of future research.

Materials and Methods:

The dry friction tests were carried out using specimens taken from an IBIS overhead conductor, which consists of an AA 1350-H19. The prefix H19 indicates that the wires were die drawn at room temperature. The nominal chemical composition of the wires are shown in table 1. The wires of overhead conductors are helically bundled to allow isotropic movements (radial direction). Due to the winding of the wires, the extracted specimens had small flections, which could not be completely removed by straightening. Such deflections negatively influence the tests results, as they avoid a plane line contact between cylinder (wire) and plate (suspension clamp). To diminish such influence, the size of the cylinder wires was limited to a length of approximately 7.8 mm. The preparation of the specimens was executed using a mounting device with a drilled retainer for the specimen (average specimen diameter 3.15 mm, diameter of the retainer 3.2 mm). To trim the cylinders on their testing length, the wires were fixed in the mounting device and both ends were grinded (abrasive paper Mesh 800) until the final length was reached. To avoid edge effects during the tests, the edges of the cylinders were manually rounded using the same grinding paper.
Most of the nowadays-used suspension clamps are made of aluminium alloy SAE 305 by mould casting. As the manufacturing process strongly influences the microstructure and the surface conditions of the devices, the used specimens were extracted from unused suspension clamps. The fabricated specimens for the cylinder-on-disc tests had a diameter of 24 mm and a height of 7.5 mm. The nominal chemical composition of the AA SAE 305 is listed in Table 1.

Table 1. Elements contained in the tested materials in %..

Materials	Al	Fe	Si	Mn	Zn	Cr	Cu	Mg
Conductor IBIS (AA 1350-H19)	99.5	0.4	0.1	0.05	0.05	0.01	0.05	-
Suspension clamps (SAE 305)	86.0	0.5	12	0.35	0.35	-	0.25	0.1

As shown in table 1, the material of the conductor can be considered as low alloyed aluminium, whereas the suspension clamp contains higher quantities of alloying elements (mainly silicon). Since the surface roughness influence the COF and the wear behaviour [10], all testes specimens were grinded and polished until they reached an equal surface condition (Ra = 1.16 µm).
The friction and wear tests were executed with a reciprocating wear tester (Optimol). A section of the testing equipment is shown in fig. 2.

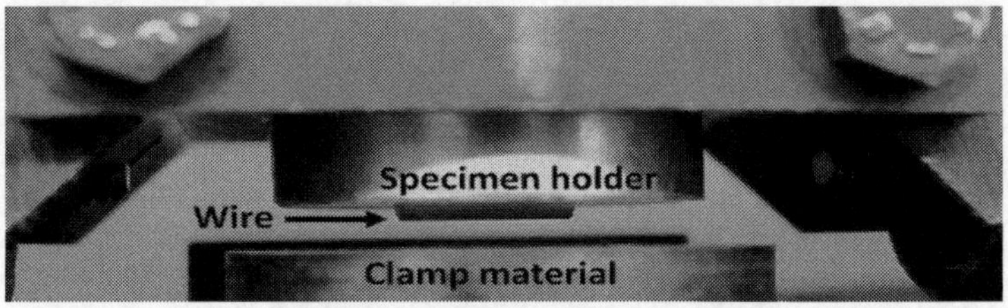

Figure 2: Wear and friction tester

The mounting of the tests were carried out in the following sequence. First, the prepared wires were clamped in the specimen holder (top side), which was fixed in the testing machine subsequently. Afterwards, the clamp material was positioned under the wire. To control if a plane contact occurred, both counterparts were slowly driven together. In case of a visible gap between the specimens, the clamp material was readjusted until no gap was noticeable. The test rig operates with unilateral movement. More precisely, the clamp material is fixed on the base (bottom) of the wear tester, while the specimen (top) is moved reciprocally to the left and the right.

The tests were carried out using a normal force of 20 N, a frequency of 200 Hz and a lateral displacement of 100 µm for 90 min. Three specimens were tested under these conditions. As the results showed a good accordance, no further tests were performed.

The COF and the vertical displacement (combined wear depth of the interfaces) was measured continuously by the testing machine. To abolish disaccords, the term vertical displacement will be explained more detailed. Before the start of the tests, the vertical displacement was set to 0. After this reset, any occurring wear resulting in a vertical displacement of the specimen holder (top side) can be recorded by the test machine. The result of this method is not the total wear volume (Due to the fact that wear occurs on both surfaces and in different lateral extensions), but it allows the detection of wear and friction effects resulting in vertical displacements. Such data is especially valuable for the investigation of the running-in phase, which cannot be analysed with post-test measurements (in the case of long lasting test cycles).

The measurement of the total wear volume (clamp and wire specimens) was realized with a confocal laser microscope (Olympus/Lext). The wear volume survey of the clamp specimens did not require further data processing. The determination of the wire wear volume required the use of a subsequently performed data processing. Fig. 3 shows the schematic drawing of the wire cross section after the wear tests. With the use of the confocal laser microscope, the dimension of the generated wear marks can be easily measured (via polygon adaptation). For the wear volume calculation, the wire wear mark was considered as rectangular (Dividing the measured wear mark area though the specimens length). Using this technique, a mean wear mark width can be calculated. In fig. 3, this theoretical wear mark surface is labelled with the letter s. With the additional assumption that the wire is perfectly circular, the wear volume can be calculated using equation 1 and 2.

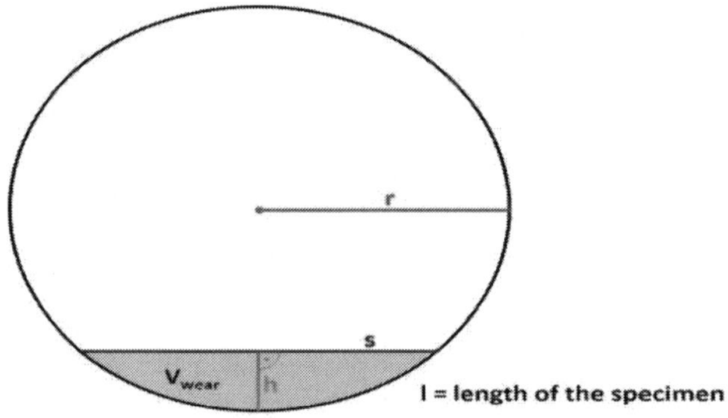

Figure 3: Schematic view of the wire cross section

- $V_{wear} = \left(r^2 \cdot \arccos\left(1 - \frac{h}{r}\right) - (r - h) \cdot \sqrt{2rh - h^2}\right) \cdot l$
- $h = r - \left(\sin\left(\arccos\left(\frac{\frac{s}{2}}{r}\right)\right) \cdot r\right)$

In the shown equations, r represents the diameter of the wire, s the normalized wear mark width, h the depth of the wear mark. To facilitate the understanding of fig. 3, the removed wear volume is shown in green.

The microstructure analysis as well as the investigation of the chemical composition were performed using a Scanning Electron Microscope (SEM) equipped with a Jeol 7500 Energy dispersive x-ray spectrometer (EDX). Beside the EDX measurements on the wear mark and the different bulk materials, a cross section scan of the wear mark (clamp) was performed. Therefore, the clamp specimens were cut in a 90 ° angle to the wear mark. Subsequently, the cut specimens was embedded and polished to scan the depth profile of the wear mark.

Results:

The used friction and wear tester measured the COF and the vertical displacement constantly during the tests. In fig. 4, the measured COF values are shown.

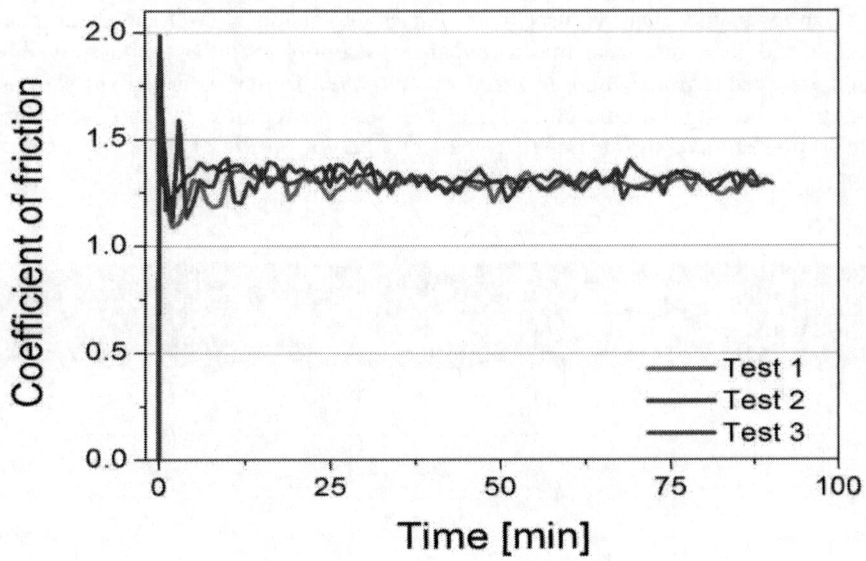

Figure 4: Coefficient of friction

The results shown in fig. 4 illustrate a clear running-in behaviour. In the first minute, the COF increased abruptly to a value of two. After this overshoot, the COF decreases to a value of approx. 1.15. This running in phase took approx. one minute. In the following eight minutes, the COF increased slightly to 1.3, where it reached its steady state value. The deviation between the results are negligible. The vertical displacement, also recorded by the wear tester, is shown in fig. 5.

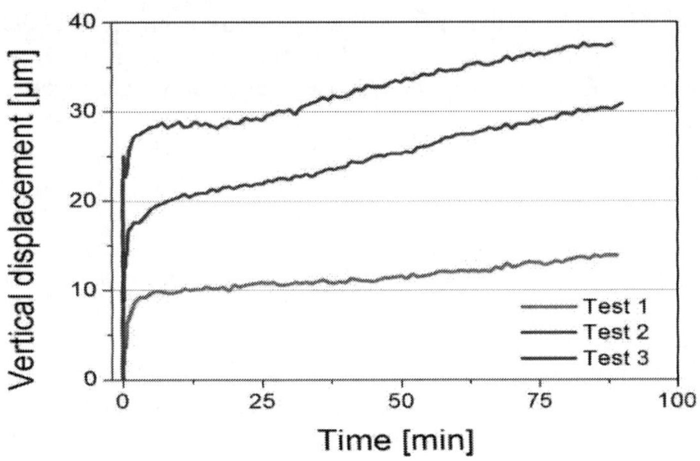

Figure 5: Vertical displacement curves

The measurement of the vertical displacement revealed strong deviation between the tested specimens. After the running-in phase, the slope of all tests stabilized. The main difference between the tests

occurred during the running-in phase. Here, the measure vertical displacement of the three specimens jumped to 10 µm, 28 µm and 18 µm, respectively.
After the wear tests, the wear volume was determined using a confocal laser microscope. The investigation of the generated wear marks revealed a strongly irregular topography. The wear marks on the clamps showed regions where material was removed from the surface, as well as areas, which were higher than the original surface level. In the built-up regions, material was deposited on the clamps. One representative colour image as well as a height profile of the same wear mark is shown in fig. 6.

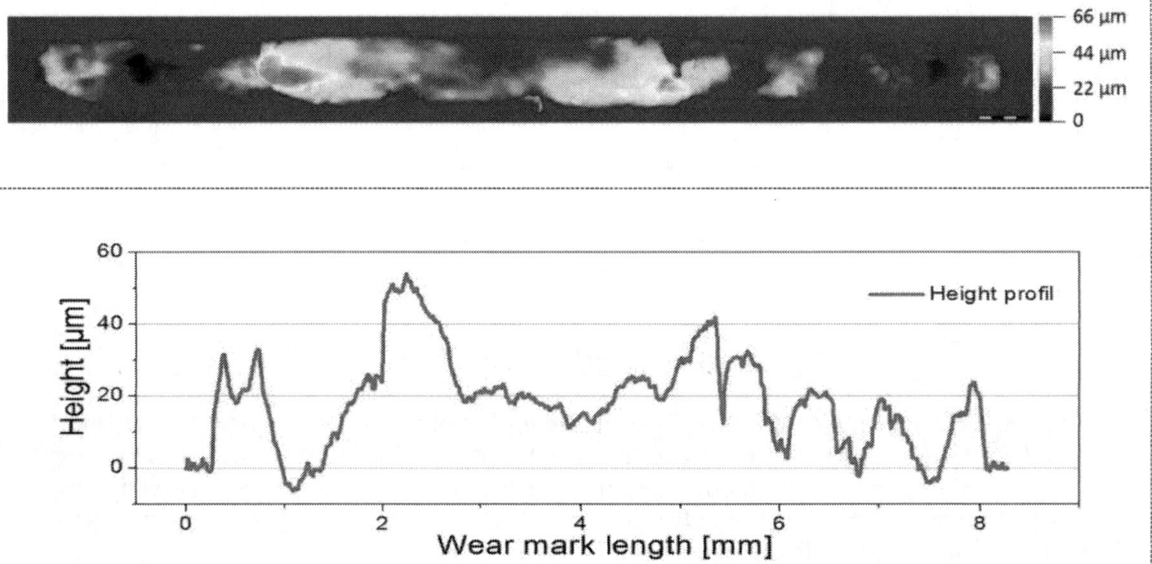

Figure 6: Wear mark and height profile of a wear mark

The colour image of fig. 6 reveals that most of the wear mark is higher than the original surface. The homogeneous violet colour distribution around the wear mark designate the unloaded surface. The darker colours indicate areas where material has been removed from the clamp surface. The brighter colours indicate areas that are high than the original surface. Here, material was transferred from the wire to the clamp. The additionally shown height profile visualizes the explained results with a topographic graph. To facilitate the understanding of the height profile, the unloaded surface level was set to 0. Due to this Y-axis adjustment, the negative values represent areas where material was removed, whereas the positive values denote a material deposition.
The software of the confocal laser microscope allows the determination of the wear volume. Therefore, an imaginary plane is visually adjusted congruently to the unloaded surface. Using this reference plane, the software is able to differentiate the deposited from the removed wear volume. According to this method, three different columns are presented in table 2. The wear volume is normally represent as a positive value. In the present study, a small amount of aluminium has been removed from the suspension clamps. On the other hand, a comparably high amount of material has been transferred to the clamps. Therefore, the total wear volume is presented with a negative sign. The column "total wear volume clamps" shows calculated values, where the removed wear volume has been subtracted from the deposited wear volume. The column "removed volume clamps" lists the amount of removed clamps material and the column "wire wear volume" represents the calculated wire wear volume (calculation method explained in the method section).

Table 2. Determined wear volume clamps and wires.

Number of test	Total wear volume clamps (106 μm^3)	Removed volume clamps (106 μm^3)	Wire wear volume
1	-64.35	0.63	64.29
2	-72.84	0.37	67.12
3	-71.22	0.15	78.88

Similar to the vertical displacement measurements, the listed total wear volume of the clamps show some differences between the tests. Comparing the measured wear volume of the clamps with the calculated wear volume of the wires, a good accordance was found. The maximum deviation between the values is 10 %. Considering that not all removed wire material was transferred to the clamps (loose abrasive particles were found on and around the wear mark), a good accordance can be assumed. The material removed from the clamps was found to be small for all tests.
The microstructural analysis of the wear marks was carried out using SEM and EDX techniques. Three representative images of one clamp/wire pair is shown in fig. 7. Fig. 7a images an overview of the wear mark on the clamp. Fig. 7b shows the wear mark on the wire and fig. 7c represents a higher resolution image of the wear mark on the wire.

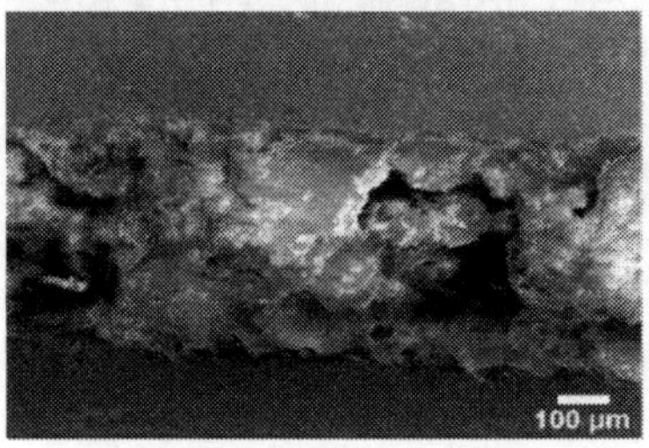

Figure 7a: Wear mark of the clamp

Figure 7b: Wear mark of the wire

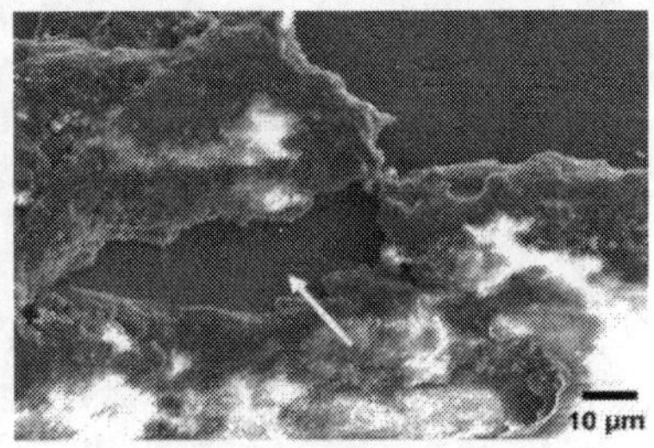

Figure 7c: Wear mark of the wire (tribolayer)

As already shown in fig. 6, a massive material transfer from the wire to the clamp took place during the tests. This phenomenon is also visible in fig. 7a, where mounds and cavities above the ground level can be recognized. In this overview picture, the wear mark can be clearly differentiated from the plane, unloaded surface. The wire wear marks show a similar topographic structure with mounds and cavities (fig. 7b). Furthermore, scale-like structures are visible on the surface of the wire wear mark. Such structures were also found on the clamp wear marks, nonetheless have not been illustrated in fig. 7a. Fig. 7c pictures a higher resolution image of the wire wear mark. Here, a tribologically unloaded surface is visible between two massive tribolayer scales (designated with a white arrow).
To gain more clarity for the interpretation of the SEM images, EDX measurements were carried out. The results are listed in table 3.

Table 3. Results of the EDX measurements

	Elements [Mass%]			
Position	Al	C	Si	O
Unloaded clamp	85.7	4.7	7.9	6.7
Unloaded wire	89.7	7.2	-	3.2
Wear mark clamp	28.8	5.7	0.3	65.1
Wear mark wire	33.8	3.6	2.5	60.1
Cross section wear mark	25.1	9.3	-	65.6

The executed EDX measurements confirm the data taken from the specification sheet of the manufacturers (table 1). The main difference between the aluminium wire and the aluminium clamps is the high amount of Si in the clamp material. The concentration of the other alloying elements shown in table 1 is too small to be detected with the used method. The main difference between the wear marks and the unloaded bulk material is the increased amount of oxygen. In this research, the increased oxidation state was used to differentiate the tribologic layer form the unloaded bulk material. The measurements carried out on the cross section of the clamp wear marks did not show a gradient in oxygen concentration or any other element, for which reason only one representative result is shown in table 3.

Discussion:

The analysis of the COF results revealed a pronounced running-in behaviour. In the first minute of the tests, the COF increased to a value of two. Such overshoots are typical for sliding wear tests with aluminium/aluminium friction partners and are contributed to the deformation of asperities [further named μd] [11, 12]. In the beginning of the tests, the contact between the two surfaces mainly occurs via few surface asperities. This means that the actual area of contact is drastically reduced. Due to the applied normal load, vertical evading is not possible, whereas wedged asperities need to deform in order to allow lateral movement. Especially the high COF values during the running-in phase are generated by the deformation of asperities [13]. In this initial phase, the biggest asperities are smoothened. Nevertheless, μd is also present after the running-in phase, as asperities are constantly built and unbuilt. The initially high friction values stand in good accordance with the found high vertical displacement (fig. 5). These results reveal that a relevant part of the wear occurs in the first seconds of the tests. The following decreased COF values are related to the smoothening of the asperities and the formation of a tribologic layer.

In the further course of the tests, the COF first decreases to approx. 1.15 and afterwards increases to its steady state value of approx. 1.3. In this transition time, surface adhesion [further named μa] plays an important role, what was not the case during the running-in phase. Due to the fact that surfaces are normally contaminated or, in the present case, oxidized, the importance of μa is little during the running-in phase [14]. μa contributes strong to COF, when the friction partners are self-mated. Removed aluminium clusters easily attach via interatomic bonds to the opposite surface when consisting of the same material [14]. After asperity deformation occurred and at least partially unoxidized clusters were detached from one surface, the integration on the friction partner or the tribolayer is easily possible.

Another friction and wear phenomena that contributes to the high COF is micro ploughing of hard asperities or abrasive particles [further named μp]. Similar to μa, μp gains more important after the running-in phase, as abrasive particles first need to be generated (present study). It is expected that abrasive wear particles are formed directly after the running-in phase. The generation of abrasive wear particles in self-mated contacts normally occurs in the following sequence: (a) asperity deformation (b) extraction of the softer sliding partner (c) oxidation of the removed or transferred material. Other research showed that material transfer occurs already after a low number of cycles [15]. In the present case, the softer wire material was transferred to the harder clamp surface. After the test specimens have been removed from the testing machine, dark abrasive particles were found on and around wear mark. Their presence proved the generation of loose abrasive particles. To sum up, the presence of all three described friction and wear phenomena could be demonstrated with the performed tests.

The results of vertical displacement curves stand in well accordance with the discussed results of the COF. The high COF values during the running-in phase are fitting well to the initially strong vertical displacement alias strong material loss between the wear partners. In literature, such severe forms of wear characterized by a macroscopic material transfer between solid surfaces is called galling [16]. Galling arises when strong plastic deformation occurs on at least one of the contact partners [17]. The tested materials have a comparably low yield strength. Therefore, plastic deformations occur even at low contact forces. Due to the reduced real contact area (asperity contact theory), even small normal loads are sufficient to generate plastic deformation. Other researchers, whose used the same experimental setup (cylinder-on-plate), found out that galling on aluminium (6061-T6) already occurred at nominal stresses 0.07 times the yield strength of the material [15]. The mentioned outcome reinforces the conclusion that the high COF and vertical displacement values in the beginning of the tests are mainly attributed to μd.

After the running-in process, the vertical displacement increased steadily until the end of the test. It is assumed that the asperity deformations in the beginning of the tests lead to a fast formation of a tribolayer. After its formation, the generated vertical displacement was found to be stable. Here, the relative movement occurs either between one of the friction partners and the tribolayer (which is attached to the opposite surface) or in between two tribolayers which are attached to their respective surfaces. As a strongly oxidized tribolayer was found on both friction partners, the latest possibly arose during the tests. Certainly, the tribolayers are further exposed to wear and asperity deformation/formation continues, whereas the vertical displacement shows a more or less stable slop. The comparison of the post-test determined wear volumes (clamp/wire) are standing in well accordance with each other as well as with the vertical displacement values. The differences between the single measurements first seem to be extraordinary. However, the microscopic views of the loaded surfaces can explain the found differences. As shown in fig. 8, the form and size of the wear marks differs strongly from each other.

Figure 8a: Wear mark on the conductor (test 3)

Figure 8b: Wear mark on the conductor (test 1)

The found differences are attributed to the not perfect line contact between the wire and clamp specimens. Even though their plane contact was visually verified, a complete contact over the entire wire length was not be reached for all tests. Comparing the image shown in fig. 8b, the right part of the wire did not have contact to the clamp surface in the beginning of the tests. After some cycles, the wear of the wire increased and the so far unloaded part of the wire was put in contact with the clamp surface. These stated differences are seen to be responsible for the deviations in vertical displacement results of the running-in phase (fig. 5).

The images taken with the SEM clearly show the formation a tribolayer (fig. 7). Different from the unloaded surface, the tribolayer has a scale-like structure, which also inhibits a higher brightness in the SEM images. Brighter reflexes indicate a stronger oxidation (present study). The scale-like textures are generated by the relative movement and the thereby generated grading of the asperities. Fig. 7c represents an amplified image of the wire wear mark edge. Here two surfaces are visible. (a), the bumpy oxidized surface of the tribolayer and (b), the smooth homogeneously flawed surface of the wire. The shown surface structure is typical for wire-drowned materials. Important to notice is that regions of unloaded wire surface are visible inside the wear mark (arrow). This shows that the lateral extension of the tribolayer cambers over the unloaded surface. The cylindrical shape of the wire facilitates this movement. It is assumed that with proceeding test cycles, parts of this overlapped material were detached from the wire and formed the found loose wear debris beside the wear mark.

The EDX analysis was used to clarify the differences between the clamp and the wire specimens as well as the formed tribolayer. As shown in table 3, the main differences between the wire and the clamp material is the amount of silicon. The carbon and the oxide detected in both materials can be related to the natural protection layer of the aluminium and negligible contaminations on the surface of the specimens. The similar results on both wear marks confirmed the visual impression that both surfaces are covered with a tribologic film. The differences in oxygen and carbon deviate slightly. Interesting to notice is that the measurement executed on the wire showed a slightly increased amount of silicon compared to the measurement on the tribolayer of clamp. As shown in fig. 6, there are some areas where the surface of the clamp material has been removed. Even though the main wear volume origins from the wire, small amounts were also removed from the clamp, for which reason silicon was also detected in the tribolayer of the wires. The cross-section of the wear mark was scanned via EDX to examine if gradual changes in the element compositions are present. Such changes could not be found, whereas only one result is shown in table 4. A comparison of the element concentration found in the cross-section with the results of the tribolayer did not reveal relevant differences.
After the straight tribologic discussion of the results, the evaluation of the relevance for the transmission line applications is done. As already described in the introduction, the performed model tests did not aim to reproduce fretting fatigue conditions (due to the missing tensile stress in the wire). However, the wear and tribologic conditions of the real application seem to be well represented by the performed tests. Similar to the in fig. 1a shown wear marks, a material transfer from the wire to the clamp was observed with the used test conditions and parameters. The lateral displacement of 100 μm stands in well accordance with the expected movement of the conductor close to the last point of contact. For investigations of the fretting fatigue behaviour of the wire, the stroke should be reduced, as the crack initiation occurs closer to the clamp centre.
Based on the detailed evaluation of the wear and friction behaviour of the aluminium/aluminium contact, the superiorly aim of tailoring contact conditions between conductor and suspension clamp can be realized with the same test procedure. Tests with surfaced that provide reduced COF and lower wear rates will be tested to delay wire damage and therefore improve the lifespan of overhead transmission lines.

Conclusion:

The wear and friction behaviour of aluminium wires used for overhead transmission lines was investigated using a cylinder-on-disc wear tester. The major findings of the carried-out experiments are the following:

- The aluminium/aluminium contact led to a sever wire damage after the applied friction/wear test.
- Adhesion between the materials, deformation of asperities and ploughing of generated abrasive particles were identified as reasons for the high COF and the high wear damage.
- After an initial smoothening process of the surfaces, a tribolayer was formed on both surfaces.
- The tribolayer mainly consist of oxidized aluminium.
- After an initially high COF values, the COF stabilized at a level of 1.3.
- The used test method was found to be adequate for the tribologic evaluation of the contact condition present in the described application.

Acknowledgement:

The author would like to acknowledge the Brazilian Program “Science Without Borders” (SWB-CNPq) for the financial support of the project 314034/2013-7. He furthermore would like to thank the Fraunhofer IWM, Freiburg for the possibility of using their wear tester.

IOP Conf. Series: Journal of Physics: Conf. Series **843** (2017) 012057 doi:10.1088/1742-6596/843/1/012057

References:

[1] Rawlins CB (1979). EPRI Transmission Line Reference Book: Wind-Induced Conductor Motion (Orange Book), *Electric Power Research Institute*.

[2] Preston B and Ramey GE (1986). Effect of suspension clamp geometry on transmission line fatigue. *Journal of Energy Engineering* 112, 168–184.

[3] Azevedo CRF, Henriques AMD, Pulino Filho AR, Ferreira JLA and Araújo JA (2009). Fretting fatigue in overhead conductors: Rig design and failure analysis of a Grosbeak aluminium cable steel reinforced conductor. *Engineering Failure Analysis* 16, 136-151.

[4] Kumar D, Biswas R, Poh LH, and Wahab MA. (2017). Fretting fatigue stress analysis in heterogeneous material using direct numerical simulations in solid mechanics. *Tribology International* 109, 124-132.

[5] Martínez JC, Vanegas Useche LV and Wahab MA (2017). Numerical prediction of fretting fatigue crack trajectory in a railway axle using XFEM. *International Journal of Fatigue* 100, Part 1, 32-49.

[6] Pereira K, Bordas S, Tomar S, Trobec R, Depolli M, Kosec G, and Abdel Wahab M. (2016). On the Convergence of Stresses in Fretting Fatigue. *Materials* 9, 639.

[7] CIGRE-Report-273 (2005). Overhead Conductor Safe Design Tension with Respect to Aeolian Vibrations. In Task Force B2.11.04. *CIGRE*.

[8] Hills DA, Nowell D and O'Connor JJ. (1988). On the mechanics of fretting fatigue. *Wear* 125, 129-146.

[9] Franco Steier V, Kalombo Badibanga R, Roberto MSC, Magalhães Nogueira M., and Araújo JA. (2014). Effect of chromium nitride coatings and cryogenic treatments on wear and fretting fatigue resistance of aluminum. *Electric Power Systems Research* 116, 322-329.

[10] Zhou ZR, Cardou A, Fiset M, and Goudreau S. (1994). Fretting fatigue in electrical transmission lines. *Wear* 173, 179-188.

[11] Bhatti NA, and Wahab MA. (2017). Finite element analysis of fretting fatigue under out of phase loading conditions. *Tribology International* 109, 552-562.

[12] Zhou ZR, Cardou A, Goudreau S, and Fiset M. (1996). Fundamental investigations of electrical conductor fretting fatigue. *Tribology International* 29, 221-232.

[13] Szolwinski MP, and Farris TN. (1996). Mechanics of fretting fatigue crack formation. *Wear* 198, 93-107.

[14] Gah, KHZ. (1998). Wear by hard particles. *Tribology International* 31, 587-596.

[15] Blau PJ and Devore CE. (1989). Interpretations of the sliding friction break-in curves of alumina-aluminum couples. *Wear* 129, 81-92.

[16] Mahdavian SM, Mai YW and Cotterel B. (1982). Friction, metallic transfer and debris analysis of sliding surfaces. *Wear* 82, 221-232.

[17] Challen JM and Oxley PLB. (1979). An explanation of the different regimes of friction and wear using asperity deformation models. *Wear* 53, 229-243.

[18] Suh NP and Sin HC. (1981). The genesis of friction. *Wear* 69, 91-114.

[19] Wycliffe P. (1993). Galling in aluminum alloys and Duralcan aluminum matrix composites. *Wear* 162, 574-579.

[20] ASTM (1989). ASTM-G98-89 Galling Resistance of Materials, Standard Test Method For. *ASTM*.

[21] Bhansali KJ and Miller AE (1982). The role of stacking fault energy on galling and wear behavior. *Wear* 75, 241-252.

An experimental study of flank wear in the end milling of AISI 316 stainless steel with coated carbide inserts

P B Odedeyi[1], K Abou-El-Hossein[2], M Liman[3]

[1] Research Assistant, Department of Mechatronics, Nelson Mandela Metropolitan University (NMMU), Port Elizabeth, South Africa

[2] Director of School of Engineering, (NMMU), Port Elizabeth, South Africa

[3] Research Assistant, Department of Mechatronics, (NMMU), South Africa

Email: khaled.abou-el-hossein@nmmu.ac.za

Abstract. Stainless steel 316 is a difficult-to-machine iron-based alloys that contain minimum of about 12% of chromium commonly used in marine and aerospace industry. This paper presents an experimental study of the tool wear propagation variations in the end milling of stainless steel 316 with coated carbide inserts. The milling tests were conducted at three different cutting speeds while feed rate and depth of cut were at (0.02, 0.06 and 01) mm/rev and (1, 2 and 3) mm, respectively. The cutting tool used was TiAlN-PVD-multi-layered coated carbides. The effects of cutting speed, cutting tool coating top layer and workpiece material were investigated on the tool life. The results showed that cutting speed significantly affected the machined flank wears values. With increasing cutting speed, the flank wear values decreased. The experimental results showed that significant flank wear was the major and predominant failure mode affecting the tool life.

1. Introduction

According to Baddo [1], the annual consumption of stainless steels has gone up to 5% for two decades. However, recent report by EN [1, 2] , showed that stainless steel production has increase globally by 8.3% in 2015 [2]. It was estimated that in 2006, a total of about 4 million tons of stainless steels were used for construction purposes globally [1, 3]. This account for about 14% of total quantity used. AISI 316 Stainless steel have become ever more main-stream materials with growing applications in the industrial, marine, chemical processing and construction industries.

Generally, Stainless steels are known to be a difficult to machine materials due to their high propensity to their toughness, work harden and moderately low thermal conductivity ability[3, 4]. Flank wear, a well-known and the most predominant types of tool wear can be seen as the rubbing of the wear land against the machined surface causes damage to the surface, flank wear occurs on the relief face which results in the formation of the wear land [5-9]. It can also be a volumetric loss at the top of the tool tip edge that are mainly caused by abrasion, it manifests at a very low operating speed [10].

According to Ciftci [11], the Surface of coated carbide cutting tools should must be able to develop resistance to abrasive wear, hardness and chemically passive to avoid the tool and the work material from interacting chemically with each other during machining. For these objectives to be achieved, in about 1970, coated carbide tools were technologically advanced and developed. This progress was a substantial development in cutting tool technology [12]. Ciftci [11] further explained that, coated carbides are essentially a cemented carbide insert coated with one or more thin layers of wear-resistant material, such as titanium carbide (TiC), titanium nitride (TiN), (TiAlN) used in this research study and/or aluminum oxide (Al_2O_3).

Further studies carried out by [13-15] revealed that that thin and hard coatings can reduce tool wear and improve tool life and productivity. This discovery has led to a situation whereby most of the carbide tools used in metal cutting industry are currently multilayered coated. According to Ciftci[11] and [16] conclusions, Higher surface roughness values at lower cutting speeds were ascribed to the high BUE formation tendency. Chipping of the cutting edges, which can seldom be validated by SEM inspections, was also discovered to be accountable for the high surface roughness values [17].

2. Methodology and Machine set up

The experimental setup was conducted on a Deckel Maho DMU 40 mono Block CNC machine of 12, 000 rpm maximum spindle speed where milling experiments were conducted with soluble castor oil as coolant. The machine parameters used can be found in Table 1. The flank wear rate on the inserts was observed using an equipped image analyzer called ZEISS stereo Microscope version 20 which provides observations with a magnification as high as 150X. It possesses three focal lenses with motorized zoom expansion and resolution adjustment. The panel combines buttons, joystick and touch screen in a compact design, allowing intelligent control of all microscope functions with real time display of main microscope parameters.

Table 1: Machine parameters

Machine conditions	**Specifications**		
	Low	**medium**	**high**
Feed rate (mm/rev)	58	243	520
Cutting speed (rpm)	2900	4050	5200
Depth of Cut (mm)	1	2	3
Coolant type	Castor oil		

2.1 Workpiece

The workpiece that was used in this experiment is AISI 316 stainless steel. This type of stainless steel belongs to the family of austenitic stainless steel 300-series which are generally difficult to machine. It is an austenitic chromium nickel stainless steel containing molybdenum Table 2 and Figure1(a) and (b). The steel contains high percentage of Nickel than 304 stainless steel.

The resultant composition of this material in Table 2 and Figures 1(a) and (b) give these steels much improved corrosion resistance in many aggressive environments. The molybdenum addition ensures more resistance to pitting and cavies' corrosion in chloride containing media, sea water and chemical environments such as sulfuric acid compounds, phosphoric and acetic acids. 316 stainless steel with the properties in Table 2 offers good strength and creep resistance and possess excellent mechanical and corrosion-resistant properties at sub-zero temperatures. When machining 316 stainless steel, unlike other austenitic steels, it alloys group machines with a rough and stringy swarf., rigidly supported tools with as heavy a cut as possible is used to prevent glazing.

Table 2: General Chemical properties of 316 Stainless steel (%):

C	Mn	P	S	Si	Cr	Ni	N	Mo
0.08max	2max	0.045max	0.03max	0.75max	16-18	10-14	0.1max	2-3

2.2 Cutting tool

The cutting tests were conducted using a 16-mm diameter Indexable end-mill with two cutting edges. Figure 2 shows a sample picture of the tool from Kennametal. KC725M Grade Inserts utilized are composed of carbide grade with a TiAIN coating. A high-performance TiAlN-PVD-coated carbide grade for milling steel, stainless steel and ductile cast iron. The good thermal shock resistance of the

IOP Conf. Series: Journal of Physics: Conf. Series **843** (2017) 012058 doi:10.1088/1742-6596/843/1/012058

substrate makes this grade ideal for both wet and dry machining. KC725M is primarily for use in general and heavy machining.

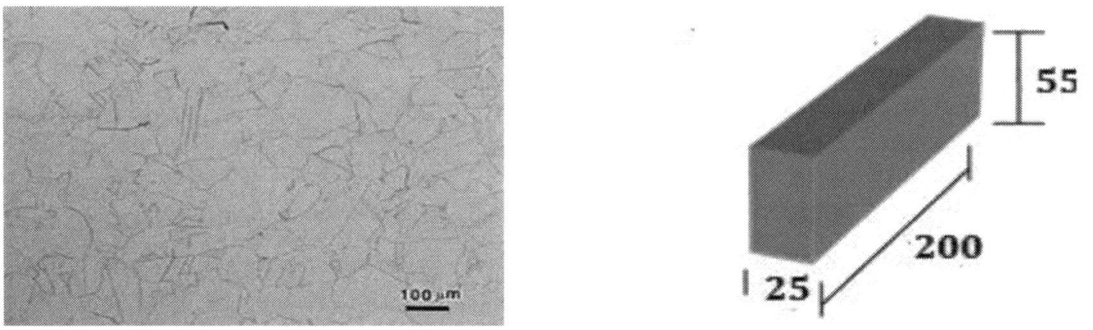

(a) (b)

Figure 1: (a) Typical microstructure of 316 SS (b) The dimension of the work piece used (200 x 55 x 25) mm.

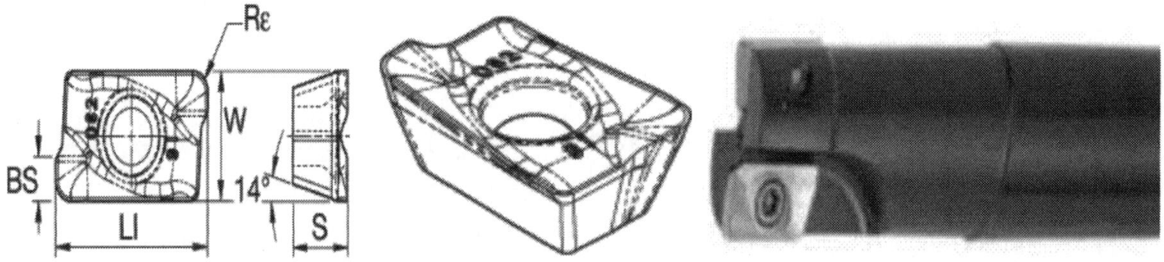

Li=12.06, S=3.7, W=6.74, Bs=1.3, Re=1.19, hm=0.082

Figure 2: Indexable tool inserts and tool holder from Kennametal

3. Analysis and Discussion

3.1 Effects of cutting Speed on Tool life

The combination of cutting speeds at constant feeds has a tremendous significant importance on the value of tool life. In the cutting tests carried out on AISI 316 stainless steel, an increase in cutting speed at a constant feed produced a visible reduction in tool life. Looking closely at Figures 3 (a) (b) and (c), an increase in cutting speed validates the reduction in tool life. It is also noticeable that even at constant feed rate of various cutting speed, a reduction in tool life is also experienced. In Figure 3(a), at high speeds of 310mm/min and feeds of 0.06 mm/rev, almost no effect was observed on the tool for the first 10 mins of tool life. Consequently, the same trend was observed in Figures (b) and (c). However, at low feed of 0.02mm/rev (Figure 3(b)), the tool life almost double at high speed of 97mm/min when compare to Figure 3(c) at feed rate of 0.06 mm/rev and speed about 310mm/min. this could due to chip load at this feed rate. To conclude on the effect of cutting speed on tool life, one can say highspeed end milling of AISI 316 should be carried out at low values of feed such as 0.02mm/rev.

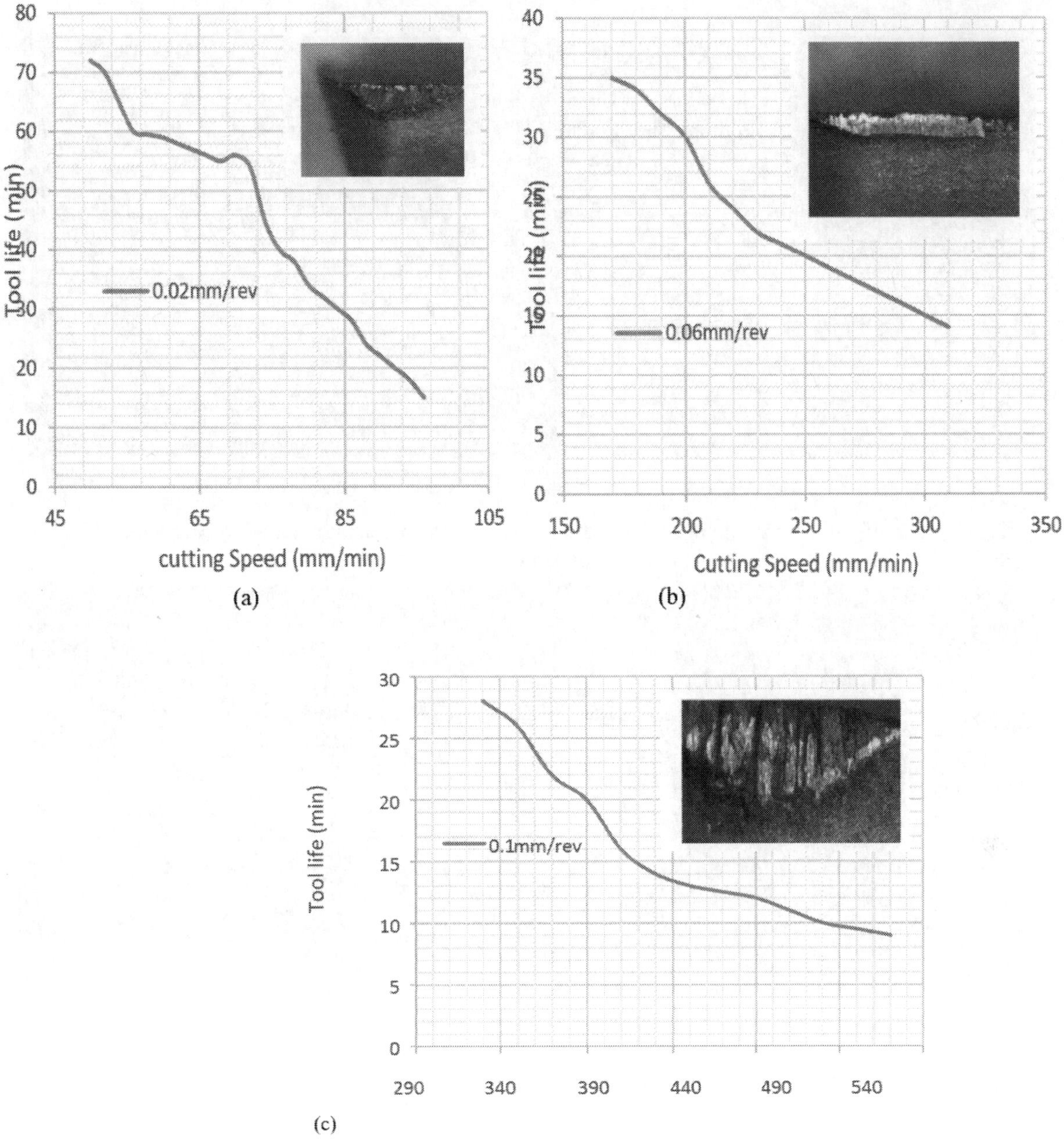

Figure 3: (a) Cutting speed versus Tool life at feed 0.02mm/rev (b) Cutting speed versus Tool life at feed 0.06mm/rev (c) Cutting speed versus Tool life at feed 0.1mm/rev

3.2 Modes of wear during machining AISI 316 stainless steel

Built-up edge, crater wear and Notch wear are the types of flank wears that are noticeable on the tool cutting edge at the location corresponding to the original surface of the machined part see Figures 4(a), 4(b) and Figures 5(a), 5(b). Flank wear is a groove development that occurs concurrently on the face and flank of the tool at the depth of cut. This may be because of Machining parts with severe (hard or oxidized) surfaces. It occurs because the original work surface is harder and more abrasive than the internal material due to work hardening from previous machining. Flank wear was noticeable as a mode of failure observed during the cutting experiments.

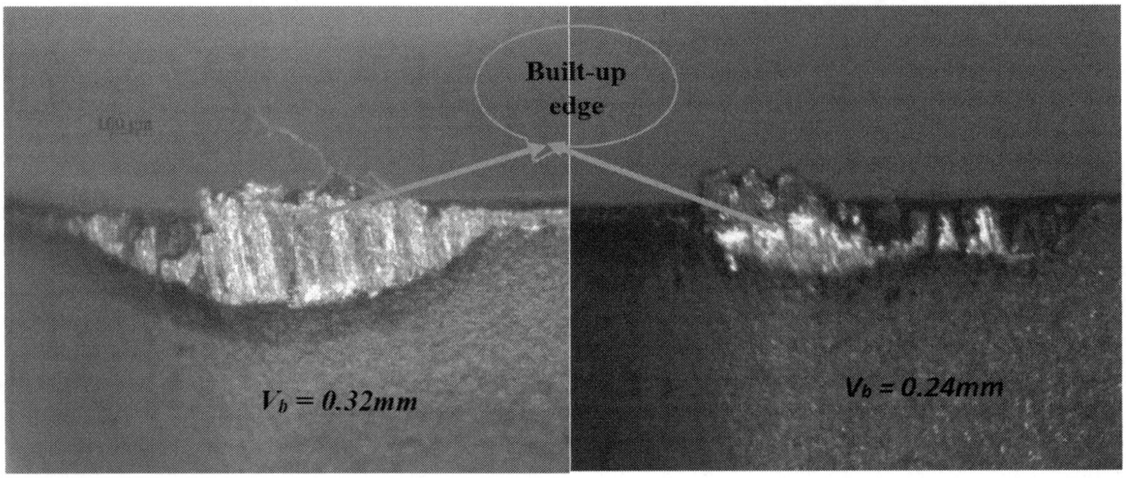

Figure 4: (a) Flank Wears at cutting speed V = 205m/min (b) Flank wear at cutting speed V = 261m/min

Figure 5:(a) Flank Wear at cutting speed V = 146m/min (b) Flank Wear at cutting speed V= 205m/min

This mode was clearly noticeable at conditions of the two highest cutting speeds (205 and 261) m/min and low feeds (0.02 mm/rev) used in this study (see Figures 4(a), 4(b) and Figures 5 (a), 5(b)). This failure is however due to the work hardening effect produced in the workpiece surface after being machined at low feeds. On the tool face, crater wear appears as a shallow trough in localized areas (Figure 5a). This patterns indicate that the tool material is diffusing into the chip. It causes very high temperatures on the tool face especially when carrying out test without coolant. Crater wear will keep rising until it reaches the cutting edge leading to chipping or fracture. Another common flank wear that was noticeable was BUE which is caused by low surface feet per minute or poor shearing action of the work piece material (Figures 5(a) and (b)). The work piece material is stick to the face of the tool due to sometimes an attraction of the work material to the insert or its coating which eventually results into chipping of the tool cutting edges.

4. Conclusion

The conclusions drawn from this study on end-milling of AISI 316 stainless steel using a newly developed multilayered coated carbide tool is that an increase in cutting speed resulted in a dramatic reduction in tool life. The tool lives of the highest three cutting speeds used in this study (146, 205 and 261 m/min) were not too far from one another while the tool life at Spindle speed = 2900rpm (cutting speed =146m/min) was almost doubled and the most predominant mode of tool failure was flank wear. It was also discovered that Built Up Edge happened at high values of cutting speeds (205 and 261 m/min) and feeds, which come to an agreement with the findings by Abou el Hossein et [17] and Sun et [16]. Crater wear was noticeable because of material being diffused into the chip at a very high temperature when machining at a low cutting speed of 146m/min. This type of wear (crater) could be worse if coolant is not applied. A low feed of 0.02m/rev will be ideal to maximize the tool life when machining AISI 316 Stainless steel.

References

[1] Baddoo N 2008 Stainless steel in construction: A review of research, applications, challenges and opportunities *J. Constr. Steel. Resrch.* **64** 1199-1206
[2] http://www.engineeringnews.co.za/print-version/global-stainless-steel-production-up-83-in-2014-2015-04-28.
[3] Nomani J, Pramanik A, Hilditch T, and Littlefair G 2013 Machinability study of first generation duplex (2205), second generation duplex (2507) and austenite stainless steel during drilling process *Wear* **304** 20-28
[4] Peckner D 1977 Handbook of stainless steels (New York : McGraw-Hill)
[5] Boud F M 1998 An investigation of the relationship between temperature, forces and tool wear in turning and drilling (Aston University)
[6] Khrais S K and Lin Y 2007 Wear mechanisms and tool performance of TiAlN PVD coated inserts during machining of AISI 4140 steel *Wear* **262** 64-69
[7] Barry J and Byrne G 2001 Cutting tool wear in the machining of hardened steels: Part I: Alumina/TiC cutting tool wear *Wear* **247** 139-151
[8] Che-Haron C 2001 Tool life and surface integrity in turning titanium alloy *J. Mat. Procs. Technol* **118** 231-237
[9] Lee M, Horne J, and Tabor D 1979 The mechanism of notch formation at the depth of cut line of ceramic tools machining nickel base superalloys *Wear. Mat* 460-469
[10] Oosthuizen G 2009 Innovative cutting materials for finish shoulder milling Ti-6A1-4V aero-engine alloys (University of Stellenbosch)
[11] Ciftci I 2006 Machining of austenitic stainless steels using CVD multi-layer coated cemented carbide tools *J. Triblgy. Inter* **39** 565-569
[12] Sarwar M, Zhang X-y, and Gillibrand D 1997 Performance of titanium nitride-coated carbide-tipped circular saws when cutting stainless steel and mild steel *J. Surf. Coatings. Technol* **94** 617-621
[13] Lim C, Lim S, and Lee K 1999 The performance of TiN-coated high speed steel tool inserts in turning *J. Triblgy. Inter* **32** 393-398
[14] Ezugwu E and Okeke C 2001 Tool life and wear mechanisms of TiN coated tools in an intermittent cutting operation *J. Mat. Procs. Technol* **116** 10-15

[15] Nordin M, Sundström R, Selinder T, and Hogmark S 2000 Wear and failure mechanisms of multilayered PVD TiN/TaN coated tools when milling austenitic stainless steel *J. Surf. Coatings. Technol* **133** 240-246
[16] Sun F, Li Z, Jiang D, and Chen B 1998 Adhering wear mechanism of cemented carbide cutter in the intervallic cutting of stainless steel *Wear* **214** 79-82
[17] Abou-El-Hossein K and Yahya Z 2005 High-speed end-milling of AISI 304 stainless steels using new geometrically developed carbide inserts *J. mat. procs. Technol* **162** 596-602

Acknowledgement
Many thanks go to the Research Capacity Development (RCD) Department of Nelson Mandela Metropolitan University for their Financial supports.

AE MONITORING OF DIAMOND TURNED RAPIDLY SOLDIFIED ALUMINIUM 443

G Onwuka[1] , K Abou-El-Hossein[2*] and Z Mkoko [3]

[1]Ultrahigh precision engineering research unit, Department of Mechatronics, Nelson Mandela Metropolitan University, Port Elizabeth, South Africa. s214006611@live.nmmu.ac.za

[2]Ultrahigh precision engineering research unit, Department of Mechatronics, Nelson Mandela Metropolitan University, Port Elizabeth, South Africa. khaled.abou-el-hossein@nmmu.ac.za

[3]Ultrahigh precision engineering research unit, Department of Mechatronics, Nelson Mandela Metropolitan University, Port Elizabeth, South Africa. s8502536@live.nmmu.ac.za

Abstract. The fast replacement of conventional aluminium with rapidly solidified aluminium alloys has become a noticeable trend in the current manufacturing industries involved in the production of optics and optical molding inserts. This is as a result of the improved performance and durability of rapidly solidified aluminium alloys when compared to conventional aluminium. Melt spinning process is vital for manufacturing rapidly solidified aluminium alloys like RSA 905, RSA 6061 and RSA 443 which are common in the industries today. RSA 443 is a newly developed alloy with few research findings and huge research potential. There is no available literature focused on monitoring the machining of RSA 443 alloys. In this research, Acoustic Emission sensing technique was applied to monitor the single point diamond turning of RSA 443 on an ultrahigh precision lathe machine. The machining process was carried out after careful selection of feed, speed and depths of cut. The monitoring process was achieved with a high sampling data acquisition system using different tools while concurrent measurement of the surface roughness and tool wear were initiated after covering a total feed distance of 13km. An increasing trend of raw AE spikes and peak to peak signal were observed with an increase in the surface roughness and tool wear values. Hence, acoustic emission sensing technique proves to be an effective monitoring method for the machining of RSA 443 alloy.

Keywords: Rapidly Solidified Aluminium (RSA), Melt spinning process, Ultra-High Precision Diamond Turning (UHPDT, Acoustic Emission(AE), Root Mean Square (RMS), Single Point Diamond Turning (SPDT), Total Integrated Scatter (TIS).

1. Introduction

Previously, diamond turned and more recently polished conventional aluminium AA6061 had been widely employed in the production of mirrors used for astronomical and terrestrial instruments. Reasons are understandably from its reflective properties, high specific stiffness and the ability maintain the coefficient of thermal expansion when subjected to severe temperature conditions [1, 2]. Furthermore, a low cost is incurred while applying AA6061 to astronomy due to sufficient available data on this material, hence selecting the material does not require new certification trajectories [1].

However, conventional AA6601 is a polycrystalline material which comprises of different crystallographic orientations creating differences in hardness and shear incompatibility in the neighbouring grains. This inherent defect creates a major drawback as the single point diamond turning of AA6061 can only produce surface roughness values in the range of 5-8nm which is only adequate for the infrared spectral range but not for the visual spectral range and near ultraviolet (NUV) spectral range [3]. Alloys with smaller grain size can produce better surface roughness values which determines the Total Integrated Scatter (TIS) [4].

Total integrated scatter from a given mirror surface is determined by the ratio of the frequency band-limited root mean square surface roughness (Rq) to the wavelength of light (λ) and it us used to measure the amount of scatter from the total reflected light by a given mirror [5]. The TIS equation (Eq. 1) below infers that the minimum surface roughness (Rq) required to generate 1% scatter at a minimum visual wavelength (λ) of 300nm will be 2.38nm which is about half the achievable surface roughness value of conventional aluminium alloy [4]. Different methods have been successful applied to achieve lower surface roughness values while machining AA6061, though not without some inherent challenges. One of these methods is plating with high purity aluminium before machining. The Alumni plating is known to have achieved a surface roughness of 2-4nm but it creates a difficulty in cleaning the machines surface due to its softness [1]. The use of nickel plating before machining also helps to achieve a surface roughness range of 1-2nm but the difference in the thermal expansion of nickel and aluminium could result in bimetallic bending at extreme temperatures [1]. Furthermore, extra cost and time is incurred with the process of plating conventional aluminium or optical polishing to achieve better surface roughness values. These conditions are not desirable in a manufacturing process.

$$TIS = \left[\frac{4\pi Rq}{\lambda}\right]^2 \quad Rq << \lambda \qquad (1)$$

From a comprehensive view, the developing trends in the use of aluminium for optical applications are viewed from two broad categories. Firstly, an improvement on the material structure observed progressively from the use of traditional aluminium to conventional aluminium 6061 and the birth of rapid solidification processes for the development of rapidly solidified aluminium grades like 905,708, and 443. Secondly, a development of the machining processes employed for the production of the optics in relation to improving on the required shapes and accuracies. As a result, instigating a rise in the need for effective process monitoring techniques. This process monitoring strategies help to study and understand the performance of the newly developed RSA alloy grades and their manufacturing processes.

With these recent development of interested researchers in RSA alloys 901, t6, 708, etcetera. There is limited available literature on AE monitoring of diamond turned RSA alloy 443.

2. Literature Survey

2.1. Rapidly Solidified Aluminium

Conventional aluminium materials are produced through powder metallurgy which cannot achieve the best uniformity of the alloying elements [2, 6]. Rapid solidification can be used to achieve the following: formation of new metastable phases, extension of the solid solubility to give precipitate strengthening and solute strengthening and refinement of the grain size of the matrix and other constituents to give fine grain strengthening [7-9]. Rapid solidification processes are now being applied to overcome the inherent challenges with conventional aluminium through refining the structure of the aluminium alloys. This process helps to manufacture uniform aluminium alloys with small grain sizes hence reducing the use of exorbitant and vulnerable plating to achieve lower surface roughness values [1]. Furthermore, rapidly solidified aluminium alloys contain fewer impurities which is advantageous as it reduces the chances of break outs occurring during subsequent polishing [2].

The fundamental principle in rapid solidification process is the removal of heat from a mass of liquid metal at a fast rate. Generally, this is made possible by controlling a thin mass of liquid metal in a particular dimension to be in good thermal contact with an efficient heat sink [10]. A range of diverse rapid solidification processes have been developed including: inert gas and water atomisation [11, 12], laser and electron beam surface melting [13], melt extraction [11], melt spinning and planar flow casting and plasma and atomised spray deposition [11]. Judging from the grades produced, melt spinning is the best rapid solidification technique [7].

The melt spinning process is a rapid solidification technique with cooling rates up to 10^6 k/s. The melt is prepared using different alloying elements which is poured through a small nozzle on a rotating copper wheel to obtain rapidly solidified ribbons. The ribbon is chopped to flakes and gathered in a vessel at high cooling rates in the range of 1Mk/s. The flakes are degassed and subjected to hot isostatic pressure at which a consolidated material can be realized in the form of billets [2]. These billets can be further extruded or forged to required dimensions. The process provides the ultimate solution for light weight high-end applications in optics, precision equipment, aerospace, medical, electronics and automotive industries [6]. Melt spinning process (MSP) has proven to be a better manufacturing method compared to powder metallurgy since it can be used to refine the grain size of the aluminum about 10-50 times thereby making the grain size of rapidly solidified aluminum alloys very small compared to the conventional aluminum. As a result, RSA alloys can be machined to smaller surface roughness values for applications in the near ultraviolet wavelength spectral range because MSP generates ultrafine and homogenous micro structures which enable better diamond turning and surface finish generation hence more suitability for a wide range of optical applications.

2.2. UHPDT of Rapidly Solidified Aluminium Alloys

With the advancement in technology and increase in material application to the development of micro-parts, moulding process has become an indispensable mass manufacturing process. Therefore, several products, micro-lenses and micro-components are being manufactured by moulding process. Furthermore, the molding process has been widely employed in plastic and glass lens production, automobile, aerospace and other manufacturing industries [14].

6061 Aluminum alloys, nickel coated steel alloys and beryllium copper alloys have been frequently used in the production of lens mold inserts because they are easily processed with single point diamond turning [15]. Of these materials, Al 6061-T6 was a preferred choice for mold inserts due to its high machinability, and low density owing to ease of assembly unlike steel which cannot be easily cut because it inflicts serious wear on the single point diamond tool and beryllium copper which require relatively costly machining tools due to its high strength[15, 16]. However, the strength and low wear resistance limitations of aluminium alloys poses a major disadvantage in molding inserts [16].

The quality of mold inserts is determined by the surface finish of the mold insert. Zhong and Gubbels et al concluded that ultra-precision machined RSA grades generated better surface finishes when compared with other moulding materials [1, 14]. Furthermore, To et al achieved a surface roughness of 2-3nm in fewer machining steps and less tool damage with RSA 6061 on the other hand, nickel plated steel generated a higher margin of surface roughness values with the same number of machining steps [17].

Furthermore, tool wear investigation on ultra-high precision diamond turning RSA 905 alloy by Abou-el-hossein et al has revealed that minimum abrasion occurred at low feed but achieved a highest value at the middle range of feed for the selected levels of machining parameters [18]. A recent study on diamond insert wear while turning RSA 443 alloy revealed that reduced cutting speed degenerated to increased tool wear and higher surface roughness occurred at lower spindle speed while feed rate effect was not as significant as the spindle speed. However, the flattened tool edge which occurred as a result of increased time and machining distance improved on the surface roughness. It was also suggested that surface roughness of RSA443 alloy will increase with the further advancement of tool wear [19].

2.3. AE Sensing Technique

Acoustic emission waves are low amplitude energy waves with a high frequency up to 50MHz, occurring as a physical phenomenon propagating through a structural lattice dislocation and atomic level disturbances [20]. Over the years this phenomenon has been proven to be invaluable in the monitoring manufacturing processes and machining conditions. Compared to other prevalent monitoring techniques applying force, vibration, pressure and temperature sensors, AE monitoring has achieved a high success rate due to its high signal to noise ratio and sensitivity even at micro scale cutting depth where other monitoring techniques are not efficient [21]. Furthermore, its immunity to low frequency noise source and environmental noise influence is a desirable consideration for monitoring techniques.

AE monitoring has been applied to monitoring several macro and micro machining operations ranging from turning and milling to the different modes of grinding but there is no available literature on AE monitoring of RSA 443 alloy grade. Abou-el-hossein et al investigated the tool wear encountered during ultra-high precision machining of RSA 905 alloy using acoustic emission sensing technique. They observed a notable change in AE raw and rms features for different conditions of tool wear and surface roughness. Furthermore, they observed higher amplitudes of raw and rms AE signals occurring at 25mm/min feed rates compared to 15mm/min feed category although the tool wear at 25mm/min was less than 15mm/min [22].

3. Experimental Design

3.1. Workpiece

RSA 443 alloy has fine microstructure and is relatively harder than traditional aluminium because of its high silicon content. The research was carried out on RSA 443 grade with 60.5 mm diameter, manufactured by RSP technology Ltd. This workpiece contains about 40% silicon, having a low thermal expansion, density, high specific thickness and high thermal conductivity. Table 1 highlights some physical and mechanical properties of RSA 443.

Table 1. Mechanical properties of RSA 443 Alloy

Composition	Al, SiO_4
Hardness	105
Density [g/cm³]	2.54
Thermal expansion [10/k]	13.6
Young's modulus [GPa]	102
Ultimate tensile strength [MPa]	245
Yield strength [MPa]	155
Elongation [%]	1.5

3.2. Parameter Selection

Three machining parameters: feed rate, cutting speed and depth of cut were varied for the experimental study. These parameters were varied at three levels as shown in (Table 2).

Table 2. Machining parameters and combination

Parameter	Combination		
Feed [mm/min]	5	15	25
Depth [µm]	5	15	25
Speed [rpm]	500	1750	3000

3.3. Machining setup

A 4 axes ultra-high precision lathe machine (Nanoform 250), manufactured by Precitech was utilised for the research (Fig. 1). The machine boasts of 0.1 nm control resolution. Diamond tool inserts by Contour Fine Tool Ltd were used in the research with a nose radius 0.5mm, rake angle -5° (mounted on a 0° tool holder) and front clearance 10° (mounted on a 0° tool holder) and standard waviness. The tool was centred using a visual display unit coupled to an integral Precitech while the coordinates of the radius and flank radius were recorded. (Fig. 2).

With help of an adaptor the workpiece was fastened and mounted on the machine's vacuum chuck while the specimen was centered according to its perimeter using an electronic dial test indicator. At 2000rpm, the chuck was rotated together with the specimen to ensure balance while kerosene coolant was used to prevent thermal damage to the work piece.

Surface roughness measurements on the workpiece were carried out using Taylor Hobson PGI Dimension XL Surface Profilometer which is a 3D nano-profiler that has 0.2 nanometre resolution (Fig. 2). Surface roughness measurements carried at different intervals. Using a sampling length of ± 4 mm, three consecutive measurements were taken on the machined surfaces and the average was calculated and recorded . The tool insert wear was measured using a JOEL JSM- 6380 Scanning Electron Microscope (SEM) as the inserts were mounted on a special stub holder for positioning in the SEM chamber. In order to measure the lenghth and width of the wear area, The X-Y and R controls axis of the SEM were manipulated to bring the required insert face (edge) to view.

Figure 1 Nanoform 250 Precison Lathe Machine

Figure 2. Taylor Hobson PGI Dimension XL Profilometer

3.4 Acoustic emission set up.

Acoustic emission data was acquired at a high sampling rate of 2million samples per second using a Kistler piezo-electric sensor. The AE sensor was attached using a magnetic clamp. The AE sensor connection comes with a built in coupler unit for signal pre-processing. Signal filtering was done with a band pass filter range (5 KHz – 700 KHz) and a gain amplification of 20db. The AE signals were further relayed to National Instrument (NI) BNC board and a 16bit NI PIC 6110 data acquisition card for further processing. AE data acquisition code for real time acquisition was developed with the use of a Graphical User Interphase (GUI), NI Virtual Instrumentation Engineering Workbench (LabVIEW).

IOP Conf. Series: Journal of Physics: Conf. Series **843** (2017) 012059 doi:10.1088/1742-6596/843/1/012059

Figure 3 AE Acquisition Set up

4. Results and Discussion

The results recorded below and the corresponding AE data from some of the experimental combinations were measured and acquired after the feed has travelled a machining distance of precisely 13km (Table 3). Only raw AE signal voltages were acquired during the course of the study. The tool numbers are classified as LH (left hand) and RH (right hand).

Table 3. Surface roughness and wear measurement results

Tool no	Cutting speed [rpm]	Feed [mm/min]	Depth [µm]	Ra[20]	Wear[20]
72LH	1750	25	25	37	8
72RH	500	5	15	24	6.5
73RH	1750	5	5	14	3
74LH	3000	5	15	13.5	2
74RH	500	15	5	43	8
75LH	1750	15	15	21	4
77LH	500	15	25	50	15
78LH	3000	25	15	28	7

The AE signals were subjected to band pass filtering and segmented to an online time segment window approximately 5 seconds. The average amplitude of the raw AE signals in all the different combinations were observed to vary between 0.02Volts and 0.05Volts for the different parameter, surface roughness and wear conditions (Fig. 4 – Fig. 9).

IOP Conf. Series: Journal of Physics: Conf. Series **843** (2017) 012059 doi:10.1088/1742-6596/843/1/012059

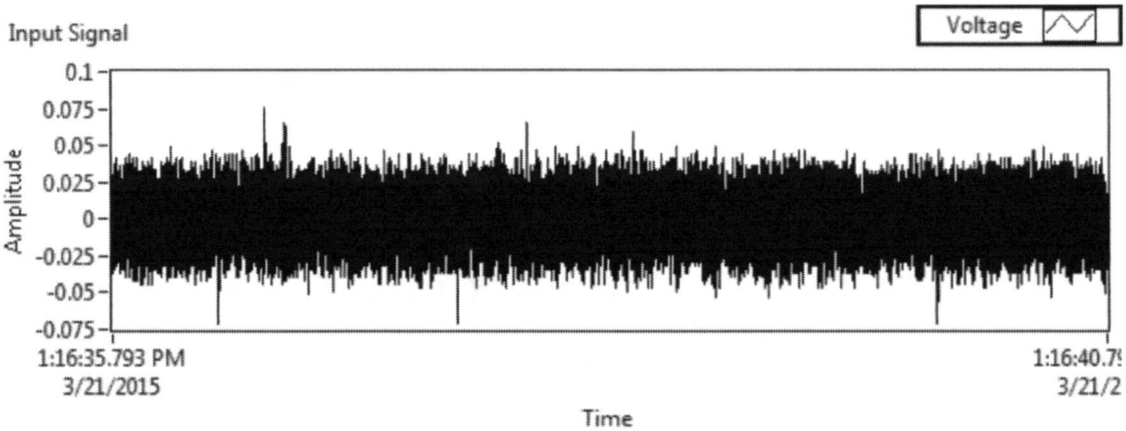

Figure 4 77LH, Ra 50nm, Wear 15nm

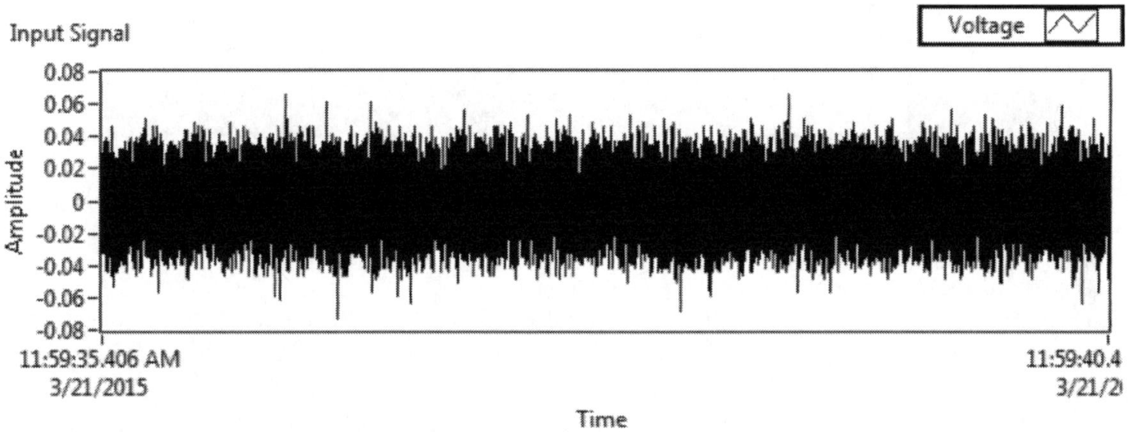

Figure 5. 74RH, Ra 43nm, Wear 8nm

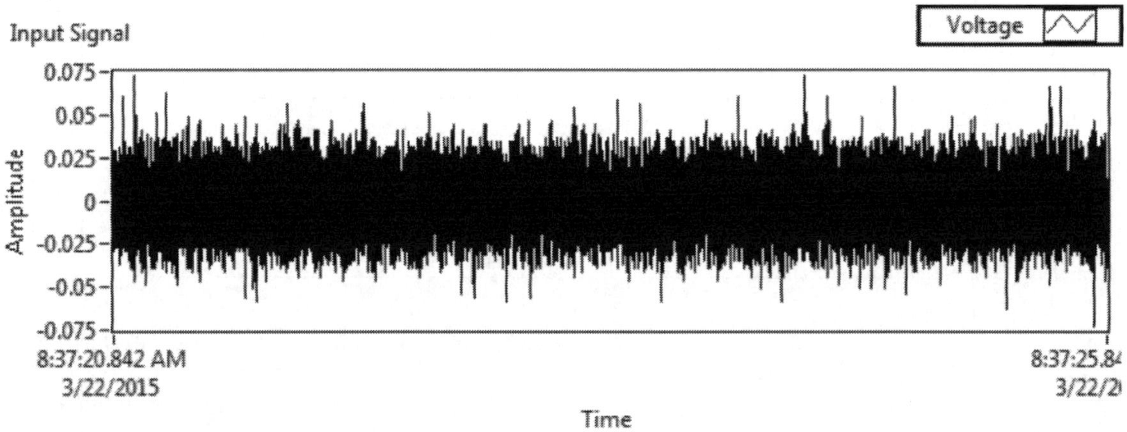

Figure 6. 78LH, Ra 28nm, Wear 7nm

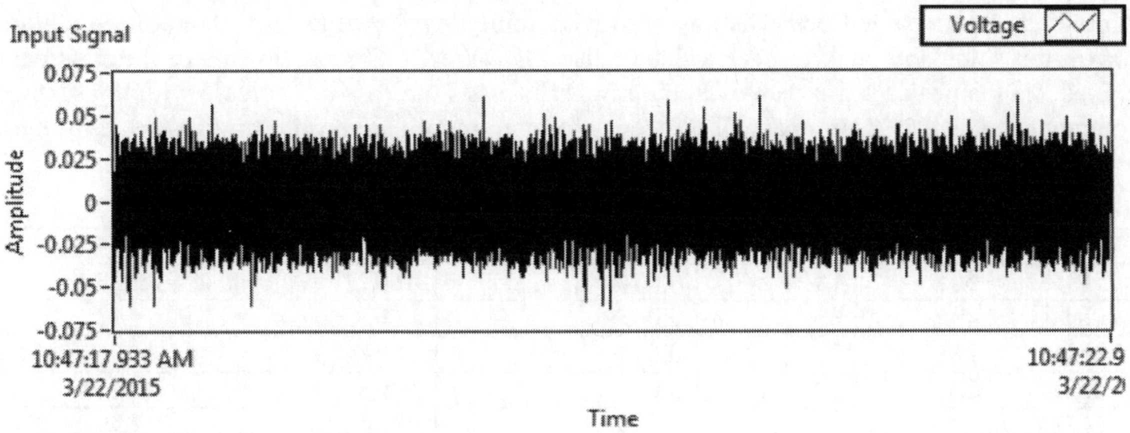

Figure 7. 72RH, Ra 24nm, Wear 6.5nm

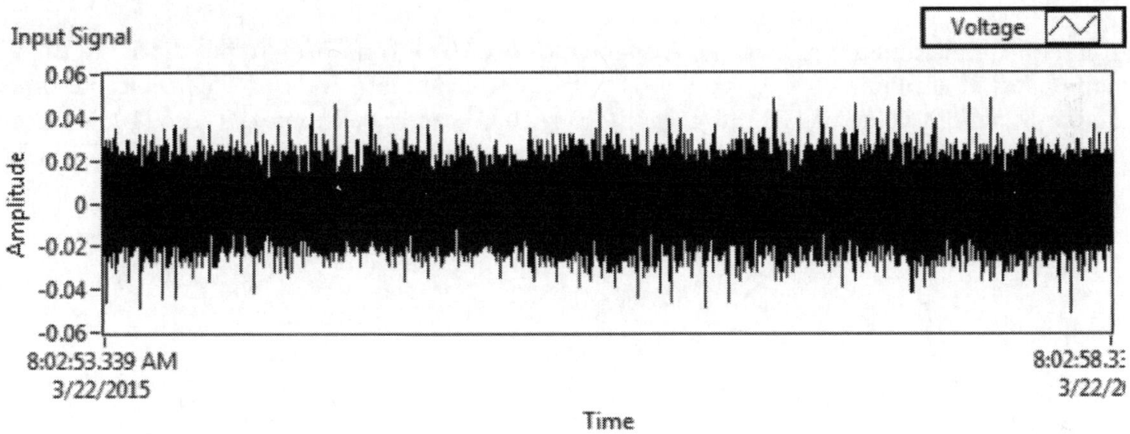

Figure 8. 75LH, Ra 21nm, Wear 4nm

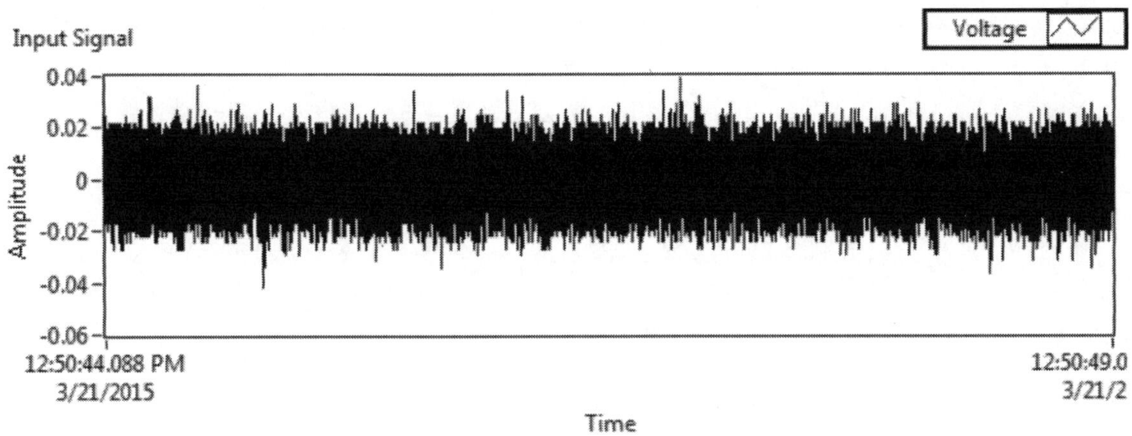

Figure 9. 74LH, Ra 13.5nm, Wear 2nm

IOP Conf. Series: Journal of Physics: Conf. Series **843** (2017) 012059 doi:10.1088/1742-6596/843/1/012059

Moreover, a noticeable trend was observed with some raw AE time series features which are maximum value and peak to peak value of the AE waves. The peak value of the dominant spikes and the peak to peak values in the raw AE signals indicate an increase with the surface roughness and wear. A summary of the trends for maximum AE amplitude (AE_{MAX}), minimum amplitude (AE_{MIN}) and peak to peak values are represented in (Table 4 and Fig. 10).

Table 4. AE feature analysis

Tool No.	Ra[20]	Wear[20]	AE_{MAX} x 10^{-2} [V]	AE_{MIN} x 10^{-2} [V]	AE_{P-P} 10^{-2} [V]
77LH	50	15	0.080	-0.073	0.153
74RH	43	8	0.068	-0.075	0.143
78LH	28	7	0.074	-0.074	0.1
72RH	24	6.5	0.060	-0.061	0.121
75LH	21	4	0.049	-0.050	0.099
74LH	13.5	2	0.039	-00.04	0.079

From table 4, maximum value of the AE signal (8.0 x 10^{-2}V) and peak to peak (15.3 x 10^{-2}V) were observed at 50nm surface roughness with a corresponding wear of 15nm. On the other hand, a minimum value of AE raw signal (4.0 x 10^{-2}V) and peak to peak (7.9 x 10^{-2}V) were observed to correspond with low values of 13.5nm surface roughness and 2nm wear (Fig. 10).

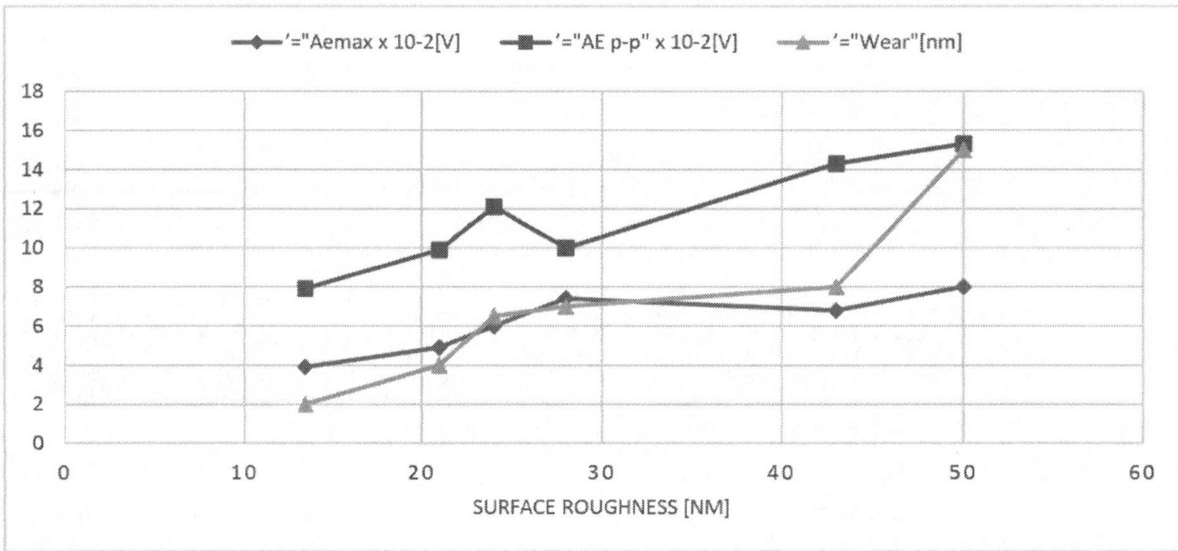

Figure 10. Increasing trend of AE features and wear with surface roughness

An exceptional case was observed at surface roughness of 28nm and wear of 7nm with AE maximum spike clearly higher than the immediate neighbouring points on the chat (Fig 10). This could be because of the high levels of machniing paranmeters (3000[rpm], 25[mm/min] and 15[μm]) used for this particular run. AE spikes are normally higher with an increased speed, feed and depth of cut.

Furthermore, previous studies show that surface roughness values increase with lower values of the cutting speed as also observed by Aboul el hossien et al. Within the range of their experimental design parameters, they also observed time series AE rms levels to be highest at a feed 15mm/min for RSA905 alloy [22] in contrast to the maximum AE spike and peak to peak observed for RSA443 alloy at 25mm/min feed rate. However, the average raw AE levels were closer to the 0.05[V] mark at a feed of 15mm/min for RSA 443 with which corresponds to high value of surface roughness and wear condition.

Abou-El-Hossein et al also noted the improved surface roughness conditions with the increase in machining time as the tool wear flattened out after the initial stage of wear and predicted a possible increase in surface roughness values as the tool wear further deteriorates [18]. Hence, from our observation so far, it is safe to posit that an increase in AE spikes which corresponds to increased surface roughness values could occur with further significant increase in machining distance covered for RSA 443 alloy.

5. Conclusion

Despite the high silicon content present in RSA443 similar AE signal behaviours were observed to be maximum at 15mm/min while more influence of the cutting speed could be noticed overall on the surface roughness, wear and AE signal spike levels. The Raw AE signals spikes and peak to peak values were observed to increase with an increase in surface roughness and tool wear respectively. Therefore, acoustic emission sensing with Raw AE signal monitoring has proved to be an efficient means of monitoring diamond turning RSA443 for optical mould inserts.

The monitoring technique can be further improved by extracting relevant direct features from the raw AE signals which could better relate directly to condition of the tool and surface roughness. Furthermore, the potential noticed with the AE sensing technique should be able to solve the dilemma of determining the transition point from initial wear stage to flattened tools and further extreme wear conditions in diamond turning of RSA443 alloys. Further analysis will be done on the raw AE signals to ascertain the frequency characteristics in relation to the surface roughness and wear.

6. References

[1] Gubbels GP, van Venrooy BW, Bosch AJ, Senden R. Rapidly solidified aluminium for optical applications. SPIE Astronomical Telescopes+ Instrumentation: International Society for Optics and Photonics; 2008. p. 70183A-A-9.

[2] ter Horst R, de Haan M, Gubbels G, Senden R, van Venrooy B, Hoogstrate A. Diamond turning and polishing tests on new RSP aluminum alloys. 2012. p. 84502M-M-11.

[3] Wouters O, Vellinga W, Van Tijum R, De Hosson JTM. On the evolution of surface roughness during deformation of polycrystalline aluminum alloys. Acta materialia. 2005;53:4043-50.

[4] Elson J, Rahn J, Bennett J. Relationship of the total integrated scattering from multilayer-coated optics to angle of incidence, polarization, correlation length, and roughness cross-correlation properties. Applied Optics. 1983;22:3207-19.
[5] Harvey JE, Schröder S, Choi N, Duparré A. Total integrated scatter from surfaces with arbitrary roughness, correlation widths, and incident angles. Optical Engineering. 2012;51:013402-1.
[6] Parsonage T. New technologies for optical systems utilizing aluminum beryllium. SPIE Astronomical Telescopes+ Instrumentation: International Society for Optics and Photonics; 2008. p. 70180M-M-9.
[7] Anantharaman T, Suryanarayana C. Rapidly solidified metals: Trans Tech Publ.; 1987.
[8] Habraken F. Consolidation, plastic yield and formability of rapidly solidified Al-Si-X alloys. 1996.
[9] Katgerman L, Dom F. Rapidly solidified aluminium alloys by meltspinning. Materials Science and Engineering: A. 2004;375:1212-6.
[10] Cantor B. Fundamentals of Rapid Solidification. In: Sahm PR, Jones H, Adam CM, editors. Science and Technology of the Undercooled Melt: Rapid Solidification Materials and Technologies. Dordrecht: Springer Netherlands; 1986. p. 3.
[11] Cantor B. Fundamentals of rapid solidification. Science and Technology of the Undercooled Melt: Springer; 1986. p. 123.
[12] Martin P, Lipsitt H, Williams J. Rapid Solidification Processing Principles and Technologies II, R. Mehrabian, BH Kear, and M Cohen, eds, Claitor's. 1980:123-28.
[13] White CW. Laser and electron beam processing of materials: Elsevier; 2012.
[14] Zhong Z, Leong M, Liu X. The wear rates and performance of three mold insert materials. Materials & Design. 2011;32:643-8.
[15] Vancoliie E, Xiangdong L, Ling Y, Annergren I. Materials Selection for Durable Optical Inserts Used in Plastic Lens Moulding. Technical Report (DD/00/009/OTC), SIMTech, Singapore; 2000.
[16] Menges G, Michaeli W, Mohren P, Menges G, Michaeli W, Mohren P. How to make injection molds: Carl Hanser Verlag GmbH & Co. KG; 2001.
[17] To S, Lee W, Cheung C. Orientation changes of aluminium single crystals in ultra-precision diamond turning. Journal of Materials Processing Technology. 2003;140:346-51.
[18] Abou-El-Hossein K, Olufayo O, Mkoko Z. Diamond tool wear during ultra-high precision machining of rapidly solidified aluminium RSA 905. Wear. 2013;302:1105-12.
[19] Mkoko Z, Abou-El-Hossein K. Diamond Insert Wear when Turning Rapidly Solidified Aluminium RSA 443 of High-Silicon Content. Materials Science Forum: Trans Tech Publ; 2015. p. 265-71.
[20] Bhaskaran J, Murugan M, Balashanmugam N, Chellamalai M. Monitoring of hard turning using acoustic emission signal. Journal of mechanical science and technology. 2012;26:609-15.
[21] Dornfeld D, Lee Y, Chang A. Monitoring of ultraprecision machining processes. The International Journal of Advanced Manufacturing Technology. 2003;21:571-8.
[22] Olufayo OA. Acoustic Emission Monitoring in Ultra-High Precision Machining of Rapidly Solidified Aluminium. COMA'132013.

Wear resistance and mechanisms of composite hardfacings at abrasive impact erosion wear

A Surzhenkov, M Viljus, T Simson, R Tarbe, M Saarna and F Casesnoves

Department of Mechanical and Industrial Engineering, Tallinn University of Technology, Ehitajate tee 5, 19086 Tallinn, Estonia

andrei.surzenkov@ttu.ee

Abstract. Tungsten carbide based hardmetal containing sprayed and melted composite hardfacings are prospective for protection against abrasive wear. For selection of abrasive wear resistant hardfacings under intensive impact wear conditions, both mechanical properties (hardness, fracture toughness, etc.) and abrasive wear conditions (type of abrasive, impact velocity, etc.) should be considered.

This study focuses on the wear (wear rate and mechanisms) of thick metal-matrix composite hardfacings with hardmetal (WC-Co) reinforcement produced by powder metallurgy technology. The influence of the hardmetal reinforcement type on the wear resistance at different abrasive impact erosion wear (AIEW) conditions was studied. An optimal reinforcement for various wear conditions is described. Based on wear mechanism studies, a mathematical model for wear prediction was drafted.

1. Introduction

Abrasive impact erosion wear (AIEW) of materials depends on their mechanical properties and on wear parameters. The dominating mechanisms at AIEW may generally be predicted on the grounds of the material / abrasive hardness ratio (H_m/H_a) and the impact angle (0...90°). Depending on the first ratio, these mechanisms are [1,2]:

- $H_m < H_a$: microcutting or plastic deformation with surface fatigue;
- $H_m \approx H_a$: deformation with microcutting and/or surface fatigue;
- $H_m > H_a$: deformation with surface fatigue and direct fracture.

Toughness of a material is another important mechanical characteristic that determines the behavior of a material at impact erosion. Depending on fracture toughness, the following mechanisms of wear may occur [3,4]:

– in brittle materials (low K_{1c}): plastic deformation with surface fatigue and great probability of direct fracture;

– in ductile materials (high K_{1c}): deformation with microcutting and/or surface fatigue [3,4].

Thick metal matrix composite (MMC) hardfacings are recommended for extreme abrasive wear conditions (i.e., high abrasivity, hardness and impact velocity of abrasive particles) due to their optimal hardness-toughness ratio. In this case, different wear mechanisms may simultaneously exist.

IOP Conf. Series: Journal of Physics: Conf. Series **843** (2017) 012060 doi:10.1088/1742-6596/843/1/012060

Such hardfacings may be produced, for example, by casting, plasma transferred arc (PTA) welding and submerged arc welding (SAW), vacuum sintering (VS), spray-fusion (SF), and high velocity spraying (HVS) [5–10].

The concept of plastic deformation and brittle fracture and a combined model of erosion were proposed to calculate the wear of composite structure materials [11]. A relatively soft metal matrix allows for use of the energetic theory of wear with the mean hardness and dimensionless specific energy parameter τ_o/e_s. In the wear calculations of hardphase, the models of plastic deformation and brittle fracture using hardness distribution and fracture probability must be taken into consideration [2,11].

The calculated and the experimental results showed that the wear rates of the Ni-based matrix composite coating with a relatively low hardness ($H_m < H_a$) have very good coincidence [12].

In the analysis of the abrasive wear resistance of HVOF-spayed, PTA-welded and PM hardfacings in different wear conditions (abrasion at rubber wheel, impact erosion) in [13], potential application areas of selected composite hardfacings were proposed.

Wear of matrix material at impact erosion is contributed from erosion by microcutting [1,2]. As not all impacts lead to the formation of a clean machined chip that is removed, at multiple impacts within the plastic strain field of the previous impacts, it may be possible for material to be removed. In [14] that kind of wear is classified as fatigue wear for only two overhapping impacts.

The aim of the present research was (a) to study wear resistance and wear mechanisms of surface damage of thick MMC hardfacings at different AIEW modes (oblique and normal impact), and (b) to propose criteria for selection of hardfacings for the above-mentioned conditions based on the mechanical properties (hardness-toughness ratio) and microstructure (reinforcement content, size and shape).

2. Experimental

2.1 Materials

Metal matrix composite hardfacings were produced by powder metallurgy (PM) technology using vacuum liquid-phase sintering. For matrix iron based self-fluxing alloy powder Höganäs 6AB with composition, wt.%: 13.72 Cr, 2.67 Si, 0.32 Mn, 2.07 C, 0.02 S, 3.40 B, 6.04 Ni, bal. Fe was used. As a reinforcement, WC-15Co hardmetal powder produced by mechanical milling using a disintegrator milling system [15], with a particle size of 1.0–2.5 mm (coarse), 0.16–0.315 mm (fine) and their mixture (50% coarse + 50% fine) was used. The amount of 50% (optimal) was determined in our previous studies [8].

Table 1 shows the composition and hardness and Figure 1 – the microstructures of the studied hardfacings. As a reference material, Hardox 400 steel was used.

Table 1. Designation, composition and hardness of studied hardfacings.

Designation	Reinforcement		Hardness [16]	
	Particle size, mm	Content, vol.%	HV30	HV1[a]
P1	–	0	870 ±30	1035 ±70
C5	Coarse angular (C) 1.0–2.5	50	1260±435	1005±40/1855±65
F5	Fine angular (F) 0.160–0.315	50	830±160	900±90/1445±135
M5	Mixture 50C + 50F	50	1715±295	815±65/1480±125
H400	Reference steel Hardox 400	–	425±25	–

[a] metal matrix/hard phase

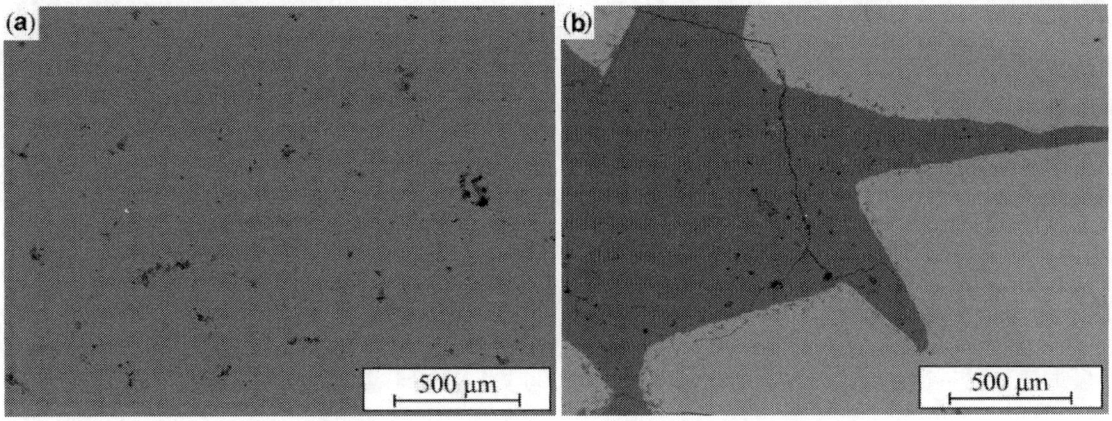

Figure 1. Microstructures of the studied hardfacings: a – P1 (unreinforced), b – C5 (50 vol.% WC-Co angular reinforcement) [16].

2.2 Abrasive wear studies

Abrasive impact erosion wear (AIEW) tests were used to determine the wear resistance of the hardfacings. At the low-energy wear test, granite and quartz sand of fraction 0.2–0.3 mm and centrifugal type tester CAK were used; at the high-energy wear test, granite gravel of fraction 3.0–5.6 mm and a disintegrator type tester DESI (Figure 3) were used.

The schemes and parameters of AIEW are given in Tables 2 and 3.

Based on the weight loss of abraded hardfacings, the volumetric wear rate (loss of volume per 1 kg of abrading material) in mm^3/kg was determined.

The relative wear resistance ε was calculated as the ratio of volumetric wear rates of the reference material (Hardox 400) to the wear rate of the studied hardfacing.

Mechanisms of the impact erosion wear were studied using a scanning electron microscope (SEM) EVO MA-15 (Carl Zeiss, Germany).

Table 2. Abrasive impact erosion wear (AIEW) modes.

	Oblique impact erosion wear $0^\circ > \alpha < 90^\circ$	Normal impact erosion wear $\alpha = 90^\circ$
Low energy	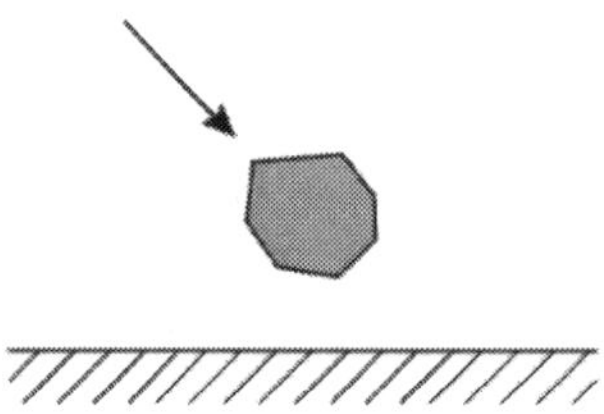	Abrasive 0.2–0.3 mm 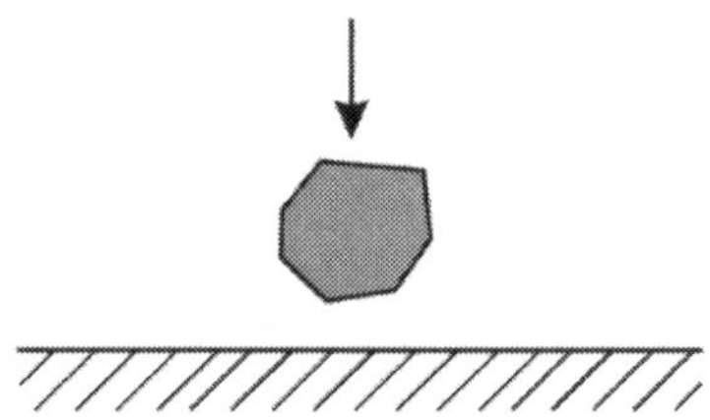 Kinetic energy E_k at v = 40 m/s 3.0×10^{-5} J at v = 80 m/s 1.2×10^{-4} J
High energy	–	Abrasive 3.0–5.6 mm 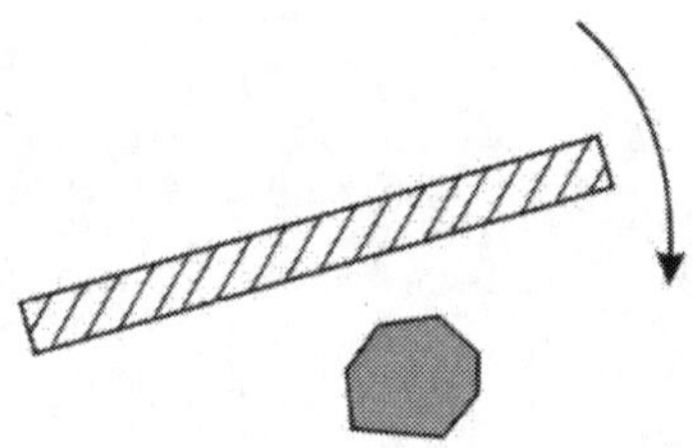 Kinetic energy E_k at v = 40 m/s 1.4×10^{-2} J at v = 80 m/s 5.6×10^{-1} J

Table 3. Abrasive impact erosion wear (AIEW) parameters (a).

Scheme of wear tester	Wear test parameters
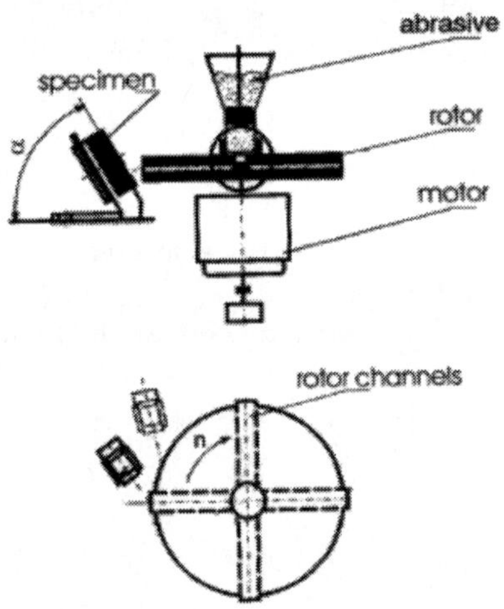 **Figure 2.** AIEW tester CAK.	Quartz sand 0.2–0.3 mm, 1000–1100 HV Granite sand 0.2–0.3 mm, 900–950 HV Impact angle $\alpha = 30^\circ$ and 90° Impact velocity v = 40 m/s and 80 m/s Quantity of abrasive Q = 6 kg

Table 3 (continues). Abrasive impact erosion (AIEW) parameters (b).

Scheme of wear tester	Wear test parameters
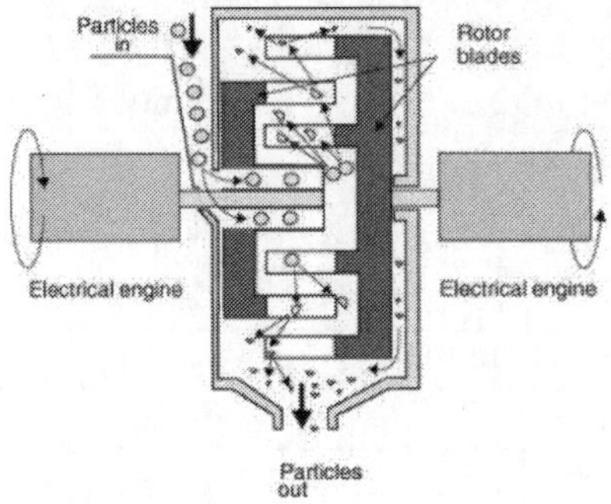 **Figure 3.** AIEW tester DESI.	Granite gravel 3.0–5.6 mm; 900–950 HV Impact angle $\alpha \approx 90°$ Impact velocity v = 40 m/s and 80 m/s Quantity of abrasive Q = 15 kg

3. Results and discussion

3.1 Wear resistance of hardfacings at AIEW

3.1.1 Influence of abrasive hardness. Results of low-energy AIEW are given in Table 4 and Figure 4. As it follows from Figure 4, the wear resistance of hardfacings is better with a softer abrasive – granite (abrasive hardness is comparable with matrix hardness); relative wear resistance exceeds that of steel Hardox 400 by about 3 times. With a harder abrasive – quartz sand ($H_a \approx 1.2\ H_m$), the wear resistance of the studied hardfacings is low; relative wear resistance is at the same level or below that in comparison with Hardox 400.

Table 4. Wear rate (mm³/kg) of hardfacings at low-energy AIEW.

Designation	Granite sand HV 900–950	Quartz sand HV 1000–1100
P1	12.0	26.9
C5	11.7	30.2
F5	30.2	90.3
M5	21.9	76.3
H400	37.0	32.8

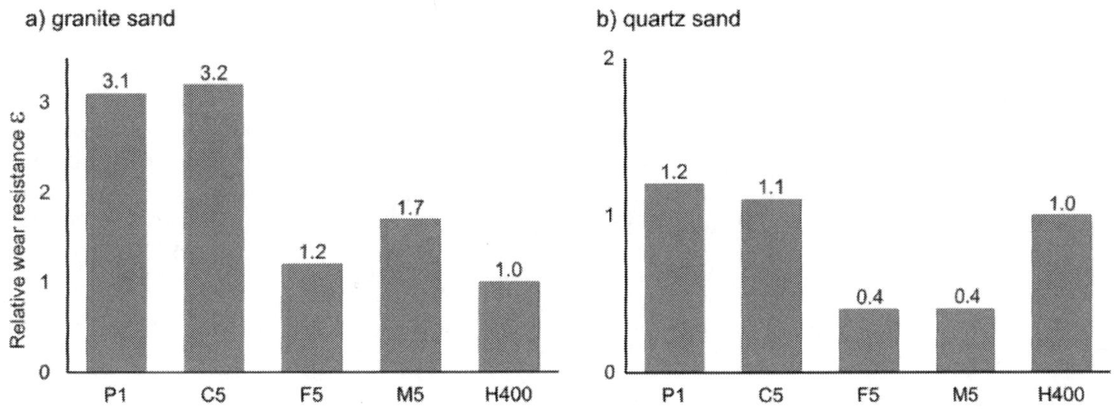

Figure 4. Relative wear resistance of hardfacings at low-energy AIEW with different abrasives (reference material – steel Hardox 400).

IOP Conf. Series: Journal of Physics: Conf. Series **843** (2017) 012060 doi:10.1088/1742-6596/843/1/012060

3.1.2 Influence of impact angle at AIEW. The results of wear resistance comparison of studied hardfacings at oblique impact ($\alpha = 30^o$) and at normal impact ($\alpha = 90^o$) show that the wear rate at straight impact is approximately 1.5 times higher (see Table 2). The relative wear resistance of the best hardfacing (C5) at $\alpha = 30^o$ was 3.2 times higher, at normal impact ($\alpha \approx 90^o$) it is only 1.9 times higher (Figure 5).

Table 5. Wear rates (mm^3/kg) of hardfacings at different impact angles at low-energy AIEW (abrasive – granite sand, impact velocity v = 80 m/s).

Designation	Impact angle, α	
	α = 30°	α= 90°
P1	12.0	21.5
C5	11.7	15.5
F5	30.2	41.0
M5	21.9	40.4
Hardox 400	37.0	30.0

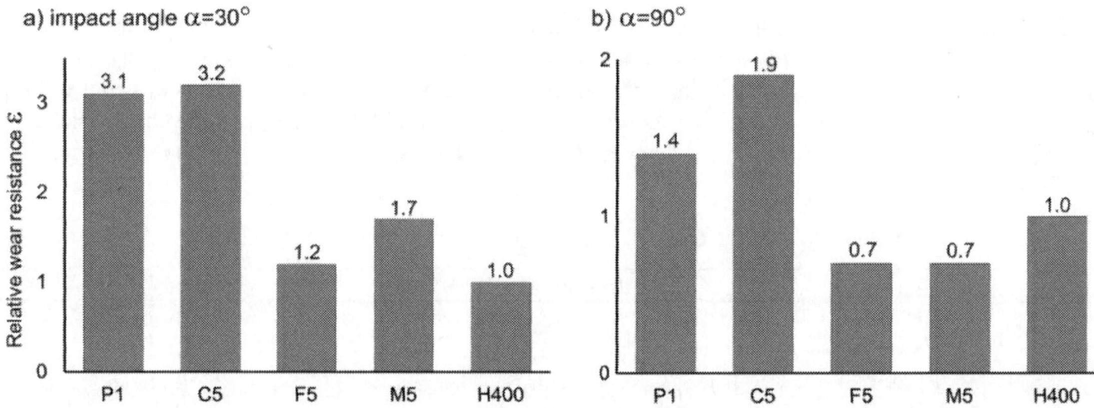

Figure 5. Relative wear resistance of hardfacings at low-energy AIEW under different impact angles (abrasive – granite sand, reference material – Hardox 400).

3.1.3 Influence of abrasive particle velocity. The results of wear rate comparison of studied hardfacings at 40 m/s and 80 m/s (Table 6 and Figure 6) show that the wear rate of reinforced hardfacings at higher velocities is 3–4 times higher when the wear of the reference steel Hardox 400 is 7–8 times higher.

Thus, at low velocity (in the range $\varepsilon = 0.3–0.8$), the relative wear resistance is lower to compare with Hardox 400; at high velocity, wear resistance of composite hardfacings is very low (in the range $\varepsilon = 0.1–0.2$). It can be explained by the differences in the wear mechanisms of hardened steel as compared with metal matrix composite hardfacings.

IOP Conf. Series: Journal of Physics: Conf. Series **843** (2017) 012060 doi:10.1088/1742-6596/843/1/012060

Table 6. Wear rates (mm^3/kg) at different velocities at low-energy AIEW (abrasive – granite sand, impact angle $\alpha = 90^o$).

Designation	Velocity of abrasive particles, v	
	v = 40 m/s	v = 80 m/s
P1	10.0	21.5
C5	4.8	15.5
F5	13.8	41.0
M5	5.6	40.4
H400	3.8	30.0

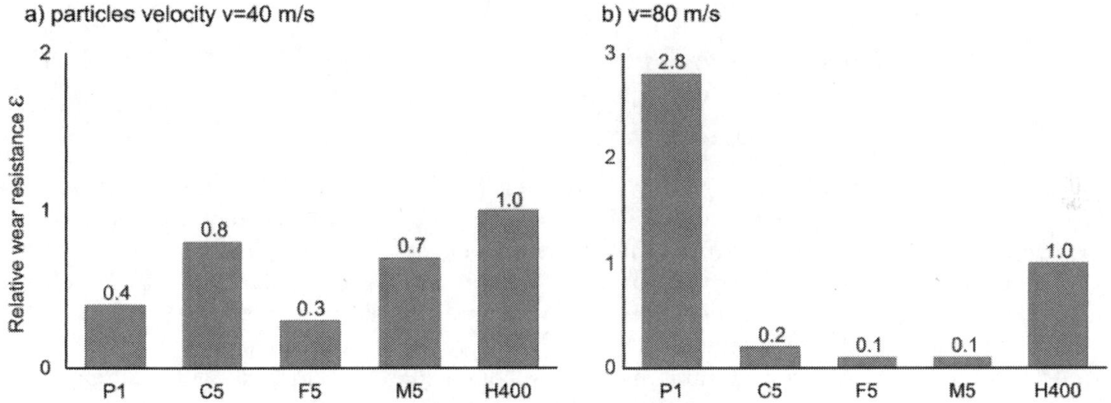

Figure 6. Relative wear resistance of hardfacings under different impact velocities (abrasive – granite sand, reference material – steel Hardox 400).

3.1.4 Influence of impact energy of abrasive particles. As the difference in studied impact energies is high (ratio of kinetic energies of high and low energy impact at v = 40 m/s is about 500 times), it influences the wear resistance significantly (see Table 7 and Figure 7). Wear rates of best hardfacing C5 differ about 23 times while the difference of the reference steel is only about 7 times.

Table 7. Wear rates (mm^3/kg) at different impact energies (abrasive – granite, v = 40 m/s, $\alpha = 90^o$).

Designation	Low energy (sand) $E_k = 3.0\times10^{-4}$ J	High energy (gravel) $E_k = 1.4\times10^{-2}$ J
P1	10.0	9.3
C5	4.8	108.6
F5	13.8	215.2
M5	5.6	348.6
Hardox 400	3.8	26.1

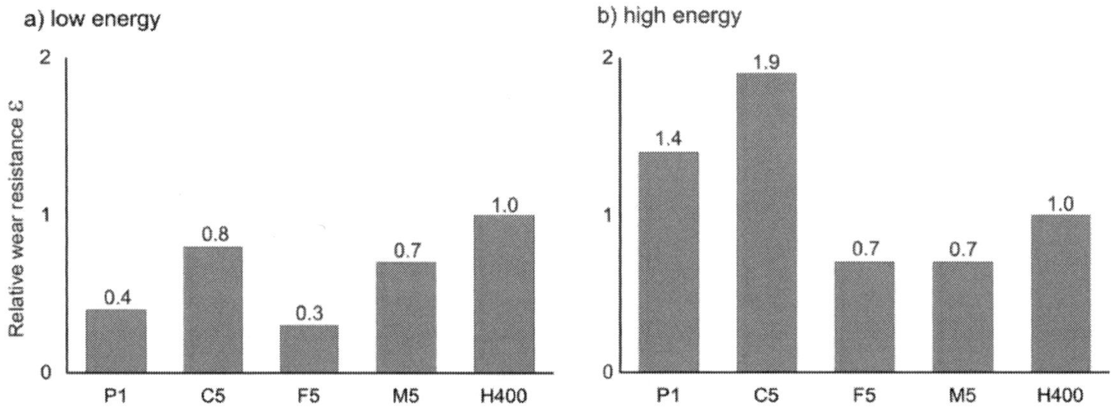

Figure 7. Relative wear resistance of hardfacings at low-energy and high-energy AIEW (abrasive – granite, reference material – steel Hardox 400).

3.2 Mechanisms of AIEW

As the unreinforced hardfacing (P1) and the hardfacing reinforced with the coarse hardmetal (C5) exhibited the lowest wear among the studied hardfacings, they were taken as the object for our analysis of wear mechanisms. Both under low-energy and under high-energy AIEW conditions, the general wear process took place in two stages: firstly, destruction of the matrix and secondly, loss of loose WC-Co particles (Figures 8 and 9). Because the wear of the matrix was much more intensive than the wear of the reinforcement, higher magnification pictures of the first are demonstrated separately.

Under the low-energy AIEW conditions, the wear of both the FeCrSiB matrix (both in unreinforced and composite hardfacing) and the WC-Co reinforcement occurred by the low-cycle fatigue mechanism (Figure 8) [2]. In the first case, it included the stages of work hardening by the impact particles, resulting in the formation and development of lateral cracks [17] and, finally, spalling of flat fragments. It is interesting to note that the wear of the FeCrSiB matrix was more extensive in the composite hardfacing, and in the proximity of the reinforcing particles (Figure 8 b,c), it became most remarkable. The most probable cause for that is the thermally induced tensile stresses at the matrix-reinforcement interface [8], which favor the removal of the material [18]. At reinforcement, the wear started with the extrusion of the binder and continued by the subsequent chipping of the exposed carbide particles, as described in [4] (not shown in Figures 8, 9).

Under high-energy AIEW conditions, the wear mechanism of the reinforcement was identical to that under the low-energy AIEW conditions. However, the wear mechanism of FeCrSiB alloy underwent some changes (Figure 9). In addition to lateral cracks, median ones [17] may be seen (Figure 9). Thus, the wear mechanism of the FeCrSiB matrix may be described as a combination of low-cycle surface fatigue and direct fracture. As at the low-energy AEIW, the wear of the FeCrSiB matrix was higher at the composite hardfacing (Figures 8 b,c) for the same reason.

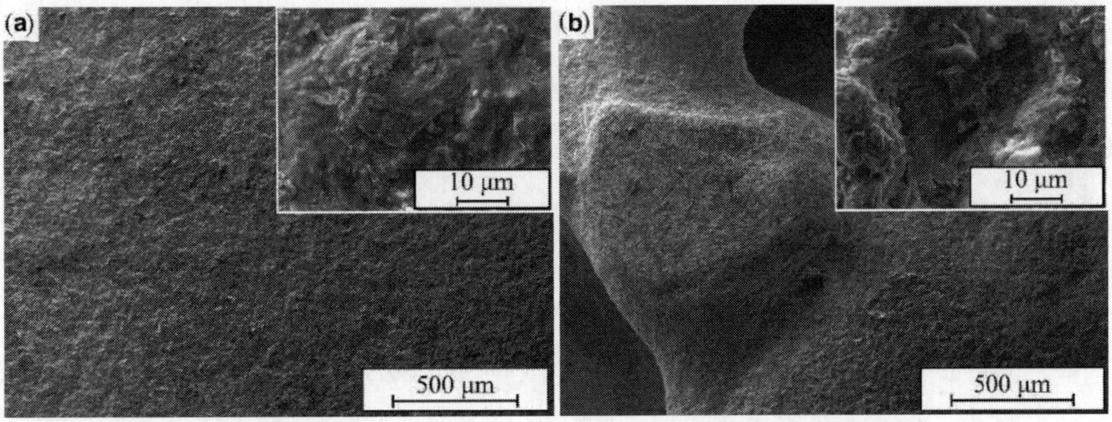

Figure 8. Worn surfaces of the hardfacings under low-energy AIEW conditions (v = 80 m/s): a – P1 (unreinforced), b – C5 (50 vol.% angular WC-Co reinforcement).

Figure 9. Worn surfaces of the hardfacings under high-energy AIEW conditions (v = 40 m/s): a – P1 (unreinforced), b – C5 (50 vol.% angular WC-Co reinforcement).

4. Conclusions

1. Wear rate of AIEW of studied composite WC-Co containing hardfacings depends first on the wear parameters (type and hardness of abrasive, velocity and kinetic energy of particles). Relative volumetric wear resistance of better hardfacings (coarse angular reinforcement, C5) exceeds that of the reference steel Hardox 400 about 3 times at low-energy AIEW with the abrasive – granite sand. At high velocity, AIEW relative wear resistance is low.
2. The dominating wear mechanism of both matrix and reinforcement at the low-energy AIEW was surface fatigue, at the high-energy AIEW – a combination of surface fatigue and direct fracture at the matrix and surface fatigue at the reinforcement.
3. The criteria for coating selection by composition and hardness-toughness ratio for hardmetals containing hardfacings are the following:

a) at oblique impact AIEW – maximal hardphase content and higher hardness of the composite;
b) at normal impact AIEW – lower hardphase content and hardness of metal matrix and higher toughness of hardphase.

5. References

[1] Hutchings I M, Winter R E and Field J E 1976 *Proc. Roy. Soc. London* A **348** 379–92
[2] Kleis I and Kulu P 2008 *Solid Particle Erosion. Occurrence, Prediction and Control* (London:

Springer Verlag) p 206
[3] Kulu P 2002 *Adv. Eng. Mater.* **4** 392–7
[4] Hussainova I, Kübarsepp J and Pirso J 2001 *Wear* **250** 818–25
[5] Rojacz H, Varga M, Kerber H and Winkelmann H 2014 *J. Mater. Process. Tech.* **214** 1285–92
[6] Kulu P, Tarbe R, Saarna M, Surženkov A, Peetsalu P and Talviste K 2015 *Int. J. Microstructure and Materials Properties* **10** 101–13
[7] Bendikiene R, Ciuplys A and Kavaliauskiene L 2016 *Proc. Est. Acad. Sci.* **65** 117–22
[8] Simson T, Kulu P, Surženkov A, Tarbe R, Viljus M, Tarraste M and Goljandin D 2016 *Proc. Est. Acad. Sci.* **65** 90–6
[9] *Reparation and wear protection technologies*. Product Catalogue, Castolin Eutectic, 2010
[10] Surzhenkov A, Vallikivi A, Mikli V, Viljus M, Vilgo T and Kulu P 2012 *Proc. 2nd Int. Conf. Manufacturing Engineering & Management 2012 (Prešov, Slovak Republic)* ed S Hloch et al. (Prešov: Udulibri) 34–7
[11] Kulu P, Veinthal R and Kaerdi H 2007 *Int. J. Mater. Prod. Tech.* **28** 425–47
[12] Veinthal R 2005 *Characterization and Modelling of Erosion Wear of Powder Composite Materials and Coatings* PhD Thesis (Tallinn: TUT Press) p 61
[13] Kulu P, Surženkov A, Simson T and Sarjas H 2016 *Proc. European Conf. Heat Treatment 2016 & 3rd Int. Conf. Heat Treatment and Surface Engineering in Automotive Applications*, 11–13 May 2016, Prague, Czech Republic (Prague: Asociace pro tepelné zpracování kovů)
[14] Andrews D R 1980 *The Erosion of Metals* PhD Thesis (London: Cambridge Press) p 156
[15] Goljandin D and Kulu P 2015 *Disintegrators and Disintegrator Treatment of Materials* (Lambert Academic Publishing) p 162
[16] Simson T, Kulu P, Surženkov A, Goljandin D, Tarbe R, Tarraste M and Viljus M 2017 *Key Eng. Mat.* **721** 351–5
[17] Sheldon G L and Finnie I 1966 *Am. Soc. Mech. Eng. Trans. J. Eng. Ind.* **88** 393–9
[18] Stewart D A, Shipway P H and McCartney D G 1998 *Surf. Coat. Tech.* **105** 13–24

Acknowledgements
This work was supported by institutional research funding IUT19-29 "Multi-scale structured ceramic-based composites for extreme applications" of the Estonian Ministry of Education and Research and by the project AR12132 "Development of advanced coatings and polymer-ceramic composites for road construction machinery wear parts (WearHard)" of Archimedes Foundation. This work was also supported by the Dora Plus activity 1.1 (Archimedes Foundation).

The self-sense monitoring of coating wear by in situ SiO_2-coated $SrAl_2O_4:Eu^{2+},Dy^{3+}$ luminescent particles

Wensheng LI* Ling He Chunxia Hu XiaoTian Yang

State Key Laboratory of Advanced Processing and Recycling of Nonferrous Metals, Lanzhou University of Technology, Lanzhou 730050, China

Wensheng-li@sohu.com

Abstract. Monitoring of coating health was achieved by embedding sensor materials $SrAl_2O_4:Eu^{2+},Dy^{3+}$ in tribological coatings, and luminescence spectra of sensor materials were used for damage sensing by monitoring the intensity of the spectra. When the light emitting from coating disappeared, it indicated the coating wearied off and the substrate exposed. But emission intensity of the $SrAl_2O_4:Eu^{2+},Dy^{3+}$ in tribological coating was quenched in preparation process which will weaken the effect of coating instructions. In this work, SiO_2-coated $SrAl_2O_4:Eu^{2+},Dy^{3+}$ were coupled with Cu-14Al-X alloy powders to form sensor composite coating by supersonic plasma spraying. The composite coatings before and after SiO_2-coating were compared of phase composition, microstructure, luminescence properties and tribological performance. The composite coating shows not only good resistance to wear but also exhibits strong green light emitting under UV light. The SiO_2-coating could prevent phosphors quenching due to high temperature in the supersonic plasma spraying process, and could repair the wear surface. The composite coating with SiO_2-coated $SrAl_2O_4:Eu^{2+},Dy^{3+}$ particles had very stable friction coefficient, When the coating ratio of SiO_2 was 10% for $SrAl_2O_4:Eu^{2+},Dy^{3+}$ particles, the luminescent properties and friction property of composite was optimal.

1. Introduction

Recently, a novel self-sensing tribological coating were developed for monitoring wear using handheld battery-operated UV lights at the operating site which can be fast and conveniently applied in periodic inspections. In some reports[1-2], the Cu-14Al-X were coupled with $SrAl_2O_4:Eu^{2+},Dy^{3+}$ phosphorescence forming the self-sensing tribological coating and the phosphors particles are incorporate into wear-resistant coating served as sensor materials for warning coating worn off and replacement. But the luminescence intensity decreased in coating and the quenching behavior of phosphors will be harmful to the indication of the coating. For improving the anti-degradation of $SrAl_2O_4:Eu^{2+}, Dy^{3+}$ during thermal spraying, the most effective technique is to coat the surface of phosphor particles with protective materials [3-6].

In this paper, the surface of $SrAl_2O_4:Eu^{2+},Dy^{3+}$ was coated with SiO_2 by Sol-gel process. Not only the microstructures and luminescent properties of the coatings but also tribological properties are investigated before and after coating SiO_2.

2. Experimental

The aluminum bronze metals (Cu-14Al-X) were melted at 1200 ~ 1260 °C, followed by water cooling atomization to form the alloys powders as listed in Table 1.

Table.1 High-aluminum bronze composition (Cu-14Al-X)

Element	Cu	Al	Fe	Ni	Balance.
wt. %	70~80	12~16	2.0~4.5	0.5~2.0	0.5~2.0

The core–shell structured $SrAl_2O_4:Eu^{2+},Dy^{3+}$ microspheres were prepared by a simple sol–gel method and the molar ratio of tetraethoxysilane, ethanol and water was 1:3:5. The tetraethyl orthosilicate (TEOS) was served as silicon source. A small amount of ammonia was dripped into these solutions as catalyst for hydrolyzing of TEOS, and the pH value was kept at 4-5; After being stirred for 2h, SiO_2-coated $SrAl_2O_4:Eu^{2+},Dy^{3+}$ phosphors with various SiO_2 addition content (5wt%, 10wt% ,15wt% and 20wt%) were obtained.

Cu-14Al-X / $SrAl_2O_4:Eu^{2+},Dy^{3+}$ composite coating doped with 10% SiO_2 coating for the phosphors were prepared on 45# carbon steel substrate by supersonic plasma spray. Wear tests were done on a pin-on-disk wear tester (MMW-1A). The wear test mode is illustrated in Fig. 1, the Phosphorescent composite coating sample acted as the disk. The test load is 35N, 70N, 110N, speed is 80r/min.

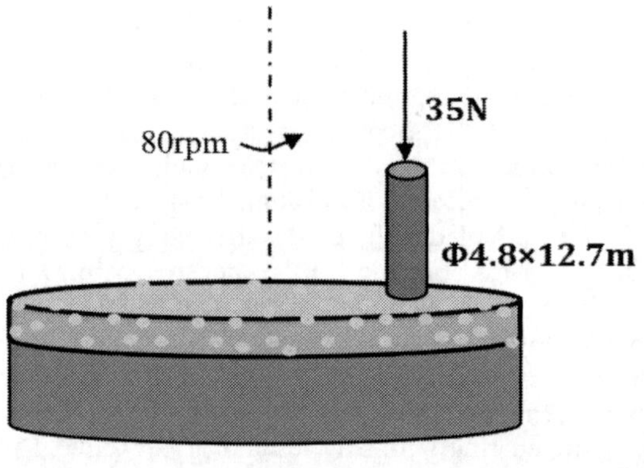

Fig. 1 Wear testing schematic diagram

3. Results and Discussion

Fig.2 SEM micrographs of phosphor before and after coating: a uncoatd; b after SiO_2 coated

SEM images of pure $SrAl_2O_4:Eu^{2+},Dy^{3+}$ (Fig.2a), and $SrAl_2O_4:Eu^{2+},Dy^{3+}$ microspheres coated by a layer of SiO_2 shell (Fig.2 b) were showed in Fig.2. The surface of the coated sample particles are relatively smooth, and uniform, complete, and with high density. The composition of the film can be characterized by EDX. The EDX (table.2) confirms the above elements and silicon (Si) in the SiO_2-coated

$SrAl_2O_4:Eu^{2+},Dy^{3+}$ samples. The results indicate that the surfaces of $SrAl_2O_4:Eu^{2+},Dy^{3+}$phosphors are successfully covered with SiO_2 thin layer via the sol-gel treatment process.

Tab.2 Element content of phosphor before and after coating

	The quality of elements（wt.%）			
Sample（Region）	Al	Sr	O	Si
（a）-1	25.18	44.22	36.60	
（b）-2	1.23	2.1	42.55	54.12

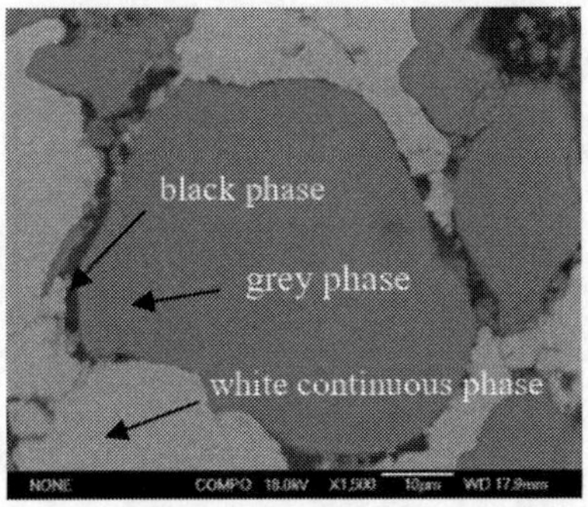

Fig. 3 The typical SEM micrographs of the composite fluorescent coatings with SiO_2 coated

Cu-14Al-X / $SrAl_2O_4:Eu^{2+},Dy^{3+}$ composite coating doped with SiO_2 coating for the phosphors were prepared on 45# carbon steel substrate by supersonic plasma spray and the typical SEM micrographs of the composite fluorescent coatings were showed in Fig.3. These coatings all composed of white continuous phase, grey phase and black phase. The white continuous phase is the Cu-14Al-X powders, thus it is concluded the composition of Cu-14Al-X powder has no changed before and after supersonic plasma spray. In Fig.3 clear interfaces exist amongst white continuous phases, grey phase is the $SrAl_2O_4:Eu^{2+},Dy^{3+}$ phosphors with the SiO_2 coating around the grey phase as black phase.

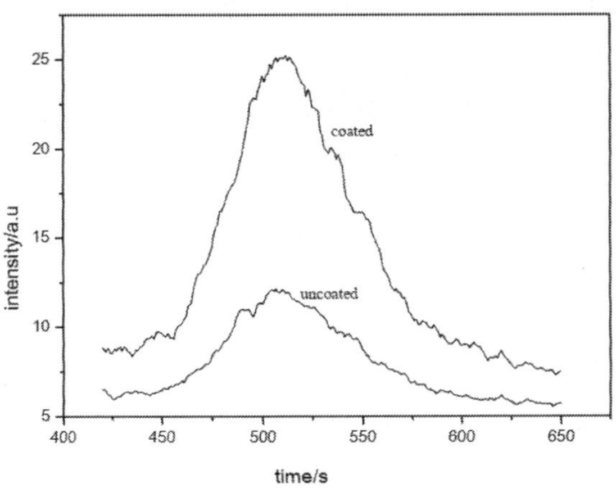

Fig.4 Emission spectrum of phosphors before and after SiO_2 coating in the composite coatings

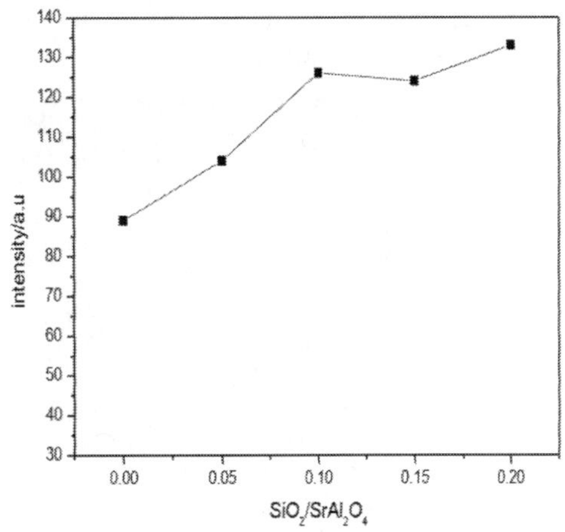

Fig.5 Emission intensity of composite coatings with different coated ratio of SiO_2

Fig.4 shows Emission spectra of coated and uncoated $SrAl_2O_4:Eu^{2+},Dy^{3+}$ phosphors after supersonic plasma spray. The emission spectra exhibit a broad band peaking at 505nm which is attributed to the emission of Eu^{2+} ions with $4f^65d$-$4f^7$ (^8s) transition and the intensity of $SrAl_2O_4:Eu^{2+}$, Dy^{3+} after SiO_2 coating is stronger than that of un-coating, which indicated quenching of

$SrAl_2O_4:Eu^{2+},Dy^{3+}$ were efficiently avoided after SiO_2 coating. This is because of SiO_2 coating showed good high temperature resistance on the surface of phosphor and segregate the phosphors from the metal ions, which can quench the luminescence. As showed in Fig.5, the luminescent intensity shows a maximum when the mass ratio of SiO_2 is 10%.

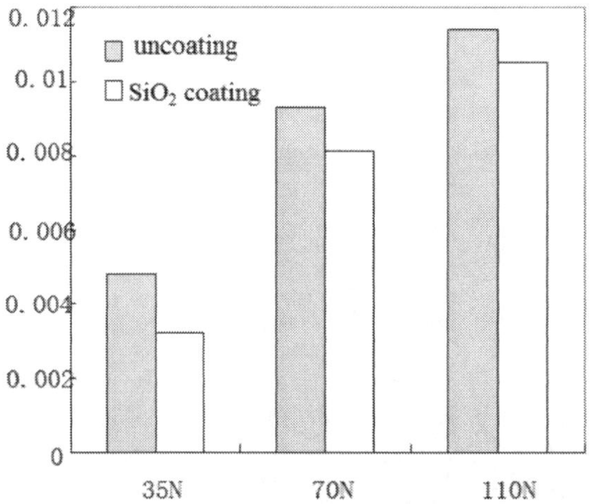

Fig.6 The wear loss of the composite coating coatings with different coated ratio of SiO_2

The wear loss of composite coatings comprised with Cu-14Al-X / $SrAl_2O_4:Eu^{2+},Dy^{3+}$ before and after SiO_2 coating have been recorded in Fig.6. Under the test loads of 35N, 70N, 110N, the composite coating after SiO_2 coating shows the lower wear loss than un-coating which means good anti-friction and anti-wear property. The reason is that SiO_2 coating can not only contribute high hardness to the $SrAl_2O_4:Eu^{2+},Dy^{3+}$ particles but also deposit in boundary film play the role of self-replenish in the worn surface. The SiO_2 coating behaves like a self-lubricant between pin and substrate, which could repair the wear surface. It is known that as a solid lubricant, the crystal structure of SiO_2 films is hexagonal layered and the mechanism behind their effective lubricating performance is attributed to easy shearing of the hexagonal crystalline structures [7].

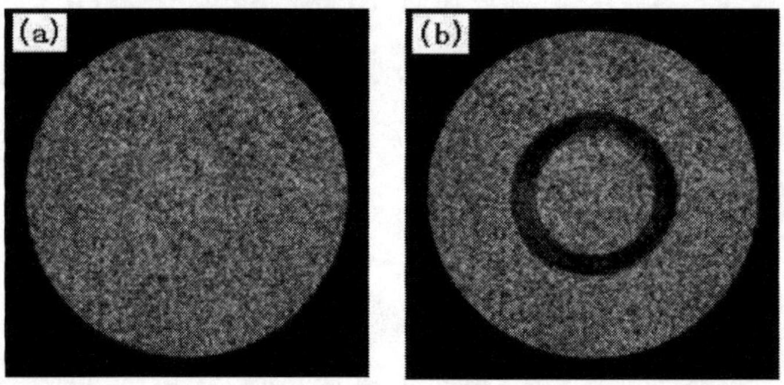

Fig. 7 Morphology of the Phosphorescent composite coatings before and after wear test
(a) Coatings before wear; (b) coatings after wear

The surface of the coating was observed under UV light before and after the wear test (Fig 7a and Fig7 b). The wear test time of the coating is 2800s.Compared with the unworn coating (Fig 7 a), the wear track is clearly visible as black circles. This demonstrates that the damage caused during wear test can be clearly observed under UV light. When the light disappeared, it indicates the coating wearied off and the substrate exposed, so the composite coating is expected as luminescence function coating for monitor wear.

4. Conclusion

The results shows that, thermal quenching of $SrAl_2O_4:Eu^{2+},Dy^{3+}$ were efficiently avoided after SiO_2 coting and the luminescent intensity shows a maximum when the mass ratio of SiO_2 is 10%. Cu-14Al-X / $SrAl_2O_4:Eu^{2+},Dy^{3+}$ composite coating doped with SiO_2 coating shows not only the good luminescence function for monitor wear but also good anti-friction and anti-wear property. Also, SiO_2 coating as additive could repair the wear surface in the wear process that can increase anti wear ability and load-carrying capacity of composite coatings.

Acknowledgements: National Natural Science Foundation of China (51674130) and The International Science and Technology Cooperation Projects (2015DFR51090) supported this research.

References

[1] Li Wensheng, WangZhipin, LuYang, et al. Preparation mechanical properties and wear behaviors of novel aluminum bronze for dies, Trans. Nonferrous Met. Soc. China, 16(2006) 607-612.

[2] Li Wensheng, Lu Yang, Yuan Kexiang, Yuan Chongguang. Effects of cerium on microstructure and bonding strength of Cu-14Al-4.5Fe bronze plasma sprayed coating, J. Rare Earth, 29 (2011) 363-369.

[3] Guo C, Chu B, Wu M, Su Q. Oxide coating for alkaline earth sulphide based Phosphor [J].J Lumin ,2003;105(12)1–6.

[4] Xue L, Kajiyoshi K, Sasaoka H, Nishimura K. *et al.* Preparation of uniform titania nanocoating on ZnS-based phosphors by a sol–gel process [J]. Thin Solid Films,2008;516(483)3–8.

[5] Do H-S, Kim E-J, Hong S-H. *et al.* Improved moisture resistance of $SrS:Eu^{2+}$ phosphors with nanoscale SiO_2 coating [J]. J Lumin, 2010,130:1400–3.

[6] Avci N, Cimieri I, Smet PF, *et al.* Stability improvement of moisture sensitive $CaS:Eu^{2+}$ micro-particles by coating with sol–gel alumina [J]. Opt Mater ,2011;33(103)2–5.

[7] M. Yazıcı, O. Çomaklı, T. Yetim, A.F. Yetim, A. Çelik, Effect of sol aging time on the wear properties of TiO_2–SiO_2 composite films prepared by a sol–gel method, [J].Tribology International, 2016, 104:175–18

IOP Conf. Series: Journal of Physics: Conf. Series **843** (2017) 012062 doi:10.1088/1742-6596/843/1/012062

Identification of Critical Operation Conditions of Industrial Gearboxes by 24/7 Monitoring of Oil Quality, Oil Aging, and Additive Consumption

M. Mauntz[1], J. Peuser[1]
[1]cmc Instruments GmbH, Rudolf Diesel Strasse 12 A, 65760 Eschborn, Germany

Abstract. The demand for wind energy grows at exponential rates. At the same time improving reliability, reduced operation and maintenance costs are the key priorities in wind tur-bine maintenance strategies [1]. This paper provides information about a novel online oil condition monitoring system to give a solution to the mentioned priorities. The presented sensor system enables damage prevention of the wind turbine gear-box by an advanced warning time of critical operation conditions and an enhanced oil exchange interval realized by a precise measurement of the electrical conductivity, the relative permittivity and the oil temperature. A new parameter, the WearSens® Index (WS_i) is introduced. The mathematical model of the WS_i combines all measured values and its gradients in one single parameter for a comprehensive monitoring to prevent wind turbines from damage. Furthermore, the WS_i enables a long-term prognosis on the next oil change by 24/7 server data logging. Corrective procedures and/or maintenance can be carried out before actual damage occurs. First WS_i results of an onshore wind turbine installation compared to traditional vibration monitoring are shown.

1. Introduction

In general, the field of maintenance can be divided into three sectors: preventive (time-based), intelligent (condition-based) and reactive maintenance (run to failure), which show different dependencies between costs and number of failures. Figure 1 shows the costs associated with the different strategies [1].

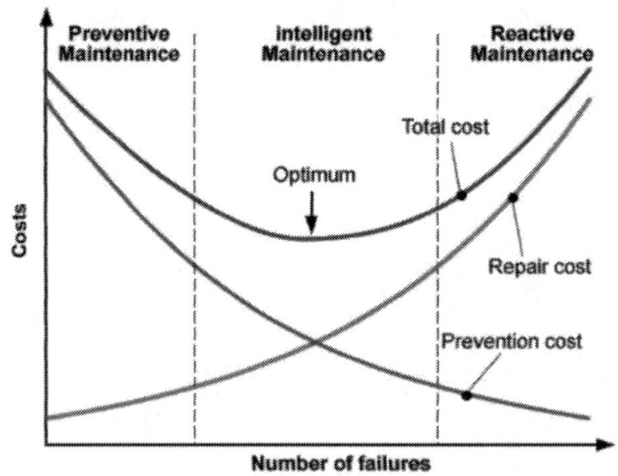

Figure 1. Costs associated with traditional maintenance strategies [1].

From this graph, the optimal point in terms of costs and number of failures can be identified within the center of the intelligent maintenance sector; intelligent maintenance can be realized with an online condition monitoring solution. Different kinds of online monitoring systems have been established over the past years: temperature [2, 3, 3], vibration [4] and particle counting [1, 5]. A vibration monitoring system analyses changes in the frequency spectrum of the observed bearing/gearbox component by Fast Fourier Transformation by converting a time-domain signal into a frequency-domain signal [6]. An optical particle-counter can detect particles larger than 4μm due to the optical resolution of the laser light source. All of these systems need a significant change in the topological contact sur-faces, which means damaged surfaces, particles and pitting.

The presented oil sensor system in contrast is already sensible at the very beginning of the damage formation stage, when the tribological layer gets depleted due to over-load conditions. At this early stage, an increase of electrical charge carriers can be identified by the oil sensor system WearSens®: the electrical conductivity, relative permittivity and temperature are measured with a high precision and low noise over a broad range to enable the detection of small changes in the oil induced by variations in the tribology of the device under test [9]. Inorganic compounds occur at con-tact surfaces from the wear of parts, broken oil molecules, acids or oil soaps. These all lead to an increase in the electrical conductivity, which correlates directly with the wear. In oils containing additives, changes in dielectric constant infer the chemical breakdown of additives. A reduction in the lubricating ability of the oils, the determination of impurities, the continuous evaluation of the wear of bearings and gears and the oil aging all together follow the holistic approach of real-time monitoring of changes in the oil-machine system [10, 11]. Abrasive (metallic) wear, ions, broken oil molecules, acids, oil soaps, etc., cause an increase of the oil conductivity κ. It rises with increasing ion concentration and mobility. The electrical conductivity of almost all impurities is high compared with the extremely low corresponding property of original pure oils. Oils are principally electrical non-conductors. The electrical residual conductivity of pure oils lies in the range below one pS/m. A direct connection between the degree of contamination of oils and the electrical conductivity is found. An in-crease of the electrical conductivity of the oil in operation can thus be interpreted as increasing wear or contamination of the lubricant. The aging of the oil is also evident in the degradation of additives, which are reflected in the relative permittivity [12, 13].

To measure the electrical conductivity and the dielectric constant the oil is passed through an electrode array, which determines the electrical resistance and the capacitance of the sensor assembly using the base oil as a resistive material and dielectric. Figure 2 shows a detail picture of the sensor electrode array with the triple plate de-sign and the schematic electronic circuit. By the high sensitivity of a time measurement method the sensor system detects critical operation conditions much earlier than existing technologies such as vibration measurement or particle counting [14].

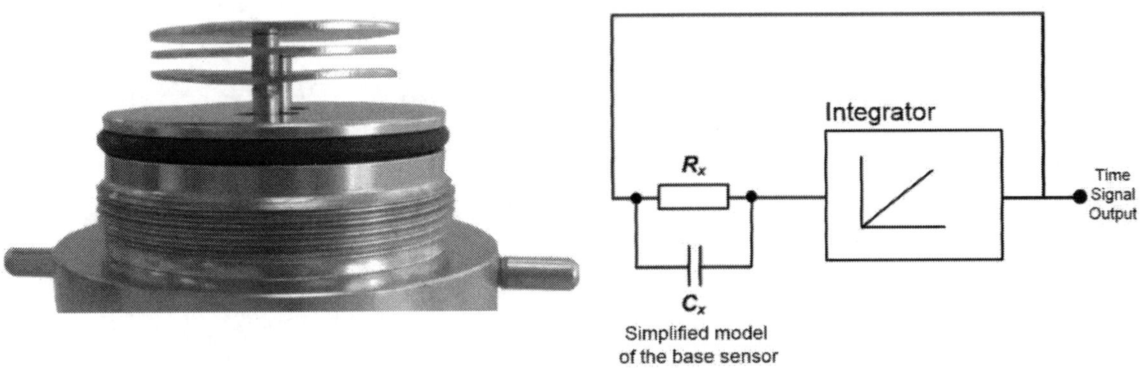

Figure 2. Detail of the triple plate design of the WearSens® base sensor and the corresponding simplified drawing of the electronic circuit.

IOP Conf. Series: Journal of Physics: Conf. Series **843** (2017) 012062 doi:10.1088/1742-6596/843/1/012062

To determine the conductivity and permittivity with a direct measurement of AC observables in a RC circuit is quite inaccurate, because oil has a very high resistance R (several GΩ) and a very low capacity C, which leads to high uncertainties, errors and a low resolution, which is necessary to follow the effects in oil. In the presented system, the electrical conductivity κ and relative permittivity ε_r are determined by a precise time measurement with a very high accuracy and repeatability based on an integrating measurement technique with a high time / bandwidth product: the measurement range for the conductivity starts from 0.1 pS/m up to 1,000,000 pS/m with a resolution of 0.01 pS/m; the relative permittivity is measured between 1 and 5 with a resolution of 1*10-6.

2. Temperature Compensation

Ion mobility and thus, electrical conductivity κ are dependent on the internal friction of the oil and therefore, also on its temperature. The conductivity κ of the oil increases with temperature. The type of contamination and its temperature dependence cannot be assumed to be known. To improve the comparability of measurements, a self-learning adaptive temperature compensation algorithm is necessary. A change of the oil quality can then be assessed by the temperature compensated conductivity value, even though the specific contamination is not determinable [16]. Calculating the electrical conductivity and the dielectric constant at the reference temperature of 40° Celsius is realized by approximating the polynomial form of the temperature dependence.

$$\kappa_{T0} = \kappa_{T0a} + \left(a\Delta T_i + b\Delta T_i^2 + c\Delta T_i^3\right) \cdot \kappa_m$$

κ_{T0} is the approximate electrical conductivity of the oil at the reference temperature T0, κ_{T0a} is the previously calculated (old) electrical conductivity at the reference temperature T0, κ_m is the non-temperature compensated measured value of the electrical conductivity, a, b and c are the coefficients of the approximating polynomial to be adaptively determined during the runtime of the sensor system.

$$\Delta T_i = T_0 - T_i$$

is the temperature difference.

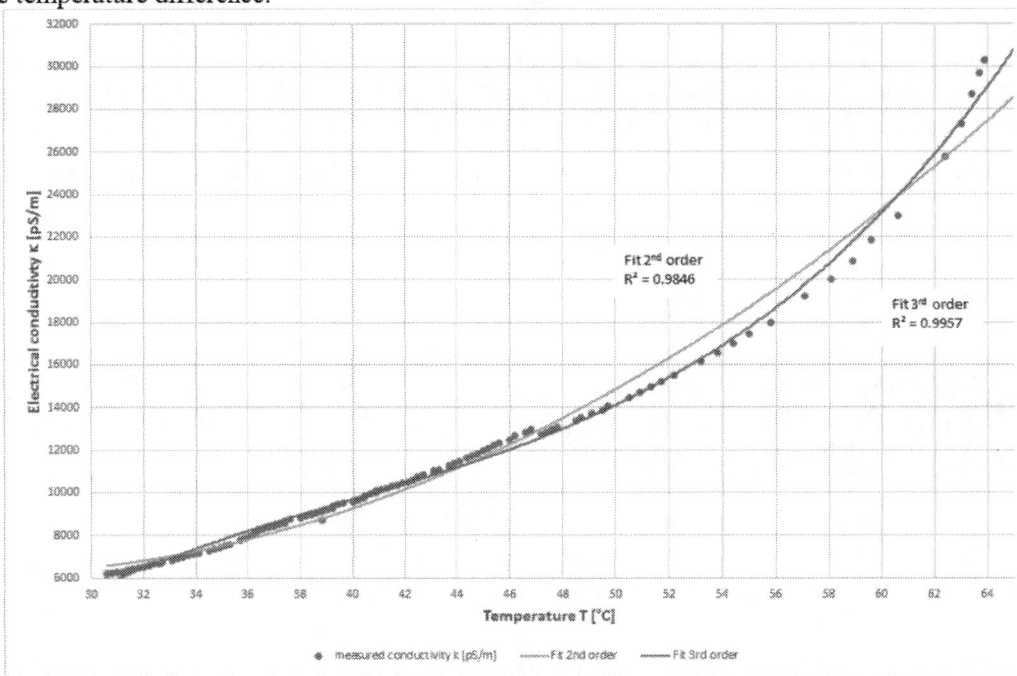

Figure 3. Electrical conductivity versus temperature with second and third order fit curves.

Figure 3 shows the dependency of the electrical conductivity with the temperature of a gearbox oil. Two different trend lines with second order fit and third order fit are plot-ted on the measured data. The approximation by a polynomial of third degree guarantees a good approximation at a reasonably low computational effort for the used microcomputer with an optimal coefficient of determination R^2. For the adaptive determination of the coefficients a, b and c of the polynomial a risk function is defined on the basis of the Gaussian method of least squares from the N measured values pairs and the approximating polynomial, whose minimization enables determination of the desired coefficients.

Figure 4 shows the effect of the adaptive temperature compensation of the electrical conductivity. The algorithm needs about 10 initial measurements to start the compensation and is getting more precise after a short time.

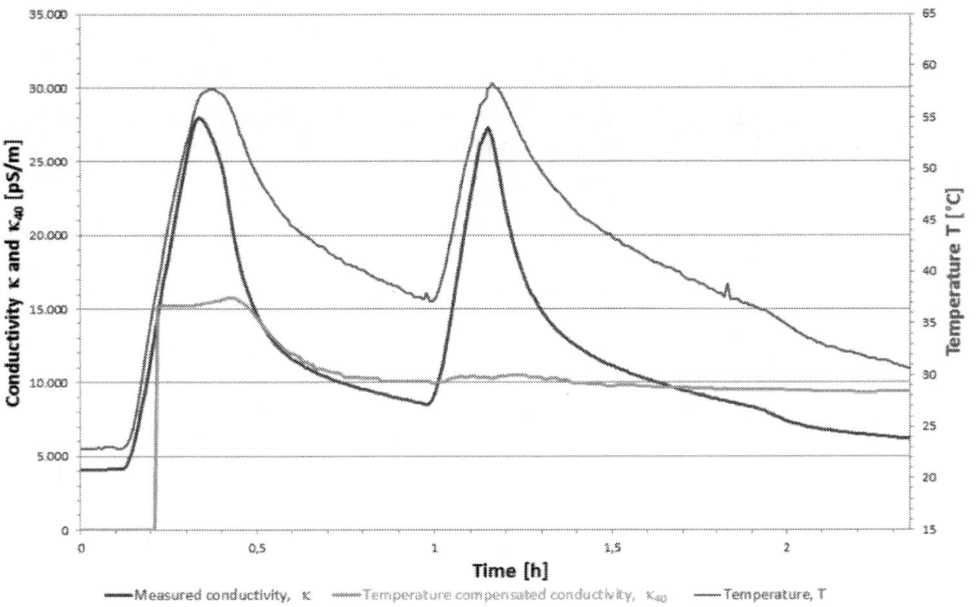

Figure 4. Graph of the measured and compensated electrical conductivity at a reference temperature of 40 °C.

While the measured conductivity κ changes significantly with temperature, the temperature compensated conductivity κ_{40} stays nearly constant. The implemented adaptive algorithm can now work in the background of the measurement procedure autonomously; it has to be reset only after an oil exchange to adapt to the new lubricant. Without the adaptive temperature compensation it is not possible to identify any critical operation conditions in the monitored system due to the high influence of the temperature on the conductivity and the relative permittivity [17].

3. Mathematical model of the WearSens® Index – WS_i

The WearSens® Index (WS_i) has originally been developed for the lubricant analysis of a wind turbine gearbox; however, it can be adapted to any other lubricated system and different oil types with individual modifications. The following description is based on the wind turbine application.

The WS_i model considers short, mid and long-term changes in the lubricant by continuous monitoring of the conductivity, relative permittivity and temperature over a time period of several years with a high time resolution of < 45 seconds. Because of the measurement sensitivity and the high time resolution critical operation conditions can be identified much earlier and a damage can be evaded in short term analysis. The stress of the lubricant and the turbine itself is based on the actual wind condition, wind fluctuation and wind turbine settings (e.g. pitch control, torque control) resulting in instantaneous changes of the conductivity and relative permittivity and their gradients. Critical operation conditions result in an increased charge carrier generation and will change the conductivity and its gradient significantly. A big change in a short time period in the measured values leads to a high WS_i Signal; for example, a significant increase in the electrical conductivity in a short time period is an indication of an abrupt high load or depending on the increase in the conductivity a critical operation condition. Frequent critical operation conditions lead to faster degradation of the oil additive complex. The hypothesis is that the consumption of the additives is directly correlated with the reduction of the relative permittivity of the oil: the relative permittivity ε_r is directly affected by the presence of polar elements; there is a high content of polar additives in gearbox oils [7]. The polar additives combine together with other polar elements (e.g. wear products, water contamination), so from this point of view the consumed additives are not polar anymore, which results in the reduction of the relative permittivity. The gradient, i.e. the time derivative, of the conductivity or the dielectric constant progression respectively represents a measure of the additive degradation and consumption. After identifying initial base κ_{i40}, ε_{ir40}, $\Delta\varepsilon_{ir40}$, $\Delta\kappa_{i40}$, T_i can be feed into the simplified WS_i model below:

$$WS_i = \int_{t1}^{t2} \Big[f(\kappa_{40}, \kappa_{i40}) + f(\varepsilon_{r40}, \varepsilon_{ir40}) + f\left(\frac{\Delta\varepsilon_{r40}}{\Delta t}, \Delta\varepsilon_{ir40}\right) + f\left(\frac{\Delta\kappa_{40}}{\Delta t}, \Delta\kappa_{i40}\right) + f(T, T_i) \Big] dt$$

A simulated wind power profile over a time of ten years is depicted in figure 5 including cyclic high and low wind conditions.

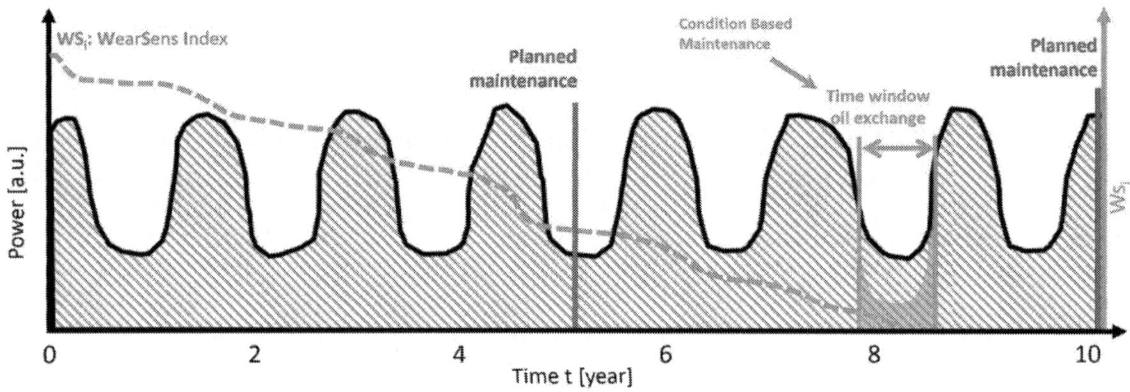

Figure 5. Time course of the simulated WearSens® Index WS_i and wind power profile over 10 years.

The corresponding continuous decrease of the WS_i due to the varying load conditions and the resulting degradation of the oil is schematically shown in the dashed green line. From this point, the benefits of a condition based oil change are clearly eminent: an offshore oil exchange is performed every 5 years independent of the actual oil quality. By using the WS_i as an indicator for the oil change on demand – condition based – the time interval of the oil exchange can be increased quite a lot due to the real condition of the oil: this will save money as a direct effect to the wind park owner, preserve environment and resources. Furthermore, the short-term analysis can avoid critical operation conditions and prevent the wind turbine from damage.

IOP Conf. Series: Journal of Physics: Conf. Series **843** (2017) 012062 doi:10.1088/1742-6596/843/1/012062

4. Sensor installation on a wind turbine: WS_i and vibration monitoring

This section demonstrates first results of the 24/7 oil condition monitoring of an on-shore wind turbine installation with WearSens® in the cooling bypass of a Südwind S77, Winergy (FAG gearbox), 3MW. Figure 6 below shows the onshore wind turbine, the wind turbine gearbox in the nacelle and the installed base sensor.

Figure 6. a) picture of the onshore wind turbine in Büddenstedt, Germany, b) preparation at the gearbox for the WearSens® base sensor installation and c) the installed base sensor in the existing cooling bypass.

The onshore wind turbine was equipped with an existing vibration monitoring system. More general information on the computation, limitations and analysis of the vibration data can be found in the literature [1, 4, 6, 8].

The time course of the vibration monitoring data and the normalized WS_i data, recorded on 14th of November 2015, was compared in figure 7; the WS_i data is plotted in green colour at low and in blue colour at high time resolution, to demonstrate the high information density available at an average update rate of less than 45 seconds. The orange curve shows the dimensionless CMI data, which trend is interpreted as abnormal shifts in the cycling frequencies of the bearing on the fast shaft; here the warning level was set at CMI=50 and the alarm level at CMI=100 from the vibration CMS manufacturer.
The data of the WS_i at low time resolution has a similar signal sequence as the vibration data, but at a different time base: the peaks in the WS_i signal are significantly in front of the vibration signal, provided from the vibration-monitoring operator. Based on an empirical approach the hypothesis is, that fluctuations from normal to critical operation conditions can be identified earlier with the presented online oil sensor system by the precise measurement of the electrical conductivity and relative permittivity, because changes in the tribological layer, an increased charge carrier generation occurs before changes in the vibration signal of gearbox components due to material fatigue. Therefore, it is possible to react much faster on events of critical conditions to prevent the gearbox from damage to enhance the overall life time. By the long-term analysis over several month and years, it is possible to perform condition-based oil change on demand to preserve the environment, to protect the oil resources and to reduce costs.

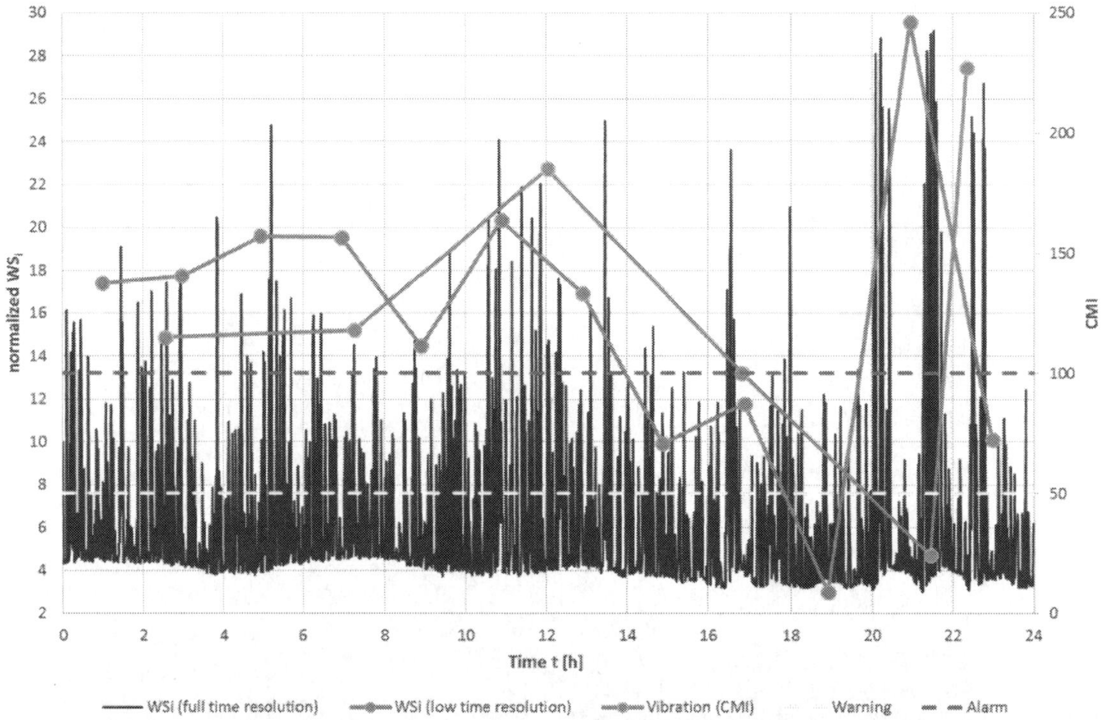

Figure 7. Time course (24 hours) of WS_i with low time resolution (green), WS_i high time resolution (blue) and the vibration signal (red).

5. Conclusion

The online diagnostics system measures components of the specific complex impedance of oils. The indication of forming stage of damage and wear is measured as an integral factor of, e.g., the degree of pollution, oil aging and acidification, water con-tent and the decomposition state of additives or abrasion of the bearings, which is correlated to the changes in the electrical conductivity and relative permittivity. By the adaptive temperature compensation of the measured values, it is possible to identify even small variation in the actual charge carrier generation and additive consumption. For an efficient machine utilization and targeted damage prevention, the WearSens® online condition monitoring system and the WearSens® index offers the prospect to carry out timely preventative maintenance on demand rather than in rigid inspection intervals. The benefits of an extended oil change interval are reduced costs, preservation of the environment and resource protection. The oil sensor system has been in-stalled into an onshore wind turbine performing short and mid-term analysis of the lubricant quality in a time period of 360 days. The direct advantages of this online oil condition online monitoring are detection of critical operation conditions and damage prevention to increase the lifetime of gearbox. The high time resolution, fast response time and accuracy of WearSens® allows earlier intervention control / optimization of the current operation in comparison to sole vibration analysis.

6. References

[1] P. Tchakoua, R. Wamkeue, M. Ouhrouche, F.S. Hasnaoui, T.A. Tameghe, G. Ekemb: Wind Turbine Condition Monitoring: State-of-the-Art Review, New Trends, and Future Challenges, Energies 2014,7, pp. 2595-2630, (2014)

[2] C.J. Hellier: Handbook of Non-destructive Evaluation, McGraw-Hill Professional Publishing, New York, USA, (2003)

[3] J.-Y. Park, J.-K. Lee, K.-Y. Oh, J.-S. Lee, B.-J. Kim: Design for 3MW Wind Turbine and its Condition Monitoring System, Proceedings of the International Multi Conference of Engineers and Computer Scientists, IMECS 2010, Kowloon, Hong Kong, 17–19 March 2010, Volume II, pp. 930–933,(2010)
[4] B.N Madsen: Condition Monitoring of Wind Turbines by Electric Signature Analysis, Master's Thesis, Technical University of Denmark, Copenhagen, Denmark, October 2011, (2011)
[5] A.C. Goncalves, J.B. Campos: Predictive maintenance of a reducer with contaminated oil under an excentrical load through vibration and oil analysis, J. Braz. Soc. Mech. Sci. Eng. 2011, 33, 1–7, (2011)
[6] M.M. Khan, M.T. Iqbal, F. Khan: Reliability and Condition Monitoring of a Wind Turbine, Proceedings of the 2005 Canadian Conference on Electrical and Computer Engineering, Saskatoon, SK, Canada, 1–4 May 2005, pp. 1978–1981, (2005)
[7] Noria Corporation: The critical role of additives in lubrication, Machinery Lubrication June 2012, (2012)
[8] A. Jablonsky, T. Barszcz, M. Bielecka: Automatic validation of vibration signals in wind farm distributed monitoring systems, Measurement 2011, 44, 1954–1967, (2011)
[9] M. Mauntz and U. Kuipers: Ölsensorsystem- Sensorsystem zur Messung von Komponenten der komplexen Impedanz elektrisch gering leitender und nichtleitender Fluide, dessen Realisierung und Anwendung, patent application no. 10 2008 047 366.9, German Patent Office, Munich, (2008)
[10] M. Mauntz and U. Kuipers: Verfahren, Schaltungsanordnung, Sensor zur Messung physikalischer Größen in Fluiden sowie deren Verwendung, European patent application no. EP 09000244, European Patent Office, Munich, (2009)
[11] M. Mauntz, J. Gegner and U. Kuipers: Ölsensorsystem zur Echtzeit-Zustandsüberwachung von technischen Anlagen und Maschinen, Technisches Messen 77, pp. 283-292, (2010)
[12] M. Mauntz, U. Kuipers and J. Gegner: New Electric Online Oil Condition Monitoring Sensor – an Innovation in Early Failure Detection of Industrial Gears, The 4th International Multi-Conference on Engineering and Technological Innovation July 19th – July 22nd, 2011, Orlando, Florida, USA 2011, Proceedings Volume I, International Institute of Informatics and Systemics, Winter Garten, FL, USA, 2011, pp. 238-242, (2011)
[13] M. Mauntz, U. Kuipers and J. Gegner: High-precision online sensor condition monitoring of industrial oils in service for the early detection of contamination and chemical aging, Sensor + Test Conf., 7.-9.6.2011, Nürnberg, AMA Service GmbH, Wunstorf, pp. 702-709, (2011)
[14] M. Mauntz, J. Gegner, S. Klingauf and U. Kuipers: Continu-ous Wear Measurement in Tribological Systems to Control Operational Wear Damage with a new Online Oil Sensor System, TAE Technische Akademie Esslingen, 19th International Colloquium Tribology, Esslingen, January 21-23, 2014, (2014) [15] A. Saeed: Online Condition Monitoring System for Wind Tur-bine, Master's Thesis, Blekinge Institute of Technology, Karlskrona, Sweden, (2008)
[16] M. Mauntz, U. Kuipers and J. Peuser: Continuous, online detection of critical operation conditions and wear damage with a new oil condition monitoring system, WearSens®, 14th In-ternational Conference on Tribology - SERIATRIB '15 Proceedings, Belgrad, Serbian Tribology Society Kragujevac, University of Belgrade, Faculty of Mechanical Engineering, Belgrade, ISBN: 978-86-7083-857-4, S. 283-288, (2015)
[17] M. Mauntz, U. Kuipers and J. Peuser: New oil condition monitoring system, WearSens® enables continuous, online detection of critical operating conditions and wear damage, Malaysian International Tribology Conference 2015 - MITC2015, Penang, Malaysia on November 16-17, 2015, Conference Proceedings, ISBN: 978-967-13625-0-1, S. 179-180, (2015)

Acknowledgments
The authors wish to thank the company Speedwind GmbH for the support and installation of the WearSens® sensor system.

IOP Conf. Series: Journal of Physics: Conf. Series **843** (2017) 012063 doi:10.1088/1742-6596/843/1/012063

Tribological behaviour of plasma nitrided cast iron D6510 and cast steel S0050A under the inclined-impact sliding condition with extremely high contact pressure

C. Zhao, J. Zhang, and X. Nie

Department of Mechanical, Automotive and Materials Engineering, University of Windsor, Windsor, Ontario N9B 3P4, Canada

Corresponding author: xnie@uwindsor.ca

Abstract. Plasma nitriding as a surface modification was applied on two substrate materials: cast iron D6510 and cast steel S0050A. After measurement of the friction coefficients of the treated samples using a pin-on-disc tribotester, an inclined impact-sliding wear tester was utilized to investigate their tribological behaviour under tilting contact with extremely high contact pressure. While numerous surface fatigue cracks, severe chipping, and peeling of the compound layer were observed for the treated cast steel sample, the treated cast iron sample had far fewer surface fatigue cracks without chipping or peeling of the compound at the same test condition. The governing mechanisms of the treated cast iron sample's superior resistance to surface fatigue failure were revealed by studying the cross-sectional hardness and nitrogen concentration profile. Energy-dispersive X-ray spectroscopy (EDS) analysis indicated that the treated cast iron sample had a smaller nitrogen concentration gradient, which led to a smaller hardness gradient as measured. The results suggest that a smaller hardness gradient between the compound layer and the diffusion zone and a thicker hardened case was able to improve the wear resistance and surface fatigue cracking resistance against high contact loads. Moreover, the smaller friction coefficient of the treated cast iron sample could also be beneficial for improving the wear resistance.

1. Introduction

The application of nitriding to substrate materials results in a hardened layer with limited distortion and dimensional variations [1–5], making it an efficient and cost-effective method for the surface modification of precise mechanical parts, like stamping dies and gear teeth.

The tribological behaviours of nitrided iron and steel have been extensively studied [2, 3, 5–7]. However, this knowledge may not be always appropriate to be used for the selection of die materials and surface modification methods in terms of industrial stamping application. For instance, a cast steel has usually been considered as a better material than a cast iron. But, this may not be true at some situations. A laboratory testing method should imitate the loading conditions of real applications.

For a stamping die used for metal forming of advanced high strength steel (AHSS), the loads in the working corner areas where the most wear may occur consist of both normal and tangential forces. Moreover, the maximum stress [8] at the radius of the die could be as high as 4000 MPa and the sliding energy density [9] could be as high as 0.4 J/mm^2. Such harsh conditions cannot be simulated by the traditional pin-on-disk apparatus, and a new kind of wear tester is needed for the tribological evaluation of the nitrided iron and nitrided steel.

In this study, plasma nitriding was used to form case hardened layers on ductile cast iron D6510 and cast steel S0050A. After measuring the friction coefficients of the treated samples, a newly developed inclined impact-sliding wear tester [10–12] was utilized to study the tribological behaviours of the treated samples. The difference in the wear resistance of the samples was explained with respect to the characteristics of their case hardened layers.

2. Sample preparation and experiments

The materials chosen for this study were two commonly used stamping die materials: ductile cast iron NAAMS-D6510 and cast steel NAAMS-S0050A. The chemical composition of the ductile iron is 3.41 wt.% C, 2.22 wt.% Si, 0.41 wt.% Mn, 0.42 wt.% Mo, 0.56 wt.% Cu, 1.12 wt.% Ni, 0.1 wt.% Cr, 0.02 wt.% S, 0.04 wt.% P, 0.05 wt.% Mg, and the balance Fe. The chemical composition of the cast steel is 0.45 wt.% C, 0.36 wt.% Si, 1.05 wt.% Mn, 0.41 wt.% Mo, 0.90 wt.% Cr, 0.03 wt.% S, 0.03 wt.% P, 0.12 wt.% V, and the balance Fe [13].

For the specimen preparation, two sets of samples were cut into 25.4 mm × 25.4 mm × 5 mm plates and subsequently machined to ensure that they had the same surface finish. The plates were cleaned, degreased, and then plasma-nitrided in a system developed by Advanced Heat Treat Corp. with a cathodic heating shield and a pulse plasma generator of 6 kHz frequency. The plasma nitriding was conducted at 570°C for 4 hours. Heating in the plasma process from room to the final temperature was performed in a mixture of 30% nitrogen and 70% hydrogen with the total pressure of 2.9 mbar. For the sake of simplicity, the treated samples will be referred to as TI for the plasma nitrided cast iron sample and TS for the plasma nitrided cast steel sample.

Metallographic sections prepared from the nitrided samples were used for measuring hardness and nitrogen concentration profiles. The surface hardness and hardness profiles of the case hardened layers of the treated samples were determined using a Vickers microhardness tester with indentation loads of 0.0981 and 0.245 N, respectively. The nitrogen concentration profiles were measured by scanning electron microscope (SEM) equipped with energy-dispersive X-ray spectroscopy (EDS) detector (FEG 2000, 13 KV). The results were the average of 16 scan frames. The elastic moduli of the case hardened layers were studied by nanoindentation with a load of 2500 μN (Hysitron).

Pin-on-disk sliding tests (sliding velocity = 0.1 m/s, normal load = 10N) were done to determine the friction coefficients of the treated samples. The wear and surface fatigue resistance were evaluated with an inclined impact-sliding wear tester.

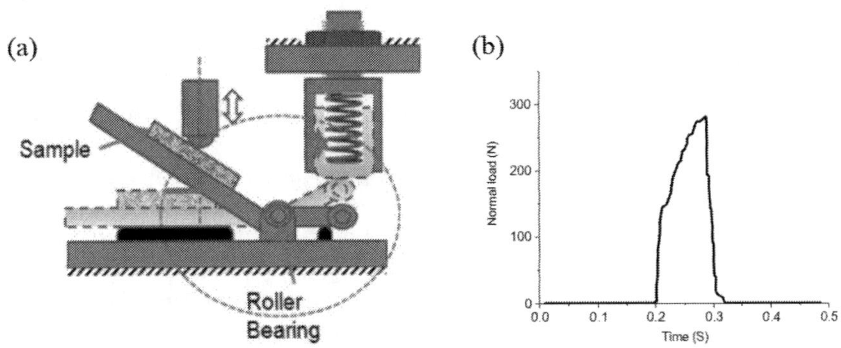

Figure 1. (a) Setup illustration of the inclined impact-sliding wear tester; (b) normal load curve for one of the test cycles.

An illustration of the setup of the inclined impact-sliding wear tester is shown in Figure 1(a). A steel ball with a diameter of 9.5 mm was used as a counterface material. The applied load curve is shown in Figure 1(b), where it can be seen that the load quickly increased to 140 N and reached the maximum force of 300 N before decreasing back to zero in each cycle. The tests were carried out in an ambient environment without lubrication. According to the Hertzian calculation, when F = 300 N, the maximum contact pressure should be around 3.60 GPa. This stress range is close to the calculated maximum stress possible at the die working corner (about 4.0 GPa) [8], while the maximum contact stress for the Pin-on-disk (POD) sliding test is only about 1.2 GPa and the purpose of the POD test was to measure coefficient of friction in this study. The period for each inclined reciprocating sliding test cycle was 0.5 s; the numbers of test cycles were 500, 1000, 1500, and 2500 for the TI sample, and 100, 250, 500, and 1000 for the TS sample. The way of choosing test cycles was the following: firstly, both the samples were tested with 500 and 1000 cycles; after observing the surface damage conditions, to increase the test cycles was decided for the TI sample and to decrease the test cycles for the TS sample. For all the tests, the balls as the counterface materials were made of SAE 52100 hardened steel (60–62 HRC).

3. Results and discussion

3.1. Tribological performance under inclined sliding wear tests

Figure 2 shows the wear volume loss and average wear volume loss per test cycle at different numbers of test cycles. Obviously, the accumulated wear volume loss increased with test cycles for both TI and TS samples. However, the average wear volume loss per test cycle is different for different samples. For the TI sample, the average wear volume loss gradually decreased from 5.92E-6 to 2.5E-6 mm^3/cycle as the number of test cycles increased from 500 to 2500; that is, the wear rate decreased as the test continued. On the other hand, the average wear volume loss of the TS sample increased sharply to 1.6E-5 mm^3/cycle at 250 test cycles and then decreased quickly to 9.38E-6 mm^3/cycle at 500 test cycles, indicating that there must be a change in the wear mechanisms.

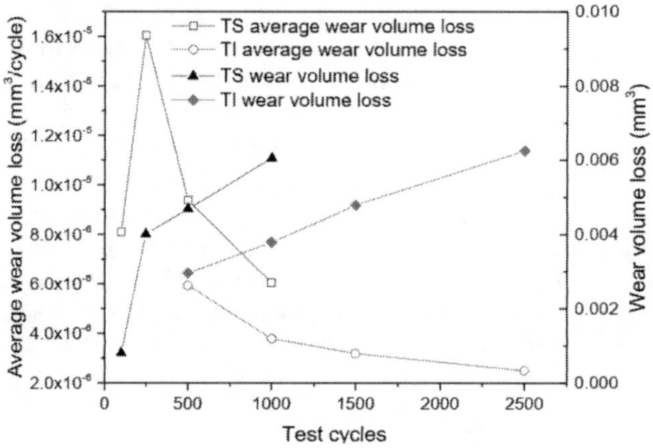

Figure 2. Wear volume loss and average wear volume loss per test cycle for the TI and TS samples at different test cycles.

The SEM morphologies of the wear track on the TS sample after 500 test cycles are presented in Figure 3; B, C, and D represent the head, middle, and tail areas of the wear track, respectively. In Figure 3(b), a large area of chipping/peeling is revealed. The inset in Figure 3(e) shows a detailed image of area E in Figure 3(b), in which materials transfer and wear debris pile-up can be seen. It can also be observed that numerous surface fatigue cracks initiated and propagated across the whole wear track.

In the middle area, as shown in Figure 3(c), severe material transferring from the ball and wear debris pile-up led to form a loose transfer layer on the surface of the wear track. The inset in Figure 3(f) shows this kind of transfer layer at a higher magnification. The small stars in Figure 3(f) represent the locations of EDS point analysis, which was applied to distinguish the compound layer, the diffusion zone and the materials transfer. The results are summarized in Table 1. It should be noted that only iron, nitrogen and oxygen were counted during the EDS analysis. Therefore, the nitrogen concentration for the nitrided layer was higher (up to 30 at. % for the compound layer and 15 at. % for the diffusion zone) than the commonly-expected value, and cannot be viewed as standard reference values. But the relative differences were sufficient to distinguish the compound layer, the diffusion zone and the materials transfer. The red star shows relatively low O and N concentrations, indicating that the compound layer has already been peeled away. The green and blue stars show relatively high O concentration without the existence of N. This indicated the materials transfer layer at the green star and the wear debris pile-up at the blue star.

In the tail area of the wear track, only slight abrasive wear was observed without surface fatigue cracks.

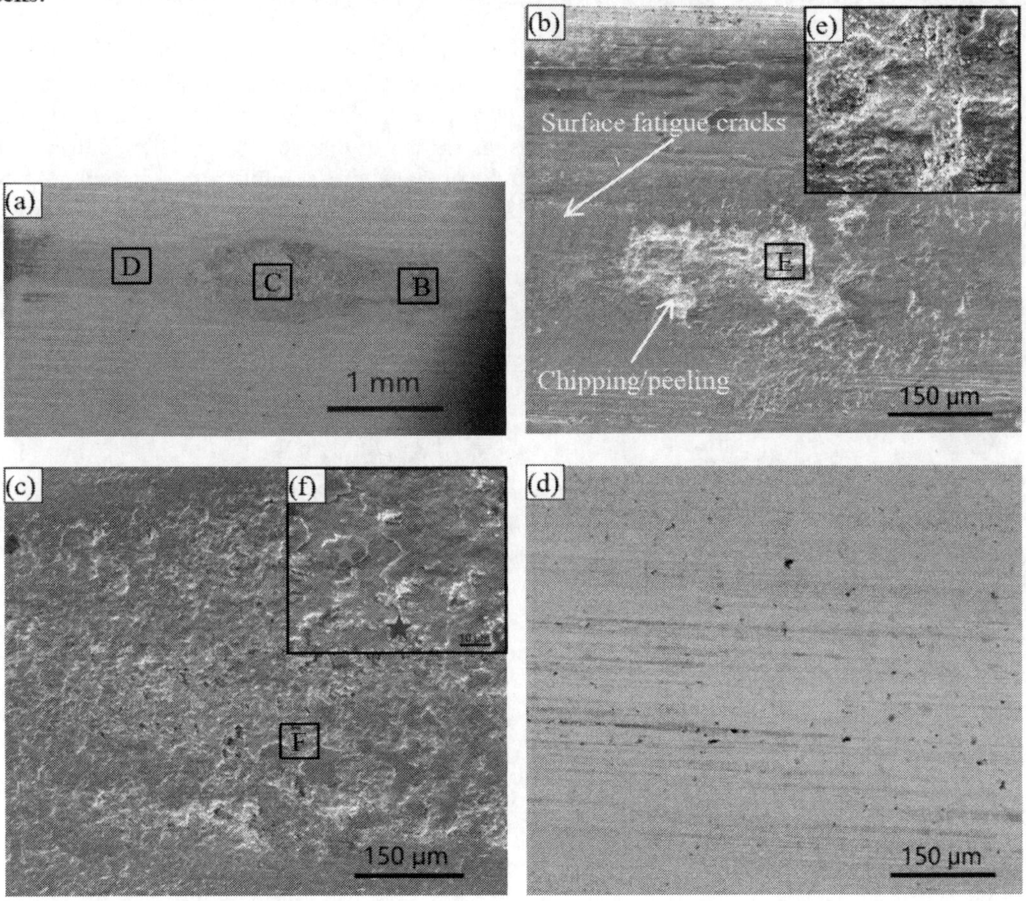

Figure 3. SEM observation of the wear track of the TS sample after 500 test cycles: (a) overview of the wear track; (b) head area of the wear track, the inset shows the detailed structure of the worn surface in the head area; (c) middle area of the wear track, the inset shows the detailed structure of the worn surface in the middle area and the arrangement of EDS point analysis; (d) tail area of the wear track.

The SEM observations revealed the change of wear mechanisms for the TS sample during the inclined impact-sliding wear test. At the beginning (50 to 100 cycles), the surface fatigue cracks were initiated and propagated in the head and middle areas. The cracking patterns were similar to those observed in scratch tests. When the crack density reached a critical value, the compound layer was chipped and peeled away. This process occurred first in the middle area (at around 200 to 250 cycles) and then in the head area (at around 500 cycles). This could explain the sharp increase of average wear volume loss during the 250 test cycles. Then, severe materials transfer from the ball and wear debris pile-up in the area of chipping and peeling could occur. This process reduced the wear volume loss measured. Meanwhile, chipping and peeling of the compound layer still happened in other areas. Thus, the wear volume loss still increased but at a much lower rate, as shown in Figure 2.

Table 1. EDS analysis results for the middle area of the TS sample's wear track after 500 test cycles

	Fe (at.%)	N (at.%)	O (at.%)
Red	80.88	9.11	10.01
Green	72.72	NA	27.28
Blue	57.61	NA	42.39

Note: The values should only be used for comparison with each other to distinguish the compound layer, the diffusion zone and the materials transfer.

Figure 4 shows the detailed SEM images of the worn surface of the TI sample after 2500 cycles of the test; B, C, and D represent the head, middle, and tail areas of the wear track, respectively. The head and middle areas of the wear track appear to have numerous surface fatigue cracks, large areas of material transfer, and a few very small chipping areas, as shown in Figure 4(b, c). The tail area of the wear track appears to have a little material transfer in the areas where graphite was originally located, as shown in Figure 4(d). Slight abrasive wear was also observed in this area without the existence of surface fatigue cracks.

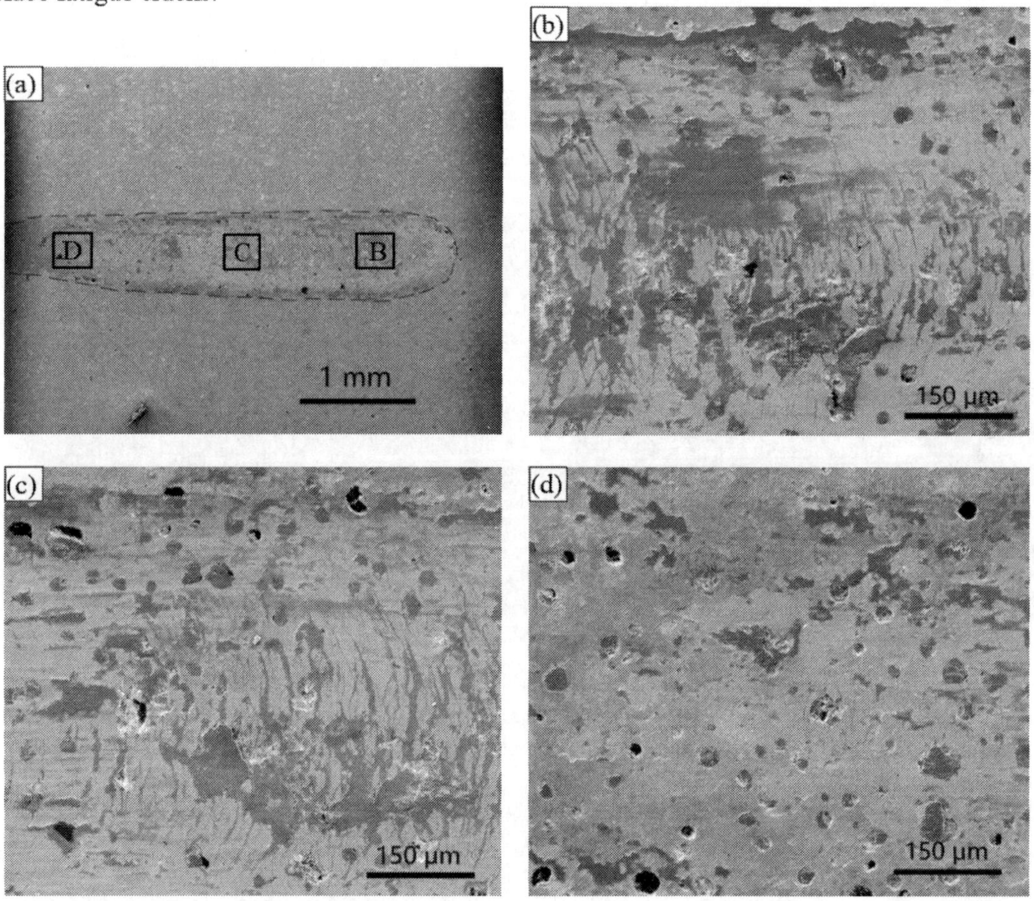

Figure 4. The worn surface of the TI sample after 2500 test cycles: (a) overview of the wear track at ×30 magnification; (b) head area, (c) middle area, and (d) tail area at ×200 magnification. The dashed curve indicates the boundary of the wear scar.

SEM observation on the worn surface of the TI sample after 500 cycles of the test revealed the existence of a porous surface layer (which was mostly worn away, with only a small area left in the wear track), as shown in Figure 5. This could explain the high average wear volume loss for the 500 test cycles. At the beginning (50 to 500 cycles), the wear volume loss was mainly contributed by the wear of this porous layer. Thus, the average wear volume loss, that is, the wear rate, was relatively high. Later on (500 cycles to 2500 cycles), surface fatigue cracks initiated and propagated in the head and middle areas, but no chipping or peeling occurred. In this period, the wear volume loss was mainly contributed by the abrasive wear of the inner compound layer with more compact structure and higher hardness. Thus the average wear volume loss per cycle, that is, the wear rate, gradually decreased. On the other hand, we also noticed that small areas of chipping and peeling started to occur after 2500 test cycles, which implies that severe surface fatigue failure may happen later (perhaps after 3000 or 5000 test cycles).

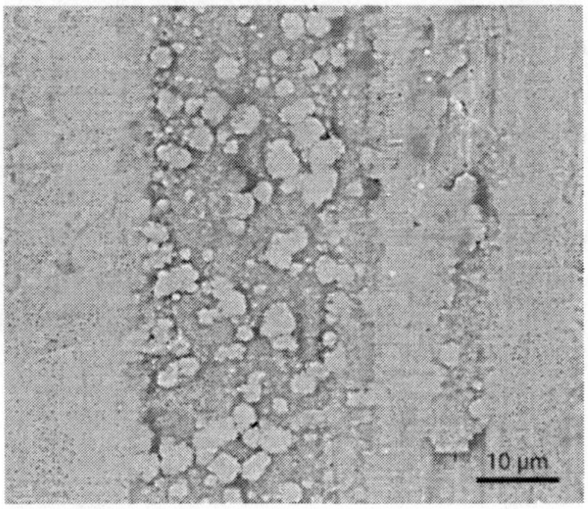

Figure 5. Worn surface of the wear track of the TI sample after 500 test cycles, showing the porous surface layer of the TI sample

3.2 Influence of the friction coefficient.

The coefficients of friction of the samples under pin-on-disk tests are plotted in Figure 6. For the treated steel sample, after an initial running-in stage with a coefficient of friction value of 0.58, the coefficient of friction reached a steady-state value of 0.74 ± 0.2. The low friction coefficient of the running-in stage was achieved through the existence of an iron oxide layer, which was confirmed by the EDS analysis.

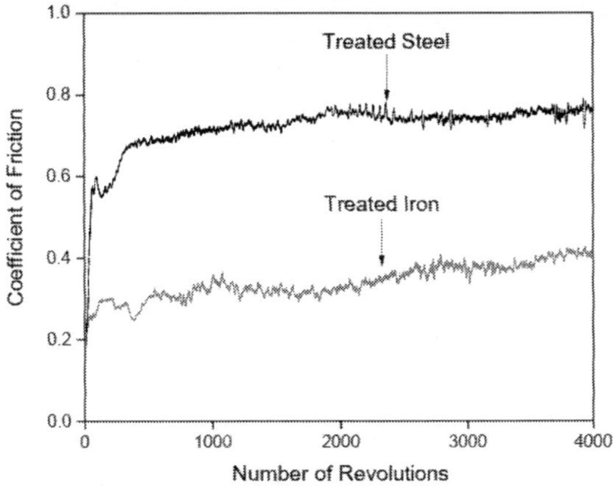

Figure 6. Recorded friction coefficient curves of the treated samples.

Compared with the TS sample, the treated iron sample had a much lower friction coefficient of 0.34 ± 0.2. The surface-distributed graphite, which could act as a solid lubricant, would be responsible for the reduction of the friction coefficient.

Since the TI sample had a much lower friction coefficient, the sliding energy density applied on the surface of the TI sample during the inclined impact-sliding wear test should be smaller. This led to the

IOP Conf. Series: Journal of Physics: Conf. Series **843** (2017) 012063 doi:10.1088/1742-6596/843/1/012063

delayed initiation and propagation of surface fatigue cracks and thus increased the resistance to surface fatigue failure. The lower friction coefficient could also lead to smaller tangential stresses during the wear test, which was beneficial for improving the resistance to surface fatigue failure.

3.3 Influence of hardness gradient and mismatch of elastic modulus

Besides the influence of the coefficient of friction, the difference in elastic modulus and hardness gradient between the compound layer and the diffusion zone of the nitriding cases could also play an important role. The sectional nitrogen concentration profiles (which would be the main cause of hardness gradient and mismatch of elastic modulus) were measured. Here the subject interested was the difference of iron nitrides in the two samples, thus only Fe and N were counted when measured the nitrogen concentration profiles. The nitrogen concentration presented thereby was thus too high. Therefore, the values could only be considered for relative comparison. The difference of the nitrogen concentration profiles are revealed in Figure 7.

Figure 7 demonstrates the sectional nitrogen concentration profiles below the surfaces of nitrided samples. It is clear that at the boundary between the compound layer and diffusion zone (at 10–15 μm from surface), the TI sample had a much smaller nitrogen concentration gradient. Moreover, at the top of the diffusion zone (which was covering a range of 15–50 μm starting from the boundary), the TI sample had a higher nitrogen concentration.

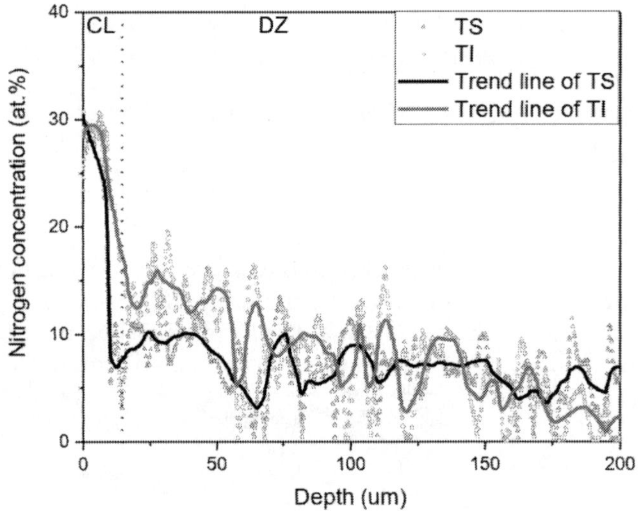

Figure 7. Sectional nitrogen concentration profiles of the nitrided samples (Y scale should be considered as arbitrary).

The Vickers hardness profiles below the surfaces of the samples are plotted in Figure 8. Each of these values is the average of five measurements. It is clear that, compared with sample TS, sample TI had a higher hardness in the top area of the diffusion zone with a lower hardness-depth gradient at the boundary between the compound layer and the diffusion zone.

Nanoindentation was utilized to study the elastic modulus of the compound layer and the diffusion zone of the TI and TS samples, respectively. The indentations were made at 5 μm and 30 μm below the surface to characterize the compound layer and the diffusion zone. It is clear that the compound layers of both samples had similar elastic modulus while their diffusion zones had significantly different elastic modulus. The mismatches of the elastic modulus of the compound layer and diffusion zone were 66.2 ± 3.2 GPa for the TS sample and 31.5 ± 2.7 GPa for the TI sample.

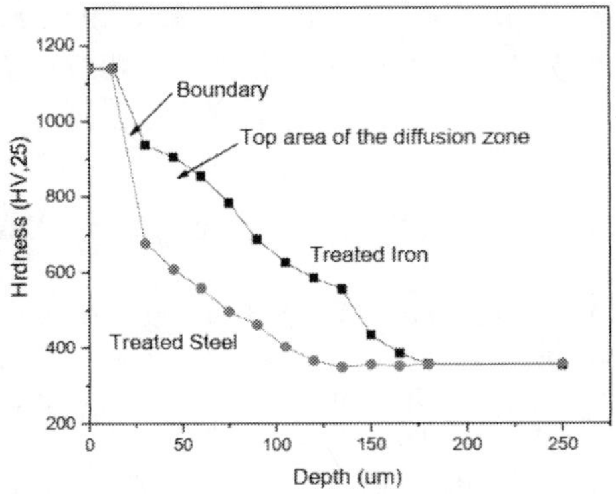

Figure 8. Cross-sectional hardness profiles of the TI and TS samples.

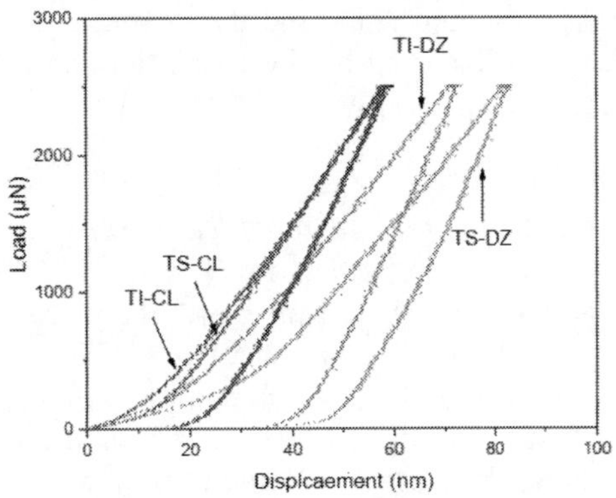

Figure 9. Nanoindentation load–displacement curves for the compound layers of the TI sample (TI-CL) and TS sample (TS-CL) at 5 μm below the surfaces, as well as for the diffusion zones of the TI sample (TI-DZ) and TS sample (TS-DZ) at 30 μm below the surfaces.

The large mismatch of the elastic modulus and the high hardness gradient at the boundary between the compound layer and the diffusion zone of the TS sample had a critical influence on the elastic/plastic deformation behaviours of the compound layer and the diffusion zone. When the tested sample was subjected to extremely high contact, such as 3.6 GPa during the inclined impact-sliding wear test in this study, the corresponding strain (elastic plus plastic) of the compound layer and the diffusion zone differed considerably. This led to a concentration of applied normal stress at the interface nearby area. Combined with the tangential stress caused by the sliding friction, the total stress was high enough to initiate surface fatigue cracks very quickly. The high combined stress also accelerated the propagation of the surface fatigue cracks and thus caused severe surface fatigue failure like chipping and peeling. On the other hand, the elastic modulus mismatch and hardness gradient of the TI sample were much smaller. Therefore, the corresponding strain and stress concentration were significantly smaller during the wear test. In addition, the tangential stress caused by the sliding friction was much smaller, leading to a smaller combined stress. The lower combined stress would cause the delayed initiation and propagation of surface fatigue cracks.

From above analysis, reducing the sliding friction of the surface and decreasing hardness gradient and mismatch of elastic modulus between the compound and diffusion layers are beneficial for increasing the surface fatigue resistance of the hardened case.

4. Conclusions

During the inclined impact-sliding wear test, numerous surface fatigue cracks formed on the nitrided samples. This was due to the extremely high contact stresses and sliding energies applied on the surfaces of the treated samples during the wear test.

On contrast to the plasma nitrided steel sample, the plasma nitrided iron sample showed higher surface fatigue resistance, resulting in delayed surface fatigue failure, which was usually initiated with chipping and peeling.

The superior surface fatigue resistance of the nitrided iron sample was attributed to the smaller elastic modulus mismatch and hardness gradient at the boundary between the compound layer and the diffusion zone. The lower coefficient of friction also contributed to the improvement in the surface fatigue resistance.

The cast iron and steel samples studied in this project were treated at the same way. The plasmas nitriding process parameters may need to be modified so that it can provide an improved wear property

IOP Conf. Series: Journal of Physics: Conf. Series **843** (2017) 012063	doi:10.1088/1742-6596/843/1/012063

for the cast steel to withstand the extremely high stressed condition (by reducing the sliding friction of surface, and lowering hardness gradient and elastic modulus mismatch between different layers, for instance).

References

[1] Cubberly W H. 1981 *Metals Handbook*, 9th edn , vol 4, Heat Treating (Metals Park, PA: American Society for Metal).

[2] Kunst H, Oster P and Elstorpff M. Investigations of load carrying capacity of nitrocarburized gears 1992 *Proc. 3rd World Congress on Gearing and Power Transmission* p 622–31

[3] Wells A and Shaw M 1985 An investigation into dry sliding wear behaviour of the ferritic nitrocarburized surface of a plain low carbon steel *Wear* **103**(1) p 29–42

[4] Tong W P 2003 Nitriding iron at lower temperatures *Science* **299**(5607) p 686–88

[5] Liang W 2003 Surface modification of AISI 304 austenitic stainless steel by plasma nitriding *Appl. Surf. Sci* **211**(1–4) 308–14

[6] Nie X, Wang L, Yao Z, Zhang L and Cheng F 2005 Sliding wear behaviour of electrolytic plasma nitrided cast iron and steel *Surf. Coat. Technol* **200**(5–6) p 1745–50

[7] Xi Y, Liu D and Han D 2008 Improvement of corrosion and wear resistances of AISI 420 martensitic stainless steel using plasma nitriding at low temperature *Surf. Coat. Technol* **202**(12) p 2577–83

[8] Pereira M P, Yan W and Rolfe B F 2010 Sliding distance, contact pressure and wear in sheet metal stamping *Wear* **268**(11–12) p 1275–84

[9] Zhou D J and Du C Q 2015 Recent lessons learned on UHSS stamping *Internal report to Auto/Steel Partnership* (Southfield, Michigan, USA)

[10] US Patent *Cyclic impact-sliding fatigue wear testing instrument* US 8402811 B2.

[11] Su J F,Yu D, Nie X and Hu H 2011 Inclined impact-sliding wear tests of TiN/Al_2O_3/TiCN coatings on cemented carbide substrates *Surf. Coat. Technol* **206**(7) 25 p 1998–2004

[12] Su J F Impact-sliding wear tests on duplex-treated die materials *Final report for Tribology Team of Auto/Steel Partnership* under Project ASP360 Project No.: 11-2538-AMP

[13] *Index for stamping dies cast materials* Retrieved 25 August 2016 from http://www.naamsstandards.org/Standards/chapters/stamping/D.pdf

[14] Holmberg K, Ronkainen H, Laukkanen A, Wallin K, Erdemir A and Eryilmaz O 2008 Tribological analysis of TiN and DLC coated contacts by 3D FEM modelling and stress simulation *Wear* **264**(9–10) p 877–84

IOP Conf. Series: Journal of Physics: Conf. Series **843** (2017) 012064 doi:10.1088/1742-6596/843/1/012064

The effect of boriding on wear resistance of cold work tool steel

Y Anzawa[1], S Koyama[2] and I Shohji[2]

[1]Graduate School of Science and Technology, Gunma University, Tenjin-cho, Kiryu, Gunma 376-8515, Japan
[2]Faculty of Scinece and Technology, Gunma University, Tenjin-cho, Kiryu, Gunma 376-8515, Japan

koyama@gunma-u.ac.jp

Abstract. Recently, boriding has attracted extensive attention as surface stiffening processing of plain steel. In this research, the influence of processing time on the formation layer of cold work tool steel (KD11MAX) by Al added fused salt bath was examined. In addition, in order to improve the abrasion resistance of KD11MAX, the effect of the treatment of boronization on the formation layer has been investigated. Boriding were performed in molten borax which contained about 10 mass% Al at processing time of 1.8 ~ 7.2 ks (processing temperature of 1303 K). As a result of the examination, the hardness of the boriding layer becomes about 1900 HV when the processing time of 3.6 ks. Also the abrasion resistance has improved remarkably. Furthermore, it was revealed that the formation layer was boronized iron from the Vickers hardness and analysis of the X-ray diffraction measurement.

1. Introduction

In recent years, machine parts manufactured by forging have been increasing. The reason is that it because processing speed of forging is extremely faster than that of cutting. Among them, the workpiece manufactured by cold forging has good dimensional accuracy and surface finish to perform a compression molding of metallic materials at room temperature with die. However, a die for cold forging have a short life because the molding pressure and the load to a die is high. Thus, to elongate the lifetime of a die, it is necessary to improve the surface hardness and wear resistance by surface reforming.

As for the cold working tool steel used for cold forging, the surface hardness and wear resistance are generally improved by quenching or nitriding [1-3]. Nevertheless, the hardness of the hardening layer to be formed by each hardening method is less than 1200 HV. On the other hand, a boron compound is an example of a compound which had high wear resistance and hardness [4, 5]. As boriding method, the molten salt immersion method is known as the method that special equipment are unnecessary, and is the simplest. However, there is very little report about boriding than nitriding and carburizing of colt working tool steels.

In this study, for the purpose of mechanical property improvement of the cold working tool steel, boriding was carried out with quenching by performing heat treatment at the time of the quenching in Al added fused salt bath. The mechanical property of the processed layer was evaluated by the surface and sectional hardness measurement and abrasion test. The identification of the processed layer was analysed by using an X-ray diffractometer (XRD).

IOP Conf. Series: Journal of Physics: Conf. Series **843** (2017) 012064 doi:10.1088/1742-6596/843/1/012064

2. Experimental details

As shown in Figure 1, the specimen used in this study comprised a 50 × 30 × 5 mm^3 plate cut from cold work tool steel (Nihon Koshuha Co., Ltd., Japan, KD11MAX) ingot. The surface to be treated of the steel was finished by grinding with emery paper (grade 1200). The Al added fused salt bath was made by dissolving a mixture in crucible made by SUS304 stainless steel after mixing 10% Al with sodium tetraborate. Boriding were performed by immersing a specimen in Al added fused salt bath while turning a specimen at rotary speed of 20 rpm with shaft. The boriding temperature was fixed at 1303 K, although the boriding time was changed ranging from 1.8 ks to 7.2 ks. After boriding, the specimen was air-cooled after taken out from a bath. Subsequently, the specimen was boiled in water for 1.8 ks to remove surface contamination. The surface and the sectional hardness was measured by using Vickers hardness tester. To determine the compound formed on the surface after the boriding, the treated surfaces were investigated using an X-ray diffractometer (XRD) with CuKα radiation working at the optimum voltage of 32 kV and anodic current of 20 mA. Furthermore, a specimen showing the vertical cross section of the boriding surface was also prepared and observed by SEM and EDX analysis at an acceleration voltage of 15 kV. This specimen was prepared by first polishing it using emery paper (grade 1200), followed by ion milling system. The abrasion test was carried out at room temperature under a load of 4.9 N (sliding distance of 5760 m) by the ball-on-disk-type abrasion testing equipment. In addition, the ball is made by ZrO_2 of a radius of 3/32 inch. The depth profile of the sliding portion after the abrasion test was measured by a laser microscope.

3. Results and discussion

3.1. Surface hardness and X-ray structure analysis

To identify influence of the processing time on the surface hardness, the surface hardness was measured by Vickers hardness tester. The measuring load is changed to investigate a crack resistance. As shown in Figure 2, surface hardness decreased with increase of the processing time without depending on measuring load. Regardless of processing time, surface hardness decreased with increase in measuring load.

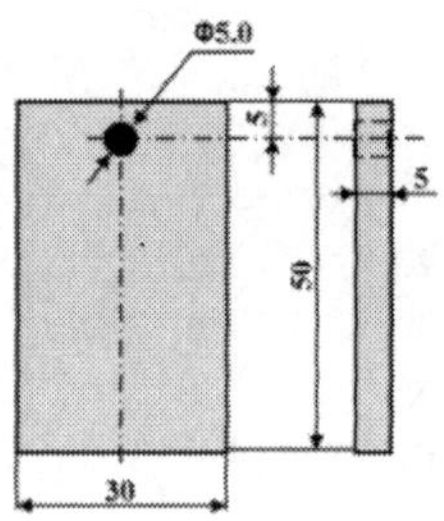

Figure 1. Schematic illustration of to-be-processed specimen (in mm).

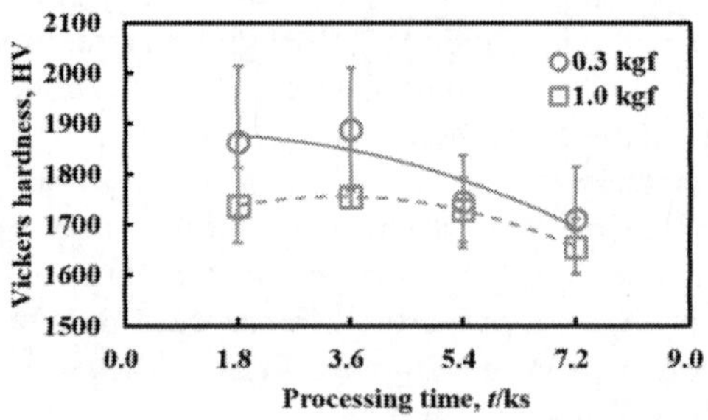

Figure 2. Surface microhardness of the specimen after boriding.

In addition, the X-ray diffraction measurement was performed to examine a metallic structure formed on the boriding surface. As shown in Figure 3, from the x-ray diffraction measurement, it was revealed that FeB and Fe_2B were formed. Moreover, with a increase in processing time, peaks shifted to the low angle side. From the X-ray diffraction measurement result, it is inferred that the boriding surface has the compression stress. It is thought that the reason why the surface hardness decreased at the processing time of 5.4 ks or more is because a crack progressed around an indentation when the hardness testing was performed. Actually, from the circumstance of the indentation, a radial crack was observed.

3.2. Cross-sectional structure and hardness distribution

To determine the effect of the processing time on the cross-sectional metallic structure and hardness distribution of a boriding layer, SEM and EDX analysis was carried out. As shown in Figure 4, a lamellar structure consist of three layers was observed in the vicinity of the surface. As a result of the EDX analysis, the difference in color of each layer was shown to depend on the B content. With increase of the processing time, the tendency that thickness of the lamellar structure increased was recognized. In addition, the detected amount of Cr was decreased in the third layer from the surface. It is inferred that this is because chromium boride was precipitated and chromium-depleted zone was formed.

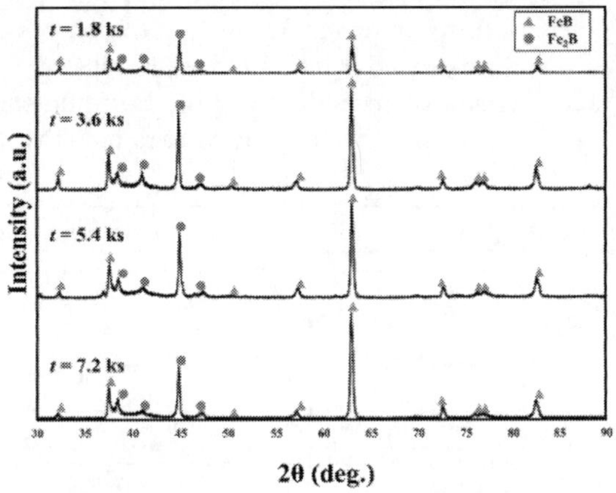

Figure 3. X-ray diffraction patterns of the surfaces of various specimens after boriding: (a) 1.8 ks, (b) 3.6 ks, (c) 5.4 ks and (d) 7.2 ks.

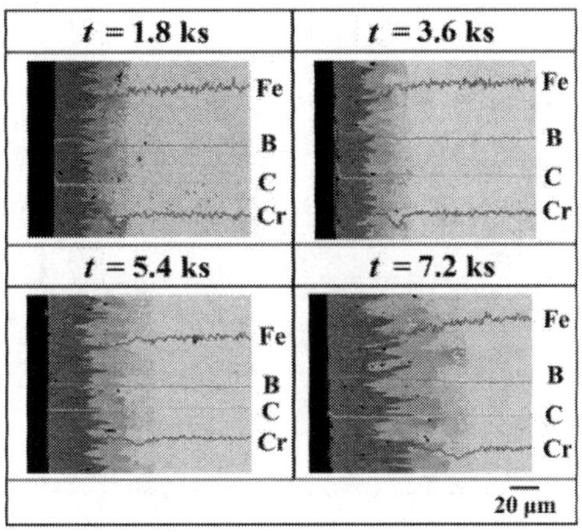

Figure 4. Cross-sectional SEM images and line analysis profiles of the surface layer after boriding: (a) 1.8 ks, (b) 3.6 ks, (c) 5.4 ks and (d) 7.2 ks.

As shown in Figure 5, with increase of the processing time, the width of the high hardness layer increased. The hardness decreased gradually following two stepped hardness distribution. As for the boronized iron, the hardness of FeB and Fe_2B is known as 1800~2000 HV and about 1600 HV, respectively. Hence, it is inferred that first layer and second layer from the top surface consisted of FeB and Fe_2B, respectively.

3.3. Effect of the processing time on the abrasion resistance

Figure 6 shows the relationship between processing time and the wear depth of the processed specimen. The wear depth can be calculated by the next equation (1) by supposing that an abrasion partner ball is worn.

$$d = \left[(2.381)^2 - \left(\frac{w}{2}\right)^2\right]^{1/2} \qquad (1)$$

In the equation (1), d and w represent a calculated value of the wear depth and a measured value of wear width.

With increase of the processing time, the wear depth increased linearly although the wear width increased quickly from the processing time of 5.4 ks. As described above, at the processing time is more than 5.4 ks, boron is dissolved and supersaturated in base metal (refer to Figure 3). Because an excessive solid-dissolution of boron into base metal promoted the spalling in FeB layer, it was revealed that the wear depth increased. As shown in Table 1, at the processing time is more than 5.4 ks, a calculated value of the wear depth exceeded a measured value. It was found from these results that the abrasion of the ball surface was caused in addition to the boriding surface. It is inferred that the cause of this disagreements are changes from mechanical wear to abrasive wear.

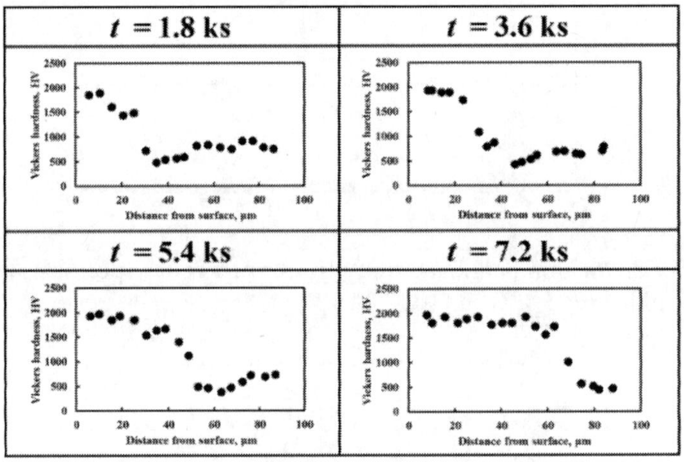

Figure 5. Microhardness profiles of the specimen after boriding (load = 0.025 kgf).

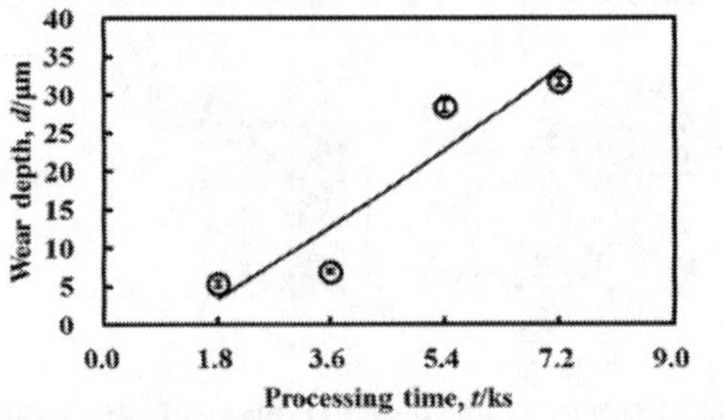

Figure 6. Influence of processing time on wear depth.

Table 1. Wear width and depth of the specimens after wear testing at various processing time.

Processing time, *t* [ks]	Wear width, *w* [μm]	Wear depth, *d* [μm]	
		measured	calculated
30	746	5.39	29.4
60	651	6.70	22.4
90	1778	28.5	172
120	1962	31.7	211

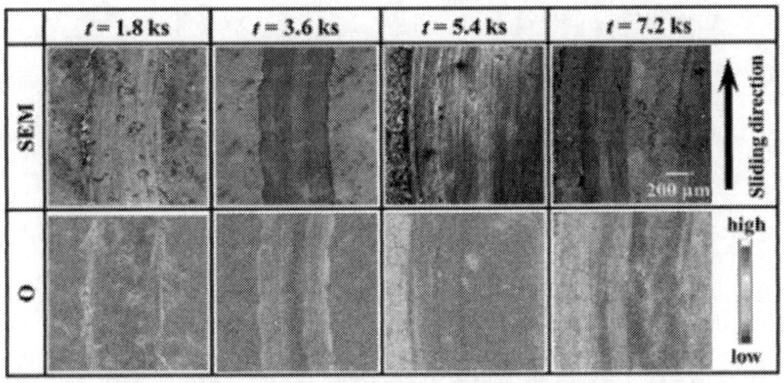

Figure 7. SEM images and EDX analysis of the surfaces after wear test.

Figure 7 shows the SEM micrographs and EDX analysis results of the sliding portion after the abrasion test. As a result of elementary analysis outside of the sliding portion, oxygen was detected from abrasion powder. Therefore, it is inferred that oxidative wear occurred in all processing time. When the processing time was more than 5.4 ks, a scratch marks came to be observed. From the investigation results of the cross sectional microstructure and hardness (refer Figure 4 and Figure 5), it is thought that abrasive wear occurred after FeB layer peeled from a processed surface. Moreover, on the basis of the depth profile of the sliding portion, it was revealed that abrasion was controlled in layers consisting of Fe_2B. From the above results, it was suggested that optimization of the processing time according to an applicable part is required.

4. Conclusions

The conclusions of this study are summarized as follows.

(1) As a result of a hardness measurement, it was found that the most suitable processing time was less than 3.6 ks to form a layer having the high surface hardness and toughness.

(2) When the processing time was more than 5.4 ks, the abrasion was controlled in Fe_2B layer, although the abrasive wear occurred after FeB layer peeled.

References

[1] Conci M D, Bozzi A C and Franco Jr. A R 2014 *Wear* **317** 188
[2] Mashreghi A R, Soleimani S M Y and Saberifar S 2013 *Mater. and Design* **46** 532
[3] Nanesa H G, Boulgakoff J and Jahazi M 2016 *J. Manufacturing Processes* **22** 115
[4] Jan Y, Huang Z, Xing J, Liu X, Sun L, Zheng B and Wang Y 2016 *Wear* **362-363** 68
[5] Carrera-Espinoza R, Figueroa-López U, Martínez-Trinidad J, Campos-Silva I, Hernández-Sánchez E and Motallebzadeh A 2016 *Wear* **362-363** 1

IOP Conf. Series: Journal of Physics: Conf. Series **843** (2017) 012065 doi:10.1088/1742-6596/843/1/012065

The influence of carbide coarsening on the friction properties of thermally affected 9–12 wt. % Cr steels

I. Velkavrh[1*], F. Kafexhiu[2], S. Klien[1], F. Ausserer[1], J. Voyer[1], A. Diem[1], B. Podgornik[2]

[1] V-Research GmbH, Stadtstrasse 33, 6850 Dornbirn, Austria

[2] Institute of Metals and Technology, Lepi pot 11, 1000 Ljubljana, Slovenia

*Corresponding author: igor.velkavrh@v-research.at

Abstract. The purpose of the present study is to evaluate the influence of tempering-induced carbide coarsening in the heat affected zone of two grades of 9–12 wt. % Cr steels, X20 and P91, on their friction properties. In order to compare the frictional properties of the selected microstructures without and with the presence of oxides on the contacting surfaces, friction tests were performed under Ar atmosphere and in air. It was observed that after tempering, coefficient of friction increases for both steels in both atmospheres. This deterioration of friction properties after tempering is probably due to the precipitation of carbides, which acted as mechanical anchors and/or increased the surface energy of steel. Since neither the number of carbides nor their size distribution showed an influence on the friction properties of the selected steels, it is presumed that the minor differences between the friction properties of the tempered X20 and P91 steels are primarily related to their oxidative properties.

1. Introduction

9–12 wt. % Cr martensitic alloy steels are used in vital parts of fossil-fuel power plants, such as turbines, headers, main steam pipes, etc., due to their good mechanical properties and corrosion resistance at elevated temperatures. Although sliding typically does not occur in these applications, components made of these steels may come into sliding contact with neighboring elements under special conditions such as overload or start-ups/shut-downs of fossil-fuel power plants. For instance, steam turbine valve seat materials require an excellent resistance to galling, wear, oxidation and thermal expansion. Namely, it was already reported that most of the turbine overspeed incidents have been caused by valves failing to close or due to the failure of the overspeed protection mechanism [1].

Welding is a major joining and repair technology for 9–12 wt. % Cr steels used for high temperature power plant structures. For applications with elevated temperatures in which martensitic 9–12 wt. % Cr steels are typically used, different temperature-induced processes occur, including coarsening of carbide particles, which significantly affect the material's mechanical properties [2–4]. In general, precipitation of carbides during tempering increases toughness and tensile strength of steels [5–7].

For Inconel 600 mill-annealed alloy, it was already reported that carbide precipitation after aging treatment strongly affected its fretting wear behavior as wear volume increased with the increase of carbides and that carbide precipitation slightly increased the coefficient of friction [8].

In a previous study of the present authors [9], it was observed that X20 steel provided better wear resistance than P91 steel, which was attributed to the less pronounced tempering-induced coarsening and the correlated smaller dissolution of carbide particles. Namely, after tempering, the number of carbide particles was typically higher in the steel X20 than in the P91 and at the same time, a higher number of medium-sized carbide particles (100–200 nm) was observed in the steel X20, while in the steel P91, a higher number of small-sized carbide particles was found (0–50 nm). It was presumed that the less pronounced coarsening, i.e. higher stability of carbide particles in X20 steel is related to its higher carbon content (Table 1).

Alongside the wear resistance, coefficient of friction is another important property to be considered for sliding conditions. However, in the available literature, no studies on the influence of carbide coarsening on the friction properties of 9–12 wt. % Cr steels can be found. The purpose of the present study is to further evaluate the potential of the X20 and P91 steels as materials for applications operating under a combined effect of mechanical wear and alternating high/low temperature conditions. The influence of tempering-induced carbide coarsening on the friction properties of the weld heat affected regions of the selected materials and microstructures is analyzed.

2. Experimental

2.1. Materials preparation

Two grades of creep-resistant 9–12 wt. % Cr steels, X20 and P91, were used in this study. In Table 1, chemical composition of both steels according to the DIN EN 10216-2 standard is given. For the heat treatment procedures, cylindrical samples with a diameter of 13 mm were used. In order to obtain the microstructure of tempered martensite, a routine heat treatment was performed. It started with austenitizing at 1050°C for 30 minutes in order to dissolve the carbide particles in austenite. Afterwards, steels were rapidly cooled (quenched) in oil, which led to a relatively brittle and hard martensitic microstructure. Finally, tempering at 810°C for 30 minutes was performed followed by air-cooling, resulting in a tempered martensitic microstructure, which represented the basis (parent metal, α) for further heat treatment processes.

Additional heat-treatment procedures with the aim to simulate two different HAZ regions, i.e., inter-critical (α+γ) and coarse-grained (γ) microstructures, were carried out. The α+γ microstructure was obtained by keeping the samples in the temperature range of $Ac_1 < T < Ac_3$, i.e. at 845°C, for 60 minutes. The γ microstructure was obtained by keeping the samples above the Ac_3 temperature, i.e. at 1000°C, for 30 minutes. After keeping the samples at the selected temperatures over the selected time periods, the samples were rapidly cooled in air.

The final part of the heat treatment process was performed in order to induce the changes of size, distribution and mutual spacing of carbide particles. It consisted of tempering of the α+γ and γ microstructures of both steels for 1 week and 1 month (168 h and 720 h) at 650°C and 750°C.

After the heat treatment, samples were longitudinally cut in order to obtain flat surfaces for microstructure characterization and wear tests. Samples were first ground with silicon carbide paper to a sufficient depth in order to remove any unrepresentative surface and then polished in several steps in order to achieve a surface roughness of $R_a < 0.01$ μm.

Table 1. Chemical composition of X20 and P91 steels according to the DIN EN 10216-2 standard.

El. (wt. %)	C	Si	Mn	P	S	Cr	Ni	Mo	V	Cu	Nb	Al	N
X20	0.17 - 0.23	max 0.5	max 1.0	max 0.03	max 0.03	10.0 - 12.5	0.30 - 0.80	0.8 - 1.2	0.25 - 0.35	-	-	-	-
P91	0.08 - 0.12	0.20 - 0.50	0.30 - 0.60	max 0.02	max 0.01	8.00 - 9.50	max 0.4	0.85 - 1.05	0.18 - 0.25	max 0.30	0.06 - 0.10	max 0.04	0.03 - 0.07

2.2. Microstructure characterisation

Microstructure changes due to different heat treatment procedures were evaluated via the analysis of micrographs obtained with JEOL JSM-6500F (JEOL Ltd., Japan) scanning electron microscope (SEM). In order to obtain the number, average spacing and size values of carbide particles, image analyses of SEM micrographs were carried out using FIJI (ImageJ) software [10]. Number density, average spacing and size distribution of carbide particles were evaluated and are presented in Table 2 for the steel X20 and in Table 3 for the steel P91. A detailed analysis of microstructural changes is presented in [9].

Table 2. Number density, average spacing and percentage size distribution of carbide particles in the steel X20 after tempering for different times at different temperatures.

X20	Tempering	No. particles per 100 μm^2	Average spacing (μm)	Percentage size distribution (%)						
				0–50 nm	50–100 nm	100–200 nm	200–300 nm	300–400 nm	400–500 nm	500–600 nm
α+γ	168 h / 650°C	1262	0.17 ± 0.06	26	25	40	9	1	0	0
	720 h / 650°C	917	0.19 ± 0.07	23	18	41	15	3	0	0
	168 h / 750°C	646	0.22 ± 0.09	20	23	36	18	3	1	0
	720 h / 750°C	505	0.24 ± 0.13	26	22	26	19	5	2	1
γ	168 h / 650°C	1033	0.18 ± 0.07	16	29	45	9	1	0	0
	720 h / 650°C	852	0.20 ± 0.08	8	24	53	12	2	0	0
	168 h / 750°C	588	0.22 ± 0.10	7	27	48	14	3	1	0.15
	720 h / 750°C	529	0.23 ± 0.12	29	17	29	16	7	2	0

Table 3. Number density, average spacing and percentage size distribution of carbide particles in the steel P91 after tempering for different times at different temperatures.

P91	Tempering	No. particles per 100 μm^2	Average spacing (μm)	Percentage size distribution (%)						
				0–50 nm	50–100 nm	100–200 nm	200–300 nm	300–400 nm	400–500 nm	500–600 nm
α+γ	168 h / 650°C	1099	0.15 ± 0.08	50	21	23	6	0	0	0
	720 h / 650°C	948	0.16 ± 0.09	44	27	24	5	0	0	0
	168 h / 750°C	594	0.21 ± 0.12	35	26	25	11	2	1	0
	720 h / 750°C	207	0.36 ± 0.25	35	25	20	12	4	2	1
γ	168 h / 650°C	1061	0.18 ± 0.08	37	23	28	10	1	0	0
	720 h / 650°C	600	0.21 ± 0.11	31	19	31	15	3	1	0
	168 h / 750°C	491	0.24 ± 0.11	20	38	25	12	3	2	0
	720 h / 750°C	377	0.26 ± 0.13	48	26	16	6	3	1	0

2.3. Friction tests

Friction tests were performed on an oscillating tribological test rig (SRV4, Optimol Instruments Prüftechnik GmbH, Germany) with a ball-on-flat testing geometry using DIN 100Cr6 steel (AISI 52100 equivalent) balls with a diameter of 10 mm, sliding against the flat side of a half-cylinder. The balls had a hardness of 62.5±2 HRC (corresponds to approximately 760 HV) and a roughness of R_a 0.02 μm. The initial hardness of the steels X20 and P91 was significantly lower than the hardness of the ball and had values of 238 HV and 228 HV, respectively.

In order to compare the frictional properties of the selected microstructures without and with the presence of oxides on the contacting surfaces, friction tests were performed under Ar atmosphere and in air. In air, metallic surfaces are covered with a thin layer of oxides which usually possess more favorable tribological properties than non-oxidized metallic surfaces [11]. The properties of nascent, non-oxidized metallic surfaces can be studied only in non-reactive atmospheres such as Ar, where the oxide layers cannot reform after removal (due to the high reactivity of nascent metallic surfaces, oxide layers normally form very rapidly – in e.g., 10^{-2} s at a pressure of 10^{-4} Torr [12]).

Before starting the friction tests, a flow of Ar gas with a gas pressure of 0.05 bar at the gas-bottle outflow was directed onto the contact for 3 minutes in order to remove any adsorbed atmospheric contaminations. After 3 minutes of flushing with Ar gas, the friction test was started and after 2 minutes of sliding, Ar gas flow was stopped so that air could come into the contact. Afterwards, the test was conducted for an additional duration of 2 minutes.

The frequency of the oscillating motion was 0.25 Hz with amplitude of 2 mm resulting in a maximum sliding velocity of 1.6 mm/s (ball movement is sinusoidal). A normal load of 30 N was applied, corresponding to 0.98 GPa and 1.46 GPa of mean and maximal Hertzian contact pressures respectively. It should be noted that the calculated Hertzian contact is significantly higher than the yield strength of both materials and doesn't correspond to the actual contact pressure. Namely, the Hertzian contact pressure considers only the elastic deformation of the material; hence when the yield strength of the material is exceeded, the actual contact pressure does not continue to increase exponentially, but remains more or less constant due to the plastic flow of the material and the related increase of the contact surface.

The coefficient of friction was calculated as the ratio of the tangential force (friction force) and the normal force. The maximum coefficient of friction occurred at the beginning of each stroke, where a break-away friction was typically observed (Figure 1). This maximum value corresponds to the so-called static coefficient of friction and was regarded as the coefficient of friction of interest. The measurement presented in Figure 1 was performed under Ar atmosphere. It should be noted that due to the high adhesion between the ball and the flat surface, after the end of the oscillating cycle, at 6 s, the ball did not yet return into its initial position although the ball holder did. Hence, due to the residual stresses between the ball holder and the ball, coefficient of friction was not zero, but gradually decreased to zero as the ball was slowly dragged into its initial position, which typically required 2–3 s after the end of an oscillating cycle. Therefore, after each oscillating cycle a waiting time of 12 s was applied.

Each microstructure was tested 3 times in order to ensure the repeatability of the measured friction curves.

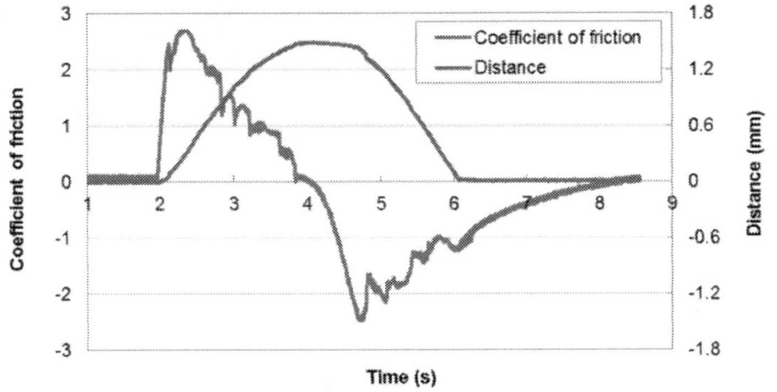

Figure 1: Typical friction signal for a single oscillating cycle.

3. Results

Figure 2 shows typical friction curves obtained in Ar and air atmospheres. Each friction point represents an average value from 3 tests. From Figure 2 it is clear that for both steels and microstructures, relatively low friction values in the range from 0.3 to 1 were initially measured. Afterwards, within the first 5 cycles, i.e. within the first 1 min of test, they increased to values higher than 2. This shows that the surfaces were initially covered with a frictionally favorable oxide layer, which was quickly removed due to sliding under high contact pressure. After the removal of the oxide layer, nascent steel surfaces, which could not re-oxidize under Ar atmosphere, came into contact, resulting in high adhesion and consequently high friction (coefficients of friction between 2.2 and 2.7). After 2 min of test, when surfaces were exposed to air, coefficients of friction started to decrease due to the re-oxidation of the surfaces and reached values between 1.5 and 2.5 at the end of the test. Friction values were generally around 20% lower in air than in Ar atmosphere.

From Figure 2 it can be also observed that for both steels and microstructures, friction values in both atmospheres were generally the lowest for the samples before tempering, i.e. in the initial state. This indicates that tempering deteriorated the friction properties of the selected materials, possibly due to a higher amount of carbide precipitates which either acted as mechanical anchors and/or increased the surface energy of the samples.

For the steel X20, both microstructures, α+γ and γ, showed similar friction curves before tempering (Figures 2a and b). On the other hand, for the steel P91, coefficient of friction was lower for γ than for α+γ microstructure, which was especially pronounced in air where the difference was around 15%. This observation suggests that in the initial state, the γ microstructure of the steel P91 had a lower adhesion and/or surface energy than the α+γ microstructure. In air, where a more pronounced difference was observed, this was probably correlated with properties of the surface oxides. On the other hand, in Ar atmosphere where the difference was minor, this could be correlated with the distribution of carbides or with the crystalline properties of the material.

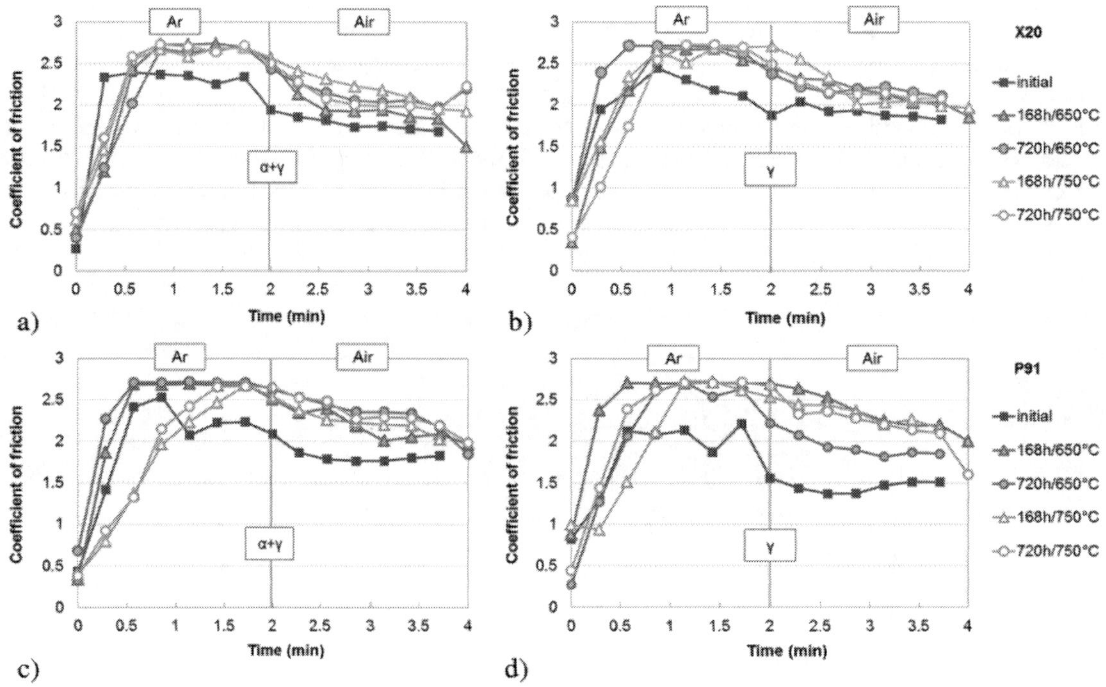

Figure 2: Typical friction curves from Ar and air atmospheres: (a) α+γ microstructure of the steel X20, (b) γ microstructure of the steel X20, (c) α+γ microstructure of the steel P91 and (d) γ microstructure of the steel P91.

After tempering, coefficient of friction increased for both steels in both atmospheres. The increase of friction due to tempering was more pronounced for the steel P91 (Figures 2c and 2d) than for the steel X20 (Figures 2a and 2b), which shows that the surface properties of the steel X20 were less affected by tempering than those of the steel P91. Furthermore, the changes in friction values due to different tempering conditions (time and temperature) were more pronounced for the P91 than for the steel X20, which was especially true for the γ microstructure.

It is interesting to note that after tempering, the initial increase of friction in Ar atmosphere was typically faster for the microstructures tempered at 650°C as compared to those tempered at 750°C, which was especially pronounced for the steel P91. Namely, for both microstructures of the steel P91 tempered at 650°C, the maximum friction was reached already after 0.5 to 0.8 min, while with the microstructures tempered at 750°C the maximum friction was reached first after around 1 to 1.5 min. This behavior could be the result of different carbide/nitride and microstructure morphology due to different precipitate coarsening kinetics at two different temperatures. For the steel X20, a similar trend was observed; however it was less pronounced when compared to the steel P91. This indicates that the surface morphology of the steel X20 was less affected by high temperature tempering as that of P91. Possibly, this may have affected also the surface oxidation of P91 samples, which would have been more pronounced due to its lower Cr content (Table 1) – generally friction is lower when more iron oxides form on the steel surfaces. The maximum values of friction in Ar atmosphere were similar for both steels and microstructures.

In air, coefficient of friction showed only minor differences depending on the tempering conditions. A notable influence of tempering time was observed only for the α+γ microstructure of the steel X20 tempered at 750°C and for the γ microstructure of the steel P91 tempered at 650°C – in both cases, friction decreased with tempering time. For all other microstructures, the influence of tempering time on friction was negligible. Similarly, the influence of tempering temperature was observed only for the α+γ microstructure of the steel X20 tempered for 168 h and for the γ microstructure of the steel P91 tempered for 720 h – in both cases, friction increased with tempering temperature. For all other microstructures, the influence of tempering temperature on friction was negligible.

Figure 3 shows coefficient of friction values measured in air in dependence of the number of carbide particles for both steels and microstructures after tempering. The number of carbides can be found in Tables 2 and 3, while the presented coefficient of friction values correspond to the values measured at 3.1 min of testing time (Figure 2). It is believed that at this time interval (3.1 min), surfaces were fully re-oxidised and the contact conditions have stabilized themselves. It is interesting to note that a linear regression of all coefficient of friction values shows an increasing trend with decreased number of carbide particles (growth of larger particles on the account of dissolution of smaller particles – both effects were promoted with tempering temperature and tempering time). Although the observed decrease was not very pronounced (within the standard deviation values), this indicates that a reduced number of carbides, which is related to carbide coarsening, increased the friction of the tempered samples. However, it should be noted that for the presented linear regression, two measurements which were significantly outside of the observed trend were omitted: α+γ microstructure of the steel P91 tempered for 720h at 650°C and γ microstructure of the steel P91 tempered for 720h at 650°C.

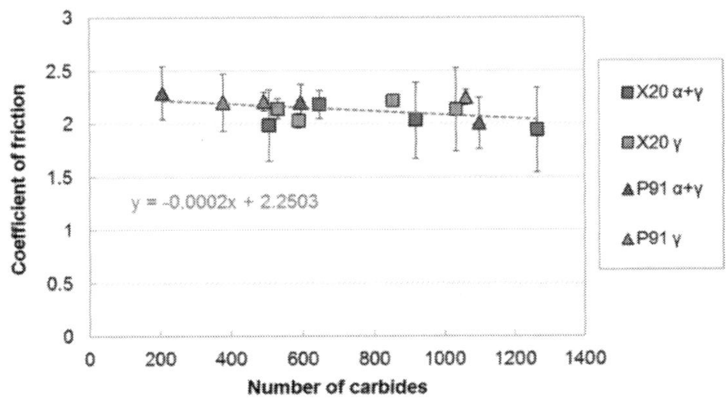

Figure 3: Linear regression of coefficient of friction values measured in air at 3.1 min of testing time for both steels and microstructures after tempering in dependence of the number of carbide particles.

4. Discussion

In the previous study of the authors [9], it was observed that the wear behavior of weld heat affected zones of the steels X20 and P91 is influenced by tempering and that the wear was typically lower for the tempered microstructures than for the microstructures in initial state, i.e. before tempering (exception was γ microstructure of the steel X20). Furthermore, lower wear occurred when a higher number of medium- and large-sized carbide particles were present in the microstructure. In the present study, it was shown that the influence of tempering on the frictional properties of these microstructures is contradictory to the influence on wear, since after tempering, the friction increased for all microstructures. This is attributed to the carbide precipitation where the carbides could either act as mechanical anchors and/or increase the surface energy of the samples. This conclusion is also supported by observations from other authors which reported that the carbide precipitation increased the friction of Inconel 600 mill-annealed alloy [8].

Although for the steel P91, coarsening and dissolution of carbide particles due to tempering was more pronounced than for the steel X20, which is probably correlated with the higher C content in the latter, it is interesting to note that for the steel P91, the increase of friction due to tempering was slightly more pronounced than for the steel X20.

Tempering at a higher temperature probably provided different carbide/nitride and microstructure morphology, resulting from higher coarsening kinetics and lower thermal stability of precipitates. Namely, for both steels, samples tempered at a higher temperature required a longer time for reaching the maximum friction in Ar atmosphere. This could be attributed to higher solid solubility of carbide/nitride forming elements at higher temperatures, and therefore higher solid solution hardening. This effect was stronger for the steel P91 as compared to the steel X20; the difference could also be correlated with a lower oxidation stability of the steel P91 due to its lower Cr content (Table 1).

While in Ar atmosphere the maximum friction values were similar for all microstructures, in air, minor differences were observed but only for some microstructures ($\alpha+\gamma$ microstructure of the steel X20 and γ microstructure of the steel P91). In air, coefficient of friction of both steels and microstructures showed a decreasing trend with an increased number of carbide particles, which indicates that the reduced number of carbides due to carbide coarsening on the account of dissolution of smaller particles, increased the friction properties of the tempered steels.

In Figure 4, the influence of the number of carbides (related to carbide coarsening) on friction and wear of the steels X20 and P91 is schematically presented. The wear curve was estimated on the basis of the results from the previous study [9]. Generally, wear was low when the number of carbide particles due to coarsening was small. However, if the number of carbide particles decreased too much due to their complete dissolution, this significantly increased wear. On the other hand, coefficient of friction showed a tendency to linearly decrease with the increased number of carbides, but this was true only for air, which indicates that changes of the surface morphology due to carbide precipitation have also affected the properties of the surface oxides of the tempered steel samples.

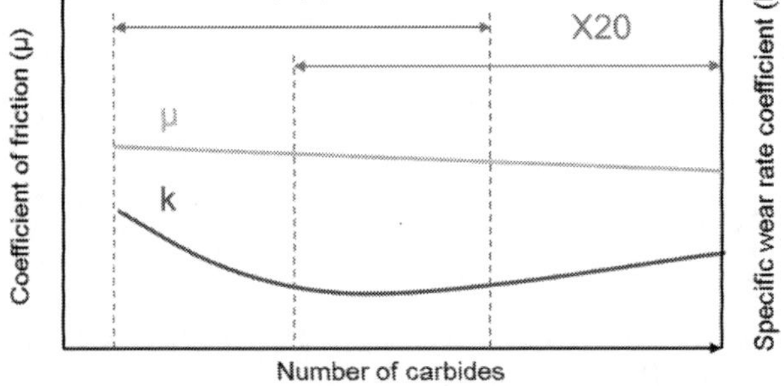

Figure 4: Schematic representation of the influence of the number of carbides (related to carbide coarsening) on friction and wear of the steels X20 and P91.

IOP Conf. Series: Journal of Physics: Conf. Series **843** (2017) 012065 doi:10.1088/1742-6596/843/1/012065

5. Conclusions

1. After tempering, coefficient of friction increased for both steels in both atmospheres (Ar and air). It is presumed that the deterioration of friction properties after tempering occurs primarily due to carbide precipitation, where carbides act either as mechanical anchors and/or increase the surface energy of the materials.
2. For both steels, samples tempered at a higher temperature required longer time for reaching the maximum friction in Ar atmosphere, which could be attributed to higher solid solubility of carbide/nitride forming elements at higher temperatures, and therefore higher solid solution hardening. This effect was stronger for P91 steel, and was possibly correlated also with the lower oxidation stability of this steel due to its lower Cr content.
3. In Ar atmosphere, the maximum friction values were similar for both steels and microstructures, while in air, more pronounced differences were observed for different microstructures and tempering conditions. Since friction is generally lower when more iron oxides form on the steel surfaces, it is possible that the changes of the surface morphology due to carbide precipitation have affected the properties of the surface oxides of the tempered steel samples.
4. Coefficient of friction showed a light tendency to linearly decrease with the increased number of carbides, indicating the that the reduced number of carbides (due to carbide coarsening on the account of dissolution of smaller particles) has affected the friction properties of the tempered steels.

Acknowledgements
The work presented was funded by the Austrian COMET Program (Project XTribology, no. 849109) and carried out at the "Excellence Centre of Tribology" (AC2T research GmbH) in cooperation with V-Research GmbH and Institute of Metals and Technology IMT.

References

[1] Mutama K R 2013 *Proc. ASME Power Conf. (July 29th–August 1st 2013, Boston, MA)* vol. 1 (New York: ASME) paper no. POWER2013-98289 pp V001T04A009
[2] Prat O, García J, Rojas D, Sanhueza J P and Camurri C 2014 *Mater. Chem. Phys.* **143** 754–64
[3] Miao K, He Y, Zhu N, Wang J, Lu X and Li L 2015 *J. Alloys Compd.* **622** 513–23
[4] Xu Y, Zhang X, Tian Y, Chen C, Nan Y, He H and Wang M 2016 *Mater. Charact.* **111** 122–7
[5] Han Y, Xue X, Zhang T, Hu R and Li J 2016 *Mater. Sci. Eng.* A **667** 391–401
[6] Liu F, Rashidi M, Johansson L, Hald J and Andrén H O 2016 *Scr. Mater.* **113** 93–6
[7] Lu Q, Xu W and van der Zwaag S 2014 *Acta Mater.* **77** 310–23
[8] Zhang H Y, Lu Y H, Ma M and Li J 2014 *Wear* **315** 58–67
[9] Velkavrh I, Kafexhiu F, Klien S, Diem A and Podgornik B 2017 *Metall. Mater. Trans.* A **48** 109-25
[10] Schindelin J, Arganda-Carreras I, Frise E, Kaynig V, Longair M, Pietzsch T, Preibisch S, Rueden C, Saalfeld S, Schmid B, Tinevez J Y, White D J, Hartenstein V, Eliceiri K, Tomancak P and Cardona A 2012 *Nat. Methods* **9** 676–82
[11] Velkavrh I, Ausserer F, Klien S, Voyer J, Ristow A, Brenner J, Forêt P and Diem A 2016 *Tribol. Int.* **98** 155–71
[12] Cabrera N and Mott N F 1949 *Rep. Prog. Phys.* **12** 163–84
[13] Vodopivec F, Jenko M and Vojvodič-Tuma J 2006 *Metalurgija* **45** 147–53
[14] Skobir D A, Vodopivec F, Jenko M, Spaić S and Markoli B 2004 *Int. J. Mater. Res.* **95**,1020–4

IOP Conf. Series: Journal of Physics: Conf. Series **843** (2017) 012066 doi:10.1088/1742-6596/843/1/012066

Wear resistance of polypropylene-SiC composite

J Abenojar, B Enciso, MA Martínez and F Velasco

Department of Materials Science and Engineering. Universidad Carlos III de Madrid. Av. Universidad, 30. 28911 – Leganés (Spain)

Email: abenojar@ing.uc3m.es

Abstract. In this work, the wear resistance of thermoplastic composites with a high amount of ceramic is evaluated. Composites made of polypropylene (PP) and silicon carbide (SiC) powder at 50 wt% were used with the final objective of manufacturing ablative materials. This is the first part of a project studying the wear resistance and the mechanical properties of those composites, to be used in applications like habitat industry. In theory, the exposure to high temperature of ablative materials involves the elimination of thermal energy by the sacrifice of surface polymer. In our case, PP will act as a heat sink, up to the reaction temperature (melting or sublimation), where endothermic chemical decomposition into charred material and gaseous products occurs. As the surface is eroded, it is formed a SiC like-foam with improved insulation performance. Composites were produced by extrusion and hot compression. The wear characterization was performed by pin-on-disk test. Wear test was carried out under standard ASTM G99. The parameters were 120 rpm speed, 15 N load, a alumina ball with 6 mm as pin and 1000 m sliding distance. The tracks were also observed by opto-digital microscope.

1. Introduction

When talking about ablative materials, it is thought in aerospace applications; for example, they are very important in missions re-entering the atmosphere. They are also used in internal areas of the nozzle as protective system to high temperature and thermal erosion [1]. According to Laub and Venkatapathy [2] when the ablative material is pyrolyzed, decomposition gases are blown in the boundary layer surrounding the capsule; then a part of the heat-flux coming from the high temperature shock layer is blocked. According to Galileo heritage, the convection heat-flux can be reduced mostly and blocked by the strong blowing. Convective blockage during earth re-entry has been discussed by Reynier [3]. There is many research about ablative materials, mainly focused on the study and simulation of heat flows, all of them based in ablation models [4,5], at surface temperatures ranging from 1700 to 2400 K.

However, an ablative material could be used in many different applications. Applications with high heat-loads and heat fluxes, like for example a fire, could imply the use of an ablative material. In this study, ablative material is planned as a construction material, for its application on floors or walls of garages, parking or public places. The work is included in a SUDOE project named "KrEaTive", which is raised to habitat industry, being this material one of the proposed to be used. For this reason, the ablative material also has to be wear and impact resistant. In this first part of the project, the wear tests carried out on the manufactured materials with thermoplastics resin and silicon carbide to 50 wt% are presented. The selection of the thermoplastic resin was made according to patent WO 2001062841

A3 [6], where thermoplastic composite materials were patented as ablative materials. It has been found that combining thermoplastic materials, reinforcing additives, a mineral, and optionally a foaming agent, produces an ablative material with good mechanical properties. With thermosetting resins, like polyester, and ceramic powder Portocarrero [7] manufactured the thermal zone of low orbit rocket nozzle.

2. Experimental Procedure

When carrying out this work, composites of polypropylene (PP) and silicon carbide (SiC) to 50 wt% were manufactured. PP was provided in the form of pellets with a diameter of 3 mm by Repsol (Madrid, Spain). SiC had a size of 43 μm, with a polygonal shape as it can see in figure 1, and it was provided by Carburos Navarro (Cuenca, Spain). Solid composites were melted, blending the polymer and silicon carbide particles in a Haake Rheomix 252P, MA, USA, at 190 °C, 50 rpm speed, for 6 min. With the feedstock obtained, precursors of 3 mm thickness were processed by hot plate press at 190 °C and 5.5 MPa during 20 min.

Figure 1. Silicon carbide of 43 μm

Three different compositions were studied: (1) PP without SiC (using it as reference); (2) PP with SiC 50 wt%; (3) PP with SiC 50 wt% and glass fibre (GF) 1 wt%. GFs used were short fibres, 6 mm long, from Feroca (Madrid, Spain).

Dry wear tests were carried out at room temperature using a pin-on-disk tribometer (Microtest, Madrid, Spain). In this test, a stationary specimen (pin) with a defined normal force is pressed against the surface of another specimen placed on the rotary disk. Normal force, temperature, wear rate, frictional coefficient, number of turns, speed and other parameters are, firstly, registered and then real time displayed. The normal force is applied over the pin or ball by means of a set of dead weights between 0 N and 60 N (in this case 15 N was applied). This way of application allows a stable force during the test. A 6 mm diameter alumina ball was used for the pin. The test conditions were 120 rpm speed, relative humidity below 30% and a friction radius of 8 mm. The sliding distance was 1000 m. These parameters have been previously optimized for other composite materials by Abenojar [8,9].

Wear was evaluated using Archard´s equation (Eq. 1). To calculate the wear and the friction coefficient, ASTM G99-05 standard was used. Measurements of the wear track radius (*R*) and the wear track width (*d*) were performed using a profiler and volume loss was calculated according to Eq. 2, where r is the pin end radius. In these tests, no significant pin wear is assumed and the wear debris was not removed from track. The mass loss was also taken into account for the calculation of volume loss, being the densities of samples measured using Archimedes' principle.

$$W = \frac{Volume\ loss}{Load\ x\ Sliding\ distance} \tag{1}$$

$$Volume\ loss = 2\pi R\left[r^2 sin^{-1}\left(d/_{2r}\right) - \left(d/_4\right)(4r^2 - d^2)^{\frac{1}{2}}\right] \quad (2)$$

After wear test, optical microscopy and profilometer measurements were performed on the track by Olympus microscopy Dsx500 (Tokyo, Japan). Three tracks were made for each material.

3. Results

The first wear results were calculated using mass loss and density. The densities measured by Archimedes' principle show that composite densities are lower than theoretical densities. The PP+SiC composites and PP+SiC+GF composites show relative densities of 90% and 92% respectively. So there is a 10 and 8% the porosity inside composite materials respectively. In table 1, it is shown the wear for the three studied materials; although the composite materials have higher porosity than plain PP, when 1% of GF is added, wear decreases a 16% and when it is not added wear increases a 50%.

Table 1. Wear of the three materials calculated by mass loss.

	Mass Loss (g)	Density (g/cm^3)	Volume Loss (mm^3)	Wear ($mm^3\ N^{-1}m^{-1}$)
PP	0.003	0.92	3.51	2.34E-04 ± 5.24E-05
PP+SiC	0.007	1.29	5.28	3.52E-04 ± 4.79E-05
PP+SiC+GF	0.004	1.28	2.95	1.97E-04 ± 1.20E-05

If wear is calculated by volume loss (equation 2), the results are enough different from that of mass loss (figure 2). The depth and width of the tracks were measured with the profilometer at three points. In total, nine measurements were carried out for each material. In this case, both composites have a higher wear than PP, although composite with SiC and GF shows lower wear than PP+SiC. This can be explained when the tracks were studied together with the wear mechanisms.

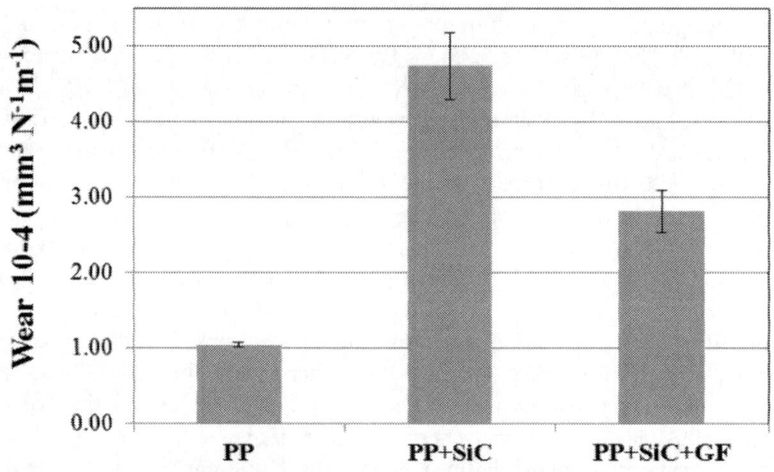

Figure 2. Wear measured by volume loss for the three materials.

The figures 3 and 4 show the wear tracks in 2D. Figure 3 corresponds to the PP track. An abrasive wear (lines on track) following an adhesive wear (higher lines) are found. PP is deformed by the alumina pin, the track is uniform and it is slightly recovered after test. For this reason the wear is smaller in figure 2 when compared to values in table 1.

Figure 3. Polypropylene wear track in 2D

Something similar takes place on the composites. In all cases (figures 4a and 4b), there are abrasive and adhesive wears. For this reason, the mass loss is lower than volume loss (table 1 and figure 2). The porosity also has influence on mass loss, as it is filled with the wear debris and mass loss is lower. According to the porosity, the PP + SiC composite would have shown less wear. This may be offset by the third body present in the track, which increases wear and track width also increases. The track width values found for PP, PP + SiC and PP + SiC + GF were 1308 ± 18 μm, 2157 ± 134 μm and 1816 ± 83 μm respectively. Depth also changes, although these changes are smaller, measuring a depth of 80 ± 14 μm for PP + SiC and 75 ± 19 μm for PP + SiC + GF.

In consequence, if volume loss is calculated from track width, bigger wear will be found because the tracks are wider. But if wear is calculated from mass loss, the wear measured is smaller due to adhesive wear.

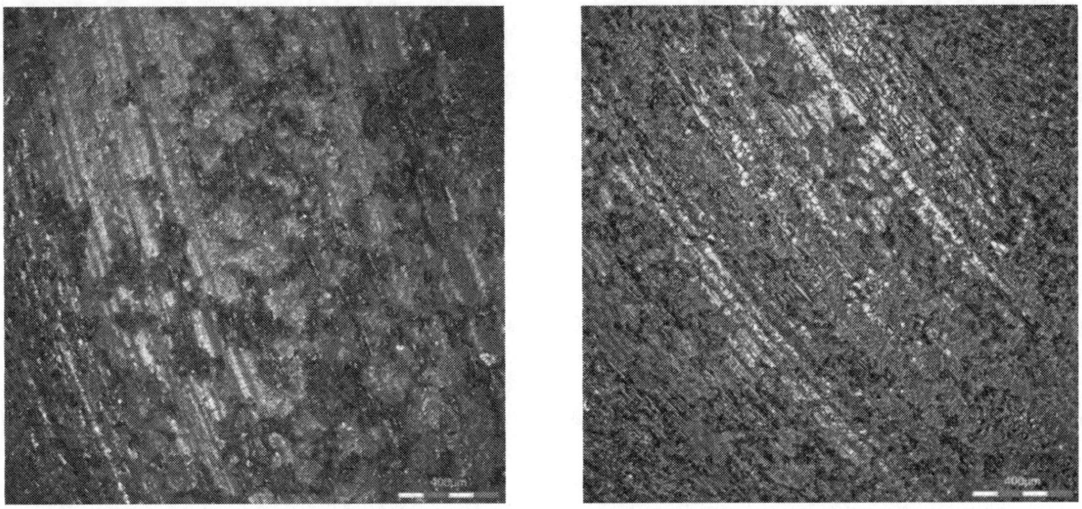

(a) **(b)**

Figure 4. Wear tracks in 2D: (a) PP + SiC; (b) PP + SiC + GF.

(a) **(b)**

(c)

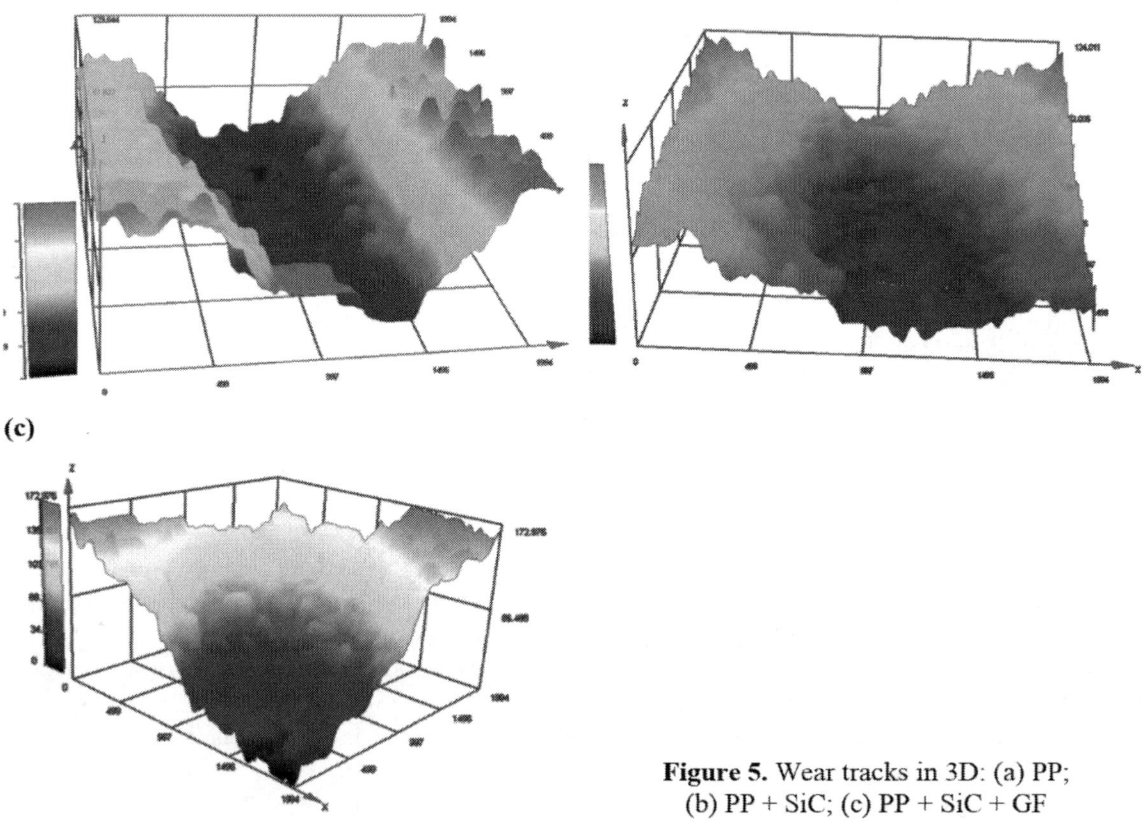

Figure 5. Wear tracks in 3D: (a) PP; (b) PP + SiC; (c) PP + SiC + GF

The figure 5 shows wear track in 3D. With these plots, the differences among materials can be clearly seen. Although defining the accumulation zones at the edge of the tracks is complicated, it can be appreciated that PP + SiC + GF track is enough different to the other tracks. Its width in the surface is similar, but it is much narrower in the deepest zone. For this reason, the wear is smaller. According to these plots, wear should be lower, as it was observed in the wear calculated by mass loss.

The effect of GF in the composite is not understood yet, but it seems that it produces higher interaction between matrix and ceramic filler. This way, better properties are expected when the GFs are added.

Due to the presence of SiC particles, the friction coefficient (figure 6) increases from 0.45 (in PP) to 0.52 (in the composite); this increment is short if it is compared to the amount of added ceramic. Difference between composites is not found. This effect informs about the join of the matrix with the particles, if SiC particles were not joined to PP, the friction coefficient would be higher.

4. Conclusions

Wear resistance changes with the method used to calculate it, as there are factors which have influence on the outcome.

Two types of wear mechanism are found in these materials: abrasive and adhesive wear.

The adhesive wear is more significant in PP than in composites, where the porosity helps to decrease mass loss by accumulation of debris powder.

Addition of 1% GF to the composite improves wear resistance.

According with wear tests, the mass loss method seems a more correct way of calculation than track width to calculate volume loss.

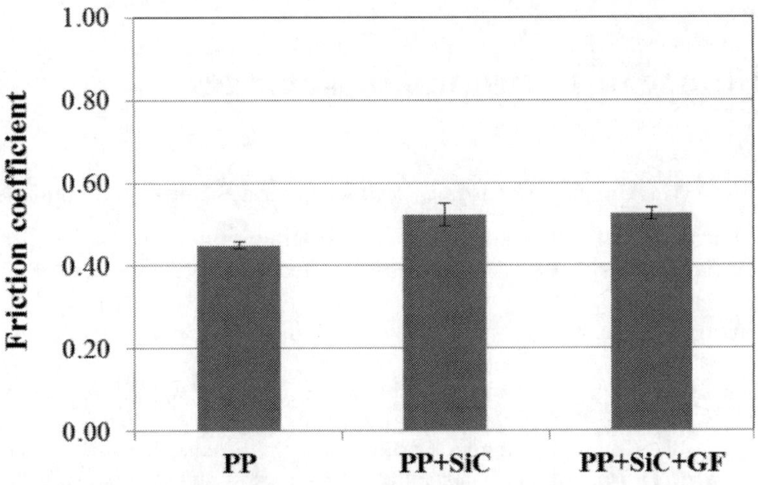

Figure 6. Friction coefficient for the studied three materials.

5. References

[1] Sutton G P and Biblarz O 2001 *Rocket Propulsion Elements* (New Yor: John Wiley & Sons, INC) p 561

[2] Laub B and Venkatapathy E 2004 Thermal protection system technology and facility needs for demanding future planetary missions *in Proc Int Workshop on Planetary Probe Atmospheric Entry and Descent Trajectory Analysis and Science* ed A. Wilson (Lisbon: ESA SP-544) pp 239-47

[3] Reynier P 2013 Survey of convective blockage for planetary entries *Acta Astronaut.* **83** 175-95

[4] Park C and Balakrishnan A 1985 Ablation of Galileo Probe heat-shield models in a ballistic range *AIAA Journal* **23** 301-8

[5] Godin D, Trépanier J Y, Reggio M,. Zhang X D and Camarero R 2000 Modelling and simulation of nozzle ablation in high-voltage circuit-breakers *J. Phys. D: Appl. Phys.* **33** 2583–90

[6] Kemmish D J and Leibfried R T 2001 Light weight ablative materials WO 2001062841 A3

[7] Portocarrero J and Maldonado J 2016 Ablative behaviour analysis of low orbit rocket nozzles based on composite materials. *XXVII International Congress of Adhesion and Adhesive in Tendencias en Adhesión y Adhesivos vol IX* ed J Narbón and S Nuere (Madrid:Universidad Politécnica de Madrid) pp 138-49

[8] Abenojar J, Tutor J, Ballesteros Y, del Real J C and Martínez M A 2015 Wear and cavitation effect in an epoxy filled with boron and silicon nanocarbides. *4th International Conference on Fracture Fatigue and Wear, FFW 2015 in International Journal of Fracture Fatigue and Wear* vol 3 ed M Abdel Wahab (Zwijnaarde: Laboratory Soete – Ghent University) pp167-73

[9] Abenojar J, Tutor J, Ballesteros Y, del Real J C and Martínez M A 2014 Friction wear behaviour of carbon based epoxy nanocomposites. *3rd International Conference on Fracture Fatigue and Wear. FFW2014 in International Journal of Fracture Fatigue and Wear* vol 2 ed M Abdel Wahab (Zwijnaarde: Laboratory Soete – Ghent University) pp 281-87

Acknowledgments

The authors would like to express their gratitude for the financial support to Interreg SUDOE, through KrEaTive Habitat project (Ref. SOE1/P1/E0307).

IOP Conf. Series: Journal of Physics: Conf. Series **843** (2017) 012067 doi:10.1088/1742-6596/843/1/012067

Wear resistance of hydrophobic surfaces

MA Martinez, J Abenojar, M Pantoja and S López de Armentia

Department of Materials Science and Engineering. Universidad Carlos III de Madrid. Avda Universidad 30, 28911-Leganes, Spain

mamc@ing.uc3m.es

Abstract. Nature has been an inspiration source to develop artificial hydrophobic surfaces. During the latest years the development of hydrophobic surfaces has been widely researched due to their numerous ranges of industrial applications. Industrially the use of hydrophobic surfaces is being highly demanded. This is why many companies develop hydrophobic products to repel water, in order to be used as coatings. Moreover, these coating should have the appropriated mechanical properties and wear resistance. In this work wear study of a hydrophobic coating on glass is carried out. Hydrophobic product used was Sika Crystal Dry by Sika S.A.U. (Alcobendas, Spain). This product is currently used on car windshield. To calculate wear resistance, pin-on-disk tests were carried out in dry and water conditions. The test parameters were rate, load and sliding distance, which were fixed to 60 rpm, 5 N and 1000 m respectively. A chamois was used as pin. It allows to simulate a real use. The friction coefficient and loss weight were compared to determinate coating resistance

1. Introduction

For buildings with large glass areas, accumulation photovoltaic modules or windshields it is very interesting to get "Easy to Clean" effect on the surfaces. Self-cleaning surface are indeed important for the above applications, since the dust, pollution, and other particles accumulation reduce the transparency [1]. On the other hand, icing is the cause of numerous air accidents. If water is not deposited on the aircraft surface, these accidents can be avoided [2]. This property is related to hydrophobic character of the surfaces.

Substrates are permanently damaged by aggressive environmental conditions and dirt from both a visual and functional standpoint. Therefore costs and time needed for cleaning activities are increased. An additional problem is the possible photocatalysis, wherein organic contaminants adsorbed on film surface are decomposed under ultraviolet light [3].

Hydrophobic coatings counterbalance these negative influences by creating a repellent and non-stick effect from aqueous and organic liquids. It considerably reduces the adherence of dirt particles on the substrate surface. When the coated surfaces are in contact with water, non-stuck dirt particles are removed by the water. Consequently, it reduces cleaning activities and provides lasting protection from the aggressive environment influences to the surface.

There are some superhydrophobic coatings based on lotus leaf effect (water resistance by rough spots with air on the surface) on the market [4]. However, that technology does not work for organic or complex liquids. Besides, if the surface that is coated is damaged or is subjected to extreme conditions, liquid drops tends to stay on the surfaces or penetrate into them, instead of slipping.

One of the reasons that wetting and surface energy are often neglected in tribological studies and engineering design models is due to the lack of an understanding about the correlation between them [5]. However, another reason may be the lack of any direct evidence for their influence on tribological performance in engineering macro contacts. Namely, the effects of surface properties, such as wetting, surface energy, oleo/hydrophobicity, and oleo/ hydrophilicity, were investigated in only a small number of tribological studies using macro contacts with engineering materials and lubricants [6,7,8,9].

However, in literature, detailed analyses of the anti-soling film in term of structural properties, transparency, robustness and cost are still missing.

2. Experimental

Flat glass samples were ultrasonically cleaned with ethanol (EtOH) and air dried. As coating, Sika Crystal Dry (supplied by SIKA SAU, Alcobendas, Spain) has been used. On the surface the product was sprayed and the film thickness was evaluated by surface imaging.

Surface hydrophobicity is determined by static contact angle For the contact angle measurements a goniometer Dataphysics Contact angle system OCA 30-2 from Data Physics Instruments (GmbH, Filderstadt, Germany) was used. The model is able to measure in a range of 1 to 180° with an accuracy of ± 0.5°. The instrument contains a 3x zoom and a software SCA 202 V.3.11.13 build 162. For these measurements, deionized water droplets volume was fixed at 3 μL, 2 drops per sample and 3 replicates were done.

Friction test was performed in a pin-on-disk tribometer (supplied by Microtest-Madrid Spain). The test was performed following ASTM G99. Abrasion test was carried out between a mobile probe fixed in a carriage with reciprocal movement and a fixed pin. The pin was manufactured by gluing a disk of 20 mm of diameter of polishing cloth Omega (supplied by ATM GMBH Gremaqny) to an aluminum plate of the same dimensions. In some cases during the test, the cloth was lubricated with water [10]. Test was fixed to 60 rpm, a turning radius of 8 mm and 500 m of total length, which corresponds to 9947 cycles of movement of the mobile surface with a load of 5 N. Tests were carried out at room conditions (25°C, 30% relative air humidity). After abrasion test the transparency of samples and weight loss were measured.

The aesthetic appearance of these coated surfaces was evaluated by transparency and color changes. The colorimetric data of the surface were calculated from the acquired spectra with the most common color space in the industry: the CIEL*a*b* color space. A ColorEyE XTH colorimeter (Neurtek Instruments S.A, Eibar, Spain) was used for D65/10◦illuminant/observer condition and without specular gloss (SCE). CIELAB system measures color in a three-dimensional space, defined by the coordinates L*, a* and b* (figure 1). The first one corresponds to luminosity: 0 is black and 100 is white; coordinate a* measures variation between red (+a) and green (-a); and b* gives changes between yellow (+b) and blue (-b). Finally, ΔE is calculated (equation. 1) in order to measure the module of the vector from the reference to the measurement.

$$\Delta E = (\Delta L^{*2} + \Delta a^{*2} + \Delta b^{*2})^{1/2} \quad (1)$$

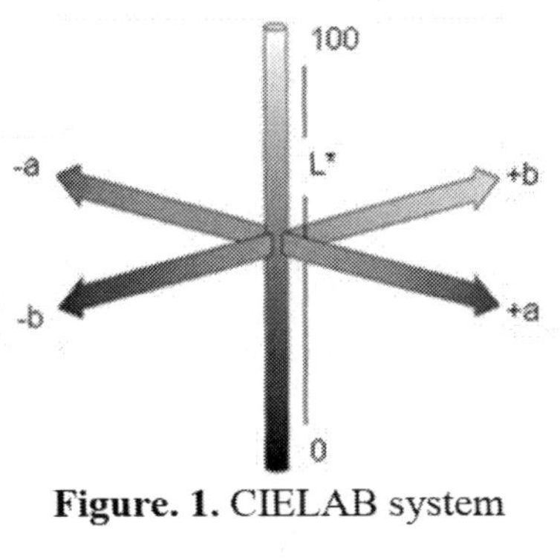

Figure. 1. CIELAB system

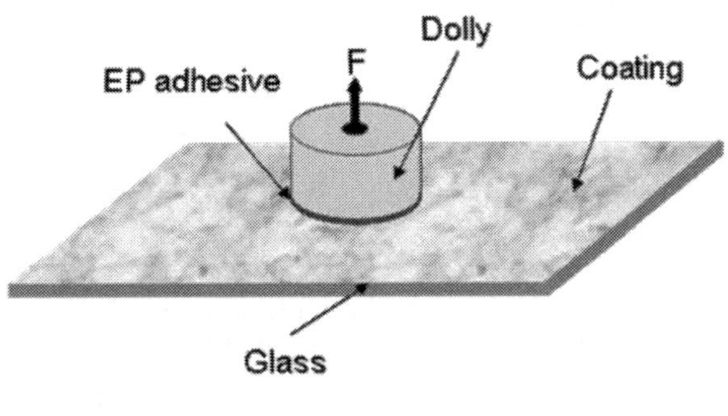

Figure. 2. The pull-off test scheme

The pull-off test method is used to determine the force required to separate aluminium dolly (20 mm diameter) fixed to the coating glass with epoxy adhesive (figure. 2). The test was performed following the ASTM D 4541 standard. The standard procedure was developed for metal substrates, but it may also be suitable for other rigid substrates such as some type of polymers and glass. The system was tested with the tensile test (relative to the plane of the adhesive bond) was performed with an universal testing machine with a 1kN load cell and 0.1 mm/min speed (Microtest, Madrid, Spain).

Characterization of surface energy, aesthetical aspect, abrasion test and adherence measurements were performed before and after ultraviolet (UV) light irradiation (15 W mercury lamp–Techlux emitting 254 nm wavelength).The samples were exposed to light irradiation, 5 cm far from the light source. Aging consisted on 120 min cycles: 102 min of radiation and 18 min immersed in deionized water without radiation.

3. Results and discussion

The purpose of this coating is to make glass hydrophobic and to improve its facility of cleaning. To check it, variations of water and olive oil contact angles between uncoated and coated glass were measured (table 1). An increase of more than 100% was observed when the glass is coated. The uncoated hydrophilic surface becomes hydrophobic.

Table 1. Contact angles of water and olive oil on the studied surfaces.

	Cicles	Water	Olive oil
Glass	--	50±4	34±1
Coated glass	0	107±5	83±2
	48	99±4	65±6
	100	93±2	64±6
	200	93±4	23±2

In table 1 it is observed that water contact angle is decresing slightly after aging cycles. The decrease is more evident in the case of olive oil.

Variation of color defined by ΔE (equation. 1) is shown in table 2. In all cases, before and after aging, it is below the threshold of visibility to the human eye.

Table 2. Variation of color parameters

	Cycles	L*	a*	b*	ΔE
Glass	--	86.60	3.56	-12.07	
Coated glass	0	87.02	3.63	-12.35	0.51
	24	86.91	3.50	-11.97	0.33
	48	87.18	3.57	-11.90	0.60
	100	86.87	3.42	-11.60	0.56
	200	86.90	3.26	-11.10	1.63

Pull-off test on glass and coated glass surfaces show epoxy-glass joint strength of 14.6 MPa, while epoxy-coated glass joint strength is 3.8 MPa, being the break through glass-coating interface.
Friction coefficient vs. time curves of the samples under friction test show two different behaviours depending on the conditions: dry or wet test (figure 3). Uncoated glass shows a friction coefficient of 2.2 when the test is carried out under dry conditions and 0.4 when it is done with water. In the case of coated glass, a reduction of the coefficient of friction of 2.4 in dry conditions (figure 5a, continuous line) and 0.3 in wet test (figure 5b, continuous line) is observed. Figure 4 shows abrasive wear marks obtained after water-submerged tests of glass and coated glass samples.

Differences on dry friction coefficient of aging plates at different number of cycles are not observed once the measured value has become stable (figure 5a). When the tests were carried out immersed in water, the curve is more similar to uncoated curve when the number of cycles is high (figure 5b).

IOP Conf. Series: Journal of Physics: Conf. Series **843** (2017) 012067 doi:10.1088/1742-6596/843/1/012067

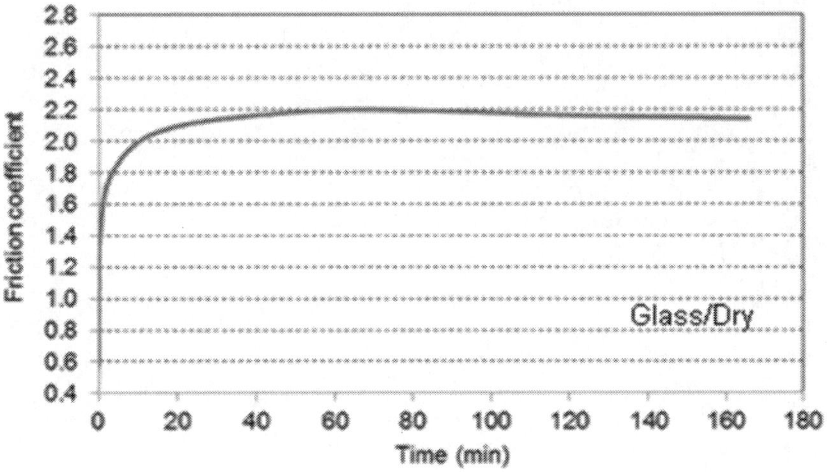

(a)

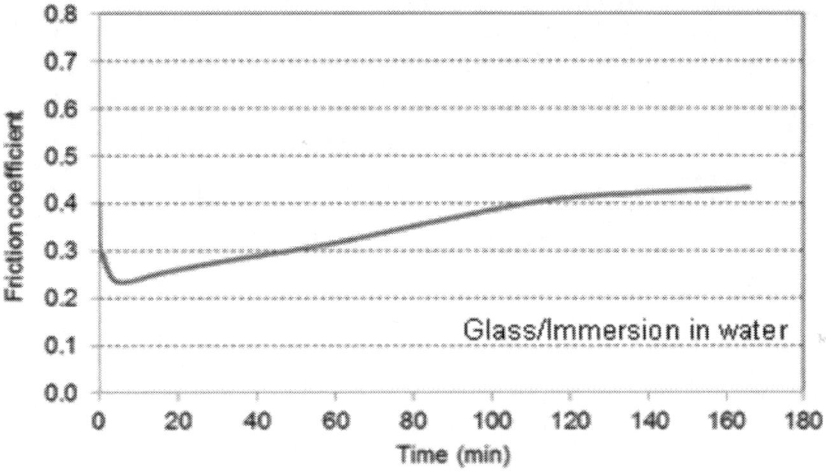

(b)

Figure 3: Friction coefficient vs. time for uncoated glass: (a) Dry; (b) Immersion in water

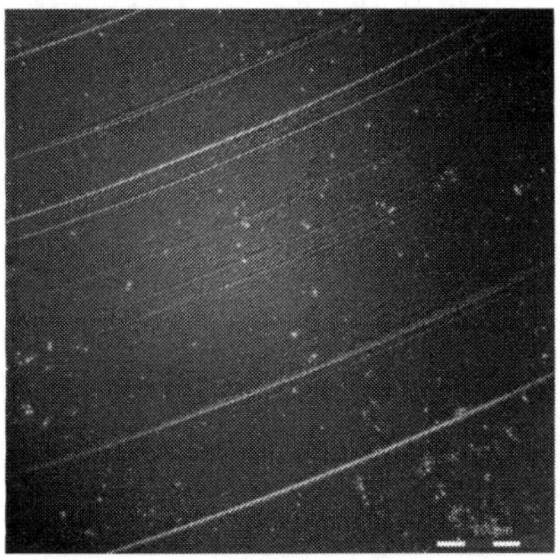

(a)

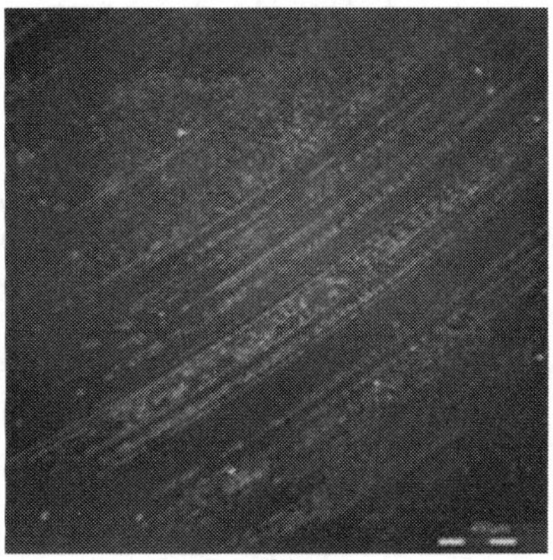

(b)

Figure 4. Wear track immersed in water (a) uncoated glass test; (b) coated glass test

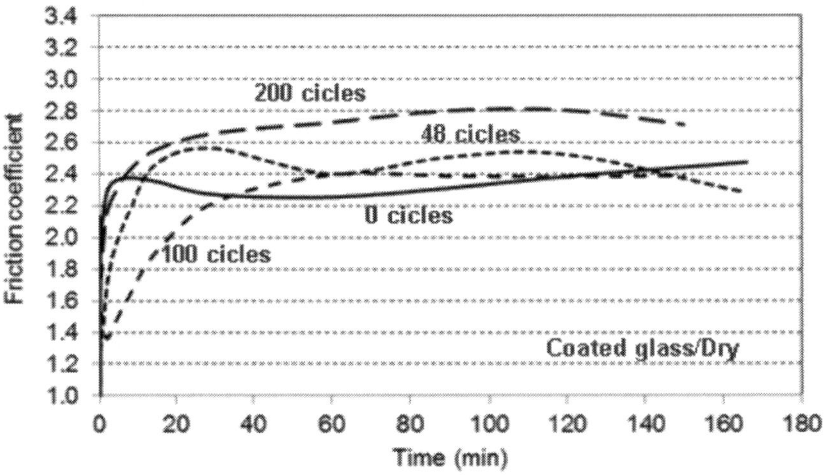

(a)

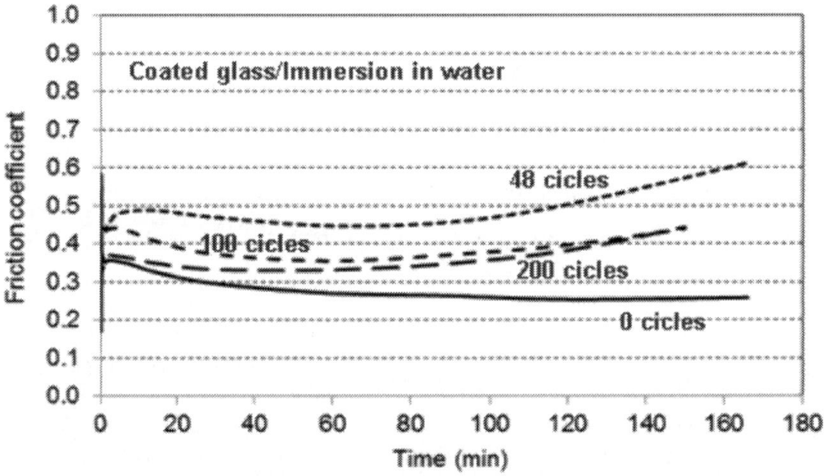

(b)

Figure 5. Tests on aging samples: Friction coefficient vs. time for coated glass. A) Dry; B) Immersion in water

Table 3 Values of the parameters obtained in friction tests

Surface	Cicles	Mass loss ($g.10^{-4}$)		Static fricction coefficient		Initial dinamic friction coefficient	
		Dry	Immersion in water	Dry	Immersion in water	Dry	Immersion in water
Glass	--	4	2	0.57	0.40	1.20	0.32
Coated glass	0	4	3	0.54	0.28	1.69	0.44
	48	8	2	0.44	0.35	1.51	0.54
	100	6	2	0.92	0.50	1.75	0.41
	200	8	6	0.84	0.43	1.46	0.33

On both dry and immersed in water tests, coated glass lost less mass than uncoated plates (table 3). It is also observed lower static friction coefficient (measured at the beginning of the test) for coated glass. Two seconds after the beginning of the test, initial dynamic friction coefficient was measured. This parameter is higher for coated glass due to the increase of surface roughness when glass is coated. The water acts as a lubricant which decreases the values of the friction coefficient and reduces the wear both in uncoated and coated glass. As the number of aging cycles increases it is possible to observe an increase in the static friction coefficient both in dry and wet tests. This same trend can be observed in the initial dynamic coefficient although it is not as clearly. Significant mass loss is observed in test dry case.

4. Conclusions

The proposed wear test is similar to the effect of the wipers on the windshields. Wear produced by these tests is abrasive on both uncoated and coated glass.

All films presented abrasion resistant property in contact with polishing cloth in dry and immersed in water It has been demonstrated that high transparency and self-cleaning have been obtained by a simple sprayed route presenting good potential to be applied on transparent super-hydrophobic surfaces.

Tests with hydrophobic surfaces revealed a reduction in friction, which may be attributed to lubricant slip against the hydrophobic surface.

5 References

[1] Ghazi S, Sayigh A and Ip K. 2014. Dust effect on flat surfaces - A review paper. *Renew. Sust. Energ. Rev.* **33** 742-51

[2] Cooper WA, Sand WR, Politovich MK and Veal DL. 1984. Effects of icing on performance of a research airplane. *J. Aircraft* **21** 708-15

[3] Fujishima A, Zhang X and Tryk DA. 2008. TiO_2 photocatalysis and related surface phenomena. *Surf. Sci. Rep.* **63** 515-82,

[4] Pantoja M, Abenojar J and Martinez MA. 2017. Influence of the type of solvent on the development of superhydrophobicity from silane-based solution containing nanoparticles. *Applied. Surface. Sci.* **397** 87-94.

[5] Kalin M, Polajnar M. 2013. The correlation between the surface energy, the contact angle and the spreading parameter, and their relevance for the wetting behaviour of DLC with lubricating oils. *Tribol. Int.* **66** 225-33

[6] Choo JH, Spikes HA, Ratoi M, Glovnea R and Forrest RA. 2007. Friction reduction in low-load hydrodynamic lubrication with a hydrophobic Surface. *Tribol. Int.* **40** 154-9

[7] Evans RD, Cogdell JD, Richter GA, and Doll GL. 2009. Traction of lubricated rolling contacts between thin-film coatings and steel. *Tribol. T.* **52** 106-13

[8] Kalin M, Velkavrh I and Vizintin J. 2009. The Stribeck curve and lubrication design for non-fully wetted surfaces. *Wear* **267** 1232-40

[9] Jahanmir S, Hunsberger AZ and Heshmat H. 2011. Load capacity and durability of H-DLC coated hydrodynamic thrust bearings. *J. Tribol.* **133** 031301-10

[10] Vinogradova OI. 1999. Slippage of water over hydrophobic surfaces. *Int. J. Miner. Process.* **56**, 31-60

Acknowledgments

The authors wish to thank SIKA SAU Spain by their collaboration in this work.

IOP Conf. Series: Journal of Physics: Conf. Series **843** (2017) 012068 doi:10.1088/1742-6596/843/1/012068

Wear-caused deflection evolution of a slide rail, considering linear and non-linear wear models

Dongwook Kim[1], Luca Quagliato[2], Donghwi Park[1], Mohanraj Murugesan[1], Naksoo Kim[1] and Seokmoo Hong[3]

[1]Department of Mechanical Engineering, Sogang University, Seoul, 04107, Republic of Korea

[2]Department of Management and Engineering, University of Padua, Vicenza, I-36100, Italy

[3]Department of Metal Mold Design Engineering, Kongju National University, Cheonan, 32588, Republic of Korea

E-mail: nskim@sogang.ac.kr

Abstract. The research presented in this paper details an experimental-numerical approach for the quantitative correlation between wear and end-point deflection in a slide rail. Focusing the attention on slide rail utilized in white-goods applications, the aim is to evaluate the number of cycles the slide rail can operate, under different load conditions, before it should be replaced due to unacceptable end-point deflection. In this paper, two formulations are utilized to describe the wear: Archard model for the linear wear and Lemaitre damage model for the non-linear wear. The linear wear gradually reduces the surface of the slide rail whereas the non-linear one accounts for the surface element deletion (i.e. due to pitting). To determine the constants to use in the wear models, simple tension test and sliding wear test, by utilizing a designed and developed experiment machine, have been carried out. A full slide rail model simulation has been implemented in ABAQUS including both linear and non-linear wear models and the results have been compared with those of the real rails under different load condition, provided by the rail manufacturer. The comparison between numerically estimated and real rail results proved the reliability of the developed numerical model, limiting the error in a ±10% range. The proposed approach allows predicting the displacement vs cycle curves, parametrized for different loads and, based on a chosen failure criterion, to predict the lifetime of the rail.

1. Introduction

A slide rail is a linear guide composed of two or more members and several balls. When the slide rail is put in motion its members move forwards and backwards, and their sliding contact with the balls cause a progressive wear of the inner surface of the rails. The amount of wear progressively increases during the operational life time of the rail and, for different average applied loads, the rail experiences a failure for considerably different numbers of forward-and-backward cycles. The failure of the slide rail is normally identified by the condition for which, due to the progressive wear, the clearance between balls and rails increase to such a point that the end-point deflection of the last member of the rail makes the product, where the rail is installed, to be unusable.

In the literature, several contribution dealt with the sliding contact between objects, as follows. Singh et al. [1] studied the importance of coating and lubrication in the sliding contact, highlighting how they can influence the lifetime of the considered product. Seo et al. [2] analyzed the different contribution of wear and fatigue in the resulting lifetime of two different railway steels, showing how different material are subjected, in a different way, to these two wearing effects. Yang et al. [3] utilized the Weibull distribution to predict the lifetime of slewing bearing by means of accelerated life testing. Maru et al. [4] studied the influence of possible contaminations in the lubricant those can make the bearing to fail earlier than its expected lifetime. In addition to that, several authors studied the influence of both vibrations and noise on the lifetime of the bearing, as here summarized. Nguyen et al. [5] used an empirical mode decomposition model for the detection of the vibration in the bearing systems which can indicated an incoming failure. Karacay et al. [6] studied the different signals produced by different defects those may occur in the bearing, developing a method which could also predict their position. Finally, by utilizing a non-linear dynamic numerical model, Singh et al. [7] analyzed the effect of the impulsive force has on the bearings and on the signals originated from their application. Concerning the lifetime prediction based on wear, Lee et al. [8] and Hwang et al. [9] utilized the accelerated life testing to predict the lifetime of roll bearing systems.

The present research work deals with the development a numerical model able to predict the cycle-based evolution of the end-point displacement of the last member of a slide rail, for different average operating loads, taking into account both linear and non-liner wear contributions, by utilizing two different wear models. Afterwards, based on a user-defined failure criterion, the number of cycles the slide rail can operate before it should be substituted can also be determined.

In order to determine the model constants to be utilized in both linear and non-linear wear model, a test machine replicating the operating conditions of the slide rail has been designed and build and simple tension test have been also conducted. Moreover, in order to validate the proposed numerical approach, the results of the numerical model have been compared with those of experiments conducted on real slide rails by the rail producer, showing the reliability of the developed approach in predicting the end-point displacement vs number of cycles, for different average loads. The developed approach is of very interest for the industrial compartment since it can be used to efficiently operate a predictive substitution of the slide rail after a certain amount of cycles, reducing the risk of critical failures of the products where they are installed.

2. The slide rail application

This study focuses on the development of a numerical model for the prediction of the end point deflection of a slide rail utilized in many white-goods application. In Figure 1 the real product, target of the study, is shown in both closed and opened configurations. In Figure 2, a schematic representation of the slide rail in opened and closed configuration and the detail of its cross-section are presented.

(a) Closed state

(b) Opened state

Figure 1. Real slide rail in (a) closed and (b) opened state

During its operational lifetime, the slide rail experience different load conditions which may affect its wear, causing an increase of the end-point deflection of the rail as well of the drawer where the rail is installed. Since excessive deflection of the end-point may result in collision or malfunctions, slide rail producers need to set up rules to define, in a reasonable and quantities way, the operational lifetime of the slide rails based on average loading conditions. For this reason, the wear phenomenon should be firstly measured, by means of material characterization experiments, and afterwards modeled in a numerical simulation, for the quick and precise estimation of the of the end-point deflection and lifetime of the slide rail.

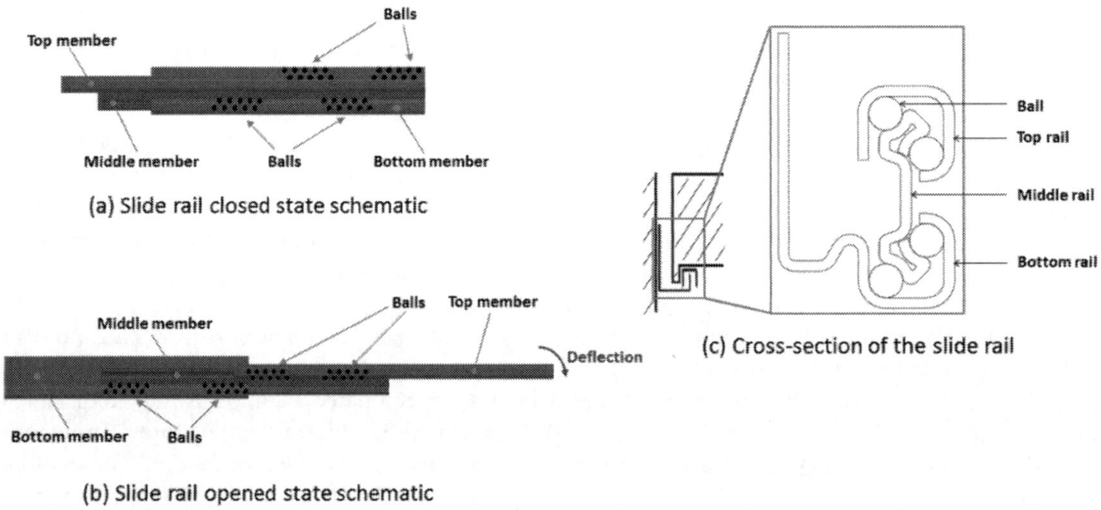

Figure 2. (a) Slide rail in closed (b) and opened state and (c) cross-section detail

3. Theoretical background

As previously anticipated, in the present research work the wear phenomena is subdivided into two different contributions, the linear one and the non-linear one. As shown in Figure 3(a), the application of the Archard model results in a continuous inward movement of the rail surface toward its thickness as a result of the progressive wear. Instead, the consequence of the utilization of the Lemaitre model accounts for the surface element deletion in case the critical damage value is reached, Figure 3(b). The meaning of "element deletion" accounts for the pitting phenomena of the rail surface which is taken into account in the numerical simulation by deleting the elements those reach the critical damage value. Since both these models are utilized in FEM simulation, for the accurate calculation of the surface update due to the wear phenomena, the theoretical backgrounds relevant for both Archard and Lemaitre model are summarized in the following two paragraphs.

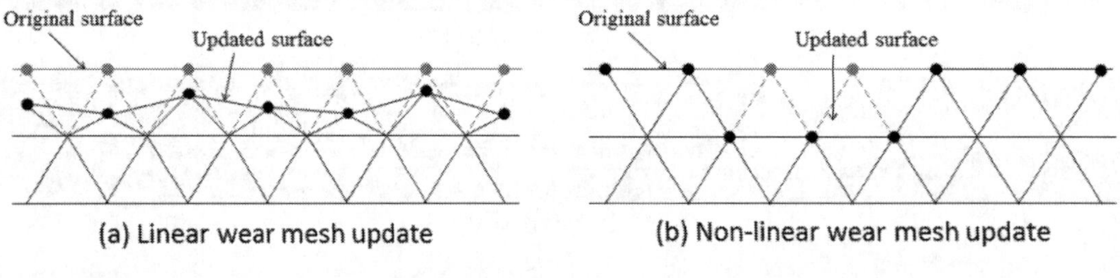

Figure 3. Schematic representation of (a) linear and (b) non-linear wear phenomena

Concerning the linear wear, a modified Archard model is utilized in order to estimate the linear wear along the thickness direction of the rail rather than the total volume loss, as in the original model.

3.1. Archard wear model

The original wear model due to Archard [10] takes into account the amount of volume loss due sliding contact between two material, Eq.(1).

$$Q = \frac{K \cdot W \cdot L}{H} \tag{1}$$

In Eq.(1), Q stands for the total amount of volume loss in the contact surface between the sliding objects; K is a model constants which is determined from the material characterization; W is the normal load applied to the sliding member, which influences the amount of wear, the higher the load the higher the effect of wear. L is the total length of the sliding movement and it is the independent variable in the calculation; as L increases the amount of wear, and accordingly the amount of volume loss, increases as well. Finally, H is the hardness of the considered material, expressed in MPa. In the considered case however, since the linear material loss in the thickness direction of the rail, δ, is meant to be calculated, the modified formulation shown in Eq. (2) is utilized.

$$\delta = \frac{K \cdot W \cdot L}{H \pi r_b l} \tag{2}$$

In Eq. (2), the remaining parameters r_b and l stand for the slide rail ball radius and ball sliding contact distance, respectively. The schematization of the contact is shown in Figure 2(c), where the red lines represent the contact length between balls and rails. According to the original Archard model, on the basis of the progressive increment of the total sliding length L, the amount of the material erosion in the thickness direction of the slide rail increases, following an L-dependent linear path. Since, in some of the analyzed cases, the linear formulation of Eq.(1) and Eq.(2) have shown not to properly fit the experimental data, a modification of the Archard model is proposed and shown in Eq.(3), where the exponent a has been added in order to better model the slight non–linearity in the curves.

$$\delta = \frac{K \cdot W \cdot L^a}{H \pi r_b l} \tag{3}$$

The impact of this modification will be shown in the result section where, especially under heavy load condition, the original Archard model fails to catch to the trend of the experimental data. The material properties K and a are derived from the material characterization experiments made by utilizing the developed test machine, and the relevant test conditions and results are presented in the following paragraph.

3.2. Lemaitre damage model

In the original work due to Lemaitre [11], a thermodynamic model for the estimation of the damage evolution for isotropic materials is proposed. Based on a simple tension test, the critical damage value, over which the material is considered to fail, can be calculated according to Eq. (4), where UTS stands for the ultimate tensile strength of the material under uniaxial loading condition and $\sigma_{break,true}$ for the uniaxial true break stress.

$$D_c = 1 - \frac{UTS}{\sigma_{break,true}} \tag{4}$$

As the damage increases, both the equivalent stress tensor and the deviatoric stress tensor vary according to evolution of the damage, as shown in following Eq. (5) and Eq. (6), respectively. The value of σ_d and s_d stand for the vectors relevant for the damaged material, whereas σ_{orig} and s_{orig} are those of the undamaged one.

$$\sigma_{orig} = \frac{\sigma_d}{1-D} \tag{5}$$

$$s_{orig} = \frac{s_d}{1-D} \tag{6}$$

In the Lemaitre model, the increment of damage value D, Eq. (7), is influenced by the equivalent plastic strain rate ε_p and by the damage strain released rate Y, the last defined as in Eq. (8).

$$D = \varepsilon_p \frac{1}{1-D}\left(\frac{-Y}{S_0}\right)^b \tag{7}$$

$$-Y = \frac{\sigma_{eq,vM}}{2E(1-D)}\left[\frac{2}{3}(1-\nu)+3(1-2\nu)\left(\frac{s_{dev}}{\sigma_{eq,vM}}\right)\right] \tag{8}$$

In Eq. (8) $\sigma_{eq,vM}$, s_{dev}, E and ν stand for von Mises equivalent stress, the deviatoric stress, the Young's modulus and the Poisson ratio, respectively. The ratio between s_{dev} and $\sigma_{eq,vM}$ stands for the triaxiality factor, which represents the amount of off-plane stress in comparison to the von Mises plane stress state. For the determination of the constants to be utilized in the Lemaitre damage model, S_0 and b respectively, simple tension test experiments on the material utilized for the realization of the rails have been conducted and are presented in the following paragraph.

3.3. Combined utilization of linear and non-linear wear models

In the present research work, since two different models are used to describe the progressive wear of the material, further explanation of how these approaches are combined together are given in this section of the paper.

Analyzing the increment of the end-point displacement for a slide rail, three different behaviors can be identified, as shown in Figure 4. In case the load applied does not make the material of the rail to reach UTS during the operational lifetime, the value of the end point displacement manifests small variation from the initial static deflection as the number of cycles increase, Figure 4 "curve 1". As it will be shown in the result section, the small variation of the end-point displacement (for high number of cycles) are well modeled by utilizing only the Archard modified model shown in previous Eq. (3).

If the average load increases, as shown by the result "curve 2" in Figure 4, the slide rail manifests two consequent different behaviors. From the static loading to most of its operational lifetime, the wear curve agrees with the Archard modified model but, from a certain point, the slope of the curve abruptly changes and the slide rail experience a critical increase of the end-point deflection in few hundred cycles. The transition between these two behaviors is explained by the fact that a huge portion of the sliding surface of the rail reaches the UTS value and the damage accumulation starts, leading soon after to an inadmissible end-point displacement. For the proper modelization of this phenomenon, an accurate material testing as well as a precise inverse calibration of the material data in the numerical simulation are essential.

Finally, the third and last observed behavior is described by "curve 3" in Figure 4, where the excessive load applied to the slide rail makes the linear wear contribution to be almost negligible and the end-point deflection to abruptly increase from the static load to the failure in few thousand cycles.

Although the number of cycles and loads for which different slide rails models experience these three kind of deflection vs cycles behavior differ, these three patterns seems to be always present and, as it will be shown in the result section, the combination of the modified Archard model and of the Lemaitre model can successfully describe them.

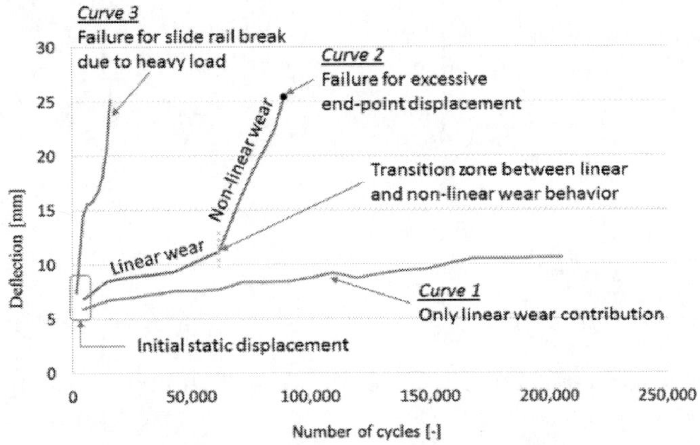

Figure 4. Different wear behaviour patterns

4. Material characterization and model constants determination

In order to properly characterize the basic mechanical properties of the material of the rail guides as well as the wear behaviour due to the interactions between balls and rails, two different tests have been carried out. First, a simple tension test under room temperature conditions for the rail material SCP10 has been conducted utilizing the ASTM E8 standard for the realization of the specimen, as shown in Figure 5. The test has been conducted at 1 mm/min.

The results of the test are shown, in terms of engineering stress-strain, in following Figure 6(a) whereas in Figure 6(b) the true stress-strain curves, together with the fitting obtained by utilizing the Swift model, are reported. The model parameters and material properties obtained from the simple tension test material characterization are reported in Table 1. The Poisson ratio is assumed to be 0.31 according to the SCP 10 (DIN specification 1.0330) material datasheet.

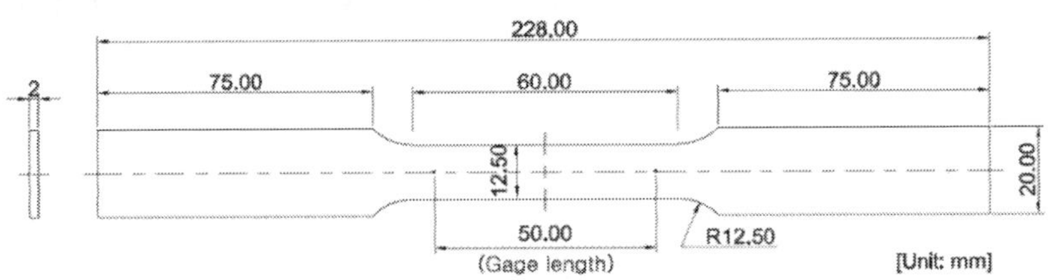

Figure 5. ASTM E8 standard tensile test specimen

In order to properly simulate the material behaviour in the numerical simulation, inverse calibration approach has been utilized to minimize the differences between experimental and numerical stress-strain curve, both in the plastic and in the necking region, as shown in Figure 6a. The constant for the Lemaitre damage model, derived from the inverse calibration of the results of the simple tension test, are summarized in Table 2.

Concerning the determination of the constants of the modified Archard model, the test machine shown in Figure 7 has been designed and built. The machine replicates the operating conditions of the slide rail and, thanks to the encoder placed on the members, speed and displacement can be recorded during the test.

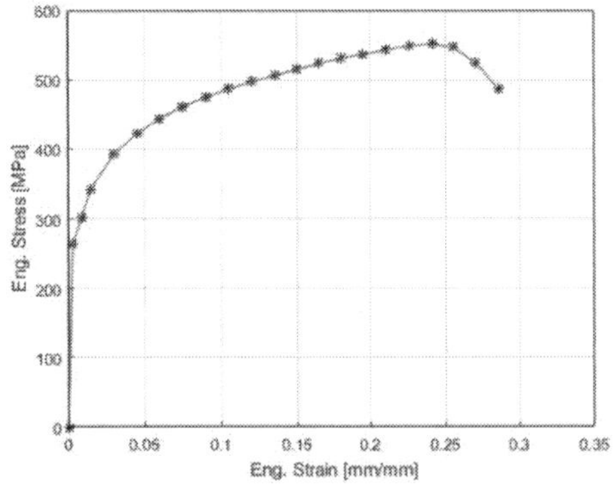

(a) Simple tension test curve

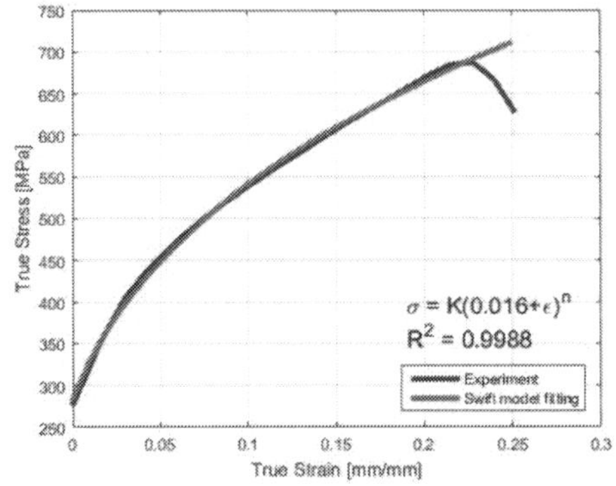

(b) Swift model fitting

Figure 6. Simple tension test result (a) and Swift model fitting (b)

Table 1. Results and fitting parameter of the simple tension test for the SCP10 material

Parameter	**Value**	**Description**
E	200 GPa	Young's modulus
σ_{ys}	301.8 MPa	Yield strength
K	690.0 MPa	Swift model constant
ε_0	0.0016	Initial strain
n	0.169	Hardening exponent

Moreover, in order to reduce the overall complexity of the sliding wear test, modified rails with only 4 balls has been realized by the slide rail producer and have been utilized in this test. In principle, the presence of only 4 balls, instead of the original 40, allows reducing to 1/10 the weight to apply on the last member of the rail thus make the experiment to be more manageable and reduces possible undesired inertia-induced failures in the components of the test machine. Afterwards, by correlating the load condition to the number of balls, it is possible to link the results of the test machine with those of the real rails. Balls radius, r_b, and sliding distance, l, there-and-back, are 6.4mm and 200mm respectively.

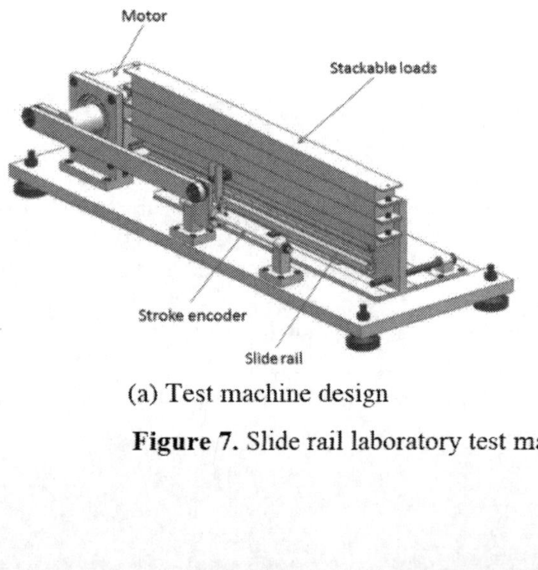

(a) Test machine design

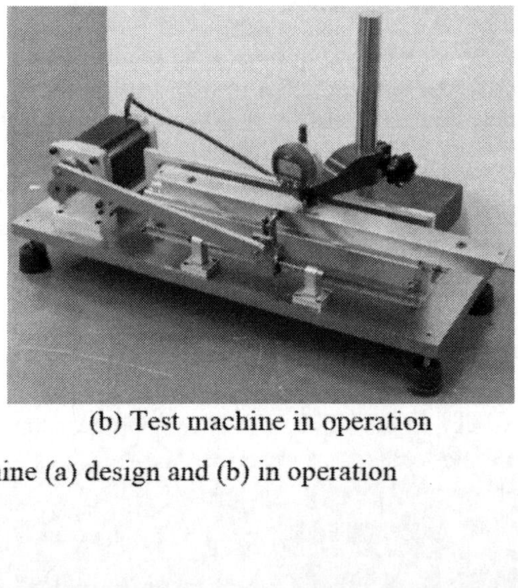

(b) Test machine in operation

Figure 7. Slide rail laboratory test machine (a) design and (b) in operation

In order to simulate all the feasible average weight conditions, experiments have been carried out considering different distributed loads on the last member of the slide rail in the test machine, in detail: 4.5kgf, 6.75kgf, 7.875kgf and 9kgf, which correspond to 45kgf, 67.5kgf, 78.75kgf and 90kgf in the real application. In the products where the slide rails are utilized, for instance a drawer, two rails are installed, whereas in the developed test machine one rail is tested and only 4 balls are installed between the members. For instance, in the real application, a weight of 45kg is distributed on two rails, each one having 40 balls, which corresponds to a 0.56kgf load for each ball. In the test machine, since only 4 balls are installed, to replicate the exact test condition, a load of 2.25kgf should be applied but, in order to conduct accelerated tests, a two times higher load has been applied, leading to a distributed load of 4.5kgf, as previously anticipated.

Table 2. Parameters for the wear models

Parameter	Value	Description
K	$1.039 \cdot 10^{-5} e^{0.1774 \cdot W}$	Archard model constant
a	$-0.0109\,W + 1.0578$	Archard model constant
S_0	36.4	Lemaitre model constant
b	0.89	Lemaitre model constant
H	81 HB	Rail material hardness

Since the high variation of loads among the experiments, the modified Archard model constants varies from case to case and, for this reason, in Table 2, the fitting equations from which to calculate them are reported. As it will be shown in the results section, for both fitting equation, the coefficient of determination (R^2) have been calculated in a range between 0.95 and 0.99, proving the reliability of the proposed approach, in comparison to the linear model. The fitting equation have been included in the numerical model for a precise estimation of the Archard model constants based on the weight applied to the slide rail. Finally, the remaining constants, to be utilized in both linear and non-linear wear models, are summarized in Table 2.

5. Numerical model implementation

The numerical model, replicating all the assembly product made of three rails and 40 balls, have been realized by utilizing ABAQUS/Explicit. In order to simulate the screws connecting the slide rail to the product where it is utilized, an encastre boundary condition have been applied to all the nodes on the side wall of the rail. For all the various load condition, a distributed load has been applied on the top surface of the last member of the slide rail, as shown in Figure 8.

The model has been meshed utilizing the C3D8R hexaedral, an 8-node linear brick, reduced integration, hourglass control element. The C3D8R allows a fast calculation while assuring precise results, as shown in the previous work of Zou et al. [12] for the roll forming process, Tajyar et al. [13] for the non-circular tube forming process, Cai et al. [14] for the roll forming process, Yang et al. [15] and Guo et al. [16] for the ring rolling process.

The mesh size varies according to different zones; the values of the mesh size in the contact are between balls and rails is set as 0.1mm whereas in the remaining zone, not relevant for the calculation, is set to 20mm.

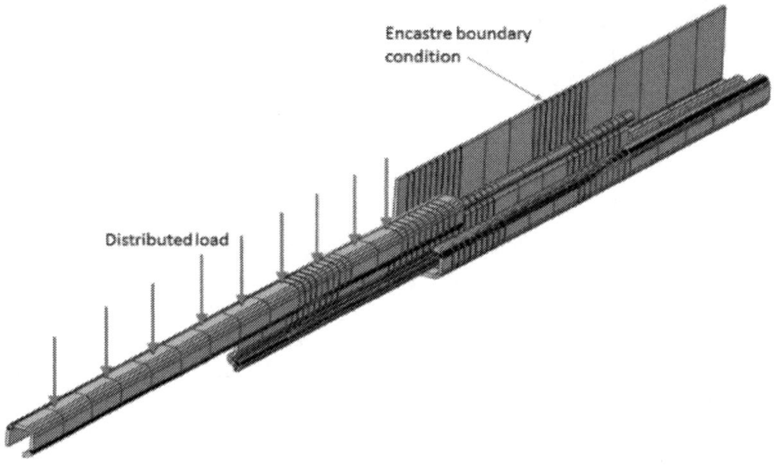

Figure 8. Slide rail numerical model (ABAQUS)

6. Results and discussion

In order to prove the reliability of the modified Archard wear model, the fitting curve relevant for the original one and for the modified one are here after reported, for each one of the four study cases, along with the relevant coefficients of determination. The wear depth amount vs number of cycles curves, along with the interpolation made with the models shown in previous Eq.(2) and Eq.(3), are shown in Figure 9, for the 4.5kg case, Figure 10, for the 6.75kg case, Figure 11 for the 7.875kg and Figure 12, for the 9kg case whereas the relevant coefficients of determination are reported in Table 3.

As it is possible to conclude from the comparison chart and from the coefficients of determination, the modified Archard model is able to better catch the trend of the experimental data and the improvement is as greater as the load applied to the slide rail increases. The reason for this is given by the fact that the higher the load the higher the non-linearity of the displacement vs cycles, resulting in an increasing error if a linear function, as the original Archard model, is used to interpolate the data.

Table 3. Coefficient of correlation comparison between original and modified Archard model

Case	R^2 modified Archard	R^2 original Archard
4.5kg	0.9901	0.8581
6.75kg	0.9516	0.3253
7.875kg	0.9632	0.1589
9kg	0.9939	0.0860

In the numerical simulation, as a consequence of the application of two different wear models, two different update of the surface of the slide rail are present. As previously anticipated, the modified Archard model progressively erodes the surface of the rail whereas the Lemaitre damage model accounts for the surface element deletion, when an element reaches the critical damage.

The critical damage stands for the value over which the material completely fails and cannot withstand any more load, resulting in the detachment of the considered portion of material from the rail. The damage accumulation starts from the UTS point, where the damage accumulation starts, and end in the break point where, according to the Lemaitre damage model, the accumulated damage reaches the unity value.

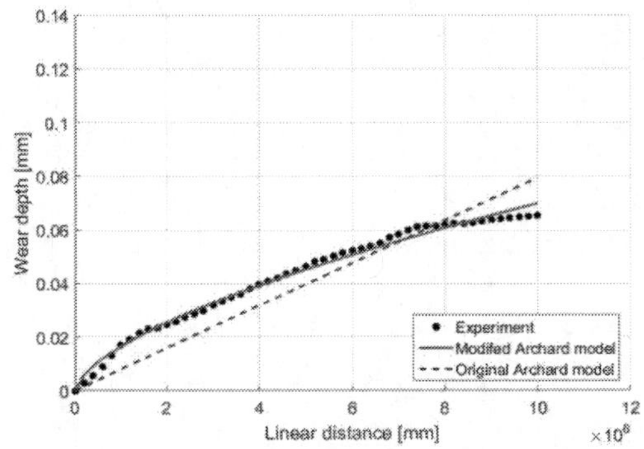

Figure 9. Wear model comparison for the 4.5kg case

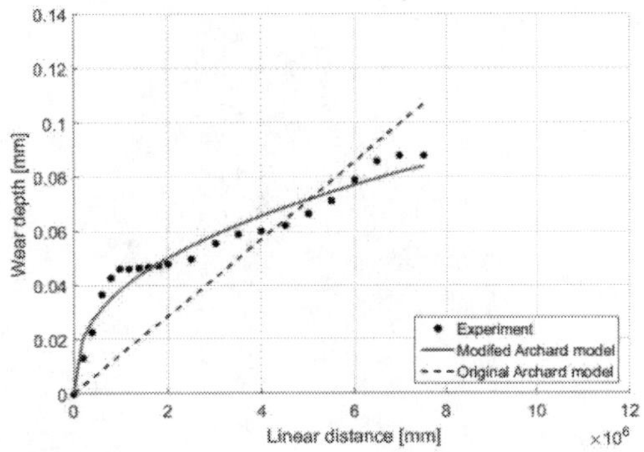

Figure 10. Wear model comparison for the 6.75kg case

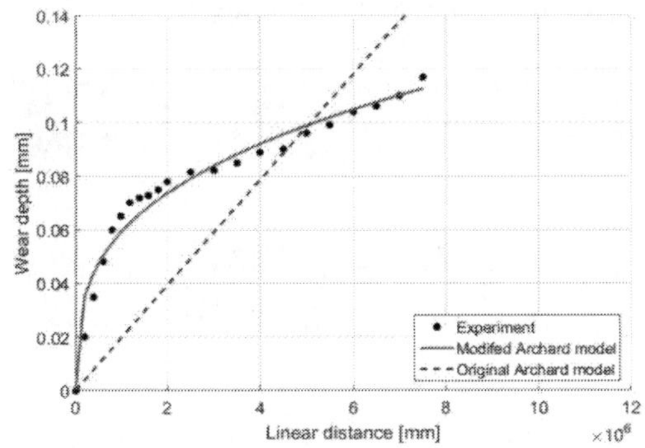

Figure 11. Wear model comparison for the 7.875kg case

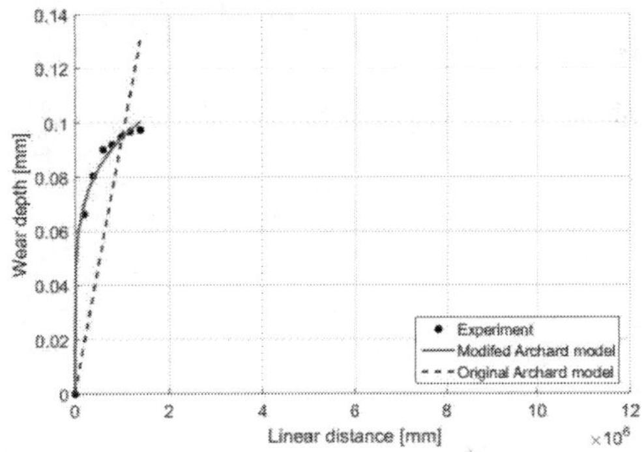

Figure 12. Wear model comparison for the 9kg case

In the numerical simulation, the linear wear contribution is described by an inward movement of the surface nodes, as the number of cycles increases, resulting in a variation of the dimension of the inner part of the rail, while the original shape of the geometry is almost kept constant. On the other hand, when an element reaches the critical damage value in the Lemaitre model, it is deleted from the mesh, resulting in a change of the geometry of the considered portion of the rail, which starts to look scarred. Both this different mesh updates are shown, along with the original surface of the inner part of the rail, in following Figure 13. As a consequence of the progressive surface update due to continuous wear, the gap between balls and rail increases and the end-point deflection of the last member increases as well. In order to show the increment of end-point deflection of the slide rail as the number of cycles increases, in the result shown in Figure 14 where the displacement scale has been doubled in order to allow a better understanding of the deflection.

Finally, in order to prove the reliability of the proposed numerical simulation, the developed model has been utilized for the estimation of the end point deflection for three different loads, 30kg, 45kg and 50kg respectively. The results of the numerical model have been compared with those provided by the slide rail producer and the comparison of the results is shown in Figure 15.

The utilization of a combined wear models approach allows a precise estimation of the sliding wear effect on the rail inner surface, and the error is limited to 9.82% in comparison to the relevant FEM simulation result. It must be noticed that the modified Archard model is able to precisely follow the trend of the experimental results whereas the Lemaitre model presents some inaccuracies, especially in the proximity of the failure.

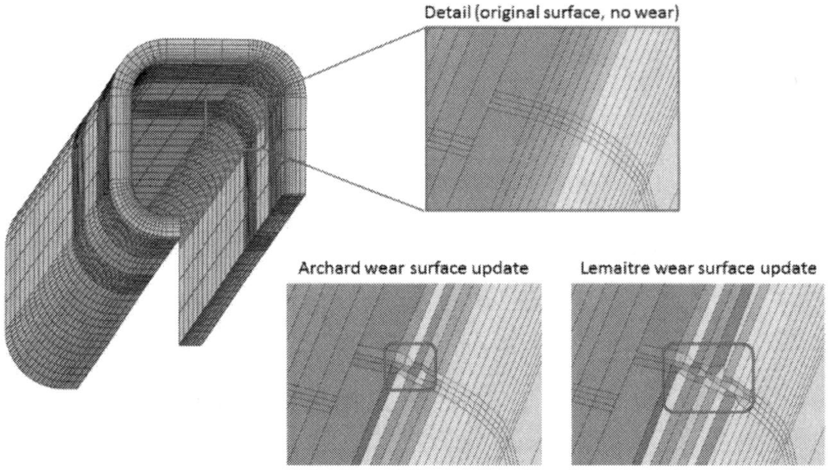

Figure 13. Rail surface update due to Archard and Lemaitre wear models

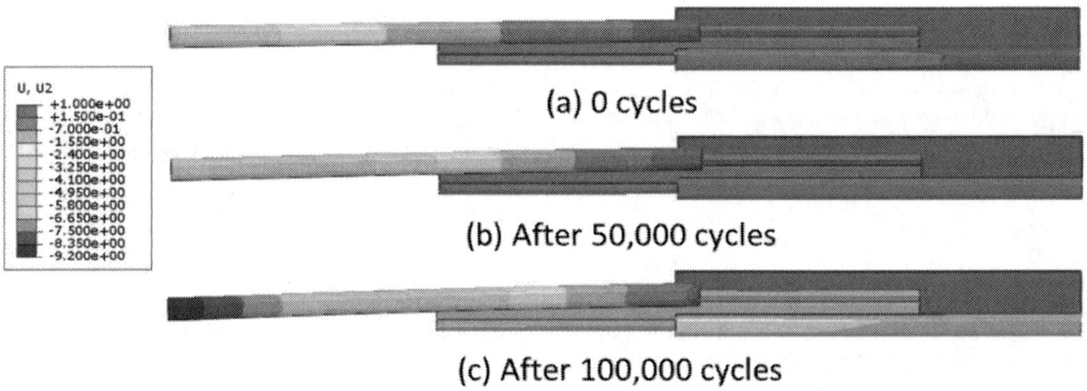

Figure 14. End-point deflection of the slide rail for (a) 0, (b) 50,000 and (c) 100,000 cycles

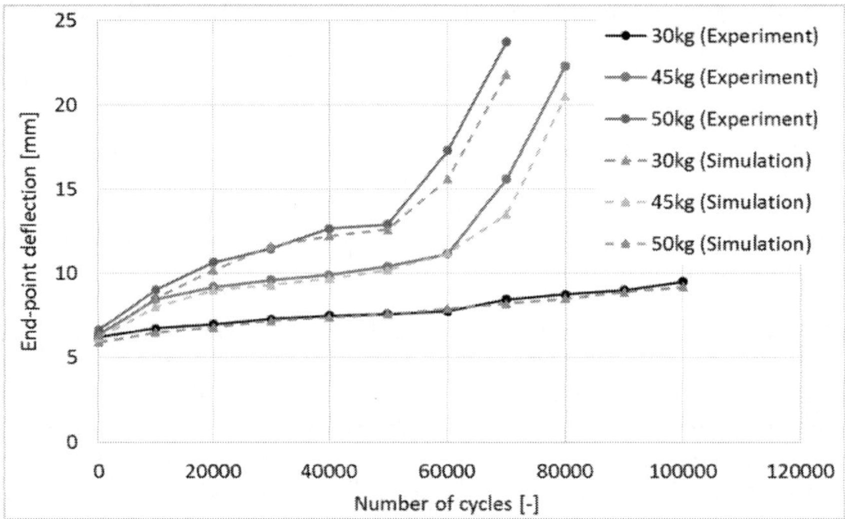

Figure 15. Comparison between simulation and experimental results of the end point deflection

7. Conclusion

In this paper a numerical model for the prediction of the end-point deflection in a slide rail utilized in several white-goods application in proposed. In order to proper account for the wear which occurs in the members of the slide rail due to the sliding contact with the balls, linear and non-linear wear formulations have been utilized and, concerning the linear one, a modified Archard model is proposed and validated, showing how it can well follow the trend of the linear wear occurring during the operational life time.

Finally, the developed model has been applied to three different study case provided by the slide rail manufacturer and its reliability in calculating the end-point deflection, as the number of cycles increase, has been proved.

The developed model is of very interest for the industry since it allows a precise estimation of the increase of wear during the operational life time of rails, and can be extended also to different typology of rails if a proper material characterization is operated.

References

[1] Singh H, Mutyala K C, Evans R D and Doll G L 2015 *Surf. & Coat. Tech.* **284** 281–289
[2] Seo J W, Jun H K, Kwon S J and Lee D H 2016 *I. J. Fat.* **83** 184–194
[3] Feng Y, Huang X, Chen J, Wang H and Hong R 2014 *Mech. Mach. Th.* **81** 94–106
[4] Maru M M, Castillo R S and Padovese L R 2007 *Trib. Int.* **40** 433-440
[5] Nguyen P, Kang M, Kim J M, Ahn B H, Ha J M and Choi B K 2015 *Exp. Sys. App.* 42 9024–32
[6] Karacay T and Akturk N 2009 Trib. Int. 42 836-843
[7] Singh S, Köpke U G, Howard C Q and Petersen D 2014 *J. Sou. Vib.* **333** 5356–5377
[8] Lee N R, Hwang S Y and Kim N S, 2015 *TechConnect W.* 6 21–24
[9] Hwang S Y, Jeong H and Kim N S 2015 *I. J. Mech. & Mechat. Eng.* **15:3** 106-111
[10] Archard J F 1953 *J. Appl. Phys.* **24** 981–988
[11] Lemaitre J 1985 *J. Eng. Mat. Tech.* **107** 83–89
[12] Zou T, Zhou N, Peng Y, Tang D and Li D 2016 *J. Phys. Conf. S.* **734** 032016
[13] Tajyar A and Abrinia K 2009 *I. J. Rec. Tr. Eng.* **1:5** 82–85
[14] Cai Z, Wang M and Li M 2014 *J. Mat. Proc. Tech.* **214** 1820–1827.
[15] Yang H, Guo L, Zhan M and Sun Z 2006 *J. Mat. Proc. Tech.* **177** 634–638
[16] Guo L and Yang H 2006 *Trans. Nonfer Met. Soc.* **16** s645-s651

Effect of initial contact surface condition on the friction and wear properties of bearing steel in cyclic reciprocating sliding contact

Y. Tanaka[1], M. Endo[2, 3], and S. Moriyama[2, 3]

[1]Graduate School of Engineering, Fukuoka University, 8-19-1 Nanakuma, Jonan-ku, Fukuoka 814-0180, Japan
[2]Department of Mechanical Engineering, Fukuoka University, 8-19-1 Nanakuma, Jonan-ku, Fukuoka 814-0180, Japan
[3]Institute of Materials Science and Technology, Fukuoka University, 8-19-1 Nanakuma, Jonan-ku, Fukuoka 814-0180, Japan

E-mail: endo@fukuoka-u.ac.jp

Abstract. Delamination failure is one of the most important engineering problems. This failure can frequently be detrimental to rolling contact machine elements such as bearings, gear wheels, etc. This phenomenon, called rolling contact fatigue, has a close relationship not only with opening-mode but also with shear-mode fatigue crack growth. The crack face interference is known to significantly affect the shear-mode fatigue crack propagation and its threshold behavior. Quantitative investigation on friction and wear at fatigue crack faces in the material is essentially impossible. Previously, thus, a novel ring-on-ring test by making use of fatigue testing machine was proposed to simulate a cyclic reciprocating sliding contact of crack surfaces. However, this test procedure had some problems. For instance, in order to achieve the uniform contact at the start of test, the rubbing of specimens must be conducted in advance. By this treatment, the specimen surfaces were already damaged before the test. In this study, an improvement of experimental method was made to perform the test using the damage-free specimens. The friction and wear properties for heat-treated high carbon-chromium bearing steel were investigated with this new method and the results were compared to the results obtained by using the initially damaged specimens.

1. Introduction

Delamination failure such as flaking, pitting, and shelling is caused by rolling contact fatigue. This ailure often causes fatal problem for the reliability of various rolling contact machine elements like bearings, gear wheels, railway rails, etc. It is known that rolling contact fatigue is intimately related to shear-mode (mode II and mode III) as well as opening-mode (mode I) fatigue crack growth [1-4]. Generally, shear-mode fatigue crack growth and its threshold condition are significantly influenced by he interaction of opposing crack faces (i.e., crack face interference). In order to establish the easonable design method based on fracture mechanics for rolling contact machine elements, a systematic research for the friction and wear at the shear-mode fatigue crack faces is strongly required. However, it is difficult to conduct this research because the real crack faces inside the material can not be observed by using the conventional tribological testing methods. The amount of relative displacement between the surfaces of a shear-mode fatigue crack can be in the sub-millimeter order at the central part and in the sub-micron order near the crack tip. The frictional properties of minimal slip distance in the micro- and nano-Newton range have extensively been investigated using an atomic force microscopy [5-7]. However, for the shear-mode fatigue cracks in the actual engineering elements such as bearings, gears, rails, etc., the knowledge about cyclic reciprocating sliding contact of surfaces with a contact area of square millimetres level under an apparent contact stress of MPa to GPa level is essential. In recent years, the material surface damage due to fretting is studied based on the finite element analysis [8-11], and this approach would surely provide useful information, but the experimental study on cyclic reciprocating sliding contact for a variety of materials under a wide range of experimental conditions is also essential to understand the mechanism of this phenomenon. In our previous study [12, 13], a new ring-on-ring testing procedure by making use of fatigue testing machine was developed. In this procedure, two end-surfaces of hollow cylinders are mutually contacted and relatively cyclically twisted. Under this condition, the cyclic reciprocating sliding contact of crack faces can be simulated. Then, using this testing method, the friction and wear properties of a high carbon-chromium bearing steel (JIS SUJ2) were investigated [13]. However, this method had the some problems. The uniform contact of specimens is indispensable at the start of this test. If the initial contact does not satisfy this condition, non-uniform contact is caused due to misalignment of specimens. In order to attain the uniform contact at the start of test, in the previous test procedure [12, 13], rubbing of the specimens was conducted before the actual test. This means that the contact surfaces had substantially been worn and damaged before the start of test. The alignment of specimens needs to accurately be adjusted without rubbing process so as to obtain the uniform contact condition and the damage-free specimens. In this study, a renovation of the testing method was conducted by developing an alignment fixture. This fixture can execute the centering and the angular adjusting of a specimen. Moreover, the friction and wear of bearing steel in cyclic reciprocating sliding contact was investigated again by using this improved method. In this paper, the effect of initial contact surface condition was newly discussed by comparing the present results without rubbing and the previous results with rubbing [13].

2. Experimental method

The material used in the test was a high carbon-chromium bearing steel (JIS SUJ2), which was held at 840 °C for 30 minutes, oil hardened and then tempered at 170 °C. Table 1 shows the chemical composition in mass%. The Vickers hardness, *HV*, measured with a load of 9.8 N was 753. Figure 1 shows the shape and dimensions of the specimen. The end-surfaces of hollow cylinders were finished by polishing with an emery paper and buffing with an alumina paste. An MTS servo-hydraulic combined axial and torsional fatigue testing machine was used to conduct the ring-on-ring test, as shown by Figure 2. This machine is designed to carry out the high cycle fatigue test in which axial force or displacement and twisting moment or angle are flexibly superimposed. The capacities are 100 kN for axial load and 1000 Nm for twisting moment. The operating frequency, f, can be applied up to 60 Hz.

In this study, the end-faces of hollow cylinders shown in Figure 1 were attached mutually and the static compressive force was applied. The specimen was compressed, then the sinusoidal displacement of constant angular amplitude, θ, was applied by rotating the specimen at the driving side under the control of angular displacement, as illustrated in Figure 2. The cyclic reciprocating relative sliding was thus generated on the contact surfaces. The tests were performed at room temperature under dry condition.

A uniform contact condition at the contact surface is important during the test as mentioned previously. In order to obtain the uniform contact condition, the alignment fixture specialized for this novel ring-on-ring test procedure was developed, by which a flexible control of both centering and angular adjusting of a specimen became possible (cf. Figure 2). The capacities of this fixture are -60 kN for static compressive force and 300 Nm for twisting moment. Figure 3(a) shows the initial contact condition for which the pressure distribution on the contact surface was measured by a pressure-sensitive paper. As shown by this figure, the initial pressure distribution was not uniform. Then, better contact condition was sought by rotating the relative position of specimens by trial and error. The contact situation became better than Figure 3(a) but usually sufficient uniform contact was not attained, as shown in Figure 3(b). Then, by adjusting the alignment of specimens using the alignment fixture shown in Figure 2, a satisfactory uniform contact was finally achieved, as shown in Figure 3(c). Accordingly, the sliding contact tests could be started with the damage-free specimens.

Table 1. Chemical composition of SUJ2 in mass%.

C	Si	Mn	P	S	Cu	Ni	Cr	Mo	O_2	Ti
1.01	0.25	0.34	0.017	0.007	0.10	0.05	1.41	0.03	5 ppm	20 ppm

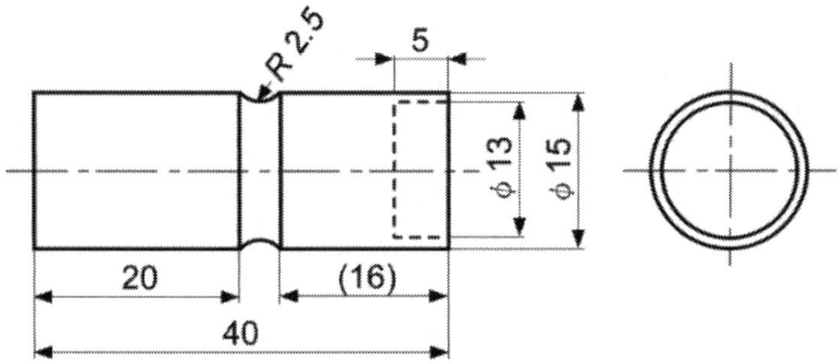

Figure 1. Shape and dimension of the specimen.

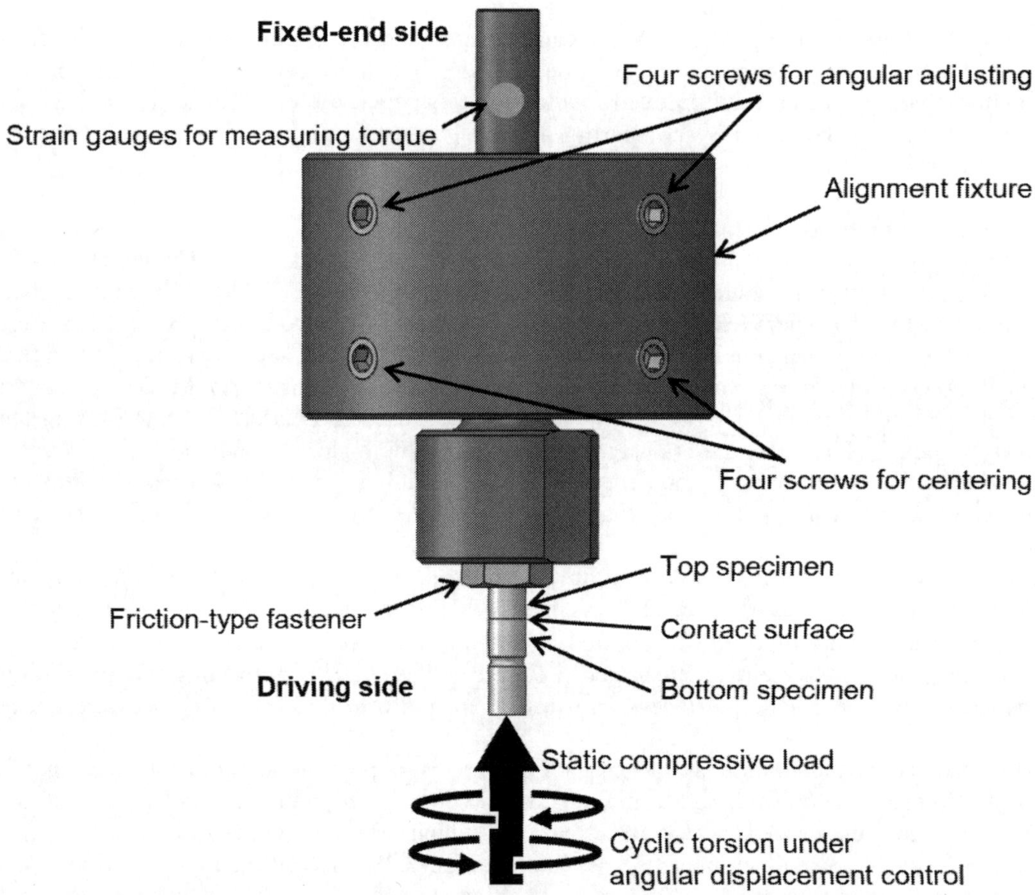

Firuge 2. Configuration of specimens and alignment fixture.

In this study, the nominal contact pressure, p, was defined as:

$$p = W / A \tag{1}$$

where W is the static compressive force, and A is the nominal contact area for which a value of 44.0 mm^2 was used for the calculation. The tangential force, F, was defined as:

$$F = T / r \tag{2}$$

where T is the twisting moment measured by the strain gages attached on a rod of the alignment fixture (cf. Figure 2), and r is the mean radius of hollow cylinder. A nominal value of 7.0 mm was used for r (cf. Figure 1). The relative displacement at the contact surfaces, S, was defined as:

$$S = S_D - S_F \tag{3}$$

In this equation, S_D and S_F are the displacements of the driving side specimen and the fixed-end side specimen, respectively. These displacements were measured by a laser displacement meter (KEYENCE: LK-H020). A mirrored thin plate made from cermet tip was used as a target of laser light, and the target was attached to the jig made by a 3D-printer, as shown by Figure 4. The material of jig was light ABS resin, so that its inertia was negligibly small. The jig was mounted on each specimen at a distance of 1 mm from the edge of hollow cylinder. A 4-point support was used to fix the jig to the specimen (cf. Figures 4 and 5).

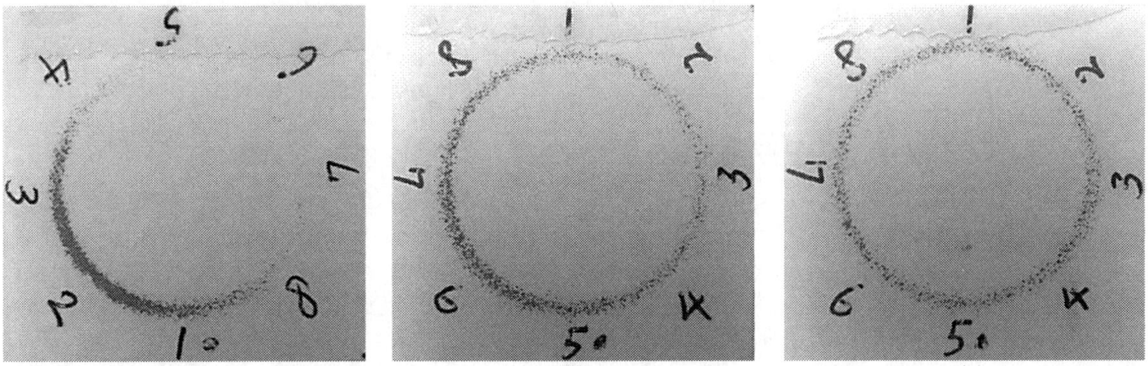

(a) (b) (c)

Figure 3. Contact conditions of the same surfaces; (a) the first contact condition, (b) the improved contact condition after rotation of the driving side specimen, and (c) the uniform contact condition after adjustment with the alignment fixture.

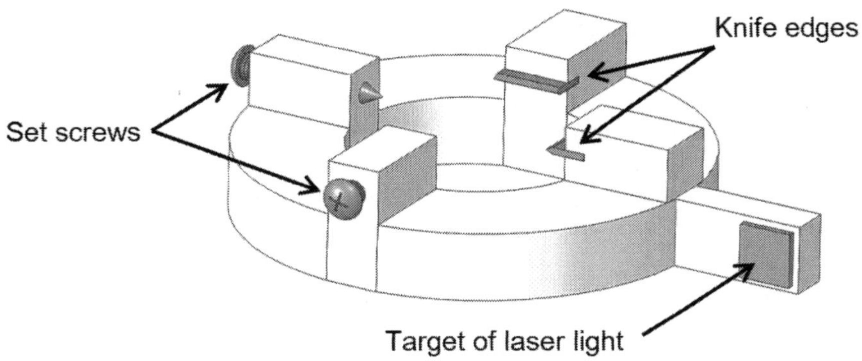

Figure 4. Shape of a jig with a target of laser light.

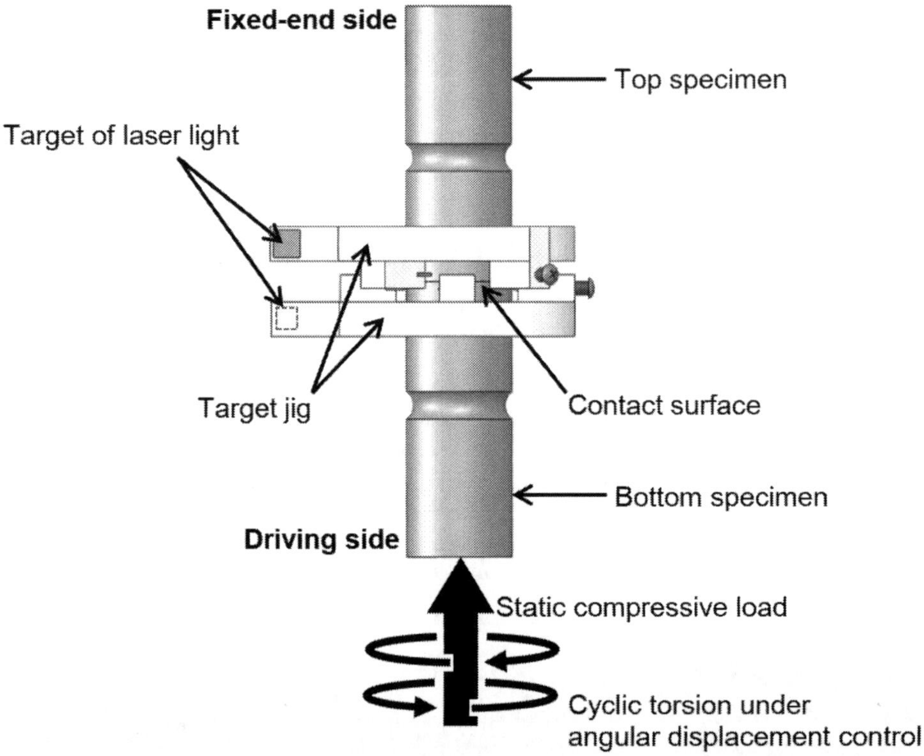

Figure 5. Configuration of specimens and measurement jigs.

3. Results and discussion

3.1. Measurement of coefficient of kinetic friction

Figure 6(a) shows the time variation of the tangential force, F, and the relative displacement, S, which were measured under p = 100 MPa, f = 1 Hz and θ = 1 deg, and at $N \approx$ 1000 cycles after the start of the test. The variation of F and S was approximately of a rectangular waveform.

Figure 6(b) shows the relationship between F and S in the dashed frame of Figure 6(a). This relationship between F and S exhibited almost a rectangular hysteresis loop. The horizontal lines surrounded by the dotted frames in Figure 6(b) indicate the regions where an entire slip happened all over the contact surfaces. An expected slope of the relationship between F and S for a hollow cylindrical specimen calculated on the assumption of full adhesion, i.e., a simple thin-wall pipe, is also shown in Figure 6(b). The slope of vertical change of the hysteresis loop was almost identical to that of calculated line. In this study, the coefficient of kinetic friction for the entire slip was defined as

$$\mu_k = F_{km} \;/\; W \tag{4}$$

where F_{km} is the mean value of tangential force during the entire slip. With this equation, $\mu_k \approx 0.8$ was obtained.

Figure 7 shows the relationships between μ_k and N that were observed under p = 100 MPa, f = 1 Hz and θ = 1 deg. The values of μ_k were measured at N = 10, 100, 1000 and 10000 cycles, respectively. As shown by Figure 7, the value of μ_k under this condition converges to about 0.8 with increase of N.

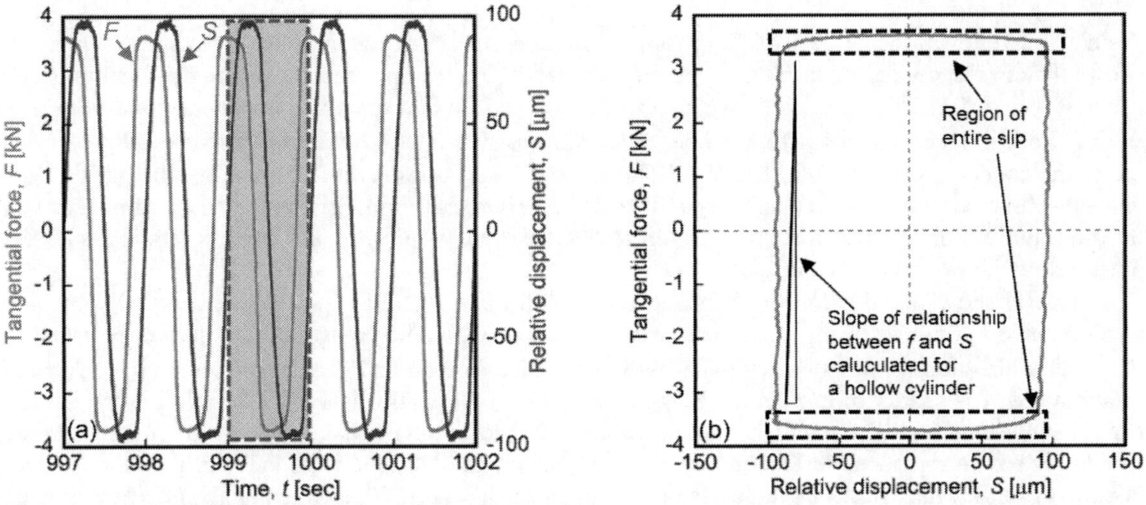

Figure 6. (a) Time variation of tangential force, F, and relative displacement, S, and (b) relationship between F and S, which were observed at $N = 1000$ cycles under $p = 100$ MPa, $f = 1$ Hz and $\theta = 1$ deg.

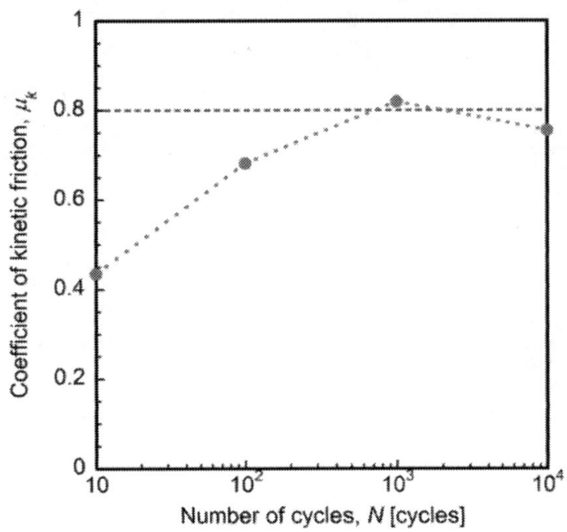

Figure 7. Relationship between μ_k and N observed under $p = 100$ MPa, $f = 1$ Hz and $\theta = 1$ deg.

3.2. Effect of initial contact surface condition

In the previous study, the effect of static compressive force on the coefficient of kinetic friction was investigated for the same material of SUJ2. Figure 8(a) shows the previous results [13]. As shown by this figure, μ_k under $p = 100$ MPa was almost constant and its value was about 0.8. On the other hand, μ_k under $p = 10$ MPa showed a larger scatter ranging from 0.4 to 1.0. The cause of this large scatter has been an open question.

Figure 9 shows the present results obtained using the damage-free specimens under the test conditions identical to the above previous test. All results show the convergence of μ_k to about 0.8 with increase of N, independently of the contact pressure, p. Compared to Figure 8, the test performed using the damage-free specimens provided a reasonable result with a small scatter. By comparing these results, it is expected that the coefficient of kinetic friction is affected by the initial contact surface condition, such as the roughness and plastic flow of the surface, etc. of base metal.

In what follows, the effects of the plastic flow and surface roughness are discussed. Figure 10 shows the cross-sectional surface observed after the test using the specimens damaged by rubbing. The rubbing was performed by the 1000 cycle reciprocating sliding of a specimen under the conditions of p = 25 MPa, f = 1 Hz and θ = 10 deg. Thereafter, the test was started and continued to 10000 cycles under the conditions of p = 100 MPa, f = 1 Hz and θ = 1 deg. As shown in Figure 10, the plastic flow was not observed at the contact surface at all. It is supposed that the large plastic flow did not happen by the initial rubbing. Therefore, the scatter of data seen in Figure 8 may have no relation with the plastic flow of initial contact surface.

Figure 11 shows the relationship between μ_k and N for p = 10 MPa, f = 1 Hz, and θ = 1 deg, which was obtained in the tests using specimens with rough surfaces. The contact surfaces were polished in the sliding direction using an emery paper with a grit size of ANSI #180. The values of μ_k converged to about 0.8. The scatter of data in the range of $N = 10^3 \sim 10^4$ cycles is a little larger than the data in Figure 9 but much smaller than the data in Figure 8. Accordingly, it is concluded that initial surface roughness has no significant effect on the value of coefficient of kinetic friction. Very recently, Pereira et al. [14] studied the effect of roughness on fretting wear by the finite element analysis. They showed that the surface asperities interaction had influence on the wear scars only for the initial thousands cycles, e.g. 1800 cycles. This result is consistent with the present result shown by Figures 9 and 11.

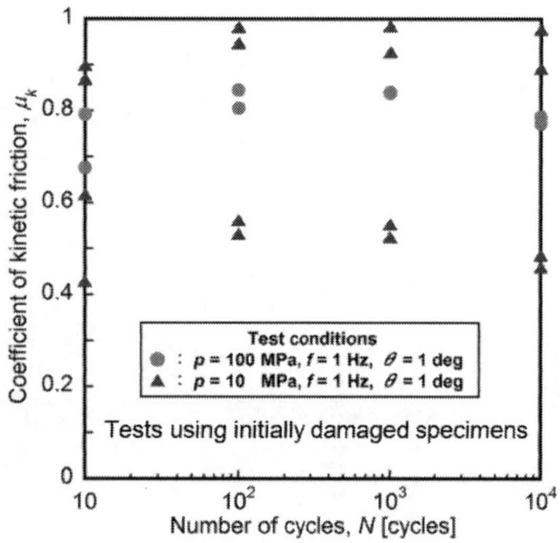

Figure 8. Results obtained in the previous study [13].

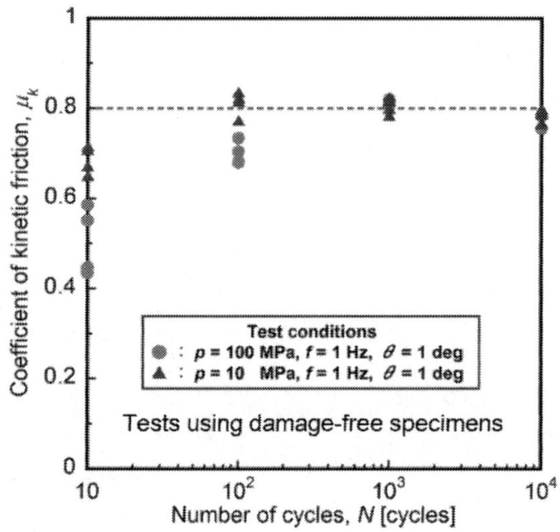

Figure 9. Results obtained in this study.

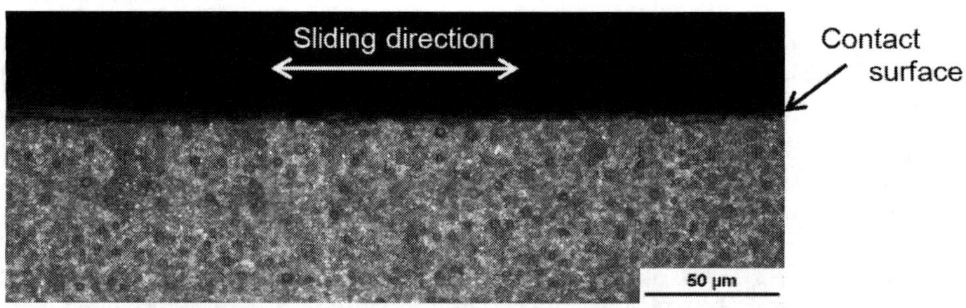

Figure 10. Cross-section of contact surface after the test with damaged specimens.

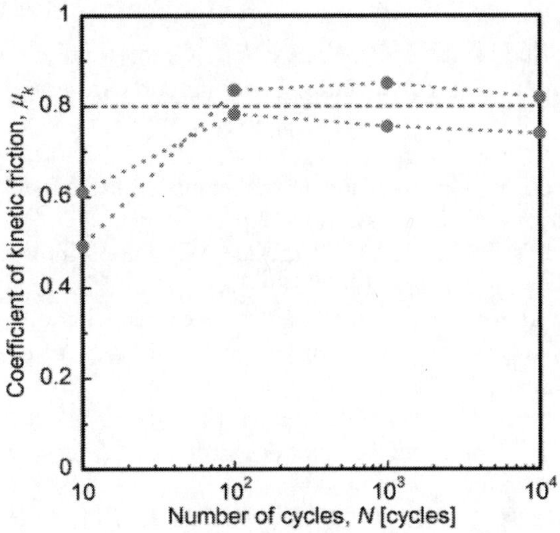

Figure 11. Relationship between μ_k and N for p = 10 MPa, f = 1 Hz and θ = 1 deg, which was obtained in the tests using the specimens polished with an emery paper (ANSI #180).

4. Conclusions

The previously proposed method of ring-on-ring test by making use of fatigue testing machine was renovated in this study. This renovated method made it possible to perform a cyclic reciprocating sliding contact test with the damage-free specimens. The coefficient of kinetic friction of JIS SUJ2 (high carbon-chromium bearing steel) was investigated by this improved method. Moreover, by comparing the results to the previous results using initially-damaged specimens, the effects of initial contact surface condition were newly discussed. The obtained results are summarized as follows:

- The renovated ring-on-ring test does not need the rubbing of specimens before the test. Therefore, the tests can be started using the damage-free specimens.
- The relationship between the tangential force (i.e., frictional force), F, and the relative displacement, S, exhibits approximately a rectangular hysteresis loop.
- The value of coefficient of kinetic friction converges to about 0.8 with increase of the number of cycles.
- Static compressive force has no influence on the coefficient of kinetic friction.
- Plastic flow does not happen by rubbing under p = 25 MPa, and therefore it is not responsible for the scatter of data.
- Surface roughness at the start of test has no significant influence on the coefficient of kinetic friction.

Acknowledgement
This work was in part supported by the JSPS KAKENHI (Grant Number: JP16K06057) and the NSK Foundation for the Advancement of Mechatronics. Their supports are gratefully appreciated.

References

[1] S. Beretta, M. Boniardi, M. Carboni and H. Desimone 2005 Mode II Fatigue Failures at Rail Butt-welds *Engineering Failure Analysis* **12** p 157-165.

[2] M. W. J. Lewis and B. Tomkins 2012 A Fracture Mechanics Interpretation of Rolling Bearing Fatigue *J. Engineering Tribology* **226** p 389-405.

[3] A. Otsuka, Y. Fujii and K. Maeda 2004 A New Testing Method to Obtain Mode II Fatigue Crack Growth Characteristics of Hard Materials *Fatigue & Fracture of Engineering Materials & Structures* **27** p 203-212.

[4] H. Matsunaga, N. Shomura, S. Muramoto and M. Endo 2010 Shear Mode Threshold for a Small Fatigue Crack in a Bearing Steel *Fatigue & Fracture of Engineering Materials & Structures* **34** p 72-82.

[5] E.-S. Yoon, S. H. Yang, H.-G. Han and H. Kong 2003 An Experimental Study on the Adhesion at a Nano-contact *Wear* **254** p 974-980

[6] S. Achanta, D. Drees and J.-P Celis 2005 Friction and Nanowear of Hard Coatings in Reciprocating Sliding at Milli-Newton Loads *Wear* **259** p 719-729

[7] V. L. Popov 2010 *Contact Mechanics and Friction: Physical Principles and Applications* Springer Heidelberg

[8] D. Kumar, R. Biswas, L. H. Poh and M. A. Wahab 2017 Fretting fatigue stress analysis in heterogeneous material using direct numerical simulations in solid mechanics *Tribology International* **109** p 124-132

[9] T. Yue and M. A. Wahab 2017 Finite element analysis of fretting wear under variable coefficient of friction and different contact regimes *Tribology International* **107** p 274-282

[10] N. A. Bhatti and M. A. Wahab 2017 Finite element analysis of fretting fatigue under out of phase loading conditions *Tribology International* **109** p 552-562

[11] T. Yue and M. A. Wahab 2014 Finite element analysis of stress singularity in partial slip and gross sliding regimes in fretting wear *Wear* **321** p 53-63

[12] M. Endo, T. Saito, S. Moriyama, S. Okazaki and H. Matsunaga 2015 Friction and Wear Properties of Heat-treated Cr-Mo Steel during Reciprocating Sliding Contact with Small Relative Motion *International Journal of Fracture Fatigue and Wear* **3** p 215-220.

[13] Y. Tanaka, M. Endo and S. Moriyama 2016 Frictional Behavior and Surface Damage in Cyclic Reciprocating Contact of Bearing Steel *M&M2016 MATERIALS AND MECHANICS CONFERENCE PROCEEDINGS* p 857-859.

[14] K. Pereira, T. Yue and M. A. Wahab 2017 Multiscale analysis of the effect of roughness on fretting wear *Tribology International* **in press** (http://dx.doi.org/10.1016/j.triboint.2017.02.024)

Study of friction and wear of thermoplastic vulcanizates: the correlation with abraded surfaces topology

E Harea[1], R Stoček[1,2], M Machovský[1]

[1] Centre of Polymer Systems, University Institute, Tomas Bata University in Zlín, třída Tomáše Bati 5678, 760 01 Zlín, Czech Republic

[2] PRL Polymer Research Lab., Nad Ovčírnou 3685, 760 01 Zlín, Czech Republic

E-mail: harea@cps.utb.cz

Abstract. The work was focused on the study of friction and wear properties of thermoplastic vulcanizates (TPV) based on polypropylene (PP), natural rubber (NR) and styrene butadiene rubber (SBR) compounds containing all common additives and curatives using ball-on-flat method. Pure materials and binary TPV blends of PP/NR, as well those of PP/SBR with the compositions 95/5, 75/25 and 50/50 (in weight %) were compounded and analysed. It is very well known that the coefficient of friction (COF), as well as wear values of pure thermoplastic matrix are significantly lower than those for pure rubber. Thus, it was found that the friction coefficient and wear of TPVs significantly increased in accordance with increased content of rubber material. Surprisingly, NR compared with SBR of similar concentrations in PP matrix, considerably affected wear of samples and the friction coefficient remained almost unaffected. Finally the topology of abraded surfaces were examined by using scanning electron microscopy (SEM) in order to understand the relationship between the COF, wear process and the composition of TPVs.

1. Introduction

Thermoplastic dynamic vulcanizates (TPV) represent the plastics and rubber blends. Such kinds of elastomers are frequently used in automotive, electric or footwear industries. They overcome other conventional materials due to combination of outstanding mechanical characteristics of elastomers and the excellent processing properties of thermoplastics. TPVs also gain substantial attention due to the simple realization of the desired physical properties by varying the blend ratios [1]. The cross-linked rubber phase in TPVs gives them a good compression behavior and a good chemical resistance against water, acids and bases and can play an essential role in recycling of "end of life" rubber products.

TPVs are produced by dynamic crosslinking of blends of thermoplastics and curable elastomers. During an intensive mixing process, fine dispersion of crosslinked, micron-size rubber particles are imbedded in a thermoplastic matrix [2]. The morphology of multiphase blends strongly depends on volume ratio, the solubility, the viscosity of components and mixing time. Usually the major phase forms the continuous matrix and governs the properties of entire blend, while the minor phase forms dispersed micro domains and contribute to certain specialized properties of blend [3].

Friction and wear characteristics of TPV materials were extensively studied for the last several decades [4-7]. Such materials require combining simultaneously, properties like durability resistance to a great number of solvents and external forces throughout the service life. The friction of products is taken in to account directly during the designing process. Thus, understanding of friction mechanisms of polymeric materials becomes very important.

Friction is governed by the surface interaction in moving contact between both of frictional partners [6]. The real contact area depends on interfacial properties and commonly vary during sliding process because of mechanical behavior of material. Usually, interfacial properties change due to tribo-film and wear debris formation and consequently roughness evolution [7], whereas the tribo-film is the well-known transfer thin layer formation on the metallic counterface, which takes place during the friction process between metal and polymer surfaces. The friction film formed on the polymer composite itself as well plays an important role in the friction and wear behavior of polymer composites [8]. Friction variation during the test is frequently related to an increase in real contact area [9].

Up to date a major interest is dedicated to friction and wear properties of micro-nanoparticles [10, 11] and ground tire rubber (GTR) filled TPVs [12]. It has been found that an increase in COF takes place as a function of the concentration of rubber powder in PP matrix for GTR modified PP blends. This result was explained through the nature of GTR powder to form an inter-material layer with enhanced COF.

In this paper studies were done on the effect of blend ratios on friction and wear of TPVs based on polypropylene (PP) + natural rubber (NR) or styrene-butadiene rubber (SBR). PP/(NR or SBR) blends were compounded as potential materials with enriched wear resistance and toughness for possible engineering applications. The main goal of the work was the investigation of tribological properties of polymeric materials with high difference of COF, to study the relation between COF, wear and the topology of abraded surfaces. Finally the wear mechanism of the TPVs was investigated.

2. Experimental

2.1. Materials

The PP used in this research was Polypropylene homopolymer (TITANPRO 6531, isotactic type) with a specified melt flow index of 3.5 g/10 minutes. It was supplied by Titan PP Polymers (M) Sdn. Bhd. NR of the type SMR 20 CV/BP1 was supplied by the Astlett Rubber Inc. SBR of the type Buna CB 24 was provided by TRINSEO. Sulphur was used as the curing agent, whereas zinc oxide and stearic acid were used as activators. Sulphur, zinc oxide, and stearic acid were supplied by Sigma-Aldrich®. Carbon black of the type N339 was used as the filler. CBS (N-cyclohexyl-2-benzothiazolesulfenamide) was employed as curing activator and IPPD (N-isopropyl-N'-phenyl-p-phenylenediamine) as antioxidant. The complete composition of the compounds is listed in the table 1.

Table1. Rubber composition.

	NR	**SBR**	**Carbon Black**	**CBS**	**Sulphur**	**IPPD**	**ZnO**	**Stearic Acid**
NR	100	-	50	2,5	1,7	1,5	3	1
SBR	-	100	50	2,5	1,7	1,5	3	1

*phr – parts per hundred rubber

2.2. Rubber compound manufacturing

The rubber blends for both NR and SBR were mixed according to ASTM D 3192 – 09 in an internal mixer Banbury (Pomini Farrel) with a total batch weight of 360 g each at 70 °C. Chemicals in proportion listed in table 1 were successively added as follows: elastomers mixed for 2 min followed by mixing of ZnO, Stearic Acid, IPPD and Carbon Black for next 5 min. Each mixing was accomplished at a rotor speed of 65 rpm. CBS and sulphur were added in the batches and further homogenized in an open two roll mixing mill (Farrel G-2603) at 60 °C for 5 min.

IOP Conf. Series: Journal of Physics: Conf. Series **843** (2017) 012070 doi:10.1088/1742-6596/843/1/012070

Each TPV batch was prepared on a Brabender Plastograph. Compounding was done at 190°C, 60 rpm and a fill factor of 80 %. PP was melted in the mixer for 2 min and then NR or SBR containing additives and curing agents were added in proportion defined in table 2 in % by weight. The mixing was continued for 3 min. Finally, the material was collected and compression molded into a 120 x 120 x 2 mm sheet. The remaining time between the collecting of the compound and molding has to be 2 minutes.

Table 2. Blend ratio of the samples

	PP	NR	SBR	PP/5NR	PP/25NR	PP/50NR	PP/5SBR	PP/25SBR	PP/50SBR
PP	100	0	0	95	75	50	95	75	50
NR	0	100	0	5	25	50	0	0	0
SBR	0	0	100	0	0	0	5	25	50

Samples of the blends were molded in an electrically heated hydraulic press upon 250 kN force. Hot-press procedures involved preheating at 200 °C for 3 min followed by compressing for 3 min at the same temperature. The final cooling of the molded sheets was performed using manual cooling press for 5 min at the maximal reached compression force.
Sheets based on pure PP were molded directly from PP granulates using aforementioned procedure. Rubber samples were cured at 150 °C in accordance with optimum curing time established by using a moving die rheometer (MDR 3000, Montech).

2.3. Testing
Dry friction tests were conducted using Bruker Universal Mechanical Tester (UMT) in a linear ball-on-flat mode at room temperature in air. The relative humidity was 28 ± 5 %. An 8 mm diameter ball, made of stainless steel, was slid against a flat sample in reciprocating mode (figure 1) under constant load F = 0,5 N. The ball traversed 300 cycles during 120 s at 125 mm/s sliding speed and 25 mm stroke length.

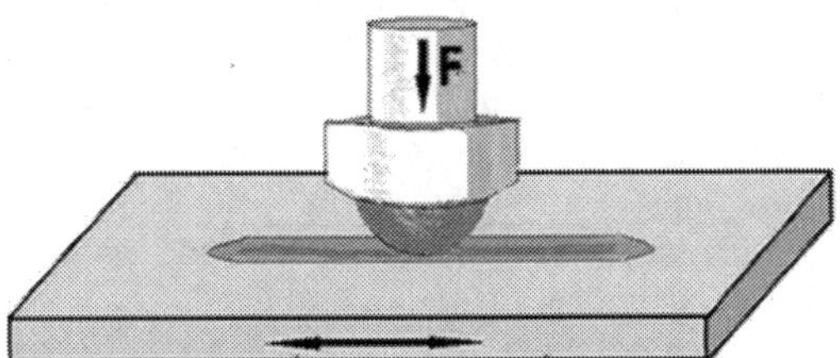

Figure 1. Scheme of ball-on –flat reciprocating friction test.

Determination of hardness was done using BAREISS HP-D - Shore D durometer. The same measuring procedure was applied for all samples including rubber samples (method SHORE, ISO 868). For rubber materials the Shore D method was used instead of the usual Shore A to maintain a consistency in comparison amongst all the samples, based on varied polymeric materials. This method measures penetration of a special shaped steel rod after a fixed time, which in the present study was 2 minutes. Frequently, hardness of materials helps to understand COF and wear evolution [13] and serves as a parameter which complements the comprehension of materials behaviors.

The data of wear were compared using the width of friction tracks measured by optical microscope Leica DVM2500.

Studies on the morphology of friction tracks were carried out using a scanning electron microscope (SEM), model Nova NanoSEM 450 (FEI) with Schottky field emission electron source.

3. Results and discussion

The homogeneity of the studied TPV blends is a very important factor, which is dependent on the compounding procedure.

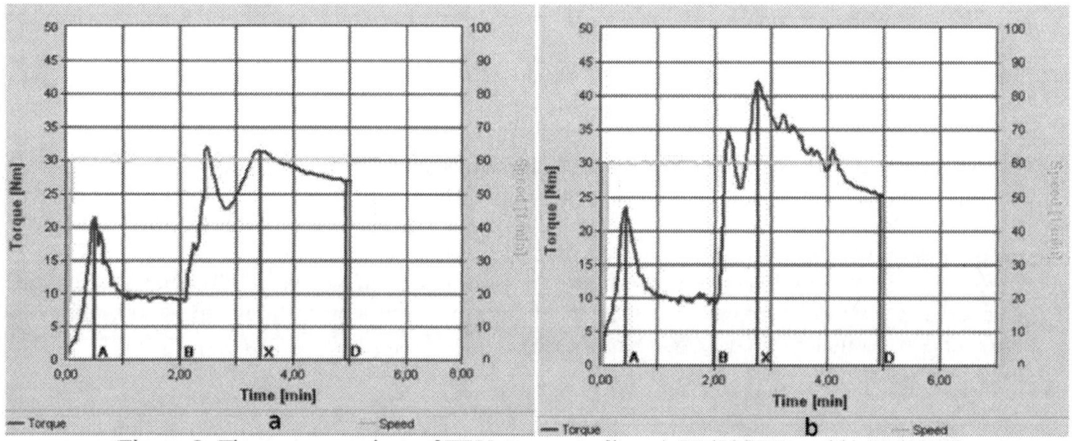

Figure 2. The torque vs time of TPV compounding a) PP/50SBR and b) PP/50NR .

In figure 2, the compounding process exemplary is shown for TPV blends, based on the highest content 50 phr of SBR (a) and NR (b). Figures 2a-b show the dependence of compounding time on torque. For all the blends, PP was inserted into the mixing chamber without preheating. The torque increased sharply due to the resistance exerted on the rotors by the un-melted PP at the first phase of the compounding process during the first 30 s, followed by a region where PP melted and got transformed from solid to liquid phase (see interval AB). After 2 min, rubber was added into the mixing chamber, which caused the next rise of torque due to the un-homogenized rubber and impregnation into PP matrix. The torque increased up to 35 Nm for NR and 32 Nm for addition of SBR. Next, the decrease in torque was due to the melting of rubber and was followed by another increase in torque due to the rubber getting cross linked at this stage as is shown on the plot. A large difference is observed between cross linking behaviours of NR and SBR. NR crosslinking (see figure 2b) started earlier than that of SBR (figure 2a) and attained the maximum in about of 30 s compared with more than 60 s for SBR (see point X). Evident fluctuation of torque during continuous decreasing on the last part of graph for NR denotes combined mechanism of crosslinking with simultaneous breaking of already existing links until final stabilization was attained. On the other hand, smooth decreasing for similar region for SBR exhibited predominantly breaking of already formed links until some kind of saturated state was reached. SBR dispersed more easily in the PP matrix compared with NR.

Figure 3 exhibits the variation of COF during the 120 s or 300 cycles for PP, pure NR and SBR samples as well as for their blends. For all the samples, a fast increase of COF can be observed in the course of the first 5-20 s of test followed by slight and continuous decrease for rubbers, increasing for pure PP and blends with lower (5%) rubber concentration in PP matrix. Blends with 25 % and 50 % rubbers represented more or less stable COF with fluctuations caused by wear debris formation. Generally, it is well known that COF in rubber-steel interaction is high [14]. The results show the COF ranged between 2,00 and 2,25 for rubbers used in the investigation (NR and SBR). Noticeable

IOP Conf. Series: Journal of Physics: Conf. Series **843** (2017) 012070 doi:10.1088/1742-6596/843/1/012070

differences in COF are observed in the first 40 s or 75 cycles of ball sliding, followed by continuous and sluggish decrease. It can be the witness about the different wear mechanisms, which at initial stage play an important role in COF values.

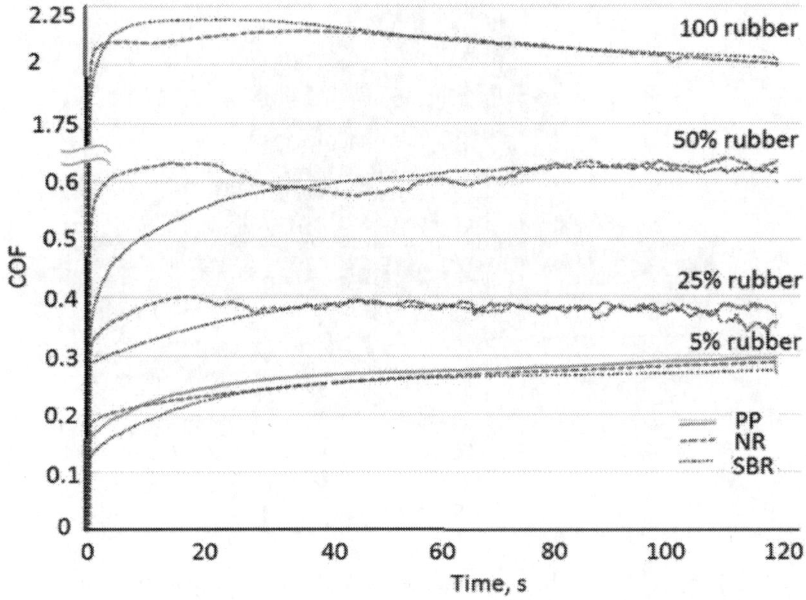

Figure 3. Evolution of COF of PP/rubber blends and their components.

Due to different mechanical properties of NR and SBR such as hardness, elastic modulus, fracture toughness etc., initial abrasion and topology inside of friction tracks are quite different, which directly affect COF through the contact surface. COF usually is governed by surface adhesion, surface roughness, surface deformation, and surface contamination. Continuously increasing number of ball passes creates a specific surface topology (figure 4) which can be compared with waves on water surface, with certain amplitude, periodicity and width for certain applied COF (load and speed) and rubber (different for NR or SBR). This particular surface kept more or less constant morphology due to permanent creation of new "waves" instead of abraded ones. COF on waved surfaces was lower than on pristine rubber surfaces.

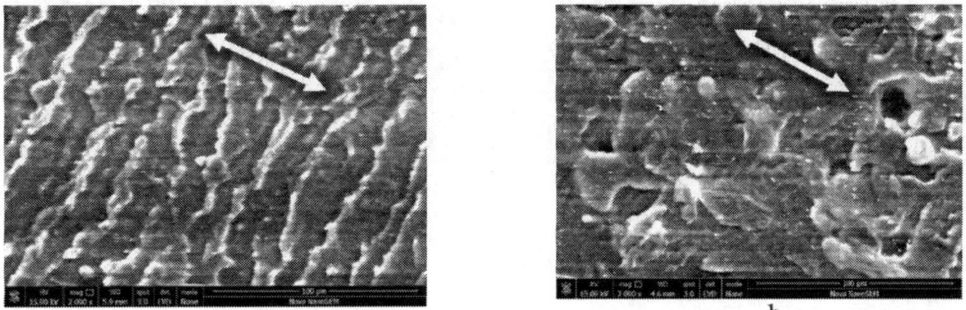

a b

Figure 4. Morphology of friction tracks: a) NR, b) SBR. Arrows show the sliding direction.

The mechanism of COF variation versus time can be illustrated assuming the instantaneous presence of two regions inside of friction track. One is the waved area (zone 1) located along the middle of track and the second (zone 2) surrounding the zone 1 and has the COF higher than that of the waved area. Abrasion process changed the ratio between the widths of the aforementioned areas

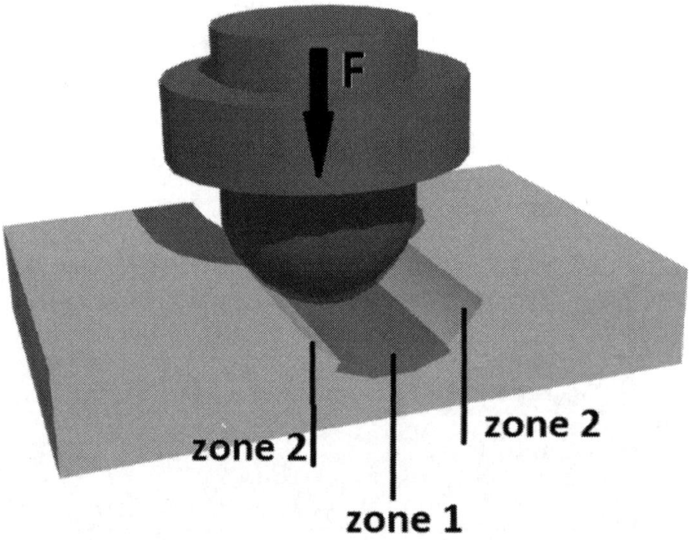

Figure 5. Scheme of bi-zonal mechanism of friction in ball-on-plate mode.

(see figure 5) and led to slight decrease in COF. On the opposite side, COF of PP slightly increased from 0,15 to 0,30 during 300 cycles (figure 3). Noticeably, it is almost 10 times lower than COF of studied rubbers. Adding of 5 % of rubber in PP matrix did not affect significantly the COF as well as the topology of the abraded surfaces (figure 6). Friction tracks topology is characterized by small cracks formed in PP matrix, which are aligned perpendicular to the sliding direction of counterbody. It was noticed that adding of small quantity of rubber (5 %) decreased the crack formation and propagation in PP matrix.

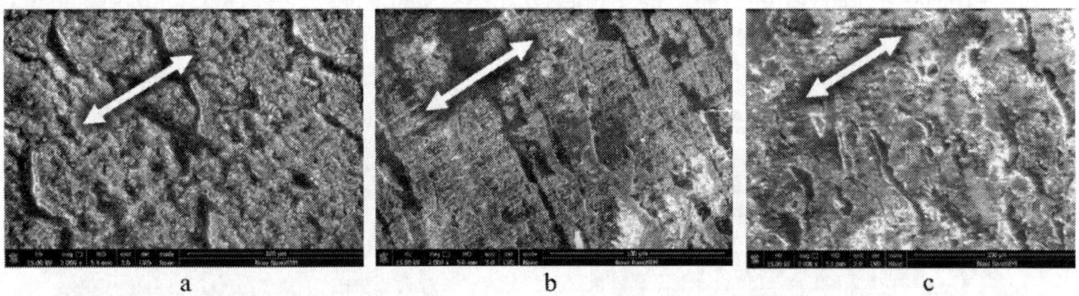

Figure 6. Topology of tracks. a) PP, b) PP/5NR, c) PP/5SBR. Arrows show the sliding direction.

A PP film was developed from rupture within the polymer and adhered to the friction track. This process as well as the film transfer to the counterbody are characteristic for most of the polymers except the highly cross linked and some glassy polymers [9]. According to article [10], the transfer film from polymers has a tendency to adhere on metal surfaces and influence the friction and wear. In the present study, PP film creation inside of tracks, decreased the influence of rubber filler on COF and wear in PP/rubber blends. Due to weaker crosslinking between rubber particles and PP matrix relative to crosslinking inside of rubber particles and inside of PP, such particles were easily rubbed or compressed comparatively with PP matrix and facilitated the PP film creation. Blends containing 25 % of rubber show COF ranged between 0.3 and 0.4. 50 % of rubber in PP matrix increased COF up to 0.63, which was still almost 3.5 times lower than COF of pure rubbers. Morphology of tracks was still dominated by PP film creation (figure 7), but was quite different for blends containing NR (figures 7 a, c) and blends containing SBR (figures 7 b, d). For ball-on-flat contact the initial conditions changed over time, due to changes in the topography, microstructure and composition of the material.

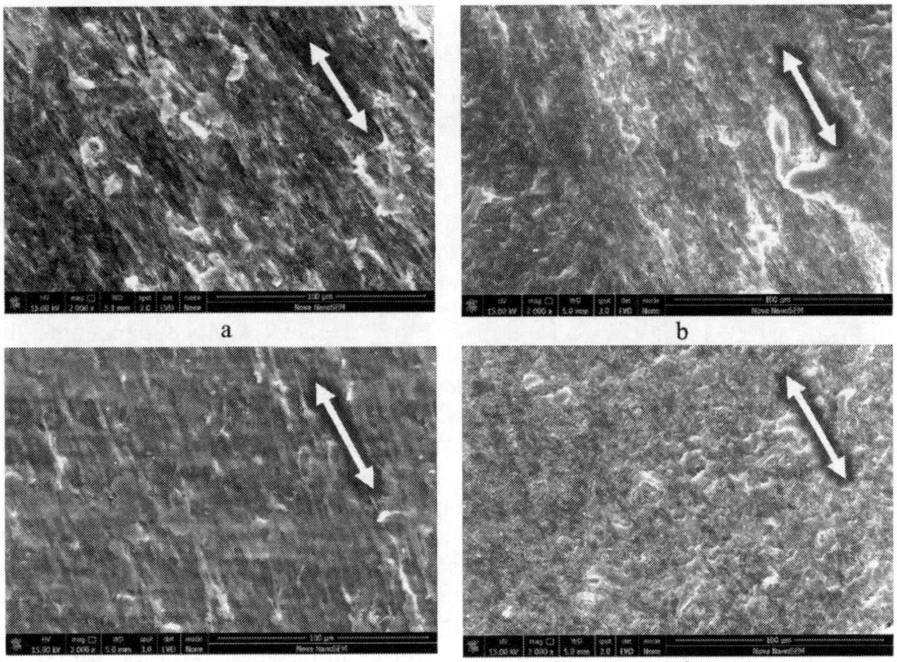

a b
c d

Figure 7. SEM images of friction tracks. a) PP/25 NR, b) PP/25 SBR, c) PP/50 NR, d) PP50/SBR. Arrows show the sliding direction.

The analyses of average value of COF during the last 5 cycles from 300 cycles exhibited a general trend of friction variation dependent on rubber concentration in TPVs as well as the nature of rubber (figure 8). An increase of rubber concentration in PP matrix showed a continuous increase of COF, the differences between NR and SBR filled samples being insignificant.

Friction wear usually is a combination of several parameters which include the contact pressure, temperature, and sliding velocity [11]. The wear of polymers is a complex process which clearly is related to the formation of the transfer film. Chemical reactions in the contact spots and additives in the polymer composition affect the tribological properties [10]. The wear of samples was analysed measuring widths of tracks in optical microscope. The obtained results are presented in figure 9. In order to compare the wear of samples, the simple width of track measurements were used on optical microscope. The determination of wear in viscoelastoplastic materials like TPVs is sometimes very difficult. If the wear is defined like a volume of material which has lost its initial properties during friction process, then the calculation of the real volume of such a material may be a great challenge. This is because besides detached material there is also present abraded surface with unknown depth. In this case, according to the authors of the present study, the most suitable method was found to be the widths of track measurement. The leading wear mechanism in the prepared blends was adhesive wear related to the formation, shearing and rupture of the adhesive junction in the contact spots and the material transfer from one surfaces to another. The film transfer from the polymer was continuously created and smeared out. With increasing of rubber concentration wear increased, but contradictory to COF, wear changed drastically depending on nature of fillers. For example, in case of 50 % rubber in PP, width of track for NR blend was two times larger than that for SBR blend.

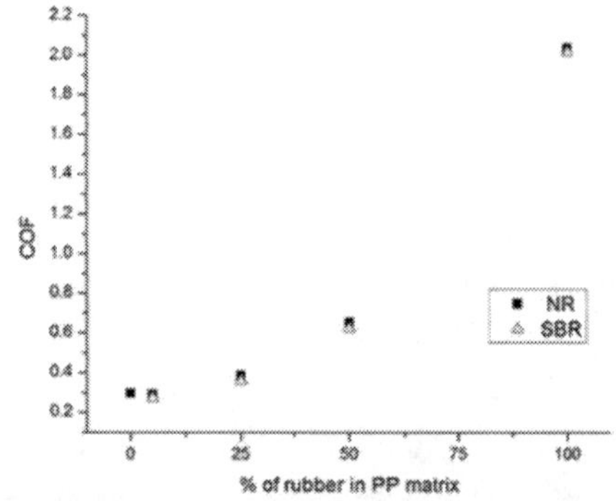

Figure 8. COF of TPVs versus rubber concentration in TPVs.

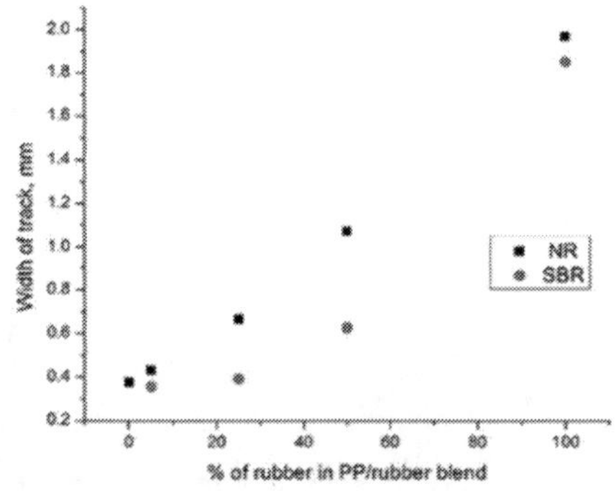

Figure 9. Width of tracks after 300 cycles vs rubber concentration in TPVs.

Hardness measurements usually supplement the tests which determine the mechanical properties of materials. Hardness is the property of a material to resist to scratching, perforation or deformation and usually expresses the deformability of materials under concentrated load action. The Shore hardness was quantified by penetration of steel rod, shaped like a truncated cone, into surface and is related to the modulus of elasticity and viscoelastic properties of the material. The adding of rubbers in PP matrix commonly modifies physical and mechanical properties of blends. There are the results of Shore D hardness measurements for investigated samples in figure10. Hardness almost linearly decreased with increasing of rubber concentration, showing the maximum of 64 Shore degree for pure PP, where the minimum was found for pure rubbers: 30 for NR and 32 for SBR. The difference of hardness between NR and SBR filled blends having the same rubber concentration was stated by slight decreasing in hardness for PP/NR blends.

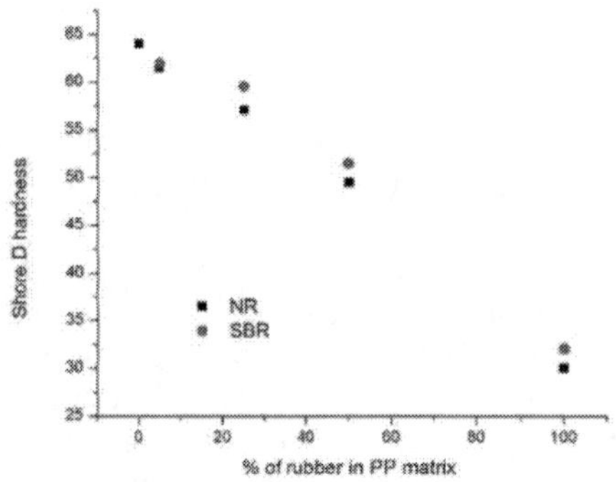

Figure 10. Shore D hardness vs rubber concentration in the PP matrix.

The relationship between hardness and wear are quite difficult to be computed. It was observed that samples having higher hardness showed lower values of wear.

4. Conclusions

Friction coefficient of PP/NR, PP/SBR TPV materials increased with increasing of rubber concentration in PP matrix.

Wear of investigated materials increased with increasing content of rubber and also depended of the rubber type.

Even a small quantity of rubber decreased the crack formation and propagation in PP matrix during the friction process.

Hardness results contributed to the understanding of wear trends in investigated materials, where samples having higher hardness showed lower values of wear.

Acknowledgements

This article was written with the support of Operational Program Research and Development for Innovations co-funded by the European Regional Development Fund (ERDF) and national budget of the Czech Republic, within the framework of the project CPS - strengthening research capacity (reg. number: CZ.1.05/2.1.00/19.0409) and supported by the Ministry of Education, Youth and Sports of the Czech Republic – Program NPU I (LO1504).

Reference

[1] Thomas S, Gupta B R and De S K 1987 *J. of Materials Science* **22** pp 209-3216
[2] Naskar K, Gohs U, Wagenknecht U and Heinrich G 2009 *eXPRESS Polymer Letters* **3** pp 677–683
[3] Radusch H-J and Pham T 1996 *Kautschuk Gummi Kunststoffe* **49** p 249
[4] Alonso I 2013 *TPE Magazine* **2** pp 106-110.
[5] Heng Yao Lin, Holmberg K and Säynätjoki M 1995 *Lubrication Science* **12** pp 39–49
[6] Dae-Hyun Cho, Bhushan B and Dyess J 2016 *Tribology International* **94** pp 165–175
[7] Bhushan B and Gupta B K 1991 *Handbook of tribology: materials, coatings, and surface treatments* (McGraw-Hill, New York) p 1168
[8] Jacko M G et al 1989 *Wear* **133** pp 23-38
[9] Bhushan B 2013 *Introduction to tribology* (Wiley, New York) p 738
[10] Brostow W, Datashvili T and Geodakyan J 2012 *Polym. Int.* **61** pp 1362–1370
[11] Karger-Kocsis J, Felhös D and Thomann R 2008 *J. Appl. Polym. Sci.* **108** pp 724–730
[12] Sivaraos, Yap T, Qumrul, Amran M, Anand T, Izamshah R and Aziz A 2013 *Procedia Engineering* **68** pp 743 – 749
[13] Harea E, Lapsker I, Laikhtman A and Rapoport L 2013 *Tribol. Lett.* **52** pp 205–212
[14] Belinski D 2007 *Archives of Civil and Mechanical Engineering* **4** pp 15-26

IOP Conf. Series: Journal of Physics: Conf. Series **843** (2017) 012071 doi:10.1088/1742-6596/843/1/012071

Effects of Material Combinations on Friction and Wear of PEEK/Steel Pairs under Oil-Lubricated Sliding Contacts

T Akagaki[1], T Nakamura[2], Y Hashimoto[2] and M Kawabata[3]

[1] Department of Industrial System Engineering, National Institute of Technology, Hachinohe College, 16-1 Uwanotai, Tamonoki, Hachinohe, Aomori 039-1192, Japan

[2] Department of Mechanical Engineering, National Institute of Technology, Hachinohe College, 16-1 Uwanotai, Tamonoki, Hachinohe, Aomori 039-1192, Japan

[3] TRIBOTEX Inc., 45-7 Yamaguchi, Nagakusa-cho, Obu, Aichi 474-0052, Japan

E mail: akagaki-m@hachinohe-ct.ac.jp

Abstract. The effects of material combinations on the friction and wear of PEEK/steel pairs are studied using blocks on a ring wear tester under oil-lubricated conditions. The rings are made of forged steel (SF540A) and a PEEK composite filled with 30 wt% carbon fibre. The surface roughness is 0.15 and 0.32 μm Ra, respectively. The blocks are also made of the same materials as the rings: the forged steel and the PEEK composite. Finished with an emery paper of #600, the surface roughness is 0.06 and 0.23 μm Ra, respectively. Sliding tests for 4 combinations of two materials are conducted. The load is increased up to 1177 N at 1 N s^{-1}. The sliding velocity is varied in the range of 10 to 19 m s^{-1}. In some cases, the ring temperature is measured with a thermocouple with a diameter of 0.5 mm, located 1 mm below the frictional surface. Results indicate that the forged steel's ring and the PEEK composite's block is the best combination among 4 combinations, because seizure does not occur under the increasing load up to 1177 N at the sliding velocity of 10–19 m s^{-1}. In contrast, seizure occurs at 15 and 19 m s^{-1} in the other three combinations. However, the PEEK composite's ring shows a lower friction coefficient as compared to the forged steel's ring, when seizure does not occur. Wear scars are observed with a scanning electron microscope (SEM). The seizure mechanisms are then discussed.

1. Introduction

Poly-ether-ether-ketone (PEEK) is one of advanced engineering polymeric materials and has excellent mechanical and chemical properties [1]. It also has been well known that PEEK has superior tribological properties. To further improve PEEK's tribological behaviors, PEEK composites filled with various fillers and solid lubricants have been developed and studied [2-5]. Although numerous papers regarding PEEK's tribological properties have been published, most of them have been conducted under un-lubricated conditions and at low sliding velocities.

PEEK composites have recently been applied to various machine elements such as a thrust bearing [6], a pump bearing [7], a seal ring, and a retainer for a rolling bearing. These PEEK composites must operate under severe conditions such as high sliding velocities and high temperatures. At a high sliding velocity, lubricants are indispensable for maintaining PEEK's low friction and low wear for a long time.

However, there has been little work regarding the lubricated friction and wear of PEEK materials at a high sliding velocity [8, 9]. Furthermore, the effects of the mating materials sliding against PEEK materials on the tribological behavior have not been studied so far. Our previous works showed that PEEK materials transit from hydrodynamic lubrication or mixed lubrication to seizure under severe oil-lubricated conditions such as high sliding velocities and high loads [10, 11]. In order to increase PEEK applications in practice and establish maintenance guidelines for PEEK sliding bearings, it is important to study PEEK's friction and wear behaviors under various experimental conditions.

In this study, the friction and wear behaviors of PEEK/steel pairs were studied from the viewpoint of material combinations under oil-lubricated sliding contacts. During the test, the steel ring's temperature was measured. Wear scars were observed using a scanning electron microscope (SEM). Based on the SEM observation results and the measurement results of steel ring's temperature, the seizure mechanisms were discussed.

2. Experimental Apparatus and Procedure

Experiments were conducted with blocks on the ring wear tester shown in figure 1 [10]. The testing materials are listed in table 1. The rings were made of forged steel (SF540A) and a PEEK composite filled with 30 wt% carbon fiber, with 130 mm diameter and 20 mm thickness; it was shaped by turning and further finished by cylindrical grinding or grinding with an emery paper. The carbon fiber was 8 μm in diameter and its length ranged from 50 to 200 μm. The ring's surface roughness was 0.15 and 0.32 μm Ra, respectively. The blocks were also made of the same materials as the rings. The blocks were 90 mm long, 10 mm wide, and 10 mm thick. They were finished with an emery paper of #600, and their surface roughness was 0.06 for the steels' block and 0.23 μm Ra for the PEEK composite's block. The experiments were conducted for four combinations of two materials at each sliding velocity. In this paper, the expression of "SF540A/PEEK composite" stands for the combination of the SF540A ring and the PEEK composite's block. The experimental conditions are listed in table 2. The sliding velocity ranged from 10.2 to 19 m s^{-1}, and the load was increased at a rate of 1 N s^{-1} up to 1177 N. During the test, a frictional torque was measured with a torque meter. The steel's ring temperature was measured with an alumel-chromel thermo-couple with a diameter of 0.5 mm, located 1mm below the frictional surface. The maximum testing load was 1177 N. However, when the frictional torque increased suddenly, the tests were ended. The lubricant used was a non-additive turbine oil (ISO VG46). It was supplied to the frictional surface at a flow rate of 1.1 ml s^{-1} with a pump. The oil temperature was kept at 30 ± 3°C with a controller.

After the test, wear scars were observed with a scanning electron microscope (SEM).

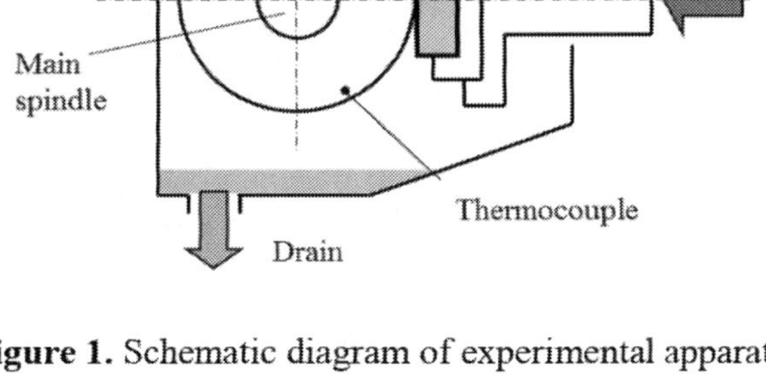

Figure 1. Schematic diagram of experimental apparatus.

Table 1. Properties of testing materials.

	Material	Hardness	Surface roughness, Ra (μm)
Ring	Forged steel (SF540A)	HV189 ± 8	0.15 ± 0.04
	PEEK composite	HRR124	0.32 ± 0.03
Block	Forged steel (SF540A)	HV189 ± 8	0.06 ± 0.02
	PEEK composite	HRR124	0.23 ± 0.04

Table 2. Experimental conditions.

Sliding velocity (m s^{-1})	10.2, 15.0, 19.0
Load (N)	Up to 1177 N at increasing rate of 1 N s^{-1}
Lubricant	Non-additive turbine oil (ISO-VG46)
Flow rate (ml s^{-1})	1.1
Oil temp. (°C)	30 ± 3

3. Results and Discussion

3.1. Friction property

Figure 2 illustrates the relationship between the friction coefficient and the load at 10.2 m s^{-1}. Figure 3 illustrates the relationship between the steel's ring temperature and the load, corresponding to figure 2. The friction coefficient decreases with the increase of the load and becomes almost constant and low values for all the material combinations. The steel's ring temperature increases linearly with the increase of the load, and reaches about 60°C at the end of testing. It does not become constant. However, the friction coefficients at the steady state become constant, and its values depend on the material combinations. Their values are 0.0109 ± 0.0022 for SF540A/SF540A, 0.007 ± 0.002 for SF540A/PEEK composite, 0.0055 ± 0.0023 for PEEK composite/SF540A, and 0.0043 ± 0.0015 for PEEK composite/PEEK composite. The PEEK composite's ring shows the lower friction coefficient than the steel's ring. Among four combinations, PEEK composite/PEEK composite shows the lowest value.

A lambda ratio [12] is calculated for discussion of the results. It is defined as $\Lambda = h_{min} / (R_{q1}^2+R_{q2}^2)^{1/2}$, where h_{min} is the minimum film thickness and R_{q1} and R_{q2} are the root mean square roughnesses of the ring and block, respectively. The minimum film thickness is calculated using the Dowson-Higginson formula [13]. Oil viscosity is calculated at the ring temperatures obtained from each test, and R_{q1} and R_{q2} are calculated from the arithmetic average roughness, R_a, where $R_q = 1.25\ R_a$ [14]. The lambda ratio is calculated from the initial surface roughness. Their values for all the material combinations are mostly more than 3 in the temperature ranges shown in figure 3, which suggests that hydrodynamic lubrication is predominant and low friction coefficients are maintained. The minimum film thickness calculated in both rings of the steel and the PEEK composite was almost the same values of about 1.0 μm under the conditions of 10.2 m s^{-1}, 1177N, and the oil temperature of 60°C. Thus the minimum oil thickness calculated cannot explain the reason why the PEEK composite's ring gives lower friction coefficients. As the width of the PEEK composite's ring is wider than that of the block, it is supposed that the ring's surface might deform elastically and could easily hold the oil between two surfaces. This result is very important, and further works are the next subject in future.

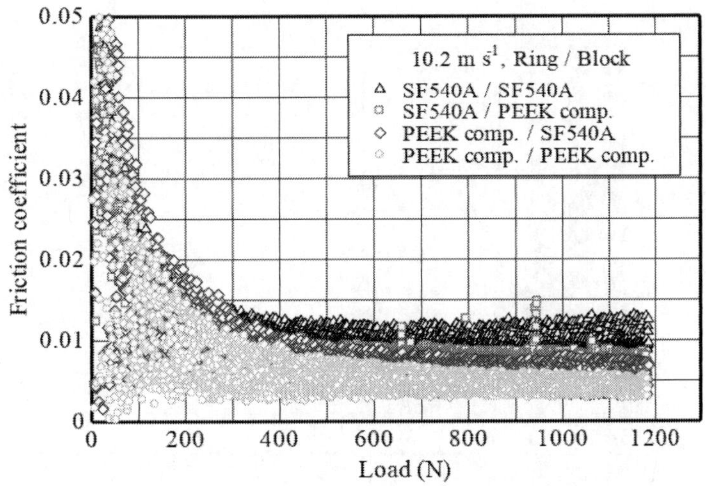

Figure 2. Relationship between friction coefficient and load at 10.2 m s^{-1}.

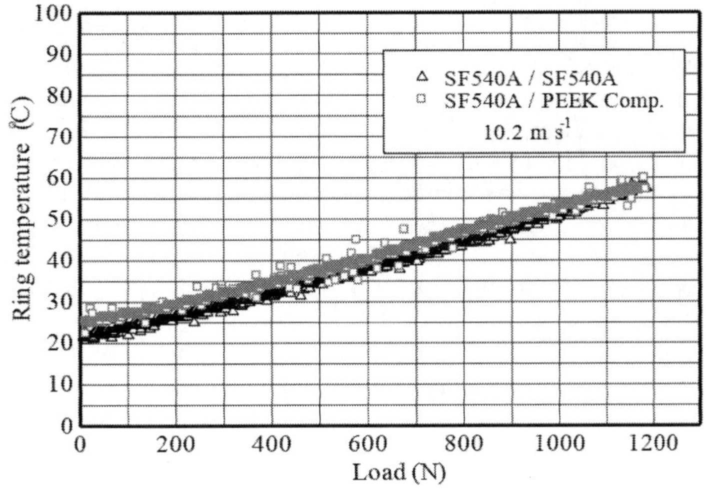

Figure 3. Relationship between ring temperature and load at 10.2 m s^{-1}

Figure 4 illustrates the relationship between the friction coefficient and the load at 19.0 m s^{-1}. It is apparent that the friction behaviors are strongly dependent on the material combinations. Except for SF540A/PEEK composite, the seizure occurs in other three material combinations as the load exceeds a certain value. The critical load is defined as the seizure load. In SF540A/SF540A, the friction coefficient rapidly increases from 0.015–0.02 to 0.27 at 270 N. In PEEK composite/SF540A and PEEK composite/PEEK composite, its values rapidly increase from 0.005–0.008 to 0.023 at 500 N and from 0.005–0.01 to 0.026 at 290 N, respectively. The friction coefficients, when the seizure occurs in the PEEK composite's ring, are significant lower values as compared to SF540A/SF540A. When the seizure occurred in the PEEK composite's ring, however, oily smokes together with noise were generated. It has been confirmed that the oily smoke is generated at the ring temperature of 140–160°C [12]. Thus it is said that the seizure in the PEEK composite' ring has much to do with the high temperature.

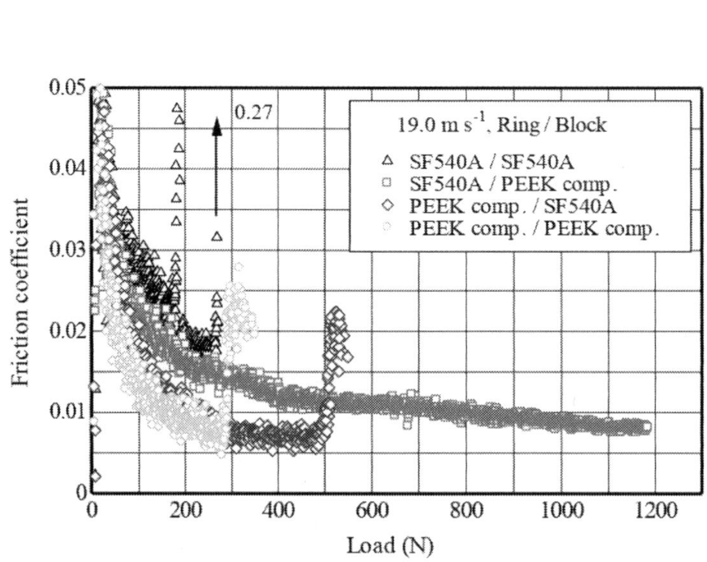

Figure 4. Relationship between friction coefficient and load at 19.0 m s^{-1}.

3.2. Comparison of seizure load

Figure 5 shows the seizure load obtained at each sliding velocity for all the material combinations. At 10.2 m s^{-1}, the seizure does not occur for all the material combinations. However, as the sliding velocity increases over 15.0 m s^{-1}, except for SF540A/PEEK composite, the seizure occurs for other three combinations. Their seizure loads decrease as the sliding velocity increases. With higher sliding velocity, there are clear differences in the seizure loads between the material combinations. The order of the seizure load at 19.0 m s^{-1} is as follows: SF540A/PEEK composite >> PEEK composite/SF540A > PEEK composite/PEEK composite, SF540A/SF540A.

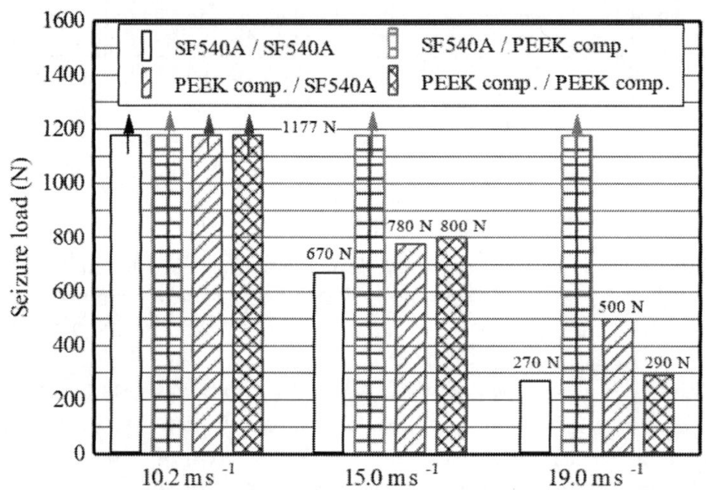

Figure 5. Seizure load at each sliding velocity for all material combinations.

In SF540A/PEEK composite, the seizure does not occur within the tested load range at 10.2–19.0 m s^{-1}. Figure 6 illustrates the relationship between the friction coefficients, the steel ring's temperature, the lambda ratio, and the load at 19 m s^{-1} for SF540A/PEEK composite. Similar to the results of 10 m s^{-1}, although the friction coefficient becomes a small and constant value of less than 0.01, the ring temperature increases linearly up to about 90°C and does not become constant. The lambda ratio is mostly more than 3, which suggests that the hydrodynamic lubrication is predominant, resulting in a low friction coefficient. Thus it is said that this pair has a high ability to maintain the hydrodynamic lubrication, leading to a superior seizure-resistant property.

The self-mated pair of SF540A has the lowest seizure load at 15.0 and 19.0 m s^{-1}, as shown in figure 5. They are about 670 and 270 N, respectively. Figure 7 illustrates the relationship between the friction coefficient, the ring temperature, the lambda ratio, and the load at 19.0 m s^{-1} in SF540A/SF540A. Although the friction coefficient increases instantaneously up to 0.1 at 30°C, eventually the seizure occurs at 40°C, and the friction coefficient rapidly increases from 0.02 to 0.27. The lambda ratio calculated at 40°C, which is the ring temperature for transiting to seizure, is much higher than 3. This value does not agree with the result at all. Considered the surface roughness after seizure, it becomes less than 1 when the ring temperature is 40°C or so. As the Hertz contact pressure between metals is higher than polymers, it is supposed that the surface roughness might become rough due to wear during a running-in process. If so, it is also expected that wear particles might be generated and remain on the surface. Even when the seizure does not occur, it has been confirmed that oil sample collected from the drain became a little dark because it contained small wear particles. Based on this idea, even if the lambda ratio calculated is high, wear particles and rough surface probably promote adhesion between metals at high sliding velocity, leading to the seizure.

When the seizure occurs in the PEEK composite's ring, oily smokes and loud noise were generated. This fact is the evidence that the surface temperature reaches or exceeds 140–160°C. The PEEK composite has a glass transition point of 143°C, above which it becomes soft. Therefore the friction coefficient increases because the contact areas between the PEEK composite's pair or the PEEK composite and the steel increases. This high temperature causes severe plastic deformation and flow of the PEEK composite, leading to severe tearing fracture [11, 12]. The seizure load at 19.0 m s^{-1} in PEEK composite/SF540A is higher than in the self-mated pair of the PEEK composite. This is probably because the mated steel block has a higher heat conductivity than the PEEK composite's block, and the increasing rate of the surface temperature is relatively low.

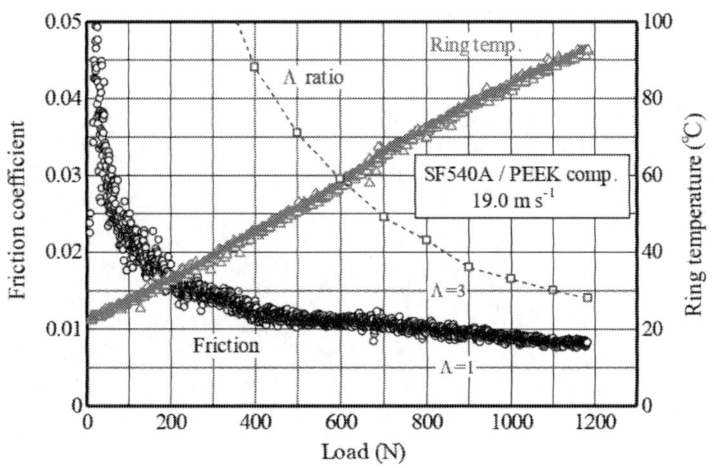

Figure 6. Relationships between friction coefficient, ring temperature, lambda ratio, and load at 19.0 m s^{-1} for SF540A/PEEK composite.

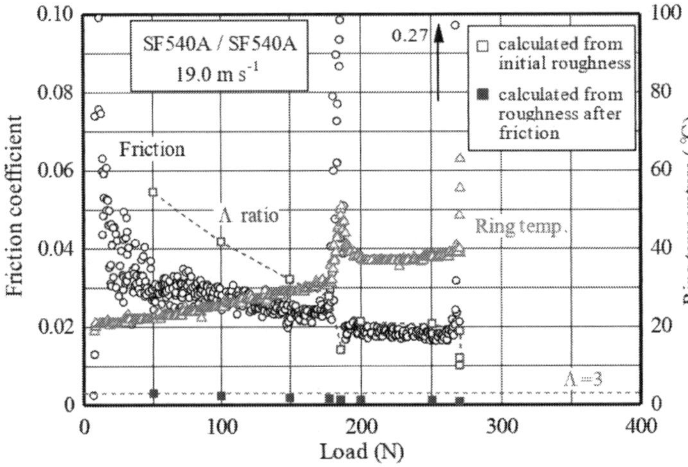

Figure 7. Relationships between friction coefficient, ring temperature, lambda ratio, and load at 19 m s^{-1} for SF540A/SF540A.

3.3. *SEM observations of wear scars*

Figure 8 shows the SEM micrographs of the ring and block's wear scars generated when the seizure occurs at 19.0 m s^{-1} in SF540A/SF540A. It is apparent that both the ring and block's surfaces suffer severe damages. The surface layers, probably formed due to the plastic flow and the transfer of materials, separate as plates less than 100 μm or so in size, as shown in figures 8 (a) and (b). It is well known that these plate-like wear particles are commonly generated in a seizure of steels [15]. Replaced the block material of SF540A with the PEEK composite in SF540A/SF540A, the low friction coefficient of less than 0.01 was maintained as shown in figures 4 and 6. Figure 9 shows the optical microscope and SEM micrographs of the wear scars in SF540A/PEEK composite. The ring's wear surface does not change appreciably and the original grinding marks remain, as shown in figure 9 (a). The block's wear scar has evidences showing slight contacts as small ripple marks are commonly observed, as shown in figures 9 (b) and (c). The original grinding grooves and carbon fiber fillers do not change and remain on the surface. These ripple marks are probably formed during the plastic flow process due to the mild plowing actions of the ring's surface asperities during the running-in period.

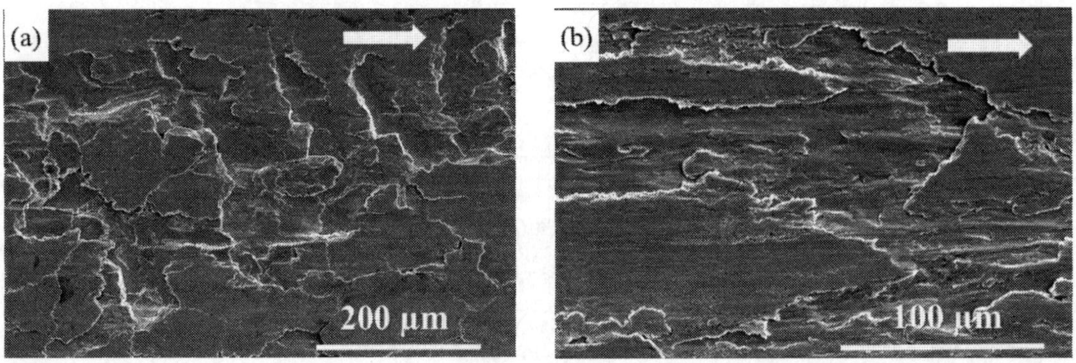

Figure 8. SEM micrographs of wear scars of ring and block generated in seizure of SF540A/SF540A at 19.0 m s^{-1}. (a) Ring; (b) Block. The arrow indicates the relative direction of motion of the counterface.

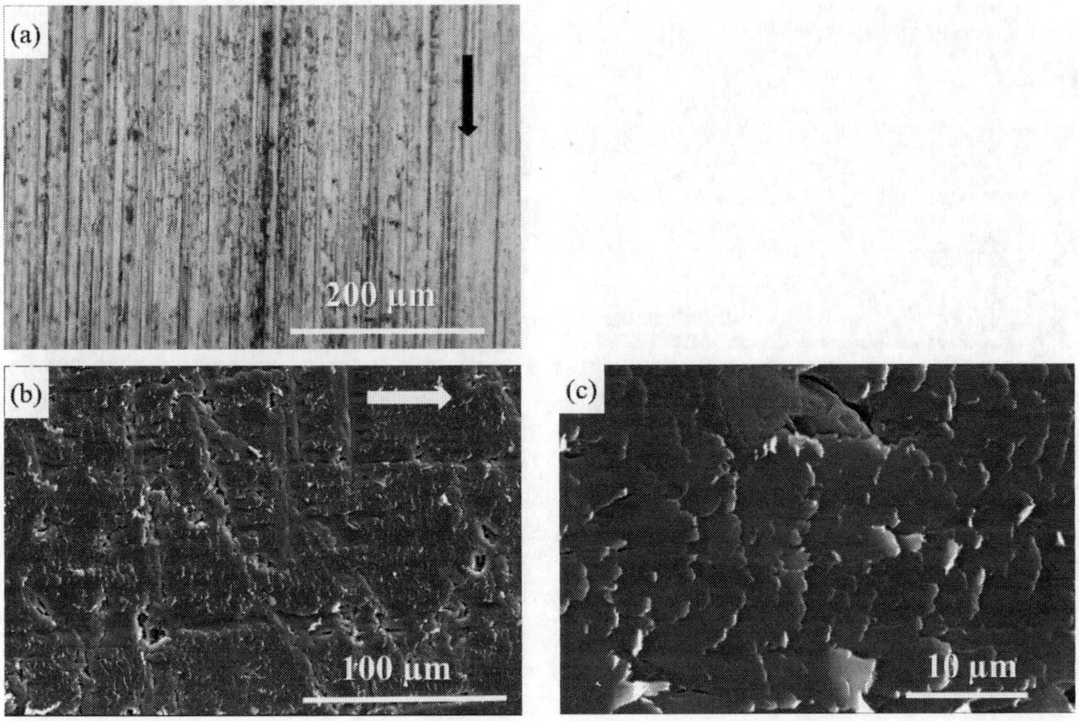

Figure 9. Optical microscope and SEM micrographs of wear scars of ring and block when low friction coefficient is maintained in SF540A/PEEK composite at 19.0 m s^{-1}. (a) Ring; (b)–(c) Block. The arrow indicates the relative direction of motion of the counterface.

Figure 10 shows the optical microscope and SEM micrographs of the ring and block's wear scars generated in the seizure of PEEK composite/SF540A at 19.0 m s^{-1}. Only the ring surface suffers the severe damages in the seizure, as shown in figures 10 (a) and (b). Regions look white in (a) are the original surface of the PEEK composite' ring. There is only a few original surface left. The severe damages are over a wide range and the worn surface looks very rough. In contrast, the block surface of SF540A does not suffer severe damages although slight and regular grooves are observed, as shown in figures 10 (c) and (d).

Figure 11 shows the optical microscope and SEM micrographs of the ring and block's wear scars generated in the seizure of PEEK composite/PEEK composite at 19.0 m s^{-1}. Both the ring and block's surfaces suffer severe damages. The ring's worn surface is very rough, and it seems like the original surface mostly disappears and the ring surface's fracture occurs, as shown in figures 11 (a) and (b). The block's surface is also rough, and severe damages are observed over a wide range, as shown in figure 11 (c). The separations of the large plate-like fragments and the exposure of carbon fibers are common features, as shown in figures 11 (c) and (d). Figures 11 (e) and (f) are the scars after the large fragments separate from the surface. Many elongated strings which are typical features of tearing fracture are commonly observed, as shown in figure 11 (e). Porous structures which is probably formed in semi-melting or melting states are observed, as shown in figure 11 (f). These observation results show that the surface temperature of the PEEK composite rises to the high value of more than the glass transition point of 143 °C.

The seizure of SF540A/SF540A shows a significant higher friction coefficient as compared to that of the seizure in the PEEK composite's ring, as shown in figure 4. Based on these observations, it is probably because the resistance due to the adhesion between two metals is significantly higher than that due to the tearing fracture of the PEEK composites at high temperature.

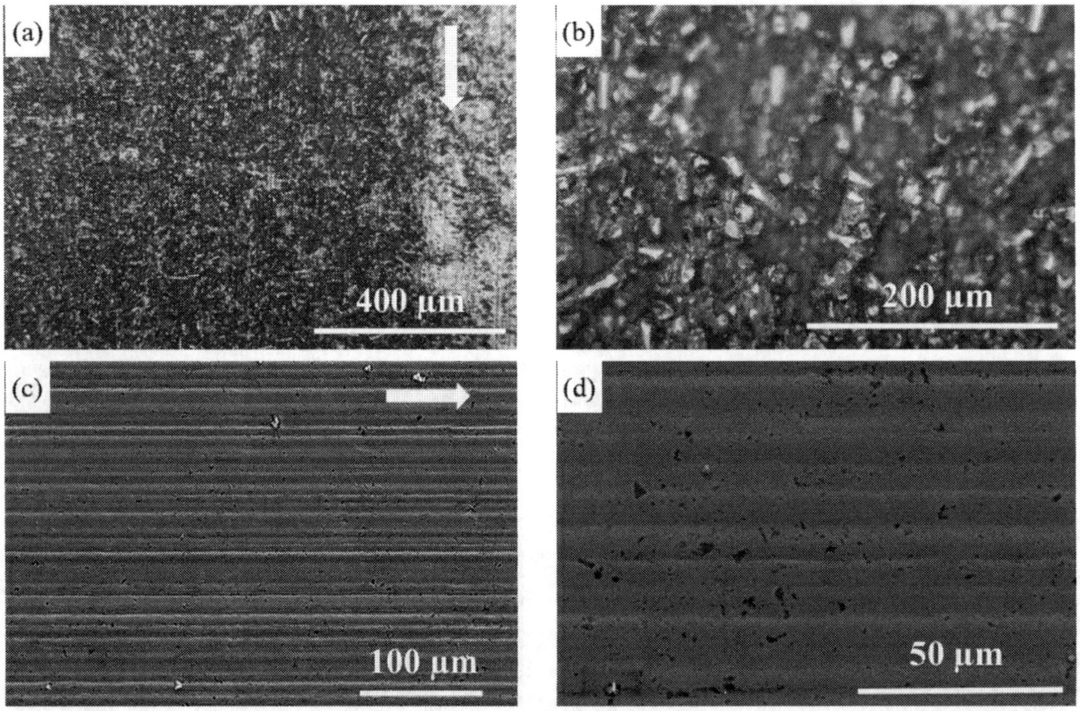

Figure 10. Optical microscope and SEM micrographs of wear scars of ring and block generated in seizure of PEEK composite/SF540A at 19.0 m s^{-1}. (a)–(b) Ring; (c)–(d) Block. The arrow indicates the relative direction of motion of the counterface.

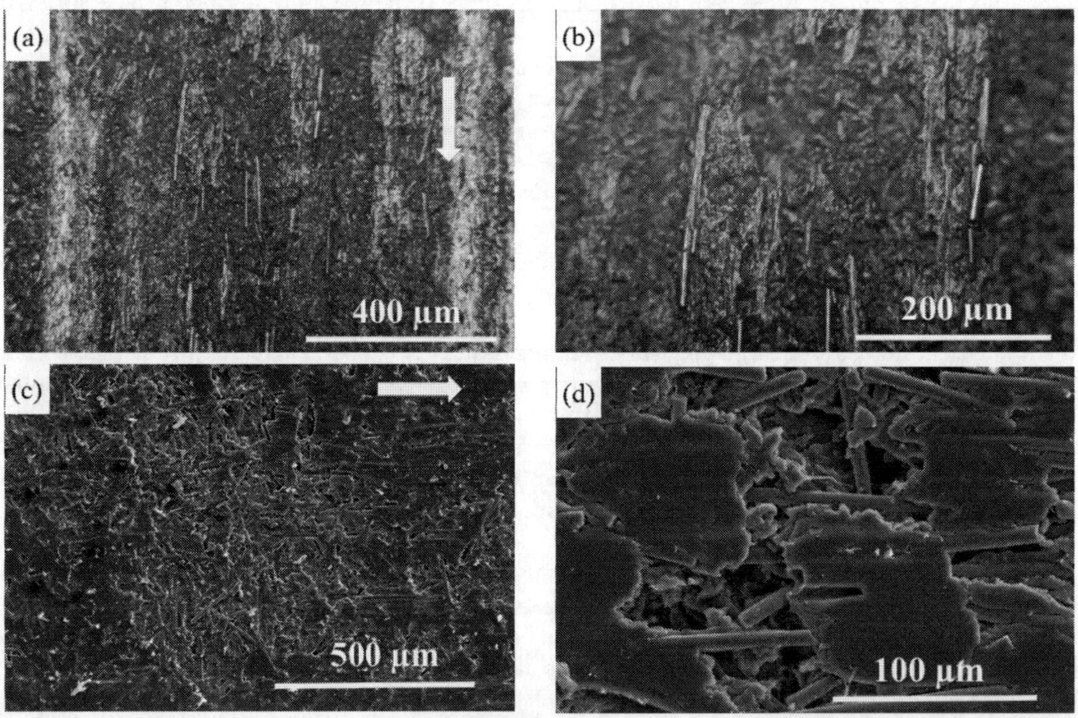

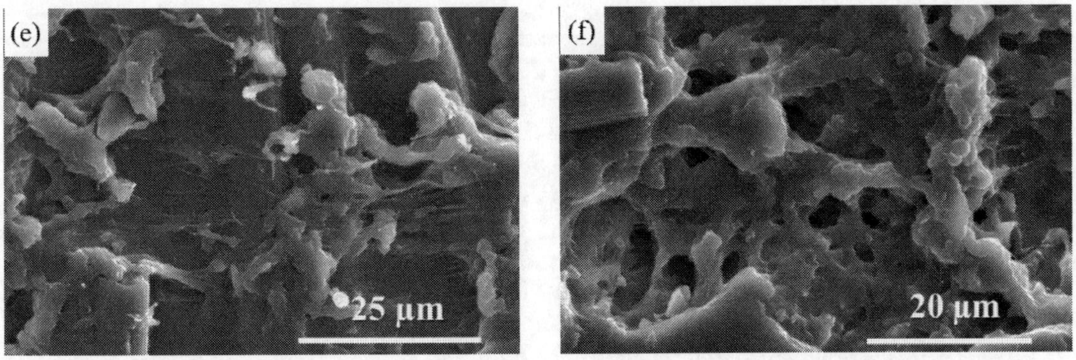

Figure 11. Optical microscope and SEM micrographs of wear scars of ring and block generated in seizure of PEEK composite/PEEK composite at 19.0 m s^{-1}. (a)–(b) Ring; (c)–(e) Block. The arrow indicates the relative direction of motion of the counterface.

In the combination of SF540A and the PEEK composite, there are clear differences in the seizure behaviors when the materials of the ring and the block are exchanged, as shown in figures 4, 5, 9, and 10. The seizure-resistant property of SF540A/PEEK composite is much better than that of PEEK composite/SF540A. The heat conductivity of SF540A is much higher than that of PEEK composite. The ring's surface area is larger than the block's one. As the heat conduction and the heat radiation performances of SF540A's ring are better than those of the PEEK composite's ring, low surface temperature seems to be maintained in SF540A/PEEK composite, leading to low friction and wear, as shown in figures 4 and 6. In PEEK composite/SF540A, the PEEK composite's ring temperature increases gradually, and the ring surface becomes soft when its temperature exceeds the glass transition point. The high temperature has been confirmed by the generation of oily smokes. As SF540A's block is harder than the PEEK composite's ring, it might be considered that SF540A's block bites into the PEEK composite's ring surface at the high temperature. As the result, the seizure occurs, and only the PEEK composite' ring suffers severe damages due to the tearing fracture, as shown in figure 10. These results indicate that the surface temperature plays an important role in the seizure behaviors of the PEEK composite.

4. Conclusions

The effects of material combinations on the friction and wear behaviors of PEEK/steel pairs were studied under oil-lubricated sliding contacts. The following conclusions were obtained.

(1) The combination of SF540A/PEEK composite (ring/block) showed the best seizure-resistant property among four combinations.

(2) The seizure load was dependent on the material combination. The combination of SF540A/SF540A was the worst.

(3) When the seizure did not occur, the PEEK composite's ring showed lower friction coefficients of less than 0.01 as compared to the forged steel's ring.

(4) The seizure, when PEEK composite's ring was used, was characterized by the ring's severe damages due to the tearing fracture of the surface layers.

References

[1] Japanese Society of Tribologists 1991 *Tribology in New Materials* (Tokyo: Yokendo Press)
[2] Lu Z P and Friedrich K 1995 *Wear* **181-183** 624
[3] Voort J V and Bahadur S 1995 *Wear* **181-183** 212
[4] Zhang Z, Breidt C, Chang L and Friedrich K 2004 *Tribology Int.* **37** 271
[5] Chang Li, Zhang Z, Ye L and Friedrich K 2007 *Tribology Int.* **40** 1170

[6] Matsueda Y, Tomizawa M, Ochiai T, Tamiya Y and Goto H 2000 *Mitsubishi-Electric Co. Technical Report* **11** 6
[7] Sugiyama K, Kimura Y, Zinno H, Takayama H and Kume S 2001 *Proc. JAST Ann. Conf. (Utsunomiya Japan)* 235
[8] Akagaki T, Kawabata M and Kato K 2002 *Proc. Int. Conf. on Hydro 2002* 61
[9] Akagaki T and Kawabata M 2016 *Tribology Online* **11** 494
[10] Akagaki T, Kataoka K and Kawabata M 2016 *Tribology Online* **11** 623
[11] Akagaki T and Kawabata M 2010 *Advanced Tribology* (Springer Berlin Heidelberg) 378
[12] Hutching I M 1992 *Tribology: Friction and Wear of Engineering Materials* (Elsevier Science & Technology)
[13] Yamamoto Y and Kaneta M 1998 *Tribology* (Tokyo: Rikogakusya Press)
[14] Hashimoto H 2006 *Tribology to learn from basics* (Tokyo: Morikita Press)
[15] Kayaba T, Kato K and Akagaki T 1986 *J. JSLE Int. Edi.* **7** 53

Wear Calculation Approach for Sliding - Friction Pairs

G Springis, J Rudzitis, J Lungevics and K Berzins

Institute of Mechanical Engineering, Riga Technical university, Viskalu street 36A, Riga, LV-1006, Latvia

Abstract. One of the most important things how to predict the service life of different products is always connected with the choice of adequate method. With the development of production technologies and measuring devices and with ever increasing precision one can get the appropriate data to be used in analytic calculations. Historically one can find several theoretical wear calculation methods but still there are no exact wear calculation model that could be applied to all cases of wear processes because of difficulties connected with a variety of parameters that are involved in wear process of two or several surfaces. Analysing the wear prediction theories that could be classified into definite groups one can state that each of them has shortcomings that might impact the results thus making unnecessary theoretical calculations. The offered wear calculation method is based on the theories of different branches of science. It includes the description of 3D surface micro-topography using standardized roughness parameters, explains the regularities of particle separation from the material in the wear process using fatigue theory and takes into account material's physical and mechanical characteristics and definite conditions of product's working time. The proposed wear calculation model could be of value for prediction of the exploitation time for sliding friction pairs thus allowing the best technologies to be chosen for many mechanical details.

1. Introduction

With the development of production technologies and materials the task concerning prediction of service life of fitted components and factors influencing it is of crucial value. Historically, several models of the wear process were developed which allow approximate prediction of the friction pairs service life. Due to the fact that the wear process is multiflorous it is affected by different parameters: the geometry of surface peaks (roughness, buckles, shape deviations, *etc.*), physically-mechanical conditions of the upper layer, material of contacting components, frictional heating during the wear process, wear conditions, etc.. These factors are impossible to take into consideration in the analytical calculation; therefore, the wear calculation development in the course of time went in different directions taking into account the main influencing factor complexes.

There are several wear calculation models based on the application of theories from several branches - the probability theory when predicting the fittings service life [6], the laws of classical physics [7], the analysis of various geometrical forms of fittings [8] and combined sliding-friction pairs calculation method [9], but none of these models contains technologically manageable surface roughness parameters.

The article deals with the wear calculation model in which the wear process is characterized by three stages seen in figure 1 [5].

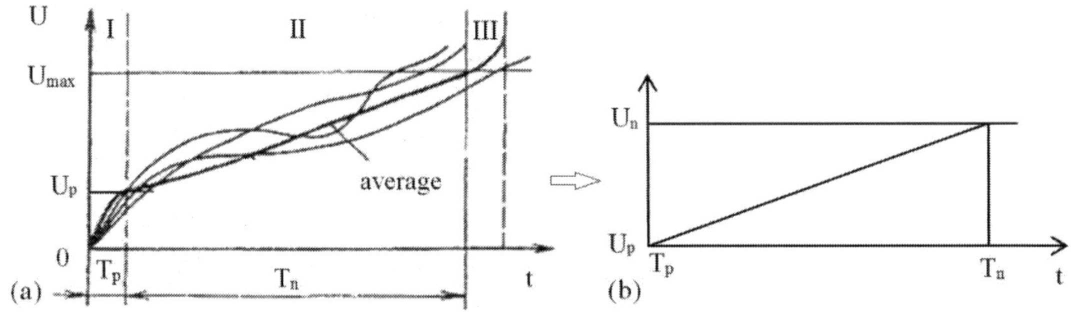

Figure 1. Wear U dependent from duration.

Figure 1 (a) shows three stage wear. I stage is the run in process, II stage is normal wear and III stage is the critical wear (b) shows the nominal wear period.

According to figure 1:

$$U_{\max} = U_p + U_n \tag{1}$$

where U_p and U_n is average wear value for run-in and normal exploitation stages.
In the given model run-in process wear (U_p) is determined experimentally by measuring it. It is so because during the run-in process there are large amount of parameters who are constantly changing and they cannot be calculated. For the offered model, run-in wear U_p will be determined experimentally and normal exploitation time wear U_n - by calculations.
That is the reason why this methodology is experimentally theoretical.
Previously mentioned facts reflects in figure 1 (b) where average wear in normal exploitation time is linear.
This process is described with equation:

$$U_n = h_{def} \frac{N_{cf}}{N_c} \tag{2}$$

where h_{def} is friction surface deformed layer height (see figure 2), N_{cf} is number of cycles that cams are exposed to, N_c is the number of cycles that surface cams can withstand before shearing off.

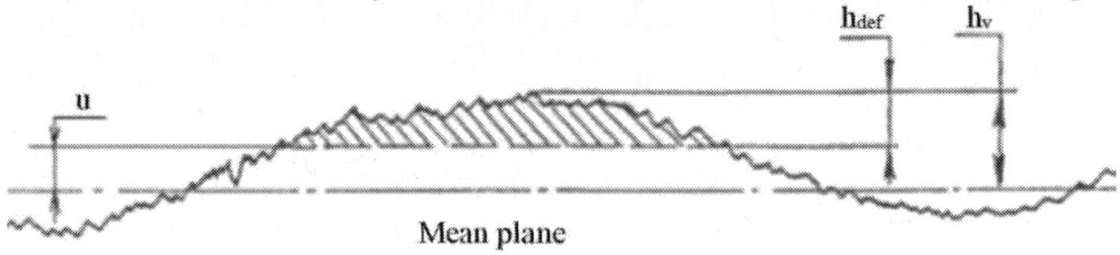

Figure 2. Potential wear particle separation from surface where u is unevenness deformation level, h_v - unevenness height.

2. Wear calculation principles

Wear calculations are based on following principles:

- Wear pair surface roughness is described with normal random field h(x,y) (Roughness characteristics).
- Wear process is conducted with the principles of the theory of fatigue (Surface fatigue wear principles).
- Roughness unevenness contact correspond to Hertz theory (Roughness contact).

2.1. Roughness characteristics

Random function is a function the value of which at each specified argument (or several arguments) value is a random size. When the random function depends on several parameters it is called a random field. The random field in case of threadbare surfaces can be normal, i.e. ordinates of such field are distributed according to the normal distribution law which is characterized by height parameter σ (standard deviation). Besides the random field can be uniform (stationary) and non-uniform. The random field is called stationary if its average size is constant and the correlation function depends on the difference between the surface points. The mean value of random field is constituted by a plane which can be called a mid-plane. Thus, to describe a normal random field the mathematical expectation of this field, standard deviation σ and correlation function should be known [1].
According to normal random field principles, surface characteristic parameters are calculated [1]:
Average arithmetic deviation Sa:

$$E\{Sa\} = \sqrt{\frac{2}{\pi}}\sigma \tag{3}$$

where $E\{Sa\}$ is mathematical expectation value.
Surface anisotropy coefficient c see figure 3 [2]:

$$c = \frac{E\{RSm_1\}}{E\{RSm_2\}} = \frac{E\{n_2(0)\}}{E\{n_1(0)\}} \tag{4}$$

Where RS_{m1} is a step perpendicular to the processing trace direction along midline and RS_{m2} is a step towards the processing trace along the midline, $n_1(0)$, $n_2(0)$ are the numbers of zeros in two mutually perpendicular directions of surface cuts x and y.

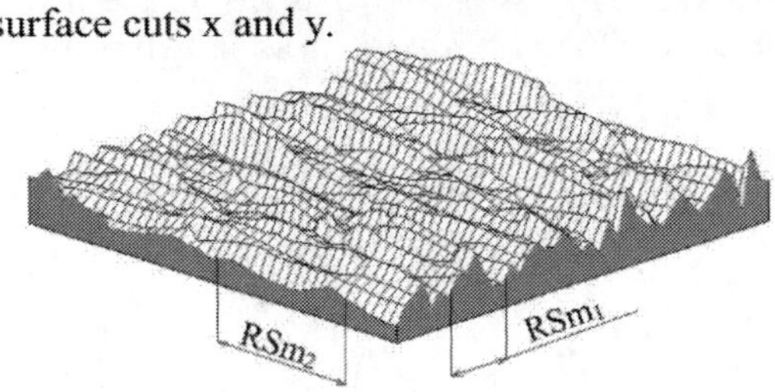

Figure 3. Irregular surface roughness spacing parameters.

Peak count per area :

$$E\{N\gamma\} \approx \frac{\pi E\{n_1(0)\}E\{\{n_2(0)\}}{2\sqrt{2\pi}} e^{-\frac{\gamma^2}{2}} \tag{5}$$

where γ is surface roughness relative deformation level ($\gamma = u/\sigma$).

2.2. Surface fatigue wear principles

Various independent researchers have shown that material fatigue is one of the main reasons that cause wear of the surfaces. Fatigue crack formation is inevitable due to the mutual movement of two body's that causes the fracture of the surface asperities. In order to better understand the process it is useful to take a look at two body contact model see figure 4.

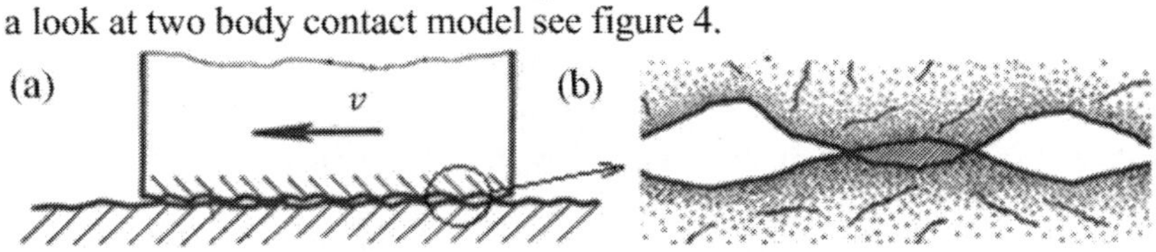

Figure 4. Two body asperity contact model. (a) is at macro scale and (b) is at micro scale.

According to Hamilton's theory [3] contacting surface asperities are exposed to tension witch is the cause of material upper layer collapse. If Huber - Mises - Genki`s criterion is used one can calculate material yield strength. The following figure shows tension field for one surface asperity where contact area ellipse semi-major axle is represented as a and the tension is represented as contact maximums pressure.

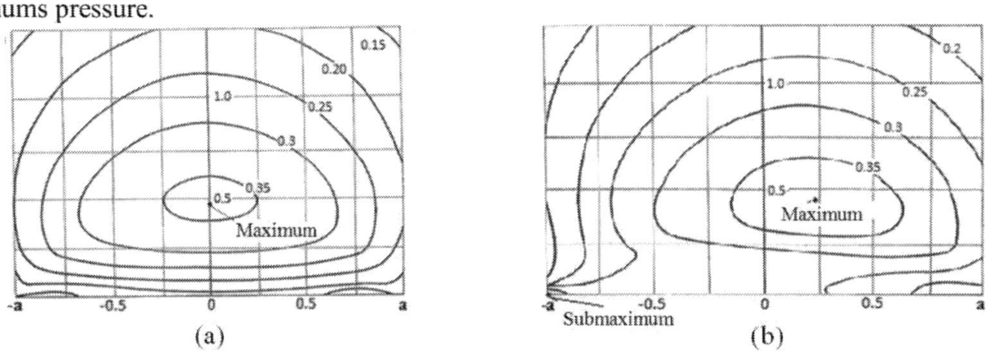

Figure 5. Surface tension field whit friction coefficient (a) , (b) .

Figure 5 shows that if coefficient of friction tension maximum is approximately located at distance 0.5a from the contact surface but if tension maximum moves towards the material surface. Besides that increase also causes formation of second (submaximum) pressure point see figure 5 (b). If contacting bodies move along each other, tension fields change their location.
As far as possibility that during wear process only two asperities are in contact is impossible one must analyse system as complex surfaces with many asperities as shown below.

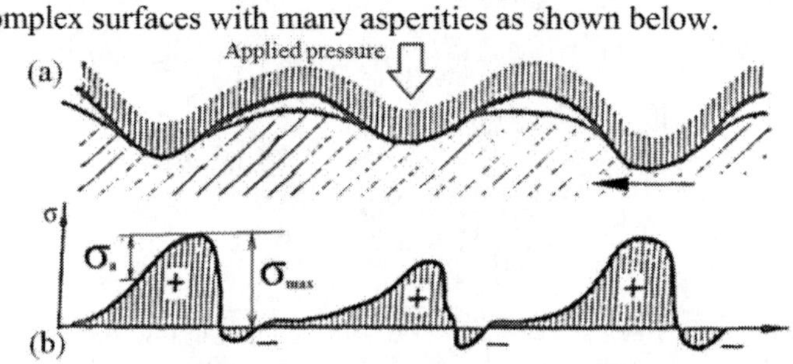

Figure 6. Irregular roughness surface where (a) asperity contact and (b) tension fields.

During such surface movement, every asperity, which is high enough to reach some other asperity from second surface, is deformed thus creating tension field. Figure 5 (b) shows that surface tension in such model changes according to asymmetric cycle. According to fatigue theory for asymmetric cycles, elastic limit σ_r is σ_0 because index r is calculated as follows:

$$r = \frac{\sigma_{\min}}{\sigma_{\max}} = \frac{0}{\sigma_{\max}} = 0 \tag{6}$$

Acquired equation allows calculating maximum amount of asperity bending cycles N_c until the material cracks [5]:

$$E\{N_c\} = \frac{N_0}{5m!} t_\sigma^m \tag{7}$$

where N_0 is number of asperity bending cycles until cracks, m is indicator of fatigue curve degree, t_σ is dimensionless tension ratio.
Parameters N_0, m and σ_0 are chosen as average values for specifically used material that receives tension.

2.3. Roughness contact.

Friction surface asperities projects as ellipses. According to Hertz theory [10]:

$$b_i = [\frac{3E(e)\theta P_i}{2(1-e^2)H}]^{\frac{1}{3}} \tag{8}$$

$$a_i = b_i(1-e^2)^{\frac{1}{2}} \tag{9}$$

$$\alpha_i = \frac{3}{2}K(e)\frac{\theta P_i}{b_i} \tag{10}$$

where, $K(e)$, $E(e)$ are first and second degree elliptical integrals, a_i, b_i are elliptical contact area semi-minor and semi-major axis, α_i is deformation of i asperity, e is eccentricity of elliptic contact area ($e = \sqrt{1-c^{\frac{8}{3}}}$), where c is anisotropy coefficient (see equation (4)),
θ – deformation level ($\theta = \frac{1-\mu^2}{\pi E}$), where E – modulus of elasticity, μ - Poisson's ratio, is average curvature of asperities:

$$H=(k1+k2)/2) \tag{11}$$

where , $k1$, $k2$ are main asperity curvatures.

3. Summary of wear calculations

Based on previously viewed principles and equation (2) linear wear can be determined:

$$E\{h_{def}\} \approx E\{V_i\}E\{N_\gamma\} \approx \frac{\sigma}{6\pi\gamma^2} \tag{12}$$

where $E\{V_i\}$ is separated volume average value for one i-asperity.

$$E\{V_i\} \approx \frac{\sigma}{\gamma^2 \pi E\{n_1(0)\}E\{n_2(0)\}} \tag{13}$$

Based on literature [4]:

$$\gamma^2 = \frac{1}{5}\left(\frac{Sa}{RSm_1}\right)^{\frac{2}{3}}\left(\frac{E}{q}\right)^{\frac{2}{3}} \tag{14}$$

where q is pressure on contacting surfaces.
Required cycles for material destruction are calculated [5]:

$$E\{N_c\} = \frac{N_0}{5m!}[\frac{\sqrt{2}}{\pi^2}\frac{\sigma_0 RSm1K^{1/2}(e)}{E \cdot Sa}]^m \tag{15}$$

Actual number of cycles N_{cf} is calculated:

$$N_{cf} = \frac{L_b}{S^a_{m2}} \tag{16}$$

where L_b is friction path length and S^a_{m2} is surface roughness average pitch in friction direction for the part than promote wear.
Inserting obtained coherences in formula (2) average linear wear can be calculated:

$$E\{U_n\} \approx k_{e-m} \cdot k_r \cdot k_{f-m} \cdot (\frac{q}{E})^{2/3} \cdot Sa \cdot \frac{L_b}{S^a_{m2}} \tag{17}$$

where k_{e-m} is coefficient dependent from surface anisotropy ($e^2 = 1 - c^{8/3}$) and fatigue curve parameters:

$$k_{e-m} = \sqrt{\pi}\frac{m!}{N_0}\left(\frac{\pi^4}{2K(e)}\right)^{m/2} \tag{18}$$

k_r is surface roughness parameter complex:

$$k_r = (S_a / RS_{m1})^{m-\frac{2}{3}} \tag{19}$$

$k_{f\text{-}m}$ is physical and mechanical parameter complex:

$$k_{f-m} = \left(\frac{E}{\sigma_0}\right)^m \tag{20}$$

Average linear wear in normal exploitation conditions can be calculated using equation (17). These conditions should be met:

- friction coefficient f ≤0.1;
- both friction surfaces must be flat.

Total wear is determined using equation (1).
Equation (17) can be rewritten when taking in count that:

$$L_b = vt \tag{21}$$

where v is friction pair speed, t – friction pair duration.
Then $E\{U_n\}$ is:

$$E\{U_n\} \approx k_{e-m} \cdot k_r \cdot k_{f-m} \cdot (\frac{q}{E})^{2/3} \cdot \frac{Sa}{S^a_{m2}} \cdot vt \tag{22}$$

4. Experimental studies.

To verify the theory an experimental study was carried out using "pin-on-disc" type tribometer. A special disk (38KHN3MA steel) with diameter 100mm and thickness of 6mm was manufactured to work as a tribometers`s disc and bronze (BrOF7-0.2) rod with a diameter of 6.5mm was used as pin. 3D surface roughness was determined with "Taylor Hobson Form Talysurf Intra 50" profilometer. Experimental study parameters are shown in table 1.

Table 1. Experiment parameters

Kinematic parameters	Physical and mechanical parameters	Roughness parameters
q=11 MPa	m=4	S_a=0.6 μm
v=5 m/s	σ_{-1}=600 MPa	RS_{m1}=0.05 mm
t_{max}=272 h	$N_0=5*10^6$	RS_{m2}^a=0.5 mm
	$E=1.05*10^5$ MPa	

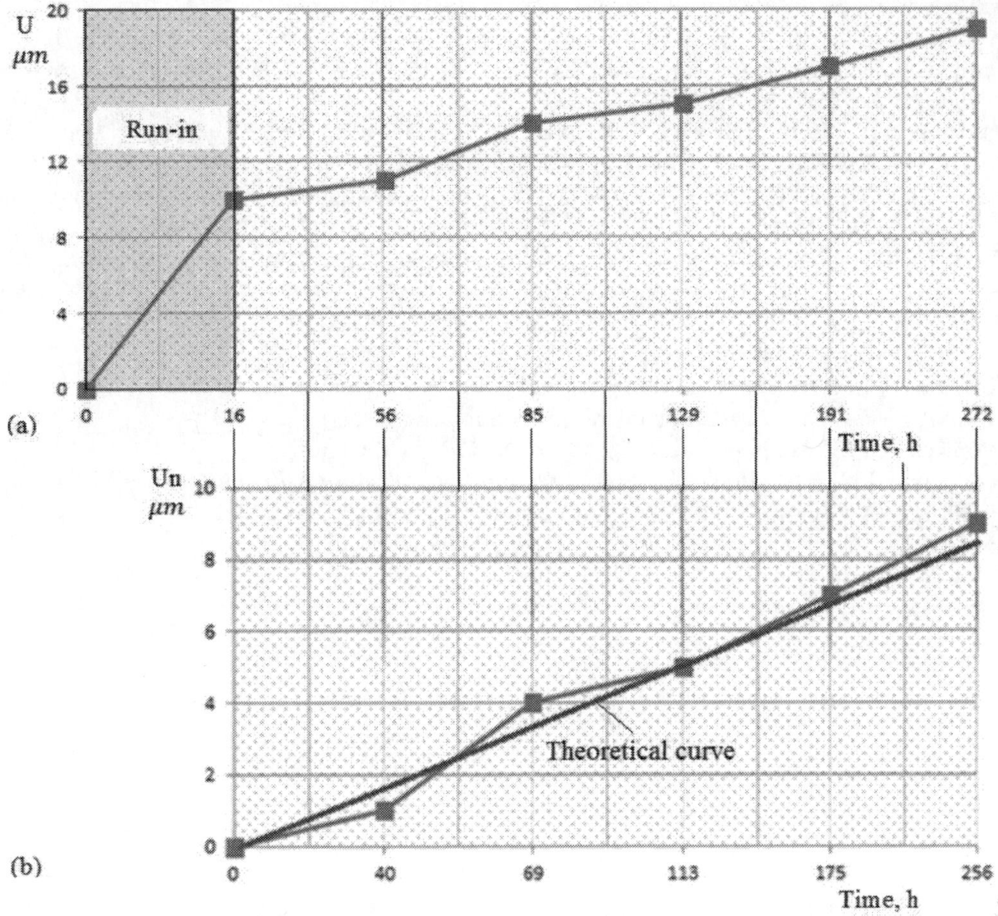

Figure 7. Linear wear experimental and theoretical data comparison where (a) total experimental data and (b) experimental and theoretical data comparison for normal exploitation stage.

Experiment results are summarized in figure 7 where (a) shows both experiment stages - run in and normal wear. Total duration time of the experiment took 272 hours. Experiment was paused after 16 hours (run-in) to measure surface roughness parameters used in further calculations. In figure 7 (b) one can see comparison of experimental data after the run-in process and theoretically calculated curve. Strong correlation is obtained for experimental and theoretical curves with result scatter less than 10%.

5. Conclusion

Paper offers experimental-theoretical methodology to determine slide frictions wear. It is carried out in a sequence:

- Determine starting data that are necessary for calculations:
  - kinematic properties: load (q), speed (v), travelled distance (L_b), duration (t);
  - material physical properties: material fatigue fracture parameters (m, σ_0, N_0);
  - material mechanical properties (E, μ)
- Determine parameters after run-in:
  - Surface roughness parameters (Sa, RSm_1, RSm_2, S_{m2}^a);
  - run-in wear (U_p) and its duration (T_p).

6. Literature

[1] Springis G, Rudzitis J, Avisane A and Leitans A 2014 Wear calculation for sliding friction pairs *Latv. J. Phys. Tech. Sci.* **51** 41-54
[2] Rudzitis J 1992 Surface Roughness topography investigations *Internationales Oberflächenkolloquium* (Chemnitz: Technische Universität Chemnitz) p 123-128
[3] Hamilton G M and Goodmen L E 1966 The stress field created by a circular sliding contact *J. Appl. Mech.* 125-131
[4] Ruzitis J 2007 Contact mechanics for surfaces 2-nd part (Riga: Riga Technical University)
[5] Springis G, Rudzitis J, Avisane A, Kumermanis M, Lungevics J 2012 Wear Calculation Possibility in Slide-Friction Pairs Using Surfaces with Nanocoatings (Riga: Mechanical Engineering and Mechanics) pp 24-28
[6] Kordonskiy H 1963 *Application of probability theory in engineering* (Moscow)
[7] Holm R 1946 *Eletric Contacts* (Stocholm: H Gebers) p 398
[8] Pronikov A 1969 *Machine's reliability and durability* (Moscow)
[9] Kragelsky I V, Alisin V V 2009 *Tribology-Lubrication, Friction and Wear* (London: Professional Engineering Publishing) chapters 3 and 4 pp 90 - 144
[10] Belyaev N 1975 *Works on the theory of plasticity elasticity* (Moscow)

Effect of abrasive particle size on friction and wear behaviour of various microstructures of 25CD4 steel

C Trevisiol[1], A Jourani[1] and S Bouvier[1]

[1] Sorbonne Universités, Université de technologie de Compiègne, CNRS, UMR 7337 Roberval, Centre de recherché Royallieu – CS 60 319 – 60 203 Compiègne cedex, France

E-mail: celine.trevisiol@utc.fr

Abstract. Many parameters, such as normal load and material bulk hardness, control the wear and friction behaviours of materials. Nonetheless, the investigation of the coupled contributions of microstructure and abrasive particle size are still lacking. A contribution is proposed by using steel pins with various microstructures with a similar macro-hardness (around 410H_V) and chemical composition. A quenched martensitic microstructure, a tempered martensitic microstructure and three ferrite-martensite dual-phase microstructures, with a similar martensite volume fraction (around 67%) and different martensite colony morphologies, are established. Friction tests are performed between these pins and abrasive papers with different sizes ranging from 15μm to 200μm. Compared to single-phase microstructures (quenched and tempered martensitic microstructures) and whatever the abrasive particle size, ferrite-martensite dual-phase microstructures reduce the friction coefficient and provide better wear resistance. For the ferrite-martensite dual-phase microstructures and unlike fine and fibrous martensite colonies, coarse and granular martensite colonies minimize the friction coefficient. In addition, characterized by a change of wear mechanisms between abrasion and adhesion, an intermediate abrasive particle around 35 μm minimizes the friction coefficient. This study also reveals that the wear rate increases with the abrasive particle size which is associated to an increase of the attack angle of abrasive grains.

1. Introduction

There is an increasing interest to reduce friction and wear in industrial applications. Excessive wear of mechanical parts requires their replacement. In addition, considerable energy losses are consumed to overcome friction which represent major economic losses [1]. For instance, a better understanding of the effect of roughness on friction and wear behaviour could contribute to reduce friction and wear [2-6]. Previous works aim to study the effect of the abrasive particle size on the tribological behaviour but without considering the material properties [2-4]. The bulk hardness is mainly used to control friction and wear resistance [7-8]. Indeed, Archard claims that as the hardness increases, wear rate decreases. However, other parameters such as the chemical compositions and the microstructures are also strongly linked to the tribological behaviour [5-6][9-16]. Compared to bainitic, pearlitic and ferritic microstructures, it is well known that steels with a martensitic microstructure, characterized by a high hardness, are mainly used to assure good wear resistance. Nonetheless, efforts are devoted to find microstructures which present better combinations between toughness and ductility and therefore less brittle structures. For example, investigations are done on tempered martensitic microstructures and ferrite-martensite dual-phase microstructures [12-15]. Mainly used in the automotive industry because of its very interesting mechanical properties, steels with dual-phase microstructures present a good combination between ductility with a soft ferrite matrix and strength with hard martensite colonies. So far, tribological behaviours of various microstructures are compared but without fixing the hardness and/ or the chemical composition. In addition, these investigations on the tribological behaviour are focused to the wear without considering the friction behaviour. Therefore, this work proposes a contribution in studying the effect of microstructures with the same hardness level and chemical composition on friction and wear mechanisms. In addition, the coupled effects of microstructure and surface roughness on the tribological behaviour of the studied low alloy steel are studied.

2. Material and methods

This work aims to investigate the tribological behaviour of a low alloy steel. Table 1 shows the chemical composition of this steel. Five cylindrical pins, with 10mm in diameter and length, are cut from a rod of this steel.

Table 1. Chemical composition of the low alloy steel.

C	Si	Mn	P	S	Cr	Mo
0.22-0.29	≤0.40	0.40-0.90	≤0.035	≤0.035	0.9-1.20	0.15-0.30

These steel pins undergo various heat treatments to generate various microstructures. These microstructures include three martensite-ferrite dual-phase microstructures (DP1, DP2 and DP3), a quenched martensitic microstructure (QM) and a tempered martensitic microstructure (TM). The procedures of the performed heat-treatments are presented in Figure 1.

Figure 1(a-c) show the different heat treatments used to develop ferrite-martensite dual-phase microstructures with various martensite colony morphologies. A step quench, an intercritical annealing and an intermediate quench heat treatments are performed to generate these microstructures. In Figure 1a, it is schematically drawn that the first dual-phase microstructure DP1 is obtained by using a step quench starting by an autenitization (at 900°C during 30 minutes) followed by an intercritical annealing (at 725°C during 3 minutes) and a water quench. In Figure 1b, it is schematically drawn that the second dual-phase microstructure DP2 is obtained by using an intercritical annealing starting by an autenitization (at 900°C during 30 minutes) and a slow cooling. Then, the specimen is intercritical annealed (at 790°C during 17 minutes) and then water quenched. In Figure 1c, it is schematically drawn that the third dual-phase microstructure DP3 is obtained by using an intermediate quench starting by an autenitisation (at 900°C during 30 minutes) and a water quench. Then, the specimen is intercritical annealed (at 790°C during 12 minutes 30) and then water quenched.

Figure 1d shows the heat treatment used to generate a quenched martensite microstructure QM. This heat treatment starts by an austenitization (at 1100°C during 2 hours) followed by a water quench.

Finally, Figure 1e shows the heat treatment used to generate a tempered martensitic microstructure TM. This heat treatment starts by an austenization (at 900°C during 2 hours) followed by a water quench and then a tempering (at 400°C during 2 hours).

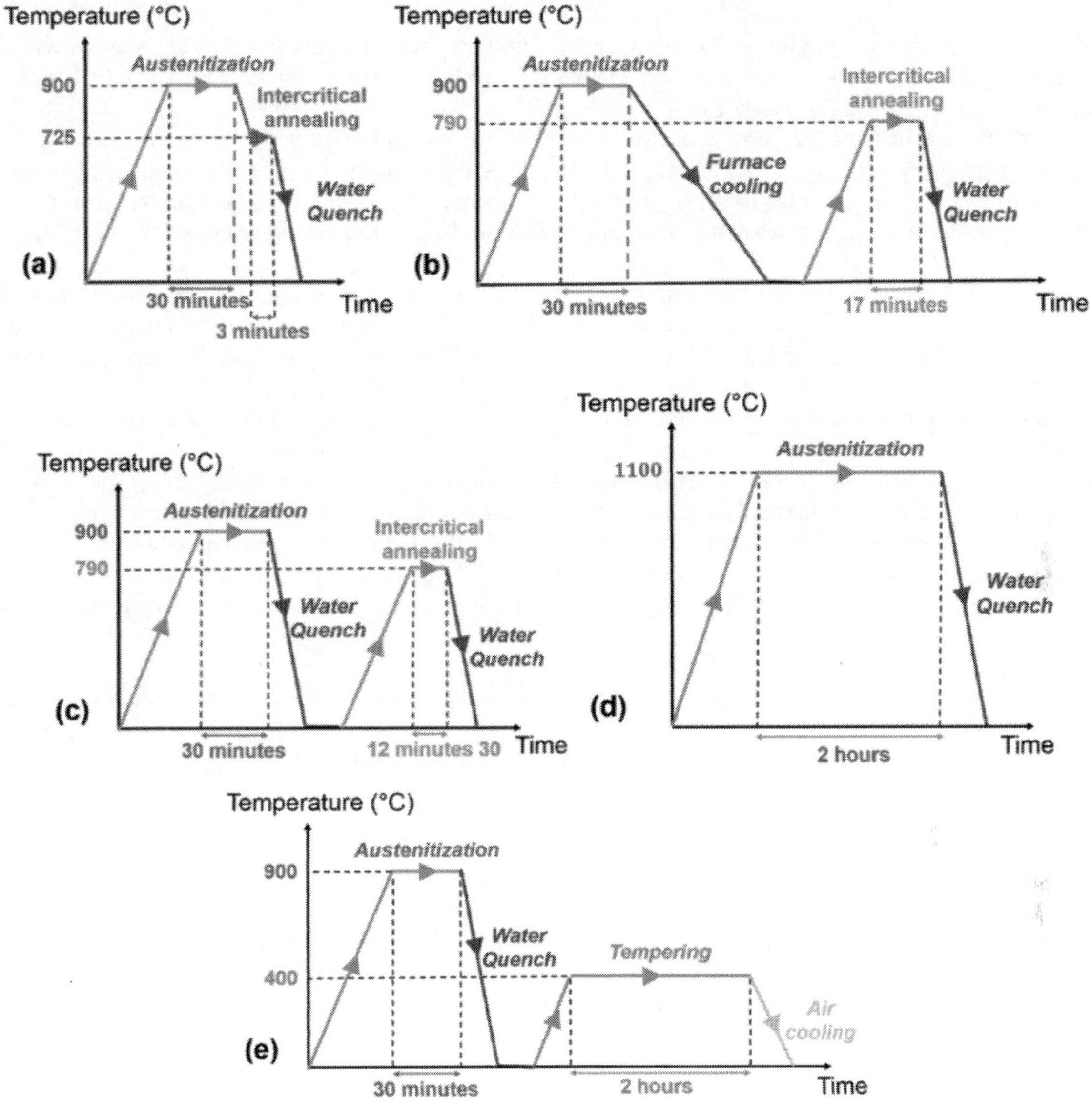

Figure 1. The schematic drawings of the heat treatments of the (a) DP1, (b) DP2, (c) DP3, (d) QM and (e) TM microstructures.

To characterize the microstructures, the heat-treated pins are ground mechanically. First, the bulk material structure is reached by using grinding papers of 80 grit to remove 1mm of material from the surface. Second, the surfaces are ground mechanically on SiC grinding papers (whose grit is ranging from 320 to 4000). Third, the surface pins are polished with diamond paste of 1µm. Finally, the pins are etched with a 1% Nital solution during 10s after having cleaned by ultrasound with ethanol during 10min. For the ferrite-martensite dual-phase microstructures, the martensite volume fractions are determined by image processing using ImageJ. Thus, the martensite volume fraction is averaged from ten micrographs done with a Field Emission Electronic Gun-Scanning Electron Microscope (FEG-SEM) ZEISS Sigma at 15keV.
Under a load of 10kg, a vickers indenter is used to characterize the macro-hardness of each microstructure. From different locations, three measurements are averaged.

Before the friction tests, the heat-treated pins are ground mechanically on SiC grinding papers with a grit ranging from 320 to 4000 grit and cleaned with ethanol by ultrasound for 10 min. Thus, the surface pins present the same initial roughness.
Then, dry friction tests are performed between the steel pin and an emery paper according to a pin-on-plate configuration and a linear reciprocating motion. The used emery papers embedded hard abrasive Struers silicon carbide particles (with particle sizes of 15μm, 35μm and 200μm). The variation of the abrasive particle size aims to study the roughness effect on the friction and wear behaviour of the heat-treated pins.
These friction tests are performed at room temperature (23°C) in air, at a relative humidity of 50-60%, under a pressure of 1.6MPa, a linear velocity of 0.06m/s and a frequency of 2Hz. The total sliding distance is 18m and the stroke length is 15mm. During these friction tests, average friction coefficients and wear rates are determined from three measurements. The friction coefficient is calculated as the ratio between the tangential force F_T and the normal force F_N as function of the test time. By using a load cell and a piezoelectric transfer transducer, F_T and F_N are measured each 0.1s. For each configuration, an average friction coefficient is determined for the steady-state. Then, after having ultrasonic cleaned the heat-treated steel pins with ethanol for 10 min, the pin mass is measured before and after the friction test. The wear rate is calculated from these pin mass losses measured by using a direct reading balance (with a minimum scale of 0.1 mg). The wear rate Q is calculated as the ratio between the cumulative lost volume V and the total sliding distance d as shown in the following equation:

$$Q = V/d \text{ (with } d = 18\text{m)} \quad (1)$$

Finally, the initial microstructures, the worn surfaces of steel pins and abrasive papers are observed to analyze the generated wear mechanisms. Scanning electron microscope (SEM) is used at 15kV for the initial microstructures and the worn steel pin surfaces and at 2kV for the abrasive paper surfaces.

3. Results and discussion

3.1. Effect of heat treatment on hardness and microstructure

Figure 2 shows the obtained microstructures and the associated hardness.
The micrographs in Figure 2(a-c) exhibit three ferrite-martensite dual-phase microstructures (DP1, DP2 and DP3). Figure 2a, 2b and 2c show dual-phase microstructures with respectively coarse and granular martensite colonies, smaller and granular martensite colonies and then fine and fibrous martensite colonies. The variation of the martensite colony morphologies depends on the initial microstructure before the intercritical annealing in the two phase region composed of ferrite and austenite.
In the case of the first dual-phase microstructure DP1 in Figure 2a, after an austenitization, the microstructure presents austenite. Then, the intercritical annealing results in transforming austenite into austenite and ferrite. Indeed, ferrite nucleates and grows at the prior austenite grain boundaries. The final water quench leads to transform the unchanged austenite into martensite and ferrite remains unchanged. The generated microstructure presents coarse martensite colonies surrounded by ferrite.
In the case of the second dual-phase microstructure DP2 in Figure 2b, the slow cooling after the autenitization generates a microstructure of hypoeutectoid ferrite and pearlite. Then, the intercritical annealing results in the nucleation of austenite at the interfaces between ferrite and carbide inside the pearlite colonies. Austenite grows into the carbides which dissolve. The final water quench leads to a microstructure composed of globular martensite colonies situated at the ferrite boundaries. In this case, the martensite colonies are smaller than for DP1.
In the case of the third dual-phase microstructure DP3 in Figure 2c, after an austenitization and a water quench, a martensitic microstructure is generated. Then, the intercritical annealing results in increasing the number of nucleation sites of autenite and ferrite at the boundaries of the prior austenite grains and martensite laths. The final water quench leads to a microstructure composed of fibrous and fine martensite colonies in a ferrite matrix

In addition, it is shown that these ferrite-martensite dual-phase microstructures present almost the same martensite volume fraction around 67±2%. Indeed, DP1, DP2 and DP3 have respectively a martensite volume fraction of 67±2%, 69±3% and 65±3%.

The micrograph in Figure 2d exhibits a microstructure composed of martensite in the shape of laths grouped into packets. This kind of microstructure is typical of a quenched martensitic microstructure QM of low and median carbon steels. During this heat treatment, the transformations of the austenite into martensite and a small amount of retained austenite occur. Nonetheless, the retained austenite is not observed in the micrographs which should be in too small amount.

The micrograph in Figure 2e exhibits a tempered martensitic microstructure TM. Indeed, the tempering heat treatment, preceded by an austenitization followed by a water quench, results in transforming a brittle and hard martensitic microstructure into a microstructure composed of martensite in the shape of laths, impoverished in carbon, and fine carbides [17]. It is shown that the prior austenite grain size increases with the time and/or the temperature of austenitization [18]. Thus, because of a lower time and temperature of austenitization, the TM microstructure presents smaller prior austenite grain sizes than the QM microstructure.

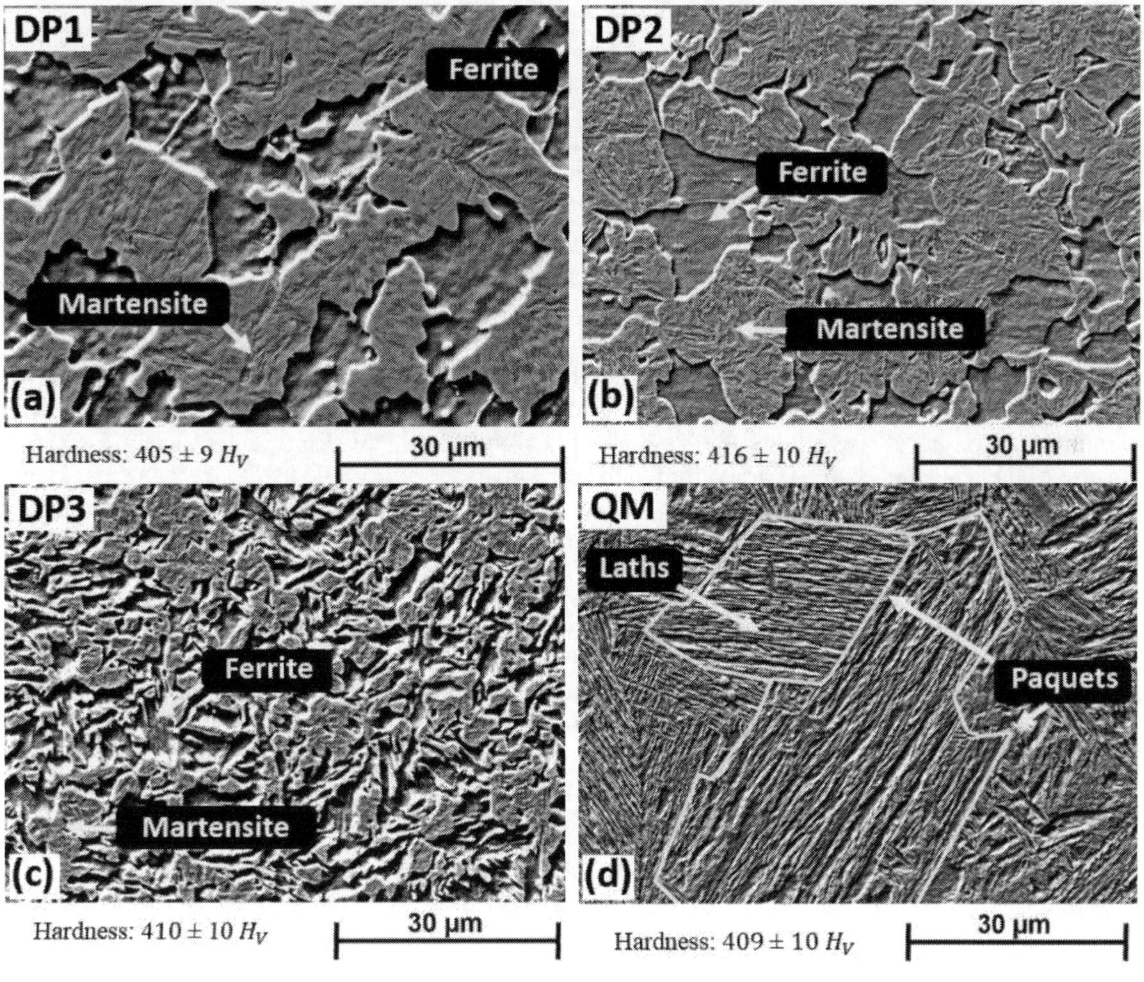

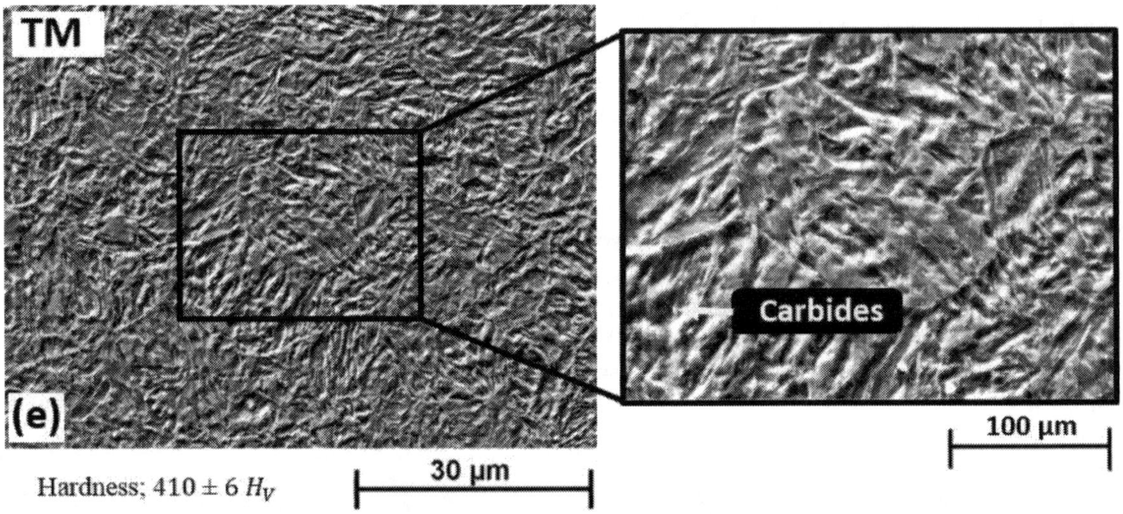

Hardness; 410 ± 6 H_V

Figure 2. Scanning electronic micrographs (SEM) of the microstructures of the (a) DP1, (b) DP2, (c) DP3, (d) QM and (e) TM.

3.2. Effect of abrasive particle size on tribological behaviour

Figure 3a shows the evolution of friction coefficient as function of abrasive particle size for the different microstructures. In Figure 3a, it is highlighted that an intermediate abrasive particle size around 35µm minimizes the friction coefficient. This minimization is due to a transition of wear mechanisms between adhesive and abrasive as shown in some works [2,5]. Bowden and Tabor [20] define the coefficient of friction μ as the sum of an adhesion component μ_a and a plowing component μ_p. The adhesion component μ_a is related to the real contact area whereas the plowing component μ_p is related to the attack angle of the asperities. This intermediate abrasive particle size leads to minimize these two components. As the abrasive particle size decreases from 35µm, the friction coefficient increases because of the real contact area increase due to the predominance of adhesion wear mechanisms. Then, as the abrasive particle size increases from 35 µm, abrasive wear mechanisms become predominant. Thus, there is an increase of the pressure and plastic deformation at each abrasive grain which results in increasing the friction coefficient [3,19,22-26].

Figure 3b shows the evolution of wear rate as function of abrasive particle size for the different microstructures. In the case of the QM, TM and DP3 microstructures, it is shown that the wear rate increases with the abrasive particle size and thus the attack angle of the abrasive particles. It is explained that as the size and the penetration depth of abrasive particles increase, the predominant wear mechanisms are plowing, followed by wedge formation and then cutting [7,9] which generate an increase of the wear rate. More particularly, for DP1 and DP2, by considering the error bars, when the abrasive particle size increases, the wear rate increases and then remains constant for an abrasive particle size higher than 35µm.

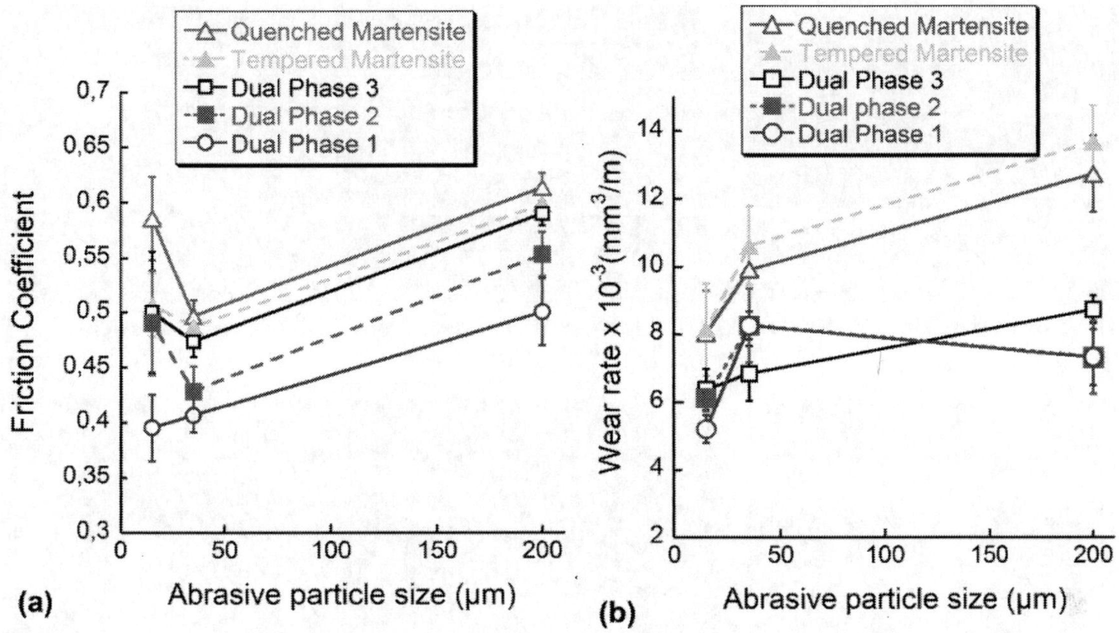

Figure 3. Evolution of (a) friction coefficient and (b) wear rate as function of abrasive particle size.

Figure 4 shows the SEM observations of the worn surfaces of the pin, the abrasive paper and the steel wear debris in the case of the QM microstructure. These surfaces are obtained after friction tests performed between a pin with a QM microstructure and an abrasive paper with particle sizes of 15µm, 35µm and 200µm.

The observations of the worn pin surfaces reveal parallel grooves in which the material is displaced to the edges. Thus, these surfaces reveal severe plastic deformations produce by the hard abrasive particles. The edges may be removed after several passages of abrasive particles on the pin surface and generate wear debris. The presence of wear debris on the worn surfaces of the abrasive paper and the pin reveal the cutting process. Thus, the predominant wear mechanisms are ploughing with severe plastic deformation and cutting.

Then, compared to fine abrasive particles, the coarse abrasive particles generate wider and deeper grooves and therefore a higher wear rate. Indeed, the wear rate increases because of the increases of the steel wear debris size and the cutting process [2-3].In the case of coarse abrasive particles, the worn abrasive paper exhibits abrasive wear mechanisms characterized by abrasive grain fractures. As the abrasive particle size increases from 35µm, the later phenomenon leads to the increase of the friction coefficient [2,5]. In the case of small abrasive particles, the worn abrasive paper reveals accumulations of steel wear debris around the abrasive particles and thus clogging. As mentionned by De Pellegrin et al. [21], as the particle size decreases, the space between abrasive grains decreases. This phenomenon is due to a facilitated accumulation of the metallic debris in smaller volumes. As the abrasive particle size decreases from 35µm, this clogging phenomenon generates adhesion and a higher friction coefficient. Thus, in the case of the intermediate abrasive particle size of 35µm, the worn abrasive paper presents a combination between adhesive and abrasive wear mechanisms which contributes to minimize the friction coefficient.

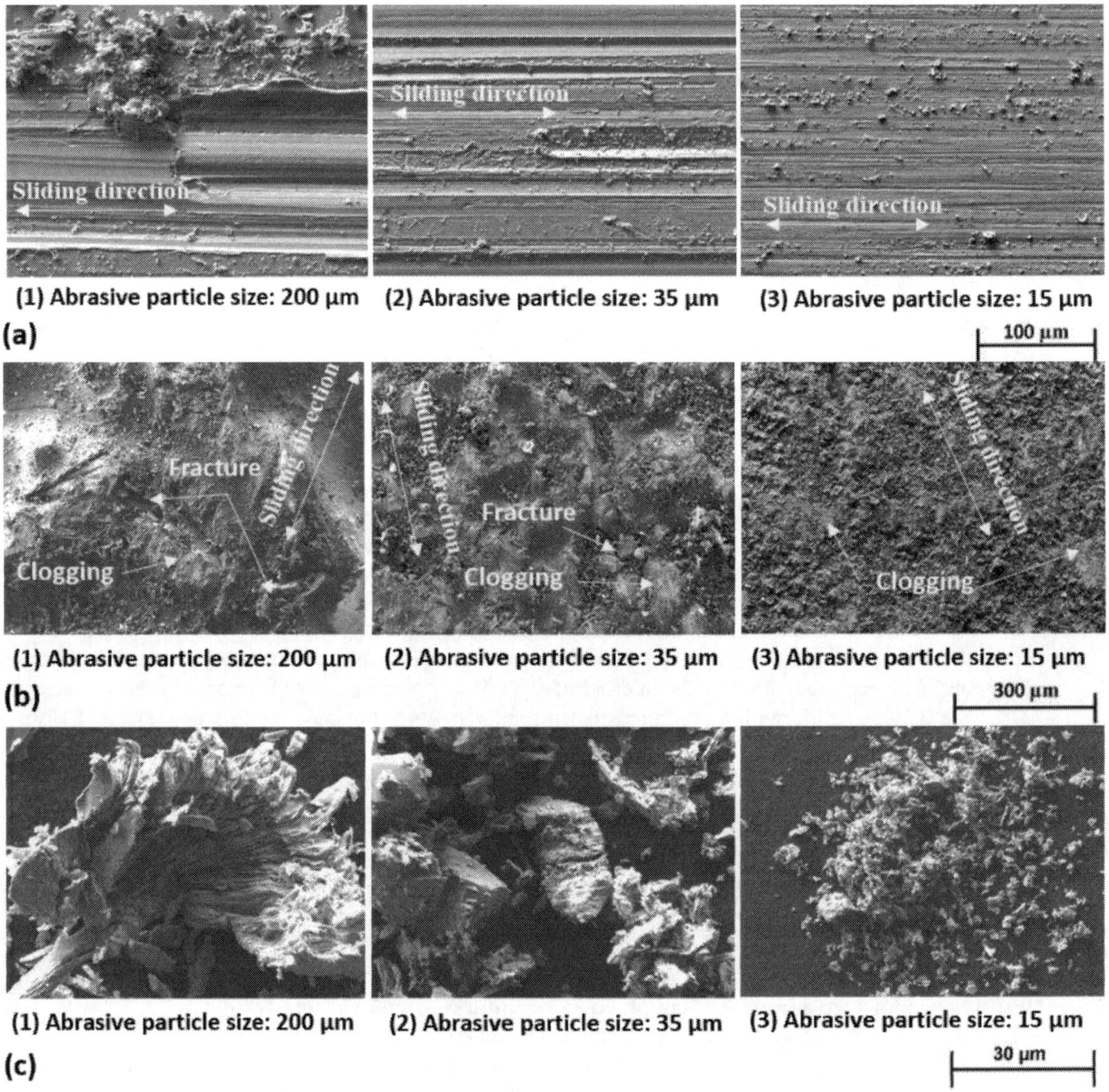

Figure 4. In the case of the QM microstructure, scanning electron micrographs (SEM) of the worn surfaces of the (a) pins, (b) abrasive papers and (c) wear debris.

3.3. Effect of microstructure on tribological behaviour

Figure 3a and 3b reveals that dual-phase microstructures DP1, DP2 and DP3, present a lower friction coefficient and wear rate than single-phase microstructures such as the QM and TM microstructures. Because of a higher carbon content, the martensite of the DP microstructures are harder than the QM and TM microstructures. Indeed, it is well known that the carbon content controls the hardness of the phases. In the case of the ferrite–martensite dual phase steels, the martensite phase, which is a metastable iron phase supersaturated in carbon, is significantly harder than the ferrite phase. On the other hand, the ferrite phase is able to dissolve a maximum carbon amount of only 0.02%wt. Based on a mixture rule as mentioned is in some works [28-29], the dual-phase steel carbon concentration C is estimated by considering the carbon concentration C_m of the isolated martensite, the carbon concentration C_f of the ferrite, the martensite volume fraction V_m and the ferrite volume fraction V_f as followed:

$$C = C_f V_f + C_m V_m = C_f(1 - V_m) + C_m V_m$$

Thus, the carbon concentration of the isolated martensite can be characterized by the following equation:

$$C_m = C_f + \frac{C - C_f}{V_m}$$

Thus, for a martensite volume fraction V_m around 67% for the three dual-phase microstructures studied, the carbon concentration C_m is estimated around 0,36% which leads to a harder martensite than for the martensite of the single-phase microstructures. Therefore, in the case of DP microstructures, the martensite carries much higher pressure and exibits lower penetration depth of the abrasive grains than the martensite of the single-phase microstructures. Thus, in the case of DP microstructures, the depth penetration of the abrasive grains on the martensite are lower. In addition, the microstructures impacts the evolution of the wear rate as function of the abrasive particle size. Indeed, for the QM, TM and DP3 microstructures, the wear rate increases with the abrasive particle size. On the other hand, for the DP1 and DP2 microstructures, when the abrasive particle size increases, the wear rate increases and then remains constant for an abrasive particle size of 35μm. It is known that for abrasive particle sizes higher than a critical particle size usually around 100μm, the wear rate remains constant [27]. Nonetheless, this critical particle size depends on the microstructure. The critical particle size for the QM, TM and DP3 is higher than for the DP1 and DP2 microstructures.

Figure 5 shows the SEM observations of the worn surface of the steel pin, the abrasive paper and the wear debris after a friction test carried out with an abrasive paper with a particle size of 200μm in the case of the QM and the DP1 microstructures.
As regards the QM microstructure, the pin presents a predominance of plowing with severe plastic deformation and cutting. On the used worn abrasive paper, adhesion of wear debris on the abrasive particle tops appear and contribute to increase the contact area and therefore the friction coefficient [2,5]. On the other hand, with coarse and granular martensite colonies, the DP1 microstructure presents a predominance of cutting followed by plowing. The DP1 microstructure also generates smaller wear debris which leads to decrease the wear rate.

Among the dual-phase microstructures, it is shown that coarse and granular martensite colonies presents lower friction coefficient than fine and fibrous martensite colonies. The coarse and granular martensite (DP1) supports the majority of load and leads to reduce the plastic deformation compared to fine and fibrous martensite (DP3) [12]. On the other hand, the effects of martensite colony morphologies on wear rate are related to the abrasive particle size.
Among the single-phase microstructures, the TM microstructure has a higher wear rate and a lower friction coefficient than the QM microstructure. Indeed, in the case of the TM microstructure, the presence of martensite laths impoverished in carbon, the carbides and the finer prior austenite grain size than the QM microstructure generates a friction coefficient decrease and a wear rate increase.

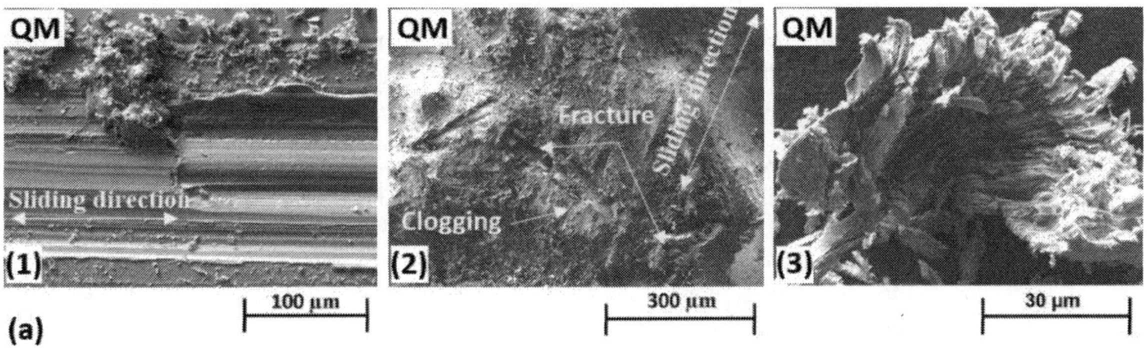

Figure 5. Scanning electron micrographs (SEM) of the worn surfaces of the (1) pins, (2) abrasive papers and (3) wear debris in the case of the (a) QM and (b) DP1 microstructures.

4. Conclusions

For a similar hardness level (around $410H_V$) and from the same low alloy steel, various microstructures are generated by using various heat treatments. The main aim is to analyze the combined effects of the abrasive particle size and the microstructure on the friction and wear behaviours. Therefore, dry friction tests are performed between these heat-treated pins and an abrasive paper with various particles sizes. By using coarse and fine abrasive particles, the effect of roughness on the tribological behaviour are investigated. The main findings are:

- Compared to single-phase microstructures (quenched martensitic and tempered martensitic microstructures) and whatever the abrasive particle size, ferrite-martensite dual-phase microstructures minimize the friction coefficient and enhance the wear resistance.
- Among the dual-phase microstructures and whatever the abrasive particle size, the morphologies of the martensite colonies influence the friction behaviour. Indeed, coarse and granular martensite colonies minimizes the friction coefficient whereas fine and fibrous martensite colonies increase the friction coefficient.
- The friction coefficient is minimized for an intermediate abrasive particle size around 35µm. This minimization is explained by a transition of wear mechanisms between abrasion and adhesion for this intermediate abrasive particle size. However, the wear rate and the wear debris size increase with the size and the attack angle of the abrasive particles. Nonetheless, the wear rate becomes constant for a critical abrasive particle size which is closely linked to the microstructures. Indeed, the dual-phase microstructures with granular martensite colony present a lower critical abrasive particle size than the others microstructures.

References

[1] Holmberg K and Erdemir A 2015 Global impact of friction on energy consumption, economy and environment, *FME Trans.* **43(3)** 181–5.
[2] Jourani A and Bouvier S 2015 Friction and wear mechanisms of 316L stainless steel in dry sliding contact: effect of abrasive particle size, *Tribol Trans* **58(1)** 131–9.
[3] Jourani A, Hagege B, Bouvier S, Bigerelle M and Zahouani H 2013 Influence of abrasive grain geometry on friction coefficient and wear rate in belt finishing, *Tribol Int* **59** 30–7.
[4] Dube N B and Hutchings I M 1999 Influence of particle fracture in the high-stress and low-stress abrasive wear of steel, *Wear* **233–235** 246–56.
[5] Trevisiol C, Jourani A and Bouvier S 2016 Effect of martensite volume fraction and abrasive particles size on friction and wear behaviour of a low alloy steel, *Tribol. Int.* In press.

[6] Trevisiol C, Jourani A and Bouvier S 2017 Effect of hardness, microstructure, normal load and abrasive size on friction and on wear behaviour of 35NCD16 steel, *Wear* Forthcoming.
[7] Hokkirigawa K, Kato K and Li Z Z 1988 The effect of hardness on the transition of the abrasive wear mechanism of steels, *Wear* **123(2)** 241–51.
[8] Khruschov M M 1974 Principles of abrasive wear, *Wear* **28(1)** 69–88.
[9] Rigney D A 1994 The roles of hardness in the sliding behavior of materials, *Wear* **175(1-2)** 63-9.
[10] Modi O P, Pandit P, Mondal D P, Prasad B K, Yegneswaran A H and Chrysanthou A 2007 High-stress abrasive wear response of 0.2% carbon dual phase steel: effects of microstructural features and experimental conditions, *Mater Sci Eng A* **458(1-2)** 303–11.
[11] Zum Gahr K.-H 1987 *Microstructure and Wear of Material*, Tribology Series, Vol 10, Elsevier, Amsterdam.
[12] Xu X, van der Zwaag S and Xu W 2016 The effect of ferrite–martensite morphology on the scratch and abrasive wear behaviour of a dual phase construction steel, *Wear* **348–349** 148–57.
[13] Xu X, van der Zwaag S and Xu W 2016 The effect of martensite volume fraction on the scratch and abrasion resistance of a ferrite–martensite dual phase steel, *Wear* **348–349** 80–8.
[14] Saghafian H and Kheirandish S 2007 Correlating microstructural features with wear resistance of dual phase steel, *Mater Lett* **61** 3059–63.
[15] Abouei V, Saghafian H and Kheirandish S 2007 Dry sliding oxidative wear in plain carbon dual phase steel, *J Iron Steel Res Int* **14** 43–8.
[16] Coronado J J, Rodríguez S A and Sinatora A 2013 Effect of particle hardness on mild–severe wear transition of hard second phase materials, *Wear* **301(1-2)** 82–8.
[17] Speich G R and Leslie W C 1972 Tempering of steel, *Metall. Trans.* **3(5)** 1043–54.
[18] Prawoto Y, Jasmawati N and Sumeru K 2012 Effect of Prior Austenite Grain Size on the Morphology and Mechanical Properties of Martensite in Medium Carbon Steel, *J. Mater. Sci. Technol.* **28(5)** 461–6.
[19] Jourani A 2015 Three-dimensional modelling of temperature distribution during belt finishing, *Int. J. Surf. Sci. Eng.* **9 (2-3)** 231-46.
[20] Bowden F P and Tabor D 1954 *The Friction and Lubrication of Solids*, Clarendon, Oxford.
[21] De Pellegrin D V, Torrance A A and Haran E 2009 Wear mechanisms and scale effects in two-body abrasion, *Wear* **266(1-2)** 13–20.
[22] Khellouki A, Rech J and Zahouani H 2013 Energetic analysis of cutting mechanisms in belt finishing of hard materials, *Proc Inst Mech Eng B: J Eng Manuf* **227(9)** 1409–13.
[23] Jourani A 2015 Effect of roughness geometries in contact mechanics, *Int. J. Mater. Prod. Technol.* ***51 (2)*** 127-138.
[24] Jourani A 2014 A new three-dimensional numerical model of rough contact: Influence of mode of surface deformation on real area of contact and pressure distribution, *J. Tribol.* **137(1)** 011401.
[25] Petit L, Bigerelle M, Jourani A, Dore E, Prelle C and Lamarque F 2010 Presentation of a new method to measure the friction coefficient using an electromagnetic digital device *Proc. Inst. Mech. Eng. Part J J. Eng. Tribol.* **224 (9)** 1019-1026.
[26] Mezghani S, Jourani A and Zahouani H 2006 The scale effect of roughness in contact problems WIT Trans. Built. Environ. **85**, 369–378.
[27] Misra A, Finnie I 1981 On the size effect in abrasive and erosive wear, Wear **65** 359–373.
[28] P. Movahed, S. Kolahgar, S.P.H. Marashi, M. Pouranvari, N. Parvin, The effect of intercritical heat treatment temperature on the tensile properties and work hardening behavior of ferrite-martensite dual phase steel sheets, Mater Sci Eng A 518 (2009) 1–6.
[29] M. Calcagnotto, Y. Adachi, D. Ponge, D. Raabe, Deformation and fracture mechanisms in fine- and ultrafine-grained ferrite/martensite dual-phase steels and the effect of aging, Acta Mater 59 (2011) 658–670.

IOP Conf. Series: Journal of Physics: Conf. Series **843** (2017) 012074 doi:10.1088/1742-6596/843/1/012074

Erosion resistance comparison of alternative surface treatments

Z Česánek[1] J Schubert[1] and Š Houdková[1]

[1]Senior researcher, Výzkumný a zkušební ústav Plzeň, Pilsen, CZ

E-mail: cesanek@vzuplzen.cz

Abstract. Erosion is a process characterized by the particle separation and the damage of component functional surfaces. Thermal spraying technology HP/HVOF (High Pressure / High Velocity Oxygen Fuel) is commonly used for protection of component surfaces against erosive wear. Alloy as well as cermet based coatings meet the requirements for high erosion resistance. Wear resistance is in many cases the determining property of required component functioning. The application suitability of coating materials is particularly influenced by different hardness. This paper therefore presents an erosion resistance comparison of alloy and cermet based coatings. The coatings were applied on steel substrates and were subjected to the erosive test using the device for evaluation of material erosion resistance working on the principle of centrifugal erodent flow. Abrasive sand Al_2O_3 with grain size 212-250 μm was selected as an erosive material. For this purpose, the specimens were prepared by thermal spraying technology HP/HVOF using commercially available powders Stellite 6, NiCrBSi, Cr_3C_2-25%NiCr, Cr_3C_2-25%CoNiCrAlY, Hastelloy C-276 and experimental coating TiMoCN-29% Ni. Erosion resistance of evaluated coatings was compared with erosive resistance of 1.4923 high alloyed steel without nitridation and in nitrided state and further with surface treatment using technology PVD. According to the evaluation, the resulting erosive resistance depends not only on the selected erodent and surface protection, but also on the erodent impact angle.

1. Introduction

Erosion is a process characterized by separation of particles and functional surface damage. Erosive wear occurs e.g. in machine parts transporting abrasive suspension (pump components, water turbines, pipes, fittings, nozzles), is caused by particles dispersed in air or gas stream (fan, cyclone, blaster, pipe components) or is generated by fluid, steam, droplets and gas flow (steam fittings, steam and gas turbine components). Regarding the erosive wear caused by particles, there can occur several ways in which particles impact on the material surface. If a particle moving askew to the surface has sufficient impact energy, it penetrates into the exposed surface resulting in material dislodge or segregation. If a particle has small impact energy or moves almost parallel with the surface, it can load only within the limits of elastic deformation, and therefore without any damage. Another case is perpendicular impact of particles on the surface resulting in elastic or plastic surface deformation. After the perpendicular impact on the surface, the particle is usually reflected back [1].

Two main methods of erosive wear evaluation are based on particle flow. Particles are moved by air or gas, or on the basis of centrifuge. In the first case, the particles are under a certain pressure accelerated by the air flow towards a specimen and impinge the surface perpendicularly or under a specific angle thanks to adjusted specimen positioning. In the second case, the centrifugal force is

IOP Conf. Series: Journal of Physics: Conf. Series **843** (2017) 012074 doi:10.1088/1742-6596/843/1/012074

introduced by the rotation and acceleration of abrasive particles from channels in the rotating disk. Deng et al. [2] investigated differences between two above mentioned methods, particularly with regard to size, shape, concentration and velocity of abrasive particles. In terms of velocity, they found that the relationship between air pressure and particles velocity is in the gas equipment almost linear for all shapes, sizes and concentrations of abrasive particles. Identical dependency was proved also for the centrifugal equipment, precisely the relationship between the angular velocity of rotating disk and the velocity of particles leaving from channels. In case of gas equipment, it was found that smaller and sharper particles are accelerated to higher velocities. Regarding the dependency between concentration and particle velocity, they found that higher concentrations reduce particles velocity as a result of increased collisions leading to slower particles with lower kinetic energy. On the contrary, the dependence between particle properties and their velocity is minimal by the centrifugal device. An important effect occurring only in gas equipments is an extension of particle flow after leaving an acceleration tube. This effect increases with increasing velocity and concentration of particles. This phenomenon was attributed to the aerodynamic clustering between particles and air stream carrying them, mutual interaction of particles and interaction between particles and acceleration tube wall. The gas stream expansion further results in the generation the so-called “dead zone” in the middle of particle flow. Moreover, smaller particles have greater tendency to copy the gas flow compared to bigger abrasive grains. Furthermore, the effect of abrasive particles properties on material erosive wear was mentioned. The shape of abrasive particles is an important factor. Angular particles cause higher wear in the form of surface cutting, especially at lower impact angles. On the other hand, oval particles cause material removal as a result of plastic deformation and scratching.

Moreover, measurements at elevated temperatures can be performing using centrifugal device, as indicated e.g. Hayashi et al. [3]. In that case, the circumferentially attached specimens are provided with a heating spiral and the entire working part is installed in a tightly sealed chamber, in which the pressure is regulated by means of a vacuum pump. From this reason, erosive tests may be performed in inert or corrosive atmosphere.

2. Experiment

Five commercially available powders were used to prepare the samples. These powders were Amperit 588.074 (Cr_3C_2-25%NiCr) with particle size distribution for HVOF (-45+15 mm), Amperit 594.074 (Cr_3C_2-25%CoNiCrAlY) with particle size distribution for HVOF (-45+15 mm), Stellite 6 - M-484.33 (CoCrWC) with particle size distribution for HVOF (-53+20 mm), M-341.33 (Hastelloy C-276) with particle size distribution for HVOF (-53+20 mm), M-771.33 (NiCrBSi) with particle size distribution for HVOF (-53+20 mm) and one experimental powder labeled T10 (TiMoCN-29%Ni). All coatings were deposited by HP/HVOF (High Pressure/High Velocity Oxygen Fuel) technology with JP-5000 torch from the company TAFA Incorporated. Already optimized spray parameters were used for preparation of each coating. Nitrided stainless steel 1.4923 (P91) was used as a competing surface treatment technology. We were also evaluated surface treatment using PVD technology. Deposited coating was TiAlN.

Construct steel 1.0421 (DIN 11 523) was used as the base material. The substrate surface was degreased and grit blasted before spraying. Brown corundum F22 with grain size (0.8 to 1.0) mm was used as abrasive medium. Before coating deposition, all specimens were blasted in order to achieve proper adhesion of coating on the substrate material. The coating thickness was in the range of 250-400 µm.

2.1. Erosion – Equipment description

The used equipment for evaluation of material erosive wear operates as follows. A given constant amount of abrasive media is led from the feeder in the upper part of the equipment through the "choke valve" and replaceable cross-sectional iris. The abrasive media continues between two firmly connected, rotating discs with four channels resulting in acceleration caused by centrifugal force. As the abrasive media leave channels, it impinges on specimens attached to the disc circumference.

Specimens are placed on predefined positions, towards both the axis of disc rotation and the axis of channels. Material weight loss of specimen was evaluated. If we know the density of test material the weight loss can be converted to volume loss, which allows quantitative comparisons.

The equipment consists of three main parts shown in Figure 1. The first part is an erodent feeder. The cover with scale to regulate erodent consumption is placed on the feeder, shown in Figure 2b. Erodent particle flow through the choke valve is regulated by a control cone, which is secured against loosening on a bar using a nut. Then the erodent continues into the second, so-called working part, through replaceable cross-sectional iris. After passing through the iris, the particles are ejected through radial rotor channels towards the specimens fixed in desired angle. The longitudinal specimen positioning correlates with the trajectory of particle flow from the rotor. The impact angle can be adjusted by tilting around the horizontal axis. This ensures that the contact area of specimen remains constant even with the different setting of impact angle. Specimens are attached using retractable plates and compression springs in a way that ensures easy attachment and removal. Erodent further falls through the middle part of the equipment, where an electric motor is placed. High speeds could lead to motor overheating and for this reason, the motor is cooled by compressed air which is regulated by mechanical valve. Engine revolutions and thus also erodent speed are controlled by frequency converter. After passing through the middle part, the erodent continues to the bottom of the equipment, into the so-called gripper (tray for abrasive). A sealing (rubber sleeve) to prevent dust particles leakage is placed between the central part and the gripper.

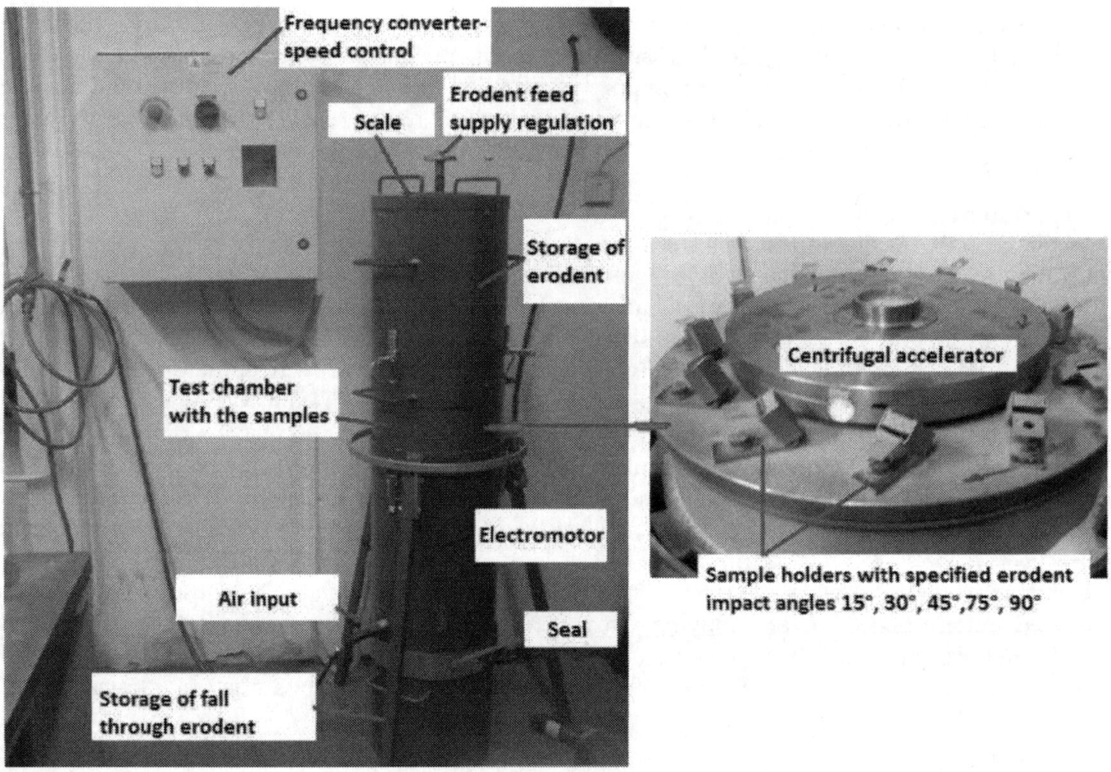

Figure 1. Equipment for evaluation of material erosion resistance working on the principle of centrifugal erodent flow.

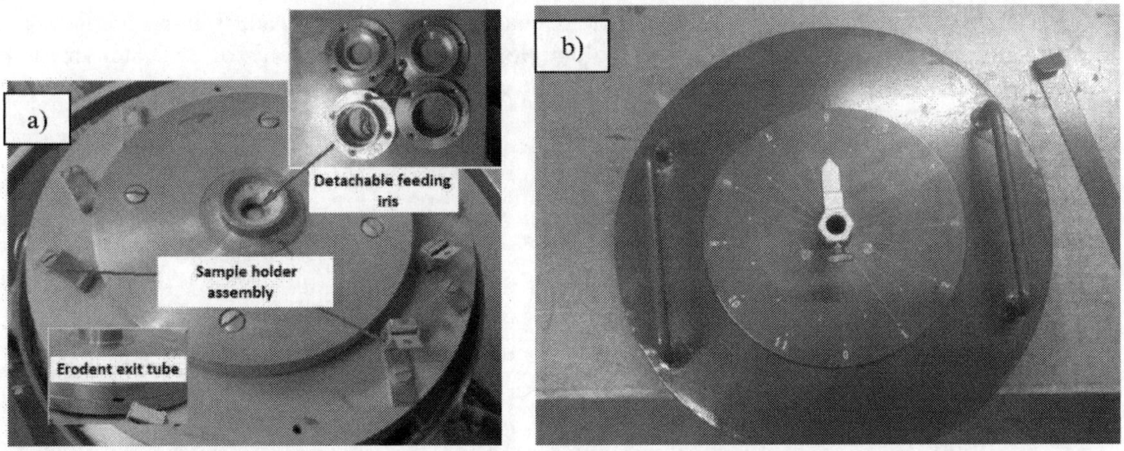

Figure 2. a) Working part of the equipment, b) Cover with scale for regulation of abrasive flow.

2.2 Measurement description

It is necessary to establish some basic parameters before starting the measurement itself. The parameters are the rotational speed of rotating disk and the amount of abrasive media depending on cone rotation (reading on the cover scale). After the finalization of these experiments, other parameters important for repeatable equipment operations were set – the dependence of erodent impact velocity and abrasive media amount on the erosive wear of tested materials.

After equipment tuning and determination of basic parameters, the evaluation of erosive wear resistance by selected thermally sprayed coatings using HVOF technology high was conducted. Optimized parameters were used for coating application. Parameters of erosive tests were following:

- Impact erodent angle - 90 °, 60 °, 30 ° and 15 °
- Evaluation time - 2 min
- Speed of rotating disc - 4500 rev / min
- Iris - 20 mm diameter
- Abrasive flow - cone rotation of 0.5 revolutions (180°)

3. Results

The resulting weight losses were converted using material density to volume losses in order to compare tested materials. Two specimens were used for each angle and the weight losses were measured three times for each specimen. Densities of evaluated materials are shown in Table 1.

Table 1. Densities of evaluated coatings.

Material	Powder indication	Density at 24.5 °C (g/cm^3)
Cr_3C_2-25% NiCr	588.074	6,685
Stellite 6	484.33	8,296
Hastelloy C-276	341.33	8,611
Cr_3C_2-25% CoNiCrAlY	594.074	6,479
NiCrBSi	771.33	7,176
TiMoCN-29% Ni	T10	5,894
Wr.Nr. 1.4923	X22 (P91)	7,7
TiAlN (PVD)		8,000 (estimation)

Weight loss measurement results of evaluated coatings under four erodent impact angles are summarized in the graph in Figure 3 and 4. Volume loss measurement results of evaluated coatings under four erodent impact angles are summarized in the graphs in Figure 5 and 6.

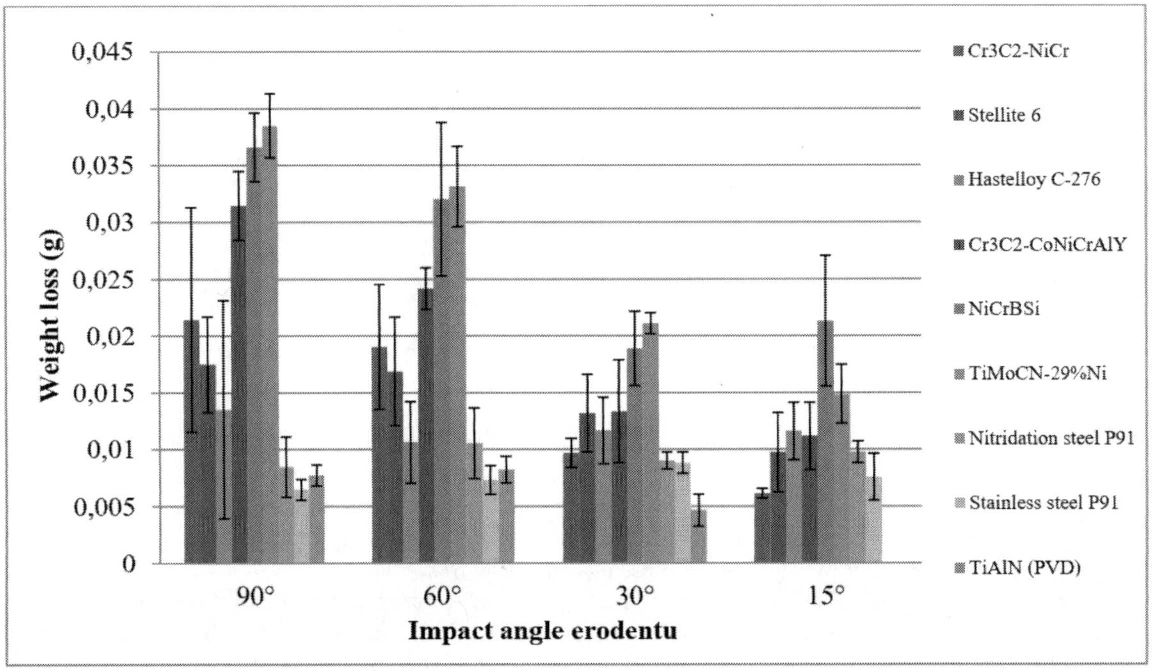

Figure 3. Weight loss of evaluated coatings under all erodent impact angles.

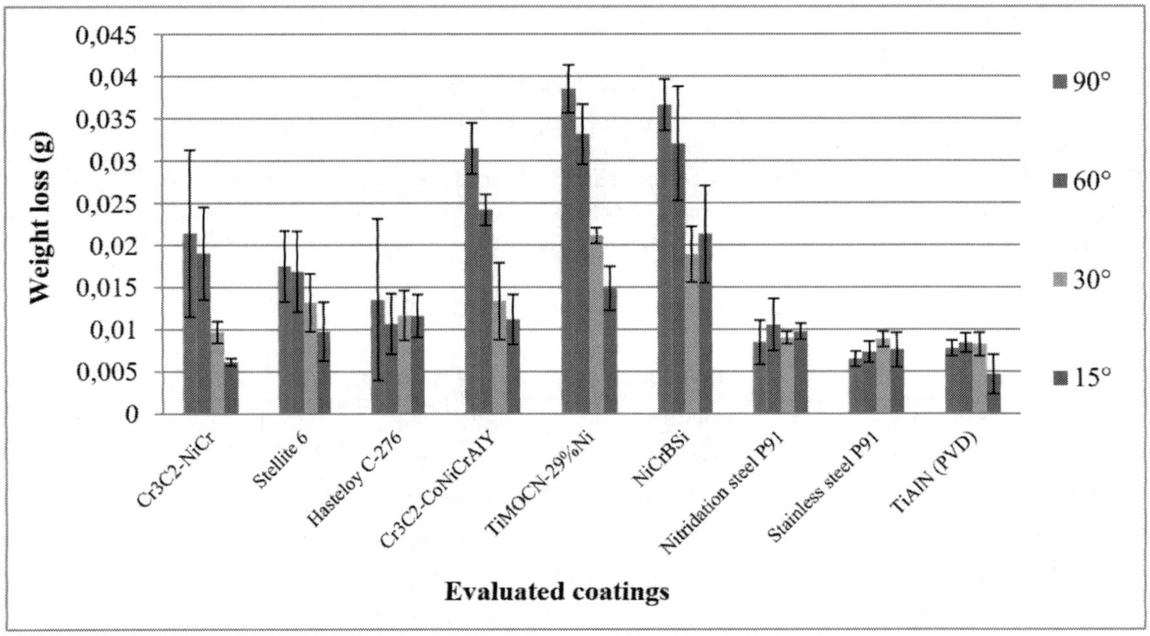

Figure 4. Weight loss of all evaluated coatings.

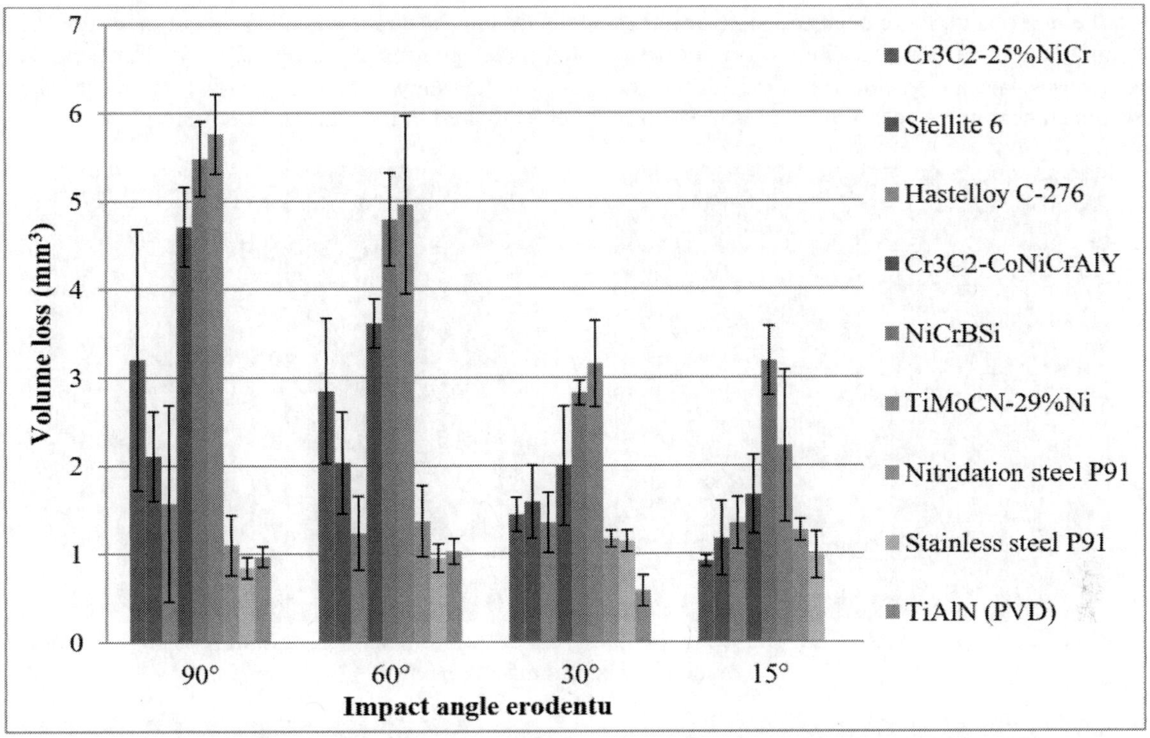

Figure 5. Volume loss of evaluated coatings under all erodent impact angles.

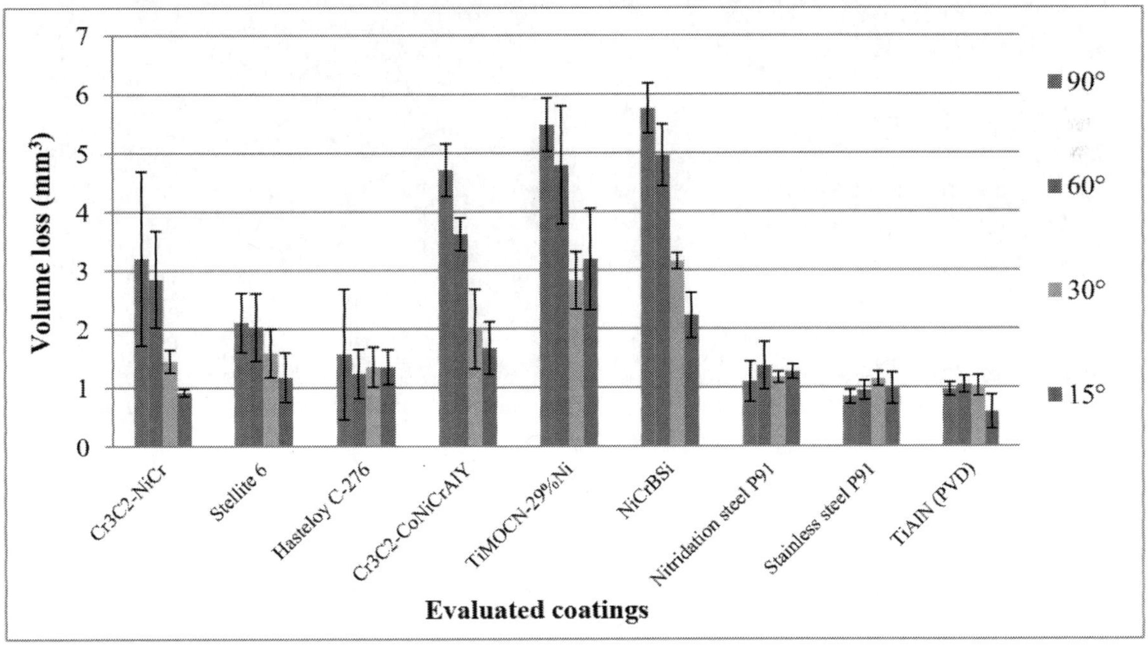

Figure 6. Volume loss of all evaluated coatings.

The wear mechanism under the erodent impact angle of 90 ° and 15 ° for Cr3C2-25%NiCr coating is illustrated in the Figure 7. Surface appearance after the erosive wear test varies significantly depending on the erodent impact angle. In case of perpendicular erodent impact angle, the traces of Al_2O_3 adhesion are visible on the surface. In case of 15 ° impact angle no erodent adhesion occurred.

On the other hand, there occurred significantly noticeable impact direction and the wear mechanism is a combination of plastic deformation and release of bigger coating parts [6], [7], [8]. Furthermore, previous splats are visible on some places. Figures 8 to 15 show traces of erosive wear for other evaluated coatings. Erosion mechanism will be further evaluated in the chapter Discussion.

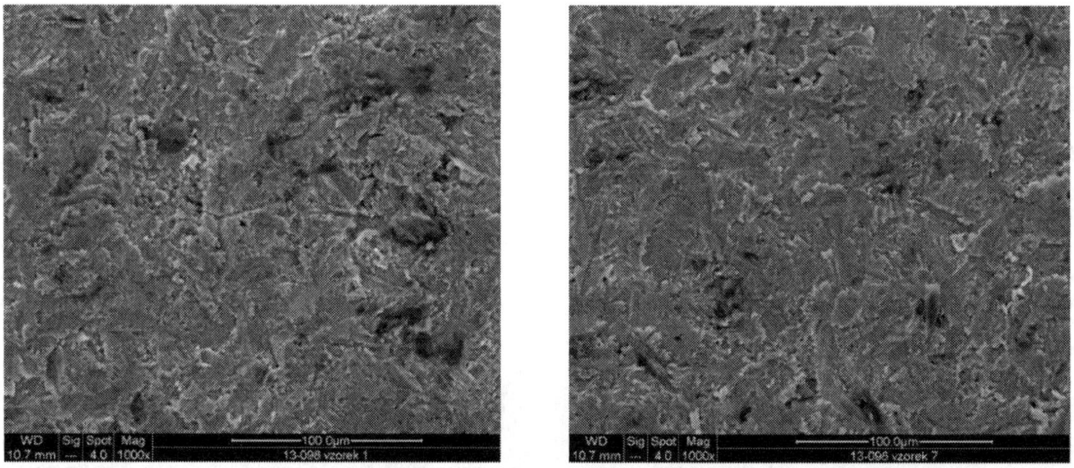

Figure 7. Erosive wear of Cr_3C_2-25% NiCr coating under 90 ° and 15 ° erodent impact angle in MIX mode (1000x magnification).

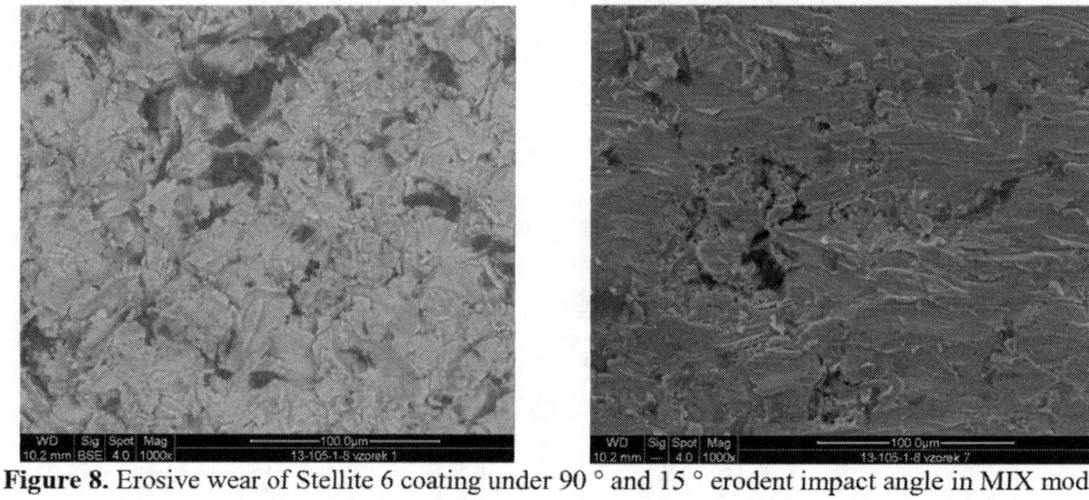

Figure 8. Erosive wear of Stellite 6 coating under 90 ° and 15 ° erodent impact angle in MIX mode (1000x magnification).

IOP Conf. Series: Journal of Physics: Conf. Series **843** (2017) 012074 doi:10.1088/1742-6596/843/1/012074

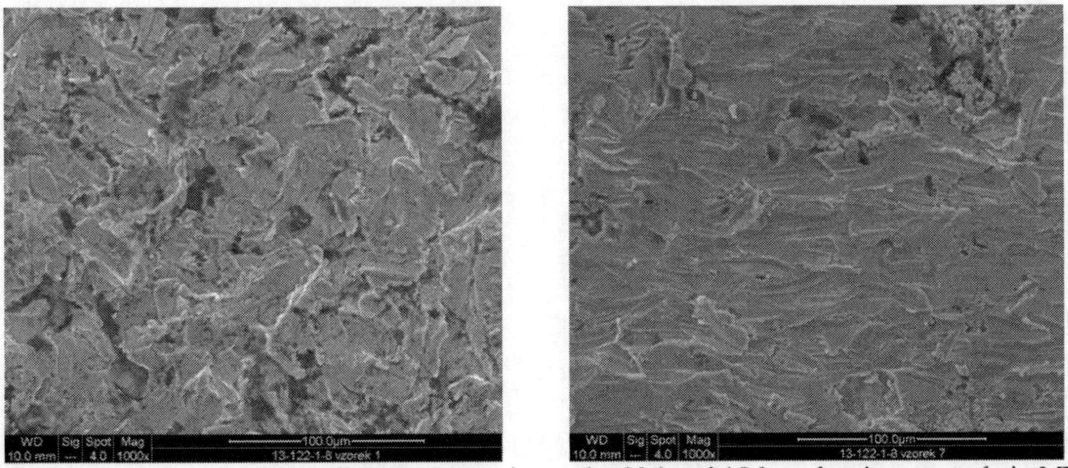

Figure 9. Erosive wear of Hastelloy C-276 coating under 90 ° and 15 ° erodent impact angle in MIX mode (1000x magnification).

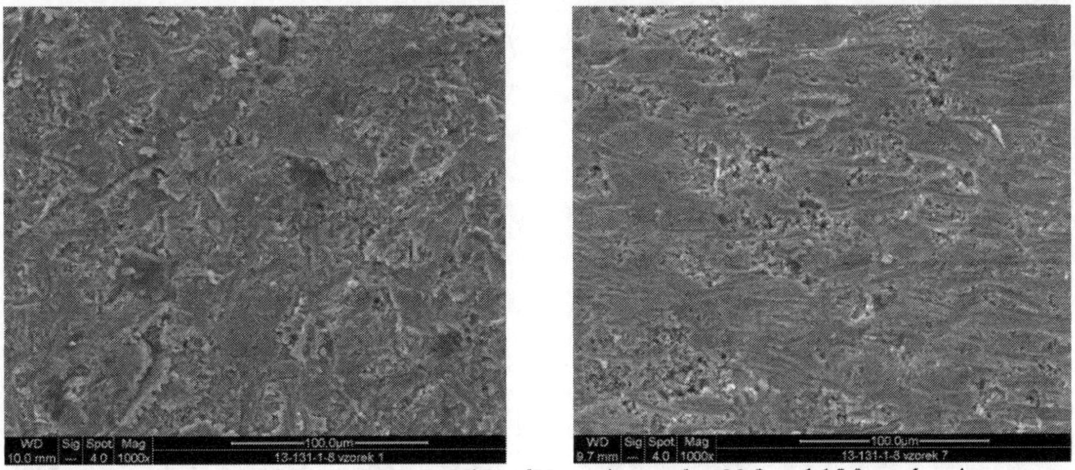

Figure 10. Erosive wear of Cr_3C_2-25% CoNiCrAlY coating under 90 ° and 15 ° erodent impact angle in MIX mode (1000x magnification).

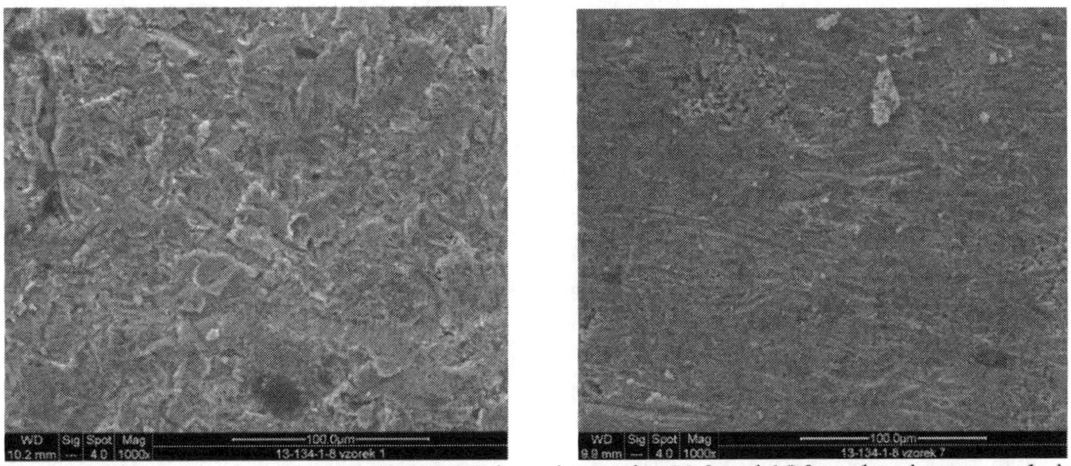

Figure 11. Erosive wear of TiMoCN-29 %Ni coating under 90 ° and 15 ° erodent impact angle in MIX mode (1000x magnification).

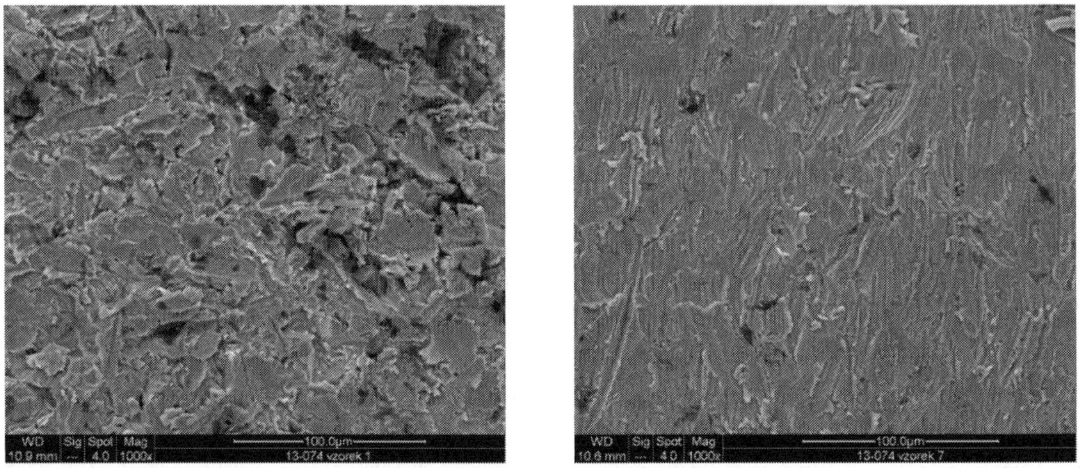

Figure 12. Erosive wear of NiCrBSi coating under 90 ° and 15 ° erodent impact angle in MIX mode (1000x magnification).

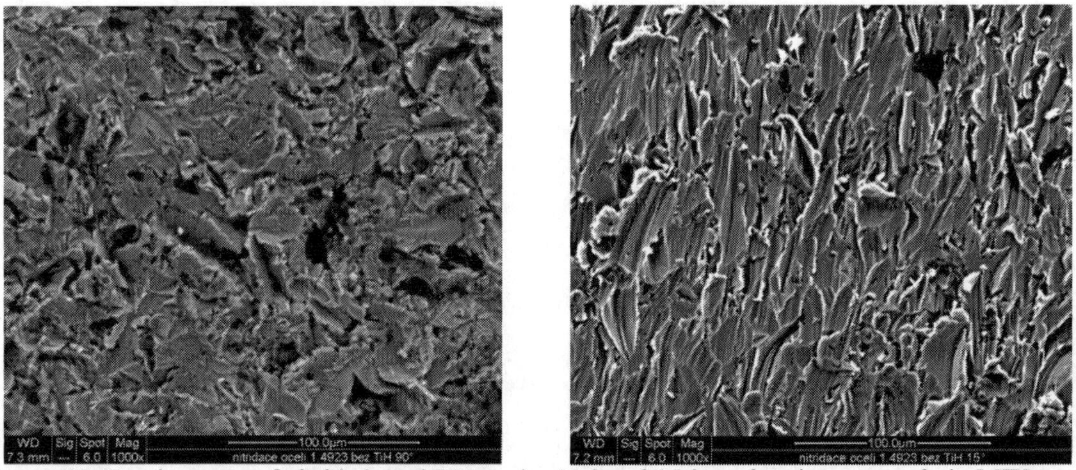

Figure 13. Erosive wear of nitrided steel P91 under 90 ° and 15 ° erodent impact angle in MIX mode (1000x magnification).

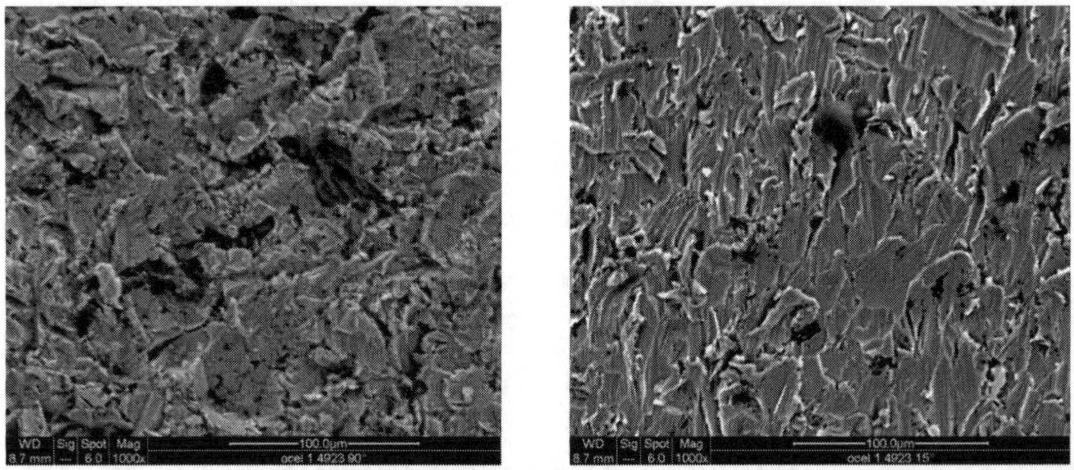

Figure 14. Erosive wear of steel P91 under 90 ° and 15 ° erodent impact angle in MIX mode (1000x magnification).

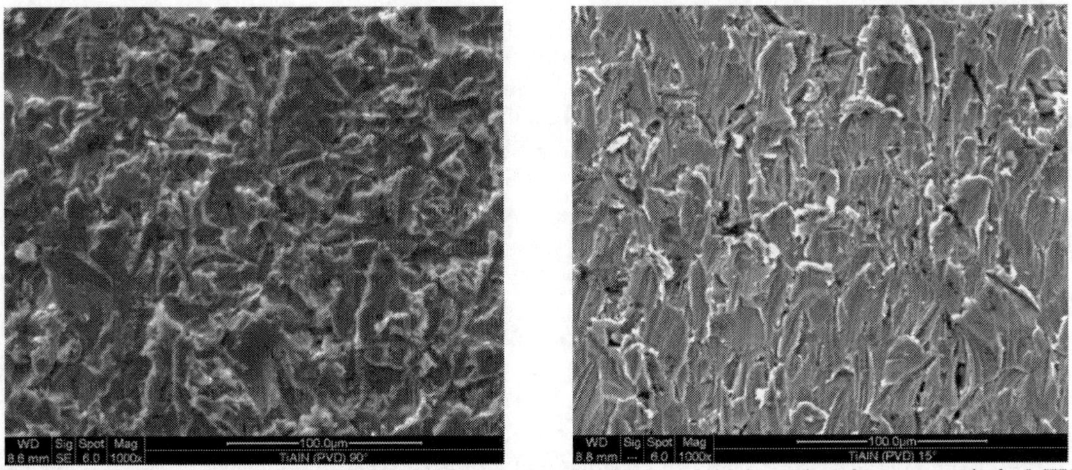

Figure 15. Erosive wear of TiAlN (PVD) coating under 90 ° and 15 ° erodent impact angle in MIX mode (1000x magnification).

4. Discussion

Cr_3C_2-25% NiCr coating exhibited significant dependence on the erodent impact angle. Under the perpendicular impact angle, its wear resistance is lower in comparison with alloy based coatings Hastelloy C-276 and Stellite 6. The wear resistance gradually increases with decreasing erodent impact angle. On the other hand, regarding the erosive wear resistance under 15 ° erodent impact angle, Cr_3C_2-25% NiCr coating shows one of the best results. A similar trend, although to lower extent, shows Stellite 6 coating. On the contrary, alloy based Hastelloy C-276 coating appears to be independent on the erodent impact angle, similarly also nitrided steels P91, TiAlN (PVD) coating and steel P91 without surface treatments.

The results can be interpreted with regard to the presumed material hardness and toughness (brittleness). While hard and abrasion-resistant Cr_3C_2-25% NiCr coating benefits under low erodent impact angle from its high hardness and its erosion wear resistance is similar to abrasive wear resistance; perpendicular erodent impact angle leads to brittle damage, release of carbide particles and overall higher weight loss. Alloy based coatings with lower hardness but higher toughness and plastic deformation ability are less dependent on impact angle. Similar results were achieved during the evaluation of erosion resistance by cermet and alloy based coatings within the project COST 523 [4] and in the publication [5].

Cr_3C_2-25% CoNiCrAlY, TiMoCN-29% Ni and NiCrBSi coatings seems to be very inappropriate as protective coatings against erosive wear for functional surfaces of components.

5. Conclusions

The aim of this study was to evaluate and compare coatings applied using thermal spraying HP/HVOF technology in terms of their erosive wear resistance with competing surface treatments. Figure 5 and Figure 6 evidently show that under the erodent impact angles of 90° and 60°, there occurred the highest erosive wear by Cr_3C_2-25% NiCr, Cr_3C_2-25% CoNiCrAlY, TiMoCN-29% Ni and NiCrBSi coating; somewhat better erosive wear resistance exhibited alloy based coating Stellite 6 and the best results among thermal sprayed coatings exhibited alloy based coating Hastelloy C-276.

Regarding the erosive wear under the erodent impact angle of 30 °, Cr_3C_2-25% NiCr, Hastelloy C-276 and Stellite 6 coatings proved to have very similar resistance. However, the situation changes under the erodent impact angle of 15 °. Due to the different wear mechanism dependent on the impact angle, there were obtained completely different results in comparison with 90° impact angle. In case of perpendicular erodent impact angle, the traces of Al_2O_3 adhesion are visible on the surface. In case of 15 ° impact angle no erodent adhesion occurred. On the other hand, there occurred significantly noticeable impact direction and the wear mechanism is a combination of plastic deformation and

release of bigger coating parts. Under this impact angle, the best erosive wear resistance showed cermet based coating Cr_3C_2-25% NiCr and the worst NiCrBSi coating.
Cr_3C_2-25% CoNiCrAlY, TiMoCN-29% Ni and NiCrBSi coatings seems to be very inappropriate as protective coatings against erosive wear for functional surfaces of components.

Wear resistance of nitridation, TiAlN (PVD) and P91 steel does not seem to be influenced by any erodent impact angle and they appear to be the most appropriate out of all evaluated surface treatments. From the economical point of view, the best solution would be the base material (steel P91), as it surprisingly showed the highest erosion wear resistance.

References

[1] Bushan B 2002 *Introduction to Tribology* John Wiley & Sons Inc. (New York)
[2] Deng T Bingley M S Bradley M S A Silva S R 2008 *A comparison of the gas-blast and centrifugal-accelerator erosion testers: The influence of particle dynamics* Wear pp 945-955
[3] Hayashi N Kagimoto Y Notomi A Takeda Y Kato K 2005 *Development of new testing method by centrifugal erosion tester at elevated temperature,* Wear pp 443-457
[4] Houdková Š Enžl R 2005 *Erozní odolnost povrchových vrstev,* VYZ 0777/05, Interní výzkumná zpráva, ŠKODA VÝZKUM s.r.o.
[5] Houdková Š Česánek Z Polach P 2015 *Solid Particle Erosion of Selected HVOF Sprayed Coatings* Applied Machanics and Materials Vol. 827 (2016) pp 39-46
[6] Yoshida M Tani K 2004 *Erosion mechanism of thermally sprayed coatings depending on shape and hardness of partickles,* ITSC 2004, May 10-12, OSAKA, Japan
[7] Legoux J-G Arsenault B Hawthorne H Immarigeon J-P 2003 *Erosion behavior of WC-10CO-4Cr HVOF Coatings*, Thermal Spray Technology, ASM International, p.405-410
[8] *Wang B Q Shui Z R 2002 The hot erosion behavior of HVOF chromium carbide-metal cermet coatings sprayed with different powders*, Wear 253), p. 550-557

Experimental verification of the coming out of the shaft for the ceramic rolls/rollers with shrink fitting system

Y Sano[1], H Sakai[1],G Zhang[1] and N-A Noda[1]

[1]Department of Mechanical and Control Engineering, Kyushu Institute of Technology Sensui-Cho 1-1 Tobata-Ku, Kitakyushu-Shi, Fukuoka, Japan

E-mail: sano.yoshikazu029@mail.kyutech.jp

Abstract. Ceramic rolls/rollers have an advantage in steel strip production because of its high wear resistance and corrosion resistance. The new rolls/rollers consist of a ceramic sleeve and two alloyed steel shafts connected to the sleeve ends only under a small shrink fitting ratio because of the brittleness. However, the alloyed steel shafts may be loosening out from the ceramic sleeve under such low shrink fitting ratio. In the some previous studies have considered the coming out mechanism was considered by the numerical simulation focusing on the displacement and the shear stress at the joint end. This phenomenon should be confirmed without experiment. In this study, therefore, in order to clarify the coming out behavior, rotating bending experiment is conducted for a small roller specimen consisting of a steel sleeve and steel shafts connected by low shrink fitting ratio. Then, it is found that the coming out can be realized in the small specimen.

1. Introduction

Steel conveying rollers are used in the heating furnace (see Figure 1) for producing high-quality steel plates for automobiles. The conventional roller (Figure 2(a)) material is steel with ceramic spray coating on the outside of sleeve. To reduce the temperature, inside of the roller is cooled by water. However the thermal expansion mismatch may exceed the adhesion strength of the ceramic layer and causes failure on the roller surface such as crack, peeling, and wearing [1]. Therefore, the life of the roller becomes quite short.

Figure 2(b) shows a new ceramic roller consisting of steel shaft at both ends and ceramic sleeve having high heat resistance, high corrosion resistance and high wear resistance [2, 3]. Several previous studies suggested that the shrink fitting system may be the most suitable joining method for cylindrical structures to reduce the maintenance time and cost for the shaft replacement [4-8]. However, since only a small shrink fitting ratio can be used because of the ceramic brittleness [9-15], the coming out of the shaft from the sleeve may happen during operation. In our previous studies, therefore, the coming out simulations are performed by using the finite element method [4, 5] for this kind of roll. Suryadi [4] et al conducted 3D model to simulate the coming out but only until N = 5 loading circles because of the large calculation time. Therefore, Xu [5] et al developed 2D model to reduce the calculation time and obtained the results until N = 40 circles or more. Those studies [4, 5] have proved that the coming out behavior can be realized in the 2D and 3D simulations.

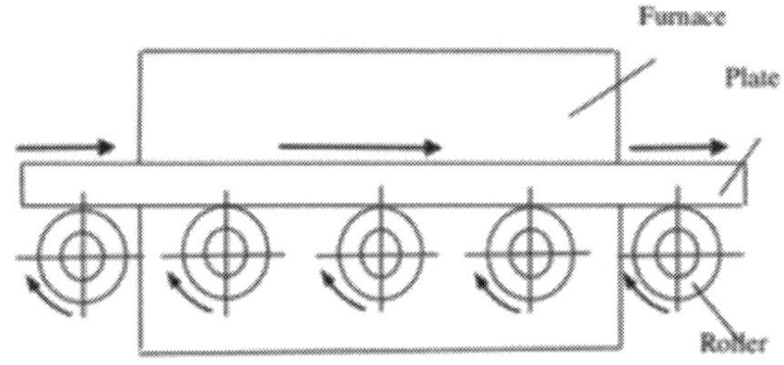

Figure 1 Layout of rollers in heating furnace

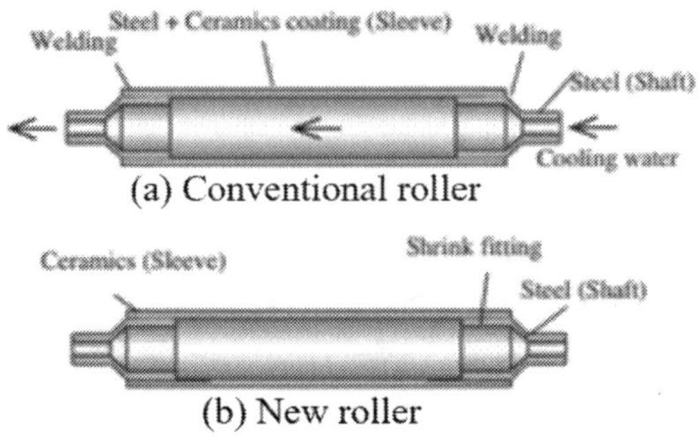

Figure 2 Roller structure

However, in the previous studies, the discussion was simply based on the numerical simulation without experiment. The real ceramic rollers and rolls are very expensive having large dimensions (see Figure 3) and therefore not suitable for confirming the coming out phenomenon.

In this study, therefore, in order to clarify the coming out behavior, rotating bending experiment will be conducted for a small roller specimen consisting of a steel sleeve and steel shafts connected by low shrink fitting ratio.

2. Coming out analysis

2.1. Analytical conditions

Figure 3 shows dimensions of the roller considered whose outer diameter D = 300 mm. Here, the roller consists of ceramic sleeve and steel shaft connected by shrink fitting. The shrink fitting ratio is defined as δ/d, where δ is the diameter difference and d is the inner diameter of the sleeve d = 240 mm. The shrink fitting connection was analyzed in the previous studies [8, 13, 14, 17-20].

In this study, the coming out of the shaft is considered under room temperature because the coming out occurs more easily. When we consider the roller in the shrink fitting ratios is considered in the range $\delta/d = 0.01 \times 10^{-3}$-$4.0 \times 10^{-3}$ at room temperature.

Figure 4 shows the roller rotation under loading where Point A located at the bottom of the shaft moves to the top after rotating 180° as shown in Figure 4(b). To simulate the coming out behavior, the roll rotation is replaced by the shifted load win the circumferential direction on the fixed roll as shown in Figure 5. The roller is subjected to distributed load w = 30 N/mm as the weight of the conveyed steel assuming the shaft ends are simply supported. As shown in Figure 5, the continuous load shifting can be replaced by discrete load shifting with load shift angle θ_0, which is usually used as a standard discretization numerical analysis. The suitable shift angle θ_0 =12° is considered with the numerical results. Here, the initial load position θ_0=0° is corresponding to the number of cycle N = 0, and θ = 360° is corresponding to the number of cycle N = 1.

Due to the symmetry, the half model considered with the total number of element 154,320. The smallest element size at the contact portion between sleeve and shaft is 1.25 mm × 1.25 mm × 6 mm. Static structural analysis is performed to the roller by using MSC Marc Mentat 2011[24] with full Newton–Raphson iterative sparse solver of multifrontal method. In this study, a three dimensional elastic FEM analysis can be applied because the loading condition does not exceed the yielding stress for the steel shaft, and the macroscopic plastic deformation does not appear for ceramics sleeve until failure.

The effect of the torsional load at the contact portion can be ignored because the shear stress $\tau_{r\theta}$ is very small compared to the shear stress τ_{rz} [14]. In contact analysis, it is known that two types of friction models, that is, stick–slip model and bilinear model have good accuracy [24]. However, since the stick–slip model needs large amount of data to determine friction force during repetitive calculation process, in this study the bilinear model is applied where the friction force is

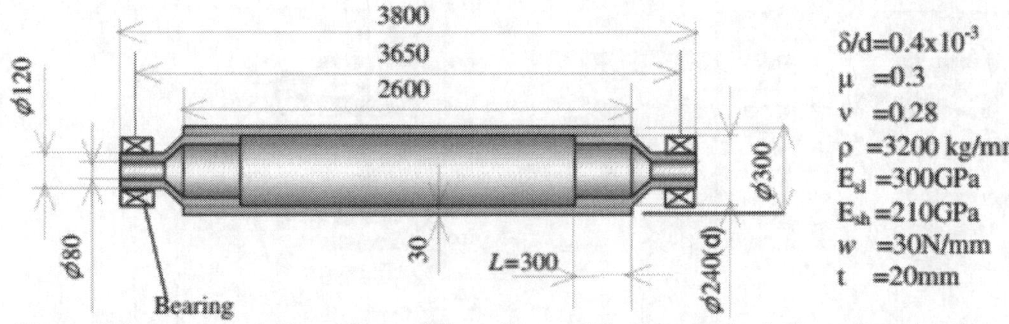

Figure 3 Structure and dimensions of the roller model with standard dimensions (mm)

Table 1 Material properties

Properties	Ceramic	Steel
Young's modulus [GPa]	300	210
Poisson's ratio	0.28	0.3
Tensile strength [MPa]	500	600
Mass density [kg/m³]	3200	7800
Thermal expansion coefficient [1/K]	0.3×10^{-5}	1.2×10^{-5}

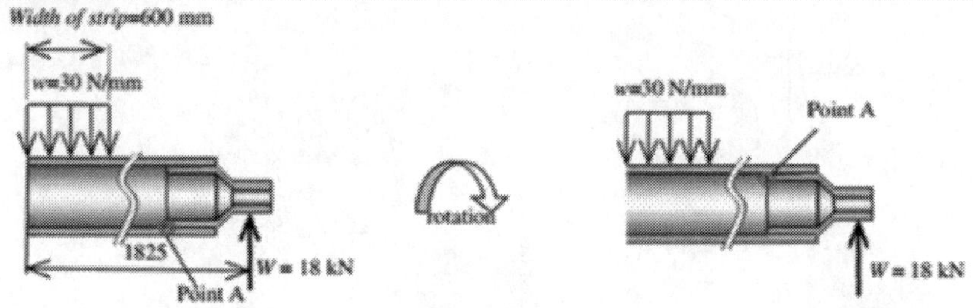

(a) Initial state (b) 180° rotation

Figure 4 Dimensions and loading condition of new roller

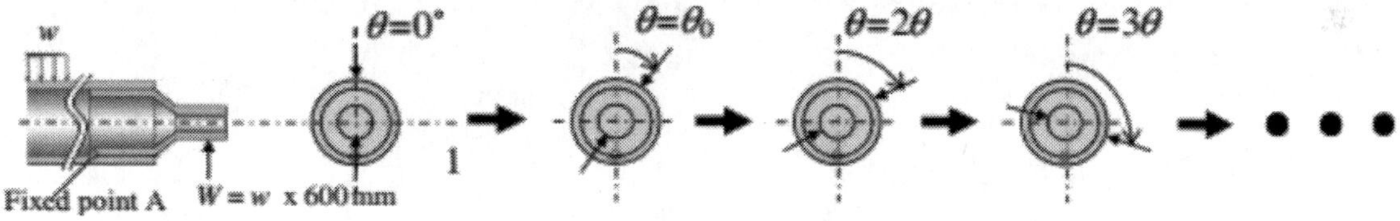

Figure 5 The rotation of the roller replaced by the shifted load at the interval of the load shift angle θ_0. When $\theta = 0°$ the number of cycle $N = 0$, and when $\theta = 360°$ the number of cycle $N = 1$.

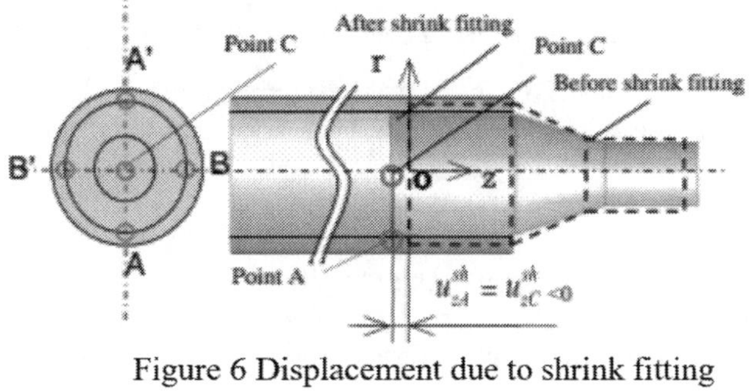

Figure 6 Displacement due to shrink fitting

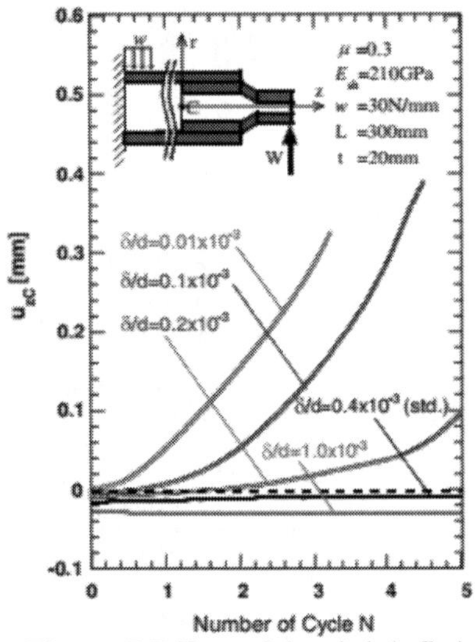

Figure 7 Effect of the shrink fitting ratio on the coming out of the shaft

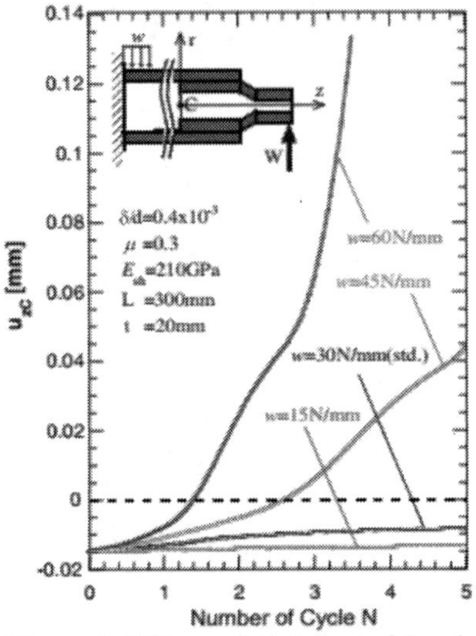

Figure 8 Effect of distributed load on the coming out of the shaft when δ/d=0.4×10^{-3}

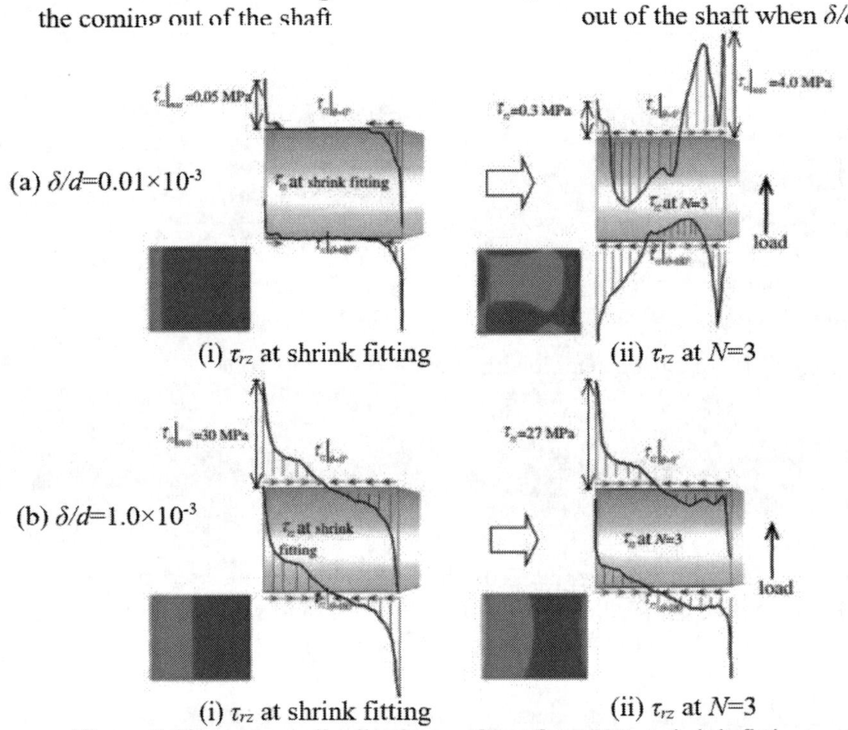

Figure 9 Shear stress distribution τ_{rz} from θ = 180° at shrink fitting and N=3

simply determined from the displacement. The friction coefficient between sleeve and shaft at joint portion is assumed as μ= 0.3. All conditions above are used as a reference condition.

2.2. *Analysis result*

2.2.1. *Effect of mechanical properties on the coming out*

Figure 6 illustrates the shaft deformation with the (r, z) coordinate defined before shrink fitting. The displacement u_{zC} in the z-direction is determined from the values at 4 points as $u_{zC}^{sh} = (u_{zA}^{sh} + u_{zA'}^{sh} + u_{zB}^{sh} + u_{zB'}^{sh})/4 = u_{zA}^{sh} < 0$ as shown in Figure 6. Then, since the shaft is under

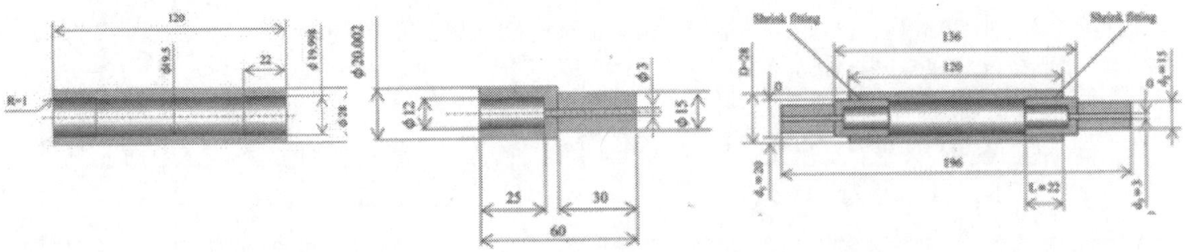

(a) Before shrink fitting (b) After shrink fitting

Figure 10 Dimensions of the sleeve and shaft with shrink fitting ratio δ/d=0.2×10^{-3} used as the test specimen (mm)

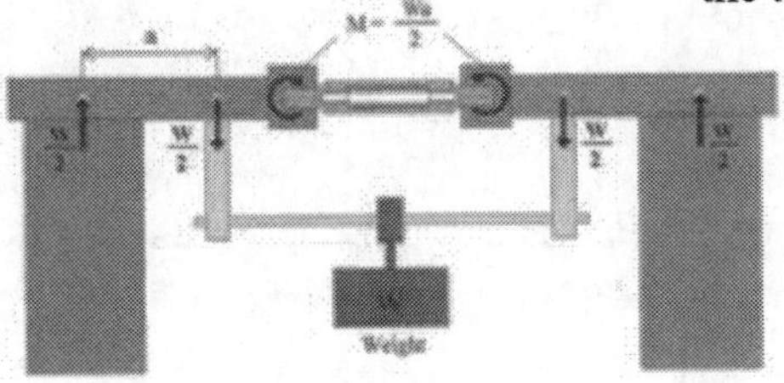

Figure 11 Schematic view of rotating bending machine

Table 2 Specification of gage

GAGE FACTOR	2.23±1.0%
GAGE LENGTH	0.2mm
GAGE REAIATANCE	120.2±0.2Ω
ADOPTABLE THERMAL EXPANSION	11.7 PPM/°C

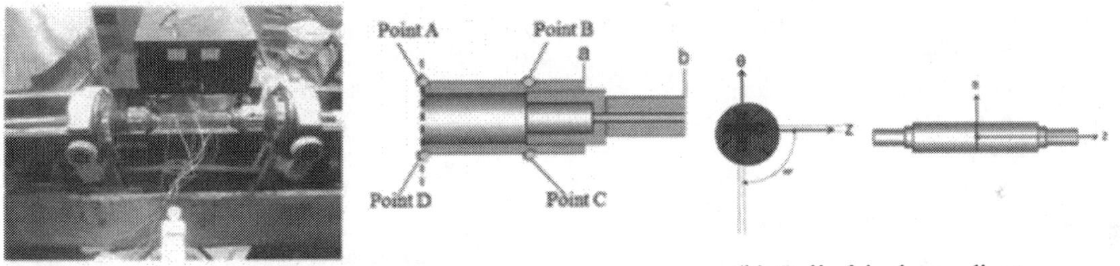

(a) Measurement position (b) Cylindrical coordinate

Figure 12 Strain gauge (KFG-1-120-D-16-11 LIM2S)

compression in the *r*-direction; we have $u_{zA}^{sh} = u_{zC}^{sh} < 0$ Figure 7 shows the results for u_{zC}. With increasing the shrink fitting ratio, the compressive stress increases causing the negative initial value of u_{zC} as shown in Figure 7 at N = 0. Under low shrink fitting ratio, the displacement u_{zC} increases significantly with increasing N.

Figure 8 shows the effects of the magnitude of load is investigated for w = 15, 45, 60 N/mm. The coming out speed significantly increases with increasing the magnitude of the load. It is seen that the coming out accelerates at N = 0–4. In other words, the coming out easily occurs when the distributed load $w \geq 45$ N/mm.

2.2.2. The coming out mechanism

The coming out of the shaft has been realized in the numerical simulation as shown in previous chapters. The results are different depending on the shrink fitting ratio and other parameters. Since it is difficult to obtain the results for large number N because of large calculation time, it is desirable that we can judge the coming out appears or not even when N is small. In this chapter, the coming out mechanism is considered by focusing on the stress distribution appearing at the contact portion.

Figure 9(a) focuses on the stresses along the lines θ = 0° and 180°. Here, $\tau_{rz}(0)|_{\theta=0°}$ and $\tau_{rz}(0)|_{\theta=180°}$ are compared between N = 0 and 3 under δ/d = 0.01 × 10^{-3}. The shear stress due to shrink fitting is

Table 3 Comparison of experiment value and FEM value

(a) Comparison of σ_z

		Point			
		A	B	C	D
σ_z [MPa]	Experiment	-101	-88	101	95
	FEM	95	80	107	72

(b) Comparison of σ_θ

		Point			
		A	B	C	D
σ_θ [MPa]	Experiment	-1.6	-4.8	1.6	1.3
	FEM	-1.3	-4.4	1.3	1.1

approximately equal to zero except at both ends when $N = 0$. Therefore, when $N = 3$, the shear stress at shrink fitting is greatly changed, and the shear stress directions are reversed at both shaft ends.

The shear stress at the left end tries to prevent the coming out of the shaft. In Figure 9(b), $\tau_{rz}(0)|_{\theta=0°}$ and $\tau_{rz}(0)|_{\theta=180°}$ are compared between $N = 0$ and $N = 3$ under $\delta/d = 1.0 \times 10^{-3}$. When N=0, it is seen the maximum value $\tau_{rz}|_{max}$= 30 MPa is about 600 times larger than the maximum value $\tau_{rz}|_{max}$= 0.05 MPa in Figure 9(a). Although not indicated, the maximum compressive stress in Figure 9(b) is about 180 times larger than the maximum compressive stress in Figure 9(a). Due to those large shrink fitted stresses, as shown in Figure 9(b), the effect of the bending load becomes smaller and stress does not change very much between $N = 0$ and $N = 3$ under $\delta/d = 1.0 \times 10^{-3}$.

3. Experiment verification to confirm and discuss by using small specimen

3.1. *Specimen shape and dimensions*

The material of small roll specimen is steel and the shafts are connected to the sleeve end by shrink fitting. Figure 10(a) shows dimensions of the sleeve and shaft before shrink fitting and Figure 10(b) shows dimensions of the roll specimen after shrink fitting. In this experiment we assumed the same value shrink fitting ratio δ/d=0.2×10^{-3}of real ceramic roll (see chapter 2.1). Figure 11 shows rotating bending machine.

3.2. *Stress due to shrink fitting and bending*

To confirm the real stress of the specimen, the stresses are measured by using the strain gauge. Figure 12 shows how to measure the stress due to bending and rotating for the test specimen. Figure 12(b) shows the positions A, B, C, D where the strains are measured. Here the stresses σ_z and σ_θ are calculated. Table 3 shows comparison between the experiment and FEM results. From Table 3, it is found that both results are good agreement within 10% error in most case.

3.3. Experiment simulation of coming out by rotating bending machine

The rotating bending test cannot be done at lower rotational speed less than 1800rpm. Therefore, in the experimental simulation, the shaft is rotated by hand to realize the coming out.

Figure13 shows the displacement u_{zc} of the shaft under rotating bending. Figure13(a) shows the results of the left shaft and Figure13(b) shows the results of the right shaft. Here, the distances l_1 and l_2 are measured as the distance between the sleeve end and the shaft end as shown Figure 14(a). Assume the average distance between the sleeve end and the shaft end is $l_{ab} = (l_1 + l_2)/2$. Here, l_{ab0} is the initial distance and l_{ab} is the distance after N cycle. Then the displacement u_{zc} is defined from l_{ab0} and l_{ab} as $u_{zc} = l'_{ab} - l_{ab}$ (see Figure 14(b)).

It is seen that the both displacement $u_{zc\text{-left}}$ and $u_{zc\text{-right}}$ increases in the z-direction with increasing the cycle N, that is, the shaft moves in the coming out direction. Therefore, it is found that the coming out are realized in the small specimen.

IOP Conf. Series: Journal of Physics: Conf. Series **843** (2017) 012075 doi:10.1088/1742-6596/843/1/012075

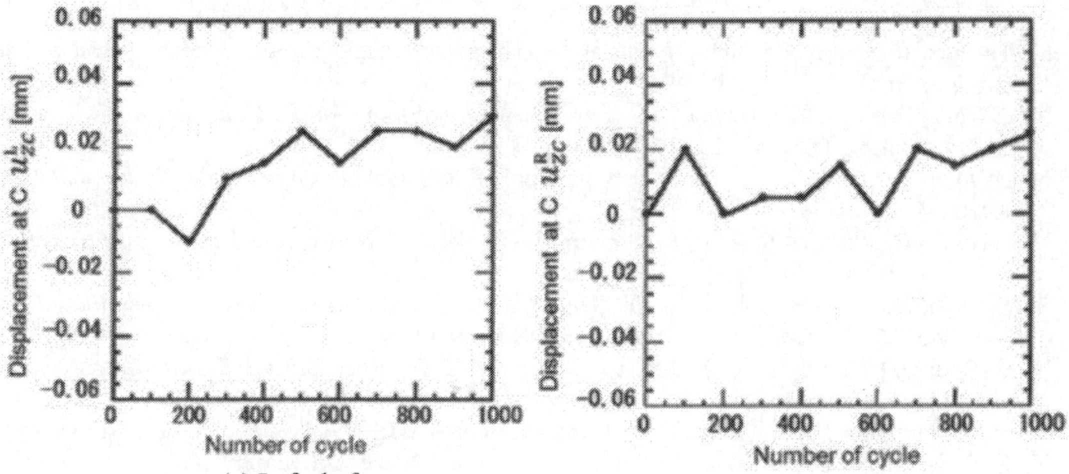

(a) Left shaft (b) Right shaft

Figure 13 The displacement u_{zC} vs. number of cycle

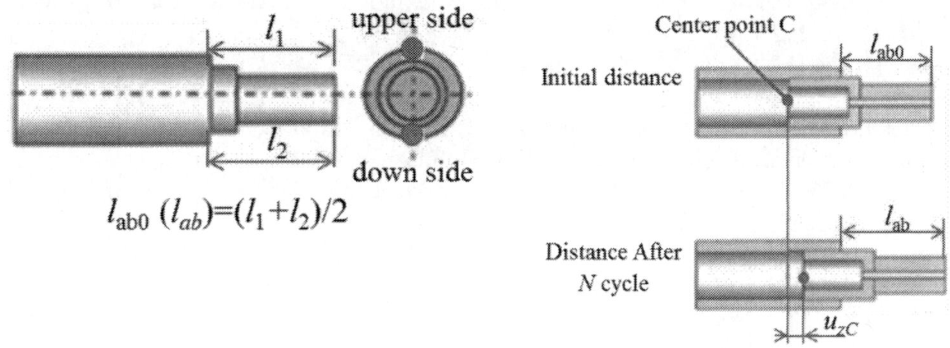

(a) Distances l_1 and l_2 (b) Displacement u_{zc}

Figure 14 Definition u_{zC} and l_{ab0}, l_{ab}

References

[1] M. Fujii, A. Yoshida, J. Ishimaru, S. Shigemura, K. Tani: *Transactions of the Japan Society of Mechanical Engineering* C, 72(2006), 716, 1354-1360.

[2] C. Liour, T. Mori, H. Kobayashi, T. Mitamura: *Journal of the Ceramic Society of Japan*, 98(1990), 4, 348-354.

[3] T. Ono: J. Jpn. Soc. Mech. Eng, 86(1983), 470.

[4] N.A. Noda, D. Suryadi, S. Kumasaki, Y. Sano, Y. Takase: *Engineering Failure Analysis*, 57(2015), 219-235.

[5] N.A Noda, Y. Xu, D. Suryadi, Y. Sano, Y. Takase: Journal of ISIJ International, 56(2016), 2, 303-310.

[6] W. Li, N-A. Noda, H. Sakai, Y. Takase: *Journal of Solid Mechanics and Materials Engineering*, 5(2011), 1, 14-24.

[7] W. Li, N.A. Noda, H. Sakai, Y. Higashi: *Key Engineering Materials*, 452-453(2011), 241-244.

[8] A. Rusin, G. Nowak, and W. Piecha: *Engineering Failure Analysis*, 34(2013), 217-227.

[9] E. Ogawa, K. Shimizu, S. Hamayoshi, N. Kumagai, Y. Ohtsubo, N.A. Noda, Y. Takase, K, Kishi, K. Shobu, T. Tabaru, E, Maeda, S. Koga, T. Matsuda: *Hitachi Metals Technical*

Review, 28(2012), 50.
[10] M. Tsuyunaru, N.A. Noda, Hendra, Y. Takase: *Transactions of the Japan Society of Mechanical Engineering* A, 74(2008), 919-925.
[11] N.A. Noda, Hendra, Y. Takase, M. Tsuyunaru: *Jounal of Solid Mechanics and Materials Engineering*, 2(2008), 11, 1410-1419.
[12] N.A. Noda, Y. Yamada, Y. Sano, S. Sugiyama, S. Kobayashi: Engineering Failure Analysis, 15(2008), 4, 261-274.
[13] N.A. Noda, Hendra, M. Oosato, K. Suzumoto, Y. Takase, W. Li: Key Engineering Materials, 462-463(2011), 1140-1145.
[14] S. Matsuda, D. Suryadi, N.A. Noda, Y. Sano, Y. Takase, S. Harada : *Transactions of the Japan Society of Mechanical Engineering* A, 79(2013), 989-999.
[15] N-A. Noda, D. Suryadi, S. Matsuda, Y. Sano, Y. Takase: *Journal of ISIJ International*, 55(2015), 11, 2416-2425.
[16] Marc Mentat team, Theory and User Information. Vol.A, MSC.Software, Tokyo, 2012.531-596.
[17] N.A. Noda, Hendra, Y. Takase, M. Tsuyunaru: *Journal of Solid Mechanics and Materials Engineering,* 2(2008), 11, 1410–1419.
[18] W. Li, N.A. Noda, H. Sakai, Y. Takase: *Journal of Solid Mechanics and Materials Engineering*, 5(2011), 1,14–24.
[19] W. Li, N.A. Noda, H. Sakai, Y. Takase: *Key Engineering Materials*, 452-453(2011), 241-244.
[20] S. Dedi, N.A. Noda, Y. Sano, Y. Takase, S. Dedi, N.A. Noda, Y. Sano, Y. Takase: *Proceeding of the 4th ACEE 2014*, pp. 221–222.
[21] N.A. Noda, Hendra, M. Oosato, K. Suzumoto, Y. Takase, W. Li: *Key Engineering Materials*, 462-463(2011), 1140-1145.
[22] S. Matsuda, D. Suryadi, N.A. Noda, Y. Sano, Y. Takase, S. Harada: *Transactions of the Japan Society of Mechanical Engineering* A, 79 (2013), 803, 989–999.
[23] A. Rusin, G. Nowak, W. Piecha, Shrink connection modelling of the steam turbine rotor, Eng. Fail. Anal. 34 (2013) 217–227.
[24] Marc Mentat team, Theory and User Information. Vol. A, MSC. Software, Tokyo, 2008. 532.

Shifted laser surface texturing for bearings applications

Š Houdková[1], P Šperka[2], M Repka[2], J Martan[1] and D Moskal[1]

[1]University of West Bohemia, new Technology Research Centre, Univerzitní 8, Plzeň, 30614

[2]Brno University of Technology, faculty of Mechanical Engineering, Technická 2, Brno, 61669

houdkov@ntc.zcu.czType the corresponding author's e-mail address here

Abstract. The laser surface texturing (LST) technologies, based on creation of micro-pattern with pre-defined geometry can positively influence both the friction and wear of tribo-elements. In practice, the integration of LST technology is often limited due to its slowness. The new method, so called shifted laser surface texturing (sLST) with increased process speed was developed to make the technology more attractive for the industrial application. In the paper, the texture created by sLST technology was applied onto the steel samples and Al-Sn-Si surface of sliding bearings. Both block-on-ring (ASTM G-77) laboratory tests of steel samples and high-loaded working application tests on Al-Sn-Si bearings surface were carried out to evaluate the influence of texture on tribological behaviour. The ASTM G-77 laboratory tests showed a positive effect of the texture on friction behaviour. Under the high-loaded testing conditions, the positive effect was observed in initial stages of the tests, decreasing the torque of textured bearings compared to the untreated one. Lately, the texture was worn out and have no influence on the overall wear of the bearings. Based on the above mentioned observations, the use of alternative bearing material with higher hardness or application of protective layer over the created texture was suggested to exploit the texture benefits.

1. Introduction

In recent years, the utilization of lasers as a tools for surface modification became popular. Apart from the most widespread technologies, such as laser hardening or cladding for wear-resistant applications, a more sophisticated surface treatment can be applied to modify the surface properties.

The laser surface texturing (LST)is a technology of creation of pattern with defined geometry on the surface of treated parts. Depending on the shape of the texture, the surface can receive a new functional properties, such as hydrophobicity [1, 2], adhesion of subsequent layers [3,4], light absorption [5,6], self-cleaning [5] or tribological performance [7].

The principle of LST can be described as a removing of material by ablation process. Using the scanning optical systems for movement of the laser beam on the workpiece on predefined trajectories, the desired texture geometry can be created [8]. Although, the LST is a promising technology increasingly used in industrial applications, there are some limitations of current LST methods, such as heat accumulation and oxidation [9-11], plasma shielding effect [12] or precision at high speed [13]. To overcome the limitations the new method, called shifted laser surface texturing (sLST) was developed, enables effective use of existing state-of-the-art high power ultrashort pulse lasers. This surface texturing process, which utilises the average power and repetition rate of the ultrashort pulsed laser source more effectively than current methods via spreading the pulses temporally across the component to be textured, enables precise laser surface texturing to be performed, at industrially relevant speeds, whilst ensuring minimal thermal degradation of the component's surface. In detail, the sLST is described in [14,15].

The tribological performance is the matter of interest of many industrial applications such as bearings [16], shafts, piston rings [17] etc. Many studies exist concerning the influence of surface texturing on the sliding performance. They are addressing the influence of texture geometry on the shape of the sliding behavior experimentally [18] [19] [20], or using numerical models [21].

From the published work, the factors playing significant role can be identified, such as the density of textured area or the diameter to depth ratio of created dimples [22] [23]. Besides them, the quality of the texture, namely the amount of re-deposited material inside [24] or around [25] the created objects, is crucial for success in effort to decrease the friction and wear.

Based on the above mentioned studies, the texture design with potential to decrease the friction in the boundary/mixed lubrication mode was chosen and applied on the surface of bearings steel samples and Al-Sn-Si bearings. The laboratory ASTM G-77 test as well as the high loaded test simulating the bearings working conditions were performed to verify the positive effect of the created texture on the sliding behavior.

2. Experimental

The texture design (Fig. 1) was chosen based on the literature review and the results of preliminary study of sliding properties [26]. The diameter of the dimples was set 78 μm, the depth varied between 6.5-7 μm. The distance between the dimples centers was 200 μm, leading to the texture area density of 12%.

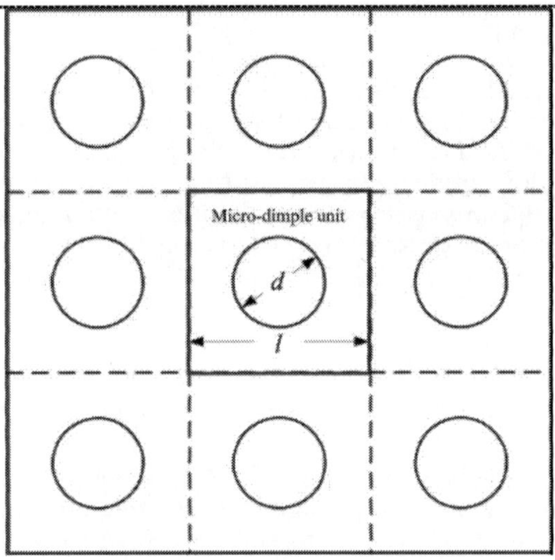

Figure 1. Texture design

The texture was applied on the outer surface of Timken A4138 rings, 35 mm outer diameter, made from the bearing steel (Fig. 2), and on the inner surface of Al-Sn-Si half bearings with inner diameter of 56 mm (Fig. 3).

The geometrical preciseness of the textured surface was evaluated by optical profilometer Bruker Contour GT X8.

Figure 2. Textured Timken A4138 ring

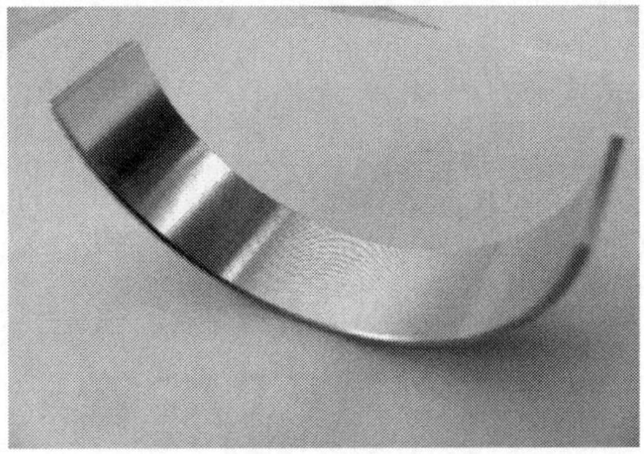

Figure 3. Textured half bearing

2.1. Texturing procedure

The Picosecond slab laser EdgeWave PX25-2-G with wavelength 532 nm, average power 14 W, pulse duration 10 ps and maximum pulse energy 100 µJ was used for texturing. The laser was equipped with scanning head with f-theta objective. The objective focal length was 255 mm, the laser spot diameter was 28 µm. Laser beam movement is produced by Scanlab software in connection with scanhead intelliSCAN® III 14. Maximal speed of laser moving is 8 m/s.

The sLST process of rings and bearings texturing was divided into several rectangular areas composed of array of dimples. Every textured rectangle was formed by sequential repetition of laser scanning processing of linear raster of individual spots – one spot per dimple in one repetition (Fig. 4). Between each repetition, the raster is slightly shifted to place next laser pulse with certain overlap. Individual spots are collected in four rings for every dimple. Whole quantity of the spots in one dimple is 127 laser pulses. Frequency of the laser pulses generation was 10 kHz, pulse energy 30 µJ and laser beam movement 2 m/s. Full process of forming of one textured rectangle contains 127 rasters (equal to the laser spots for one dimple) multiplied by 6 repetitions. After full finishing one of the textured rectangle the bearing was rotated to the next non-textured surface area for providing the next sLST processing. Rotations of the bearings were provided with step equal to the textured rectangular width 2 – 4 mm, depending on the bearing geometry. Speed of one movement between neighbor rectangles was set 50 degree per second.

Correct connection of the first and last textured rectangles was controlled by optical microscopy on Hirox 3D microscope. Total quantity of rectangles for full textured surface on ring with diameter 3.5 cm was 77 movements.

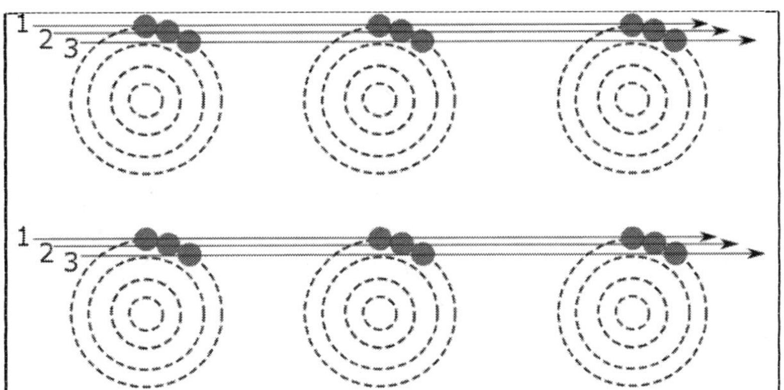

Figure 4. Schematic image of shifted Laser Surface Texturing (sLST)

IOP Conf. Series: Journal of Physics: Conf. Series **843** (2017) 012076 doi:10.1088/1742-6596/843/1/012076

2.2. Laboratory Block-on-Ring testing

The RTEC (RTEC Instruments, USA) universal tribometer was used to compare the sliding behavior of textured and non-textured Timken A4138 rings samples. The Block-on-Ring test, according to ASTM G-77 was carried out on non-textured sample (twice) and 3 samples with texture. The testing parameters are summarized in the Table1. Dip lubrication of the rings in mineral base oil was used and temperature of oil in the bath was recorded. All tests were started at similar temperature of the oil in the bath. During testing, the coefficient of friction (COF) was recorded. The surface topography was evaluated before and after the test.

Table 1. Testing parameters of "block-on-ring" testing

Item	Dimensions	Unit
Samples	Timken A4183, ø 35	mm
Counterparts	Block, 100Cr6, Ra 0.4	µm
Load	40	N
Speed (RPM)	10,20,50,100,200,400,700,1000,1500	min^{-1}
Intervals for each speed	10 (6 for 1000 and 1500 min^{-1})	s
Temperature	25 ± 2	°C
Lubricant	Mineral oil R834/90 (dyn. viscosity 0.115 Pas)	
Repetition	3 times from low seed to highest and vice versa	

2.3. Harsh condition testing

The testing system, developed in Daido Metal CO., Ltd. –The European Technical Center, was used to evaluate the behavior of Al-Sn-Si textured half bearings. The system is composed of a driving part where the shaft rotation can achieve up to 10 000 rpm, with loading and lubrication systems. For overloading conditions, the clutch is integrated. The loading system is applied by a hydraulic system able to simulate static and dynamic testing with force amplitude 100 KN and frequency 60 Hz. Bearings were tested for their wear resistance and ant seizure properties.

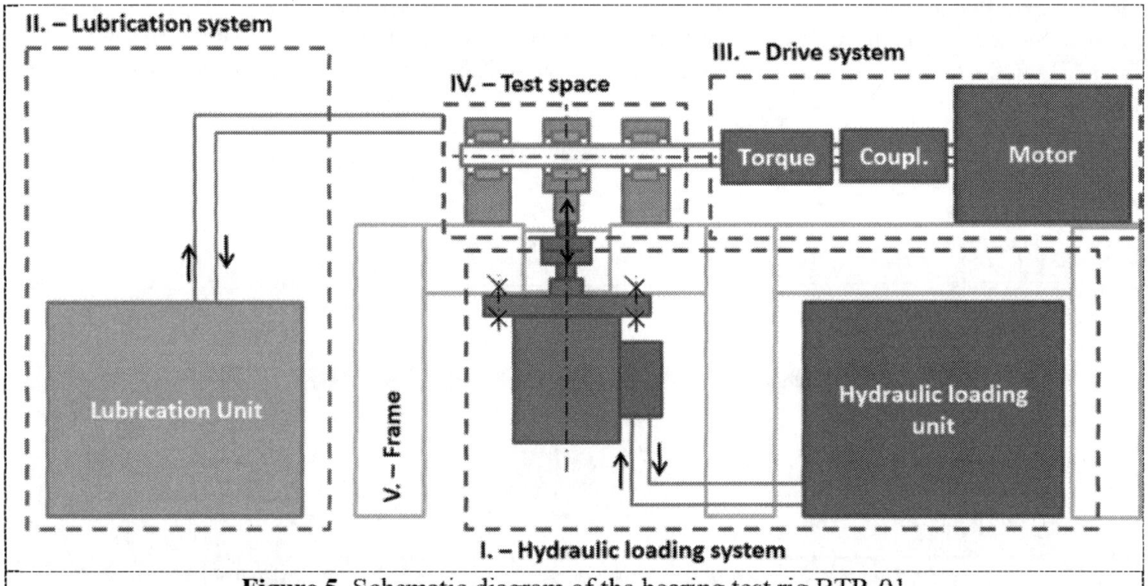

Figure 5. Schematic diagram of the bearing test rig BTR-01

2.3.1. Wear Test. Wear test conditions are described below in Table 2. The test machine was set for repeated start/stop cycling conditions for rotation of shaft to 1 m/s similar to hybrid vehicle or start-stop engines to be reproduced, i.g. with frequent start-ups when the oil film is sparse to promote bearing wear. The test mode is shown in Figure 6. Constant static load was set to the specific load (or nominal pressure) 5 MPa with limited oil flow supply of lubricant at room temperature. The wear volume was calculated from the measuring of the weight and the dimensional checks.

Table 2. Testing parameters of wear test

Item	Dimensions	Unit
Samples	Al-Sn-Si, ø 35W15xT1.5	mm
Counterparts	Shaft, Carbon steel, Rz ≤ 0.1	µm
Test specific load	5	MPa
Sliding speed	1	m/s
Test duration	20	hour
Lubricant	VG05	
Oil inlet temperature	Room temperature	
Oil flow	2.0	ml/min
Test mode	Start-stop	

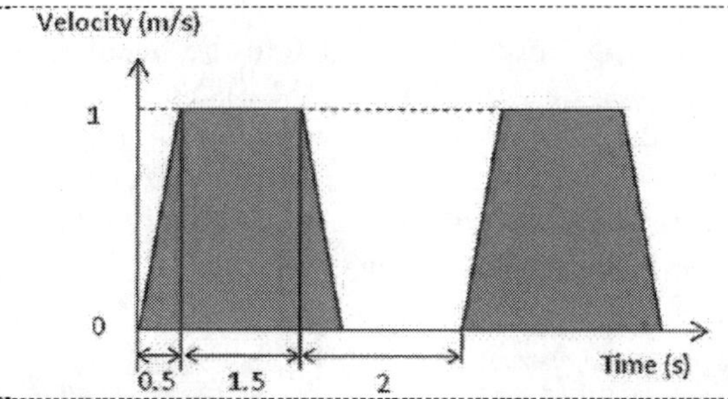

Figure 6. Schematic diagram showing the wear test mode

3. Results and discussion

3.1. Laboratory Block-on-Ring testing

The results of the Block-on-Ring testing summarized in the Figure 7. The average value of coefficient of friction and standard deviation is calculated from 6 measured values (each test was performed 3 times, the rotational speed was changed up and down).

From the measured results it can be seen, that the two rings (Textured 1 and Textured 2) showed lower friction in the area of lower speeds (up to 1000 min^{-1}). For higher speeds, the textured surfaces has the tendency to increase the friction compare to non-textured surface. According to the shape of the Stribeck curve and measured values, it can be said that the textured surface improves the friction in the mixed lubrication mode.

The results of Textured 2 sample showed opposite manner, but it has to be mentioned, that the behavior of the Textured 2 samples was not standard at all. According to optical profilometers measurements it can be estimated, that the sample differ from the others namely in the cylindricity of the basic geometry.

The same data are replotted against lubrication parameter in Fig. 8. Lubrication parameter is a ratio of lubrication layer thickness and the composite surface roughness. For the given testing configuration and materials, using the Hertz contact theory, film thickness prediction formula [27] and composite roughness (Rq 0.45 µm for the block; Rq 0.3 µm for the ring), the lowest measured COF was reached at the lubrication parameter in the range 3.5-4. It is known, that to full separation of the surfaces the value of lubrication parameter higher than 3 is needed [28].

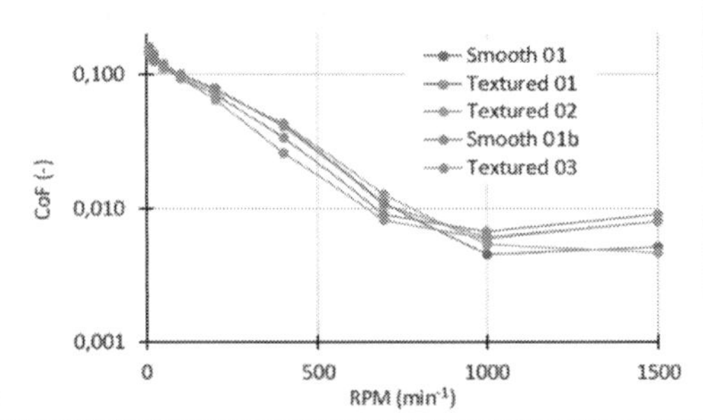

Figure 7. Comparison of COF for all measured Timken A4183 samples in dependence on the rotation speed

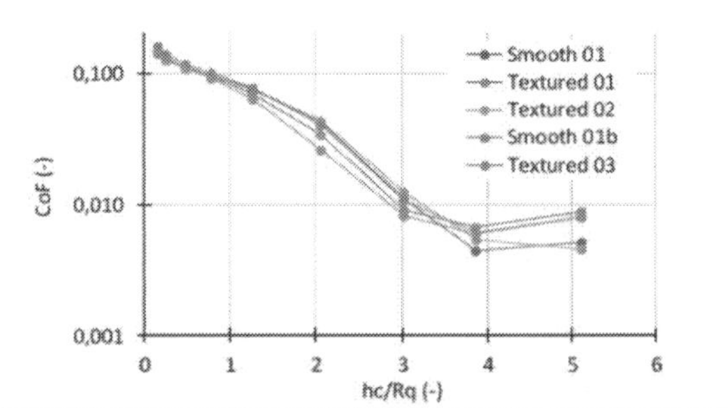

Figure 8. Comparison of COF for all measured Timken A4183 samples in dependence on the lubrication parameter

Discussing the choice of testing approach, it has to be mentioned, that the block-on-ring configuration, with line contact, is not the same as it is in the case of hydro-dynamically lubricated plain bearings. According to the Hertz theory, for the given testing configuration, the contact area is 72 µm wide. The value is smaller than the diameter of dimples in the texture (75-85 µm). Therefore, to improve performance of textures, a modification lead to wider contact area would be appreciated– e.g. use of a part of inner cylinder diameter as a counterpart geometry.

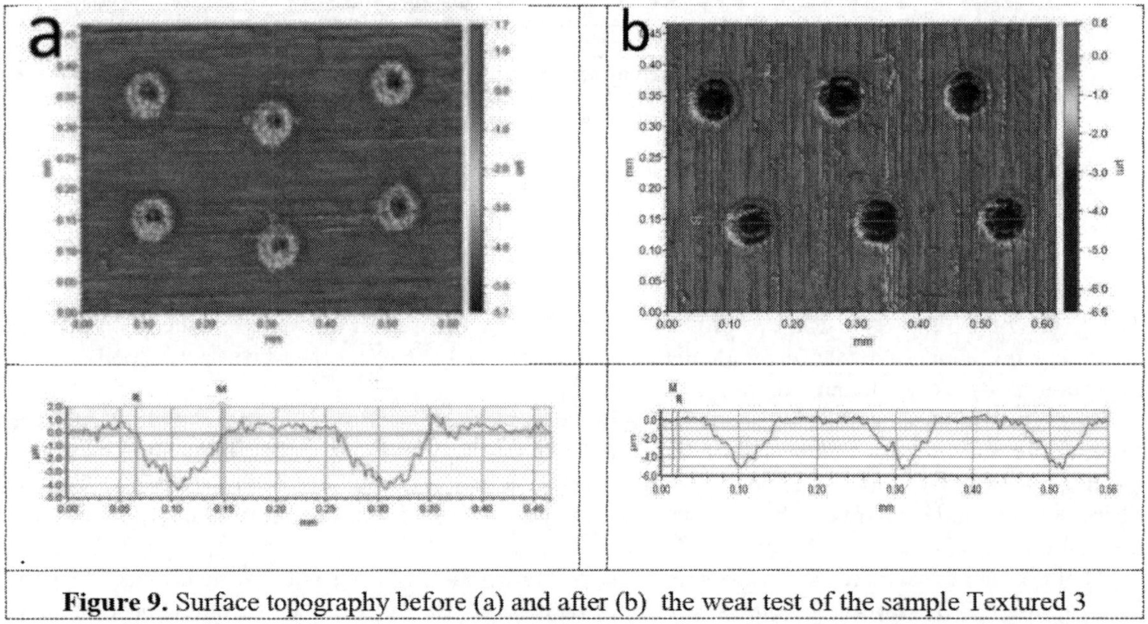

Figure 9. Surface topography before (a) and after (b) the wear test of the sample Textured 3

No wear of textured surfaces was recording during the block-on-ring tests. In the Figure 9, the comparison of the Textured 3 surface profile before (Fig. 9a) and after (Fig. 9b) test can be seen. Similar results were obtained also for the others samples.

3.2. Harsh condition testing

3.2.1. Wear test results. From the wear tests results (Fig. 10) it is obvious, that the texture was worn-out from the surface quite soon after the running of the tests, since the depth of textures varied between 6.5-7 μm. There is no significant difference in wear amount between the textured bearings and the untreated bearings.

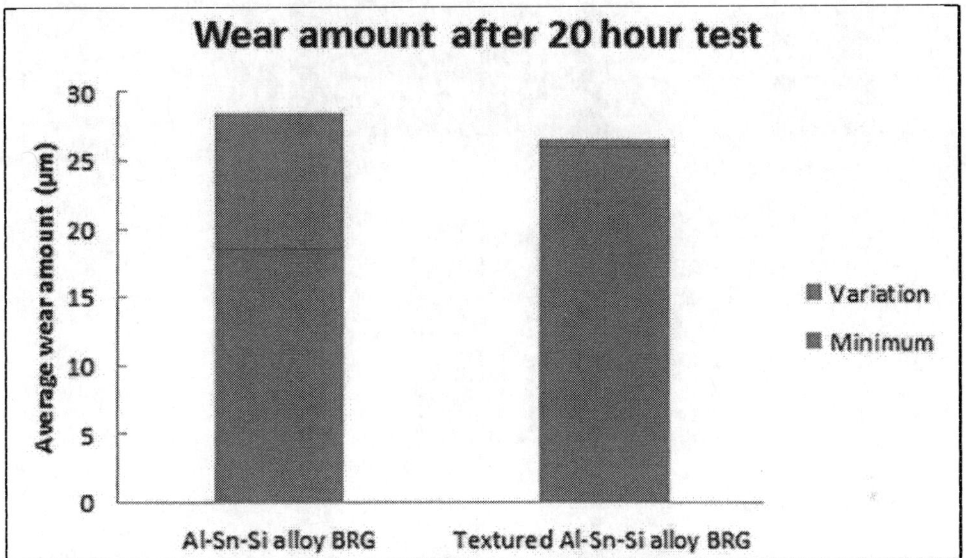

Figure 10. Wear amount after 20 hour wear test determined from the dimensional checks (two separate test bearings)

The wear occurs mostly around the main loaded area. The dimples in this area were completely removed after the wear test. The remaining textured surfaces were measured for surface topography to observe the evolution of the wear (Figure 11). The scans were carried out near the radial edge of the bearing, as there is less wear around that and the interface can be observed. Closer to the main area, no remaining texture was observed.

During the test, the torque on shaft was recorded. In the Figure 12, the comparison between the un-treated and textured bearings is shown in initial stage of stages of the test (Fig 12a) and after one hour of testing (Fig. 12b). While at the initial stages of the test the positive effect of the texture is obvious, just a little difference in torque between non-textured and textured bearings was found after one hour of testing due to worn out of the texture.

4. Conclusions

The efficient sLST method proves its suitability to produce the texture on the parts surface, which can positively influence the sliding behavior. The potential of the texture to decrease the friction in boundary lubrication mode was proved by both laboratory testing as well as by the testing under harsh conditions, where the torque of the shaft was decreased without the negative influence on the bearing wear. The lower measured torque in initial stages of star-stop test could be explained by faster oil formation during running. Although the positive effect of the texture was recorded namely in the initial stages of the test due to the fast worn-out, the very harsh condition of the test has to be taken into account.

IOP Conf. Series: Journal of Physics: Conf. Series **843** (2017) 012076 doi:10.1088/1742-6596/843/1/012076

Under real working conditions, the time to worn out of the texture is expected to be much longer. If the sLST method will be applied on more wear resistant bearing material that the Al-Sn-Si alloy, the lifetime of the textured surface can be also increased.

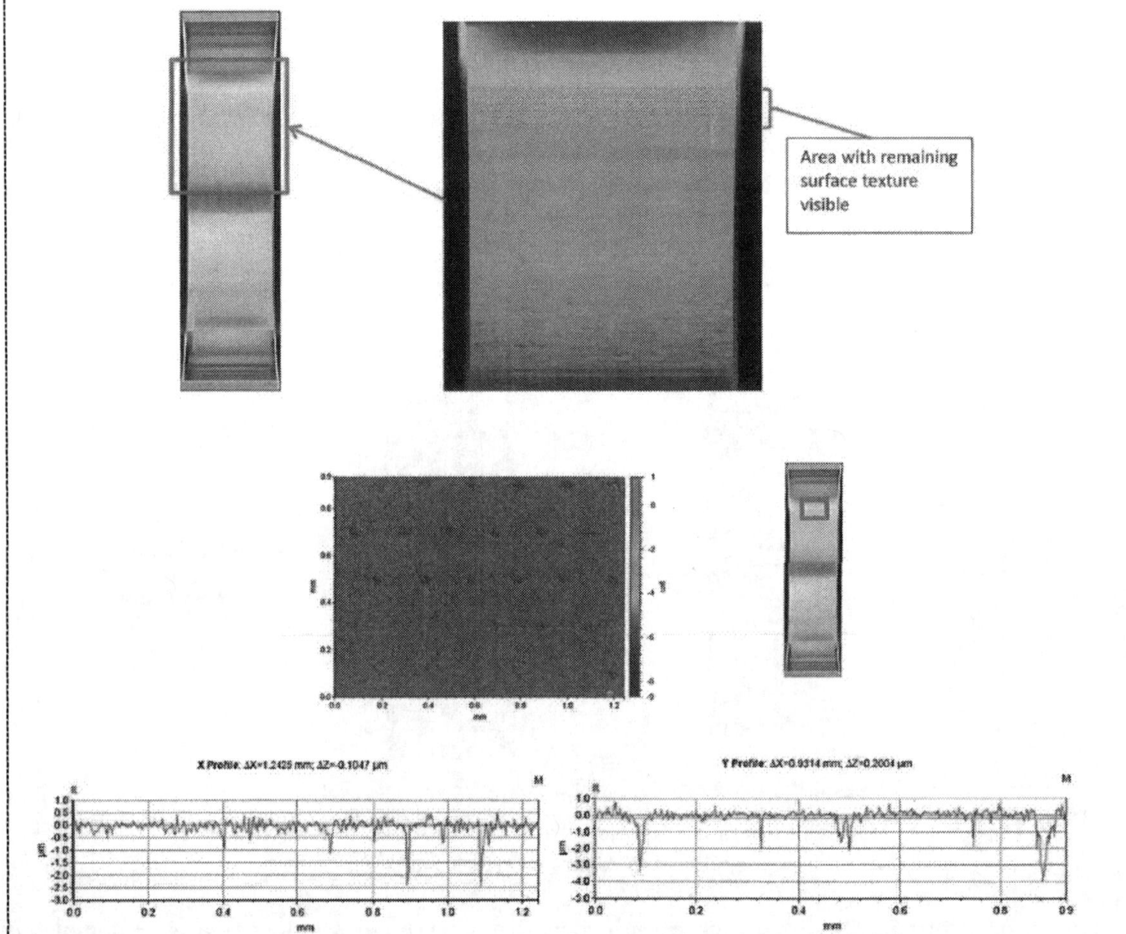

Figure 11. Dimpled area after the wear test

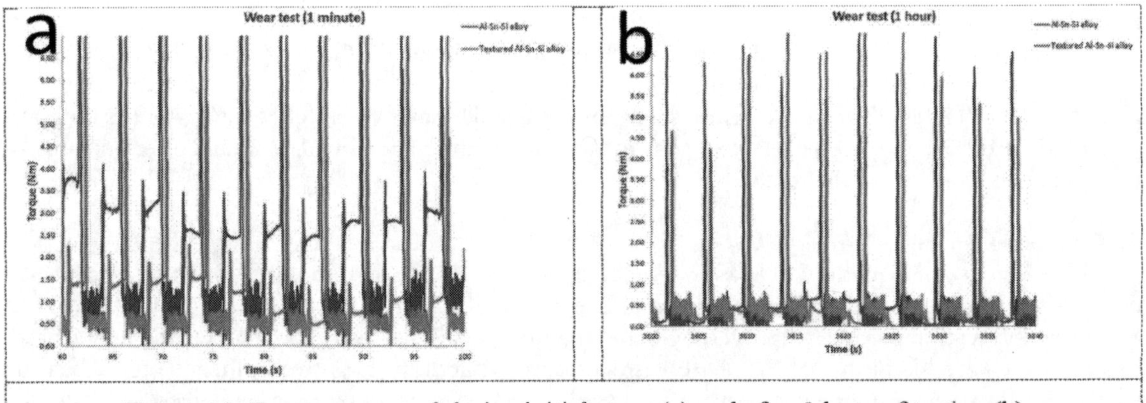

Figure 12. Torque measured during initial stage (a) and after 1 hour of testing (b)

Acknowledgement
The result was developed within the CENTEM project, reg. no. CZ.1.05/2.1.00/03.0088, co-funded by the ERDF as part of the Ministry of Education, Youth and Sports of the Czech Republic OP RDI program and, in the follow-up sustainability stage, supported through CENTEM PLUS (LO1402) by financial means from the Ministry of Education, Youth and Sports under the "National Sustainability Program I" and project SGS-2016-005.

The authors thank to Daido Metal Co-Ltd.-The European Technical Center for use of their testing rig BTR-01.

References
[1] Yilbas B S, Khaled M, Abu-Dheir N, Aqeeli N and Furquan S Z 2013 Laser texturing of alumina surface for improved hydrophobicity *Appl. Surf. Sci.* **286** 161–70
[2] Shum P W, Zhou Z F and Li K Y 2014 To increase the hydrophobicity, non-stickiness and wear resistance of DLC surface by surface texturing using a laser ablation process *Tribol. Int.* **78** 1–6
[3] Kromer R, Costil S, Cormier J, Berthe L, Peyre P and Courapied D 2016 Laser Patterning Pretreatment before Thermal Spraying: A Technique to Adapt and Control the Surface Topography to Thermomechanical Loading and Materials *J. Therm. Spray Technol* **25** 401–10
[4] Kromer R. et al. 2015 Laser surface patterning to enhance adhesion of plasma sprayed coatings *Surf. Coatings Technol* **278** 171–82
[5] Vorobyev A Y and Guo C 2015 Multifunctional surfaces produced by femtosecond laser pulses *J. Appl. Phys.* **33103** 6–11
[6] Zheng B, Wang W, Jiang G and Mei X 2016 Fabrication of broadband antireflective black metal surfaces with ultra-light-trapping structures by picosecond laser texturing and chemical fluorination *Appl. Phys. B* **122** 180
[7] Etsion I 2005 State of the Art in Laser Surface Texturing *J. Tribol. Trans. ASME* **127** 248
[8] Jaeggi B, Neuenschwander B, Zimmermann M, Loor R D and Penning L 2014 High throughput ps-laser micro machining with a synchronized polygon line scanner *8th International Conference on Photonic Technologies LANE 2014* Bayerisches Laserzentrum GmbH Germany1–8
[9] Weber R et al 2014 Heat accumulation during pulsed laser materials processing *Opt. Express* **22** 11312–24
[10] Bauer F, Michalowski A, Kiedrowski T. and Nolte S 2015 Heat accumulation in ultra-short pulsed scanning laser ablation of metals *Opt. Express* **23** 1035–43
[11] Schonlau T, Hebel R, Pause S and Mayer G 2014 Machining strategies for versatile ultra-short pulse laser applications. *8th International Conference on Photonic Technologies LANE 2014* Bayerisches Laserzentrum GmbH Germany1–4
[12] Bulgakova N M et al 2014 Impacts of ambient and ablation plasmas on short- and ultrashort-pulse laser processing of surfaces *Micromachines* **5** 1344–72
[13] Moskal D, Kučera M, Smazalová E, Houdková Š and Kromer R 2015 Application of Shifted Laser Surface Texturing *METAL 2015 - 24th International Conference on Metallurgy and Materials* Tanger Ltd. Czech Republic 1016–21
[14] Moskal D, Martan J, Kučera M, Houdková Š, Kromer R, 2016 Picosecond Laser Surface Cleaning of AM1 Superalloy *Physics Procedia* **83** 249-57
[15] Martan J, Moskal D, Kučera M 2017 Laser surface texturing for large area functional surfaces, *Nature Communications*, Preprint
[16] Brizmer V, Kligerman Y and Etsion I 2003 A Laser Surface Textured Parallel Thrust Bearing *Tribol. Trans* **46** 397–403
[17] Kligerman Y, Etsion I and Shinkarenko A 2005 Improving Tribological Performance of Piston

Rings by Partial Surface Texturing *J. Tribol.* **127** 632

[18] Galda L, Pawlus P, and Sep J 2009 Dimples shape and distribution effect on characteristics of Stribeck curve *Tribol. Int.* **42** 1505–12

[19] Wan Y and Xiong D S 2008 The effect of laser surface texturing on frictional performance of face seal *J. Mater. Process. Technol.* **197** 96–100

[20] Li K, Yao Z, Hu Y and Gu W 2014 Friction and wear performance of laser peen textured surface under starved lubrication *Tribol. Int.* **77** 97–105

[21] Gropper D, Wang L and Harvey T J 2016 Hydrodynamic lubrication of textured surfaces: A review of modeling techniques and key findings *Tribol. Int.* **94** 509–29

[22] Mishra S P and Polycarpou A A 2011 Tribological studies of unpolished laser surface textures under starved lubrication conditions for use in air-conditioning and refrigeration compressors *Tribol. Int.* **44** 1890–1901

[23] Braun D, Greiner C, Schneider J and Gumbsch P 2014 Efficiency of laser surface texturing in the reduction of friction under mixed lubrication *Tribol. Int.* **77** 142–47

[24] Zuhlke C A, Anderson T P and Alexander D R 2013 Formation of multiscale surface structures on nickel via above surface growth and below surface growth mechanisms using femtosecond laser pulses *Opt. Express.* **21** 8460–73

[25] Dumitru G, Romano V, Weber H P, Haefke H, Gerbig Y and Pflüger E 2000 Laser microstructuring of steel surfaces for tribological applications *Appl. Phys. A Mater. Sci. Process.* **70** 485–87

[26] Houdková Š, Martan M, Moskal D and Muzika L 2017 Laser technologies for modification of functional surface properties *Manufacturing technology* Preprint

[27] Pan P, Hamrock B J 1989 Simple Formulas for Performance Parameters Used in Elastohydrodynamically Lubricated Line Contacts *J. Tribol.* **111** 246-51.

[28] Tallian T E 1967 On Competing Failure Modes in Rolling Contact *ASLE Transactions* **10** 418-39.

Multi technical analysis of wear mechanisms in axial piston pumps

G Schuhler[1,2,3]**, A Jourani**[1]**, S Bouvier**[1] **and J-M Perrochat**[2]

[1] Sorbonne University, Université de technologie de Compiègne, CNRS, UMR 7337 Roberval, Centre de recherche Royallieu - CS 60 319 - 60 203 Compiègne cedex , France

[2] Actuation & Propeller Systems UTC AEROSPACE SYSTEMS 106, rue Fourny - 78530 Buc France

E-mail: guillaume.schuhler@utc.fr

Abstract. Axial piston pumps convert a motor rotation motion into hydraulic or pneumatic power. Their compactness and efficiency of approximately 0.9 make them suitable for actuation applications especially in aeronautics. However, they suffer a limited life due to the wear of their components. In the literature, studies of axial piston pumps deal with contact between its different elements under lubrication conditions. Nevertheless, they are more focused on analytic or numerical approaches. This study consists in an experimental analysis of worn pump components to highlight and understand wear mechanisms. Piston shoes are central components in the axial piston pump since they are involved in three tribological contacts. These three contacts are thereby studied: piston shoes/swashplate, piston shoes/pistons and piston shoes/ shoes hold down plate (SHDP). To perform this analysis, helicopter hydraulic pumps after different operating times have been studied. The wear damage mechanisms and wear debris are analysed using SEM observations. 3D surface roughness measurements are then used to characterize worn surfaces. The observations reveal that in the contact between shoes and swashplate, the main wear mechanism is three-body abrasive wear due to coarse carbides removal. Between shoes and pistons, wear occurs in a less severe way and is mainly due to the debris generated in the first contact and conveyed by the lubricating fluid. In the third contact, the debris are also the prime cause of the abrasive wear and the generation of deep craters in the piston shoes.

1. Introduction

Axial piston pumps convert the mechanical power generated by a motor rotation into hydraulic power. The motor rotational motion is transmitted to a barrel by a shaft. Pistons within the barrel are held on a tilted plate - the swashplate. Consequently, the barrel rotation drives the pistons to translate in their bore and pump the fluid. Then a two-holed plate - the valve plate - distributes the fluid between two conducts (inlet and outlet). The hydraulic fluid also plays a lubricating role since the carter is filled with the fluid. In regular cases, piston shoes (or shoes) are added to improve pistons driving on the swashplate and create a plane on plane contact. To improve lubrication in the contact between shoes and swashplate, a small cavity in each piston/shoe assembly provides pressurized fluid. Shoes are kept in contact with the swashplate by the action of a retainer: here a shoe-hold-down-plate (SHDP) and a spring. The mechanism and kinematics of axial piston pumps are detailed by many authors [1–5].

[3] Corresponding author

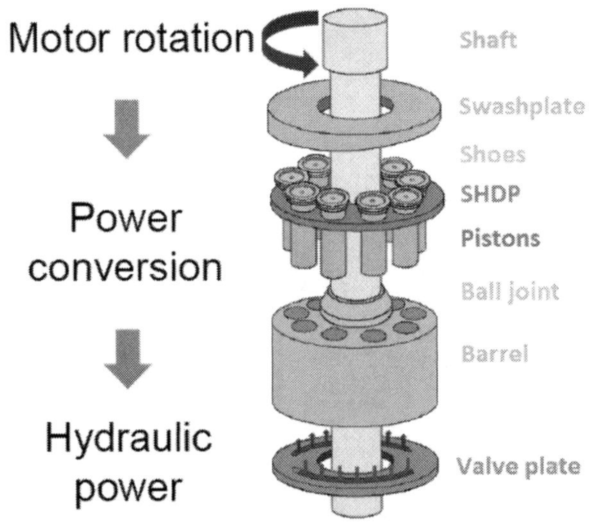

Figure 1. Axial piston pump exploded view.

The hydraulic fluid flows in the carter and in the circuitry, and thus lubricates all contacts between components. Deeken [6] simulated the combined hydrodynamic and tribological phenomena in valve plate/barrel, pistons/barrel and shoes/swashplate contacts. Some authors combined simulations and measurements for several contacts in an axial piston pump [5, 7, 8]. The lubricating film in shoe/swashplate contact is the most simulated or studied in literature [9–16]. Despite this lubrication of every contact, each component undergoes wear at different levels and. By generating wider gaps and greater leakage flows, this loss of material results in declining performances and reduced life. Wear can also lead to catastrophic failure. Consequently, it is necessary to study and understand wear mechanisms in axial piston pump.

Wear is a complex phenomenon and is rarely simulated in axial piston pumps or analogue systems. The above-mentioned simulations and experiments are focused on lubrication film thickness, contact pressures, heat generation, elastic and plastic deformations due to the fluid-structure interactions. These quantities are linked to wear volume but the dependence on wear mechanism is complex to establish. Indeed, laws of wear have been developed to predict wear and correlate experimental results [17–21].

For example, Archard and Holm proposed a simple law of wear [22, 23], but it is initially limited to dry contact and the constant K is not well understood [24]. Stolarski and Zou et al. developed abrasive wear models for lubricated contact for prediction [25, 26].

Ma et al. adapted Zou's model for the contact between shoes and swashplate [27]. They compared simulations results to experimental wear rate evaluations that consisted in measuring a variation of shoe thickness before and after testing. Wang et al. proposed a statistical approach to predict the remaining useful life of an axial piston pump, and compared result to experiments by monitoring return oil flow (sum of leakage in four contacts) as a consequence of internal wear [28]. Some experimental studies are performed on test machines instead of test rigs with adapted setup to reproduce the pumps contacts [29, 30]. It allows a simpler control of operating conditions but in accordance with pump contact specifics.

Kalin et al. observed the influence on wear resistance of DLC-coated shoes. They followed wear evolution with optical microscope and leakage monitoring [31]. This type of qualitative study can be found on other systems than axial piston pumps. Mucchi et al. determined lubrication regimes in variable displacement vane pumps by measuring roughness to quantify

IOP Conf. Series: Journal of Physics: Conf. Series **843** (2017) 012077 doi:10.1088/1742-6596/843/1/012077

wear [32]. Archard's law K takes typical values depending on the lubrication regime and has been identified with the experimental wear volumes. Brandao et al. and Maruda et al. did complete analyses of wear of respectively gears and a cutting tool for different operating conditions [33, 34]. These studies are close to the one conducted in this paper. Indeed, surfaces are analysed after normal use with techniques such as optical microscopy, SEM, X-rays micro analyses or 3D optical topography, and qualitative conclusions are drawn.

This literature review shows that few wear analyses are performed on real components of axial piston pump. More generally, there is a lack of qualitative analyses aiming to understand wear mechanisms. A better knowledge of these mechanisms enables to choose more accurately wear prediction models. From an engineering point of view, it also permits to propose tailored wear resistance improvement solutions for each contact. This study consists in an experimental analysis of worn pump components to highlight and understand wear mechanisms. It is conducted on helicopter hydraulic pumps after different operating times. The wear damage mechanisms and wear debris are analysed using SEM observations. 3D surface roughness measurements are then used to characterize worn surfaces. The study provides information on wear mechanisms and chronology by using a multi-technical analysis.

2. Material and methods

2.1. Material - The pumps

In this paper, the study focuses on helicopter axial piston pumps. Five pumps of the same model have been sampled after different operating times, but similar operating conditions. The components and subsystems are detailed in the introduction but some extra information is required to perform this study:

The swashplate and pistons are made of the same high carbon martensitic stainless steel. The bars that are machined into swashplate and pistons did not receive the same treatment during steelmaking. As specified by the norms AIR 9160 and AIR0819, classes are distinguished by a limit size of inclusions. Class 2 (bigger inclusions) is manufactured in a usual steelmaking furnace. Whereas class 3 is refined in a vacuum furnace after the usual steelmaking furnace. Consequently, class 3 has less impurities and finer carbides. Once the components are machined, they are thermally treated to reach a hardness higher than 58 HRc and a good thermal stability. This treatment is identical for classes 2 and 3. The SHDP is made of nitrided steel and the shoes of a copper alloy.

2.2. Measurement

The aim of this study is to understand wear mechanisms and their evolution during time in the three following contacts: shoes/swashplate, shoes/pistons and shoes/SHDP. For each contact, the contacting surfaces are analysed.

Fives pumps are observed and analysed after 0h, 300h 550h, 870h, and 1050h real operating times. Then, they are disassembled and the concerned components for the different analyses. Scanning electronic microscope (SEM) observations are performed to obtain a first understanding of the microstructure evolution and the wear phenomena that occurred in the different contacts. Chemical analysis completes this approach by giving insight on adhesion phenomena, nature of debris. 3D surface roughness measurements characterize worn surfaces in order to determine abrasive wear mechanisms and areal roughness parameters of the worn surfaces.

2.3. Method for roughness analysis

Roughness enlightens wear chronology and severity compared to the initial state. However, a good choice of parameters is mandatory. The arithmetic mean deviation of the surface (Sa) will be studied, because it provides a first general information on the surface roughness. However,

Sa alone is not enough because several surfaces with a similar Sa can have totally different topologies. Hence the need for more functional parameters [35, 36]. Various authors use Abbott-Firestone bearing curves and their derived parameters to monitor wear [37–42]. The Abbott-Firestone curve is the cumulative density function of surface heights.

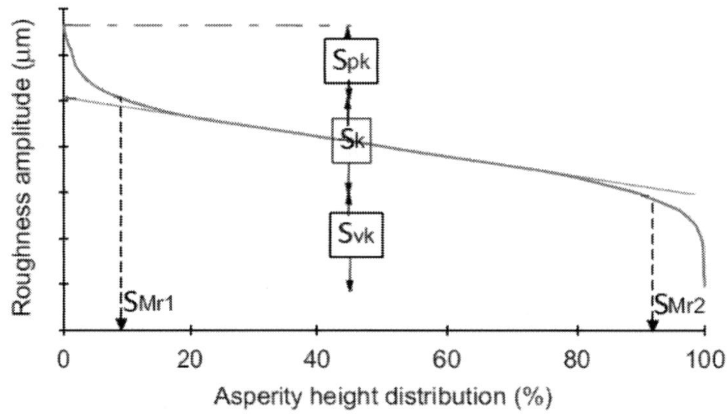

Figure 2. Abbott-Firestone curve and derived parameters from [37], adaptated for surfaces.

As a complement to Sa, the five parameters on the Abbott-Firestone curve (fig 2) will be studied (ISO 13565-2):

- Spk: reduced summit height
- Svk: reduced valley depth
- Sk: core roughness depth
- SMr1: Percentage of surface considered as peaks
- SMr2: Percentage of surface considered higher than valleys

To make SMr2 easier to compare to SMr1, its complementary to 100, SMr2'= 100-SMr2, will be used instead.

All these roughness parameters are given in a normalized form. Parameters in micrometers (Sa, Sk, Spk, Svk) are normalized with respect to Sa. SMr1 and SMr2' (in %) are normalized with respect to SMr1.

3. Results and discussion

The results of the analysis are detailed below. First, a table with the measured Vickers hardness is presented.

Table 1. Measured Hv5 hardness of the different materials.

Material	Pump components	Vickers Hardness (Hv5)
Stainless steel	Swashplate, pistons	903.3 ± 42.7
Copper alloy	Shoes	249.8 ± 1.2
Nitrided steel	Shoes hold down plate	1159.5 ± 58.4

3.1. Swashplate - Shoes contact

The shoes are made of a copper alloy and the swashplate of a thermally treated stainless steel. This treatment results in a martensitic matrix sprinkled with fine and coarse chromium carbides. The pistons-shoes assembly is designed to ensure lubrication in the swashplate-shoes contact. However, this contact is open, which enables the lubricant to flow out of it. Accordingly, a higher stress concentration occurs on the shoes border. The following results show how this configuration influence wear mechanisms.

The SEM observations of the swashplates surfaces (figure 3 A) show a combination of motor-rotation-oriented and randomly oriented grooves of approximately 5 micrometers large. This highlights two-body and three-body abrasive wear. The same conclusions can be drawn from the SEM observations of the shoes surfaces (figure 3 B) with larger (until 15 micrometers), more randomly oriented and more numerous grooves.

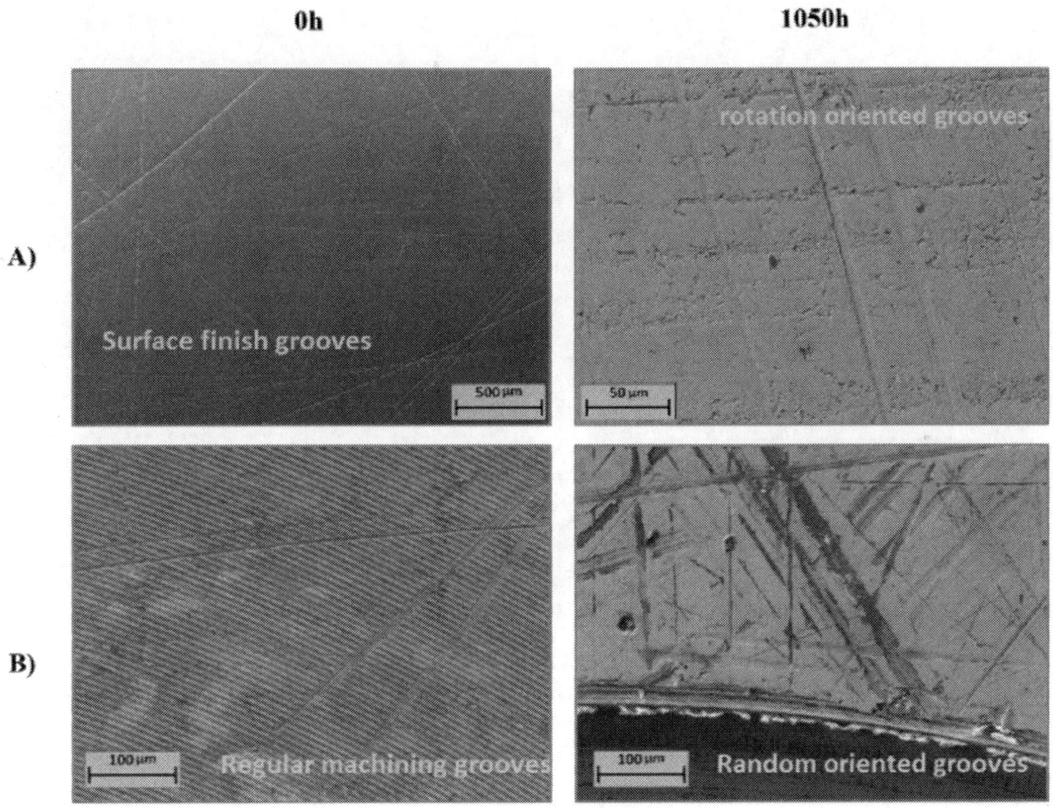

Figure 3. SEM pictures: A) Swashplate, B) Shoe in contact with the swashplate after 0h and 1050h functioning time. Generation of grooves.

The presence of copper smearing on the swashplate cannot be confirmed by EDX, because of the too small thickness of the deposit. Therefore, hypothesis are made after optical microscopy. A copper-colored deposit is observed on the swashplate (figure 4). This confirms the soft adhesive wear, and the limit lubrication regime.

The EDX analysis reveals the third-body is composed of chromium carbides (figure 5 left). These carbides are removed from the swashplate stainless steel surface (figure 5 right) and generate grooves on the contacting surfaces. They also get stuck in shoes surfaces (figure 5 left) and scratch the swashplate.

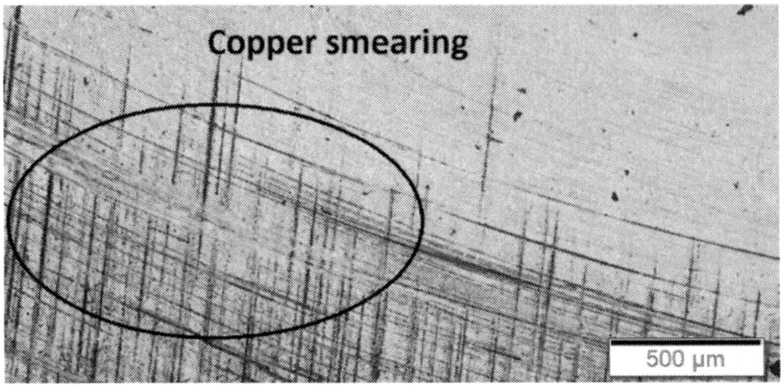

Figure 4. Copper smearing on the swashplate after 1050h functioning hours.

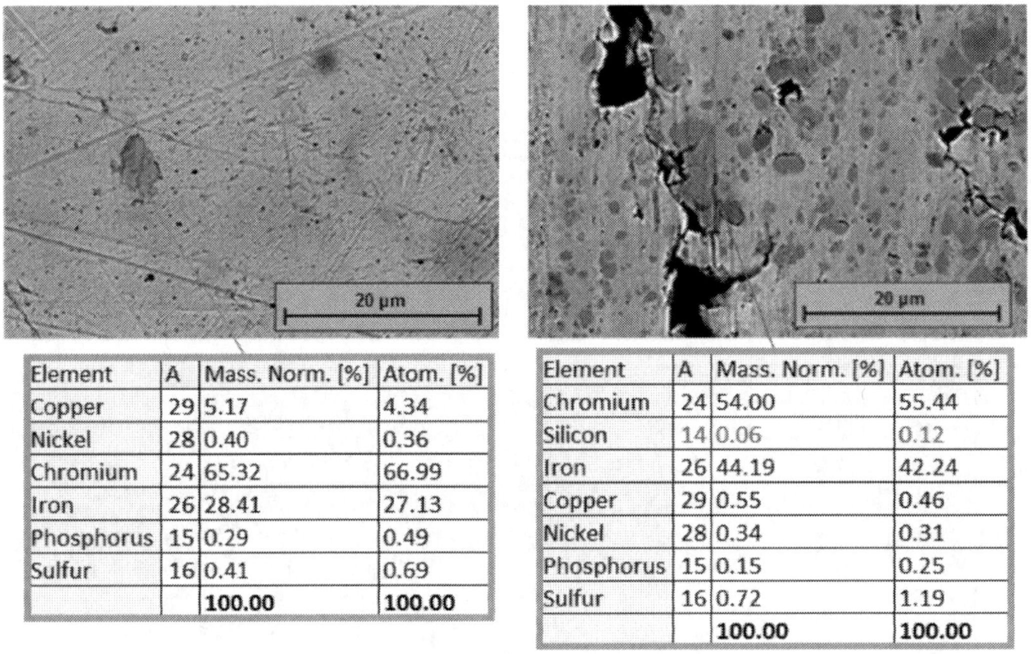

Element	A	Mass. Norm. [%]	Atom. [%]
Copper	29	5.17	4.34
Nickel	28	0.40	0.36
Chromium	24	65.32	66.99
Iron	26	28.41	27.13
Phosphorus	15	0.29	0.49
Sulfur	16	0.41	0.69
		100.00	**100.00**

Element	A	Mass. Norm. [%]	Atom. [%]
Chromium	24	54.00	55.44
Silicon	14	0.06	0.12
Iron	26	44.19	42.24
Copper	29	0.55	0.46
Nickel	28	0.34	0.31
Phosphorus	15	0.15	0.25
Sulfur	16	0.72	1.19
		100.00	**100.00**

Figure 5. SEM pictures with EDX pointing on chromium carbides. (left) chromium carbide stuck in a shoe surface, (right) swashplate surface with removed carbides.

These removed carbides seem to be the coarse carbides as shown in the following microstructure (figure 6). According to various authors they are undissolved M_7C_3 eutectic carbides [43–47]. This explains the presence of a carbide composed third-body. The other component of the third-body is a mix of strain hardened debris from the different parts of the pump.

Between 0 and 300 hours (figure 7), the swashplate is worn by running-in (Sa and Sk decrease). Peaks that are higher than the bearing surface are more numerous and conserv their height whereas valley that are lower than the bearing surface. It can be explained by important losses under the mean line, such as carbides removal and grooves (three-body abrasive

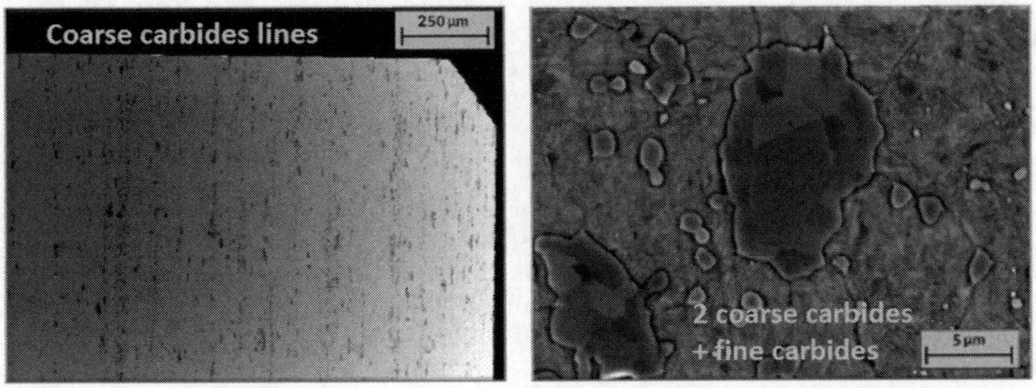

Figure 6. (left) SEM pictures of swashplate microstructure with coarse carbides lines (right) zoom on a coarse carbide.

wear). Meanwhile, the whole shoes surfaces are more severely worn. The small variation of SMr parameters means an almost constant quantity of high peaks and deep valleys. However, peaks and and valleys that are not wide enough to be counted as peaks and valleys increased, hence a higher Sk. This variation corresponds to a three-body abrasive wear mechanism. Between 300 and 870 hours, the same mechanisms occur on the swashplate but with less intensity. Less carbides are removed and a two-body abrasive wear begins. In the meantimes, a third-body randomly pollutes the fluid and generate random grooves. The shoes also experiment theis combined mechanism, but the rise of Svk and MMr2 displays the generation of more grooves. During the last period, shoes surface seem to be more polished but with more peaks eventough their height decreases. Carbides and particles getting stuck in shoes surfaces explain this result and the increase in swashplate parameters : the particles that are stuck in the shoes scratch the swashplate.

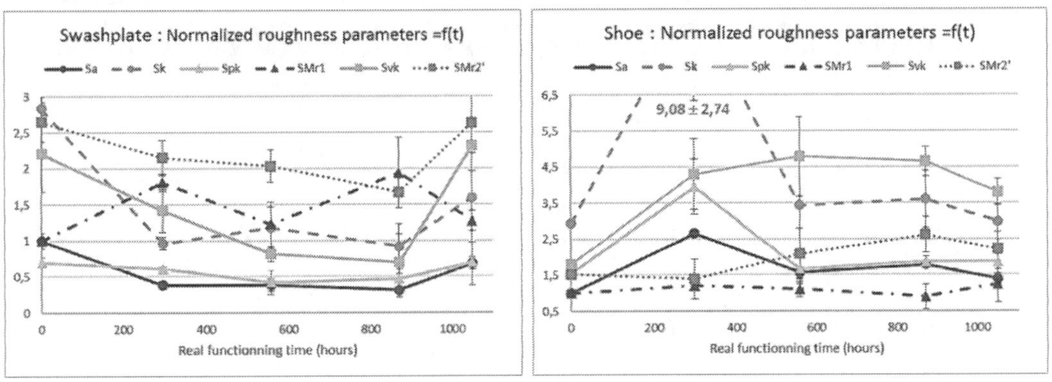

Figure 7. Evolution of different normalized roughness parameters for swashplate-shoes contact.

While studying failure and cracks in blades made of 440C, Neri and Colas, concluded that the failure was due to the presence of brittle coarse carbides [48]. Indeed, they initiated crack or provided easy crack path. The brittleness of coarse carbides might explains their removal from the martensitic matrix. Meurling et al found the same result for several steel grades [49].

Carbide have a direct influence on matrix as well. Indeed, coarse carbides are due to a non-optimal austenizing temperature and can result in a lower carbon concentration in the matrix and a lower strength [47]. The following sections show that the impact of this carbide removal is not limited to this contact.

3.2. Pistons - Shoes contact

The cyclic motions of the pistons in the shoes bring the lubricant in the contact. Since this ball-joint is a semi-closed contact the fluid is then trapped in it. Hence, the high stress concentration due to sphere-on-sphere contact is avoided by the constant presence of lubricant. The absence of copper (optical microscopy) on the piston sphere tends to confirm the effective lubrication in the ball-joint. Besides external particles and debris pollution are partially blocked by this semi-closed contact.

The SEM observations of the pistons surfaces (figure 8 A)) show rare grooves and no carbides removal. The shoes' initial machining grooves seems to have been polished and replaced by randomly oriented grooves (figure 8 B)). This mechanism is a combination of two and three-body abrasive wear.

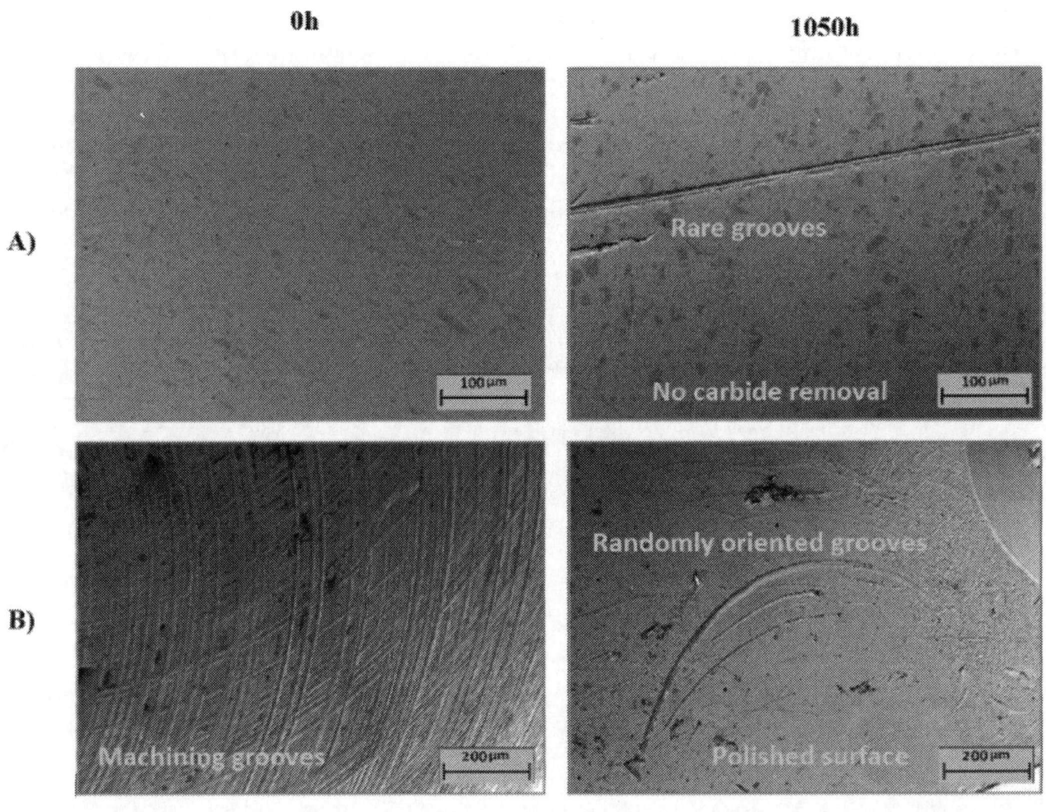

Figure 8. SEM pictures: A) Piston, B) Shoe after 0h and 1050h functioning time.

The micrographies show no carbides removal from the piston surface. It is indeed prevented by the lubrication conditions. In addition, less coarse carbides lines are observed in pistons microstructure (figure 9), hence less carbides can be removed. This confirms the effect of secondary steelmaking made on the class 3 high carbon martensitic stainless steel.

IOP Conf. Series: Journal of Physics: Conf. Series **843** (2017) 012077 doi:10.1088/1742-6596/843/1/012077

Figure 9. SEM pictures - microstructure of a piston: lines of coarse carbides.

The small probability of carbide removal in pistons is also due to their machining process. Swashplate and pistons are machined from steel bars. The coarse carbides lines are in the same direction as the bars. Consequently, there are more coarse carbides in contact with shoes on the swashplate than on the piston sphere.

However, there are random grooves due to a third-body in the shoes surfaces (figure 10). This contact is thus polluted by debris and carbides coming from other contacts, especially between swashplate and shoes.

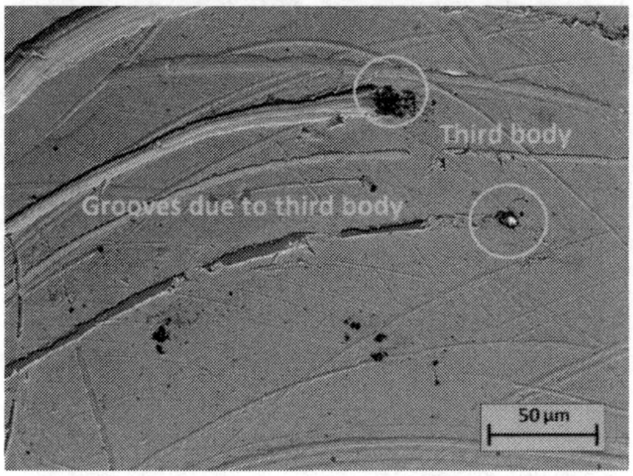

Figure 10. Grooves in a 1050h old shoe and remaining third-body.

According to the roughness parameters analysis (figure 11), the piston sphere is slightly polished until the appearance of grooves between 870 and 1050 hours of functionning. The shoes spheres are polished too, but also suffer a three-body abrasive wear due to polluting particles (Spk variations and Svk decrease with constant SMr). Between 870h and 1050h, new grooves are generated and new particles are stuck in the surface and will scratch pistons' spheres.

3.3. SHDP - Shoes contact

When the edge of shoes enter in contact with the SHDP, a stress concentration occurs. The SHDP is made of nitrided steel and has a hardness of Hv5 1159,5±58,4 against Hv5 249,8±1,2 for the copper alloy shoes (table 1). The surface of SHDP in contact with the shoes seems to be lightly polished (figure 12 A)), whereas the shoes seem serverely worn (figure 12 B)). The surface is impacted and worn by two-body abrasion. Although, wear is not regular on the whole surface and some zones with the initial machining grooves can still be observed.

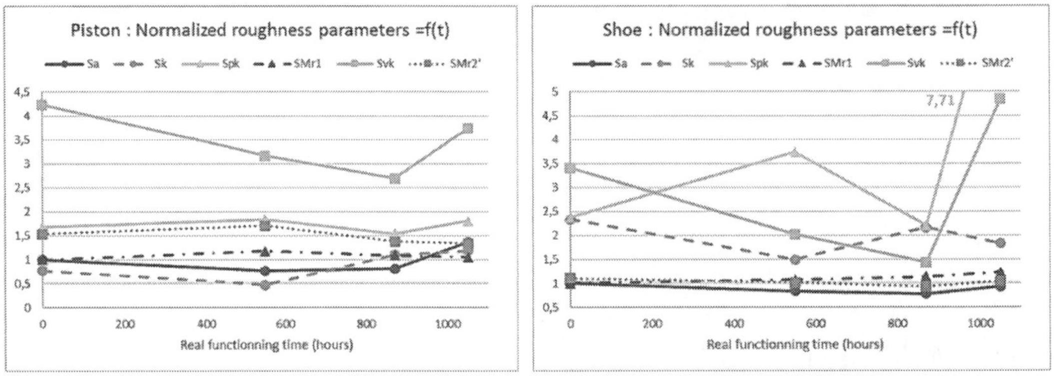

Figure 11. Evolution of different normalized roughness parameters for pistons-shoes contact.

Figure 12. SEM pictures: A) SHDP, B) Shoe after 0h and 1050h functioning time.

Observation of sliced SHDPs (figure 13) confirms that wear is limited on their surfaces. The nitriding combination layer (1 to 2 micrometers thick) resisted 1050 functionning hours.

EDX analysis of the impacts on the shoes surfaces explain their origin. Carbides and debris that pollute the fluid are trapped between the surfaces and pressed in the shoes because of the hardness difference between SHDP and shoes (figure 14 left). It generates edge formations and a remaining crater (figure 14).

IOP Conf. Series: Journal of Physics: Conf. Series **843** (2017) 012077 doi:10.1088/1742-6596/843/1/012077

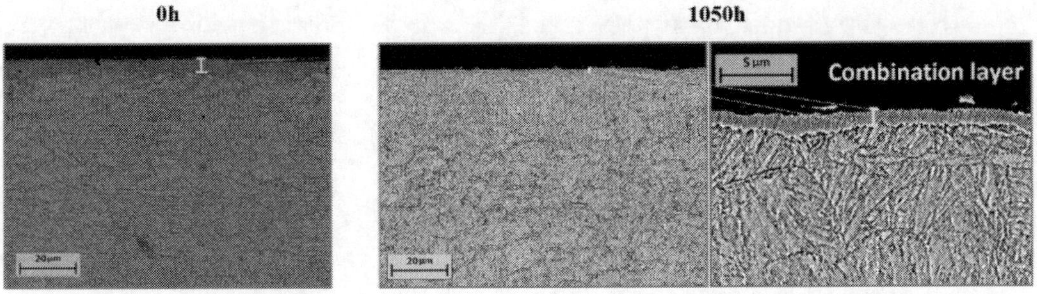

Figure 13. SEM pictures of SHDPs slices after 0h and 1050h.

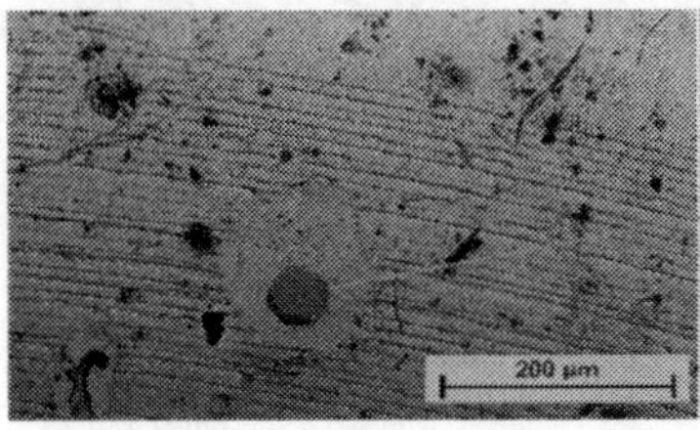

Element	A	Mass. Norm. [%]	Atom. [%]
Aluminium	13	0.99	2.03
Copper	29	6.49	5.65
Nickel	28	0.47	0.45
Iron	26	91.01	90.10
Chromium	24	0.16	0.17
Silicon	14	0.25	0.50
Phosphorus	15	0.62	1.10
		100.00	**100.00**

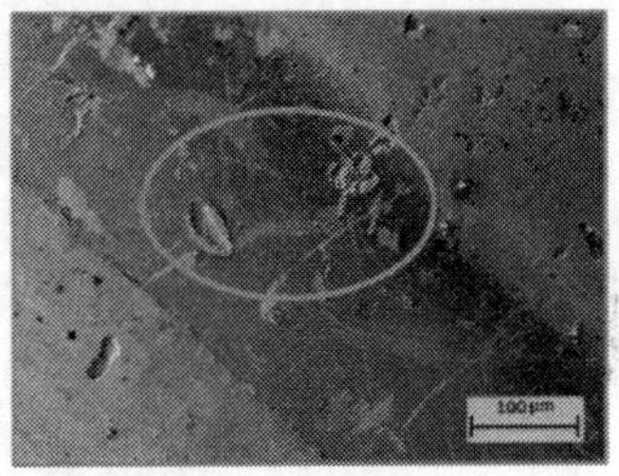

Figure 14. SEM pictures (left) with EDX analysis of a debris stuck in the shoes surface, (right) remaining crater and edge formations.

In addition, the edges of the shoes enter in contact with the SHDP, the load is concentrated and a plastic deformation occurs (figure 15).

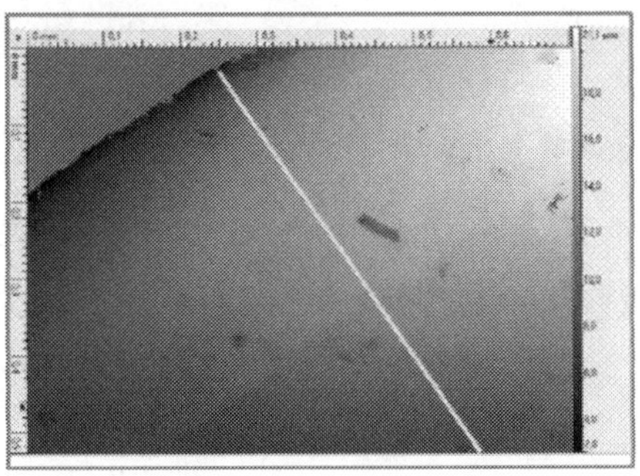

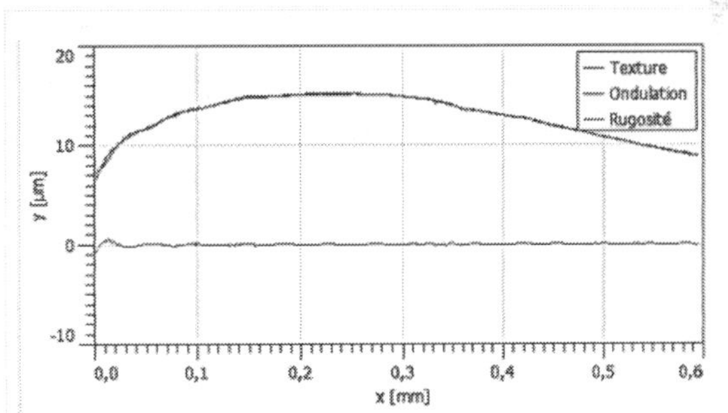

Figure 15. Plastic deformation on the shoe edge - 3D optic profile and an extracted 2D profile.

Due to these combined wear mechanisms this surface of the shoes is reported to be the most worn surface of the pump. The decrease of roughness parameters on the SHDP reflects a polishing without other signs of wear (no groove or impact) : running-in (0 to 550h) and soft abrasive wear. On the shoes, between 0 and 550h, particles and debris are getting stuck in the surface (less valleys, more peaks, both wider). After 550h, the particles seem to have left the shoes with deep valleys (grooves and craters) that are progressively removed by two-body abrasive wear (figure 16).

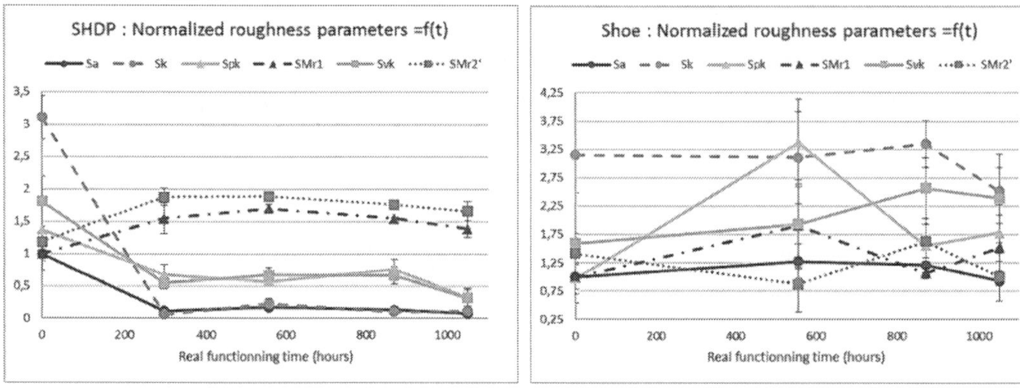

Figure 16. Evolution of different normalized roughness parameters for SHDP-shoes contact.

The fluid pollution with carbides and other particles is a form of fluid contamination. In their overview about fluid contamination in hydraulic systems, Frith and Scott concluded that an appropriate filtration is compulsory to limit contamination effects. However, they inferred that improving wear resistance of component is also necessary [50]. It is in accordance with the quasi no wear of the SHDP surface opposed to the important wear of shoes which are made of the softest material in the assembly. Kalin's study of DLC-coated shoes on a laboratory hydraulic system displays similar conclusions : shoes surfaces with only two-body abrasive wear in the rotation direction [31]. The DLC coating increased shoes' hardness and resistance to third-body. He also reported less wear debris generation.

4. Conclusions

Wear mechanisms in axial piston pumps have been investigated using a multitechnical analysis. The conditions of contact are different for the three tribological contacts and influence the severity of wear.

- The swashplate is a thermally treated stainless steel. with a martensitic matrix and a combination of fine and coarse chromium carbides. As a consequence of the uncertain lubrication and the stress concentration on shoes borders, carbides are removed from the matrix and act as a third body. This ensuing important abrasive wear of both the surfaces is typical of dry friction or boundary lubrication cases. The removed carbides pollute the fluid and the and the other two contacts.
- The pistons' stainless steel is the same as the swashplate's with rarer coarse carbides. This contact is a ball joint which traps the fluid and offers a better lubrication. It also limits the intrusion of external particles and debris. These combined effetcs of the ball -joint result in a less severe wear of the antagonist surfaces. Indeed, the wear of the piston surface is very low. Additionally, the lower quantity and size of coarse carbides also prevents carbides removal from the piston. However, deep grooves and holes due to carbides are observed on the shoes. The three -body abrasive wear is thus mainly due to external particles. Carbides are removed in the first contact and conveyed by the fluid.
- The SHDP is made of a nitrided martensitic steel. The contact with shoes is open and immersed in the carter -pressurized fluid. Lubricant can be exhausted from the contact, which lead to a limited lubrication. When the edges of shoes enter in contact with the SHDP, the stress concentration results in plastic deformation. Debris and carbides get stuck

between the two surfaces, are pressed in the shoes surface, which generates edge formations and craters. These deformations are then gradually removed by two body abrasive wear. In regard of the 3 different contacts, the shoes in this contact are the most worn surfaces, whereas the SHDP is just slightly polished.

In the three studied contacts, wear is due to carbide removal and to the comparatively low shoes' hardness.

References

[1] Merritt H E 1967 *Hydraulic control systems* (John Wiley & Sons)
[2] Milestone W D, Solomon A and Olsen G 1983 *Mechanism and Machine Theory* **18** 475–9
[3] Manring N D 1996 *Torque on the cylinder block of an axial-piston swash-plate type hydrostatic pump* PhD Iowa State University
[4] Khalil M B, Svoboda J and Bhat R 2004 *Journal of mechanical design*
[5] Ivantysynova M 2012 *Ventil (Ljubljana)* **18** 374–86
[6] Deeken M 2004 *ASME 2004 International Mechanical Engineering Congress and Exposition* (American Society of Mechanical Engineers) pp 71–5
[7] Manring N D 1999 *Journal of dynamic systems, measurement, and control* **121** 531–7
[8] Manco, Nervegna, Lettini and Gilardino 2002
[9] Wohlers A and Murrenhoff H 2007 *Bath Workshop on Power Transmission and Motion Control, PTMC* pp 129–41
[10] Rokala M, Calonius O, Koskinen K T and Pietola M 2008 *Proceedings of the JFPS International Symposium on Fluid Power* **2008** 91–4
[11] Bergada J and Watton J 2010 *Meccanica* **45** 585–602
[12] Schenk A and Ivantysynova M 2011 *Symposium on Fluid Power* vol 2011
[13] Sun Y and Jiang J 2011 (Okinawa)
[14] Lin S and Hu J 2015 *Applied Mathematical Modelling* **39** 548–58
[15] Schenk A and Ivantysynova M 2015 *Journal of Tribology* 031701–1–10
[16] Manring, Wray and Dong 2004 *Journal of Tribology* **126** 511–8
[17] Jourani A 2015 *International Journal of Surface Science and Engineering* **9** 231–46
[18] Jourani A 2015 *International Journal of Materials and Product Technology* **51** 127–138
[19] Jourani A 2014 *Journal of Tribology* **137**
[20] Petit L, Bigerelle M, Jourani A, Dore E, Prelle C and Lamarque F 2010 *Proceedings of the Institution of Mechanical Engineers, Part J: Journal of Engineering Tribology* **224** 1019–26
[21] Mezghani S, Jourani A and Zahouani H 2006 *WIT Transactions on the Built Environment* **85** 369–78
[22] Holm R 1946 *Electrical contact handbook* (Stockholm)
[23] Archard 1980 *Wear Control Handbook* ed Winer W O and Peterson M B (American society of mechanical engineers)
[24] *Wear* **181**
[25] Stolarski T A 1996 *Lubrication Science* **8** 315–51
[26] Zou Q, Huang P and When S 1996 *Wear* **196** 72–6
[27] Ma J, Chen J, Li J, Li Q and Ren C 2015 *Tribology International* **90** 467–72
[28] Wang X, Lin S, Wang S, He Z and Zhang C 2016 *Chinese Journal of Aeronautics* **29** 779–88
[29] Lee S Y and Hong Y S 2007 *Surface and Coatings Technology* **202** 1129–34
[30] Kapsiz M, Durat M and Ficici F 2011 *Advances in Engineering Software* **42** 595–603
[31] Kalin M, Majdi F, Viintin J, Pezdirnik J and Velkavrh I 2007 *J. Tribol.* **130** 011013–1–8
[32] Mucchi E, Agazzi A, D'Elia G and Dalpiaz G 2013 *Wear* **306** 36–46
[33] Brandao J A, Cerqueira P, Seabra J H O and Castro M J D 2016 *Tribology International*
[34] Maruda R W, Krolczyk G M, Feldshtein E, Nieslony P, Tyliszczak B and Pusavec F 2017 *Wear* **372-373** 54–67
[35] Zecchino M 2003 *Advanced materials & processes* **161** 25–8
[36] Gadelmawla E S, Koura M M, Maksoud T M A, Elewa I M and Soliman H H 2002 *Journal of Materials Processing Technology* **123** 133–45
[37] Bigerelle M and Iost A 2007 *Tribology International* **40** 1319–34
[38] Schmhling J and Hamprecht F A 2007 *Wear* **262** 1360–71
[39] Petropoulos G, Pandazaras C and Stamos I *Tribology in industry* **23**
[40] Torrance A A 1997 *Tribology International* **30** 239–44

IOP Conf. Series: Journal of Physics: Conf. Series **843** (2017) 012077 doi:10.1088/1742-6596/843/1/012077

[41] Stewart M 1990 *A new approach to the use of Bearing Area Curve* (International Honing Technologies and Applications: Society of Manufacturing Engineers)
[42] Ripa M, Tomescu L, Hapenciuc M and Crudu I 2003 *ANNALS* **24** 26
[43] Girodin and Dudragne 2000 (3rd symposium Ascometal, Arles)
[44] Yang J R, Yu T H and Wang C H 2006 *Materials Science and Engineering: A* **438440** 276 – 80
[45] Salleh S H 2009 *International Journal of Mechanical and Materials Engineering* **4** 21–4
[46] Arparjirasakul, Patcharawit and Kitkamthorn 2014 (Singapore)
[47] Reddy A V 2004 *Investigation of Aeronautical and Engineering Component Failures* (CRC Press) ISBN 978-0-203-49209-3
[48] Neri M A and Colas R 2001 *Materials characterization* **47** 283–9
[49] Meurling F, Melander A, Tidesten M and Westin L 2001 *International Journal of Fatigue* **23** 215–24
[50] Frith R H and Scott W 1993 *Wear* **165** 69–74

The effect of specimen temperature on the fretting wear behaviour of a non-nitrided against nitrided high strength alloy steel contact pair

N Arun Prakash[1] and C J Bennett[2]
[1,2] Gas Turbine and Transmission Research Centre, Faculty of Engineering, The University of Nottingham, University Park, Nottingham NG7 2RD, United Kingdom
[1]Email: Arun.Natarajan@nottingham.ac.uk, Tel: +44 (0)11584 67683
[2]Email: Chris.Bennett@nottingham.ac.uk, Tel: +44 (0)11584 68460

Abstract. The fretting wear performance of the heterogeneous non-nitrided against nitrided contact pair of the high strength alloy steel, SCMV was investigated in the gross slip regime at ambient and elevated temperatures (120 and 150 °C). The tests were performed with displacement amplitudes of 25 and 100 μm at an applied normal load of 250 N using a cylinder-on-flat configuration. As the fretting contact temperature increases, the energy dissipation and total wear volume decreases for the same applied normal load irrespective of the displacement amplitude. The energy dissipation increases with increase in displacement amplitude for the same applied normal load. During fretting at ambient temperature, a substantial loss of material as wear debris was observed in the contact, whereas the overall loss of material from the contact significantly reduces with an increase in temperature. The variation in the tangential force coefficient and the total wear volume at elevated temperatures may be attributed to the sintering of oxide particles in the contact to form a protective debris bed. The worn surface morphology of the fretted specimen investigated using a scanning electron microscope shows the presence of wear debris in the wear track, oxide patches, delamination cracks, delamination cavities and discontinuities, scale like topography and plate-like wear debris. The abrasive action of the particles in the contact interface promoted wear at ambient temperature whereas at elevated temperatures, the compaction of the oxide particles reduced the wear.

Keywords: Fretting wear, high strength steel, wear mechanism, temperature

1. Introduction

Fretting wear is a surface degradation phenomenon, predominant in quasi-static loaded assemblies of machine elements and engineering structures subjected to small amplitude relative oscillatory motion occurring between two normally loaded contacting solid surfaces as a result of cyclic loading [1]. The influence of elevated temperature on the friction and wear performance of different types of steels in an unlubricated environment under fretting loading conditions have been examined by previous researchers and it has been reported that the fretting wear decreases with an increase in temperature [2-5]. Researchers have observed an abrupt decrease in wear rate once the temperature is increased beyond 200 °C, commonly referred as the transition temperature for austenitic stainless steels [2] and mild steels [3, 4] which can be attributed to the difference in the properties of the oxide formed at the contact interface. In the fretting wear process, the adhesion wear mechanism becomes more effective with increasing interface contact temperature [2-4]. Experimental investigations by Hurricks [3, 4] on mild steel showed that the decrease in the wear rate with temperature may be attributed to the transformation in oxide structure where a spinel Fe_3O_4 structure becomes more thermodynamically stable above 200 °C than a rhombohedral Fe_2O_3 (α- Fe_2O_3) structure and that Fe_3O_4 is more effective in preventing wear compared to α- Fe_2O_3; as Fe_3O_4 adheres more strongly to the surface, then acts as a solid lubricant. The formation of oxide at the contact interface and its separation from the surface are the two main processes of oxidative wear. Oxidation is a time-dependent process that is accelerated at high temperature [2] and is also affected by the partial pressure of the oxygen. It is also described [5] that the properties of the oxide generated at the contact interface influences the fretting wear performance more significantly than the hardness of the material.

The quantitative modelling of fretting wear is difficult due to the complex relationship between fretting wear and slip regime (stick/partial/sliding), coupled with highly the nonlinear stress–strain relationship and the evolution and release of debris at the fretting interface. Finite element analysis of the fretting wear process has been studied by a number of authors to solve the evolution of stress and strain distribution for the gross sliding and partial slip regimes using various techniques [6 – 8]. Jin et al. [9] studied the effect of fretting frequency and the contact temperature due to the dissipation of frictional power in a stainless steel contact and a simple analytical model was developed to predict this behaviour. It was reported that at a fretting frequency of 200 Hz with a bulk specimen temperature of 150 °C, due the frictional power dissipation the contact temperature at the interface was raised around 70 °C and hence they suggested that there is a strong dependence of fretting behaviour upon fretting frequency at this temperature. Ding et al. [10] proposed a methodology to evaluate the fretting wear life of highly loaded assemblies based on the simple laboratory tests by developing a fretting wear simulation and predictive tool. Ding developed a fretting contact model using finite element analysis including the presence and evolution of the debris layer with increasing number of cycles and reported that with the introduction of the debris layer a smaller wear scar width and slightly greater wear depth was predicted when compared to the model without the debris layer. The model also predicts a shift in the maximum contact pressure from the centre of the contact to the contact edges and additionally predicts the retention and ejection rate of the wear debris as a function of number of cycles. Yue and Wahab [11] developed a two dimensional fretting wear model with a debris layer using the finite element method to study the effect of debris on fretting wear behaviour and reported, in contrast to Ding et al. [10], that the wear volume with debris layers was slightly higher compared to the model without the debris layer. Ding et al. [10] modelled the debris accumulation on the fretting contact interface as a layer structure, by defining the mechanical properties using an anisotropic elastic-plastic material model, whereas Yue and Wahab [11] described the material behaviour of the debris layer as elastic and homogeneous. The contact interface between the cylindrical pad and debris layer was assumed as a rigid connection by Ding et al. [10] whereas Yue and Wahab [11] assumed both the top and bottom contact interface as implemented contacts. The assumption and the material model considered by Ding et al. [10] and Yue and Wahab [11] are different which may be attributed to difference in the predicted wear volume value by these two approaches. Yue and Wahab [12] developed a two dimensional contact model by FE simulations using the ABAQUS commercial finite element software to study the effect of variable coefficient of friction with different fretting contact regimes. It is reported that the variable coefficient of friction in gross sliding conditions had a marginal influence in the prediction of wear volume at the end of the steady state condition, however in the partial slip regime or the running in stages of the gross sliding condition, the predicted wear volume by FE models with variable coefficient of friction are closer to the experimental results reported by McColl et al. [13]. Yue and Wahab [14] also studied the effect of the stress singularity in the simulation of the fretting wear using cylinder on flat contact using a two-dimensional FE contact model based on the fretting parameters such as applied displacement, coefficient of friction and number of cycles, and reported that the stress singularity has a close relationship to the fretting regime based on the stress singularity method. They also reported that the sensitivity of the element size at the contact interface is important and that the element size required for a converged solution should be determined.

In this work, the fretting wear behaviour of the high strength alloy steel, SCMV against nitrided SCMV (SCMVN) was investigated at ambient and, for the first time, elevated temperatures using a cylinder-on-flat configuration at 250 N applied normal load with displacement amplitudes of 25 and 100 μm under dry sliding conditions. The basis for this investigation is the need for high strength materials with enhanced mechanical and tribological properties in highly fatigue loaded components in order to achieve a longer service life at ambient and elevated temperatures. Creating an understanding of the fretting wear performance of SCMV-SCMVN material pair at the ambient and elevated temperatures will allow the development of the new materials with enhanced surface properties on the top surface layer, used in highly fatigue loaded components and it will provide a platform to better understanding the role of the contact conditions, surface hardening, temperature on the mechanisms controlling the fretting wear process.

2. Methodology

A series of fretting wear tests were performed on the high strength chromium–molybdenum–vanadium steel, SCMV and SCMVN at ambient and elevated temperatures using a crossed cylinder-on-flat configuration, as shown in Figure1, which results in a 10 mm line contact. SCMV steel contains 3 – 3.5 Cr, 0.8 – 1.10 Mo, 0.15 – 0.25 V, 0.35 – 0.43 C, 0.1 – 0.35 Si, 0.4 – 0.7 Mn, < 0.007 P, < 0.002 S, < 0.3 Ni and balance Fe by weight percentage [15]. The ultimate tensile strength and yield strength of the SCMV material determined by tensile test was 1247 MPa and 1697 MPa respectively. The test material was initially cut into specimen blanks providing a machining allowance on all the sides. Once the heat treatment process was completed on the specimens, the raw material blanks were further machined into the required final dimensions of the flat and cylindrical specimens confirming that the decarburised layer formed by the heat treatment process had been completely removed from the specimen surface. The cylindrical fretting wear specimens were pulsed plasma gas nitrided in an atmosphere of 4 parts nitrogen and 1 part hydrogen for 30 hours with a nitriding temperature of 495 ± 5 °C. A minimum surface hardness of 7.85 GPa with a case depth of 0.2 to 0.4 mm at a hardness of +100 HV greater than the core was obtained in the nitrided specimen. The Vickers hardness of the heat treated SCMV material machined surface was 485 ± 10 HV30. The average surface hardness of the nitrided SCMV specimen was 1010 ± 20 HV0.5.

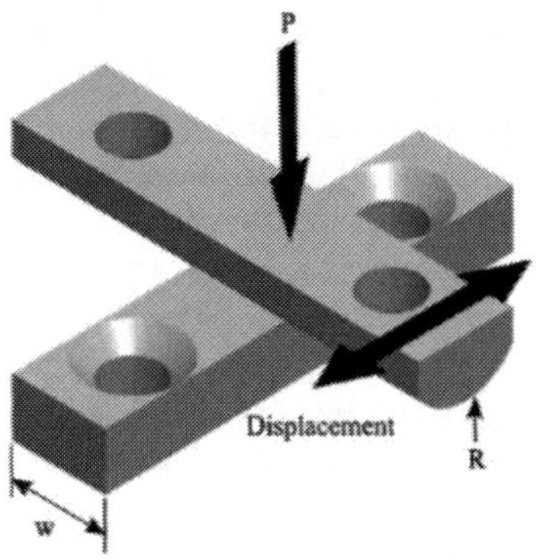

Figure 1. Cylinder-on-flat specimen configuration used in the fretting experiments, where W = 10 mm, R = 6 mm, P is the applied normal load.

The flat specimens were manufactured with a width of 10 mm while the cylindrical specimens had a radius of 6 mm. A brief description of the fretting wear test rig is presented here for ease of understanding although the full details of the setup is given elsewhere [16]. All fretting specimens were demagnetised, thoroughly degreased using detergent, rinsed with industrial methylated spirit (IMS) and acetone and finally dried using a hot air dryer prior to the experiment. The flat specimen was attached on a stationary lower specimen mounting block (LSMB) while the cylindrical specimen was fixed on a moving upper specimen mounting block (USMB). Fretting wear tests were conducted at an applied normal load of 250 N with displacement amplitudes of 25 and 100 μm at ambient and elevated temperatures (120 and 150 °C) with a relative humidity of 35 ± 5 % under a constant frequency of 20 Hz. Cartridge heaters were used to heat the specimen to the required temperature and independent control loops exists for the heaters in the USMB and LSMB. In order to monitor the temperature near the fretting contact, additional thermocouples were spot welded to the surface of the flat and cylindrical specimens. The deviation between the surface temperature and the heater block temperature was relatively small, within ± 5 °C for all the fretting experiments. Table 1 presents the fretting wear test conditions used in the present work. A linear variable differential transformer (LVDT) was used to measure the relative displacements between the upper and lower specimen mounting blocks and the tangential force at the contact interface was measured using piezoelectric load cell. The normal loads were applied by dead weights using the lever arm acting on the USMB. The control and data acquisition system programmed in LabVIEW was used to log the tangential force (Q) and relative displacement (δ), continuously. To evaluate the surface topography, the wear scar on both the flat and cylindrical specimens were scanned individually using an ALICONA G5 "Infinite Focus" non-contact measuring instrument which has a vertical resolution of ~ 10 nm and a lateral resolution of 4.4 µm. Prior to scanning, the specimens were rinsed with acetone to remove the loose wear debris that was not adhered to the surface. A Philips XL 30 SEM scanning electron microscope was used to characterize the worn surface morphology of the fretted surfaces.

Table 1. Summary of the fretting wear test conditions.

Material combination (Flat/Round)	SCMV-SCMVN
Normal load, P (N)	250
Displacement amplitude, δ (µm)	25 and 100
Number of cycles, N	100 x 10^3
Radius of the cylindrical specimen, R (mm)	6
Frequency, f (Hz)	20
Temperature, T (°C)	Ambient, 120 and 150

3. Results and discussion

The ratio of tangential force amplitude to the applied normal load is defined as the tangential force coefficient under sliding fretting conditions. The variation of tangential force coefficient with cycles elapsed for SCMV fretted against SCMVN at 250 N applied normal load with displacement amplitudes of 25 and 100 µm under the ambient and elevated temperatures is shown in Figure 2. At the fretting contact area, the tangential force is considered to be formed either by means of direct interlocking of surface asperities or through entrapment of wear debris in between the surface asperities. The tangential force coefficient is low at the beginning of the fretting action and after the completion of a few hundred cycles, the tangential force coefficient increases rapidly and reaches a more steady state value for the ambient conditions investigated. During the fretting process at elevated temperatures, the tangential force coefficient increases for few thousand cycles and then decreases and reaches a more steady state value with increasing fretting cycles. The initial increase in tangential force coefficient for all the conditions investigated under ambient and elevated temperatures is attributed to the strong adhesion at the contact interface after the removal of the natural lubricant layer by spalling and micro fragmentation, which is a weak oxide and provides minimal protection to wear; this initial case is a two-body wear system [17]. Once the preliminary oxide layer is removed, there is no creation/formation of the oxide layer and the metal surface remains reactive, resulting in high adhesion at the contact interface. This phenomenon not only results in high tangential frictional forces, but in some conditions (displacement amplitudes) it can also lead to seizure [18]. In the ambient temperature condition, as the fretting cycles advance, more and more wear debris is produced at the mating interface due to the repeated sliding action. This debris may remain stuck between the two mating interface thereby forming a three-body

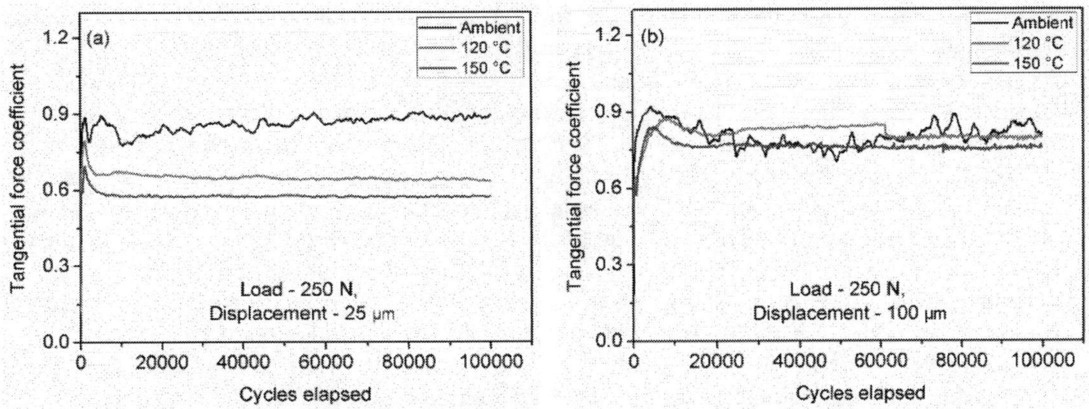

Figure 2. Variation of tangential force coefficient with cycles elapsed for the SCMV-SCMVN material pair at different displacement amplitudes with temperature, (a) 250 N, 25 μm and (b) 250 N, 100 μm.

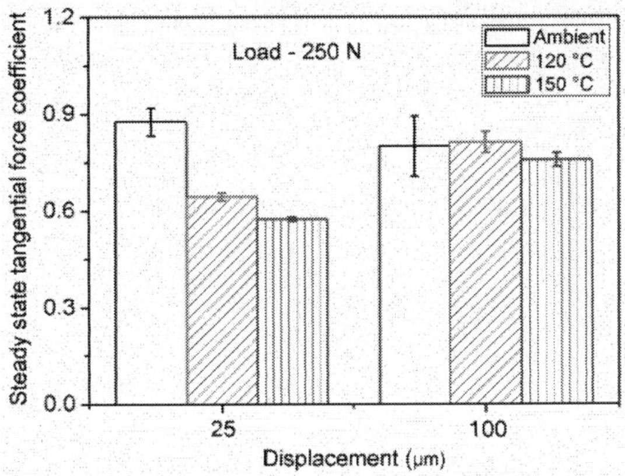

Figure 3. Variation of steady state tangential force coefficient with cycles elapsed for SCMV-SCMVN material pair at different displacement amplitudes with temperature.

wear system. This third-body (wear debris) acts as a natural barrier by splitting the direct interaction between the two mating interface, consequently the local stress field is lowered, which prevents further crack nucleation and propagation and thereby the tangential force coefficient reaches a steady state value. The experimental investigation by Kayaba and Iwabuchi [2] reported that, during the initial period of testing, the contact interface suffers plastic deformation and subsequently oxidation is facilitated causing the formation of an oxide layer. Fretting wear decreases with increasing temperature and there exists a transition temperature for all the materials in fretting contact, which influences the oxidation characteristics of the material. At low displacement amplitude (25 μm) with increasing temperature, the tangential force coefficient decreases drastically, which may be attributed to the formation of the surface oxides which adhere strongly to the substrate without being easily removed from the contact interface, thereby able to agglomerate and compact and thus reducing metal to metal contact. It is also evident from the wear scar depth profile that material transfer takes places during fretting at high temperature at low displacement amplitude, which is not the case at a displacement amplitude of 100 μm. At high displacement amplitude, the retention of the wear debris formed at the contact interface is minimal as it is easily ejected from the contact surface due to the large displacement, this free movement of the debris in the contact region increases the adhesion and the tangential force coefficient. To enable further comparison between the different fretting conditions and temperatures, the average values of the tangential force coefficient between 40 x 10^3 to 100 x 10^3 cycles was calculated and is presented in

Figure 3. At elevated temperatures, the tangential force coefficient reaches a stabilized value (i.e. steady state) more quickly compared to at ambient temperature. The steady state tangential force coefficient was highest at ambient temperature and decreases monotonically with increasing temperature at low displacement amplitude, whereas at high displacement amplitude, there is little variation in the value of the steady state tangential force coefficient with temperature. Figure 4 (a) shows the influence of applied normal load and displacement amplitude on total energy dissipated, which increases with increasing displacement amplitude for the same applied normal load at both ambient and elevated temperatures, as a gross slip condition exists for the material combinations investigated. The higher value of total energy dissipated at ambient and elevated temperature may be due to the following reasons: at ambient temperature, due to shearing of asperities, the maximum frictional energy is needed for the removal of the material and at high temperature maximum frictional energy is needed due to the delamination of the tribo-oxide layer. Figure 4(b) shows the influence of applied normal load and displacement amplitude on the total wear volume of the material pair examined at ambient and elevated temperatures. At ambient and elevated temperature, irrespective of the displacement amplitude, the total wear volume decreases with increasing temperature. Figure 5 shows the wear depth profile of the fretting wear scar of the flat SCMV specimen tested at 250 N load with temperature and displacement amplitudes of (a) 25 µm and (b) 100 µm. At 25 µm displacement amplitude, the wear volume of the fretted specimens are very small as the applied displacement is accommodated by elastic deformation of the asperities. As the displacement amplitude increases, the damage mechanism is governed by abrasive wear which causes both the depth and width of the wear scars to grow monotonically as illustrated in Figures 5 (a and b). A similar wear behaviour was observed during the fretting of annealed 304 stainless steel against itself [2] and AISI 301 stainless steel fretted against AISI 52100 steel [19]. Figure 6 shows the worn surface morphology of the SCMV fretted against SCMVN at 250 N, 100 µm at ambient and

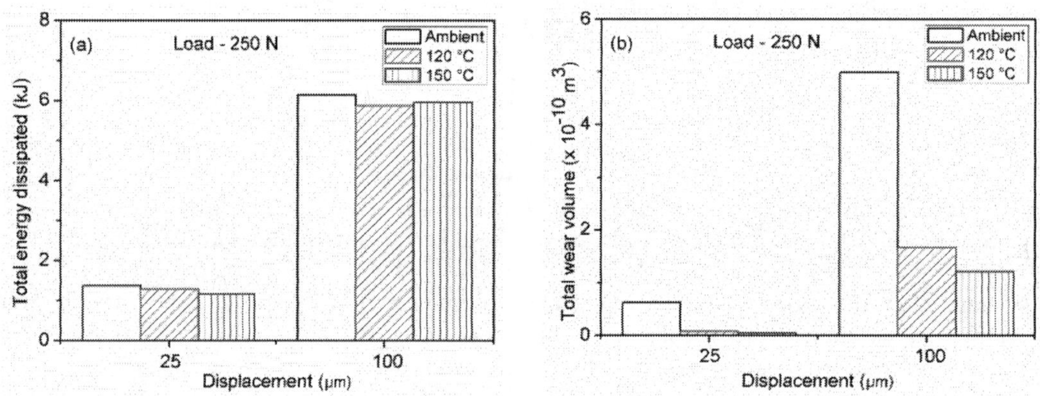

Figure 4. Influence of applied normal load and displacement amplitudes for SCMV-SCMVN material pair with temperature on (a) total energy dissipated and (b) total wear volume.

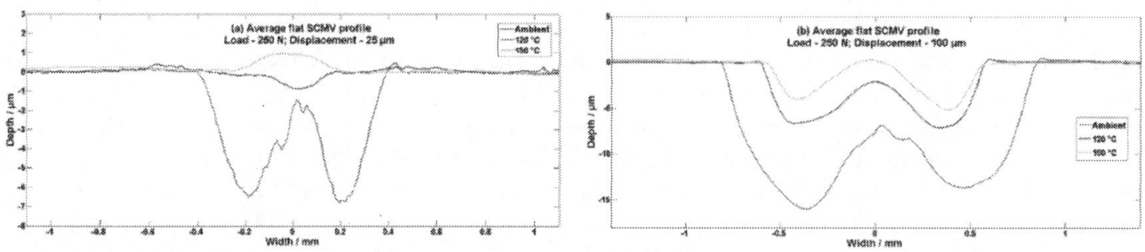

Figure 5. Wear depth profile of the flat SCMV specimen for the test conducted at 250 N with temperature (a) 25 µm and (b) 100 µm.

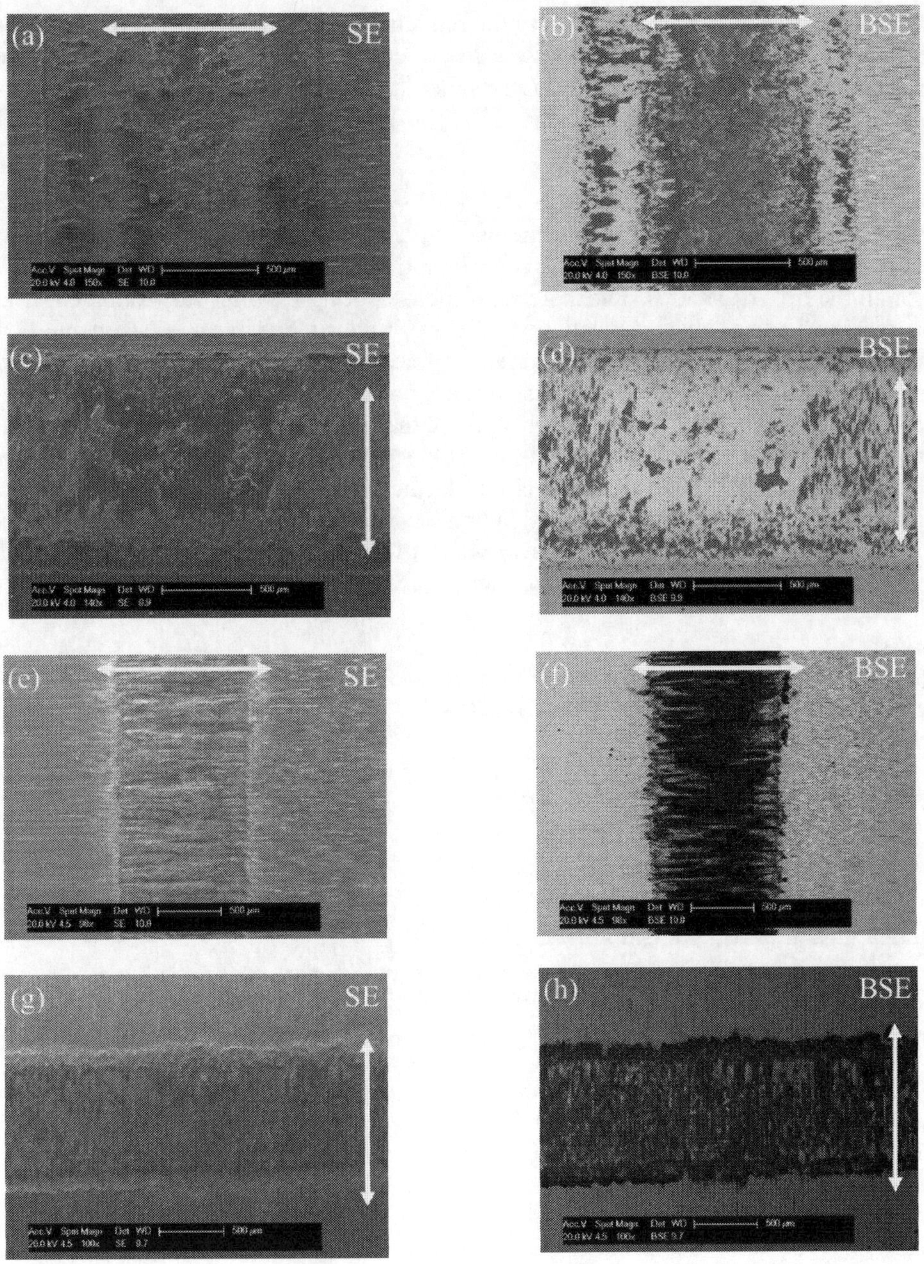

Figure 6. Worn surface morphology of the SCMV fretted against SCMVN specimen at 250 N, 100 μm (i) (a-d) – Ambient temperature; (ii) (e-h) –150 °C. (a, b, e, f) – flat SCMV specimen and (c, d, g, h) – round SCMVN specimen.

150 °C temperature. From the BSE images, it is evident that at ambient temperature, patches of oxides debris are present; whereas at high temperature, the wear scar is fully covered by the oxide debris, illustrating the bed of compacted oxide debris on the wear surface. At ambient temperature, delamination, adhesive wear and abrasion were found to be the most predominant mechanisms of failure at the contact interface in the high strength alloy steel under the fretting conditions investigated. The oxidation of the wear particles is accelerated with the increase in temperature and ploughing by wear particles are observed with increasing temperature. The adhesive transfer layer seemed to protect the

surface with the increase in temperature due to the reduction in the interaction of the worn surface and the total wear volume decreases with increasing temperature. Ploughing, abrasion and abrasive wear are the most predominant wear mechanisms at high temperatures in the high strength alloy steel under the fretting conditions examined.

4. Conclusions

The influence of temperature on the fretting wear performance of SCMV against SCMVN was investigated using a cylinder-on-flat specimen configuration. Experiments were conducted in gross slip condition for all the fretting studies. The material response during fretting was examined by analysis of the friction force response, characterization of the wear debris and wear measurement. At 25 μm displacement amplitude, the steady state tangential force coefficient decreases monotonically with increasing temperature whereas at 100 μm displacement amplitude there is little variation between the temperatures examined. The total dissipated energy and the total wear volume decreases with increasing temperature irrespective of the displacement amplitude. The reduction in wear and tangential force coefficient with the increase in temperature is due to the retention of wear debris within the contact interface, leading to the formation of a stable loading bearing bed and progressive formation of a glaze oxide layer. At ambient temperature, the abrasive action of the particles promoted the wear at the contact interface whereas compaction of the oxide particles reduces the wear at elevated temperatures.

5. References

[1] Arun Prakash N, Gnanamoorthy R and Kamaraj M 2012 *Wear* **294–295** 427–37.
[2] Kayaba T and Iwabuchi A 1981 *Wear* **74** 229–45.
[3] Hurricks P L 1972 *Wear* **19** 207–29.
[4] Hurricks P L 1970 *Wear* **15** 389–09.
[5] Kayaba T and Iwabuchi A 1981 *Wear* **66** 27–41.
[6] Ding J, Leen S B and McColl I R 2004 *International Journal of Fatigue* **26** 521–31.
[7] Fouvry S, Liskiewicz T and Paulin C 2007 *Wear* **263** 518–31.
[8] Mohd Tobi A L, Sun W and Shipway P H 2016 *Tribology International* (In Press: http://dx.doi.org/10.1016/j.triboint.2016.12.009).
[9] Jin X, Shipway P H and Sun W 2015 *Wear* **330–331** 103–11.
[10] Ding J, McColl I R, Leen S B and Shipway P H 2007 *Wear* ***263*** *481–91.*
[11] Yue T and Abdel Wahab M 2016 *Materials* **9** 597.
[12] Yue T and Abdel Wahab M 2017 *Tribology International* **107** 274–82.
[13] McColl I R, Ding J and Leen S B 2004 *Wear* **256** 1114-27.
[14] Yue T and Abdel Wahab M 2014 *Wear* **321** 53–63.
[15] Leen S B, Hyde T H, Ratsimba C H H, Williams E J and McColl I R 2002 *Journal of Strain Analysis for Engineering Design* **37** 565–83.
[16] Pearson S R, Shipway P H, Abere J O and Hewitt R A A 2013 *Wear* **303** 622–31.
[17] Sarkar A D 1980 *Friction and Wear* Academic Press, London.
[18] Chaudhry V and Satish V Kailas 2013 *Wear* **301** 524–39.
[19] Hirsch M R and Neu R W 2013 *Tribology International* **68** 77–84.

Acknowledgements
This work is part of a Collaborative R&T Project, SILOET II Project 6 High Temperature Capability - Compressor and Discs supported by the Technology Strategy Board and carried out by The University of Nottingham and Rolls-Royce plc.

The effect of plasma nitriding and post oxidation on fretting wear behaviour of a high strength alloy steel

N Arun Prakash[1] and C J Bennett[2]
[1,2] Gas Turbine and Transmission Research Centre, Faculty of Engineering, The University of Nottingham, University Park, Nottingham NG7 2RD, United Kingdom
[1]Email: Arun.Natarajan@nottingham.ac.uk, Tel: +44 (0)11584 67683
[2]Email: Chris.Bennett@nottingham.ac.uk, Tel: +44 (0)11584 68460

Abstract. The fretting wear performance of the non-nitrided, nitrided and nitrided-post oxidized high strength alloy steel, W460 were investigated in the gross slip regime at ambient condition. Fretting wear tests were performed with an applied normal load of 250 and 650 N at a displacement amplitude of 100 μm using a cylinder-on-flat configuration. X-ray analysis (XRD) revealed the formation of the iron-nitrided Fe_3N and Fe_4N during plasma nitriding and iron oxide phases of hematite (Fe_2O_3) and magnetite (Fe_3O_4) during post-oxidation of the cylindrical steel samples. The steady state tangential force coefficient decreases when the nitrided and post-oxidized samples were fretted against the non-nitrided steel material when compared to the non-nitrided steel contact pair. The steady state tangential force coefficient decreased with an increase in applied normal load across all of the fretting conditions. The total dissipated energy and the total wear volume increased with an increase in applied normal load with total wear volume of the non-nitrided vs nitrided and non-nitrided vs nitrided post-oxidized sample pairs, showing a reduction in the wear volume of approximately 50% compared to the non-nitrided vs non-nitrided combination under the fretting conditions examined. The worn surface morphology of the fretted samples examined using a scanning electron microscope showed the presence of loose wear debris in the wear track, fragmented wear debris, delamination cracks, delamination with large discontinuities, plate-like wear debris, oxide patches and formation of large cavities.

Keywords: Fretting wear, high strength steel, plasma nitriding, plasma nitriding and post oxidation

1. Introduction

Surface strengthening of metallic materials is achieved by various methods: modifying the surface chemistry, introducing compressive residual stresses and surface hardening. Various mechanical and thermo-chemical surface modification techniques have been developed to strengthen the surfaces of engineering materials, as many machine components experience fatigue loading in service, and mostly, fail due to fatigue related mechanisms. To overcome the tribological issues of any metallic material, surface strengthening of metallic materials are continually being developed for use with specific applications such as increased surface hardness, reduced coefficient of friction and improved mechanical and tribological properties [1]. One of the most economical ways of improving the tribological and corrosion properties of high strength steels is through the use of plasma nitriding, a thermochemical process which involves the diffusion of nitrogen atoms into the steel surface at elevated temperature to produce a compound layer and diffusion layer. A post-nitriding oxidation treatment forms superficial oxide layer over the compound layer on the surface of the metallic materials. Plasma nitriding and post-nitriding oxidation treatments find increasing application on low alloy steels, stainless steels, cast iron and tool steels due to the ability to use a lower process temperature, shorter duration of the treatment, minimal distortion to the surface and low use of energy compared to conventional surface modifications [2-3].

Fretting is characterised by small amplitude relative oscillatory motion in the order of a few tens of microns, occurring between the two normally loaded contacting solid surfaces as a result of cyclic loading and is predominant in quasi-static loaded assemblies of machine elements and engineering structures. Although the fretting wear is complicated in nature, the two most important parameters affecting the fretting mechanism are the displacement amplitude and applied normal load [4]. Mashreghi et al. [5] investigated the effect of plasma nitriding temperature and duration on the microstructure, surface hardness, wear and corrosion resistance of a quenched and tempered cold work tool steel. It was reported that plasma nitriding resulted in a significant improvement in the surface hardness, wear and corrosion resistance of the tool steel and the samples with a higher surface hardness and thicker compound layer with epsilon iron nitride (ε:Fe_3N) phase showed better wear and corrosion resistance. It was stated by Abedi et al. [6] that plasma nitriding and plasma nitriding along with post oxidation surface treatment on AISI 316 austenitic stainless steels reduced the friction coefficient and surface roughness whereas as the oxidation treatment time increases, the friction coefficient and surface roughness increased and the wear resistance decreased. The wear resistance decreased after the oxidation treatment at high applied normal loads and the authors described that this may be attributed to the reduction in the nitride layer thickness [6]. Ramesh and Gnanamoorthy [7] investigated the influence of post-oxidation on the fretting wear behaviour of liquid nitrided structural steel, En 24 against the hardened and tempered En 31 steel and conveyed that the sample liquid nitrided and post-oxidized for 60 minutes showed a lower coefficient of friction and reduced wear volume and hence exhibited a higher resistance to fretting wear compared to the sample liquid nitrided and liquid nitrided and post-oxidized for 15 minutes due to a thicker oxide layer.

BÖHLER W460 is a recently developed ultra high strength steel, manufactured by a double vacuum melted production route, leading to an ultra clean steel with excellent mechanical properties which finds usage in motorsport and advanced engineering applications, particularly highly fatigue-loaded components such as camshafts, gears, connecting rods, crankshafts, bolts and drive shafts, all of which are susceptible to fretting wear and fretting-fatigue failure when the contact conditions prevail. The high range of tempering temperatures for W460 allows for a comprehensive range of surface treatments. However, very few research articles have been published on the friction and wear characteristics of surface modified metallic materials, especially on high strength steels under fretting conditions. In this work, the fretting wear performance of the non-nitrided (quenched and tempered), nitrided and nitrided post-oxidized W460 steel were studied using a cylinder-on-flat samples configuration under various applied normal loads and constant displacement amplitude in dry sliding conditions. The basis for this investigation is the need for high strength materials with enhanced mechanical and tribological properties, providing mechanical integrity for longer life in the engineering applications. Creating an understanding of the behaviour of the material pair at the ambient temperature will allow new materials to be used in highly fatigue loaded components and will provide a platform to better understand the role of contact conditions on the mechanisms controlling the fretting wear process.

2. Methodology

Fretting wear studies were performed on the non-nitrided, nitrided and nitrided post-oxidized BÖHLER W460 high strength chromium–molybdenum–vanadium steel at ambient temperature using a crossed cylinder-on-flat sample configuration. BÖHLER W460 steel contains 4.55 Cr, 3 Mo, 0.75 V, 0.5 C, 0.2 Si, 0.45 Mn and balance Fe by weight percentage. The flat quenched and tempered W460 samples were fretted against the cylindrical non-nitrided, nitrided and nitrided post-oxidized W460 samples. Initially, the test materials were cut into sample blanks providing an allowance for heat treatment on all the sides. The heat treatment consisted of a preheat at 560 °C for 120 mins, 650 °C for 30 mins and 860 °C for 30 mins followed by normalising at 1050 °C for 30 mins after which samples were oil quenched; subsequently, samples were tempered at three different tempering temperatures with air cooling after each successive temperature, 550 °C for 180 mins, 580 °C for 180 mins and 580 °C for 120 mins. The flat and cylindrical samples were further machined into the required dimensions from the blanks once the heat treatment process was completed, removing the decarburised layer formed by the heat treatment process from the sample surface. The machined cylindrical samples were subjected to nitriding and nitriding post-oxidation treatment. The cylinder-on-flat test sample configuration used for the fretting studies, which results in a 10 mm line contact, is shown in Figure 1. The cylindrical samples were manufactured with a radius of 6 mm and the flat samples had a width of 10 mm. The surface hardness of the quenched and tempered W460 material was measured as 605 ± 15 HV30. The elastic modulus and Poison's ratio of the W460 material were 206 GPa and 0.28 respectively. X-ray analysis (XRD) revealed the formation of the iron-nitrided Fe_3N and Fe_4N during plasma nitriding and iron oxide phases of hematite (Fe_2O_3) and magnetite (Fe_3O_4) during post-oxidation of the cylindrical steel sample.

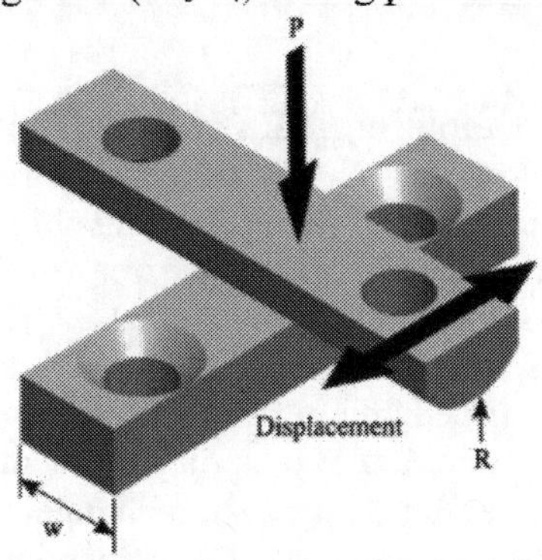

Figure 1. Cylinder-on-flat sample configuration used in the fretting experiments, where $W = 10$ mm, $R = 6$ mm, P is the applied normal load.

A detailed description of the fretting wear test setup is given elsewhere [8, 9] although a brief explanation of the test rig is presented here for ease of understanding. Initially all the fretting samples were demagnetised, thoroughly degreased using detergent, rinsed with industrial methylated spirit (IMS) and acetone and finally dried using a hot air dryer to remove the surfactant present in the samples before the start of the experiment. On the stationary lower specimen mounting block (LSMB) the flat sample was mounted, while on the moving upper specimen mounting block (USMB) the cylindrical sample was mounted. The fretting studies were conducted with applied normal loads of 250 and 650 N, with a displacement amplitude of 100 μm, at a constant frequency of 20 Hz, with relative humidity of 35 ± 5 % at ambient temperature. Table 1 shows the fretting wear test parameters used in the current study. The main components of fretting wear test rig are upper and lower specimen mounting blocks, electromagnetic vibrator, sensors and data acquisition systems. The relative displacement between the flat and cylindrical samples is produced by an electromagnetic vibrator which develops a force that is axially guided by a leaf spring. The relative displacement between the upper and lower specimen mounting block was measured using a linear variable differential transformer (LVDT). The tangential force at the contact interface are measured using the piezoelectric load cell. Normal loads were exerted on the sample by dead weights using the lever arm to the USMB. The tangential force (Q) and relative displacement (δ) during the running condition was continuously logged using the control and data acquisition system programmed in LabVIEW. Both the load sensor and displacement sensor were calibrated externally and in-situ under static conditions. To assess the surface topography of the flat and cylindrical samples a Taylor-Hobson Talysurf CL1 1000 tactile profilometer with a fine diamond stylus was used to scan the wear scar individually. Prior to scanning, the samples were rinsed with acetone to remove the loose wear debris that was not adhered to the surface. The worn surface morphology of the fretted flat and cylindrical surfaces was characterised using a Philips XL 30 SEM scanning electron microscope to reveal more information about the nature of contact and associated damage mechanisms exhibited at the contact interface.

Table 1. Summary of the fretting wear test parameters.

Material combination (Flat/Round)	W460 – W460, W460 – W460N, W460 – W460NO
Normal load, P (N)	250 and 650
Displacement amplitude, δ (μm)	100
Number of cycles, N	100 x 10^3
Radius of the cylindrical sample, R (mm)	6
Frequency, f (Hz)	20
Temperature, T (°C)	Ambient, 120 and 150

3. Results and discussion

The tangential force coefficient, or coefficient of friction, is defined as the ratio of tangential force amplitude *(Q)* to the applied normal load *(P)* under fretting sliding conditions. As pure sliding motion occurs between the two contacting surface in the reciprocating sliding wear process, the coefficient of friction is used to describe the friction at the interfaces whereas in fretting wear process as there is no pure sliding motion between the two members at the contact interface, tangential force coefficient is used to define friction at the contacts. The variation of the tangential force coefficient with number of cycles elapsed for the non-nitrided, nitrided and nitrided post-oxidized W460 high strength alloy steel under the applied normal loads of 250 and 650 N with displacement amplitude of 100 μm is shown in Figure 2. During the initial stages of the fretting process, the tangential force coefficient increases with the number of cycles for all the conditions examined. The tangential force coefficient for the quenched and tempered material pair (non-nitrided) is low at the beginning of the test and after the completion of a few hundred cycles, the tangential force coefficient increases rapidly and reaches a more steady state value under the conditions investigated. The initial increase in the tangential force coefficient is mainly attributed to the strong adhesion at the contact interface after the removal of the initial protective oxide films by the mechanical action. Once the initial oxide layer is interrupted there is no further formation of the oxide layer and the metal surface remains reactive resulting in high adhesion at the contact interface. This phenomenon not only results in high tangential frictional forces but in some conditions (displacement amplitudes) it can also lead to seizure [10]. Seizure is usually defined as the arrest of the relative motion between the contacting bodies as a result of the adhesive interaction of the rubbing surfaces as well as the debris entrapment. The probability of seizure reduces at high displacement amplitude due to larger relative motion between the contacting bodies as they are accommodated by elastic deformation and large variation in the tangential force coefficient with applied normal load and cycles elapsed has been observed. The tangential force coefficient of the nitrided and nitrided post-oxidized material pair at an applied normal load of 250 N and displacement amplitude of 100 μm increases initially as the cycles progressed (< 15,000) and then fluctuates for a period of cycles (~15,000 to 55,000) and reaches a more steady state value (> 55,000) under the conditions investigated. At an applied normal load of 650 N, the nitrided and nitrided post-oxidized material pair showed a sharp drop in the tangential force coefficient during the initial cycles with a gradual increase in the tangential force coefficient with an increase in number of cycles and as the fretting cycles progressed the tangential force coefficient fluctuates.

The steady state tangential force coefficient under fretting conditions represents the friction coefficient between the materials where the various parameters influencing the friction process attain a steady state value after initial perturbations. A steady state condition depends on the development of the wear debris, deformation and ejection of the wear debris from the contact interface by wearing down of tangentially interlocked protrusions and depressions. The average value of the tangential force

IOP Conf. Series: Journal of Physics: Conf. Series **843** (2017) 012079 doi:10.1088/1742-6596/843/1/012079

coefficient between 40 x 10^3 to 100 x 10^3 cycles was calculated for all the conditions examined to enable further comparison of the behaviour of the material pairs under different fretting conditions and is presented in Figure 3 (a). The steady state tangential force coefficient decreases with an increase in applied normal loads across all of the fretting conditions. The lower value of the steady state tangential force coefficient in the nitrided and nitrided post-oxidized W460 samples are attributed to the presence of the white iron nitride Fe_3N and Fe_4N layer on the surface formed during plasma nitriding and iron oxide phases of hematite (Fe_2O_3) and magnetite (Fe_3O_4) during post-oxidation. Figure 3 (b) shows the influence of the applied normal load and displacement amplitude on the total energy dissipated, it can be seen that the total energy dissipation increases with an increase in the normal load for the material pairs and displacement amplitude examined. The increase in total energy dissipated with applied normal load is due to the shearing of asperities at the contact interface and the frictional energy needed for the removal of the material at high loads. There is very little change in the total energy dissipation due to nitriding and post-oxidization of W460 steel. Figure 4 shows the influence of the applied normal load and displacement amplitude on the total wear volume and the individual wear volume of the flat and the cylindrical samples examined. The total wear volume decreases with the nitrided and nitrided post-oxidized samples compared to the non-nitrided test due to (i) the presence of a dense compound layer of iron nitride and iron oxide phases on the surface, (ii) an increased surface hardness that allows

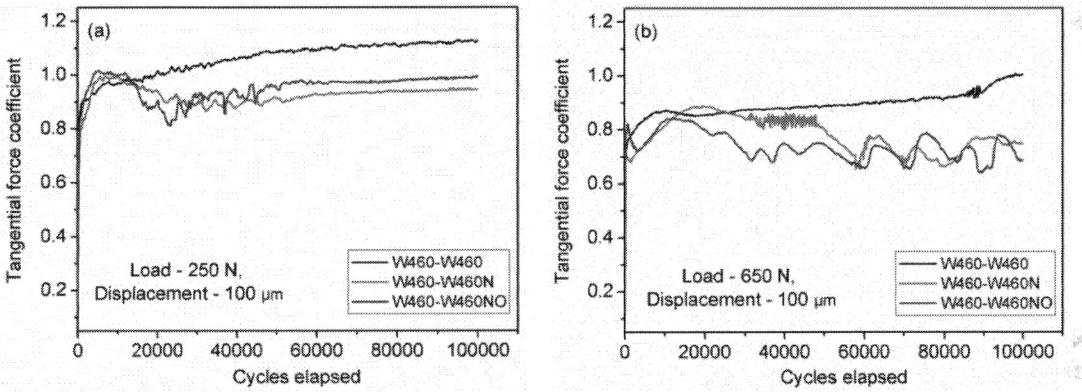

Figure 2. Variation of tangential force coefficient with cycles elapsed for different material contact pairs at (a) 250 N, 100 μm and (b) 650 N, 100 μm.

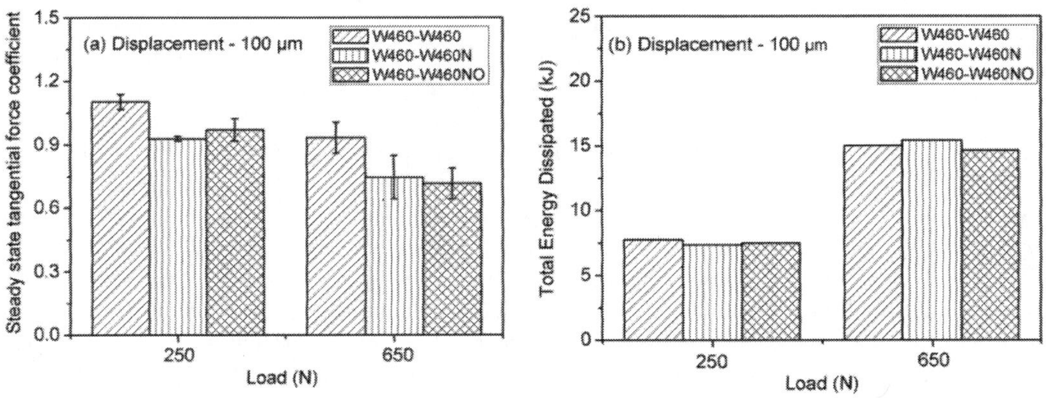

Figure 3. Influence of normal load on (a) steady state tangential force coefficient and (b) total energy dissipated.

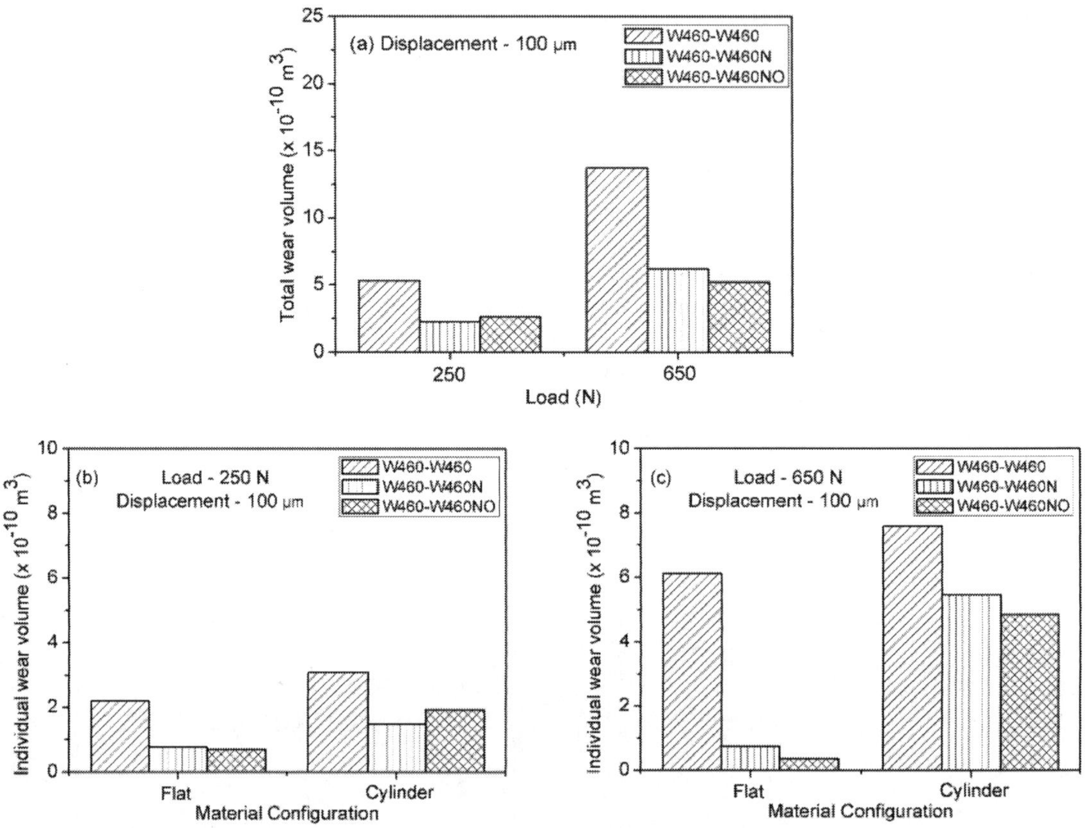

Figure 4. Influence of applied normal loads and displacement amplitude on (a) total wear volume, (b) individual wear at 250 N, 100 μm and (c) individual wear at 650 N, 100 μm.

only limited plastic deformation on the surface and (iii) a reduction in the tangential force coefficient. Similar behaviour of a decrease in the wear volume of liquid nitrided steels were reported by Ramesh and Gnanamoorthy [11] and Kato et al. [12], they suggested that the improvement in the fretting and sliding wear resistance of the nitrided steels are attributed to the presence of a thin mono-phase epsilon iron nitride layer on En24 steel [11] and En41B steel [12]. It is also reported by Ramesh and Gnanamoorthy [11] that the load at which the transition between the slip regimes occurs depends on the surface composition of the contact pair and not on the hardness of the material pair studied. At an applied normal load of 650 N, the total wear volume of the nitrided post-oxidized samples was lower than the nitrided and non-nitrided samples which is attributed to the reduction in the nitride layer thickness and prolonged oxidation time; similar behaviour was reported by Abedi et al. [6]. The oxide film formed by the post-oxidation treatment may act as a solid lubricant reducing the tangential force coefficient. At high load the post-oxidized sample had a lower tangential force coefficient and wear compared to the non-nitrided and nitrided samples, whereas at low load the tangential force coefficient and wear volume are greater. The individual wear volume of the flat samples were lower than the round W460 samples for all the conditions examined. As the applied normal load increases for the same displacement amplitude (i.e., 100 μm), the wear volume increases as the fretting contact area increases thereby increasing the total energy required to wear the fretted surface. Figure 5 shows the wear depth profile of the flat sample with an applied normal load of 250 and 650 N. The profile of the wear scars was trough-like in nature, showing U-shape for the non-nitrided samples and a W-shaped for nitrided and nitrided post-oxidized samples, indicating bulk material removal. The magnitude of the fretting damage increases with increase in applied normal load for the non-nitrided material pair as illustrated in Figures 5 (a and b), whereas there is not much increase in the fretting damage for the nitrided and nitrided

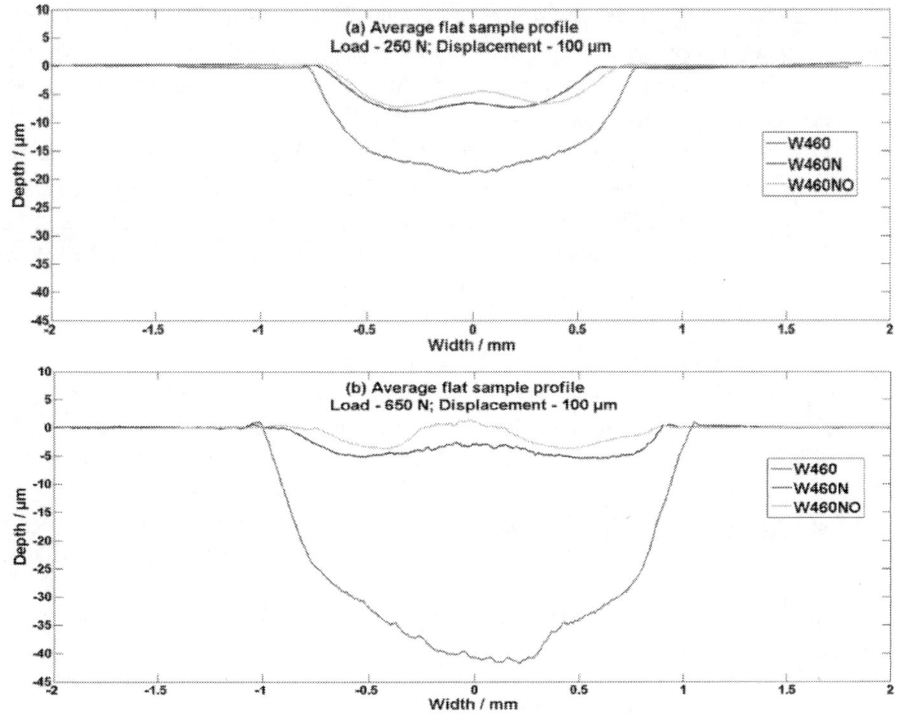

Figure 5. Wear depth profile of the flat sample for the test conducted at 100 μm displacement amplitude for different applied normal loads (a) 250 N and (b) 650 N.

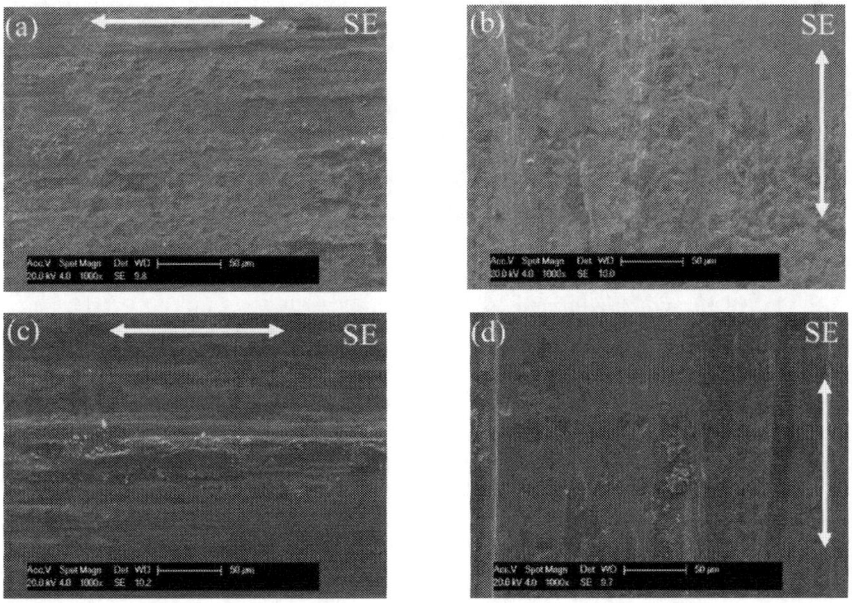

Figure 6. Worn surface morphology of the fretted sample at 250 N, 100 μm (i) W460 – W460N and (ii) W460 – W460NO; (a, c) – flat W460 sample, (b) – round W460N sample and (d) – round W460NO sample.

post-oxidized samples which are attributed to the presence of the iron-nitride and iron oxide phases. At the edge of the contact, an increase in the height of the wear depth was observed indicating either plastic deformation or embedding of wear debris taking place during the tests for all the conditions examined. Figure 6 shows the worn surface morphology of the W460 fretted against W460N and W460NO at 250 N, 100 μm. The worn surface morphology of the fretted samples showed the presence of loose wear debris in the wear track, fragmented wear debris, delamination cracks, delamination with large discontinuities, plate-like wear debris, oxide patches and formation of large cavities. Abrasion was found to be the most predominant mechanism of failure in non-nitrided, nitrided and nitrided-post oxidized W460 high strength alloy steel under the fretting conditions investigated.

4. Conclusions

The fretting wear performance of the non-nitrided, nitrided and nitrided-post oxidized W460 high strength alloy steel was investigated in the gross slip regime using a cylinder-on-flat configuration. For the conditions studied, the steady state tangential force coefficient of the nitrided and nitrided post-oxidized W460 samples was lower than the non-nitrided W460 samples. The wear volume of the nitrided and nitrided post-oxidized W460 samples was reduced by almost 50 percent compared to the non-nitrided samples due to the presence of the iron-nitrided Fe_3N and Fe_4N during plasma nitriding and iron oxide phases of hematite (Fe_2O_3) and magnetite (Fe_3O_4) during post-oxidation of W460. As the applied normal load increased for the same displacement amplitude, the wear volume increased as the fretting contact area increased thereby increasing the total energy required to wear the fretted surface. The worn surface morphology of the fretted samples showed the presence of loose wear debris in the wear track, fragmented wear debris, delamination cracks, delamination with large discontinuities, plate-like wear debris, oxide patches and formation of large cavities. Abrasion was found to be the most predominant mechanism of failure in non-nitrided, nitrided and nitrided-post oxidized W460 high strength alloy steel under the fretting conditions investigated.

5. References

[1] Arun Prakash N, Gnanamoorthy R and Kamaraj M 2010 *Material Science and Engineering B* **168** 176–81.
[2] Alsaran A, Altun H, Karakan M and Celik A 2004 *Surface and Coatings Technology* **176** 344–48.
[3] Li Y, Wang L, Zhang D and Shen L 2012 *Applied Surface Science* **256** 4149–52.
[4] Arun Prakash N, Gnanamoorthy R and Kamaraj M 2012 *Surface and Coatings Technology* **207** 450–60.
[5] Mashreghi A R, Soleimani S M Y and Saberifar S 2013 *Materials and Design* **46** 532–38.
[6] Abedi H R, Salehi M and Yazdkhasti M 2010 *Materials Letters* **64** 698–701.
[7] Ramesh R and Gnanamoorthy R 2006 *Surface and Coatings Technology* **200** 3926–32.
[8] Warmuth A R, Pearson S R, Shipway P H and Sun W 2013 *Wear* **301** 491–500.
[9] Pearson S R, Shipway P H, Abere J O and Hewitt R A A 2013 *Wear* **303** 622–31.
[10] Chaudhry V and Satish V Kailas 2013 *Wear* **301** 524–39.
[11] Ramesh R and Gnanamoorthy R 2006 *Journal of Materials Processing Technology* **171** 61–7.
[12] Kato H, Eyre T S and Ralph B 1994 Surface Engineering **10** 65–75.

Acknowledgements
This work is part of a Collaborative R&T Project, SILOET II Project 6 High Temperature Capability - Compressor and Discs supported by the Technology Strategy Board and carried out by The University of Nottingham and Rolls-Royce plc.

IOP Conf. Series: Journal of Physics: Conf. Series **843** (2017) 012080 doi:10.1088/1742-6596/843/1/012080

Development of new duplex treatments on 100Cr6steel combining Thermochemical Treatments, Laser Shock Peening and Physical Vapour Deposition

J. Osés[1], G. G. Fuentes[1], S. Santo Domingo[2], I. Miguel[2], S. Gimeno[2], L. Carreras[3], P. Peyre[4], C. Gorny[4]

[1] Asociación de la Industria Navarra, AIN, Ingeniería de Superficies, Carretera Pamplona n°1, Cordovilla 31191, Spain

[2] FERSA INNOVA, Calle Bari 18, Parque empresarial PLA-ZA, Zaragoza 50197, Spain

[3] Tratamientos Térmicos Carreras, Passatge Enginyer Playa 38-42 Sabadell, 08205, Barcelona, Spain

[4] PIMM - Procédés et Ingénierie en Mécanique et Matériaux, Arts et Métiers ParisTech - 151 bd de l'Hôpital, 75013 Paris, France

joses@ain.es

Abstract. 100Cr6 steel (AISI 52100) is one of the most used steel grades in the manufacturing of through hardening bearings mainly due to its properties: controlled impurities during steel making process, high hardenability and well known mechanical properties such as wear and fatigue resistance on clean environments. These characteristics play an important role on the performance of a bearing together with the bearing design, loads and environment. However, there is an increasing set of demanding applications where the above mentioned steel does not fulfil the required needs and thus, bearing manufacturers continuously work on the development of technologies to improve the bearing performance.

Nowadays thermochemical treatments (TCT), such as carbonitriding are being applied to this steel in order to enhance the performance of such pieces in contaminated environment, where particles can produce defects on the raceway, increasing the onset of defects that eventually lead to premature fail. These treatments induce the formation of carbides and nitrides which are directly related to the enhancement of the wear resistance and also to increasing the amount of Retained Austenite (RA) in the surface which may have a beneficial effect as it delays the crack propagation on subsurface regions, then increasing bearing fatigue life.

In this work, different TCTs have been applied to 100Cr6 steel flat samples. Using a tribometer (ball-on-disc configuration) and a grinding machine, surface and in-depth wear resistance measurements have been carried out, obtaining wear resistance profiles that have been correlated with the microstructure, microhardness profiles and RA content.

The most promising TCT has been combined either with Laser Shock Peening (LSP) treatments or carbonaceous Physical Vapour Deposition (PVD) coatings with the aim of improving not only the wear resistance but also the CoF of the duplex treated sample.

The results obtained on flat samples are promising; the combination of treatments produces long-lasting low CoF and a reduction of 60% in the wear rate. However, the treatments should be applied on real pieces and tested in a test bench in order to obtain more appropriate data about the lifespan of duplex treated bearings.

1. Introduction

Rolling bearings are one of the most critical components in any mechanical system, especially in automotive and industrial applications where they usually work under high cyclic loads and high speeds requirements, not only in clean but also in contaminated environments where metallic particles are present [1].

Nowadays load bearing capacity requirements for bearings are becoming more demanding, as well as longer service life and higher reliability [2,3]. These requirements are aimed to develop new high-end applications and to reduce maintenance costs.

To meet these demands, tribologists and material experts are exploring new approaches in order to enhance existing materials, such as SAE 52100 steel for bearings, studied in this work [3,4].

Failures in bodies working under rolling or sliding contact, such as bearings, are normally due to Rolling Contact Fatigue (RCF). Cyclic loads lead to crack generation and propagation, what ends in material removal through pitting/delamination of the near-surface area [4,5]. Is therefore clear that surface characteristics have a great influence in RCF resistance [6].

Different material and surface modifications have been analyzed to enhance life performance and wear resistance, as well as friction reduction, by preventing the appearance of RCF.

One of the modifications analyzed are thermochemical treatments, specifically Carbonitriding [7]. It involves the diffusion of C and N into the steel substrate, producing fine carbides and nitrides, compressive surface stress and a higher concentration of Retained Austenite than through hardening treatment. The result is higher hardness and enhanced fatigue life up to 10-15 times longer due to crack propagation reduction [4].

Other modifications include surface treatments such as Laser Peening. It is an industrial surface treatment featuring the formation of shock waves on the surface through laser pulses that produce compressive stress on it [8,9]. For that reason mechanical properties of the material are improved, namely higher hardness, wear and friction reduction under certain conditions, and improved fatigue strength by delaying crack propagation [8,10]. Besides one of the advantages of this technique is that compressive stress from Laser Peening process penetrate deeper than those from Shot Peening [10].

The other surface treatments tested are Diamond-Like Carbon (DLC) coatings. DLC are amorphous materials with different %C (variants: a-C:C) containing significant fractions of sp3 bonds. It is deposited through Physical Vapour Deposition or Plasma Assisted Chemical Vapour Deposition techniques in the form of a sputtered film composed of several layers, and improves coefficient of friction and wear rate by generating compressive stress and high microhardness [11,12]. DLC thickness, which is in the range of 0.75 – 1μm, is a critical factor because is directly related to internal residual stress within the coating. Thick DLC layers are associated to high failure rates in the first steps of RCF [5].

2. Experimental

2.1. Materials

A 100Cr6 steel bar of 50 mm in diameter was cut into slices of 5 mm each. The bar had been previously annealed following this process: 815°C for 3 hours, 735°C for 4 hours, 675°C for 3 hours, slow cool down to 540°C and finally air cooling down to room temperature.

2.2. Thermochemical treatments (TCT)

The samples were thermally/thermo-chemically treated in an Eros T4 with 2 chambers transfer oven.

The reference samples were heated at 830 °C in a controlled atmosphere during 1 h and then were oil quenched.

For the TCT, four different temperatures between 800 and 925 °C were used: T1, T2, T3 and T4. Other parameters set in the TCT were: C potential, NH3 intake and treatment time (see Table I). The samples were oil quenched and split in two groups. The first group was kept in that state and the second group was subjected to sub-zero (SZ) treatment, performed with the aim of optimizing the retained austenite.

After the different thermal/thermochemical treatments, all the samples were tempered at 180 °C during 2 h.

Table I: thermal/thermochemical treatment conditions

Sample	Temperature (°C)	Time (h)	Atmosphere	Sub-zero
Ref	830	1	1% C	No
T1	T1	4	1,1 %C + NH_3	No
T1SZ	T1	4	1,1 %C + NH_3	-45 °C
T2	T2	4	1,1 %C + NH_3	No
T2SZ	T2	4	1,1 %C + NH_3	-45 °C
T3	T3	4	1,1 %C + NH_3	No
T3SZ	T3	4	1,1 %C + NH_3	-45 °C
T4	T4	4	1,1 %C + NH_3	No

2.3. Grinding process

A Chevalier CNC flat grinding machine was used for removing the material for the in-depth wear tests. Borazon extra-fine grinding wheels and low feed rate were used with the aim of minimizing the distortion and phase transformation during this process.

Different thicknesses of material were removed from the samples: 100, 150, 200 and 250 µm. These depths were chosen taking into account the manufacturing process of the bearings and to study wear resistance behaviour of the structure in each layer.

The surface of the samples was polished using cloths and polycrystalline diamond suspension of 9, 3 and 1 µm for the final polishing stage.

2.4. Laser Shock Peening (LSP)

LSP treatments were carried out using a Thales Gaia HP Laser. The configuration used in this work produces pulses of 10 ns and 14 J, with a wavelength of 532 nm, and they are able to produce plastic deformation in the material and induce a compressive layer in the first mm of the surface. Three different configurations were used: LC1, LC2 and LC3.

2.5. Physical Vapour Deposition (PVD)

PVD coatings were deposited using a Metaplas Ionon MZR 323. The equipment has three magnetron sputtering sources, being two of them balanced and one unbalanced.

WC:C coatings of about 1 µm were deposited on the samples.

2.6. Metallography

The treated samples were cut using a Buehler Isomet 4000 precision saw and then hot-mounted in phenolic resin using a Buehler Simplimet 1000 mounting press. The cross sections were first polished using SiC polishing papers (P320, P600, P1200) and then using diamond suspensions of 9, 3 and 1 µm for a final mirror polish. In order to reveal the microstructure of the material and the effect of the TCT, the samples were chemically etched with nital 3 % and Picral.

The optical inspection was carried out using a Leica DMI 5000M metallography microscope. The microstructure, the in-surface depth of the TCT's and the grain size, following the ASTM E112, in the area affected by the TCT were analysed.

2.7. Microhardness

A micrometer Buehler Micromet 2103 was used for determining the hardness profiles of the material. Polished cross sections were tested using a Knoop indenter and a load of 50 gf. Five indents were performed on each studied depth.

2.8. X-Ray Diffraction
Quantitative XRD analysis was used to determine the fraction of retained austenite. For these experiments, samples were polished up to Ra = 10 nm. X-ray diffraction measurements were performed with a Bruker D8 Discover diffractometer equipped with a Co X-ray tube, a Lynx-eye position sensitive detector and Goebel mirror optics to obtain a parallel and monochromatic X-ray beam. A current of 30 mA and a voltage of 40 keV were used. Operational conditions were selected to obtain X-ray diffraction data of sufficiently high quality, e.g., sufficient counting statistics and narrow peak widths. XRD data were collected over a 2θ range of 35 -116° in steps of 0.01° and 0.3 s/step. To minimize the effect of texture, the volume fraction of retained austenite was calculated from the integrated intensities of (111), (200), (220), and (311) austenite peaks and the ferrite (110), (002), (112) planes. For this goal, a calibration curve plotted from the data of three standard reference materials certified by the National Bureau Standards with specific amounts of austenite (5, 15 and 30%) was used.

In-depth residual stress profiles were obtained using the same equipment. The surfaces were measured at different angles and the position of the diffraction peeks was studied, obtaining the surface stress. The measurements were repeated several times after removing layers of material by electropolishing. The results allowed the obtention of the stress profiles.

2.9. Wear tests
The wear tests were performed on a ball-on-disc CSM HT Tribometer with a WC 6 mm in diameter ball as a counterbody, 10 N of normal load and duration of 50.000-100.000 cycles at 10cm/s and track radius from 4 to 8 mm.

The obtained wear tracks were measured by means of an interferometric profilometer Wyko RST 500 TM using the Vertical Scanning Interferometry (VSI) mode. The loss volume was used to determine the wear coefficient of each sample.

3. Results & discussion

3.1. Thermochemical treatments
In the first part of the study, flat samples were treated with the aim of studying the effects of the TCT on the material and discard the treatments that are not suitable to be scaled up to real parts.

3.1.1. Microstructure. Optical microscopy was used to determine the penetration of the TCT. There are two factors affecting the measurement. Firstly, the diffusion of nitrogen and carbon into the material and secondly, the effect of sub-zero quench that changes the susceptibility of the material to be chemically etched, so the thickness of the carbonitrided layer is not clear (See figure 1). Besides, sub-zero treatment transform part of the retained austenite, white and clearly visible in the optical analysis, into martensite, that is similar to the rest of the bulk material. Due to this, the determination of the treated layer by optical methods was not accurate. Results are collected in table II.

3.1.2. Grain size. The optical analysis of the samples allowed the determination of the grain size in the carbonitrided surface. The temperature and time of the TCT are higher than in the conventional quenching process, so the grain size is expected to be bigger (figure 2).

As it can be seen in Table II there are big differences in the depth observed by optical microscopy due to the effect of sub-zero treatment.

The grain size of the material in the area close to the surface has been increased due to the thermochemical treatment (see figure 2). TCT performed at T3 and T4 were dismissed due to the excessive growth of the grain size. It must be considered that the study is focused on the development of treatments intended for bearings, where ASTM grain size number under 9 is not allowed.

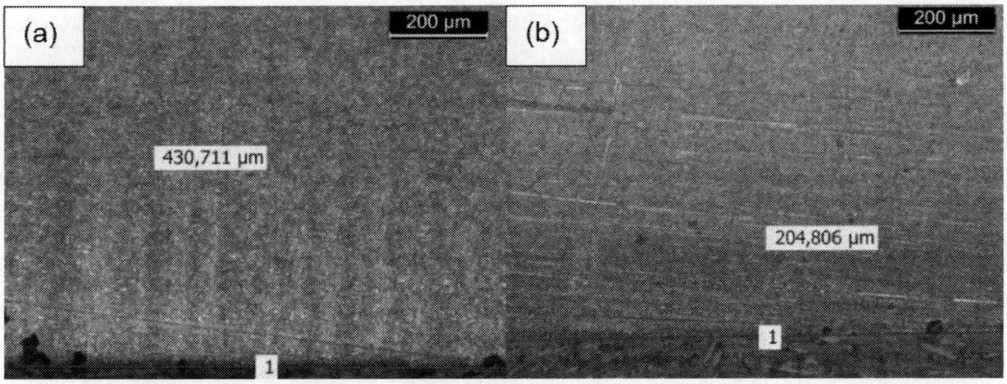

Figure 1: Optical micrographs (100X) of samples T4 (a) and T3SZ (b) after nital etching and estimated treatment depth.

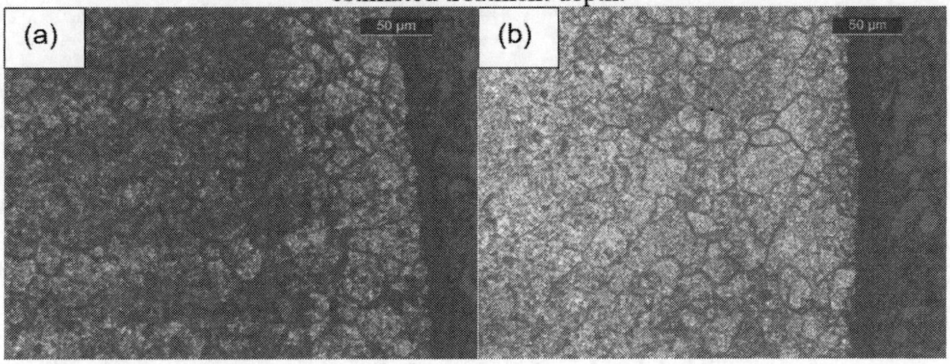

Figure 2: Optical micrographs (200 X) of samples T2 (a) and T3 (b) after Picral etching

Table II: treatment depth, grain size and retained austenite surface content of the different treatments

Sample	Treatment depth (μm)	Surface grain size ASTM	RA content (%)
Ref	---	9-10	7.5
T4	380-450	7	32.6
T3SZ	200-220	7	26.5
T3	300-400	8	29.6
T2SZ	200-220	9	29.7
T2	260-290	9	42.8
T1SZ	180-210	9	22.5
T1	200-220	9	37.0

3.1.3. XRD surface measurements. Surface XRD measurements showed an important increase in the RA content on the surfaces of the treated samples (table II). The RA content of the quenched sample is 7.5 %, but the carbonitriding process increases this amount due to the stabilization effect that nitrogen & carbon induces on austenite.

A drop in the RA due to sub-zero treatment can be observed in all samples. The RA content of sample T3 is lower than expected, which can be ascribed to problems during surface preparation of the sample. Due to this, the difference between the samples T3 and T3SZ is less than 3 %.

Surface RA content higher than 30% is not allowed due to potential distortions during operation in real pieces, so subzero treatments were preferred versus non subzero treatments.

3.2. In-depth measurements

After the first analysis, the number of suitable treatments was reduced to two: T1SZ and T2SZ.
After this first stage of the study, two sets of samples of each treatment were prepared, then grinded at different depths and polished with the aim of evaluating the in-depth properties of the treatments and obtain the information needed for scaling-up the process to be applied on real bearings.

Different characterization analyses were performed on the samples: optical microscopy, RA measurement by XRD and wear resistance measurements by means of Pin-on-disc tribology analysis. The results are gathered on table III.

The microstructure study (figure 3) shows that, at T1, the cementite content of the surface is increased, producing carbide networks. These networks could represent a problem in the manufacturing process of real parts, so must be taken into account.

In addition to this, microhardness profiles were measured in cross sections of samples T1SZ and T2SZ as it can be seen in figure 4.

The analysis of the curves in figure 4 allows determining the depth at which the wear is minimum in light of the tribology results, which is a goal in this study. This point is 200 μm for T1SZ and 150 μm for T2SZ and the obtained value can be considered similar to the wear rate of the quenched (Ref) sample. These points correspond with RA contents of 8.8 % in the treatment at T1 and 22.4 % in the one at T2. This difference could be explained for the different content in nitrides, carbides and carbonitrides, which was not analysed.

Regarding the microhardness, figures 4 and 5 show how a high RA content affects strongly the surface hardness of sample T2SZ, lowering it, especially in the first 100 μm. The lower content of RA observed in sample T1SZ, combined with the presence of cementite, allow the increase of the hardness with respect to the Ref sample. The hardness has been increased in the first 400-500 μm of the material, to decrease later to bulk values.

Figure 3: Optical micrographs (200 X) of samples T1SZ (a) and T2SZ (b) after nital etching. Cementite segregates in grain boundary in the first 100 μm of the samples T1SZ

After this stage, T1SZ was selected to be combined with LSP and PVD treatments. Is expected that treatment T1SZ will be able to improve the fatigue resistance of the material without losing surface hardness and wear resistance.

Table III: samples obtained after grinding process and measured properties

Sample	Removed	RA (%)	K wear * E16

IOP Conf. Series: Journal of Physics: Conf. Series **843** (2017) 012080 doi:10.1088/1742-6596/843/1/012080

	material (µm)		(m3/Nm)
Ref	0	7.5	5.2 ± 3.0
T1SZ	0	22.5	23.2 ± 4.6
T1SZ100	100	15.2	19.9 ± 4.4
T1SZ150	150	13.6	17.4 ± 5.5
T1SZ200	200	8.8	10.1 ± 5.1
T1SZ250	250	8.2	12.8 ± 5.9
T2SZ	0	29.7	26.1 ± 6.9
T2SZ100	100	27.4	17.4 ± 5.0
T2SZ150	150	22.4	9.8 ± 5.6
T2SZ200	200	20.3	13.7 ± 8.8
T2SZ250	250	19.3	18.0 ± 9.6

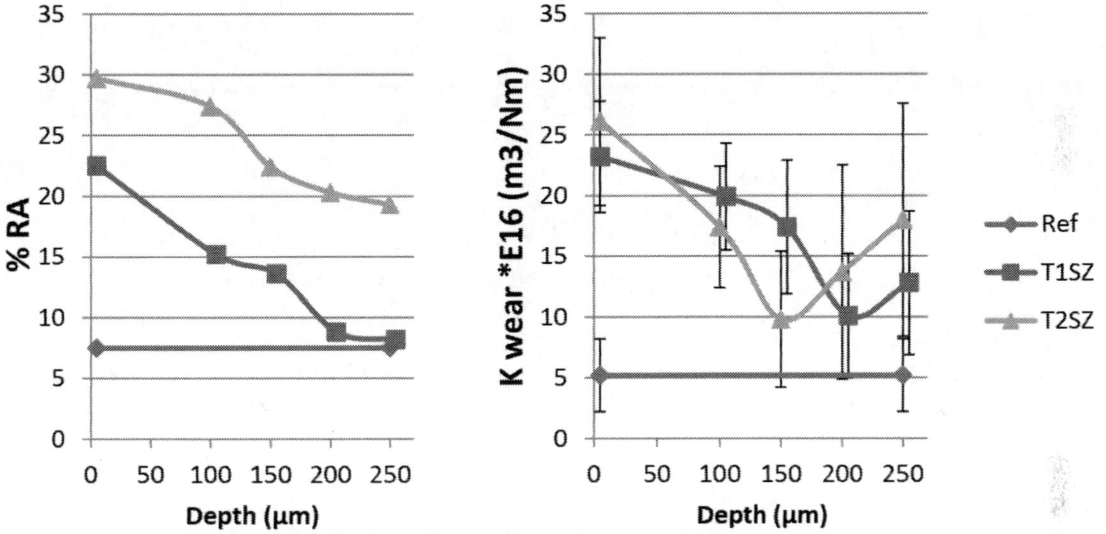

Figure 4: Graphs of RA content and Kwear profiles of Ref, T1SZ and T2SZ

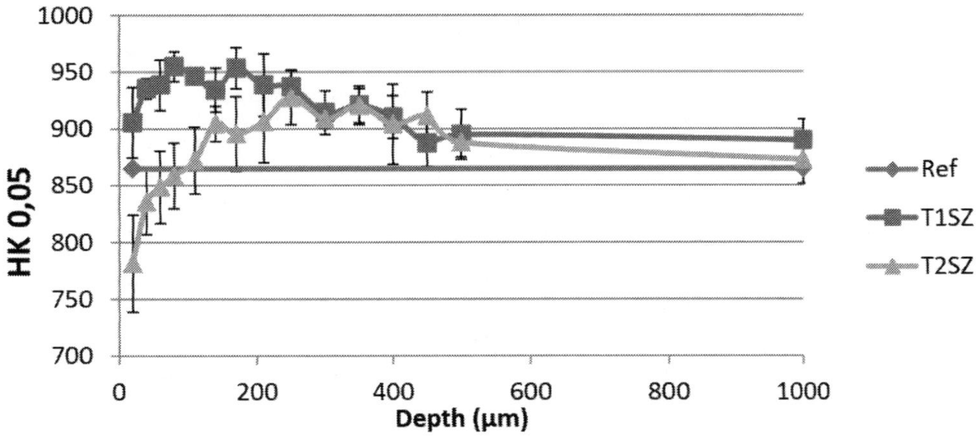

Figure 5: Microhardness profiles of samples Ref, T1SZ and T2SZ

3.3. Laser Shock Peening (LSP)

Stress profiles were studied with the aim of determining the effect of LSP on the 100Cr6 treated samples. Figure 6 shows the profiles obtained after the application of 3 different LSP treatments. All of them induce a compressive layer of about 400 MPa at 50 μm that decreases to 200 MPa at 600 μm of depth.

LSP1 was the best of the treatments taking into account the profile produced and the time consumption.

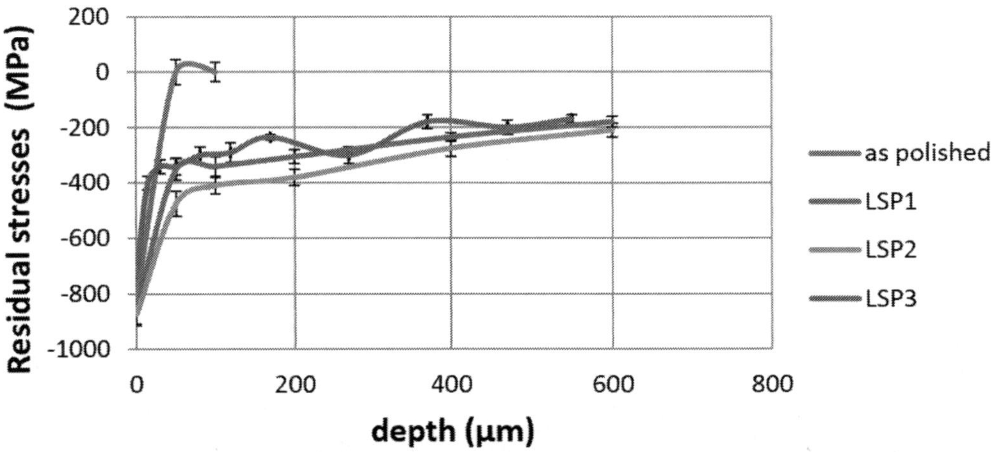

Figure 6: Stress profiles produced by LSP treatments

3.4. Duplex treatments

With the aim of characterizing the duplex treatments, samples with T1SZ treatment were grinded, 200 μm of material were then removed, and finally the samples were treated either by LSP (Treatment LSP1) or by PVD (WC:C coating).

Wear resistance of the duplex treated samples were tested and the results compared with the reference and the single treated sample.

As can be seen in figure 7, PVD coating improves the wear resistance of the sample with respect to the reference, so it is a suitable treatment for the required application. The improvement on the wear resistance of 100Cr6 provided by LSP treatment is not clear and the obtained measured wear rate is higher than the reference one.

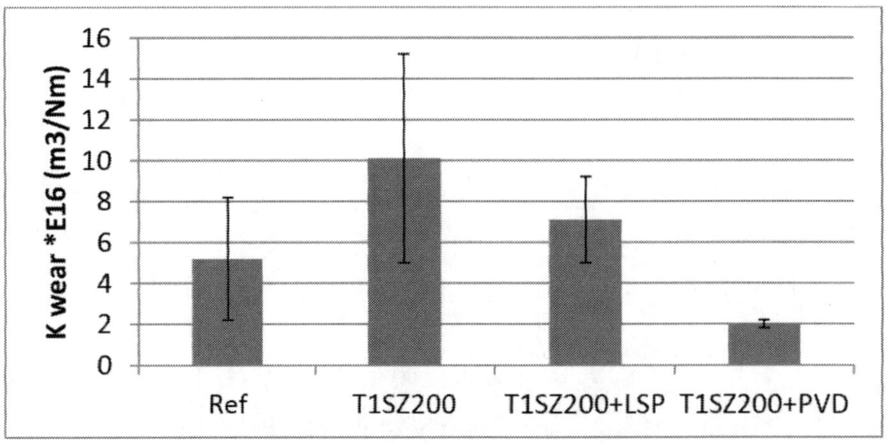

Figure 7: Kwear of the different duplex treatments

IOP Conf. Series: Journal of Physics: Conf. Series **843** (2017) 012080 doi:10.1088/1742-6596/843/1/012080

Table IV: Kwear of the duplex treated samples

Sample	Removed material (µm)	K wear * E16 (m^3/Nm)
Ref	0	5.2 ± 3.0
T1SZ200	0	10.1 ± 5.1
T1SZ200+LSP	200	7,1 ± 2.1
T1SZ200+PVD	200	2.0 ± 0.2

4. Conclusions

The application of TCT on 100Cr6 intended for improving the performance of bearings must be controlled due to several characteristics of the treatment.

The grain size of the material grows in contrast to the quenched samples due to the temperature of the process, and the longest times at these temperatures (4 h Vs. 1h). Samples treated at temperatures over T2 have suffered an unacceptable growth of its grain size, from 9-10 to 7-8 in ASTM Grain Number.

The diffusion of C and N in the material is hardly observable by optical microscopy due to changes in the chemical etching susceptibility of the material.

The diffusion on N into the steel matrix stabilizes the austenite, increasing the RA content from 7.5 %, present in the quenched sample, to a maximum of 42.8 %, measured in the surface of the sample T2.

At T1 the low diffusion of the C produces enrichment in the cementite content of the surface, generating harmful carbides networks which can affect the machinability of the material. The depth of the affected layer is less than 100 µm.

The in-depth wear resistance of the material is affected by the RA content and the presence of particles of carbonitrides, nitrides and carbides, producing profiles that present a minimum in the wear rate curve. In the studied treatments this minimum was found at 200 µm in the sample T1SZ and at 150 µm in the case of sample T2Z. In both cases this parameter is slightly higher than the one of the quenched sample.

Hardness profiles show that the treatment can produce an increase in the hardness of the first 300-400 µm for the sample T1SZ. The low RA content of this treatment and the presence of cementite in the surface are responsible of this result.

However, the sample T2SZ is affected by the high RA content, so the hardness is lower than the reference in the first 200 µm and slightly higher between 200 and 400 µm.

The combination of TCT and LSP does not produce a clear improvement in the wear resistance of the material. On the contrary, TCT + PVD coating enhance the properties of the surface, reducing up to 60 % the wear rate.

Acknowledgements
The authors gratefully acknowledge the support of the CDTI, EUREKA and EUROSTARS offices for supporting this work through the project E!8323 ULC Bearings.

References
[1] Bhadeshia H.K.D.H. 2012 Steel for bearings Progress in Materials Science 57 268-435.
[2] Ganguly S. et al 2002 Ultra clean steel for anti-friction bearing applications Bearing Steel Technology ASTM STP1419 47-68.
[3] Shibata M. et al 2002 Development of long life rolling bearings for use in the extreme conditions Bearing Steel Technology ASTM STP1419 285-296.
[4] Rajan K. 2013 Effect of carbonitriding on endurance life of ball bearing produced from sae 52100 bearing steels Journal of Surface Engineered Materials and Advanced Technology 3 172-177
[5] Stewart S. and Ahmed R. 2002 Rolling contact fatigue of surface coatings—a review Wear 253 1132-1144

[6] Dommarco, R. C. et al 2004 Residual stresses and retained austenite evolution in sae 52100 steel under non-ideal rolling contact loading Wear 257 1081-1088
[7] Lampman S. 1990 Introduction to surface hardening of steels ASM Handbook 4 559-267.
[8] Yakimets I. et al 2004 Laser peening processing effect on mechanical and tribological properties of rolling steel 100Cr6 Wear 256 311-320
[9] Ocaña J.L. et al 2004 Numerical simulation of surface deformation and residual stresses fields in laser shock processing experiments Applied Surface Science 238 242-248.
[10] Liao Y. et al 2012 The mechanisms of thermal engineered laser shock peening for enhanced fatigue performance Acta Materalia 60 4997-5009.
[11] Grill A. 1999 Diamond-like carbon: state of the art" Diamond and Relates Materials 8 428-434.
[12] Vetter J. 2014 60 years of DLC coatings: historical highlights and technical review of cathodic arc processes to synthesize various DLC types, and their evolution for industrial applications Surface and Coatings Technology 257 213-240.

IOP Conf. Series: Journal of Physics: Conf. Series **843** (2017) 012081 doi:10.1088/1742-6596/843/1/012081

Abrasive wear of ceramic wear protection at ambient and high temperatures

M. Varga*[1], K. Adam[2], M. Tumma[3], K.O. Alessio[3]

1 AC2T research GmbH, Wiener Neustadt, Austria
2 voestalpine Stahl GmbH, Linz, Austria
3 Kalenborn Kalprotect GmbH, Vettelschoss, Germany

Corresponding author: markus.varga@ac2t.at

Abstract. Ceramic wear protection is often applied in abrasive conditions due to their excellent wear resistance. This is especially necessary in heavy industries conveying large amounts of raw materials, e.g. in steel industry. Some plants also require material transport at high temperatures and velocities, making the need of temperature stable and abrasion resistant wear protection necessary.
Various types and wear behaviour of ceramic protection are known. Hence, the goal of this study is to identify the best suitable ceramic materials for abrasive conditions in harsh environments at temperatures up to 950°C and severe thermal gradients. Chamottes, known for their excellent thermal shock resistance are compared to high abrasion resistant ceramic wear tiles and a cost efficient cement-bounded hard compound. Testing was done under high-stress three-body abrasion regime with a modified ASTM G65 apparatus enabling for investigations up to ~950°C. Thereto heated abrasive is introduced into the wear track and also preheated ceramic samples were used and compared to ambient temperature experiments.
Results indicate a significant temperature influence on chamottes and the hard compound. While the chamottes benefit from temperature increase, the cement-bounded hard compound showed its limitation at abrasive temperatures of 950°C. The high abrasion resistant wear tiles represented the materials with the best wear resistance and less temperature influence in the investigated range.

1. Introduction

1.1. Application: cokery

Coke is produced from coal in the cokery. The hard coal is heated to ~1260°C in low oxygen atmosphere in evacuated chambers for ~24 hours. After the coking process the coke is pressed out from the chamber onto the coke quenching car. To prevent it from burning down, it is quenched with water. Afterwards the coke slides onto the coke wharf at temperatures of ~120-150°C. Sometimes there are hot spots of unquenched coke inside the pile, which reach ~650-750°C. These burning spots are not treated and usually extinguish before the coke gets transported further on a rubber conveyor belt. During windy weather conditions this initial burning spots can spread to local fire on the coke wharf, destroying the coke and causing high thermal stress on the coke wharf. In such cases manual water quenching is conducted. After cooling down the coke is removed by a plough onto a conveyor belt for further transportation. The surface temperature on the coke wharf is quickly dropping to environmental temperature after exposure to ambient conditions. Hence, thermal shock of the wharf flooring takes place and strains the flooring out of ceramic plates. The tiles get cracks and may start chipping or breaking into pieces. Besides thermal stress, high wear loss is caused by the abrasive coke. The mechanical load imparted by the coke depends on the slope of the wharf (~30-45°), with a bulk density of ~0.5 t/m³ a surface load of ~0.24 t/m² is applied on the coke wharf. Hence, the static mechanical load is relatively small compared to the thermal and abrasive load. It is therefore neglected in the following investigation. A third component which stresses the applied materials in the coke wharf is chemical attack of corrosive elements in the surrounding environment. After quenching the coke, quenching cars still contain a significant amount of quenching water which is spread onto the coke wharf. This water contains corrosive components from the coke, e.g. Ca, Cl, SO_4, NO_3. [1,2] Nevertheless, this investigation does not deal with chemical effects.

1.2. Ceramics for wear protection

Abrasion is a serious issue in heavy industries like mining or steel production. Transportation of goods is a main reason for serious wear loss on plant components. In addition to the material costs for worn parts, production losses due to component failure must be kept in mind. In the face of worldwide competitive markets this is intolerable. To avoid excessive maintenance expenses wear protection is inevitable. Here, a beneficial material choice is very important in relation to the tribosystem's properties of the application. The tribosystem describes the interplay of surfaces including lubrication, abrasive particles and wear particles as well as operating parameters such as load, velocity, temperature, etc. Conveying systems are so called open tribosystems, as no encapsulation is possible due the nature of the application. This normally renders the reduction of friction and wear by lubrication impossible. Hence, other forms of wear protection must be utilised in the form of wear resistant surfaces. [3-4]

To meet the application's requirements, e.g. the wearing processes and the thermal shock, it is fundamental to investigate prospective materials and compare materials under close-to-real application conditions in the laboratory. Lab-scale tests must ensure the simulation of equal wear processes as taken from failure analysis and application measurements.

Non-metallic components or hard materials, e.g. ceramics have special properties, like high hardness, low density and high crack resistance. These properties are based on the chemical bond of ceramics and non-metallic components which are mostly covalent. These special properties entail excellent wear-protection characteristics in the field of abrasion. Almost all ceramics which are used in coke wharfs have a heterogeneous microstructure with large phases and a significant amount of porosity. [5]

Also the application's temperature, especially the temperature of the medium is of high relevance in the wear process. Ceramics usually have a very high melting point and are stable over a wide temperature range. This does not hold true for many SiO_2-types, which feature transition points already below 600°C. This effect on high temperature wear is not yet fully scientifically established up to now. Certain ceramics or non-metallic components have a very good thermal shock resistance and temperature resistance. [5] In most cases the main problem is not the maximum temperature but rather the temperature fluctuations, so called thermal shock. This is also a crucial factor for coke wharfs.

The aim of this work is to cast light on the high temperature abrasive wear of commercially available ceramic wear protection for the thermal shock regime. In view of the application coke wharf the currently used chamotte should be compared with high wear resistant materials in a high temperature abrasive wear test. Results should point out temperature limitations of certain material groups and present wear mechanisms in order to aid material selection for the application.

2. Experimental

2.1. Materials

Different ceramic material groups can be applied for high temperature applications. Typical commercially available materials were chosen and are given in Table 1. Usually chamottes are used for highest temperatures. Chamottes are heterogeneous mineral aggregates, which are hydraulically pressed before the sintering process. The compositions of chamotte 1-3 is based on the binary system SiO_2-Al_2O_3. Whether the chamotte is fire resistant or chemically resistant depends on the Al_2O_3- and SiO_2-content.

The thermal shock resistance can be customised by the particle size distribution, the particle shape as well as the size and shape of pores.

For heavy wear regimes, especially in abrasive environments, very dense and hard ceramics are used, which usually impairs thermal shock resistance. The wear tiles 1-2 are based on the ternary system Al_2O_3-ZrO_2-SiO_2. By the addition of zirconium oxide the material reaches a harder structure. Wear tile 1 is prepared by a melt-cast process and wear tile 2 is made by vibro-casting followed by a subsequent sinter-process. Wear tile 2 thereby is optimised for high thermal shock resistance, like the chamottes, with additional high wear resistance.

The hard compound is a low cost wear lining material based on cement and hard material which can be applied by the customer as a monolithic liner on-site. In Table 1 the examined materials are listed and some basic physical properties are given.

Table 1. Chemical composition and main parameters of the materials investigated.

Name	Chamotte 1	Chamotte 2	Chamotte 3	Wear tile 1	Wear tile 2	Hard compound
Chemical composition	69 % SiO_2 25 % Al_2O_3 1 % Fe_2O_3	69 % SiO_2 25 % Al_2O_3 1 % Fe_2O_3	46 % Al_2O_3 1,5 % Fe_2O_3 SiO_2 bal.	51 % Al_2O_3 33 % ZrO_2 13 % SiO_2	82 % Al_2O_3 + ZrO_2 13 % SiO_2	70 % Bauxite SiO_2 bal.
Density [g/cm³]	2.2	2.2	2.3	3.5	3.2	2.7
Compressive strength [MPa]	150	150	>110	400	400	135
Max. application temperature	1000°C	1000°C		1000°C	1250°C	1200°C
Thermal shock resistance	300°C/h	300°C/h		150°C/h	300°C/h	300°C/h

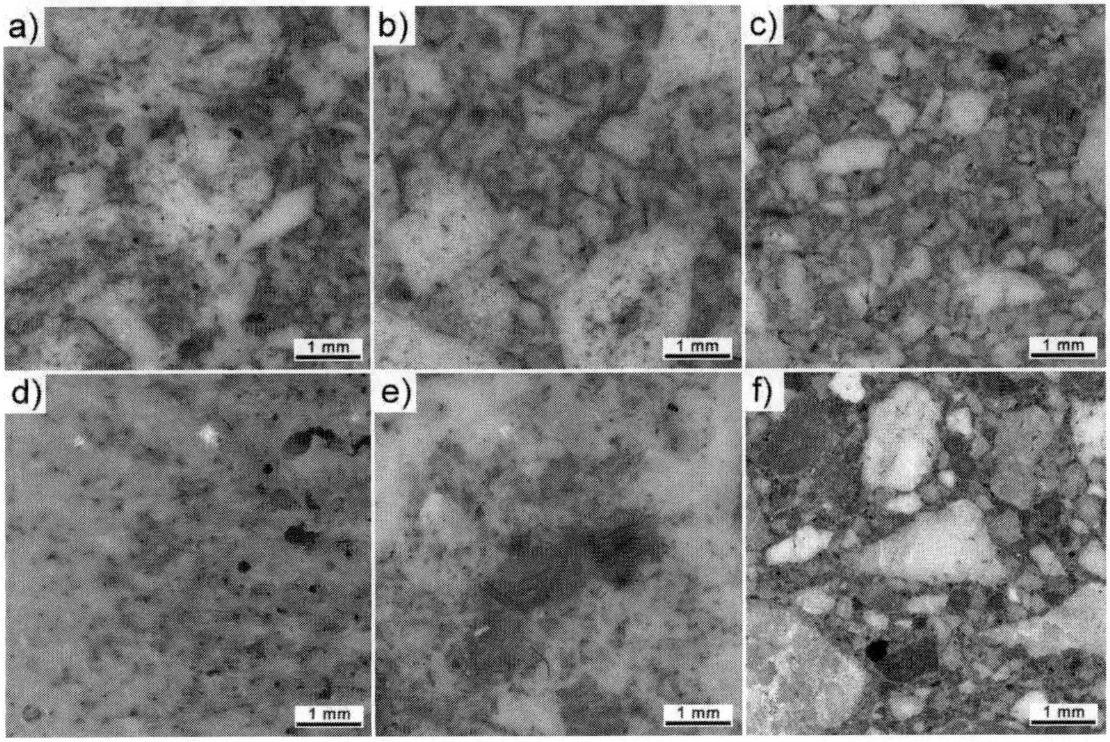

Figure 1. Microstructures of the materials investigated: a) Chamotte 1, b) Chamotte 2, c) Chamotte 3, d) Wear tile 1, e) Wear tile 2, f) Hard compound.

The microstructures as seen by optical microscopy (OM) are given in Figure 1. Chamottes 1-2 are of inhomogeneous microstructure with intergranular porosity. The grains are covered with a glassy phase. The grain size of the large SiO_2 grains is of several millimetres, with some smaller Al_2O_3 grains (white in Figure 1a, b). Chamotte 3 (Figure 1c) shows an inhomogeneous microstructure with significant inter- and intra-granular porosity. The Al_2O_3-content is higher compared to the chamottes 1-2, the grain size is smaller.

Wear tile 1 is of homogenous microstructure with shrinkage cavity in the centre of the tiles. The microstructure grows finer and less porous to the border of the tile, as can be seen in Figure 1d on the left side. Wear tile 2 is of inhomogeneous microstructure with a low percentage of porosity. Al_2O_3 hardphases are of large size of several millimetres (Figure 1e).

The hard compound is of heterogeneous microstructure with a hydraulic bounding. Particle size is of several millimetres as can be seen in Figure 1f.

2.2. *Abrasive wear testing - High temperature continuous abrasion test.*

The high temperature continuous abrasion test (HT-CAT) is based on a rotating wheel where a sample is pressed against (Figure 2a,b). An abrasive flow is introduced between the contacting bodies. To simulate a typical high temperature application, where hot abrasive is transported, the abrasive can be heated. Furthermore, to simulate also enhanced operation temperatures ceramic samples can be preheated and placed into the test rig with a certain start temperature. Reference tests were done with all components cold: i.e. sample, abrasive and wheel at room temperature (RT). This was compared to tests with heated abrasive (950°C) and cold sample and wheel, as well as preheated samples (750°C, 1 h), heated abrasive (950°C) and cold wheel.

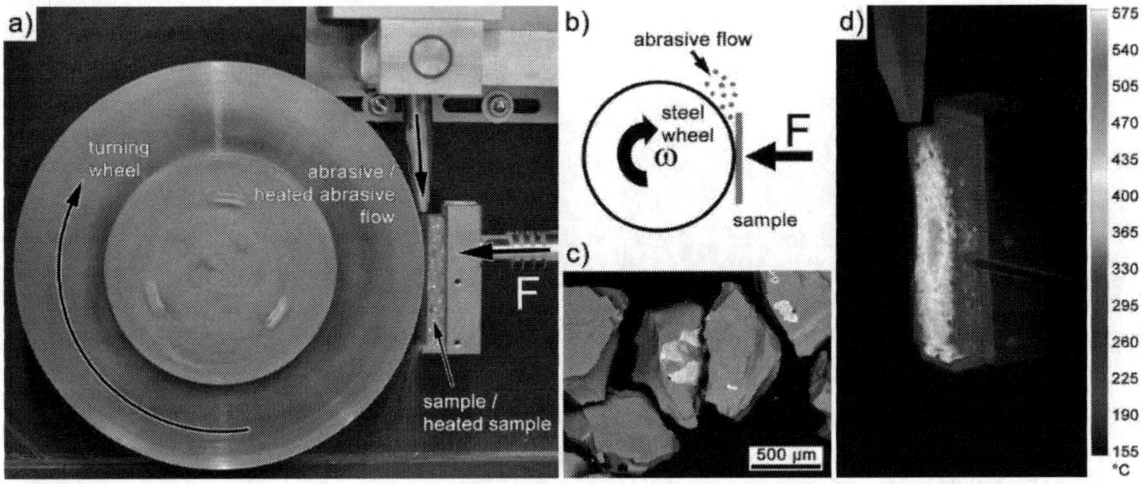

Figure 2. High temperature continuous abrasion test: a) test chamber; b) test principle; c) SiC abrasive; d) temperature distribution example after hot abrasive-hot sample test (hard compound).

A Hardox® 400 steel wheel is used as counter body (ø232×12 mm, 360±12 HV10 [6]), the load was set to 10 N and the speed was 1 m/s. Two test distances were used, in order to cope with the different abrasion resistance of the materials investigated: 114 m for the chamottes and the hard compound and 664 m for the wear tiles. The longer tests were necessary for accurate determination of the wear loss. As abrasive fused silicon carbide (SiC) was used in broken condition (angular), which is much more temperature stable as commonly used quartz abrasive for similar tests (e.g. ASTM G65 [7]), but also the angularity and higher hardness entails higher abrasivity of the abrasive [cf. 8]. The abrasive is displayed in Figure 2c. For the two high temperature tests it was heated in a tube furnace to 950°C and introduced into the wear zone by a pipe system at a flow rate of 180 g/min. To guarantee high temperature conditions the first 10 s of the abrasive flow are not used, after that the sample is brought into contact and begins to heat-up and abrade the sample.

Three samples were tested for each condition. Wear quantification was done via determination of the mass loss. The wear volume was than calculated via material's density and set in correlation to the wear distance. This resulted in the wear rate [mm³/m]. Furthermore the temperature evolution during testing is recorded by thermal imaging. Thereto a FLIR® SC7600 MB infrared camera with InSb-detector was utilised. The sample's surface is captured right after the end of the test and turn-off of the abrasive. Thereby the temperature distribution and maximum temperature of the wear track can be measured. An example is given in Figure 2d. After testing exemplarily wear tracks were also measured by OM and 3D topography measurements (Alicona® InfinteFocus G5).

3. Results and discussion

3.1. Abrasive wear results

The results of the abrasive wear tests at the different temperature conditions are displayed in Figure 3a. The RT condition shows three wear levels: highest wear rates were measured at the chamottes with ~3.8-4.1 mm³/m, with no significant difference between the types 1-3; the second level of about half wear rate with 1.8 mm³/m features the cement bounded hard compound; lowest wear rates were found for the wear tile types. They both have ~0.6 mm³/m wear rate.

The enhanced temperature tests were done first with heated abrasive (950°C) and second with both heated, abrasive (950°C) and sample (750°C). Temperature influence is diverse for the materials investigated: chamottes and wear tiles show a decrease in wear rate with increasing temperature input. The hard compound on the other hand shows considerably increased wear rates.

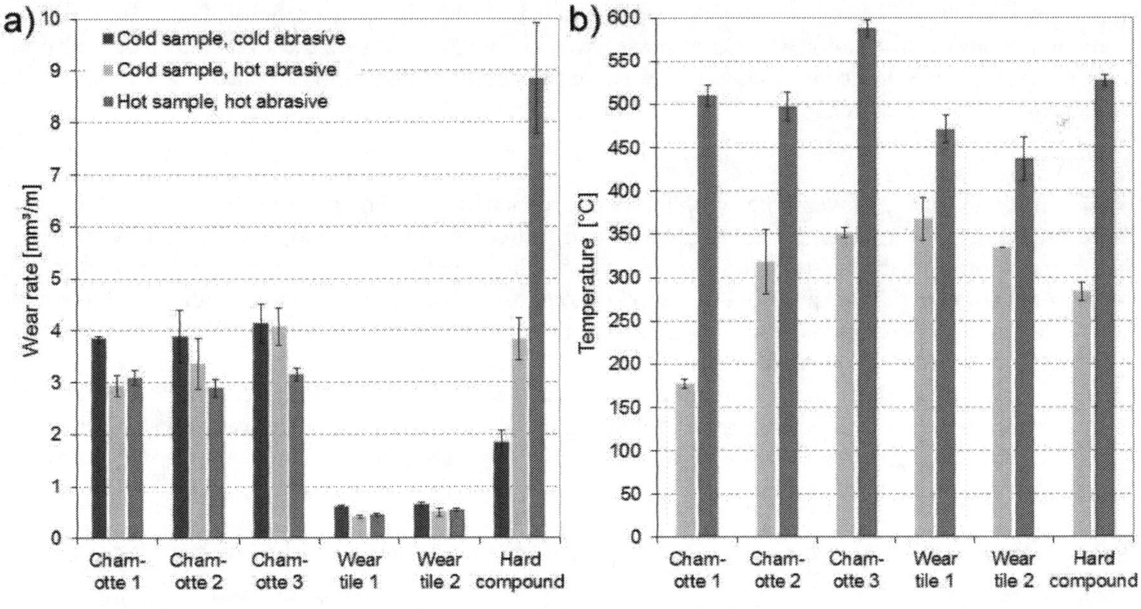

Figure 3. Results of HT-CAT testing: a) wear rates; b) temperature in the wear zone after testing.

For chamotte 1 wear decreases already at testing with heated abrasive from ~3.8 mm³/m to 2.9 mm³/m, which stays in the same range also with additional preheated sample. Chamotte 3 stays at the higher RT level also at heated abrasive testing, and shows a decrease to 3.1 mm³/m at preheated samples. Chamotte 2 shows wear rates between those of the two other chamottes, with a medium decrease at heated abrasive and further decrease at heated abrasive and heated sample to the lowest value of the chamottes measured of 2.9 mm³/m.

The wear tiles show the lowest wear rates in the temperature regime investigated. Like chamotte 1 already at heated abrasive testing the wear rate significantly decreases, and then stays constant also at preheated sample testing. Wear rates reduce from ~0.6 mm³/m for wear tile 1, RT testing, to ~0.45 mm³/m at HT testing. For wear tile 2 the reduction is from ~0.65 mm³/m to 0.55 mm³/m.

The cement bounded hard compound shows diverging behaviour. While featuring a relative good wear resistance at RT with ~1.8 mm³/m, this more than doubles at hot abrasive testing to ~3.8 mm³/m and reaches the same level as the chamottes. For further temperature increase with preheated sample, the wear rate peaks at to ~8.8 mm³/m, meaning a 4.8-fold wear increase compared to RT testing.

3.2. *Temperature evolution*

For exact reproduction of temperature conditions of the application, testing was not done at isothermal condition (e.g. in a closed oven), but with heated abrasive. This entails inhomogeneous temperature distribution, both locally and temporally: heating of the sample starts at the contact area with the abrasive (see Figure 2d), and the sample heats-up with time for the test configuration of cold sample and hot abrasive. Figure 3b shows the maximum temperature of the contact area (wear zone) directly after testing and turn-off of the heated abrasive flow. Temperature increase after 2 min testing time was lowest for the chamotte 1, resulting at ~175°C followed by the hard compound with ~280°C. The other values are close together in a range from 320-370°C. With the highest value for the wear tile 1. Here must be kept in mind, that the wear tiles were tested 4-times longer than the other materials and hence had more time to heat-up.

For experiments with preheated samples, the temperature lowers over the test duration: Samples are introduced into the test with ~750°C, gradually cooling down in the open environment. Opposite to this cooling stays the concomitant heating-up by the hot abrasive. This entails ~500°C after the test for the chamottes 1-2. Chamotte 3 shows ~690°C and seems to have a higher heat capacity. The wear tiles show lower temperature of 430-460°C, which most probably is the result of the higher test duration and thereby longer time for cooling-down. The hard compound shows ~530°C and lies between the temperature values of the chamottes.

3.3. *Wear mechanisms*

To elucidate the change of wear rates with temperature increase more detailed investigations of the wear zones are required. Therefore for each material and test condition the 3D-topgraphy was measured. A 5×5 mm section from the centre of the wear zone is displayed in Figure 4. Here the first line corresponds to the cold sample-cold abrasive condition, the second to cold sample-hot abrasive and the third to hot sample-hot abrasive. Furthermore, the roughness parameters according to ISO 25178 were derived from topography data (Table 2). Generally roughness levels are high after high-stress abrasion testing of the ceramics. Only the hard and more homogenous material wear tile 1 shows significantly lower roughness values, which is also clearly visible in the topography images in Figure 4. The roughness of chamotte types decreases at higher temperature testing, leading to the assumption that breakage, which induces high roughness, takes place less often.

Table 2. Roughness parameters of the centre of the wear zone after wear testing at various conditions.

		Chamotte 1	Chamotte 2	Chamotte 3	Wear tile 1	Wear tile 2	Hard compound
Cold sample-cold abrasive	Sa [µm]	44.9	60.6	75.7	17.6	56.7	41.7
	Sz [µm]	517.3	664.0	697.0	160.1	479.2	353.7
Cold sample-hot abrasive	Sa [µm]	39.2	53.4	52.1	19.4	64.3	42.9
	Sz [µm]	535.1	617.0	435.8	344.9	467.7	370.8
Hot sample-hot abrasive	Sa [µm]	41.3	35.8	34.2	23.1	43.5	36.4
	Sz [µm]	337.4	408.5	339.3	345.6	325.9	405.3

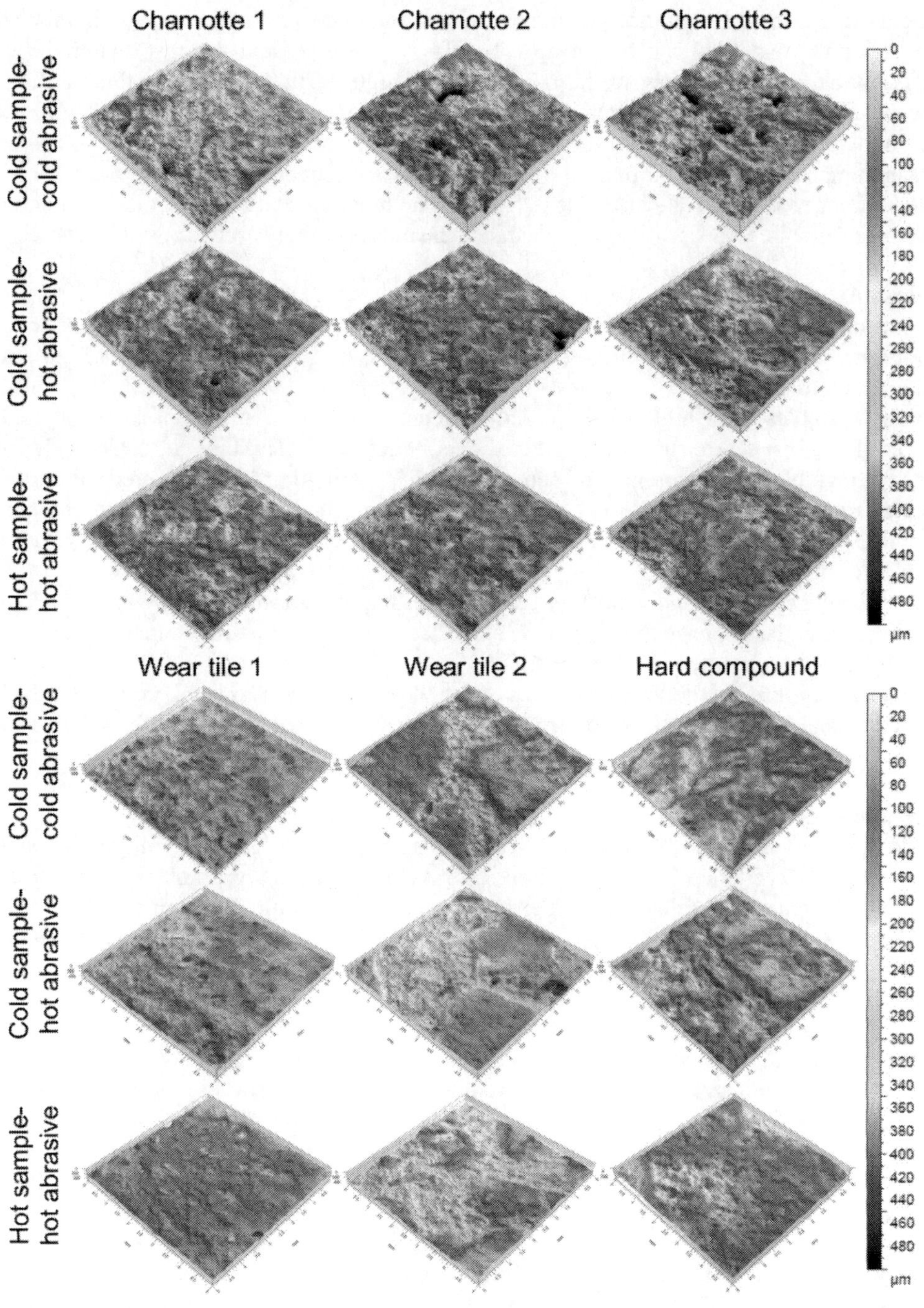

Figure 4. 3D-topography in the centre of the wear zone (5×5 mm) after the various test conditions.

The roughness of wear tile 1 stays in a similar range over the temperatures investigated. Sz values are not very meaningful for that material, as there are many small pores present. Wear tile 2 and the hard compound have very course microstructure, hence the measured area only shows a small insight on the wear process. The hard Al_2O_3 particles remain more wear resistant and the matrix between them is removed, which is similar to metal matrix compounds wear [cf. 9,10], where the metallic matrix is removed more easily than the hard phases.

Taking these information into account a hypothesis about the wear rate drop at high temperatures for the chamottes and wear tiles can be set up. The glassy phase present in these materials is not very temperature stable, and especially in the range >500°C significant softening of this phase can be assumed. Temperatures in this range were approved for the preheated samples, but also with just heated abrasive, the temperature transfer on the microscale over the hot abrasive particles likely entail local high contact temperatures. This softening of the glassy phase reduces the brittleness of the compounds leading to an improved wear resistance. Nevertheless, one must keep in mind that the further softening at even higher temperature will significantly destabilise the microstructure and wear will definitely increase.

4. Conclusions

In order to study the abrasion resistance of ceramics and its temperature influence, six commercially available ceramics for high temperature (HT) application were investigated under a high-stress three-body abrasion regime.

The two wear tiles with high Al_2O_3 + ZrO_2 content showed least wear loss at all temperatures investigated. Here, especially the wear tile 2 shows beneficial behaviour, as it has a thermal shock resistance comparable to chamottes. The cement bounded hard compound featured the next best wear resistance at room temperature (RT), but showed wear increase when exposed to HT-environment (950°C abrasive temperature in this investigation) due to its chemical bounding. Close together on a third wear level lie the chamottes.

The three chamotte-types and wear tiles show decreasing wear rates at HT, which was put down to the softening of the glassy phase therein, increasing their ductility. This limits the dominant brittle wear mechanism present at room temperature testing.

For practical application the low wear rates of the wear tiles are promising, compared to chamottes. As long as the mechanism of abrasive wear is dominant, their use will significantly reduce maintenance expenses.

Acknowledgments

This work was funded by the Austrian COMET Program (Project K2 XTribology, no. 849109) and carried out at the "Excellence Centre of Tribology". Acknowledged are Frank Schütze for performing the wear tests, Tetyana Khmelevska for topography analysis, Ewald Badisch and Manel Rodriguez Ripoll for fruitful discussions and Wanggo Gummitechnik GmbH for active research cooperation.

References

[1] Varga M, Winkelmann H, Badisch E, Adam K 2010 *Einsatzbedingungen und Schadensanalyse im Bereich der Kokerei* (Göttingen: Proceedings GfT-Tribologie Fachtagung 27.-29.9.2010)
[2] Varga M, Winkelmann H, Badisch E, Adam K 2012 *Lebenszyklus-Optimierung mittels tribo-systematischer Analyse des Realsystems und den daraus abgeleiteten werkstofflichen und numerischen Ansätzen* (Göttingen: Proceedings GfT-Tribologie Fachtagung 24.-26.9.2012)
[3] Pigors O 2001 *Werkstoffe in der Tribotechnik. Reibung, Schmierung und Verschleißbeständigkeit von Werkstoffen und Bauteilen* (Weinheim: Wiley-Vch)
[4] Habig KH 1980 *Verschleiß und Härte von Werkstoffen* (München: Hanser Fachbuchverlag)
[5] Salmang H, Scholze H 2007 *Keramik* (Berlin, Heidelberg: Springer)
[6] Varga M, Rojacz H, Winkelmann H, Mayer H, Badisch E: Wear reducing effects and temperature dependence of tribolayer formation in harsh environment, *Tribol Int* **65** (2013) 190-199
[7] ASTM G65: Standard Test Method for Measuring Abrasion Using the Dry Sand/Rubber Wheel Apparatus, ASTM (1980)
[8] Kelly DA, Hutchings IM: A new method for measurment of particle abrasivity, *Wear* **250** (2001) 76-80
[9] Rojacz H, Varga M, Kerber H, Winkelmann H: Processing and wear of cast MMCs with cemented carbide scrap, *J Mater Process Tech* **214/6** (2014) 1285-1292
[10] Winkelmann H, Varga M, Badisch E: Influence of secondary precipitations in Fe-based MMCs on high temperature wear behaviour, *Tribol Lett* **43** (2011) 229-234